ATLANTIS STUDIES IN MATHEMATICS FOR ENGINEERING AND SCIENCE

VOLUME 7

SERIES EDITOR: **C.K. CHUI**

Atlantis Studies in Mathematics for Engineering and Science

Series Editor:

C. K. Chui, Stanford University, USA

(ISSN: 1875-7642)

Aims and scope of the series

The series 'Atlantis Studies in Mathematics for Engineering and Science' (AMES) publishes high quality monographs in applied mathematics, computational mathematics, and statistics that have the potential to make a significant impact on the advancement of engineering and science on the one hand, and economics and commerce on the other. We welcome submission of book proposals and manuscripts from mathematical scientists worldwide who share our vision of mathematics as the engine of progress in the disciplines mentioned above.

All books in this series are co-published with World Scientific.

For more information on this series and our other book series, please visit our website at:

www.atlantis-press.com/publications/books

ATLANTIS PRESS

AMSTERDAM – PARIS

World Scientific

Boundary Element Methods with Applications to Nonlinear Problems

Goong Chen

Professor of Mathematics and Aerospace Engineering, Texas A&M University
College Station, Texas 77843, USA

and

Distinguished Professor of Mathematics
National Taiwan University
Taipei, Taiwan, Republic of China

Jianxin Zhou

Professor of Mathematics
Texas A&M University
College Station, Texas 77843, USA

ATLANTIS PRESS

AMSTERDAM – PARIS

World Scientific

Atlantis Press

29, avenue Laumière
75019 Paris, France

For information on all Atlantis Press publications, visit our website at: *www.atlantis-press.com*

Atlantis Studies in Mathematics for Engineering and Science
Volume 1: Continued Fractions: Volume 1: Convergence Theory – L. Lorentzen, H. Waadeland
Volume 2: Mean Field Theories and Dual Variation – T. Suzuki
Volume 3: The Hybrid Grand Unified Theory – V. Lakshmikantham, E. Escultura, S. Leela
Volume 4: The Wavelet Transform – R.S. Pathak
Volume 5: Theory of Causal Differential Equations – V. Lakshmikantham, S. Leela, Z. Drici
Volume 6: The Omega Problem of all Members of the United Nations – E.N. Chukwu

ISBN: 978-90-78677-31-4
ISSN: 1875-7642

Dedicated to
Professor David L. Russell
on the occasion of his 70th birthday

and

Professor George C. Hsiao
on the occasion of his 75th birthday

Preface to the 1st edition

Boundary element methods (BEM) have undergone rapid advancement in recent years. Because of their numerous advantages, such as ease of coding, small memory requirement and computational efficiency, they have become a *major* numerical tool. The analysis and rigor of BEM have been strengthened through the work of several mathematicians. It is now possible for the authors to fit together the many scattered contributions into a comprehensive account. In this book, we present mathematical formulations of *boundary integral equations* (BIE) for several of the most important linear elliptic *boundary value problems* (BVP), discuss their computational algorithms and the accuracy of their solutions, illustrate the numerical solutions and show some applications. We wrote this monograph as a reference source for researchers who are concerned with numerical solutions of *partial differential equations* (PDE), and as a graduate text for a course in this subject.

It has been known for a long time that it is possible to formulate BIE with singular kernels to solve BVP of the linear elliptic type. The unknown boundary data in the solution depend on the prescribed boundary data through those BIE. Once all the boundary data (or the Cauchy data) become available, the solution of the given PDE is obtained by an application of the Green's formula. Complex variable techniques for solving BIE on planar complex contours were perfected by the Russian school (see Muskhelishvili [137]) in the early 1950s, and are still widely used today for *two-dimensional* potential and elasticity problems. A monumental achievement, developing a *general theory* of BIE using the *fundamental solution* rather than the complex Cauchy kernel, is presented in V.D. Kupradze's book [117], published in 1965. In that decade, with electronic computers gaining larger power and faster speed, and being in wider use in the U.S., T.A. Cruse, F.J. Rizzo [58, 160, 59] and others started to compute solutions to many BIE arising in continuum mechanics and obtained excellent numerical results. The BIE they used at that time were primarily formulated from the Green's formula, which is commonly referred to as *the direct approach* in the literature.

Actually, for elliptic BVP an alternative way of formulating simpler systems of BIE is possible. In a paper [72] published in 1961 G. Fichera used ideas from *potential theory* to represent the solution as a *simple-layer potential*. This *ansatz* leads to greatly simplified systems of BIE whose solutions, as a consequence, are much easier to compute. Nevertheless, Fichera's paper remained largely obscure until more than a decade later, when in 1973 G.C. Hsiao and R.C. MacCamy [95] adapted his ideas and lucidly exemplified them for the Poisson equation on an exterior domain, and for the second biharmonic problem in two-dimensional linear elasticity. The arguments in [95] are now standard in proving the existence and uniqueness of solutions of BIE, as are shown at several places in our book.

When BIE are approximated numerically, the questions of *convergence* and *error analysis* need to be addressed. The mathematical foundation of such an analysis was laid down in 1977 in a paper [96] by G.C. Hsiao and W.L. Wendland. There, they used the *strong ellipticity* of the simple-layer boundary operator to derive sharp error estimates for the solutions of the *Galerkin approximations* of BIE. Their proof works in *arbitrary space dimensions* for any strongly elliptic pseudodifferential equations arising from elliptic BVP.

In practice, however, BIE are more easily approximated by *point collocation*. Error analysis questions concerning collocation approximations were essentially resolved by D.N. Arnold, J. Saranen and W.L. Wendland in [9, 165, 10] in the case of *two-dimensional* problems, where in the culminating paper [10] the main technique was a delicate Fourier series analysis. Nevertheless, as of this writing, no analogous success has been achieved for collocation analysis of three-dimensional problems.

The authors' interests in BEM stem from numerical solutions to various applied problems in science and engineering. In our pursuit of numerical solutions of PDE, we marvelled at the efficacy of the BEM and we were gratified by the accuracy of the numerical results. During those years we have also received growing demands from graduate students in many engineering departments asking us to teach courses containing BEM material. While teaching those courses, we began to realize the needs of developing a graduate level text and monograph that is sufficiently *self-contained*, emphasizes more *mathematical rigor*, and includes many useful *examples of applications and illustrations* to stimulate the audience's interests. Indeed, these are the three major goals we have been striving to achieve throughout the volume—we hope with some success.

The organization and selection of the material carry the strong bias of the authors. There are 10 chapters in the book. The first five consist of preparatory material:

(1) Chapter 1, an introduction, gives readers some quick ideas as how BEM work and how they compare with other numerical PDE methods.
(2) Sobolev spaces are fundamental in the modern theory of PDE. In Chapter 2, we explain some of their basic properties that are essential in understanding the solvability of BIE.
(3) Hypersingular integrals occur naturally in boundary element computations. The main objective of Chapter 3 is to enable readers to understand how to regularize such integrals based upon the theory of distributions.
(4) Boundary integral operators can be studied in an elegant way using the theory of pseudodifferential and Fredholm operators. In Chapter 4, we give a concise account of this approach and its applications to BIE.
(5) The mathematical theory of finite elements is a prerequisite for BEM. In Chapter 5, we describe the commonly used finite element spaces and their basic properties.

The subsequent chapters deal with the corpus of BEM for PDE:

(6) Chapter 6 studies BIE for the potential equation.
(7) Chapter 7 studies BIE for the Helmholtz equation.
(8) Chapter 8 studies BIE for the thin plate equation.
(9) Chapter 9 studies BIE for the linear elastostatic systems.
(10) Finally, Chapter 10 contains error estimates for Galerkin and collocation solutions of general BIE.

In Chapters 6–9, for each specific type of PDE treated, the physical background is first described, and from there follow the different layer formulations for various types of boundary conditions. Numerical examples and computer graphics are then presented to illustrate the theory.

From our computational experience, our favorite approach is the (augmented) *simple-layer representation collocated by piecewise constant boundary elements*. Among different *admissible* layer representations and discretization schemes, this approach is certainly the easiest to program on a computer. It yields rather smooth profiles of solutions due to the smoothing property of the simple-layer boundary operator. In the discussion of the computational aspects of the examples in Chapters 6–9, we have avoided minute technical details of discretization and quadrature, as readers can easily improvise on their own, also since such material is already available in existing BEM books (cf. Brebbia's [23, 24], for example). We strongly encourage readers to develop computer packages themselves.

We have tried to present the material as straightforwardly as possible, even though the exposition may occasionally be repetitive and may not be the optimal one. For many beginning engineering graduate students, some sections involving Sobolev spaces and pseudodifferential operators in Chapters 2 and 4 may still appear somewhat technical. Our suggestion for instructors and students alike in such a mathematics/engineering course is to accept the statements of the theorems but to put off the details of the mathematical justification for the first reading. It has been our experience that this will not severely hamper study of the material in Chapters 6–9. Instructors may also assign numerous numerical examples in those chapters as computer projects for students. The computer programs for Chapter 6 are relatively simple. It normally takes about a week to develop a computer program for an example. But the time taken to do this increases with the chapter numbers. Computer programs for the examples in Chapter 9 generally require month-long efforts because of the three-dimensional geometry involved.

We have left out many interesting topics, among them crack and fracture problems, domains with corners, Maxwell's equations in electromagnetism, the linear Stokes flow and time-dependent evolution equations. We hope to be able to treat some of these in a separate volume. Readers may also consult a forthcoming monograph [99] by Hsiao and Wendland for some advanced material on BEM.

The Bibliography contains only those references most pertinent to the content of the book. Based upon the information therein, we hope that interested readers can trace the literature on their own. We apologize in advance for any inadvertent omissions of references, citations and acknowledgements.

Any corrections, comments and suggestions for improvement from readers will be greatly appreciated.

Goong CHEN
Jianxin ZHOU

College Station, Texas June 1991

Dedication

Professor David L. Russell is the former Ph.D. thesis advisor of the first author (G.C.) at the University of Wisconsin-Madison. His work and personality have had a profound influence in shaping both authors' career interests and in developing their professionalism.

Professor George C. Hsiao is a founder of the mathematical theory of boundary element methods. We have benefitted greatly from reading his papers. He also generously consulted with us and provided assistance on numerous occasions.

We have tremendous admiration for both individuals. To them we dedicate this monograph.

Acknowledgements for the 1st edition

Our work in partial differential equations and boundary element methods was supported by National Science Foundation Grants DMS 84-01297 and 87-18510, and Air Force Office of Scientific Research Grants 85-0253, 87-0334, 88-0091 and 91-0097. The Program Managers were Drs John E. Lagnese, Andre Manitius and Michael P. Polis at NSF, and Drs John A. Burns, James M. Crowley and Marc Jacobs at AFOSR. Their interest in our work, their helpful suggestions and generous support are gratefully acknowledged. Dr Andre Manitius was instrumental in arranging a SUN 3/60 Workstation for our research through NSF Grant 87-18510. This equipment proves to benefit us immensely in our computer graphics work.

Supercomputer time was allocated to us by the former John von Neumann Center at Princeton, New Jersey, the National Center for Supercomputing Applications (NCSA) at Urbana-Champaigne, Illinois and the Texas A&M University Supercomputer Center. We thank them for their support.

The two authors are extremely fortunate to be able to work closely together at the same institution. In this regard we are particularly indebted to Professors Steven G. Krantz and Jon Pitts, who helped us move to Texas A&M University in 1987.

At various stages of our research and the writing of this book we have benefitted from discussions with many colleagues and collaborators. This list includes Professors Thomas J. Bridges, Michel C. Delfour, Stephen A. Fulling, Nilotpal Ghosh, D. Joe Hartfiel, Wendell H. Mills, E. Bruce Lee, Kaitai Li, Walter Littman, Philip J. Morris, Michael Pedersen, Shunhua Sun and Joseph D. Ward. Professors Andrew Kurdila, Jeff Morgan and Michael Pedersen, Dr Frank Molzahn, Messrs Chih-Hsuan Chen, Yuanhua Deng, Zhonghai Ding and Christopher Peoples have read portions of the manuscript, spotted many errors and offered invaluable suggestions.

The material in § 4.4 was adapted from notes communicated to us from Professor Michael

Pedersen of the Technical University of Denmark, while a large portion of the material in § 5.5 was adapted from Dr Wendell H. Mills' unpublished lecture notes on finite element methods taught at Pennsylvania State University. We are grateful to their contributions.

The computer work on linear elastostatics in Chapter 9 was obtained using the computer packages from Ms Link Ji's Ph.D. thesis directed by the first author at Texas A&M University. Ms Ji has also helped us in many computer-related matters.

Many segments of the material have been taught at Pennsylvania State University and Texas A&M University. During the summer of 1990, by invitation of Professor Sze-Bi Hsu, Director of the Institute of Applied Mathematics, National Tsing Hua University at Hsinchu, Taiwan, Republic of China, the first author gave a short course on boundary element methods. The organization of the book was significantly reshaped after that lecture series. We wish to thank Professor Hsu, all the attendees and students, and the respective institutions for the teaching and lecturing support.

Messrs Kevin Chen and J. Brian Campbell spent many hours in drawing figures and tables. Ms Jean Chen and Robin Campbell patiently typed the long manuscript. We truly admire the excellent quality of their work.

Finally, we deeply thank the Editor, Professor John R. Whiteman, and the editorial staff, Mr Andrew Carrick, Misses Kate Brewin and Fiona Murphy, for all of their efforts and assistance that have made the publication of this volume possible.

Preface to the 2nd edition

Boundary element methods (BEM) continue making the progress since the first edition of this book was published in 1992. The theory and numerical estimates have become more complete, with also an ever increasing number of applications in science and engineering.

Although the advances made in BEM are many-faceted and, in fact, too numerous for us to describe, the authors' main interest in making this updated edition regarding BEM is in adding the theory and applications of BEM to *nonlinear problems*. This is reflected in the new adjusted title of the book: "Boundary Element Methods, with Applications to Nonlinear Problems". Since the last quarter of the 20th Century, beginning during the 1970s, there has been a great shift of interests and emphases in the research and development from the linear to the nonlinear theory, methods and models in all branches of science and engineering. The linear theory of the majority of subjects under study is now widely regarded as reaching an asymptotically complete state of development and has become classical. Nonlinear science and technology are definitely at the frontier of scientific endeavors of the 21st Century. Nonlinear phenomena, such as the lack of uniqueness and stability, the emergence of pattern formations and chaos, etc., now attract the major attention of the scientific community. Therefore, it is imperative for scholars to learn and study the basics of nonlinear science and mathematics. It is based on this conception that, in bringing out the revised edition, we provide two new chapters, Chapters 11 and 12 in this book, to expressly address the theory and applications of BEM to nonlinear elliptic boundary value problems and systems. Therein, we describe the development of various *iterative algorithms* to be coupled with the linear BEM methodology in order to capture solutions of nonlinear equations and systems. Many graphics are accompanied to help visualize the solutions to the nonlinear equations with the hope to aid in the understanding of the physics of those systems.

In making this 2nd edition, we have also tried our best to pick out a number of typographical errors. A majority of the graphics in the original edition of the book has been refurbished or improved, and we hope they are more eye-pleasing. One section in Chapter 8, the former Section 8.9, has been removed as it is not tightly related to the other parts of the book.

Many of our former collaborators, in particular Dr. Yuanhua Deng, Prof. Zhonghai Ding and Prof. Wei-Ming Ni, have made direct or indirect contributions to the newly added Chapters 11 and 12 in this volume. Ms. Nancy G. Chen and Mr. Changchun Wang have helped reproduce and improve many graphics and photos. Ms. Robin Campbell helped the editing and typesetting of the book. To them, we wish to register our sincerest thanks.

Finally, we express our gratitude to Dr. Keith Jones as Publisher and Zeger as Director at Atlantis Press, for their patience and constant encouragements in making the publication of this volume happen.

Goong CHEN
Jianxin ZHOU

College Station, Texas and Taipei, Taiwan
College Station, Texas
2010

Notation and Abbreviations

$\forall$	for all
$\exists$	there exists
$\equiv$	defined by, or identically equal to
$\perp$	orthogonal to
$\|\ \|$	absolute value of a real number, or the length of a finite-dimensional vector
$\langle\ ,\ \rangle$	the inner product of two vectors in a Hilbert space: if $x = (x_1, \ldots, x_N)$ and $y = (y_1, \ldots, y_N)$ are two vectors in $\mathbb{R}^N$, we use either $\langle x, y\rangle$ or $x \cdot y$ to denote $\sum_{j=1}^{N} x_j y_j$
$\\|\ \\|$	the norm of a vector in a Banach space or a Hilbert space
∇	gradient
$\overline{\nabla}$	surface gradient
Δ	Laplacian
$\nabla\cdot$	divergence
Δ^2	biharmonic operator
$\otimes$	tensor product
$\oplus$	direct sum
#	the cardinal number (of a set)
$\hookrightarrow$	continuous injection (imbedding)
dx	the infinitesimal volume element $dx_1\, dx_2 \ldots dx_N$
$d\sigma_x$	the infinitesimal surface element (with respect to the x variable)
∞	infinity
$*$	for two functions f and g, $f * g$ is the convolution of f with g
a.e.	almost everywhere
BC	boundary condition(s)

BEM	boundary element method(s)
BIE	boundary integral equation(s)
BVP	boundary value problem(s)
c.c.	completely continuous
codim	codimension (of a subspace)
dim	the dimension (of a space)
ess. sup	the essential supremum (of a function)
FDM	finite difference method(s)
FEM	finite element method(s)
iff	if and only if
LHS	the left-hand side (of an equation)
PDE	partial differential equation(s)
RHS	the right hand side (of an equation)
s.t.	such that
w.r.t.	with respect to

arg	the argument of a complex number
$\mathbb{C}$	the set of all complex numbers
$C_0^\infty(\Omega)$	the set of all infinitely differentiable functions with compact support contained in Ω
C^k	k-times continuously differentiable
$C^{k,\alpha}(\overline{\Omega})$	the set of all k-times continuously differentiable functions whose k th derivatives are Hölder-continuous with exponent α on $\overline{\Omega}$
$\mathscr{C}_1, \mathscr{C}_2$	the Calderón projectors
Coker $\mathscr{L}$	the cokernel of a linear operator $\mathscr{L}$
$\mathscr{D}, \mathscr{D}(\mathbb{R}^N)$	the space of test functions $(= C_0^\infty(\mathbb{R}^N))$
$\mathscr{D}'$	the space of distributions (on $\mathbb{R}^N$)
D^α	the partial differential operator $\left(\frac{\partial}{\partial x_1}\right)^{\alpha_1}\left(\frac{\partial}{\partial x_2}\right)^{\alpha_2}\cdots\left(\frac{\partial}{\partial x_N}\right)^{\alpha_N}$, for a given multi-index $\alpha = (\alpha_1, \ldots, \alpha_N)$, where α_j is a nonnegative integer for $j = 1, 2, \ldots, N$
$\frac{\partial}{\partial n}, \partial n, \frac{\partial}{\partial n_x}$	the normal derivative (at x or with respect to the x-variable)
$\delta(x)$	the Dirac delta distribution
$E(x,\xi)$	the fundamental solution (of a partial differential operator)
$\mathscr{F}$	the Fourier transform, $\mathscr{F}(f)(\xi) = \widehat{f}(\xi) = \int_{\mathbb{R}^N} e^{-2\pi i\langle x,\xi\rangle} f(x)dx$
$\mathscr{F}^{-1}, \overline{\mathscr{F}}$	the inverse (and the conjugate) Fourier transform,

$$\mathscr{F}^{-1}(g)(x) = \overline{\mathscr{F}}(g)(x) = \int_{\mathbb{R}^N} e^{2\pi i\langle x,\xi\rangle} g(\xi)d\xi$$

$\Gamma(x)$	the gamma function
$H_\alpha^{(1)}, H_\alpha^{(2)}$	Hankel functions of the first and the second kinds of order α
$H^s(\Omega)$	the Sobolev space of order s on a domain Ω
$H^s_{\text{loc}}(\mathbb{R}^N)$	the space of all functions or distributions f on $\mathbb{R}^N$ such that $f \in H^s(K)$ for every compact set $K \subset \mathbb{R}^N$
Im	the imaginary part (of a complex number)
Ind	the index (of a Fredholm operator)
inf	infimum
J_α	Bessel function of order α
ln	the natural logarithm function
$L^p(\Omega)$	the space of all p th $(p \geqslant 1)$ power Lebesgue-integrable functions on a domain Ω
$L^p_{\text{loc}}(\mathbb{R}^N)$	the space of all functions whose p^{th} $(p \geqslant 1)$ power is absolutely Lebesgue-integrable on every compact set of $\mathbb{R}^N$
$\mathscr{L}(X,Y)$	the space of all bounded linear operators from space X into space Y
max	the maximum (of a given function or functional): when the supremum is attained (over a given set), we write max instead of sup
meas A	the measure (or volume) of a given set A
min	the minimum (of a given function or functional): when the infimum is attained (over a given set), we write min instead of inf
$\mathbb{N}$	the set of all *positive* integers
n, $n(x)$, n_x	the unit normal vector pointing outward on the boundary $\partial\Omega$ of a bounded domain Ω (at $x \in \partial\Omega$)
N_α	Neumann function of order α
$\mathscr{N}(\mathscr{L})$	the null space (or the kernel) of a linear operator $\mathscr{L}$
Ω	an open domain in $\mathbb{R}^N$: Ω is usually simply connected and bounded with smooth boundary $\partial\Omega$; its topological closure is denoted by $\overline{\Omega}$
Ω^c	the complement of $\overline{\Omega}$ in $\mathbb{R}^N$, i.e. $\mathbb{R}^N \backslash \overline{\Omega}$, which is an exterior open domain
$\overline{\Omega^c}$	the closure of Ω^c in $\mathbb{R}^N$
PF	pseudofunction
P.V.	the principal value (of an integral with Cauchy kernel)
$\mathbb{R}$	the set of all real numbers

$\mathbb{R}_+$	the set of all positive real numbers
$\mathbb{R}_-$	the set of all negative real numbers
$\mathbb{R}^N$	the N-dimensional real Euclidean space
$\mathbb{R}^N_+$	the set consisting of all $x = (x_1, \dots, x_N) \in \mathbb{R}^N$ where $x_N > 0$
$\mathbb{R}^N_-$	the set consisting of all $x = (x_1, \dots, x_N) \in \mathbb{R}^N$ where $x_N < 0$
Re	the real part of a complex number
$\mathscr{R}(\mathscr{L})$	the range of a linear operator $\mathscr{L}$
$S_1 \setminus S_2$	the set consisting of all the elements belonging to set S_1 but not to set S_2
$\mathscr{S}$	the Schwartz space of all rapidly decreasing C^∞ functions (on $\mathbb{R}^N$)
$\mathscr{S}'$	the Schwartz space of distributions
sgn	signature, the sign of a real number
span	the linear span (of a set of vectors)
sup	the supremum (of a given function or functional)
supp ϕ	the support of a function or a distribution ϕ
M^{Tr}	the transpose of matrix M
$\mathrm{tr}(M)$	the trace of a matrix M
χ_A	the characteristic function of a set A
$\mathbb{Z}$	the set of all integers
$\mathbb{Z}^+$	the set of all nonnegative integers

Contents

Chapter 1

Introduction

1.1 How boundary element methods work

To explain what boundary element methods (BEM) are and how they work, let us consider a motivating example. Let Ω be an open domain in $\mathbb{R}^N$ with smooth boundary $\partial\Omega$. Consider the boundary value problem (BVP)

$$\left.\begin{aligned} \Delta w(x) &= 0 \quad \text{on } \Omega, \\ w(x) &= g(x) \ \text{ on } \partial\Omega, \end{aligned}\right\} \tag{1.1}$$

where g is called the Dirichlet data. Besides BEM, many analytical and numerical methods exist for solving this problem, such as

(i) separation of variables in curvilinear coordinates;
(ii) the use of integral transforms;
(iii) finite-difference methods (FDM);
(iv) finite-element methods (FEM);
(v) spectral methods;
(vi) the use of the Green's functions;
(vii) complex variables.

Each method has its own advantages and disadvantages, and these methods are all related to one another in some way.

We assume the reader is sufficiently familiar with methods (i) and (ii). For methods (iii) (FDM), the Laplacian Δ in (1.1) is replaced, e.g., by a central finite-difference operator such as

$$\begin{aligned} \frac{\partial^2 w}{\partial x_1^2} &\sim \frac{1}{h^2}\left(w_{n+1,m} - 2w_{n,m} + w_{n-1,m}\right), \qquad w_{n,m} \equiv w(nh, mk) \\ \frac{\partial^2 w}{\partial x_2^2} &\sim \frac{1}{k^2}\left(w_{n,m+1} - 2w_{n,m} + w_{n,m-1}\right), \end{aligned}$$

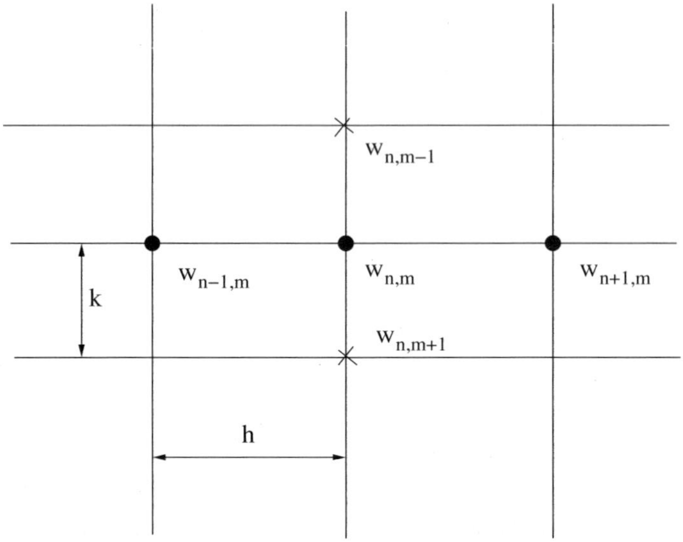

Figure 1.1: A finite-difference grid.

see Fig. 1.1. This method works best for *rectangular geometries*.

Methods (iv) (FEM) are based upon the (Galerkin) *variational principle*, where a family of finite-dimensional approximation spaces V_h is used to approximate the solution w of (1.1) as follows. First, find a (sufficiently smooth) function F on $\overline{\Omega}$ such that on $\partial\Omega$, F is equal to g, i.e.,

$$F(x) = g(x) \quad \text{for } x \in \partial\Omega.$$

Write

$$v(x) = w(x) - F(x).$$

Then

$$\Delta v(x) = \Delta w(x) - \Delta F(x) = -\Delta F(x) \qquad \text{on } \Omega,$$

$$v(x) = 0 \qquad \text{on } \partial\Omega.$$

Let $\{\phi_i^{(h)} \mid 1 \leqslant i \leqslant m(h)\}$ be a basis of V_h, where each function $\phi_i^{(h)}$ satisfies the zero boundary condition. We approximate $v(x)$ by

$$v(x) = \sum_{i=1}^{m(h)} \alpha_i \phi_i^{(h)}(x), \qquad x \in \Omega,$$

and use

$$\int_{\Omega} [\Delta v(x)]\, \phi_j^{(h)}(x)\, dx = -\int_{\Omega} \nabla v(x) \cdot \nabla \phi_i^{(h)}(x)\, dx$$
$$= -\int_{\Omega} \Delta F(x) \phi_j^{(h)}(x)\, dx \quad \forall 1 \leqslant j \leqslant m(h)$$

to determine the coefficients α_i, with $1 \leqslant i \leqslant m(h)$. Usually, the basis $\{\phi_i^{(h)}(x) \mid 1 \leqslant i \leqslant m(h)\}$ is chosen to be locally supported piecewise-continuous polynomials so that the corresponding linear matrix system above has the pleasant properties of being banded and sparse. Such basis functions will also be used in BEM computations. A survey of FEM is provided in Chapter 5.

Spectral methods (v) use *smooth polynomials or exponential functions*, either in the form of Galerkin approximation as FEM or collocation, to approximate the solutions of PDE (see [82]).

The Green's function method (vi) has the closest relationship to the development of boundary integral equations (BIE) or BEM. So let us briefly describe it. Assume that we can find a function $E(x, \xi)$ depending on x and ξ such that

$$\Delta_{\xi} E(x, \xi) = -\delta(x - \xi), \quad \forall x, \xi \in \mathbb{R}^N. \tag{1.2}$$

Here $\Delta_{\xi} \equiv \partial^2/\partial \xi_1^2 + \cdots + \partial^2/\partial \xi_N^2$ denotes the Laplacian with respect to the ξ variables, and δ is the Dirac delta function concentrated at ξ (or at x, depending on how one looks at it.) Later, we will show that such an $E(x, \xi)$ exists and

$$E(x, \xi) = \begin{cases} -(2\pi)^{-1} \ln|x - \xi|, & N = 2, \\ \dfrac{\Gamma\left(\frac{1}{2}N\right)}{2\pi^{N/2}(N-2)|x - \xi|^{N-2}}, & N > 2, \end{cases} \tag{1.3}$$

where Γ is the gamma function. E is the *fundamental solution* for the Laplacian, and is defined everywhere except at $x = \xi$, where it is singular. This E is unique up to a harmonic function of ξ. We know that the Dirac delta function is not a pointwise-defined function in the classical sense, but is rather a *generalized function* or a *distribution*, and as E satisfies (1.2) *in the sense of distributions*, E itself should be regarded as a generalized function.

Let $x \in \Omega$. From the defining property (see [76, p. 1]) of the delta function, we have

$$w(x) = \int_{\Omega} \delta(x - \xi) w(\xi)\, d\xi. \tag{1.4}$$

Using (1.1), (1.2) and (1.4), we obtain

$$w(x) = \int_{\Omega} \{[\Delta w(\xi)] E(x, \xi) - [\Delta_{\xi} E(x, \xi)]\, w(\xi)\}\, d\xi. \tag{1.5}$$

Integration by parts (see Appendix A.1) yields

$$w(x) = \int_{\partial\Omega} \left[\frac{\partial w(\xi)}{\partial n} E(x,\xi) - \frac{\partial E(x,\xi)}{\partial n_\xi} w(\xi) \right] d\sigma_\xi, \quad \forall x \in \Omega. \tag{1.6}$$

The integrand on the right-hand side (RHS) of (1.6) does not contain singularities when $x \in \Omega$; therefore the step from (1.5) to (1.6) causes no problem. Equation (1.6) shows that the value of w at an interior point of Ω can be determined in terms of the boundary data w, and $\partial w/\partial n$ on the boundary $\partial\Omega$.

From equation (1.6), to solve for w on the domain Ω we need to know *both* the Dirichlet data w and the Neumann data $\partial w/\partial n$. Only $w = g$ on $\partial\Omega$ is given in (1.1). Therefore something must be done, either

(a) to eliminate the presence or dependence of $\partial w/\partial n$ in (1.6), or
(b) to find $\partial w/\partial n$ so that we can use it to obtain w from (1.6).

The method of Green's functions takes approach (a). Assume we can find a function $G(x,\xi)$ of two variables x and ξ (called the Green's function) such that for each $x \in \Omega$

$$\left.\begin{aligned} &\Delta_\xi G(x,\xi) = -\delta(x-\xi), && \xi \in \Omega, \\ &G(x,\xi) = 0, && \xi \in \partial\Omega. \end{aligned}\right\} \tag{1.7}$$

Then repeating the similar procedures (1.4)–(1.6) for w with G replacing E yields

$$\begin{aligned} w(x) &= \int_{\partial\Omega} \left[\frac{\partial w(\xi)}{\partial n} G(x,\xi) - \frac{\partial G(x,\xi)}{\partial n_\xi} w(\xi) \right] d\sigma_\xi \\ &= -\int_{\partial\Omega} \frac{\partial G(x,\xi)}{\partial n_\xi} g(\xi)\, d\sigma_\xi. \end{aligned} \tag{1.8}$$

If a $G(x,\xi)$ is available that satisfies (1.7) then $\partial G/\partial n_\xi$ is known. Thus w is solved by (1.8).

When Ω has special geometries such as that of a half-space or a sphere, a "reflection principle" can be applied and G can be readily obtained ([171, pp. 169–172]): we have the following:

(1) *The half space case*

$$G(x,\xi) = \frac{1}{4\pi} \left(\frac{1}{|x-\xi|} - \frac{1}{|\widetilde{x}-\xi|} \right) \tag{1.9}$$

where

$$x = (x_1, x_2, x_3), \quad \xi = (\xi_1, \xi_2, \xi_3), \quad \widetilde{x} = (-x_1, x_2, x_3)$$

$$\Omega = \left\{ x = (x_1, x_2, x_3) \in \mathbb{R}^3 | x_1 > 0 \right\}.$$

By (1.8) and (1.9), the solution $w(x)$ to (1.1) is given by

$$w(x) = \frac{x_1}{2\pi} \int_{-\infty}^{\infty} \int_{-\infty}^{\infty} \frac{g(\xi_2, \xi_3)}{\left[x_1^2 + (x_2 - \xi_2)^2 + (x_3 - \xi_3)^2\right]^{3/2}} d\xi_2 d\xi_3, \quad x \in \Omega. \tag{1.10}$$

(2) *The spherical case*

$$G(x,\xi) = \frac{1}{4\pi} \left[\frac{1}{|x-\xi|} - \frac{a/|x|}{|\xi - (a/|x|)^2 x|} \right] \tag{1.11}$$

where

$$\begin{aligned} x &= (x_1, x_2, x_3), \quad \xi = (\xi_1, \xi_2, \xi_3) \\ \Omega &= \left\{ x = (x_1, x_2, x_3) \in \mathbb{R}^3 \mid |x| < a, a > 0 \right\} \\ &= \text{the sphere with radius } a \text{ centered at the origin.} \end{aligned}$$

By (1.8), the solution $w(x)$ to (1.1) is given by

$$w(x) = w(r,\theta,\phi) = \frac{a(a^2 - |x|^2)}{4\pi}$$

$$\int_0^{2\pi} \int_0^{\pi} \frac{\widetilde{g}(\eta, \psi) \sin \eta}{\{a^2 + |x|^2 - 2a|x|[\cos\theta \cos\eta + \sin\theta \sin\eta \cos(\phi - \psi)]\}^{1/2}} d\eta \, d\psi, \tag{1.12}$$

where (r,θ,ϕ) and (ρ,η,ψ) are the spherical coordinates of x and ξ respectively, $r = |x| < a$, $\rho = a$, $\widetilde{g}(\eta,\psi) \equiv g(a,\eta,\psi)$, $0 \leqslant \theta, \eta \leqslant \pi, 0 \leqslant \phi, \psi \leqslant 2\pi$.

Exercise 1.1. Show that the Green's functions (1.9) and (1.11) satisfy (1.7). Take their normal derivatives and verify (1.10) and (1.12). □

Exercise 1.2. Derive the Green's function for the Neumann BVP on $\Omega = \{x = (x_1, x_2) \in \mathbb{R}^2 \mid x_1 > 0\}$, i.e., the open right half plane, and use it to derive the representation formula for w.

Show that for the Neumann BVP on $\Omega = \{x = (x_1, x_2) \in \mathbb{R}^2 \mid |x| < a, a > 0\}$, i.e., the open disk with radius a centered at the origin, there does not exist a Green's function. □

It is interesting to note that, although formulas (1.10) and (1.12) are perfectly well defined for interior points $x \in \Omega$, they are not so if x is a boundary point on $\partial\Omega$. First, the integrals in (1.10) and (1.12) may be divergent. Secondly, the factors in front of the integrals vanish when $x \in \partial\Omega$, seeming to give $w(x) = 0$ on $\partial\Omega$, which is obviously incorrect. (As a matter of fact, the boundary data $w = g$ can be recovered by the application of (1.10) or (1.12) in each case only through a limiting process.)

The Green's function method requires knowledge of $G(x,\xi)$ from solving the BVP (1.7), which in general is more difficult than (1.1) unless Ω has very special geometries such as a half space or a sphere as discussed above. We thus get stuck.

Boundary element methods take the second approach (b). Let us return to (1.6) and examine each term carefully. For simplicity, assume that the space dimension N is two. Let Ω be bounded with smooth boundary $\partial\Omega$. Let $w(x)$ be a harmonic function whose Dirichlet and Neumann boundary data w and $\partial w/\partial n$ are C^0 continuous on $\partial\Omega$. Then, from (1.6), the integral of the first product

$$\int_{\partial\Omega} \frac{\partial w(\xi)}{\partial n} E(x,\xi)\, d\sigma_\xi$$

exists even for $x \in \partial\Omega$, because $\ln|x-\xi|$ is absolutely integrable on $\partial\Omega$ with respect to the measure $d\sigma_\xi$ for any given x. For the second product in (1.6), we have

$$\frac{\partial E(x,\xi)}{\partial n_\xi} = -\frac{1}{2\pi}\frac{(\xi - x)\cdot n_\xi}{|x-\xi|^2} = \mathcal{O}(|x-\xi|^{-1}), \qquad x \in \Omega, \tag{1.13}$$

where it appears as though, when x tends to a point on $\partial\Omega$ (from Ω), the integral

$$-\int_{\partial\Omega} \frac{\partial E(x,\xi)}{\partial n_\xi} w(\xi)\, d\sigma_\xi \tag{1.14}$$

may be divergent. It turns out, somewhat surprisingly, that this is not so, i.e., by letting

$$v(x) = -\int_{\partial\Omega} \frac{\partial E(x,\xi)}{\partial n_\xi} w(\xi)\, d\sigma_\xi, \quad x \in \Omega, \tag{1.15}$$

it holds that

$$\lim_{x \to x_0 \in \partial\Omega} v(x) \tag{1.16}$$

exists (though it is worth noting that *this integral does not converge absolutely*). In fact, it will be shown (see Lemma 6.3) that v can be extended to a continuous function on $\overline{\Omega}$. For $x \in \partial\Omega$, the kernel $\partial E(x,\xi)/\partial n_\xi$ does not become singular:

$$\frac{\partial E(x,\xi)}{\partial n_\xi} = \mathcal{O}(1), \quad x,\, \xi \in \partial\Omega, \tag{1.17}$$

which is completely distinctive from (1.13), cf. Lemma 6.1 in Chapter 6. Therefore the boundary integral

$$V(x) = -\int_{\partial\Omega} \frac{\partial E(x,\xi)}{\partial n_\xi} w(\xi)\, d\sigma_\xi, \quad x \in \partial\Omega, \tag{1.18}$$

is well defined. (Actually, V is a function in $C^\infty(\partial\Omega)$ if $\partial\Omega$ is C^∞!) From (1.16), we wonder whether

$$\lim_{x \to x_0 \in \partial\Omega} v(x) = V(x_0)$$

holds. It turns out that the answer is no. What we actually have is

$$\lim_{x \to x_0 \in \partial\Omega} v(x) = \frac{1}{2} w(x_0) + V(x_0), \tag{1.19}$$

i.e., a *jump* of magnitude $\frac{1}{2}w(x_0)$ takes place in the limit. From (1.6) and (1.19), we obtain the BIE

$$w(x_0) = \frac{1}{2}w(x_0) + \int_{\partial\Omega} \left[E(x_0,\xi)\frac{\partial w(\xi)}{\partial n} - \frac{\partial E(x_0,\xi)}{\partial n_\xi} w(\xi) \right] d\sigma_\xi$$
$$\forall x_0 \in \partial\Omega. \qquad (1.20)$$

We can also arrive at the above equation more intuitively as follows. From (1.5) and (1.6) we have, for $x \in \mathbb{R}^N \setminus \partial\Omega$,

$$\int_{\partial\Omega} \left[\frac{\partial w(\xi)}{\partial n} E(x,\xi) - \frac{\partial E(x,\xi)}{\partial n_\xi} w(\xi) \right] d\sigma_\xi = \begin{cases} w(x), & x \in \Omega, \\ 0, & x \in \Omega^c. \end{cases}$$

For $x \in \Omega^c$, the integral on the LHS is zero because the singularity $\xi = x$ is not enclosed in $\partial\Omega$. What happens if $x \in \partial\Omega$? It seems reasonable that we will get the "average" of the RHS, i.e.,

$$\int_{\partial\Omega} \left[\frac{\partial w(\xi)}{\partial n} E(x_0,\xi) - \frac{\partial E(x_0,\xi)}{\partial n_\xi} w(\xi) \right] d\sigma_\xi = \frac{1}{2} w(x_0), \qquad x_0 \in \partial\Omega, \qquad (1.21)$$

which is just the equation we have obtained at (1.21). Replacing the dummy variable x_0 by x, simplifying and rearranging, we obtain

$$\int_{\partial\Omega} E(x,\xi)\frac{\partial w(\xi)}{\partial n} d\sigma_\xi = \frac{1}{2}w(x) + \int_{\partial\Omega} \frac{\partial E(x,\xi)}{\partial n_\xi} w(\xi)\, d\sigma_\xi \qquad \forall x \in \partial\Omega. \qquad (1.22)$$

This is a boundary integral equation (*of the first kind*) relating the unknown Neumann data $\partial w/\partial n$ to the given Dirichlet data $w = g$ (cf. (1.1)). On the other hand, if at the outset the Neumann data $\partial w/\partial n$ is prescribed in the second equation of (1.1) instead, then we obtain

$$\frac{1}{2}w(x) + \int_{\partial\Omega} \frac{\partial E(x,\xi)}{\partial n_\xi} w(\xi)\, d\sigma_\xi = \int_{\partial\Omega} E(x,\xi)\frac{\partial w(\xi)}{\partial n} d\sigma_\xi \qquad \forall x \in \partial\Omega, \qquad (1.23)$$

which is a BIE *of the second kind* for the unknown Dirichlet data w. In principle, if we know the Dirichlet data w, we can use (1.22) to solve the Neumann data $\partial w/\partial n$, and vice versa. Once w and $\partial w/\partial n$ on the boundary are both known, we can substitute them into (1.6) to obtain $w(x)$ at any point $x \in \Omega$. So the BVP (1.1) is solved. The BIE (1.22) generally does not admit closed form solutions. Therefore numerical approximations must be performed. A common procedure is to discretize the relevant function spaces for w and $\partial w/\partial n$ on $\partial\Omega$ and the equation (1.22) by finite elements or point collocation. This gives the approach the name of the boundary element method.

Those familiar with complex analysis (method (vii)) may note some analogy between the integral in (1.15) and the Cauchy integral

$$F(z) = \frac{1}{2\pi i}\oint_{\partial\Omega} \frac{f(\zeta)}{\zeta - z} d\zeta, \qquad z \in \mathbb{C} \setminus \partial\Omega, \qquad (1.24)$$

where f is a Hölder-continuous function of a certain exponent $\alpha > 0$, i.e., $f \in C^{0,\alpha}(\partial\Omega)$. Then $F(z)$ is analytic on both Ω and $\Omega^c = \mathbb{C} \setminus \overline{\Omega}$. The kernels of both integrals (1.15) and (1.24) (for $z \in \Omega$) have the same degree of singularity. Indeed, the analogy between these two integrals is not accidental, because we know that every harmonic function is the real part of an analytic function. The complex contour integral (1.24) (integrated in a counterlockwise sense) fails to be defined if $z \in \partial\Omega$. However, in complex analysis, it is possible to define the *Cauchy principal value* for a divergent integral when $z \in \partial\Omega$ in (1.24). The *principal value* (P.V.) of (1.24) is defined as

$$(\text{P.V.})\frac{1}{2\pi i}\oint_{\partial\Omega}\frac{f(\zeta)}{\zeta - z}\,d\zeta = \lim_{\varepsilon\downarrow 0}\frac{1}{2\pi i}\int_{\partial\Omega\setminus B_\varepsilon(z)}\frac{f(\zeta)}{\zeta - z}\,d\zeta \quad z \in \partial\Omega \tag{1.25}$$

where

$$B_\varepsilon(z) = \{\zeta \in \mathbb{C} \mid |\zeta - z| < \varepsilon\},$$
$$\partial\Omega \setminus B_\varepsilon(z) = \{\zeta \in \partial\Omega \mid |\zeta - z| \geqslant \varepsilon\},$$

i.e., we truncate a small portion from the contour $\partial\Omega$ evenly around the singular point z and shrink it to zero, as shown in Fig. 1.2.

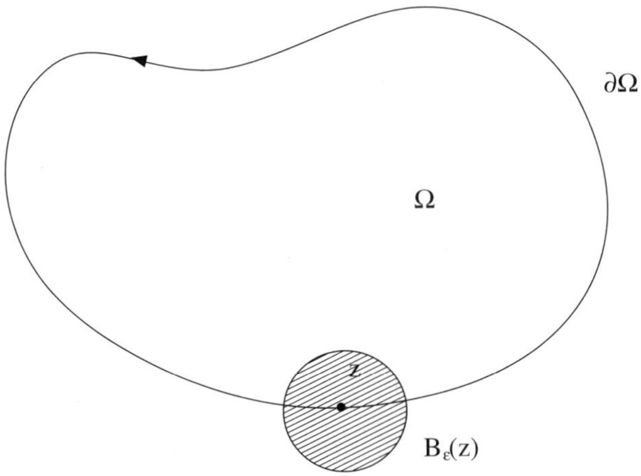

Figure 1.2: Exclusion of disk $B_\varepsilon(z)$ from the contour $\partial\Omega$ in evaluating the Cauchy principal value.

Then it is not difficult to show that the principal value (1.25) is well defined for any $z \in \partial\Omega$. Call it $\Phi(z)$. One can even show that $\Phi \in C^{0,\alpha}(\partial\Omega)$ if $f \in C^{0,\alpha}(\partial\Omega)$ and $0 < \alpha < 1$.

The well-known Plemelj formulas [137, p. 42] state

$$\lim_{\substack{z\in\Omega \\ z\to z_0\in\partial\Omega}} F(z) = \frac{1}{2}f(z_0) + (\text{P.V.})\frac{1}{2\pi i}\oint_{\partial\Omega}\frac{f(\zeta)}{\zeta - z_0}\,d\zeta, \tag{1.26}$$

$$\lim_{\substack{z\to\Omega^c \\ z\to z_0\in\partial\Omega}} F(z) = -\frac{1}{2}f(z_0) + (\text{P.V.})\frac{1}{2\pi i}\oint_{\partial\Omega}\frac{f(\zeta)}{\zeta - z_0}\,d\zeta, \tag{1.27}$$

i.e., approaching $\partial\Omega$ from inside and outside, the analytic function $F(z)$ suffers a jump, respectively, of $f(z_0)/2$ and $-f(z_0)/2$, if the Cauchy integral is now interpreted as a Cauchy principal value. Again, (1.26) and (1.27) are BIE of the second kind for the unknown function f.

Exercise 1.3. Let Ω be the open disk with radius $a > 0$ centered at the origin, and let

$$f(z) = z + \gamma, \quad \text{where } \gamma \in \mathbb{C} \text{ is a constant.}$$

Compute the principal value (1.25) and verify (1.26) and (1.27). □

The first Plemelj formula (1.26) looks similar to (1.19), because the RHS of both equations have a jump term of the same magnitude. But this is where the similarity ends, since we know that the kernel $1/(\zeta - z_0)$ in (1.26) causes a *divergent integral*, while (1.18) is a *convergent integral* from (1.17). In the subsequent chapters, we will be faced with many different types of integrals whose sense of existence must be carefully interpreted.
The preceding discussion explains in a relatively simple manner how the boundary element method works. There are many fundamental questions we can think of that remain to be investigated:

(Q1) Is the BIE (1.22) always solvable? For example, in (1.22) given the Dirichlet data w, can we uniquely obtain the Neumann data $\partial w/\partial n$, or vice versa?

(Q2) Are there ways to formulate a simpler BIE for the Dirichlet problem (1.1)? This would save a large amount of coding and computing work.

(Q3) Can we determine the smoothness of the Neumann data $\partial w/\partial n$ from that of the Dirichlet data w, or vice versa, from the BIE (1.22)?

(Q4) How do we formulate the BIE for the Dirichlet problem when Ω is an exterior domain (i.e., $\mathbb{R}^N \setminus \Omega$ is a bounded domain)? How do we solve them?

(Q5) What are the BIE formulations for many other partial differential equations in physics and engineering?

(Q6) In the BIE (1.26), the kernel contains a singularity $\mathcal{O}(|\zeta - z_0|^{-1})$. But when treating the BIE for other PDE, we may encounter kernels, called *hypersingular kernels*, whose singularities are even worse. How do we handle them?

(Q7) A basic assumption made during the derivation of the BIE (1.22) or (1.23) was that the boundary $\partial\Omega$ was smooth. What adjustment do we need to make if $\partial\Omega$ has corners?

(Q8) What are effective ways to discretize the BIE and to obtain numerical solutions?

(Q9) How can we utilize the BIE and BEM to solve nonlinear BVPs?

We will try to answer some of these questions in this book.

1.2 An example of implementation

Let us look at some practical aspects of solving the BIE (1.22) and (1.23). Consider, respectively, the Dirichlet and Neumann BVP on a two-dimensional bounded smooth open domain Ω:

$$\text{(DBVP)} \qquad \begin{cases} \Delta w(x) = 0 & \text{on } \Omega, \\ w(x) = g_1(x), & x \in \partial\Omega, \end{cases}$$

$$\text{(NBVP)} \qquad \begin{cases} \Delta w(x) = 0 & \text{on } \Omega, \\ \dfrac{\partial w(x)}{\partial n} = g_2(x), & x \in \partial\Omega. \end{cases}$$

We discuss them separately below.

(I) *Neumann Case*: To solve (NBVP), we use the (BIE) (1.23), where we solve for the *unknown* Dirichlet data $w(x) = g_1(x)$ on $\partial\Omega$ from

$$\frac{1}{2}g_1(x) + \int_{\partial\Omega} \frac{\partial E(x,\xi)}{\partial n_\xi} g_1(\xi)\, d\sigma = \int_{\partial\Omega} E(x,\xi) g_2(\xi)\, d\sigma_\xi, \qquad x \in \partial\Omega. \tag{1.28}$$

First, note that for $(NBVP)$ to be properly posed, the given data g_2 must satisfy the compatibility condition

$$0 = \int_\Omega \Delta w(x)\, dx = \int_{\partial\Omega} \frac{\partial w(x)}{\partial n}\, d\sigma = \int_{\partial\Omega} g_2(x)\, d\sigma. \tag{1.29}$$

Now we discretize (1.28) as follows. Let $\partial\Omega$ be subdivided into m piecewise arcs Γ_j, $1 \leqslant j \leqslant m$, with x_j as the midpoint of Γ_j. On each Γ_j, let $g_1(x)$ be approximated by a constant:

$$\widetilde{g_1}(x) = w_j, \quad x \in \Gamma_j, \quad j = 1, 2, \ldots, m. \tag{1.30}$$

In other words, $\widetilde{g_1}$ is a piecewise-constant approximation to g_1 on $\partial\Omega$. Then (1.28) gives the $m \times m$ linear system

$$\frac{1}{2}w_j + \sum_{i=1}^{m} w_i \int_{\Gamma_i} \frac{\partial E(x_j,\xi)}{\partial n_\xi}\, d\sigma_\xi = \int_{\partial\Omega} E(x_j,\xi) g_2(\xi)\, d\sigma_\xi, \quad 1 \leqslant j \leqslant m. \tag{1.31}$$

With $\zeta = \xi = (\xi_1, \xi_2) = \xi_1 + i\xi_2 \in \mathbb{C}$, $z = x = (x_1, x_2) = x_1 + ix_2 \in \mathbb{C}$,

$$\frac{1}{2\pi}\ln(z-\zeta) = \frac{1}{2\pi}[\ln|x-\xi| + i\arg(x-\xi)] \tag{1.32}$$

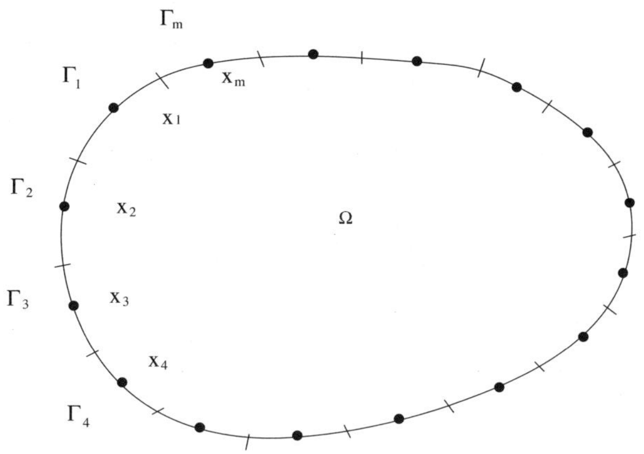

Figure 1.3: Subdividing $\partial\Omega$ into piecewise arcs Γ_j, $\partial\Omega = \bigcup_{j=1}^{m} \Gamma_j$, where x_j is the midpoint of Γ_j.

is a holomorphic function. By the Cauchy–Riemann equations, if $u + iv$ is holomorphic then

$$\frac{\partial u}{\partial x} = \frac{\partial v}{\partial y}, \qquad \frac{\partial u}{\partial y} = -\frac{\partial v}{\partial x}. \tag{1.33}$$

Since

$$\frac{\partial}{\partial n_\xi} = n_1(\xi)\frac{\partial}{\partial \xi_1} + n_2(\xi)\frac{\partial}{\partial \xi_2}, \tag{1.34}$$

$$\frac{\partial}{\partial \sigma_\xi} = -n_2(\xi)\frac{\partial}{\partial \xi_1} + n_1(\xi)\frac{\partial}{\partial \xi_2}, \tag{1.35}$$

(the counterclockwise tangential derivative), (1.36)

from (1.32) we have

$$-\frac{1}{2\pi}\frac{\partial}{\partial n_\xi}\ln|x-\xi| = -\frac{1}{2\pi}\frac{\partial}{\partial \sigma_\xi}\arg(x-\xi). \tag{1.37}$$

As $\partial E(x,\xi)/\partial n_\xi$ is equal to the LHS of (1.37) by (1.3), we get

$$\int_{\Gamma_i} \frac{\partial E(x_j,\xi)}{\partial n_\xi}\, d\sigma_\xi = -\frac{1}{2\pi}\int_{\Gamma_i} \frac{\partial \arg(x_j-\xi)}{\partial \sigma_\xi}\, d\sigma_\xi = -\frac{1}{2\pi}\theta_{ji} \tag{1.38}$$

where for a convex domain Ω, θ_{ji} is the angle subtended by the arc Γ_i from x_j, see Fig. 1.4. Substituting (1.38) into the LHS of (1.31), we obtain

$$\frac{1}{2}w_j - \frac{1}{2\pi}\sum_{i=1}^{m}\theta_{ji}w_i = \eta_j, \quad j = 1, 2, \ldots, m, \tag{1.39}$$

where

$$\eta_j \equiv \int_{\partial\Omega} E(x_j,\xi)g_2(\xi)\,d\sigma_\xi \tag{1.40}$$

can be evaluated by quadrature. Denoting

$$\vec{W} = \begin{bmatrix} w_1 \\ w_2 \\ \vdots \\ w_m \end{bmatrix}, \quad K \equiv \frac{1}{\pi}\begin{bmatrix} \theta_{11} & \theta_{12} & \cdots & \theta_{1m} \\ \theta_{21} & \theta_{22} & \cdots & \theta_{2m} \\ \vdots & \vdots & \ddots & \vdots \\ \theta_{m1} & \theta_{m2} & \cdots & \theta_{mm} \end{bmatrix}, \quad \vec{Z} = \begin{bmatrix} 2\eta_1 \\ 2\eta_2 \\ \vdots \\ 2\eta_m \end{bmatrix},$$

we can write (1.39) as

$$(I-K)\vec{W} = \vec{Z}. \tag{1.41}$$

This linear system is *noninvertible*, since it can be easily verified that

$$\sum_{j=1}^{m} \theta_{ij} = \pi, \quad i = 1,2,\ldots,m, \tag{1.42}$$

implying

$$\vec{W}_0 \equiv [1,1,\ldots,1]^{\mathrm{Tr}} \in \mathcal{N}(I-K). \tag{1.43}$$

By the Fredholm alternative (see Chapter 4), (1.41) is solvable if and only if

$$\vec{Z} \perp \mathcal{N}(I-K)^*. \tag{1.44}$$

This compatibility condition is related to the compatibility condition (1.29) of the original problem (NBVP) thus its occurrence is rather natural. Even if (1.44) is satisfied, the solution $\vec{W}$ of (1.41) still lacks uniqueness, as $\vec{W} + \alpha\vec{W}_0$ is a solution for any $\alpha \in \mathbb{R}$, whenever $\vec{W}$ itself is. This is the case for the original (NBVP), since $w + \alpha$ is a solution whenever w is, for any $\alpha \in \mathbb{R}$. Therefore, an *accessory condition must be prescribed* to fix this one degree of freedom α.

For instance, we can assume that a pointwise value

$$w(x_1) = c \tag{1.45}$$

is prescribed at $x_1 \in \Gamma_1$. We can let $w_1 = c$ in (1.39) and drop the first equation in (1.39). As the rank of the matrix $I-K$ is $m-1$, it is expected that the submatrix of $I-K$ obtained by deleting the first column and row from $I-K$ will have rank $m-1$, thus the linear system

$$\frac{1}{2}w_i - \frac{1}{2\pi}\sum_{j=2}^{m}\theta_{ij}w_j = \eta_i + \frac{1}{2\pi}\theta_{i1}c, \quad i = 2,\ldots,m, \tag{1.46}$$

is expected to be uniquely solvable.

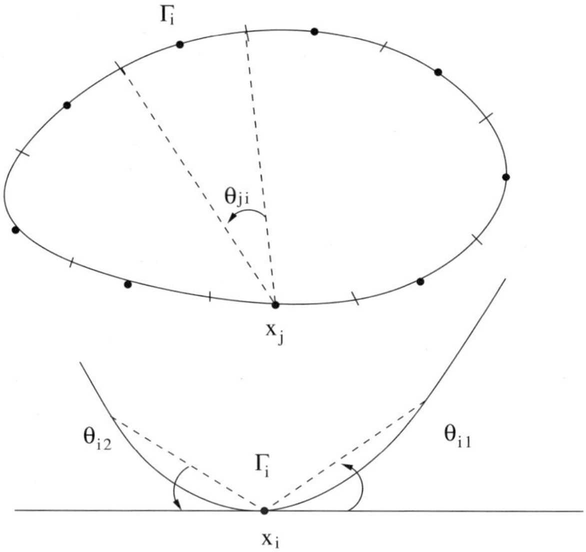

Figure 1.4: The angle θ_{ji} subtended by the arc Γ_i from x_j. Note that $\theta_{ii} = \theta_{i1} + \theta_{i2}$ and $\sum_{i=1}^{m} \theta_{ji} = \pi$.

For example, when Ω is a disk with circumference $\partial\Omega$ divided uniformly into m equal arcs, $\theta_{ij} = \pi/m$, $1 \leqslant i, j \leqslant m$, and (1.39) gives

$$\frac{1}{2}w_i - \frac{1}{2\pi}\frac{\pi}{m}\sum_{j=1}^{m} w_j = \eta_i, \quad 1 \leqslant i \leqslant m. \tag{1.47}$$

Now $w_1 = c$ is given, so

$$\sum_{j=1}^{m} w_j = m(w_1 - 2\eta_1) \equiv A$$

is known. From (1.47), we get

$$w_i = \frac{1}{m}A + 2\eta_i, \quad i = 2, \ldots, m.$$

So the piecewise-constant approximation (1.30) to g_1 has been computed. The approximate solution for (NBVP) is then

$$w(x) = \int_{\partial\Omega} \left[E(x,\xi)g_2(\xi) - \frac{\partial E(x,\xi)}{\partial n_\xi}\widetilde{g}_1(\xi) \right] d\sigma_\xi, \quad x \in \Omega.$$

Readers may now use the above procedures to write a simple computer program to see how the BEM work.

(II) *Dirichlet Case:* To solve (DBVP), we wish to use BIE (1.22) to obtain the unknown Neumann data $\partial w(x)/\partial n = g_2(x)$ on $\partial\Omega$:

$$\int_{\partial\Omega} E(x,\xi)g_2(\xi)\,d\sigma_\xi = \frac{1}{2}g(x) + \int_{\partial\Omega} \frac{\partial E(x,\xi)}{\partial n_\xi} g_1(\xi)\,d\sigma_\xi, \quad x \in \partial\Omega. \tag{1.48}$$

Although for the original (DBVP), g_2 always exists and is unique (provided that g_1 is sufficiently smooth), unfortunately for (1.48) g_2 is not necessarily uniquely solvable by this boundary integral equation, as can be observed as follows. Let Ω be the unit disk in $\mathbb{R}^2$. Set

$$g_2(\xi) \equiv 1 \text{ on } \partial\Omega.$$

Using polar coordinates

$$(x,\xi) = (re^{i\theta}, \rho e^{i\phi}) = (e^{i\theta}, e^{i\phi}) \text{ on } \partial\Omega,$$

we have

$$\begin{aligned}\int_{\partial\Omega} E(x,\xi)g_2(\xi)\,d\sigma_\xi &= \int_{\partial\Omega} E(x,\xi)\,d\sigma_\xi \\ &= -\frac{1}{2\pi}\int_0^{2\pi} \ln\left|e^{i\theta} - e^{i\phi}\right| d\phi \\ &= -\frac{1}{2\pi}\int_0^{2\pi} \ln\{2[1-\cos(\theta-\phi)]\}^{1/2}\,d\phi \\ &= -\frac{1}{4\pi}\left\{2\pi\ln 2 + \int_0^{2\pi} \ln[1-\cos(\theta-\phi)]d\phi\right\} \\ &\equiv 0 \quad \text{for } \theta \in [0,2\pi].\end{aligned}$$

Therefore we see that the integral operator on the LHS of (1.48) has the constant function 1 in its kernel, implying that the operator is not invertible. Thus any approximate operator is either noninvertible or very ill conditioned.

This situation will be remedied later (Theorem 6.33 in Chapter 6).

1.3 Comparison between BEM and FEM

We conclude this chapter by briefly describing a comparison of advantages and disadvantages between BEM and FEM.

The finite element methods mentioned in § 1.1, based upon the variational principle, can handle almost any linear, nonlinear or time-dependent PDE on domains with a curved boundary. They are unquestionably the most general and powerful methods for solving PDE. Nevertheless, difficulties do arise, which involve the following:

(i) The PDE must be discretized over the entire domain Ω. The amount of computer program coding is rather large and laborious.

(ii) For unbounded domains, artificial constraints must normally be imposed (near ∞) to truncate the domain to a bounded one, resulting in loss of numerical accuracy. In some cases, spurious or incorrect solutions may occur.

(iii) For PDE of higher order ($\geqslant 4$), the most convenient piecewise-linear finite elements must usually be given up. Higher-order elements must be employed. The regularity conforming requirements become increasingly severe. For example, for *plate* or *shell* equations of order 6, 8 or higher, conforming requirements can cause work so tedious as to render the use of finite elements impractical.

(iv) FEM yield the displacement data most accurately. But for force data (such as bending moments and shear) corresponding to higher-order derivatives, which are of primary interest in structural mechanics models, FEM are rather ineffective.

In contrast, the advantages of the BEM are as follows:

(i) The BIE are formulated on the boundary thus there is a significant reduction of dimensionality. Consequently, the discretization is much simpler. The computer programs require much less memory space, and are easier to develop.

(ii) For unbounded domains, the BVP are again reformulated as BIE on the boundary. The resulting solutions naturally satisfy the boundary conditions at ∞ (i.e., the radiation conditions.) Better still, computer programs for interior BVP will also work for the exterior BVP after making rather simple adjustments. (In contrast, FEM for interior and exterior BVP involve completely different discretization and quadrature work.)

(iii) BEM are highly effective in the computation of forces in elasticity and structural problems. Such forces usually appear as boundary conditions of higher-order.

(iv) BEM allow us to compute certain point values and values over a subdomain of special interest, through the application of formulas like (1.5), without having to compute the solution $w(x)$ everywhere on Ω. This is not possible by FEM or FDM. Such an advantage becomes apparent for distributed parameter structural control problems, e.g., where the engineering model has *point sensors* distributed inside the domain. To treat such point sensors would entail a large amount of work by other methods.

(v) For engineering applications, an important motivation for using BEM is its amenability for "singular geometry" such as cracks.

The formulation of BIE from given boundary value problems depends on the availability of an explicit fundamental solution $E(x,\xi)$. For linear PDE with variable coefficients or for nonlinear PDE, such fundamental solutions are usually not available in explicit form. Therefore, BEM are in general not applicable. Research is under way to partially overcome some of these difficulties by coupling the BEM with various iteration methods.

Chapter 2

Some Basic Properties of Sobolev Spaces

In the modern theory of partial differential equations, the smoothness of solutions is commonly described in terms of Sobolev spaces. They have nearly become the basic language of PDE. Detailed studies of Sobolev spaces can be found in [2, 138], for example. In this section, we will briefly introduce what these spaces are and some of their basic properties, to the extent that they are sufficient for the subsequent material in this book. The most important thing here is to understand the content of Theorems 2.1, 2.2 and 2.8 in this chapter. Their proofs, although not difficult to comprehend, do appear a little technical for an inexperienced reader and, thus, can be put off to the second reading.

2.1 Definition and imbedding theorems

Let Ω be a domain in $\mathbb{R}^N$ with boundary $\partial\Omega$. Assume that the domain locally lies on only one side of $\partial\Omega$, and $\partial\Omega$ does not have cusps. Thus the situations in Fig. 2.1 are ruled out.

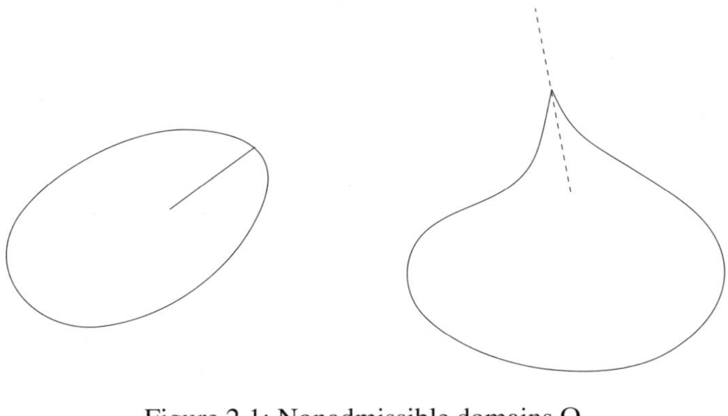

Figure 2.1: Nonadmissible domains Ω.

Let f be a real- or complex-valued function defined on Ω. Let $\alpha = (\alpha_1, \alpha_2, \ldots, \alpha_N) \in$

$(\mathbb{Z}^+)^N$ be a *multi-index* of nonnegative integers. We write

$$D^\alpha f = \left(\frac{\partial}{\partial x_1}\right)^{\alpha_1} \left(\frac{\partial}{\partial x_2}\right)^{\alpha_2} \cdots \left(\frac{\partial}{\partial x_N}\right)^{\alpha_N} f \tag{2.1}$$

to denote a mixed partial derivative of f of order

$$|\alpha| \equiv \alpha_1 + \alpha_2 + \cdots + \alpha_N.$$

In (2.1), we actually want to take the derivatives of f *in the sense of distributions* (see Chapter 3). But for now we may regard the derivative in (2.1) as the classical derivative: the one about which we have learned in advanced calculus. Let $1 \leqslant p \leqslant \infty$, and $k \in \mathbb{Z}^+$. The $\|f\|_{k,p}$ norm of f is defined by

$$\|f\|_{k,p} = \begin{cases} \left(\sum_{|\alpha|\leqslant k} \int_\Omega |D^\alpha f|^p \, dx\right)^{1/p}, & 1 \leqslant p < \infty, \\ \operatorname{ess.sup}_{x\in\Omega} \sum_{|\alpha|\leqslant k} |D^\alpha f(x)|, & p = \infty, \end{cases} \tag{2.2}$$

if the RHS is finite. Note that we take all the mixed partial derivatives of order $\leqslant k$ in (2.2). The integrals in (2.2) are Lebesgue integrals (as are those throughout the rest of the book), but for all practical purposes in this book they may be regarded as Riemann integrals. We define the *Sobolev space*

$$W^{k,p}(\Omega) = \{f \colon \Omega \to \mathbb{C} \text{ or } \mathbb{R} \mid \|f\|_{k,p} < \infty\}. \tag{2.3}$$

$W^{k,p}(\Omega)$ becomes a normed linear space with the usual sum of functions and scalar multiplication. $W^{k,p}(\Omega)$ is actually a *Banach space* (see Appendix A.2) if, as we indicated earlier, the derivatives are taken in the sense of distributions and the integrals are the Lebesgue integrals. (Otherwise, by using the classical dervatives and the Riemann integrals, $W^{k,p}(\Omega)$ is definitely not a Banach space!)

For later use, we also define

$$|f|_{k,p} = \begin{cases} \left(\sum_{|\alpha|=k} \int_\Omega |D^\alpha f|^p \, dx\right)^{1/p}, & 1 \leqslant p < \infty, \\ \operatorname{ess.sup}_{x\in\Omega} \sum_{|\alpha|=k} |D^\alpha f(x)|, & p = \infty. \end{cases} \tag{2.4}$$

Then $|\cdot|_{k,p}$ is a seminorm on $W^{k,p}(\Omega)$.

When $p = 2$, $W^{k,2}(\Omega)$ becomes a *Hilbert space* (see Appendix A.4) with respect to the inner product

$$\langle f, g \rangle_{W^{k,2}} \equiv \sum_{|\alpha|\leqslant k} \int_\Omega D^\alpha f(x) \overline{D^\alpha g(x)} \, dx.$$

We denote

$$H^k(\Omega) \equiv W^{k,2}(\Omega). \tag{2.5}$$

We refer to $H^k(\Omega)$ as a *Sobolev space of order k.*

When Ω is an N-dimensional compact differentiable manifold in $\mathbb{R}^{N+1}$ such as the *surface* of a ball or torus in $\mathbb{R}^3$, we can also define Sobolev spaces on Ω. Such a surface is a two-dimensional manifold, on which every point x has a neighborhood $\mathcal{N}_x$ homeomorphic to the unit open disk D in $\mathbb{R}^2$.

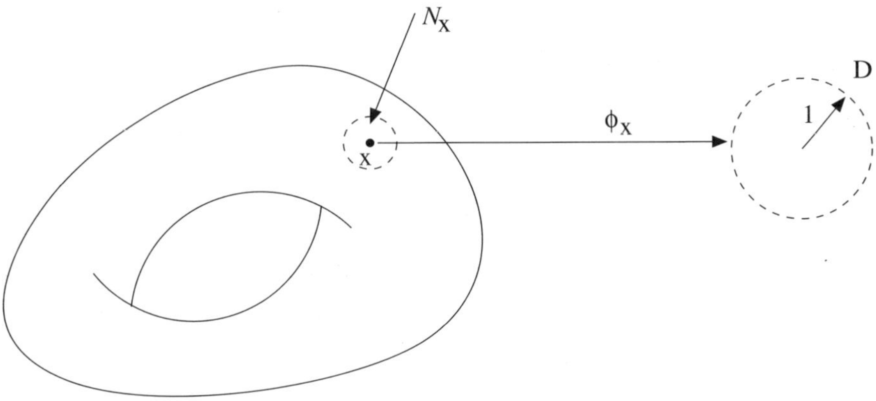

$\phi_x : \mathcal{N}_x \longrightarrow D$ is C^∞, 1-1 and onto

$\phi_x^{-1} : D \longrightarrow \mathcal{N}_x$ is C^∞, 1-1 and onto

Figure 2.2: A two dimensional differentiable manifold.

In general, for each $x \in \Omega$, there exists a neighborhood $\mathcal{N}_x$ of x in Ω and a diffeomorphism ϕ_x from $\mathcal{N}_x$ to B_1, where B_1 is the unit open ball in $\mathbb{R}^N$. Then ϕ_x establishes a local coordinate system on $\mathcal{N}_x$, called a *chart* on $\mathcal{N}_x$. Because $\{\mathcal{N}_x | x \in \Omega\}$ forms an open covering of Ω (i.e., $\Omega = \cup_{x \in \Omega} \mathcal{N}_x$) and because Ω is compact (see Appendix A.5), there exists a finite subcovering $\{\mathcal{N}_{x_i} \,|\, x_i \in \Omega, i = 1, 2, \ldots, M\}$ such that

$$\Omega = \bigcup_{i=1}^{M} \mathcal{N}_{x_i}. \tag{2.6}$$

Corresponding to the finite covering $\{\mathcal{N}_{x_i} \,|\, 1 \leqslant i \leqslant M\}$, we can find a set of smooth functions $\{\theta_i\colon \Omega \to \mathbb{R} \,|\, 1 \leqslant i \leqslant M\}$ such that

$$\begin{aligned} &0 \leqslant \theta_i(x) \leqslant 1 \qquad && \forall x \in \Omega, \\ &\operatorname{supp} \theta_i \subset \mathcal{N}_{x_i}, \\ &\sum_{i=1}^{M} \theta_i(x) = 1 && \forall x \in \Omega. \end{aligned} \tag{2.7}$$

This set of functions $\{\theta_i \mid 1 \leqslant i \leqslant M\}$ is called a *partition of unity* for Ω (see [116, p. 96]). For any function

$$f\colon \Omega \to \mathbb{C},$$

the function

$$F_i \equiv \theta_i(\varphi_{x_i}^{-1}(\cdot)) \cdot f(\varphi_{x_i}^{-1}(\cdot))\colon B_1 \longrightarrow \mathbb{C}$$

is defined on domain B_1. We now define

$$\|f\|_{k,p} = \left(\sum_{i=1}^{M} \sum_{|\alpha| \leqslant k} \int_{B_1} |D^\alpha F_i|^p \, dx \right)^{1/p},$$

if the RHS is finite. It can be shown that different coordinate charts $\{\varphi_{x_i}^{(1)} \mid 1 \leqslant i \leqslant M_1\}$, $\{\varphi_{x_j}^{(2)} \mid 1 \leqslant j \leqslant M_2\}$ and their associated partitions of unity $\{\theta_i^{(1)} \mid 1 \leqslant i \leqslant M_1\}$, $\{\theta_j^{(2)} \mid 1 \leqslant j \leqslant M_2\}$ yield equivalent norms $\|\cdot\|_{k,p}^{(1)}$, $\|\cdot\|_{k,p}^{(2)}$, i.e., there exists $C > 1$ such that

$$C^{-1} \|f\|_{k,p}^{(2)} \leqslant \|f\|_{k,p}^{(1)} \leqslant C \|f\|_{k,p}^{(2)}$$

for all f such that $\|f\|_{k,p}^{(1)} < \infty$.
Therefore we can define

$$W^{k,p}(\Omega) = \{f\colon \Omega \to \mathbb{C} \;\big|\; \|f\|_{k,p} < \infty\}$$

whenever Ω is an N-dimensional compact differentiable manifold in $\mathbb{R}^{N+1}$. Similarly, we write $H^k(\Omega)$ for $W^{k,p}(\Omega)$ when $p = 2$.

Throughout our discussions so far, k is restricted to be a nonnegative integer. k is the largest order of derivatives that functions in $W^{k,p}(\Omega)$ all admit. Can we define $W^{k,p}(\Omega)$ for *non-integral* values of k? Obviously, such a definition would suggest the idea of "fractional-order" derivatives of functions.

To investigate this, we first look at the case when $\Omega = \mathbb{R}^N$, the entire space. For a complex-valued function f defined on $\mathbb{R}^N$, we define the (multivariate) Fourier transform of f by

$$\widehat{f}(\xi) \equiv \mathscr{F}(f)(\xi) = \int_{\mathbb{R}^N} f(x) e^{-2\pi i \langle x, \xi \rangle} \, dx, \quad \xi \in \mathbb{R}^N, \tag{2.8}$$

if such an integral exists in a certain sense. Obviously, $\widehat{f}(\xi)$ at least exists if $f \in L^1(\mathbb{R}^N)$.

The well-known Fourier inversion formula gives

$$\begin{aligned} f(x) = \overline{\mathscr{F}}(\widehat{f})(x) &= \int_{\mathbb{R}^N} \widehat{f}(\xi) e^{2\pi i \langle x,\xi\rangle}\, d\xi \\ &= \overline{\mathscr{F}}(\mathscr{F}(f))(x) = \mathscr{F}^{-1}(\widehat{f})(x), \end{aligned} \tag{2.9}$$

(for continuous functions with compact support, or for functions in $L^2(\mathbb{R}^N)\cap C(\mathbb{R}^N)$, for example), from which we also have

$$f(-x) = \mathscr{F}(\widehat{f})(x) = \widehat{\widehat{f}}(x). \tag{2.10}$$

Remark 2.1. There are several ways to define the Fourier transform. Perhaps the most popular is

$$\widehat{u}(\xi) = (\mathscr{F}u)(\xi) = \int_{\mathbb{R}^N} e^{-i\langle x,\xi\rangle} u(x) \frac{dx}{(2\pi)^{N/2}}. \tag{2.11}$$

Then the inverse transform is again (the adjusted) $\overline{\mathscr{F}}$, and the inversion formula becomes

$$u(x) = \left(\overline{\mathscr{F}}\widehat{u}\right)(x) = \int_{\mathbb{R}^N} e^{i\langle x,\xi\rangle}\widehat{u}(\xi)\frac{d\xi}{(2\pi)^{N/2}}. \qquad \square$$

Assume that f is a well-behaved function at ∞. From (2.8), integration by parts, we have

$$\begin{aligned} \mathscr{F}\left(\frac{\partial}{\partial x_j} f\right) &= \int e^{-2\pi i\langle x,\xi\rangle}\frac{\partial f(x)}{\partial x_j}\, dx \\ &= 2\pi i\,\xi_j \int e^{-2\pi i\langle x,\xi\rangle} f(x)\, dx \\ &= 2\pi i\,\xi_j \mathscr{F}(f). \end{aligned}$$

Repeating this procedure several times, we get

$$\begin{aligned} \mathscr{F}(D^\alpha f) &= \mathscr{F}\left(\frac{\partial^{|\alpha|}}{\partial x_1^{\alpha_1}\partial x_2^{\alpha_2}\ldots\partial x_N^{\alpha_N}} f\right) \\ &= (2\pi i)^{|\alpha|}\xi_1^{\alpha_1}\xi_2^{\alpha_i}\ldots\xi_N^{\alpha_N}\mathscr{F}(f) \\ &\equiv (2\pi i)^{|\alpha|}\xi^\alpha \mathscr{F}(f), \end{aligned} \tag{2.12}$$

where

$$\xi^\alpha \equiv \xi_1^{\alpha_1}\xi_2^{\alpha_2}\ldots\xi_N^{\alpha_N} \quad \text{for the multi-index} \quad \alpha = (\alpha_1,\alpha_2,\ldots,\alpha_N).$$

Similarly, we have

$$\begin{aligned} D^\alpha(\mathscr{F}f) &= \frac{\partial^{|\alpha|}}{\partial\xi_1^{\alpha_1}\partial\xi_2^{\alpha_2}\ldots\partial\xi_N^{\alpha_N}}(\mathscr{F}f) \\ &= (-2\pi i)^{|\alpha|}\mathscr{F}(x_1^{\alpha_1}x_2^{\alpha_2}\ldots x_N^{\alpha_N} f) \\ &= (-2\pi i)^{|\alpha|}\mathscr{F}(x^\alpha f). \end{aligned} \tag{2.13}$$

The Bessel–Parseval formula says that

$$\langle f,g\rangle_{L^2(\mathbb{R}^N)} = \langle \widehat{f},\widehat{g}\rangle_{L^2(\mathbb{R}^N)} \qquad \forall f,g \in L^2(\mathbb{R}^N), \tag{2.14}$$

so

$$\|f\|_{L^2(\mathbb{R}^N)} = \|\widehat{f}\|_{L^2(\mathbb{R}^N)} \quad \forall f \in L^2(\mathbb{R}^N).$$

We say that $\mathscr{F} : L^2(\mathbb{R}^N) \longrightarrow L^2(\mathbb{R}^N)$ is an isometric isomorphism, with inverse $\overline{\mathscr{F}}$.

For two functions u and v, their convolution is defined by

$$(u*v)(x) = \int_{\mathbb{R}^N} u(x-y)v(y)\,dy.$$

Then the Fourier transform satisfies

$$\begin{aligned} \mathscr{F}(u*v)(\xi) &= \widehat{u}(\xi)\widehat{v}(\xi), \\ \mathscr{F}(uv)(\xi) &= (\widehat{u}*\widehat{v})(\xi). \end{aligned} \tag{2.15}$$

The justification of (2.12)–(2.15) can be found in [161, Chapter 9], for example.

Using equations (2.12) and (2.14), we can now deduce that

$$\begin{aligned} \|f\|^2_{H^k(\mathbb{R}^N)} &= \int_{\mathbb{R}^N} \sum_{|\alpha|\leqslant k} |D^\alpha f|^2\,dx \\ &= \sum_{|\alpha|\leqslant k} \|D^\alpha f\|^2_{L^2(\mathbb{R}^N)} \\ &= \sum_{|\alpha|\leqslant k} \|\mathscr{F}(D^\alpha f)\|^2_{L^2(\mathbb{R}^N)} \\ &= \sum_{|\alpha|\leqslant k} (2\pi)^{2|\alpha|} \|\xi^\alpha \mathscr{F}(f)\|^2_{L^2(\mathbb{R}^N)} \\ &\leqslant C_1 \int_{\mathbb{R}^N} (1+|\xi|^2)^k |\widehat{f}(\xi)|^2\,d\xi, \end{aligned}$$

where C_1 is a positive constant depending only on k. Similarly, we can prove

$$\|f\|^2_{H^k(\mathbb{R}^N)} \geqslant C_2 \int_{\mathbb{R}^N} (1+|\xi|^2)^k \|\widehat{f}(\xi)\|^2\,d\xi$$

for some positive C_2 depending on k only. Therefore

$$\begin{aligned} C^{-1} \int_{\mathbb{R}^N} (1+|\xi|^2)^k |\widehat{f}(\xi)|^2\,d\xi &\leqslant \|f\|^2_{H^k(\mathbb{R}^N)} \\ &\leqslant C \int_{\mathbb{R}^N} [1+|\xi|^2]^k |\widehat{f}(\xi)|^2\,d\xi \end{aligned} \tag{2.16}$$

for some sufficiently large $C > 1$, for all $f \in H^k(\mathbb{R}^N)$.

The inequalities (2.16) say that

$$\left[\int_{\mathbb{R}^N} (1+|\xi|^2)^k |\widehat{f}(\xi)|^2\,d\xi\right]^{1/2} \tag{2.17}$$

defines an equivalent norm in $H^k(\mathbb{R}^N)$. Associated with this norm, we can define a natural inner product

$$\langle f,g\rangle_{H^k(\mathbb{R}^N)} = \int_{\mathbb{R}^N} (1+|\xi|^2)^k \widehat{f}(\xi)\overline{\widehat{g}(\xi)}\,d\xi \quad \forall f,g \in H^k(\mathbb{R}^N). \tag{2.18}$$

Note that the expressions in both (2.17) and (2.18) are well defined for any *real number k*, if $\widehat{f}$ and $\widehat{g}$ are well behaved, i.e., k does not have to be a nonnegative integer. This enables us to define

$$\|f\|_{s,2} = \left[\int_{\mathbb{R}^N} (1+|\xi|^2)^s |\widehat{f}(\xi)|^2\,d\xi\right]^{1/2}, \tag{2.19}$$

$$H^s(\mathbb{R}^N) = \{f | f \text{ is a "generalized function"} \tag{2.20}$$
$$\text{(see Chapter 3) such that } \|f\|_{s,2} < \infty\},$$

for any $s \in \mathbb{R}$. Note that elements of $H^s(\mathbb{R}^N)$ are not necessarily functions that are pointwise-defined almost everywhere. We admit generalized functions as well, whose definition will be made more precise in the next chapter.

Exercise 2.1. The Dirac delta function $\delta(x)$ concentrated at the origin is a generalized function whose defining property is

$$\int_{\mathbb{R}^N} f(x)\delta(x)\,dx = f(0)$$

for any sufficiently smooth function f. According to this property,

$$\widehat{\delta}(\xi) = \mathscr{F}(\delta)(\xi) = \int_{\mathbb{R}^N} \delta(x)e^{-2\pi i\langle x,\xi\rangle}\,dx = e^{-2\pi i\langle 0,\xi\rangle}$$
$$= e^0 = 1.$$

(Thus the Fourier transform of the delta function is the constant 1. Taking the inverse Fourier transform of 1,

$$\overline{\mathscr{F}}(1) = \int_{\mathbb{R}^N} 1\cdot e^{2\pi i\langle x,\xi\rangle}\,d\xi,$$

we realize that the integral does not exist in the classical sense. Nevertheless, we can *define*

$$\overline{\mathscr{F}}(1) = \delta(x)$$

so that formally the Fourier inversion formula is made to hold:

$$\overline{\mathscr{F}}(\mathscr{F}(\delta)) = \delta.$$

These can all be made rigorous as in the next chapter.)
For what s does $\delta(x)$ belong to $H^s(\mathbb{R}^N)$ according to definition (2.20)? □

We can now reexamine the definition of $H^k(\Omega)$, $k \in \mathbb{N}$, for Ω a proper subdomain of $\mathbb{R}^N$. If we redefine

$$H^k(\Omega) = \left\{ f\colon \Omega \longrightarrow \mathbb{C} \mid \text{there exists an } F \in H^k(\mathbb{R}^N) \text{ s.t. } F|_\Omega = f \right\}, \tag{2.21}$$

$$\|f\|_{H^k(\Omega)} = \inf\{ \|F\|_{H^k(\mathbb{R}^N)} \mid F|_\Omega = f \}, \tag{2.22}$$

it turns out this definition of $H^k(\Omega)$ is completely equivalent to the earlier definition given in (2.3). But the new definitions (2.21) and (2.22) have the advantage that k *does not have to be a nonnegative integer*; it can be *any nonnegative real number*. From now on, we will use (2.21) and (2.22) as our definitions for the Sobolev space $W^{k,2}(\Omega) = H^k(\Omega)$ on a domain Ω, for $k \geqslant 0$.

Note that if Ω is a "pathological" domain such as those depicted in Fig. 2.1, then the new definition (2.21) is not equivalent to the old one in (2.3). So we must be rather careful about the way in which the Sobolev spaces are defined.

We now state and prove some basic properties of Sobolev spaces.

Theorem 2.1. *If $s > \frac{1}{2}N$, then the injection*

$$H^s(\mathbb{R}^N) \longrightarrow C^0(\mathbb{R}^N)$$

is continuous.

Proof. We want to show that there exists a constant $C > 0$ such that

$$\sup_{x \in \mathbb{R}^N} |u(x)| \leqslant C\|u\|_{H^s(\mathbb{R}^N)} \quad \forall u \in H^s(\mathbb{R}^N).$$

It is easy to verify that the inverse Fourier transform

$$\overline{\mathscr{F}}\colon L^1(\mathbb{R}^N) \longrightarrow C^0(\mathbb{R}^N) \tag{2.23}$$

is continuous [161, Chapter 9]. Since

$$u = \overline{\mathscr{F}}\widehat{u} \quad (\widehat{u} = \mathscr{F}u),$$

if we can show $\widehat{u} \in L^1(\mathbb{R}^N)$ the proof will be complete. But

$$\widehat{u}(\xi) = \left[(1+|\xi|^2)^{s/2}\widehat{u}(\xi) \right] \left[\frac{1}{(1+|\xi|^2)^{s/2}} \right]. \tag{2.24}$$

The first factor on the right is in $L^2(\mathbb{R}^N)$ because $u \in H^s(\mathbb{R}^N)$. The second factor is in $L^2(\mathbb{R}^N)$ provided that $s > \frac{1}{2}N$, because

$$\int_{\mathbb{R}^N} \frac{d\xi}{(1+|\xi|^2)^s} < \infty \quad \text{iff } 2s > N.$$

Hence $\widehat{u}$ is the product of two $L^2(\mathbb{R}^N)$ functions, and is in $L^1(\mathbb{R}^N)$ by the Cauchy–Schwarz inequality. □

Corollary 2.1. *If $s > \frac{1}{2}N + k$ for some integer k, then the injection*

$$H^s(\mathbb{R}^N) \longrightarrow C^k(\mathbb{R}^N)$$

is continuous. □

Exercise 2.2. Let (a,b) be a finite open interval in $\mathbb{R}$ and let f be a piecewise continuous function with finitely many discontinuities on (a,b), with each segment a C^∞ function on a subinterval. Show that

$$f \in H^{1/2-\varepsilon}(a,b)$$

for any $\varepsilon > 0$. □

Theorem 2.2. *If $s = \frac{1}{2}N + \alpha$ for some $\alpha, 0 < \alpha < 1$, then the injection*

$$H^s(\mathbb{R}^N) \longrightarrow C^{0,\alpha}(\mathbb{R}^N)$$

is continuous.

Proof. For any $y \in \mathbb{R}^N$, define the translated function

$$u_y(x) = u(x+y).$$

Then

$$\begin{aligned}
u_y(x) &= \overline{\mathscr{F}}[\widehat{u}(\xi)e^{2\pi i\langle y,\xi\rangle}](x),\\
u_y(x) - u(x) &= \overline{\mathscr{F}}[\widehat{u}(\xi)(e^{2\pi i\langle y,\xi\rangle} - 1)](x),\\
|u_y(x) - u(x)| &\leqslant \left[\int_{\mathbb{R}^N} (1+|\xi|^2)^s |\widehat{u}(\xi)|^2 \, d\xi\right]^{1/2}\\
&\quad \cdot \left[\int_{\mathbb{R}^N} \frac{|e^{2\pi i\langle y,\xi\rangle} - 1|^2}{(1+|\xi|^2)^s} \, d\xi\right]^{1/2}\\
&\equiv (T_1)^{1/2}(T_2)^{1/2}.
\end{aligned}$$

Now if $|y| \leqslant 1$,

$$\begin{aligned}
T_2 &= \int_{\mathbb{R}^N} |e^{2\pi i\langle y,\xi\rangle} - 1|^2 (1+|\xi|^2)^{-s} \, d\xi\\
&\leqslant C_1 \int_{|\xi|\leqslant|y|^{-1}} |y|^2|\xi|^2(1+|\xi|^2)^{-s} \, d\xi + 4\int_{|\xi|\geqslant|y|^{-1}} (1+|\xi|^2)^{-s} \, d\xi\\
&= C_2|y|^2 \int_0^{|y|^{-1}} \frac{r^2}{(1+r^2)^s} r^{N-1} \, dr + 4\int_{|y|^{-1}}^{\infty} \frac{r^{N-1}}{(1+r^2)^s} \, dr,
\end{aligned}$$

where we have made the change of variables to spherical coordinates. Since

$$\int_0^{|y|^{-1}} \frac{r^{N+1}}{(1+r^2)^s}\,dr \leqslant \int_0^{|y|^{-1}} \frac{r^{N+1}}{r^{2s}}\,dr = \int_0^{|y|^{-1}} r^{N+1-2s}\,dr$$
$$= \frac{1}{N+2-2s}(|y|^{-1})^{N+2-2s} \equiv C_3|y|^{2s-(N+2)}$$

due to the contributions by $r = |y|^{-1}$ and $r = 0$, and

$$\int_{|y|^{-1}}^{\infty} \frac{r^{N-1}}{(1+r^2)^s}\,dr = \int_1^{\infty} \frac{|y|^{1-N}\rho^{N-1}}{(1+|y|^{-2}\rho^2)^s}\frac{1}{|y|}\,d\rho$$
$$\leqslant C_4|y|^{2s-N}\int_1^{\infty}\frac{\rho^{N-1}}{\rho^{2s}}\,d\rho$$
$$\equiv C_5|y|^{2s-N} \quad \text{if } 0 < 2s-N.$$

(Note that when $2s - N \leqslant 0$, the integral above is divergent.) So

$$\int_{\mathbb{R}^N} |e^{2\pi i\langle y,\xi\rangle} - 1|^2(1+|\xi|^2)^s\,d\xi \leqslant C_6|y|^{-N+2s},$$

implying

$$\frac{|u(x+y)-u(x)|}{|y|^{(-N+2s)/2}} \leqslant C_6^{1/2}\|u\|_{H^s}.$$

If $|y| \geqslant 1$, use of the triangle inequality will suffice. So we have proved

$$\sup_{x,y\in\mathbb{R}^N} \frac{|u(x+y)-u(x)|}{|y|^{s-N/2}} \leqslant C_7\|u\|_{H^s}.$$

□

Corollary 2.2. *If* $s = \frac{1}{2}N + k + \alpha$ *for some* $k \in \mathbb{Z}^+$ *and* $\alpha, 0 < \alpha < 1$, *then the injection*

$$H^s(\mathbb{R}^N) \longrightarrow C^{k,\alpha}(\mathbb{R}^N)$$

is continuous.

Proof. One first shows that $H^s(\mathbb{R}) \longrightarrow C^k(\mathbb{R}^N)$ is continuous. This follows from Corollary 2.1. Then one can utilize the fact that all the distributional derivatives of order up to k are the same as the classical derivatives of order up to k in $C^k(\mathbb{R}^N)$ (the proof is not given here). Finally, one uses Theorem 2.2 to show that the injection

$$H^{s-k}(\mathbb{R}^N) \longrightarrow C^{0,\alpha}(\mathbb{R}^N)$$

is continuous. □

We can now state some consequences of these theorems for bounded smooth domains Ω.

Theorem 2.3. *If* $s = \frac{1}{2}N + k + \alpha$ *for some* $k \in \mathbb{Z}^+$ *and* $\alpha, 0 < \alpha < 1$, *then the injection*

$$H^s(\Omega) \longrightarrow C^{k,\alpha}(\overline{\Omega}) \tag{2.25}$$

is continuous.

Proof. This follows as a consequence of (2.22) and Corollary 2.2 □

Remark 2.2. Actually, the injection (or imbedding) (2.25) is *compact*: every bounded sequence in $H^s(\Omega)$ contains a convergent subsequence in $C^{k,\alpha}$ provided Ω is a bounded domain with sufficiently smooth boundary. □

We state a few further properties of Sobolev spaces without proof.
Let $u \in H^s(\mathbb{R}^N)$ and $v \in H^{-s}(\mathbb{R}^N)$, $s \geqslant 0$. Define the "pairing product"

$$\langle u,v\rangle_{H^s,H^{-s}} \equiv \int_{\mathbb{R}^N} \widehat{u}(\xi)\overline{\widehat{v}}(\xi)\,d\xi. \tag{2.26}$$

This is finite, because by Hölder's inequality

$$\left|\int_{\mathbb{R}^N} \widehat{u}(\xi)\overline{\widehat{v}}(\xi)\,d\xi\right| \leqslant \left[\int_{\mathbb{R}^N}(1+|\xi|^2)^s|\widehat{u}(\xi)|^2\,d\xi\right]^{1/2}\cdot\left[\int_{\mathbb{R}^N}(1+|\xi|^2)^{-s}|\widehat{v}(\xi)|^2\,d\xi\right]^{1/2} < \infty,$$

by (2.20). $H^{-s}(\mathbb{R}^N)$ is the dual of $H^s(\mathbb{R}^N)$ in the following sense:

Theorem 2.4. *Define $\langle u,v\rangle$ as above for $u \in H^s(\mathbb{R}^N)$ and $v \in H^{-s}(\mathbb{R}^N)$. Then for some $C>0$*

(1)
$$|\langle u,v\rangle| \leqslant \|u\|_{H^s(\mathbb{R}^N)}\cdot\|v\|_{H^{-s}(\mathbb{R}^N)}, \quad \forall u \in H^s(\mathbb{R}^N),\ v \in H^{-s}(\mathbb{R}^N); \tag{2.27}$$

(2) *let $T\colon\ H^s(\mathbb{R}^N) \longrightarrow \mathbb{C}$ be a linear functional such that*

$$|T(u)| \leqslant C\|u\|_{H^s(\mathbb{R}^N)} \quad \forall u \in H^s(\mathbb{R}^N). \tag{2.28}$$

Then there exists a unique $v \in H^{-s}(\mathbb{R}^N)$ such that

$$T(u) = \langle u,v\rangle \tag{2.29}$$

and

$$\|v\|_{H^{-s}(\mathbb{R}^N)} = \|T\| = \sup_{\substack{u\neq 0\\ u\in H^s(\mathbb{R}^N)}} \frac{|T(u)|}{\|u\|_{H^s(\mathbb{R}^N)}}; \tag{2.30}$$

(3) *for any fixed $v \in H^{-s}(\mathbb{R}^N)$, the mapping*

$$T\colon H^s(\mathbb{R}^N) \longrightarrow \mathbb{C},\ \ T(u) = \langle u,v\rangle \quad \forall u \in H^s(\mathbb{R}^N) \tag{2.31}$$

is a continuous linear functional, and we have

$$\|v\|_{H^{-s}(\mathbb{R}^N)} = \|T\|. \tag{2.32}$$

□

Using the Fourier transform, one can further show that

$$u \in H^{-k}(\mathbb{R}^N) \text{ for } k > 0,\ k \in \mathbb{N}, \text{ if and only if}$$
$$u = \sum_{|\alpha| \leqslant k} D^\alpha f_\alpha \text{ for some (nonunique) } f_\alpha \in L^2(\mathbb{R}^N),$$

where the derivatives are taken *in the sense of distributions*. See § 3.1.
Let M be a compact differentiable manifold without boundary in a Euclidean space. For $s \geqslant 0$, we can define $H^s(M)$ through partition of unity as mentioned at the beginning of this section. Then the dual of $H^s(M)$ in the sense similar to (2.27)–(2.32) will be denoted by $H^{-s}(M)$.

Theorem 2.5 (Compact imbedding). *Let Ω be a bounded domain with smooth boundary in $\mathbb{R}^N$ or a compact differentiable manifold without boundary in $\mathbb{R}^{N+1}$. For any $s_1, s_2 \in \mathbb{R}$, $s_1 > s_2$, the injection*

$$H^{s_1}(\Omega) \longrightarrow H^{s_2}(\Omega)$$

is completely continuous. □

Theorem 2.6 (Interpolation inequality). *Let Ω be a bounded domain with sufficiently smooth boundary in $\mathbb{R}^N$ and $k \in \mathbb{N}$. Then for any $\varepsilon > 0$, there exists $C(\varepsilon) > 0$ such that*

$$\|f\|_{H^k(\Omega)} \leqslant \varepsilon \sum_{|j|=k+1} \|D^j f\|^2_{L^2(\Omega)} + C(\varepsilon)\|f\|^2_{L^2(\Omega)}, \quad \forall f \in H^{k+1}(\Omega). \tag{2.33}$$

□

The L^p properties of Sobolev spaces are given in the following.

Theorem 2.7. *Let Ω be an open set in $\mathbb{R}^N$. Then the imbeddings*

$$H^s(\Omega) \hookrightarrow L^p(\Omega), \qquad \text{where } \frac{1}{p} = \frac{1}{2} - \frac{s}{N} > 0,$$

and

$$H^s(\Omega) \hookrightarrow L^p_{\text{loc}}(\Omega) \quad \forall p, 1 \leqslant p < \infty, \text{ when } \frac{1}{2} - \frac{s}{N} = 0,$$

are continuous. Further, if Ω is bounded, then the imbeddings are completely continuous, for $H^s(\Omega) \hookrightarrow L^q(\Omega)$, $1 \leqslant q < p$. □

We refer the reader to [2] for proofs of these theorems.

2.2 The trace theorems

Given a function u defined on $\mathbb{R}^N$, consider the restriction of u to an $(N-1)$-dimensional subspace $\mathbb{R}^{N-1} = \{x = (x_1, x_2, \ldots, x_N) \in \mathbb{R}^N | x_N = 0\}$. The restriction of u on $\mathbb{R}^{N-1}$ is called the *trace* of u. It is similar to taking a boundary value of u. The following is the important trace theorem.

Theorem 2.8. *The trace operator*

$$u \longmapsto u|_{\mathbb{R}^{N-1}}$$

is a continuous linear mapping from $H^s(\mathbb{R}^N)$ *into* $H^{s-1/2}(\mathbb{R}^{N-1})$ *for* $s > \frac{1}{2}$.

Proof. First, let $u \in \mathscr{D}(\mathbb{R}^N)$, the space of $\mathbb{C}^\infty$ functions with compact support on $\mathbb{R}^N$. Write

$$x_{\widehat{N}} = (x_1, x_2, \ldots, x_{N-1}) \in \mathbb{R}^{N-1}.$$

Then

$$\begin{aligned} v(x_{\widehat{N}}) &= u(x_{\widehat{N}}, 0) \\ &\equiv \int_{\mathbb{R}} e^{2\pi i(0\cdot\xi_N)} \left[\int_{\mathbb{R}} u(x) e^{-2\pi i x_N \xi_N}\, dx_N\right] d\xi_N \\ &= \int_{\mathbb{R}} \left[\int_{\mathbb{R}} u(x) e^{-2\pi i x_N \xi_N}\, dx_N\right] d\xi_N. \end{aligned} \tag{2.34}$$

So

$$(\mathscr{F}_{\widehat{N}} v)(\xi_{\widehat{N}}) = \int_{\mathbb{R}} (\mathscr{F} u)(\xi)\, d\xi_N, \tag{2.35}$$

where $\mathscr{F}_{\widehat{N}}$ is the Fourier transform in $\mathbb{R}^{N-1}$ (with the ξ_N variable excluded). We want to prove that

$$(1 + |\xi_{\widehat{N}}|^2)^{(1/2)(s-1/2)} (\mathscr{F}_{\widehat{N}} v)(\xi_{\widehat{N}}) \in L^2(\mathbb{R}^{N-1}).$$

From (2.35),

$$\begin{aligned} |(\mathscr{F}_{\widehat{N}} v)(\xi_{\widehat{N}})|^2 &\leqslant \left[\int_{\mathbb{R}} |(\mathscr{F} u)(\xi)|\, d\xi_N\right]^2 \\ &\leqslant \int_{\mathbb{R}} (1 + |\xi_{\widehat{N}}|^2 + \xi_N^2)^s |(\mathscr{F} u)(\xi)|^2\, d\xi_N \cdot \\ &\quad \int_{\mathbb{R}} \frac{d\xi_N}{(1 + |\xi_{\widehat{N}}|^2 + \xi_N^2)^s}. \end{aligned} \tag{2.36}$$

In the final integral, we make the change of variable

$$\xi_N = (1 + |\xi_{\widehat{N}}|^2)^{1/2} t$$

to get

$$\int_{\mathbb{R}} \frac{(1+|\xi_{\widehat{N}}|^2)^{1/2}}{(1+|\xi_{\widehat{N}}|^2)^s(1+t^2)^s}\,dt \equiv C(1+|\xi_{\widehat{N}}|^2)^{-s+1/2}, \tag{2.37}$$

which is finite if $s > \frac{1}{2}$. Substituting the above into (2.36) and integrating over $\mathbb{R}^{N-1}$, we get

$$\begin{aligned}
&\int_{\mathbb{R}^{N-1}} (1+|\xi_{\widehat{N}}|^2)^{s-1/2}|(\mathscr{F}_{\widehat{N}}v)(\xi_{\widehat{N}})|^2\,d\xi_{\widehat{N}} \\
&\leqslant C\int_{\mathbb{R}^{N-1}}\int_{\mathbb{R}} (1+|\xi_{\widehat{N}}|^2+\xi_N^2)^s|(\mathscr{F}u)(\xi)|^2\,d\xi_N\,d\xi_{\widehat{N}} \\
&= C\int_{\mathbb{R}^N} (1+|\xi|^2)^s|(\mathscr{F}u)(\xi)|^2\,d\xi.
\end{aligned} \tag{2.38}$$

Since C is independent of u, for any $u \in H^s(\mathbb{R}^N)$, we can approximate it by a sequence of functions in $\mathscr{D}(\mathbb{R}^N)$ and prove the desired consequence for $H^s(\mathbb{R}^N)$ functions. □

Theorem 2.9 (Trace theorem). *Let Ω be a bounded open domain with smooth boundary $\partial\Omega$. If $s > \frac{1}{2}$, then the trace operator*

$$u \longmapsto u|_{\partial\Omega}$$

is a continuous mapping from $H^s(\Omega)$ onto $H^{s-1/2}(\partial\Omega)$.

Proof. Since $\overline{\Omega}$ is compact, let $\{\theta_i\colon \overline{\Omega} \to \mathbb{R} \mid i = 1,2,\ldots,M\}$ be a partition of unity for a chart $\{\Omega_i \mid \Omega_i \subseteq \overline{\Omega}$ is open, $i = 1,2,\ldots,M\}$. The chart consists of the diffeomorphisms from Ω_i to B_1 as given in (2.6)–(2.7). For each $u \in H^s(\Omega)$,

$$u = \sum_{i=1}^{M} \theta_i u; \tag{2.39}$$

therefore we can assume that u has support on Ω_j for some j because it is a sum of such functions. Consider an Ω_j, such that $\Omega_j \cap \partial\Omega$ is relatively open in $\partial\Omega$. We can choose the Ω_j such that there exists a smooth mapping

$$\begin{aligned}
&f_j\colon \Omega_j \longrightarrow \mathbb{R}^N, \\
&f_j\colon \Omega_j \cap \partial\Omega \longrightarrow \mathbb{R}^{N-1} = \{x = (x_1,x_2,\ldots,x_{N-1},0) \in \mathbb{R}^N\}.
\end{aligned}$$

(This procedure is called "flattening out" the boundary $\partial\Omega$.) Now apply Theorem 2.8, and use (2.39) to sum on $H^{s-1/2}(\Omega_j \cap \partial\Omega)$ for those j such that $\Omega_j \cap \partial\Omega$ is relatively open in $\partial\Omega$. The proof of surjectivity involves additional work and is omitted. □

Corollary 2.3. *Let $s - \frac{1}{2}k > 0$ and $u \in H^s(\mathbb{R}^N)$. Then*

$$u(x_1,x_2,\ldots,x_{N-k},0,\ldots,0) \in H^{s-k/2}(\mathbb{R}^{N-k})$$

and the mapping

$$u \longmapsto u\Big|_{\mathbb{R}^{N-k}}$$

is continuous from $H^s(\mathbb{R}^N)$ *onto* $H^{s-k/2}(\mathbb{R}^{N-k})$. □

Corollary 2.4. *Let* $s - j - \frac{1}{2} > 0$ *and* $u \in H^s(\mathbb{R}^N)$. *Then*

$$\left(\frac{\partial}{\partial x_N}\right)^j u(x_1,\ldots,x_N)\Big|_{x_N=0} \in H^{s-j-1/2}(\mathbb{R}^{N-1}), \quad j = 0, 1, 2, \ldots,$$

and the mapping

$$u \longmapsto \left(\frac{\partial}{\partial x_N}\right)^j u\Big|_{\mathbb{R}^{N-1}}$$

is continuous from $H^s(\mathbb{R}^N)$ *onto* $H^{s-j-\frac{1}{2}}(\mathbb{R}^{N-1})$. □

Corollary 2.5. *Let* Ω *be a bounded open domain with smooth boundary* $\partial\Omega$. *If* $s - \frac{1}{2} - j > 0$ *then the mapping*

$$u \longmapsto \left(\frac{\partial}{\partial n}\right)^j u\Big|_{\partial\Omega}$$

is continuous from $H^s(\Omega)$ *onto* $H^{s-j-1/2}(\partial\Omega)$. □

Definition 2.1. The completion of $C_0^\infty(\Omega)$ in $H^m(\Omega)$, $m \in \mathbb{Z}^+$, will be denoted by $H_0^m(\Omega)$. □

By the trace theorem, we have $u = \partial u/\partial n = \cdots = (\partial/\partial n)^{m-1} u = 0$ on $\partial\Omega$ for $u \in H_0^m(\Omega)$, where higher-order normal derivatives $(\partial/\partial n)^k$ are defined by extending $n(x)$ in a C^∞ way to a neighborhood of $\partial\Omega$ through a Gaussian normal coordinate system (see [64, 23.48.4] and [107, §5.2]); cf. (6.41). So we often write

$$H_0^m(\Omega) = \left\{ u \in H^m(\Omega) \;\middle|\; u = \frac{\partial}{\partial n} u = \cdots = \left(\frac{\partial}{\partial n}\right)^{m-1} u = 0 \text{ on } \partial\Omega \right\}. \tag{2.40}$$

Chapter 3

Theory of Distributions

3.1 Test functions and generalized functions. Regularization of divergent integrals

In the study of boundary integral equations, the theory of distributions naturally comes into use. This theory began with the use of the Dirac delta function by the British physicist P.A.M. Dirac during the 1930s and 1940s. It was found to be extremely useful in solving ordinary and partial differential equations and became very popular, but was rejected by many mathematicians because it was not a classical function and its usage lacked mathematical rigor. In 1950–51, the French mathematician L. Schwartz published *Théorie des Distributions* [168], making rigorous the theory concerning the usage of the delta function and other distributions. Today, this theory is fundamental in the study of partial differential equations. For a detailed account of the theory of distributions, we refer the reader to [27, 76, 107].

The space $\mathscr{D}(\mathbb{R}^n)$, or briefly $\mathscr{D}$, where

$$\mathscr{D} \equiv C_0^\infty(\mathbb{R}^N),$$

in Chapter 2, is also called the *test function space*. In $\mathscr{D}$, we define the topology of uniform convergence. We write $\phi_n \longrightarrow 0$ in $\mathscr{D}$ if

(i) there exists a bounded open set G in $\mathbb{R}^N$ such that $\operatorname{supp}\phi_n \subseteq G$ for all $n = 1, 2, \ldots$

(ii) For each multi-index α, $0 \leqslant |\alpha| < \infty$,

$$\lim_{n\longrightarrow\infty} D^\alpha \phi_n(x) = 0 \text{ uniformly in } \mathbb{R}^N.$$

We write $\phi_n \longrightarrow \phi$ in $\mathscr{D}$ if $\phi_n - \phi \longrightarrow 0$ in $\mathscr{D}$.

Definition 3.1. A continuous linear functional T on $\mathscr{D}$ is a mapping from $\mathscr{D}$ to $\mathbb{C}$, denoted by $\langle T, \phi\rangle$ for $\phi \in \mathscr{D}$, satisfying

(i) $\langle T, c_1\phi_1 + c_2\phi_2\rangle = c_1\langle T,\phi_1\rangle + c_2\langle T,\phi_2\rangle\ \forall c_1, c_2 \in \mathbb{C}, \phi_1, \phi_2 \in \mathscr{D}$,

(ii) $\phi_n \to 0$ in $\mathscr{D}$ implies that $\langle T, \phi_n\rangle \to 0$ in $\mathbb{C}$.

We also call such a continuous linear functional a *distribution* or a *generalized function*. The space of all distributions will be denoted by $\mathscr{D}'(\mathbb{R}^n)$, or $\mathscr{D}'$. □

The Dirac delta function δ is now well defined as a distribution because

$$\langle \delta, \phi\rangle \equiv \phi(0)$$

is a continuous linear functional on $\mathscr{D}(\mathbb{R}^N)$.
We may add distributions or multiply them by C^∞ functions to form new distributions. However, the product of two distributions is not well defined in general.
Every $L^1_{\text{loc}}(\mathbb{R}^N)$ function f defines a distribution via

$$\langle f, \phi\rangle \equiv \int_{\mathbb{R}^N} f(x)\phi(x)\,dx, \quad \forall \phi \in \mathscr{D}. \tag{3.1}$$

This suggests the notation

$$\langle T, \phi\rangle \equiv \int_{\mathbb{R}^N} T(x)\phi(x)\,dx \tag{3.2}$$

for a continuous linear functional T even when T is not an L^1_{loc} function.
Distributions such as $L^1_{\text{loc}}(\mathbb{R}^N)$ functions can not be assigned values at isolated points. Thus, one can not say that "a distribution f is equal to zero at x_0". However, one may clearly make a statement that "a distribution f is equal to zero in a neighborhood $\mathscr{N}$ of x_0". This will mean that for every $\phi \in \mathscr{D}$ with support in $\mathscr{N}$, we have $\langle f, \phi\rangle = 0$. Thus, for instance, a distribution f corresponding to an $L^1_{\text{loc}}(\mathbb{R}^N)$ function $f(x)$ vanishes in a neighborhood $\mathscr{N}$ of x_0 if $f(x)$ itself vanishes a.e. in $\mathscr{N}$. A point x_0 is called an *essential point* of a distribution f if there does not exist a neighborhood $\mathscr{N}$ of x_0 in which f vanishes. The closure of the set of all essential points is defined as the *support* of f and denoted by $\operatorname{supp} f$.
Let us now define the important differentiation operation on distributions. For any $T \in \mathscr{D}$, we define $D^\alpha T$ to be a linear functional such that

$$\langle D^\alpha T, \phi\rangle = (-1)^{|\alpha|}\langle T, D^\alpha \phi\rangle \quad \forall \phi \in \mathscr{D}, \tag{3.3}$$

where $\alpha \in (\mathbb{Z}^+)^N$ is a given multi-index. It is easy to check that $D^\alpha T$ itself is now a continuous linear functional, i.e., a distribution.
When T is a function such that $D^\beta T \in L^1_{\text{loc}}(\mathbb{R}^N)$ for all $|\beta| \leqslant |\alpha|$, then from the notation (3.2) and (3.3), the definition (3.3) amounts to no more than integration by parts. But when T (is a classical function that) does not admit classical derivatives, (3.3) says that we can still shift all the "burden of differentiability" from T to ϕ and define a new distribution.

Example 3.1. In $\mathbb{R}$, the Heaviside function

$$H(x) = \begin{cases} 0, & x \leqslant 0, \\ 1, & x > 0, \end{cases}$$

is in L^1_{loc}, so it defines a distribution

$$\langle H, \phi \rangle = \int_{-\infty}^{\infty} H(x)\phi(x)\,dx = \int_0^{\infty} \phi(x)\,dx \quad \forall \phi \in \mathscr{D}.$$

H is differentiable everywhere with pointwise derivative 0, except at the point $x = 0$, where it is nondifferentiable in the classical sense.

The distributional derivative H' of H, by definition, satisfies

$$\begin{aligned}\langle H', \phi \rangle &= -\langle H, \phi' \rangle = -\int_0^{\infty} \phi'(x)\,dx = -[\phi(\infty) - \phi(0)] \\ &= \phi(0).\end{aligned}$$

Therefore, we have

$$H'(x) = \delta(x). \tag{3.4}$$

□

Next, we consider the distribution

$$x_+^{-\lambda} = \begin{cases} 0, & x \leqslant 0, \\ x^{-\lambda}, & x > 0, \end{cases} \tag{3.5}$$

for some $0 < \lambda < 1$. It is easy to check that $x_+^{\lambda} \in L^1_{\text{loc}}(\mathbb{R})$. Let us take the classical derivative of $x_+^{-\lambda}$:

$$\frac{d}{dx}(x_+^{-\lambda}) = \begin{cases} 0, & x < 0, \\ \text{nondifferentiable}, & x = 0, \\ -\lambda x^{-(\lambda+1)}, & x > 0. \end{cases} \tag{3.6}$$

The distributional derivative of $x_+^{-\lambda}$ satisfies

$$\langle (x_+^{-\lambda})', \phi \rangle = -\langle x_+^{-\lambda}, \phi' \rangle = -\int_0^{\infty} \frac{\phi'(x)}{x^{\lambda}}\,dx. \tag{3.7}$$

Can we find a more "direct" relationship between the classical derivative (3.6) and the distributional derivative defined through (3.7)? We would pretty much like to see that the classical derivative (3.6) yields a distribution corresponding to the distributional derivative (3.7). However (after ignoring the point $x = 0$), $(x_+^{-\lambda})'$ as given by (3.6) is not in $L^1_{\text{loc}}(\mathbb{R})$. Therefore it does not generate an ordinary distribution. We are thus in a somewhat awkward situation: Every L^1_{loc} function is identifiable with a distributional derivative, but the classical derivative of an L^1_{loc} function may not be identical with the distributional derivative.

To remedy this, we will adopt a procedure called *regularization of divergent integrals*. In the computation of boundary integral equations in later chapters, we will encounter many "hypersingular integrals" that are divergent and need regularization. In this chapter, we will provide many examples of integrals that contain singularities similar to those of the hypersingular integrals.

Let us first write (3.7) as

$$\langle (x_+^{-\lambda})', \phi \rangle = -\lim_{\varepsilon \downarrow 0} \int_\varepsilon^\infty \frac{\phi'(x)}{x^\lambda} dx,$$

and integrate by parts, yielding

$$\langle (x_+^{-\lambda}), \phi \rangle = \lim_{\varepsilon \downarrow 0} \left[\frac{\phi(\varepsilon)}{\varepsilon^\lambda} - \lambda \int_\varepsilon^\infty \frac{\phi(x)}{x^{\lambda+1}} dx \right]. \tag{3.8}$$

We apply the mean value theorem to ϕ on the interval $(0,x)$ to get

$$\phi(x) = \phi(0) + x\phi'(\theta x) \equiv \phi(0) + x\psi(x), \tag{3.9}$$

where $\theta = \theta(x)$ depends on x and $0 < \theta < 1$, and $\psi(x) \equiv \phi'(\theta x)$. Thus

$$\frac{\phi(\varepsilon)}{\varepsilon^\lambda} = \frac{\phi(0)}{\varepsilon^\lambda} + \varepsilon^{1-\lambda}\phi'(\theta\varepsilon)$$

and

$$\varepsilon^{1-\lambda}\phi'(\theta\varepsilon) \longrightarrow 0 \quad \text{as} \quad \varepsilon \downarrow 0.$$

So (3.8) becomes

$$\langle (x_+^{-\lambda})', \phi \rangle = \lim_{\varepsilon \downarrow 0} \left[\frac{\phi(0)}{\varepsilon^\lambda} - \lambda \int_\varepsilon^\infty \frac{\phi(x)}{x^{\lambda+1}} dx \right]. \tag{3.10}$$

It is easy to check that the function $\psi(x)$ defined by (3.9) is C^∞ everywhere except perhaps at $x = 0$. At $x = 0$, we can easily define $\psi(0)$ so that $\psi(x)$ is continuous at $x = 0$. Let $R > 0$ be sufficiently large that $\phi(x) = 0$ for $x > R$. Then, from (3.9) we have

$$\begin{aligned} \lim_{\varepsilon \downarrow 0} \lambda \int_\varepsilon^R \frac{\phi(x)}{x^{\lambda+1}} dx &= \lim_{\varepsilon \downarrow 0} \left[\lambda \int_\varepsilon^R \frac{\phi(0) + x\psi(x)}{x^{\lambda+1}} dx \right] \\ &= \lim_{\varepsilon \downarrow 0} \left[\frac{\phi(0)}{\varepsilon^\lambda} - \frac{\phi(0)}{R^\lambda} + \lambda \int_\varepsilon^R \frac{\psi(x)}{x^\lambda} dx \right]. \end{aligned} \tag{3.11}$$

From (3.11), we have

$$\lim_{\varepsilon \downarrow 0} \left[\frac{\phi(0)}{\varepsilon^\lambda} - \lambda \int_\varepsilon^R \frac{\phi(x)}{x^{\lambda+1}} dx \right] = \lim_{\varepsilon \downarrow 0} \left[\frac{\phi(0)}{R^\lambda} - \lambda \int_\varepsilon^R \frac{\psi(x)}{x^\lambda} dx \right], \tag{3.12}$$

where the RHS is always finite for any $\phi \in \mathscr{D}$. We let $R \longrightarrow \infty$ to get

$$\langle (x_+^{-\lambda})', \phi \rangle = \lim_{\varepsilon \downarrow 0} \left[\frac{\phi(0)}{\varepsilon^\lambda} - \lambda \int_\varepsilon^\infty \frac{\phi(x)}{x^{\lambda+1}} dx \right] = \text{RHS of (3.12)}. \tag{3.13}$$

This says that the RHS above is the finite limit of the difference of two terms, each of which tends to infinity. We call this the *finite part*, in the sense of Hadamard, of the divergent integral

$$\int_0^\infty [(-\lambda)x^{-(\lambda+1)}]\phi(x)\,dx$$

and write it as

$$(\text{F.P.})\int_0^\infty [(-\lambda)x^{-(\lambda+1)}]\phi(x)\,dx = \lim_{\varepsilon\downarrow 0}\left[\frac{\phi(0)}{\varepsilon^\lambda} - \lambda\int_\varepsilon^\infty \frac{\phi(x)}{x^{\lambda+1}}\,dx\right].$$

It defines a distribution called the *pseudofunction*

$$\mathbf{PF}(-\lambda x_+^{-(\lambda+1)}).$$

Thus we can write

$$\frac{d}{dx}(x_+^{-\lambda}) = \mathbf{PF}[-\lambda x_+^{-(\lambda+1)}], \text{ in the sense of distributions.} \tag{3.14}$$

In general, let f be a function that is L^1_{loc} everywhere except at x_0, such that the integral

$$\int_{-\infty}^\infty f(x)\phi(x)\,dx, \quad \phi\in\mathscr{D},$$

is generally divergent. If we can redefine a functional such that for all test functions ϕ vanishing in a neighborhood at x_0, the functional has the value given by the above, then we say that such a functional is a *regularization* of the divergent integral $\int_{\mathbb{R}} f\phi\,dx$. Taking the *finite part* of a divergent integral is a special way of regularization that is very "natural". (In [76], Gelfand and Shilov use an *analytic continuation* (of the complex variable λ) approach instead of taking the finite part, and arrive at the very same expressions (3.10), (3.14), etc. They call such a way of regularizing the "canonical regularization" [76, pp. 44–48]. Compare also §3.3.)

It is also easy to see that if, instead of (3.5), we consider the distribution

$$x_-^{-\lambda} = \begin{cases} |x|^{-\lambda}, & x<0, \\ 0, & x\geqslant 0, \end{cases} \quad 0<\lambda<1, \tag{3.15}$$

then we have similarly

$$(\text{F.P.})\int_{-\infty}^0 \left(\lambda|x|^{-(\lambda+1)}\right)\phi(x)\,dx = \lim_{\varepsilon\downarrow 0}\left(-\frac{\phi(0)}{\varepsilon^\lambda} + \lambda\int_{-\infty}^{-\varepsilon}\frac{\phi(x)}{|x|^{\lambda+1}}\,dx\right). \tag{3.16}$$

The RHS of (3.16) defines a distribution

$$\frac{d}{dx}(x_-^{-\lambda}) = \mathbf{PF}(\lambda x^{-(\lambda+1)}). \tag{3.17}$$

The definitions of $x_+^{-\lambda}$ and $x_-^{-\lambda}$ through (3.13) and (3.16) can be easily extended to $\lambda\in\mathbb{C}$, $0<\operatorname{Re}\lambda<1$ using

$$|x|^\lambda = |x|^{\operatorname{Re}\lambda}\{\cos[\operatorname{Im}\lambda\ln|x|] + i\sin[\operatorname{Im}\lambda\ln|x|]\}.$$

From now on, we will assume that $\lambda \in \mathbb{C}$.

From (3.5) and (3.15), for $0 < \operatorname{Re}\lambda < 1$, we define

$$|x|^{-\lambda} = \left\{ \begin{array}{c} |x|^{-\lambda},\ x<0 \\ 0, \quad x=0 \\ |x|^{-\lambda},\ x>0 \end{array} \right\} \tag{3.18}$$

$$= x_-^{-\lambda} + x_+^{-\lambda},$$

$$|x|^{-\lambda}\operatorname{sgn} x = \left\{ \begin{array}{c} -|x|^{-\lambda},\ x<0 \\ 0, \quad x=0 \\ |x|^{-\lambda},\ x>0 \end{array} \right\} \tag{3.19}$$

$$= -x_-^{-\lambda} + x_+^{-\lambda},$$

then for $0 < \lambda < 1$, we have

$$\frac{d}{dx}|x|^{-\lambda} = \frac{d}{dx}(x_-^{-\lambda} + x_+^{-\lambda}) = \mathbf{PF}(\lambda x_-^{-(\lambda+1)} - \lambda x_+^{-(\lambda+1)}) \tag{3.20}$$

$$= -\lambda |x|^{-(\lambda+1)}\operatorname{sgn} x,$$

$$\frac{d}{dx}|x|^{-\lambda}\operatorname{sgn} x = \frac{d}{dx}(-x_-^{-\lambda} + x_+^{-\lambda}) \tag{3.21}$$

$$= \mathbf{PF}(-\lambda x^{-(\lambda+1)} - \lambda x_+^{-(\lambda+1)}) \tag{3.22}$$

$$= -\lambda |x|^{-(\lambda+1)}$$

in the sense of distribution. Indeed, the above procedures can be continued for $\operatorname{Re}\lambda$ not in $(0,1)$. Let

$$n < \operatorname{Re}\lambda < n+1, \quad \lambda \neq 1,2,\ldots,n, \tag{3.23}$$

$$[\operatorname{Re}\lambda] = \text{ the integral part of } \operatorname{Re}\lambda = n.$$

For any test function $\phi(x)$,

$$\int_{-\infty}^{\infty} \phi(x)H(x)|x|^{-\lambda}\,dx = \lim_{\varepsilon\downarrow 0}\int_{\varepsilon}^{\infty} x^{-\lambda}\Big[\phi(0) + x\phi'(0) + \cdots + \frac{x^{n-1}}{(n-1)!}\phi^{(n-1)}(0) + \frac{x^n}{n!}\phi^{(n)}(\theta x)\Big]\,dx,$$

where the Taylor expansion with remainder

$$\left. \begin{array}{r} \phi(x) = \phi(0) + x\phi'(0) + \cdots + \dfrac{x^{n-1}}{(n-1)!}\phi^{(n-1)}(0) + \dfrac{x^n}{n!}\phi^{(n)}(\theta x), \\ 0<\theta<1, \theta \text{ depends on } x. \end{array} \right\} \tag{3.24}$$

is used. Therefore

$$(\text{F.P.}) \int_{-\infty}^{\infty} \phi(x)H(x)|x|^{-\lambda}\,dx = \lim_{\varepsilon\downarrow 0}\int_{\varepsilon}^{\infty} x^{-\lambda}\frac{x^n}{n!}\phi^{(n)}(\theta x)\,dx$$

$$= \int_0^{\infty} x^{-(\lambda-n)}\frac{\phi^{(n)}(\theta x)}{n!}\,dx$$

$$= \int_0^{\infty} x^{-\lambda}\left[\phi(x)-\phi(0)-x\phi'(0)-\cdots-\frac{x^{n-1}}{(n-1)!}\phi^{(n-1)}(0)\right]dx. \tag{3.25}$$

Using the identity

$$\int_1^{\infty} x^{-\mu}\,dx = \frac{1}{\mu-1} \quad \text{for } \mu\in\mathbb{C},\ \ \operatorname{Re}\mu>1, \tag{3.26}$$

we can further rewrite (3.25) as

$$\begin{aligned}\langle x_+^{-\lambda},\phi\rangle \equiv (\text{F.P.}) \int_0^{\infty} x^{-\lambda}\phi(x)\,dx \\ = \int_0^1 x^{-\lambda}\Big[\phi(x)-\phi(0)-x\phi'(0)-\cdots \\ -\frac{x^{n-1}}{(n-1)!}\phi^{(n-1)}(0)\Big]dx \\ +\int_1^{\infty} x^{-\lambda}\phi(x)\,dx+\sum_{k=1}^{n}\frac{\phi^{(k-1)}(0)}{(k-1)!(k-\lambda)}.\end{aligned} \tag{3.27}$$

Remark 3.1. We can interpret (3.27) as follows: when $\langle x_+^{-\lambda},\phi\rangle$ is treated as a function of λ, it has simple poles at $\lambda=1,2,\ldots$, and its residue at $\lambda=k\in\mathbb{N}$ is $\phi^{(k-1)}(0)/(k-1)!$. Since

$$\phi^{(k-1)}(0) = (-1)^{(k-1)}\langle\delta^{(k-1)}(x),\phi\rangle,$$

we may say that the functional $x_+^{-\lambda}$ itself has simple poles at $\lambda=k$, and the residues there are

$$\frac{(-1)^{k-1}}{(k-1)!}\delta^{(k-1)}(x), \quad k=1,2,\ldots$$

□

Exercise 3.1. For $\lambda\leqslant 0$, it is easy to define $x_+^{-\lambda}$. Show that

$$\frac{d}{dx}x_+^{-\lambda} = -\lambda x_+^{-(\lambda+1)}, \quad \lambda\in\mathbb{R},\lambda\notin\mathbb{N}, \tag{3.28}$$

in the sense of distributions, where for $\lambda>0$, $\lambda\notin\mathbb{N}$, $x_+^{-\lambda}$ is the pseudofunction defined through (3.27) (or, equivalently, by (3.25)). □

The distribution corresponding to

$$x_-^{-\lambda} = \begin{cases} |x|^{-\lambda}, & x < 0, \\ 0, & x \geqslant 0, \end{cases}$$

can be treated in the same way as $x_+^{-\lambda}$, by making a reflection $x \to -x$:

$$\begin{aligned}
\langle x_-^{-\lambda}, \phi\rangle &= \int_0^\infty x^{-\lambda}\phi(-x)\,dx = -\langle x_+^{\lambda}, \phi(-x)\rangle \\
&= \int_0^1 x^{-\lambda}\left[\phi(-x) - \phi(0) + x\phi'(0) - \cdots - \frac{(-x)^{n-1}}{(n-1)!}\phi^{(n-1)}(0)\right] dx \\
&\quad + \int_1^\infty x^{-\lambda}\phi(-x)\,dx + \sum_{k=1}^{n} \frac{(-1)^{k-1}\phi^{(k-1)}(0)}{(k-1)!(-\lambda+k)} \\
&= \int_0^\infty x^{-\lambda}\left[\phi(-x) - \phi(0) + x\phi'(0) - \cdots - \frac{(-x)^{n-1}}{(n-1)!}\phi^{(n-1)}(0)\right] dx
\end{aligned} \tag{3.29}$$

for λ satisfying (3.23). We thus can interpret the pseudofunction $x_-^{-\lambda}$ as having simple poles at $\lambda = 1, 2, \ldots$ and residues $\delta^{(k-1)}(x)/(k-1)!$ at $\lambda = k$.

Now let us consider combination of $x_+^{-\lambda}$ and $x_-^{-\lambda}$ as in (3.18) and (3.19), but without $\mathrm{Re}\,\lambda$ being restricted to $(0,1)$:

$$\left.\begin{aligned} |x|^{-\lambda} &= x_+^{-\lambda} + x_-^{-\lambda}, \\ |x|^{-\lambda}\operatorname{sgn} x &= x_+^{-\lambda} - x_-^{-\lambda}. \end{aligned}\right\} \tag{3.30}$$

Let us find the singularities of these distributions (i.e., pseudofunctions). Recall from Remark 3.1 and (3.29) above that at $\lambda = k$ both $x_+^{-\lambda}$ and $x_-^{-\lambda}$ have poles, with respective residues

$$\frac{(-1)^{(k-1)}}{(k-1)!}\delta^{(k-1)}(x) \text{ and } \frac{1}{(k-1)!}\delta^{(k-1)}(x).$$

Thus $|x|^{-\lambda}$ has poles only at $\lambda = 1, 3, 5, \ldots$, with residues

$$\frac{2}{(2m)!}\delta^{(2m)}(x) \quad \text{at } \lambda = 2m+1, m \in \mathbb{Z}^+.$$

At $\lambda = 2m$, $m \in \mathbb{Z}^+$, the distribution $|x|^{-\lambda}$ is well defined, and for these values of λ we shall naturally write x^{-2m} instead of $|x|^{-2m}$.

Similarly, $|x|^{-\lambda}\operatorname{sgn} x$ has poles at $\lambda = 2, 4, \ldots$, with residues

$$-\frac{2}{(2m-1)!}\delta^{(2m-1)}(x) \quad \text{at } \lambda = 2m, m \in \mathbb{N}.$$

We shall naturally write the distribution $x^{-(2m+1)}$ instead of $|x|^{-(2m+1)}\operatorname{sgn} x$.

Therefore the distributions x^{-n} are now defined for all $n \in \mathbb{N}$. Their explicit expression can be derived. Using $\lambda = 2m$ in (3.27) and (3.29) and adding, we obtain

$$\langle x^{-2m}, \phi \rangle = \int_0^\infty x^{-2m} \Big\{ \phi(x) + \phi(-x) \\ - 2\left[\phi(0) + \frac{x^2}{2!}\phi(0) + \cdots + \frac{x^{2m-2}}{(2m-2)!}\phi^{(2m-2)}(0) \right] \Big\} dx.$$

Similarly, using $\lambda = 2m+1$ in (3.27) and (3.29) and subtracting, we obtain

$$\langle x^{-(2m+1)}, \phi \rangle = \int_0^\infty x^{-(2m+1)} \Big\{ \phi(x) - \phi(-x) \\ - 2\left[x\phi'(0) + \frac{x^3}{3!}\phi'''(0) + \cdots + \frac{x^{2m-1}}{(2m-1)!}\phi^{(2m-1)}(0) \right] \Big\} dx. \tag{3.31}$$

For example

$$\langle x^{-2}, \phi \rangle = \int_0^\infty \frac{\phi(x) + \phi(-x) - 2\phi(0)}{x^2}\, dx, \tag{3.32}$$

$$\langle x^{-1}, \phi \rangle = \int_0^\infty \frac{\phi(x) - \phi(-x)}{x}\, dx. \tag{3.33}$$

This last expression is identical with the Cauchy principal value of the integral of $\phi(x)/x$ (cf. (1.25)):

$$\text{(P.V.)} \int_{-\infty}^\infty \frac{\phi(x)}{x}\, dx = \lim_{\varepsilon \downarrow 0} \left\{ \int_{-\infty}^{-\varepsilon} \frac{\phi(x)}{x}\, dx + \int_\varepsilon^\infty \frac{\phi(x)}{x}\, dx \right\}.$$

By Exercise 3.1, on differentiating (in the sense of distributions) $|x|^{-\lambda}$ and $|x|^{-\lambda}\operatorname{sgn} x$, we obtain

$$\frac{d}{dx}|x|^{-\lambda} = -\lambda |x|^{-(\lambda+1)} \operatorname{sgn} x,$$

$$\frac{d}{dx}|x|^{-\lambda} \operatorname{sgn} x = -\lambda |x|^{-(\lambda+1)}.$$

In particular, for $\lambda = n \in \mathbb{Z}^+$, we have

$$\frac{d}{dx}x^{-n} = -nx^{-(n+1)},$$

which agrees with the classical derivative.

3.2 The pseudofunctions x_+^{-n}, $n = 1, 2, 3, \ldots$

We note from Exercise 3.1 that the derivative formula (3.28) does not apply to x_+^{-n}. Similarly, we note in the last section that we have not defined distributions corresponding to

$$\left. \begin{aligned} x_+^{-(2m+1)} + x_-^{-(2m+1)}, \\ x_+^{-2m} - x_-^{-2m}, \end{aligned} \right\} \tag{3.34}$$

when $m \in \mathbb{N}$. Indeed, the derivative formula (3.28) is not valid for x_+^{-n}. The regularization of x_+^{-n} has the special feature that the delta function and its derivatives appear when $n \geqslant 2$. For $n = 1$, we regularize the integral

$$\int_{-\infty}^{\infty} \phi(x) x_+^{-1}\, dx = \int_0^{\infty} \frac{\phi(x)}{x}\, dx \tag{3.35}$$

by taking its finite part:

$$\begin{aligned}\lim_{\varepsilon \downarrow 0} \int_{\varepsilon}^{\infty} \frac{\phi(x)}{x}\, dx &= \lim_{\varepsilon \downarrow 0} \int_{\varepsilon}^{1} \frac{1}{x}\left[\phi(0) + x\phi'(\theta x)\right] dx + \int_1^{\infty} \frac{\phi(x)}{x}\, dx \\ &\quad \text{(cf. (3.24), with } n = 1\text{)} \\ &= \lim_{\varepsilon \downarrow 0} [-\phi(0) \ln \varepsilon] + \int_0^1 \phi'(\theta x)\, dx + \int_1^{\infty} \frac{\phi(x)}{x}\, dx. \end{aligned} \tag{3.36}$$

Therefore the finite part is

$$\begin{aligned}\text{(F.P.)} \int_0^{\infty} \frac{\phi(x)}{x}\, dx &= \int_0^1 \phi'(\theta x)\, dx + \int_1^{\infty} \frac{\phi(x)}{x}\, dx \\ &= \int_0^1 \frac{\phi(x) - \phi(0)}{x}\, dx + \int_1^{\infty} \frac{\phi(x)}{x}\, dx. \end{aligned} \tag{3.37}$$

Integrating by parts on the RHS, we get

$$\begin{aligned}\text{(F.P.)} \int_0^{\infty} \frac{\phi(x)}{x}\, dx &= \int_0^1 [\phi(x) - \phi(0)] (\ln x)'\, dx + \int_1^{\infty} \phi(x)(\ln x)'\, dx \\ &= \lim_{\varepsilon \downarrow 0} [\phi(\varepsilon) - \phi(0)] \ln \varepsilon - \int_0^{\infty} \phi'(x) \ln x\, dx. \end{aligned} \tag{3.38}$$

The limit term on the RHS is 0 because $\phi(\varepsilon) - \phi(0) = \mathscr{O}(\varepsilon)$. Combining (3.35)–(3.38), we obtain

$$\text{(F.P.)} \int_{-\infty}^{\infty} \phi(x) x_+^{-1}\, dx = -\int_0^{\infty} \phi'(x) \ln x\, dx.$$

Hence

$$x_+^{-1} = \frac{d}{dx} \mathbf{PF}(\ln x)_+, \tag{3.39}$$

where the pseudofunction $(\ln x)_+$ is an L^1_{loc} function defined by

$$(\ln x)_+ = \begin{cases} 0, & x \leqslant 0, \\ \ln x, & x > 0. \end{cases}$$

For $n = 2$, we write

$$\int_{-\infty}^{\infty} \phi(x) x_+^{-2}\, dx = \int_0^{\infty} \phi(x) x^{-2}\, dx.$$

But

$$\int_{\varepsilon}^{\infty} \phi(x)x^{-2}\,dx = \int_{\varepsilon}^{1} \frac{\phi(0)+x\phi'(0)+x^2\phi''(\theta x)/2}{x^2}\,dx + \int_{1}^{\infty} \frac{\phi(x)}{x^2}\,dx$$

(cf. (3.24), with $n = 2$)

$$= \frac{1}{2}\int_{\varepsilon}^{1} \phi''(\theta x)\,dx + \phi(0)\left(\frac{1}{\varepsilon}-1\right) - \phi'(0)\ln\varepsilon + \int_{1}^{\infty} \frac{\phi(x)}{x^2}\,dx$$

$$= \left[\frac{1}{2}\int_{\varepsilon}^{1} \phi''(\theta x)\,dx + \int_{1}^{\infty} \frac{\phi(x)}{x^2}\,dx - \phi(0)\right] + \left[\frac{\phi(0)}{\varepsilon} - \phi'(0)\ln\varepsilon\right]$$

Therefore the finite part is

$$\text{(F.P.)} \int_{-\infty}^{\infty} \phi(x)x_+^{-2}\,dx = \lim_{\varepsilon\downarrow 0} \frac{1}{2}\int_{\varepsilon}^{1} \phi(\theta x)\,dx + \int_{1}^{\infty} \frac{\phi(x)}{x^2}\,dx - \phi(0)$$

$$= \int_{0}^{1} \frac{\phi(x)-\phi(0)-x\phi'(0)}{x^2}\,dx + \int_{1}^{\infty} \frac{\phi(x)}{x^2}\,dx - \phi(0). \tag{3.40}$$

To find the derivative relation, we integrate by parts for the RHS and obtain

$$\text{(F.P.)} \int_{-\infty}^{\infty} \phi(x)x_+^{-2}\,dx = -\int_{0}^{1} [\phi(x)-\phi(0)-x\phi'(0)]\left(\frac{1}{x}\right)'\,dx$$

$$-\int_{1}^{\infty} \phi(x)\left(\frac{1}{x}\right)'\,dx - \phi(0)$$

$$= -\left\{\phi(1)-\phi(0)-\phi'(0) - \lim_{\varepsilon\downarrow 0}\frac{[\phi(\varepsilon)-\phi(0)-\varepsilon\phi'(0)]}{\varepsilon}\right\}$$

$$+\int_{0}^{1} \frac{\phi'(x)-\phi'(0)}{x}\,dx + \phi(1) + \int_{1}^{\infty} \frac{\phi'(x)}{x}\,dx$$

$$= \int_{0}^{1} \frac{\phi'(x)-\phi'(0)}{x}\,dx + \int_{1}^{\infty} \frac{\phi'(x)}{x}\,dx + \lim_{\varepsilon\downarrow 0}\frac{\phi(\varepsilon)-\phi(0)}{\varepsilon}. \tag{3.41}$$

But for the sum of the first two terms on the RHS, we have

$$\text{(F.P.)} \int_{-\infty}^{\infty} \phi'(x)x_+^{-1}\,dx = \int_{0}^{1} \frac{\phi'(x)-\phi'(0)}{x}\,dx + \int_{1}^{\infty} \frac{\phi'(x)}{x}\,dx$$

$$\text{by (3.37)}, \tag{3.42}$$

and for the third term,

$$\lim_{\varepsilon\downarrow 0}\frac{\phi(\varepsilon)-\phi(0)}{\varepsilon} = \phi'(0). \tag{3.43}$$

Therefore (3.41)–(3.43) give

$$\text{(F.P.)} \int_{-\infty}^{\infty} \phi(x)x_+^{-2}\,dx = -\text{(F.P.)} \int_{-\infty}^{\infty} [-\phi'(x)]x_+^{-1} - \int_{-\infty}^{\infty} \phi(x)\delta'(x)\,dx. \tag{3.44}$$

Hence

$$x_+^{-2} = -\frac{d}{dx}(x_+^{-1}) - \delta'(x). \tag{3.45}$$

Exercise 3.2. Prove by induction that

$$\langle x_+^{-m}, \phi \rangle = \int_0^\infty x^{-m} \left[\phi(x) - \phi(0) - x\phi'(0) - \cdots - \frac{x^{m-2}}{(m-2)!}\phi^{(m-2)}(0) \right.$$
$$\left. - \frac{x^{m-1}}{(m-1)!}\phi^{(m-1)}(0) H(1-x) \right] dx \tag{3.46}$$

$$x_+^{-(m+1)} = -\frac{d}{dx}\left(\frac{1}{m} x_+^{-m}\right) + \frac{(-1)^m}{m!}\frac{\delta^{(m)}(x)}{m}, m \in \mathbb{Z}^+. \tag{3.47}$$

□

3.3 The distributions $(x \pm i0)^{-\lambda}$

We consider the *Heisenberg distributions* (see [107, p. 30, Example 5] and [76, §§ I.3.6, I.4.4 and I.4.5])

$$\langle \delta^{\pm}(x), \phi(x) \rangle \equiv \mp \lim_{\varepsilon \downarrow 0} \frac{1}{2\pi i} \int_{-\infty}^{\infty} \frac{\phi(x)}{x \pm i\varepsilon} dx, \quad \phi \in \mathscr{D}. \tag{3.48}$$

We assume that supp $\phi \subset [-A, A]$ for some $A > 0$ and use (3.9). Then for each $\varepsilon > 0$,

$$\int_{-\infty}^{\infty} \frac{\phi(x)}{x+i\varepsilon} dx = \int_{-A}^{A} \frac{x - i\varepsilon}{x^2+\varepsilon^2}\phi(x)\, dx$$
$$= \phi(0) \int_{-A}^{A} \frac{x - i\varepsilon}{x^2+\varepsilon^2} dx + \int_{-A}^{A} \frac{x - i\varepsilon}{x^2+\varepsilon^2} x\psi(x)\, dx$$
$$= -2i\phi(0) \tan^{-1}\frac{A}{\varepsilon} + \int_{-A}^{A} \frac{x - i\varepsilon}{x^2+\varepsilon^2} x\psi(x)\, dx. \tag{3.49}$$

Thus,

$$\lim_{\varepsilon \downarrow 0} \frac{1}{2\pi i} \int_{-\infty}^{\infty} \frac{\phi(x)}{x+i\varepsilon} dx = \frac{1}{2\pi i}\left[-2i\phi(0)\frac{\pi}{2}\right] + \frac{1}{2\pi i}\int_{-A}^{A} \psi(x)\, dx. \tag{3.50}$$

But

$$\int_{-A}^{A} \psi(x)\, dx = \lim_{\varepsilon \downarrow 0} \left[\left(\int_{-A}^{-\varepsilon} + \int_{\varepsilon}^{A} \right) \psi(x)\, dx \right]$$
$$= \lim_{\varepsilon \downarrow 0} \left[\left(\int_{-A}^{\varepsilon} + \int_{\varepsilon}^{A} \right) \left(\frac{\phi(0)}{x} + \psi(x) \right) dx \right]$$
$$= \lim_{\varepsilon \downarrow 0} \int_{|x|>\varepsilon} \frac{\phi(x)}{x} dx = \langle \mathbf{PF}(x^{-1}), \phi \rangle,$$

so (3.50) becomes

$$\lim_{\varepsilon \downarrow 0} \frac{1}{2\pi i} \int_{-\infty}^{\infty} \frac{\phi(x)}{x+i\varepsilon} dx = -\frac{1}{2}\phi(0) + \frac{1}{2\pi i}\langle \mathbf{PF}(x^{-1}), \phi \rangle. \tag{3.51}$$

Similarly,

$$\lim_{\varepsilon \downarrow 0} \frac{1}{2\pi i} \int_{-\infty}^{\infty} \frac{\phi(x)}{x-i\varepsilon} dx = \frac{1}{2}\phi(0) + \frac{1}{2\pi i}\langle \mathbf{PF}(x^{-1}), \phi \rangle. \tag{3.52}$$

Equations (3.51) and (3.52) are called Plemelj formulas; cf. (1.26) and (1.27). We can write (3.51) and (3.52) symbolically as

$$\frac{1}{2\pi i}\frac{1}{x\pm i0} \equiv \lim_{\varepsilon\downarrow 0}\frac{1}{2\pi i}\frac{1}{x\pm i\varepsilon} = \mp\frac{1}{2}\delta(x) + \frac{1}{2\pi i}\mathbf{PF}(x^{-1}). \tag{3.53}$$

From (3.48),

$$\left.\begin{aligned} \delta^{+}(x) &= \frac{1}{2}\delta(x) - \frac{1}{2\pi i}\mathbf{PF}(x^{-1}),\\ \delta^{-}(x) &= \frac{1}{2}\delta(x) + \frac{1}{2\pi i}\mathbf{PF}(x^{-1}), \end{aligned}\right\} \tag{3.54}$$

so that

$$\delta(x) = \delta^{+}(x) + \delta^{-}(x) = \lim_{\varepsilon\downarrow 0}\frac{1}{\pi}\frac{\varepsilon}{x^2+\varepsilon^2},$$

$$\frac{1}{\pi i}\mathbf{PF}(x^{-1}) = \delta^{-}(x) - \delta^{+}(x) = \lim_{\varepsilon\downarrow 0}\frac{1}{\pi}\frac{x}{x^2+\varepsilon^2}.$$

The Heisenberg distributions can also be obtained by differentiating the regular distribution

$$\ln(x+i0) = \lim_{\varepsilon\downarrow 0}\ln(x+i\varepsilon),$$

because

$$\ln(x+i\varepsilon) = \ln|x+i\varepsilon| + i\arg(x+i\varepsilon)$$

$$\begin{aligned} \lim_{\varepsilon\downarrow 0} i\arg(x+i\varepsilon) &= i\lim_{\varepsilon\downarrow 0}\tan^{-1}\frac{\varepsilon}{x}\\ &= \begin{cases} 0, & x>0\\ i\pi, & x<0 \end{cases}\\ &= i\pi H(-x). \end{aligned}$$

Therefore

$$\lim_{\varepsilon\downarrow 0}\ln(x+i\varepsilon) = \ln|x| + i\pi H(-x), \tag{3.55}$$

and

$$\begin{aligned} \left\langle \frac{d}{dx}\ln(x+i0),\phi\right\rangle &= \lim_{\varepsilon\downarrow 0}\left\langle \frac{1}{x+i\varepsilon},\phi\right\rangle\\ &= \left\langle \mathbf{PF}\left(\frac{1}{x}\right) - i\pi\delta(x),\phi\right\rangle, \quad \phi\in\mathscr{D}. \end{aligned}$$

We can also define $(x\pm i0)^{-\lambda}$ with exponent $-\lambda$, $\lambda\in\mathbb{C}$. If $\operatorname{Re}\lambda > -1$, we can simply define

$$\begin{aligned} (x\pm i0)^{-\lambda} &= \lim_{\varepsilon\downarrow 0}(x^2+\varepsilon^2)^{-\lambda/2}e^{-i\lambda\arg(x\mp i\varepsilon)}\\ &= x_+^{-\lambda} + e^{\pm i\pi\lambda}x_-^{-\lambda}. \end{aligned} \tag{3.56}$$

For other complex values of λ, we can analytically continue (3.56) to make it defined for those values [76, loc. cit.].

It can be shown [76, p. 94] that for $m \in \mathbb{N}$,

$$(x \pm i0)^{-m} = x_+^{-m} + (-1)^m x_-^{-m} \mp \frac{(-1)^{m-1}(i\pi)}{(m-1)!}\delta^{(m-1)}(x). \tag{3.57}$$

Thus, in particular, for $m = 2$,

$$\lim_{y\to 0^+}\int_{-\infty}^{\infty}\frac{\phi(x)}{(x+iy)^2}\,dx = \left\langle \frac{1}{(x+i0)^2}, \varphi \right\rangle \tag{3.58}$$

$$= \int_0^{\infty}\frac{\phi(x)+\phi(-x)-2\phi(0)}{x^2}\,dx - i\pi\phi'(0),$$

$$\lim_{y\to 0^+}\int_{-\infty}^{\infty}\frac{\phi(x)}{(x-iy)^2}\,dx = \left\langle \frac{1}{(x-i0)^2}, \phi \right\rangle \tag{3.59}$$

$$= \int_0^{\infty}\frac{\phi(x)+\phi(-x)-2\phi(0)}{x^2}\,dx + i\pi\phi'(0).$$

Compare (3.58) and (3.59) with (3.32). For $m = 1$, (3.57) gives

$$\lim_{y\to 0^+}\int_{-\infty}^{\infty}\frac{\phi(x)}{(x\pm iy)}\,dx = \left\langle \frac{1}{x\pm i0}, \phi \right\rangle$$

$$= \int_0^{\infty}\frac{\phi(x)-\phi(-x)}{x}\,dx \mp i\pi\phi(0). \tag{3.60}$$

Compare (3.60) with (3.33).

3.4 Regularizing divergent integrals in $\mathbb{R}^N$ with an algebraic singularity

A function $f(x)$ defined for $x \in \mathbb{R}^N$ is said to have an algebraic singularity at $a \in \mathbb{R}^N$ if $f \notin L^1$ in a neighborhood of a but $f(x)|x-a|^m \in L^1$ in a neighborhood of a, for some positive power m. If a function $f(x)$ has at most finitely many algebraic singularities in any bounded subdomain of $\mathbb{R}^N$, then the divergent integral $\int_{\mathbb{R}^N} f(x)\phi(x)\,dx$ can be regularized. This is due to the following theorem. For simplicity, we assume that there is only one singularity $a = 0$.

Theorem 3.1. *If there exists $m \in \mathbb{Z}^+$ such that $f(x)|x|^m$ is in $L^1_{loc}(\mathbb{R}^N)$, then the divergent integral $\int_{\mathbb{R}^N} f(x)\phi(x)\,dx$, $\phi \in \mathscr{D}$, can be regularized.*

Proof. We can construct a regularization of f by

$$\langle f,\phi\rangle = \int_{\mathbb{R}^N} f(x)\Bigg\{\phi(x) - \Bigg[\phi(0) + \sum_{i=1}^{N}\frac{\partial\phi(0)}{\partial x_i}x_i + \frac{1}{2!}\sum_{i,j=1}^{N}\frac{\partial^2\phi(0)}{\partial x_i\partial x_j}x_ix_j$$

$$+\cdots+\sum_{|\alpha|=m-1}D^{\alpha}\phi(0)x^{\alpha}\Bigg]H\left(1-\frac{|x|}{\varepsilon}\right)\Bigg\}\,dx, \tag{3.61}$$

where H is the Heaviside function and $\varepsilon > 0$ is small.

The expression in square brackets is the Taylor expansion of degree $m-1$, so

$$\phi(x) - \left[\phi(0) + \sum_{i=1}^{N} \frac{\partial \phi(0)}{\partial x_i} x_i + \cdots + \sum_{|\alpha| \leqslant m-1} D^\alpha \phi(0) x^\alpha \right] = \mathscr{O}(|x|^m).$$

Hence the integral (3.61) is finite for all $\phi \in \mathscr{D}$. It is not hard to verify that (3.61) is a continuous linear functional on $\mathscr{D}$.

If $\phi \in \mathscr{D}$ vanishes in a neighborhood of the origin,then

$$D^\alpha \phi(0) = 0 \quad \text{for} \quad 0 \leqslant |\alpha| \leqslant m-1,$$

and (3.61) becomes

$$\langle f, \phi \rangle = \int_{\mathbb{R}^N} f(x)\phi(x)\, dx.$$

Therefore, except at the origin, the functional f coincides with $f(x)$, and (3.61) is a regularization. □

Because the positive constant ε in (3.61) can be chosen arbitrarily, the divergent integral $\langle f, \phi \rangle$ has many regularizations. The following theorem somewhat clarifies the situation.

Theorem 3.2. *Let $t_0 \in \mathscr{D}'$ regularize the divergent integral (3.61). Then any $t \in \mathscr{D}'$ regularizing (3.61) is obtainable by adding to t_0 a linear combination of the delta function and its derivatives concentrated at the origin (where the algebraic singularity of $f(x)$ is located).*

Proof. Let g be a linear combination of the delta function and its derivatives at $x = 0$. Then for any $\phi \in \mathscr{D}$ vanishing in a neighborhood of the origin,

$$\langle t_0 + g, \phi \rangle = \langle t_0, \phi \rangle + \langle g, \phi \rangle = \langle t_0, \phi \rangle = \int_{\mathbb{R}^N} f(x)\phi(x)\, dx,$$

so $t_0 + g$ is also a regularization. Conversely, if t is any regularization then for all such ϕ, we have

$$\begin{aligned}
\langle t - t_0, \phi \rangle &= \langle t, \phi \rangle - \langle t_0, \phi \rangle \\
&= \int_{\mathbb{R}^N} f(x)\phi(x)\, dx - \int_{\mathbb{R}^N} f(x)\phi(x)\, dx \\
&= 0,
\end{aligned}$$

so $t - t_0$ is concentrated at $x = 0$. Any distribution concentrated at $x = 0$ is a finite linear combination of the delta function and its derivatives (why?). Therefore $t - t_0 = g$ for some g described earlier in the proof. □

Remark 3.2. If the function $f(x)$ has a singularity (say, at $x = 0$) *of infinite order*, e.g.,

$$f(x) = e^{1/|x|^2},$$

and satisfies

$$|f(x)| > A_m/|x|^m$$

in a neighborhood of the origin for each $m = 0, 1, 2, \ldots$ then $f(x)$ does not admit a regularization in the way given by Theorem 3.1.

This does not mean that in generalized functional analysis, one may never consider functions with singularities of infinite order. It is still possible to define test function spaces other than $\mathscr{D}$ on which functions with singularities of any kind can be given meaning as functionals. □

Theorems 3.1 and 3.2 still have not told us exactly how to find the finite part of a divergent integral. Indeed, no such general formulas are available in $\mathbb{R}^N$ when $N > 1$. Fortunately, if f satisfies certain nice properties such as (3.62) below, as it does in many applications, we can take the finite part nearly as easily as in the one-dimensional case.

Let us assume that

$$f(x) = |x|^{-\lambda} g(x), \quad \lambda \geqslant N, g \in C^\infty(\mathbb{R}^N). \tag{3.62}$$

For such $f, \langle f, \phi \rangle$ defines a divergent integral. We note that for $\phi \in \mathscr{D}$,

$$\langle f, \phi \rangle = \int_{\mathbb{R}^N} |x|^{-\lambda} g(x)\phi(x)\, dx,$$

which can be expressed in hyperspherical coordinates as

$$\langle f, \phi \rangle = \int_0^\infty r^{-\lambda} \left[\int_{\omega_N} g(r\omega)\phi(r\omega)\, d\omega \right] r^{N-1}\, dr,$$

where $r = |x|$, ω_N is the hypersurface of the unit ball and $d\omega$ is the hypersurface element on the unit sphere. We further denote

$$\int_{\omega_N} g(r\omega)\phi(r\omega)\, d\omega \equiv |\omega_N| S_{g\phi}(r),$$

where

$\lvert\omega_N\rvert = \dfrac{2\pi^{N/2}}{\Gamma(\frac{1}{2}N)}$	is	the area of the hypersurface ω_N,
$S_{g\phi}(r)$	is	the mean value of $g(x)\phi(x)$ on the spherical surface of radius r.

Thus, we arrive at

$$\langle f, \phi \rangle = |\omega_N| \int_0^\infty r^{N-(\lambda+1)} S_{g\phi}(r)\, dr. \tag{3.63}$$

We claim that $S_{g\phi}(\cdot) \in C_0^\infty(\mathbb{R}^+)$, and all of its derivatives of odd order vanish at $r = 0$.
It is obvious that $S_{g\phi}(\cdot)$ is infinitely differentiable for $r > 0$. For sufficiently large r, $g(x)\phi(x)$ vanishes, so the average $S_{g\phi}(r)$ has compact support.
To show that $S_{g\phi}(r)$ is also infinitely differentiable at $r = 0$, we expand $\psi(x) = g(x)\phi(x)$ as a Taylor series and obtain

$$|\omega_N| S_{g\phi}(r) = \int_{\omega_N} \left[\psi(0) + \sum_{j=1}^{N} \frac{\partial \psi(0)}{\partial x_j} x_j + \cdots + \frac{1}{m!} \sum_{|\alpha|=m} D^\alpha \psi(0) x^\alpha + \cdots \right] d\omega.$$

After integrating with respect to the angular variables ω, the odd-power terms x^α disappear, and only even-power terms remain. They must be functions of r only; thus, we have

$$S_{g\phi}(r) = \psi(0) + a_1 r^2 + a_2 r^4 + \cdots + a_k r^{2k} + o(r^{2k}).$$

Therefore $S_{g\phi}(r)$ has derivatives of arbitrary order at $r = 0$, and all its derivatives of odd order vanish. $S_{g\phi}(r)$ is an even function if we naturally extend it to negative r values.
We can now interpret (3.63) as

$$\begin{aligned} \langle f, \phi \rangle &\equiv |\omega_N| \int_0^\infty x^{N-(\lambda+1)} S_{g\phi}(x)\, dx \\ &\equiv |\omega_N| \langle x_+^{N-(\lambda+1)}, S_{g\phi} \rangle, \end{aligned} \tag{3.64}$$

where the pseudofunction $x_+^{N-(\lambda+1)}$ has been defined as a distribution in §3.1 and 3.2, for which we know how to take the finite part. Thus, (3.64) gives us the *finite part*, or the *canonical regularization* of the divergent integral $\langle f, \phi \rangle$.

Exercise 3.3. Let $\phi \in \mathscr{D}(\mathbb{R})$. Give the finite part of the divergent integral

$$\int_0^\infty x^{-3} \phi(x)\, dx = \langle x_+^{-3}, \phi \rangle.$$ □

3.5 The Fourier transform of tempered distributions

We have defined the classical Fourier transform (FT) for functions in Chapter 2. Now we study the FT for distributions.
Consider $\mathbb{R}^1$ first. In an attempt to define the FT of a distribution t, it would be natural to just let

$$\widehat{t}(\xi) = \mathscr{F}(t)(\xi) = \int_{-\infty}^{\infty} t(x) e^{-2\pi i x \xi}\, dx.$$

However, for each given $\xi \in \mathbb{R}$, $e^{-2\pi i x\xi}$ is not a test function in $\mathscr{D}$; thus

$$\widehat{t}(\xi) = \mathscr{F}(t)(\xi) = \langle t, e^{-2\pi i x\xi} \rangle$$

is not defined. We could try the Bessel–Parseval formula (2.14), now rewritten as

$$\langle \widehat{f}, g \rangle_{L^2(\mathbb{R}^N)\times L^2(\mathbb{R}^N)} = \langle f, \widehat{g} \rangle_{L^2(\mathbb{R}^N)\times L^2(\mathbb{R}^N)},$$

i.e.,

$$\int_{-\infty}^{\infty} \widehat{f}(x) g(x)\, dx = \int_{-\infty}^{\infty} f(x)\widehat{g}(x)\, dx,$$

connecting the FT of two functions f and g. That is, we might define

$$\langle \widehat{t}, \phi \rangle = \langle t, \widehat{\phi} \rangle \quad \forall \phi \in \mathscr{D}.$$

But this does not work either, because $\widehat{\phi}$ may not be a test function even though ϕ itself is. To make things work, we define the Schwartz space $\mathscr{S}$ and the notion of tempered distributions.

Definition 3.2. The (Schwartz) space $\mathscr{S}$ of *rapidly decreasing test functions* consist of functions $\phi\colon\ \mathbb{R}^N \to \mathbb{C}$ satisfying

(1) $\phi \in C^\infty(\mathbb{R}^N)$;

(2) ϕ, as well as all of its derivatives, decay faster than the reciprocal of any polynomial at ∞, i.e., for any nonnegative integral multi-indices m and n, there exists $C_{mn} > 0$ depending only on ϕ, m and n such that

$$|x^m D^n \phi(x)| < C_{mn} \qquad \forall x \in \mathbb{R}^N. \tag{3.65}$$

A sequence $\{\phi_p \mid p = 1, 2, 3, \ldots\}$ is said to converge to ϕ_0 in $\mathscr{S}'$ if $D^n\phi_p \to D^n\phi_0$ uniformly on all compact sets K, and

$$|x^m D^n [\phi_p(x) - \phi_0(x)]| < C_{mn}(K) \quad \forall x \in \mathbb{R}^N,\ \forall p.$$

When $\phi_0(x) \equiv 0$, the sequence $\{\phi_p\}$ is called a *null sequence.* □

Definition 3.3. A continuous linear functional t on $\mathscr{S}$ is called a *tempered distribution*, i.e., t satisfies

(1) $$\langle t, c_1\phi_1 + c_2\phi_2 \rangle = c_1\langle t, \phi_1 \rangle + c_2\langle t, \phi_2 \rangle \quad \forall c_1, c_2 \in \mathbb{C},\ \ \phi_1, \phi_2 \in \mathscr{S};$$

(2) $\lim_{p\to\infty} \langle t, \phi_p \rangle = 0$ for every null sequence $\{\phi_p\}$. We denote by $\mathscr{S}'$ the space of all tempered distributions. A sequence $\{t_p \mid p = 1, 2, \ldots\}$ is said to converge to t_0 in $\mathscr{S}'$ if for every $\phi \in \mathscr{S}$,

$$\lim_{p\to\infty} \langle t_p, \phi \rangle = \langle t_0, \phi \rangle.$$ □

Definition 3.4. Let f be a continuous function on $\mathbb{R}^N$. If

$$|f(x)| = \mathscr{O}(|x|^\alpha), \quad |x| \text{ large},$$

for some real α, we call f a *tempered function*. □

The space $\mathscr{D}$ is *dense* in $\mathscr{S}$. To see this, let us use a *cutoff function* $\theta \in C_0^\infty(\mathbb{R})$:

$$\theta(x) = \begin{cases} 1 & \text{for } |x| \leqslant 1, \\ 0 & \text{for } |x| \geqslant 2. \end{cases}$$

Then, for any $\phi \in \mathscr{S}$, the sequence of functions

$$\phi_p(x) \equiv \theta(|x|/p)\phi(x), \quad p = 1, 2, 3, \ldots, \tag{3.66}$$

is in $\mathscr{D}$ with the limit ϕ in the sense of convergence in $\mathscr{S}$. Therefore

$$\mathscr{D} \subset \mathscr{S}.$$

This yields the reversed inclusion

$$\mathscr{S}' \subset \mathscr{D}'.$$

Derivatives of tempered distributions can be defined in a similar way as in (2.3).

Exercise 3.4. Show that every tempered function f defines a tempered distribution by

$$\langle f, \phi \rangle = \int_{\mathbb{R}^N} f(x)\phi(x)\, dx, \quad \forall \phi \in \mathscr{S}.$$ □

The most pleasing property of the test function space $\mathscr{S}$ is its invariance under the Fourier transform.

Theorem 3.3. *The Fourier transform $\mathscr{F}$ and its inverse are continuous, linear 1–1 mappings of $\mathscr{S}$ onto itself.*

Proof. The Fourier transform

$$\widehat{\phi}(\xi) = \int_{\mathbb{R}^N} \phi(x) e^{-2\pi i \langle x, \xi \rangle}\, dx, \quad \phi \in \mathscr{S}, \tag{3.67}$$

and its mixed derivatives

$$D^n \widehat{\phi(\xi)} = (-2\pi i)^{|n|} \int_{\mathbb{R}^N} x^n \phi(x) e^{-2\pi i \langle x, \xi \rangle}\, dx \tag{3.68}$$

(n is a nonnegative integral multi-index)

are absolutely convergent integrals, thanks to the rapid decreasing property (3.65). Also, for any nonnegative integral multi-index m, by (2.12) and (2.13),

$$\left| \xi^m D^n \widehat{\phi}(\xi) \right| = \left| (-2\pi i)^{|n|} \int_{\mathbb{R}^N} x^n \phi(x) (-2\pi i)^{-|m|} \left(D_x^m e^{-2\pi i \langle x, \xi \rangle} \right) dx \right|. \tag{3.69}$$

Integration by parts $|m|$ times in the RHS gives

$$\left|\xi^m D^n \widehat{\phi}(\xi)\right| = \left|(-2\pi i)^{|n|-|m|} \int_{\mathbb{R}^N} D^m \left[x^n \phi(x)\right] e^{-2\pi i\langle x,\xi\rangle}\, dx\right|$$
$$\leqslant (2\pi)^{|n|-|m|} \int_{\mathbb{R}^N} |D^m [x^n \phi(x)]|\, dx. \tag{3.70}$$

But $D^m[x^n\phi(x)]$ is a sum of terms of the form

$$x^\alpha \left[D^\beta \phi(x)\right],$$

which are all absolutely integrable by (3.65). Hence $\widehat{\phi} \in \mathscr{S}$. This gives

$$\mathscr{F}(\mathscr{S}) \subseteq \mathscr{S}. \tag{3.71}$$

The inverse transform $\mathscr{F}^{-1}$ has the same property (just drop the negative signs before $2\pi i$ in (3.67)–(3.70)), so

$$\mathscr{F}^{-1}(\mathscr{S}) \subseteq \mathscr{S}. \tag{3.72}$$

Because

$$\mathscr{F}\mathscr{F}^{-1}\phi = \phi, \quad \forall \phi \in \mathscr{S}, \tag{3.73}$$

we have $\mathscr{F}(\mathscr{S}) = \mathscr{S}$ and $\mathscr{F}$ is 1–1 and onto. □

We can now give the following.

Definition 3.5. Let $t \in \mathscr{S}'$ be a tempered distribution. Then its Fourier transform $\widehat{t}$ is provided by the Bessel–Parseval formula

$$\langle \widehat{t}, \phi\rangle = \langle t, \widehat{\phi}\rangle \quad \forall \phi \in \mathscr{S}. \tag{3.74}$$

□

We note that the RHS of (3.74) is a well-defined continuous linear functional, thanks to Theorem 3.3. Therefore, $\widehat{t}$ is a well-defined tempered distribution in $\mathscr{S}'$. Similarly, we can define the inverse FT of a tempered distribution by

$$\langle \mathscr{F}^{-1}t, \phi\rangle = \langle t, \mathscr{F}^{-1}\phi\rangle, \quad \forall \phi \in \mathscr{S}. \tag{3.75}$$

It is straightforward from (3.73)–(3.75) that for any $t \in \mathscr{S}'$, we have

$$\mathscr{F}\left(\mathscr{F}^{-1}t\right) = \mathscr{F}^{-1}(\mathscr{F}t) \;=\; t, \tag{3.76}$$
$$\mathscr{F}\left(D^\alpha t\right) = (2\pi i)^{|\alpha|} \xi^\alpha \widehat{t}, \tag{3.77}$$
$$D^\alpha \left(\mathscr{F}t\right) = (-2\pi i)^{|\alpha|} \mathscr{F}(x^\alpha t), \tag{3.78}$$

which are, in a sense, generalizations of (2.9), (2.12) and (2.13) respectively.

3.6 Examples of Fourier transforms of tempered distributions

The singular distributions $\delta(x)$ and its derivatives, x_+^λ, $|x|^\lambda$, etc. treated in the earlier sections are also tempered distributions, whose classical Fourier transforms mostly were not well defined. Now we can compute them in the following examples.

Example 3.2. The delta function

$$\begin{aligned}\langle \widehat{\delta}, \phi \rangle = \langle \delta, \widehat{\phi} \rangle &= \langle \delta(x), \int_{\mathbb{R}^N} \phi(y) e^{-i2\pi\langle y,x \rangle}\, dy \rangle \\ &= \int_{\mathbb{R}^n} \phi(y)\, dy = \langle 1, \phi \rangle, \quad \forall \phi \in \mathscr{S};\end{aligned}$$

thus

$$\widehat{\delta} = 1. \tag{3.79}$$

By (3.76), we get

$$\widehat{1} = \widehat{\widehat{\delta}}(x) = \delta(-x) = \delta(x), \tag{3.80}$$

because δ is an even distribution:

$$\langle \delta(-x), \phi \rangle = \phi(0) = \langle \delta(x), \phi \rangle, \quad \forall \phi \in \mathscr{S}.$$

□

Example 3.3. The Heaviside function (in $\mathbb{R}^1$) satisfies

$$H'(x) = \delta(x), \qquad \text{by (3.4).}$$

By (3.79) and (3.77),

$$1 = \widehat{H'}(\xi) = (i2\pi)\xi \widehat{H}(\xi).$$

The solution to this equation is not

$$\widehat{H}(\xi) = \frac{1}{(2\pi i)} \frac{1}{\xi}$$

in $\mathscr{S}'$, but rather

$$\widehat{H}(\xi) = \frac{1}{2\pi i} \frac{1}{\xi} + c\delta(\xi) \tag{3.81}$$

for some constant $c \in \mathbb{C}$, because $\xi \delta(\xi) \equiv 0$ in $\mathscr{S}'$. (Can you rigorize the argument in (3.81)?) To determine the constant c, we note

$$[H(-x)]^\wedge(\xi) = c\delta(\xi) - \frac{1}{2\pi i} \frac{1}{\xi}. \tag{3.82}$$

By (3.80),

$$\begin{aligned}\widehat{1} = \delta(\xi) &= \mathscr{F}[H(-x) + H(x)](\xi) \\ &= 2c\delta(\xi), \qquad \text{by (3.81) and (3.82).}\end{aligned}$$

or $c = \frac{1}{2}$. Therefore

$$\widehat{H}(\xi) = \frac{1}{2}\delta(\xi) + \frac{1}{2\pi i}\frac{1}{\xi}. \tag{3.83}$$

Since

$$\begin{aligned}\widehat{H}(\xi) &= \int_0^\infty e^{-2\pi i x\xi}\,dx \\ &= \int_0^\infty \cos(2\pi x\xi)\,dx - i\int_0^\infty \sin(2\pi x\xi)\,dx,\end{aligned}$$

separating the real and imaginary parts gives

$$\begin{aligned}\int_0^\infty \cos(2\pi x\xi)\,dx &= \frac{1}{2}\delta(\xi), \\ \int_0^\infty \sin(2\pi x\xi)\,dx &= \frac{1}{2\pi}\frac{1}{\xi}.\end{aligned} \tag{3.84}$$

Also, since

$$\operatorname{sgn} x = H(x) - H(-x), \tag{3.85}$$

from (3.81) and (3.82),

$$[\operatorname{sgn} x]^\wedge(\xi) = \widehat{H}(\xi) - [H(-x)]^\wedge(\xi) = \frac{1}{i\pi}\frac{1}{\xi}. \tag{3.86}$$

Thus

$$\left(\frac{1}{\xi}\right)^\wedge(x) = i\pi\,\operatorname{sgn}(-x) = -i\pi\,\operatorname{sgn} x. \tag{3.87}$$

□

Example 3.4. The Heisenberg delta distributions. From (3.54),

$$\delta^+(x) = \frac{1}{2}\delta(x) - \frac{1}{2\pi i}\frac{1}{x},$$

we have

$$\begin{aligned}\widehat{\delta}^+(\xi) &= \frac{1}{2}\widehat{\delta}(\xi) - \frac{1}{2\pi i}\left(\frac{1}{x}\right)^\wedge(\xi) \\ &= \frac{1}{2} + \frac{i\pi}{2\pi i}\operatorname{sgn}\xi \quad \text{(cf. (3.87) and (3.79))} \\ &= \frac{1}{2}(1 + \operatorname{sgn}\xi) = H(\xi).\end{aligned} \tag{3.88}$$

Similarly,

$$\widehat{\delta^-}(\xi) = H(-\xi). \tag{3.89}$$

□

Example 3.5. The pseudofunction $|x|^{-1}$ in $\mathbb{R}^1$. From § 3.1, in particular (3.30), we define the pseudofunction $|x|^{-1}$ by

$$\mathbf{PF}|x|^{-1} = x_+^{-1} + x_-^{-1},$$

i.e.,

$$\langle |x|^{-1}, \phi \rangle = \int_{|x|<1} \frac{\phi(x)-\phi(0)}{|x|}\,dx + \int_{|x|>1} \frac{\phi(x)}{|x|}\,dx, \quad \forall \phi \in \mathscr{S}.$$

We have

$$\begin{aligned}
\langle \mathscr{F}(|x|^{-1}), \phi \rangle &= \langle |x|^{-1}, \widehat{\phi}(x) \rangle \\
&= \int_{-1}^{1} \frac{\widehat{\phi}(x)-\widehat{\phi}(0)}{|x|}\,dx + \int_{|x|>1} \frac{\widehat{\phi}(x)}{|x|}\,dx \\
&= \int_{-1}^{1} \frac{1}{|x|}\left[\int_{-\infty}^{\infty} \phi(u)(e^{-2\pi i x u}-1)\,du\right]dx + \int_{|x|>1} \frac{1}{|x|}\left[\int_{-\infty}^{\infty} \phi(u)e^{-2\pi i x u}\,du\right]dx \\
&= 2\int_0^1 \int_{-\infty}^{\infty} \phi(u)\frac{\cos(2\pi x u)-1}{x}\,du\,dx + 2\int_1^{\infty}\int_{-\infty}^{\infty} \phi(u)\frac{\cos(2\pi x u)}{x}\,du\,dx \\
&= \frac{2}{2\pi}\int_0^1 \int_{-\infty}^{\infty} \phi\left(\frac{v}{2\pi}\right)\frac{\cos xv - 1}{x}\,dv\,dx + \frac{2}{2\pi}\int_1^{\infty}\int_{-\infty}^{\infty} \phi\left(\frac{v}{2\pi}\right)\frac{\cos xv}{x}\,dv\,dx \\
&= \frac{1}{\pi}\int_{-\infty}^{\infty} \phi\left(\frac{v}{2\pi}\right)\left(\int_0^{|v|}\frac{\cos y - 1}{y}\,dy + \int_{|v|}^{\infty}\frac{\cos y}{y}\,dy\right)dv \\
&= \frac{1}{\pi}\int_{-\infty}^{\infty} \phi\left(\frac{v}{2\pi}\right)\left(\int_0^{1}\frac{\cos y - 1}{y}\,dy + \int_1^{|v|}\frac{\cos y - 1}{y}\,dy\right. \\
&\qquad\qquad \left. + \int_1^{\infty}\frac{\cos y}{y}\,dy - \int_1^{|v|}\frac{\cos y}{y}\,dy\right)dv \\
&= \frac{1}{\pi}\int_{-\infty}^{\infty} \phi\left(\frac{v}{2\pi}\right)\left(\int_0^{1}\frac{\cos y - 1}{y}\,dy + \int_1^{\infty}\frac{\cos y}{y}\,dy - \int_1^{|v|}\frac{dy}{y}\right)dv. \qquad (3.90)
\end{aligned}$$

But it is known that

$$\int_0^1 \frac{1-\cos y}{y}\,dy - \int_1^{\infty}\frac{\cos y}{y}\,dy = \gamma,$$

where γ is the Euler constant

$$\begin{aligned}
\gamma &\equiv \lim_{n\to\infty}\left(1+\frac{1}{2}+\frac{1}{3}+\cdots+\frac{1}{n}-\ln n\right) \\
&= -\int_0^{\infty} e^{-t}\ln t\,dt = -\Gamma'(1). \qquad (3.91)
\end{aligned}$$

Continuing, we get

$$\text{RHS}\,(3.90) = \frac{1}{\pi}\int_{-\infty}^{\infty}\phi\left(\frac{v}{2\pi}\right)(-\gamma - \ln|v|)\,dv \qquad (3.92)$$

$$= -2\int_{-\infty}^{\infty}\phi(u)(\gamma + \ln|2\pi u|)\,du.$$

Hence

$$(|x|^{-1})^{\wedge}(\xi) = -2\left[\ln|\xi| + (\gamma + \ln 2\pi)\right]. \qquad (3.93)$$

□

Example 3.6. The pseudofunction $x_+^{-\lambda}$ in $\mathbb{R}^1$. We first consider

$$0 < \operatorname{Re}\lambda < 1.$$

We take advantage of the fact that

$$x_+^{-\lambda} = x^{-\lambda}H(x) = \lim_{\tau\downarrow 0}\left[e^{-2\pi\tau x}x^{-\lambda}H(x)\right],$$

to get

$$(x_+^{-\lambda})^{\wedge}(\xi) = \int_{-\infty}^{\infty} x_+^{-\lambda}e^{-2\pi i x\xi}\,dx \qquad (3.94)$$

$$= \lim_{\tau\downarrow 0}\int_0^{\infty} x^{-\lambda}e^{2\pi i(-\xi + i\tau)x}\,dx$$

$$= \lim_{\tau\downarrow 0+}\int_0^{\infty} x^{-\lambda}e^{-2\pi i s x}\,dx, \quad \text{where } s = \xi - i\tau.$$

Since $\tau = \text{Im}$, $s < 0$, we have $0 < -\arg s < \pi$. We make a change of variables $2\pi i s x = \zeta$, or $x = -i\zeta/(2\pi s)$, $dx = -id\zeta/(2\pi s)$, and obtain

$$(3.94) = \lim_{\tau\downarrow 0}\int_L \left(\frac{-i}{2\pi s}\right)^{-\lambda}\zeta^{-\lambda}e^{-\zeta}\left(\frac{-i}{2\pi s}\right)d\zeta \qquad (3.95)$$

$$= \lim_{\tau\downarrow 0}\left(\frac{-i}{2\pi s}\right)^{-(\lambda-1)}\int_L \zeta^{-\lambda}e^{-\zeta}\,d\zeta,$$

where the contour of integration is L as shown in Fig. 3.1 for the case $\xi > 0$. By an application of the Cauchy integral theorem, we can show that

$$\int_L \zeta^{-\lambda}e^{-\zeta}\,d\zeta = \int_0^{\infty}\zeta^{-\lambda}e^{-\zeta}\,d\zeta = \Gamma(-\lambda+1),$$

where Γ is the gamma function. So (3.95) gives

$$(x_+^{-\lambda})^{\wedge}(\xi) = \lim_{\tau\downarrow 0+}\left(\frac{-i}{2\pi s}\right)^{-(\lambda-1)}\Gamma(-\lambda+1)$$

$$= \lim_{\tau\downarrow 0+}\left[\left(\frac{-i}{2\pi}\right)^{-(\lambda-1)}(\xi - i\tau)^{\lambda-1}\Gamma(-\lambda+1)\right]$$

$$= (2\pi)^{\lambda-1}e^{\pi i(\lambda-1)/2}\Gamma(-\lambda+1)(\xi - i0)^{\lambda-1}. \qquad (3.96)$$

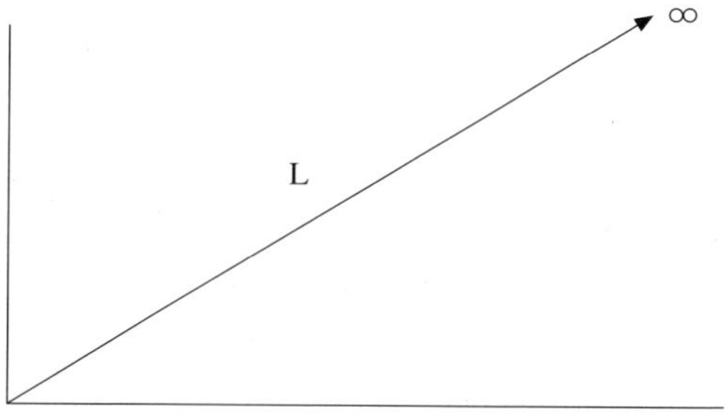

Figure 3.1: Contour L for the integration of (6.17).

But by (3.56),

$$(\xi - i0)^{\lambda-1} = \xi_+^{\lambda-1} + e^{-i\pi(\lambda-1)}\xi_-^{\lambda-1}. \tag{3.97}$$

Combining (3.96) and (3.97), we obtain

$$\begin{aligned}(x_+^{-\lambda})^\wedge(\xi) &= (2\pi)^{\lambda-1}\Gamma(-\lambda+1)\left[e^{\pi i(\lambda-1)/2}\xi_+^{\lambda-1} + e^{-\pi i(\lambda-1)/2}\xi_-^{\lambda-1}\right] \\ &= (2\pi)^{\lambda-1}\Gamma(-\lambda+1)e^{\pi i(\lambda-1)(\operatorname{sgn}\xi)/2}|\xi|^{\lambda-1},\end{aligned} \tag{3.98}$$

for $\lambda \in \mathbb{C}$, $0 < \text{Re}\,\lambda < 1$. By analytic continuation, this remains valid for all $\lambda \in \mathbb{C}$, except at $\lambda = 1, 2, 3, \ldots$, where $\Gamma(-\lambda+1)$ becomes singular. □

Exercise 3.5. Mimic the above derivations and show that

$$\begin{aligned}(x_-^{-\lambda})^\wedge(\xi) &= (2\pi)^{\lambda-1}e^{-\pi i(\lambda-1)/2}\Gamma(-\lambda+1)(\xi+i0)^{\lambda-1} \\ &= (2\pi)^{\lambda-1}\Gamma(-\lambda+1)e^{-\pi i(\lambda-1)(\operatorname{sgn}\xi)/2}|\xi|^{\lambda-1}.\end{aligned} \tag{3.99}$$

□

Example 3.7.

(i) $|x|^{-\lambda} = x_+^{-\lambda} + x_-^{-\lambda}$, $\lambda \neq 1, 2, 3, \ldots$

$$\begin{aligned}(|x|^{-\lambda})^\wedge(\xi) &= (x_+^{-\lambda})^\wedge(\xi) + (x_-^{-\lambda})^\wedge(\xi) \\ &= (2\pi)^{\lambda-1}\Gamma(-\lambda+1)e^{\pi i(\lambda-1)(\operatorname{sgn}\xi)/2}|\xi|^{\lambda-1} \\ &\quad + (2\pi)^{\lambda-1}\Gamma(-\lambda+1)e^{-\pi i(\lambda-1)(\operatorname{sgn}\xi)/2}|\xi|^{\lambda-1} \\ &= -i(2\pi)^{\lambda-1}\Gamma(-\lambda+1)[(e^{\pi i\lambda/2}\xi_+^{\lambda-1} - e^{-\pi i\lambda/2}\xi_-^{\lambda-1}) \\ &\quad - (e^{-\pi i\lambda/2}\xi_+^{\lambda-1} - e^{\pi i\lambda/2}\xi_-^{\lambda-1})] \\ &= 2(2\pi)^{\lambda-1}\Gamma(-\lambda+1)\sin(\frac{\lambda\pi}{2})|\xi|^{\lambda-1}.\end{aligned} \tag{3.100}$$

(ii)

$$|x|^{-\lambda}\operatorname{sgn} x = x_+^{-\lambda} - x_-^{-\lambda}, \quad \lambda \neq 1, 2, 3, \ldots$$
$$(|x|^{-\lambda}\operatorname{sgn} x)^\wedge = (x_+^{-\lambda})^\wedge(\xi) - (x_-^{-\lambda})^\wedge(\xi).$$

So in (3.100), if we do subtraction instead of addition, we obtain

$$[|x|^{-\lambda}\operatorname{sgn} x]^\wedge(\xi) = -2i(2\pi)^{\lambda-1}\Gamma(-\lambda+1)\cos(\lambda\pi/2)|\xi|^{\lambda-1}\operatorname{sgn}\xi.$$

(iii) $x_+^{-\lambda}\ln x_+$, $\lambda \neq 1, 2, 3, \ldots$ We differentiate (3.100) with respect to λ and obtain

$$\begin{aligned}(-x_+^{-\lambda}\ln x_+)^\wedge(\xi) = {} & (2\pi)^{\lambda-1}e^{\pi i(\lambda-1)/2}\Big[(\ln 2\pi)\Gamma(-\lambda+1)(\xi-i0)^{\lambda-1} \\ & + \frac{\pi i}{2}\Gamma(-\lambda+1)(\xi-i0)^{\lambda-1} - \Gamma'(-\lambda+1)(\xi-i0)^{\lambda-1} \\ & + \Gamma(-\lambda+1)(\xi-i0)^{\lambda-1}\ln(\xi-i0)\Big]. \end{aligned} \tag{3.101}$$

In particular, when $\lambda = 0$, we obtain

$$\begin{aligned}&(\ln x_+)^\wedge(\xi) \\ &= \frac{i}{2\pi}\left\{\left[\ln 2\pi + \frac{\pi i}{2} - \Gamma'(1)\right](\xi-i0)^{-1} + (\xi-i0)^{-1}\ln(\xi-i0)\right\}.\end{aligned} \tag{3.102}$$

(iv) $x_-^{-\lambda}\ln x_-$, $\lambda \neq 1, 2, 3, \ldots$ Similarly, we differentiate (3.99) and obtain

$$\begin{aligned}(-x_-^{-\lambda}\ln x_-)^\wedge(\xi) = {} & (2\pi)^{\lambda-1}e^{-\pi i(\lambda-1)/2}[(\ln 2\pi)\Gamma(-\lambda+1)(\xi+i0)^{\lambda-1} \\ & - \frac{1}{2}\pi i\Gamma(-\lambda+1)(\xi+i0)^{\lambda-1} - \Gamma'(-\lambda+1)(\xi+i0)^{\lambda-1} \\ & + \Gamma(-\lambda+1)(\xi+i0)^{\lambda-1}\ln(\xi+i0)]. \end{aligned} \tag{3.103}$$

In particular, when $\lambda = 0$, the above gives

$$\begin{aligned}&(\ln x_-)^\wedge(\xi) \\ &= -\frac{i}{2\pi}\left\{\left[\ln 2\pi - \frac{1}{2}\pi i - \Gamma'(1)\right](\xi+i0)^{-1} + (\xi+i0)^{-1}\ln(\xi+i0)\right\}.\end{aligned} \tag{3.104}$$

(v) $\ln|x|$. Adding (3.103) and (3.104), we obtain

$$\begin{aligned}(\ln|x|)^\wedge(\xi) &= (\ln x_+)^\wedge(\xi) + (\ln x_-)^\wedge(\xi) \\ &= \text{RHS of (3.103)} + \text{RHS of (3.104)}.\end{aligned} \tag{3.105}$$

We further simplify the above by using (3.55) and (3.57). Consider first $\xi > 0$. Then continuing from (3.105) we have

$$\begin{aligned}(\ln|x|]^\wedge(\xi) = {} & \frac{i}{2\pi}\Big\{\Big[\ln 2\pi + \frac{1}{2}\pi i - \Gamma'(1)\Big][\pi i\delta(\xi) + \xi^{-1}] \\ & + [\pi i\delta(\xi) + \xi^{-1}]\ln\xi\Big\} - \frac{i}{2\pi}\Big\{\Big[\ln 2\pi - \frac{1}{2}\pi i - \Gamma'(1)\Big]\cdot \\ & [-\pi i\delta(\xi) + \xi^{-1}] + [-\pi i\delta(\xi) + \xi^{-1}]\ln\xi\Big\} \\ = {} & -[\ln 2\pi - \Gamma'(1)]\delta(\xi) - \frac{1}{2}\xi^{-1}, \quad \xi > 0,\end{aligned} \tag{3.106}$$

where we have used (3.55) and (3.57). Similarly, for $\xi < 0$, (3.106) gives

$$\begin{aligned}(\ln|x|)^\wedge(\xi) &= \frac{i}{2\pi}\Big\{\Big[\ln 2\pi + \frac{1}{2}\pi i - \Gamma'(1)\Big][\pi i\delta(\xi) + \xi^{-1}] \\ &\quad + [\pi i\delta(\xi) + \xi^{-1}](\ln|\xi| - i\pi)\Big\} \\ &\quad - \frac{i}{2\pi}\Big\{\Big[\ln 2\pi - \frac{1}{2}\pi i - \Gamma'(1)\Big][-\pi i\delta(\xi) + \xi^{-1}] \\ &\quad + [-\pi i\delta(\xi) + \xi^{-1}](\ln|\xi| + i\pi)\Big\} \\ &= -[\ln 2\pi - \Gamma'(1)]\delta(\xi) + \frac{1}{2}\xi^{-1}, \quad \xi < 0. \end{aligned} \tag{3.107}$$

Thus (3.106) and (3.107) combined give

$$(\ln|x|)^\wedge(\xi) = -(\ln 2\pi + \gamma)\delta(\xi) - \frac{1}{2}|\xi|^{-1}, \quad \text{by (3.91)}. \tag{3.108}$$

The reader may find that (3.108) and (3.93) are completely consistent with each other under the Fourier inversion formula (3.76). □

Example 3.8. $r^{-\lambda} = |x|^{-\lambda} = \left(\sum_{i=1}^N x_i^2\right)^{-\lambda/2}$ in $\mathbb{R}^N$.
Consider first $\lambda \in \mathbb{C}$, $0 <$ Re $\lambda < N$. Write

$$g_\lambda(\xi) = (r^{-\lambda})^\wedge(\xi) = \int_{\mathbb{R}^N} r^{-\lambda} e^{-2\pi i\langle x,\xi\rangle}\,dx. \tag{3.109}$$

We show that $g_\lambda(\xi)$ is a homogeneous function of degree $\lambda - N$ of the variable ξ, i.e.,

$$g_\lambda(t\xi) = t^{\lambda-N} g_\lambda(\xi), \quad t > 0, \tag{3.110}$$

since

$$\begin{aligned} g_\lambda(t\xi) &= \int_{\mathbb{R}^N} r^{-\lambda} e^{-2\pi i\langle x,t\xi\rangle}\,dx \\ &= \int_{\mathbb{R}^N} |x|^{-\lambda} e^{-2\pi i\langle tx,\xi\rangle}\,dx \\ &= t^{\lambda-N}\int_{\mathbb{R}^N} |tx|^{-\lambda} e^{-2\pi i\langle tx,\xi\rangle}\,d(tx) \\ &= t^{\lambda-N}\int_{\mathbb{R}^N} |y|^{-\lambda} e^{-2\pi i\langle y,\xi\rangle}\,dy \\ &= t^{\lambda-N} g_\lambda(\xi). \end{aligned}$$

Therefore g_λ must be a function of power $\lambda - N$:

$$g_\lambda(\xi) = C_\lambda|\xi|^{\lambda-N}, \qquad \xi \in \mathbb{R}^N. \tag{3.111}$$

To determine C_λ, we use the Bessel–Parseval relation (2.14)

$$\langle f,\phi\rangle = \langle \widehat{f},\widehat{\phi}\rangle, \quad f \in \mathscr{S}', \phi \in \mathscr{S}. \tag{3.112}$$

For $\phi(x) = e^{-2\pi^2 r^2} \in \mathscr{S}$, the following result is easy to obtain (and well known):

$$\begin{aligned}\widehat{\phi}(\xi) &= \int_{\mathbb{R}^N} e^{-|2\pi x|^2/2} e^{-2\pi i\langle x,\xi\rangle}\, dx \\ &= (2\pi)^{-N/2} e^{-|\xi|^2/2}.\end{aligned}$$

For $f = r^{-\lambda}$ and such a ϕ, (3.111) and (3.112) give

$$\int_{\mathbb{R}^N} r^{-\lambda} e^{-2\pi^2 r^2}\, dx = (2\pi)^{-N/2} C_\lambda \int_{\mathbb{R}^N} |\xi|^{\lambda-N} e^{-|\xi|^2/2}\, d\xi. \tag{3.113}$$

We evaluate the integrals on both sides in spherical coordinates:

$$dx = r^{N-1}\, dr\, d\omega_N, \quad d\xi = \rho^{N-1}\, d\rho\, d\omega_N'.$$

So

$$\begin{aligned}\int_{\mathbb{R}^N} r^{-\lambda} e^{-2\pi^2 r^2}\, dx &= |\omega_N| \int_0^\infty r^{-\lambda+N-1} e^{-2\pi^2 r^2}\, dr \\ &= 2^{(\lambda-N-2)/2} \pi^{\lambda-N} \Gamma\left(\frac{-\lambda+N}{2}\right) |\omega_N|\end{aligned}$$

and

$$\begin{aligned}\int_{\mathbb{R}^N} |\xi|^{\lambda-N} e^{-|\xi|^2/2}\, d\xi &= |\omega_N| \int_0^\infty \rho^{\lambda-1} e^{-\rho^2/2}\, d\rho \\ &= 2^{-\frac{1}{2}(-\lambda+2)} \Gamma\left(\frac{\lambda}{2}\right) |\omega_N|.\end{aligned}$$

Thus, (3.113) yields

$$\begin{aligned}C_\lambda &= (2\pi)^{N/2} 2^{(-\lambda+2)/2} \left[\Gamma\left(\frac{1}{2}\lambda\right)\right]^{-1} 2^{\frac{1}{2}(\lambda-N-2)} \pi^{\lambda-N} \Gamma\left(\frac{1}{2}(-\lambda+N)\right) \\ &= \pi^{\lambda-N/2} \Gamma\left(\frac{1}{2}(-\lambda+N)\right) / \Gamma\left(\frac{1}{2}\lambda\right).\end{aligned} \tag{3.114}$$

Consequently, from (3.111) we have

$$(r^{-\lambda})^\wedge(\xi) = \pi^{\lambda-N/2} \frac{\Gamma(\frac{1}{2}(-\lambda+N))}{\Gamma(\frac{1}{2}\lambda)} |\xi|^{\lambda-N}. \tag{3.115}$$

By analytic continuation, this formula is valid for all $\lambda \in \mathbb{C}$, except at the singularities of the gamma function, i.e.,

$$\begin{aligned}&\frac{1}{2}(-\lambda+N) \neq -k \quad \text{for} \quad k = 1, 2, 3, \ldots, \\ &\frac{1}{2}\lambda \neq -k \qquad\qquad \text{for} \quad k = 1, 2, 3, \ldots\end{aligned}$$

For example,

$$(r^{-(N-2)})^\wedge(\xi) = \frac{\pi^{N/2-2}}{\Gamma(\frac{1}{2}(N-2))} |\xi|^{-2}. \tag{3.116}$$

But

$$|\omega_N| = \frac{2\pi^{N/2}}{\Gamma(\frac{1}{2}N)}, \quad \Gamma\left(\frac{1}{2}N\right) = \left(\frac{1}{2}N - 1\right)\Gamma\left(\frac{1}{2}N - 1\right);$$

so (3.116) gives

$$\begin{aligned}\overline{\mathscr{F}}\left(|\xi|^{-2}\right)(x) &= \frac{\Gamma(\frac{1}{2}N)}{\pi^{N/2-2}}\frac{2}{N-2}r^{-(N-2)} \\ &= \frac{1}{|\omega_N|}\frac{4\pi^2}{N-2}r^{-(N-2)},\end{aligned}$$

which for $N = 3$, becomes

$$\overline{\mathscr{F}}(|\xi|^{-2}) = \pi r^{-1}. \tag{3.117}$$

□

Chapter 4

Pseudodifferential Operators and Their Fredholm Properties

Pseudodifferential operators are a natural extension of linear integral and partial differential operators. The theory of such operators grew out of the study of singular integral operators by Giraud, Mikhlin, Calderòn and Zygmund, among others. It developed rapidly after 1965 with the systematic studies of Kohn and Nirenberg [114], Hörmander, and other researchers. This theory has found many fields of application. In particular, all the boundary integral operators corresponding to the elliptic boundary value problems studied in this book are such operators. By using such a theory, the analysis of boundary integral equations and boundary element methods can be either greatly simplified or presented in a more general and elegant form.

In this chapter, we present some basic facts and properties of pseudo-differential operators for our subsequent applications. The interested reader can find a general account of the theory in Grubb [85], Kumano-go [116], Taylor [177], and Treves [179, 180], for example. A concise account given in Gilkey [80] is also helpful.

4.1 Symbol class S^m

A linear partial differential operator is a polynomial expression

$$P(x,D) = \sum_{|\alpha|\leqslant m} a_\alpha(x) D_x^\alpha, \quad \alpha\text{s are multi-indices.} \tag{4.1}$$

The *symbol* of P is denoted by

$$\sigma P = p(x,\xi) = \sum_{|\alpha|\leqslant m} a_\alpha(x)(i\xi)^\alpha. \tag{4.2}$$

Since

$$u(x) = \left[\mathscr{F}(\mathscr{F}u)\right](x) = \int_{\mathbb{R}^N} e^{2\pi i\langle x,\xi\rangle}\widehat{u}(\xi)\,d\xi,$$

we have

$$D_x^\alpha u(x) = \int_{\mathbb{R}^N} e^{2\pi i\langle x,\xi\rangle} \left[(2\pi i\xi)^\alpha \widehat{u}(\xi)\right] d\xi,$$

thus

$$\begin{aligned} P(x,D)u(x) &= \int_{\mathbb{R}^N} e^{2\pi i\langle x,\xi\rangle} \left[\sum_{|\alpha|\leqslant m} a_\alpha(x)(2\pi i\xi)^\alpha\right] \widehat{u}(\xi)\, d\xi \\ &= \int_{\mathbb{R}^N} e^{2\pi i\langle x,\xi\rangle} p(x,2\pi\xi)\widehat{u}(\xi)\, d\xi. \end{aligned} \tag{4.3}$$

We use this expression to define the action of a pseudodifferential operator (ψDO) for a larger class of functions $p(x,\xi)$ than polynomials.

Remark 4.1. Had we defined the Fourier transform by (2.8), then formula (4.3) would now be

$$P(x,D)u(x) = \int_{\mathbb{R}^N} e^{i\langle x,\xi\rangle} p(x,\xi)\widehat{u}(\xi) \frac{d\xi}{(2\pi)^{N/2}}, \tag{4.4}$$

appearing somewhat more elegant than (4.3). Using the definition of the Fourier transform as in (2.8) throughout all our discussions, the only adjustment we need to make is to use $p(x,2\pi\xi)$ instead of $p(x,\xi)$, noting the correspondence between (4.3) and (4.4). □

Definition 4.1. $p(x,\xi)$ is said to be a symbol of order m, $m \in \mathbb{R}$, denoted by $p \in S^m$, of a pseudodifferential operator $P(x,D)$ defined through (4.6), if it satisfies

(i) $p(x,\xi)$ is C^∞ in x and ξ;
(ii) $p(x,\xi)$ has compact x-support, i.e., $p(x,\xi)$ is zero for all sufficiently large x; and
(iii) for all multi-indices α, β, there is a constant $C_{\alpha\beta}$ such that

$$\left|D_x^\alpha D_\xi^\beta p(x,\xi)\right| \leqslant C_{\alpha\beta} \left(1+|\xi|\right)^{m-|\beta|}. \tag{4.5}$$

□

Let $a_\alpha \in C_0^\infty(\mathbb{R}^N)$ in (4.1). Then it is easy to check that the symbol of $P(x,D)$ is in S^m if it is given by (4.1).

Theorem 4.1. *Let $p \in S^m$ for some $m \in \mathbb{R}$. Then*

$$P\colon\ H^s(\mathbb{R}^N) \longrightarrow H^{s-m}(\mathbb{R}^N)$$

defined by

$$(Pu)(x) = \int_{\mathbb{R}^N} e^{2\pi i\langle x,\xi\rangle} p(x,2\pi\xi)\widehat{u}(\xi)\, d\xi \tag{4.6}$$

is a continuous linear operator for all $s \in \mathbb{R}$.

Proof. From (4.3),

$$\widehat{Pu}(\eta) = \iint e^{2\pi i\langle x,\xi-\eta\rangle} p(x,2\pi\xi)\widehat{u}(\xi)\,d\xi\,dx. \tag{4.7}$$

Define

$$\widetilde{p}(\lambda,\xi) = \int e^{-2\pi i\langle x,\lambda\rangle} p(x,2\pi\xi)\,dx. \tag{4.8}$$

Then

$$\widehat{Pu}(\eta) = \int \widetilde{p}(\eta-\xi,\xi)\widehat{u}(\xi)\,d\xi. \tag{4.9}$$

Now let $\phi \in H^{m-s}(\mathbb{R}^N)$. We have

$$\begin{aligned}\int \widehat{Pu}(\eta)\widehat{\phi}(\eta)\,d\eta &= \iint \widetilde{p}(\eta-\xi,\xi)\widehat{\phi}(\eta)\widehat{u}(\xi)\,d\xi\,d\eta \\ &= \iint \widetilde{p}(\eta-\xi,\xi)(1+|\xi|)^{-s}(1+|\eta|)^{s-m}\widehat{\phi}(\eta)\cdot \\ &\qquad (1+|\eta|)^{m-s}\widehat{u}(\xi)(1+|\xi|)^{s}\,d\xi\,d\eta. \end{aligned} \tag{4.10}$$

Define

$$K(\xi,\eta) = \widetilde{p}(\eta-\xi,\xi)(1+|\xi|)^{-s}(1+|\eta|)^{s-m}. \tag{4.11}$$

We claim that K satisfies

$$\int |K(\xi,\eta)|\,d\xi \leqslant C, \qquad \int |K(\xi,\eta)|\,d\eta \leqslant C, \tag{4.12}$$

for some constant C independent of $\xi,\eta \in \mathbb{R}^N$. From (4.5), $p \in S^m$ implies

$$|D_x^\alpha p| \leqslant C_1(1+|\xi|)^m, \quad \text{for any multi-index } \alpha.$$

Since p has compact x-support,

$$\begin{aligned}\left|\widetilde{D_x^\alpha p}(\lambda,\xi)\right| &= \left|\int e^{-2\pi i\langle x,\lambda\rangle} D_x^\alpha p(x,2\pi\xi)\,dx\right| \\ &\leqslant C_2(1+|\xi|)^m. \end{aligned} \tag{4.13}$$

Integrating by parts $|k|$ times for the above integral, we obtain

$$\left|(-2\pi i\lambda)^k \int e^{-2\pi i\langle x,\lambda\rangle} D_x^{\alpha-k} p(x,2\pi\xi)\,dx\right| \leqslant C_2(1+|\xi|)^m.$$

Therefore

$$\left|\int e^{-2\pi i\langle x,\lambda\rangle} D_x^{\alpha-k} p(x,2\pi\xi)\,dx\right| \leqslant C_3(1+|\xi|)^m(1+|\lambda|)^{-|k|}. \tag{4.14}$$

Since the RHS above is independent of α, we obtain

$$\left|\widetilde{D_x^\alpha p}(\lambda,\xi)\right| \leqslant C_4(k)(1+|\xi|)^m(1+|\lambda|)^{-|k|}. \tag{4.15}$$

From (4.11) and (4.15), with $\alpha = (0,\dots,0)$ in (4.15), we get

$$|K(\xi,\eta)| = \left|\widetilde{p}(\eta-\xi,\xi)(1+|\xi|)^{-s}(1+|\eta|)^{s-m}\right|$$
$$\leqslant C_5(k)\frac{(1+|\xi|)^{m-s}}{(1+|\eta|)^{m-s}}(1+|\xi-\eta|)^{-|k|}. \tag{4.16}$$

We apply the inequality

$$\frac{1+|a|}{1+|b|} \leqslant 1+|a-b| \tag{4.17}$$

to (4.16) and get

$$|K(\xi,\eta)| \leqslant C_5(k)(1+|\xi-\eta|)^{-|k|+m-s}.$$

By choosing $|k|$ sufficiently large, we obtain

$$\int |K(\xi,\eta)|\,d\xi \leqslant C_5(k)\int (1+|\xi-\eta|)^{-|k|+m-s}\,d\xi < \infty,$$
$$\int |K(\xi,\eta)|\,d\eta \leqslant C_5(k)\int (1+|\xi-\eta|)^{-|k|+m-s}\,d\xi < \infty,$$

so (4.12) has been verified.

We continue from (4.10):

$$\left|\int \widehat{Pu}(\eta)\widehat{\phi}(\eta)\,d\eta\right| = \left|\iint K(\xi,\eta)(1+|\eta|)^{m-s}\widehat{\phi}(\eta)(1+|\xi|)^{s}\widehat{u}(\xi)\,d\xi\,d\eta\right|$$
$$\leqslant \left\{\iint |K(\xi,\eta)|\left[(1+|\eta|)^{m-s}|\widehat{\phi}(\eta)|\right]^2 d\xi\,d\eta\right\}^{1/2}$$
$$\cdot\left\{\iint |K(\xi,\eta)|\,[(1+|\xi|)^{s}|\widehat{u}(\xi)|]^2\,d\xi\,d\eta\right\}^{1/2}$$
$$\leqslant C\|\phi\|_{H^{m-s}}\|u\|_{H^s}, \quad \text{by (4.12)}.$$

Since H^{m-s} is dual to H^{s-m}, this implies that

$$\|Pu\|_{H^{s-m}} \leqslant C\|u\|_{H^s}.$$

Therefore P is a bounded operator from H^s into H^{s-m}. □

Exercise 4.1. Prove the inequality (4.16). □

Example 4.1. By Definition 4.1,

$$p(x,\xi) = c(x)(1+|\xi|^2)^{m/2} \in S^m, \tag{4.18}$$

where

$$c \in C_0^\infty(\mathbb{R}^N);$$

m here is not necessarily an integer. Then the ψDO P associated with p satisfies

$$P\colon\ H^s(\mathbb{R}^N) \longrightarrow H^{s-m}(\mathbb{R}^N), \quad \forall s \in \mathbb{R}.$$

For example, when $m=-2, P\colon\ H^s \to H^{s+2}$; thus P makes functions smoother, which is not the case with differential operators. □

Remark 4.2. In (ii) of Definition 4.1, the requirement that $p(x,\xi)$ has compact x–support is very restrictive. For example, $p(x,\xi) = (1+|\xi|^2)^{m/2}$ does not have compact x–support. This is somewhat inconvenient. Actually, we can remove the requirement (ii) in Definition 4.1, and prove that the operator P defined by (4.6) satisfies

$$P\colon\ H^s_{\text{loc}}(\mathbb{R}^N) \to H^{s-m}_{\text{loc}}(\mathbb{R}^N)$$

is continuous. See [179]. From now on, we will deemphasize the restriction of compact x-support of symbols in S^m. □

4.2 Products and adjoints of pseudo-differential operators. Asymptotic expansions of a symbol

As an example, let us consider the following ψDOs:

$$\left.\begin{aligned} &P = \frac{1}{i}a(x)\frac{\partial}{\partial x}, \quad Q = b(x),\, x\in\mathbb{R}, \quad a,b,\in C_0^\infty(\mathbb{R}),\\ &\sigma P = a(x)\xi, \qquad \sigma Q = b(x).\end{aligned}\right\} \tag{4.19}$$

For $u \in \mathscr{S}(\mathbb{R})$,

$$\begin{aligned}(PQ)u = P(Qu) &= \frac{1}{i}a(x)\frac{\partial}{\partial x}[b(x)u(x)]\\ &= \frac{1}{i}a(x)\left[b'(x)u(x)+b(x)u'(x)\right]\\ &= \left(\frac{1}{i}ab\frac{\partial}{\partial x}+\frac{1}{i}ab'\right)u.\end{aligned}$$

Thus

$$PQ = \frac{1}{i}\,ab\,\frac{\partial}{\partial x}+\frac{1}{i}\,ab'$$

and

$$\begin{aligned}\sigma(PQ) = ab\xi + \frac{1}{i}ab' &= \sigma P\cdot\sigma Q + i^{-1}D_\xi(\sigma P)D_x(\sigma Q)\\ &= \sum_{\alpha=0,1} i^{-|\alpha|}\left[D_\xi^\alpha(\sigma P)\right][D_x^\alpha(\sigma Q)].\end{aligned} \tag{4.20}$$

Let P^* denote the adjoint of P with respect to the L^2 inner product, i.e.,

$$\langle P^*u,v\rangle = \langle u,Pv\rangle. \tag{4.21}$$

Let us compute P^* for the P in (4.19):

$$\begin{aligned}\int_{\mathbb{R}}\left\{\left[\frac{1}{i}a(x)\frac{\partial}{\partial x}\right]^* u\right\}\overline{v}\,dx &= \int u\overline{\left[\frac{1}{i}a(x)\frac{\partial v}{\partial x}\right]}\,dx\\ &= i\int u\overline{av'}\,dx\\ &= -i\int(\overline{a}u)'\overline{v}\,dx \quad \text{(integration by parts).}\end{aligned}$$

Therefore

$$\left(\frac{1}{i}a\frac{\partial}{\partial x}\right)^* u = \frac{1}{i}(\overline{a}u)'$$
$$\left(\frac{1}{i}a\frac{\partial}{\partial x}\right)^* = \frac{1}{i}a^*\frac{\partial}{\partial x} + \frac{1}{i}(a^*)', \quad a^* \equiv \overline{a}.$$

So

$$\sigma(P^*) = a^*\xi + \frac{1}{i}(a^*)' \tag{4.22}$$
$$= (\sigma P)^* + i^{-1}D_\xi D_x(\sigma P)^* \tag{4.23}$$
$$= \sum_{\alpha=0,1} i^{-|\alpha|}D_\xi^\alpha D_x^\alpha(\sigma P)^*. \tag{4.24}$$

Next, we consider how to compute $\sigma(PQ)$ and $\sigma(P^*)$ for general ψDO's P and Q. In the following discussion, the corresponding ψDO of a symbol denoted by a particular lower-case letter will be denoted by the corresponding capital letter. We introduce an equivalence relation on the class of symbols:

$$a \sim b \;\text{ iff }\; a - b \in S^m \quad \forall\, m \in \mathbb{R}. \tag{4.25}$$

Consequently,

$$A - B\colon\ H^s \longrightarrow H^{s'} \quad \forall s, s' \in \mathbb{R}$$
$$A - B\colon\ H^s \longrightarrow \cap_{s'>0} H^{s'}.$$

Then a consequence of the Sobolev imbedding theorem (Corollary 2.1) gives

$$A - B\colon\ H^s \longrightarrow C^\infty. \tag{4.26}$$

(Note that C^∞ is equipped with the topology of uniform convergence of any finite-order derivatives on compact sets.) Hence $A - B$ is an infinitely smoothing operator, whose effect can be ignored for our purposes later on and whose symbol is thus quotiented out in the equivalence relation.

Definition 4.2. Given symbols a, $\{a_j\}_{j=0}^\infty$, we write

$$a \sim \sum_{j=0}^\infty a_j \tag{4.27}$$

provided that for any $m \in \mathbb{R}^+$ there exists $N_0 \in \mathbb{Z}^+$ such that

$$a - \sum_{j=0}^n a_j \in S^{-m} \quad \text{for } n \geqslant N_0. \tag{4.28}$$

That is, a can be approximated up to an arbitrarily smoothing symbol by taking finitely many terms in $\{a_j\}_{j=0}^\infty$. □

Let K be a compact set in $\mathbb{R}^N$. We denote by Ψ_K the class of ψDOs restricted to (act on) functions with support in K. We first prove a technical lemma.

Lemma 4.1. *Let $r(x,y,\xi)$ be of compact x– and y–support in a compact set K and satisfy the condition that for all multi-indices α, β, γ, there exists a constant $C_{\alpha\beta\gamma} > 0$ such that*

$$\left|D_x^\alpha D_y^\beta D_\xi^\gamma r(x,y,\xi)\right| \leqslant C_{\alpha\beta\gamma}(1+|\xi|)^{m-|\gamma|} \tag{4.29}$$

for some fixed $m \in \mathbb{R}$. If

$$(Ru)(x) \equiv \iint e^{2\pi i\langle x-y,\xi\rangle} r(x,y,2\pi\xi)u(y)\,dy\,d\xi, \tag{4.30}$$

then $R \in \Psi_K$ and

$$\sigma R \sim \sum_\alpha \frac{i^{-|\alpha|}}{\alpha!} D_y^\alpha D_\xi^\alpha r(x,y,\xi)\,|_{y=x}. \tag{4.31}$$

Proof.

$$\int e^{-2\pi i\langle y,\xi\rangle} r(x,y,2\pi\xi)u(y)\,dy = \mathscr{F}_y\left(r(x,\cdot,2\pi\xi)u(\cdot)\right)(\xi), \tag{4.32}$$

where $\mathscr{F}_y$ is the Fourier transform with respect to the y variable. By (2.15)

$$\int e^{-2\pi i\langle y,\xi\rangle} r(x,y,2\pi\xi)u(y)\,dy = ([\mathscr{F}_y r(x,\cdot,2\pi\xi)] * \widehat{u})(\xi).$$

Therefore

$$(Ru)(x) = \iint e^{2\pi i\langle x,\xi\rangle}\,[\mathscr{F}_y r(x,\cdot,2\pi\xi)](\xi-\eta)\widehat{u}(\eta)\,d\eta\,d\xi. \tag{4.33}$$

Because $r(x,y,\xi)$ and $u(y)$ have compact y-support, performing an estimation such as in (4.8)–(4.17), we obtain

$$\left.\begin{aligned} &\left|[\mathscr{F}_y r(x,\cdot,2\pi\xi)](\xi-\eta)\right| \leqslant C_1(k)(1+|\xi|)^m(1+|\xi-\eta|)^{-k},\\ &|\widehat{u}(\eta)| \leqslant C_2(k)(1+|\eta|)^{-k}, \end{aligned}\right\} \tag{4.34}$$

for any $k \in \mathbb{N}$, where $C_1(k)$ and $C_2(k)$ depend only on k. So we can apply Fubini's theorem to the RHS of (4.33) and obtain

$$(Ru)(x) = \iint e^{2\pi i\langle x,\xi-\eta\rangle}\,[\mathscr{F}_y r(x,\cdot,2\pi\xi)](\xi-\eta)\widehat{u}(\eta)\,d\xi\,d\eta. \tag{4.35}$$

Now define

$$p(x,2\pi\eta) = \int e^{2\pi i\langle x,\xi-\eta\rangle}\,[\mathscr{F}_y r(x,\cdot,2\pi\xi)](\xi-\eta)\,d\xi.$$

Then

$$(Ru)(x) = \int e^{2\pi i\langle x,\eta\rangle} p(x,2\pi\eta)\widehat{u}(\eta)\,d\eta. \tag{4.36}$$

Note that $p(x,\eta) \in S^m$ because of (4.29) and (4.34), from which, (4.36) and the property of r it follows that $R \in \Psi_K$. In the RHS of (4.33), if we change the variable ξ to $\xi + \eta$, then

$$p(x,2\pi\eta) = \int e^{2\pi i\langle x,\xi\rangle} \left[\mathscr{F}_y r(x,\cdot,2\pi(\xi+\eta))\right](\xi)\,d\xi.$$

But $\mathscr{F}_y r$ allows a Taylor series expansion

$$\left[\mathscr{F}_y r(x,\cdot,2\pi(\xi+\eta))\right](\xi) = \sum_{|\alpha|\leqslant k} \frac{1}{\alpha!} D_\eta^\alpha \left[\mathscr{F}_y r(x,\cdot,2\pi\eta)\right](\xi)(2\pi\xi)^\alpha + \text{remainder}.$$

Thus

$$\begin{aligned}
p(x,2\pi\eta) &= \sum_{|\alpha|\leqslant k} \int e^{2\pi i\langle x,\xi\rangle} \cdot \frac{1}{\alpha!} D_\eta^\alpha \left[\mathscr{F}_y r(x,\cdot,2\pi\eta)\right](\xi)(2\pi\xi)^\alpha\,d\xi \\
&\quad + \text{remainder} \\
&= \sum_{|\alpha|\leqslant k} D_\eta^\alpha \left\{ \int e^{2\pi i\langle x,\xi\rangle} \frac{1}{\alpha!} \left[\mathscr{F}_y r(x,\cdot,2\pi\eta)\right](\xi)(2\pi\xi)^\alpha\,d\xi \right\} \\
&\quad + \text{remainder} \\
&= \sum_{|\alpha|\leqslant k} D_\eta^\alpha \left\{ \left(\frac{1}{i}\right)^\alpha \frac{1}{\alpha!} D_y^\alpha \int e^{2\pi i\langle y,\xi\rangle} \left[\mathscr{F}_y r(x,\cdot,2\pi\eta)\right](\xi)\,d\xi \Big|_{y=x} \right\} \\
&\quad + \text{remainder} \\
&= \sum_{|\alpha|\leqslant k} D_\eta^\alpha \left(\frac{1}{i} D_y\right)^\alpha \frac{1}{\alpha!} r(x,y,2\pi\eta)\Big|_{y=x} + \text{remainder}.
\end{aligned}$$

It is easy to show that the remainder belongs to S^{m-k}. Hence (4.31) has been proved. □

Using the technical lemma, we can now prove the following crucial theorem.

Theorem 4.2. *For $P \in \Psi_K$, let P^* denote its adjoint defined by*

$$\langle P^* u, v\rangle = \langle u, Pv\rangle \quad \forall u,v \in C_0^\infty(K).$$

Then

(i)

$$P^* \in \Psi_K \text{ and } \sigma P^* \sim \sum_\alpha \frac{i^{-|\alpha|} D_\xi^\alpha D_x^\alpha p^*}{\alpha!}, \tag{4.37}$$

where $p^*(x,\xi) = \overline{p(x,\xi)}$;

(ii)

$$P, Q \in \Psi_K \text{ implies } PQ \in \Psi_K \text{ and } \sigma(PQ) \sim \sum_\alpha i^{-|\alpha|} D_\xi^\alpha p D_x^\alpha \frac{q}{\alpha!}. \tag{4.38}$$

Proof. **(i)**

$$\begin{aligned}\langle u, Pv\rangle &= \iint u(x)e^{-2\pi i\langle x,\xi\rangle}\overline{p}(x,2\pi\xi)\overline{\widehat{v}}(\xi)\,d\xi\,dx\\ &= \iiint e^{-2\pi i\langle x-y,\xi\rangle}u(x)\overline{v}(y)\overline{p}(x,2\pi\xi)\,dy\,dx\,d\xi.\end{aligned}$$

We approximate $p(x,2\pi\xi)$ by functions of compact ξ-support as in (3.66), and apply Fubini's theorem in this limit process to obtain

$$\langle u, Pv\rangle = \iiint e^{-2\pi i\langle x-y,\xi\rangle}u(x)\overline{v}(y)\overline{p}(x,2\pi\xi)\,dx\,d\xi\,dy.$$

Now define

$$(P^*u)(y) = \iint e^{-2\pi i\langle x-y,\xi\rangle}\overline{p}(x,2\pi\xi)u(x)\,dx\,d\xi.$$

Then

$$\langle P^*u, v\rangle = \langle u, Pv\rangle.$$

Choose $\phi \in C_0^\infty(\mathbb{R}^N)$ such that $\phi = 1$ on K. Then

$$(P^*u)(x) = \iint e^{-2\pi i\langle y-x,\xi\rangle}\phi(x)\overline{p}(y,2\pi\xi)u(y)\,dy\,d\xi, \tag{4.39}$$

and

$$r(x,y,\xi) \equiv \phi(x)\overline{p}(y,\xi)$$

satisfies the conditions of Lemma 4.1 so that $P^* \in \Psi_K$ and

$$\begin{aligned}\sigma P^* &\sim \sum_\alpha \frac{i^{-|\alpha|}}{\alpha!}D_\xi^\alpha D_y^\alpha\,[\phi(x)\overline{p}(y,\xi)]\,|_{y=x}\\ &= \sum_\alpha \frac{i^{-|\alpha|}}{\alpha!}D_\xi^\alpha D_y^\alpha\overline{p}(y,\xi)\,|_{y=x},\end{aligned}$$

so (4.37) is proved.

(ii) From (4.39),

$$(Q^*u)(x) = \iint e^{-2\pi i\langle y-x,\xi\rangle}\overline{q}(y,2\pi\xi)u(y)\,dy\,d\xi.$$

Thus

$$\begin{aligned}(Qu)(x) &= [(Q^*)^*u]\,(x)\\ &= \iint e^{-2\pi i\langle y-x,\xi\rangle}\widetilde{q}(y,2\pi\xi)u(y)\,dy\,d\xi,\end{aligned} \tag{4.40}$$

where

$$\widetilde{q}(y,\xi) = \overline{\sigma Q^*}. \tag{4.41}$$

Observe that the RHS of (4.40) above is simply $\overline{\mathscr{F}}_\xi[\int e^{-2\pi i\langle y,\xi\rangle}\widetilde{q}(y,2\pi\xi)u(y)dy]$; therefore

$$\widehat{Qu}(\xi) = \int e^{-2\pi i\langle y,\xi\rangle}\widetilde{q}(y,2\pi\xi)u(y)dy.$$

We thus get

$$\begin{aligned}[(PQ)u](x) &= P[Qu(x)] \\ &= \int e^{2\pi i\langle x,\xi\rangle} p(x,2\pi\xi)\widehat{Qu}(\xi)\,d\xi \\ &= \iint e^{2\pi i\langle x-y,\xi\rangle} p(x,2\pi\xi)\widetilde{q}(y,2\pi\xi)u(y)\,dy\,d\xi.\end{aligned}$$

But now

$$r(x,y,\xi) \equiv p(x,\xi)\widetilde{q}(y,\xi)$$

satisfies condition (4.29) of Lemma 4.1; therefore $PQ \in \Psi_K$ and

$$\begin{aligned}\sigma(PQ) &\sim \sum_\alpha \frac{i^{-|\alpha|}}{\alpha!} D_\xi^\alpha D_y^\alpha r(x,y,\xi)|_{y=x} \\ &= \sum_\alpha \frac{i^{-|\alpha|}}{\alpha!} D_\xi^\alpha D_y^\alpha \left[p(x,\xi)\widetilde{q}(y,\xi)\right]|_{y=x} \\ &= \sum_\alpha \frac{i^{-|\alpha|}}{\alpha!} D_\xi^\alpha \left[p(x,\xi) D_x^\alpha \widetilde{q}(x,\xi)\right].\end{aligned} \tag{4.42}$$

The Leibnitz rule says that

$$D_\xi^\alpha(fg) = \sum_{\alpha_1+\alpha_2=\alpha} \frac{\alpha!}{(\alpha_1!)(\alpha_2!)}(D_\xi^{\alpha_1} f)(D_\xi^{\alpha_2} g). \tag{4.43}$$

Applying this to (4.42), with $f = p(x,\xi)$ and $g = D_x^\alpha \widetilde{q}(x,\xi)$, we obtain

$$\begin{aligned}\sigma(PQ) &\sim \sum_{\alpha_1+\alpha_2=\alpha} \frac{i^{-|\alpha|}\alpha!}{\alpha_1!\alpha_2!\alpha!}[D_\xi^{\alpha_1} p(x,\xi)][D_\xi^{\alpha_2} D_x^{\alpha_1} D_x^{\alpha_2}\widetilde{q}(x,\xi)] \\ &= \sum_{\alpha_1} \frac{i^{-|\alpha_1|}}{\alpha_1!}[D_\xi^{\alpha_1} p(x,\xi)] D_x^{\alpha_1}\left[\sum_{\alpha_2} \frac{i^{-|\alpha_2|}}{\alpha_2!} D_\xi^{\alpha_2} D_x^{\alpha_2}\widetilde{q}(x,\xi)\right] \\ &\sim \sum_{\alpha_1} \frac{i^{-|\alpha_1|}}{\alpha_1!}[D_\xi^{\alpha_1} p(x,\xi)][D_x^{\alpha_1} q(x,\xi)] \quad \text{(by (4.43) and (4.37))} \\ &= \sum_\alpha \frac{i^{-|\alpha|}}{\alpha!}[D_\xi^\alpha p(x,\xi)][D_x^\alpha q(x,\xi)].\end{aligned} \tag{4.44}$$

□

Thus, we see that (4.20) and (4.22) are special consequences of Theorem 4.2.

We are now in a position to prove the asymptotic expansion of a symbol.

Theorem 4.3. *Let $p_j \in S^{m_j}$, $j = 0, 1, 2, \ldots$, where $m_j \in \mathbb{R}$ decreases monotonically to $-\infty$, and the x-support of p_j is contained in some fixed compact set for all j. Then there is a unique $p \in S^{m_0}$ (up to an infinitely smoothing symbol) such that*

$$p \sim \sum_{j=0}^{\infty} p_j. \tag{4.45}$$

Proof. Let $\psi \colon \mathbb{R}^N \to [0,1]$ be C^∞ such that $\psi \equiv 0$ on $|\xi| < 1$ and $\psi \equiv 1$ on $|\xi| > 2$. That is, $1 - \psi(x)$ is a cutoff function as used earlier in (3.66).

Choose a sequence $t_j \in \mathbb{R}_+$ such that $t_j \uparrow \infty$.

We first observe that the sum $\sum_{j=0}^{\infty} \psi(\xi/t_j) p_j(x,\xi)$ is well-defined, since for any fixed (x,ξ) there are only finitely many nonzero terms.

Let $j \in \mathbb{Z}^+$. Then

$$\begin{aligned} \left|p_j(x,\xi)\right| &\leqslant C(1+|\xi|)^{m_j} \\ &= C(1+|\xi|)^{m_j-m_0}(1+|\xi|)^{m_0} \\ &\leqslant \frac{1}{2^j}(1+|\xi|)^{m_0} \quad \text{for sufficiently large } |\xi|, \\ &\qquad\qquad \text{because } m_j - m_0 < 0. \end{aligned}$$

Also,

$$\begin{aligned} \left|D_\xi^\alpha p_j(x,\xi)\right| &\leqslant C(1+|\xi|)^{m_j-|\alpha|} \\ &= C(1+|\xi|)^{m_j-m_0}(1+|\xi|)^{m_0-|\alpha|} \\ &\leqslant \frac{1}{2^j}(1+|\xi|)^{m_0-|\alpha|} \quad \text{for sufficiently large } \xi. \end{aligned}$$

Using a subsequence $\{t_{01}, t_{02}, t_{03}, \ldots\}$ of the t_js we can assume that

$$\left|D_x^\alpha D_\xi^\beta \left[\psi(\xi/t_{0j}) p_j(x,\xi)\right]\right| \leqslant \frac{1}{2^j}(1+|\xi|)^{m_0-|\beta|} \quad \forall \xi \in \mathbb{R}^N.$$

Therefore

$$\begin{aligned} \left|D_x^\alpha D_\xi^\beta \left[\sum_{j=0}^{\infty} \psi(\xi/t_{0j}) p_j(x,\xi)\right]\right| &\leqslant (1+|\xi|)^{m_0-|\beta|} \sum_{j=0}^{\infty} \frac{1}{2^j} \\ &\leqslant C(1+|\xi|)^{m_0-|\beta|} \quad \text{for some } C > 0. \end{aligned}$$

This shows that

$$\sum_{j=0}^{\infty} \psi(\xi/t_{0j}) p_j(x,\xi) \in S^{m_0}.$$

Inductively, one can show that for any $k \in \mathbb{Z}^+$, we have

$$\sum_{j=k}^{\infty} \psi(\xi/t_{kj}) p_j(x,\xi) \in S^{m_k}$$

for a suitably chosen subsequence $\{t_{kj} \mid j = 1, 2, \ldots\}$ of $\{t_{k-1,j} \mid j = 1, 2, \ldots\}$.

Now define

$$p(x,\xi) = \sum_{j=0}^{\infty} \psi(\xi/t_{jj}) p_j(x,\xi).$$

We see that $p \in S^{m_0}$ by the way it was constructed. Let $\phi = \psi - 1$. For each j, $\psi(\xi/t_{jj})p_j(x,\xi) - p_j(x,\xi)$ has compact ξ-support and therefore lies in S^k for all k, so it is infinitely smoothing. Now

$$\begin{aligned} p(x,\xi) - p_0(x,\xi) &= \phi(\xi/t_{00})p_0(x,\xi) + \sum_{j=1}^{\infty} \psi(\xi/t_{jj})p_j(x,\xi) - p_0(x,\xi) \\ &\sim \sum_{j=1}^{\infty} \psi(\xi/t_{jj})p_j(x,\xi) \in S^{m_1}, \end{aligned}$$

and

$$\begin{aligned} p(x,\xi) - \sum_{j=0}^{k} p_k(x,\xi) &= \sum_{j=0}^{k} \phi(\xi/t_{jj})p_j(x,\xi) + \sum_{j=k+1}^{\infty} \psi(\xi/t_{jj})p_j(x,\xi) \\ &\sim \sum_{j=k+1}^{\infty} \psi(\xi/t_{jj})p_j(x,\xi) \in S^{m_{k+1}}. \end{aligned}$$

This proves (4.45). □

4.3 Elliptic operators

Most of the boundary integral operators in the boundary element methods in this book are elliptic pseudodifferential operators. We give the following important definition.

Definition 4.3. Let $p(x,\xi) \in S^m$.

(i) p is said to be *elliptic* of order m if there exist $R > 0$ and $C > 0$ such that

$$|p(x,\xi)| \geqslant C(1+|\xi|)^m, \quad \forall\, |\xi| \geqslant R. \tag{4.46}$$

(ii) p is said to be *strongly elliptic* of order m if there exist $R > 0$ and $C > 0$ such that

$$\operatorname{Re} p(x,\xi) \geqslant C(1+|\xi|)^m, \quad \forall\, |\xi| \geqslant R. \tag{4.47}$$

A ψDO is said to be (strongly) elliptic if its symbol p is (strongly) elliptic. □

Example 4.2.

(i) The negative Laplacian $-\Delta$ on functions in $\mathbb{R}^N$,

$$-\Delta f = -\sum_{i=1}^{N} \frac{\partial^2 f}{\partial x_i^2}, \quad \sigma(-\Delta) = |\xi|^2,$$

is strongly elliptic of order 2.

(ii) The operator $\partial/\partial\bar{z}$ on functions in $\mathbb{C}$

$$\frac{\partial f}{\partial \bar{z}} = \frac{1}{2}\left(\frac{\partial f}{\partial x} + i\frac{\partial f}{\partial y}\right), \quad \sigma\left(\frac{\partial}{\partial \bar{z}}\right) = \frac{1}{2}(i\xi - \eta)$$

is elliptic of order 1.

(iii) The biharmonic operator Δ^2 on functions in $\mathbb{R}^N$,

$$\Delta^2 f = \left(\sum_{i=1}^{N} \frac{\partial^2}{\partial x_i^2}\right)^2 f, \quad \sigma(\Delta^2) = |\xi|^4,$$

is strongly elliptic of order 4. □

Given an elliptic ψDO, we want to construct an *approximate inverse*, called a *parametrix*, of this ψDO.

Lemma 4.2. *Let $p(x,\xi) \in S^m$ be elliptic of order m. Then there exists $q(x,\xi) \in S^{-m}$ such that $pq \sim qp \sim 1$.*

Proof. Let $R > 1$ be such that

$$|p(x,\xi)| \geqslant C(1+|\xi|)^m \qquad \text{for } |\xi| \geqslant R.$$

Choose a C^∞ function $\psi\colon \mathbb{R}^N \to [0,1]$ such that $\psi \equiv 0$ for $|\xi| < R-1$ and $\psi \equiv 1$ for $|\xi| \geqslant R$. Define a C^∞ function $q(x,\xi)$ such that

$$q(x,\xi) = \begin{cases} 0 & \text{for } |\xi| < R-1, \\ \psi(\xi)p(x,\xi)^{-1} & \text{for } |\xi| \geqslant R. \end{cases} \tag{4.48}$$

Then $q(x,\xi)$ has compact x-support because p does, and both $pq-1$ and $qp-1$ are nonzero at most on a compact subset of $|\xi| \leqslant R$. Therefore, $pq-1$ and $qp-1$ are C^∞ smoothing symbols. It remains to show that $q \in S^{-m}$.
Clearly

$$|q(x,\xi)| \leqslant C(1+|\xi|)^{-m} \ \forall\, \xi \in \mathbb{R}^N, \text{ for some } C > 0,$$

and

$$D_\xi^\alpha q(x,\xi) = \begin{cases} 0 & \text{for } |\xi| < R-1, \\ \dfrac{D_\xi \psi}{p} - \dfrac{\psi(D_\xi p)}{p^2} = -\dfrac{D_\xi p}{p^2} & \text{for } |\xi| \geqslant R, \ |\alpha| = 1. \end{cases}$$

But $D_\xi p \in S^{m-1}$ and $p^2 \in S^{2m}$, so

$$\frac{D_\xi p}{p^2} \in S^{m-1-2m} = S^{-(m+1)}.$$

Therefore $D_\xi q \in S^{-(m+1)}$. Similarly, one can show that $D_x D_\xi q(x,\xi) \in S^{-(m+1)}$, i.e., differentiating with respect to x leaves the order unchanged. Computing higher-order derivatives inductively yields

$$\left| D_x^\alpha D_\xi^\beta q(x,\xi) \right| \leqslant C_{\alpha\beta}(1+|\xi|)^{-m-|\beta|}. \tag{4.49}$$

□

Theorem 4.4. *If P is an elliptic ψDO then there is a unique (up to an infinitely smoothing operator) ψDO Q such that*

$$PQ - I \sim QP - I \sim 0.$$

Proof. By modifying $p(x,\xi)$ on $|\xi| \leqslant R$, we can assume that $p(x,\xi)^{-1}$ exists and is C^∞ for all ξ. We construct Q by a series as follows:

$$\sigma Q = q = q_0 + q_1 + \cdots, \quad q_i \in S^{-m-i}, \quad i \in \mathbb{N},$$

where

$$q_0 = \frac{1}{p} \in S^{-m} \quad \text{by Lemma 4.2.}$$

Assume that q_j has been defined for $j < k$. We define

$$q_k = -\frac{1}{p} \sum_{\substack{|\alpha|+j=k \\ j<k}} \frac{i^{-|\alpha|}}{\alpha!} D_\xi^\alpha p D_x^\alpha q_j \in S^{-m-k}.$$

By (4.38), we have

$$\begin{aligned}
\sigma(PQ) &\sim \sum_{\alpha,j} \frac{i^{-|\alpha|}}{\alpha!} D_\xi^\alpha p D_x^\alpha q_j \\
&= \sum_{k=0}^{\infty} \sum_{|\alpha|+j=k} \frac{i^{-|\alpha|}}{\alpha!} D_\xi^\alpha p D_x^\alpha q_j \\
&= pq + \text{lower-order terms} \\
&\approx 1.
\end{aligned}$$

Therefore, $PQ - I \sim 0$. Similarly, one can construct $\widetilde{Q}$ such that $\widetilde{Q}P - I \sim 0$. But

$$\widetilde{Q} \sim \widetilde{Q}(PQ) \sim (\widetilde{Q}P)Q \sim Q,$$

so Q is unique up to an infinitely smoothing operator. □

The following lemma builds up the machinery for proving the Gårding inequality.

Lemma 4.3. *Let $p \in S^0$ be strongly elliptic of order 0, i.e.,*

$$\operatorname{Re} p(x,\xi) \geqslant C > 0 \quad \forall\, x,\xi, \text{ for some } C.$$

Then there exists B such that $b \in S^0$ and

$$\sigma\,(\operatorname{Re} P - B^*B) \in S^{-\infty}, \tag{4.50}$$

where $\operatorname{Re} P = \frac{1}{2}(P+P^*)$.

Proof. We construct B's symbol $b \sim \sum_{j=0}^{\infty} b_j$, $b_j \in S^{-j}$, inductively:

$$b_0(x,\xi) \equiv [\operatorname{Re} p(x,\xi)]^{1/2}.$$

It is straightforward to verify that $b_0 \in S^0$. The formulas for adjoints and products (Theorem 4.2) show that

$$\sigma(\operatorname{Re} P - B_0^* B_0) \equiv r_1 \in S^{-1}.$$

Assume that we have $b_0, b_1, \ldots, b_j$. We want $b_{j+1} \in S^{-(j+1)}$ such that

$$\begin{aligned}&\sigma\{\operatorname{Re} P - [(B_0^* + B_1^* + \cdots + B_j^*) + B_{j+1}^*]\cdot\\ &[(B_0 + B_1 + \cdots + B_j t) + B_{j+1}]\} \equiv r_{j+1} \in S^{-(j+1)}.\end{aligned} \tag{4.51}$$

But the LHS is just

$$r_j - \sigma\left[(B_0^* + \cdots + B_j^*)B_{j+1} + B_{j+1}^*(B_0 + B_1 + \cdots + B_j) + B_{j+1}^* B_{j+1}\right]. \tag{4.52}$$

Let us choose

$$B_{j+1} = -\frac{1}{2}(B_0^*)^{-1} R_j.$$

Then

$$b_{j+1} \in S^{-j}, \tag{4.53}$$

$$B_{j+1}^* = -\frac{1}{2} R_j (B_0)^{-1} \qquad (\text{because } R_j = R_j^*), \tag{4.54}$$

so

$$R_j - (B_0^* B_{j+1} + B_{j+1}^* B_0) = 0, \tag{4.55}$$

and from (4.51)–(4.55) we obtain

$$\begin{aligned}r_{j+1} &= \sigma[-(B_1^* + \cdots + B_j^*)B_{j+1} - B_{j+1}^*(B_1 + \cdots + B_j) - B_{j+1}^* B_{j+1}]\\ &\in S^{-(j+1)}.\end{aligned}$$

The proof is complete. □

We define operators

$$\Lambda^{\sigma} u = \int (1+|\xi|^2)^{\sigma/2} e^{2\pi i\langle x,\xi\rangle} \widehat{u}(\xi)\, d\xi, \qquad \sigma \in \mathbb{R}. \tag{4.56}$$

By Theorem 4.1 and the Fourier inversion formula,

$$\Lambda^{\sigma}:\ H^s(\mathbb{R}^N) \longrightarrow H^{s-\sigma}(\mathbb{R}^n) \quad \text{isomorphically.} \tag{4.57}$$

Theorem 4.5 (Gårding's inequality). *Let $p \in S^m$ be strongly elliptic of order m. Then for any $s \in \mathbb{R}$ and for any bounded $\Omega \subset \mathbb{R}^N$, there are $C_1 > 0$, $C_2 > 0$ such that*

$$\mathrm{Re}\langle Pu,u\rangle \geqslant C_1 \|u\|^2_{H^{m/2}} - C_2 \|u\|^2_{H^s}, \ \ \forall u \in C_0^{\infty}(\Omega). \tag{4.58}$$

Sketch of Proof. By Theorem 4.2, it is easy to verify that

$$\sigma(\Lambda^{-m/2} P \Lambda^{-m/2}) \in S^0 \text{ is strongly elliptic of order } 0.$$

Without loss of generality, we can replace P by $\Lambda^{-m/2} P \Lambda^{-m/2}$ and assume that $m = 0$. By making a C^{∞} modification of p, we can assume that

$$\mathrm{Re}\, p(x,\xi) \geqslant C_0 > 0 \quad \forall x,\xi, \text{ for some } C_0.$$

We apply Lemma 4.3 to $q(x,\xi) \equiv \mathrm{Re}\, p(x,\xi) - \frac{1}{2}C_0$ to yield $b \in S^0$ satisfying

$$\sigma(Q - B^*B) = r \in S^{-\infty};$$

hence

$$\begin{aligned} \mathrm{Re}\,\langle Pu,u\rangle - \frac{1}{2}C_0\langle u,u\rangle &= \langle B^*Bu,u\rangle + \mathrm{Re}\,\langle Ru,u\rangle \\ &= \langle Bu,Bu\rangle + \mathrm{Re}\,\langle Ru,u\rangle, \\ \mathrm{Re}\,\langle Pu,u\rangle &= \frac{1}{2}C_0\langle u,u\rangle + \langle Bu,Bu\rangle + \mathrm{Re}\,\langle Ru,u\rangle \\ &\geqslant \frac{1}{2}C_0\|u\|^2 + \mathrm{Re}\,\langle Ru,u\rangle. \end{aligned}$$

Since $r \in S^{-\infty}$ and R is infinitely smoothing, we can apply an interpolation inequality in the form (cf. (2.33))

$$|\mathrm{Re}\,\langle Ru,u\rangle| \leqslant \varepsilon\|u\|^2 + C(\varepsilon,s)\|u\|^2_{H^s}$$

to all u such that $\mathrm{supp}\, u$ is compactly contained in Ω. Hence

$$\begin{aligned} \mathrm{Re}\,\langle Pu,u\rangle &\geqslant \frac{1}{2}(C_0 - 2\varepsilon)\|u\|^2 - C(\varepsilon,s)\|u\|^2_{H^s} \\ &\equiv C_1\|u\|^2 - C_2\|u\|^2_{H^s}. \end{aligned}$$

□

Let M be a compact manifold without boundary in $\mathbb{R}^N$. By changing the $L^2(\Omega)$–inner product $\langle\cdot,\cdot\rangle$ on the LHS of (4.58) to the $H^j(M)$–inner product $\langle\cdot,\cdot\rangle_{H^j}$ and by refining the proof of Theorem 4.5, one can further show that

$$\operatorname{Re}\langle Pu,u\rangle_{H^j} \geqslant C_0\|u\|^2_{H^{j+m/2}} - \langle Ku,u\rangle_{H^j} \;\; \forall u \in H^{j+m/2}(M), \tag{4.59}$$

where $K\colon\; H^{j+m/2}(M) \to H^{j-m/2}(M)$ is a compact operator (thus $\langle Ku,u\rangle_{H^j}$ is considered as a duality pairing in a sense similar to Theorem 2.4). The above is the form of Gårding's inequality we will refer to in the future.

4.4 Calculation of the principal symbols of boundary integral operators representing multiple-layer potentials

Boundary element methods are based upon the use of boundary integral operators representing multiple-layer potentials. The regularity properties of these operators are important in the later development of the theory of boundary integral equations and error estimates. It is known [170] that these boundary integral operators are all pseudodifferential operators. In this section, we will calculate the principal symbols of several multilayer potentials of the Laplacian, thereby determining the regularity properties of these operators.

Let $E(x,\xi)$ be the fundamental solution of the Laplacian as given by (1.3). Let Ω be a bounded open domain in $\mathbb{R}^N$ with C^∞ boundary $\partial\Omega$. For $\phi \in C^\infty(\partial\Omega)$, define boundary layer potentials

$$V_1\phi(x) = \int_{\partial\Omega} E(x,y)\phi(y)\,d\sigma_y, \tag{4.60}$$

$$V_2\phi(x) = \int_{\partial\Omega} \frac{\partial E(x,y)}{\partial n_y}\phi(y)\,d\sigma_y, \tag{4.61}$$

$$V_3\phi(x) = \frac{\partial}{\partial n_x}\int_{\partial\Omega} \frac{\partial E(x,y)}{\partial n_y}\phi(y)\,d\sigma_y \tag{4.62}$$

for $x \in \Omega$, where the normal vector field n_y has been extended in a C^∞ way to a neighborhood of $\partial\Omega$ through a Gaussian normal coordinate system (see [64, 23.48.4], and [107, §5.2], as in § 2.2). Then $V_1\phi$, $V_2\phi$ and $V_3\phi$ are in $C^\infty(\overline{\Omega})$. Define three boundary integral operators by the traces

$$A_1\phi = V_1\phi|_{\partial\Omega}, \tag{4.63}$$

$$A_2\phi = V_2\phi|_{\partial\Omega}, \tag{4.64}$$

$$A_3\phi = V_3\phi|_{\partial\Omega}. \tag{4.65}$$

The pseudodifferential operators considered so far are defined on the free space $\mathbb{R}^N$. In this section, we will study instead pseudodifferential operators on Ω. Thus we first define the class of symbols $S^m_{1,0}(\Omega, \mathbb{R}^N)$ to consist of C^∞ functions $p(x,\xi)$ satisfying the estimates

$$|D_x^\alpha D_\xi^\beta p(x,\xi)| \leqslant C_{\alpha\beta}(x)(1+|\xi|)^{m-|\beta|}, \qquad \forall (x,\xi) \in \Omega \times \mathbb{R}^N$$

where $C_{\alpha\beta}(x)$ is a continuous function of $x \in \Omega$. $S^m(\Omega, \mathbb{R}^N)$, a subclass of $S^m_{1,0}(\Omega, \mathbb{R}^N)$, is defined to consist of *polyhomogeneous symbols*, which are symbols possessing an asymptotic expansion as a sum of symbols $p_{m-l}(x,\xi)$, $l \in \mathbb{N}$, such that p_{m-l} is *homogeneous* in ξ, for $|\xi| \geqslant 1$, and

$$p - \sum_{l<k} p_{m-l} \in S^{m-k}_{1,0}(\Omega, \mathbb{R}^N) \ \forall \ k \in \mathbb{N}.$$

For a pseudodifferential operator P on $\mathbb{R}^N$, we restrict it to Ω by defining

$$P_\Omega u = r^+ P e^+ u$$

for $u \in \mathscr{D}'(\Omega)$, where e^+ is the "extension by zero" operator,

$$e^+ : \ \mathscr{D}'(\Omega) \to \mathscr{D}'(\mathbb{R}^N),$$

$$e^+ u(x) = \begin{cases} u(x) & \text{for } x \in \Omega, \\ 0 & \text{for } x \in \mathbb{R}^N \setminus \Omega, \end{cases}$$

and r^+ is the "restriction" operator,

$$r^+ : \ \mathscr{D}'(\mathbb{R}^N) \to \mathscr{D}'(\Omega),$$

$$r^+ u = u|_\Omega \in \mathscr{D}'(\Omega).$$

We are now in a position to study boundary integral operators. Denote

$$\begin{aligned} \mathbb{R}^N_+ = \{x \in \mathbb{R}^N \mid x = (x_{\widehat{N}}, x_N) \text{ such that} \\ x_{\widehat{N}} = (x_1, \ldots, x_{N-1}) \in \mathbb{R}^{N-1}, x_N > 0\}, \\ \mathbb{R}^N_- = \mathbb{R}^N \setminus \overline{\mathbb{R}}^N_+. \end{aligned}$$

We can locally identify the smooth boundary $\partial\Omega$ with $(x_{\widehat{N}}, 0)$ in the sense illustrated in Fig. 4.1.

The neighborhood $\mathscr{N}(y_0)$ of a point $y_0 \in \partial\Omega$ is mapped into $\mathbb{R}^N$ by a local diffeomorphism v such that y_0 is mapped into the origin of $\mathbb{R}^N$ and $\mathscr{N}(y_0) \cap \overline{\Omega}$ is mapped into $\overline{\mathbb{R}}^N_+$ such that the directions $n(x)$ for $x \in \mathscr{N}(y_0) \cap \partial\Omega$ are mapped into the direction $\vec{e}_N = (0, \ldots, 0, 1)^{\mathrm{Tr}}$. In this fashion, $\partial\Omega$ is locally identified with (part of) $\mathbb{R}^{N-1}$, and v induces a normal coordinate variable x_N.

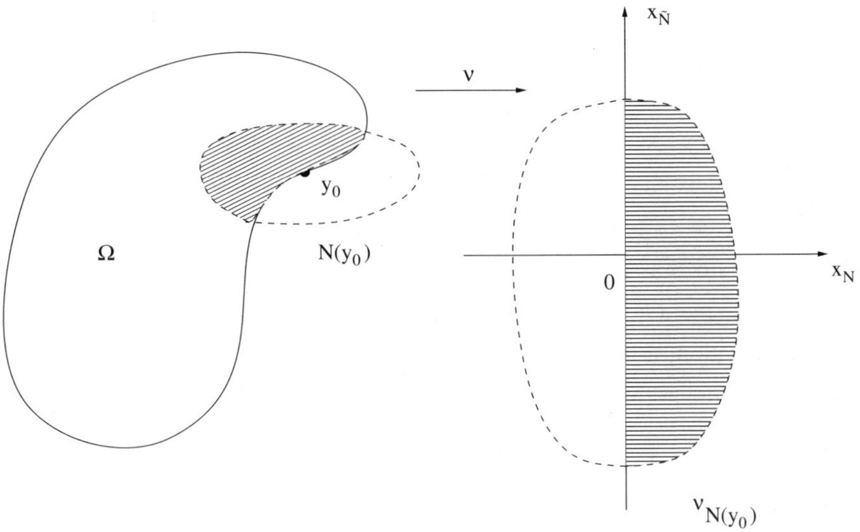

Figure 4.1: Flattening out $\partial\Omega$ locally.

Let γ_0 be the boundary trace operator:

$$\gamma_0\psi = \psi|_{\partial\Omega} \quad \forall\psi \in C^\infty(\overline{\Omega}). \tag{4.66}$$

Then γ_0 can be extended such that $\gamma_0\colon\ H^s(\Omega) \to H^{s-1/2}(\partial\Omega)$ is continuous for $s > \frac{1}{2}$, by Theorem 2.9.

Now, we study the operators A_1, A_2 and A_3 in (4.63)–(4.65) sequentially. First note that the fundamental solution E of the Laplacian satisfies

$$\Delta_x E(x,0) = -\delta(x), \quad x \in \mathbb{R}^N.$$

Using the Fourier transform, we obtain

$$\begin{aligned}|2\pi\xi|^2\widehat{E}(\xi,0) = 1,&\\ E(x,0) &= \mathscr{F}^{-1}\left(\frac{1}{|2\pi\xi|^2}\right)\\ &= \mathscr{F}^{-1}\left(\frac{1}{4\pi^2}\frac{1}{|\xi_{\widehat{N}}|^2+\xi_N^2}\right),\end{aligned}$$

The operator A_1 defined in (4.63) and (4.60) is a convolution of E with ϕ on $\partial\Omega$. Therefore A_1, as a pseudodifferential operator on $\mathbb{R}^{N-1} \leftrightarrow \partial\Omega$, through the correspondence shown in Fig. 4.1, can be written as

$$A_1\phi(x_{\widehat{N}}) = \gamma_0 r^+ \int_{\mathbb{R}^{2N}} e^{2\pi i\langle x-y,\xi\rangle}\frac{1}{4\pi^2}\frac{\phi(y_{\widehat{N}})\delta(y_N)}{|\xi_{\widehat{N}}|^2+\xi_N^2}\,dy\,d\xi \tag{4.67}$$

for $x \in \mathbb{R}^N_+$, by the Fourier convolution formula (2.15). Now writing

$$\frac{1}{|\xi_{\hat{N}}|^2+\xi_N^2} = \frac{1}{2|\xi_{\hat{N}}|}\left(\frac{1}{|\xi_{\hat{N}}|+i\xi_N} + \frac{1}{|\xi_N|-i\xi_N}\right), \tag{4.68}$$

we see that the above symbol is split into two parts: the first part has singularities at $\xi_N = i|\xi_{\hat{N}}|$ and is thus analytic in the lower half plane $\operatorname{Im}\xi_N < 0$, and the second part has singularities at $\xi_N = -i|\xi_{\hat{N}}|$ and is analytic in the upper half plane $\operatorname{Im}\xi_N > 0$, where these two parts are regarded as functions of the complex variable ξ_N. From the distribution of singularities $-i|\xi_{\hat{N}}|$ and $i|\xi_{\hat{N}}|$ on the RHS of (4.68), it is known (see e.g., [191, Chapt. VI, §4]) by an application of the inverse Fourier transform with respect to the y variable and the Paley–Wiener theorem that we have

$$\operatorname{supp}\left[\frac{1}{4\pi^2}\int_{\mathbb{R}^N} e^{2\pi i\langle x,\xi\rangle}\left(\frac{1}{2|\xi_{\hat{N}}|}\frac{1}{|\xi_{\hat{N}}|+i\xi_N}\right)(\mathscr{F}\phi)(\xi_{\hat{N}})\,d\xi\right] \subseteq \overline{\mathbb{R}^N_+}$$

and

$$\operatorname{supp}\left[\frac{1}{4\pi^2}\int_{\mathbb{R}^N} e^{2\pi i\langle x,\xi\rangle}\left(\frac{1}{2|\xi_{\hat{N}}|}\frac{1}{|\xi_{\hat{N}}|-i\xi_N}\right)(\mathscr{F}\phi)(\xi_{\hat{N}})\,d\xi\right] \subseteq \overline{\mathbb{R}^N_-}.$$

Therefore (4.67) gives

$$\begin{aligned} A_1\phi(x_{\hat{N}}) &= \gamma_0 r^+\left[\frac{1}{4\pi^2}\int_{\mathbb{R}^N} e^{2\pi i\langle x,\xi\rangle}\left(\frac{1}{2|\xi_{\hat{N}}|}\frac{1}{|\xi_{\hat{N}}|+i\xi_N}\right)(\mathscr{F}\phi)(\xi_{\hat{N}})\,d\xi\right] \\ &= \int_{\mathbb{R}^{N-1}} e^{2\pi i\langle x_{\hat{N}},\xi_{\hat{N}}\rangle} p_1(2\pi\xi_{\hat{N}})(\mathscr{F}\phi)(\xi_{\hat{N}})\,d\xi_{\hat{N}},\ \ x\in\mathbb{R}^N_+, \end{aligned}$$

where $p_1(\xi_{\hat{N}})$ is the symbol of A_1, given by

$$p_1(2\pi\xi_{\hat{N}}) = (2\pi)^{-2}\gamma_0 r^+\int_{\mathbb{R}} e^{2\pi i x_N\xi_N}\left(\frac{1}{2|\xi_{\hat{N}}|}\frac{1}{|\xi_{\hat{N}}|+i\xi_N}\right)d\xi_N.$$

Since γ_0 imposes the restriction to $x_N = 0$, the above is just

$$p_1(2\pi\xi_{\hat{N}}) = (2\pi)^{-2}\oint_{C_+}\frac{1}{2|\xi_{\hat{N}}|}\frac{1}{|\xi_{\hat{N}}|+i\xi_N}\,d\xi_N, \tag{4.69}$$

where $\oint_{C_+}$ is a contour integral along a counterclockwise closed path C_+ in the upper half plane enclosing the singularity $\xi_N = i|\xi_{\hat{N}}|$. We need only calculate the residue at the singularity $\xi_N = i|\xi_{\hat{N}}|$. Since

$$\frac{1}{2|\xi_{\hat{N}}|}\frac{1}{|\xi_{\hat{N}}|+i\xi_N} = \frac{1}{2|\xi_{\hat{N}}|i}\left(\frac{1}{\xi_N - i|\xi_{\hat{N}}|}\right), \tag{4.70}$$

from (4.69) we get

$$p_1(2\pi\xi_{\hat{N}}) = \frac{1}{2\pi}\frac{1}{2\pi i}\oint_{C+}\frac{1}{2|\xi_{\hat{N}}|}\frac{1}{\xi_N - i|\xi_{\hat{N}}|}\,d\xi_N = \frac{1}{2}\frac{1}{|2\pi\xi_{\hat{N}}|}. \tag{4.71}$$

Hence the principal symbol p_1 is found to be

$$p_1(\xi_{\hat{N}}) = \frac{1}{2}|\xi_{\hat{N}}|^{-1}, \tag{4.72}$$

which, after mapping from $\mathbb{R}^N$ back to Ω, implies that the principal symbol of A_1 is

$$\sigma(A_1) = \frac{1}{2}|\xi|^{-1}, \qquad \xi \in \partial\Omega, \tag{4.73}$$

where ξ is the dual variable of x for x restricted to $\partial\Omega$, i.e., the "cotangent vector" on $\partial\Omega$, and, as a consequence,

$$A_1\colon\ H^s(\partial\Omega) \to H^{s+1}(\partial\Omega) \quad \forall s \in \mathbb{R}. \tag{4.74}$$

We proceed to study the principal symbols of A_2 and A_3. We note that normal differentiation $\partial/\partial n$ on $\partial\Omega$ corresponds to $\partial/\partial x_N$ on $\mathbb{R}^{N-1} = \partial\mathbb{R}^N_+$, and, after the Fourier transformation, $\partial/\partial n_y$ and $\partial/\partial n_x$ become respectively the multiplication by $2\pi i\xi_N$ and $-2\pi i\xi_N$ of the integrand of (4.67), i.e., for $x \in \mathbb{R}^N_+$,

$$A_2\phi(x_{\widehat{N}}) = \gamma_0 r^+ \left[\int_{\mathbb{R}^{2N}} e^{2\pi i\langle x-y,\xi\rangle} \frac{1}{4\pi^2} \frac{2\pi i\xi_N}{|\xi_{\widehat{N}}|^2 + \xi_N^2} \phi(y_{\widehat{N}})\delta(y_N)\,dy\,d\xi\right], \tag{4.75}$$

$$A_3\phi(x_{\widehat{N}}) = \gamma_0 r^+ \left[-\int_{\mathbb{R}^{2N}} e^{2\pi i\langle x-y,\xi\rangle} \frac{1}{4\pi^2} \frac{(2\pi i\xi_N)^2}{|\xi_{\widehat{N}}|^2 + \xi_N^2} \phi(y_{\widehat{N}})\delta(y_N)\,dy\,d\xi\right]. \tag{4.76}$$

Therefore, our primary task is to find a splitting for

$$\frac{i\xi_N}{|\xi_{\widehat{N}}|^2 + \xi_N^2} \tag{4.77}$$

and

$$\frac{-\xi_N^2}{|\xi_{\widehat{N}}|^2 + \xi_N^2} \tag{4.78}$$

such as (4.68). But this is straightforward, since

$$\frac{(i\xi_N)^k}{|\xi_{\widehat{N}}|^2 + \xi_N^2} = \frac{(i\xi_N)^k}{2|\xi_{\widehat{N}}|}\left(\frac{1}{|\xi_{\widehat{N}}| + i\xi_N} + \frac{1}{|\xi_{\widehat{N}}| - i\xi_N}\right), \quad k = 1, 2,$$

for (4.77) and (4.78). The parts

$$\frac{(i\xi_N)^k}{|\xi_{\widehat{N}}| + i\xi_N} \quad \text{and} \quad \frac{(i\xi_N)^k}{|\xi_{\widehat{N}}| - i\xi_N}$$

have singularities at $\xi_N = i|\xi_{\widehat{N}}|$ and $\xi_N = -i|\xi_{\widehat{N}}|$, respectively, and thus are analytic on the lower and upper half planes respectively. By the same arguments as in the preceding paragraphs, the calculations of the symbols $p_2(\xi_{\widehat{N}})$ and $p_3(\xi_{\widehat{N}})$ for A_2 and A_3 respectively boil down to the following contour integrals:

$$p_k(2\pi\xi_{\widehat{N}}) = (2\pi)^{-2}(-1)^k \oint_{C_+} \frac{(2\pi i\xi_N)^{k-1}}{2|\xi_{\widehat{N}}|} \frac{1}{|\xi_{\widehat{N}}| + i\xi_N}\,d\xi_N, \quad k = 2, 3. \tag{4.79}$$

We need only calculate the residues. Since

$$\begin{aligned}\frac{2\pi i\xi_N}{2|\xi_{\hat{N}}|}\frac{1}{|\xi_{\hat{N}}|+i\xi_N} &= \frac{\pi\xi_N}{|\xi_{\hat{N}}|}\frac{1}{\xi_N - i|\xi_{\hat{N}}|}\\ &= \frac{\pi}{|\xi_{\hat{N}}|}\left(1+\frac{i|\xi_{\hat{N}}|}{\xi_N - i|\xi_{\hat{N}}|}\right),\end{aligned} \tag{4.80}$$

the coefficient of the singular term $(\xi_N - i|\xi_{\hat{N}}|)^{-1}$ is πi. Thus, from (4.79) and (4.80),

$$\begin{aligned}p_2(2\pi\xi_{\hat{N}}) &= \frac{1}{(2\pi)^2}\oint_{C_+}(\pi i)\frac{1}{\xi_N - i|\xi_{\hat{N}}|}\,d\xi_N\\ &= -\frac{1}{2\pi i}\oint_{C_+}\frac{1}{2}\frac{1}{\xi_N - i|\xi_{\hat{N}}|}\,d\xi_N\\ &= -\frac{1}{2}.\end{aligned} \tag{4.81}$$

Therefore the leading symbol of A_2 is

$$\sigma(A_2) = -\frac{1}{2}. \tag{4.82}$$

A_2 is a pseudodifferential operator on $\partial\Omega$ of order 0, satisfying

$$A_2\colon H^s(\partial\Omega) \to H^s(\partial\Omega) \quad \text{continuously}, \quad \forall s \in \mathbb{R}.$$

Finally,

$$\begin{aligned}\frac{(2\pi i\xi_N)^2}{2|\xi_{\hat{N}}|}\frac{1}{|\xi_{\hat{N}}|+i\xi_N} &= -\frac{2\pi^2}{i|\xi_{\hat{N}}|}\left(\xi_N + \frac{i\xi_N|\xi_{\hat{N}}|}{\xi_N - i|\xi_{\hat{N}}|}\right)\\ &= -\frac{2\pi^2}{i|\xi_{\hat{N}}|}\left[\xi_N + i|\xi_{\hat{N}}|\left(1+\frac{i|\xi_{\hat{N}}|}{\xi_N - i|\xi_{\hat{N}}|}\right)\right];\end{aligned}$$

so the coefficient of the singular term $(\xi_N - i|\xi_N|)^{-1}$ is $-i2\pi^2|\xi_{\hat{N}}|$, yielding

$$\begin{aligned}p_3(2\pi\xi_{\hat{N}}) &= -\frac{1}{(2\pi)^2}\oint_{C_+}(-i2\pi^2)|\xi_{\hat{N}}|\frac{1}{\xi_N - i|\xi_{\hat{N}}|}\,d\xi_N\\ &= -\frac{1}{2\pi i}\oint_{C_+}\frac{|2\pi\xi_{\hat{N}}|}{2}\frac{1}{\xi_N - i|\xi_{\hat{N}}|}\,d\xi_N\\ &= -\frac{1}{2}|2\pi\xi_{\hat{N}}|.\end{aligned}$$

Therefore the leading symbol of A_3 is

$$\sigma(A_3) = p_3(\xi) = -\frac{1}{2}|\xi|, \tag{4.83}$$

where ξ is the dual variable of x for x restricted to $\partial\Omega$. Consequently,

$$A_3\colon H^s(\partial\Omega) \to H^{s-1}(\partial\Omega) \quad \text{is continuous } \forall s \in \mathbb{R}. \tag{4.84}$$

The operators A_1, A_2 and A_3 will be shown to be Fredholm, see § 4.6 and Chapter 6.

4.5 The Calderòn projector

There are some simple relationships between the boundary integral operators representing various layer potentials. They are obtainable through an elegant algebraic operator called *the Calderòn projector* [29, 64].

As before, let Ω be a bounded open domain in $\mathbb{R}^N$ with C^∞ boundary $\partial\Omega$, with outward unit normal vector field n on $\partial\Omega$. Let u_1 and u_2 be respectively C^∞ functions on $\overline{\Omega}$ and $\overline{\Omega^c}$. We first extend $n(x)$ to a C^∞ vector field on an open neighborhood of $\partial\Omega$ through a Gaussian normal coordinate system off $\partial\Omega$ and then define the kth normal derivative trace operators

$$\gamma_k u_i = \left[\left(\frac{\partial}{\partial n} \right)^k u_i \right] \bigg|_{\partial\Omega}, \qquad i = 1, 2.$$

Let u be a C^∞ function on $\mathbb{R}^N \setminus \partial\Omega$ such that

$$u(x) = \begin{cases} u_1(x) \text{ on } \Omega, \\ u_2(x) \text{ on } \Omega^c, \end{cases} \tag{4.85}$$

where $u_1 \in C^\infty(\overline{\Omega})$ and $u_2 \in C^\infty(\overline{\Omega^c})$. Define the jump of $\gamma_k u$ across $\partial\Omega$ by

$$[\gamma_k u] \equiv \gamma_k u_2 - \gamma_k u_1 \quad \text{on } \partial\Omega, \tag{4.86}$$

as well as the jump vector (of order $2m, m \in \mathbb{N}$) across $\partial\Omega$ by

$$[\gamma u] \equiv ([\gamma_0 u], [\gamma_1 u], \ldots, [\gamma_{2m-1} u])^{\mathrm{Tr}}.$$

Note that the above definitions also are meaningful if u is a vector-valued function.

Let P be a given linear partial differential operator with C^∞ coefficients. Then we can write P as

$$P = \sum_{j=0}^{2m} P_j \partial_n^j, \tag{4.87}$$

where

$$P_j = \sum_{|l| \leqslant 2m-j} a_{jl}(x) D^l \qquad (x \in \partial\Omega) \tag{4.88}$$

is a partial differential operator of order $2m - j$ involving only tangential derivatives on $\partial\Omega$ [64, 23.48.13.3] (or more generally, on hypersurfaces parallel to $\partial\Omega$ through a local coordinate system on a neighborhood of $\partial\Omega$). Naturally, we may also consider P that are $q \times q$ matrices of operators acting on vector-valued functions u.

Let χ denote the characteristic function of Ω^c, i.e.,

$$\chi(x) = \begin{cases} 0, \ x \in \overline{\Omega}, \\ 1, \ x \in \Omega^c. \end{cases}$$

From § 3.1, it is easy to verify that

$$\partial_n \chi = \frac{\partial}{\partial n}\chi = \delta_{\partial\Omega} \tag{4.89}$$

in the sense of distributions, where $\delta_{\partial\Omega}$ is the Dirac delta distribution concentrated on $\partial\Omega$ defined by

$$\int_{\mathbb{R}^N} \delta_{\partial\Omega}(x)\phi(x)\,dx = \int_{\partial\Omega} \phi|_{\partial\Omega}\,d\sigma \quad \forall\phi \in \mathscr{D}(\mathbb{R}^N).$$

Let $v \in L^1(\partial\Omega)$. We define the tensorial distribution $v \otimes \delta_{\partial\Omega} \in \mathscr{D}'(\mathbb{R}^N)$ by

$$\int_{\mathbb{R}^N} (v \otimes \delta_{\partial\Omega})(x)\phi(x)\,dx = \int_{\partial\Omega} v\phi|_{\partial\Omega}\,d\sigma \quad \forall\phi \in \mathscr{D}(\mathbb{R}^N). \tag{4.90}$$

Lemma 4.4 ([64, 23.48.13.2]). , *Let u be C^∞ on $\mathbb{R}^N \setminus \partial\Omega$ and on $\overline{\Omega^c}$, and let P_0 be an lth-order operator of the form*

$$P_0 = D^j \partial_n^k, \qquad j+k=l,$$

where D^j is a tangential operator of order j. Then

$$P_0(u\chi) = (P_0 u)\chi + \sum_{i=0}^{k-1} D^j \partial_n^i \left[(\partial_n^{k-i-1} u)\big|_{\partial\Omega^+} \otimes \delta_{\partial\Omega} \right], \tag{4.91}$$

where in the above χ is the characteristic function of Ω^c and $(\partial_n^{k-i-1}u)\big|_{\partial\Omega^+}$ denotes the $(k-i-1)$th-order normal derivative trace taken from Ω^c.

Proof. First, we easily verify that

$$\partial_n(u\chi) = (\partial_n u)\chi + u(\partial_n \chi) = (\partial_n u)\chi + u|_{\partial\Omega} \otimes \delta_{\partial\Omega},$$

by (4.89) and (4.90). Repeating the above, we have

$$\begin{aligned}
\partial_n^2(u\chi) &= \partial_n[(\partial_n u)\chi + u|_{\partial\Omega} \otimes \delta_{\partial\Omega}] \\
&= (\partial_n^2 u)\chi + (\partial_n u)|_{\partial\Omega} \otimes \delta_{\partial\Omega} + \partial_n(u|_{\partial\Omega} \otimes \delta_{\partial\Omega}) \\
&\ \vdots \\
\partial_n^k(u\chi) &= (\partial_n^k u)\chi + \sum_{i=0}^{k-1} \partial_n^i[(\partial_n^{k-i-1}u)|_{\partial\Omega} \otimes \delta_{\partial\Omega}].
\end{aligned}$$

Since $u|_{\partial\Omega} \in C^\infty(\partial\Omega)$, for a tangential operator D^j we have

$$D^j\partial_n^k(u\chi) = (D^j\partial_n^k u)\chi + \sum_{i=0}^{k-1} D^j\partial_n^i[(\partial_n^{k-i-1}u)|_{\partial\Omega} \otimes \delta_{\partial\Omega}],$$

and (4.91) has been verified. □

Let $u \in C^\infty(\mathbb{R}^N \setminus \partial\Omega)$ satisfy (4.85) with $u_1 \in C^\infty(\overline{\Omega})$ and $u_2 \in C^\infty(\overline{\Omega^c})$. On $\mathbb{R}^N \setminus \partial\Omega$, let

$$Pu(x) = f(x) = \begin{cases} f_1(x), \; x \in \Omega, \\ f_2(x), \; x \in \Omega^c, \end{cases} \tag{4.92}$$

define f. Then, for P given by (4.87) (subject to (4.88)), from Lemma 4.4 we have

$$\begin{aligned} Pu &= P[u_2\chi + u_1(1-\chi)] \\ &= (Pu_2)\chi + (Pu_1)(1-\chi) \\ &\quad + \sum_{j=0}^{2m}\sum_{i=0}^{j-1}\{P_j\partial_n^i[(\partial_n^{j-i-1}u)|_{\partial\Omega^+} \otimes \delta_{\partial\Omega}] \\ &\quad - P_j\partial_n^i[(\partial_n^{j-i-1}u)|_{\partial\Omega} \otimes \delta_{\partial\Omega}]\} \\ &= f + \sum_{j=0}^{2m}\sum_{i=0}^{j-1}[P_j\partial_n^i([\partial_n^{j-i-1}u] \otimes \delta_{\partial\Omega})]. \end{aligned} \tag{4.93}$$

By making the change of indices $j-i-1 = l$, the above becomes

$$Pu = f + \sum_{j=0}^{2m-1}\sum_{l=0}^{2m-j-1} P_{l+i+1}\partial_n^i([\partial_n^l u] \otimes \partial_{\partial\Omega}),$$

which is equation (1.3) of [55, p. 38]. Each term $\partial_n^i([\partial_n^l u] \otimes \delta_{\partial\Omega})$ is a distribution (also called an *i-fold layer*), satisfying

$$\int_{\mathbb{R}^N} ([\partial_n^l u] \otimes \delta_{\partial\Omega})\phi(x)\,dx = \int_{\partial\Omega} [\partial_n^l u](\phi|_{\partial\Omega})\,d\sigma \quad \forall\phi \in \mathscr{D}(\mathbb{R}^N),$$

for $i = 0$. For $i = 1$, letting $\psi_l \equiv [\partial_n^l u]$ on $\partial\Omega$, noting that $n(x)$ has been extended outside $\partial\Omega$, we have

$$\begin{aligned} \int_{\mathbb{R}^N} [\partial_n([\partial_n^l u] \otimes \delta_{\partial\Omega})]\phi(x)\,dx &= \int_{\mathbb{R}^N} \left[\sum_{i=1}^N n_i(x)\frac{\partial}{\partial x_i}(\psi_l \otimes \delta)\right]\phi(x)\,dx \\ &= -\int_{\mathbb{R}^N} (\psi_l \otimes \delta)(x)\sum_{i=1}^N \frac{\partial}{\partial x_i}[n_i(x)\phi(x)]\,dx \\ &= -\int_{\mathbb{R}^N} (\psi_l \otimes \delta)\left[\frac{\partial}{\partial n}\phi + (\operatorname{div} n)\phi\right]\,dx \\ &= -\int_{\partial\Omega} \psi_l[(\partial_n\phi) - 2H(\phi|_{\partial\Omega})]\,d\sigma, \\ &= -\int_{\partial\Omega} [\partial_n^l u](\partial_n\phi - 2H\phi|_{\partial\Omega})\,d\sigma, \end{aligned}$$

where it is known [107, p. 112] that for $x \in \partial\Omega$,

$$H(x) = -\frac{1}{2}\operatorname{div}\, n(x) \tag{4.94}$$

is the mean curvature of the hypersurface $\partial\Omega$ at x. Similarly, for $i = 2$,

$$\int_{\mathbb{R}^N} [\partial_n^2([\partial_n^l u] \otimes \delta_{\partial\Omega})]\phi(x)\,dx = \int_{\partial\Omega} [\partial_n^l u]\{\partial_n^2\phi - 4H\cdot\partial_n\phi + [-2\partial_n(H) + 4H^2](\phi|_{\partial\Omega})\}\,d\sigma,$$

etc. Following the notation of [55], we define the distribution

$$v \otimes (\partial_n^i \delta_{\partial\Omega}) \equiv \partial_n^i (v \otimes \delta_{\partial\Omega}).$$

Then, for any test function $\phi \in \mathscr{D}(\mathbb{R}^N)$, we have

$$\begin{aligned}\langle v \otimes \partial_n^i \delta_{\partial\Omega}, \phi\rangle &= \int_{\mathbb{R}^N} (v \otimes \partial_n^i \delta_{\partial\Omega})\phi\,dx \\ &= \int_{\mathbb{R}^N} [\partial_n^i (v \otimes \delta_{\partial\Omega})]\phi\,dx \\ &= \int_{\partial\Omega} v[(\partial_n^i)^*\phi]\,d\sigma,\end{aligned}$$

where $(\partial_n^i)^* = (\partial_n^*)^i$ is the adjoint operator of ∂_n^i satisfying

$$\partial_n^* = -\partial_n + 2H, \tag{4.95}$$

$$\partial_n^{*^2} = (-\partial_n + 2H)^2 = \partial_n^2 - 2H\partial_n + [4H^2 - 2(\partial_n H)],$$

$$\vdots$$

$$(\partial_n^*)^k = (-1)^k \partial_n^k + \text{lower-order terms.} \tag{4.96}$$

Equation (4.93) can now be rewritten as

$$Pu = f + \sum_{i=0}^{2m-1} \sum_{l=0}^{2m-i-1} P_{l+i+1}[\partial_n^l u] \otimes (\partial_n^i \delta_{\partial\Omega}). \tag{4.97}$$

To explain Calderòn's ideas [29], we now assume that the operator P in (4.87) is elliptic with a fundamental solution G, i.e., G satisfies

$$GP\psi = PG\psi = \psi \tag{4.98}$$

for all distributions $\psi \in \mathscr{D}'(\mathbb{R}^N)$ with compact support. It is known [170] that G is a pseudodifferential operator of order $-2m$ and as a consequence has a locally integrable kernel $G(x,\xi)$, $x,\xi \in \mathbb{R}^N$, which is C^∞ for $x \neq \xi$ and

$$(G\phi)(x) = \int_{\mathbb{R}^N} G(x,\xi)\phi(\xi)\,d\xi, \quad \phi \in \mathscr{D}(\mathbb{R}^N).$$

The application of G to an i-fold layer $\phi \otimes \partial_n^i \delta_{\partial\Omega}$ gives rise to an i-fold layer potential

$$V_{i+1}\phi \equiv G(\phi \otimes \partial_n^i \delta_{\partial\Omega}). \tag{4.99}$$

For $x \notin \partial\Omega$, (4.99) gives

$$(V_{i+1}\phi)(x) = \int_{\partial\Omega} \partial_{n(\xi)}^{*^i} G(x,\xi)\phi(\xi)\,d\sigma_\xi, \quad \phi \in \mathscr{D}(\mathbb{R}^N). \tag{4.100}$$

It is known [64, §23.53]

$$V_{i+1}\phi \in H_{\mathrm{loc}}^{-1/2+2m-i-\varepsilon}(\mathbb{R}^N) \quad \forall \varepsilon > 0, \phi \in H_{\mathrm{loc}}^{0}(\mathbb{R}^N). \tag{4.101}$$

Now, applying G to (4.97) and utilizing (4.98), we obtain a general representation formula for the solution of the linear PDE (4.92) on $\mathbb{R}^N \setminus \partial\Omega$,

$$u = Gf + \sum_{\substack{i+l+1\leqslant 2m\\ i,l\geqslant 0}} V_{i+1}(P_{l+i+1}[\partial_n^l u]), \tag{4.102}$$

for u, which is equal to respectively u_1 on Ω and u_2 on Ω^c, with $u_1 \in C_0^\infty(\overline{\Omega})$ and $u_2 \in C_0^\infty(\overline{\Omega^c})$, and where $f = Pu$ on $\mathbb{R}^N \setminus \partial\Omega$.
Let $u_2 \equiv 0$ on Ω^c. Then (cf. (4.86))

$$[\gamma u] = -(\gamma_0 u_1, \gamma_1 u_1, \ldots, \gamma_{2m-1} u_1)^{\mathrm{Tr}}. \tag{4.103}$$

The RHS above is called the *Cauchy data* of u_1. Substituting (4.103) into (4.102) and then taking the Cauchy data of u in (4.102), we obtain an identity for the Cauchy data of u_1:

$$\begin{aligned}\gamma u_1 &= \gamma G f_1 - \sum_{\substack{i+l+1\leqslant 2m\\ i,l\geqslant 0}} \gamma[(V_{i+1}(P_{l+i+1}\gamma_l u_1))|_\Omega] \qquad (4.104)\\ &\equiv \gamma G f_1 + \mathscr{C}_1(\gamma u_1) \quad \forall u_1 \in C_0^\infty(\overline{\Omega}).\end{aligned}$$

The operator $\mathscr{C}_1$ is *the Calderòn projector*. More precisely, define the matrices of operators

$$\begin{aligned}\mathscr{P} &\equiv [\mathscr{P}_{il}] = [P_{i+l+1}]_{i,l=0,\ldots,2m-1}, \qquad (4.105)\\ &\text{with } P_t = 0 \text{ for } t > 2m,\\ \mathscr{K}_j &\equiv [\mathscr{K}_{ik}^{(j)}] = [\gamma_i V_{k+1}^{(j)}]_{i,k=0,\ldots,2m-1}, \qquad (4.106)\end{aligned}$$

with

$$V_{k+1}^{(j)}(\phi) = \begin{cases} V_{k+1}(\phi)|_{\overline{\Omega}}, & j=1,\\ V_{k+1}(\phi)|_{\overline{\Omega^c}}, & j=2.\end{cases}$$

Then

$$\mathscr{C}_1 = -\mathscr{K}_1\mathscr{P}. \tag{4.107}$$

Similarly, setting $u_1 \equiv 0$ in (4.102), we obtain

$$\gamma u_2 = Gf_2 + \mathscr{C}_2(\gamma u_2) \tag{4.108}$$

for any $u_2 \in C_0^\infty(\overline{\Omega^c})$, where $f_2 = Pu|_{\Omega^c}$ and

$$\mathscr{C}_2 = \mathscr{K}_2\mathscr{P}. \tag{4.109}$$

Note that all the operators $\mathscr{P}, \mathscr{K}_j$ and $\mathscr{C}_j$, $j = 1, 2$, in (4.107) and (4.109) are well defined if $[\gamma u]$, γu_1 and γu_2 respectively are replaced by an arbitrary (vector-valued) function $\phi \in C^\infty(\partial\Omega; \mathbb{C}^{2m})$. According to [64, 23.4.7.3.1 and 23.48.14.3], it is known that $\mathscr{K}_j$ and $\mathscr{C}_j$ are pseudodifferential operators with

$$\begin{aligned} \text{order } \mathscr{K}_j &= [\text{order } K_{ik}^{(j)}] \\ &= [i+k-2m+1]_{i,k=0,\ldots,2m-1}, \end{aligned} \tag{4.110}$$

$$\begin{aligned} \text{order } \mathscr{C}_j &= [\text{order } (\mathscr{C}_j)_{ik}] \\ &= [i-k]_{i,k=0,\ldots,2m-1}. \end{aligned} \tag{4.111}$$

with also the explicit formulas for the principal symbols given therein. A purely algebraic method for computing the principal symbols of $\mathscr{C}_j$ in terms of the symbol of P can be found in [86, Appendix].

Now we prove the following proposition, which dates back to Seeley [169].

Proposition 4.1 ([55, Lemma 1.1], [169, Lemma 5]). *Denote*

$$[\mathscr{K}] \equiv \mathscr{K}_2 - \mathscr{K}_1.$$

Then

$$[\mathscr{K}] = \mathscr{P}^{-1}, \tag{4.112}$$

where $\mathscr{P}^{-1}$ is the inverse matrix of tangential operators to $\mathscr{P}$. Equivalently, for the Calderòn operators,

$$\mathscr{C}_1 + \mathscr{C}_2 = I. \tag{4.113}$$

Proof. By (4.105), the matrix operator $\mathscr{P}$ is given by

$$\mathscr{P} = \begin{bmatrix} P_1 & P_2 & P_3 & \ldots & P_{2m} \\ P_2 & P_3 & \ldots & P_{2m} & 0 \\ P_3 & \vdots & & \vdots & \vdots \\ \vdots & P_{2m} & \ldots & \ldots & 0 \\ P_{2m} & 0 & \ldots & \ldots & 0 \end{bmatrix}.$$

It is upper-triangular with respect to the transverse diagonal, every entry of which is equal to P_{2m}. Since P_{2m} is of order zero and hence is just a function, P_{2m}^{-1} exists due to the ellipticity of P. Consequently, it is straightforward to verify that $\mathscr{P}$ is invertible, with $\mathscr{P}^{-1}$ being a lower-triangular (with respect to the transverse diagonal) matrix of tangential differential operators.

Thus, it suffices to show that $[\mathscr{K}]$ is a right inverse to $\mathscr{P}$. Let $v \in C^\infty(\partial\Omega)$ and $i \in \{0,1,\ldots,2m-1\}$. Define

$$u = V_{i+1}(v) = G(v \otimes \partial_n^i \delta_{\partial\Omega}).$$

Then u is C^∞ on $\overline{\Omega}$ and $\overline{\Omega^c}$, and by (4.97)

$$Pu = v \otimes \partial_n^i \delta_{\partial\Omega}.$$

Since $v \otimes \partial_n^i \delta_{\partial\Omega}$ is concentrated on $\partial\Omega$, we have

$$f = Pu|_{\mathbb{R}^N \setminus \partial\Omega} \equiv 0.$$

Apply (4.97) to the above u, yielding

$$v \otimes \partial_n^i \delta_{\partial\Omega} = \sum_{k=0}^{2m-1} \sum_{l=0}^{2m-k-1} P_{k+l+1}[\gamma_l u] \otimes \partial_n^k \delta_{\partial\Omega}. \tag{4.114}$$

Noting that for $\phi_k \in C^\infty(\partial\Omega)$ that do not vanish identically, the distributions

$$\{\phi_k \otimes \partial_n^k \delta_{\partial\Omega} \mid k = 0, \ldots, 2m-1\}$$

are linearly independent, we obtain from (4.114) that

$$\sum_{l=0}^{2m-k-1} P_{k+l+1}[\gamma_l V_{i+1}(v)] = \begin{cases} v, & k = i, \\ 0, & k \neq i. \end{cases} \tag{4.115}$$

Since v is arbitrary, (4.115) gives

$$\mathscr{P}[\mathscr{K}] = I.$$

The equivalence between (4.112) and (4.113) is easily obtained after observing (4.107) and (4.109), and noting that $[\mathscr{K}]$ is a two-sided inverse of $\mathscr{P}$. □

The Calderòn projectors $\mathscr{C}_1$ and $\mathscr{C}_2$ satisfy the following useful properties as shown in the next theorem. We see that they are indeed the *projection operators into the space of Cauchy data* of solutions on Ω and Ω^c respectively.

Theorem 4.6. *For any $v = (v_0, v_1, \ldots, v_{2m-1})^{Tr} \in [C^\infty(\partial\Omega)]^{2m}$, we have*

$$\mathscr{C}_1\mathscr{C}_2 v = 0, \quad \mathscr{C}_2\mathscr{C}_1 v = 0.$$

Consequently, $\mathscr{C}_1$ and $\mathscr{C}_2$ satisfy

$$\left.\begin{aligned} \mathscr{C}_1 + \mathscr{C}_2 = I, \quad \mathscr{C}_1^2 = \mathscr{C}_1, \\ \mathscr{C}_2^2 = \mathscr{C}_2, \mathscr{C}_1\mathscr{C}_2 = \mathscr{C}_2\mathscr{C}_1 = 0. \end{aligned}\right\} \tag{4.116}$$

Proof. For any given $v = (v_0, v_1, \ldots, v_{2m-1}) \in [C^\infty(\partial\Omega)]^{2m}$, define u on $\mathbb{R}^N \setminus \partial\Omega$ by

$$\begin{aligned} u &= \sum_{\substack{k+l+1\leqslant 2m \\ k,l\geqslant 0}} V_{k+1}(P_{k+l+1}v_l) \\ &= \sum_{\substack{k+l+1\leqslant 2m \\ k,l\geqslant 0}} G((P_{k+l+1}v_l)\otimes \partial_n^k \delta_{\partial\Omega}). \end{aligned} \tag{4.117}$$

Define

$$u_1 = u|_\Omega, \quad u_2 = u|_{\Omega^c}.$$

Then u_1 and u_2 can be extended respectively to be C^∞ on $\overline{\Omega}$ and $\overline{\Omega^c}$. Using (4.98), we obtain, on $\mathbb{R}^N \setminus \partial\Omega$,

$$Pu = \sum_{\substack{k+l+1\leqslant 2m \\ k,l\geqslant 0}} P_{k+l+1}v_l \otimes \partial_n^k \delta_{\partial\Omega}.$$

Hence $Pu_1 = 0$ and $Pu_2 = 0$ on Ω and Ω^c respectively. Now extend u_1 and u_2 by 0 on Ω^c and Ω respectively, and denote the extended functions as $\widetilde{u}_1$ and $\widetilde{u}_2$. We can apply (4.102), (4.104) and (4.108) to get

$$\widetilde{u}_1 = -\sum_{\substack{k+l+1\leqslant 2m \\ k,l\geqslant 0}} V_k(P_{l+k+1}\gamma_l u_1), \tag{4.118}$$

$$\widetilde{u}_2 = \sum_{\substack{k+l+1\leqslant 2m \\ k,l\geqslant 0}} V_k(P_{l+k+1}\gamma_l u_2), \tag{4.119}$$

$$\gamma(\widetilde{u}_1|_\Omega) = \gamma u_1 = \mathscr{C}_1(\gamma u_1), \tag{4.120}$$

$$\gamma(\widetilde{u}_2|_{\Omega^c}) = \gamma u_2 = \mathscr{C}_2(\gamma u_2). \tag{4.121}$$

It is obvious that

$$\mathscr{C}_2(\gamma u_1) = \mathscr{C}_2\mathscr{C}_1(\gamma u_1) = \mathscr{C}_2(\gamma(\widetilde{u}_1|_{\Omega^c})) = 0, \tag{4.122}$$

$$\mathscr{C}_1(\gamma u_2) = \mathscr{C}_1\mathscr{C}_2(\gamma u_2) = \mathscr{C}_1(\gamma(\widetilde{u}_2|_\Omega)) = 0. \tag{4.123}$$

Since u_1 and u_2 are defined from u in (4.117), by (4.102) we have

$$\begin{aligned} u &= \sum_{\substack{k+l+1\leqslant 2m \\ k,l\geqslant 0}} G((P_{k+l+1}v_l)\otimes \partial_n^k \delta_{\partial\Omega}) \\ &= \sum_{\substack{k+l+1\leqslant 2m \\ k,l\geqslant 0}} G((P_{k+l+1}[\gamma_l u]\otimes \partial_n^k \delta_{\partial\Omega}) \\ &= \sum_{\substack{k+l+1\leqslant 2m \\ k,l\geqslant 0}} G((P_{k+l+1}(\gamma_l u_2 - \gamma_l u_1))\otimes \partial_n^k \delta_{\partial\Omega}). \end{aligned}$$

Using the same arguments as in the proof of Proposition 4.1, we obtain

$$P_{k+l+1}v_l = P_{k+l+1}(\gamma_l u_2 - \gamma_l u_1) \quad \forall k,l \geqslant 0 \text{ with } k+l+1 \leqslant 2m.$$

Choosing $k+l+1 = 2m$, we have

$$P_{2m}v_l = P_{2m}(\gamma_l u_2 - \gamma_l u_1).$$

For the same reason as before, we get

$$v_l = \gamma_l u_2 - \gamma_l u_1, \text{ for } l = 0,1,\dots,2m-1.$$

Hence

$$v = \gamma u_2 - \gamma u_1 = \mathscr{C}_2(\gamma u_2) - \mathscr{C}_1(\gamma u_1),$$

by (4.121) and (4.120). Therefore, by (4.120)–(4.123),

$$\begin{aligned}
\mathscr{C}_1 v &= \mathscr{C}_1\mathscr{C}_2(\gamma u_2) - \mathscr{C}_1^2(\gamma u_1)\\
&= 0 - \mathscr{C}_1^2(\gamma u_1)\\
&= -\gamma u_1,\\
\mathscr{C}_2 v &= \mathscr{C}_2\mathscr{C}_2(\gamma u_2) - \mathscr{C}_2\mathscr{C}_1(\gamma u_1)\\
&= \gamma u_2,\\
\mathscr{C}_1\mathscr{C}_2 v = \mathscr{C}_1(\gamma u_2) = 0, \mathscr{C}_2\mathscr{C}_1 v = \mathscr{C}_2(\gamma u_1) &= 0.
\end{aligned}$$

The proof is complete. □

It is easy to see that for $w \in [C^\infty(\partial\Omega)]^{2m}$ the following are equivalent

(i) $w = \gamma u$ for some u satisfying $Pu = 0$ on $\mathbb{R}^N \setminus \partial\Omega, u$ being sufficiently smooth on $\overline{\Omega}$ and $\overline{\Omega^c}$;

(ii) $\mathscr{C}_j w = w, j = 1$ or 2;

(iii) $w = \mathscr{C}_j g$ for some $g \in [C^\infty(\partial\Omega)]^{2m}, j = 1$ or 2.

We now illustrate an application of the above theory to the Laplacian [55].

Example 4.3. The Calderòn projectors for Δ.

Let Ω be a bounded open domain in $\mathbb{R}^N$ with C^∞ boundary $\partial\Omega$. Consider $P = -\Delta = -\sum_{i=1}^N \partial^2/\partial x_i^2$. It is known (see e.g., [107, p. 124, last line]) that P can be written as

$$P = -\Delta = -\Delta_0 + 2H\partial_n - \partial_n^2, \text{ with } \mathscr{P} = \begin{bmatrix} 2H & -1 \\ -1 & 0 \end{bmatrix},$$

where H is mean curvature as in (4.94), and Δ_0 is the Laplace–Beltrami operator on $\partial\Omega$. (For example, when Ω is a sphere with center at the origin, we have

$$P = -\Delta = -\frac{1}{r^2 \sin\theta}\left[\frac{\partial}{\partial\theta}\left(\sin\theta\frac{\partial}{\partial\theta}\right) + \frac{1}{\sin\theta}\frac{\partial^2}{\partial\phi^2}\right] - \frac{2}{r}\frac{\partial}{\partial r} - \frac{\partial^2}{\partial r^2},$$

where $\partial/\partial r = \partial/\partial n$ is the normal derivative).

From (4.95)

$$\partial_n^* = -\partial_n + 2H,$$

and the layer potentials (4.100) take the forms

$$\text{(simple layer)} \qquad V_1\phi(x) = \int_{\partial\Omega} E(x,\xi)\phi(\xi)\,d\sigma_\xi,$$

$$\text{(double layer)} \qquad V_2\phi(x) = -\int_{\partial\Omega}\left[\frac{\partial}{\partial n_\xi}E(x,\xi)\right]\phi(\xi)\,d\sigma_\xi + 2\int_{\partial\Omega} H(\xi)E(x,\xi)\phi(\xi)\,d\sigma_\xi,\ x \in \mathbb{R}^N \setminus \partial\Omega,$$

where $E(x,\xi)$ is given by (1.3).

In comparison with the classical double-layer potential (4.61) to be used and studied later on, V_2 above is actually the sum of the classical double-layer potential with a simple-layer potential due to the form of ∂_n^*. To obtain the classical double-layer potential and, more generally, the classical multiple-layer potentials

$$\widetilde{V}_{k+1}\phi \equiv G(\phi \otimes \partial_n^{*k}\delta_{\partial\Omega}) = \int_{\partial\Omega}\left(\frac{\partial}{\partial n_\xi}\right)^k E(x,\xi)\phi(\xi)\,d\sigma_\xi,$$

we can interchange the roles of ∂_n and ∂_n^* in (4.100) and (4.87):

$$P = \Sigma \widetilde{P}_j \partial_n^{*j}.$$

For the special case $P = -\Delta$, there is an easier and more straightforward way. We have, from (4.92),

$$\begin{aligned} -\Delta u &= f + (2H[\gamma_0 u] - [\gamma_1 u]) \otimes \delta_{\partial\Omega} + (-[\gamma_0 u]) \otimes \partial_n\delta_{\partial\Omega} \\ &= f - [\gamma_1 u] \otimes \delta_{\partial\Omega} + [\gamma_0 u] \otimes \partial_n^*\delta_{\partial\Omega}. \end{aligned}$$

Thus (4.102) and the above give the direct Green's representation formula

$$u(x) = \int_{\mathbb{R}^N} E(x,\xi)f(\xi)\,d\xi + \int_{\partial\Omega}\frac{\partial E(x,\xi)}{\partial n_\xi}[\gamma_0 u](\xi)\,d\sigma_\xi - \int_{\partial\Omega} E(x,\xi)[\gamma_1 u](\xi)\,d\sigma_\xi.$$

As will be shown later (cf. Theorem 6.3, with a slight change of notation), the jump conditions of potentials yield

$$\gamma_0 u_1 = \frac{1}{2}(1 - 2\widetilde{V}_2)(\gamma_0 u_1) + V_1(\gamma_1 u_1), \tag{4.124}$$

$$\gamma_1 u_1 = -\widetilde{V}_2^d(\gamma_0 u_1) + \frac{1}{2}(1 + 2\widetilde{V}_2^*)(\gamma_1 u_1), \tag{4.125}$$

where $u_1 = u|_\Omega$, and for $x \in \partial\Omega$,

$$V_1\phi(x) = \int_{\partial\Omega} E(x,\xi)\phi(\xi)\,d\sigma_\xi,$$

$$\widetilde{V}_2\phi(x) = \int_{\partial\Omega} \left[\frac{\partial}{\partial n_\xi} E(x,\xi)\right] \phi(\xi)\,d\sigma_\xi,$$

$$\widetilde{V}_2^*\phi(x) = \int_{\partial\Omega} \left[\frac{\partial}{\partial n_x} E(x,\xi)\right] \phi(\xi)\,d\sigma_\xi,$$

$$\widetilde{V}_2^d\phi(x) = \int_{\partial\Omega} \left[\frac{\partial^2}{\partial n_x \partial n_\xi} E(x,\xi)\right] \phi(\xi)\,d\sigma_\xi,$$

i.e., V_1, $\widetilde{V}_2$, $\widetilde{V}_2^*$ and $\widetilde{V}_2^d$ are respectively, the simple-layer, double-layer, the adjoint of the double-layer, and the normal derivative of the double-layer potential operators. Hence the first Calderòn operator, which projects into the space of Cauchy data of $Pu = 0$ on Ω, by (4.124) and (4.125), is

$$\mathscr{C}_1 = \begin{bmatrix} \frac{1}{2} - \widetilde{V}_2 & V_1 \\ -\widetilde{V}_2^d & \frac{1}{2} + \widetilde{V}_2^* \end{bmatrix}.$$

Similarly, the second Calderòn projector is

$$\mathscr{C}_2 = \begin{bmatrix} \frac{1}{2} + \widetilde{V}_2 & -V_1 \\ \widetilde{V}_2^d & \frac{1}{2} - \widetilde{V}_2^* \end{bmatrix}.$$

Thus, we indeed have

$$\mathscr{C}_1 + \mathscr{C}_2 = \begin{bmatrix} 1 & 0 \\ 0 & 1 \end{bmatrix} = I$$

proved in Proposition 4.1. From the projection property $\mathscr{C}_1^2 = \mathscr{C}_1$ (or, identically, $\mathscr{C}_2^2 = \mathscr{C}_2$) in Theorem 4.6,

$$\mathscr{C}_1^2 = \begin{bmatrix} \frac{1}{4} - \widetilde{V}_2 + (\widetilde{V}_2)^2 - V_1\widetilde{V}_2^d & V_1 - \widetilde{V}_2V_1 + V_1\widetilde{V}_2^* \\ -\widetilde{V}_2^d + \widetilde{V}_2^d\widetilde{V}_2 - \widetilde{V}_2^*\widetilde{V}_2^d & -\widetilde{V}_2^dV_1 + \frac{1}{4} + \widetilde{V}_2^* + (\widetilde{V}_2^*)^2 \end{bmatrix}$$

$$= \mathscr{C}_1 = \begin{bmatrix} \frac{1}{2} - \widetilde{V}_2 & V_1 \\ -\widetilde{V}_2^d & \frac{1}{2} + \widetilde{V}_2^* \end{bmatrix},$$

we obtain the following simple relations between these layer potential operators:

$$-\frac{1}{4} + (\widetilde{V}_2)^2 - V_1\widetilde{V}_2^d = 0, \tag{4.126}$$

$$V_1\widetilde{V}_2^* - \widetilde{V}_2V_1 = 0, \tag{4.127}$$

$$\widetilde{V}_2^d\widetilde{V}_2 - \widetilde{V}_2^*\widetilde{V}_2^d = 0, \tag{4.128}$$

$$-\frac{1}{4} + (\widetilde{V}_2^*)^2 - \widetilde{V}_2^dV_1 = 0. \tag{4.129}$$

Note that (4.128) and (4.129) are respectively the adjoint relations of (4.127) and (4.126). □

4.6 Fredholm operators

Let V and H be Hilbert spaces such that

$$V \subset H, \quad V \quad \text{is dense in} \quad H, \tag{4.130}$$

and the injection is continuous. The *dual space* of all continuous linear functionals l on V satisfying

$$l\colon\ V \to \mathbb{C}, \quad |l(v)| \leqslant M\|v\|_V \ \text{ for some } M > 0,\ \forall\, v \in V,$$

is V'. We write

$$l(v) = (v, l)_{V \times V'}.$$

Assume that $l_0 \in V'$ also satisfies

$$|l_0(v)| \leqslant M_0\|v\|_H \ \text{ for some } M_0 > 0, \forall\, v \in H.$$

Then, by the closed graph theorem (see Appendix A.3(ii) or [161]), l_0 can be uniquely extended as a bounded linear functional to all of H. We still call it l_0. Then $l_0 \in H'$ and we have

$$(v, l_0)_{V \times V'} = (v, l_0)_{H \times H'}. \tag{4.131}$$

We can identify H' with H through the Riesz representation theorem (see Appendix A.4). For any $l \in H \equiv H'$, through the pivotal relation (4.131), we define

$$l(v) \equiv (v, l)_{H \times H'},$$

so

$$|l(v)| \leqslant M_0\|v\|_H \leqslant M_1\|v\|_V \ \text{ for some } M_0, M_1 > 0, \forall\, v \in V,$$

by the dense imbedding of V in H. Therefore $l \in V'$ and (algebraically)

$$V \subset H(\equiv H') \subset V'. \tag{4.132}$$

It is not difficult to verify that the injection from H into V' is also dense and continuous, so the inclusions in (4.132) are also topological.

Proposition 4.2. *Let X and Y be Banach spaces, and let $A \in \mathscr{L}(X, Y)$. Then*

$$[\mathscr{R}(A)]^\perp = \mathscr{N}(A^*). \tag{4.133}$$

Proof. Let $y^* \in \mathscr{N}(A^*)$ and $y \in \mathscr{R}(A)$. Then $y = Ax$ for some $x \in X$. We have

$$(y, y^*)_{Y \times Y'} = (Ax, y^*) = (x, A^* y^*) = 0,$$

so $\mathscr{N}(A^*) \subset [\mathscr{R}(A)]^{\perp}$.

Now let $y^* \in [\mathscr{R}(A)]^{\perp}$. Then

$$(Ax, y^*) = 0 \quad \forall x \in X.$$

So

$$(x, A^* y^*) = 0 \quad \forall x \in X.$$

Hence $[\mathscr{R}(A)]^{\perp} \subset \mathscr{N}(A^*)$. □

The dual of Proposition 4.2 is

Proposition 4.3. *Let X and Y be Banach spaces, and let $A \in \mathscr{L}(X,Y)$. Let $\mathscr{R}(A)$ be closed. Then*

$$\mathscr{R}(A^*) = [\mathscr{N}(A)]^{\perp}. \tag{4.134}$$

Proof. Let $x^* \in \mathscr{R}(A^*)$. Then $x^* = A^* y^*$ for some $y^* \in Y'$. Let $x \in \mathscr{N}(A)$. Then

$$(x, x^*)_{X \times X'} = (x, A^* y^*) = (Ax, y^*) = 0.$$

Thus $x^* \in [\mathscr{N}(A)]^{\perp}$. It follows that $\mathscr{R}(A^*) \subset [\mathscr{N}(A)]^{\perp}$.

To prove the reverse inclusion, more work is required. For each $x \in X$, let $[x] = x + \mathscr{N}(A)$, the equivalence class containing x in the quotient space (see Appendix A.6) $\overline{X} \equiv X/\mathscr{N}(A)$. Then A induces a 1–1 mapping

$$\overline{A}\colon\ X/\mathscr{N}(A) \longrightarrow \mathscr{R}(A), \qquad \overline{A}[x] = Ax.$$

This 1–1 mapping has an inverse $\overline{A}^{-1} \in \mathscr{L}(\mathscr{R}(A), \overline{X})$ because $\mathscr{R}(A)$ is closed (and therefore is a Banach space with the subspace topology of Y). Therefore there is a constant $M > 0$ such that for any $y \in \mathscr{R}(A)$ such that $y = Ax$ for some $x \in X$,

$$\|[x]\|_{\overline{X}} \leqslant M \|y\|_Y \quad \forall x \in X. \tag{4.135}$$

Now let $x^* \in [\mathscr{N}(A)]^{\perp}$. For $y \in \mathscr{R}(A)$ and $x \in X$ satisfying $Ax = y$, define

$$f(y) = (\xi, x^*)_{X \times X'} \quad \forall \xi \in [x]. \tag{4.136}$$

Then for any $a_1, a_2 \in \mathbb{C}$, $y_1, y_2 \in \mathscr{R}(A)$, we have

$$\begin{aligned} f(a_1 y_1 + a_2 y_2) &= (a_1 \xi_1 + a_2 \xi_2 + \mathscr{N}(A), x^*) \\ &= (a_1 \xi_1 + a_2 \xi_2, x^*) \\ &= a_1 f(y_1) + a_2 f(y_2), \end{aligned}$$

for any $x_1, x_2 \in X$ such that $y_1 = Ax_1$, $y_2 = Ax_2$, and $\xi_1 \in [x_1]$, $\xi_2 \in [x_2]$. Therefore f is well defined and linear. f is bounded because

$$|f(y)| = |([x], x^*)| \leqslant \|x^*\|_{X'} \|[x]\|_{\overline{X}} \leqslant M\|x^*\|_{X'} \|y\|_Y,$$
$$\text{for } x \text{ such that } Ax = y, \text{by (4.135) and (4.136)}.$$

We now invoke the closed graph theorem (see Appendix A.3(ii)). Extend f to a linear functional $F \in Y^*$. From

$$\begin{aligned}(x, A^*F^*) &= (Ax, F^*) \\ &= (y, F^*) \\ &= (x, x^*) \quad \forall x \in X,\end{aligned}$$

it follows that $A^*F^* = x^*$, and thus $\mathscr{R}(A^*) \supset [\mathscr{N}(A)]^\perp$. □

Definition 4.4. Let X and Y be two Banach spaces. Let A be a bounded linear operator from X into Y, with $\mathscr{R}(A)$ closed in Y. We define the *cokernel* of A to be

$$\text{Coker}\, A = \{y^* \in Y' \mid (Ax, y^*) = 0\ \forall x \in X\}.$$

we say that A is a *Fredholm operator* if the dimension of its null space (or kernel) and the codimension of its range (i.e., dim Coker A) are both finite. For a Fredholm operator A, the number

$$\text{Ind}\, A = \dim \mathscr{N}(A) - \text{ codim}\, \mathscr{R}(A).$$

is called the *index* of A. □

Theorem 4.7. *Let X and Y be reflexive Banach spaces, and let $A \in \mathscr{L}(X,Y)$. If A is Fredholm then so is A^*, and* Ind $A = -$ Ind A^*.

Proof. By Propositions 4.2 and 4.3 , because of reflexivity and $A = A^{**}$, we have the symmetric relations

$$\dim \mathscr{N}(A^*) = \text{codim}\, \mathscr{R}(A), \tag{4.137}$$

$$\dim \mathscr{N}(A) = \text{codim}\, \mathscr{R}(A^*). \tag{4.138}$$

$\mathscr{R}(A^*)$ is closed in X' because $\mathscr{R}(A)$ is closed in X due to the closed range theorem of Banach [191, p. 205]. □

Corollary 4.1. *Let H be a Hilbert space, and suppose $A \in \mathscr{L}(H)$ is self-adjoint, i.e., $A = A^*$. If A is Fredholm then* Ind $A = 0$.

Definition 4.5. Let X and Y be Banach spaces, and let $A \in \mathscr{L}(X,Y)$. We say that A is *of finite rank* if $\dim \mathscr{R}(A)$ is finite. □

Proposition 4.4. *Suppose that X and Y are reflexive Banach spaces and that $A \in \mathscr{L}(X,Y)$. Then the following two properties are equivalent:*

(i) *A is Fredholm;*

(ii) *There is $B \in \mathscr{L}(Y,X)$ such that $AB - I_Y$ and $BA - I_X$ both have finite rank.*

If the above hold then the operator B in (ii) is also Fredholm. We can choose B such that

$$A \circ B \circ A = A, \qquad B \circ A \circ B = B. \tag{4.139}$$

Proof. First assume that (i) holds. Let $d = \operatorname{codim} \mathscr{R}(A)$. Then by (4.133) we can select a basis $f_1^*, f_2^*, \ldots, f_d^*$ in $\mathscr{N}(A^*)$, and d vectors $f_1, f_2, \ldots, f_d$ in Y, such that $(f_k, f_j^*) = \delta_{kj}$. Define

$$P\colon\ Y \to Y$$

$$Py = y - \sum_{j=1}^{d} (y, f_j^*) f_j.$$

Then $(Py, f_j^*) = 0$ for all $y \in Y$ and $j = 1, 2, \ldots, d$. Therefore the range of P is contained in $A(X)$. The restriction of P to $A(X)$ is the identity. P is a continuous projection (see Appendix A.7) of Y onto $A(X)$.

Let X_0 be a closed subspace of X such that X is the direct sum (see Appendix A.7) of X_0 and $\mathscr{N}(A)$: $X = X_0 \oplus \mathscr{N}(A)$. We restrict A to X_0 by defining

$$A_0 : X_0 \longrightarrow A(X),$$

$$A_0(x_0) = Ax \quad \text{if } x = x_0 + n \text{ for some } n \in \mathscr{N}(A).$$

Then A_0 is a continuous linear isomorphism, so its inverse $A_0^{-1}\colon\ A(X) \to X$ exists. Now define

$$B = A_0^{-1} \circ P.$$

The range of B is X_0, and it can easily be checked that $A \circ B = P$. The operator $I_Y - P$ $(= I_Y - A \circ B)$ is a continuous projection of Y into span $\{f_1, f_2, \ldots, f_d\}$ and thus has finite rank.

Similarly, $PA = A$, so $BA = A_0^{-1}A$ is a continuous projection of X onto X_0 whose kernel in $\mathscr{N}(A)$. Hence $I_X - BA$ is a continuous projection of X onto $\mathscr{N}(A)$; it also has finite rank. It is easy to check that (4.139) holds.

Suppose that (ii) holds. By (4.139), since $\mathscr{N}(A)$ is contained in the range of $I_X - BA$, it is finite-dimensional. Let $Y_0 = \mathscr{R}(I_Y - AB)$. From (4.139), it is clear that

$$Y_0 + A(X) = Y.$$

Therefore $\operatorname{codim} A(X) \leqslant \dim Y_0$. The fact that $A(X)$ is closed in Y is obvious from (4.139).

□

Theorem 4.8. *Let X and Y be reflexive Banach spaces. Then $A \in \mathscr{L}(X,Y)$ is Fredholm if and only if there exists $B \in \mathscr{L}(Y,X)$ such that both $I_Y - AB$ and $I_X - BA$ are compact operators.*

Proof. The "only if" part follows directly from Proposition 4.4 because every linear operator with finite rank is compact.

Now consider the "if" part. Define $K_X = I_X - BA$ and $K_Y = I_Y - AB$. Then K_X and K_Y are compact operators on X and Y respectively. We have

$$BA = I_X - K_X.$$

On $\mathscr{N}(I_X - K_X)$,

$$x = K_X x, \quad \text{for } x \in \mathscr{N}(I_X - K_X),$$

i.e., the identity operator is a compact operator on $\mathscr{N}(I_X - K_X)$. This is possible when and only when $\mathscr{N}(I_X - K_X)$ is finite-dimensional, implying $\dim \mathscr{N}(BA) < \infty$. Since

$$\mathscr{N}(A) \subset \mathscr{N}(BA),$$

so $\mathscr{N}(A)$ has finite dimension.

On the other hand, K_Y being compact implies that K_Y^* is compact [191, p. 282], so

$$B^*A^* = (AB)^* = (I_Y - K_Y)^* = I_{Y^*} - K_Y^*,$$

with K_Y^* being compact. We again have $\dim \mathscr{N}[(AB)^*] < \infty$, giving

$$\mathscr{N}(A^*) \subset \mathscr{N}(B^*A^*) = \mathscr{N}[(AB)^*],$$
$$\dim \mathscr{N}(A^*) = \dim[\mathscr{R}(A)]^{\perp} \leqslant \dim \mathscr{N}[(AB)^*] < \infty.$$

So $\mathscr{R}(A)$ has finite codimension. Now

$$Y \supset \mathscr{R}(A) \supset \mathscr{R}(AB) = [\mathscr{R}(I_Y - K_Y)].$$

But $\mathscr{R}(I_Y - K_Y)$ is closed in Y (see Exercise 4.2), whose algebraic complement in Y is finite dimensional. Therefore $\mathscr{R}(A)$ is also closed in Y. Hence A is Fredholm. □

Exercise 4.2. Let X be a Banach space, and let $K \in \mathscr{L}(X)$ be compact. Show that $\mathscr{R}(I-K)$ is closed in X. □

Exercise 4.3.

(i) Let X_j and Y_j be Banach spaces and let $A_j\colon\ X_j \to Y_j$ be bounded linear with closed range, for $j = 1,2$. Define

$$A_1 \oplus A_2\colon\ X_1 \oplus X_2 \longrightarrow Y_1 \oplus Y_2$$

by

$$(A_1 \oplus A_2)(x_1 \oplus x_2) = A_1x_1 \oplus A_2x_2.$$

Let A_1 and A_2 be Fredholm. Show that $A_1 \oplus A_2$ is also Fredholm, and

$$\operatorname{Ind}(A_1 \oplus A_2) = \operatorname{Ind}A_1 + \operatorname{Ind}A_2.$$

(ii) Let

$$X \xrightarrow{A} Y \xrightarrow{B} Z,$$

where X, Y and Z are Banach spaces and A, B are Fredholm. Then $B \circ A$ is also Fredholm, and

$$\operatorname{Ind}(B \circ A) = \operatorname{Ind}A + \operatorname{Ind}B. \qquad \square$$

The following theorem and its corollaries contain further properties of the index of Fredholm operators. Their proofs are not difficult, and are based upon the above discussion. See also [179, Chap. II, § 2.1], for example.

Theorem 4.9. *Let X and Y be Banach spaces. If $A \in \mathscr{L}(X,Y)$ is Fredholm and $K \in \mathscr{L}(X,Y)$ is compact, then $A+K$ is Fredholm. Furthermore,* $\operatorname{Ind}(A+K) = \operatorname{Ind}A$. □

Corollary 4.2. *Let X be a Banach space and let $K \in \mathscr{L}(X)$ be compact. Then*

$$\operatorname{Ind}(I_X + K) = 0. \qquad \square$$

Corollary 4.3. *Let X and Y be Banach spaces. The following two properties of an operator $A \in \mathscr{L}(X,Y)$ are equivalent.*

(i) *A is Fredholm and* $\operatorname{Ind}A = 0$*;*

(ii) *There is a compact operator $K \in \mathscr{L}(X,Y)$ such that $A+K$ is invertible.* □

Now we are able to present the following theorems, which will be important in the study of boundary element methods.
Let X and Y be Banach spaces and let $A\colon\ X \to Y$ be Fredholm. Assume that

$$m = \dim \mathscr{N}(A) = \text{ codim } \mathscr{R}(A) \in \mathbb{N}, \tag{4.140}$$

so Ind $A = 0$. We want to augment A to become an operator $\widetilde{A}\colon\ \mathbb{R}^m \oplus X \to \mathbb{R}^m \oplus Y$ that is *invertible*. The augmented operator $\widetilde{A}$ then takes the form

$$\widetilde{A} = \begin{bmatrix} R & T \\ S & A \end{bmatrix} : \ \mathbb{R}^m \oplus X \to \mathbb{R}^m \oplus Y, \tag{4.141}$$

where

$$R\colon \mathbb{R}^m \to \mathbb{R}^m,\ \ S\colon \mathbb{R}^m \to Y,\ \ T\colon\ X \to \mathbb{R}^m. \tag{4.142}$$

Obviously, R is an $m \times m$ constant matrix. From (4.142), we also see that S and T are both operators of finite rank, admitting representations

$$S = \sum_{j=1}^{m} \sigma_j \otimes e_j, \quad T = \sum_{j=1}^{m} e_j \otimes \tau_j^*, \tag{4.143}$$

where

$$\{\sigma_j \mid 1 \leqslant j \leqslant m\} \subset Y,\ \ \{\tau_j^* \mid 1 \leqslant j \leqslant m\} \subset X^*,$$
$$e_j = (0,0,\ldots,0,1,0,\ldots,0) \in \mathbb{R}^m, \text{ the } j\text{th unit vector}, \ \ 1 \leqslant j \leqslant m,$$

and

$$\left.\begin{aligned} S\vec{a} &= \sum_{j=1}^{m} a_j\sigma_j, && \text{for } \vec{a} = (a_1,\ldots,a_m) \in \mathbb{R}^m, \\ T\vec{f} &= \vec{\alpha} = (\alpha_1,\ldots,\alpha_m), && \text{with } \alpha_j = (f_j,\tau_j^*)_{X\times X^*},\ \ 1 \leqslant j \leqslant m. \end{aligned}\right\} \tag{4.144}$$

We state and prove the following.

Theorem 4.10 ([40]). *Let X and Y be Banach spaces, and let $A \in \mathscr{L}(X,Y)$ be a Fredholm operator with index zero, with m given by (4.140). Let $\{k_j \mid 1 \leqslant j \leqslant m\}$ and $\{k_j^* \mid 1 \leqslant j \leqslant m\}$ be, respectively, the bases of $\mathscr{N}(A)$ and $\mathscr{N}(A^*)$. Then an augmented operator $\widetilde{A}$ as given by (4.141), with S and T represented by (4.143), is invertible if and only if*

$$\det\left[\langle k_i, \tau_j^*\rangle\right]_{1\leqslant i,j\leqslant m} \neq 0, \quad \det\left[\langle \sigma_i, k_j^*\rangle\right]_{1\leqslant i,j\leqslant m} \neq 0. \tag{4.145}$$

Proof. First note that

$$\widetilde{A} = \begin{bmatrix} R & T \\ S & 0 \end{bmatrix} + \begin{bmatrix} 0 & 0 \\ 0 & A \end{bmatrix} \equiv \widetilde{\mathscr{A}_1} + \widetilde{\mathscr{A}_2},$$

and that $\widetilde{A}_1$ is of finite rank, and thus compact. We see by Theorem 4.9 that $\widetilde{A}$ is Fredholm, with index

$$\begin{aligned}
\operatorname{Ind}\widetilde{A} &= \operatorname{Ind}\begin{bmatrix} 0 & 0 \\ 0 & A \end{bmatrix} \\
&= \dim\left\{\mathcal{N}\begin{bmatrix} 0 & 0 \\ 0 & A \end{bmatrix}\right\} - \dim\left\{\mathbb{R}^m \oplus Y \Big/ \mathcal{R}\left(\begin{bmatrix} 0 & 0 \\ 0 & A \end{bmatrix}\right)\right\} \\
&= m + \dim \mathcal{N}(A) - m + \dim[Y/\mathcal{R}(A)] \\
&= \dim \mathcal{N}(A) - \dim[Y/\mathcal{R}(A)] \\
&= \operatorname{Ind} A = 0.
\end{aligned}$$

By Theorem 4.9, $\widetilde{A}$ is invertible if and only if $\mathcal{N}(\widetilde{A}) = \{0\}$. Let $(\vec{a}, f) \in \mathbb{R}^m \oplus X$ be such that $\widetilde{A}(\begin{bmatrix} \vec{a} \\ f \end{bmatrix}) = 0$. Thus

$$\left.\begin{aligned} R\vec{a} + Tf = 0 \in \mathbb{R}^m, \\ S\vec{a} + Af = 0 \in Y. \end{aligned}\right\} \tag{4.146}$$

We get

$$\begin{aligned}
0 &= \langle S\vec{a} + Af, k_j^* \rangle_{Y \times Y^*} \\
&= \sum_{i=1}^{m} a_i \langle \sigma_i, k_j^* \rangle + \langle f, A^* k_j^* \rangle \\
&= \sum_{i=1}^{m} a_i \langle \sigma_i, k_j^* \rangle \quad \text{for } 1 \leqslant j \leqslant m, \text{ by (4.144)}.
\end{aligned} \tag{4.147}$$

If the second relation in (4.145) is satisfied then $\vec{a} = (a_1, \ldots, a_m) = 0$. Thus (4.146) gives

$$Tf = 0 \text{ in } \mathbb{R}^m, \quad Af = 0 \text{ in } Y. \tag{4.148}$$

So $f \in \mathcal{N}(A)$, implying

$$f = \sum_{i=1}^{m} c_i k_i \in X, \quad c_i \in \mathbb{R}. \tag{4.149}$$

Substituting (4.149) into the first equation of (4.148), we get

$$0 = Tf = \sum_{i=1}^{m} c_i \sum_{j=1}^{m} \langle k_i, \tau_j^* \rangle e_j.$$

Therefore

$$\sum_{i=1}^{m} c_i \langle k_i, \tau_j^* \rangle = 0, \quad \text{for } j = 1, 2, \ldots, m. \tag{4.150}$$

By the first condition in (4.145), we get $c_i = 0$ for $1 \leqslant i \leqslant m$. Therefore $f = 0$ in X. Therefore, if (4.145) holds, $\mathscr{N}(A) = \{0\}$.
To show that (4.145) is also necessary for $\mathscr{N}(A) = \{0\}$, we set, separately, $\vec{a} = 0$ and $f = 0$ in (4.146) and conclude that (4.150) and (4.147) can admit only trivial solutions $c_i = 0$ and $a_i = 0$ for $1 \leqslant i \leqslant m$. Hence (4.145) must hold. □

We now show that the augmented system will always provide a solution to the original problem, if $\widetilde{A}$ is invertible. The original equation is

$$Au = g, \tag{4.151}$$

where the m compatibility conditions

$$\langle g, k_j^* \rangle_{Y \times Y^*} = 0, \qquad 1 \leqslant j \leqslant m, \tag{4.152}$$

are satisfied. Let us solve an augmented equation

$$\widetilde{A} \begin{bmatrix} \vec{a} \\ f \end{bmatrix} = \begin{bmatrix} \vec{a} \\ g \end{bmatrix} \in \mathbb{R}^m \oplus Y \tag{4.153}$$

for any given $\vec{\alpha} \in \mathbb{R}^m$. If $\widetilde{A}$ is invertible, we get a unique solution $(\vec{a}, f) \in \mathbb{R}^m \oplus X$, satisfying

$$\left.\begin{aligned} R\vec{a} + Tf = \vec{\alpha}, \\ S\vec{a} + Af = g. \end{aligned}\right\} \tag{4.154}$$

From the second equation above, using (4.152), we get

$$\langle S\vec{a}, k_j^* \rangle + \langle Af, k_j^* \rangle = \langle g, k_j^* \rangle, \quad 1 \leqslant j \leqslant m,$$
$$\sum_{i=1}^{m} a_i \langle \sigma_i, k_j^* \rangle + 0 = 0, \quad 1 \leqslant j \leqslant m,$$

by (4.152), (4.144) and $A^* k_j^* = 0$.

By (4.145), we have

$$\vec{a} = (a_1, \ldots, a_m) = 0.$$

Hence the second equation in (4.154) gives

$$Af = g.$$

Thus, the second component f in the solution of the augmented system (4.153) can always be used to serve as the solution u in (4.151). The extra m equations will always be consistent with the original equation (4.151).

Remark 4.3. The presence of R (cf. (4.141)) in $\widetilde{A}$ is immaterial in the statement of Theorem 4.10; therefore it is convenient to just set $R = 0$. □

Theorem 4.10 will be applied to various types of BVP in the next section.

The following two theorems are fundamental in the analysis of boundary integral equations, where the compact differentiable manifold M is just $\partial\Omega$, the boundary of some bounded open set Ω in $\mathbb{R}^N$.

Theorem 4.11. *Let M be a compact differentiable manifold and let $p(x,\xi) \in S^m$ be elliptic of order m. Then*

$$P\colon\ H^s(M) \to H^{s-m}(M), \tag{4.155}$$

is Fredholm for any $s \in \mathbb{R}$.

Proof. By applying Theorem 4.1, and a partition of unity on M, we find that the mapping P in (4.155) is linear continuous. Since P is elliptic, by Theorem 4.4, we can let $q(x,\xi) = \sigma(Q) \in S^{-m}$ be such that

$$Q\colon\ H^{s-m}(M) \to H^s(M)$$

and $I - QP$ is regularizing, i.e., Q is a parametrix of P. So

$$I - QP\colon {}^s(M) \to C^\infty(M) \subset H^s(M). \tag{4.156}$$

But $C^\infty(M) = \cap_{\tau \geqslant s} H^\tau(M)$, and the injection $H^\tau(M) \to H^s(M)$ is compact if $\tau > s$, by Theorem 2.5. So $I - QP \in \mathscr{L}(H^s(M))$ is compact. By Theorem 4.8, P is Fredholm. □

Corollary 4.4. *Under the assumptions of Theorem 4.11, we have*

$$\mathscr{N}(P) \subset C^\infty(M).$$

Proof. Let $u \in \mathscr{N}(P)$. By (4.156), we have

$$(I - QP)u = Ru, \quad \text{where } R\colon\ H^s(M) \to C^\infty(M) \ \text{ is infinitely smoothing.}$$

Since $Pu = 0$, we get

$$u = Ru \in C^\infty(M). \tag{4.157}$$

□

4.7 Applications to BIE of elliptic BVP with Neumann boundary conditions

An elliptic partial differential operator satisfies the Gårding inequality (4.59), which leads to a Fredholm opeorator when the boundary conditions are also properly posed. In our subsequent applications, this Fredholm operator has zero index. Therefore m orthogonality (or compatibility) conditions must be satisfied by the data, and the solutions have m degrees of freedom—they are nonunique. In order to fix these m degrees of freedom, m accessory linear conditions are usually prescribed. This leads to an augmented system like (4.141). In this section, we apply the theory of § 4.6, particularly Theorem 4.10, to elliptic BVP whose boundary conditions involve higher-order derivatives, i.e., Neumann-type boundary conditions. We will use Theorem 4.10 to determine when the augmented linear system is uniquely solvable.

The main advantages of the invertibility of a Fredholm operator by augmentation are the following.

(i) *It produces a unique solution* satisfying the given PDE, boundary data and the m accessory conditions. Further, our characterization of the augmented operator indicates what types of accessory conditions may be used for solving a boundary value problem uniquely.

(ii) *It bypasses the verification of the compatibility conditions.* When the compatibility conditions are satisfied, the augmented system will automatically provide a solution to the original system. This is particularly useful in the numerical solutions for such BVP, because numerically the discretized compatibility conditions usually do not correspond to the compatibility conditions for the matrix equation obtained by discretizing the PDE and boundary conditions.

We mention two applications: one to the Neumann boundary value problem of the Laplacian, and the other to the traction value problem of linear elastostatics. The plate equation will also be discussed in Example 4.7.

Throughout (except in Example 4.7) we let Ω be a simply connected bounded domain in $\mathbb{R}^3$ with smooth boundary $\partial\Omega$. We want to solve the BVP by boundary integral equations.

Example 4.4 (The Neumann boundary value problem of the Laplace equation).
Consider the following problem:

$$\left.\begin{array}{l}\Delta u(x) = 0 \quad \text{on} \quad \Omega,\\ \dfrac{\partial u(x)}{\partial n} = g(x) \in H^r(\partial\Omega), r \in \mathbb{R},\\ (n \equiv \text{unit exterior normal on } \partial\Omega).\end{array}\right\} \tag{4.158}$$

From elliptic regularity (cf. § 6.7), we know there exists a solution $u \in H^{r+3/2}(\Omega)$ satisfying (4.158), and, by (6.157),

$$\inf \|u\|_{H^{r+3/2}(\Omega)} \leqslant C\|g\|_{H^r(\partial\Omega)}$$

for some $C>0$ independent of g, with the infimum taken over all such u satisfying (4.158), if and only if the compatibility condition

$$\int_{\partial\Omega} g(x)\,d\sigma = 0 \tag{4.159}$$

is satisfied. We note that in the above, the integral should be interpreted as a duality pairing $\langle 1,g\rangle$ in $H^{-r}(\partial\Omega)\times H^{r}(\partial\Omega)$ if $r<0$. From now on throughout this section, we will assume such an interpretation for all similar situations without further mention. □

Let

$$E(x,\xi)=\frac{1}{4\pi}\frac{1}{|x-\xi|},\qquad x,\,\xi\in\mathbb{R}^3, \tag{4.160}$$

be the fundamental solution of the Laplacian satisfying

$$\Delta_\xi E(x,\xi)=-\delta(x-\xi),\quad \Delta_\xi\equiv\sum_{i=1}^{3}\frac{\partial^2}{\partial\xi_i^2},$$

The simple-layer potential solution for (4.158) is based on the *ansatz* (see [72, 95] and Theorem 6.20)

$$u(x)=\int_{\partial\Omega}E(x,\xi)f(\xi)\,d\sigma_\xi,\quad x\in\overline{\Omega}. \tag{4.161}$$

Taking the normal derivation, from the jump property of the boundary layer potentials (see Theorem 6.3), we get

$$\frac{\partial u(x)}{\partial n}=\frac{1}{2}f(x)+\int_{\partial\Omega}\frac{\partial E(x,\xi)}{\partial n_x}f(\xi)\,d\sigma_\xi,\quad x\in\partial\Omega, \tag{4.162}$$

where the integral on the RHS of (4.162) is interpreted as a weakly singular integral for $f\in C^\infty(\partial\Omega)$. Define boundary integral operators

$$(\mathscr{L}_1 f)(x)=\int_{\partial\Omega}E(x,\xi)f(\xi)\,d\sigma_\xi, \tag{4.163}$$

$$(\mathscr{L}_2 f)(x)=\frac{1}{2}f(x)+\int_{\partial\Omega}\frac{\partial E(x,\xi)}{\partial n_x}f(\xi)\,d\sigma_\xi. \tag{4.164}$$

According to the theory in §4.4 and §6.8, it is known that the following hold:

(1) $\mathscr{L}_1$ is a strongly elliptic pseudodifferential operator of order -1, mapping continuously

$$\mathscr{L}_1:\ H^r(\partial\Omega)\to H^{r+1}(\partial\Omega). \tag{4.165}$$

$\mathscr{L}_1$ is positive definite and invertible, satisfying

$$\langle\mathscr{L}_1 f,f\rangle_{H^0(\partial\Omega)}=\int_{\partial\Omega}|\nabla V(f)|^2\,d\sigma\geqslant\gamma\|\mathscr{L}_1 f\|^2_{H^{1/2}(\partial\Omega)},$$

$$\gamma>0,\ \forall f\in H^0(\partial\Omega), \tag{4.166}$$

where

$$V(f)(x) \equiv \int_{\partial\Omega} E(x-\xi)f(\xi)\,d\sigma_\xi, \quad x \in \mathbb{R}^3,$$

$$C_r^{-1}\|f\|_{H^r(\partial\Omega)} \leqslant \|\mathscr{L}_1 f\|_{H^{r+1}(\partial\Omega)} \tag{4.167}$$

$$\leqslant C_r\|f\|_{H^r(\partial\Omega)}, \tag{4.168}$$

$$C_r > 0 \text{ independent of } f \in H^r(\partial\Omega).$$

(2) $\mathscr{L}_2$ is a Fredholm operator with zero index on $H^r(\partial\Omega), r \in \mathbb{R}$. It is a pseudodifferential operator of order 0, with principal symbol equal to $\frac{1}{2}$. The integral operator on the RHS of (4.164) is a compact operator on $H^r(\partial\Omega)$. Further,

$$\mathscr{N}(\mathscr{L}_2) = \text{span}\{k \in C^\infty(\partial\Omega) \mid \mathscr{L}_1 k = 1 \text{ on } \partial\Omega\} \tag{4.169}$$

$$\text{Coker } \mathscr{L}_2 = \mathscr{N}(\mathscr{L}_2^*) = \text{span}\{1 \text{ on } \partial\Omega\}, \tag{4.170}$$

$$(\mathscr{L}_2^* f)(x) = \frac{1}{2}f(x) + \int_{\partial\Omega} \frac{\partial}{\partial n_\xi} E(x,\xi)f(\xi)\,d\sigma_\xi, \quad x \in \partial\Omega, \tag{4.171}$$

$$\mathscr{L}_2^* : H^{-r}(\partial\Omega) \to H^{-r}(\partial\Omega),\ r \in \mathbb{R},$$

is also Fredholm with zero index.

To solve the BVP (4.158) subject to (4.159), we use the simple-layer solution (4.161). In order to fix the solution of (4.158) and (4.159) uniquely, we consider appending an extra condition such as

$$\int_{\partial\Omega} x_j^n u(x)\,d\sigma_x = \alpha \in \mathbb{R}, \quad j = 1, 2 \text{ or } 3 \text{ (exclusively)} \tag{4.172}$$

$$n \text{ is a given nonnegative integer,}$$

or

$$u(x^{(0)}) = \alpha \in \mathbb{R}, \quad x^{(0)} \in \partial\Omega \text{ given.} \tag{4.173}$$

Condition (4.172) means that the nth moment of the solution u in the x_j-direction on $\partial\Omega$ must be equal to a specified number α. (Note that only one pair (j,n) is allowed here.) Condition (4.173) means that the solution takes a specified value α at a given boundary point $x^{(0)}$.

Let us amend the boundary integral equation

$$\frac{\partial u(x)}{\partial n} = g(x) = \frac{1}{2}f(x) + \int_{\partial\Omega} \frac{\partial E(x,\xi)}{\partial n_x} f(\xi)\,d\xi \tag{4.174}$$

$$= (\mathscr{L}_2 f)(x), \quad x \in \partial\Omega,$$

by (4.172) or (4.173). We treat these two cases separately.

(a) *Extra condition (4.172):* $\int_{\partial\Omega} x_j^n u(x)\,d\sigma_x = \alpha$.

We first note that, corresponding to the simple-layer representation (4.161), (4.173) gives

$$\begin{aligned}\int_{\partial\Omega} x_j^n \int_{\partial\Omega} E(x,\xi)f(\xi)\,d\sigma_\xi\,d\sigma_x &= \int_{\partial\Omega}\left[\int_{\partial\Omega} x_j^n E(x,\xi)\,d\sigma_x\right] f(\xi)\,d\sigma_\xi \\ &\equiv \int_{\partial\Omega} \tau^*(\xi) f(\xi)\,d\sigma\ (=\langle f,\tau^*\rangle).\end{aligned} \tag{4.175}$$

From the smoothing property of the simple-layer potential (4.74), we see that $\tau^* \in C^\infty(\partial\Omega)$. Let $k \in \mathscr{N}(\mathscr{L}_1)$ be its basis element satisfying

$$\int_{\partial\Omega} E(x,\xi)k(\xi)\,d\sigma_\xi = 1 \quad \forall x \in \partial\Omega, \text{ cf. (4.169)}. \tag{4.176}$$

We let the augmented system be

$$\begin{bmatrix} 0 & \tau^* \\ 1 & \mathscr{L}_2 \end{bmatrix}\begin{bmatrix} a \\ f \end{bmatrix} = \begin{bmatrix} \alpha \\ g \end{bmatrix} \quad \text{in } \mathbb{R}\oplus H^r(\partial\Omega). \tag{4.177}$$

i.e.,

$$\left.\begin{aligned} &\langle f,\tau^*\rangle = \int_{\partial\Omega} f(\xi)\tau^*(\xi)\,d\sigma = \alpha, \\ &a + (\mathscr{L}_2 f)(x) = g(x). \end{aligned}\right\}$$

According to Theorem 4.10, (4.177) is invertible if and only if

$$\langle k,\tau^*\rangle \neq 0, \quad \langle 1,k^*\rangle \neq 0. \tag{4.178}$$

The second condition above is always satisfied, because by (4.170), $k^* \equiv 1$ and

$$\begin{aligned}\langle 1,k^*\rangle_{H^r(\partial\Omega)\times H^{-r}(\partial\Omega)} = \langle 1,k^*\rangle_{L^2(\partial\Omega)} &= \int_{\partial\Omega} 1\cdot 1\,d\sigma \\ &= \text{area of } \partial\Omega \\ &> 0.\end{aligned}$$

The first condition in (4.145) requires that

$$\begin{aligned} 0 \neq \langle k,\tau^*\rangle_{H^r(\partial\Omega)\times H^{-r}(\partial\Omega)} &= \langle k,\tau^*\rangle_{L^2(\partial\Omega)} \\ &= \int_{\partial\Omega} k(\xi)\left[\int_{\partial\Omega} x_j^n E(x,\xi)\,d\sigma_x\right] d\sigma_\xi \\ &= \int_{\partial\Omega} x_j^n \left[\int_{\partial\Omega} E(x,\xi)k(\xi)\,d\sigma_\xi\right] d\sigma_x \\ &= \int_{\partial\Omega} x_j^n\,d\sigma_x \qquad (\text{cf.}(4.176)). \end{aligned} \tag{4.179}$$

It is strictly positive if $n = 0, 2, 4, \ldots$, but can vanish if n is odd (just let Ω be the unit ball and use the spherical coordinate representation for x_1, x_2 or x_3). Therefore, the augmented system (4.177) is invertible if and only if

$$\int_{\partial\Omega} x_j^n \, d\sigma_x \neq 0 \quad \text{for given } (j,n).$$

It is obvious that the invertibility of the augmented system

$$\widetilde{A} = \begin{bmatrix} R & \tau^* \\ S & \mathscr{L}_2 \end{bmatrix} \tag{4.180}$$

in the present case of (4.177) will remain unchanged even if R and S in (4.180) are chosen differently from (4.175), because every harmonic function (with sufficiently regular boundary data on $\partial\Omega$) on $\Omega \subset \mathbb{R}^3$ is representable by a simple-layer potential, as a consequence of the invertibility of the operator $\mathscr{L}_1$.

Therefore, we conclude the following.

Corollary 4.5. *The Neumann BVP*

$$\left.\begin{aligned} \Delta u(x) &= 0 \quad \text{on } \Omega \\ \frac{\partial u(x)}{\partial n} &= g(x) \in H^r(\partial\Omega), \quad r \in \mathbb{R}, \end{aligned}\right\} \tag{4.181}$$

with the compatibility condition $\int_{\partial\Omega} g \, d\sigma = 0$ *has a* unique *solution* $u \in H^{r+3/2}(\Omega)$ *also satisfying the accessory condition*

$$\int_{\partial\Omega} x_j^n u(x) \, d\sigma_x = \alpha, \quad j = 1, 2 \text{ or } 3, \tag{4.182}$$

with $n =$ *a nonnegative integer and* $\alpha \in \mathbb{R}$ *both given, if and only if*

$$\int_{\partial\Omega} x_j^n \, d\sigma_x \neq 0. \tag{4.183}$$

□

Furthermore, the augmented boundary integral equation system (4.177) based on the simple-layer potential (4.161) corresponding to (4.181) and (4.182) has a unique solution $(a, f) = (0, \widetilde{f}) \in \mathbb{R} \times H^r(\partial\Omega)$ satisfying

$$\langle \widetilde{f}, \tau^* \rangle = \alpha = \int_{\partial\Omega} x_j^n u(x) \, d\sigma_x,$$

where τ^* is defined by (4.175).

(b) *Extra condition (4.173)*: $u(x^{(0)}) = \alpha, x^{(0)} \in \partial\Omega$

First, note that u is defined pointwise at $x^{(0)} \in \partial\Omega$, provided that $g \in H^r(\partial\Omega)$ with $r > \frac{1}{2}$ and, thus, $u \in H^{3/2+r}(\Omega)$. Also, its trace satisfies

$$u|_{\partial\Omega} \in H^{3/2+r-1/2}(\partial\Omega) = H^{1+r}(\partial\Omega) \subset H^{1+\varepsilon}(\partial\Omega) \subset C^0(\partial\Omega),$$
$$\text{for } r > 1/2, \text{ and for } \varepsilon\colon\ 0 < \varepsilon \leqslant \frac{1}{2},$$

since $\partial\Omega$ is two-dimensional (by the Sobolev imbedding theorem; cf. Theorem 2.2). Corresponding to the simple-layer representation (4.161), (4.173) implies that

$$\int_{\partial\Omega} E(x^{(0)},\xi) f(\xi)\, d\sigma_\xi = \alpha. \tag{4.184}$$

So we let

$$\int_{\partial\Omega} E(x^{(0)},\xi) f(\xi)\, d\sigma_\xi \equiv \int_{\partial\Omega} f(\xi)\tau^*(\xi)\, d\sigma = \langle f, \tau^* \rangle_{L^2}. \tag{4.185}$$

Note that the distribution $\tau^* \in L^{2-\varepsilon_1}(\partial\Omega) \cap H^{-\varepsilon_2}(\partial\Omega)$ for all $\varepsilon_1,\ \varepsilon_2 > 0$.
Again, we let the augmented system be (4.177), where $\tau^* = E(x^{(0)},\cdot)$ is used in (4.177) instead. The system (4.177) is invertible if and only if

$$\langle k, \tau^* \rangle_{H^r(\partial\Omega)\times H^{-r}(\partial\Omega)} = \langle k, \tau^* \rangle_{L^2(\partial\Omega)} \neq 0,$$

i.e.,

$$\int_{\partial\Omega} E(x^{(0)},\xi) k(\xi)\, d\sigma \neq 0. \tag{4.186}$$

But we know that k satisfies (4.176). So $1 =$ LHS of (4.186) $\neq 0$ is always satisfied. We conclude the following.

Corollary 4.6. *The Neumann BVP*

$$\left.\begin{aligned} &\Delta u(x) = 0 \\ &\frac{\partial u(x)}{\partial n} = g(x) \in H^r(\partial\Omega),\ r > \tfrac{1}{2}, \end{aligned}\right\} \tag{4.187}$$

with the compatibility condition $\int_{\partial\Omega} g\, d\sigma = 0$ *has a* unique *solution* $u \in H^{r+3/2}(\Omega)$ *also satisfying the accessory condition*

$$u(x^{(0)}) = \alpha, \quad \textit{with } x^{(0)} \in \partial\Omega \textit{ and } \alpha \in \mathbb{R} \textit{ given.} \tag{4.188}$$

□

Furthermore, the augmented boundary integral equation system (4.177) based on the simple-layer potential (4.161) corresponding to (4.187) and (4.188) has a unique solution $(a, f) = (0, \widetilde{f}) \in \mathbb{R} \times H^r(\partial\Omega)$ satisfying

$$\langle \widetilde{f}, \tau^* \rangle = \alpha = u(x^{(0)}),$$

where τ^* is defined by (4.185).

Example 4.5. (The traction boundary value problem in three-dimensional linear elastostatics (Chapter 9)). Consider

$$\left\{\begin{array}{l} \mu\Delta\vec{u}(x)+(\lambda+\mu)\nabla[\nabla\cdot\vec{u}(x)]=0 \text{ on } \Omega, \\ \vec{\tau}(\vec{u})(x)=\vec{g}(x)\in[H^r(\partial\Omega)]^3, \qquad r\in\mathbb{R}, \end{array}\right\} \tag{4.189}$$

where $\vec{u}(x)=(u_1(x),u_2(x),u_3(x))$ represents the displacement field of a homogeneous isotropic Hookean solid at $x\in\Omega$; $\lambda,\ \mu>0$ are the Lamé constants; $\vec{\tau}(\vec{u})(x)$ represents the traction vector given by

$$\vec{\tau}(\vec{u})(x)=\lambda[\nabla\cdot\vec{u}(x)]n(x)+2\mu\frac{\partial\vec{u}(x)}{\partial n}+\mu\vec{n}(x)\times[\nabla\times\vec{u}(x)] \tag{4.190}$$

at $x\in\partial\Omega$. The BVP (4.189) behaves very similarly to the Neumann BVP in Example 4.4, except that now it is a *system* and more compatibility conditions must be satisfied: (4.189) has a solution $\vec{u}$ if and only if

$$\int_{\partial\Omega}\vec{g}(x)\cdot\vec{m}_j(x)\,d\sigma=0, \quad j=1,2,\ldots,6, \tag{4.191}$$

are satisfied, where $\vec{m}_j(x)$ represents the j-th column vectors of the rigid-body motion matrix

$$M(x)=\begin{bmatrix} 1 & 0 & 0 & -x_2 & 0 & x_3 \\ 0 & 1 & 0 & x_1 & -x_3 & 0 \\ 0 & 0 & 1 & 0 & x_2 & -x_1 \end{bmatrix}. \tag{4.192}$$

(The first three column vectors represent translations along the three coordinate directions, while the last three represent angular rotations.) □

Again, let us solve (4.189) by the simple-layer potential method. Write

$$\vec{u}(x)=\int_{\partial\Omega}E(x,\xi)\vec{f}(\xi)\,d\sigma_\xi, \quad x\in\overline{\Omega}, \tag{4.193}$$

where the fundamental 3×3 matrix solution

$$E(x,\xi)=\frac{\lambda+3\mu}{8\pi(\lambda+2\mu)}\left[\frac{1}{|x-\xi|}I_3+\frac{\lambda+\mu}{\lambda+3\mu}\frac{1}{|x-\xi|^3}(x-\xi)(x-\xi)^{\mathrm{Tr}}\right] \tag{4.194}$$

satisfies

$$\mu\Delta_\xi E(x,\xi)+(\lambda+\mu)\nabla_\xi[\nabla_\xi\cdot E(x,\xi)]=-\delta(x-\xi)I_3,$$

with I_3 being the 3×3 identity matrix. Taking the traction vector of the simple-layer displacement field (4.193), similarly to (4.162), we get

$$\begin{aligned}\vec{\tau}(\vec{u})(x)&=\frac{1}{2}\vec{f}(x)+\int_{\partial\Omega}\vec{\tau}_x(E)(x,\xi)\vec{f}(\xi)\,d\sigma_\xi \\ &=\vec{g}(x), \quad x\in\partial\Omega. \end{aligned} \tag{4.195}$$

where $\vec{\tau}_x(E)(x,\xi)$ is given by (9.99). Again, we define boundary integral operators

$$\mathscr{L}_1(\vec{f})(x) = \int_{\partial\Omega} E(x,\xi)\vec{f}(\xi)\,d\sigma_\xi, \qquad x \in \partial\Omega, \tag{4.196}$$

$$\mathscr{L}_2(\vec{f})(x) = \frac{1}{2}\vec{f}(x) + \int_{\partial\Omega} \vec{\tau}_x(E)(x,\xi)\vec{f}(\xi)\,d\sigma_\xi, \quad x \in \partial\Omega. \tag{4.197}$$

The singularity of the kernel $E(x,\xi)$ is essentially similar to that of the kernel (4.160) for the Laplacian. Using the same ideas in Example 4.4, it is known (see [98]) the following.

(1) $\mathscr{L}_1$ is a strongly pseudodifferential operator of order -1 mapping continuously from

$$\mathscr{L}_1:\ [H^r(\partial\Omega)]^3 \longrightarrow \left[H^{r+1}(\partial\Omega)\right]^3, \quad r \in \mathbb{R}. \tag{4.198}$$

$\mathscr{L}_1$ is positive definite and invertible, satisfying

$$\begin{aligned}\langle \mathscr{L}_1\vec{f},\vec{f}\rangle_{[H^0(\partial\Omega)]^3} &= \int_{\mathbb{R}^3}\left\{\lambda[\nabla\cdot\vec{V}(\vec{f})]^2 + \frac{\mu}{2}\sum_{j,k=1}^{3}\left[\frac{\partial}{\partial x_j}V_k(\vec{f}) + \frac{\partial}{\partial x_k}V_j(\vec{f})\right]^2\right\}dx \\ &\geqslant \gamma\|\mathscr{L}_1\vec{f}\|^2_{[H^{1/2}(\partial\Omega)]^3}, \quad \gamma>0,\ \forall\,\vec{f}\in[H^0(\partial\Omega)]^3,\end{aligned} \tag{4.199}$$

where

$$\vec{V}(\vec{f})(x) \equiv \int_{\partial\Omega} E(x,\xi)\vec{f}(\xi)\,d\sigma_\xi, \quad x \in \mathbb{R}^3, \tag{4.200}$$

and

$$\begin{aligned} c_r^{-1}\|\vec{f}\|_{[H^r(\partial\Omega)]^3} &\leqslant \|\mathscr{L}_1\vec{f}\|_{[H^{r+1}(\partial\Omega)]^3} \\ &\leqslant c_r\|\vec{f}\|_{[H^r(\partial\Omega)]^3}, \quad c_r>0 \text{ independent of } \vec{f}.\end{aligned} \tag{4.201}$$

(2) $\mathscr{L}_2$ is Fredholm with zero index on $[H^r(\partial\Omega)]^3, r\in\mathbb{R}$. $\mathscr{L}_2$ is a pseudodifferential operator whose principal matrix symbol at each point $x\in\partial\Omega$ is similar to a symmetric symbol matrix with a diagonal part $\frac{1}{2}I_3$, with some nonzero off-diagonal entries (cf. [98, p. 49]). The integral operator on the RHS of (4.164) represents a Cauchy principal value. The overall principal matrix symbol of $\mathscr{L}_2$ is strongly elliptic of order 0 at each $x\in\partial\Omega$, and therefore Fredholm with zero index on $[H^r(\partial\Omega)]^3$ by Corollary 4.2. Further,

$$\mathscr{N}(\mathscr{L}_2) = \operatorname{span}\left\{\vec{k}_j \in [C^\infty(\partial\Omega)]^3 \mid \mathscr{L}_1\vec{k}_j = \vec{m}_j \text{ on } \partial\Omega, 1\leqslant j\leqslant 6\right\} \tag{4.202}$$

$$\operatorname{Coker}\mathscr{L}_2 = \mathscr{N}(\mathscr{L}_2^*) = \operatorname{span}\{\vec{m}_j \text{ on } \partial\Omega \mid 1\leqslant j\leqslant 6\}, \tag{4.203}$$

$$(\mathscr{L}_2^* f)(x) = \frac{1}{2}\vec{f}(x) + \int_{\partial\Omega}\vec{\tau}_\xi(E)(x,\xi)\vec{f}(\xi)\,d\sigma_\xi, \quad x\in\partial\Omega.$$

$\mathscr{L}_2^*$ is also a pseudodifferential operator of order 0, with principal matrix symbol at each point $x\in\partial\Omega$ similar to a symmetric symbol matrix whose diagonal part is $\frac{1}{2}I_3$, and some of whose off-diagonal entries are nonzero, and with an overall principal matrix symbol strongly elliptic of order 0. Thus $\mathscr{L}_2^*$ is also a Fredholm operator with zero index on $[H^r(\partial\Omega)]^3$. In order to uniquely solve the solution of (4.189) satisfying (4.191), we need to *fix six additional constants*. We may consider, e.g.,

$$\int_{\partial\Omega} \vec{u}(x) \cdot \vec{m}_j(x)\, d\sigma = \alpha_j, \quad j = 1, 2, \ldots, 6, \tag{4.204}$$

or

$$\vec{u}(x^{(1)}) = \vec{\beta}^{(1)}, \ \ \vec{u}(x^{(2)}) = \vec{\beta}^{(2)} \tag{4.205}$$
$$\text{where } x^{(1)}, x^{(2)} \in \overline{\Omega}, \vec{\beta}^{(1)}, \vec{\beta}^{(2)} \in \mathbb{R}^3 \text{ are given,}$$

or six additional restrictive conditions on the simple layer density $\vec{f}$:

$$\int_{\partial\Omega} \vec{m}_j(x) \cdot \vec{f}(x)\, d\sigma = \alpha_j \in \mathbb{R}, \tag{4.206}$$
$$1 \leqslant j \leqslant 6, \ \ \alpha_j\text{s are given.}$$

Any of (4.204)–(4.206) would provide six extra conditions to possibly help obtain a unique solution $\vec{u}(x)$. Note that (4.205) is analogous to (4.173), but we now use two points $x^{(1)}$ and $x^{(2)}$ (providing six data) instead.

We use the simple-layer solution (4.193). We treat cases (4.204), (4.205) and (4.206) separately below.

(a) *Six extra conditions (4.204):* $\int_{\partial\Omega} \vec{u}(x) \cdot \vec{m}_j(x)\, d\sigma = \alpha_j, j = 1, 2, \ldots, 6.$

Note that, corresponding to (4.193), conditions (4.204) imply

$$\begin{aligned}
\int_{\partial\Omega}\int_{\partial\Omega} [E(x,\xi)\vec{f}(\xi)] \cdot \vec{m}_j(x)\, d\sigma_\xi\, d\sigma_x = \alpha_j \\
&= \int_{\partial\Omega} \vec{f}(x) \cdot \left[\int_{\partial\Omega} E(x,\xi)\vec{m}_j(\xi)\, d\sigma_\xi\right] d\sigma_x \\
&\equiv \int_{\partial\Omega} \vec{f}(x) \cdot \vec{\tau}_j^*(x)\, d\sigma_x \\
&= \langle \vec{f}, \vec{\tau}_j^* \rangle_{[L^2(\partial\Omega)]^3}, \ \ \tau_j^* \in C^\infty(\partial\Omega).
\end{aligned} \tag{4.207}$$

From (4.195), (4.197) and (4.207), we form an augmented system by

$$\begin{aligned}
\widetilde{A}\begin{bmatrix} \vec{a} \\ \vec{f} \end{bmatrix} &\equiv \begin{bmatrix} 0 & \sum_{j=1}^{6} e_j \otimes \vec{\tau}_j^* \\ \sum_{j=1}^{6} \vec{m}_j \otimes e_j & \mathscr{L}_2 \end{bmatrix} \begin{bmatrix} \vec{a} \\ \vec{f} \end{bmatrix} \\
&= \begin{bmatrix} \vec{\alpha} \\ \vec{g} \end{bmatrix}, \ \text{ in } \mathbb{R}^6 \oplus [H^r(\partial\Omega)]^3.
\end{aligned} \tag{4.208}$$

Then, $\widetilde{A}$ is invertible if and only if

$$\det[\langle \vec{k}_i, \vec{\tau}_j^*\rangle] \neq 0, \quad \det[\langle \vec{m}_i, k_j^*\rangle] \neq 0. \tag{4.209}$$

We have

$$\begin{aligned}\langle \vec{k}_i, \vec{\tau}_j^*\rangle &= \int_{\partial\Omega} \vec{k}_i(x) \cdot \int_{\partial\Omega} E(x,\xi)\vec{m}_j(\xi)\, d\sigma_\xi\, d\sigma_x \\ &= \int_{\partial\Omega} \vec{m}_j(x) \cdot \int_{\partial\Omega} E(x,\xi)\vec{k}_i(\xi)\, d\sigma_\xi\, d\sigma_x \\ &= \int_{\partial\Omega} \vec{m}_j(x) \cdot \vec{m}_i(x)\, d\sigma. \end{aligned}\tag{4.210}$$

Because $\vec{m}_i(x), i = 1,\dots,6$, are linearly independent functions in $L^2(\partial\Omega)$, we see that the first condition in (4.145) is always satisfied. The second condition is also satisfied, because by (4.203), $\langle \vec{m}_i, \vec{k}_j^*\rangle = \langle \vec{m}_i, \vec{m}_j\rangle$, which form entries of an invertible matrix because $\vec{m}_i, 1 \leqslant i \leqslant 6$, are linearly independent in $[L^2(\Omega)]^3$.

Therefore we conclude the following.

Corollary 4.7. *The traction BVP*

$$\left.\begin{aligned} &\mu\Delta\vec{u} + (\lambda+\mu)\nabla(\nabla\cdot\vec{u}) = 0 && \text{on } \Omega \\ &\vec{\tau}(\vec{u})(x) = \vec{g}(x) \in [H^r(\partial\Omega)]^3, && r \in \mathbb{R}, \end{aligned}\right\}\tag{4.211}$$

with compatibility conditions $\int_{\partial\Omega} \vec{g}\cdot\vec{m}_j\, d\sigma = 0,\ 1 \leqslant j \leqslant 6$, *has a unique solution* $\vec{u} \in [H^{r+3/2}(\Omega)]^3$ *if six accessory conditions*

$$\int_{\partial\Omega} \vec{u}(x)\cdot\vec{m}_j(x)\, d\sigma = \alpha_j, \tag{4.212}$$

$$j = 1, 2, \dots, 6, \text{ with } \alpha_j \in \mathbb{R} \text{ given,}$$

are specified. □

Furthermore, the augmented boundary integral equation system (4.208) based on the simple-layer potential (4.193) corresponding to (4.211) and (4.212) has a unique solution $(\vec{a}, \vec{f}) = (0, \vec{\widetilde{f}}) \in \mathbb{R}^6 \oplus [H^r(\partial\Omega)]^3$ satisfying

$$\left\langle \vec{\widetilde{f}}, \sum_{j=1}^{6} e_j \otimes \tau_j^* \right\rangle = \vec{\alpha} = \left[\int_{\partial\Omega} \vec{u}(x)\cdot\vec{m}_j(x)\, d\sigma\right]_{j=1}^{6},$$

where $\tau_j^*,\ 1 \leqslant j \leqslant 6$, are defined by (4.207).

(b) *Six extra conditions (4.205):* $\vec{u}(x^{(1)}) = \vec{\beta}^{(1)}$ and $\vec{u}(x^{(2)}) = \vec{\beta}^{(2)}$

As in Example 4.4, case (b), in order for $\vec{u}$ to be defined pointwise at $x^{(1)}, x^{(2)} \in \partial\Omega$, we require the traction data $\vec{g}$ in (4.188) to satisfy

$$\vec{g} \in [H^r(\partial\Omega)]^3, \quad r > \frac{1}{2}.$$

Then $\vec{u} \in [H^{r+3/2}(\Omega)]^3$, and by the trace theorem and the Sobolev imbedding theorem (Theorem 2.2),

$$\vec{u}\,|_{\partial\Omega} \in [H^{1+r}(\partial\Omega)]^3 \subset [H^{1+\varepsilon}(\partial\Omega)]^3 \subset [C^0(\partial\Omega)]^3,$$
$$\text{for } r > \frac{1}{2}, \; 0 < \varepsilon \leqslant \frac{1}{2}.$$

Denote $\vec{\alpha} = (\vec{\beta}^{(1)}, \vec{\beta}^{(2)}) \in \mathbb{R}^6$. The conditions

$$\vec{u}(x^{(i)}) = \vec{\beta}^{(i)}, \quad i = 1, 2,$$

require that

$$\int_{\partial\Omega} E(x^{(i)}, \xi)\vec{f}(\xi)\,d\sigma_\xi = \vec{\beta}^{(i)},$$

so we define τ_j^* by

$$\langle \vec{f}, \vec{\tau}_j^* \rangle = \int_{\partial\Omega} \vec{f}(\xi) \cdot \vec{\tau}_j^*(\xi)\,d\sigma$$

where

$$\left.\begin{aligned} &\vec{\tau}_j^* \text{ is the } j\text{th row vector of } E(x^{(1)}, \xi), \\ &\qquad \text{transposed for } j = 1, 2, 3; \\ &\vec{\tau}_j^* \text{ is the } (j-3)\text{th row vector of } E(x^{(2)}, \xi), \\ &\qquad \text{transposed for } j = 4, 5, 6. \end{aligned}\right\} \tag{4.213}$$

Similarly, we now form the augmented system (4.208), where $\vec{\tau}_j^*$ defined by (4.213) are used instead in (4.208). $\vec{\tau}_j^*$ here has the same type of singularity as the $\vec{\tau}^*$ in (4.185), so $\vec{\tau}_j^* \in [L^{2-\varepsilon_1}(\partial\Omega)]^3 \cap [H^{-\varepsilon_2}(\partial\Omega)]^3$ for any $\varepsilon_1, \varepsilon_2 > 0$. We now check the two conditions in (4.145). The second condition in (4.145) is easily verified as before, so we need only check the first condition in (4.145). We have

$$\begin{aligned} \langle \vec{k}_i, \vec{\tau}_j^* \rangle &= \int_{\partial\Omega} \vec{k}_i(\xi) \cdot \vec{E}_j(x^{(1)}, \xi)\,d\sigma_\xi \\ &= \text{the } j\text{th component of } \vec{m}_i(x^{(1)}), \quad 1 \leqslant j \leqslant 3, \\ \langle \vec{k}_i, \vec{\tau}_j^* \rangle &= \int_{\partial\Omega} \vec{k}_i(\xi) \cdot \vec{E}_{j-3}(x^{(2)}, \xi)\,d\sigma_\xi \\ &= \text{the } (j-3)\text{th component of } \vec{m}_i(x^{(2)}), \quad 4 \leqslant j \leqslant 6, \end{aligned}$$

where $\vec{E}_l(x, \xi)$ denotes the lth row vector of $E(x, \xi)$, transposed. Therefore the matrix of $\langle \vec{k}_i, \vec{\tau}_j^* \rangle$ is assembled, yielding

$$\det \begin{bmatrix} 1 & 0 & 0 & 1 & 0 & 0 \\ 0 & 1 & 0 & 0 & 1 & 0 \\ 0 & 0 & 1 & 0 & 0 & 1 \\ -x_2^{(1)} & 0 & x_3^{(1)} & -x_2^{(2)} & 0 & x_3^{(2)} \\ x_1^{(1)} & -x_3^{(1)} & 0 & x_1^{(2)} & -x_3^{(2)} & 0 \\ 0 & x_2^{(1)} & -x_1^{(1)} & 0 & x_2^{(2)} & -x_1^{(2)} \end{bmatrix} = 0. \tag{4.214}$$

So the first condition of (4.145) is violated. Because every solution of linear elastostatics is representable by the simple-layer potential (4.193), we see that the six extra conditions (4.205) will not yield an augmented system that is invertible. We conclude the following.

Corollary 4.8. *Consider the traction BVP*

$$\left.\begin{aligned} &\mu\Delta\vec{u} + (\lambda+\mu\nabla)(\nabla\vec{u}) = 0, \\ &\vec{\tau}(\vec{u}) = \vec{g} \in [H^r(\partial\Omega)]^3, \quad r > \tfrac{1}{2}, \end{aligned}\right\} \tag{4.215}$$

satisfying the compatibility conditions $\int_{\partial\Omega} \vec{g}(x)\cdot\vec{m}_j(x)\,d\sigma = 0$ *for* $j = 1, 2, \ldots, 6$. *Then adding the six extra accessory conditions*

$$\vec{u}(x^{(1)}) = \vec{\beta}^{(1)}, \quad \vec{u}(x^{(2)}) = \vec{\beta}^{(2)}, \tag{4.216}$$

where $x^{(1)},\ x^{(2)} \in \overline{\Omega}$ *and* $\vec{\beta}^{(1)},\ \vec{\beta}^{(2)} \in \mathbb{R}^3$ *are given, will not yield a unique solution* $\vec{u}$ *satisfying (4.215) and (4.216).* □

Physically, Corollary 4.1 implies that *holding the displacements of two points of a (3D) solid fixed will not yield a unique equilibrium configuration when a traction force is applied to the surface of the solid.*

(c) *Six extra conditions (4.206):* $\int_{\partial\Omega} \vec{f}(x)\cdot\vec{m}_j(x)\,d\sigma = \alpha_j,\ 1 \leqslant j \leqslant 6$

Obviously, we choose

$$\vec{\tau}_j^*(x) = \vec{m}_j(x), \qquad j = 1, 2, \ldots, 6, \tag{4.217}$$

in (4.208).

The second condition in (4.209) is fulfilled in the same way as before. To check the first condition in (4.209), we note that

$$\langle \vec{k}_i, \vec{\tau}_j^*\rangle = \langle \mathscr{L}_1\vec{m}_i, \vec{m}_j\rangle, \qquad 1 \leqslant i, j \leqslant 6. \tag{4.218}$$

Because $\mathscr{L}_1$ is positive definite (cf. (4.199)) and invertible (cf. (4.201)), the matrix with entries (4.218) is positive definite. So

$$\det[\langle \vec{k}_i, \vec{\tau}_j^*\rangle] > 0.$$

Thus Theorem 4.10 is satisfied. We state the following.

Corollary 4.9. *The augmented boundary integral equation system*

$$\begin{bmatrix} 0 & \sum\limits_{j=1}^{6} e_j\otimes\vec{m}_j \\ \sum\limits_{j=1}^{6} \vec{m}_j\otimes e_j & \mathscr{L}_2 \end{bmatrix} \begin{bmatrix} \vec{a} \\ \vec{f} \end{bmatrix} = \begin{bmatrix} \vec{\alpha} \\ \vec{g} \end{bmatrix} \tag{4.219}$$

corresponding to the simple-layer potential (4.193) for the traction BVP (4.189), where the simple-layer density $\vec{f}$ is subject to the constraints (4.206), is uniquely solvable with solution

$$(\vec{a}, \vec{f}) = (0, \vec{\tilde{f}}) \in \mathbb{R}^m \oplus [H^r(\partial\Omega)]^3. \qquad \square$$

Remark 4.4.

(1) Example 4.4 can be immediately generalized to the Neumann problem in an N-dimensional space, $N \geqslant 3$.

(2) Both Examples 4.4 and 4.5, after suitable modification, also work for a two-dimensional domain Ω. In the 2 D case, because of the logarithmic potential, the operators $\mathscr{L}_1$ in (4.163) and (4.196) may not be invertible, and the arguments are slightly more complicated. □

Example 4.6. Corollaries 4.5–4.9 above are most useful when the boundary integral equation approach is used for these boundary value problems. We now show *a shortcut to directly determine whether a given set of accessory conditions ensures the uniqueness of the solution*. Consider the Neumann BVP (4.181) subject to (4.182):

$$\int_{\partial\Omega} x_j^n u(x)\, d\sigma = \alpha, \quad \text{with } \alpha \in \mathbb{R} \text{ given.} \tag{4.220}$$

□

First let $u_0(x)$ be any particular solution to (4.181). (u_0 exists if the compatibility condition $\int_{\partial\Omega} g(x)\, d\sigma = 0$ is satisfied.) Write the general solution as

$$u(x) = u_0(x) + c, \tag{4.221}$$

where the constant $c \in \mathbb{R}$ is yet to be determined. Substituting (4.221) into (4.220), we get

$$\int_{\partial\Omega} x_j^n [u_0(x) + c]\, d\sigma = \alpha.$$

So

$$c \int_{\partial\Omega} x_j^n\, d\sigma = \alpha - \int_{\partial\Omega} x_j^n u_0(x)\, d\sigma,$$

where the RHS above is known. Thus c is uniquely determinable if and only if

$$\int_{\partial\Omega} x_j^n\, d\sigma \neq 0.$$

This is exactly condition (4.183) in Corollary 4.5.

When the Neumann BVP (4.181) is subject to (4.173),

$$u(x^{(0)}) = \alpha, \quad \text{with } x^{(0)} \in \partial\Omega \text{ and } \alpha \in \mathbb{R} \text{ given,} \tag{4.222}$$

let us again substitute (4.221) into (4.222), and see that the constant c is always uniquely determinable:

$$c = \alpha - u_0(x^{(0)}).$$

Note that $u_0(x)$ is pointwise-defined at $x^{(0)} \in \partial\Omega$ once $g \in H^r(\partial\Omega)$ with $r > 0$. So uniqueness of solution u is always assured, just as Corollary 4.6 states.

Next, we consider the elastostatic traction BVP (4.189), where the traction data $\vec{g}$ satisfies the compatibility conditions (4.191). Let the six accessory conditions be imposed as (4.204). Let $\vec{u}^{\,0}$ be a particular solution of (4.189). Then any general solution $\vec{u}$ can be represented as

$$\vec{u}(x) = \vec{u}^{\,0}(x) + \sum_{i=1}^{6} c_i \vec{m}_i(x), \quad x \in \Omega. \tag{4.223}$$

Substituting (4.223) into (4.204), we get

$$\sum_{i=1}^{6} c_i \int_{\partial\Omega} \vec{m}_i(x) \cdot \vec{m}_j(x)\, d\sigma = \alpha_j - \int_{\partial\Omega} \vec{u}^{\,0}(x) \cdot \vec{m}_j(x)\, d\sigma, \quad j = 1, 2, \ldots, 6. \tag{4.224}$$

Because of the linear independence of $\{\vec{m}_i \mid 1 \leqslant i \leqslant 6\}$ in $[L^2(\partial\Omega)]^3$, c_i, $1 \leqslant i \leqslant 6$, are uniquely solvable. Therefore $\vec{u}(x)$ is uniquely given by (4.223), which agrees with what is stated in Corollary 4.7.

If instead, the six constraints (4.205) are imposed,

$$\vec{u}(x^{(1)}) = \vec{\beta}^{\,(1)}, \quad \vec{u}(x^{(2)}) = \vec{\beta}^{\,(2)}, \qquad \text{with } x^{(1)}, x^{(2)} \in \overline{\Omega} \text{ and } \vec{\beta}^{\,(1)}, \vec{\beta}^{\,(2)} \in \mathbb{R}^3 \text{ given,} \tag{4.225}$$

we substitute (4.223) into (4.225), getting

$$\begin{bmatrix} \vec{m}_1(x^{(1)})\ \vec{m}_2(x^{(1)}) \ldots \vec{m}_6(x^{(1)}) \\ \vec{m}_1(x^{(2)})\ \vec{m}_2(x^{(2)}) \ldots \vec{m}_6(x^{(2)}) \end{bmatrix} \begin{bmatrix} c_1 \\ c_2 \\ \vdots \\ c_6 \end{bmatrix} = \begin{bmatrix} \vec{\beta}^{\,(1)} - \vec{u}^{\,0}(x^{(1)}) \\ \vec{\beta}^{\,(2)} - \vec{u}^{\,0}(x^{(2)}) \end{bmatrix} \in \mathbb{R}^6. \tag{4.226}$$

The determinant of the matrix on the LHS of (4.226) is zero for the same reason as (4.214). Therefore $c_1, \ldots, c_6$ are not uniquely determinable, and the solution lacks uniqueness, just as stated in Corollary 4.8.

The uniqueness of solution in Corollary 4.9 is *not obtainable* by this shortcut argument, because the six accessory conditions are formulated in terms of the simple-layer density $\vec{f}$ rather than the solution $\vec{u}$. (Recall that the simple-layer density $\vec{f}$ still has the physical meaning of being twice the jump discontinuity of the traction field across the layer surface.) □

Example 4.7. In the same spirit of Example 4.6, we now look at a fourth-order (biharmonic) elastostatic Kirchhoff pure bending plate model (Chapter 8):

$$\left.\begin{aligned} &\Delta^2 u(x) = 0, \qquad \Delta^2 = \left(\frac{\partial^2}{\partial x_1^2} + \frac{\partial^2}{\partial x_2^2}\right)^2, \\ &\qquad\qquad x = (x_1, x_2) \in \Omega \subset \mathbb{R}^2, \\ &B_1 u(x) = g_1(x),\ x \in \partial\Omega, \\ &B_2 u(x) = g_2(x),\ x \in \partial\Omega, \end{aligned}\right\} \tag{4.227}$$

where

$$B_1 u \equiv \frac{\partial}{\partial n}\Delta u + (1-\nu)\frac{\partial}{\partial s}\left[(n_1^2 - n_2^2)\frac{\partial^2 u}{\partial x_1 \partial x_2} - n_1 n_2\left(\frac{\partial^2 u}{\partial x_1^2} - \frac{\partial^2 u}{\partial x_2^2}\right)\right],$$

$$B_2 u \equiv \nu\Delta u + (1-\nu)\left(n_1^2\frac{\partial^2 u}{\partial x_1^2} + n_2^2\frac{\partial^2 u}{\partial x_2^2} + 2n_1 n_2\frac{\partial^2 u}{\partial x_1 \partial x_2}\right),$$

are, respectively, the shear and bending moment operators on $\partial\Omega$; $n = (n_1, n_2) = (n_1(x), n_2(x))$ and $s = (-n_2, n_1)$ are respectively the unit exterior normal vector and the unit counterclockwise tangent vector on $\partial\Omega$; $\partial/\partial s$ denotes the counterclockwise tangential derivative; $\nu\colon\ 0 < \nu < \frac{1}{2}$ denotes the Poisson ratio; and $u(x)$ is the vertical displacement of the thin plate. □

The variational formulation, Sobolev space setting and ellipticity of the BVP (4.227) can be found in §§ 8.1–8.3. Assume that the boundary data g_1 and g_2 in (4.227) are sufficiently regular. It is not hard to verify that (4.227) has a solution if and only if the three compatibility conditions

$$\int_{\partial\Omega} [g_1(x)\phi_1(x) + g_2(x)\phi_2(x)]\, d\sigma = 0 \tag{4.228}$$

are satisfied, where

$$\begin{bmatrix}\phi_1\\ \phi_2\end{bmatrix} \in \left\{ \begin{bmatrix}1\\0\end{bmatrix}, \begin{bmatrix}x_1\\ -n_1(x)\end{bmatrix}, \begin{bmatrix}x_2\\ -n_2(x)\end{bmatrix} \,\middle|\, x \in \partial\Omega \right\}.$$

Once (4.228) is met, the solution has three degrees of freedom:

$$\begin{aligned} &u(x) = u_0(x) + a_0 + a_1 x_1 + a_2 x_2, \\ &a_0,\ a_1,\ a_2 \in \mathbb{R},\ \ x = (x_1, x_2) \in \Omega, \end{aligned} \tag{4.229}$$

where u_0 is some particular solution to (4.227).

In order to fix the three constants in (4.229), we may impose one of the two sets of commonly used accessory conditions in the following forms: either

$$u(x^{(0)}) = \alpha_0, \qquad \frac{\partial u(x^{(0)})}{\partial n} = \alpha_1, \qquad \frac{\partial u(x^{(0)})}{\partial s} = \alpha_2, \tag{4.230}$$
$$\text{with } x^{(0)} \in \partial\Omega \text{ and } \alpha_j \in \mathbb{R},\ 0 \leqslant j \leqslant 2 \text{ given,}$$

or

$$u(x^{(1)}) = \beta_1, \quad u(x^{(2)}) = \beta_2, \quad u(x^{(3)}) = \beta_3, \tag{4.231}$$
$$\text{with } x^{(1)},\ x^{(2)},\ x^{(3)} \in \overline{\Omega} \text{ and } \beta_j \in \mathbb{R},\ 1 \leqslant j \leqslant 3, \text{ given.}$$

We see that (4.231) is analogous to (4.173) or (4.205) treated earlier. We study (4.230) and (4.231) separately below:

(a) *Three extra conditions (4.230)*

First of all, to ensure that the solution u and first-order derivatives are pointwise well defined in (4.230), it is sufficient that

$$g_1 \in H^{-3/2+\varepsilon_1}(\partial\Omega), \ g_2 \in H^{-1/2+\varepsilon_2}(\partial\Omega), \ \forall \varepsilon_1,\ \varepsilon_2 > 0. \tag{4.232}$$

(Thus $u \in H^{2+\varepsilon}(\Omega)$, $\varepsilon = \min(\varepsilon_1, \varepsilon_2)$, must be interpreted as a weak solution.) We substitute the general solution (4.229) into (4.230), yielding

$$\left.\begin{aligned} a_0 + a_1 x_1^{(0)} + a_2 x_2^{(0)} &= \alpha_0 - u_0(x^{(0)}), \\ a_1 n_1(x^{(0)}) + a_2 n_2(x^{(0)}) &= \alpha_1 - \frac{\partial}{\partial n} u_0(x^{(0)}), \\ -a_1 n_2(x^{(0)}) + a_2 n_1(x^{(0)}) &= \alpha_2 - \frac{\partial}{\partial s} u_0(x^{(0)}), \\ \text{with } x^{(0)} &= (x_1^{(0)}, x_2^{(0)}) \in \partial\Omega. \end{aligned}\right\} \tag{4.233}$$

Since the determinant satisfies

$$\det \begin{bmatrix} 1 & x_1^{(0)} & x_2^{(0)} \\ 0 & n_1(x^{(0)}) & n_2(x^{(0)}) \\ 0 & -n_2(x^{(0)}) & n_1(x^{(0)}) \end{bmatrix} = 1,$$

a_0, a_1 and a_2 in (4.229) are uniquely solvable. Therefore *uniqueness of the solution $u(x)$ for (4.227) is always assured.*

(b) *Three extra conditions (4.231)*

First, assume regularities

$$g_1 \in H^{-5/2+\varepsilon_1}(\partial\Omega), \ g \in H^{-3/2+\varepsilon_2}(\partial\Omega), \quad \varepsilon_1, \varepsilon_2 > 0$$

to assure that $u \in H^{1+\varepsilon}(\Omega)$, $\varepsilon = \min(\varepsilon_1, \varepsilon_2) > 0$, with trace $u \in H^{1/2+\varepsilon}(\partial\Omega) \subset C^0(\partial\Omega)$. Next, substitute the general solution (4.229) into (4.231), yielding

$$\left\{ \begin{aligned} a_0 + a_1 x_1^{(1)} + a_2 x_2^{(1)} &= \beta_1 - u_0(x^{(1)}), \\ a_0 + a_1 x_1^{(2)} + a_2 x_2^{(2)} &= \beta_2 - u_0(x^{(2)}), \\ a_0 + a_1 x_1^{(3)} + a_2 x_2^{(3)} &= \beta_3 - u_0(x^{(3)}), \\ \text{with } x^{(j)} &= (x_1^{(j)}, x_2^{(j)}) \in \overline{\Omega}, \;\; 1 \leqslant j \leqslant 3. \end{aligned} \right\} \tag{4.234}$$

The determinant of the coefficient matrix on the LHS of (4.234), evaluated,

$$\begin{aligned} &\det \begin{bmatrix} 1 & x_1^{(1)} & x_2^{(1)} \\ 1 & x_1^{(2)} & x_2^{(2)} \\ 1 & x_1^{(3)} & x_2^{(3)} \end{bmatrix} \\ &= \left(x_1^{(2)} - x_1^{(1)}\right)\left(x_2^{(3)} - x_2^{(1)}\right) - \left(x_1^{(3)} - x_1^{(1)}\right)\left(x_2^{(2)} - x_2^{(1)}\right) \neq 0, \end{aligned}$$

if and only if $x^{(1)}$, $x^{(2)}$ and $x^{(3)}$ are not colinear. Therefore, we see that *the three extra displacement data (4.231) assure uniqueness of the solution of the plate BVP (4.227) if and only if the three locations* $x^{(1)}$, $x^{(2)}$ *and* $x^{(3)}$ *are chosen to be non colinear*. Physically, this implies that *holding (the displacements of) three points fixed on a plate subject to boundary shear and bending will yield a unique equilibrium configuration if and only if the three points are not colinear.* □

Note that the conclusions in Example 4.7 are also obtainable by a boundary integral equation method.

Chapter 5

Finite-Element Methods: Spaces and Properties

Finite-element method(s) (FEM) were first introduced by R. Courant [56] in 1943. Since the 1950s, the method has found wide applications in civil, mechanical and aerospace engineering. A rigorous mathematical theory also began to emerge in the late 1960s. Many software packages for applying these methods to various types of partial differential equations have been developed by scientists and engineers. Today, the FEM are probably the most important and powerful tools for obtaining numerical solutions of linear and nonlinear partial differential equations; see § 1.3. The main references for this chapter are [173, 49]. In discretizing boundary integral equations, finite elements are used to collocate the equations at nodal points. In this chapter, we will introduce the commonly used finite-elements in one, two and three space dimensions, and prove some of the basic properties of finite-element approximation spaces, which are essential in understanding the collocation error estimates for BEM in Chapter 10.

5.1 Minimization of a quadratic functional. The Ritz variational formulation

Many elliptic partial differential equations can be formulated using a coercive bilinear form approach. This also provides a formal setting for the finite-element method.
Let H be a Hilbert space, with inner product $\langle \cdot, \cdot \rangle$ and norm $\|\cdot\|$. Let $a(\cdot,\cdot)$ be a bilinear form on H satisfying the following properties:

$$\left.\begin{aligned} &a\colon H \times H \to \mathbb{R} \quad \text{is continuous,}\\ &a(u, v_1+v_2) = a(u,v_1)+a(u,v_2), \ \forall u,\, v_1,\, v_2 \in H,\\ &a(u_1+u_2, v) = a(u_1,v)+a(u_2,v), \ \forall u_1,\, u_2,\, v \in H,\\ &a(ku,v) = a(u,kv) = ka(u,v), \qquad \forall k \in \mathbb{R}. \end{aligned}\right\} \tag{5.1}$$

We say that a is *symmetric* if

$$a(u,v) = a(v,u), \quad \forall u, v \in H, \tag{5.2}$$

and *coercive* if there exists a $c > 0$ such that

$$a(u,u) \geqslant c\|u\|^2, \quad \forall u \in H. \tag{5.3}$$

Note that the conditions in (5.1) imply that there exists $c' > 0$ such that

$$|a(u,v)| \leqslant c'\|u\|\,\|v\| \quad \forall u,v \in H. \tag{5.4}$$

Let θ be a linear form on H satisfying

$$\theta:\ H \to \mathbb{R},$$

and there exists $c'' > 0$ such that

$$|\theta(u)| \leqslant c''\|u\| \forall u \in H, \tag{5.5}$$

$$\theta(ku_1 + u_2) = k\theta(u_1) + \theta(u_2) \forall k \in \mathbb{R}, \forall u_1, u_2 \in H. \tag{5.6}$$

Now we consider the problem of minimizing a quadratic functional:

$$\inf_{v\in H} J(v) \equiv \frac{1}{2}a(v,v) - \theta(v). \tag{5.7}$$

Theorem 5.1. *Assume $a(\cdot,\cdot)$ is a symmetric coercive bilinear form on H and θ is a linear form on H. Then problem (5.7) has a unique solution $\widehat{u}$ such that*

$$J(\widehat{u}) = \min_{v\in H} J(v). \tag{5.8}$$

Furthermore, $\widehat{u}$ is characterized by the variational equation

$$a(\widehat{u}, v) = \theta(v) \quad \forall v \in H. \tag{5.9}$$

Proof. We first assume that a solution $\widehat{u}$ to (5.8) exists. Let us derive (5.9). Since $\widehat{u}$ satisfies (5.8), we have

$$\lim_{t\to 0+} \frac{1}{t}\,[J(\widehat{u}+tv) - J(\widehat{u})] \geqslant 0 \quad \forall v \in H.$$

$$\begin{aligned}\frac{1}{t}\,[J(\widehat{u}+tv) - J(\widehat{u})] &= \frac{1}{t}\{[\frac{1}{2}a(\widehat{u}+tv, \widehat{u}+tv) - \theta(\widehat{u}+tv)] \\ &\quad - [\frac{1}{2}a(\widehat{u},\widehat{u}) - \theta(\widehat{u})]\} \\ &= \frac{1}{t}\{ta(\widehat{u},v) + \frac{1}{2}t^2 a(v,v) - t\theta(v)\}.\end{aligned}$$

Let $t \to 0+$. We get

$$a(\widehat{u}, v) - \theta(v) \geqslant 0 \quad \forall v \in H. \tag{5.10}$$

Since (5.10) is valid for all $v \in H$, we use $-v$ for v and obtain

$$a(\widehat{u}, -v) - \theta(-v) \geqslant 0 \quad \forall v \in H,$$

or

$$a(\widehat{u},v)-\theta(v)\leqslant 0. \tag{5.11}$$

The inequalities (5.10) and (5.11) combined give us (5.9).

To prove the uniqueness of $\widehat{u}$, assume that both $\widehat{u}_1$ and $\widehat{u}_2$ satisfy (5.9). Then

$$a(\widehat{u}_1,v)=\theta(v)\quad \forall v\in H,$$
$$a(\widehat{u}_2,v)=\theta(v)\quad \forall v\in H.$$

Subtracting, we get

$$a(\widehat{u}_1-\widehat{u}_2,v)=0\quad \forall v\in H. \tag{5.12}$$

Using $v=\widehat{u}_1-\widehat{u}_2$ in (5.12), we get

$$a(\widehat{u}_1-\widehat{u}_2,\widehat{u}_1-\widehat{u}_2)=0.$$

By coercivity (5.3),

$$0=a(\widehat{u}_1-\widehat{u}_2,\widehat{u}_1-\widehat{u}_2)\geqslant c\|\widehat{u}_1-\widehat{u}_2\|^2\geqslant 0.$$

This implies $\widehat{u}_1=\widehat{u}_2$.

We now show that if $\widehat{u}$ satisfies (5.9), then $\widehat{u}$ is also the solution of (5.8). For any $v\in H$, we have

$$\begin{aligned} J(v)-J(\widehat{u}) &= \left[\frac{1}{2}a(v,v)-\theta(v)\right]-\left[\frac{1}{2}a(\widehat{u},\widehat{u})-\theta(\widehat{u})\right]\\ &= \frac{1}{2}a(v-\widehat{u},v-\widehat{u})+\{a(\widehat{u},v-\widehat{u})-\theta(v-\widehat{u})\}\\ &= \frac{1}{2}a(v-\widehat{u},v-\widehat{u})\geqslant 0, \end{aligned} \tag{5.13}$$

where the term in { } vanishes because of (5.9). Therefore $J(v)\geqslant J(\widehat{u})$, and $\widehat{u}$ solves (5.8). According to the Lax–Milgram theorem 5.2 to be proved in the next section, the solution $\widehat{u}$ to the variational equation (5.9) uniquely exists. The proof is complete. □

Let us give some examples of elliptic BVP in mathematical physics that can be formulated as a minimization problem in the setting of Theorem 5.1.

Example 5.1. The solution u to the Dirichlet problem

$$\left.\begin{array}{l} \Delta u(x)=f(x),\ x\in\Omega, \text{bounded open in } \mathbb{R}^N, x=(x_1,\ldots,x_N),\\ u(x)\mid_{\partial\Omega}=0,\quad \Delta=\dfrac{\partial^2}{\partial x_1^2}+\cdots+\dfrac{\partial^2}{\partial x_N^2},\quad f\in L^2(\Omega), \end{array}\right\} \tag{5.14}$$

is also the unique solution to the problem

$$\min_{v\in H_0^1(\Omega)} J(v)=\frac{1}{2}a(v,v)-\theta(v), \tag{5.15}$$

where

$$a(v_1,v_2) \equiv \int_\Omega \nabla v_1(x)\cdot\nabla v_2(x)\,dx, \nabla = \text{ grad } = \left(\frac{\partial}{\partial x_1},\cdots,\frac{\partial}{\partial x_N}\right),$$

$$\theta(v) \equiv -\int_\Omega f(x)v(x)\,dx.$$

To show that (5.14) and (5.15) are equivalent, we use Theorem 5.1. By Poincaré's inequality, for all v satisfying $v\,|_{\partial\Omega}=0$, there is $c>0$ (independent of v) such that

$$a(v,v) = \int_\Omega |\nabla v(x)|^2\,dx \geqslant c\left\{\int_\Omega \left[|\nabla v(x)|^2 + v^2(x)\right]\,dx\right\} = c\|v\|^2_{H^1(\Omega)}.$$

Therefore a is coercive, and a is bilinear and symmetric. For $f\in L^2(\Omega)$, θ defined from f is a linear form on $H^1_0(\Omega)$. Therefore Theorem 5.1 is applicable. We see that (5.15) has a unique solution $\widehat{u}$, characterized by

$$a(\widehat{u},v) = \theta(v) \quad \forall v\in H^1_0(\Omega), \tag{5.16}$$

i.e.,

$$\int_\Omega \nabla\widehat{u}(x)\cdot\nabla v(x)\,dx = -\int_\Omega f(x)v(x)\,dx.$$

We apply Green's theorem and integrate by parts for the left hand term, getting

$$\begin{aligned}\int_\Omega \nabla\widehat{u}(x)\cdot\nabla v(x)\,dx &= \int_{\partial\Omega}\frac{\partial\widehat{u}(x)}{\partial n}v(x)d\sigma - \int_\Omega [\Delta\widehat{u}(x)]\,v(x)\,dx\\ &= -\int_\Omega f(x)v(x)\,dx,\\ &\int_\Omega [\Delta\widehat{u}(x)-f(x)]v(x)\,dx = 0 \quad \forall v\in H^1_0(\Omega).\end{aligned}$$

Since $H^1_0(\Omega)$ is dense in $L^2(\Omega)$, the above is valid if and only if

$$\Delta\widehat{u}(x) - f(x) = 0 \quad \text{on } \Omega.$$

Therefore $\widehat{u}$ satisfies (5.14). (The Dirichlet boundary condition $u\,|_{\partial\Omega}=0$ is satisfied because $\widehat{u}$ is taken from $H^1_0(\Omega)$.) □

Example 5.2. Consider the elliptic problem with Neumann condition

$$\left.\begin{aligned}&-\Delta u(x) + \alpha u(x) = f(x), \quad \alpha>0, f\in L^2(\Omega),\\ &\frac{\partial u(x)}{\partial n}\,|_{\partial\Omega}=0.\end{aligned}\right\} \tag{5.17}$$

It has a unique solution, which is also the unique solution of

$$\min_{v\in H^1(\Omega)} J(v) = \frac{1}{2}a(v,v) - \theta(v), \tag{5.18}$$

where

$$a(v_1, v_2) \equiv \int_\Omega [\nabla v_1(x) \cdot \nabla v_2(x) + \alpha v_1(x) v_2(x)]\, dx,$$
$$\theta(v) = \int_\Omega f(x) v(x)\, dx.$$

The solution u satisfies

$$a(u, v) = \theta(v) \quad \forall v \in H^1(\Omega). \tag{5.19}$$

□

Remark 5.1. There is a considerable difference in handling the Dirichlet boundary condition in Example 5.1 and the Neumann boundary condition in Example 5.2. In Example 5.1, the Dirichlet condition must be imposed on the space of admissible functions (i.e., $H_0^1(\Omega)$) for the minimization problem. In Example 5.2, we do not impose any boundary conditions for the minimization problem (5.15). But the solution will automatically satisfy the boundary condition $\partial u/\partial n\,|_{\partial\Omega}= 0$. Thus, the first boundary condition is called an *essential boundary condition*, whereas the second boundary condition is called a *natural boundary condition*. □

Example 5.3. Let

$$\vec{v}(x) = (v_1(x), v_2(x), v_3(x)), \quad x \in \Omega \subseteq \mathbb{R}^3,$$

denote the displacement vector at x of an elastic body undergoing deformation. The body is assumed to be an isotropic homogeneous Hookean solid. Define

$$V = \{\vec{v} = (v_1, v_2, v_3) \in \left(H^1(\Omega)\right)^3, v_i\,|_{\Gamma_0}= 0, i = 1, 2, 3\},$$

where $\partial\Omega = \Gamma_0 \cup \Gamma_1$. Let

$$\varepsilon_{ij}(\vec{v}) = \frac{1}{2}\left(\frac{\partial v_i}{\partial x_j} + \frac{\partial v_j}{\partial x_i}\right), \quad 1 \leqslant i, j \leqslant 3,$$
$$\sigma_{ij}(\vec{v}) = \lambda \left[\sum_{i=1}^{3} \varepsilon_{ii}(\vec{v})\right] \delta_{ij} + 2\mu\varepsilon_{ij}(\vec{v}), \quad 1 \leqslant i, j \leqslant 3$$

(δ_{ij} is the Kronecker delta)

denote respectively the linear strain tensor and the linear stress tensor; λ and μ are Lamé constants.

We define

$$a(\vec{v}^1,\vec{v}^2) \equiv \int_\Omega \sum_{i,j=1}^{3} \sigma_{ij}(\vec{v}^1)\varepsilon_{ij}(\vec{v}^2)\,dx$$

$$\left(= \int_\Omega \left[\lambda(\nabla\cdot\vec{v}^1)(\nabla\cdot\vec{v}^2) + 2\mu \sum_{i,j=1}^{3} \varepsilon_{ij}(\vec{v}^1)\varepsilon_{ij}(\vec{v}^2) \right] dx \right),$$

$$\theta(v) \equiv \int_\Omega \vec{f}\cdot\vec{v}\,dx + \int_{\Gamma_1} \vec{g}\cdot\vec{v}\,d\sigma,$$

where

$$\vec{f} = (f_1,f_2,f_3) \in \left[L^2(\Omega)\right]^3,$$

$$\vec{g} = (g_1,g_2,g_3) \in \left[L^2(\Gamma_1)\right]^3.$$

One can show that the solution $\vec{u}$ to the problem

$$\min_{v\in V} J(\vec{v}) = \frac{1}{2}a(\vec{v},\vec{v}) - \theta(\vec{v}), \tag{5.20}$$

or equivalently

$$a(\vec{u},\vec{v}) = \theta(\vec{v}) \quad \forall \vec{v}\in V, \tag{5.21}$$

is also the solution to the boundary value problem

$$\left.\begin{aligned} &-\mu\Delta\vec{u} - (\lambda+\mu)\nabla(\nabla\cdot\vec{u}) = \vec{f} \quad \text{on } \Omega,\\ &\vec{u}\,|_{\Gamma_0} = 0,\\ &\textstyle\sum_{j=1}^{3} \sigma_{ij}(\vec{u})_j n_j\,|_{\Gamma_1} = g_i, \quad i=1,2,3. \end{aligned}\right\} \tag{5.22}$$

See Fig. 5.1 □

The minimization problems (5.15), (5.18) and (5.20) are called the *Ritz variational formulations* for, respectively, the partial differential equations (5.14), (5.17) and (5.22), while (5.16), (5.19) and (5.21) are the *Galerkin variational formulations*.

5.2 Error bounds for internal approximations

Let H be a Hilbert space as given in § 5.1.

We consider a parameter $h \in \mathbb{R}$, the discretization parameter, which will tend to 0. There is no loss of generality in assuming $0 < h < 1$. For every h, there is a space V_h satisfying

$$V_h \subset H \quad \text{(i.e., it is an internal approximation)}, \tag{5.23}$$

$$V_h \text{ is finite-dimensional}, \tag{5.24}$$

$$V_h \text{ "converges" to } H, \text{i.e., } \forall v\in H, \text{there exists } v_h\in V_h$$

$$\text{such that } v_h \to v \text{ in } H \text{ as } h\to 0; \text{ equivalently},$$

$$\lim_{h\to 0} \inf_{v_h\in V_h} \|v - v_h\| = 0. \tag{5.25}$$

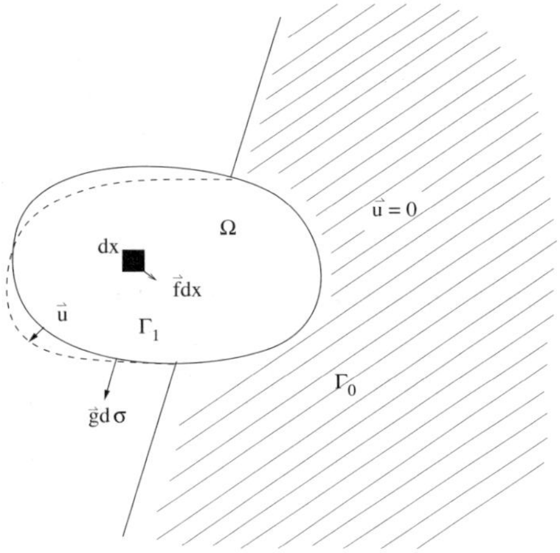

Figure 5.1: A solid is fixed at Γ_0 but has traction force applied at Γ_1.

We want to construct approximations $\widehat{u}_h$ of the solution $\widehat{u}$ of (5.9). We define $\widehat{u}_h$ as the solution of

$$\left.\begin{aligned} &a(\widehat{u}_h, v_h) = \theta(v_h) \quad \forall v_h \in V_h, \\ &\widehat{u}_h \in V_h. \end{aligned}\right\} \tag{5.26}$$

The above problem has a unique solution $\widehat{u}_h$. This is a consequence of the Lax–Milgram theorem below.

Theorem 5.2 (The Lax–Milgram theorem). *Let H be a Hilbert space and let $a(\cdot,\cdot)$ be a symmetric, coercive and continuous bilinear form on $H \times H$ satisfying (5.1)–(5.3). Let $\theta(\cdot)$ be a continuous linear form on H satisfying (5.5) and (5.6). Then the variational problem*

$$\text{“Find } u \in H \text{ satisfying } a(u,v) = \theta(v) \quad \forall v \in H\text{”}, \tag{5.27}$$

has a unique solution $\widehat{u}$ satisfying

$$\|\widehat{u}\|_H \leqslant \frac{1}{c}\|\theta\|, \tag{5.28}$$

where c is the bounding constant in (5.3), and $\|\theta\|$ is the norm of θ in the dual space of H.

Proof. Because $a(\cdot,\cdot)$ is symmetric and satisfies

$$c\|v\|^2 \leqslant a(v,v) \leqslant c'\|v\|^2,$$

$a(\cdot,\cdot)$ defines an inner product and generates a norm on H which is equivalent to the norm $\|\cdot\|$. With respect to this new norm, $\theta(\cdot)$ is still a continuous linear form. By the Riesz representation theorem (Appendix A.4), there exists a unique $\widehat{u}$ such that

$$a(\widehat{u},v) = \theta(v) \quad \forall v \in H.$$

Also

$$c\|\widehat{u}\|^2 \leqslant a(\widehat{u},\widehat{u}) = \theta(\widehat{u}) \leqslant \|\theta\|\,\|\widehat{u}\|$$

gives (5.28). □

Theorem 5.3 (Céa's lemma). *Let a and θ satisfy the assumptions (5.1)–(5.6). Let $\widehat{u}$ and $\widehat{u}_h$ denote respectively solutions to (5.27) and (5.26). Then*

$$\|\widehat{u}-\widehat{u}_h\| \leqslant \frac{c'}{c} \inf_{v_h \in V_h} \|\widehat{u}-v_h\|. \tag{5.29}$$

Proof. From (5.27), $\widehat{u}$ satisfies

$$a(\widehat{u},v_h) = \theta(v_h) \quad \forall v_h \in V_h. \tag{5.30}$$

Subtracting (5.26) from (5.30), we get

$$a(\widehat{u}-\widehat{u}_h, v_h) = 0 \quad \forall v_h \in V_h. \tag{5.31}$$

So

$$\begin{aligned} c\|\widehat{u}-\widehat{u}_h\|^2 &\leqslant a(\widehat{u}-\widehat{u}_h,\widehat{u}-\widehat{u}_h) \\ &= a(\widehat{u}-\widehat{u}_h,\widehat{u}-v_h) + a(\widehat{u}-\widehat{u}_h, v_h-\widehat{u}_h) \\ &= a(\widehat{u}-\widehat{u}_h,\widehat{u}-v_h) \qquad (a(\widehat{u}-\widehat{u}_h, v_h-\widehat{u}_h) = 0 \text{ by } (5.31)) \\ &\leqslant c'\|\widehat{u}-\widehat{u}_h\|\,\|\widehat{u}-v_h\| \quad \forall v_h \in V_h. \end{aligned}$$

Dividing by $\|\widehat{u}-\widehat{u}_h\|$ on both sides, we get

$$\|\widehat{u}-\widehat{u}_h\| \leqslant \frac{c'}{c}\|\widehat{u}-v_h\| \quad \forall v_h \in V_h.$$

Hence

$$\|\widehat{u}-\widehat{u}_h\| \leqslant \frac{c'}{c} \inf_{v_h \in V_h} \|\widehat{u}-v_h\|.$$ □

An immediate consequence of the above theorem is the *convergence of the approximation* (5.26), because, by (5.25) and (5.29), we have

$$\lim_{h\to 0} \|\widehat{u}-\widehat{u}_h\| = 0.$$

Remark 5.2. In the proof of the previous theorem, we never used the symmetry of $a(\cdot,\cdot)$. Under the assumption of symmetry, we can improve (5.29) to

$$\|\widehat{u}-\widehat{u}_h\| \leqslant \left(\frac{c'}{c}\right)^{1/2} \inf_{v_h \in V_h} \|\widehat{u}-v_h\|. \tag{5.32}$$

since $(c'/c)^{1/2} < c'/c$ generally holds. We prove (5.32) as follows: Write

$$\begin{aligned} a(\widehat{u}-v_h,\widehat{u}-v_h) &= a\left((\widehat{u}-\widehat{u}_h)+(\widehat{u}_h-v_h),(\widehat{u}-\widehat{u}_h)+(\widehat{u}_h-v_h)\right) \\ &= a(\widehat{u}-\widehat{u}_h,\widehat{u}-\widehat{u}_h)+a(\widehat{u}_h-v_h,\widehat{u}_h-v_h) \\ &\quad + a(\widehat{u}_h-v_h,\widehat{u}-\widehat{u}_h)+a(\widehat{u}-\widehat{u}_h,\widehat{u}_h-v_h). \end{aligned}$$

The last two terms are zero by (5.31) and the symmetry of a. Therefore

$$\begin{aligned} a(\widehat{u}-v_h,\widehat{u}-v_h) &= a(\widehat{u}-\widehat{u}_h,\widehat{u}-\widehat{u}_h)+a(\widehat{u}_h-v_h,\widehat{u}_h-v_h), \\ c\|\widehat{u}-\widehat{u}_h\|^2 &\leqslant a(\widehat{u}-\widehat{u}_h,\widehat{u}-\widehat{u}_h) \leqslant \inf_{v_h\in V_h} a(\widehat{u}-v_h,\widehat{u}-v_h) \\ &\leqslant c' \inf_{v_h\in V_1} \|\widehat{u}-v_h\|^2. \end{aligned}$$

Therefore, (5.32) is proved. □

For later use, we will include a more general form of the variational principle, due to I. Babuska [14, Chap. 5].

Theorem 5.4. *Let H_1 and H_2 be two real Hilbert spaces with inner products $(\cdot,\cdot)_1$, $(\cdot,\cdot)_2$ and norms $\|\cdot\|_1, \|\cdot\|_2$, respectively. Let $a(\cdot,\cdot)$ be a bilinear form on $H_1\times H_2$ such that*

$$|a(u,v)| \leqslant C_1\|u\|_1\|v\|_2, \quad u\in H_1, v\in H_2, \tag{5.33}$$

$$\inf_{\substack{u\in H_1\\ \|u\|_1=1}} \sup_{\substack{v\in H_2\\ \|v\|_2=1}} |a(u,v)| \geqslant C_2 > 0, \tag{5.34}$$

$$\inf_{\substack{v\in H_2\\ \|v\|_2=1}} \sup_{\substack{u\in H_1\\ \|u\|_1=1}} |a(u,v)| \geqslant C_3 > 0. \tag{5.35}$$

Let θ be a continuous linear form on H_2. Then there exists a unique $\widehat{u}\in H_1$ such that

$$a(\widehat{u},v) = \theta(v) \quad \forall v\in H_2,$$

and

$$\|\widehat{u}\|_1 \leqslant \frac{1}{C_2}\|\theta\|.$$

Proof. For a given $u\in H_1$,

$$\Phi_u(v) \equiv a(u,v)$$

defines a linear functional on H_2, with norm

$$\|\Phi_u\| = \sup_{\substack{v\in H_2\\ \|v\|_2=1}} |a(u,v)| \leqslant C_1\|u\|_1. \tag{5.36}$$

By the Riesz representation theorem (see Appendix A.4), there exists a unique $z \in H_2$ such that

$$(z,v)_2 = a(u,v) \quad \forall v \in H_2. \tag{5.37}$$

This defines a mapping

$$A\colon H_1 \to H_2,\ a\ Au = z, \text{ with } u \text{ and } z \text{ associated by (5.37).}$$

It is straightforward to verify that A is linear. A is also bounded:

$$\|A\| \leqslant C_1,$$

by (5.36).

We show that $A(H_1)$, the range of A in H_2, is a closed subspace. Because

$$\begin{aligned}\|Au\|_{H_2} &= \sup_{\substack{v\in H_2\\ \|v\|_2=1}} |(Au,v)_2| \\ &= \sup_{\substack{v\in H_2\\ \|v\|_2=1}} |a(u,v)| \geqslant C_2\|u\|_1, \quad \text{by (5.34).}\end{aligned} \tag{5.38}$$

Thus, if $\{Au_n\}_{n=1}^{\infty}$ is a Cauchy sequence in H_2, then $\{u_n\}_{n=1}^{\infty}$ is a Cauchy sequence in H_1, because

$$\|Au_n - Au_m\|_{H_2} = \|A(u_n - u_m)\|_{H_2} \geqslant C_2\|u_n - u_m\|_{H_1}.$$

Hence $A(H_1)$ is closed.

As a matter of fact, the range of A consists of the entire H_2. Let $v_0 \in H_2$, $v_0 \neq 0$ be orthogonal to $A(H_1)$ in H_2:

$$(Au, v_0)_2 = 0 \quad \forall u \in H_1.$$

It follows that

$$\sup_{\substack{u\in H_1\\ \|u\|_1=1}} |(Au,v_0)_2| = \sup_{\substack{u\in H_1\\ \|u\|_1=1}} |a(u,v_0)| = 0,$$

contradicting (5.35).

By (5.38) and the surjectivity of A, we see that A is invertible from H_1 onto H_2. The inverse satisfies

$$\|A^{-1}\| \leqslant \frac{1}{C_2}$$

by (5.38). For any continuous linear form θ on H_2, by the Riesz representation theorem,

$$\theta(v) = (\zeta, v)_2 \quad \forall v \in H_2,$$

for a unique $\zeta \in H_2$. ζ satisfies

$$\|\zeta\|_2 = \|\theta\|,$$

where $\|\theta\|$ is the norm of θ in H_2'. By choosing $\widehat{u} = A^{-1}\zeta$, we have

$$(A\widehat{u}, v)_2 = a(\widehat{u}, v) = (\zeta, v)_2 = \theta(v) \quad \forall v \in H_2,$$
$$\|\widehat{u}\|_1 \leqslant \|A^{-1}\| \, \|\zeta\|_2 \leqslant \frac{1}{C_2}\|\zeta\|_2 = \frac{1}{C_2}\|\theta\|.$$

$\widehat{u}$ is unique because ζ is. The proof follows. □

Remark 5.3. The condition (5.35) can be weakened to

$$\sup_{\substack{u \in H_1 \\ \|u\|_1 = 1}} |a(u, v)| > 0 \quad \text{for } v \neq 0, \tag{5.39}$$

as given in [14, (5.2.3), p. 112]. □

The approximation property based upon the variational formulation of Theorem 5.4 is given below; see [14, pp. 186–188].

Theorem 5.5. *Assume the hypotheses of Theorem 5.4. Let M_1 and M_2 be closed subspaces of H_1 and H_2, respectively, satisfying*

$$\inf_{\substack{u \in M_1 \\ \|u\|_1 = 1}} \sup_{\substack{v \in M_2 \\ \|v\|_2 = 1}} |a(u, v)| \geqslant d(M_1, M_2) > 0, \tag{5.40}$$

$$\inf_{\substack{v \in M_2 \\ \|v\|_2 = 1}} \sup_{\substack{u \in M_1 \\ \|u\|_1 = 1}} |a(u, v)| \geqslant d'(M_1, M_2) > 0. \tag{5.41}$$

Let $\widehat{u}$ be the unique solution in H_1 satisfying

$$a(\widehat{u}, v) = \theta(v) \quad \forall v \in H_2,$$

and let $\widetilde{u}$ be the unique solution in M_1 satisfying

$$a(\widetilde{u}, v) = \theta(v) \qquad \forall v \in M_2. \tag{5.42}$$

Then

$$\|\widehat{u} - \widetilde{u}\|_1 \leqslant \left[1 + \frac{C_1}{d(M_1, M_2)}\right] \inf_{\mu \in M_1} \|\widehat{u} - \mu\|_1, \tag{5.43}$$

where C_1 is the same as in (5.36).

Proof. Let $P_i\colon\ H_i \to H_i$, $i = 1,2$ be the orthogonal projection operator from H_i into M_i:

$$P_i^2 = P_i,\ \ P_i(I-P_i) = 0,\ \ P_i = P_i^*, \tag{5.44}$$
$$w = P_i w + (I-P_i)w,\ \ P_i w \perp (I-P_i)w \text{ in } H_i,\ \ i = 1,2.$$

Let $A\colon\ H_1 \to H_2$ be the linear operator constructed in the proof of Theorem 5.4. Let $Q \equiv P_2 A\colon\ H_1 \to M_2$. Further, define

$$\widetilde{A}\colon\ M_1 \to M_2$$
$$(\widetilde{A}u, v)_2 = a(u,v) \quad \forall u \in M_1, v \in M_2.$$

The mapping $\widetilde{A}$ can be constructed in the same way as A in the proof of Theorem 5.4. We show that

$$QP_1 u = \widetilde{A}u \quad \forall u \in M_1. \tag{5.45}$$

This means that we must verify

$$(P_2 A P_1 u, v)_2 = a(u,v) \quad \forall u \in M_1, v \in M_2. \tag{5.46}$$

But for $u \in M_1$, $v \in M_2$,

$$\begin{aligned}(P_2AP_1u,v)_2 &= (P_2Au,v)_2 = (Au,P_2^*v)_2 = (Au,P_2v)_2, \quad \text{by (5.44)},\\ &= (Au,v)_2 = a(u,v).\end{aligned}$$

So (5.46) holds.

Now let $\widetilde{\zeta} \in M_2$ be such that

$$(\widetilde{\zeta}, v)_2 = \theta(v) \quad \forall v \in M_2.$$

(Here, $\widetilde{\zeta}$ plays the same role as ζ in the proof of Theorem 5.4.) It is easy to check that

$$\widetilde{\zeta} = P_2 A\widehat{u}.$$

Thus, the solution $\widetilde{u}$ in M_1 for the variational problem (5.42) is

$$\widetilde{u} = \widetilde{A}^{-1} P_2 A \widehat{u} \in M_1.$$

For any $\mu \in M_1$, write $\widehat{u} = \mu + \widehat{z}$. ($\widehat{z}$ is dependent on μ.) Then

$$\begin{aligned}\widetilde{u} - \widehat{u} &= \widetilde{A}^{-1}P_2A\widehat{u} - \widehat{u}\\ &= \widetilde{A}^{-1}P_2A\mu + \widetilde{A}^{-1}P_2A\widehat{z} - \mu - \widehat{z}\\ &= (\widetilde{A}^{-1}P_2A\mu - \mu) + \widetilde{A}^{-1}P_2A\widehat{z} - \widehat{z}\\ &= -\widehat{z} + \widetilde{A}^{-1}P_2A\widehat{z}, \qquad \text{by (5.45)}.\end{aligned}$$

Therefore

$$\begin{aligned}\|\widetilde{u}-\widehat{u}\|_1 &\leqslant \left\|I-\widetilde{A}^{-1}P_2A\right\| \|\widehat{z}\| \\ &\leqslant \left[1+\frac{C_1}{d(M_1,M_2)}\right] \|\widehat{z}\| \\ &= \left[1+\frac{C_1}{d(M_1,M_2)}\right] \|\widehat{u}-\mu\|.\end{aligned}$$

So (5.43) is proved. □

Similarly to Remark 5.3, condition (5.41) can be weakened to

$$\sup_{\substack{u\in M_1\\ \|u\|_1=1}} |a(u,v)| > 0 \quad \text{for any } v\in M_2, v\neq 0.$$

5.3 Finite-element computation of BVP: an example

The implementation of the finite-element method to solve boundary value problems can be roughly divided into the following steps:

(i) Usually, a boundary value problem consists of an elliptic partial differential operator of even order $2m$, with m boundary conditions. (The most common ones are the Laplace and the biharmonic equations.) The weak (i.e., Galerkin) formulation requires the computation of partial derivatives of order up to m. Among the m boundary conditions, some correspond *essential* boundary conditions, and some to *natural* boundary conditions. (A rule of thumb is that if the order of the boundary condition is less than or equal to $m-1$ then it is an essential boundary condition. Otherwise it is a natural boundary condition.)

(ii) Discretize the spatial domain into "finite elements".

(iii) Choose appropriate "trial functions" which are piecewise polynomials with sufficient smoothness to approximate the solution.

To show how the above works, we give a very simple example.

Example 5.4. The boundary value problem

$$\left.\begin{aligned}-u''(x) &= 1,\\ u(0) &= u(1)=0,\end{aligned}\right\} \tag{5.47}$$

has exact solution $u(x) = -\frac{1}{2}x^2+\frac{1}{2}x$. Its weak formulations is

$$\left.\begin{aligned}&\int_0^1 u'(x)v'(x)\,dx = \int_0^1 1\cdot v(x)\,dx \quad \forall v(x), v\in H_0^1((0,1)),\\ &u\in H_0^1((0,1)).\end{aligned}\right\} \tag{5.48}$$

Let us divide the interval $[0,1]$ into five equal subintervals and define four "roof functions"

$$\varphi_i(x) = \begin{cases} \dfrac{x-x_{i-1}}{x_i-x_{i-1}}, \; x_{i-1} \leqslant x \leqslant x_i, \\ & i=1,2,3,4, \\ \dfrac{x_{i+1}-x}{x_{i+1}-x_i}, \; x_i \leqslant x \leqslant x_{i+1}, \end{cases} \tag{5.49}$$

which are locally supported on the interval $[x_{i-1}, x_{i+1}]$ and equal to 0 elsewhere. See Fig. 5.2.

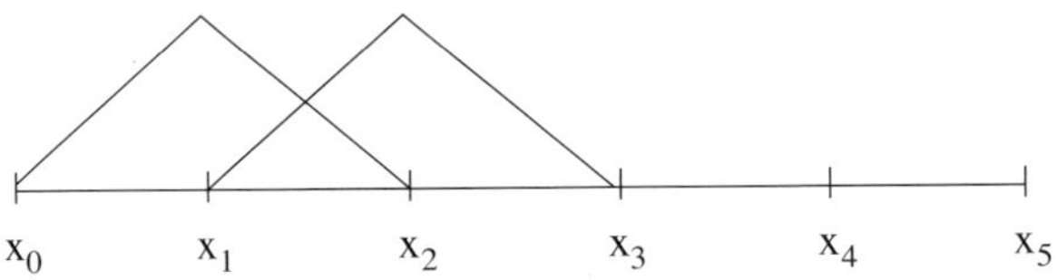

Figure 5.2: The interval $[0,1]$ is divide into five subintervals with roof functions.

We write the approximate solution as

$$\widehat{u}_h(x) = \alpha_1\varphi_1(x) + \alpha_2\varphi_2(x) + \alpha_3\varphi_3(x) + \alpha_4\varphi_4(x).$$

Then $\widehat{u}_h(x)$ is determined by the equations

$$\int_0^1 \widehat{u}_h'(x)\varphi_j'(x)\,dx = \int_0^1 \varphi_j(x)\,dx, \qquad j=1,2,3,4,$$

which can be written as a matrix equation

$$\begin{bmatrix} 2/h & -1/h & 0 & 0 \\ -1/h & 2/h & -1/h & 0 \\ 0 & -1/h & 2/h & -1/h \\ 0 & 0 & -1/h & 2/h \end{bmatrix} \begin{bmatrix} \alpha_1 \\ \alpha_2 \\ \alpha_3 \\ \alpha_4 \end{bmatrix} = \begin{bmatrix} h \\ h \\ h \\ h \end{bmatrix}.$$

We can see that

$$\frac{-\left(\dfrac{\alpha_3-\alpha_2}{h} - \dfrac{\alpha_2-\alpha_1}{h}\right)}{h} = 1,$$

for example, which corresponds to a second-order finite-difference approximation for $-u''$. One can expect that the error for $u - \widehat{u}_h$ is $\mathcal{O}(h^2)$. □

5.4 (t,m)-systems of approximating subspaces

In the approximation of a variational problem, using the formulation (5.26) or Theorem 5.5, spaces of *trial functions* (cf. V_h in (5.26) and M_2 in Theorem 5.5) are chosen, which are generally piecewise-polynomial spline functions. Such functions can form approximating subspaces satisfying nice properties that are useful in estimating the errors of the approximate solutions in Sobolev spaces. We give the following important definition.

Definition 5.1. Let Ω be a bounded domain with piecewise-smooth boundary in $\mathbb{R}^N$. Let t,m be integers such that $0 \leqslant m \leqslant t-1$. A collection of linear spaces $\{S_h(\Omega) \mid 0 < h < 1\}$ of functions defined on Ω is called a (t,m)-*system* in $H^m(\Omega)$ if

(i) $S_h(\Omega) \subset H^m(\Omega)$; and

(ii) for each integer $r \geqslant 0$, each $g \in H^r(\Omega)$, and each integer s, $0 \leqslant s \leqslant \min(r,m)$, there exists $v \in S_h(\Omega)$ such that

$$\|g-v\|_{H^s(\Omega)} \leqslant Ch^{\mu}\|g\|_{H^r(\Omega)}, \qquad \mu = \min(t-s, r-s), \tag{5.50}$$

where C is a constant independent of g and h. □

In (5.50), h usually denotes the mesh size of a partition of Ω. This inequality supplies convergence behavior and discretization error as $h \to 0$.

In the following sections, we present a variety of commonly used (t,m)-systems in one-, two- and three-dimensional spaces. They are extremely useful in solving boundary value problems.

5.5 Polynomial splines in one dimension

Let $a = x_0 < x_1 < \cdots < x_N = b$ be a partition of $[a,b]$ with mesh length $h = \max\limits_{1 \leqslant j \leqslant N}(x_j - x_{j-1})$. We define polynomial spline spaces as follows. Let m, t be integers.

(1) For $0 < m < t$,

$$\begin{aligned} S^{t,m}(x_0,x_N) = \{f \in C^{m-1}([x_0,x_N]) \mid\ & f \text{ restricted} \\ & \text{to each subinterval } (x_j,x_{j+1}),\ j=0,1,\ldots,N-1, \\ & \text{is a polynomial of degree } \leqslant t-1\}; \end{aligned} \tag{5.51}$$

(2) For $0 = m < t$,

$$\begin{aligned} S^{t,0}(x_0,x_N) = \{f \in L^2((x_0,x_N)) \mid\ & f \text{ restricted} \\ & \text{to each subinterval } (x_j,x_{j+1}),\ j=0,\,1,\ldots,N-1, \\ & \text{is a polynomial of degree } \leqslant t-1\}. \end{aligned} \tag{5.52}$$

Remark 5.4.

(i) $S^{t,m}(x_0,x_N)$ is a finite-dimensional linear space of dimension $N(t-m)+m$.

(ii) If $m = t > 0$, then $S^{t,m}(x_0,x_N)$ can be defined to be just $P_{t-1}(a,b)$, the set of *all* polynomials on (a,b) of degree $\leqslant t-1$.

(iii) For $m > 0$, functions in $S^{t,m}(x_0,x_N)$ are continuous at the interior mesh points $x_1, x_2, \ldots, x_{N-1}$. For $m = 0$, they may not be.

(iv) If $m = t-1$, $S^{t,m}(x_0,x_N)$ is called the space of polynomial B-splines. □

Of utmost importance to the finite-element method is the following approximation theoretical result concerning $S^{t,m}(x_0,x_N)$.

Theorem 5.6. *Let $t, m \in \mathbb{Z}^+$ such that $0 \leqslant m \leqslant t-1$ and let*

$$\{S_h^{t,m}(x_0,x_N) \mid 0 < h < 1\}$$

be a collection of $S^{t,m}(x_0,x_N)$ polynomial spline spaces each defined on a partition having mesh length h. If the mesh ratios of the partitions

$$R_h = \frac{\max_{1\leqslant j\leqslant N} |x_j - x_{j-1}|}{\min_{1\leqslant j\leqslant N} |x_j - x_{j-1}|}, \quad 0 < h < 1,$$

are bounded independent of h, then this collection is a (t,m)-system in $H^m(x_0,x_N)$. □

The proof of this theorem will become evident by Theorem 5.9 given later on in this chapter. We now list most of these spline spaces below. The first four types ((5.53)–(5.60)) given are for arbitrary partitions, and the last two ((5.61), and (5.62)) are for uniform partitions.

(1) $S^{2,1}(x_0,x_N)$ (C^0 piecewise-linear): (2,1)-system in $H^1((x_0,x_N))$ (see Fig. 5.3).

This has dimension $N+1$. The standard basis consists of one hill (roof) function ϕ_{i+1} for each mesh point x_i, $0 \leqslant i \leqslant N$.

$$\phi_{i+1}(x) = \begin{cases} \dfrac{x - x_{i-1}}{x_i - x_{i-1}}, & x_{i-1} \leqslant x \leqslant x_i, \\ \dfrac{x - x_{i+1}}{x_i - x_{i+1}}, & x_i \leqslant x \leqslant x_{i+1}, \\ 0, & \text{elsewhere}. \end{cases} \tag{5.53}$$

(2) $S^{3,1}(x_0,x_N)$ (C^0 piecewise-quadratic): (3,1)-system in $H^1((x_0,x_N))$ (see Fig. 5.4).

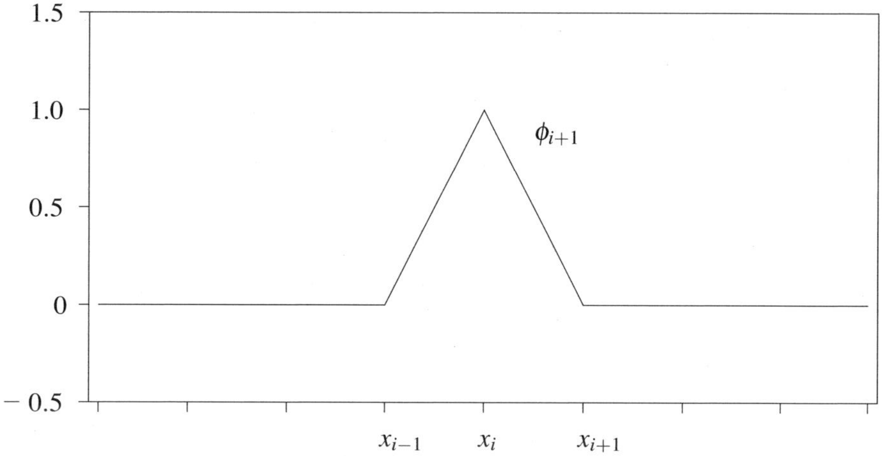

Figure 5.3: A basis element of $S^{2,1}$.

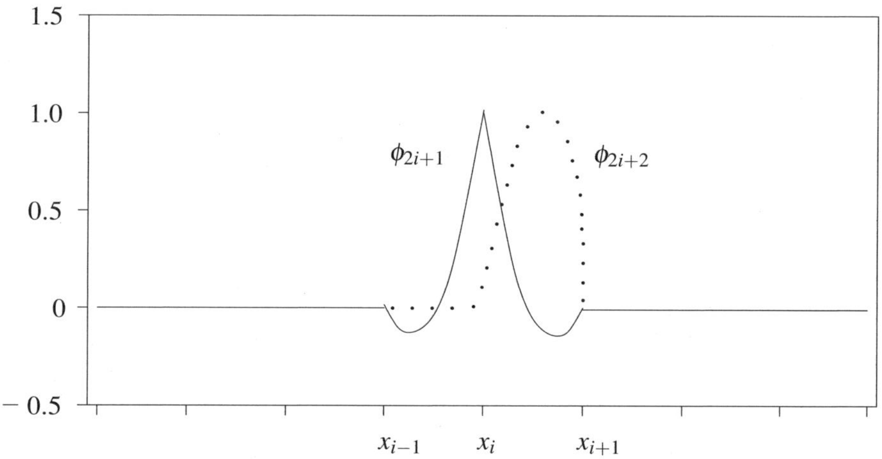

Figure 5.4: Basis elements of $S^{3,1}$.

This has dimension $2N+1$. The standard basis consists of the $2N+1$ hill functions $\phi_{2i+1}(x)$, $0 \leqslant i \leqslant N$, and $\phi_{2i+2}(x)$, $0 \leqslant i \leqslant N-1$.

$$\phi_{2i+1}(x) = \begin{cases} \dfrac{(2x - x_i - x_{i-1})(x - x_{i-1})}{(x_i - x_{i-1})^2}, & x_{i-1} \leqslant x \leqslant x_i, \\ \dfrac{(2x - x_i - x_{i+1})(x - x_{i+1})}{(x_{i+1} - x_i)^2}, & x_i \leqslant x \leqslant x_{i+1}, \\ 0, & \text{elsewhere}, \end{cases} \tag{5.54}$$

$$\phi_{2i+2(x)} = \begin{cases} \dfrac{-4(x - x_i)(x - x_{i+1})}{(x_{i+1} - x_i)^2}, & x_i \leqslant x \leqslant x_{i+1}, \\ 0, & \text{elsewhere}. \end{cases} \tag{5.55}$$

(3) $S^{4,1}(x_0,x_N)$ (C^0 piecewise-cubic): (4,1)-system in $H^1((x_0,x_N))$ (see Fig. 5.5).

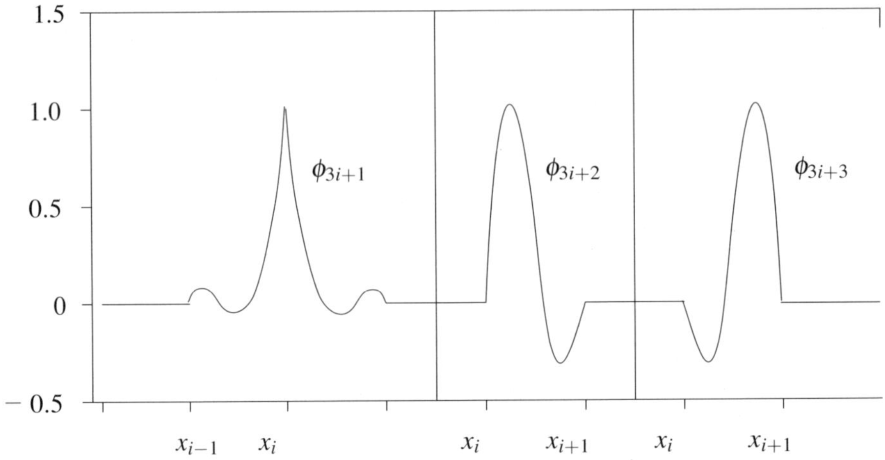

Figure 5.5: Basis elements of $S^{4,1}$.

This has dimension $3N+1$. The standard basis consists of the $3N+1$ hill functions $\phi_{3i+1}(x)$, $0 \leqslant i \leqslant N$, $\phi_{3i+2}(x)$, $0 \leqslant i \leqslant N-1$, and $\phi_{3i+3}(x)$, $0 \leqslant i \leqslant N-1$.

$$\phi_{3i+1}(x) = \begin{cases} \dfrac{(x-x_{i-1})(3x-2x_{i-1}-x_i)(3x-2x_i-x_{i-1})}{2(x_i-x_{i-1})^3}, & x_{i-1} \leqslant x \leqslant x_i, \\ \dfrac{(x-x_{i-1})(3x-2x-x_{i+1})(3x-2x_{i+1}-x_i)}{2(x_i-x_{i+1})^3}, & x_i \leqslant x \leqslant x_{i+1}, \\ 0, & \text{elsewhere}, \end{cases} \tag{5.56}$$

$$\phi_{3i+2}(x) = \begin{cases} \dfrac{9(x-x_i)(3x-2x_{i+1}-x_i)(x-x_{i+1})}{2(x_{i+1}-x_i)^3}, & x_i \leqslant x \leqslant x_{i+1}, \\ 0, & \text{elsewhere}, \end{cases} \tag{5.57}$$

$$\phi_{3i+3}(x) = \begin{cases} \dfrac{9(x-x_i)(3x-2x_i-x_{i+1})(x-x_{i+1})}{2(x_i-x_{i+1})^3}, & x_i \leqslant x \leqslant x_{i+1}, \\ 0, & \text{elsewhere}. \end{cases} \tag{5.58}$$

(4) $S^{4,2}(x_0,x_N)$ (C^1 piecewise-cubic): (4,2)-system in $H^2((x_0,x_N))$ (see Fig. 5.6).

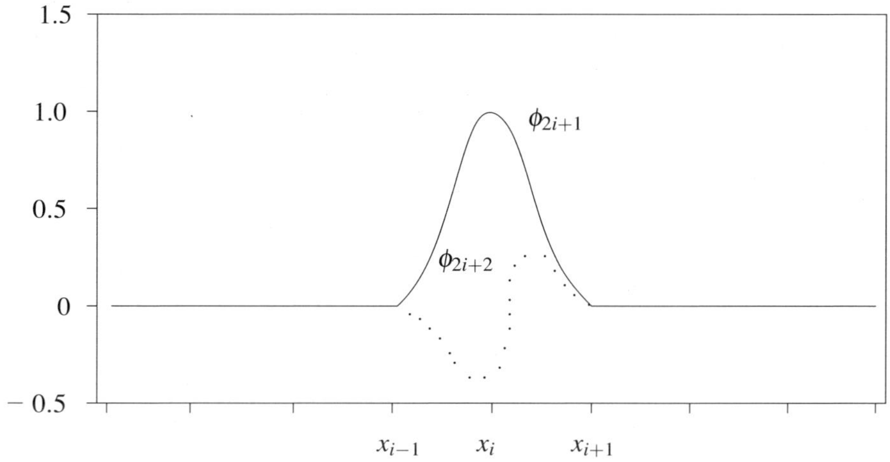

Figure 5.6: Basis elements of $S^{4,2}$.

This has dimension $2N+2$. The standard basis consists of two hill functions $\phi_{2i+1}(x)$ and $\phi_{2i+2}(x)$ for each mesh point x_i, $0 \leqslant i \leqslant N$.

$$\phi_{2i+1}(x) = \begin{cases} \dfrac{(x-x_{i-1})^2(3x_i - x_{i-1} - 2x)}{(x_i - x_{i-1})^3}, & x_{i-1} \leqslant x \leqslant x_i, \\ \dfrac{(x_{i+1}-x)^2(x_{i+1} - 3x_i + 2x)}{(x_{i+1} - x_i)^3}, & x_i \leqslant x \leqslant x_{i+1}, \\ 0, & \text{elsewhere}, \end{cases} \tag{5.59}$$

$$\phi_{2i+2}(x) = \begin{cases} \dfrac{(x-x_{i-1})^2(x - x_i)}{(x_i - x_{i-1})^2}, & x_{i-1} \leqslant x \leqslant x_i, \\ \dfrac{(x-x_i)(x - x_{i+1})^2}{(x_{i+1} - x_i)^2}, & x_i \leqslant x \leqslant x_{i+1}, \\ 0, & \text{elsewhere}. \end{cases} \tag{5.60}$$

(5) $S^{3,2}(x_0,x_N)$ (C^1 piecewise-quadratic): (3,2)-system in $H^2((x_0,x_N))$ (see Fig. 5.7).

This has dimension $N+2$ and is a cardinal spline space. Let the partition be equally spaced with mesh length $h = x_{i+1} - x_i$, and let $x_{-2} < x_{-1} < x_0$ and $x_N < x_{N+1} < x_{N+2}$ be four additional equally spaced points outside the partition. The standard basis consists of one

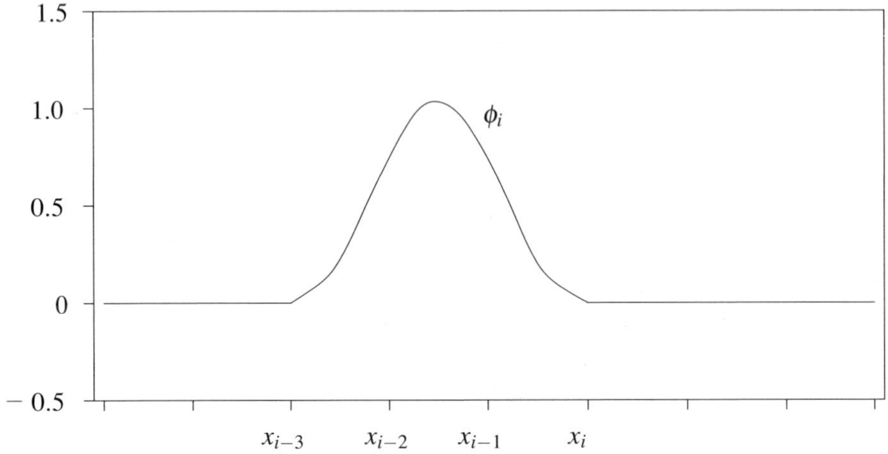

Figure 5.7: A basis element of $S^{3,2}$.

hill function $\phi_i(x)$ for each mesh point $x_1,\ldots,x_{N+2}$.

$$\phi_i(x) = \frac{4}{3}\begin{cases} \dfrac{(x-x_{i-3})^2}{2h^2}, & x_{i-3} \leqslant x \leqslant x_{i-2}, \\ \dfrac{(x-x_{i-2})(x_{i-1}-x)}{h^2} + \dfrac{1}{2}, & x_{i-2} \leqslant x \leqslant x_{i-1}, \\ \dfrac{(x_i-x)^2}{2h^2}, & x_{i-1} \leqslant x \leqslant x_i, \\ 0, & \text{elsewhere}. \end{cases} \tag{5.61}$$

(6) $S^{4,3}(x_0,x_N)$ (C^2 piecewise-cubic): (4,3)-system in $H^3((x_0,x_N))$ (see Fig. 5.8).

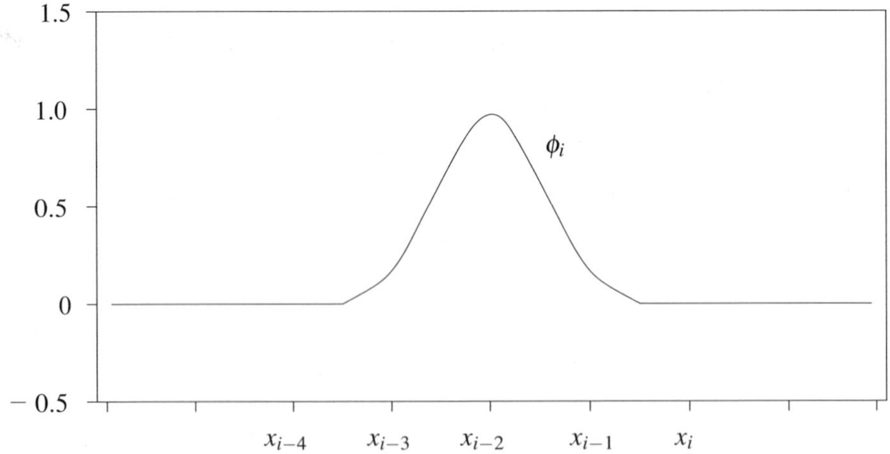

Figure 5.8: A basis element of $S^{4,3}$.

This has dimension $N+3$ and is a cardinal spline space. Let the partition be equally spaced with mesh length $h = x_{i+1} - x_i$ and let $x_{-3} < x_{-2} < x_{-1} < x_0$ and $x_N < x_{N+1} < x_{N+2} < x_{N+3}$ be six additional equally spaced points outside the partition. The standard basis consists of one hill function $\phi_1(x)$ for each of the mesh points x_i, $i = 1, \ldots, N+3$.

$$\phi_i(x) = \frac{1}{6}\begin{cases} \frac{(x-x_{i-4})^3}{h^3}, & x_{i-4} \leqslant x \leqslant x_{i-3}, \\ 1 + \frac{3(x-x_{i-3})}{h} + \frac{3(x-x_{i-3})^2}{h^2} - \frac{3(x-x_{i-3})^3}{h^3}, & x_{i-3} \leqslant x \leqslant x_{i-2}, \\ 1 - \frac{3(x-x_{i-1})}{h} + \frac{3(x-x_{i-1})^2}{h^2} + \frac{3(x-x_{i-1})^3}{h^3}, & x_{i-2} \leqslant x \leqslant x_{i-1}, \\ \frac{(x_i-x)^3}{h^3}, & x_{i-1} \leqslant x \leqslant x_i, \\ 0, & \text{elsewhere.} \end{cases} \tag{5.62}$$

Exercise 5.1. Consider the eigenvalue problem

$$\begin{aligned} &a^4\phi^{(4)}(x) + \lambda^2\phi(x) = 0, \; -1 \leqslant x \leqslant 1, \\ &\phi(-1) \;=\; \phi'(-1) = 0, \\ &a^4\phi'''(1) - k_0^2\lambda\phi(1) = 0, \quad k_0^2 \geqslant 0, \\ &a^4\phi''(1) + k_1^2\lambda\phi'(1) = 0, \quad k_1^2 \geqslant 0. \end{aligned}$$

(i) Formulate the above as a variational problem.

(ii) Let $h = \frac{2}{3}$ and divide $[-1,1]$ into three subintervals of equal length. Use an $S^{4,2}$–system to discretize the variational problem. Show that the discretized matrix equation is of the form

$$\lambda^2 M_0 \vec{q} + \lambda M_1 \vec{q} + M_2 \vec{q} = 0,$$
$$\left(\vec{q} = (q_1, q_2, \ldots, q_6)^{\mathrm{Tr}}, \;\; \phi(x) = \sum_{j=1}^{6} q_i \phi_j(x)\right)$$

where M_0, M_1 and M_2 are 6×6 symmetric matrices given by

$$M_0 = \begin{bmatrix} \frac{26}{35}h & 0 & \frac{9}{70}h & \frac{-13}{420}h^2 & 0 & 0 \\ & \frac{2}{105}h^3 & \frac{13}{420}h^2 & \frac{-1}{140}h^3 & 0 & 0 \\ & & \frac{26}{35}h & 0 & \frac{9}{70}h & \frac{-13}{420}h^2 \\ & & & \frac{2}{105}h^3 & \frac{13}{420}h^2 & \frac{-1}{140}h^3 \\ & & & & \frac{13}{35}h & \frac{-11}{210}h^2 \\ & & & & & \frac{1}{105}h^3 \end{bmatrix},$$

$$M_1 = \begin{bmatrix} 0 \cdots & & 0 \\ \ddots & & \vdots \\ & 0 & 0 \\ & k_0^2 & 0 \\ 0 & 0 & k_1^2 \end{bmatrix},$$

$$M_2 = a^4 \begin{bmatrix} 24/h^3 & 0 & -12/h^3 & 6/h^2 & 0 & 0 \\ & 8/h & -6/h^2 & 2/h & 0 & 0 \\ & & 24/h^3 & 0 & -12/h^3 & 6/h^2 \\ & & & 8/h & -6/h^2 & 2/h \\ & & & & 12/h^3 & -6/h^2 \\ & & & & & 4/h \end{bmatrix}. \tag{5.63}$$

□

5.6 Barycentric coordinates

To describe finite elements in higher dimensions, it is necessary that we establish a local coordinate system defined on each element and beyond, based upon the vertices of each element. Such a system is called a barycentric coordinate system.

Let $A_j = (a_{ij})_{i=1}^N$, $j = 1, 2, \ldots, N+1$, be a set of $N+1$ points in the N-dimensional space $\mathbb{R}^N$. Assume that the determinant of the $(N+1) \times (N+1)$ matrix

$$A = \begin{bmatrix} A_1 & A_2 & \cdots & A_{N+1} \\ 1 & 1 & \cdots & 1 \end{bmatrix} = \begin{bmatrix} a_{11} & a_{12} & \cdots & a_{1,N+1} \\ a_{21} & a_{22} & \cdots & a_{2,N+1} \\ \vdots & \vdots & & \vdots \\ a_{N1} & a_{N2} & \cdots & a_{N,N+1} \\ 1 & 1 & \cdots & 1 \end{bmatrix}$$

is nonzero. (This means that these $N+1$ points $A_1, A_2, \ldots, A_{N+1}$ do not fall on a hyperplane in $\mathbb{R}^N$.) Then

$$K = \left\{ x \in \mathbb{R}^N \,\middle|\, x = \sum_{j=1}^{N+1} \lambda_j A_j; 0 \leqslant \lambda_j \leqslant 1, 1 \leqslant j \leqslant N+1, \sum_{j=1}^{N+1} \lambda_j = 1 \right\}$$

is said to be an N-simplex. Note that a 1-simplex is a line segment, a 2-simplex is a triangle, a 3-simplex is a tetrahedron. See Fig. 5.9.

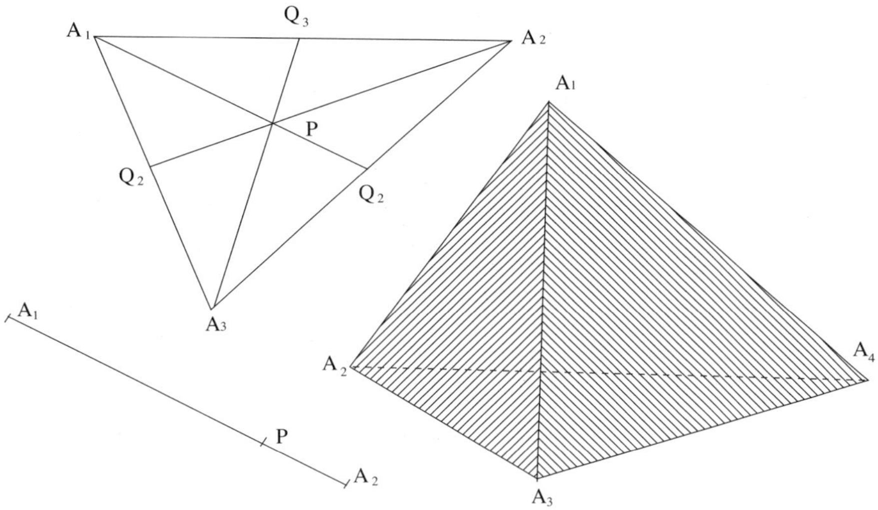

Figure 5.9: 1-, 2- and 3-simplices.

$$P = \lambda_1 A_1 + \lambda_2 A_2,$$
$$\frac{A_1P}{PA_2} = \frac{\lambda_2}{\lambda_1},$$

$$P = \lambda_1 A_1 + \lambda_2 A_2 + \lambda_3 A_3,$$
$$\lambda_1 = \frac{\text{area } \Delta PA_2A_3}{\text{area } \Delta A_1A_2A_3}, \quad \frac{A_1Q_3}{Q_3A_2} = \frac{\lambda_2}{\lambda_1},$$
$$\lambda_2 = \frac{\text{area } \Delta PA_1A_3}{\text{area } \Delta A_1A_2A_3}, \quad \frac{A_1Q_2}{Q_2A_3} = \frac{\lambda_3}{\lambda_1},$$
$$\lambda_3 = \frac{\text{area } \Delta PA_1A_2}{\text{area } \Delta A_1A_2A_3}, \quad \frac{A_2Q_1}{Q_1A_3} = \frac{\lambda_3}{\lambda_2}.$$

The magnitudes of λ_i have simple geometrical meanings as shown above.

For any $x \in K$, we can combine the two relations

$$x = \sum_{j=1}^{N+1} \lambda_j A_j, \quad x = \begin{bmatrix} x_1 \\ \vdots \\ x_N \end{bmatrix},$$
$$\sum_{j=1}^{N+1} \lambda_j = 1,$$

into a matrix equation

$$\begin{bmatrix} x \\ 1 \end{bmatrix} = \lambda_1 \begin{bmatrix} A_1 \\ 1 \end{bmatrix} + \lambda_2 \begin{bmatrix} A_2 \\ 1 \end{bmatrix} + \cdots + \lambda_{N+1} \begin{bmatrix} A_{N+1} \\ 1 \end{bmatrix}$$

$$= \begin{bmatrix} A_1 & A_2 & \cdots & A_{N+1} \\ 1 & 1 & \cdots & 1 \end{bmatrix} \begin{bmatrix} \lambda_1 \\ \lambda_2 \\ \vdots \\ \lambda_{N+1} \end{bmatrix},$$

or

$$\begin{bmatrix} x \\ 1 \end{bmatrix} = A\lambda, \quad \lambda = \begin{bmatrix} \lambda_1 \\ \lambda_2 \\ \vdots \\ \lambda_{N+1} \end{bmatrix}.$$

Since $\det A \neq 0$, A is invertible and we have

$$\lambda = A^{-1} \begin{bmatrix} x \\ 1 \end{bmatrix}. \tag{5.64}$$

$\lambda = (\lambda_1, \ldots, \lambda_{N+1})$ are called the *barycentric coordinates* (or area coordinates, or sometimes called triangular coordinates by engineers) of the point x.

Definition 5.2. Let $k \geqslant 0$ be an integer. We use P_k to denote the space of all polynomials of degree $\leqslant k$ in $x_1, x_2, \ldots, x_N$. □

Let $\overline{\Omega}$ be a bounded domain in $\mathbb{R}^N$. Assume that $\overline{\Omega}$ can be divided (i.e., triangulated) into a finite collection of N-simplexes $\{\sigma_j\}_{j=1}^L$. We call $\mathscr{T} \equiv \{\sigma_j\}_{j=1}^L$ a *triangulation* of $\overline{\Omega}$. Let V_h be a finite-dimensional space of trial functions on $\overline{\Omega}$. For any $K \in \mathscr{T}$, we define

$$P(K) \equiv \{v|_K \mid v \in V_h\}. \tag{5.65}$$

5.7 Finite elements in two dimensions

Finite elements in two dimensions are used most often in boundary-element computations, because the boundary of a three-dimensional domain is two dimensional.

In this section, we mention the major types of finite elements–triangles and rectangles in $\mathbb{R}^2$. Triangular elements are the ones used most widely, because

(i) their shape is very simple;

(ii) they can approximate the boundary of any two-dimensional domain very well; combinations of planar triangular pieces can also approximate two-dimensional surfaces very accurately;

(iii) "complete polynomials" may be used on the entire triangle and/or on each side of the triangle (finite elements of other shapes do not have this property).

Rectangular elements are most useful when the domain itself is rectangular, or has portions of its boundary parallel to the coordinate axes. They are easy to construct, using high degrees of polynomials. But they do not fit a curved boundary well. They can be used in conjunction with other types of elements to simplify computations or to improve accuracy. We present these finite elements below.

(FE_1) *C^0-linear Lagrangian triangular elements (see Fig. 5.10).*

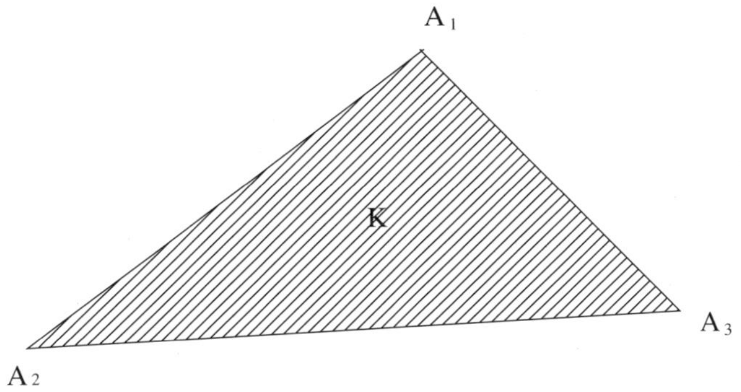

Figure 5.10: Nodes of an (FE_1) element $(N = 2)$.

Let K be an N-simplex. We use P_1, the space of linear polynomials in N variables and define

$$\Sigma_K = \{p(A_i) \mid 1 \leqslant i \leqslant N+1, A_i \text{ is a vertex of } K, p \in P_1\}, \tag{5.66}$$

which is the *set of degrees of freedom* for $p \in P_1$. Then the set Σ_K will determine every polynomial $p \in P_1$ uniquely as noted below. First, note that $\dim P_1 = N+1$. Consider the barycentric coordinate functions $\lambda_1(x), \ldots, \lambda_{N+1}(x) \in P_1$. They satisfy the property that

$$\lambda_i(A_j) = \delta_{ij} = \begin{cases} 1, & i = j, \\ 0, & i \neq j. \end{cases} \tag{5.67}$$

Therefore any linear combination $\sum_{k=1}^{N+1} \alpha_k \lambda_k(x) = 0$ implies that $\alpha_k = 0$ for $k = 1, 2, \ldots, N+1$. Thus $\lambda_1(x), \ldots, \lambda_{N+1}(x)$ form a basis for P_1. For any $p \in P_1$, let us write

$$p(x) = \sum_{i=1}^{N+1} \alpha_i \lambda_i(x).$$

Then

$$p(A_j) = \sum_{i=1}^{N+1} \alpha_i \lambda_i(A_j) = \sum_{i=1}^{N+1} \alpha_i \delta_{ij} = \alpha_j.$$

Thus

$$p(x) = \sum_{i=1}^{N+1} p(A_i)\lambda_i(x). \tag{5.68}$$

Note that we use (5.64) to find the basis functions $\lambda_1(x),\ldots,\lambda_{N+1}(x)$. When $N=2$, e.g., we have the following three basis functions

$$\begin{bmatrix} \lambda_1(x) \\ \lambda_2(x) \\ \lambda_3(x) \end{bmatrix} = \begin{bmatrix} a_{11} & a_{12} & a_{13} \\ a_{21} & a_{22} & a_{23} \\ 1 & 1 & 1 \end{bmatrix}^{-1} \begin{bmatrix} x_1 \\ x_2 \\ 1 \end{bmatrix}$$

$$= \frac{1}{\det A} \begin{bmatrix} a_{22}-a_{23} & -(a_{12}-a_{13}) & (a_{12}a_{23}-a_{22}a_{13}) \\ -(a_{21}-a_{23}) & a_{11}-a_{13} & -(a_{11}a_{23}-a_{21}a_{13}) \\ a_{21}-a_{22} & -(a_{11}-a_{12}) & a_{11}a_{22}-a_{21}a_{12} \end{bmatrix} \begin{bmatrix} x_1 \\ x_2 \\ 1 \end{bmatrix},$$

where

$$\det A = \det \begin{vmatrix} a_{11} & a_{12} & a_{13} \\ a_{21} & a_{22} & a_{23} \\ 1 & 1 & 1 \end{vmatrix} = \det \begin{vmatrix} a_{12}-a_{11} & a_{13}-a_{11} \\ a_{22}-a_{21} & a_{23}-a_{21} \end{vmatrix}$$

$$= (a_{12}-a_{11})(a_{23}-a_{21}) - (a_{22}-a_{21})(a_{13}-a_{11}).$$

(FE_2) C_0-quadratic Lagrangian triangular elements (see Fig. 5.11).

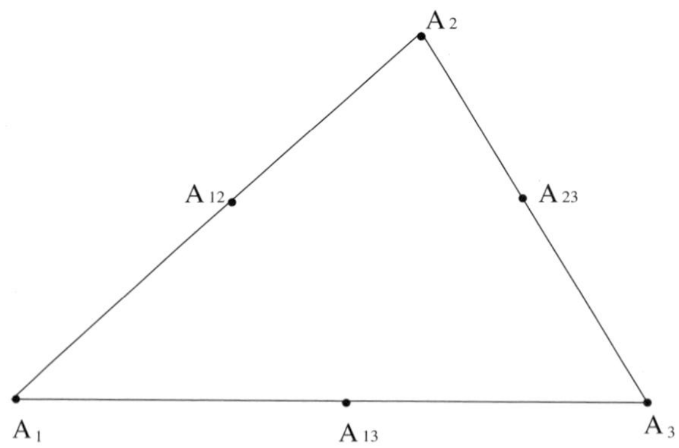

Figure 5.11: Nodes of an (FE_2) element $(N=2)$.

Let K be an N-simplex with vertices $\{A_1,\ldots,A_{N+1}\}$. Let

$$A_{ij} = \frac{1}{2}(A_i + A_j), \quad i < j,$$

be the midpoints of $\overline{A_iA_j}$.

We use P_2. Let the set of degrees of freedom be

$$\Sigma_K = \{p(A_i) \mid 1 \leqslant i \leqslant N+1\} \cup \left\{p(A_{ij}) \mid 1 \leqslant i < j \leqslant N+1\right\} \text{ for } p \in P_2. \tag{5.69}$$

The basis functions for P_2 are

$$\{\lambda_i(x)[2\lambda_i(x)-1] \mid 1 \leqslant i \leqslant N+1\} \cup \{\lambda_i(x)\lambda_j(x) \mid 1 \leqslant i < j \leqslant N+1\}.$$

Since

$$\lambda_i(A_j) = \delta_{ij}, \lambda_i(A_{kj}) = \begin{cases} \frac{1}{2}, \ i = k \text{ or } i = j, \\ 0, \text{ otherwise,} \end{cases}$$

we see that these basis functions are linearly independent in P_2. Write

$$p(x) = \sum_{i=1}^{N+1} \alpha_i \lambda_i(x)\,[2\lambda_i(x)-1] + \sum_{1 \leqslant i<j \leqslant N+1} \beta_{ij}\lambda_i(x)\lambda_j(x).$$

Then

$$\begin{aligned} p(A_k) &= \sum_{i=1}^{N+1} \alpha_i \delta_{ik}(2\delta_{ik}-1) = \alpha_k, \\ p(A_{kl}) &= \sum_{i=1}^{N+1} \alpha_i \left[2\lambda_i^2(A_{kl}) - \lambda_i(A_{kl})\right] + \sum_{1 \leqslant i<j \leqslant N+1} \beta_{ij}\lambda_i(A_{kl})\lambda_j(A_{kl}). \end{aligned} \tag{5.70}$$

But

$$\begin{aligned} \lambda_i(A_{kl}) &= 0 \ \text{ or } \ \frac{1}{2}, \\ \lambda_i(A_{kl})\lambda_j(A_{kl}) &= \begin{cases} \frac{1}{4}, \ (i,j) = (k,l) \text{ or } (l,k), \\ 0, \ \text{otherwise.} \end{cases} \end{aligned}$$

Hence the first sum on the RHS of (5.70) is 0, and

$$\begin{aligned} p(A_{kl}) &= \frac{1}{4}\beta_{kl}, \\ p(x) &= \sum_{i=1}^{N+1} p(A_i)\lambda_i(x)\,[2\lambda_i(x)-1] \\ &\quad + \sum_{1 \leqslant i<j \leqslant N+1} 4p(A_{ij})\lambda_i(x)\lambda_j(x) \ \ \forall p \in P_2. \end{aligned} \tag{5.71}$$

(FE_3) C_0-cubic Lagrangian triangular elements (see Fig. 5.12).

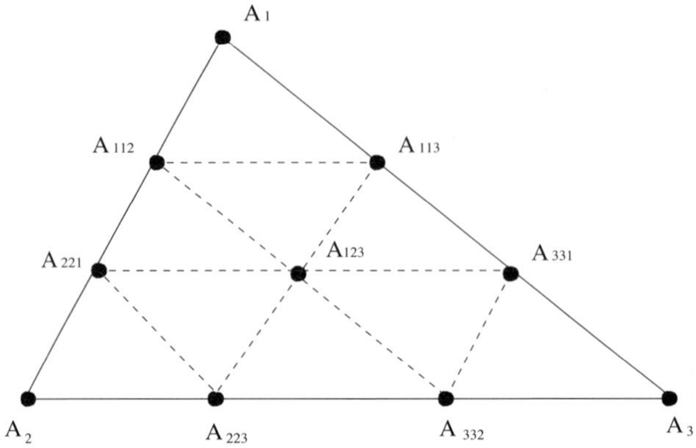

Figure 5.12: Nodes of an (FE_3) element $(N=2)$.

Let K be an N-simplex with vertices $\{A_1,\dots,A_{N+1}\}$. Let

$$A_{iij} = \frac{1}{3}(2A_i + A_j), \qquad i \neq j$$
$$A_{ijk} = \frac{1}{3}(A_i + A_j + A_k), \quad i < j < k.$$

We use P_3. Let the set of degrees of freedom be

$$\Sigma_K = \{p(A_i) \mid 1 \leqslant i \leqslant N+1\} \cup \{p(A_{iij}) \mid 1 \leqslant i \neq j \leqslant N+1\} \cup \{p(A_{ijk}) \mid 1 \leqslant i < j < k \leqslant N+1\}.$$

Then, as before, one can verify that the following set forms a basis for P_3:

$$\begin{aligned} &\{\lambda_i(x)[3\lambda_i(x)-1][3\lambda_i(x)-2] \mid 1 \leqslant i \leqslant N+1\} \\ &\cup \{\lambda_i(x)\lambda_j(x)[3\lambda_i(x)-1] \mid 1 \leqslant i \neq j \leqslant N+1\} \\ &\cup \{\lambda_i(x)\lambda_j(x)\lambda_k(x) \mid 1 \leqslant i < j < k \leqslant N+1\}. \end{aligned} \tag{5.72}$$

For any $p \in P_3$, we have

$$\begin{aligned} p(x) = &\sum_{i=1}^{N+1} \frac{1}{2} p(A_i)\lambda_i(x)[3\lambda_i(x)-1][3\lambda_i(x)-2] \\ &+ \sum_{1\leqslant i\neq j\leqslant N+1} \frac{9}{2} p(A_{iij})\lambda_i(x)\lambda_j(x)[3\lambda_i(x)-1] \\ &+ \sum_{1\leqslant i<j<k\leqslant N+1} 27 p(A_{ijk})\lambda_i(x)\lambda_j(x)\lambda_k(x). \end{aligned}$$

Using the same ideas as above, we can easily construct C^0 polynomials of degree n for any positive n for any Lagrangian N-simplex element.

We now show how to use the previous local finite-element basis to define the space V_h.

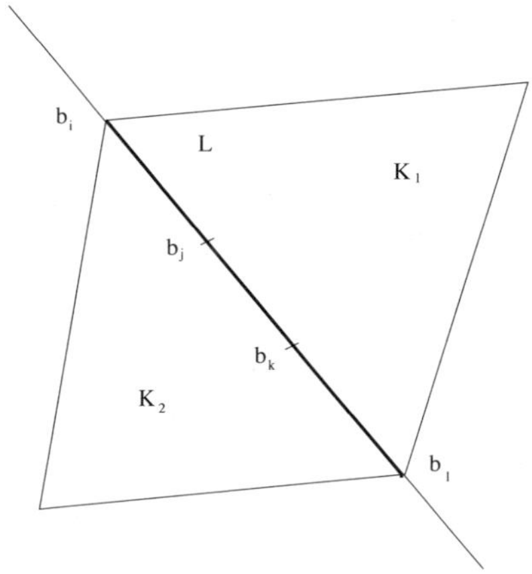

Figure 5.13: Two neighboring elements K_1 and K_2.

We require that $V_h \subset C^0(\overline{\Omega})$. For each function $v_h \in V_h$, v_h is continuous on the interior of each element. Let K_1 and K_2 be two neighboring triangles (here $N = 2$) sharing a common side L, as shown in Fig. 5.13.

For example, let us consider C^0-cubics as in (5.72). Let t be a coordinate variable along the side $L = \overline{b_i b_l}$, where b_i, b_j, b_k, b_l are nodal points for C^0-cubics. Then $v_h|_{K_1}$ and $v_h|_{K_2}$ (the restriction of v_h to K_1 and K_2 respectively) on L become cubic polynomials of the variable t, and $v_h|_{K_1}$, $v_h|_{K_2}$ agree at four distinct points b_i, b_j, b_k, and b_l. Therefore

$$v_h|_{K_1} \equiv v_h|_{K_2} \quad \text{on } L.$$

Hence v_h is continuous on $\overline{\Omega}$.

When C_0-cubics are used, each basis function $\phi_h \in V_h$ has one of the following *three types of supports* as shown in Fig. 5.14.

For example, $\varphi_{(i)}$, $\varphi_{(j)}$, and $\varphi_{(k)}$ satisfy

$$\begin{aligned}
&\varphi_{(l)}(b_m) = \delta_{lm} \quad \text{for any node } b_m,\ l = i, j \text{ or } k,\\
&\varphi_{(l)} \text{ is continuous on } \overline{\Omega}, l = i,\ j \text{ or } k, \text{etc.},\\
&\varphi_{(l)} \equiv 0 \text{ outside support } \omega_l, \text{for } l = i,\ j \text{ or } k, \text{etc.}
\end{aligned}$$

Exercise 5.2. Assume that $\overline{\Omega}$ is an irregular pentagon that is triangulated as shown in Fig. 5.15. Assume that $V_h \subset C^0(\overline{\Omega})$ consists of C^0-cubics.

(i) How many nodal points are there for the above triangulation?

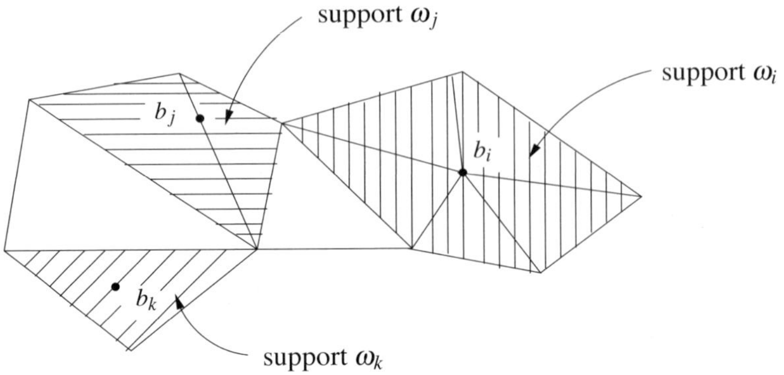

Figure 5.14: The supports of three basis functions $\phi_{(i)}$, $\phi_{(j)}$ and $\phi_{(k)}$.

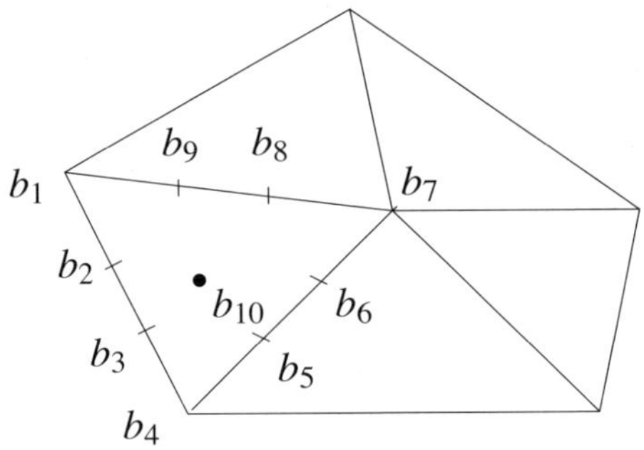

Figure 5.15: A triangulated pentagon.

(ii) How many basis functions are there in V_h?

(iii) Let $b_1, b_2, \ldots, b_9$ and b_{10} be given nodal points as indicated in the figure. Determine the supports of basis functions $\varphi_{(j)} \in V_h$ satisfying

$$\varphi_{(j)}(b_i) = \delta_{ij} \quad \text{for nodes } b_i, \; i = 1, 2, \ldots, 10; \; j = 1, 2, \ldots, 10.$$

(iv) Write out explicitly the function $\varphi_{(9)}(x)$. □

(FE_4) *The cubic Z_3 Hermite triangular elements (see Fig. 5.16).*

Let $K \subset \mathbb{R}^2$ be a triangle with vertices $\{A_1, A_2, A_3\}$. Let $\lambda_i(x), k = 1, 2, 3$, be the barycentric coordinate functions. Then one can check that any polynomial $p \in P_3$ can be expanded as

$$\begin{aligned} p(x) = & \sum_{i=1}^{3} p(A_i)\left[-2\lambda_i^3(x)+3\lambda_i^2(x)-7\lambda_1(x)\lambda_2(x)\lambda_3(x)\right] \\ & + 27p(A_{123})\lambda_1(x)\lambda_2(x)\lambda_3(x) \\ & + \sum_{i=1}^{3}\sum_{\substack{j=1 \\ j\neq i}}^{3} \left[Dp(A_i)(A_j-A_i)\right]\lambda_i(x)\lambda_j(x)\left[2\lambda_i(x)+\lambda_j(x)-1\right], \\ & x=(x_1,x_2), \end{aligned}$$

where $A_{123} = \frac{1}{3}(A_1+A_2+A_3)$ is the centroid and $Dp(A_i)(A_j-A_i)$ denotes the directional derivative of $p(x)$ at A_i in the direction of A_j-A_i, i.e.,

$$Dp(A_i)(A_j-A_i) = \left[(a_{1j}-a_{1i})\frac{\partial}{\partial x_1}p(x)+(a_{2j}-a_{2i})\frac{\partial}{\partial x_2}p(x)\right]\Bigg|_{x=A_i}.$$

Thus,

$$\Sigma_K = \{p(A_i) \mid 1 \leqslant i \leqslant 3\} \cup \{Dp(A_i)(A_j-A_i) \mid 1 \leqslant i \neq j \leqslant 3\} \cup \{p(A_{123})\}.$$

is the set of (10) degrees of freedom. Note that the above set can be replaced by

$$\Sigma_K' = \{p(A_i) \mid 1 \leqslant i \leqslant 3\} \cup \left\{\frac{\partial p}{\partial x_1}(A_i), \frac{\partial p}{\partial x_2}(A_i) \,\middle|\, 1 \leqslant i \leqslant 3\right\} \cup \{p(A_{123})\}.$$

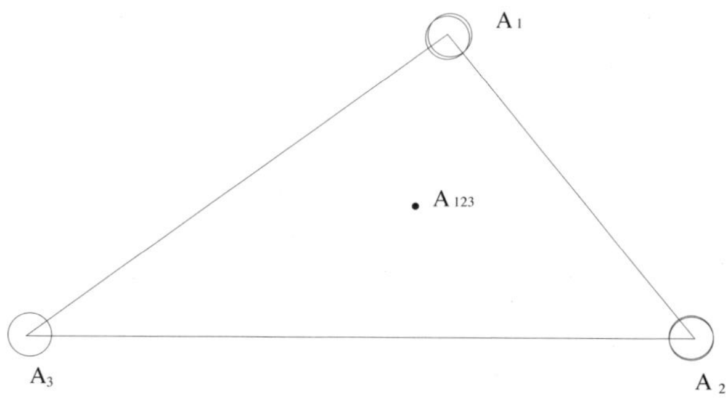

Figure 5.16: Nodes of a cubic $Z_3(FE_4)$.

(FE_5) The Zienkiewicz triangles (Hermite elements).

The centroid A_{123} in Fig. 5.16 and (FE_4) is rather inconvenient. It can be eliminated by imposing the constraint

$$6p(A_{123}) - 2\sum_{i=1}^{3} p(A_i) + \sum_{i=1}^{3} Dp(A_i)(A_i - A_{123}) = 0, \quad p \in P_3. \tag{5.73}$$

The new set of (nine) degrees of freedom becomes

$$\Sigma_K = \left\{ p(A_i),\ \frac{\partial p}{\partial x_1}(A_i),\ \frac{\partial p}{\partial x_2}(A_i) \,\middle|\, i = 1, 2, 3, p \in P_3 \text{ satisfying (5.73)} \right\}.$$

Note that any $p \in P_2$ satisfies (5.73). This is the idea of a serendipity family. See (FE_{10}) later.

(FE_6) The Argyris triangles (Hermite elements) (see Fig. 5.17).

This is also known as the *21-degree-of-freedom triangle*. We use P_5 and

$$\begin{aligned}\Sigma_K = \Bigg\{ & p(A_i),\ \frac{\partial}{\partial x_1}p(A_i),\ \frac{\partial^2}{\partial x_1^2}p(A_i),\ \frac{\partial}{\partial x_2}p(A_i),\ \frac{\partial^2}{\partial x_2^2}p(A_i),\\ & \frac{\partial^2}{\partial x_1 \partial x_2}p(A_i) \,\Bigg|\, i = 1, 2, 3 \Bigg\} \cup \left\{ \frac{\partial}{\partial n}p(A_{ij}) \,\middle|\, 1 \leqslant i < j \leqslant 3 \right\}\\ & \left(\frac{\partial}{\partial n} \text{ is the normal derivative} \right).\end{aligned}$$

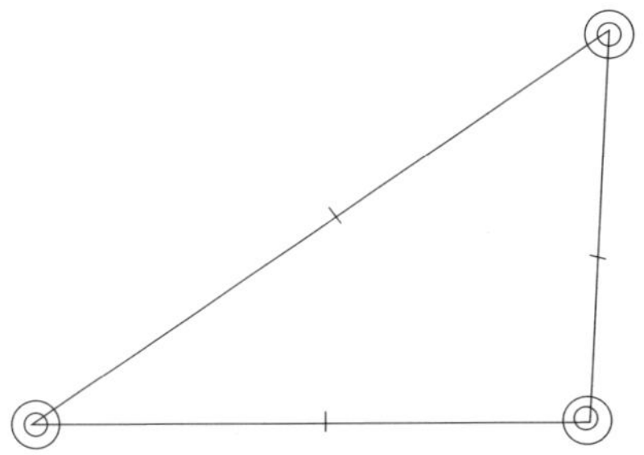

Figure 5.17: Nodes of an Argyris triangle.

It can be shown that Σ_K uniquely determines a $p \in P_5$, for any $p \in P_5$.
We now show that from Σ_K one can construct $V_h \subset C^1(\overline{\Omega})$. Consider two adjacent Argyris triangles K_1 and K_2 with common side L, along which s is an abscissa; see Fig. 5.18.
Let $v \in V_h$. Now $v|_{K_1}, v|_{K_2}$ are polynomials of degree five in t along L and they are required to agree, along with their first and second derivatives at the endpoints. This gives six conditions for six unknown coefficients in this polynomial of degree five in t. So v is continuous across L.
Now $(\partial/\partial n)v|_{K_1}$ and $(\partial/\partial n)v|_{K_2}$ along L are polynomials of degree 4. They are required to agree at the midpoint of L, as well as at the two endpoints. We further require that

$$\frac{\partial}{\partial s}\left(\frac{\partial v}{\partial n}|_{K_1} \right) = \frac{\partial}{\partial s}\left(\frac{\partial v}{\partial n}|_{K_2} \right) \quad \text{at the two endpoints.}$$

This imposes five conditions on five unknown coefficients of this fourth degree polynomial. Thus

$$\left.\begin{aligned}\frac{\partial}{\partial n}(v|_{K_1}) &= \frac{\partial}{\partial n}(v|_{K_2})\\ \frac{\partial}{\partial s}(v|_{K_1}) &= \frac{\partial}{\partial s}(v|_{K_2})\end{aligned}\right\}\text{ on } L.$$

Hence $v \in C^1(\overline{\Omega})$ and $V_h \subset C^1(\overline{\Omega})$.

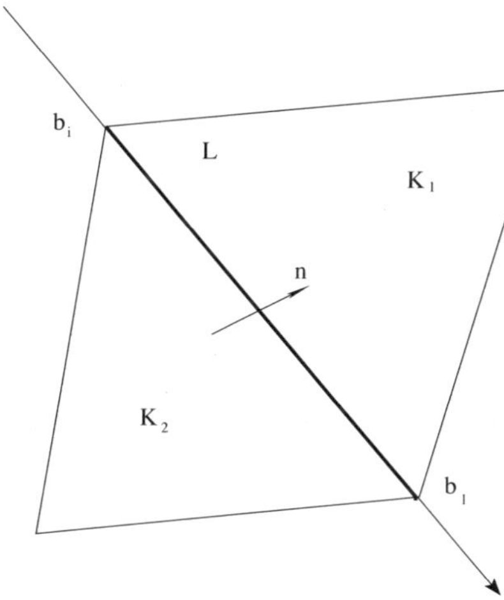

Figure 5.18: C_1 continuity of Argyris triangles across and along a common side.

(FE_7) *Bell's 18 degree-of-freedom triangles (Hermite elements)*

Let V_h consist of those piecewise polynomials v such that on each triangle $K \subset \mathbb{R}^2, v|_K$ is a polynomial of degree five, and along each side of K, the normal derivative is a polynomial of degree $\leqslant 3$. In other words, the leading coefficient of what would otherwise be a quartic polynomial (in the abscissa variable s along that edge) is made to vanish. With this constraint, the nodal values of $\partial v/\partial n$ and $\partial^2 v/\partial s \partial n$ determine the cubic $\partial v/\partial n$ along that edge, and $v \in C^1(\overline{\Omega})$.

The complete quintic is lost, reducing the accuracy of $\|u - v_h\|$ from $\mathcal{O}(h^6)$ to $\mathcal{O}(h^5)$. At the same time, however, the dimension of V_h is significantly reduced by a factor of 9:6, and this more than compensates for the loss of three quintic terms.

The set of degrees of freedom is

$$\Sigma_K = \left\{ p(A_i),\ \frac{\partial}{\partial x_1}p(A_i),\ \frac{\partial^2}{\partial x_1^2}p(A_i),\ \frac{\partial}{\partial x_2}p(A_i),\ \frac{\partial^2}{\partial x_2^2}p(A_i),\ \frac{\partial^2}{\partial x_1 \partial x_2}p(A_i) \,\middle|\, i = 1,2,3 \right\}.$$

Note that P_4 is a subspace of $\{v|_K \mid v \in V_h\}$.

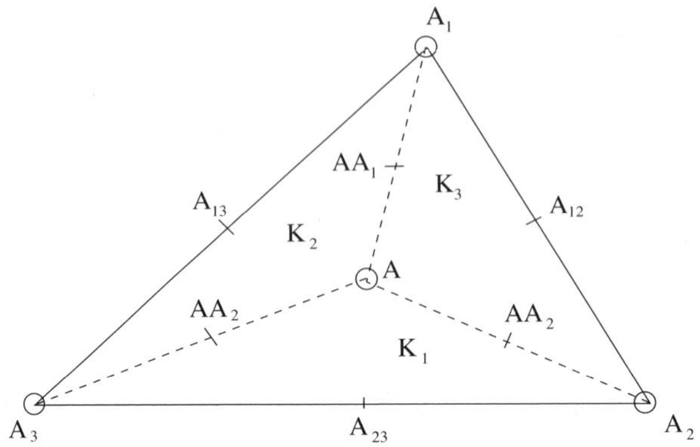

Figure 5.19: Nodes of an HCT triangle.

(FE_8) The HCT (Hsieh–Clough–Tocher) triangles (Hermite elements) (see Fig. 5.19)

Let K be a triangle with vertices A_1, A_2 and A_3. Let A be an interior point subdividing the triangle into three triangles K_1, K_2 and K_3. We choose piecewise polynomials

$$p \in C^1(K), \;\; p|_{K_i} \in P_3, \;\; \text{for } i = 1,2,3.$$

Obviously these polynomials contain P_3 as a subset.
We have to determine three polynomials $p_i = p|_{K_i}$, each of degree $\leqslant 3$. We need to determine a total of 30 coefficients. The following 30 conditions are used.

(i) $p_k(A_i)$, $(\partial/\partial x_1)p_k(A_i)$, $(\partial/\partial x_2)p_k(A_i)$; $p_k(A_j), (\partial/\partial x_1)p_k(A_j)$, $(\partial/\partial x_2)p_k(A_j); (\partial/\partial n)p_k(A_{ij})$, on each triangle K_k for $k \neq i, k \neq j$. Thus 7 conditions on each K_k are given and $3 \times 7 = 21$ conditions are obtained.

(ii) $p_1(A) = p_2(A) = p_3(A)$ gives 2 conditions.

(iii) $(\partial/\partial x_i)p_1(A) = (\partial/\partial x_i)p_2(A) = (\partial/\partial x_i)p_3(A)$ for $i = 1,2$ gives 4 more conditions.

(iv) $\partial p_1/\partial n = \partial p_2/\partial n$ along $\overline{AA_3}$ and two other similar conditions give 3 conditions.

Thus, we have 30 conditions to determine 30 coefficients.
In the foregoing discussions, we have covered many useful triangular elements. Now, let us study *rectangular elements*.
We first define

$$Q_k = \left\{ p(x_1, \ldots, x_N) \;\middle|\; p(x_1, \ldots, x_N) = \sum_{\substack{0 \leqslant \alpha_i \leqslant k \\ 1 \leqslant i \leqslant N}} a_{\alpha_1, \alpha_2, \ldots, \alpha_n} x_1^{\alpha_1} x_2^{\alpha_2} \cdots x_N^{\alpha_N} \right\} \tag{5.74}$$

to be the space of N-multivariate polynomials. It is easy to see that

$$\dim Q_k = (k+1)^N,$$

$$P_k \subset Q_k \subset P_{Nk}.$$

Let K be the unit hypercube in $\mathbb{R}^N$:

$$K = [0,1]^N = \{(x_1,\ldots,x_N) \mid 0 \leqslant x_i \leqslant 1,\ i = 1,2,\ldots,N\}.$$

Let Ξ_k be a subset of K defined by

$$\Xi_k = \left\{ x = \left(\frac{i_1}{k}, \frac{i_2}{k}, \ldots, \frac{i_N}{k}\right) \in \mathbb{R}^N \;\middle|\; i_j \in \{0, 1, \ldots, k\},\ 1 \leqslant j \leqslant N \right\}.$$

Then, we have

$$p = \sum_{\substack{0 \leqslant i_j \leqslant k \\ 1 \leqslant j \leqslant N}} \left[\prod_{j=1}^{N} \left(\prod_{\substack{i'_j = 0 \\ i'_j \neq i_j}}^{k} \frac{kx_j - i'_j}{i_j - i'_j} \right) \right] p\left(\frac{i_1}{k}, \frac{i_2}{k}, \ldots, \frac{i_N}{k}\right) \quad \forall p \in Q_k.$$

(FE_9) C^0-bilinear rectangles, C^0-biquadratic rectangles, C^0-bicubic rectangles, etc. (Lagrangian elements) (see Figs. 5.20–F5.22)

Figure 5.20: Nodes of a bilinear rectangle.

Let $N = 2$. Then the following sets of degrees of freedom determine C^0 bilinear, biquadratic and bicubic elements.

(*Bilinear* Q_1) $\Sigma_K = \{p(A_i) \mid 1 \leqslant i \leqslant 4\}$; see Fig. 5.20.

$$P(K) = Q_1, \quad \dim P(K) = 4.$$

(*Biquadratic* Q_2) $\Sigma_K = \{p(A_i) \mid 1 \leqslant i \leqslant 9\}$; see Fig. 5.21.

$$P(K) = Q_2, \quad \dim P(K) = 9.$$

(*Bicubic* Q_3) $\Sigma_K = \{p(A_i) \mid 1 \leqslant i \leqslant 16\}$; see Fig. 5.22.

$$P(K) = Q_3, \quad \dim P(K) = 16.$$

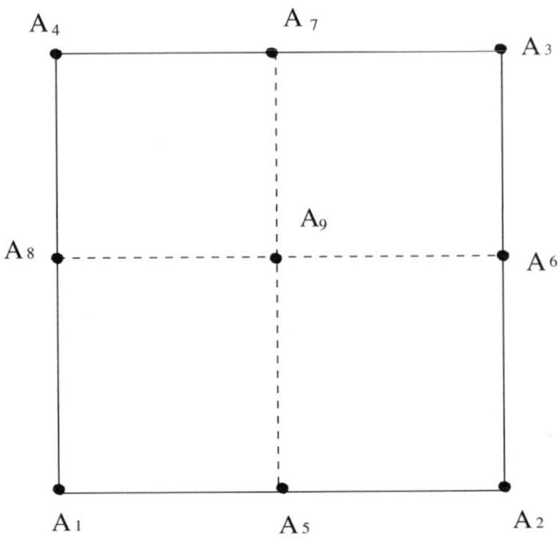

Figure 5.21: Nodes of a biquadratic rectangle.

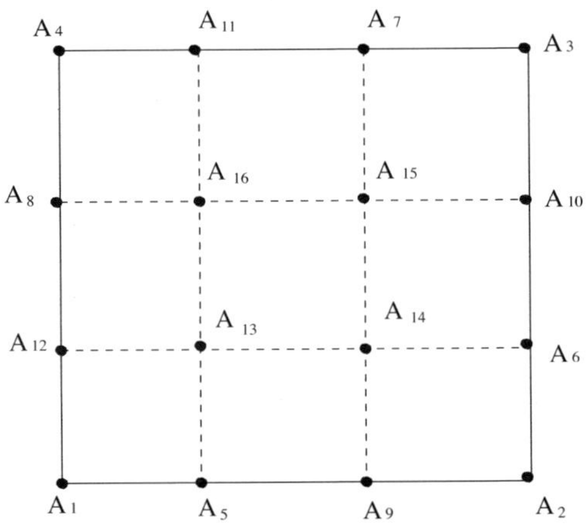

Figure 5.22: Nodes of a bicubic rectangle.

(FE_{10}) *The serendipity family (Lagrangian rectangular elements).*

The central node A_9 of the biquadratic rectangular element in Fig. 5.21 is not convenient in coordinate transformations. It is quite distinct from other nodes since it does not lie on the edges of K.

We wish to eliminate this node. We use the following idea.

For any $p(x_1,x_2) \in P_2$, we have

$$4p(A_9) + \sum_{i=1}^{4} p(A_i) - 2\sum_{i=5}^{8} p(A_i) = 0 \tag{5.75}$$

This can be verified for the following unit rectangle (Fig. 5.23):

$$p(x_1,x_2) = \alpha_0 + \alpha_1 x_1 + \alpha_2 x_2 + \alpha_3 x_1^2 + \alpha_4 x_1 x_2 + \alpha_5 x_2^2;$$

$$[4] \times p(A_9) = p\left(\frac{1}{2},\frac{1}{2}\right) = \alpha_0 + \alpha_1 \frac{1}{2} + \alpha_2 \frac{1}{2} + \alpha_3 \left(\frac{1}{2}\right)^2 + \alpha_4 \left(\frac{1}{2}\right)^2 + \alpha_5 \left(\frac{1}{2}\right)^2;$$

$$[1] \times \begin{cases} p(A_1) \\ = p(0,0) = \alpha_0, \\ p(A_2) \\ = p(1,0) = \alpha_0 + \alpha_1 \qquad + \alpha_3, \\ p(A_3) \\ = p(1,1) = \alpha_0 + \alpha_1 + \alpha_2 + \alpha_3 + \alpha_4 + \alpha_5, \\ p(A_4) \\ = p(0,1) = \alpha_0 \qquad + \alpha_2 \qquad + \alpha_5; \end{cases}$$

$$[-2] \times \begin{cases} p(A_5) \\ = p(\tfrac{1}{2},0) = \alpha_0 + \alpha_1(\tfrac{1}{2}) \qquad + \alpha_3(\tfrac{1}{2})^2, \\ p(A_6) \\ = p(1,\tfrac{1}{2}) = \alpha_0 + \alpha_1 \quad + \alpha_2(\tfrac{1}{2}) + \alpha_3 \quad + \alpha_4(\tfrac{1}{2}) + \alpha_5(\tfrac{1}{2})^2, \\ p(A_7) \\ = p(\tfrac{1}{2},1) = \alpha_0 + \alpha_1(\tfrac{1}{2}) + \alpha_2 \quad + \alpha_3(\tfrac{1}{2})^2 + \alpha_4(\tfrac{1}{2}) + \alpha_5 1, \\ p(A_8) \\ = p(0,\tfrac{1}{2}) = \alpha_0 \qquad + \alpha_2(\tfrac{1}{2}) \qquad + \alpha_5(\tfrac{1}{2})^2; \end{cases}$$

$$+)(\text{summing the above}) = 0.$$

So (5.75) is verified.

Now define a subspace Q_2' of Q_2 by

$$Q_2' = \left\{ p \in Q_2 \;\middle|\; 4p(A_9) + \sum_{i=1}^{4} p(A_i) - 2\sum_{i=5}^{8} p(A_i) = 0 \right\}.$$

Then $P_2 \subset Q_2'$, and every polynomial $p \in Q_2'$ can be determined from the following set of degrees of freedom:

$$\Sigma_K = \{p(A_i) \mid 1 \leqslant i \leqslant 8\}.$$

Q_2' is called the *serendipity biquadratic family*.

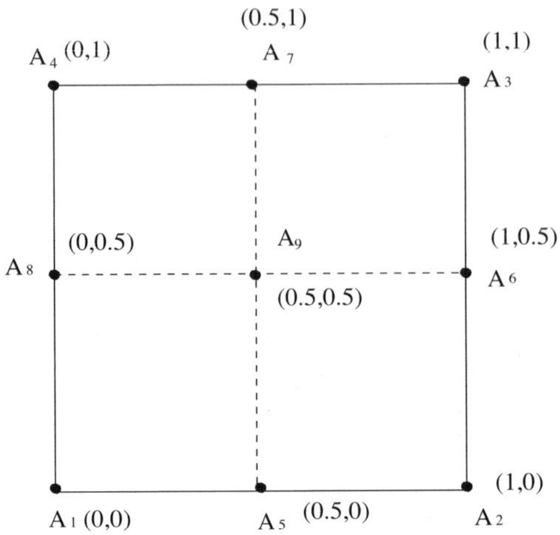

Figure 5.23: Eliminating the central node of a C^0-biquadratic rectangle.

Next, we consider the elimination of the interior nodes A_{13}, A_{14}, A_{15} and A_{16} in Fig. 5.22. It can be shown that for all $p \in P_3$, the following identities hold:

$$\Lambda_1(p) \equiv 9p(A_{13}) + 4p(A_1) + 2p(A_2) + p(A_3) + 2p(A_4) - 6p(A_5) \\ -3p(A_6) - 3p(A_{11}) - 6p(A_{12}) = 0,$$

$$\Lambda_2(p) \equiv 9p(A_{14}) + 4p(A_2) + 2p(A_3) + p(A_4) + 2p(A_1) - 6p(A_6) \\ -3p(A_7) - 3p(A_{12}) - 6p(A_9) = 0,$$

$$\Lambda_3(p) \equiv 9p(A_{15}) + 4P(A_3) + 2p(A_4) + p(A_1) + 2p(A_2) - 6p(A_7) \\ -3p(A_8) - 3p(A_9) - 6p(A_{10}) = 0,$$

$$\Lambda_4(p) \equiv 9p(A_{16}) + 4p(A_4) + 2p(A_1) + p(A_2) + 2p(A_3) - 6p(A_8) \\ -3p(A_5) - 3p(A_{10}) - 6p(A_{11}) = 0.$$

Note that Λ_2, Λ_3 and Λ_4 are obtainable by successive cyclic permutations

$$\begin{pmatrix} 1\,2\,3\,4 \\ 2\,3\,4\,1 \end{pmatrix}, \quad \begin{pmatrix} 5\,6\,7\,8 \\ 6\,7\,8\,5 \end{pmatrix}, \quad \begin{pmatrix} 9 & 10 & 11 & 12 \\ 10 & 11 & 12 & 9 \end{pmatrix}, \quad \begin{pmatrix} 13 & 14 & 15 & 16 \\ 14 & 15 & 16 & 13 \end{pmatrix}.$$

We define

$$Q_3' = \{p \in Q_3 \mid \Lambda_i(p) = 0 \text{ for } i = 1,2,3,4\}.$$

Then any $p \in Q_3'$ is uniquely determined by the following set of degrees of freedom:

$$\Sigma_K = \{p(A_i) \mid 1 \leqslant i \leqslant 12\}.$$

Q_3' is called the *serendipity bicubic family*.

This idea can be easily generalized to dimension three.

(FE_{11}) *The Bogner–Fox–Schmit rectangles (Hermite elements) (see Fig. 5.24).*

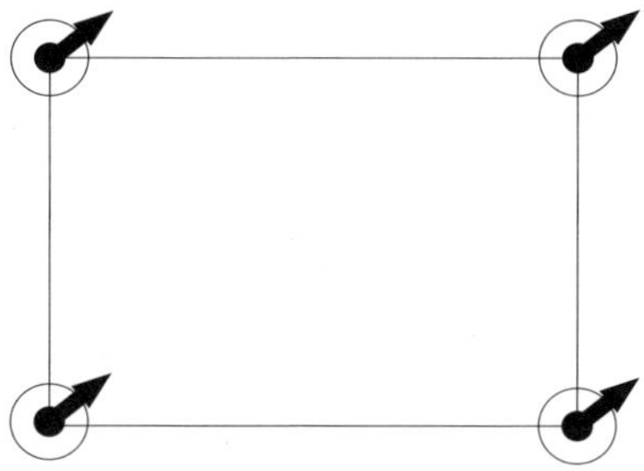

Figure 5.24: Nodes of a Bogner-Fox-Schmit rectangle.

Let $P(K) = Q_3$ and let the set of degrees of freedom be

$$\Sigma_K = \left\{ p(A_i), \frac{\partial p}{\partial x_1}(A_i), \frac{\partial p}{\partial x_2}(A_i), \frac{\partial^2 p}{\partial x_1 \partial x_2}(A_i) \middle| 1 \leqslant i \leqslant 4 \right\}.$$

Then Σ_K uniquely determines a polynomial $p \in Q_3$ (a double-dotted arrow indicates that the mixed second derivative is a degree of freedom). In this case $V_h \subset C^1(\overline{\Omega})$.

5.8 Finite elements in three dimensions

Three-dimensional elements are generally much more complicated than two-dimensional ones. In this section we introduce some commonly used types that have simple representations.

(FE_{12}) *C^0-linear, quadratic or cubic Lagrangian tetrahedral elements (see Fig. 5.25).*

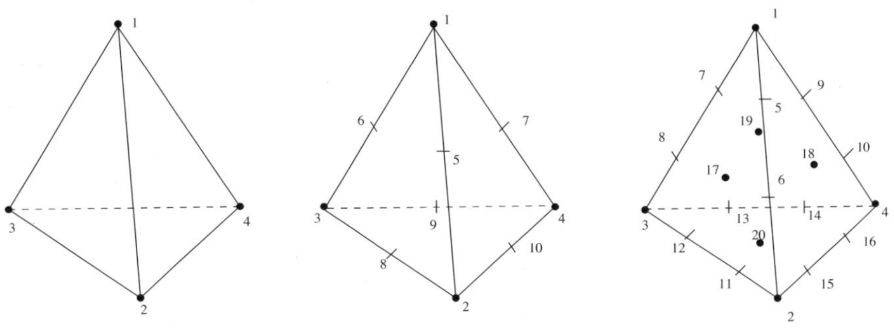

Figure 5.25: Nodes for C^0-linear, quadratic and cubic tetrahedral elements.

These elements have been obtained in (5.67), (5.69) and (5.72) respectively, using $N = 3$. (Recall that a tetrahedron is a 3-simplex.)

We just note that a complete polynomial of degree n in 3 variables has $C\begin{bmatrix} n+3 \\ 3 \end{bmatrix} = (n+1)(n+2)(n+3)/6$ independent coefficients, i.e., $C\begin{bmatrix} n+3 \\ 3 \end{bmatrix}$ degrees of freedom. Thus the correspondence between the degree of the polynomial and the number of degrees of freedom is

$$
\begin{aligned}
&n = 1 \text{ (linear)} && 4, \\
&n = 2 \text{ (quadratic)} && 10, \\
&n = 3 \text{ (cubic)} && 20, \\
&n = 4 \text{ (quartic)} && 35, \text{etc.}
\end{aligned}
$$

The nodes corresponding to $n = 1, 2$ and 3 on a tetrahedron are indicated in Fig. 5.25.

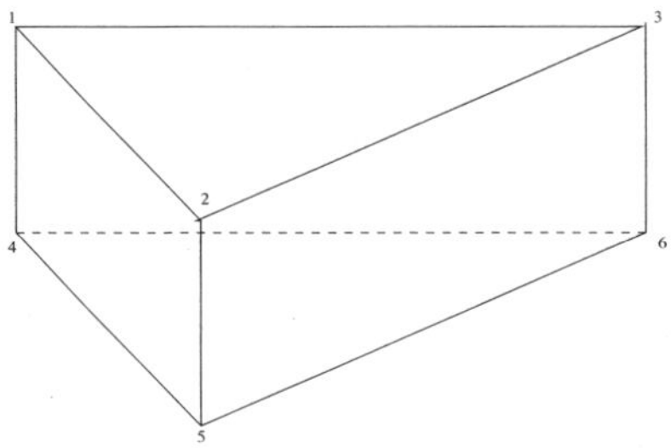

Figure 5.26: Nodes of a bilinear prismatic element.

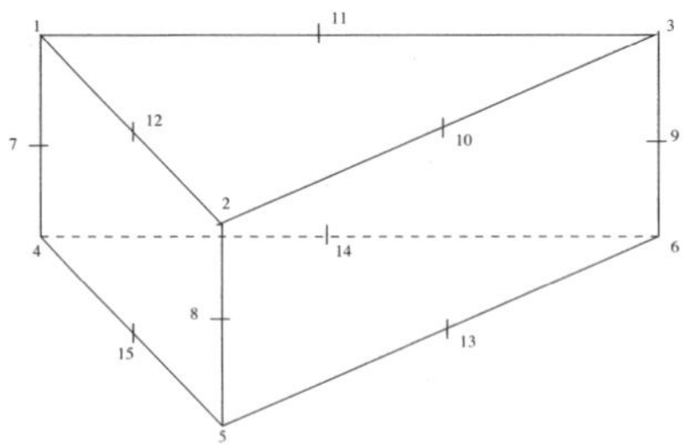

Figure 5.27: Nodes of a biquadratic prismatic element.

(FE_{13}) *C^0-bilinear, biquadratic and bicubic Lagrangian prismatic elements (see Figs. 5.26 and 5.27)*

For a right prism, we can establish a local coordinate system formed by the barycentric coordinates on the triangular facets and the vertical coordinate axis. Assume that the vertical axis is parallel to the x_3 (i.e., z) direction. Let the upper and lower facets be parallel to the x_1x_2 (i.e., xy) plane, with x_3 coordinates z_{i+1} and z_i respectively. On the upper and lower facets, the barycentric coordinates are $(\lambda_1, \lambda_2, \lambda_3)$. Transform the x_3 coordinate according to

$$\zeta = \frac{2}{z_{i+1} - z_i} x_3 - \frac{z_{i+1} + z_i}{z_{i+1} - z_i}.$$

Then every point in this prism has coordinate representation $(\lambda_1, \lambda_2, \lambda_3, \zeta)$. (Note that $\zeta = 1$ and -1 correspond respectively to the upper and lower facets.)

There are six nodal points on a *bilinear prismatic* element; see Fig. 5.26. The basis functions are

$$p_i(\lambda_1, \lambda_2, \lambda_3, \zeta) = \frac{1}{2}\lambda_{i_0}(1 + \zeta_i \zeta), \quad i = 1, 2, \ldots, 6,$$

where

$$\zeta_i = \begin{cases} 1, & i = 1, 2, 3, \\ -1, & i = 4, 5, 6, \end{cases} \tag{5.76}$$

$$i_0 = \begin{cases} i, & i = 1, 2, 3, \\ i - 3, & i = 4, 5, 6. \end{cases} \tag{5.77}$$

For a *biquadratic prismatic* element, there are 15 nodal points, with basis functions

$$p_i(\lambda_1, \lambda_2, \lambda_3, \zeta) = \frac{1}{2}\lambda_{i_0}(2\lambda_{i_0} - 1)(1 + \zeta_i \zeta) - \frac{1}{2}\lambda_{i_0}(1 - \zeta^2)$$

$$i = 1, 2, \ldots, 6, \text{corresponding to the six vertices } 1, 2, \ldots, 6,$$

$$\text{cf. (5.77) and (5.76) for } i_0 \text{ and } \zeta_i.$$

Corresponding to the midpoints of each side on the upper and lower facets, the basis functions are

$$p_i(\lambda_1, \lambda_2, \lambda_3, \zeta) = 2\lambda_m \lambda_l (1 + \zeta_i \zeta),$$

where

$$\zeta_i = \begin{cases} 1, & i = 10, 11, 12, \\ -1, & i = 13, 14, 15, \end{cases}$$

$$i_0 = \begin{cases} i - 9, & i = 10, 11, 12, \\ i - 12, & i = 13, 14, 15, \end{cases}$$

and permute (i_0, m, l) by $(1, 2, 3)$ symmetrically. See Fig. 5.27.

For the remaining three nodes, which are the midpoints of three vertical sides, we have

$$p_i(\lambda_1, \lambda_2, \lambda_3, \zeta) = \lambda_{i_0}(1 - \zeta^2), \quad i_0 = i - 6.$$

One can similarly construct bicubic prismatic elements.

(FE_{14}) C^0-trilinear, triquadratic and tricubic Lagrangian right hexagonal elements (see Fig. 5.28).

Any hexagon can be transformed into a cube by a linear mapping. Let the length of each side of the cube be 2. We establish coordinates (ξ, η, ζ) on the cube: $-1 \leqslant \xi, \eta, \zeta \leqslant 1$. For a *trilinear* right hexagonal element, there are eight nodal points, with basis functions linear in each variable:

$$p_i(\xi, \eta, \zeta) = \frac{1}{8}(1 + \xi_i\xi)(1 + \eta_i\eta)(1 + \zeta_i\zeta), \quad 1 \leqslant i \leqslant 8,$$

where (ξ_i, η_i, ζ_i) are the coordinates of the ith vertex; see Fig. 5.28(a).
For a *triquadratic* right hexagonal element, there are 20 nodal points, with basis functions quadratic in each variable:

(a) $p_i(\xi, \eta, \zeta) = \frac{1}{8}(1 + \xi_i\xi)(1 + \eta_i\eta)(1 + \zeta_i\zeta)(\xi_i\xi + \eta_i\eta + \zeta_i\zeta - 2)$, corresponding to $(\xi_i, \eta_i, \zeta_i) = (\pm 1, \pm 1, \pm 1)$, the coordinates of the ith vertex;
(b) $p_i(\xi, \eta, \zeta) = \frac{1}{4}(1 - \xi^2)(1 + \eta_i\eta)(1 + \zeta_i\zeta)$, corresponding to $(\xi_i, \eta_i, \zeta_i) = (0, \pm 1, \pm 1)$, the mid-points 9, 10, 11, 12 in Fig. 5.28(b). Then permute the variables ξ, η, ζ in (b), etc.

For a *tricubic* right hexagonal element, there are 32 nodal points, with basis functions cubic in each variable:

(a') $p_i(\xi, \eta, \zeta) = \frac{1}{64}(1 + \xi_i\xi)(1 + \eta_i\eta)(1 + \zeta_i\zeta) \cdot [9(\xi^2 + \eta^2 + \zeta^2) - 19]$, corresponding to $(\xi_i, \eta_i, \zeta_i) = (\pm 1, \pm 1, \pm 1)$, the coordinates of the ith vertex;
(b') $p_i(\xi, \eta, \zeta) = \frac{9}{16}(1 - \xi^2)(1 + 9\xi_i\xi)(1 + \eta_i\eta)(1 + \zeta_i\zeta)$, corresponding to nodes $(\xi_i, \eta_i, \zeta_i) = (\pm\frac{1}{3}, \pm 1, \pm 1)$, the points $9, 10, \ldots, 16$ in Fig. 5.28(c). Then permute the variables ξ, η, ζ in (b').

(FE_{15}) C^1-cubic Hermite tetrahedral elements.

Such elements have eight nodal points, consisting of the four vertices and the four centroids of the facets. The degrees of freedom are specified as follows:

4 degrees of freedom corresponding to function values at 4 vertices;
12 degrees of freedom corresponding to first-order derivatives in 3 axial directions at each of the 4 vertices;
4 degrees of freedom corresponding to function values at 4 centroids.

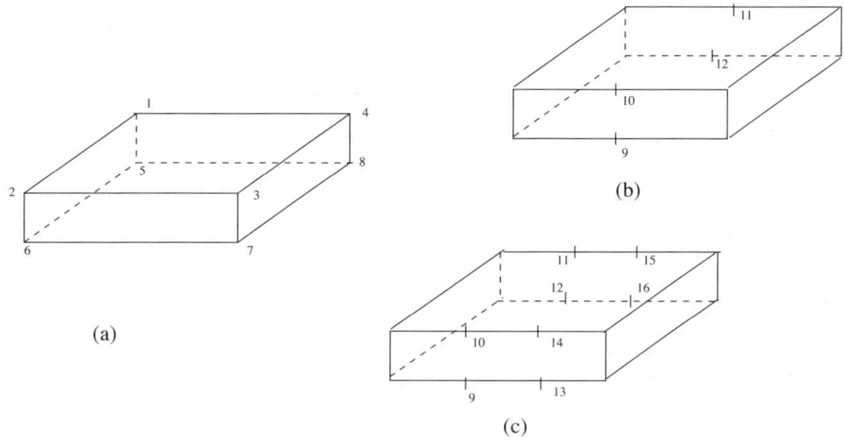

Figure 5.28: Nodes of trilinear, triquadratic and tricubic right hexagonal element.

Thus there are a total of 20 degrees of freedom, agreeing with the degrees of freedom of complete polynomials of degree 3 ($C\begin{bmatrix}6\\3\end{bmatrix}=20$).

Let $\alpha=(\alpha_1,\alpha_2,\alpha_3)$ be a multi-index as defined in § 2.1. We use the superscript α in $p^{(\alpha)}$ to denote $(\partial/\partial x_1)^{\alpha_1}(\partial/\partial x_2)^{\alpha_2}(\partial/\partial x_3)^{\alpha_3}p$. Then, the basis functions corresponding to the four vertices are

$$p_i^{(0,0,0)}(\lambda_1,\lambda_2,\lambda_3,\lambda_4)=3\lambda_i^2-2\lambda_i^3-7\sum_{1\leqslant i\leqslant j\leqslant k\leqslant 4}\lambda_i\lambda_j\lambda_k,\ i=1,2,3,4.$$

The basis functions corresponding to the four centroids are

$$p_{i+4}^{(0,0,0)}(\lambda_1,\lambda_2,\lambda_3,\lambda_4)=27\lambda_1\lambda_2\lambda_3\lambda_4/\lambda_i,\quad i=1,2,3,4.$$

Let the four vertices of the tetrahedron have coordinates (ξ_i,η_i,ζ_i), $i=1,2,3,4$. Define a determinant

$$\triangle=\begin{vmatrix}1&1&1&1\\ \xi_1&\xi_2&\xi_3&\xi_4\\ \eta_1&\eta_2&\eta_3&\eta_4\\ \zeta_1&\zeta_2&\zeta_3&\zeta_4\end{vmatrix}.$$

Let $\mathscr{C}_1$, $\mathscr{C}_2$, $\mathscr{C}_3$ and $\mathscr{C}_4$ be the cofactors corresponding to the i-th entry of the first row of $\triangle$. The entries in the matrix of $\mathscr{C}_j$ are denoted by $a_{mk}^{(j)}$ $(m,k=1,2,3,\ j=1,2,3,4)$. The cofactors of $a_{mk}^{(j)}$ in $\mathscr{C}_j$ are denoted by $A_{mk}^{(j)}$. Then the basis functions corresponding to the first-order derivatives at the vertices are

$$p_i^{(\alpha_i)}(\lambda_1,\lambda_2,\lambda_3,\lambda_4)=\frac{\triangle}{\mathscr{C}_i}\Big\{\lambda_i^2(A_{lj}^{(i)}\lambda_j+A_{lm}^{(i)}\lambda_m+A_{lk}^{(i)}\lambda_k)$$
$$-\lambda_i\Big[(A_{lj}^{(i)}+A_{lm}^{(i)})\lambda_j\lambda_m+(A_{lj}^{(i)}+A_{lk}^{(i)})\lambda_j\lambda_k+(A_{lm}^{(i)}+A_{lk}^{(i)})\lambda_m\lambda_k\Big]\Big\},$$

where

$$(\alpha_i) = (1,0,0),(0,1,0) \text{ or } (0,0,1)$$

corresponds respectively to $l = 1,2,3$, and the permutations of (i,j,m,k) are taken by $(1,2,3,4)$.

Finally, we note that if p is a piecewise polynomial such that $p \in C^{k-1}(\overline{\Omega})$, then $p \in H^k(\Omega)$, for any positive integer k. For example, Argyris triangular element (FE_6) are in $C^1(\overline{\Omega})$, hence they are in $H^2(\Omega)$. Thus they can be used to compute the plate problem, cf. Chapter 8, whose variational formulation requires $H^2(\Omega)$ regularity. (Nevertheless, in general, $C^{k-1}(\overline{\Omega}) \subseteq H^k(\Omega)$ is not true.)

5.9 Computation of element matrices

The variational equation

$$a(\widehat{u},v) = \theta(v) \quad \forall v \in V,$$

for many boundary value problems is approximated by finite elements. For example, for the Dirichlet problem in Example 5.1 for a two-dimensional domain Ω, one must compute

$$a(v,v) = \int_\Omega \nabla v \cdot \nabla v \, dx.$$

Let

$$v = \sum_i q_i \zeta_i, \qquad \zeta_i \in V_h.$$

Then

$$a(v,v) = q^{\mathrm{Tr}} K q, \qquad \text{a bilinear form in } q,$$

where the entries of the square matrix K are $K_{ij} = a(\zeta_i, \zeta_j)$. In practice, the calculation of these entries is reduced to the evaluation of $a(v,v)$ over each element e:

$$a_e(v,v) = q_e^{\mathrm{Tr}}(k_e)q_e. \tag{5.78}$$

One then assembles these element matrices k_e into the overall matrix K:

$$a(v,v) = q^{\mathrm{Tr}} K q = \sum_e q_e^{\mathrm{Tr}} k_e q_e. \tag{5.79}$$

Each entry k_{ij} in k_e is of the form

$$k_{ij} = \iint_e \psi_i(x_1,x_2)\psi_j(x_1,x_2)\, dx_1\, dx_2. \tag{5.80}$$

First, one should always find the connection matrix H between the nodal parameters q and the polynomial coefficients a:

$$a = Hq. \tag{5.81}$$

(Refer to Example 5.5 below.) Then one gets

$$a_e(v,v) = a^{\mathrm{Tr}} N \mathrm{a} = q^{\mathrm{Tr}} (H^{\mathrm{Tr}} NH) q,$$

for some matrix N whose entries are integrals of the form

$$\iint_e x_1^r x_2^s \, dx_1 \, dx_2.$$

We mention some methods below.

(1) *Use some local coordinate system* (due to Cowper, Kosko, Lindberg and Olson [57]). In the local system as shown in Fig. 5.29, the convenient formula

$$\iint_e \xi^r \eta^s \, d\xi \, d\eta = c^{s+1} [a^{r+1} - (-b)^{r+1}] \frac{r!s!}{(r+s+2)!} \tag{5.82}$$

can be used.

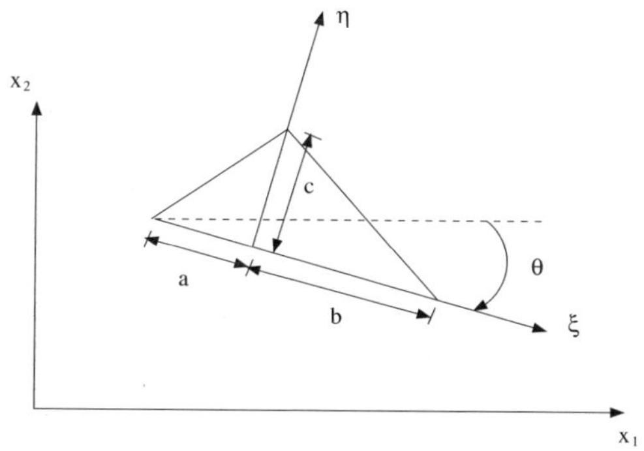

Figure 5.29: A local coordinate system for the integration formula (5.82).

(2) *Use barycentric coordinates* The barycentric coordinate system $(\zeta_1, \zeta_2, \zeta_3)$ on the triangle satisfies the transformation

$$\begin{bmatrix} x_1 \\ x_2 \\ 1 \end{bmatrix} = \begin{bmatrix} \alpha_1 & \alpha_2 & \alpha_3 \\ \beta_1 & \beta_2 & \beta_3 \\ 1 & 1 & 1 \end{bmatrix} \begin{bmatrix} \zeta_1 \\ \zeta_2 \\ \zeta_3 \end{bmatrix} = A \begin{bmatrix} \zeta_1 \\ \zeta_2 \\ \zeta_3 \end{bmatrix}; \text{ see Fig. 5.30.}$$

Thus

$$\iint_e x_1^r x_2^s \, dx_2 \, dx_2 = \iint_e (\alpha_1 \zeta_1 + \alpha_2 \zeta_2 + \alpha_3 \zeta_3)^r (\beta_1 \zeta_1 + \beta_2 \zeta_2 + \beta_3 \zeta_3)^s \, dx_1 \, dx_2. \tag{5.83}$$

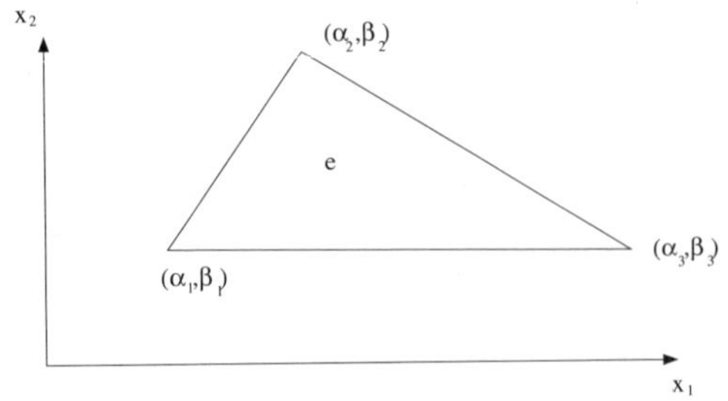

Figure 5.30: Barycentric coordinates for the integration formula (5.83).

One expands the above and uses

$$\iint_e \zeta_1^m \zeta_2^n \zeta_3^p \, dx_1 \, dx_2 = \frac{m!n!p!}{(m+n+p+2)!} \det A \tag{5.84}$$

to get the value of (5.83).

Example 5.5. Assume that we use a system of $S^{4,2}$ splines (cf. (5.59) and (5.60)). On a typical subinterval $[0,h]$, there are four basis functions as shown in Fig. 5.31.

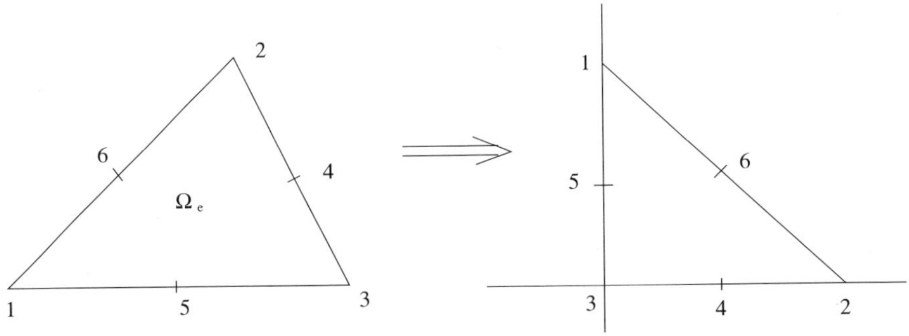

Figure 5.31: Four basis functions on a subinterval $(0,h)$.

If we write

$$\begin{aligned} u_h(x) &= q_0\phi_1(x) + q_0'\phi_2(x) + q_1\phi_3(x) + q_1'\phi_4(x), \quad 0 \leqslant x \leqslant h, \\ &= a_0 + a_1 x + a_2 x^2 + a_3 x^3, \end{aligned}$$

where

$$q_0 = u(0), \; q'(0) = u'(0), \; q_1 = u(h), \; q_1' = u'(h),$$

then (a_0, a_1, a_2, a_3) is related to (q_0, q_0', q_1, q_1') through

$$\begin{bmatrix} a_0 \\ a_1 \\ a_2 \\ a_3 \end{bmatrix} = \begin{bmatrix} 1 & 0 & 0 & 0 \\ 0 & 1 & 0 & 0 \\ -3/h^2 & -2/h & 3/h^2 & -1/h \\ 2/h^2 & 1/h^2 & -2/h^3 & 1/h^2 \end{bmatrix} \begin{bmatrix} q_0 \\ q_0' \\ q_1 \\ q_1' \end{bmatrix} \equiv H \begin{bmatrix} q_0 \\ q_0' \\ q_1 \\ q_1' \end{bmatrix},$$

which is the connection relation (5.81). □

5.10 Curved boundary and isoparametric transformations

Consider the variational equation: Find $\widehat{u} \in V$ such that

$$a(\widehat{u}, v) = \theta(v) \quad \forall v \in V,$$

derived from a boundary value problem. This problem is approximated by finite-elements. We say that the finite-element space V_h is *conforming* if

$$V_h \subset V. \tag{5.85}$$

Otherwise the space V_h is said to be *nonconforming*.

Example 5.6. For the Dirichlet problem in Example 5.1, $V = H_0^1(\Omega)$. For a finite-dimensional trial function space V_h consisting of piecewise polynomials to satisfy (5.85), we must have

$$V_h \subset C^0(\overline{\Omega}),$$

for V_h to be conforming. □

Example 5.7. Consider the variational formulation of the plate equation subject to the boundary conditions in (8.9), cf. Theorem 8.1 in Chapter 8. In this case, for a finite-element space consisting of piecewise polynomials to be conforming, one must have

$$V_h \subset C^1(\overline{\Omega}). \tag{5.86}$$

□

Consider, e.g., a region $\Omega \subset \mathbb{R}^3$ that is subdivided into finite elements. The elements are not of the proper shape: they may have one or more curved sides, or they may be nonrectangular quadrilaterals. By changing (x_1, x_2, x_3) variables into (ξ_1, ξ_2, ξ_3) variables, the elements can be given the desirable shape. The element matrices are then evaluated by integrations in the new variables, leading to a finite-element solution $u_h(\xi_1, \xi_2, \xi_3)$, which can then be transformed back to x_1, x_2 and x_3.

The transformation rule is, e.g.,

$$\iiint_{e_i} p(x_1,x_2,x_3)v_{x_1}^2(x_1,x_2,x_3)\,dx_1\,dx_2\,dx_3$$
$$=\iiint_E p(x(\xi))\left(v_{\xi_1}\xi_{1_{x_1}}+v_{\xi_2}\xi_{2_{x_1}}+v_{\xi_3}\xi_{3_{x_1}}\right)^2 J\,d\xi_1\,d\xi_2\,d\xi_3, \tag{5.87}$$

$$J = \text{Jacobian} = \frac{\partial(x_1,x_2,x_3)}{\partial(\xi_1,\xi_2,\xi_3)}. \tag{5.88}$$

In this coordinate change, one must be careful about the following.

(i) If the energy integral involves mth derivatives then the coordinate change must be of class C^{m-1} in order to *preserve the conforming property*, i.e., conforming elements in (ξ_1,ξ_2) will be conforming in (x_1,x_2).

(ii) For two elements e_i and e_j, a point common to e_i and e_j must not split into two separate points when $e_i \to E_i$ and $e_j \to E_j$.

(iii) The Jacobian (5.88) should be bounded away from 0, as $h \to 0$, forcing all angles to be bounded away from 0 and π.

(iv) The coordinate change and its derivatives must be conveniently computable.

The above requirements can be met by the use of *isoparametric technique*, which consists in choosing piecewise polynomials to define the coordinate transformation. If the transformation

$$x_1(\xi_1,\xi_2,\xi_3),\ x_2(\xi_1,\xi_2,\xi_3),\ x_3(\xi_1,\ \xi_2,\xi_3) \tag{5.89}$$

uses the same type of polynomial functions as the trial functions themselves, we say that it is *isoparametric*. If only a subset of lower-degree polynomials are used for (5.89), we say that the coordinate transformation is *subparametric*.

Volume integrals are transformed according to (5.87), for example. To evaluate the *boundary integrals* on ∂e_i, the boundary of the element e_i, we can pick two variables (η,μ) in the isoparametric coordinate system (ξ_1,ξ_2,ξ_3) to parametrize the surface ∂e_i, say

$$x_1 = x_1(\eta,\mu),\ x_2 = x_2(\eta,\mu),\ x_3 = x_3(\eta,\mu),\ (x_1,x_2,x_3) \in \partial e_i.$$

Then

$$\int_{\partial e_i}(\cdots)\,d\sigma = \int_S(\cdots)(EG-F^2)^{1/2}\,d\eta\,d\mu, \tag{5.90}$$

where S is a flat two-dimensional domain corresponding to ∂e_i, and E, F and G are defined as in (6.28) in Chapter 6.

When e_i is a two-dimensional element, ∂e_i is a curve, a line segment which sometimes can be parametrized by the abscissa s, for example; see Fig. 5.18. Then

$$\int_{\partial e_i} (\cdots)\, d\sigma = \int_L (\cdots) \left(\frac{dx_1}{ds}^2 + \frac{dx_2}{ds}^2 \right)^{1/2} ds, \tag{5.91}$$

where L ($\subset \mathbb{R}$) is an interval with endpoints corresponding to the curve ∂e_i.

Let us look at some examples.

Example 5.8. (Bilinear transformation from a square to a quadrilateral [173, p. 157]; see Fig. 5.32). The coordinate change between S and Q is given by

$$\left.\begin{aligned} x_1(\xi_1,\xi_2) &= \alpha_1 + (\beta_1 - \alpha_1)\xi_1 + (\delta_1 - \alpha_1)\xi_2 + (\gamma_1 - \delta_1 - \beta_1 + \alpha_1)\xi_1\xi_2, \\ x_2(\xi_1,\xi_2) &= \alpha_2 + (\beta_2 - \alpha_2)\xi_1 + (\delta_2 - \alpha_2)\xi_2 + (\gamma_2 - \delta_2 - \beta_2 + \alpha_2)\xi_1\xi_2. \end{aligned}\right\} \tag{5.92}$$

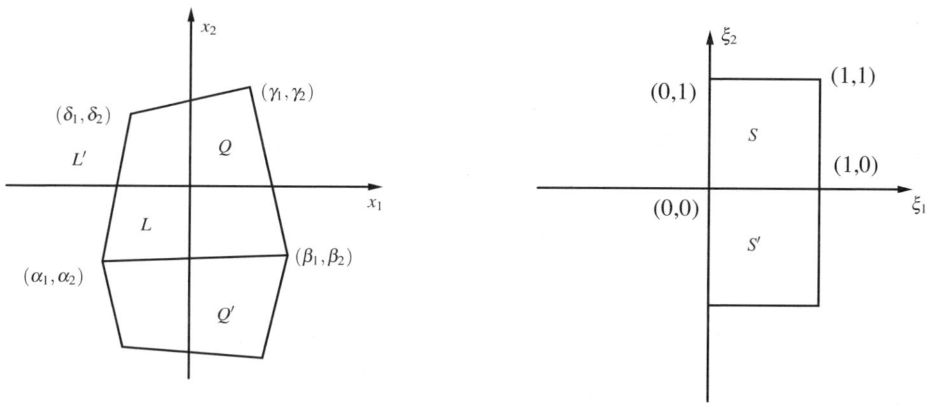

Figure 5.32: The mapping from S to Q in Example 5.8.

Each edge of the square S goes linearly into the corresponding edge of Q. This guarantees that any conforming element in ξ_1 and ξ_2—say a bilinear element for which the mapping is isoparametric—will still be conforming in x_1 and x_2.

We now check that the transformation (5.92) is invertible. This can be achieved by showing that the Jacobian is nonvanishing inside S:

$$J = x_{1_{\xi_1}} x_{2_{\xi_2}} - x_{1_{\xi_2}} x_{2_{\xi_1}} = \begin{vmatrix} \beta_1 - \alpha_1 + A\xi_2 & \delta_1 - \alpha_1 + A\xi_1 \\ \beta_2 - \alpha_2 + B\xi_2 & \delta_2 - \alpha_2 + B\xi_1 \end{vmatrix}$$

$$\begin{pmatrix} A = \gamma_1 - \delta_1 - \beta_1 + \alpha_1, \\ B = \gamma_2 - \delta_2 - \beta_2 + \alpha_2. \end{pmatrix}$$

At $(\xi_1,\xi_2)=(0,0)$, e.g.,

$$J=(\beta_1-\alpha_1)(\delta_2-\alpha_2)-(\beta_2-\alpha_2)(\delta_1-\alpha_1)$$

$$=\begin{vmatrix} 1 & 1 & 1 \\ \alpha_1 & \beta_1 & \delta_1 \\ \alpha_2 & \beta_2 & \delta_2 \end{vmatrix}$$

$\neq 0$ iff the angle formed by L and L' is not equal to 0 or π.

Hence, J is nonvanishing if and only if the quadrilateral Q is convex. □

Example 5.9 (Cubic isoparametric transformation for the triangle [134, pp. 93–94]). Let $\varphi_1(\xi_1,\xi_2)$, $\varphi_2(\xi_1,\xi_2),\ldots,\varphi_{10}(\xi_1,\xi_2)$ be the 10 basis trial functions in (ξ_1,ξ_2) for the triangle with 10 nodes in (ξ_1,ξ_2) plane; cf. (FE_3) in § 5.7. These functions $\varphi_i(\xi_1,\xi_2)$ are cubic polynomials. The cubic isoparametric transformation is to determine c_i, d_i, $1\leqslant i\leqslant 10$, as functions of the coordinates of P_i, $1\leqslant i\leqslant 10$, such that

$$x_1=\sum_{i=1}^{10} c_i\varphi_i(\xi_1,\xi_2), \qquad x_2=\sum_{i=1}^{10} d_i\varphi_i(\xi_1,\xi_2).$$

Look at Fig. 5.34. For example, when $\widehat{P_4P_7}$ and $\widehat{P_7P_1}$ are straight sides,

$$x_1=\xi_1+\frac{9}{2}\xi_1\xi_2(6\alpha_{10}-\alpha_2-\alpha_3-1)+\frac{27}{2}\xi_1^2\xi_2(\alpha_2-2\alpha_{10})+\frac{27}{2}\xi_1\xi_2^2\left(\alpha_3-2\alpha_{10}+\frac{1}{3}\right),$$

$$x_2=\xi_2+\frac{9}{2}\xi_1\xi_2(6\beta_{10}-\beta_2-\beta_3-1)+\frac{27}{2}\xi_1^2\xi_2(\beta_2-2\beta_{10})+\frac{27}{2}\xi_1\xi_2^2\left(\beta_3-2\beta_{10}+\frac{1}{3}\right).$$

Note here that the replacement curve passing through the points P_1, P_2, P_3 and P_4 is a *cubic curve*. □

Exercise 5.3 (Quadrilateral with one curved side [134, pp. 98–99]). Given a quadrilateral with one curved side as shown in Fig. 5.33, find isoparametric transformation (in two stages) to map it into a standard biquadratic rectangle (cf. (FE_9)) as shown in Fig. 5.33(c). Note the two stages:

(1) first, map Fig. 5.33(a);

(2) second, map Fig. 5.33(b) to (c). □

5.11 Accuracy of finite-element approximations

The material throughout the rest of the chapter is now standard in FEM. It can be found, e.g., in [49, 12, 129]. We collect some basic theorems and proofs here to make our study sufficiently self-contained.

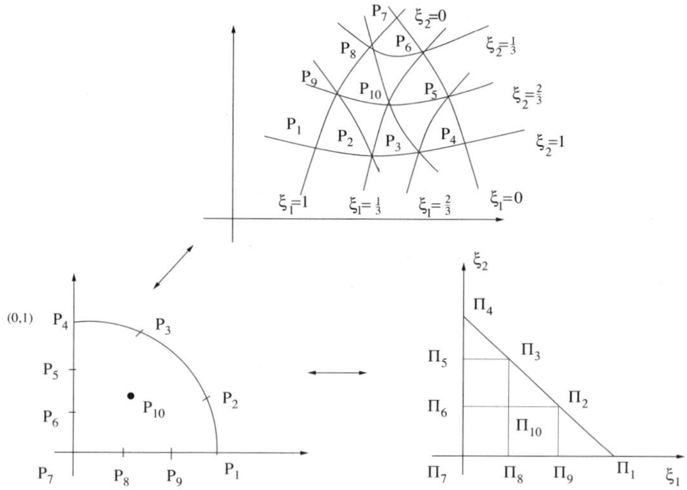

Figure 5.33: Mapping a quadrilateral with one curved side to a standard biquadratic rectangle in Exercise 5.3.

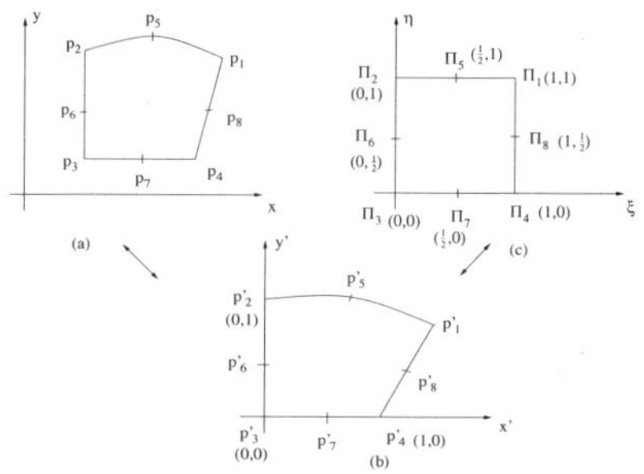

Figure 5.34: Mapping of domains in Example 5.9.

We first present some preliminary lemmas. Let Ω be a bounded convex domain in $\mathbb{R}^N$. Let k be a positive integer. Recall the definitions of the Sobolev space $W^{k,p}(\Omega)$, (2.3), the norm $\|\cdot\|_{k,p}$, (2.2) and the seminorm $|\cdot|_{k,p}$, (2.4). Here $p \in [1,\infty]$.
We first note that if $w \in W^{k,p}(\Omega)$ and $|w|_{k,p} = 0$ then w is a.e. equal to a polynomial of degree $\leqslant k-1$ on Ω. This is a known result in the theory of distributions.

Lemma 5.1. *Let $P_{k-1}(\Omega)$ be the space of all polynomials of degree $\leqslant k-1$ in N variables $x_1, x_2, \ldots, x_N$. Let $m = \dim P_{k-1}(\Omega)$. Assume that l_j, $1 \leqslant j \leqslant m$, are continuous linear functionals on $W^{k,p}(\Omega)$ such that $\{l_j\}_{j=1}^m$ is linearly independent on the subspace $P_{k-1}(\Omega)$. For*

$w \in W^{k,p}(\Omega)$, define

$$\|w\|' = \left[\sum_{j=1}^{m} |l_j(w)|^p + |w|_{k,p}^p\right]^{1/p}. \tag{5.93}$$

($p = \infty$ is interpreted by letting $p \to \infty$.)

Then $\|\cdot\|'$ defines an equivalent norm on $W^{k,p}(\Omega)$: there exist $c_1,\ c_2 > 0$ such that

$$c_1\|w\|' \leqslant \|w\|_{k,p} \leqslant c_2\|w\|', \quad \forall w \in W^{k,p}(\Omega). \tag{5.94}$$

Proof. Because $\{l_j\}_{j=1}^m$ are continuous linear functionals, the left inequality in (5.94) is obvious. We now prove the right inequality. Assume that the right inequality does not hold. Then there exists a sequence $\{w_j \in W^{k,p}(\Omega) \mid j = 1, 2, \ldots\}$ such that

$$\|w_j\|_{k,p} = 1, \tag{5.95}$$

$$\left[\sum_{i=1}^{m} |l_i(w_j)|^p + |w_j|_{k,p}^p\right]^{1/p} \leqslant \frac{1}{j}. \tag{5.96}$$

By (5.95), $\{w_j\}$ is bounded in $W^{k,p}(\Omega)$. Since $W^{k,p}(\Omega)$ is compactly imbedded in $W^{k-1,p}(\Omega)$ (cf. Theorem 2.5, e.g., where $p = 2$), $\{w_j\}_{j=1}^\infty$ has a convergent subsequence (still denoted as $\{w_j\}$ for convenience) such that for any $\varepsilon > 0$, there exists $J_1 > 0$ such that

$$\|w_{j_1} - w_{j_2}\|_{k-1,p} \leqslant \frac{\varepsilon}{2} \quad \text{for all } j_1,\ j_2 \geqslant J_1. \tag{5.97}$$

Also, from (5.96), there exists $J_2 > 0$ such that for $j_1,\ j_2 \geqslant J_2$,

$$\left|w_{j_1} - w_{j_2}\right|_{k,p} \leqslant |w_{j_1}|_{k,p} + |w_{j_2}|_{k,p} \leqslant \frac{1}{j_1} + \frac{1}{j_2} < \varepsilon/2. \tag{5.98}$$

Thus, when $j_1, j_2 \geqslant \max(J_1, J_2)$ by (5.97) and (5.98),

$$\|w_{j1} - w_{j2}\|_{k,p} \leqslant \|w_{j1} - w_{j2}\|_{k-1,p} + |w_{j1} - w_{j2}|_{k,p} < \varepsilon.$$

Hence $\{w_j\}$ is a Cauchy sequence in $W^{k,p}(\Omega)$. It has a limit w such that

$$\lim_{j\to\infty} \|w_j - w\|_{k,p} = 0.$$

From the continuity of $\{l_i | 1 \leqslant i \leqslant m\}$ and the norm $\|\cdot\|_{k,p}$, by (5.95) and (5.96), we have

$$\|w\|_{k,p} = 1, \ \ |w|_{k,p} = 0, \ \ l_i(w) = 0, \quad 1 \leqslant i \leqslant m. \tag{5.99}$$

Therefore $D^\alpha w = 0$ for all $\alpha : |\alpha| = k$, implying that w is a polynomial of degree $\leqslant k-1$. But from the third relation in (5.99) and the linear independence of $\{l_i | 1 \leqslant i \leqslant m\}$, we have $w = 0$ a.e. on Ω, contradicting $\|w\|_{k,p} = 1$. □

Let $P_k(\Omega)$ or, briefly P_k, denote the space of all polynomials of degree $\leqslant k$, k a positive integer on Ω. The quotient space (cf. Appendix (A.6)) $W^{k+1,p}(\Omega)/P_k$ is a space of equivalent classes $\widehat{v}$, where

$$\widehat{v} = \{w \in W^{k+1,p}(\Omega) \mid w - v \in P_k\},$$

with norm

$$\|\widehat{v}\|_{k+1,p/P_k} \equiv \inf_{\substack{p_k \in P_k \\ v \in \widehat{v}}} \|v + p_k\|_{k+1,p}. \tag{5.100}$$

We also define the seminorm

$$|\widehat{v}|_{k+1,p/P_k} \equiv \inf_{\substack{p_k \in P_k \\ v \in \widehat{v}}} |v + p_k|_{k+1,p}, \quad \widehat{v} \in W^{k+1,p}(\Omega)/P_k \tag{5.101}$$

Obviously,

$$|\widehat{v}|_{k+1,p/P_k} = |v|_{k+1,p} \tag{5.102}$$

and

$$|\widehat{v}|_{k+1,p/P_k} \leqslant \|\widehat{v}\|_{k+1,p/P_k}, \tag{5.103}$$

for all $\widehat{v} \in W^{k+1,p}(\Omega)/P_k$.

Lemma 5.2 ([63]). *In the quotient space $W^{k+1,p}(\Omega)/P_k$, the quotient norm $\|\cdot\|_{k+1,p/P_k}$ is equivalent to the seminorm $|\cdot|_{k+1,p}$, i.e., there exists $c = c(\Omega) > 0$ such that*

$$c^{-1}|v|_{k+1,p} \leqslant \|\widehat{v}\|_{k+1,p/P_k} \leqslant c|v|_{k+1,p}, \quad \forall v \in W^{k+1,p}(\Omega). \tag{5.104}$$

Proof. By (5.102), (5.103), we need only prove the right inequality in (5.104). Let P_k have dimension m and let $\{\phi_j \mid 1 \leqslant j \leqslant m\}$ be the basis of P_k. Define linear functionals $\{l_i \mid 1 \leqslant i \leqslant m\}$ on P_k such that

$$l_i(\phi_j) = \delta_{ij}, \quad 1 \leqslant i, j \leqslant m.$$

Extend each l_i to a continuous linear functional on $W^{k+1,p}(\Omega)$ by the Hahn–Banach theorem (see Appendix A.3(iv)). From Lemma 5.1, we have the equivalence of the two norms

$$\|w\|_{k+1,p}, \quad \left[\sum_{i=1}^{m} |l_i(w)|^p + |w|_{k+1,p}^p\right]^{1/p}.$$

For any $w \in W^{k+1,p}(\Omega)$, we can find $p_w \in P_k$ such that

$$l_i(p_w + w) = 0 \quad \text{for } i = 1, 2, \ldots, m.$$

(Indeed, $p_w = -\sum_{j=1}^{m} l_j(w)\phi_j$ will do.) Thus

$$\begin{aligned}
\|\widehat{w}\|_{k+1,p/P_k} &= \inf_{p_k\in P_k} \|w+p_k\|_{k+1,p} \\
&\leqslant c \inf_{p_k\in P_k} \left[\sum_{i=1}^{m} |l_i(w+p_k)|^p + |w+p_k|_{k+1,p}\right]^{1/p} \\
&\leqslant c \left[\sum_{i=1}^{m} |l_i(w+p_w)|^p + |w+p_w|_{k+1,p}\right]^{1/p} \\
&= c\,|w+p_w|_{k+1,p} \quad = \quad c|w|_{k+1,p} \\
&= c|\widehat{w}|_{k+1,p}.
\end{aligned}$$

The proof is complete. □

Now we are in a position to find error estimates in finite-element approximations. This is done through the *interpolation errors* $|v-\Pi_K v|_{m,p}$ and $\|v-\Pi_K v\|_{m,p}$, where Π_K is an interpolation operator on a finite element K.

In finite-element computations, we need to map an element into a "standard reference element". For example, for a triangular element Ω_e as shown in Fig. 5.35(a), we can map it onto $\widehat{\Omega}$ in (b), a 45° right triangle with base length 1, a standard element, such that nodal points on Ω_e and $\widehat{\Omega}$ correspond to each other. This can done through an affine transformation

$$x = F(\widehat{x}) = A\widehat{x} + b, \quad x \in \Omega_e, \quad \widehat{x} \in \widehat{\Omega}, \quad \Omega_e, \widehat{\Omega} \subset \mathbb{R}^N, \tag{5.105}$$

where A is an $N \times N$ constant matrix, and b is an $N \times 1$ (column) vector.

In general, let Ω and $\widehat{\Omega}$ be two open subsets of $\mathbb{R}^N$. We say that Ω and $\widehat{\Omega}$ are *affine-equivalent* if there exists an invertible affine mapping (5.105) such that $\Omega = F(\widehat{\Omega})$.

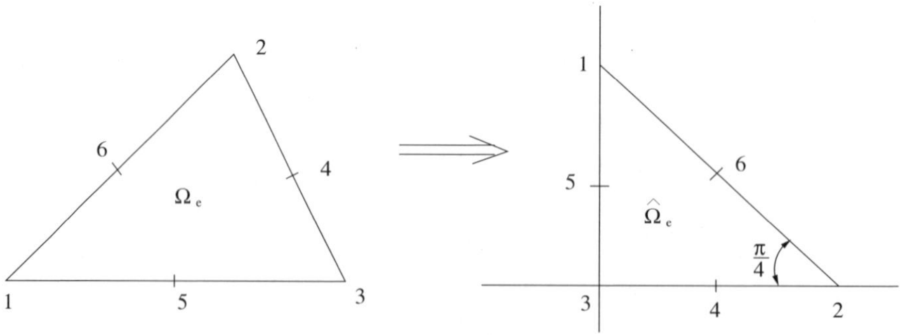

Figure 5.35: Two affine-equivalent triangular elements.

Lemma 5.3. *Let Ω_e and $\widehat{\Omega}$ be affine-equivalent through (5.105). Let*

$$H = d(\widehat{\Omega}) = \sup_{\widehat{x},\widehat{y}\in\widehat{\Omega}} |\widehat{x}-\widehat{y}|, \tag{5.106}$$

$$h_e = d(\Omega_e) = \sup_{x,y\in\Omega_e} |x-y|, \tag{5.107}$$

$$R = \sup\{d(S)|S \text{ is a sphere contained in } \widehat{\Omega}\}, \tag{5.108}$$

$$\rho_e = \sup\{d(S)|S \text{ is a sphere contained in } \Omega_e\}. \tag{5.109}$$

Let $||A||$ be the matrix norm of A in (5.105) induced by the Euclidean norm in $\mathbb{R}^N$. Then

$$||A|| = \sup_{|\widehat{x}|=R} |A\widehat{x}|/R, \tag{5.110}$$

$$||A|| \leqslant h_e/R, \tag{5.111}$$

$$||A^{-1}|| \leqslant H/\rho_e. \tag{5.112}$$

Proof. Let $z = R\widehat{x}/|\widehat{x}|$ for $\widehat{x}\neq 0$, $\widehat{x}\in\mathbb{R}^N$. Then

$$||A|| = \sup_{\widehat{x}\neq 0}\frac{|A\widehat{x}|}{|\widehat{x}|} = \sup_{|z|=R}\frac{|Az|}{R}.$$

Thus (5.110) is obtained. To prove (5.111), let $\widehat{x}_1, \widehat{x}_2 \in \overline{\widehat{\Omega}}$ satisfy $R = |\widehat{x}_1 - \widehat{x}_2|$. Then

$$\begin{aligned} ||A|| &= \frac{1}{R}\sup|A(\widehat{x}_1-\widehat{x}_2)| \\ &= \frac{1}{R}\sup|(A\widehat{x}_1+b)-(A\widehat{x}_2+b)| \\ &= \frac{1}{R}\sup|x_1-x_2| \leqslant \frac{h_e}{R}, \end{aligned}$$

so (5.111) is verified. The inequality (5.112) can be verified in the same manner by the inverse affine transformation F^{-1}. □

Lemma 5.4. *Let Ω and $\widehat{\Omega}$ be two affine-equivalent open subsets in $\mathbb{R}^N$ (through (5.105)). Let $u\in W^{m,p}(\Omega)$, where m is a nonnegative integer and $p\in[1,\infty]$. Then the composition function $\widehat{u} = u\circ F$ belongs to $W^{m,p}(\widehat{\Omega})$. Further, there exists $C>0$ independent of u such that*

$$|\widehat{u}|_{m,p,\widehat{\Omega}} \leqslant C||A||^m|\det A|^{-1/p}|u|_{m,p,\Omega} \quad \forall u\in W^{m,p}(\Omega), \tag{5.113}$$

$$|u|_{m,p,\Omega} \leqslant C||A^{-1}||^m|\det A|^{1/p}|\widehat{u}|_{m,p,\widehat{\Omega}} \quad \forall \widehat{u}\in W^{m,p}(\widehat{\Omega}). \tag{5.114}$$

Proof. Because $C^m(\overline{\Omega})$ is densely contained in $W^{m,p}(\Omega)$, we need only prove (5.113) for $u\in C^m(\overline{\Omega})$.

Given m directions $\vec{\xi}_1, \vec{\xi}_2, \dots, \vec{\xi}_m \in \mathbb{R}^N$, denote the mth-order derivatives in these directions by

$$D^m\widehat{u}(\widehat{x}) \cdot \left(\vec{\xi}_1, \vec{\xi}_2, \dots, \vec{\xi}_m\right) = \left[\prod_{j=1}^{m}\left(\xi_{j1}\frac{\partial}{\partial \widehat{x}_1} + \xi_{j2}\frac{\partial}{\partial \widehat{x}_2} + \cdots + \xi_{jN}\frac{\partial}{\partial \widehat{x}_N}\right)\right]\widehat{u}(\widehat{x}), \quad (5.115)$$

where ξ_{jk} is the kth component of the vector $\vec{\xi}_j$. Then for any multi-index α, $|\alpha| = m$, we have

$$D^\alpha\widehat{u}(\widehat{x}) = D^m\widehat{u}(\widehat{x}) \cdot (\vec{e}_{\alpha 1}, \vec{e}_{\alpha 2}, \dots, \vec{e}_{\alpha_m}),$$

where $\vec{e_{\alpha_j}}$, $1 \leqslant j \leqslant m$, are the α_jth unit vectors in $\mathbb{R}^N$. We have

$$|D^\alpha\widehat{u}(\widehat{x})| \leqslant \sup_{\substack{|\vec{\xi}_j| \leqslant 1 \\ 1 \leqslant j \leqslant m}} \left|D^m\widehat{u}(\widehat{x}) \cdot (\vec{\xi}_1, \vec{\xi}_2, \dots, \vec{\xi}_m)\right| \equiv \|D^m\widehat{u}(\widehat{x})\|, \ \widehat{x} \in \widehat{\Omega}.$$

Therefore

$$|\widehat{u}|_{m,p,\widehat{\Omega}} = \left[\int_{\widehat{\Omega}} \sum_{|\alpha|=m} |D^\alpha\widehat{u}(\widehat{x})|^p \, d\widehat{x}\right]^{1/p} \leqslant C_1 \left[\int_{\widehat{\Omega}} \|D^m\widehat{u}(\widehat{x})\|^p \, d\widehat{x}\right]^{1/p}, \quad (5.116)$$

where $C_1 > 0$ can be chosen as the total number of multi-indices α such that $|\alpha| = m$. By the chain rule,

$$D^m\widehat{u}(\widehat{x}) \cdot (\vec{\xi}_1, \dots, \vec{\xi}_m) = D^m u(x) \cdot (A\vec{\xi}_1, \dots, A\vec{\xi}_m),$$

so

$$\|D^m\widehat{u}(\widehat{x})\| \leqslant \|D^m u(x)\| \, \|A\|^m,$$

and

$$\int_{\widehat{\Omega}} \|D^m\widehat{u}(\widehat{x})\|^p \, d\widehat{x} \leqslant \|A\|^{mp} \int_{\widehat{\Omega}} \|D^m u(F(\widehat{x}))\|^p \, d\widehat{x}.$$

By the formula of change of variables in multiple integrals, we have

$$\int_{\widehat{\Omega}} \|D^m u(F(\widehat{x}))\|^p \, d\widehat{x} = |\det(A^{-1})| \int_{\Omega} \|D^m u(x)\|^p \, dx. \quad (5.117)$$

For any $x \in \Omega$, since there exists a constant $C_2 > 0$ such that

$$\|D^m u(x)\| \leqslant C_2 \max_{|\alpha|=m} |D^\alpha u(x)|, \quad (5.118)$$

from (5.116)–(5.118), we obtain (5.113) for $p \in [1, \infty)$. The case $p = \infty$ can be obtained by letting $p \to \infty$, for if $u \in W^{m,\infty}(\Omega)$ then $u \in W^{m,p}(\Omega)$ for all p, $1 \leqslant p < \infty$, and

$$|u|_{m,\infty,\Omega} = \lim_{p \to \infty} |u|_{m,p,\Omega}.$$

The inequality (5.114) can be proved in a similar fashion. □

The rest of the lemmas and theorems in this section can be found, e.g., in [49, Chap 3].

Lemma 5.5. *Let k, m be nonnegative integers, $p,q \in [1,\infty]$. Assume that the Sobolev space $W^{k+1,p}(\Omega)$ is continuously imbedded in the Sobolev space $W^{m,q}(\Omega)$. Let $\Pi \in \mathscr{L}(W^{k+1,p}(\Omega), W^{m,q}(\Omega))$ be a linear operator such that*

$$\Pi p = p, \quad \forall p \in P_k(\Omega). \tag{5.119}$$

Then there exists a constant $C > 0$ such that

$$|u - \Pi u|_{m,q} \leqslant C\|I - \Pi\|_{\mathscr{L}(W^{k+1,p}(\Omega),W^{m,q}(\Omega))}|u|_{k+1,p} \quad \forall u \in W^{k+1,p}(\Omega), \tag{5.120}$$

where in (5.120), I is the imbedding operator from $W^{k+1,p}(\Omega)$ into $W^{m,q}(\Omega)$.

Proof. By (5.119),

$$u - \Pi u = (I - \Pi)(u + p_k), \quad \forall u \in W^{k+1,p}(\Omega), \quad \forall p_k \in P_k(\Omega). \tag{5.121}$$

From the continuity of the imbedding and (5.121), we have

$$|u - \Pi u|_{m,q} \leqslant \|I - \Pi\|_{\mathscr{L}(W^{k+1,p}(\Omega),W^{m,q}(\Omega))} \inf_{p_k \in P_k(\Omega)} \|u + p_k\|_{k+1,p}$$

From Lemma 5.2, (5.120) follows. □

Let S_h be a finite-dimensional space of piecewise-continuous polynomials (i.e., a finite-element subspace). For any function $u\colon \Omega \subset \mathbb{R}^N \to \mathbb{R}$, sufficiently smooth, we can write the interpolant of u in S_h as

$$u_h = \sum_{\alpha,\triangle} u^{\triangle}\alpha\phi_{\triangle}^{(\alpha)},$$

where $\{\phi_{\triangle}^{(\alpha)}\}$ are the basis functions of S_h, and $\{u^{\triangle}\alpha\}$ are function values and the derivatives of u at nodal points. Define

$$u_h = \Pi_h u.$$

We call Π_h the *interpolation operator*. The interpolant $\Pi_h u$ restricted to a particular finite element Ω_e is denoted by

$$u_h\,|_{\Omega_e} = \Pi_h u\,|_{\Omega_e} = \Pi_e u.$$

Theorem 5.7. *Let k, m be nonnegative integers, $p,q \in [1,\infty]$. Let the imbedding*

$$W^{k+1,p}(\widehat{\Omega}) \subset W^{m,q}(\widehat{\Omega}) \quad \textit{be continuous.} \tag{5.122}$$

Assume that $\widehat{\Pi} \in \mathscr{L}(W^{k+1,p}(\widehat{\Omega}), W^{m,q}(\widehat{\Omega}))$ satisfies

$$\widehat{\Pi}\widehat{p} = \widehat{p} \quad \forall \widehat{p} \in P_k(\widehat{\Omega}). \tag{5.123}$$

Then

(1) *there exists $C > 0$, such that*

$$|\widehat{u}-\widehat{\Pi}\widehat{u}|_{m,q,\widehat{\Omega}} \leqslant C|\widehat{u}|_{k+1,p,\widehat{\Omega}}, \quad \forall \widehat{u} \in W^{k+1,p}(\widehat{\Omega}). \tag{5.124}$$

(2) *Let Ω be any open subset affine-equivalent to $\widehat{\Omega}$. Define a (projection) operator Π_Ω by*

$$(\Pi_\Omega u)^\wedge = \widehat{\Pi}\widehat{u} \quad \forall \widehat{u} \in W^{k+1,p}(\widehat{\Omega}), \tag{5.125}$$

where $u = \widehat{u} \circ F^{-1}$ through (5.105). Then there exists a constant $C > 0$ such that

$$|u-\Pi_\Omega u|_{m,q,\Omega} \leqslant C[\text{meas}(\Omega)]^{1/q-1/p} h^{k+1} \rho^{-m} |u|_{k+1,p,\Omega} \quad \forall u \in W^{k+1,p}(\Omega), \tag{5.126}$$

where h and ρ are defined by (5.107) and (5.109), with $h = h_e$ and $\rho = \rho_e$.

Proof. From (5.123) we have

$$\widehat{u}-\widehat{\Pi}\widehat{u} = (\widehat{I}-\widehat{\Pi})(\widehat{u}+\widehat{p}) \quad \forall \widehat{u} \in W^{k+1,p}(\widehat{\Omega}), \widehat{p} \in P_k(\widehat{\Omega}), \tag{5.127}$$

where $\widehat{I}$ is the imbedding operator in (5.122). We apply Lemma 5.5 to (5.127) and obtain (5.124).

By (5.125), we have

$$\widehat{u}-\widehat{\Pi}\widehat{u} = (u-\Pi_\Omega u)^\wedge.$$

By (5.124) and Lemma 5.4, we have

$$\begin{aligned}
|u-\Pi_\Omega u|_{m,q,\Omega} &\leqslant C\|A^{-1}\|^m |\det A|^{1/q} |\widehat{u}-\widehat{\Pi}\widehat{u}|_{m,q,\widehat{\Omega}} \\
&\leqslant C\|A^{-1}\|^m |\det A|^{1/q} |\widehat{u}|_{k+1,p,\widehat{\Omega}} \\
&\leqslant C\|A^{-1}\|^m |\det A|^{1/q} \|A\|^{k+1} |\det(A^{-1})|^{1/p} |u|_{k+1,p,\Omega}.
\end{aligned}$$

Now, using (5.111) and (5.112) in the above, we have

$$|u-\Pi_\Omega u|_{m,q,\Omega} \leqslant C|\det A|^{1/q-1/p} h^{k+1} \rho^{-m} |u|_{k+1,p,\Omega}.$$

Because $\det A = \text{meas}(\Omega)/\text{meas}(\widehat{\Omega})$, we obtain (5.126). □

Let Ω be a polygonal domain such that $\overline{\Omega}$ is a union of polygonal finite elements K such as those we have introduced previously:

$$\overline{\Omega} = \bigcup_{K \in \mathscr{T}_h} K.$$

For each K, let $\mathscr{P}_K$ be a space of spline functions defined on K. They correspond to those piecewise polynomials obtained earlier. Let Σ_K denote the total degrees of freedom of K. Equivalently, this means that Σ_K consists of the set of nodal points of K if K is a Lagrangian element, and Σ_K consists of the nodal points and the specified order of derivatives at nodal points if K is a Hermite element. Let

$$\dim \mathscr{P}_K = N.$$

We assume that $\sharp\Sigma_K = N$ and Σ_K is $\mathscr{P}_K$*-unisolvent*, namely, given any set of N numbers $\{\alpha_i \mid 1 \leqslant i \leqslant N\}$, there exists a unique $p \in \mathscr{P}_K$ such that

$p(x_i) = \alpha_i \;\; \forall x_i \in \Sigma_K$, if K is a Lagrangian element;
$p^{(k)}(x_j) = \alpha_{(jk)}$, for a node x_j, with the kth derivative evaluated, where $\{\alpha_{(jk)}\}$ is an (arbitrary) permutation of $\{\alpha_i\}$.

The triple $(K, \mathscr{P}_K, \Sigma_K)$ contains all the information about the finite element K. We also say that $(K, \mathscr{P}_K, \Sigma_K)$ is a finite element.

Theorem 5.8. *Let $(\widehat{\Omega}, \mathscr{P}_{\widehat{\Omega}}, \Sigma_{\widehat{\Omega}})$ be a reference finite element. Let s denote the highest order of the partial derivatives appearing in the definition of $\Sigma_{\widehat{\Omega}}$. Assume that for some integers $m \geqslant 0, k \geqslant 0$ and for some $p, q \in [1, \infty]$, the following inclusions hold:*

$$\left.\begin{array}{ll} W^{k+1,p}(\widehat{\Omega}) \subset C^s(\widehat{\Omega}), & \text{continuous imbedding,} \qquad (5.128) \\ W^{k+1,p}(\widehat{\Omega}) \subset W^{m,q}(\widehat{\Omega}), & \text{continuous imbedding,} \qquad (5.129) \\ P_k(\widehat{\Omega}) \subset \mathscr{P}_{\widehat{\Omega}} \subset W^{m,q}(\widehat{\Omega}). & \qquad (5.130) \end{array}\right\}$$

Then there exists a constant $C > 0$, depending on $(\widehat{\Omega}, \mathscr{P}_{\widehat{\Omega}}, \Sigma_{\widehat{\Omega}})$ only, such that for all affine-equivalent finite elements $(K, \mathscr{P}_K, \Sigma_K)$ and for all
$v \in W^{k+1,p}(K)$, we have

$$|v - \Pi_K v|_{m,q,K} \leqslant C(\text{meas}\,(\Omega))^{1/q-1/p} h_K^{k+1} \rho_K^{-m} |v|_{k+1,p,K}, \tag{5.131}$$

where $\Pi_K v$ denotes the $\mathscr{P}_K$-interpolant of v, and h_K and ρ_K are defined as in (5.107) and (5.109).

Proof. The inclusion $P_k(\widehat{\Omega}) \subset \mathscr{P}_{\widehat{\Omega}}$ and the $\mathscr{P}_{\widehat{\Omega}}$-unisolvency of $\Sigma_{\widehat{\Omega}}$ imply (how?) that the interpolation operator $\widehat{\Pi}$ satisfies

$$\widehat{\Pi}\widehat{p} = \widehat{p}, \quad \forall \widehat{p} \in P_k(\widehat{\Omega}).$$

(The domain of $\widehat{\Pi}$ is defined to be $C^s(\widehat{\Omega})$.) Let $\widehat{v} \in W^{k+1,p}(\widehat{\Omega})$. Since in most of our examples of finite elements, s is at most 2, we let $s = 2$ for definiteness. Then the $\mathscr{P}_{\widehat{\Omega}}$-interpolant of $\widehat{v}$ takes the form

$$\widehat{\Pi}\widehat{v} = \sum_i \widehat{v}(\widehat{a}_i)\widehat{p}_i^0 + \sum_{i,j} [D\widehat{v}(\widehat{a}_i^1)(\widehat{\xi}_{ij}^1)]\widehat{p}_{ij}^1 + \sum_{i,j,k} [D^2\widehat{v}(\widehat{a}_i^2)(\widehat{\xi}_{ij}^2, \widehat{\xi}_{ik}^2)]\widehat{p}_{ijk}^2.$$

(cf. (5.115) for directional derivatives $D\widehat{v}$ and $D^2\widehat{v}$).

The linear mapping $\widehat{\Pi}\colon\ W^{k+1,p}(\widehat{\Omega}) \to W^{m,q}(\widehat{\Omega})$ is continuous because of (5.130) and

$$\begin{aligned}
\|\widehat{\Pi}\widehat{v}\|_{m,q,\widehat{\Omega}} &= \sum_i |\widehat{v}(\widehat{a}_i^\circ)|\,\|\widehat{p}_i^0\|_{m,q,\widehat{\Omega}} + \sum_{i,j} |[D\widehat{v}(\widehat{a}_i^1)(\widehat{\xi}_{i,j}^1)]|\,\|\widehat{p}_{ij}^1\|_{m,q,\widehat{\Omega}} \\
&\quad + \sum_{i,j,k} |[D^2\widehat{v}(\widehat{a}_i^2)(\xi_{i,j}^2, \xi_{ik}^2)]|\,\|\widehat{p}_{ijk}^2\|_{m,q,\widehat{\Omega}} \\
&\leqslant C\|\widehat{v}\|_{c^2(\overline{\Omega})} \quad (C>0 \text{ depends on } \|\widehat{p}_i^0\|_{m,q,\widehat{\Omega}}, \|\widehat{p}_{ij}^1\|_{m,q,\widehat{\Omega}}, \\
&\quad \|\widehat{p}_{ijk}^2\|_{m,q,\widehat{\Omega}} \text{and } |\widehat{\xi}_{ij}^1|, |\widehat{\xi}_{ij}^2|, |\widehat{\xi}_{ik}^2|) \\
&\leqslant C\|\widehat{v}\|_{k+1,p,\widehat{\Omega}} \quad (\text{by (5.128) with } s=2).
\end{aligned}$$

Since the $\mathscr{P}_K$ and $\mathscr{P}_{\widehat{\Omega}}$-interpolation operators are related through

$$(\Pi_K v)^\wedge = \widehat{\Pi}\widehat{v} \quad \forall v \in D(\Pi_K) \ \ (\text{cf. (5.125)}),$$

we can now apply Theorem 5.7 and conclude (5.131). □

The dependence on the parameter ρ_K in the estimate (5.131) can be eliminated if the finite elements do not become "flat" in the limit. We say that a family of triangulations $\mathscr{T}_h$ with corresponding finite elements $(K, \mathscr{P}_K, \Sigma_K)$ on a polygonal domain $\overline{\Omega}$ is *regular* if

(i) there exists a constant $\sigma > 0$ such that

$$\frac{h_K}{\rho_K} \leqslant \sigma \quad \forall K \in \bigcup_h \mathscr{T}_h; \tag{5.132}$$

(ii) the diameters tend to zero, i.e.,

$$h = \max_{K \in \mathscr{T}_h} h_K \to 0. \tag{5.133}$$

The following important theorem gives us criteria for a $(k+1,m)$-system (cf. Definition 5.1).

Theorem 5.9. *Let $\mathscr{T}_h$ be a regular family of triangulations with corresponding finite elements $(K, \mathscr{P}_K, \Sigma_K)$ on a closed polygonal domain $\overline{\Omega}$. Assume that all the finite elements $(K, \mathscr{P}_K, \Sigma_K), K \in \bigcup_h \mathscr{T}_h$, are affine-equivalent to a fixed reference finite element $(\widehat{\Omega}, \mathscr{P}_{\widehat{\Omega}}, \Sigma_{\widehat{\Omega}})$. Furthermore, assume that*

$$P_k(\widehat{\Omega}) \subset \mathscr{P}_{\widehat{\Omega}} \subset H^l(\widehat{\Omega}), \tag{5.134}$$

$$H^{k+1}(\widehat{\Omega}) \subset C^s(\widehat{\Omega}) \text{ is a continuous imbedding,} \tag{5.135}$$

where s is the highest order of partial derivatives occurring in the definition of the set $\Sigma_{\widehat{\Omega}}$. *Then there exists a constant* $C > 0$ *independent of h such that for any* $v \in H^{k+1}(\Omega)$,

$$\left(\sum_{K \in \mathscr{T}_h} \|v - \Pi_h v\|^2_{H^m(K)} \right)^{1/2} \leqslant Ch^{k+1-m} |v|_{k+1,2,\Omega}, \tag{5.136}$$

$$0 \leqslant m \leqslant \min(k+1, l),$$

where Π_h *is the (global) interpolation operator satisfying*

$$\Pi_h v|_K = \Pi_K v \quad \text{on } K \in \mathscr{T}_h, \text{for every } K \text{ and } h.$$

Proof. Applying Theorem 5.8 with $p = q = 2$, and using (5.132), we obtain

$$|v - \Pi_K v|_{m,2,K} \leqslant Ch_K^{k+1-m} |v|_{k+1,2,K}.$$

For any integer $j : 0 \leqslant j < m$, we also have

$$|v - \Pi_K v|_{j,2,K} \leqslant Ch_K^{k+1-j} |v|_{k+1,2,K} \leqslant Ch_K^{k+1-m} |v|_{k+1,2,K}.$$

After summing j from 0 to m, we get

$$\|v - \Pi_K v\|_{H^m(K)} \leqslant Ch_K^{k+1-m} |v|_{k+1,2,K}.$$

Summing the above over all $K \in \mathscr{T}_h$, we obtain

$$\left(\sum_{K \in \mathscr{T}_h} \|v - \Pi_K v\|^2_{H^m(K)} \right)^{1/2} \leqslant Ch^{k+1-m} \left(\sum_{K \in \mathscr{T}_h} |v|^2_{k+1,2,K} \right)^{1/2}$$

$$= Ch^{k+1-m} |v|_{k+1,2,K} \tag{5.137}$$

for $0 \leqslant m \leqslant \min(k+1, l)$. Note that the LHS of (5.137) is just the LHS of (5.136). □

In the inequality (5.136), if $\Pi_h v \in H^m(\Omega)$ for some (particular) v then we actually have

$$\|v - \Pi_h v\|_{H^m(\Omega)} \leqslant Ch^{k+1-m} |v|_{k+1,2,\Omega}$$

for this particular v. However, for most of the finite elements we have defined earlier, $\Pi_h v \in H^m(\Omega)$ does not hold for $m > 1$ because of the smoothness of the combined (i.e., global) piecewise-polynomials across ∂K. When $m = 0$ and $m = 1$, we can adjoin piecewise polynomial basis functions $\mathscr{P}_K$ across boundaries ∂K *by continuity* and form a global finite-element space V_h with those functions as basis functions. (See Exercise 5.2, for example. This, a priori, forbids triangulations such as that shown in Fig. 5.36; namely, the intersection of two elements must consist of an entire edge of each element.) Then we can have $V_h \subset C^0(\overline{\Omega})$.

Theorem 5.10. *Under the same conditions as Theorem 5.9, assume further that* $l \geqslant 1$, $m = 0$ *or 1, and* V_h, *the global finite-element space on* $\overline{\Omega}$, *satisfies* $V_h \subset C^0(\overline{\Omega})$. *Then*

$$\|v - \Pi_h v\|_{H^m(\Omega)} \leqslant Ch^{k+1-m} |v|_{k+1,2,\Omega}, \quad 0 \leqslant m \leqslant \min(1, l) \tag{5.138}$$

for some $C > 0$ *independent of h, for all* $v \in H^m(\Omega)$.

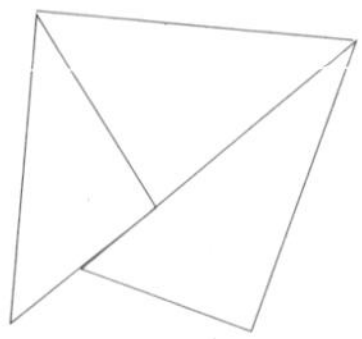

Figure 5.36: A nonadmissible triangulation.

Proof. Since $\mathscr{P}_{\widehat{\Omega}} \subset H^1(\widehat{\Omega})$ and $V_h \subset C^0(\overline{\Omega})$, we see from Theorem 2.1 in [49], p. 39] that $\Pi_h v \in V_h \subset H^1(\Omega)$. So (5.100) follows as a consequence of Theorem 5.9. □

For the various types of finite elements $(FE_1), \ldots, (FE_{15})$ we have given earlier in two and three dimensions, their accuracies as indicated by Theorem 5.10 are given in Fig. 5.37 (cf. also [49], p. 125]).

Remark 5.5. Theorems 5.9 and 5.10 are stated for polygonal domains Ω. For domains Ω with curved boundaries, we can define isoparametrically equivalent (i.e., instead of affine-equivalent) finite elements and prove analogous versions of Theorems 5.9 and 5.10. See [49, Chap. 4]. □

5.12 The Aubin–Nitsche lemma

By Theorem 5.10, we know that the best approximation

$$\|v - v_h\|_{H^1(\Omega)} = \inf_{\phi \in V_h} \|v - \phi\| = \mathscr{O}(h^k)$$

holds if the finite-element space V_h satisfies the given assumptions. Therefore $\|v - v_h\|_{H^0(\Omega)}$ is at least of accuracy $\mathscr{O}(h^k)$. By using the Aubin–Nitsche lemma (also called the Nitsche trick), we can actually show that $\|v - v_h\|_{H^0(\Omega)} = \mathscr{O}(h^{k+1})$.

Let V and H be two Hilbert spaces such that V is dense in H and the imbedding $V \subset H$ is continuous. Let V' be the dual of V and let the dual of H be identified with itself. Then we have

$$V \subset H \subset V' \quad \text{(cf. (4.130) and (4.131))}$$

where H is dense in V' and the imbedding $H \subset V'$ is also continuous. (We may take $V = H_0^1(\Omega), H = L^2(\Omega)$ and $V' = H^{-1}(\Omega)$, for example.) Let $a(\cdot,\cdot)$ be a bilinear form on V satisfying (5.1)–(5.4) (with H taken to be V). For any $g \in H$, the inner product $(g,v)_H$ for $v \in V$ defines a linear functional on $v \in V$. We denote this by

$$(g,v)_H = g(v), \quad \forall v \in V.$$

$\lVert V - \Pi_K V \rVert_{H^m(K)}$	$\mathscr{O}(h^{2-m})$ $0 \leqslant m \leqslant 2$ $k=1$	$\mathscr{O}(h^{3-m})$ $0 \leqslant m \leqslant 3$ $k=2$		$\mathscr{O}(h^{4-m})$ $0 \leqslant m \leqslant 4$ $k=3$			$\mathscr{O}(h^{5-m})$ $0 \leqslant m \leqslant 5$ $k=4$	$\mathscr{O}(h^{6-m})$ $0 \leqslant m \leqslant 6$ $k=5$
Regularity of v	$H^2(K)$	$H^3(K)$		$H^4(K)$			$H^5(K)$	$H^6(K)$
Upper bound on space dimension N to ensure $H^{k-1}(K) \subset C^{\ell}(K)$	$N \leqslant 3$ $(\ell = 0)$	$N \leqslant 5$ $(\ell = 0)$	$N \leqslant 3$ $(\ell = 1)$	$N \leqslant 7$ $(\ell = 0)$	$N \leqslant 5$ $(\ell = 1)$	$N \leqslant 3$ $(\ell = 2)$	$N \leqslant 7$ $(\ell = 1)$	$N \leqslant 9$ $(\ell = 1)$
Simplicial Finite Elements	FE_4 $FE_{12,linear}$	FE_4 $FE_{12,quadratic}$	FE_6	FE_6 $FE_{12,cubic}$	FE_4 FE_{15} FE_6		FE_7	FE_6
Rectangular and/or Prismatic Finite Elements	$FE_{9,bilinear}$ $FE_{13,bilinear}$ $FE_{14,trilinear}$	$FE_{9,biquadratic}$ $FE_{13,biquadratic}$ $FE_{14,triquadratic}$	$FE_{10,biquadratic}$ $FE_{10,triquadratic}$	$FE_{9,bicubic}$ $FE_{14,tricubic}$	$FE_{10,bicubic}$ $FE_{10,tricubic}$	FE_{11}		

Figure 5.37: Accuracies of various types of finite elements.

Let $M > 0$ be the constant such that

$$|a(u,v)| \leqslant M\|u\|_V\|v\|_V, \quad \forall\, u,v \in V.$$

Theorem 5.11 (The Aubin–Nitsche lemma). *Let $\theta \in V'$ and $V_h \subset V$, and let $u \in V$ and $u_h \in V_h$ be the solutions of the variational problems*

$$a(u,v) = \theta(v), \quad \forall v \in V, \tag{5.139}$$

$$a(u_h,v_h) = \theta(v_h), \quad \forall v_h \in V_h. \tag{5.140}$$

Then

$$\|u-u_h\|_H \leqslant M\|u-u_h\|_V \left[\sup_{g\in H}\left(\frac{1}{\|g\|_H}\inf_{v_h\in V_h}\|\phi_g - v_h\|_V\right)\right], \tag{5.141}$$

where for any $g \in H$, $\phi_g \in V$ is the unique solution to the variational problem

$$a(v,\phi_g) = (g,v)_H, \quad \forall v \in V. \tag{5.142}$$

Proof. First, note that

$$\|u-u_h\|_H = \sup_{g\in H}\frac{|(g,u-u_h)_H|}{\|g\|_H}. \tag{5.143}$$

Using the some ideas as in the proof of the Lax–Milgram theorem 5.2, we know that for any $g \in H$, there exists a unique $\phi_g \in V$ as a solution of (5.142). Since $u-u_h \in V$, we have, in particular,

$$a(u-u_h,\phi_g) = (g,u-u_h)_H. \tag{5.144}$$

But by (5.139) and (5.140),

$$a(u-u_h,v_h) = 0, \quad \forall v_h \in V_h. \tag{5.145}$$

From (5.144) and (5.145), we have

$$(g,u-u_h)_H = a(u-u_h,\phi_g - v_h), \quad \forall v_h \in V_h.$$

Therefore

$$|(g,u-u_h)_H| \leqslant M\|u-u_h\|\inf_{v_h\in V_h}\|\phi_g - v_h\|.$$

Using the above in (5.143), we have proved (5.141). □

5.13 Inverse inequalities

In this section, we discuss inverse inequalities, which are also important in the finite-element and boundary-element estimation of errors.

Theorem 5.12 ([49, pp. 140–142]). *Let $(\widehat{\Omega}, \mathscr{P}_{\widehat{\Omega}}, \Sigma_{\widehat{\Omega}})$ be a standard reference finite element in $\mathbb{R}^N$ and let $\{\mathscr{T}_h\}_h$ be a regular family of triangulations for a polygonal domain $\overline{\Omega} \subseteq \mathbb{R}^N$ such that $(K, \mathscr{P}_K, \Sigma_K)$ is affine-equivalent to $(\widehat{\Omega}, \mathscr{P}_{\widehat{\Omega}}, \Sigma_{\widehat{\Omega}})$ for each element $K \in \mathscr{T}_h$. Furthermore, let $\{\mathscr{T}_h\}_h$ satisfy (an inverse assumption)*

$$\frac{h}{h_K} \leqslant \nu \quad \text{for some } \nu > 0, \quad \forall\, K \in \cup_K \mathscr{T}_h, \tag{5.146}$$

where $h = \max_{K \in \mathscr{T}_h} h_K$, and $h_K/\rho_K \leqslant \sigma$ for all $K \in \mathscr{T}_h$ and all h, for some $\sigma > 0$, with ρ_K defined by (5.109). If there are two pairs (l,r) and (m,q) with (integers) $l, m \geqslant 0$, $l \leqslant m$ and $(r,q) \in [1,\infty]$ such that

$$\mathscr{P}_{\widehat{\Omega}} \subset W^{l,r}(\widehat{\Omega}) \cap W^{m,q}(\widehat{\Omega}),$$

then there exists a constant $C > 0$ such that

$$\left(\sum_{K \in \mathscr{T}_h} |v_h|^q_{m,q,K} \right)^{1/q} \leqslant \frac{C}{(h^N)^{\max(0,1/r-1/q)} h^{m-l}} \left(\sum_{K \in \mathscr{T}_h} |v_h|^r_{l,r,K} \right)^{1/r} \quad \forall v_h \in V_h, \tag{5.147}$$

for $q, r < \infty$, and with

$$\max_{K \in \mathscr{T}_h} |v_h|_{m,\infty,K} \ \textit{replacing} \ \left(\sum_{K \in \mathscr{T}_h} |v_h|^q_{m,q,K} \right)^{1/q} \ \textit{in (5.147) if } q = \infty,$$

$$\max_{K \in \mathscr{T}_h} |v_h|_{l,\infty,K} \ \textit{replacing} \ \left(\sum_{K \in \mathscr{T}_h} |v_h|^r_{l,r,K} \right)^{1/q} \ \textit{in (5.147) if } r = \infty.$$

Proof. Given a function $v_h \in V_h$ and an element $K \in \mathscr{T}_h$, we have, by Lemmas 5.3 and 5.4,

$$\left|(v_h|_K)^\wedge\right|_{l,r\widehat{\Omega}} \leqslant C \|A_K\|^l |\det A_K|^{-1/r} |v_h|_{l,r,K} \tag{5.148}$$

$$|v_h|_{m,q,K} \leqslant C \|A_K^{-1}\|^m |\det A_K|^{1/q} |(v_h|_K)^\wedge|_{m,q,\widehat{\Omega}}. \tag{5.149}$$

Define the space

$$\widehat{N} = \left\{ \widehat{p} \in \mathscr{P}_{\widehat{\Omega}} \mid |\widehat{p}|_{l,r,\widehat{\Omega}} = 0 \right\}.$$

Then we have

$$\widehat{N} = \begin{cases} \{0\}, & l = 0, \\ \mathscr{P}_{\widehat{\Omega}} \cap P_{l-1}(\widehat{\Omega}), & l \geqslant 1. \end{cases}$$

Since $l \leqslant m$, $\widehat{p} \in \widehat{N}$ implies $|\widehat{p}|_{m,q,\widehat{\Omega}} = 0$. Therefore the mapping

$$\widehat{p}^* = \{\widehat{p}\} \in \mathscr{P}_{\widehat{\Omega}}/\widehat{N}, \quad \|\widehat{p}^*\|_{m,q,\widehat{\Omega}} = \inf_{\widehat{\eta}\in\widehat{N}} |\widehat{p}-\widehat{\eta}|_{m,q,\widehat{\Omega}} \tag{5.150}$$

is a norm on the quotient space $\mathscr{P}_{\widehat{\Omega}}/\widehat{N}$. Since this quotient space is finite-dimensional, the norm (5.150) is equivalent to the quotient norm $\|\cdot\|_{l,r,\widehat{\Omega}}$. Therefore, there exists a constant $C > 0$ such that

$$|\widehat{p}|_{m,q,\widehat{\Omega}} = \|\widehat{p}^*\|_{m,q,\widehat{\Omega}} \leqslant C\|\widehat{p}^*\|_{l,r,\widehat{\Omega}} = C\|\widehat{p}\|_{l,r,\widehat{\Omega}}. \tag{5.151}$$

From (5.148), (5.149), (5.151), (5.146), Lemma 5.3, and the fact that

$$|\det A_K| = \frac{\text{meas } K}{\text{meas } \widehat{\Omega}},$$

we have

$$|v_h|_{m,q,K} \leqslant C\frac{(h^N)^{1/q-1/r}}{h^{m-l}}|v_h|_{l,r,K}. \tag{5.152}$$

Consider first the case $q = \infty$. By (5.152), there exists a finite element $K_0 \in \mathscr{T}_h$ such that

$$\max_{K\in\mathscr{T}_h} |v_h|_{m,\infty,K} = |v_h|_{m,\infty,K_0} \leqslant C\frac{(h^N)^{-\frac{1}{r}}}{h^{m-l}}|v_h|_{l,r,K_0} \leqslant C\frac{(h^N)^{-\frac{1}{r}}}{h^{m-l}}|v_h|_{l,r,K_0}.$$

If $q < \infty$ then from (5.152) we have

$$\left(\sum_{K\in\mathscr{T}_h} |v_h|^q_{m,q,k}\right)^{1/q} \leqslant C\frac{(h^N)^{1/q-1/r}}{h^{m-l}}\left(\sum_{K\in\mathscr{T}_h} |v_h|^q_{l,r,K}\right)^{1/q}.$$

We consider the three possibilities:

(i) $r \leqslant q$. Using the inequality

$$\sum_j a_j^{\frac{q}{r}} \leqslant \left(\sum_j a_j\right)^{q/r} \quad \text{for any finite set } \{a_j \mid a_j \geqslant 0\},$$

we have

$$\left(\sum_{K\in\mathscr{T}_h} |v_h|^q_{l,r,K}\right)^{1/q} \leqslant \left(\sum_{K\in\mathscr{T}_h} |v_h|^r_{l,r,K}\right)^{1/r}.$$

(ii) $q < r < \infty$. By Hölder's inequality,

$$\left(\sum_{K\in\mathscr{T}_h} |v_h|^q_{l,r,K}\right)^{1/q} \leqslant \mu_h^{1/q-1/r}\left(\sum_{K\in\mathscr{T}_h} |v_h|^r_{l,r,K}\right)^{1/r}.$$

where

$$\mu_h \equiv \sharp\mathscr{T}_h \leqslant Ch^{-N} \quad \text{for some } C > 0, \quad \forall h.$$

(iii) $r = \infty$. We have

$$\left(\sum_{K \in \mathscr{T}_h} |v_h|_{l,\infty,K}^q \right)^{1/q} \leqslant \mu_h^{1/q} \max_{K \in \mathscr{T}_h} |v_h|_{l,\infty,K}.$$

Therefore, (5.147) is proved for all cases. □

Remark 5.6. When $q = r = 2$ and $V_h \subset H^m(\Omega)$, (5.147) gives

$$|v_h|_{m,2,\Omega} \leqslant Ch^{-(m-l)} |v_h|_{l,2,\Omega}$$

for $l = 0, 1, 2, \ldots, m$. Since

$$|v_h|_{j,2,\Omega} \leqslant Ch^{-(j-l)} |v_h|_{l,2,\Omega} \leqslant Ch^{-(m-l)} |v_h|_{l,2,\Omega}$$

for $j = l, l+1, \ldots, m$ and $0 < h < 1$, and

$$|v_h|_{j,2,\Omega} \leqslant Ch^{-(m-l)} \|v_h\|_{H^l(\Omega)}$$

for $j = 0, 1, \ldots, l-1$, we have

$$\|v_h\|_{H^m(\Omega)} = \left(\sum_{j=0}^{m} |v_h|_{j,2,\Omega}^2 \right)^{1/2} \leqslant Ch^{-(m-l)} \|v_h\|_{H^l(\Omega)} \tag{5.153}$$

for some $C > 0$, for all $v_h \in V_h$. □

Chapter 6

The Potential Equation

The potential or Poisson equation

$$\Delta w(x) = f(x) \tag{6.1}$$

is perhaps the most studied equation in the partial differential equation literature. When the inhomogeneous term $f(x)$ is identically zero, (6.1) is called the *Laplace equation*, or the *harmonic equation*, and the solution w is said to be a harmonic function. The potential equation is the simplest type of elliptic partial differential equations studied in this book, but because of the basic properties of the boundary integral operators involved, most of the theory and methodology developed here will actually be applicable to all second-order elliptic equations or systems. Therefore, the development in this chapter will also provide the essential mathematical reasoning and procedures for treating all the other PDE in remaining chapters.

6.1 The occurrence of the potential equation

The potential equation (6.1) is the mathematical model for many important physical phenomena. We will not actually give a derivation of the equation for the many relevant physical situations. But since we are studying the BEM with a view to applications, it is useful to know the nature of the problem and the areas of physics and engineering in which the field equations can be reduced to the potential equation. We mention a few of them below.

(a) Gravitation

Consider a region Ω with mass distribution of density ρ. The gravitational force $\vec{F}$ both inside and outside Ω can be expressed in terms of a gravitational potential ϕ by

$$\vec{F} = \nabla\phi.$$

Inside Ω, ϕ satisfies

$$\Delta\phi(x) = -4\pi\rho(x), \quad x \in \Omega,$$

the potential equation. Outside Ω, assume that the space is empty. Then ϕ satisfies the Laplace equation

$$\Delta\phi(x) = 0, \quad x \in \Omega^c.$$

Instead of having mass distributed over a volume, we can also consider mass distributed over a surface. Let σ be the surface density of the matter. Then the potential function ϕ satisfies

$$\Delta\phi(x) = 0 \quad x \in \Omega \text{ or } x \in \Omega^c.$$

On $\partial\Omega$, the surface, we have

$$\phi_-(x) = \phi_+(x), \quad \frac{\partial\phi_+(x)}{\partial n} - \frac{\partial\phi_-(x)}{\partial n} = -4\pi\sigma(x),$$

where ϕ_- and ϕ_+ are the limiting values of ϕ on respectively the inside and outside of the surface.

(b) Irrotational perfect fluid

The velocity field $\vec{q}$ of a perfect fluid in irrotational motion can be expressed in terms of a velocity potential

$$\vec{q} = -\nabla\phi.$$

If there are no sources or sinks in the fluid then ϕ satisfies the Laplace equation

$$\Delta\phi = 0.$$

If the boundary is a rigid surface moving with velocity $\vec{U}_0(x)$ at x then

$$\left[\vec{q}(x) - \vec{U}_0(x)\right] \cdot n(x) = 0,$$

so the (inhomogeneous)Neumann boundary condition

$$\frac{\partial\phi}{\partial n} = -\vec{U}_0 \cdot n$$

is satisfied at the boundary.

(c) Electromagnetism

In an electrostatic field with the presence of electric charge density ρ, the electric vector $\mathbb{E}$ can be expressed in terms of a potential ϕ by

$$\mathbb{E} = -\nabla\phi.$$

The presence of charge leads to the Poisson equation

$$\Delta\phi = -4\pi\rho.$$

In a magnetostatic field, the magnetic vector $\mathbb{H}$ can be expressed in terms of a magnetic potential ψ by

$$\mathbb{H} = -\nabla\psi.$$

Let μ be the magnetic permeability. Then ψ satisfies

$$\nabla\cdot(\mu\nabla\psi) = 0.$$

When μ is constant, the above reduces to the Laplace equation.

(d) Steady-state heat flow

Let $u(x,t)$ be the temperature at point x at time t in a linear heat conduction phenomenon. Then u satisfies

$$\rho c\frac{\partial u(x,t)}{\partial t} - \nabla\cdot(k\nabla u(x,t)) = f(x,t),$$

where it is assumed that the material density ρ, specific heat c and thermal conductivity k are all constant throughout the medium. When there is no heat source $(f(x,t)\equiv 0)$ and when the heat flow has reached a steady state, the temperature does not vary with time, so $u(x,t)$ depends only on x and

$$\Delta u = 0.$$

On the boundary, if there is no heat flux then

$$\frac{\partial u}{\partial n} = 0$$

is satisfied. If the temperature is kept at $g(x)$, then the Dirichlet boundary condition

$$u(x,t) = g(x), \quad x\in\partial\Omega,$$

is satisfied. If there is radiation into a medium at a constant temperature u_0, then the Robin boundary condition

$$\frac{\partial u}{\partial n} + h(u-u_0) = 0, \quad h>0 \text{ is a constant},$$

is satisfied at the boundary.

6.2 The fundamental solution of the Laplace equation

Let $P = \sum_{|\alpha| \leqslant m} a_\alpha D^\alpha$ be a general linear partial differential operator of order m with constant coefficients. A *fundamental solution* for L with singularity at ξ is a distribution $E(x;\xi)$ satisfying

$$PE(x;\xi) = \delta(x-\xi). \tag{6.2}$$

Physically, E is the response to a concentrated unit source located at $x = \xi$. Knowing E enables us to solve the inhomogeneous equation

$$Pu(x) = q(x), \quad \text{with } q \text{ given}, \tag{6.3}$$

because if we take

$$u(x) = \int_{\mathbb{R}^N} q(\xi) E(x;\xi)\, d\xi, \tag{6.4}$$

assuming that the integral exists as a function of x, then formally

$$\begin{aligned}(Pu)(x) &= \int_{\mathbb{R}^n} q(\xi) P_x E(x;\xi)\, d\xi \\ &= \int_{\mathbb{R}^n} q(\xi)\delta(x-\xi)\, d\xi \\ &= q(x),\end{aligned} \tag{6.5}$$

so (6.3) is solved. (In (6.5), we write P_x to emphasize that the partial differentiations are taken with respect to the x variable.) The use of the fundamental solution is indeed very powerful.

Since P has constant coefficients, we can make a simple translation $x - \xi \to x$ to obtain

$$PE(x-\xi;0) = \delta(x-\xi). \tag{6.6}$$

So a solution E of (6.2) has the property

$$E(x;\xi) = E(x-\xi;0).$$

Thus, we can actually write

$$E(x;\xi) = E(x-\xi), \tag{6.7}$$

and (6.4) becomes a convolution integral

$$u(x) = (E * q)(x).$$

(However, obviously, if P does not have constant coefficients then (6.4) can no longer be written as a convolution.)

Now letting $\xi = 0$ and $P = -\Delta$ in (6.6), we consider the equation

$$-\Delta E(x) = \delta(x), \quad x \in \mathbb{R}^N, \tag{6.8}$$

in the free space $\mathbb{R}^N$. Since the Laplacian Δ is invariant under coordinate rotations, we will look for a solution E that depends only on $r = |x|$, the radial variable. For $r > 0$, E satisfies

$$-\Delta E(x) = 0, \qquad |x| = r > 0. \tag{6.9}$$

Consider

$$E(r) \equiv E(x).$$

We have

$$\begin{aligned}
\frac{\partial E}{\partial x_i} &= \frac{\partial E}{\partial r}\frac{\partial r}{\partial x_i} = \frac{x_i}{r}\frac{\partial E}{\partial r}, \\
\frac{\partial^2 E}{\partial x_i^2} &= \frac{x_i^2}{r^2}\frac{\partial^2 E}{\partial r^2} - \frac{x_i^2}{r^3}\frac{\partial E}{\partial r} + \frac{1}{r}\frac{\partial E}{\partial r}, \\
\Delta E &= \sum_{i=1}^{N} \frac{\partial^2 E}{\partial x_i^2} = \frac{\partial^2 E}{\partial r^2} - \frac{1}{r}\frac{\partial E}{\partial r} + \frac{N}{r}\frac{\partial E}{\partial r} \\
&= \frac{\partial^2 E}{\partial r^2} + \frac{N-1}{r}\frac{\partial E}{\partial r}.
\end{aligned}$$

Thus for $x \neq 0$, (6.9) can now be written as

$$\Delta E = \frac{\partial^2 E}{\partial r^2} + \frac{N-1}{r}\frac{\partial E}{\partial r} = \frac{1}{r^{N-1}}\frac{\partial}{\partial r}\left(r^{N-1}\frac{\partial E}{\partial r}\right) = 0, \tag{6.10}$$

which has the general solution

$$E(r) = \begin{cases} A \ln r + B, & N = 2, \\ C r^{-(N-2)} + D, & N > 2. \end{cases}$$

The constants B and D usually have no significance in the occasions when the fundamental solution E is useful in solving partial differential equations (indeed, their presence could add extra work), so we set $B = 0$, $D = 0$. We now determine A and C by the divergence theorem. Integrating (6.8) over B_ε, a ball of radius ε centered at the origin, we get

$$-\int_{B_\varepsilon} \Delta E \, dx = \int_{B_\varepsilon} \delta(x) \, dx = 1.$$

But

$$\int_{B_\varepsilon} \Delta E \, dx = -\int_{\varepsilon\omega_N} \left(\frac{\partial E}{\partial r}\right)_{r=\varepsilon} d\sigma,$$

where ω_N, as in (6.11) below, is the surface of the unit sphere in $\mathbb{R}^N$. Thus we get

(i) for $N = 2$,

$$d\sigma = r d\theta,$$

$$-\int_{\varepsilon\cdot\omega_N} \left(\frac{\partial E}{\partial r}\right)_{r=\varepsilon} d\sigma = -A\int_{\varepsilon\omega_2} \frac{1}{r}\Big|_{r=\varepsilon} d\sigma$$

$$= -\frac{A}{\varepsilon} 2\pi\varepsilon \;=\; 1, \tag{6.11}$$

$$A = -1/2\pi;$$

(ii) for $N > 2$,

$$\int_{\varepsilon\omega_N} d\sigma = \varepsilon^{N-1} \times \text{area of } \omega_N = \varepsilon^{N-1} \frac{2\pi^{N/2}}{\Gamma(\frac{1}{2}N)} \tag{6.12}$$

(Γ is the gamma function),

$$-\int_{\varepsilon\omega_N} \left(\frac{\partial E}{\partial r}\right)_{r=\varepsilon} d\sigma = -C\int_{\varepsilon\omega_N} [-(N-2)]r^{-(N-1)}\,|_{r=\varepsilon}\, d\sigma$$

$$= C(N-2)\varepsilon^{-(N-1)} \left[\varepsilon^{N-1} \frac{2\pi^{N/2}}{\Gamma(\frac{1}{2}N)}\right] = 1,$$

$$C = \frac{\Gamma(\frac{1}{2}N)}{2\pi^{N/2}(N-2)}.$$

Therefore, we have obtained the fundamental solution

$$E(x-\xi) = \begin{cases} -\dfrac{1}{2\pi} \ln|x-\xi|, & N = 2, \\ \dfrac{\Gamma(N/2)}{2\pi^{N/2}(N-2)} |x-\xi|^{-(N-2)}, & N > 2. \end{cases} \tag{6.13}$$

For $N = 2$, the graph of the fundamental solution $E(x-\xi)|_{\xi=0}$ is plotted in Fig. 6.1. Another way of deriving (6.13) is by the Fourier transform. We leave this as an exercise.

Exercise 6.1. Use the Fourier transform to derive the fundamental solution (6.13) for $N > 2$. □

6.3 The volume and boundary potentials

Consider the potential equation

$$\left.\begin{aligned} \Delta w(x) &= f(x) \;\text{ on } \Omega, \\ Bw(x) &= g(x) \;\text{ on } \partial\Omega, \end{aligned}\right\} \tag{6.14}$$

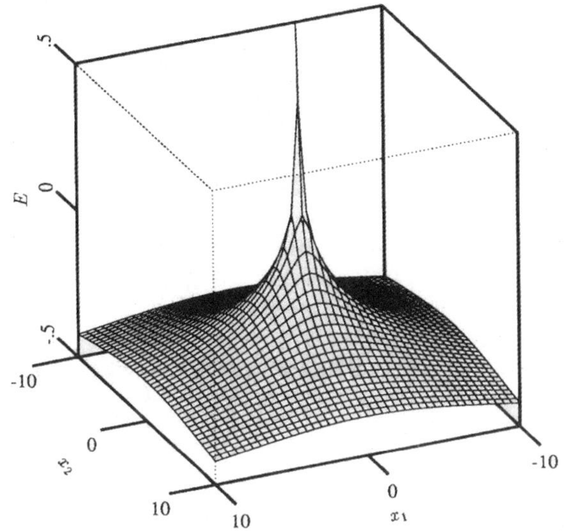

Figure 6.1: The fundamental solution $E(x) = -(2\pi)^{-1}\log|x|$ in $\mathbb{R}^2$, with $\xi = 0$.

where f and g are given, and B is a boundary operator corresponding to the Dirichlet, Neumann or Robin boundary condition on $\partial\Omega$:

$$\begin{aligned} &\text{(Dirichlet)} && w(x) = g(x), \\ &\text{(Neumann)} && \frac{\partial w(x)}{\partial n} = g(x), \\ &\text{(Robin)} && \frac{\partial w(x)}{\partial n} + hw(x) = g(x), \qquad h > 0, \end{aligned}$$

as mentioned in the problems in §6.1. We wish to solve (6.14).

In the integral equation approach, the first step is to reduce the partial differential equation itself to a homogeneous one. We let

$$\begin{aligned} v(x) &= V(f)(x) \\ &= -(E * f)(x) \\ &= -\int_\Omega E(x-\xi)f(\xi)\,d\xi. \end{aligned} \tag{6.15}$$

Then on Ω, v satisfies

$$\Delta v(x) = f(x),$$

as explained in (6.5). The operator V is linear, producing a volume potential v. According to the theory of elliptic boundary value problems [127, Chap. 2, Theorem 8.3], V has the following property.

Theorem 6.1. *Let Ω be bounded with smooth boundary $\partial\Omega$. Then*

$$V:\ H^r(\Omega) \longrightarrow H^{r+2}(\Omega)$$

is continuous for any $r \geqslant -1$. Thus V increases the order of smoothness by 2 on Ω for $r \geqslant -1$. □

Now define

$$\widetilde{w}(x) = w(x) - v(x). \tag{6.16}$$

Then $\widetilde{w}$ satisfies the homogeneous equation

$$\Delta\widetilde{w}(x) = \Delta w(x) - \Delta v(x) = 0.$$

On $\partial\Omega$, $\widetilde{w}$ satisfies

$$\text{(Dirichlet case)}\quad \widetilde{w}(x) = g(x) - v(x)\,|_{\partial\Omega} \equiv \widetilde{g}(x), \tag{6.17}$$

$$\text{(Neumann case)}\quad \frac{\partial\widetilde{w}(x)}{\partial n} = g(x) - \frac{\partial v(x)}{\partial n}\Big|\partial\Omega \equiv \widetilde{g}(x), \tag{6.18}$$

$$\text{(Robin case)}\quad \frac{\partial\widetilde{w}(x)}{\partial n} + h\widetilde{w}(x) = g(x) - \left[\frac{\partial v(x)}{\partial n} + hv(x)\right]\Big|_{\partial\Omega} \equiv \widetilde{g}(x). \tag{6.19}$$

Let us now examine the regularity of the boundary data $\widetilde{g}$ for each case. In (6.14), let

$$f \in H^r(\Omega), \qquad r \geqslant -1,$$
$$g \in H^{s_j}(\partial\Omega), \quad s_j \in \mathbb{R}, \quad j = 1, 2, 3,$$

where $j = 1, 2, 3$ respectively for Dirichlet, Neumann and Robin conditions. Then by the trace theorem in Chapter 2 and Theorem 6.1,

$$v\Big|_{\partial\Omega} \in H^{r+3/2}(\partial\Omega),$$
$$\frac{\partial v}{\partial n}\Big|_{\partial\Omega} \in H^{r+1/2}(\partial\Omega),$$
$$\left(\frac{\partial v}{\partial n} + hv\right)\Big|_{\partial\Omega} \in H^{r+1/2}(\partial\Omega).$$

Therefore, in (6.17), (6.18) and (6.19), we have

$$\left.\begin{aligned} &\widetilde{g} \in H^{\min(s_1, r+3/2)}(\partial\Omega) \ \text{ for (6.17)},\\ &\widetilde{g} \in H^{\min(s_2, r+1/2)}(\partial\Omega) \ \text{ for (6.18)},\\ &\widetilde{g} \in H^{\min(s_3, r+1/2)}(\partial\Omega) \ \text{ for (6.19)}. \end{aligned}\right\} \tag{6.20}$$

Rewriting $\widetilde{w}$ as w and $\widetilde{g}$ as g, we have a simplified problem

$$\left.\begin{aligned} \Delta w(x) &= 0 \qquad \text{on } \Omega,\\ Bw(x) &= g(x) \ \text{ on } \partial\Omega, \end{aligned}\right\} \tag{6.21}$$

where it is understood that the regularity of the boundary data g in (6.21) is related to the original data f and g in (6.14) through (6.17)–(6.19).

We can now derive a boundary integral equation as explained in (1.13)–(1.22) in Chapter 1. Acknowledging the correctness of (1.23), we would have for $x \in \partial\Omega$,

$$\frac{1}{2}w(x) + \int_{\partial\Omega} \frac{\partial E(x-\xi)}{\partial n_\xi} w(\xi)\, d\sigma_\xi = \int_{\partial\Omega} E(x-\xi) \frac{\partial w(\xi)}{\partial n}\, d\sigma_\xi, \tag{6.22}$$

an integral equation relating the Dirichlet and Neumann data w and $\partial w/\partial n$. It remains to justify all the intermediate steps in Chapter 1 that lead to (6.22).

Since the potential equation contains the basic features of most of the other second-order elliptic PDEs such as the Helmholtz equation and the equations of linear elasticity to be discussed in subsequent chapters, the mathematical procedures and reasoning in treating these equations are essentially similar. Therefore it is important for us to understand the basic properties of the surface potentials (which are the integrals appearing) in (6.22). Our presentation of the classical theory of potentials follows largely along the lines of the excellent exposition given in [53, Chap. 2]. In the following section, let us study some preliminary material required to establish those properties.

6.4 Geometry of hypersurfaces

Let S be an N-dimensional smooth hypersurface contained in $\mathbb{R}^{N+1}$. Every interior point $x_0 \in S$ has a parametrization

$$x(u) = x(u_1, u_2, \ldots, u_N), \quad u \in \mathbb{R}^N, \quad |u_j| \leqslant \delta_j, \ \delta_j > 0$$

valid for all x in a small neighborhood of x_0, where

$$x_0 = x(0) = x(u)|_{u=0}. \tag{6.23}$$

The vector

$$\left.\frac{\partial x(u)}{\partial u_j}\right|_{u=0}, \tag{6.24}$$

is a tangent vector touching S at x_0. Letting $j = 1, 2, \ldots, N$ in (6.24), we obtain a basis of N linearly independent tangent vectors to S at x_0.

A curve C on S is a set parametrized by a representation $x(u(s))$, where $s \in I$ for some interval $I \subset \mathbb{R}$. Let (6.23) hold, and $u(0) = 0$. Then

$$\tau = \left.\frac{d}{ds} x(u(s))\right|_{s=0} = \sum_{j=1}^{N} \left.\frac{\partial x}{\partial u_j} \frac{du_j(s)}{ds}\right|_{s=0}$$

is a tangent vector to C at x_0. If s is chosen to be the arclength along C then

$$ds^2 = \left\langle \sum_{j=1}^{N} \frac{\partial x}{\partial u_j} du_j, \sum_{k=1}^{N} \frac{\partial x}{\partial u_k} du_k \right\rangle \tag{6.25}$$

$$= \sum_{j,k=1}^{N} g_{jk} du_j du_k \qquad \left(g_{jk} = \left\langle \frac{\partial x}{\partial u_j}, \frac{\partial x}{\partial u_k} \right\rangle \right)$$

gives the (square of the) arclength element on C, called the *first fundamental form* in differential geometry. The matrix $\mathbb{G} = [g_{jk}]$ is a second-order covariant *tensor*, called the first fundamental tensor.

Let $N = 2$. Then (6.25) is often written as

$$ds^2 = E\,du_1^2 + 2F\,du_1\,du_2 + G\,du_2^2. \tag{6.26}$$

In this case, the surface element becomes (the magnitude of the cross vector product)

$$d\sigma = \left| \left(\frac{\partial x}{\partial u_1} du_1 \right) \times \left(\frac{\partial x}{\partial u_2} du_2 \right) \right|. \tag{6.27}$$

Since

$$\left| \frac{\partial x}{\partial u_1} \times \frac{\partial x}{\partial u_2} \right| = \left| \frac{\partial x}{\partial u_2} \right| \left| \frac{\partial x}{\partial u_2} \right| \sin\theta$$

and

$$\left\langle \frac{\partial x}{\partial u_1}, \frac{\partial x}{\partial u_2} \right\rangle = \left| \frac{\partial x}{\partial u_1} \right| \left| \frac{\partial x}{\partial u_2} \right| \cos\theta,$$

from (6.25) and (6.26),

$$\left| \frac{\partial x}{\partial u_1} \right|^2 = \left\langle \frac{\partial x}{\partial u_1}, \frac{\partial x}{\partial u_1} \right\rangle = E,$$

$$\left| \frac{\partial x}{\partial u_2} \right|^2 = \left\langle \frac{\partial x}{\partial u_2}, \frac{\partial x}{\partial u_2} \right\rangle = G,$$

$$\left\langle \frac{\partial x}{\partial u_1}, \frac{\partial x}{\partial u_2} \right\rangle = F,$$

we get

$$\cos\theta = \frac{F}{(EG)^{1/2}}$$

and thus

$$\sin\theta = \left(\frac{EG - F^2}{EG} \right)^{1/2}.$$

Hence from (6.27),

$$d\sigma = \sqrt{EG - F^2}\,du_1\,du_2. \tag{6.28}$$

The unit normal vector n_x to S at x is given by

$$n_x = \left(\frac{\partial x}{\partial u_1} \times \frac{\partial x}{\partial u_2}\right) \Big/ \left|\frac{\partial x}{\partial u_1} \times \frac{\partial x}{\partial u_2}\right| = \frac{1}{\sqrt{EG-F^2}} \frac{\partial x}{\partial u_1} \times \frac{\partial x}{\partial u_2}. \tag{6.29}$$

Since $\langle n_x, n_x\rangle = 1$, we have

$$\frac{\partial}{\partial u_j}\langle n_x, n_x\rangle = 0 = 2\left\langle \frac{\partial n_x}{\partial u_j}, n_x \right\rangle, \quad j = 1,2.$$

Thus $\partial n_x/\partial u_j$ must be a tangent vector. We write

$$\frac{\partial n_x}{\partial u_j} = -\left(b_j^1 \frac{\partial x}{\partial u_1} + b_j^2 \frac{\partial x}{\partial u_2}\right), \quad j = 1,2. \tag{6.30}$$

This leads to a matrix $\mathbb{B} = [b_j^k]$, that is also a tensor. The trace and the determinant

$$\left.\begin{aligned} H &= \frac{1}{2}(b_1^1 + b_2^2) \\ K &= b_1^1 b_2^2 - b_2^1 b_1^2 \end{aligned}\right\} \tag{6.31}$$

of $\mathbb{B}$ are scalars independent of the choice of coordinates and are called respectively the *mean curvature* and the *Gaussian curvature* of the surface.

Because $\partial x/\partial u_1$ and $\partial x/\partial u_2$ are linearly independent at every point of S, the matrix $\mathbb{G}$ is invertible, with entries

$$\mathbb{G}^{-1} = [g^{jk}], \quad \sum_{k=1}^{2} g^{jk} g_{kr} = \delta_r^j, \quad j, r = 1, 2, \tag{6.32}$$

which are components of a second-order contravariant tensor. For any scalar differentiable function ϕ defined on S, define a surface gradient

$$\overline{\nabla}\phi = \sum_{j,k=1}^{2} g^{jk} \frac{\partial \phi}{\partial u_j} \frac{\partial x}{\partial u_k}. \tag{6.33}$$

Then $\overline{\nabla}\phi$ is tangent to S and is defined independently of the choice of the coordinate system. For any direction $\vec{\alpha}$ tangent to S, the directional derivative in the direction of $\vec{\alpha}$ is given by

$$D_{\vec{\alpha}}\phi = \langle \vec{\alpha}, \overline{\nabla}\phi\rangle = \sum_{j=1}^{2} \frac{\partial \phi}{\partial u_j} D_{\vec{\alpha}} u_j.$$

We state a variant of Stokes' theorem on a surface. Its proof is adopted from [53, pp. 34–35].

Theorem 6.2. *Let $S \subset \partial\Omega$ be a connected surface with C^2-smooth boundary ∂S, where $\Omega \subset \mathbb{R}^3$. Let n_0 denote the exterior unit normal vector to the boundary curve ∂S that is perpendicular to the surface normal n on $\partial\Omega$. Let $\tau = n \times n_0$ denote the unit tangent vector along ∂S; see Fig. 6.2. Let ϕ be a sufficiently smooth function defined on $\overline{S}$. Then*

$$\int_S \overline{\nabla}\phi \, d\sigma = \int_{\partial S} \phi n_0 \, ds - 2\int_S \phi H n \, d\sigma, \tag{6.34}$$

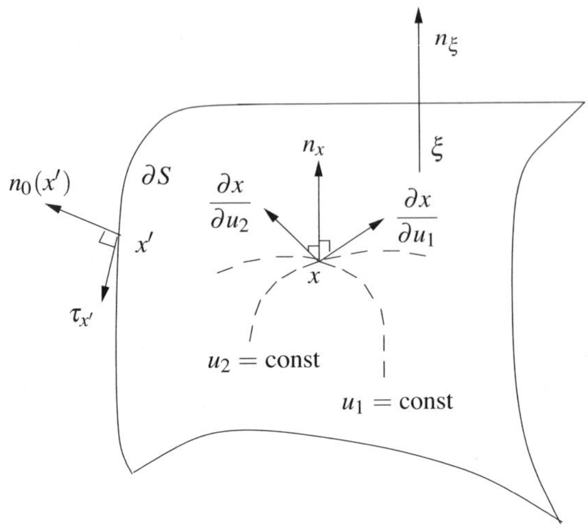

Figure 6.2: The surface S with parametrization (u_1, u_2) and tangent vectors $\partial x/\partial u_1$, $\partial x/\partial u_2$. At $x' \in \partial S$, $\tau_{x'}$ is the unit tangent vector to ∂S at x', and $n_0(x')$ is orthogonal to both $n_{x'}$ and $\tau_{x'}$.

and

$$\int_S n \times \overline{\nabla}\phi \, d\sigma = \int_{\partial S} \phi \tau \, ds, \tag{6.35}$$

where in the above H is the mean curvature as defined in (6.31), and ds is the infinitesimal arclength.

Proof. We first assume that S is a surface element:

$$S = \left\{ x(u) \in \mathbb{R}^3 \mid u \in U \subseteq \mathbb{R}^2 \text{ for some open set } U \right\}.$$

Then

$$\begin{aligned}
2\int_S \phi H n \, d\sigma &= \int_U \phi (b_1^1 + b_2^2) \left(\frac{\partial x}{\partial u_1} \times \frac{\partial x}{\partial u_2} \right) du_1 \, du_2 \\
&= \int_U \phi \left[\left(\frac{\partial n}{\partial u_2} \times \frac{\partial x}{\partial u_1} \right) - \left(\frac{\partial n}{\partial u_1} \times \frac{\partial x}{\partial u_2} \right) \right] du_1 \, du_2 \\
&\qquad \text{(by (6.30))} \\
&= \int_{\partial U} \phi \left(\frac{\partial x}{\partial u_1} \times n \, du_1 + \frac{\partial x}{\partial u_2} \times n \, du_2 \right) \\
&\quad - \int_U \left[\frac{\partial \phi}{\partial u_2} \left(n \times \frac{\partial x}{\partial u_1} \right) - \frac{\partial \phi}{\partial u_1} \left(n \times \frac{\partial x}{\partial u_2} \right) \right] du_1 \, du_2 \\
&\qquad \text{(integration by parts)}
\end{aligned} \tag{6.36}$$

It is easy to verify that on ∂U

$$\frac{\partial x}{\partial u_1}\times n\,du_1+\frac{\partial x}{\partial u_2}\times n\,du_2=(\tau\times n)\,ds=n_0\,ds$$

and on U

$$\frac{\partial \phi}{\partial u_2}\left(n\times\frac{\partial x}{\partial u_1}\right)-\frac{\partial \phi}{\partial u_1}\left(n\times\frac{\partial x}{\partial u_2}\right)=\sum_{j,k=1}^{2} g^{1/2}g^{jk}\frac{\partial \phi}{\partial u_j}\frac{\partial x}{\partial u_k}=g^{1/2}\overline{\nabla}\phi.$$

We conclude from (6.36) that

$$2\int_S \phi H n\,d\sigma=\int_{\partial S}\phi n_0\,ds-\int_S \overline{\nabla}\phi\,d\sigma.$$

Similarly, we have

$$\begin{aligned}\int_S (n\times\overline{\nabla}\phi)\,d\sigma&=\sum_{j,k=1}^{2}\int_U g^{jk}\frac{\partial \phi}{\partial u_j}\left[\left(\frac{\partial x}{\partial u_1}\times\frac{\partial x}{\partial u_2}\right)\times\frac{\partial x}{\partial u_k}\right]du_1\,du_2\\&=\int_U\left(\frac{\partial \phi}{\partial u_1}\frac{\partial x}{\partial u_2}-\frac{\partial \phi}{\partial u_2}\frac{\partial x}{\partial u_1}\right)du_1\,du_2\\&=\int_{\partial U}\phi\left(\frac{\partial x}{\partial u_1}du_1+\frac{\partial x}{\partial u_2}du_2\right)\\&=\int_{\partial S}\phi\tau\,ds.\end{aligned}$$

Therefore (6.34) and (6.35) have been proved for a surface element S. For a general surface S, we can patch the surface elements together and note that the integrals along the neighboring boundary of surface elements cancel □

This theorem will be useful later on in the layer potential estimates.

Lemma 6.1. *If $\partial\Omega$ is a C^2 surface then there exists $M>0$ such that*

$$|\langle n(y),x-y\rangle|\leqslant M|x-y|^2, \tag{6.37}$$

$$|n(x)-n(y)|\leqslant M|x-y| \tag{6.38}$$

for all $x,y\in\partial\Omega$.

Proof. We only give a brief sketch of the proof. For details, see [53, pp. 35–37]. For given $y\in\partial\Omega$, define

$$f_{(y)}(x)=\langle n(y),x-y\rangle,\quad x\in\partial\Omega.$$

Then

$$f_{(y)}(y)=0.$$

Also

$$\overline{\nabla}f_{(y)}(y)=\langle n(y),\sigma(y)\rangle\sigma(y)=0$$

because $\sigma(y) \equiv \overline{\nabla}(x-y)|_{x=y} = \overline{\nabla}x|_{x=y}$ is a tangent vector at y.
Using the Taylor expansion of $f_{(y)}$ locally at y, we then have

$$\left|f_{(y)}(x)\right| \leqslant M_{(y)}|y-x|^2.$$

Use the smoothness of $\partial\Omega$ to show that $\sup_{y\in\partial\Omega} M_{(y)}$ is finite.
For (6.38), apply the mean value theorem. □

6.5 Regularity of the layer potentials and the jump property

For a given function f defined on $\partial\Omega$ and $x \in \mathbb{R}^N$ (as in (4.60) and (4.61) earlier on), we define surface (i.e., boundary) potentials

$$V_1(f)(x) = \int_{\partial\Omega} E(x-\xi)f(\xi)\,d\sigma_\xi, \tag{6.39}$$

$$V_2(f)(x) = \int_{\partial\Omega} \frac{\partial E(x-\xi)}{\partial n_\xi} f(\xi)\,d\sigma_\xi. \tag{6.40}$$

Such potentials are well defined for $x \in \mathbb{R}^N \setminus \partial\Omega$ because the kernels are C^∞-smooth when $\xi \in \partial\Omega, x \notin \partial\Omega$. For $x \in \partial\Omega$, (6.39) and (6.40) are also well defined because of Lemma 6.1, provided that f is sufficiently smooth.
Historically, V_1 and V_2 are called respectively the *simple-* (*or single-*) and *double-layer potentials*. The reason for this will be obvious later on.
As boundary element computations are mostly performed in $\mathbb{R}^3$ or $\mathbb{R}^2$, in what follows let us limit our discussion primarily to $N = 3$. The corresponding results for $N = 2$ should be more or less obvious; see § 6.6.
For a smooth surface $\partial\Omega$, we can generate locally parallel surfaces $\partial\Omega_h$ by the representation

$$x = z + hn(z), \quad z \in \partial\Omega, \tag{6.41}$$

where h denotes the distance from x to z. The unit outward normal vector at x in (6.41) is easily seen to be $n(z)$. (z,h) establishes a local coordinate system near $\partial\Omega$ for $z \in \partial\Omega$ and h sufficiently small.

Theorem 6.3. *Let $\Omega \subseteq \mathbb{R}^3$ be bounded with smooth $\partial\Omega$. Let $f \in C^\infty(\partial\Omega)$. Denote*

$$V_i^+(f)(x) = V_i(f)(x) \quad \text{if } x \in \Omega^c,$$

$$V_i^-(f)(x) = V_i(f)(x) \quad \text{if } x \in \Omega.$$

For $x \in \partial\Omega$, define

$$V_i^+(f)(x) = \lim_{\substack{y\to x\\ y\in\Omega^c}} V_i^+(f)(y), \quad V_i^-(f)(x) = \lim_{\substack{y\to x\\ y\in\Omega}} V_i^-(f)(y), \tag{6.42}$$

where the limits are taken in the pointwise sense. Then for $x \in \partial\Omega$, we have

$$V_1^+(f)(x) = V_1^-(f)(x) = V_1(f)(x), \tag{6.43}$$

$$V_2^+(f)(x) = \frac{1}{2}f(x) + V_2(f)(x), \tag{6.44}$$

$$V_2^-(f)(x) = -\frac{1}{2}f(x) + V_2(f)(x), \tag{6.45}$$

$$\frac{\partial}{\partial n}V_1^+(f)(x) = -\frac{1}{2}f(x) + \int_{\partial\Omega} \frac{\partial E(x-\xi)}{\partial n_x} f(\xi)\, d\sigma_\xi, \tag{6.46}$$

$$\frac{\partial}{\partial n}V_1^-(f)(x) = \frac{1}{2}f(x) + \int_{\partial\Omega} \frac{\partial E(x-\xi)}{\partial n_x} f(\xi)\, d\sigma_\xi, \tag{6.47}$$

$$\frac{\partial}{\partial n}V_2^+(f)(x) = (\text{F.P.}) \int_{\partial\Omega} \frac{\partial^2 E(x-\xi)}{\partial n_x \partial n_\xi} f(\xi)\, d\sigma_\xi, \tag{6.48}$$

$$\frac{\partial}{\partial n}V_2^-(f)(x) = (\text{F.P.}) \int_{\partial\Omega} \frac{\partial^2 E(x-\xi)}{\partial n_x \partial n_\xi} f(\xi) d\sigma_\xi, \tag{6.49}$$

$$\nabla V_1^+(f)(x) = \int_{\partial\Omega} \nabla_x E(x-\xi) f(\xi)\, d\sigma_\xi - \frac{1}{2}f(x)n(x), \tag{6.50}$$

$$\nabla V_1^-(f)(x) = \int_{\partial\Omega} \nabla_x E(x-\xi) f(\xi)\, d\sigma_\xi + \frac{1}{2}f(x)n(x), \tag{6.51}$$

$$\nabla V_2^+(f)(x) = -\int_{\partial\Omega} \left\{ \nabla_x E(x-\xi) \times [\overline{\nabla} f(\xi) \times n(\xi)] \right\} d\sigma_\xi + \frac{1}{2}\overline{\nabla} f(x), \tag{6.52}$$

$$\nabla V_2^-(f)(x) = -\int_{\partial\Omega} \left\{ \nabla_x E(x-\xi) \times [\overline{\nabla} f(\xi) \times n(\xi)] \right\} d\sigma_\xi - \frac{1}{2}\overline{\nabla} f(x). \tag{6.53}$$

Note that in (6.46)–(6.49), the normal derivatives are interpreted as

$$\frac{\partial V_i^\pm(x)}{\partial n} = \lim_{h\to 0\pm} \frac{\partial V_i(x+hn(x))}{\partial n(x)},$$

while in (6.50)–(6.53), the integrals on the RHS exist as Cauchy principal values.

Let $f \in H^\alpha(\partial\Omega), \alpha \geqslant 0$. Then f can be approximated by a sequence $\{f_m\}$ in $C^\infty(\partial\Omega)$. On $\partial\Omega$, after taking the limit (i.e., closure) of $\{f_m\}$ with respect to the Sobolev space norm, we have

$$V_1^+(f) = V_1^-(f) = V_1(f) \quad \text{in } H^{\alpha+1}(\partial\Omega), \tag{6.54}$$

$$V_2^+(f) = \frac{1}{2}f + V_2(f) \quad \text{in } H^\alpha(\partial\Omega), \tag{6.55}$$

$$V_2^-(f) = -\frac{1}{2}f + V_2(f) \quad \text{in } H^\alpha(\partial\Omega), \tag{6.56}$$

$$\frac{\partial}{\partial n}V_1^+(f) = -\frac{1}{2}f + \int_{\partial\Omega} \frac{\partial E(\cdot-\xi)}{\partial n_x} f(\xi)\, d\sigma_\xi \quad \text{in } H^\alpha(\partial\Omega), \tag{6.57}$$

$$\frac{\partial}{\partial n}V_1^-(f) = \frac{1}{2}f + \int_{\partial\Omega} \frac{\partial E(\cdot-\xi)}{\partial n_x} f(\xi)\, d\sigma_\xi \quad \text{in } H^\alpha(\partial\Omega), \tag{6.58}$$

$$\frac{\partial}{\partial n}V_2^+(f) = (\text{F.P.})\int_{\partial\Omega}\frac{\partial^2 E(\cdot-\xi)}{\partial n_x \partial n_\xi}f(\xi)\,d\sigma_\xi \quad \text{in } H^{\alpha-1}(\partial\Omega), \tag{6.59}$$

$$\frac{\partial}{\partial n}V_2^-(f) = (\text{F.P.})\int_{\partial\Omega}\frac{\partial^2 E(\cdot-\xi)}{\partial n_x \partial n_\xi}f(\xi)\,d\sigma_\xi \quad \text{in } H^{\alpha-1}(\partial\Omega), \tag{6.60}$$

$$\nabla V_1^+(f) = \int_{\partial\Omega}\nabla_x E(\cdot-\xi)f(\xi)\,d\sigma_\xi - \frac{1}{2}f(\cdot)n_x \quad \text{in } [H^{\alpha}(\partial\Omega)]^3, \tag{6.61}$$

$$\nabla V_1^-(f) = \int_{\partial\Omega}\nabla_x E(\cdot-\xi)f(\xi)\,d\sigma_\xi + \frac{1}{2}f(\cdot)n_x \quad \text{in } [H^{\alpha}(\partial\Omega)]^3, \tag{6.62}$$

$$\nabla V_2^+(f) = -\int_{\partial\Omega}\{\nabla_x E(\cdot-\xi)\times[\overline{\nabla}f(\xi)\times n_\xi]\}\,d\sigma_\xi + \frac{1}{2}\overline{\nabla}f(\cdot) \quad \text{in } [H^{\alpha-1}(\partial\Omega)]^3, \tag{6.63}$$

$$\nabla V_2^-(f) = -\int_{\partial\Omega}\{\nabla_x E(\cdot-\xi)\times[\overline{\nabla}f(\xi)\times n_\xi]\}\,d\sigma_\xi - \frac{1}{2}\overline{\nabla}f(\cdot) \quad \text{in } [H^{\alpha-1}(\partial\Omega)]^3. \tag{6.64}$$

In each equation above, the mapping represented by the RHS of the equation is continuous (into the respective Sobolev space). For example, in (6.61), the mapping

$$\mathscr{L}\colon f \longrightarrow -\frac{1}{2}f(\cdot)n_x + \int_{\partial\Omega}\nabla_x E(\cdot-\xi)f(\xi)\,d\sigma_\xi, \tag{6.65}$$
$$\mathscr{L}\colon H^{\alpha}(\partial\Omega)\to[H^{\alpha}(\partial\Omega)]^3$$

is a continuous linear operator.

Furthermore, the regularity properties in (6.54)-(6.64) remain valid for all $\alpha\in\mathbb{R}$, since for $\alpha<0$, they can be established by a duality transposition argument. □

Theorem 6.3 contains nearly all the jump and regularity properties of simple- and double-layer potentials. We will prove (6.43)–(6.64) through a sequence of lemmas. We assume at the beginning that the density f is sufficiently smooth (or regular). The regularity of the layer potentials and their derivatives will become obvious when we regard the integral operators in (6.54)–(6.64) as pseudodifferential operators and determine the orders of their symbols according to § 4.4.

We treat (6.43) first.

Lemma 6.2. *Let $f\in L^\infty(\partial\Omega)$. Then $V_1(f)$ is continuous and uniformly bounded throughout $\mathbb{R}^3$.*

Proof. The continuity of $V_1^\pm(f)$ on Ω^c and Ω is rather obvious. We need only consider their continuity on or near $\partial\Omega$. For given $x\in\partial\Omega$, let

$$\Gamma_\varepsilon = \{z\in\partial\Omega \mid |z-x|\leqslant\varepsilon\}.$$

Then

$$V_1^{\pm}(f)(y) = \int_{\partial\Omega \setminus \Gamma_\varepsilon} E(y-\xi) f(\xi)\, d\sigma_\xi + \int_{\Gamma_\varepsilon} E(y-\xi) f(\xi)\, d\sigma_\xi.$$

As $y \to x$, the first integral on the right is continuous, while the second tends to

$$\lim_{y \to x} \int_{\Gamma_\varepsilon} E(y-\xi) f(\xi)\, d\sigma_\xi = \int_{\Gamma_\varepsilon} \frac{1}{4\pi} \frac{1}{|x-\xi|} f(\xi)\, d\sigma_\xi,$$

which is a convergent integral such that

$$\lim_{\varepsilon \downarrow 0} \int_{\Gamma_\varepsilon} \frac{1}{4\pi} \frac{1}{|x-\xi|} f(\xi)\, d\sigma_\xi = 0.$$

Therefore

$$\lim_{\substack{y \to x \\ x \in \partial\Omega}} V_1^{\pm}(f)(y) = \int_{\partial\Omega} E(x-\xi) f(\xi)\, d\sigma_\xi = V_1(f)(x).$$

$V_1(f)$ is uniformly bounded because $V_1(f)$ is continuous on any bounded domain, and near infinity we have the decay rate

$$\begin{aligned}
|V_1(f)(x)| &= \left| \int_{\partial\Omega} E(x-\xi) f(\xi)\, d\sigma_\xi \right| \\
&= \frac{1}{4\pi} \left| \int_{\partial\Omega} \frac{1}{|x-\xi|} f(\xi)\, d\sigma_\xi \right| \\
&\leqslant \frac{1}{4\pi} \frac{1}{|x|} \int_{\partial\Omega} \frac{|x|}{|x-\xi|} |f(\xi)|\, d\sigma_\xi \\
&\leqslant \frac{1}{5\pi|x|} \|f\|_{L^\infty} \text{ if } |x| \geqslant 4 \sup\{|\xi| \mid \xi \in \partial\Omega\}.
\end{aligned} \tag{6.66}$$

□

The continuity of the simple layer can be strengthened to Hölder continuity. To show this, we first prove a general theorem due to Giraud [81] for an integral equation with a weakly singular kernel. Our presentation follows that of [53, pp. 41–44].

Theorem 6.4. *Let $G \subseteq \mathbb{R}^3$ be a closed domain containing $\partial\Omega$ in its interior. Assume that the kernel function $K(x,\xi)$ is continuous for all $x \in G$, $\xi \in \partial\Omega$, $x \neq \xi$, and assume there exist positive constants M and $\alpha \in (0,2]$ such that for all $x \in G$, $\xi \in \partial\Omega$, $x \neq \xi$, we have*

$$|K(x,\xi)| \leqslant M|x-\xi|^{\alpha-2}. \tag{6.67}$$

Furthermore, assume there exists $m \in \mathbb{Z}^+$ such that

$$|K(x_1,\xi) - K(x_2,\xi)| \leqslant M \sum_{j=1}^{m} |x_1-\xi|^{\alpha-2-j} |x_1-x_2|^j \tag{6.68}$$

for all $x_1, x_2 \in G$, $\xi \in \partial\Omega$ *with* $2|x_1 - x_2| \leqslant |x_1 - \xi|$. *Then the potential*

$$u(x) \equiv \int_{\partial\Omega} K(x,\xi)\phi(\xi)\,d\sigma_\xi, \quad x \in G, \tag{6.69}$$

with density $\phi \in C(\partial\Omega)$ *satisfies*

$$u \in C^{0,\beta}(G) \quad \forall \begin{cases} \beta \in (0,\alpha] & \text{if } 0 < \alpha < 1, \\ \beta \in (0,1) & \text{if } \alpha = 1, \\ \beta \in (0,1] & \text{if } 1 < \alpha < 2, \end{cases} \tag{6.70}$$

and

$$\|u\|_{C^{0,\beta}(G)} \leqslant C(\beta)\|\phi\|_{C(\partial\Omega)} \tag{6.71}$$

for some $C(\beta) > 0$ *depending only on* β.

Proof. For given $\alpha > 0$, the weak singularity (6.67) makes u in (6.69) a well-defined improper integral. See the arguments between (6.72) and (6.78) below.
Select $h_0 > 0$ sufficiently small. Define parallel surfaces $\partial\Omega_h$ for all $|h| \leqslant h_0$ by (6.41), and define

$$\Omega_{h_0} \equiv \{x \in \mathbb{R}^3 \mid x = z + hn(z),\ z \in \partial\Omega,\ |h| \leqslant h_0\}.$$

We first show there exists $C > 0$ such that

$$|u(x)| \leqslant C_0\|\phi\|_\infty, \quad \forall\, x \in \Omega_{h_0} \cup \Omega, \tag{6.72}$$

where $\|\phi\|_\infty \equiv \|\phi\|_{C(\partial\Omega)}$. If $x \in \Omega \setminus \Omega_R$ for some small $R > 0$, $R \leqslant h_0$, then

$$|u(x)| \leqslant \left| \int_{\partial\Omega} K(x,\xi)\phi(\xi)\,d\sigma_\xi \right| \leqslant MR^{\alpha-2}\|\phi\|_\infty. \tag{6.73}$$

Now let $x \in \Omega_R$. Then

$$x = z + hn(z) \text{ for some } z \in \partial\Omega,\ 0 \leqslant h \leqslant R \leqslant h_0. \tag{6.74}$$

Noting that

$$\langle n(z), n(\xi)\rangle = 1 - \langle n(z), n(z) - n(\xi)\rangle,$$

by (6.38) in Lemma 6.1, we can choose $R > 0$ such that

$$\langle n(z), n(\xi)\rangle \geqslant \frac{1}{2}\ \forall\, \xi,\ z \in \partial\Omega \text{ with } |\xi - z| \leqslant R. \tag{6.75}$$

Further, we assume R is small enough that

$$S_{z,R} \equiv \{\xi \in \partial\Omega \mid |\xi - z| \leqslant R,\ z \text{ and } x \text{ are related through (6.74)}\}$$

is connected for each $x \in \partial\Omega$. Then (6.73) implies $S_{z,R}$ can be bijectively projected into the tangent plane passing $\partial\Omega$ at z. We establish polar coordinates (ρ, θ) in the tangent plane with origin at z. Then

$$d\sigma_\xi = \frac{\rho\, d\rho\, d\theta}{\langle n(\xi), n(z)\rangle} \leqslant 2\rho\, d\rho\, d\theta, \text{ on } S_{z,R},$$

and

$$\begin{aligned}\left|\int_{S_{z,R}} K(x,\xi)\phi(\xi)\, d\sigma_\xi\right| &\leqslant M\|\phi\|_\infty \int_{S_{z,R}} |x-\xi|^{\alpha-2} d\sigma_\xi \\ &\leqslant 2M\|\phi\|_\infty \frac{3}{4}\int_0^{2\pi}\int_0^R \rho^{\alpha-2}\rho\, d\rho\, d\theta \\ &= 3\pi M\|\phi\|_\infty R^\alpha/\alpha, \end{aligned} \tag{6.76}$$

where we have used

$$|x-\xi| \geqslant \frac{3}{4}|z-\xi| = \frac{3}{4}\rho, \quad \text{if } R \text{ is sufficiently small,} \tag{6.77}$$

and the projection of $S_{z,R}$ into the tangent plane is contained in the disk of radius R with origin at z. In addition,

$$\begin{aligned}\left|\int_{\partial\Omega \setminus S_{z,R}} K(x,\xi)\phi(\xi)\, d\sigma_\xi\right| &\leqslant M\|\phi\|_\infty \int_{\partial\Omega\setminus S_{z,R}} R^{\alpha-2} d\sigma_\xi \\ &\leqslant M\|\phi\|_\infty R^{\alpha-2} \text{ meas } \partial\Omega.\end{aligned} \tag{6.78}$$

Combining (6.73), (6.76) and (6.78), we have established (6.72).

To prove the uniform Hölder continuity (6.70), we need only consider $x_1, x_2 \in \Omega_{h_0}$, where $x_j = z_j + h_j n(z_j)$, $j = 1, 2$. Let

$$0 < |x_1 - x_2| < \frac{1}{4}R, \qquad R \text{ is the same as in (6.74)}$$

Using (6.38) in Lemma 6.1, we have

$$\frac{1}{2}|x_1 - x_2| \leqslant |z_1 - z_2| \leqslant 2|x_1 - x_2|,$$

provided R and h_0 are chosen small enough (depending on the M in (6.38)). Now letting

$$r = 4|x_1 - x_2|,$$

and arguing as in (6.76), we have

$$\begin{aligned}&\left|\int_{S_{z_1,r}} [K(x_1,\xi) - K(x_2,\xi)]\,\phi(\xi)\, d\sigma_\xi\right| \\ \leqslant& M\|\phi\|_\infty \left\{\int_{S_{z_1,r}} |x_1-\xi|^{\alpha-2} d\sigma_\xi + \int_{S_{z_2,3r/2}} |x_2-\xi|^{\alpha-2} d\sigma_\xi\right\} \\ \leqslant& C_1\|\phi\|_\infty |x_1-x_2|^\alpha \end{aligned} \tag{6.79}$$

for some $C_1 > 0$ depending only on M and α. In the above, we have used $S_{z_1,r} \subset S_{z_2,3r/2}$ because

$$\begin{aligned}|y-z_2| &\leqslant |y-z_1|+|z_1-z_2|\\ &\leqslant r+2|x_1-x_2|\\ &\leqslant r+\frac{1}{2}r=\frac{3}{2}r.\end{aligned}$$

Using the condition (6.68), we have

$$\begin{aligned}&\left|\int_{S_{z_1,R}\setminus S_{z_1,r}}[K(x_1,\xi)-K(x_2,\xi)]\,\phi(\xi)\,d\sigma_\xi\right|\\ \leqslant& M\|\phi\|_\infty\sum_{j=1}^{m}|x_1-x_2|^j\int_{S_{z_1,R}\setminus S_{z_1,r}}|x_1-\xi|^{\alpha-2-j}\,d\sigma_\xi\\ \leqslant& 4\pi M\|\phi\|_\infty\sum_{j=1}^{m}|x_1-x_2|^j\int_{r/4}^{R}\rho^{\alpha-1-j}\,d\rho,\end{aligned}\tag{6.80}$$

where we have used the fact that the projection of $S_{z_1,R}\setminus S_{z_1,r}$ into the tangent plane at z_1 is contained in the annulus with radius $\frac{1}{4}r$ and R. But

$$\int_{r/4}^{R}\rho^{\alpha-1-j}\,d\rho \leqslant \begin{cases}\dfrac{1}{j-\alpha}|x_1-x_2|^{\alpha-j}, & j>\alpha,\\ \ln\left[\dfrac{R}{|x_1-x_2|}\right], & j=\alpha,\\ \dfrac{1}{\alpha-j}R^{\alpha-j}, & j<\alpha,\end{cases}\tag{6.81}$$

and for $\beta\in(0,1)$, $|x_1-x_2|<1$, we have

$$|x_1-x_2|\ln|x_1-x_2|^{-1}\leqslant\frac{1}{1-\beta}|x_1-x_2|^{\beta}.$$

Hence

$$\begin{aligned}&\left|\int_{S_{z_1,R}\setminus S_{z_1,r}}[K(x_1,\xi)-K(x_2,\xi)]\,\phi(\xi)\,d\sigma_\xi\right|\\ \leqslant& C_2\|\phi\|_\infty\begin{cases}|x_1-x_2|^{\alpha}, & \alpha<1,\\ |x_1-x_2|^{\beta}, & \alpha=1,\ 0<\beta<1,\\ |x_1-x_2|, & \alpha>1,\end{cases}\end{aligned}\tag{6.82}$$

for some $C_2>0$ depending only on m,M,R,α and β.

Using (6.68) again, we have

$$\begin{aligned}&\left|\int_{\partial\Omega\setminus S_{z_1,R}}[K(x_1,\xi)-K(x_2,\xi)]\,\phi(\xi)\,d\sigma_\xi\right|\\ \leqslant& M\|\phi\|_\infty\sum_{j=1}^{m}|x_1-x_2|^j\int_{\partial\Omega\setminus S_{z_1,R}}|x_1-\xi|^{\alpha-2-j}\,d\sigma_\xi\\ \leqslant& C_3\|\phi\|_\infty|x_1-x_2|,\end{aligned}\tag{6.83}$$

for some $C_3 > 0$ depending only on m, M, R, α and meas $\partial\Omega$.

Combining (6.79), (6.82) and (6.83), we obtain

$$|u(x_1) - u(x_2)| \leqslant (C_1 + C_2 + C_3)|x_1 - x_2|^{\beta} \|\phi\|_{\infty}$$

for all $x_1, x_2 \in \Omega_{h_0}$ with $|x_1 - x_2| \leqslant \frac{1}{4}R$. If $|x_1 - x_2| \geqslant \frac{1}{4}R$, we can simply use (6.72) to obtain

$$|u(x_1) - u(x_2)| \leqslant 2C_0(4/R)^{\beta}|x_1 - x_2|^{\beta} \|\phi\|_{\infty}.$$

Therefore

$$|u(x_1) - u(x_2)| \leqslant C_4|x_1 - x_2|^{\beta} \|\phi\|_{\infty}, \quad \forall\, x_1, x_2 \in \Omega_{h_0}.$$

Hence (6.71) is proved. □

Using Theorem 6.4, we can now easily prove the following.

Theorem 6.5. *Let $f \in L^{\infty}(\partial\Omega)$. Then $V_1(f) \in C^{0,\alpha}(\mathbb{R}^3)$ for all $\alpha \in (0,1)$ and*

$$\|V_1(f)\|_{C^{0,\alpha}(\mathbb{R}^3)} \leqslant C(\alpha)\|f\|_{L^{\infty}(\partial\Omega)} \tag{6.84}$$

for some $C(\alpha) > 0$ depending only on α.

Proof. The fundamental solution $E(x - \xi)$ as a kernel satisfies

$$|E(x - \xi)| \leqslant \frac{1}{4\pi|x - \xi|}, \tag{6.85}$$

corresponding to $\alpha = 1$ in (6.67). Also

$$\begin{aligned} |E(x_1 - \xi) - E(x_2 - \xi)| &= (4\pi)^{-1}\left|\frac{1}{|x_1 - \xi|} - \frac{1}{|x_2 - \xi|}\right| \\ &\leqslant (4\pi)^{-1}\frac{|x_1 - x_2|}{|x_1 - \xi||x_2 - \xi|} \\ &\leqslant (4\pi)^{-1}\frac{2|x_1 - x_2|}{|x_1 - \xi|^2} \end{aligned} \tag{6.86}$$

for $x_1, x_2 \in \mathbb{R}^3$ satisfying $2|x_1 - x_2| \leqslant |x_2 - \xi|$,

$$|x_2 - \xi| \geqslant |x_1 - \xi| - |x_1 - x_2| \geqslant |x_1 - \xi| - \frac{1}{2}|x_2 - \xi|,$$

$$|x_1 - \xi| \leqslant \frac{1}{2}|x_2 - \xi|.$$

So (6.86) corresponds to (6.68) with $m = 1$. Hence (6.84) is proved. □

We proceed to establish (6.44) and (6.45).

Lemma 6.3. *Let $f \in C(\partial\Omega)$. Then $V_2^{+}(f)$ and $V_2^{-}(f)$ can be extended respectively from Ω^c to $\overline{\Omega^c}$ and from Ω to $\overline{\Omega}$, such that*

$$\lim_{y \to x} V_2^{\pm}(f)(y) = \pm\frac{1}{2}f(x) + \int_{\partial\Omega} \frac{\partial E(x - \xi)}{\partial n_{\xi}} f(\xi)\, d\sigma_{\xi}, \quad x \in \partial\Omega \tag{6.87}$$

where $y \to x$ from Ω^c to $\partial\Omega$ for "+", and from Ω to $\partial\Omega$ for "−", and the integral above exists as an improper integral.

Proof. We prove the case $N = 3$. Since

$$\frac{\partial E(x-\xi)}{\partial n_\xi} = \frac{\langle n_\xi, x-\xi\rangle}{4\pi|x-\xi|^3}, \tag{6.88}$$

by (6.38), we have

$$\left|\frac{\partial E(x-\xi)}{\partial n_\xi}\right| \leqslant \frac{M}{4\pi|x-\xi|}, \quad x, \xi \in \partial\Omega, x \neq \xi. \tag{6.89}$$

Hence the integral in (6.87) exists as an improper integral.

We first prove (6.87) for $f(x) \equiv 1$. From the Gauss divergence theorem, we have

$$V_2(1)(y) = \begin{cases} 0, & y \in \Omega^c, \\ -1, & y \in \Omega. \end{cases} \tag{6.90}$$

Now for $x \in \partial\Omega$, we determine

$$W(x) \equiv \int_{\partial\Omega} \frac{\partial E(x-\xi)}{\partial n_\xi} d\sigma_\xi.$$

For $x \in \partial\Omega$ and $r > 0$, let

$$H_{x,r} \equiv \{\xi \in \Omega \mid |\xi - x| = r\}.$$

Then, using the divergence theorem and a limiting process, we have

$$\begin{aligned}\int_{\partial\Omega} \frac{\partial E(x-\xi)}{\partial n_\xi} d\sigma_\xi &= \int_\Omega \Delta_\xi E(x-\xi)\, d\sigma_\xi \\ &= \int_{B(x,r)} \Delta_\xi E(x-\xi)\, d\sigma_\xi \\ &= \int_{H_{x,r}} \frac{\partial E(x-\xi)}{\partial n_\xi} d\sigma_\xi + \int_{\Gamma_r} \frac{\partial E(x-\xi)}{\partial n_\xi} d\sigma_\xi, \end{aligned} \tag{6.91}$$

where $\Gamma_r \equiv B(x,r) \cap \partial\Omega$. See Fig. 6.3. The contribution from the integral on Γ_r tends to zero as $r \to 0$, because the measure of Γ_r tends to zero. On $H_{x,r}$, we have

$$\langle n_\xi, x-\xi\rangle = -|x-\xi| = -r.$$

So

$$\frac{\partial E(x-\xi)}{\partial n_\xi} = -\frac{1}{4\pi r^2}.$$

We get

$$\begin{aligned}\int_{\partial\Omega} \frac{\partial E(x-\xi)}{\partial n_\xi} d\sigma_\xi &= \lim_{r\to 0} \int_{H_{x,r}} \frac{\partial E(x-\xi)}{\partial n_\xi} d\sigma_\xi \\ &= \lim_{r\to 0} \left(-\frac{1}{4\pi r^2}\right) \int_{H_{x,r}} d\sigma_\xi. \end{aligned}$$

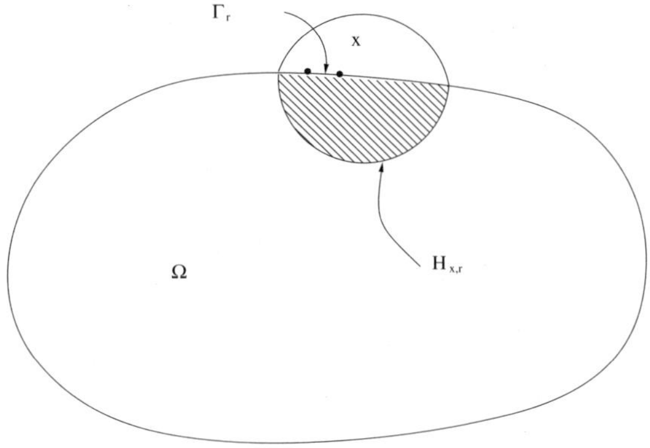

Figure 6.3: The surfaces $H_{x,r}$ and Γ_r contained in $\overline{\Omega}$.

But by (6.38), using local polar coordinates on the tangent plane to $\partial\Omega$ at x, we have

$$\begin{aligned} d\sigma_\xi &= \frac{\rho\, d\rho\, d\theta}{\langle n_x, n_\xi\rangle} = \frac{\rho\, d\rho\, d\theta}{\langle n_x, n_x\rangle + \langle n_x, n_\xi - n_x\rangle} \\ &= \frac{\rho\, d\rho\, d\theta}{1 + \langle n_x, n_\xi - n_x\rangle} = \rho\, d\rho\, d\theta + \mathscr{O}(\rho)\rho\, d\rho\, d\theta, \end{aligned}$$

hence

$$\int_{H_{x,r}} d\sigma_\xi = 2\pi r^2 + \mathscr{O}(r^3)$$

uniformly for all $x \in \partial\Omega$. Thus

$$\int_{\partial\Omega} \frac{\partial E(x-\xi)}{\partial n_\xi}\, d\sigma_\xi = -\frac{1}{2}, \quad x \in \partial\Omega.$$

From (6.90), we see that

$$\begin{aligned} \lim_{y\to x} V_2^{\pm}(1)(y) &= \begin{cases} 0 & \text{for “+”}, \\ -1 & \text{for “−”}, \end{cases} \\ &= \pm\frac{1}{2} + \int_{\partial\Omega} \frac{\partial E(x-\xi)}{\partial n_\xi}\, d\sigma_\xi, \quad x \in \partial\Omega. \end{aligned} \tag{6.92}$$

Therefore, (6.87) has been proved for $f(x) \equiv 1$.

Now let $f \in C^0(\partial\Omega)$ be arbitrary. Let $y \in \Omega_{h_0} \setminus \partial\Omega$ and $x \in \partial\Omega$. For each $y \in \Omega_{h_0} \setminus \partial\Omega$, we have a unique representation

$$y = x + h(y)n_x, \qquad -h_0 < h(y) < h_0. \tag{6.93}$$

Write

$$\begin{aligned} v(y) &\equiv \int_{\partial\Omega} \frac{\partial E(y-\xi)}{\partial n_\xi} f(\xi)\, d\sigma_\xi \\ &= f(x)\int_{\partial\Omega} \frac{\partial E(y-\xi)}{\partial n_\xi}\, d\sigma_\xi + \int_{\partial\Omega} \frac{\partial E(y-\xi)}{\partial n_\xi}[f(\xi) - f(x)]\, d\sigma_\xi \\ &\equiv f(x)w(y) + u(y), \quad y \in \Omega_{h_0} \setminus \partial\Omega, \end{aligned} \tag{6.94}$$

where in the definition of $u(y)$, we have inverted the relation of (6.93) to obtain the dependence of u on y only. We wish to show that u can be extended to the entire Ω_{h_0} such that

$$\lim_{y\to x} u(y) = \lim_{h(y)\to 0} u(x+h(y)n_x) = u(x), \qquad x \in \partial\Omega, \tag{6.95}$$

uniformly on $\partial\Omega$. If this is true then, from (6.92) and (6.93), we will have

$$\lim_{y\to x} v(y) = f(x) \left\{ \begin{array}{l} \dfrac{1}{2} + \displaystyle\int_{\partial\Omega} \frac{\partial E(x-\xi)}{\partial n_\xi} d\sigma_\xi \ (y \to x \text{ from } \Omega^c) \\ -\dfrac{1}{2} + \displaystyle\int_{\partial\Omega} \frac{\partial E(x-\xi)}{\partial n_\xi} d\sigma_\xi \ (y \to x \text{ from } \Omega) \end{array} \right\} + u(x)$$

$$= \pm\frac{1}{2} f(x) + f(x) \int_{\partial\Omega} \frac{\partial E(x-\xi)}{\partial n_\xi} d\sigma_\xi + \int_{\partial\Omega} \frac{\partial E(x-\xi)}{\partial n_\xi} [f(\xi) - f(x)]\, d\sigma_\xi$$

$$= \pm\frac{1}{2} f(x) + \int_{\partial\Omega} \frac{\partial E(x-\xi)}{\partial n_\xi} f(\xi)\, d\sigma_\xi, \quad x \in \partial\Omega,$$

and (6.87) will be proved. So we now try to prove (6.95).

Using Lemma 6.1, we have

$$\begin{aligned} |y-\xi|^2 &= |x-y|^2 + 2\langle y-x, x-\xi\rangle + |x-\xi|^2 \\ &= |x-y|^2 + 2h(y)\langle n_x, x-\xi\rangle + |x-\xi|^2 \\ &\geqslant |x-y|^2 - 2Mh_0|x-\xi|^2 + |x-\xi|^2 \\ &\geqslant \frac{1}{2}[|x-y|^2 + |x-\xi|^2], \text{ if } h_0 \text{ is sufficiently small.} \end{aligned} \tag{6.96}$$

We write

$$\frac{\partial E(y-\xi)}{\partial n_\xi} = \frac{1}{4\pi}\left(\frac{\langle n_\xi, x-\xi\rangle}{|y-\xi|^3} + \frac{\langle n_\xi, y-x\rangle}{|y-\xi|^3} \right)$$

By (6.96), we have

$$\begin{aligned} \frac{1}{4\pi}\left| \frac{\langle n_\xi, x-\xi\rangle}{|y-\xi|^3} \right| &\leqslant \frac{2^{3/2}}{4\pi} \frac{|\langle n_\xi, x-\xi\rangle|}{[|x-y|^2 + |x-\xi|^2]^{3/2}} \\ &\leqslant \frac{M}{2^{1/2}\pi} \frac{|x-\xi|^2}{|x-\xi|^3} \\ &\leqslant \frac{C_1}{|x-\xi|} \qquad \forall\, \xi \in \Omega_{h_0}. \end{aligned}$$

We also get

$$\frac{1}{4\pi}\left| \frac{\langle n_\xi, y-x\rangle}{|y-\xi|^3} \right| \leqslant \frac{2^{3/2}}{4\pi} \frac{|y-x|}{(|x-y|^2 + |x-\xi|^2)^{3/2}} \quad \text{(by (6.96))}.$$

Therefore by projecting $S_{x,r} \equiv \{\xi \in \partial\Omega \mid |\xi - x| < r\}$ into the tangent plane at x, we have

$$\begin{aligned} \int_{S_{x,r}} \left| \frac{\partial E(y-\xi)}{\partial n_\xi} \right| d\sigma_\xi &\leqslant C_1 \int_0^{2\pi}\int_0^r \frac{1}{\rho}\rho\, d\rho\, d\theta + C_2|y-x| \int_0^{2\pi}\int_0^\infty \frac{\rho\, d\rho\, d\theta}{(\rho^2 + |x-y|^2)^{3/2}} \\ &= 2\pi C_1 r + C_2 2\pi \equiv C_3. \end{aligned} \tag{6.97}$$

From the mean value theorem and (6.96), we have

$$\left|\frac{\partial E(y-\xi)}{\partial n_\xi}-\frac{\partial E(x-\xi)}{\partial n_\xi}\right|=\frac{1}{4\pi}\left|\left[(-3)\frac{\langle n_\xi,\bar{x}-\xi\rangle}{|\bar{x}-\xi|^5}(\bar{x}-\xi)+\frac{n_\xi}{|\bar{x}-\xi|^3}\right]\cdot(y-x)\right|, \tag{6.98}$$

for some $\bar{x}=\alpha y+(1-\alpha)x$, $\alpha\in(0,1)$. Thus for y such that $2|y-x|\leqslant|x-\xi|$,

$$|\bar{x}-\xi|=|(x-\xi)+\alpha(y-x)|\geqslant|x-\xi|-\alpha|y-x|\geqslant\frac{1}{2}|x-\xi|,$$

and similarly,

$$|\bar{x}-\xi|\leqslant\frac{3}{2}|x-\xi|.$$

Using the above in (6.98), we have

$$\left|\frac{\partial E(y-\xi)}{\partial n_\xi}-\frac{\partial E(x-\xi)}{\partial n_\xi}\right|\leqslant C_4\frac{|y-x|}{|x-\xi|^3}.$$

Therefore

$$\int_{\partial\Omega\setminus S_{x,r}}\left|\frac{\partial E(y-\xi)}{\partial n_\xi}-\frac{\partial E(x-\xi)}{\partial n_\xi}\right|d\sigma_\xi\leqslant C_5\frac{|x-y|}{r^3}. \tag{6.99}$$

We combine (6.97) and (6.99) to obtain

$$\begin{aligned}|u(y)-u(x)|&=\left|\int_{\partial\Omega}\left\{\frac{\partial E(y-\xi)}{\partial n_\xi}[f(\xi)-f(x)]-\frac{\partial E(x-\xi)}{\partial n_\xi}[f(\xi)-f(x)]\right\}d\sigma_\xi\right|\\&\leqslant 2\int_{S_{x,r}}\left|\frac{\partial E(y-\xi)}{\partial n_\xi}\right|d\sigma_\xi\sup_{\xi\in S_{x,r}}|f(\xi)-f(x)|\\&\quad+2\int_{\partial\Omega\setminus S_{x,r}}\left|\frac{\partial E(y-\xi)}{\partial n_\xi}-\frac{\partial E(x-\xi)}{\partial n_\xi}\right|d\sigma_\xi\sup_{x\in\partial\Omega}|f(x)|\\&\leqslant C_6\left[\sup_{|\xi-x|\leqslant r}|f(\xi)-f(x)|+\frac{|x-y|}{r^3}\right].\end{aligned}$$

Given $\varepsilon>0$, we can choose $r>0$ such that

$$|f(\xi)-f(x)|\leqslant\frac{\varepsilon}{2C_6}\quad\forall\,\xi,x\in\partial\Omega,|\xi-x|<r,$$

because f is uniformly continuous on $\partial\Omega$. Take $\delta<\varepsilon(2C_6)^{-1}r^3$. Then for all $y,x\in\partial\Omega$, $|y-x|<\delta$, we have

$$|u(x)-u(y)|\leqslant\varepsilon.$$

Therefore the proof is complete. □

Exercise 6.2. Use the ideas given in the proof of Lemma 6.3 to show that if $\partial\Omega$ is not smooth at x_0 but forms an (interior) angle θ with vertex at x_0, then (1.20) is replaced by

$$\frac{\theta}{2\pi}w(x_0) = \int_{\partial\Omega}\left[E(x_0-\xi)\frac{\partial w(\xi)}{\partial n} - \frac{\partial E(x-\xi)}{\partial n_\xi}w(\xi)\right]d\sigma_\xi$$

in $\mathbb{R}^2$. □

The integral $V_2(f)(x)$ for $x \in \partial\Omega$, interpreted as an improper integral, actually has a stronger continuity property on $\partial\Omega$. This is given in the following.

Theorem 6.6. *Let $f \in C(\partial\Omega)$ and $x \in \partial\Omega$. Then*

$$V_2(f)(x) = \int_{\partial\Omega}\frac{\partial E(x-\xi)}{\partial n_\xi}f(\xi)\,d\sigma_\xi, \qquad x \in \partial\Omega,$$

is in $C^{0,\alpha}(\partial\Omega)$, satisfying

$$\|V_2(f)\|_{C^{0,\alpha}(\partial\Omega)} \leqslant C(\alpha)\|f\|_{C(\partial\Omega)}$$

for any $\alpha \in (0,1)$, for some $C(\alpha) > 0$ depending only on α.

Proof. We wish to use Theorem 6.4 with $\alpha = 1$, $m = 2$. We first note that the condition (6.67) is verified in (6.89) for $x, \xi \in \partial\Omega$, for the kernel $\partial E(x-\xi)/\partial n_\xi$. To verify (6.68), we have

$$\begin{aligned}&4\pi\left[\frac{\partial E(x_1-\xi)}{\partial n_\xi} - \frac{\partial E(x_2-\xi)}{\partial n_\xi}\right]\\ &= \left(\frac{1}{|x_1-\xi|^3} - \frac{1}{|x_2-\xi|^3}\right)\langle x_2-\xi, n_\xi\rangle + \frac{\langle x_1-x_2, n_\xi\rangle}{|x_1-\xi|^3}.\end{aligned} \tag{6.100}$$

For $x_1, x_2, \xi \in \partial\Omega$ satisfying

$$2|x_1-x_2| < |x_1-\xi|, \tag{6.101}$$

we have

$$\begin{aligned}\frac{1}{2}|x_1-\xi| &\leqslant -|x_1-x_2| + |\xi-x_1|\\ &\leqslant |x_2-\xi| \leqslant |x_1-x_2| + |\xi-x_1|\\ &\leqslant \frac{3}{2}|x_1-\xi|\end{aligned} \tag{6.102}$$

We estimate the first term on the RHS of (6.100) using the mean value theorem:

$$\begin{aligned}\left|\frac{1}{|x_1-\xi|^3} - \frac{1}{|x_2-\xi|^3}\right| &= \left|-3|\bar{x}-\xi|^{-5}(\bar{x}-\xi)\cdot(x_1-x_2)\right|\\ &\qquad (\bar{x} = \alpha x_2 + (1-\alpha)x_1 \text{ for some } 0<\alpha<1)\\ &\leqslant 3(1-\frac{1}{2}\alpha)^{-5}(1+\frac{1}{2}\alpha)|x_1-\xi|^{-4}|x_1-x_2|,\\ &\qquad \text{by } (6.101).\end{aligned} \tag{6.103}$$

By (6.101)–(6.103), we have

$$\left|\left(\frac{1}{|x_1-\xi|^3}-\frac{1}{|x_2-\xi|^3}\right)\langle x_2-\xi,n_\xi\rangle\right| \leqslant C_0|x_1-\xi|^{-2}|x_1-x_2|.$$

For the second term on the RHS of (6.100), we have

$$\begin{aligned}|\langle x_1-x_2,n_\xi\rangle| &\leqslant |\langle x_1-x_2,n_\xi-n_{x_1}\rangle|+|\langle x_1-x_2,n_{x_1}\rangle|\\ &\leqslant C_1(|x_1-x_2||\xi-x_1|+|x_1-x_2|^2),\end{aligned}$$

by Lemma 6.1. Thus

$$\left|\frac{\langle x_1-x_2,n_\xi\rangle}{|x_1-\xi|^3}\right| \leqslant C_2\left(|x_1-\xi|^{-2}|x_1-x_2|+|x_1-\xi|^{-3}|x_1-x_2|^2\right). \tag{6.104}$$

□

The smoothness of the double-layer potential $V_2(f)$ on the closed exterior and interior domains $\overline{\Omega^c}$ and $\overline{\Omega}$ can now be given.

Theorem 6.7. *Let $f\in C^{0,\alpha}(\partial\Omega)$, $0<\alpha<1$. Then the double-layer potential $V_2(f)$ is in $C^{0,\alpha}(\overline{\Omega^c})$ and in $C^{0,\alpha}(\overline{\Omega})$, satisfying*

$$\|V_2(f)\|_{C^{0,\alpha}(\overline{\Omega^c})} \leqslant C(\alpha)\|f\|_{C^{0,\alpha}(\partial\Omega)}, \tag{6.105}$$

$$\|V_2(f)\|_{C^{0,\alpha}(\overline{\Omega})} \leqslant C(\alpha)\|f\|_{C^{0,\alpha}(\partial\Omega)} \tag{6.106}$$

for some constant $C(\alpha)>0$ depending only on $\partial\Omega$ and α.

Proof. Return to the decomposition (6.94). It is easy to observe that the first term on the RHS of (6.94) is in $C^{0,\alpha}$ on $\overline{\Omega^c}$ and $\overline{\Omega}$. For the second term, we apply the following Lemma 6.4. Just verify that the kernel $\partial E(x-\xi)/\partial n_\xi$ satisfies conditions (6.107) and (6.108) with $m=1$ for $x\in\mathbb{R}^3$ and $\xi\in\partial\Omega$. Also, from (6.99), we see that (6.109) is satisfied. Therefore the second term on the RHS of (6.94) is $C^{0,\alpha}$ on Ω_{h_0}. That $V_2(f)$ is $C^{0,\alpha}$ on $\overline{\Omega^c}$ and $\overline{\Omega}$ and satisfies (6.105) and (6.106) now easily follows from the property of the kernel $\partial E(x-\xi)/\partial n_\xi$. □

The following lemma was also originally due to Giraud [81] (see [53, pp. 44–45]).

Lemma 6.4. *Let $K(x,\xi)$ be a kernel defined and continuous for all $x\in\Omega_{h_0}$, $\xi\in\partial\Omega$, $x\neq\xi$, such that there exists $M>0$ satisfying*

$$|K(x,\xi)| \leqslant M|x-\xi|^{-2}, \quad \forall\, x\in\Omega_{h_0},\ \xi\in\partial\Omega,\ x\neq y. \tag{6.107}$$

Furthermore, assume that there exists $m\in\mathbb{Z}^+$ such that K satisfies

$$|K(x_1,\xi)-K(x_2,\xi)| \leqslant M\sum_{j=1}^{m}|x_1-\xi|^{-2-j}|x_1-x_2|^j \tag{6.108}$$

for all $x_1, x_2 \in \Omega_{h_0}$, $\xi \in \partial\Omega$, *such that* $2|x_1 - x_2| \leqslant |x_1 - \xi|$, *and that*

$$\left| \int_{\partial\Omega \setminus S_{z,r}} K(x,\xi)\, d\sigma_\xi \right| \leqslant M \tag{6.109}$$

for all $z \in \partial\Omega$, $x = z + hn(z) \in \Omega_{h_0}$, *and all* $0 < r < R$. *Let* $f \in C^{0,\alpha}(\partial\Omega)$ *for some* $0 < \alpha < 1$, *and define*

$$u(x) = \int_{\partial\Omega} K(x,\xi)\,[f(\xi) - f(z)]\, d\sigma_\xi, \quad x \in \Omega_{h_0}, \tag{6.110}$$

where

$$x = z + hn(z). \tag{6.111}$$

Then $u \in C^{0,\alpha}(\Omega_{h_0})$ *and*

$$\|u\|_{C^{0,\alpha}(\Omega_{h_0})} \leqslant C \|f\|_{C^{0,\alpha}(\partial\Omega)} \tag{6.112}$$

for some $C > 0$ *independent of* f.

Proof. From (6.107) and (6.111), for $x \in \Omega_{h_0}$ and $\xi, z \in \partial\Omega$, we have

$$|K(x,\xi)\,[f(\xi) - f(z)]| \leqslant M \cdot 2^\alpha \|f\|_{C^{0,\alpha}(\partial\Omega)} |x - \xi|^{\alpha-2}$$

hence (6.110) exists as an improper integral.
We mimic the proof in Theorem 6.4 to obtain

$$\left| \int_{S_{z_1,r}} \{K(x_1,\xi)[f(y) - f(z_1)] - K(x_2,z)[f(\xi) - f(z_2)]\}\, d\sigma_\xi \right|$$
$$\leqslant C_1 \|f\|_{C^{0,\alpha}(\partial\Omega)} |x_1 - x_2|^\alpha,$$
$$\left| \int_{S_{z_1,R} \setminus S_{z_1,r}} [K(x_1,z) - K(x_2,\xi)]\,[f(\xi) - f(z_2)]\, d\sigma_\xi \right|$$
$$\leqslant C_2 \|f\|_{C^{0,\alpha}(\partial\Omega)} |x_1 - x_2|^\alpha,$$

and

$$\left| \int_{\partial\Omega \setminus S_{z_1,R}} [K(x_1,z) - K(x_2,\xi)]\,[f(\xi) - f(z_2)]\, d\sigma_\xi \right|$$
$$\leqslant C_3 \|f\|_{C^{0,\alpha}(\partial\Omega)} |x_1 - x_2|,$$

where $C_1, C_2, C_3 > 0$ are independent of f. The only difference from the proof of Theorem 6.4 is that the logarithmic term will not appear in (6.81) because $0 < \alpha < 1$. The rest of the proof can be completed just as that of Theorem 6.4. □

Let us now study the differentiability properties of layer potentials (6.46)–(6.54). We first give two theorems concerning the derivatives of simple-layer potentials.

Theorem 6.8. *Let* $f \in C^{0,\alpha}(\partial\Omega)$ *with* $0 < \alpha < 1$*. Then* $\nabla V_1(f)$ *can be extended from* Ω^c *to* $\overline{\Omega^c}$ *and from* Ω *into* $\overline{\Omega}$*, with limiting values on* $\partial\Omega$*:*

$$\nabla V_1^{\pm}(f)(x) = \int_{\partial\Omega} \nabla_x E(x-\xi) f(\xi)\, d\sigma_\xi \mp \frac{1}{2} f(x) n(x), \quad x \in \partial\Omega, \tag{6.113}$$

where the integral exists as a Cauchy principal value. Furthermore,

$$\|\nabla V_1^{+}(f)\|_{C^{0,\alpha}(\overline{\Omega^c})} \leqslant C(\alpha) \|f\|_{C^{0,\alpha}(\partial\Omega)}, \tag{6.114}$$

$$\|\nabla V_1^{-}(f)\|_{C^{0,\alpha}(\overline{\Omega})} \leqslant C(\alpha) \|f\|_{C^{0,\alpha}(\partial\Omega)}, \tag{6.115}$$

for $C(\alpha) > 0$ *depending only on* $\partial\Omega$ *and* α*.*

Proof. We first consider the special case $f = 1$. Because

$$\nabla_x E(x,\xi) = -\nabla_\xi E(x-\xi), \tag{6.116}$$

we have

$$\nabla V_1(1)(x) = -\int_{\partial\Omega} \nabla_\xi E(x-\xi)\, d\sigma_\xi \equiv w_1(x) + w_2(x),$$

where

$$w_1(x) = -\int_{\partial\Omega} \overline{\nabla}_\xi E(x-\xi)\, d\sigma_\xi,$$

$$w_2(x) = -\int_{\partial\Omega} n_\xi \frac{\partial E(x-\xi)}{\partial n_\xi}\, d\sigma_\xi, \tag{6.117}$$

and

$$\nabla_\xi = \overline{\nabla}_\xi + n_\xi \frac{\partial}{\partial n_\xi}, \tag{6.118}$$

$\overline{\nabla}$ is the gradient in the tangential directions on $\partial\Omega$, cf. (6.33),

have been utilized. Since w_2 is a double-layer potential with density (vector) $n_\xi = n(\xi)$, $n \in C^\infty(\partial\Omega)$, we can apply Lemma 6.3 and Theorem 6.6 to obtain

$$w_2^{\pm}(x) = -\int_{\partial\Omega} n_\xi \frac{\partial E(x-\xi)}{\partial n_\xi}\, d\sigma_\xi \mp \frac{1}{2} n(x), \quad x \in \partial\Omega, \tag{6.119}$$

$$\|w_2\|_{C^{0,\alpha}(\partial\Omega)} \leqslant C \|n\|_{C^{0,\alpha}(\partial\Omega)}.$$

To handle w_1 in (6.117), we use (6.34), noting that $\partial(\partial\Omega) = \emptyset$, the empty set, obtaining

$$w_1(x) = 2\int_{\partial\Omega} H(\xi) n_\xi E(x-\xi)\, d\sigma_\xi, \qquad x \in \mathbb{R}^3 \setminus \partial\Omega, \tag{6.120}$$

i.e., w_1 is a simple layer with density $2Hn$. Thus we can apply Theorem 6.5 to obtain the uniform Hölder continuity of w_1, and also the limit

$$w_1^{\pm}(x) = 2\int_{\partial\Omega} H(\xi)n_\xi E(x-\xi)\,d\sigma_\xi, \quad x \in \partial\Omega. \tag{6.121}$$

Combining (6.119) and (6.121), we see that the proof will be complete (for $f = 1$) if we can show

$$\lim_{r\downarrow 0}\int_{\partial\Omega\setminus S_{x,r}} \overline{\nabla}_\xi E(x-\xi)\,d\sigma_\xi = -2\int_{\partial\Omega} H(\xi)n_\xi E(x-\xi)\,d\sigma_\xi, \; x \in \partial\Omega. \tag{6.122}$$

But from (6.34),

$$\begin{aligned}\int_{\partial\Omega\setminus S_{x,r}} \overline{\nabla}_\xi E(x-\xi)\,d\sigma_\xi &= \int_{\substack{|x-\xi|=r\\ \xi\in\partial\Omega}} E(x-\xi)n_0(\xi)\,ds_\xi \\ &\quad -2\int_{\partial\Omega\setminus S_{x,r}} H(\xi)n_\xi E(x-\xi)\,d\sigma_\xi,\end{aligned}$$

where ds_ξ is the infinitesimal arclength on the curve $\{\xi \mid |\xi - x| = r\} \cap \partial\Omega$. We have

$$\begin{aligned}\int_{\substack{|x-\xi|=r\\ \xi\in\partial\Omega}} E(x-\xi)n_0(\xi)\,ds_\xi &= \frac{1}{4\pi r}\int_{\substack{|x-\xi|=r\\ \xi\in\partial\Omega}} n_0(\xi)\,ds_\xi \\ &= -\frac{1}{2\pi r}\int_{S_{x,r}} H(\xi)n_\xi\,d\sigma_\xi,\end{aligned}$$

and

$$\left|\int_{S_{x,r}} H(\xi)n_\xi\,d\sigma_\xi\right| \leqslant Cr^2$$

for some C depending only on $\partial\Omega$. Therefore (6.113) is proved for $f \equiv 1$.

For arbitrary density function f, we use the decomposition

$$\nabla V_1(f)(x) = f(z)\nabla V_1(1)(x) + \int_{\partial\Omega} \nabla_x E(x-\xi)[f(\xi) - f(z)]\,d\sigma_\xi$$

and apply Lemma 6.4. □

Theorem 6.9. *Let $f \in C(\partial\Omega)$. Then*

$$\frac{\partial}{\partial n}V_1^{\pm}(f)(x) = \mp\frac{1}{2}f(x) + \int_{\partial\Omega}\frac{\partial}{\partial n_x}E(x,\xi)f(\xi)\,d\sigma_\xi, \quad x \in \partial\Omega, \tag{6.123}$$

where the normal derivative

$$\frac{\partial}{\partial n}V_1^{\pm}(f)(x) \equiv \lim_{h\downarrow 0}\frac{\partial}{\partial n_x}V_1(f)(x \pm hn_x)$$

exists uniformly for $x \in \partial\Omega$, and the integral in (6.123) exists as an improper integral.

Proof. We represent $y \in \Omega_{h_0} \setminus \partial\Omega$ as

$$y = x + hn_x, \quad x \in \partial\Omega, -h_0 < h < h_0.$$

Then, we have

$$\begin{aligned}\langle n_x, \nabla V_1^{\pm}(f)(y)\rangle + V_2^{\pm}(f)(y) &= \int_{\partial\Omega} \langle n_x, \nabla_y E(y,\xi)\rangle \, d\sigma_\xi \\ &+ \int_{\partial\Omega} \langle n_\xi, \nabla_\xi E(y,\xi)\rangle \, d\sigma_\xi \\ &= \int_{\partial\Omega} \langle n_\xi - n_x, \nabla_\xi E(y,\xi)\rangle \, d\sigma_\xi \\ &(\nabla_y E(y,\xi) - -\nabla_x E(x,\xi)). \end{aligned} \tag{6.124}$$

We can extend the vector field n_x from $\partial\Omega$ to Ω_{h_0} such that for the parallel surface passing y, the unit normal vector at y is $n_y = n_x$. Then the kernel

$$k(y,\xi) = \langle n_\xi - n_y, \nabla_\xi E(y,\xi)\rangle$$

satisfies the conditions in Giraud's Theorem 6.4 with $\alpha = 1$ and $m = 1$. Therefore the integral in the last end of (6.124) is Hölder-continuous with respect to y, yielding

$$\lim_{h\downarrow 0} \int_{\partial\Omega} \langle n_\xi - n_y, \nabla_\xi E(y,\xi)\rangle \, d\sigma_\xi = \int_{\partial\Omega} \langle n_\xi - n_x, \nabla_\xi E(x,\xi)\rangle \, d\sigma_\xi.$$

Therefore, from (6.124) we have

$$\begin{aligned}&\frac{\partial V_1^{\pm}(f)(x)}{\partial n} \pm \frac{1}{2} f(x) + \int_{\partial\Omega} \frac{\partial E(x-\xi)}{\partial n_\xi} f(\xi)\, d\sigma_\xi \\ &= \int_{\partial\Omega} \langle n_\xi - n_x, \nabla_\xi E(x,\xi)\rangle f(\xi)\, d\sigma_\xi \\ &= \int_{\partial\Omega} \frac{\partial E(x-\xi)}{\partial n_\xi} f(\xi)\, d\sigma_\xi - \int_{\partial\Omega} \langle n_x, \nabla_\xi E(x,\xi)\rangle f(\xi)\, d\sigma_\xi, \\ &= \int_{\partial\Omega} \frac{\partial E(x-\xi)}{\partial n_\xi} f(\xi)\, d\sigma_\xi + \int_{\partial\Omega} \frac{\partial E(x-\xi)}{\partial n_x} f(\xi) d\sigma_x,\end{aligned}$$

resulting in

$$\frac{\partial V_1^{\pm}(f)(x)}{\partial n} = \mp \frac{1}{2} f(x) + \int_{\partial\Omega} \frac{\partial E(x-\xi)}{\partial n_\xi} f(\xi)\, d\sigma_\xi, x \in \partial\Omega.$$

□

Remark 6.1. If $f \in C^{0,\alpha}(\partial\Omega)$ then (6.123) is a simple consequence of Theorem 6.8. But if $f \in C(\partial\Omega)$ only then there are counterexamples in Günter [87] that the gradient of the simple-layer potential may not exist (in the classical sense), so Theorem 6.8 cannot be applied to Theorem 6.9. □

Next, we present three theorems concerning the derivatives of double-layer potentials.

Theorem 6.10. *Let $f \in C(\partial\Omega)$. Then the normal derivative of the double-layer potential $V_2(f)$ is continuous across $\partial\Omega$, i.e.,*

$$\frac{\partial}{\partial n} V_2^+(f)(x) = \frac{\partial}{\partial n} V_2^-(f)(x), \quad x \in \partial\Omega, \tag{6.125}$$

in the sense that

$$\lim_{h\downarrow 0} \left[\frac{\partial}{\partial n_x} V_2(f)(x+hn_x) - \frac{\partial}{\partial n_x} V_2(f)(x-hn_x) \right] = 0$$

uniformly for $x \in \partial\Omega$.

Proof. Denote $x \pm hn_x = x_\pm$. We first note that

$$\frac{\partial^2 E(x-\xi)}{\partial n_x \partial n_\xi} = \frac{1}{4\pi} \left(\frac{\langle n_x, n_\xi \rangle}{|x-\xi|^3} - \frac{3}{|x-\xi|^5} \langle n_\xi, x-\xi \rangle \langle n_x, x-\xi \rangle \right). \tag{6.126}$$

Also, using (6.90), we have

$$\begin{aligned} &\frac{1}{4\pi} \int_{\partial\Omega} \left(\frac{\langle n_x, n_\xi \rangle}{|x-\xi|^3} - \frac{3}{|x-\xi|^5} \langle n_\xi, x-\xi \rangle \langle n_x, x-\xi \rangle \right) f(x)\, d\sigma_\xi \\ &= f(x) \frac{\partial}{\partial n_x} \int_{\partial\Omega} \frac{\partial E(x-\xi)}{\partial n_\xi} d\sigma_\xi \\ &= 0, \quad \text{for } x \in \Omega^c \text{ or } \Omega. \end{aligned} \tag{6.127}$$

Thus from (6.126) and (6.127), we can write

$$\begin{aligned} &\frac{\partial}{\partial n_x} V_2(f)(x_+) - \frac{\partial}{\partial n_x} V_2(f)(x_-) \\ &= \frac{1}{4\pi} \int_{\partial\Omega} [f(\xi) - f(x)] \left\{ \left(\frac{1}{|x_+-\xi|^3} - \frac{1}{|x_- - \xi|^3} \right) \langle n_x, n_\xi \rangle \right. \\ &\quad \left. -3 \left[\frac{\langle n_\xi, x_+ - \xi \rangle \langle n_x, x_+ - \xi \rangle}{|x_+ - \xi|^5} - \frac{\langle n_\xi, x_- - \xi \rangle \langle n_x, x_- - \xi \rangle}{|x_- - \xi|^5} \right] \right\} d\sigma_\xi . \end{aligned} \tag{6.128}$$

From

$$|x_+ - \xi|^2 - |x_- - \xi|^2 = 4h \langle n_x, x - \xi \rangle$$

and

$$\frac{1}{|x_+ - \xi|^3} - \frac{1}{|x_- - \xi|^3} = \frac{4h\langle n_x, \xi - x \rangle (|x_+ - \xi|^2 + |x_+ - \xi||x_- - \xi| + |x_- - \xi|^2)}{|x_+ - \xi|^3 |x_- - \xi|^3 (|x_+ - \xi| + |x_- - \xi|)},$$

we can use Lemma 6.1 to obtain

$$\left| \frac{1}{|x_+ - \xi|^3} - \frac{1}{|x_- - \xi|^3} \right| \leqslant C \frac{h}{|x_+ - \xi|^3} \tag{6.129}$$

for some $C > 0$ independent of x and ξ, for $h > 0$ sufficiently small. Similarly, noting that

$$\begin{aligned}&\langle n_\xi, x_+ - \xi\rangle\langle n_x, x_+ - \xi\rangle - \langle n_\xi, x_- - \xi\rangle\langle n_x, x_- - \xi\rangle \\ &= 2h\left(\langle n_\xi, n_x\rangle\langle n_x, x - \xi\rangle + \langle n_\xi, x - \xi\rangle\right),\end{aligned}$$

the same type of estimate holds for the term inside the square brackets in the integrand of (6.128). We can now establish the uniform convergence of the limit similarly to the proof of Lemma 6.3 and complete the remainder of the proof. □

Exercise 6.3. Give details of the argument in the proof of Theorem 6.10, beginning at step (6.129). □

Theorem 6.11. *Let $f \in C^{0,\alpha}(\partial\Omega)$, $0 < \alpha < 1$. Then on $\partial\Omega$, $V_2(f) \in C^{1,\alpha}(\partial\Omega)$ such that*

$$\|\overline{\nabla} V_2(f)\|_{C^{0,\alpha}(\partial\Omega)} \leqslant C(\alpha)\|f\|_{C^{0,\alpha}(\partial\Omega)}$$

for some $C(\alpha) > 0$ depending only on α and $\partial\Omega$.

Proof. Let $x \in \partial\Omega$ and let $\tau(x)$ be a unit vector tangent to $\partial\Omega$ at x. Choose a C^2 curve $\alpha(t)$ on $\partial\Omega$ passing through x such that the unit tangent vector to α at x is $\tau(x)$. Let x_h be a point on the curve α in a neighborhood of x such that the arclength between x_h and x is $|h|$. Then

$$x_h = x + h\tau(x) + \mathscr{O}(h^2).$$

We have

$$\begin{aligned}\frac{\partial}{\partial\tau(x)} V_2(f)(x) &= \lim_{h\downarrow 0}\left[V_2(f)(x_h) - V_2(f)(x)\right]/h \\ &= \lim_{h\downarrow 0}\int_{\partial\Omega}\frac{1}{h}\left[\frac{\partial E(x_h - \xi}{\partial n_\xi} - \frac{\partial E(x-\xi)}{\partial n_\xi}\right] f(\xi)\,d\sigma_\xi.\end{aligned}$$

But

$$\begin{aligned}&4\pi\frac{1}{h}\left[\frac{\partial E(x_h - \xi)}{\partial n_\xi} - \frac{\partial E(x-\xi)}{\partial n_\xi}\right] \\ &= \frac{1}{h}\left(\frac{\langle n_\xi, x_h - \xi\rangle}{|x_h - \xi|^3} - \frac{\langle n_\xi, x - \xi\rangle}{|x-\xi|^3}\right) \\ &= \frac{\langle \tau(x), n_\xi\rangle}{|x-\xi|^3} - 3\frac{\langle n_\xi, x-\xi\rangle\langle\tau(x), x-\xi\rangle}{|x-\xi|^5} + \mathscr{O}\left(\frac{|h|}{|x-\xi|^3}\right).\end{aligned}$$

So $[\partial/\partial\tau(x)]V_2(f)(x)$, if it exists, should be equal to

$$\frac{\partial V_2(f)(x)}{\partial\tau(x)} = \frac{1}{4\pi}\int_{\partial\Omega}\left[\frac{\langle\tau(x), n_\xi\rangle}{|x-\xi|^3} - 3\frac{\langle n_\xi, x-\xi\rangle\langle\tau(x), x-\xi\rangle}{|x-\xi|^5}\right] f(\xi)\,d\sigma_\xi.$$

The integrand on the RHS has a rather strong singularity. We utilize (6.127) to rewrite it as

$$u(x) \equiv \frac{1}{4\pi}\int_{\partial\Omega}[f(\xi)-f(x)]\left[\frac{\langle\tau(x),n_\xi\rangle}{|x-\xi|^3}-3\frac{\langle n_\xi,x-\xi\rangle\langle\tau(x),x-\xi\rangle}{|x-\xi|^5}\right]d\sigma_\xi$$

and consider the limit

$$v(x;h) \equiv \frac{1}{h}[V_2(f)(x_h)-V_2(f)(x)]-u(x).$$

Following the same ideas as in the proofs of Theorem 6.4 and Lemma 6.4, we can show

$$v(x;h)=\mathscr{O}(|h|^\alpha).$$

Thus $\overline{\nabla}V_2(f)$ exists and lies in $C^{0,\alpha}(\partial\Omega)$. Details of the arguments are omitted. □

Theorem 6.12. *Let $f\in C^{1,\alpha}(\partial\Omega)$, $0<\alpha<1$. Then $\nabla V_2^\pm(f)$ can be extended from Ω^c into $\overline{\Omega^c}$ and from Ω into $\overline{\Omega}$ such that on $\partial\Omega$,*

$$\nabla V_2^\pm(f)(x) = -\int_{\partial\Omega}\nabla_x E(x-\xi)\times\left[\overline{\nabla}f(\xi)\times n_\xi\right]d\sigma_\xi \pm\frac{1}{2}\overline{\nabla}f(x),$$

$$x\in\partial\Omega, \tag{6.130}$$

where the integral exists as a Cauchy principal value. Furthermore,

$$\|\nabla V_2^+(f)\|_{C^{0,\alpha}(\Omega^c)} \leqslant C(\alpha)\|f\|_{C^{1,\alpha}(\partial\Omega)}$$
$$\|\nabla V_2^-(f)\|_{C^{0,\alpha}(\overline{\Omega})} \leqslant C(\alpha)\|f\|_{C^{1,\alpha}(\partial\Omega)}$$

for some $C(\alpha)>0$ depending only on $\partial\Omega$ and α.

Proof. We first use (6.116) to rewrite

$$V_2^\pm(f)(x) = -\nabla\cdot\int_{\partial\Omega}E(x-\xi)f(\xi)n_\xi\,d\sigma_\xi,\quad x\in\mathbb{R}^3\setminus\partial\Omega, \tag{6.131}$$

and then use $\nabla\times\nabla\times A=-\Delta A+\nabla(\nabla\cdot A)$ to obtain

$$\nabla V_2(f)(x) = -\nabla\times\nabla\times\int_{\partial\Omega}E(x-\xi)f(\xi)n_\xi\,d\sigma_\xi,\quad x\in\mathbb{R}^3\setminus\partial\Omega. \tag{6.132}$$

Next, we use

$$\nabla\times(\phi A)=\phi\nabla\times A+(\nabla\phi)\times A$$

and (6.116) and (6.35) to get

$$\nabla\times\int_{\partial\Omega}E(x-\xi)f(\xi)n_\xi\,d\sigma_\xi = \int_{\partial\Omega}n_\xi\times[f(\xi)\overline{\nabla}_\xi E(x-\xi)]\,d\sigma_\xi$$
$$= -\int_{\partial\Omega}E(x-\xi)\left[n_\xi\times\overline{\nabla}f(\xi)\right]d\sigma_\xi,\ x\in\mathbb{R}^3\setminus\partial\Omega.$$

Therefore

$$\nabla V_2(f)(x) = \int_{\partial\Omega}\nabla_x E(x-\xi)\times[n_\xi\times\overline{\nabla}f(x)]\,d\sigma,\ x\in\mathbb{R}^3\setminus\partial\Omega.$$

We now quote Theorem 6.4 and Lemma 6.4 to complete the proof. □

Exercise 6.4. Let F be a vector-valued density function, sufficiently smooth and defined on $\partial\Omega$. Define the simple-layer vector potential

$$A^{\pm}(x) = \int_{\partial\Omega} E(x-\xi)F(\xi)\,d\sigma_\xi, \quad \begin{cases} \text{“}+\text{” for } x \in \Omega^c, \\ \text{“}-\text{” for } x \in \Omega. \end{cases} \tag{6.133}$$

Show that $A^{\pm}(x)$ are extendable from Ω^c to $\overline{\Omega^c}$ and from Ω to $\overline{\Omega}$ and satisfy the jump conditions

$$\nabla \times A^{\pm}(x) = \int_{\partial\Omega} \nabla_x \times [E(x-\xi)F(\xi)]\,d\sigma_\xi \mp \frac{1}{2}[n_x \times F(x)], \quad x \in \partial\Omega, \tag{6.134}$$

$$\nabla \cdot A^{\pm}(x) = \int_{\partial\Omega} \nabla_x \cdot [E(x-\xi)F(\xi)]\,d\sigma_\xi \mp \frac{1}{2}\langle n_x, F(x)\rangle, \quad x \in \partial\Omega, \tag{6.135}$$

where the integrals exist as Cauchy principal values. □

Exercise 6.5. According to the following form of Gauss' divergence theorem [136],

$$\int_{\partial\Omega} \phi(\overline{\nabla} \cdot U)\,d\sigma + \int_{\partial\Omega} \langle \overline{\nabla}\phi, U\rangle\,d\sigma = 0 \tag{6.136}$$

holds for a scalar function ϕ and a tangential vector function U sufficiently smooth on $\partial\Omega$. Let A be defined as in (6.133). Use (6.136), (6.116) and (6.118) to prove that

$$\nabla \cdot A(x) = \int_{\partial\Omega} E(x-\xi)\overline{\nabla} \cdot F(\xi)\,d\sigma_\xi \quad \text{for } x \in \mathbb{R}^3 \setminus \partial\Omega.$$

□

Corollary 6.1. *Let the scalar density function $f(\cdot)$ and the vector density function $F(\cdot)$ be sufficiently smooth on $\partial\Omega$. Then the following jump conditions hold on $\partial\Omega$:*

(i) $\dfrac{\partial}{\partial n}V_1^+(f) - \dfrac{\partial}{\partial n}V_1^-(f) = -f;$
(ii) $V_2^+(f) - V_2^-(f) = f;$
(iii) $\nabla V_1^+(f) - \nabla V_1^-(f) = -fn;$
(iv) $\dfrac{\partial}{\partial n}V_2^+(f) - \dfrac{\partial}{\partial n}V_2^-(f) = 0;$
(v) $\nabla V_2^+(f) - \nabla V_2^-(f) = \overline{\nabla} f;$
(vi) $\nabla \times A^+ - \nabla \times A^- = -n \times F$ *(cf. (6.133) for A);*
(vii) $\nabla \cdot A^+ - \nabla \cdot A^- = \langle -n, F\rangle.$

Proof. These are consequences of (6.123), (6.87), (6.113), (6.125), (6.130), (6.134) and (6.135) given earlier. □

Therefore, all the equations (6.43)–(6.54) have been proved in the sense of classical pointwise continuity for a sufficiently smooth density f given on $\partial\Omega$.

The Sobolev space regularity properties (6.54)–(6.64) of these layer potentials are now simple consequences of §4.4. We let the potential density $f \in C^\infty(\partial\Omega)$. Then the simple-layer potential $V_1(f)$ corresponds to the boundary operator A_1 defined in (4.74), whose leading symbol is (4.73). Since for $\alpha \geqslant 0$, any $f \in H^\alpha(\partial\Omega)$ can be approximated by a sequence of $\{f_n\} \in C^\infty(\partial\Omega)$, taking the closure with respect to $H^\alpha(\partial\Omega)$, we obtain (6.54). For $\alpha < 0$, apply a duality transposition argument to get the same regularity result.

The double-layer potential $V_2^-(f)(x)$, $x \in \Omega$, corresponds to the boundary operator A_2 in (4.75). Its leading symbol is (4.82), i.e., $-\frac{1}{2}I$, in (6.56), with mapping property $H^\alpha(\partial\Omega) \to H^\alpha(\partial\Omega)$ continuous. The residual symbol of A_2 is of order -1, such that

$$A_2 - (-\frac{1}{2}I)\colon\ H^\alpha(\partial\Omega) \to H^{\alpha+1}(\partial\Omega) \text{ is continuous,} \tag{6.137}$$

or

$$A_2 - (-\frac{1}{2}I)\colon\ H^\alpha(\partial\Omega) \to H^{\alpha}(\partial\Omega) \text{ is compact.} \tag{6.138}$$

Therefore (6.56) holds. Equation (6.55) can be treated in a similar manner. The boundary operators $(\partial/\partial n)V_1^+$ and $(\partial/\partial n)V_1^-$ can be treated similarly to V_2^+ and V_2^-. In fact, (6.57) and (6.58) represent, as operators, the adjoints of (6.56) and (6.55) respectively. Therefore we need only quote Theorem 4.2 (i) and deduce the Sobolev space regularity (6.57) and (6.58).

The triple-layer $(\partial/\partial n)V_2^-$ in (6.60) corresponds to the boundary operator A_3 in (4.76). Its leading symbol is given by (4.83), so

$$A_3\colon\ H^\alpha(\partial\Omega) \to H^{\alpha-1}(\partial\Omega) \text{ is continuous.} \tag{6.139}$$

Hence (6.60) holds. Equation (6.59) follows as a limiting result of (6.125).

To obtain (6.61)–(6.64), symbols of various boundary integral operators represented there need to be computed as in §4.4. Since they will not be used in subsequent discussions, the work is omitted.

6.6 The two-dimensional case

In two-dimensional space, the fundamental solution of the Laplacian is

$$E_2(x-\xi) = -\frac{1}{2\pi}\ln|x-\xi|. \tag{6.140}$$

The various Hölder-continuity and jump properties of surface layer potentials of the previous section continue to hold after making some appropriate adjustment in the proofs. We examine them below.

First, we note in the key Theorem 6.4 that for a bounded domain $\Omega \subseteq \mathbb{R}^2$, the kernel E now satisfies

$$|E_2(x-\xi)| \leqslant M|x-\xi|^{-\delta} \quad \text{for some } M>0,\ \ \forall\, x,\xi \in G, x \neq \xi,\ \forall\, \delta > 0, \tag{6.141}$$

where G is a bounded domain containing $\partial\Omega$ in its interior. Also, it is easy to check that

$$|E_2(x_1-\xi) - E_2(x_2-\xi)| \leqslant M|x_1-\xi|^{-1}|x_1-x_2| \tag{6.142}$$

for all $x_1, x_2 \in G, \xi \in \partial\Omega$ with $2|x_1-x_2| \leqslant |x_1-\xi|$ for M chosen sufficiently large. With (6.67) and (6.68) replaced by (6.141) and (6.142), respectively, the potential

$$u(x) \equiv \int_{\partial\Omega} E_2(x-\xi)\phi(\xi)\,d\sigma_\xi, \quad x \in G,$$

with density $\phi \in C(\partial\Omega)$ satisfies

$$u \in C^{0,1-\delta}(G) \quad \text{for any } \delta > 0,$$
$$\|u\|_{C^{0,1-\delta}(G)} \leqslant C_\delta \|\phi\|_{C(\partial\Omega)}, C_\delta \to \infty \text{ as } \delta \downarrow 0.$$

Thus, Theorem 6.5 remains valid in the two-dimensional case.

Next, examine the double-layer potential

$$V_2(f)(x) = \int_{\partial\Omega} \frac{\partial E(x-\xi)}{\partial n_\xi} f(\xi)\,d\sigma_\xi, \quad x \in \partial\Omega. \tag{6.143}$$

The kernel satisfies

$$\frac{\partial E(x-\xi)}{\partial n_\xi} = +\frac{1}{2\pi}\frac{\langle n_\xi, x-\xi\rangle}{|x-\xi|^2} = \mathcal{O}(1), \tag{6.144}$$

by (6.37). Therefore (6.143) exists pointwise for any $f \in L^p(\partial\Omega)$, $p \geqslant 1$. According to (6.137), we have

$$V_2(f) \in H^{\alpha+1}(\partial\Omega) \ \text{ for } f \in H^\alpha(\partial\Omega), \text{ for any } \alpha \in \mathbb{R}. \tag{6.145}$$

We now show that for a two-dimensional smooth domain Ω, (6.145) can actually be strengthened to

$$V_2(f) \in C^\infty(\partial\Omega) \ \text{ for } f \in H^\alpha(\partial\Omega), \text{ for any } \alpha \in \mathbb{R},$$

in other words, V_2 is C^∞-smoothing and

$$V_2\colon\ H^\alpha(\partial\Omega) \longrightarrow C^\infty(\partial\Omega) \text{ is continuous.} \tag{6.146}$$

Let $(x_1(t), x_2(t))$, $0 \leqslant t \leqslant 2\pi$, be a C^∞-parametrization of $\partial\Omega$, and

$$x = (x_1(t), x_2(t)) \in \partial\Omega,$$
$$\xi = (x_1(\tau), x_2(\tau)) \in \partial\Omega.$$

Then

$$\begin{aligned}
d\sigma_\xi &= [\dot{x}_1^2(\tau) + \dot{x}_2^2(\tau)]^{1/2} d\tau, \\
n_\xi &= -[\dot{x}_1^2(\tau) + \dot{x}_2^2(\tau)]^{-1/2}(-\dot{x}_2(\tau), \dot{x}_1(\tau)), \\
&\frac{1}{2\pi}\int_{\partial\Omega} \frac{\langle n_\xi, x-\xi\rangle}{|x-\xi|^2} f(\xi)\, d\sigma_\xi, \\
&= \frac{1}{2\pi}\int_0^{2\pi} \frac{-\dot{x}_2(\tau)[x_1(t)-x_1(\tau)] + \dot{x}_1(\tau)[x_2(t)-x_2(\tau)]}{[x_1(t)-x_1(\tau)]^2 + [x_2(t)-x_2(\tau)]^2} f(x(\tau))\, d\tau.
\end{aligned}$$

The above kernel,

$$k(t,\tau) \equiv \frac{1}{2\pi}\frac{-\dot{x}_2(\tau)[x_1(t)-x_1(\tau)] + \dot{x}_1(\tau)[x_2(t)-x_2(\tau)]}{[x_1(t)-x_1(\tau)]^2 + [x_2(t)-x_2(\tau)]^2}, \tag{6.147}$$

is C^∞ with respect to t and τ. Obviously, $k(t,\tau)$ is C^∞ off the diagonal $t = \tau$. When t approaches τ, we have

$$\begin{aligned}
k(t,\tau) = \frac{1}{2\pi}\Bigg[&-\dot{x}_2(\tau)\frac{(x_1(t)-x_1(\tau)) - \dot{x}_1(\tau)(t-\tau)}{(t-\tau)^2} \\
&+ \dot{x}_1(\tau)\frac{(x_2(t)-x_2(\tau)) - \dot{x}_2(\tau)(t-\tau)}{(t-\tau)^2}\Bigg] \\
&\times\left\{\left[\frac{x_1(t)-x_1(\tau)}{t-\tau}\right]^2 + \left[\frac{x_2(t)-x_2(\tau)}{t-\tau}\right]^2\right\}^{-1}.
\end{aligned} \tag{6.148}$$

Using

$$x_i(t) = x_i(\tau) + \dot{x}_i(\tau)(t-\tau) + \frac{1}{2}\ddot{x}_i(\tau)(t-\tau)^2 + \mathscr{O}((t-\tau)^3), \quad i = 1, 2,$$

we obtain from (6.148) that

$$\lim_{t\to\tau} k(t,\tau) = \frac{1}{4\pi}\frac{-\dot{x}_2(\tau)\ddot{x}_1(\tau) + \dot{x}_1(\tau)\ddot{x}_2(\tau)}{[\dot{x}_1(\tau)]^2 + [\dot{x}_2(\tau)]^2}.$$

Thus we see that k is C^∞ on $[0,2\pi]\times[0,2\pi]$. (In fact, $k(\tau,\tau)$ is the curvature of $\partial\Omega$ at $x(\tau)$.) Therefore (6.146) has been verified. Similarly, the mapping

$$V_2^*:\ f(\cdot) \longmapsto \int_{\partial\Omega} \frac{\partial E(\cdot - \xi)}{\partial n_x} f(\xi)\, d\sigma_\xi \tag{6.149}$$

is C^∞-smooth, i.e.,

$$V_2^*:\ H^\alpha(\partial\Omega) \to C^\infty(\partial\Omega) \text{ is continuous.} \tag{6.150}$$

The results (6.146) and (6.150) constitute the most salient features of two-dimensional problems. (In the language of pseudodifferential operators, we say that the pseudodifferential operators representing (6.143) and (6.149) have *vanishing leading symbols*, so they become C^∞-smoothing.)

To conclude this section, we derive an extra relation between the simple- and double-layer potentials, valid in two dimensions only, in addition to the relations (4.126)–(4.129), through the Calderón projector.

Note that

$$\begin{aligned}\frac{\partial E(x-\xi)}{\partial n_\xi} &= \langle n_\xi, \nabla_\xi E(x-\xi)\rangle = -\langle n_\xi, \nabla_x E(x-\xi)\rangle \\ &= -s_2(\xi)\frac{\partial E(x-\xi)}{\partial x_1} + s_1(\xi)\frac{\partial}{\partial x_2}E(x-\xi), \qquad (6.151)\end{aligned}$$

where $\sigma(\xi) = (s_1(\xi), s_2(\xi)) = (-n_2(\xi), n_1(\xi))$ is the unit counterclockwise tangential vector at ξ. Using

$$\Delta_x E(x-\xi) = 0 \quad \text{for} \quad x \neq \xi,$$

from (6.151) we obtain

$$\nabla_x\left(\frac{\partial E(x-\xi)}{\partial n_\xi}\right) = \left(\frac{\partial}{\partial x_2}\langle \nabla_x E(x-\xi), \sigma(\xi)\rangle, -\frac{\partial}{\partial x_1}\langle \nabla_x E(x-\xi), \sigma(\xi)\rangle\right).$$

Therefore

$$\nabla \int_{\partial\Omega} \frac{\partial E(x-\xi)}{\partial n_\xi} f(\xi)\, d\sigma_\xi = \left(\frac{\partial W(x)}{\partial x_2}, -\frac{\partial W(x)}{\partial x_1}\right),$$

where

$$\begin{aligned}W(x) &\equiv \int_{\partial\Omega} \langle \nabla_x E(x-\xi), \sigma(\xi)\rangle f(\xi)\, d\sigma_\xi \\ &= -\int_{\partial\Omega} \frac{\partial E(x-\xi)}{\partial \sigma_\xi} f(\xi)\, d\sigma_\xi \\ &= \int_{\partial\Omega} E(x-\xi)\frac{\partial f(\xi)}{\partial \sigma}\, d\sigma_\xi,\end{aligned}$$

and where we have integrated by parts along $\partial\Omega$. Hence

$$\begin{aligned}\frac{\partial}{\partial n}V_2(f)(x) &= \langle n_x, \nabla V_2(f)(x)\rangle \\ &= \left\langle n_x, \left(\frac{\partial W(x)}{\partial x_2}, -\frac{\partial W(x)}{\partial x_1}\right)\right\rangle \\ &= n_1(x)\frac{\partial W(x)}{\partial x_2} - n_2(x)\frac{\partial W(x)}{\partial x_1} \\ &= \frac{\partial}{\partial \sigma}\int_{\partial\Omega} E(x-\xi)\frac{\partial f(\xi)}{\partial \sigma}\, d\sigma_\xi.\end{aligned}$$

The above is commonly written as

$$\frac{\partial}{\partial n}V_2(f) = \frac{\partial}{\partial \sigma}V_1\left(\frac{\partial f}{\partial \sigma}\right). \qquad (6.152)$$

6.7 Regularity of solutions of the potential BVP

We state some basic Sobolev space regularity properties of the interior and exterior Poisson equations subject to the Dirichlet, Neumann and Robin boundary conditions. First, we need a more precise interpretation of what constitutes a "solution". To fix ideas, let us consider the Neumann BVP

$$\begin{aligned} \Delta w(x) &= f(x), \qquad && x \in D, \\ \frac{\partial w(x)}{\partial n} &= g(x), \qquad && x \in \partial D, \end{aligned}$$

where D is either a bounded or unbounded open domain in $\mathbb{R}^N$. In the above, w and f are *distributions* supported on $\overline{D}$, g is a distribution supported on ∂D, and all the derivatives are taken in the sense of distributions. A distribution w satisfying $\Delta w = f$ on D and $\partial w/\partial n = g$ on ∂D in the sense of distributions is called a *weak* (or *generalized*) solution of the above Neumann BVP. Equivalently, this means that for any test function $\phi \in C_0^\infty(\mathbb{R}^N)$, we have

$$\int_D (\Delta w)\phi\, dx = \int_{\partial D} \frac{\partial w}{\partial n}\phi\, d\sigma - \int_D \nabla w \cdot \nabla \phi\, dx = \int_D f\phi\, dx,$$

i.e.,

$$\int_D \nabla w \cdot \nabla \phi\, dx = \int_{\partial D} g\phi\, d\sigma - \int_D f\phi\, dx.$$

The LHS defines an (energy) bilinear form $a(\cdot,\cdot)$ on $H^1_{\text{loc}}(D)$ as in §5.1, while the RHS defines a linear form $\theta(\cdot)$ on $C_0^\infty(\partial D)$. When $f \in H^r_{\text{loc}}(D)$ and $g \in H^s_{\text{loc}}(\partial D)$ for $r \geqslant -1$ and $s \geqslant -\frac{1}{2}$, θ can be extended to a linear form on $H^1_{\text{loc}}(D)$. (Indeed, all the solutions u of $a(u,v) = \theta(v)$ considered in Chap. 5 are weak solutions.)

We further call w a *strong* (or *classical*) solution if $f \in H^r_{\text{loc}}(D)$ for some $r \geqslant 0$, and $g \in H^s_{\text{loc}}(\partial D)$ for some $s \geqslant 0$, such that there exists a sequence $w_j \in C^\infty(D)$ satisfying

$$\begin{aligned} w_j &\longrightarrow w \qquad && \text{in } L^2_{\text{loc}}(D), \\ \Delta w_j \equiv f_j &\longrightarrow \qquad && \text{in } L^2_{\text{loc}}(D), \end{aligned}$$

and

$$\frac{\partial w_j}{\partial n} \equiv g_j \longrightarrow g \quad \text{in} \quad L^2_{\text{loc}}(\partial D).$$

For other types of BVP considered in this book, readers can easily define the corresponding meanings of weak and strong solutions by following the above as an example. (In particular, if D is bounded, then all the subscripts "loc" become unnecessary.)

For the existence and uniqueness of solutions to interior problems, the proofs can be found in [127, § 2.5] for "strong" solutions, provided that s below is sufficiently large ($s \geqslant \frac{3}{2}$ in

(6.153) and $s \geqslant \frac{1}{2}$ in (6.155) and (6.158)). For smaller values of s, the solutions are "weak", see [85, (1.2.33), p. 28] and [64].

Theorem 6.13. *Let Ω be a bounded open domain in $\mathbb{R}^N$ with smooth boundary $\partial\Omega$. Then the Dirichlet BVP*

$$\left.\begin{aligned} &\Delta w = f \in H^r(\Omega), \qquad r \geqslant -1, \\ &w|_{\partial\Omega} = g \in H^s(\partial\Omega),\ s \in \mathbb{R}, \end{aligned}\right\} \tag{6.153}$$

has a unique solution $w \in H^\alpha(\Omega)$, with $\alpha = \min\left(r+2, s+\frac{1}{2}\right)$, satisfying

$$\|w\|_{H^\alpha(\Omega)} \leqslant C(\|f\|_{H^r(\Omega)} + \|g\|_{H^s(\partial\Omega)}) \tag{6.154}$$

for some $C > 0$ depending only on Ω. □

Theorem 6.14. *Let Ω be a bounded open domain in $\mathbb{R}^N$ with smooth boundary $\partial\Omega$. Then the Neumann BVP*

$$\left.\begin{aligned} &\Delta w = f \in H^r(\Omega), \qquad r \geqslant -1, \\ &\left.\frac{\partial w}{\partial n}\right|_{\partial\Omega} = g \in H^s(\partial\Omega),\ s \in \mathbb{R}, \end{aligned}\right\} \tag{6.155}$$

has a solution $w \in H^\alpha(\Omega)$, with $\alpha = \min\left(r+2, s+\frac{3}{2}\right)$, if and only if the compatibility condition

$$\int_{\partial\Omega} g\, d\sigma = \int_\Omega f\, dx \quad \left(\equiv \langle 1, f\rangle_{H^{-r}(\Omega), H^r(\Omega)}\right) \tag{6.156}$$

is satisfied. If (6.156) is satisfied and $w(\cdot)$ solves (6.155) then $w(\cdot) + c$ is a solution to (6.155) for any $c \in \mathbb{R}$. The estimate

$$\inf_{c\in\mathbb{R}} \|w(\cdot) + c\|_{H^\alpha(\Omega)} \leqslant C(\|f\|_{H^r(\Omega)} + \|g\|_{H^s(\partial\Omega)}) \tag{6.157}$$

holds for some $C > 0$ depending only on Ω. □

Theorem 6.15. *Let Ω be a bounded open domain in $\mathbb{R}^N$ with smooth boundary $\partial\Omega$. Then the Robin BVP*

$$\left.\begin{aligned} &\Delta w = f \in H^r(\Omega), \qquad\qquad r \geqslant -1, \\ &\left.\left(\frac{\partial w}{\partial n} + hw\right)\right|_{\partial\Omega} = g \in H^s(\partial\Omega),\ s \in \mathbb{R}, h > 0, \end{aligned}\right\} \tag{6.158}$$

has a unique solution $w \in H^\alpha(\Omega)$, with $\alpha = \min\left(r+2, s+\frac{3}{2}\right)$, satisfying

$$\|w\|_{H^\alpha(\Omega)} \leqslant C(\|f\|_{H^r(\Omega)} + \|g\|_{H^s(\partial\Omega)}) \tag{6.159}$$

for some $C > 0$ depending only on Ω. □

For the exterior BVP, the questions of uniqueness is more subtle as shown by the following example.

Example 6.1. Let Ω be the unit open ball in $\mathbb{R}^3$. The Dirichlet problem

$$\begin{aligned} \Delta w(x) &= 0 \quad \text{on } \Omega^c, \\ w(x) &= 1 \quad \text{on } \partial\Omega, \end{aligned}$$

has at least two solutions $w_1(x) \equiv 1$ and $w_2(x) \equiv |x|^{-1}$. □

Such nonuniqueness can be avoided if some decay conditions (or radiation conditions) are imposed near ∞. The following theorems can be proved by the maximum principle. See e.g., [132].

Theorem 6.16. *For the exterior Dirichlet BVP*

$$\left.\begin{aligned} \Delta w &= 0 && \text{on } \Omega^c \subset \mathbb{R}^N,\ N = 2 \text{ or } 3, \\ w|_{\partial\Omega} &= g \in H^s(\partial\Omega), && s \in \mathbb{R}, \end{aligned}\right\} \tag{6.160}$$

there exists a unique solution $w \in H^{s+1/2}_{\text{loc}}(\Omega^c)$ *satisfying*

$$w(x) = \begin{cases} \mathcal{O}(1), & N = 2, \\ o(1), & N = 3, \end{cases} \quad \text{for large } |x|. \tag{6.161}$$

□

Theorem 6.17. *Consider the exterior Neumann BVP*

$$\left.\begin{aligned} &\Delta w = 0 && \text{on } \Omega^c \subset \mathbb{R}^N,\ N = 2 \text{ or } 3, \\ &\left.\frac{\partial w}{\partial n}\right|_{\partial\Omega} = g \in H^s(\partial\Omega),\ s \in \mathbb{R}. \end{aligned}\right\} \tag{6.162}$$

Assume that (6.161) holds. Then the BVP (6.162) has a unique solution $w \in H^{s+3/2}_{\text{loc}}(\Omega^c)$ *when* $N = 3$. *When* $N = 2$, *the BVP has solutions of the form* $w_p(\cdot) + c$ *for any* $c \in \mathbb{R}$, *for a particular solution* $w_p(\cdot) \in H^{s+3/2}_{\text{loc}}(\Omega^c)$, *provided that the boundary data* g *satisfies*

$$\int_{\partial\Omega} g\, d\sigma = 0. \tag{6.163}$$

□

Theorem 6.18. *The exterior Robin BVP*

$$\left.\begin{aligned} &\Delta w = 0 \text{ on } \Omega^c \subset \mathbb{R}^N, && N = 2 \text{ or } 3, \\ &\left.\left(\frac{\partial w}{\partial n} - hw\right)\right|_{\partial\Omega} = g \in H^s(\partial\Omega),\ s \in \mathbb{R}, h > 0, \end{aligned}\right\} \tag{6.164}$$

has a unique solution $w \in H^{s+3/2}_{\text{loc}}(\Omega^c)$ *satisfying (6.161).* □

Remark 6.2. All the partial differential operators corresponding to the BVP considered in this book are *Fredholm operators of zero index* on properly posed Sobolev spaces. However, cases of nonzero index do occur. A notable example is the *oblique derivative BVP*. Let Ω be a bounded simply connected domain in $\mathbb{R}^2$ with smooth boundary $\partial\Omega$. Consider

$$\left.\begin{aligned} \Delta w &= f \in L^2(\Omega), \\ \partial_\nu w &= g \in H^{1/2}(\partial\Omega), \end{aligned}\right\} \tag{6.165}$$

where $\partial_\nu w = \nu_1 w_{x_1} + \nu_2 w_{x_2}$ and $\nu = (\nu_1, \nu_2)$ is a C^∞ vector field on $\partial\Omega$. Let $l = l(\nu)$ be the *winding number* of ν when $z = x_1 + ix_2$ goes around $\partial\Omega$ in a counterclockwise direction. For example, when Ω is a disk centered at the origin, $\nu(z) = \nu_1(z) + i\nu_2(z) = z^m$, $m \in \mathbb{Z}$, has winding number $l = m$. Let Δ_ν denote the bounded operator corresponding to the BVP (6.165):

$$\Delta_\nu : \; H^2(\Omega) \to L^2(\Omega) \times H^{1/2}(\partial\Omega),$$

where

$$\Delta_\nu w \equiv (\Delta w, \partial_\nu w|_{\partial\Omega}).$$

Then it is known (see [116, pp. 191–192]) that

$$\operatorname{Ind}\Delta_\nu = 2(1 - l).$$

For Ω a disk centered at the origin and $\nu = z^l$ as mentioned above, we have, in particular,

$$\dim \mathscr{N}(\Delta_\nu) = \begin{cases} 1, & l > 0, \\ 2(1-l), & l \leqslant 0, \end{cases}$$

$$\dim \operatorname{Coker}\Delta_\nu = \begin{cases} 2l - 1, & l > 0, \\ 0, & l \leqslant 0. \end{cases}$$

When $l = 1$, the case $\nu = z$ corresponds to the Neumann BVP, with $\operatorname{Ind}\Delta_z = 0$.
The index of partial differential operators, or of bounded integral operators (which are pseudodifferential operators) discussed in this book, can be computed by Noether's formula; see [116, Chap. 5, § 2]. □

6.8 Simple-layer representations for interior BVP in $\mathbb{R}^3$

In this section, we will formulate boundary integral equations for the interior and exterior Dirichlet, Neumann and Robin elliptic boundary value problems in (6.14), and prove the existence and uniqueness of solutions. The proof consists of three important ingredients:

estimation of the harmonic potential at ∞;

using the fact that the boundary integral operators and their adjoints are Fredholm of zero index;
utilization of jump discontinuities of surface layer potentials.

These ideas were first given by Fichera in [72] and adapted by Hsiao and MacCamy [95], and later became the standard techniques in the mathematical theory of boundary elements. Let Ω be a nonempty bounded simply connected domain in $\mathbb{R}^3$ with smooth boundary $\partial\Omega$. We state and prove theorems in three dimensions first. The two-dimensional case may differ slightly from the three-dimensional one and will be discussed afterwards. We first prove a fundamental uniqueness lemma for the exterior harmonic boundary value problem.

Lemma 6.5. *Assume that* $w \in H^1_{\text{loc}}(\Omega^c)$ *is a weak solution of*

$$\Delta w(x) = 0, \quad x \in \Omega^c \subset \mathbb{R}^N, \quad N = 2 \text{ or } 3, \tag{6.166}$$

subject to one of the following boundary conditions on $\partial\Omega$*:*

$$w = 0, \tag{6.167}$$

$$\frac{\partial w}{\partial n} = 0, \quad \text{and } w \text{ is not a nonzero constant on } \partial\Omega, \tag{6.168}$$

$$\frac{\partial w}{\partial n} - hw = 0, \quad h > 0, \tag{6.169}$$

such that

$$\left| w(x)\frac{\partial w(x)}{\partial r} \right| = \begin{cases} o(r^{-2}), & N = 3; \\ o(r^{-1}), & N = 2; \end{cases} \qquad r = |x|. \tag{6.170}$$

Then $w(x) = 0$ *throughout* $x \in \Omega^c$.

Proof. First consider $\mathbb{R}^3$. Let w satisfy (6.166) subject to (6.167) or (6.168). Applying the integration by parts procedure on $\Omega^c_R \equiv \Omega^c \cap \{x \in \mathbb{R}^3 \mid |x| \leqslant R\}$ for sufficiently large R, we obtain

$$\begin{aligned} a_R(w,w) &\equiv \int_{\Omega^c_R} |\nabla w|^2 \, dx = -\int_{\partial\Omega} \frac{\partial w}{\partial n} w \, d\sigma + \int_{|x|=R} \frac{\partial w}{\partial r} w \, d\sigma \\ &\leqslant 0 + \int_0^{2\pi} \int_0^{\pi} C \frac{\varepsilon(R)}{R^2} R^2 \sin\theta \, d\theta \, d\phi \quad (\varepsilon(R) \to 0 \text{ as } R \to \infty), \\ &= C' \varepsilon(R) \to 0 \text{ as } R \to \infty. \end{aligned}$$

Since $a_R(w,w)$ is a norm for w up to a constant function, as $R \to \infty$, we conclude that w must be the constant under the boundary constraint (6.167) or (6.168). This constant is zero in the case of (6.168).

For the Robin boundary condition (6.169), we integrate by parts to obtain

$$a_R(w,w) = \int_{\Omega_R^c} |\nabla w|^2\, dx + h \int_{\partial\Omega} w^2\, d\sigma$$

instead and conclude the same result. The $\mathbb{R}^2$ case can be dealt with similarly. □

We also need a uniqueness theorem for interior BVP.

Theorem 6.19. *Let Ω be an open domain in $\mathbb{R}^N$, $N > 1$, with boundary $\partial\Omega$. If $w \in H^1(\Omega)$ is a weak solution to the boundary value problem*

$$\Delta w(x) = 0 \quad \text{on} \quad \Omega, \tag{6.171}$$

subject to the Dirichlet or Robin boundary condition

$$w|_{\partial\Omega} = 0 \quad \text{or} \quad \left(\frac{\partial w}{\partial n} + hw\right)\Big|_{\partial\Omega} = 0, \quad h > 0,$$

then $w(x) \equiv 0$ on Ω. If w satisfies (6.171) and the Neumann boundary condition

$$\frac{\partial w}{\partial n}\Big|_{\partial\Omega} = 0,$$

then $w(x) \equiv c$ for some constant c on Ω.

Proof. Apply integration by parts to obtain

$$0 = \int_{\Omega} |\nabla w(x)|^2\, dx + h \int_{\partial\Omega} w^2(x)\, d\sigma.$$

□

Let us first consider the interior inhomogeneous Dirichlet problem

$$\begin{aligned} \Delta w(x) &= 0 \quad \text{on } \Omega \\ w(x) &= g(x), \ \ x \in \partial\Omega. \end{aligned}$$

An important idea of the boundary element method is to represent the solution as a simple-layer potential [72, 95]

$$w(x) = \int_{\partial\Omega} E(x-\xi) f(\xi)\, d\sigma_\xi \tag{6.172}$$

with some density function f on $\partial\Omega$ and then solve for f. □

From now on throughout the rest of the book, except in § 6.14 where multiconnected domains will be treated, we make the important stipulation that the domain Ω is simply connected.

Theorem 6.20 (Simple-layer representation for the interior Dirichlet BVP in a $\mathbb{R}^3$). *Let $\Omega \subseteq \mathbb{R}^3$ be bounded with smooth boundary $\partial\Omega$. The mapping*

$$\mathscr{L}_1(f)(x) \equiv \int_{\partial\Omega} E(x-\xi)f(\xi)\,d\sigma_\xi, \quad x \in \partial\Omega \tag{6.173}$$

is linear continuous, mapping isomorphically from $H^r(\partial\Omega)$ onto $H^{r+1}(\partial\Omega)$ for any $r \in \mathbb{R}$, satisfying

$$\dim \mathscr{N}(\mathscr{L}_1) = 0, \quad \dim \mathrm{Coker} \mathscr{L}_1 = 0.$$

Its adjoint operator $\mathscr{L}_1^*$ maps $H^{-(r+1)}(\partial\Omega)$ isomorphically onto $H^{-r}(\partial\Omega)$ and is given by

$$\mathscr{L}_1^* = \mathscr{L}_1, \quad \textit{as in (6.173)},$$

with

$$\dim \mathscr{N}(\mathscr{L}_1^*) = 0, \quad \dim \mathrm{Coker}\, \mathscr{L}_1^* = 0.$$

Consequently, the solution $w(x)$ to

$$\left.\begin{aligned} &\Delta w(x) = 0 && \text{on } \Omega, \\ &w(x) = g(x) \in H^r(\partial\Omega), && r \in \mathbb{R}, \end{aligned}\right\} \tag{6.174}$$

is uniquely given by

$$w(x) = \int_{\partial\Omega} E(x-\xi)f(\xi)\,d\sigma_\xi \in H^{r+1/2}(\Omega), \quad x \in \Omega \tag{6.175}$$

where the potential density $f \in H^{r-1}(\partial\Omega)$ uniquely solves the BIE

$$\mathscr{L}_1(f) = g. \tag{6.176}$$

Proof. The regularity property of $\mathscr{L}_1$ is a consequence of Theorem 6.3. That $\mathscr{R}(\mathscr{L}_1)$ is a closed subspace of $H^{r+1}(\partial\Omega)$ follows from Theorem 4.11 and § 4.6, the Fredholm property of $\mathscr{L}_1$.

The integral operator $\mathscr{L}_1$ with kernel

$$K(x,\xi) = E(x-\xi)$$

has an adjoint operator $\mathscr{L}_1^*$ whose kernel is

$$K^*(x,\xi) = K(\xi,x) = E(\xi - x) = E(x-\xi) = K(x,\xi).$$

So $\mathscr{L}_1^*$ is formally self-adjoint to $\mathscr{L}_1$. This formal adjointness becomes the true adjointness because of the inner products and duality pairing involved. Hence $\mathscr{L}_1^*$ mapping from $H^{-(r+1)}(\partial\Omega)$ into $H^{-r}(\partial\Omega)$ has canonical restrictions and extensions that map from $H^s(\partial\Omega)$ into $H^{s+1}(\partial\Omega)$ for any $s \in \mathbb{R}$.

Since $\mathscr{L}_1$ now is (truly, not just formally) self-adjoint, if we can prove that

$$\dim \mathscr{N}(\mathscr{L}_1) = 0 \tag{6.177}$$

then the rest of the proof will follow from the Fredholm alternative. Assume $\widetilde{f} \in \mathscr{N}(\mathscr{L}_1)$. Then $\widetilde{f} \in C^\infty(\partial\Omega)$ by Corollary 4.4. We want to show that $\widetilde{f} \equiv 0$ on $\partial\Omega$. For $x \in \Omega$, consider

$$\widetilde{w}^-(x) \equiv \int_{\partial\Omega} E(x-\xi)\widetilde{f}(\xi)\,d\sigma_\xi. \tag{6.178}$$

Then $\widetilde{w}^-$ satisfies

$$\begin{aligned} \Delta\widetilde{w}^-(x) &= 0 \quad \text{on } \Omega, \\ \widetilde{w}^- &\equiv 0 \qquad \text{on } \partial\Omega. \end{aligned}$$

From the uniqueness of solutions of the Dirichlet problem (Theorem 6.19), we have

$$\widetilde{w}^-(x) \equiv 0 \quad \text{on } \Omega.$$

We extend (6.178) to $x \in \overline{\Omega^c}$ by defining

$$\widetilde{w}^+(x) = \int_{\partial\Omega} E(x-\xi)\widetilde{f}(\xi)\,d\sigma_\xi, \quad x \in \Omega^c.$$

Then for large $|x|, x \in \Omega^c$, we have the asymptotic expansion

$$\begin{aligned} \widetilde{w}^+(x) &= \frac{1}{4\pi}\int_{\partial\Omega} |x|^{-1}\left[1 + \frac{(-2x\cdot\xi)+|\xi|^2}{|x|^2}\right]^{-1/2} \widetilde{f}(\xi)\,d\sigma_\xi \qquad (6.179)\\ &= \frac{1}{4\pi}\left[\frac{1}{|x|}\int_{\partial\Omega}\widetilde{f}(\xi)\,d\sigma_\xi - \frac{1}{2}\frac{1}{|x|^3}\int_{\partial\Omega}(-2x\cdot\xi + |\xi|^2)\widetilde{f}(\xi)\,d\sigma_\xi + \cdots\right] \\ &= \frac{1}{4\pi}\frac{1}{|x|}\int_{\partial\Omega}\widetilde{f}(\xi)\,d\sigma_\xi + \mathscr{O}(|x|^{-2}), \end{aligned}$$

so

$$\widetilde{w}^+(x) = o(1) \quad \text{and} \quad \Delta\widetilde{w}^+(x) \equiv 0 \quad \text{on} \quad \Omega^c.$$

On $\partial\Omega$, $\widetilde{w}^+$ is identically zero because $\widetilde{f} \in \mathscr{N}(\mathscr{L}_1)$. We now apply Theorem 6.16 to conclude that $\widetilde{w}^+ \equiv 0$ on $\overline{\Omega^c}$. By the jump property, Corollary 6.1 (i), we have

$$\begin{aligned} \widetilde{f} &= \frac{\partial\widetilde{w}^+}{\partial n} - \frac{\partial\widetilde{w}^-}{\partial n} \\ &= 0 - 0 = 0 \quad \text{on } \partial\Omega. \end{aligned}$$

Therefore, (6.177) is proved and the work is complete. □

The linear operator $\mathscr{L}_1$ defined in (6.173) is the restriction of the simple-layer potential V_1 (6.39) to $\partial\Omega$. $\mathscr{L}_1$ has an important positive definitiveness property, proved by Hsiao and Wendland [96] in 1977, which forms the basis of boundary element error estimates.

Theorem 6.21. *Let $\Omega \subset \mathbb{R}^N$ be a bounded open domain with smooth boundary, with $N \geqslant 3$, and let $f \in L^2(\partial\Omega)$. Let $\mathscr{L}_1$ be defined as in (6.173). Then*

$$\langle f, \mathscr{L}_1(f)\rangle_{L^2(\partial\Omega)} \geqslant C\|\mathscr{L}_1(f)\|^2_{H^{1/2}(\partial\Omega)},$$

where C is a positive constant independent of f.

Proof. Consider only $N = 3$. ($N > 3$ can be done in the same way.) From $f \in L^2(\partial\Omega)$, we have $V_1^-(f) \in H^{3/2}(\Omega)$ and $V_1^+(f) \in H^1(\Omega^c) \cap H^{3/2}_{\text{loc}}(\Omega^c)$, so

$$\begin{aligned}\int_{\mathbb{R}^3} |\nabla V_1(f)|^2\,dx &= \int_{\Omega^c} |\nabla V_1^+(f)|^2\,dx + \int_\Omega |\nabla V_1^-(f)|^2\,dx \\ &= -\int_{\Omega^c} [\Delta V_1^+(f)]V_1^+(f)\,dx - \int_{\partial\Omega} \left[\frac{\partial V_1^+(f)}{\partial n}\right][V_1^+(f)]\,d\sigma \\ &\quad - \int_\Omega [\Delta V_1^-(f)]V_1(f)\,dx + \int_{\partial\Omega} \left[\frac{\partial V_1^-(f)}{\partial n}\right][V_1(f)]\,d\sigma.\end{aligned}$$

Because $\Delta V_1(f) = 0$ for $x \in \mathbb{R}^3 \setminus \partial\Omega$,

$$\int_{\mathbb{R}^3} |\nabla V_1(f)|^2\,dx = \int_{\partial\Omega} \left[\frac{\partial V_1^-(f)}{\partial n} - \frac{\partial V_1^+(f)}{\partial n}\right] V_1(f)\,d\sigma.$$

We know by Corollary 6.1 (i) that on $\partial\Omega$,

$$\frac{\partial}{\partial n} V_1^\pm(f)(x) = \mp\frac{1}{2}f(x) + \int_{\partial\Omega} \frac{\partial E(x-\xi)}{\partial n_x} f(\xi)\,d\sigma_\xi,$$

thus

$$\frac{\partial}{\partial n}V_1^-(f)(x) - \frac{\partial}{\partial n}V_1^+(f)(x) = f(x),$$

and consequently

$$\int_{\mathbb{R}^3} |\nabla V_1(f)|^2\,dx = \int_{\partial\Omega} [V_1(f)(x)]f(x)\,d\sigma = \langle f, \mathscr{L}_1(f)\rangle_{H^0(\partial\Omega)}.$$

We want to show that there exists $C > 0$ such that

$$\begin{aligned}\int_{\mathbb{R}^3} |\nabla V_1(f)|^2\,dx &\geqslant C\|V_1(f)\|^2_{H^1(\Omega)} \\ &= C\left[\int_\Omega |\nabla V_1(f)|^2\,dx + \int_\Omega |V_1(f)|^2\,dx\right]\end{aligned}$$

for all $f \in L^2(\partial\Omega)$.

Assume the contrary. Then there exists a sequence $\{f_n\} \subset L^2(\partial\Omega)$,

$$\|V_1(f_n)\|_{H^1(\Omega)} = 1, \tag{6.180}$$

such that

$$\int_\Omega |\nabla V_1(f_n)|^2\,dx \leqslant \int_{\mathbb{R}^3} |\nabla V_1(f_n)|^2\,dx \leqslant \frac{1}{n}, \quad n = 1, 2, \ldots. \tag{6.181}$$

From the assumption, $\{V_1(f_n)\}_{n=1}^\infty$ is a bounded sequence in $H^1(\Omega) \overset{\text{c.c.}}{\hookrightarrow} L^2(\Omega) = H^0(\Omega)$. The compact imbedding implies that $\{V_1(f_n)\}_{n=1}^\infty$ has a convergent subsequence in $H^0(\Omega)$. Without loss of generality, we still denote this subsequence as $\{V_1(f_n)\}_{n=1}^\infty$, which converges to some w in $H^0(\Omega)$; that is,

$$V_1(f_n) \xrightarrow{\text{weakly}} w \text{ in } H^1(\Omega),$$

$$V_1(f_n) \xrightarrow{\text{strongly}} w \text{ in } H^0(\Omega).$$

Since

$$\|\nabla(V_1(f_n)-V_1(f_m))\|_{H^0(\Omega)} \leqslant \frac{1}{n}+\frac{1}{m},$$

we have

$$V_1(f_n)\xrightarrow{\text{strongly}} w \text{ in } H^1(\Omega),$$

and by (6.181)

$$\lim_{n\to\infty} V_1(f_n) = \text{ const } = C \text{ on } \Omega.$$

Similarly, from Lemma 5.5,

$$\lim_{n\to\infty}\int_{\Omega^c}|\nabla V_1(f_n)|^2\,dx \leqslant \inf_{c\in\mathbb{R}}\|V_1(f_n)-C\|_{H^1(\Omega_r^c)} \leqslant C\|\nabla V_1(f_n)\|_{L^2(\Omega_r^c)},$$

we have $\{c_n\}$ such that

$$\lim_{n\to\infty} V_1(f_n)^{-c_n} = 0 \text{ on } \overline{\Omega_r^c}.$$

Since $V_1(f_n)$ is continuous across $\partial\Omega$, we have $C=C'$.

But by the property of the simple-layer potential $V_1(f)$ in $\mathbb{R}^N$, $N\geqslant 3$, we have, by (6.179),

$$V_1(f_n)(x) = \mathscr{O}\left(\frac{1}{|x|}\right) \text{ for large } |x|.$$

So $C=0$, implying

$$\lim_{n\to\infty}\|V_1(f_n)\|_{H^1(\Omega)} = 0.$$

From (6.180), we obtain a contradiction. Hence

$$\langle f,\mathscr{L}_1(f)\rangle_{H^0(\partial\Omega)} \geqslant C''\|V_1(f)\|^2_{H^1(\Omega)}. \tag{6.182}$$

Furthermore, the trace theorem implies,

$$\|V_1(f)\|^2_{H^1(\Omega)} \geqslant C'''\|V_1(f)\|^2_{H^{1/2}(\partial\Omega)}, \tag{6.183}$$

so we combine (6.182) with (6.183) to obtain

$$\langle f,\mathscr{L}_1(f)\rangle_{H^0(\partial\Omega)} \geqslant C'''\|V_1(f)\|^2_{H^{1/2}(\partial\Omega)}.$$

□

Remark 6.3. The above theorem also holds for $\Omega\subset\mathbb{R}^2$ provided that the diameter of Ω is less than 1. For any star-shaped domain Ω, we can always make a scaling transformation to reduce the diameter of the transformed domain to less than 1. See Fig. 6.4. □

Corollary 6.2. *Let $\Omega\subseteq\mathbb{R}^3$ be bounded with smooth boundary $\partial\Omega$. Then $\mathscr{L}_1:\ H^r(\partial\Omega)\to H^{r+1}(\partial\Omega)$, $r\in\mathbb{R}$, satisfies*

$$C_r^{-1}\|f\|_{H^r(\partial\Omega)} \leqslant \|\mathscr{L}_1 f\|_{H^{r+1}(\partial\Omega)} \leqslant C_r\|f\|_{H^r(\partial\Omega)} \quad \forall\, f\in H^r(\partial\Omega),$$

for some constant $C_r>1$ depending only on r and Ω.

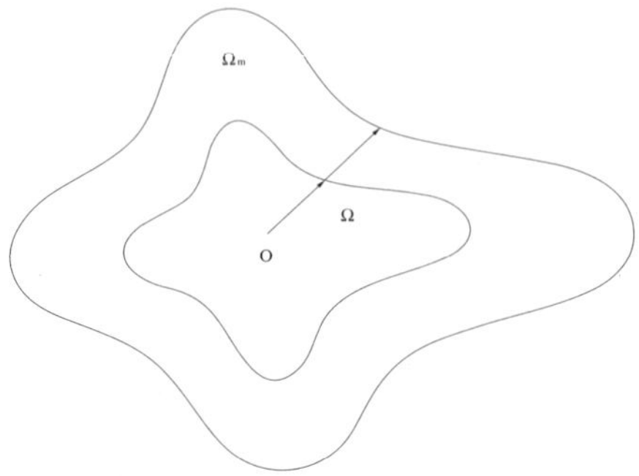

Figure 6.4: The star-shaped domain Ω and a magnified domain Ω_m.

Proof. This is an immediate consequence of the continuity and invertibility of

$$\mathscr{L}_1:\ H^r(\partial\Omega) \to H^{r+1}(\partial\Omega)$$

and

$$\mathscr{L}_1^{-1}:\ H^{r+1}(\partial\Omega) \to H^r(\partial\Omega)$$

guaranteed in Theorem 6.20. □

The boundary integral equation (6.176) for the Dirichlet problem is a *Fredholm integral equation* (with a weakly singular kernel) *of the first kind*. Next, let us consider a simple-layer representation for the Neumann boundary value problem, which leads to an *integral equation of the second kind*.

Theorem 6.22 (Simple-layer representation for the interior Neumann BVP in $\mathbb{R}^3$). *Let $\Omega \subseteq \mathbb{R}^3$ be bounded with smooth boundary $\partial\Omega$. The mapping*

$$\mathscr{L}_2(f)(x) \equiv \int_{\partial\Omega} \frac{\partial E(x-\xi)}{\partial n_x} f(\xi)\, d\sigma_\xi + \frac{1}{2} f(x), \quad x \in \partial\Omega, \tag{6.184}$$

is Fredholm:

$$\mathscr{L}_2:\ H^r(\partial\Omega) \longrightarrow H^r(\partial\Omega) \tag{6.185}$$

for any $r \in \mathbb{R}$, with

$$\dim \mathscr{N}(\mathscr{L}_2) = 1, \quad \dim \operatorname{Coker} \mathscr{L}_2 = 1, \tag{6.186}$$

and hence zero index. Its adjoint operator

$$\mathscr{L}_2^*:\ H^{-r}(\partial\Omega) \longrightarrow H^{-r}(\partial\Omega) \tag{6.187}$$

is given by

$$\mathscr{L}_2^*(\phi)(x) = \int_{\partial\Omega} \frac{\partial E(x-\xi)}{\partial n_\xi} \phi(\xi)\, d\sigma_\xi + \frac{1}{2}\phi(x) \tag{6.188}$$

on $\partial\Omega$, with

$$\dim \mathscr{N}(\mathscr{L}_2^*) = 1, \quad \dim \operatorname{Coker} \mathscr{L}_2^* = 1, \tag{6.189}$$

$$\mathscr{N}(\mathscr{L}_2^*) = \{c \mid c \in \mathbb{R},\ on\ \partial\Omega\} \ \ (= \operatorname{Coker} \mathscr{L}_2), \tag{6.190}$$

$$\begin{aligned} \operatorname{Coker} \mathscr{L}_2^* = \ & \operatorname{span}\{\lambda_0(\cdot) \mid \lambda_0(\cdot)\ \ satisfies\ \mathscr{L}_1(\lambda_0) \equiv 1 \\ & on\ \partial\Omega\} \ \ (= \mathscr{N}(\mathscr{L}_2)). \end{aligned} \tag{6.191}$$

$\mathscr{L}_2^*$ satisfies

$$\mathscr{L}_2^* :\ H^s(\partial\Omega) \to H^s(\partial\Omega) \quad \text{continuously,}$$

for any $s \in \mathbb{R}$.

Consequently, the solution $w(x)$ to the interior Neumann problem

$$\left.\begin{aligned} &\Delta w(x) = 0 && on\ \Omega, \\ &\frac{\partial w(x)}{\partial n} = g(x) \in H^r(\partial\Omega), && r \in \mathbb{R}, \end{aligned}\right\} \tag{6.192}$$

can be given in the form

$$w(x) = \int_{\partial\Omega} E(x-\xi) f(\xi)\, d\sigma_\xi + c \in H^{r+3/2}(\Omega),\ \ x \in \Omega, \tag{6.193}$$

for some $f \in H^r(\partial\Omega)$ and for any given constant $c \in \mathbb{R}$, where f is a (i.e., any nonunique) solution to the BIE

$$\mathscr{L}_2(f) = g, \tag{6.194}$$

provided that the compatibility condition

$$\langle g, 1 \rangle_{L^2(\partial\Omega)} = \int_{\partial\Omega} g(x)\, d\sigma = 0.$$

is satisfied, i.e., $g \perp \operatorname{Coker} \mathscr{L}_2$.

Proof. The regularity properties (6.185) and (6.187) of $\mathscr{L}_2$ and $\mathscr{L}_2^*$ are consequences of § 4.4 and Theorem 4.2. $\mathscr{L}_2$ is the sum of $\frac{1}{2}I$ and an integral operator with a weakly singular kernel

$$K_1(x,\xi) = \frac{\partial E(x-\xi)}{\partial n_x}. \tag{6.195}$$

Thus the adjoint operator $\mathscr{L}_2^*$ is the sum of $\frac{1}{2}I$ and an integral operator with weakly singular kernel

$$K_2(x,\xi) = K_1^*(\xi, x) = \frac{\partial E(\xi - x)}{\partial n_\xi}. \tag{6.196}$$

Therefore (6.188) is obtained.

We now show the following.

(i) $\dim \mathscr{N}(\mathscr{L}_2) = 1$.

Choose any constant $c_0 \in \mathbb{R}$, $c_0 \neq 0$. According to Theorem 6.20, there exists a unique $\widetilde{f}_0 \in C^\infty(\partial\Omega)$ such that

$$\int_{\partial\Omega} E(x-\xi)\widetilde{f}_0(\xi)\,d\sigma_\xi = c_0. \tag{6.197}$$

For $x \in \Omega$, define

$$\widetilde{w}_0(x) = \int_{\partial\Omega} E(x-\xi)\widetilde{f}_0(\xi)\,d\sigma_\xi - c_0.$$

Then $\widetilde{w}_0$ satisfies

$$\begin{aligned} \Delta\widetilde{w}_0(x) &= 0 \quad \text{on } \Omega, \\ \widetilde{w}_0(x) &\equiv 0 \quad \text{on } \partial\Omega. \end{aligned}$$

By the uniqueness Theorem 6.19, we have

$$\widetilde{w}_0(x) \equiv 0 \quad \text{on } \overline{\Omega}.$$

Taking the normal derivative, by (6.46),

$$0 = \frac{\partial \widetilde{w}_0(x)}{\partial n} = \frac{1}{2}\widetilde{f}_0(x) + \int_{\partial\Omega} \frac{\partial E(x-\xi)}{\partial n_x}\widetilde{f}_0(\xi)\,d\sigma_\xi,$$

we obtain $\widetilde{f}_0 \in \mathscr{N}(\mathscr{L}_2)$. Obviously $\widetilde{f}_0 \not\equiv 0$ because $c_0 \neq 0$ in (6.197). Therefore $\dim \mathscr{N}(\mathscr{L}_2) \geqslant 1$.

If $\widetilde{f} \in \mathscr{N}(\mathscr{L}_2)$ and $\widetilde{f} \neq 0$ then $\widetilde{f} \in C^\infty(\partial\Omega)$ by Corollary 4.4. Define

$$\widetilde{w}(x) = \int_{\partial\Omega} E(x-\xi)\widetilde{f}(\xi)\,d\sigma_\xi, \quad x \in \Omega.$$

Then

$$\frac{\partial \widetilde{w}(x)}{\partial n} = \frac{1}{2}\widetilde{f}(x) + \int_{\partial\Omega} \frac{\partial E(x-\xi)}{\partial n_x}\widetilde{f}(\xi)\,d\sigma_\xi = 0, \quad x \in \partial\Omega,$$

because $\widetilde{f} \in \mathscr{N}(\mathscr{L}_2)$. Thus $\widetilde{w}$ satisfies

$$\begin{aligned} \Delta\widetilde{w} &= 0 \quad \text{on } \Omega, \\ \frac{\partial \widetilde{w}}{\partial n} &= 0 \quad \text{on } \partial\Omega. \end{aligned}$$

By the uniqueness Theorem 6.19, we have

$$\widetilde{w}(x) \equiv c_1, \quad \text{for some} \quad c_1 \in \mathbb{R}.$$

This constant c_1 is nonzero because $\dim \mathscr{N}(\mathscr{L}_1) = 0$. Therefore

$$\int_{\partial\Omega} E(x-\xi)\left[\widetilde{f}(\xi) - \frac{c_1}{c_0}\widetilde{f}_0(\xi)\right]d\sigma_\xi = 0, \quad x \in \partial\Omega,$$

where $\widetilde{f}_0$ satisfies (6.197). Hence

$$\widetilde{f} = \frac{c_1}{c_0}\widetilde{f}_0, \quad \widetilde{f}_0 = \mathscr{L}_1^{-1}(c_0),$$

by Theorem 6.21.

(ii) dim $\mathscr{N}(\mathscr{L}_2^*) = 1$, $\mathscr{N}(\mathscr{L}_2^*) = \text{span}\{1\}$

Let w be sufficiently smooth, satisfying $\Delta w = 0$ on Ω. By (1.22), for $x \in \partial\Omega$ we have

$$\frac{1}{2}w(x) = \int_{\partial\Omega}\left[-\frac{\partial E(x-\xi)}{\partial n_\xi}w(\xi) + E(x-\xi)\frac{\partial w(\xi)}{\partial n}\right]d\sigma_\xi.$$

Pick $w(x) \equiv 1$ on Ω. Then

$$\mathscr{L}_2(1)(x) = \frac{1}{2}w(x) + \int_{\partial\Omega}\frac{\partial E(x-\xi)}{\partial n_\xi}w(\xi)\,d\sigma_\xi = 0.$$

Therefore $1 \in \mathscr{N}(\mathscr{L}_2^*)$.
It immediately follows that

$$\dim \mathscr{N}(\mathscr{L}_2^*) = 1, \quad \mathscr{N}(\mathscr{L}_2^*) = \text{span}\{1 \text{ on } \partial\Omega\},$$

because $\mathscr{L}_2$ (and $\mathscr{L}_2^*$), as a sum of an invertible operator $\frac{1}{2}I$ with a compact integral operator having kernel (6.195) (and (6.196), respectively), is Fredholm of index zero by Theorem 4.9 and Corollary 4.2. Thus

$$\dim \mathscr{N}(\mathscr{L}_2^*) = \dim \text{ Coker } \mathscr{L}_2$$

must be equal to 1.
The rest of the conclusion is obvious. □

The operator $\mathscr{L}_2$ is not invertible, since $\mathscr{N}(\mathscr{L}_2)$ has dimension 1. This causes the nonuniqueness of the solution f to the BIE (6.194), which is inconvenient both theoretically and numerically. But since we have complete information about $\mathscr{N}(\mathscr{L}_2)$ (i.e., (6.191)), and because $\mathscr{L}_2$ is Fredholm of zero index, we can apply the theory in §4.7 to solve the Neumann BVP involving $\mathscr{L}_2$. See the examples given in §4.7.

Exercise 6.6. Let $\widetilde{f}_0$ satisfy (6.197) for some $c_0 \neq 0$. Re-denote $\widetilde{f}_0$ by $\widetilde{\lambda}_0$. Show that $\widetilde{\lambda}_0$ is of one sign on $\partial\Omega$, i.e., $\widetilde{\lambda}_0(x) > 0$ or $\widetilde{\lambda}_0(x) < 0$ for all $x \in \partial\Omega$. (Hint: Use the fact that $\partial\Omega$ is an equipotential surface for $\widetilde{w}_0$, and use the jump property $\widetilde{\lambda}_0 = -\left(\partial\widetilde{w}_0^+/\partial n - \partial\widetilde{w}_0^-/\partial n\right)$.) □

Now let us give the theorem for the inhomogeneous Robin boundary condition

$$\frac{\partial w(x)}{\partial n} + hw(x) = g(x) \quad \text{on} \quad \partial\Omega, \; h > 0.$$

Theorem 6.23 (Simple-layer representation for the interior Robin BVP in $\mathbb{R}^3$). *Let $\Omega \subseteq \mathbb{R}^3$ be bounded with smooth boundary $\partial\Omega$, and let $h > 0$. The mapping*

$$\mathscr{L}_3(f)(x) \equiv \int_{\partial\Omega} \left[\frac{\partial E(x-\xi)}{\partial n_x} + hE(x-\xi)\right] f(\xi)\,d\sigma_\xi + \frac{1}{2}f(x), \quad x \in \partial\Omega, \tag{6.198}$$

is an isomorphism:

$$\mathscr{L}_3 \colon\ H^r(\partial\Omega) \to H^r(\partial\Omega), \tag{6.199}$$

for any $r \in \mathbb{R}$, with

$$\dim \mathscr{N}(\mathscr{L}_3) = 0, \quad \dim \operatorname{Coker} \mathscr{L}_3 = 0.$$

Its adjoint operator

$$\mathscr{L}_3^* \colon\ H^{-r}(\partial\Omega) \to H^{-r}(\partial\Omega) \tag{6.200}$$

is given by

$$\mathscr{L}_3^*(\phi)(x) = \int_{\partial\Omega} \left[\frac{\partial E(x-\xi)}{\partial n_\xi} + hE(x-\xi)\right] \phi(\xi)\,d\sigma_\xi + \frac{1}{2}\phi(x) \tag{6.201}$$

on $\partial\Omega$, with

$$\dim \mathscr{N}(\mathscr{L}_3^*) = 0 \quad \dim \operatorname{Coker} \mathscr{L}_3^* = 0.$$

$\mathscr{L}_3^*$ also maps isomorphically

$$\mathscr{L}_3^* \colon\ H^s(\partial\Omega) \to H^s(\partial\Omega) \tag{6.202}$$

for any $s \in \mathbb{R}$.

Consequently, the solution $w(x)$ to

$$\begin{aligned} \Delta w(x) &= 0 \qquad\qquad \text{on } \Omega \\ \frac{\partial w(x)}{\partial n} &+ hw(x) = g(x) \in H^r(\partial\Omega), \quad r \in \mathbb{R}, \end{aligned}$$

can be given in the form

$$w(x) = \int_{\partial\Omega} E(x-\xi) f(\xi)\,d\sigma_\xi$$

for unique $f \in H^r(\partial\Omega)$ solving the BIE

$$\mathscr{L}_3(f) = g. \tag{6.203}$$

Proof. The regularity properties (6.199), (6.200) and (6.202) are consequences of §4.4. We show the following.

(i) $\dim \mathscr{N}(\mathscr{L}_3) = 0.$

Let $\widetilde{f} \in \mathscr{N}(\mathscr{L}_3)$. Then $\widetilde{f} \in C^\infty(\partial\Omega)$. Define

$$\widetilde{w}^{\pm}(x) = \int_{\partial\Omega} E(x-\xi)\widetilde{f}(\xi)\,d\sigma_\xi, \quad \text{with} \quad \begin{cases} + \text{ for } x \in \Omega^c, \\ - \text{ for } x \in \Omega. \end{cases}$$

Then

$$\Delta\widetilde{w}^{\pm}(x) = 0 \quad \text{on } \Omega^c \text{ and on } \Omega.$$

Let x tend to a point on $\partial\Omega$. Then we have

$$\begin{aligned}\frac{\partial \widetilde{w}^-(x)}{\partial n} + h\widetilde{w}^-(x) &= \frac{1}{2}\widetilde{f}(w) + \int_{\partial\Omega}\left[\frac{\partial E(x-\xi)}{\partial n_x} + hE(x-\xi)\right]\widetilde{f}(\xi)\,d\sigma_\xi \\ &= \mathscr{L}_3(\widetilde{f})(x) \ \equiv \ 0.\end{aligned}$$

By the uniqueness Theorem 6.19, we conclude that

$$\widetilde{w}^-(x) \equiv 0 \quad \text{on } \Omega. \tag{6.204}$$

By the continuity of the simple layer, on $\partial\Omega$ we have

$$\widetilde{w}^+(x) = \widetilde{w}^-(x) = 0 \quad \forall\, x \in \partial\Omega.$$

Further, it is easy to check that $\widetilde{w}^+$ satisfies the growth condition (6.170). Therefore, by Lemma 6.5, we have

$$\widetilde{w}^+(x) \equiv 0 \quad \text{on } \overline{\Omega^c}. \tag{6.205}$$

Now, from the jump condition

$$\widetilde{f} = \frac{\partial \widetilde{w}^-}{\partial n} - \frac{\partial \widetilde{w}^+}{\partial n}$$

and (6.204) and (6.205), we conclude

$$\widetilde{f} \equiv 0.$$

(ii) $\dim \mathscr{N}(\mathscr{L}_3^*) = 0$

Let $\widetilde{\phi} \in \mathscr{N}(\mathscr{L}_3^*)$. Then $\widetilde{\phi} \in C^\infty(\partial\Omega)$ and

$$\begin{aligned}\mathscr{L}_3^*(\widetilde{\phi}(x)) &= \frac{1}{2}\widetilde{\phi}(x) + \int_{\partial\Omega}\left[\frac{\partial E(x-\xi)}{\partial n_\xi} + hE(x-\xi)\right]\widetilde{\phi}(\xi)\,d\sigma_\xi \\ &= 0 \quad \text{on } \partial\Omega.\end{aligned}$$

Define

$$\widetilde{w}^{\pm}(x) = \int_{\partial\Omega} \left[\frac{\partial E(x-\xi)}{\partial n_\xi} + hE(x-\xi) \right] \widetilde{\phi}(\xi)\, d\sigma_\xi$$

$$\text{with} \begin{cases} + \text{ for } x \in \Omega^c, \\ - \text{ for } x \in \Omega. \end{cases}$$

Then $\widetilde{w}^{\pm}$ can be extended to $\partial\Omega$ (but are discontinuous across $\partial\Omega$). We have

$$\widetilde{w}^{+}(x) = \mathscr{L}_3^*(\widetilde{\phi}(x)) = 0 \quad \text{on } \partial\Omega.$$

Since $\widetilde{w}^{+}$ satisfies the growth condition (6.170), by Lemma 6.5 we obtain

$$\widetilde{w}^{+}(x) \equiv 0 \quad \text{on } \Omega^c, \tag{6.206}$$

thus

$$\frac{\partial}{\partial n}\widetilde{w}^{+}(x) = -\frac{h}{2}\widetilde{\phi}(x) + \int_{\partial\Omega} \left[\frac{\partial^2 E(x-\xi)}{\partial n_x \partial n_\xi} + h\frac{\partial E(x-\xi)}{\partial n_x} \right] \widetilde{\phi}(\xi)\, d\sigma_\xi = 0 \text{ on } \partial\Omega.$$

But on $\partial\Omega$, it is easy to check that

$$0 = \frac{\partial \widetilde{w}^{+}(x)}{\partial n} + h\widetilde{w}^{+}(x) = \frac{\partial \widetilde{w}^{+}(x)}{\partial n} + h(\mathscr{L}_3^*\widetilde{\phi})(x) = \frac{\partial \widetilde{w}^{-}(x)}{\partial n} + h\widetilde{w}^{-}(x).$$

We see that $\widetilde{w}^{-}$ satisfies

$$\begin{aligned} &\Delta \widetilde{w}^{-}(x) = 0 && \text{on } \Omega, \\ &\frac{\partial \widetilde{w}^{-}(x)}{\partial n} + h\widetilde{w}^{-}(x) = 0 \text{ on } \partial\Omega. \end{aligned}$$

By the uniqueness Theorem 6.19, we obtain

$$\widetilde{w}^{-}(x) \equiv 0 \quad \text{on } \Omega. \tag{6.207}$$

By the jump condition

$$\widetilde{\phi} = \widetilde{w}^{+} - \widetilde{w}^{-}$$

and (6.206) and (6.207), we conclude that

$$\widetilde{\phi} \equiv 0.$$

The rest of the proof follows as a consequence of the Fredholm alternative. □

Remark 6.4. The boundary integral equations (6.194) and (6.203) for the Neumann and Robin BVP are weakly *singular Fredholm integral equations of the second kind.* □

Next, we study the boundary integral equations for the exterior BVP of Dirichlet, Neumann and Robin types in $\mathbb{R}^3$.

6.9 Simple-layer representations for exterior BVP in $\mathbb{R}^3$

We limit our consideration to the class of *regular* solutions.

Definition 6.1. Let Ω be a bounded domain in $\mathbb{R}^3$ or $\mathbb{R}^2$. A function f on Ω^c is said to be *regular at* ∞ if

(i) $f \in H^1_{\text{loc}}(\Omega^c)$,
(ii) $f(x) = o(1)$ and $\nabla f(x) = o(|x|^{-1})$ for large $|x|$.

A function $f(x)$ is said to be *generalized regular at* ∞ if $f(x) - c$ is regular at ∞ for some constant c. □

Lemma 6.6. *Let w be a weak solution of $\Delta w = 0$ on Ω^c such that w is generalized regular at ∞. Then*

$$w(x) = -\int_{\partial\Omega}\left[E(x-\xi)\frac{\partial w(\xi)}{\partial n} - \frac{\partial E(x-\xi)}{\partial n_\xi}w(\xi)\right]d\sigma_\xi + c, x \in \Omega^c, \tag{6.208}$$

where c is a constant such that $w - c$ is regular at ∞.

Proof. Define

$$v(x) = w(x) - c.$$

Since v is regular at ∞, the Green's formula holds for v on Ω^c:

$$v(x) = -\int_{\partial\Omega}\left[E(x-\xi)\frac{\partial v(\xi)}{\partial n} - \frac{\partial E(x-\xi)}{\partial n_\xi}v(\xi)\right]d\sigma_\xi, \quad x \in \Omega^c.$$

But

$$\frac{\partial v}{\partial n} = \frac{\partial w}{\partial n} \qquad \text{on } \partial\Omega,$$

$$\int_{\partial\Omega}\frac{\partial E(x-\xi)}{\partial n_\xi}d\sigma_\xi = 0, \quad x \in \Omega^c.$$

Therefore (6.208) follows for a classical solution w. Taking the limit, we see that (6.208) holds for $w \in H^1_{\text{loc}}(\Omega^c)$. □

Lemma 6.7. *Let w and v be weak solutions of $\Delta u = 0$ for $u = w$ and v, and assume that w and v are generalized regular at ∞. Then*

$$\int_{\Omega^c}\nabla w \cdot \nabla v\, dx = -\int_{\partial\Omega} w\frac{\partial v}{\partial n}d\sigma + \int_{\partial\Omega} c\frac{\partial v}{\partial n}d\sigma, \tag{6.209}$$

where c is a constant such that $w - c$ is regular at ∞. Consequently, we also have

$$\int_{\Omega^c}|\nabla w|^2\, dx = -\int_{\partial\Omega} w\frac{\partial w}{\partial n}d\sigma + c\int_{\partial\Omega}\frac{\partial w}{\partial n}d\sigma. \tag{6.210}$$

Proof. Let $\Omega_R^c = \Omega^c \cap \{x \mid |x| < R\}$ for some sufficiently large $R > 0$. Then integration by parts on Ω_R^c gives

$$\int_{\Omega_R^c} \nabla w \cdot \nabla v \, dx = \int_{|x|=R} w \frac{\partial v}{\partial n} d\sigma - \int_{\partial\Omega} w \frac{\partial v}{\partial n} d\sigma, \tag{6.211}$$

where on $|x| = R$ the unit normal n points outside of Ω_R^c. Since

$$v_0(x) \equiv v(x) - c', \quad x \in \Omega^c, \text{ for some constant } c'$$

is regular at ∞, we may represent v_0 by Lemma 6.6 as

$$v_0(x) = -\int_{\partial\Omega} E(x-\xi) \frac{\partial v(\xi)}{\partial n} d\sigma_\xi + \int_{\partial\Omega} \frac{\partial E(x-\xi)}{\partial n_\xi} v(\xi) \, d\sigma_\xi, x \in \Omega^c. \tag{6.212}$$

Now substituting (6.212) into (6.211), we have

$$\begin{aligned} I &\equiv \int_{|x|=R} w(x) \frac{\partial v(x)}{\partial n} d\sigma \\ &= -\int_{|x|=R} \int_{\partial\Omega} w(x) \frac{\partial E(x-\xi)}{\partial n_x} \frac{\partial v(\xi)}{\partial n} d\sigma_\xi \, d\sigma_x \\ &\quad + \int_{|x|=R} \int_{\partial\Omega} w(x) \frac{\partial^2 E(x-\xi)}{\partial n_x \partial n_\xi} v(\xi) \, d\sigma_\xi d\sigma_x. \\ &= \int_{|x|=R} \int_{\partial\Omega} [c + \mathcal{O}(|x|^{-1})] \left[-\frac{\partial E(x-\xi)}{\partial n_x} \frac{\partial v(\xi)}{\partial n} \right. \quad \text{(by (6.208))} \\ &\quad \left. + \frac{\partial^2 E(x-\xi)}{\partial n_x \partial n_\xi} v(\xi) \right] d\sigma_\xi \, d\sigma_x. \end{aligned} \tag{6.213}$$

We can represent the constant 1 by

$$1 = -\int_{|x|=R} \frac{\partial E(x-\xi)}{\partial n_x} d\sigma_x, \quad |\xi| < R. \tag{6.214}$$

Substituting the above into (6.213) and interchanging the order of integration, we obtain

$$\begin{aligned} I &= \int_{\partial\Omega} c \frac{\partial v}{\partial n} d\sigma + \int_{\partial\Omega} \left[\int_{|x|=R} c \frac{\partial^2 E(x-\xi)}{\partial n_x \partial n_\xi} d\sigma_x \right] v(\xi) \, d\sigma_\xi \\ &\quad + \mathcal{O}(|x|^{-1}) \\ &= \int_{\partial\Omega} c \frac{\partial v}{\partial n} d\sigma + \int_{\partial\Omega} \frac{\partial}{\partial n_\xi} \left[\int_{|x|=R} \frac{\partial E(x-\xi)}{\partial n_x} c \, d\sigma_x \right] v(\xi) \, d\sigma_\xi \\ &\quad + \mathcal{O}(|x|^{-1}) \\ &= \int_{\partial\Omega} c \frac{\partial v}{\partial n} d\sigma + \mathcal{O}(|x|^{-1}) \quad \text{(by (6.214))}. \end{aligned}$$

Letting $|x| \to \infty$, we obtain (6.209). Using $v = w$ in (6.209), we obtain (6.210). □

As a consequence of Lemma 6.7, we see that regular solutions to $\Delta w = 0$ subject to the Dirichlet, Neumann or Robin BVP are *unique*.

Theorem 6.24 (Simple-layer representation for the exterior Dirichlet BVP in $\mathbf{R}^3$). *Let $\Omega \subseteq \mathbb{R}^3$ be bounded with smooth boundary $\partial\Omega$. The mapping*

$$\mathscr{L}_{1e}(f)(x) \equiv \int_{\partial\Omega} E(x-\xi) f(\xi)\, d\sigma_\xi, \quad x \in \partial\Omega, \tag{6.215}$$

is an isomorphism from $H^r(\partial\Omega)$ onto $H^{r+1}(\partial\Omega)$ for any $r \in \mathbb{R}$, satisfying

$$\dim \mathscr{N}(\mathscr{L}_{1e}) = 0, \quad \dim \operatorname{Coker} \mathscr{L}_{1e} = 0.$$

Its adjoint operator $\mathscr{L}_{1e}^*$ maps $H^{-(r+1)}(\partial\Omega)$ onto $H^{-r}(\partial\Omega)$ and is formally given by

$$\mathscr{L}_{1e}^* = \mathscr{L}_{1e}, \quad \text{as in (6.215)},$$

with

$$\dim \mathscr{N}(\mathscr{L}_{1e}^*) = 0, \quad \dim \operatorname{Coker} \mathscr{L}_{1e}^* = 0.$$

$\mathscr{L}_{1e}^*$ also maps isomorphically

$$\mathscr{L}_{1e}^*\colon\ H^s(\partial\Omega) \xrightarrow{\text{onto}} H^{s+1}(\partial\Omega)$$

for any $s \in \mathbb{R}$.

The solution w, regular at ∞, to the exterior Dirichlet problem

$$\begin{aligned} \Delta w(x) &= 0 && \text{on } \Omega^c, \\ w(x) &= g(x) \in H^{r+1}(\partial\Omega), && r \in \mathbb{R}, \end{aligned}$$

is given uniquely by

$$w(x) = \int_{\partial\Omega} E(x-\xi) f(\xi)\, d\sigma_\xi \in H^{r+3/2}_{\mathrm{loc}}(\Omega^c), \quad x \in \Omega^c,$$

where the potential density $f \in H^r(\partial\Omega)$ uniquely solves the BIE

$$\mathscr{L}_{1e}(f) = g. \qquad \square$$

The proof, based upon Lemma 6.7, is identical to that of Theorem 6.20 so it is left out.

Theorem 6.25 (Simple-layer representation for the exterior Neumann BVP in $\mathbb{R}^3$). *Let $\Omega \subset \mathbb{R}^3$ be bounded with smooth boundary $\partial\Omega$. The mapping*

$$\mathscr{L}_{2e}(f)(x) \equiv \int_{\partial\Omega} \frac{\partial E(x-\xi)}{\partial n_x} f(\xi)\, d\sigma_\xi - \frac{1}{2} f(x), \quad x \in \partial\Omega,$$

is an isomorphism,

$$\mathscr{L}_{2e}\colon\ H^r(\partial\Omega) \longrightarrow H^r(\partial\Omega),$$

for any $r \in \mathbb{R}$, with

$$\dim \mathscr{N}(\mathscr{L}_{2e}) = 0, \quad \dim \operatorname{Coker} \mathscr{L}_{2e} = 0.$$

Its adjoint operator

$$\mathscr{L}_{2e}^{*}: H^{-r}(\partial\Omega) \longrightarrow H^{-r}(\partial\Omega)$$

is given by

$$\mathscr{L}_{2e}^{*}(\phi)(x) = \int_{\partial\Omega} \frac{\partial E(x-\xi)}{\partial n_{\xi}} \phi(\xi)\, d\sigma_{\xi} - \frac{1}{2}\phi(x)$$

on $\partial\Omega$, with

$$\dim \mathscr{N}(\mathscr{L}_{2e}^{*}) = 0, \quad \dim \operatorname{Coker} \mathscr{L}_{2e}^{*} = 0.$$

$\mathscr{L}_{2e}^{*}$ also maps isomorphically

$$\mathscr{L}_{2e}^{*}: H^{s}(\partial\Omega) \longrightarrow H^{s}(\partial\Omega),$$

for any $s \in \mathbb{R}$.

There is a unique solution w, regular at ∞, to the exterior Neumann problem

$$\Delta w(x) = 0 \qquad \text{on } \Omega^{c},$$
$$\frac{\partial w(x)}{\partial n} = g(x) \in H^{r}(\partial\Omega),\ r \in \mathbb{R},$$

given by

$$w(x) = \int_{\partial\Omega} E(x-\xi) f(\xi)\, d\sigma_{\xi} \in H_{\text{loc}}^{r+1/2}(\Omega^{c}), \quad x \in \Omega^{c},$$

for unique $f \in H^{r}(\partial\Omega)$, where f uniquely solves the BIE

$$\mathscr{L}_{2e}(f) = g. \tag{6.216}$$

Proof. We need only check that $\dim \mathscr{N}(\mathscr{L}_{2e}) = 0$ and $\dim \mathscr{N}(\mathscr{L}_{2e}^{*}) = 0$.

First, let

$$\widetilde{f} \in \mathscr{N}(\mathscr{L}_{2e}) \subset C^{\infty}(\partial\Omega). \tag{6.217}$$

Define

$$\widetilde{w}^{\pm}(x) = \int_{\partial\Omega} E(x-\xi)\widetilde{f}(\xi)\, d\sigma_{\xi}, \quad \text{with} \quad \begin{cases} + \text{ for } x \in \Omega^{c}, \\ - \text{ for } x \in \Omega. \end{cases}$$

Then both $\widetilde{w}^{\pm}$ are extendable to $\partial\Omega$ and are continuous across $\partial\Omega$. By (6.217),

$$\frac{\partial \widetilde{w}^{+}(x)}{\partial n} = -\frac{1}{2}\widetilde{f}(x) + \int_{\partial\Omega} \frac{\partial E(x-\xi)}{\partial n_{x}} \widetilde{f}(\xi) d\sigma_{\xi} = 0 \quad \text{on } \partial\Omega.$$

Since $\widetilde{w}^{+}$ is regular at ∞ and $\Delta\widetilde{w}^{+} = 0$ on Ω^{c}, by Lemma 6.7, we conclude that

$$\widetilde{w}^{+}(x) \equiv 0 \quad \text{on } \Omega^{c}.$$

Since $\widetilde{w}^{-}(x) = \widetilde{w}^{+}(x) = 0$ on $\partial\Omega$ and $\Delta\widetilde{w}^{-}(x) = 0$ on Ω, by the uniqueness Theorem 6.19, we get

$$\widetilde{w}^{-}(x) \equiv 0 \quad \text{on } \Omega.$$

Hence on $\partial\Omega$,

$$\widetilde{f} = \frac{\partial \widetilde{w}^{-}}{\partial n} - \frac{\partial \widetilde{w}^{+}}{\partial n} = 0.$$

The fact that $\dim \mathscr{N}(\mathscr{L}_{2e}^{*}) = 0$ can be proved in the same way. □

Remark 6.5. For the exterior Neumann BVP, the compatibility condition

$$\int_{\partial\Omega} g\,d\sigma = 0, \quad \text{where} \quad g = \frac{\partial w}{\partial n}, \tag{6.218}$$

does not hold in general, because

$$0 = \int_{\Omega^c\cap\{|x|\leqslant R\}} \Delta w\,dx = \int_{|x|=R} \frac{\partial w}{\partial r}\,d\sigma - \int_{\partial\Omega} g\,d\sigma,$$

thus $\int_{\partial\Omega} g\,d\sigma$ is balanced by the flux at ∞ when $R \uparrow \infty$. If it is known a priori that $\partial w/\partial r = \mathscr{O}(|r|^{-2})$ then g must satisfy (6.218). □

Theorem 6.26 (Simple-layer representation for the exterior Robin problem in $\mathbb{R}^3$). *Let $\Omega \subseteq \mathbb{R}^3$ be bounded with smooth boundary $\partial\Omega$. The mapping*

$$\mathscr{L}_{3e}(f)(x) \equiv \int_{\partial\Omega} \left[\frac{\partial E(x-\xi)}{\partial n_x} - hE(x-\xi)\right] f(\xi)\,d\sigma_\xi - \frac{1}{2}f(x),$$

$$x \in \partial\Omega, h > 0,$$

is an isomorphism,

$$\mathscr{L}_{3e}\colon\ H^r(\partial\Omega) \to H^r(\partial\Omega),$$

for any $r \in \mathbb{R}$, with

$$\dim \mathscr{N}(\mathscr{L}_{3e}) = 0, \quad \dim \operatorname{Coker} \mathscr{L}_{3e} = 0.$$

Its adjoint operator is given by

$$\mathscr{L}^*_{3e}\colon\ H^{-r}(\partial\Omega) \longrightarrow H^{-r}(\partial\Omega)$$

$$\mathscr{L}^*_{3e}(\phi)(x) = \int_{\partial\Omega} \left[\frac{\partial E(x-\xi)}{\partial n_\xi} - hE(x-\xi)\right] \phi(\xi)\,d\sigma_\xi - \frac{1}{2}\phi(x)$$

on $\partial\Omega$, with

$$\dim \mathscr{N}(\mathscr{L}^*_{3e}) = 0, \quad \dim \operatorname{Coker} \mathscr{L}^*_{3e} = 0.$$

$\mathscr{L}^*_{3e}$ also maps isomorphically,

$$\mathscr{L}^*_{3e}\colon\ H^s(\partial\Omega) \longrightarrow H^s(\partial\Omega)$$

for any $s \in \mathbb{R}$.

There exists a unique solution w, regular at ∞, to the exterior Robin problem

$$\Delta w(x) = 0 \quad \text{on } \Omega^c,$$

$$\frac{\partial w(x)}{\partial n} - hw(x) = g(x) \in H^r(\partial\Omega), \quad h > 0,\ r \in \mathbb{R},$$

given by

$$w(x) = \int_{\partial\Omega} E(x-\xi) f(\xi)\,d\sigma_\xi \in H^{r+3/2}_{\mathrm{loc}}(\Omega^c),\ x \in \Omega^c$$

for some (unique) $f \in H^r(\partial\Omega)$, which is the unique solution to the BIE

$$\mathscr{L}_{3e}(f) = g. \tag{6.219}$$

Proof. Exercise. □

6.10 Double-layer representations for interior BVP in $\mathbb{R}^3$

Let us use BIE to solve the interior and exterior Dirichlet, Neumann and Robin BVP for harmonic functions based upon the representation

$$w(x) = \int_{\partial\Omega} \frac{\partial E(x-\xi)}{\partial n_\xi} f(\xi)\, d\sigma_\xi.$$

We will state theorems for problems of each type. Since their proofs are similar to those given in the last section, we will leave them as exercises.

Theorem 6.27 (Double-layer representation for the interior Dirichlet BVP in $\mathbb{R}^3$). *Let $\Omega \subseteq \mathbb{R}^3$ be bounded with smooth boundary $\partial\Omega$. The mapping*

$$\mathscr{L}_4(f)(x) \equiv \int_{\partial\Omega} \frac{\partial E(x-\xi)}{\partial n_\xi} f(\xi)\, d\sigma_\xi - \frac{1}{2} f(x)$$

is an isomorphism from $H^r(\partial\Omega)$ onto $H^r(\partial\Omega)$ for any $r \in \mathbb{R}$, satisfying

$$\dim \mathscr{N}(\mathscr{L}_4) = 0, \quad \dim \text{Coker } \mathscr{L}_4 = 0.$$

Its adjoint operator

$$\mathscr{L}_4^*:\ H^{-r}(\partial\Omega) \longrightarrow H^{-r}(\partial\Omega), \quad r \in \mathbb{R}$$

is given by

$$\mathscr{L}_4^*(\phi)(x) = \int_{\partial\Omega} \frac{\partial E(x-\xi)}{\partial n_x} \phi(\xi)\, d\sigma_\xi - \frac{1}{2}\phi(x)$$

on $\partial\Omega$, with

$$\dim \mathscr{N}(\mathscr{L}_4^*) = 0, \quad \dim \text{Coker } \mathscr{L}_4^* = 0.$$

$\mathscr{L}_4^*$ also maps isomorphically,

$$\mathscr{L}_4^*:\ H^s(\partial\Omega) \to H^s(\partial\Omega)$$

for any $s \in \mathbb{R}$.
Consequently, the solution $w(x)$ to the interior Dirichlet BVP

$$\begin{cases} \Delta w(x) = 0 & on\ \Omega, \\ w(x) = g(x) \in H^r(\partial\Omega), & r \in \mathbb{R}, \end{cases}$$

is given by

$$w(x) = \int_{\partial\Omega} \frac{\partial E(x-\xi)}{\partial n_\xi} f(\xi)\, d\sigma_\xi \in H^{r+1/2}(\Omega),$$

where the density $f \in H^r(\partial\Omega)$ is the unique solution to the BIE

$$\mathscr{L}_4(f) = g. \qquad \square$$

Theorem 6.28 (Double-layer representation for the interior Neumann BVP in $\mathbb{R}^3$). *Let $\Omega \subseteq \mathbb{R}^3$ be bounded, with smooth boundary $\partial\Omega$. The mapping*

$$\mathscr{L}_5(f)(x) \equiv \int_{\partial\Omega} \frac{\partial^2 E(x-\xi)}{\partial n_x \partial n_\xi} f(\xi)\, d\sigma_\xi, \quad x \in \partial\Omega, \tag{6.220}$$

is Fredholm,

$$\mathscr{L}_5 \colon\ H^{r+1}(\partial\Omega) \to H^r(\partial\Omega), \quad r \in \mathbb{R},$$

where the integral in (6.220) is interpreted in the sense of the finite part, with

$$\dim \mathscr{N}(\mathscr{L}_5) = 1, \quad \dim \text{Coker}\, \mathscr{L}_5 = 1.$$

Its adjoint operator $\mathscr{L}_5^*$ maps $H^{-r}(\partial\Omega)$ onto $H^{-(r+1)}(\partial\Omega)$ and is formally given by

$$\mathscr{L}_5^* = \mathscr{L}_5, \quad \text{as in (6.220)},$$

with

$$\dim \mathscr{N}(\mathscr{L}_5^*) = 1, \quad \dim \text{Coker}\, \mathscr{L}_5^* = 1.$$
$$\mathscr{N}(\mathscr{L}_5) = \mathscr{N}(\mathscr{L}_5^*) = \text{Coker}\, \mathscr{L}_5 = \text{Coker}\, \mathscr{L}_5^* = \text{span}\{1\}.$$

$\mathscr{L}_5^*$ also maps continuously,

$$\mathscr{L}_5^* \colon\ H^{s+1}(\partial\Omega) \to H^s(\partial\Omega)$$

for any $s \in \mathbb{R}$.
Consequently, the solution $w(x)$ to the interior Neumann BVP

$$\Delta w(x) = 0 \quad \text{on } \Omega,$$
$$\frac{\partial w(x)}{\partial n} = g(x) \in H^r(\partial\Omega), \quad r \in \mathbb{R},$$

is given by

$$w(x) = \int_{\partial\Omega} \frac{\partial E(x-\xi)}{\partial n_\xi} f(\xi)\, d\sigma_\xi + c \in H^{r+3/2}(\Omega), \quad x \in \Omega,$$

for arbitrarily given $c \in \mathbb{R}$, where the potential density $f \in H^r(\partial\Omega)$ is a solution of the BIE

$$\mathscr{L}_5(f) = g$$

provided that the compatibility condition

$$\int_{\partial\Omega} g\, d\sigma = 0$$

is satisfied, i.e., $g \perp \text{Coker}\, \mathscr{L}_5$. □

Theorem 6.29 (Double-layer representation for the interior Robin BVP in $\mathbb{R}^3$). *Let Ω be a bounded domain in $\mathbb{R}^3$ with smooth boundary $\partial\Omega$. The boundary integral operator*

$$\mathscr{L}_6(f)(x) \equiv \int_{\partial\Omega}\left[\frac{\partial^2 E(x-\xi)}{\partial n_x \partial n_\xi} + h\frac{\partial E(x-\xi)}{\partial n_\xi}\right] f(\xi)\,d\sigma_\xi - \frac{h}{2}f(x),$$
$$x \in \partial\Omega,\ h > 0, \tag{6.221}$$

is an isomorphism,

$$\mathscr{L}_6\colon\ H^{r+1}(\partial\Omega) \to H^r(\partial\Omega) \quad \text{for any} \quad r \in \mathbb{R},$$

with

$$\dim \mathscr{N}(\mathscr{L}_6) = 0, \quad \dim\,\text{Coker}\,\mathscr{L}_6 = 0.$$

Its adjoint $\mathscr{L}_6^*$,

$$\mathscr{L}_6^*\colon\ H^{-r}(\partial\Omega) \to H^{-(r+1)}(\partial\Omega),$$

is given by

$$\mathscr{L}_6^*(\phi)(x) = \int_{\partial\Omega}\left[\frac{\partial^2 E(x-\xi)}{\partial n_x \partial n_\xi} + h\frac{\partial E(x-\xi)}{\partial n_x}\right] \phi(\xi)\,d\sigma_\xi - \frac{h}{2}\phi(x),$$
$$x \in \partial\Omega. \tag{6.222}$$

$\mathscr{L}_6^*$ maps isomorphically,

$$\mathscr{L}_6^*\colon\ H^{s+1}(\partial\Omega) \to H^s(\partial\Omega)$$

for any $s \in \mathbb{R}$. (Note that the first integrals in (6.221) and (6.222) are taken in the sense of finite parts.)

Consequently, the solution w to the interior Robin BVP

$$\begin{cases} \Delta w(x) = 0 & \text{on } \Omega, \\ \dfrac{\partial w(x)}{\partial n} + hw(x) = g(x) \in H^r(\partial\Omega), & r \in \mathbb{R}, \end{cases}$$

can be given by

$$w(x) = \int_{\partial\Omega} \frac{\partial E(x-\xi)}{\partial n_\xi} f(\xi)\,d\sigma_\xi \in H^{r+3/2}(\Omega), \quad x \in \Omega,$$

where $f \in H^{r+1}(\partial\Omega)$ is the unique solution to the BIE

$$\mathscr{L}_6(f) = g. \qquad \square$$

6.11 Double-layer representations for exterior BVP in $\mathbb{R}^3$

Now, let us consider the exterior boundary value problems by the doubler-layer representation. For the exterior problem, the double-layer representation has behavior $\mathscr{O}(r^{-2})$ at infinity:

$$\begin{aligned} w(x) &= \int_{\partial\Omega} \frac{\partial E(x-\xi)}{\partial n_\xi} f(\xi)\, d\sigma_\xi, \quad x \in \Omega^c, \\ &= \frac{1}{4\pi} \frac{x}{|x|^3} \cdot \int_{\partial\Omega} n_\xi f(\xi)\, d\sigma_\xi + \mathscr{O}(|x|^{-3}), \\ &= \mathscr{O}(|x|^{-2}), \end{aligned} \tag{6.223}$$

whereas in many applications (e.g., fluid mechanics), the required potential function is generally $\mathscr{O}(|r|^{-1})$ at infinity. This deficiency of the double-layer representation (6.223) can be circumvented by introducing a term $c|x|^{-1}$ into (6.223):

$$w(x) = \frac{c}{|x|} + \int_{\partial\Omega} \frac{\partial E(x-\xi)}{\partial n_\xi} f(\xi)\, d\sigma_\xi, \quad x \in \Omega^c, \tag{6.224}$$

in view of the appearance of the leading term of the simple layer at ∞. See (6.179).

Theorem 6.30 (Double-layer representation for the exterior Dirichlet BVP in $\mathbb{R}^3$). *Let $\Omega \subseteq \mathbb{R}^3$ be a bounded domain containing the origin and let $\partial\Omega$ be smooth. Define the operator*

$$\mathscr{L}_{4e}\colon\ \mathbb{R} \oplus H^r(\partial\Omega) \to \mathbf{R} \oplus H^r(\partial\Omega),\ r \in \mathbb{R}, \tag{6.225}$$

by

$$\mathscr{L}_{4e}\left(\begin{bmatrix} a \\ f \end{bmatrix}\right)(x) = \begin{bmatrix} \int_{\partial\Omega} |\xi|^{-1} f(\xi)\, d\sigma \\ \frac{a}{|x|} + \int_{\partial\Omega} \frac{\partial E(x-\xi)}{\partial n_\xi} f(\xi)\, d\sigma_\xi + \frac{1}{2} f(x) \end{bmatrix}, \quad x \in \partial\Omega.$$

Then $\mathscr{L}_{4e}$ is an isomorphism.
Its adjoint operator

$$\mathscr{L}_{4e}^*\colon\ \mathbb{R} \oplus H^{-r}(\partial\Omega) \to \mathbb{R} \oplus H^{-r}(\partial\Omega) \quad \text{is continuous,}$$

and is given by

$$\mathscr{L}_{4e}^*\left(\begin{bmatrix} b \\ \phi \end{bmatrix}\right)(x) = \begin{bmatrix} \int_{\partial\Omega} |\xi|^{-1} \phi(\xi)\, d\sigma \\ \int_{\partial\Omega} \frac{\partial E(x-\xi)}{\partial n_x} \phi(\xi)\, d\sigma_\xi + \frac{1}{2}\phi(x) + \frac{b}{|x|} \end{bmatrix}, \quad x \in \partial\Omega. \tag{6.226}$$

$\mathscr{L}_{4e}^*$ maps isomorphically,

$$\mathscr{L}_{4e}^*\colon\ \mathbb{R} \oplus H^s(\partial\Omega) \to \mathbb{R} \oplus H^s(\partial\Omega)$$

for any $s \in \mathbb{R}$.

Consequently, if w *is the solution of the exterior Dirichlet BVP regular at* ∞

$$\left.\begin{aligned} &\Delta w(x) = 0 \quad \text{on } \Omega^c \\ &w(x) = g(x) \in H^r(\partial\Omega), \quad r \in \mathbb{R}, \end{aligned}\right\} \tag{6.227}$$

then

$$w(x) = \frac{a}{|x|} + \int_{\partial\Omega} \frac{\partial E(x-\xi)}{\partial n_\xi} f(\xi)\, d\sigma_\xi \in H^{r+1/2}_{\text{loc}}(\Omega),\ x \in \Omega^c,$$

where $(a, f) \in \mathbb{R} \oplus H^r(\partial\Omega)$ *is the unique solution to the BIE*

$$\mathscr{L}_{4e}\left(\begin{bmatrix} a \\ f \end{bmatrix}\right) = \begin{bmatrix} A \\ g \end{bmatrix} \in \mathbb{R} \oplus H^r(\partial\Omega) \tag{6.228}$$

for some given constant $A \in \mathbb{R}$. *(The constant* A *can be chosen arbitrarily, but the resulting* (a, f) *depends on the choice of* A. *Here* a *remains the same for any choice of* A.*)*

Proof. It is straightforward to verify by transposition that $\mathscr{L}_{4e}^*$ is given by (6.226). We will verify that $\mathscr{N}(\mathscr{L}_{4e}) = \{0\}$ and $\mathscr{N}(\mathscr{L}_{4e}^*) = \{0\}$.

(i) $\mathscr{N}(\mathscr{L}_{4e}) = \{0\}$

Let $(a, f) \in \mathscr{N}(\mathscr{L}_{4e})$. Then $f \in C^\infty(\partial\Omega)$. Define

$$w^\pm(x) = \int_{\partial\Omega} \frac{\partial E(x-\xi)}{\partial n_\xi} f(\xi)\, d\sigma_\xi, \quad \text{with} \quad \begin{cases} + \text{ for } x \in \Omega^c, \\ - \text{ for } x \in \Omega. \end{cases} \tag{6.229}$$

On Ω^c, $\widetilde{w}^+ \equiv w^+(x) - a/|x|$, and

$$\begin{aligned} &\Delta \widetilde{w}^+ = 0 \quad \text{on } \Omega^c, \\ &\widetilde{w}^+ = 0 \quad \text{on } \partial\Omega, \ \text{because } (a, f) \in \mathscr{N}(\mathscr{L}_{4e}), \\ &\widetilde{w}^+(x) = \mathscr{O}(|x|^{-1}), \quad \nabla \widetilde{w}^+(x) = \mathscr{O}(|x|^{-2}), \quad \text{large } |x|. \end{aligned}$$

Therefore

$$\int_{\Omega^c} (\Delta \widetilde{w}^+) \widetilde{w}^+\, dx = -\int_{\Omega^c} |\nabla \widetilde{w}^+|^2\, dx = 0,$$

and $\widetilde{w}^+ \equiv 0$ on Ω^c. So $a = 0$. Since the normal derivative of the double-layer potential is continuous across $\partial\Omega$, we have

$$\frac{\partial w^-}{\partial n} = \frac{\partial w^+}{\partial n} = 0 \quad \text{on } \partial\Omega.$$

Therefore, w^- satisfies

$$\begin{aligned} &\Delta w^- = 0 \ \text{ on } \Omega, \\ &\frac{\partial w^-}{\partial n} = 0 \text{ on } \partial\Omega. \end{aligned}$$

By the uniqueness theorem of elliptic BVP, we have

$$w^-(x) \equiv c_1, x \in \Omega, \text{ for some constant } c_1. \tag{6.230}$$

On $\partial\Omega$, since $a = 0$, we obtain from (6.44), (6.229) and (6.230) that

$$\int_{\partial\Omega} \frac{\partial E(x-\xi)}{\partial n_\xi} f(\xi)\, d\sigma_\xi - \frac{1}{2} f(x) = c_1 \quad \forall x \in \partial\Omega. \tag{6.231}$$

But from Theorem 6.25, we see that (6.231) has the unique solution

$$f(x) \equiv -c_1 \quad \text{on } \partial\Omega.$$

Since f satisfies

$$\int_{\partial\Omega} |\xi|^{-1} f(\xi)\, d\sigma = 0 = -c_1 \int_{\partial\Omega} |\xi|^{-1}\, d\sigma,$$

we get $f(x) = -c_1 = 0$ on $\partial\Omega$.

(ii) $\mathscr{N}(\mathscr{L}_{4e}^*) = \{0\}$

Let $(b, \phi) \in \mathscr{N}(\mathscr{L}_{4e}^*)$. Define

$$v^\pm(x) = \int_{\partial\Omega} E(x-\xi)\phi(\xi)\, d\sigma_\xi, \quad \text{with} \quad \begin{cases} + \text{ for } x \in \Omega^c, \\ - \text{ for } x \in \Omega. \end{cases}$$

Then v^- is the solution to the interior Neumann BVP

$$\begin{aligned} \Delta v^- &= 0 && \text{on } \Omega, \\ \frac{\partial v^-(x)}{\partial n} &= -\frac{b}{|x|} && \text{on } \partial\Omega. \end{aligned}$$

v^- is unique up to a constant. v^- exists iff the compatibility condition

$$\int_{\partial\Omega} \frac{\partial v^-}{\partial n}\, d\sigma = 0 = -\int_{\partial\Omega} \frac{b}{|x|}\, d\sigma$$

is satisfied. Thus b must be equal to 0. We therefore have

$$v^-(x) \equiv c_2 \quad \text{on } \overline{\Omega}, \text{ for some constant } c_2.$$

On Ω^c, we apply Lemmas 6.2 and 6.5 to $w(x) \equiv v^+(x) - c_2$. We obtain

$$w(x) \equiv 0 \quad \text{on } \Omega^c.$$

Hence

$$\phi(x) = \frac{\partial v^-(x)}{\partial n} - \frac{\partial v^+(x)}{\partial n} = \frac{\partial v^-(x)}{\partial n} - \frac{\partial w(x)}{\partial n} = 0 \quad \text{on } \partial\Omega.$$

□

Theorem 6.31 (Double-layer representation for the exterior Neumann BVP in $\mathbb{R}^3$). *Let $\Omega \subseteq \mathbb{R}^3$ be a bounded domain with smooth boundary $\partial\Omega$ containing the origin in its interior. Define the operator*

$$\mathscr{L}_{5e}\colon\ \mathbb{R}\oplus H^{r+1}(\partial\Omega) \to \mathbb{R}\oplus H^{r}(\partial\Omega), \quad r\in\mathbb{R},$$

$$\mathscr{L}_{5e}\left(\begin{bmatrix} a \\ f \end{bmatrix}\right)(x) = \begin{bmatrix} \int_{\partial\Omega} f(\xi)\,d\sigma \\ -\frac{a}{|x|^3}(n_x,x) + \int_{\partial\Omega} \frac{\partial^2 E(x-\xi)}{\partial n_x \partial n_\xi} f(\xi)\,d\sigma_\xi \end{bmatrix}, \quad x\in\partial\Omega.$$

Then $\mathscr{L}_{5e}$ is an isomorphism. Its adjoint operator

$$\mathscr{L}_{5e}^*\colon\ \mathbb{R}\oplus H^{-r}(\partial\Omega) \to \mathbb{R}\oplus H^{-(r+1)}(\partial\Omega),$$

given by

$$\mathscr{L}_{5e}^*\left(\begin{bmatrix} b \\ \phi \end{bmatrix}\right)(x) = \begin{bmatrix} -\int_{\partial\Omega} |\xi|^{-3}(n_\xi\cdot\xi)\phi(\xi)\,d\sigma_\xi \\ \int_{\partial\Omega} \frac{\partial^2 E(x-\xi)}{\partial n_x \partial n_\xi} \phi(\xi)\,d\sigma_\xi + b \end{bmatrix}, \quad x\in\partial\Omega,$$

is also continuous and invertible. Further, $\mathscr{L}_{5e}^*$ maps isomorphically,

$$\mathscr{L}_{5e}^*\colon\ \mathbb{R}\oplus H^{s+1}(\partial\Omega) \to \mathbb{R}\oplus H^{s}(\partial\Omega)$$

for any $s\in\mathbb{R}$.

Consequently, if w is the solution of the exterior Neumann BVP regular at ∞:

$$\left.\begin{array}{ll} \Delta w(x) = 0 & \text{on } \Omega^c \\ \frac{\partial w(x)}{\partial n} = g(x) \in H^r(\partial\Omega), & r\in\mathbb{R}, \\ w(x) = o(1), & \text{for large } |x|, \end{array}\right\} \tag{6.232}$$

then

$$w(x) = \frac{a}{|x|} + \int_{\partial\Omega} \frac{\partial E(x-\xi)}{\partial n_\xi} f(\xi)\,d\sigma_\xi \in H^{r+3/2}_{\mathrm{loc}}(\overline{\Omega}^c), \ x\in\overline{\Omega}^c,$$

where $(a,f)\in\mathbb{R}\oplus H^{r+1}(\partial\Omega)$ is the unique solution to the BIE

$$\mathscr{L}_{5e}\begin{bmatrix} a \\ f \end{bmatrix} = \begin{bmatrix} A \\ g \end{bmatrix} \in \mathbb{R}\oplus H^r(\partial\Omega),$$

for some given constant $A\in\mathbb{R}$. (The constant A can be chosen arbitrarily, but the resulting (a,f) depends on the choice of A.) □

Finally, we state the following.

Theorem 6.32 (Double-layer representation for the exterior Robin BVP in $\mathbb{R}^3$).
Let $\Omega \subseteq \mathbb{R}^3$ be bounded with smooth boundary $\partial\Omega$ containing the origin in the interior. Define the operator

$$\mathscr{L}_{6e}\colon\ \mathbb{R}\oplus H^{r+1}(\partial\Omega) \to \mathbb{R}\oplus H^{r}(\partial\Omega), \quad r\in\mathbb{R}, \tag{6.233}$$

$$\mathscr{L}_{6e}\left(\begin{bmatrix} a \\ f \end{bmatrix}\right) = \begin{bmatrix} \displaystyle\int_{\partial\Omega} f(\xi)\,d\sigma \\ \displaystyle\left(\frac{\partial}{\partial n} - h\right)\frac{a}{|x|} + \int_{\partial\Omega}\left[\frac{\partial^2 E(x-\xi)}{\partial n_x \partial n_\xi} - h\frac{\partial E(x-\xi)}{\partial n_\xi}\right] f(\xi)\,d\sigma_\xi - \frac{1}{2}hf(x) \end{bmatrix},$$
$$x\in\partial\Omega,\ h>0$$

Then $\mathscr{L}_{6e}$ is an isomorphism. Its adjoint $\mathscr{L}^*_{6e}$ is also an isomorphism,

$$\mathscr{L}^*_{6e}\colon\ \mathbb{R}\oplus H^{-r}(\partial\Omega) \to \mathbb{R}\oplus H^{-(r+1)}(\partial\Omega), r\in\mathbb{R}, \tag{6.234}$$

$$\mathscr{L}^*_{6e}\left(\begin{bmatrix} b \\ \phi \end{bmatrix}\right)(x) = \begin{bmatrix} \displaystyle\int_{\partial\Omega}\left[\left(\frac{\partial}{\partial n_\xi} - h\right)\frac{1}{|\xi|}\right]\phi(\xi)d\sigma_\xi \\ \displaystyle\int_{\partial\Omega}\left[\frac{\partial^2 E(x-\xi)}{\partial n_x \partial n_\xi} - h\frac{\partial E(x-\xi)}{\partial n_\xi}\right]\phi(\xi)\,d\sigma_\xi + \frac{1}{2}h\phi(x) + b \end{bmatrix}, \quad x\in\partial\Omega.$$

(Note that in (6.233) and (6.234), the integrals are taken in the sense of finite parts.) Consequently, the solution $w(x)$ to the exterior Robin BVP regular at ∞

$$\begin{cases} \Delta w(x) = 0 \quad \text{on } \Omega^c, \\ \dfrac{\partial w(x)}{\partial n} - hw(x) = g(x) \in H^r(\partial\Omega), \ \ r\in\mathbb{R}, \end{cases}$$

is given by

$$w(x) = \frac{a}{|x|} + \int_{\partial\Omega}\frac{\partial E(x-\xi)}{\partial n_\xi} f(\xi)\,d\sigma_\xi \in H^{r+3/2}_{\text{loc}}(\Omega^c), \ \ x\in\Omega^c, \tag{6.235}$$

where $(a,f)\in\mathbb{R}\oplus H^{r+1}(\partial\Omega)$ is the unique solution to the BIE

$$\mathscr{L}_{6e}\left(\begin{bmatrix} a \\ f \end{bmatrix}\right) = \begin{bmatrix} A \\ g \end{bmatrix}. \tag{6.236}$$

(The constant A in (6.236) can be chosen arbitrarily, but the resulting (a,f) depends on the choice of A.) □

Remark 6.6. Comparing the various operators introduced from §6.8 up to this point, we have

$$\mathscr{L}_4 = \mathscr{L}_{2e}, \qquad \mathscr{L}_4^* = \mathscr{L}_{2e}^*.$$

In terms of these operators, we can rewrite the Calderòn projectors from Example 4.3 as

$$\mathscr{C}_1 = \begin{bmatrix} -\mathscr{L}_4 & \mathscr{L}_1 \\ -\mathscr{L}_5 & \mathscr{L}_2 \end{bmatrix}, \quad \mathscr{C}_2 = \begin{bmatrix} \mathscr{L}_2^* & -\mathscr{L}_1 \\ \mathscr{L}_5 & -\mathscr{L}_4 \end{bmatrix}.$$

Since $\mathscr{C}_i^2 = \mathscr{C}_i$, $i = 1,2$, we get the identities

$$\begin{aligned}
\mathscr{L}_4^2 - \mathscr{L}_1\mathscr{L}_5 &= -\mathscr{L}_4, \\
-\mathscr{L}_4\mathscr{L}_1 + \mathscr{L}_1\mathscr{L}_2 &= \mathscr{L}_1, \\
\mathscr{L}_5\mathscr{L}_4 - \mathscr{L}_2\mathscr{L}_5 &= -\mathscr{L}_5, \\
-\mathscr{L}_5\mathscr{L}_1 + \mathscr{L}_2^2 &= \mathscr{L}_2; \\
\mathscr{L}_2^{*2} - \mathscr{L}_1\mathscr{L}_5 &= \mathscr{L}_2^*, \\
-\mathscr{L}_2^*\mathscr{L}_1 + \mathscr{L}_1\mathscr{L}_4 &= -\mathscr{L}_1, \\
\mathscr{L}_5\mathscr{L}_2^* - \mathscr{L}_4\mathscr{L}_5 &= \mathscr{L}_5, \\
-\mathscr{L}_5\mathscr{L}_1 + \mathscr{L}_4^2 &= -\mathscr{L}_4.
\end{aligned}$$

□

6.12 Simple-layer representations for BVP in $\mathbb{R}^2$

In principle, the two-dimensional Laplace equation $\Delta w(x) = 0$ subject to various boundary conditions can be treated in the same way as the three-dimensional case using boundary integral equations methods. Nevertheless, the theorems in the previous sections for the $\mathbb{R}^3$ case must be slightly modified to accommodate the fact that now the fundamental solution is

$$E_2(x-\xi) = -\frac{1}{2\pi}\ln|x-\xi|, \quad x,\ \xi \in \mathbf{R}^2,$$

which has logarithmic growth at ∞. Let us first consider the simple-layer method for the interior Dirichlet problem

$$\begin{cases} \Delta w(x) = 0 & \text{on } \Omega, \\ w(x) = g(x) \in H^r(\partial\Omega),\ r \in \mathbb{R}, & \text{on } \partial\Omega, \end{cases}$$

where $\Omega \subseteq \mathbb{R}^2$ is a bounded open domain with smooth boundary $\partial\Omega$. If we represent our solution by the simple layer

$$w(x) = \int_{\partial\Omega} E_2(x-\xi) f(\xi)\, d\sigma_\xi, \quad x \in \Omega,$$

then the density f must satisfy the BIE

$$\mathscr{L}_1(f)(x) \equiv \int_{\partial\Omega} E_2(x-\xi)f(\xi)\,d\sigma_\xi = g(x), \quad x \in \partial\Omega,$$

similarly to the $\mathbb{R}^3$ case (6.176). Nevertheless, unlike the 3D case, $\mathscr{N}(\mathscr{L}_1)$ may now be nontrivial. We argue as follows [103, pp. 52–54]. Let Ω be star-shaped with respect to an interior point, say the origin.

Since the operator $\mathscr{L}_1$ is self-adjoint and Fredholm from $H^r(\partial\Omega)$ into $H^{r+1}(\partial\Omega)$ for any $r \in \mathbb{R}$, it has zero index and

$$\mathscr{N}(\mathscr{L}_1) \neq \{0\} \quad \text{if and only if} \quad \operatorname{Coker} \mathscr{L}_1 \neq \{0\}.$$

Assume first that $\mathscr{N}(\mathscr{L}_1) = \{0\}$. Then the BIE

$$\mathscr{L}_1(f_0) = 1$$

is uniquely solvable with density $f_0 \in C^\infty(\partial\Omega)$, i.e., f_0 satisfies

$$\int_{\partial\Omega} E_2(x-\xi)f_0(\xi)\,d\sigma_\xi = -\frac{1}{2\pi}\int_{\partial\Omega} \ln|x-\xi|\,f_0(\xi)\,d\sigma_\xi = 1,$$
$$\forall\, x \in \partial\Omega; \text{ cf. Fig. 6.4.}$$

Now consider the similar domain Ω_m that is obtained by magnifying Ω by a scale m in the radial direction:

$$\Omega_m \equiv \{mx \mid x \in \Omega\},$$

see Fig. 6.4. Further, let f_{om} be a density on $\partial\Omega_m$ defined by

$$f_{om}(\xi) = f_0(\xi/m) \quad \text{for} \quad \xi \in \partial\Omega_m.$$

Then for $x = m\tilde{x} \in \partial\Omega_m$,

$$\begin{aligned}
\int_{\partial\Omega_m} E_2(x-\xi)f_{om}(\xi)\,d\sigma_\xi &= \int_{\partial\Omega} E_2(m(\tilde{x}-\xi))f_0(\xi)\cdot m\,d\sigma_\xi \\
&= -\frac{1}{2\pi}\int_{\partial\Omega} [m\ln(m|\tilde{x}-\xi|)]f_0(\xi)\,d\sigma_\xi \\
&= -\frac{1}{2\pi}(m\ln m)\int_{\partial\Omega} f_0(\xi)\,d\sigma_\xi \\
&\quad + m\int_{\partial\Omega} E_2(\tilde{x}-\xi)f_0(\xi)\,d\sigma_\xi \\
&= -\frac{1}{2\pi}(m\ln m)\int_{\partial\Omega} f_0(\xi)\,d\sigma_\xi + m.
\end{aligned}$$

By Exercise 6.6 (which can be easily seen to remain valid in $\mathbb{R}^2$), we know that f_0 is of one sign, so we have

$$\int_{\partial\Omega} f_0(\xi)\,d\sigma_\xi = \alpha_0 \neq 0.$$

The equation

$$-\frac{1}{2\pi}(m\ln m)\alpha_0 + m = 0$$

has a solution

$$m = e^{2\pi/\alpha_0}.$$

Therefore $\mathcal{N}(\mathcal{L}_1) \neq \{0\}$ for Ω_m.

The reader can easily check that for *circular* domains Ω_r with radius r, we have

$$\mathcal{N}(\mathcal{L}_1) = \begin{cases} \{0\} & \text{if } r \neq 1, \ r > 0; \\ \text{span}\{1\} & \text{if } r = 1. \end{cases}$$

Exercise 6.7. Show that $\mathcal{N}(\mathcal{L}_1)$ has at most dimension 1 if Ω is simply connected. □

Let us reformulate the simple- and double-layer representations for interior and exterior 2-D BVP. We will be brief in the statements of theorems. In what follows, Ω will always denote a *simply connected bounded domain in* $\mathbb{R}^2$ *with smooth boundary* $\partial\Omega$.

Theorem 6.33 (Simple-layer representation for the interior Dirichlet BVP in $\mathbb{R}^2$). *We have*

$$\mathcal{L}_1' \colon \ \mathbb{R} \oplus H^r(\partial\Omega) \to \mathbb{R} \oplus H^{r+1}(\partial\Omega), \ r \in \mathbb{R}, \quad \text{isomorphically,}$$

where

$$\mathcal{L}_1'\left(\begin{bmatrix} a \\ f \end{bmatrix}\right)(x) \equiv \begin{bmatrix} \int_{\partial\Omega} f(\xi)\, d\sigma \\ \int_{\partial\Omega} E_2(x-\xi) f(\xi)\, d\sigma_\xi + a \end{bmatrix}, \quad x \in \partial\Omega \tag{6.237}$$

and

$$\mathcal{L}_1'^{*} \colon \ \mathbb{R} \oplus H^{-(r+1)}(\partial\Omega) \to \mathbb{R} \oplus H^{-r}(\partial\Omega), \quad \text{isomorphically,}$$

with

$$\mathcal{L}_1'^{*}\left(\begin{bmatrix} b \\ \phi \end{bmatrix}\right)(x) = \begin{bmatrix} \int_{\partial\Omega} \phi(\xi)\, d\sigma_\xi \\ \int_{\partial\Omega} E_2(x-\xi)\phi(\xi)\, d\sigma_\xi + b \end{bmatrix}, \quad x \in \partial\Omega.$$

The canonical restriction or extension of $\mathcal{L}_1'^{*}$ to $\mathbb{R} \oplus H^r(\partial\Omega)$ makes $\mathcal{L}_1'^{*} = \mathcal{L}_1'$, i.e., $\mathcal{L}_1'$ is self-adjoint. The solution w to the interior Dirichlet BVP

$$\left.\begin{aligned} &\Delta w(x) = 0 \quad \text{on } \Omega, \\ &w(x) = g(x) \in H^{r+1}(\partial\Omega), \ r \in \mathbb{R}, \end{aligned}\right\} \tag{6.238}$$

is given by

$$w(x) = \int_{\partial\Omega} E_2(x-\xi) f(\xi)\, d\sigma_\xi + a \in H^{r+3/2}(\Omega), \quad x \in \Omega, \tag{6.239}$$

where $(a, f) \in \mathbb{R} \oplus H^r(\partial\Omega)$ *uniquely solves the BIE*

$$\mathscr{L}_1' \left(\begin{bmatrix} a \\ f \end{bmatrix} \right) = \begin{bmatrix} A \\ g \end{bmatrix}, \tag{6.240}$$

for some (but arbitrarily) given $A \in \mathbb{R}$. □

Remark 6.7. If somehow it is known that the simple-layer equation

$$V_1(f)(x) = \int_{\partial\Omega} E_2(x-\xi) f(\xi)\, d\sigma_\xi = 0, \quad x \in \partial\Omega,$$

does not have a nontrivial solution (or equivalently, $V_1(f)(x) \equiv 1$ has a unique solution f on $\partial\Omega$), then instead of solving the BIE (6.240), one can solve the simpler BIE

$$V_1(f)(x) = g(x), \quad x \in \partial\Omega. \tag{6.241}$$

The solution w to the harmonic Dirichlet BVP (6.238) will then be given by

$$w(x) = \int_{\partial\Omega} E_2(x-\xi) f(\xi)\, d\sigma_\xi, \quad x \in \Omega.$$

Numerically, the equation (6.241) (after discretization) could become very ill-conditioned when $\partial\Omega$ is close to a contour Γ_0 for which

$$V_{1,\Gamma_0}(f)(x) = \int_{\Gamma_0} E_2(x-\xi) f(\xi)\, d\sigma_\xi = 0$$

admits a nontrivial solution f, that is, $\mathscr{N}(V_{1,\Gamma_0}) \neq \{0\}$. □

Theorem 6.34 (Simple-layer representation for the interior Neumann BVP in $\mathbb{R}^2$). *This is the same as Theorem 6.22 but with* $E(x-\xi)$ *changed to* $E_2(x-\xi)$ *throughout the statements of that theorem, except at (6.191), where it is changed to*

$$\begin{aligned} \mathscr{N}(\mathscr{L}_2) = \text{Coker}\, \mathscr{L}_2^* \; = \; & \text{span}\, \{\lambda_0(\cdot) \mid \lambda_0(\cdot) \neq 0 \ \textit{satisfies} \\ & \int_{\partial\Omega} E_2(x-\xi)\lambda_0(\xi)\, d\sigma_\xi = c \in \mathbb{R} \ \textit{on} \ \partial\Omega, \\ & \textit{for some} \ c \in \mathbb{R},\ c \ \textit{may be} \ 0.\} \end{aligned} \tag{6.242}$$

□

Theorem 6.35 (Simple-layer representation for the interior Robin BVP in $\mathbb{R}^2$). *We have, for* $h > 0$,

$$\mathscr{L}_3\colon\ \mathbb{R} \oplus H^r(\partial\Omega) \to \mathbb{R} \oplus H^r(\partial\Omega), \ r \in \mathbb{R}, \quad \text{isomorphically},$$

where

$$\mathscr{L}_3\left(\begin{bmatrix} a \\ f \end{bmatrix}\right)(x)$$
$$= \begin{bmatrix} \int_{\partial\Omega} f(\xi)\,d\sigma \\ \frac{1}{2}f(x) + \int_{\partial\Omega}\left[\frac{\partial E_2(x-\xi)}{\partial n_x} + hE_2(x-\xi)\right] f(\xi)\,d\sigma_\xi + ha \end{bmatrix}, \quad x \in \partial\Omega.$$

Also

$$\mathscr{L}_3^*:\ \mathbb{R}\oplus H^{-r}(\partial\Omega) \to \mathbb{R}\oplus H^{-r}(\partial\Omega), \quad \text{isomorphically,}$$

where

$$\mathscr{L}_3^*\left(\begin{bmatrix} b \\ \phi \end{bmatrix}\right)$$
$$= \begin{bmatrix} \int_{\partial\Omega} \phi(\xi)\,d\sigma \\ \frac{1}{2}\phi(x) + \int_{\partial\Omega}\left[\frac{\partial E_2(x-\xi)}{\partial n_\xi} + hE(x-\xi)\right] \phi(\xi)\,d\sigma_\xi + hb \end{bmatrix}, \ a\, x \in \partial\Omega.$$

The solution w to the interior Robin BVP

$$\begin{cases} \Delta w(x) = 0 \quad \text{on } \Omega \\ \dfrac{\partial w(x)}{\partial n} + hw(x) = g(x) \in H^r(\partial\Omega), \quad r \in \mathbb{R}, \end{cases}$$

is given by

$$w(x) = \int_{\partial\Omega} E_2(x-\xi) f(\xi)\,d\sigma_\xi + a \in H^{r+3/2}(\Omega), \quad r \in \mathbb{R},$$

where $f \in H^r(\partial\Omega)$, and $a \in \mathbb{R}$ solves the BIE

$$\mathscr{L}_3\left(\begin{bmatrix} a \\ f \end{bmatrix}\right) = \begin{bmatrix} A \\ g \end{bmatrix}$$

for some (but arbitrarily given) $A \in \mathbb{R}$. □

Remark 6.8. The boundary integral operator

$$T_1(f)(x) \equiv \frac{1}{2}f(x) + \int_{\partial\Omega}\left[\frac{\partial E_2(x-\xi)}{\partial n_x} + hE_2(x-\xi)\right] f(\xi)\,d\sigma_\xi, \quad x \in \partial\Omega,$$

has a nontrivial kernel if the integral operator

$$T_2(f)(x) \equiv \int_{\partial\Omega} E_2(x-\xi) f(\xi)\,d\sigma_\xi$$

has a nontrivial kernel. □

As for the simple-layer representations for exterior BVP, we note that for large $|x|$,

$$\begin{aligned}
w(x) &= \int_{\partial\Omega} E_2(x-\xi)f(\xi)\,d\sigma_\xi \\
&= -\frac{1}{2\pi}\int_{\partial\Omega} \ln\left(|x|^2\left|1-\frac{2x\cdot\xi-|\xi|^2}{|x|^2}\right|\right)^{1/2} f(\xi)\,d\sigma_\xi \\
&= -\frac{1}{2\pi}\ln|x|\int_{\partial\Omega} f(\xi)\,d\sigma_\xi \\
&\quad + \frac{1}{4\pi|x|^2}\int_{\partial\Omega} 2(x\cdot\xi)f(\xi)\,d\sigma_\xi + \mathscr{O}(|x|^{-2}).
\end{aligned} \tag{6.243}$$

The leading term of the type $A\ln|x|$ (large $|x|$) was found to be useful in the singular perturbation theory of potential flows [95, 93]. We first state an existence and uniqueness result of exterior BVP with solutions of order of magnitude $A\ln|x| + \mathscr{O}(1)$ at infinity, that is an easy consequence of Theorem 6.16.

Theorem 6.36. *Let w be a solution of the exterior Dirichlet problem*

$$\begin{cases} \Delta w(x) = 0 \quad \text{on } \Omega^c, \\ w(x) = g(x) \in H^r(\partial\Omega), \quad r \in \mathbb{R}, \\ w(x) = A\ln|x| + \mathscr{O}(1) \quad \text{for large} \quad |x| \quad \text{for some given} \quad A \in \mathbb{R}. \end{cases}$$

Then $w \in H^{r+1/2}_{\text{loc}}(\Omega^c)$ exists and is unique. □

Using this fact and similar ideas in § 6.9, it is straightforward to prove the following.

Theorem 6.37. (Simple-layer representation for the exterior Dirichlet BVP in $\mathbb{R}^2$ with $\mathscr{O}(\ln|x|)$ growth). *We have*

$$\mathscr{L}_{1e}\colon\ \mathbb{R}\oplus H^r(\partial\Omega) \to \mathbb{R}\oplus H^{r+1}(\partial\Omega), \quad r \in \mathbb{R}, \text{ isomorphically,}$$

where

$$\mathscr{L}_{1e}\left(\begin{bmatrix} a \\ f \end{bmatrix}\right)(x) = \begin{bmatrix} \int_{\partial\Omega} f(\xi)\,d\sigma \\ \int_{\partial\Omega} E_2(x-\xi)f(\xi)\,d\sigma_\xi + a \end{bmatrix}, \quad x \in \partial\Omega,$$

with adjoint

$$\mathscr{L}^*_{1e}\colon\ \mathbb{R}\oplus H^{-(r+1)}(\partial\Omega) \to \mathbb{R}\oplus H^{-r}(\partial\Omega), \quad \text{isomorphically,}$$

where

$$\mathscr{L}^*_{1e}\left(\begin{bmatrix} b \\ \phi \end{bmatrix}\right)(x) = \begin{bmatrix} \int_{\partial\Omega} \phi(\xi)\,d\sigma \\ \int_{\partial\Omega} E_2(x-\xi)\phi(\xi)\,d\sigma_\xi + b \end{bmatrix}, \quad x \in \partial\Omega.$$

The canonical restriction or extension of $\mathscr{L}_{1e}^*$ to $\mathbb{R}\oplus H^r(\partial\Omega)$ makes $\mathscr{L}_{1e}^* = \mathscr{L}_{1e}$, i.e., $\mathscr{L}_{1e}$ is self-adjoint. The solution w to the exterior Dirichlet BVP

$$\begin{aligned}
&\Delta w(x) = 0 \quad \text{on } \Omega^c,\\
&w(x) = g(x) \in H^{r+1}(\partial\Omega), \quad r \in \mathbb{R},\\
&w(x) = A\ln|x| + \mathscr{O}(1) \quad \text{for large } |x|, \text{ with } A \text{ given},
\end{aligned}$$

is uniquely given by

$$w(x) = \int_{\partial\Omega} E_2(x-\xi) f(\xi)\, d\sigma_\xi + a \in H_{\text{loc}}^{r+3/2}(\partial\Omega),$$

where $(a,f) \in \mathbb{R}\oplus H^r(\partial\Omega)$ uniquely solves the BIE

$$\mathscr{L}_{1e}\left(\begin{bmatrix} a \\ f \end{bmatrix}\right) = \begin{bmatrix} A \\ g \end{bmatrix}.$$

□

Subject to a $\mathscr{O}(1)$ growth condition at ∞ (cf. Theorem 6.16), a different existence and uniqueness of solutions to exterior Dirichlet BVP can be given.

Assuming the validity of the existence and uniqueness result subject to $\mathscr{O}(1)$ growth at ∞ in Theorem 6.16, we can formulate a different existence and uniqueness theorem for simple-layer solutions of the exterior Dirichlet BVP. Let $\lambda_0(\cdot) \in C^\infty(\partial\Omega)$, $\lambda_0 \not\equiv 0$, be a solution of the simple-layer BIE

$$\int_{\partial\Omega} E(x-\xi)\lambda_0(\xi)\, d\sigma_\xi = c, \quad x \in \partial\Omega, \tag{6.244}$$

where $c \in \mathbb{R}$ is a constant. This constant c is either zero (cf. e.g., (6.242)) or nonzero. (When c is nonzero, we can simply choose $c = 1$.) The existence of such a solution λ_0 is guaranteed by the unique solvability of the BIE (6.240), wherein we simply pick $A \neq 0$, $g = 0$, and $c = -a$ for the second component of the BIE (6.240). According to Exercise 6.6, $\lambda_0(x)$ is either strictly positive or strictly negative on $\partial\Omega$, so

$$\int_{\partial\Omega} \lambda_0(x)\, d\sigma \neq 0.$$

We define a linear operator

$$\mathscr{M}_1 : \; H^r(\partial\Omega) \to H^r(\partial\Omega),$$

by

$$\mathscr{M}_1(g) = g - \frac{\langle \lambda_0, g\rangle}{\langle \lambda_0, 1\rangle} = g - \frac{\int_{\partial\Omega}\lambda_0(\xi)g(\xi)\, d\sigma}{\int_{\partial\Omega}\lambda_0(\xi)\, d\sigma}. \tag{6.245}$$

It is easy to verify that $\mathscr{M}_1$ satisfies

$$\mathscr{M}_1 : \; H^r(\partial\Omega) \to H^r(\partial\Omega) \cap \left[\{\lambda_0\}^\perp \text{ in } H^0(\partial\Omega)\right] \tag{6.246}$$

and

$$
\begin{aligned}
\langle \mathscr{M}_1(g), \lambda_0 \rangle &= \left\langle g - \frac{(\lambda_0, g)}{(\lambda_0, 1)}, \lambda_0 \right\rangle_{H^0(\partial\Omega)} \\
&= \langle g, \lambda_0 \rangle - \frac{\langle \lambda_0, g \rangle}{\langle \lambda_0, 1 \rangle} \langle 1, \lambda_0 \rangle \\
&= 0 \quad \forall g \in H^r(\partial\Omega).
\end{aligned} \tag{6.247}
$$

We are in a position to formulate the second simple-layer solutions of exterior Dirichlet BVP.

Theorem 6.38. (Simple-layer representation for the exterior Dirichlet BVP in $\mathbf{R}^2$ with $\mathscr{O}(1)$ growth solutions). *Let $\mathscr{M}_1$ be defined as in (6.245). Define $V_1 : H^r(\partial\Omega) \to H^{r+1}(\partial\Omega), r \in \mathbb{R}$, as in (6.54). Then for any given $g \in H^{r+1}(\partial\Omega)$, there exists a unique $f \in H^r(\partial\Omega)$ such that f uniquely solves the BIE*

$$
\left.
\begin{aligned}
V_1(f) &= \mathscr{M}_1(g), \\
\langle f, 1 \rangle_{\partial\Omega} &= 0.
\end{aligned}
\right\} \tag{6.248}
$$

The solution w to the exterior Dirichlet BVP

$$
\left.
\begin{aligned}
&\Delta w(x) = 0 \quad \text{on } \Omega^c, \\
&w(x) = g(x) \in H^{r+1}(\partial\Omega), \quad r \in \mathbb{R}, \\
&w(x) = \mathscr{O}(1) \quad \text{for large } |x|,
\end{aligned}
\right\} \tag{6.249}
$$

is given uniquely by

$$
w(x) = V_1^+(f)(x) + \frac{\langle \lambda_0, g \rangle}{\langle \lambda_0, 1 \rangle} \in H_{\mathrm{loc}}^{r+3/2}(\Omega^c), \tag{6.250}
$$

where f is the unique solution to (6.248).

Equivalently and more simply, the solution to (6.249) is given by

$$
w(x) = \int_{\partial\Omega} E_2(x - \xi) f(\xi)\, d\sigma_\xi + a, x \in \Omega^c, \quad w \in H_{\mathrm{loc}}^{r+3/2}(\Omega^c), \tag{6.251}
$$

where $(a, f) \in \mathbb{R} \oplus H^r(\partial\Omega)$ is the unique solution to the BIE (6.240), with $A = 0$ therein.

Proof. If for $c = 1$, (6.244) admits a solution λ_0, then Remark 6.7 is applicable, so the BIE

$$
V_1(f) = \mathscr{M}_1(g)
$$

is uniquely solvable with $f \in H^r(\partial\Omega)$. We need only show that $\langle f, 1 \rangle_{\partial\Omega} = 0$.

$$
\begin{aligned}
\langle f, 1 \rangle_{H^0(\partial\Omega)} &= \langle f, V_1(\lambda_0) \rangle = \langle V_1(f), \lambda_0 \rangle \\
&= \langle \mathscr{M}_1(g), \lambda_0 \rangle = 0 \quad \text{(by (6.247))}.
\end{aligned}
$$

Since $f \perp 1$, by the asymptotic expansion (6.243), we have

$$\begin{aligned} w(x) &= V_1^+(f)(x) + \frac{\langle \lambda_0, g\rangle}{\langle \lambda_0, 1\rangle} \quad \text{(from (6.239))} \\ &= \frac{\langle \lambda_0, g\rangle}{\langle \lambda_0, 1\rangle} + \frac{1}{4\pi|x|^2}\int_{\partial\Omega} 2(x\cdot\xi)f(\xi)\,d\sigma_\xi + \mathscr{O}(|x|^{-2}) \\ &= \mathscr{O}(1) \quad \text{when } |x| \text{ is large.} \end{aligned}$$

On the other hand, if for $c = 0$, (6.244) admits a nontrivial solution λ_0, then by Exercise 6.7 and the Fredholm alternative, the BIE

$$V_1(f) = G \tag{6.252}$$

has a solution f if and only if $G \perp \mathscr{N}(V_1)$. (Note here that V_1 is the $\mathscr{L}_1$ in Exercise 6.7.) Taking $G = \mathscr{M}_1(g)$, we have

$$\langle G, \lambda_0\rangle = \langle \mathscr{M}_1(g), \lambda_0\rangle = 0, \quad \text{by (6.247).}$$

The solutions f to (6.252) are characterized by $f + k\lambda_0$ for any $k \in \mathbb{R}$. We choose k such that

$$\int_{\partial\Omega} [f(\xi) + k\lambda_0(\xi)]\,d\sigma = 0,$$

i.e.,

$$k = -\frac{\displaystyle\int_{\partial\Omega} f(\xi)\,d\sigma}{\displaystyle\int_{\partial\Omega} \lambda_0(\xi)\,d\sigma}, \quad \text{because } \int_{\partial\Omega} \lambda_0(\xi)d\sigma \neq 0.$$

The rest of the proof follows easily. □

Remark 6.9. By setting $A = 0$ in (6.252), we see that the constant a in (6.251) is

$$a = \frac{\langle \lambda_0, g\rangle}{\langle \lambda_0, 1\rangle} \quad \text{(with } bA = 0 \text{ in (6.240)).}$$ □

Next, we consider the exterior Neumann problem

$$\left.\begin{aligned} \Delta w(x) &= 0 \qquad \text{on } \Omega^c, \\ \frac{\partial w(x)}{\partial n} &= g(x) \;\; \text{on } \partial\Omega. \end{aligned}\right\} \tag{6.253}$$

If we consider simple-layer solutions as candidates

$$w(x) = \int_{\partial\Omega} E_2(x-\xi)f(\xi)\,d\sigma_\xi, \tag{6.254}$$

then $w(x)$ will have a magnitude of growth $A \ln |x|$ for large $|x|$. See (6.243). Since

$$\begin{aligned}
0 &= \int_{\Omega^c \cap B_R} \Delta w(x)\, dx = \int_{|x|=R} \frac{\partial w(x)}{\partial r}\, d\sigma - \int_{\partial\Omega} \frac{\partial w}{\partial n} d\sigma \\
&= \int_0^{2\pi} \left[A\frac{1}{R} + \mathscr{O}\left(\frac{1}{R^2}\right)\right] R\, d\theta - \int_{\partial\Omega} g(x)\, d\sigma \\
&\to 2\pi A - \int_{\partial\Omega} g(x)\, d\sigma \quad \text{as} \quad R \to \infty,
\end{aligned}$$

we see that (6.253) implies an asymptotic expansion of the form

$$w(x) = \frac{1}{2\pi}\left[\int_{\partial\Omega} g(\xi)\, d\sigma\right] \ln|x| + c + \mathscr{O}(|x|^{-1}), \quad \text{for large } |x|. \tag{6.255}$$

The constant c is irrelevant to g in (6.253) because it disappears when we take the normal derivative. Therefore, we can just set $c = 0$.

Theorem 6.39 (Simple-layer representation for the exterior Neumann BVP in $\mathbb{R}^2$). *Define*

$$\mathscr{L}_{2e}\colon\ H^r(\partial\Omega) \to \mathbb{R} \oplus H^r(\partial\Omega), \quad r \in \mathbb{R},$$
$$\mathscr{M}_2\colon\ H^r(\partial\Omega) \to \mathbb{R} \oplus H^r(\partial\Omega), \quad r \in \mathbb{R},$$

respectively by

$$\mathscr{L}_{2e}(f)(x) = \begin{bmatrix} \int_{\partial\Omega} f(\xi)\, d\sigma \\ -\frac{1}{2} f(x) + \int_{\partial\Omega} \frac{\partial E_2(x-\xi)}{\partial n_x} f(\xi)\, d\sigma_\xi \end{bmatrix},$$

$$\mathscr{M}_2(g)(x) = \begin{bmatrix} -\int_{\partial\Omega} g(\xi)\, d\sigma \\ g(x) \end{bmatrix}.$$

Then $\mathscr{L}_{2e}$ and $\mathscr{M}_2$ are continuous and Fredholm. Furthermore,

$$\mathscr{L}_{2e}\colon\ H^r(\partial\Omega) \to \mathscr{R}(\mathscr{M}_2)$$

is an isomorphism.

The solution to the exterior Neumann problem

$$\left.\begin{aligned}
&\Delta w(x) = 0 \quad \text{on } \Omega^c \\
&\frac{\partial w(x)}{\partial n} = g(x) \in H^r(\partial\Omega), \quad r \in \mathbb{R}, \\
&w(x) - \left(\frac{1}{2\pi}\int_{\partial\Omega} g(\xi)\, d\sigma\right) \ln|x| = \mathscr{O}(|x|^{-1}), \quad \text{large } |x|,
\end{aligned}\right\} \tag{6.256}$$

is given by

$$w(x) = \int_{\partial\Omega} E_2(x-\xi) f(\xi)\, d\sigma_\xi \in H^{r+3/2}_{\text{loc}}(\Omega^c), \tag{6.257}$$

where f is the unique solution to

$$\mathscr{L}_{2e}(f) = \mathscr{M}_2(g). \tag{6.258}$$

Consequently, if instead of the $A\ln|x| + \mathscr{O}(|x|^{-1})$ *growth of the solution in (6.256), we consider*

$$\left.\begin{aligned} &\Delta w(x) = 0 \quad \text{on } \Omega^c,\\ &\frac{\partial w(x)}{\partial n} = g(x) \in H^r(\partial\Omega), \quad r \in \mathbb{R},\\ &w(x) = \mathscr{O}(1), \quad \text{large } |x|, \end{aligned}\right\} \tag{6.259}$$

then (6.259) has a solution w if and only if $\int_{\partial\Omega} g(\xi)\,d\sigma = 0$. *The solutions are nonunique, given by*

$$w(x) = \int_{\partial\Omega} E_2(x-\xi) f(\xi)\,d\sigma_\xi + c \in H^{r+3/2}_{\text{loc}}(\Omega^c) \tag{6.260}$$

for any constant $c \in \mathbb{R}$, where $f \in H^r(\partial\Omega)$ uniquely solves the BIE

$$\mathscr{L}_{2e}(f) = \begin{bmatrix} 0 \\ g \end{bmatrix}. \tag{6.261}$$

Proof. The BIE

$$-\frac{1}{2} f(x) + \int_{\partial\Omega} \frac{\partial E_2(x-\xi)}{\partial n_x} f(\xi)\,d\sigma_\xi = g(x) \in H^r(\partial\Omega) \tag{6.262}$$

is uniquely solvable with solution $f \in H^r(\partial\Omega)$ by Theorem 6.27 (which also holds in $\mathbb{R}^2$) because the operator on the LHS corresponds to $\mathscr{L}_4^*$, which has $\mathscr{N}(\mathscr{L}_4^*) = \{0\}$, $\operatorname{Coker}\mathscr{L}_4^* = \{0\}$. Integrating both sides of (6.262) on $\partial\Omega$, we obtain

$$-\frac{1}{2}\int_{\partial\Omega} f(x)\,d\sigma + \int_{\partial\Omega} f(\xi)\left[\int_{\partial\Omega} \frac{\partial E_2(x-\xi)}{\partial n_x}\,d\sigma_x\right] d\sigma_\xi = \int_{\partial\Omega} g(x)\,d\sigma.$$

But

$$\int_{\partial\Omega} \frac{\partial E_2(x-\xi)}{\partial n_x}\,d\sigma_x = -\frac{1}{2},$$

so

$$-\int_{\partial\Omega} f(x)\,d\sigma = \int_{\partial\Omega} g(x)\,d\sigma.$$

Hence (6.258) is satisfied. Using Lemma 6.5, it is easy to show that the solution satisfying (6.256) is unique. Therefore w is uniquely given by (6.257). □

Theorem 6.40 (Simple-layer representation for the exterior Robin BVP in $\mathbb{R}^2$). *Let $h > 0$. We have*

$$\mathscr{L}_{3e}\colon\ \mathbb{R} \oplus H^r(\partial\Omega) \to \mathbb{R} \oplus H^r(\partial\Omega), \quad r \in \mathbb{R}, \quad \textit{isomorphically,}$$

where

$$\mathscr{L}_{3e}\left(\begin{bmatrix} a \\ f \end{bmatrix}\right)(x)$$

$$= \begin{bmatrix} \int_{\partial\Omega} f(\xi)\,d\sigma \\ -\frac{1}{2}f(x) + \int_{\partial\Omega}\left[\frac{\partial E_2(x-\xi)}{\partial n_x} - hE_2(x-\xi)\right] f(\xi)d\sigma_\xi - ha \end{bmatrix}, \; x \in \partial\Omega,$$

with adjoint

$$\mathscr{L}_{3e}^*\colon\ \mathbb{R}\oplus H^{-r}(\partial\Omega) \to \mathbb{R}\oplus H^{-r}(\partial\Omega), \text{ isomorphically,}$$

given by

$$\mathscr{L}_{3e}^*\left(\begin{bmatrix} b \\ \phi \end{bmatrix}\right)(x)$$

$$= \begin{bmatrix} \int_{\partial\Omega} \phi(\xi)\,d\sigma \\ -\frac{1}{2}\phi(x) + \int_{\partial\Omega}\left[\frac{\partial E_2(x-\xi)}{\partial n_\xi} - hE_2(x-\xi)\right] \phi(\xi)d\sigma_\xi - ha \end{bmatrix}, \; x \in \partial\Omega.$$

The solution to the exterior Robin BVP

$$\begin{cases} \Delta w(x) = 0 \quad on\ \Omega^c, \\ \dfrac{\partial w(x)}{\partial n} - hw(x) = g(x) \in H^r(\partial\Omega), \quad r \in \mathbb{R}, \\ w(x) = \mathscr{O}(1), \quad large\ |x|, \end{cases}$$

is given by

$$w(x) = \int_{\partial\Omega} E_2(x-\xi)f(\xi)\,d\sigma_\xi + a \in H^{r+3/2}_{\mathrm{loc}}(\Omega^c),$$

where $(a, f) \in \mathbb{R}\oplus H^r(\partial\Omega)$ *is the unique solution to the BIE*

$$\mathscr{L}_{3e}\left(\begin{bmatrix} a \\ f \end{bmatrix}\right) = \begin{bmatrix} 0 \\ g \end{bmatrix}. \qquad \square$$

6.13 Double-layer representations for two-dimensional BVP

If we represent the solution of a two-dimensional interior BVP by a double-layer potential

$$w(x) = \int_{\partial\Omega} \frac{\partial E_2(x-\xi)}{\partial n_\xi} f(\xi)\,d\sigma_\xi, \quad x \in \mathbb{R}^2 \backslash \partial\Omega, \tag{6.263}$$

then Theorems 6.27, 6.28 and 6.29 can be easily shown to remain valid for respectively the Dirichlet, Neumann and Robin BVP. We therefore simply state the following.

Theorem 6.41. *Let* Ω *be a bounded domain in* $\mathbb{R}^2$ *with smooth boundary* $\partial\Omega$. *Then Theorem 6.27 remains valid for the interior Dirichlet BVP.*

Theorem 6.42. *Let Ω be a bounded domain in $\mathbb{R}^2$ with smooth boundary $\partial\Omega$. Then Theorem 6.28 remains valid for the interior Neumann BVP.* □

Theorem 6.43. *Let Ω be a bounded domain in $\mathbb{R}^2$ with smooth boundary $\partial\Omega$. Then Theorem 6.29 remains valid for the interior Robin BVP.* □

For exterior BVP, we first note that the representation (6.263) has growth

$$w(x) = \frac{1}{2\pi}\int_{\partial\Omega}\frac{\langle n_\xi, x-\xi\rangle}{|x-\xi|^2}f(\xi)\,d\sigma_\xi = \mathscr{O}(|x|^{-1}) \tag{6.264}$$

when $|x|$ is large. This will correspond to a regular solution at ∞. In order to obtain a generalized regular solution at ∞, the representation (6.263) must be modified to

$$w(x) = a + \int_{\partial\Omega}\frac{\partial E_2(x-\xi)}{\partial n_\xi}f(\xi)\,d\sigma_\xi, \quad x\in\Omega^c,\ a\in\mathbb{R}. \tag{6.265}$$

We have the following.

Theorem 6.44. (Double-layer representation for the exterior Dirichlet BVP with $\mathscr{O}(1)$ growth in $\mathbb{R}^2$). *Let $\Omega\subseteq\mathbb{R}^2$ be bounded with smooth boundary $\partial\Omega$. The linear operator*

$$\mathscr{L}'_{4e}\colon\ \mathbb{R}\oplus H^r(\partial\Omega)\to\mathbb{R}\oplus H^r(\partial\Omega),\quad r\in\mathbb{R},$$

where

$$\mathscr{L}'_{4e}\left(\begin{bmatrix}a\\ f\end{bmatrix}\right)(x) = \begin{bmatrix}\displaystyle\int_{\partial\Omega}f(\xi)\,d\sigma_\xi\\ \displaystyle\frac{1}{2}f(x)+\int_{\partial\Omega}\frac{\partial E_2(x-\xi)}{\partial n_\xi}f(\xi)\,d\sigma_\xi + a\end{bmatrix},\ x\in\partial\Omega, \tag{6.266}$$

is an isomorphism and its adjoint operator

$$\mathscr{L}'^{*}_{4e}\colon\ \mathbb{R}\oplus H^{-r}(\partial\Omega)\to\mathbb{R}\oplus H^{-r}(\partial\Omega),$$

given by

$$\mathscr{L}'^{*}_{4e}\left(\begin{bmatrix}b\\ \phi\end{bmatrix}\right)(x) = \begin{bmatrix}\displaystyle\int\phi(\xi)\,d\sigma_\xi\\ \displaystyle\frac{1}{2}\phi(x)+\int_{\partial\Omega}\frac{\partial E_2(x-\xi)}{\partial n_x}\phi(\xi)\,d\sigma_\xi + b\end{bmatrix},\ x\in\partial\Omega, \tag{6.267}$$

is also an isomorphism. If w is the solution to the exterior Dirichlet BVP

$$\left.\begin{aligned}&\Delta w(x) = 0 \quad \text{on } \Omega^c\\ &w(x) = g(x)\in H^r(\partial\Omega),\quad r\in\mathbb{R},\\ &w(x) = \mathscr{O}(1)\quad \text{large } |x|,\end{aligned}\right\} \tag{6.268}$$

then

$$w(x) = a + \int_{\partial\Omega}\frac{\partial E_2(x-\xi)}{\partial n_\xi}f(\xi)\,d\sigma_\xi,\qquad x\in\Omega^c,$$

where $(a,f)\in\mathbb{R}\oplus H^r(\partial\Omega)$ is the unique solution to the BIE

$$\mathscr{L}'_{4e}\left(\begin{bmatrix}a\\ f\end{bmatrix}\right) = \begin{bmatrix}0\\ g\end{bmatrix}.$$

□

Remark 6.10. The reader can easily check that a regular solution (rather than a *generalized regular* solution) at ∞ exists if and only if the Dirichlet data g satisfies

$$\int_{\partial\Omega} g(x)\widetilde{\lambda}_0(x)\, d\sigma = 0,$$

where $\widetilde{\lambda}_0$ is given in Exercise 6.6. □

Similarly, we have the following.

Theorem 6.45. (Double-layer representation for the exterior Neumann BVP with $\mathscr{O}(1)$ or $o(1)$ growth in $\mathbb{R}^2$). *Let $\Omega \subseteq \mathbb{R}^2$ be bounded with smooth boundary $\partial\Omega$. The linear operator*

$$\mathscr{L}'_{5e}\colon\ \mathbb{R}\oplus H^{r+1}(\partial\Omega) \to \mathbb{R}\oplus H^{r}(\partial\Omega), \qquad r\in\mathbb{R},$$

given by

$$\mathscr{L}'_{5e}(f)(x) = \int_{\partial\Omega} \frac{\partial^2 E_2(x-\xi)}{\partial n_x \partial n_\xi} f(\xi)\, d\sigma_\xi, \quad x\in\partial\Omega, \tag{6.269}$$

is Fredholm of index zero, with

$$\mathscr{N}(\mathscr{L}'_{5e}) = \operatorname{span}\{\widetilde{\lambda}_0(\cdot)\} \quad \text{(cf. Exercise 6.6)},$$
$$\operatorname{Coker}(\mathscr{L}'_{5e}) = \mathscr{N}(\mathscr{L}'_{5e}).$$

$\mathscr{L}'_{5e}$ is self-adjoint with ${\mathscr{L}'_{5e}}^{*} = \mathscr{L}'_{5e}$.
The solution to the exterior Neumann BVP

$$\left.\begin{aligned} &\Delta w(x) = 0 \quad \text{on } \Omega^c,\\ &\frac{\partial w(x)}{\partial n} = g(x)\in H^r(\partial\Omega), \quad r\in\mathbb{R},\\ &w(x) = \mathscr{O}(1) \text{ or } o(1), \quad \text{large } |x|, \end{aligned}\right\} \tag{6.270}$$

exists if and only if $\int_{\partial\Omega} g(x)\,d\sigma = 0$. If this compatibility condition is satisfied then

$$w(x) = a + \int_{\partial\Omega} \frac{\partial E_2(x-\xi)}{\partial n_\xi} f(\xi)\, d\sigma_\xi, \quad x\in\Omega^c,\ w\in H^{r+3/2}_{\mathrm{loc}}(\Omega^c),$$

for an arbitrary constant $a\in\mathbb{R}$ ($a=0$ if $w(x)=o(1)$ for large $|x|$), where $f\in H^{r+1}(\partial\Omega)$ is a (nonunique) solution to the BIE

$$\mathscr{L}'_{5e}(f) = g.$$ □

Remark 6.11. Comparing the exterior Neumann BVP (6.232) and (6.270) in $\mathbb{R}^3$ and $\mathbb{R}^2$, we note that there is a significant difference: for two-dimensional problems, the Neumann data must satisfy the compatibility condition

$$\int_{\partial\Omega} g(x)\, d\sigma = 0.$$ □

Finally, we state the following.

Theorem 6.46. (Double-layer representation for the exterior Robin BVP with $\mathscr{O}(1)$ growth in $\mathbb{R}^2$). *Let $\Omega \subseteq \mathbb{R}^2$ be bounded with smooth boundary $\partial\Omega$, and let $h > 0$. The linear operator*

$$\mathscr{L}'_{6e}\colon\ \mathbb{R}\oplus H^{r+1}(\partial\Omega) \to \mathbb{R}\oplus H^{r}(\partial\Omega), \quad r\in\mathbb{R},$$

given by

$$\mathscr{L}'_{6e}\left(\begin{bmatrix} a \\ f \end{bmatrix}\right)(x) = \begin{bmatrix} \int_{\partial\Omega} f(\xi)\,d\sigma \\ \int_{\partial\Omega}\left[\frac{\partial^2 E_2(x-\xi)}{\partial n_x \partial n_\xi} - h\frac{\partial E_2(x-\xi)}{\partial n_\xi}\right] f(\xi)\,d\sigma_\xi \\ -\frac{h}{2}f(x) - ha \end{bmatrix}, \quad x\in\partial\Omega,$$

is an isomorphism, and its adjoint

$$\mathscr{L}'^{*}_{6e}\colon\ \mathbb{R}\oplus h^{-r}(\partial\Omega) \to \mathbb{R}\oplus H^{-(r+1)}(\partial\Omega),$$

given by

$$\mathscr{L}'^{*}_{6e}\left(\begin{bmatrix} b \\ \phi \end{bmatrix}\right)(x) = \begin{bmatrix} \int_{\partial\Omega} \phi(\xi)\,d\sigma \\ \int_{\partial\Omega}\left[\frac{\partial^2 E_2(x-\xi)}{\partial n_x \partial n_\xi} - h\frac{\partial E_2(x-\xi)}{\partial n_x}\right] \phi(\xi)\,d\sigma_\xi \\ -\frac{h}{2}\phi(x) - hb \end{bmatrix}, \quad x\in\partial\Omega,$$

is also an isomorphism.

The solution to the exterior Robin BVP

$$\left.\begin{aligned} &\Delta w(x) = 0 \quad \text{on } \Omega^c, \\ &\frac{\partial w(x)}{\partial n} - hw(x) = g(x) \in H^r(\partial\Omega), \quad r\in\mathbb{R}, \\ &w(x) = \mathscr{O}(1), \quad \text{large } |x|, \end{aligned}\right\}$$

is given by

$$w(x) = a + \int_{\partial\Omega} \frac{\partial E_2(x-\xi)}{\partial n_\xi} f(\xi)\,d\sigma_\xi,\ x\in\Omega^c,\ \ w\in H^{r+3/2}_{\text{loc}}(\Omega^c),$$

where $(a,f)\in\mathbb{R}\oplus H^{r+1}(\partial\Omega)$ is the unique solution to the BIE

$$\mathscr{L}'_{6e}\left(\begin{bmatrix} a \\ f \end{bmatrix}\right) = \begin{bmatrix} 0 \\ g \end{bmatrix}.$$

□

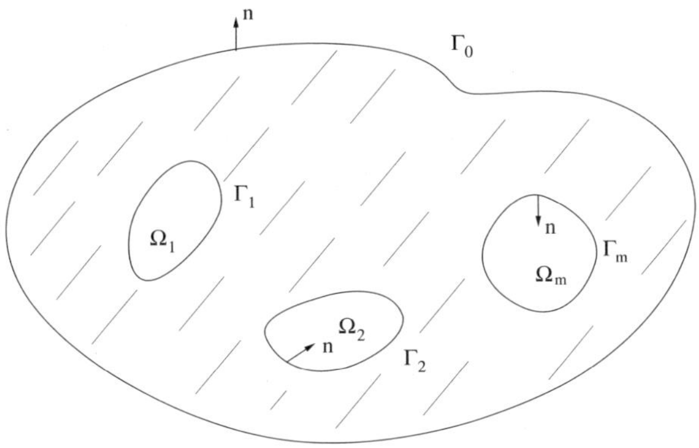

Figure 6.5: A multiconnected domain with m holes (for interior BVP).

6.14 Multiconnected domains

In § 6.8–6.13, the theorems on the solvability of BIE are stated for *simply connected domains*. What if the domain is *multiconnected*?

Let us examine the two-dimensional case. For an *interior* BVP, we assume that Ω is a multiconnected bounded open domain in $\mathbb{R}^2$ with smooth boundary $\partial\Omega$ s.t.

$$\partial\Omega = \Gamma_0 \dot{\cup} \Gamma_1 \dot{\cup} \cdots \dot{\cup} \Gamma_m \tag{6.271}$$

as shown in Fig. 6.5, where Γ_0 is the outer boundary and $\Gamma_j, 1 \leqslant j \leqslant m$, is the boundary of a "hole" Ω_j. Each $\Gamma_j, 0 \leqslant j \leqslant m$, is a smooth simple closed (i.e., Jordan) curve. For an *exterior* BVP, we assume that Ω is a disjoint union of simply connected bounded open domains:

$$\Omega = \Omega_1 \dot{\cup} \Omega_2 \dot{\cup} \cdots \dot{\cup} \Omega_m, \tag{6.272}$$

where each Ω_j has smooth boundary Γ_j, as in Fig. 6.6.

We study the validity of the BIE theorems in §§ 6.12 and 6.13 successively.

After carefully checking the arguments of the proofs, readers can see that all theorems based upon the (augmented) simple-layer representation, Theorems 6.33–6.35 and 6.37–6.40, remain valid for multiconnected domains.

However, when the double-layer representation is used, all the theorems in § 6.13 need to be modified. We first state a modified Theorem 6.41 as follows.

Theorem 6.47. *Let $\Omega \subseteq \mathbb{R}^2$ be a bounded multiconnected domain with smooth boundary, as shown in Fig. 6.5. The mapping*

$$\mathscr{L}_4(f)(x) \equiv \int_{\partial\Omega} \frac{\partial E_2(x-\xi)}{\partial n_\xi} f(\xi)\, d\sigma_\xi - \frac{1}{2} f(x), \quad x \in \partial\Omega,$$

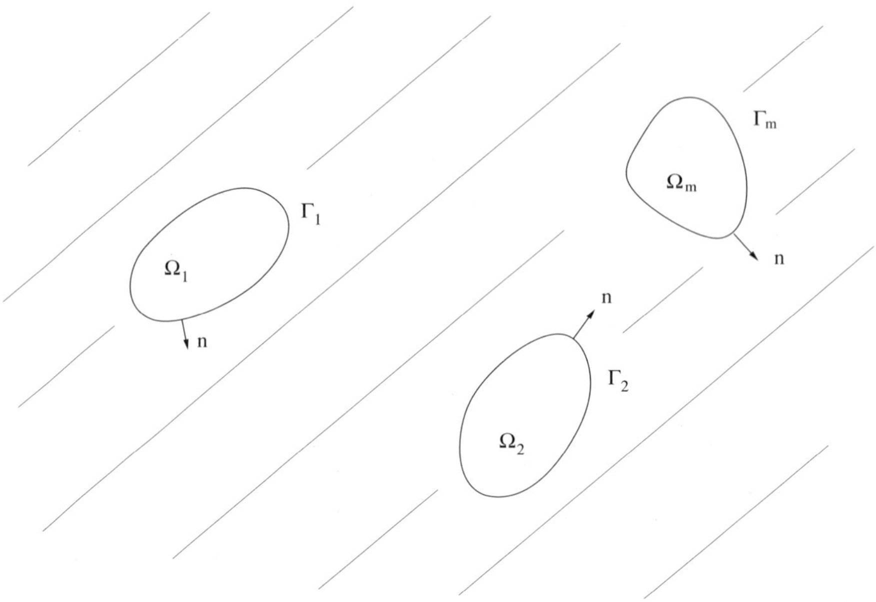

Figure 6.6: A multiconnected domain with m holes (for exterior BVP).

is Fredholm of index zero from $H^r(\partial\Omega)$ onto $H^r(\partial\Omega)$ for any $r \in \mathbb{R}$, satisfying

$$\dim \mathcal{N}(\mathcal{L}_4) = m, \qquad \dim \operatorname{Coker} \mathcal{L}_4 = m.$$

Its adjoint operator

$$\mathcal{L}_4^* :\ H^{-r}(\partial\Omega) \to H^{-r}(\partial\Omega), \qquad r \in \mathbb{R},$$

given by

$$\mathcal{L}_4^*(\phi)(x) = \int_{\partial\Omega} \frac{\partial E_2(x-\xi)}{\partial n_x} \phi(\xi)\, d\sigma_\xi - \frac{1}{2}\phi(x), \quad x \in \partial\Omega,$$

is also Fredholm, with

$$\begin{aligned} \mathcal{N}(\mathcal{L}_4^*) = {} & \operatorname{Coker} \mathcal{L}_4 \\ = {} & \operatorname{span} \Bigg\{ \lambda_j \in C^\infty(\partial\Omega) \Big| \lambda_j(\cdot) \equiv 0 \ \ \mathit{on}\ \Gamma_i,\ \ \mathit{for}\ i \neq j, \\ & \lambda_j(\cdot) \not\equiv 0 \ \ \mathit{on}\ \Gamma_j, \quad \int_{\partial\Omega} E_2(x-\xi)\lambda_j(\xi)\, d\sigma_\xi = c_j \in \mathbb{R}, \\ & c_j \ \ \mathit{may\ be}\ 0, 1 \leqslant j \leqslant m,\ \ 0 \leqslant i \leqslant m \Bigg\}, \end{aligned} \tag{6.273}$$

and

$$\begin{aligned} \operatorname{Coker} \mathcal{L}_4^* &= \mathcal{N}(\mathcal{L}_4) \\ &= \operatorname{span} \{ \chi_{\Gamma_j}(\cdot) \in C^\infty(\partial\Omega) \mid 1 \leqslant j \leqslant m \}. \end{aligned} \tag{6.274}$$

□

The verification of Theorem 6.47 is straightforward, so it is omitted.

Obviously, the solution to the Dirichlet problem

$$\left.\begin{aligned} \Delta w(x) &= 0 \quad \text{on } \Omega, \\ w(x) &= g(x) \text{ on } \partial\Omega, \end{aligned}\right\} \tag{6.275}$$

on a multiconnected domain Ω still uniquely exists, for any $g \in H^r(\partial\Omega)$, for some $r \in \mathbb{R}$. The implication of Theorem 6.47, however, is that now the solution of (6.275) is no longer representable as a double-layer potential

$$w(x) = \int_{\partial\Omega} \frac{\partial E(x-\xi)}{\partial n_\xi} f(\xi)\, d\sigma_\xi, \qquad x \in \Omega, \tag{6.276}$$

because by (6.275) and (6.276), we have

$$w(x)|_{\partial\Omega} = -\frac{1}{2} f(x) + \int_{\partial\Omega} \frac{\partial E(x-\xi)}{\partial n_\xi} f(\xi)\, d\sigma_\xi = g(x), \quad x \in \partial\Omega.$$

But the above BIE is solvable iff

$$g \perp \operatorname{Coker} \mathscr{L}_4,$$

which generally fails because the Dirichlet data g in (6.275) does not have to satisfy any compatibility conditions.

There are several possible ways to remedy this situation. For example, one might apply Theorem 4.10 to this situation by defining some augmented double-layer potentials. Nevertheless, a very convenient way is to just redefine the integral kernel as

$$K(x,\xi) = \frac{\partial E(x-\xi)}{\partial n_\xi} + \chi(x,\xi), \tag{6.277}$$

where

$$\chi(x,\xi) = \begin{cases} 1 \text{ if } x, \xi \in \Gamma_j \text{ for } j = 1,2,\ldots,m, \\ 0 \text{ otherwise.} \end{cases}$$

Exercise 6.8. Verify that the boundary integral operator

$$\mathscr{L}(f) \equiv -\frac{1}{2} f(x) + \int_{\partial\Omega} K(x,\xi) f(\xi)\, d\sigma_\xi \tag{6.278}$$

with the kernel K defined above is an isomorphism on $H^r(\partial\Omega)$ for any $r \in \mathbb{R}$. □

Using the above exercise, we can now represent the solution of (6.275) uniquely as

$$w(x) = \int_{\partial\Omega} K(x,\xi) f(\xi)\, d\sigma_\xi, \qquad x \in \Omega,$$

where $f \in H^r(\partial\Omega)$ is the unique solution to the BIE

$$\mathscr{L}(f) = g,$$

for $g \in H^r(\partial\Omega)$, where $\mathscr{L}$ is defined in (6.278).

We ask readers to check the rest of the theorems as exercises below.

Exercise 6.9. Let Ω be a bounded multiconnected domain in $\mathbb{R}^2$ with smooth boundary $\partial\Omega$ as shown in Fig. 6.5. Consider the boundary integral operator $\mathscr{L}_5$ (cf. (6.220)) in Theorem 6.42. Show

$$\begin{aligned}\mathscr{N}(\mathscr{L}_5) &= \operatorname{Coker}\mathscr{L}_5 = \mathscr{N}(\mathscr{L}_5^*) = \operatorname{Coker}\mathscr{L}_5^* \\ &= \operatorname{span}\{\chi_{\Gamma_j} \in C^\infty(\partial\Omega) \mid j = 0,1,2,\ldots,m\}.\end{aligned}$$

(Hence $\dim\mathscr{N}(\mathscr{L}_5) = m+1$ and $\operatorname{Ind}\mathscr{L}_5 = 0$.) □

Exercise 6.10. Under the same conditions on Ω as in Exercise 6.9, consider the boundary integral operator $\mathscr{L}_6$ (cf. (6.221)) in Theorem 6.43. Show that

$$\begin{aligned}\mathscr{N}(\mathscr{L}_6) &= \operatorname{Coker}\mathscr{L}_6^* \\ &= \operatorname{span}\{\chi_{\Gamma_j} \in C^\infty(\partial\Omega) \mid 1 \leqslant j \leqslant m\}\end{aligned}$$

and

$$\begin{aligned}\mathscr{N}(\mathscr{L}_6^*) = {} & \operatorname{Coker}\mathscr{L}_6 \\ = {} & \operatorname{span}\{\lambda_j \in C^\infty(\partial\Omega) \mid \lambda_j(\cdot) \equiv 0 \text{ on } \Gamma_i \text{ for } i \neq j, \\ & \lambda_j(\cdot) \text{ satisfies (6.279) on } \Gamma_j, \text{ for } 1 \leqslant j \leqslant m, 0 \leqslant i \leqslant m\},\end{aligned}$$

where on Γ_j

$$\left.\begin{aligned}&\lambda_j(\cdot) \not\equiv 0, \\ &\frac{1}{2}\lambda_j(x) + \int_{\Gamma_j}\left[\frac{\partial E(x-\xi)}{\partial n_\xi} + hE(x-\xi)\right]\lambda_j(\xi)\,d\sigma_\xi = 1 \text{ or } 0 \;\; (\text{exclusively})\end{aligned}\right\}. \tag{6.279}$$

(Hence $\dim\mathscr{N}(\mathscr{L}_5) = m$ and $\operatorname{Ind}\mathscr{L}_5 = 0$.) □

Exercise 6.11. Let Ω be as shown in Fig. 6.6. Reconsider Theorem 6.44 on the multiconnected exterior domain Ω^c. Show that $\mathscr{L}'_{4e}$ defined in (6.266) satisfies

$$\begin{aligned}&\mathscr{N}(\mathscr{L}'_{4e}) = \operatorname{Coker}\mathscr{L}'^*_{4e} \\ &= \left\{\begin{bmatrix} 0 \\ c_1\chi_{\Gamma_1} + \cdots + c_m\chi_{\Gamma_m}\end{bmatrix} \in [C^\infty(\partial\Omega)]^2 \,\middle|\, \begin{matrix}(c_1,\cdots,c_m) \in \mathbb{R}^m \\ \text{satisfies (6.280)}\end{matrix}\right\},\end{aligned}$$

where

$$\sum_{j=1}^m c_j \operatorname{meas}\Gamma_j = 0, \tag{6.280}$$

and

$$\mathscr{N}(\mathscr{L}_{4e}^{\prime *}) = \text{Coker}\,\mathscr{L}_{4e}^{\prime}$$
$$= \left\{ \begin{bmatrix} 0 \\ d_1\lambda_1(\cdot)+\cdots+d_m\lambda_m(\cdot) \end{bmatrix} \in [C^\infty(\partial\Omega)]^2 \,\middle|\, \begin{array}{l} (d_1,\ldots,d_m)\in\mathbb{R}^m \text{ and } (\lambda_1(\cdot),\ldots,\lambda_m(\cdot)) \\ \in [C^\infty(\partial\Omega)]^2 \text{ satisfy (6.281)} \end{array} \right\},$$

where

$$\left.\begin{array}{l} \lambda_j(\cdot)\equiv 0 \quad \text{on } \Gamma_i, \text{ for } i\neq j, \\ \lambda_j \not\equiv 0 \quad \text{on } \Gamma_j, \\ \displaystyle\int_{\Gamma_j} E_2(x-\xi)\lambda_j(\xi)\,d\sigma_\xi = 1 \text{ or } 0 \text{ (exclusively)}, \\ \displaystyle\sum_{j=1}^m d_j \int_{\Gamma_j} \lambda_j(\xi)\,d\sigma = 0. \end{array}\right\} \tag{6.281}$$

(Hence $\dim\mathscr{N}(\mathscr{L}_{4e}^{\prime}) = m-1$ and $\text{Ind}\,\mathscr{L}_{4e}^{\prime} = 0$.) □

Exercise 6.12. Let Ω satisfy the same condition as in Exercise 6.11. Reconsider Theorem 6.45 on the exterior multiconnected domain Ω^c. Show that the boundary integral operator $\mathscr{L}_{5e}^{\prime}$ as defined in (6.269) satisfies

$$\dim\mathscr{N}(\mathscr{L}_{5e}^{\prime}) = \dim\text{Coker}\,\mathscr{L}_{5e}^{\prime} = m,$$
$$\dim\mathscr{N}(\mathscr{L}_{5e}^{\prime *}) = \dim\text{Coker}\,\mathscr{L}_{5e}^{\prime *} = m.$$

Use Exercise 6.9 to determine $\mathscr{N}(\mathscr{L}_{5e}^{\prime}), \mathscr{N}(\mathscr{L}_{5e}^{\prime *}), \text{Coker}\,\mathscr{L}_{5e}^{\prime}$ and $\text{Coker}\,\mathscr{L}_{5e}^{\prime *}$. □

Exercise 6.13. Let Ω satisfy the same conditions as in Exercise 6.11. Reconsider Theorem 6.46 on the exterior multiconnected domain Ω^c. Determine $\mathscr{N}(\mathscr{L}_{6e}^{\prime}), \mathscr{N}(\mathscr{L}_{6e}^{\prime *}), \text{Coker}\,\mathscr{L}_{6e}^{\prime}$ and $\text{Coker}\,\mathscr{L}_{6e}^{\prime *}$ and show that their dimensions are all equal to $m-1$. (Hint: the arguments are similar to Exercise 6.10.) □

6.15 Direct formulation of BIE based upon Green's formula

Assume that w is sufficiently smooth, satisfying

$$\Delta w(x) = 0 \quad \text{on } \Omega,$$

where Ω is a bounded domain in $\mathbb{R}^2$ or $\mathbb{R}^3$ with smooth boundary $\partial\Omega$. Then integration by parts (i.e., Green's formula) gives

$$w(x) = \int_{\partial\Omega} E(x-\xi)\frac{\partial w(\xi)}{\partial n}\,d\sigma_\xi - \int_{\partial\Omega} \frac{\partial E(x-\xi)}{\partial n_\xi} w(\xi)\,d\sigma_\xi$$
$$= V_1^-\left(\frac{\partial w}{\partial n}\right)(x) - V_2^-(w)(x), \quad x\in\Omega, \tag{6.282}$$

according to the notation of Theorem 6.3. Letting x tend to a boundary point nontangentially, by (6.43) and (6.44) we have

$$\frac{1}{2}w(x) = \int_{\partial\Omega} \left[E(x-\xi)\frac{\partial w(\xi)}{\partial n} - \frac{\partial E(x-\xi)}{\partial n_\xi} w(\xi) \right] d\sigma_\xi, \quad x \in \partial\Omega.$$

Therefore for the Dirichlet problem

$$\left.\begin{aligned} \Delta w(x) &= 0 \quad \text{on } \Omega, \\ w(x) &= g_1(x), \quad x \in \partial\Omega, \end{aligned}\right\} \tag{6.283}$$

we solve the Neumann data $\partial w(x)/\partial n = f_1(x)$ by the BIE of the first kind

$$\int_{\partial\Omega} E(x-\xi) f(\xi)\, d\sigma_\xi = \frac{1}{2} g_1(x) + \int_{\partial\Omega} \frac{\partial E(x-\xi)}{\partial n_\xi} g_1(\xi)\, d\sigma_\xi, \ x \in \partial\Omega. \tag{6.284}$$

For a three-dimensional problem, the above BIE can be written as

$$\mathscr{L}_1(f) = \frac{1}{2} g_1 + \int_{\partial\Omega} \frac{\partial E(\cdot-\xi)}{\partial n_\xi} g_1(\xi)\, d\sigma_\xi, \tag{6.285}$$

according to the notation of Theorem 6.20. Consequently, it is always solvable with a unique solution $f_1 \in H^{r-1}(\partial\Omega)$, provided that $g_1 \in H^r(\partial\Omega)$. But if the space dimension is two then (6.284) is not necessarily uniquely solvable. Instead, one should augment (6.284) as

$$\begin{aligned} &\begin{bmatrix} \int_{\partial\Omega} f_1(\xi)\, d\sigma \\ \int_{\partial\Omega} E(x-\xi) f_1(\xi)\, d\sigma_\xi + a \end{bmatrix} \equiv \mathscr{L}_1' \left(\begin{bmatrix} a \\ f_1 \end{bmatrix} \right)(x) \\ &= \begin{bmatrix} A \\ \frac{1}{2} g_1(x) + \int_{\partial\Omega} \frac{\partial E(x-\xi)}{\partial n_\xi} g_1(\xi)\, d\sigma_\xi \end{bmatrix}, \end{aligned} \tag{6.286}$$

where $\mathscr{L}_1'$ is defined by (6.237), and solve it for some (arbitrary) $A \in \mathbb{R}$. By Theorem 6.33, we will obtain a unique $(a, f_1) \in \mathbb{R} \oplus H^{r-1}(\partial\Omega)$ provided that $g_1 \in H^r(\partial\Omega)$. Then the solution w to (6.283) is obtained as

$$w(x) = \int_{\partial\Omega} \left[E(x-\xi) f_1(\xi) - \frac{\partial E(x-\xi)}{\partial n_\xi} g(\xi) \right] d\sigma_\xi + a, \ x \in \Omega. \tag{6.287}$$

The reader can see that the BIE (6.285) and (6.286) have more complicated RHS than the corresponding BIE (6.176) and (6.240), respectively, if simple-layer representations of solutions have been used. Also, inside the domain Ω, computation of the solution w according to (6.282) or (6.287) is more complicated than the evaluation of the simple-layer potential (6.175) or (6.239), respectively.

For the Neumann problem

$$\begin{cases} \Delta w(x) = 0 \quad \text{on } \Omega, \\ \dfrac{\partial w(x)}{\partial n} = g_2(x), \ x \in \partial\Omega, \qquad \langle g_2, 1 \rangle_{H^0(\partial\Omega)} = 0, \end{cases}$$

we solve the Dirichlet data $w(x) = f_2(x)$ by the BIE of the second kind

$$\frac{1}{2} f_2(x) + \int_{\partial\Omega} \frac{\partial E(x-\xi)}{\partial n_\xi} f_2(\xi)\, d\sigma = \int_{\partial\Omega} E(x-\xi) g_2(\xi)\, d\sigma_\xi, \ x \in \partial\Omega. \tag{6.288}$$

In the notation of Theorems 6.22 and 6.34, the above can be written as

$$\mathscr{L}_2^*(f_2) = \int_{\partial\Omega} E(\cdot - \xi) g_2(\xi)\, d\sigma_\xi.$$

The operator $\mathscr{L}_2^*$ has a nontrivial kernel of dimension 1. Therefore, when the compatibility condition $\langle g_2, 1 \rangle = 0$ is satisfied, the above equation has a unique solution $f_2 \in H^{r+1}(\partial\Omega)$ (when $g_2 \in H^r(\partial\Omega)$) if we apply the augmentation procedure as described in § 4.7.

By comparison with the simple-layer representation (6.193), we see that the BIE (6.288) has a more complicated RHS than the corresponding BIE (6.194). Also, to obtain the solution by (6.282), more work is involved than in (6.193).

The reader can easily compare the various types of BIE derived by the direct formulation of Green's formula for the remaining cases of BVP, including the Robin BVP and the exterior BVP, with those obtained by the simple-layer and double-layer representations stated in the theorems of § 6.8–6.13. Invariably, the resulting BIE obtained by Green's formula are more complicated. *It appears that the only advantage of solving BIE using the direct formulation is that it may provide solutions (of the BIE) that correspond to the physical data, such as the Dirichlet, Neumann or Robin data, for the given equation, whereas using the simple- and double-layer representations, the solutions of the BIE (such as f in (6.176)) commonly lack physical meaning.*

6.16 Numerical example (I): an interior Dirichlet problem in $\mathbb{R}^2$

Now we are ready to apply the theorems obtained in the previous sections to compute numerical solutions. Three examples will be given. Only two-dimensional BVP will be treated here for the reasons that computer graphics are easier to illustrate, and that collocation convergence theorems are also available (see Chapter 10).

Example 6.2. Let Ω be the unit open disk in $\mathbb{R}^2$. Consider the Poisson equation with Dirichlet data

$$\left.\begin{aligned} &\Delta w(x) = g_1(x) \text{ on } \Omega, \quad x = (x_1, x_2), \\ &w(x) = g_2(x) \quad \text{on } \partial\Omega. \end{aligned}\right\} \tag{6.289}$$

It is easy to verify that the above BVP has the exact (benchmark) solution

$$w(x) = e^{x_1} + \cos 2x_2, \quad x \in \Omega, \tag{6.290}$$

for given

$$\left.\begin{aligned} g_1(x) &= e^{x_1} - 4\cos 2x_2 \text{ on } \Omega \\ g_2(x) &= e^{x_1} + \cos 2x_2 \quad \text{on } \partial\Omega. \end{aligned}\right\} \tag{6.291}$$

□

We compute numerical solutions step by step as follows.

Step 1: *Compute the volume potential*

$$v(x) = -\int_{\Omega} E(x,\xi) g_1(\xi)\, d\xi, \quad x \in \Omega. \tag{6.292}$$

Here, numerical integration must be performed over the entire domain Ω. This involves subdividing (i.e., triangulating) Ω into subdomains Ω_i, $1 \leqslant i \leqslant m$, such that

$$\Omega = \bigcup_{i=1}^{m} \Omega_i.$$

This triangulation can be done in much the same way as in finite-element methods. Because of the singularity of $E(x,\xi)$, it is not clear what is the most effective quadrature for evaluating $\int_{\Omega_i} E(\tilde{x},\xi) g_1(\xi)\, d\xi$ when $\tilde{x} \in \Omega_i$. Nevertheless, such integrals are *absolutely convergent* (for sufficiently regular g_1), and in space dimension two the logarithmic singularity of E is rather mild. In practice, it is useful to place more quadrature points near the singularity $\xi = \tilde{x}$ in evaluating $\int_{\Omega_i} E(\tilde{x},\xi) g_1(\xi)\, d\sigma$. This adaptive measure usually yields more accurate results. When $\tilde{x} \notin \Omega_i$, the integrand $E(\tilde{x},\xi) g_1(\xi)$ does not contain any singularity (for continuous g_1), thus the evaluation of $\int_{\Omega_i} E(\tilde{x},\xi) g_1(\xi)\, d\xi$ is much easier. Quadrature rules for certain standard regions Ω_i such as N-dimensional cubes can be found, e.g., in [174].

Step 2: *Use BIE to solve the BVP where the only inhomogeneity appears in the boundary condition.*

Define

$$\left.\begin{aligned} w_0(x) &= w(x) - v(x), \; x \in \Omega, \\ \tilde{g}_1(x) &= v(x), \qquad\qquad x \in \partial\Omega. \end{aligned}\right\} \tag{6.293}$$

Then w_0 satisfies

$$\left.\begin{aligned} \Delta w_0(x) &= 0 \qquad\qquad\quad \text{on } \Omega, \\ w_0(x) &= g_2(x) - \tilde{g}_1(x) \text{ on } \partial\Omega. \end{aligned}\right\} \tag{6.294}$$

Thus the inhomogeneity appears only in the boundary condition. Denote

$$g_0(x) \equiv g_2(x) - \tilde{g}_1(x) = e^{x_1} + \cos 2x_2 - \tilde{g}_1(x), \quad x \in \partial\Omega.$$

We now use the modified simple-layer potential (6.239) to solve (6.294) by Theorem 6.33. (This approach leads to the simplest and most effective BIE formulation.) Thus

$$w_0(x) = a + \int_{\partial\Omega} E(x,\xi) f(\xi) d\sigma, \quad x \in \Omega, \tag{6.295}$$

where $a \in \mathbb{R}$ and $f(\cdot)$ are solved by

$$\mathscr{L}_1' \left(\begin{bmatrix} a \\ f \end{bmatrix} \right) = \begin{bmatrix} A \\ g_0 \end{bmatrix}, \quad A \in \mathbb{R}, \tag{6.296}$$

i.e.,

$$\left. \begin{aligned} &\int_{\partial\Omega} f(\xi) d\sigma = A, \\ &a + \int_{\partial\Omega} E(x,\xi) f(\xi) d\sigma_\xi = g_0(x), \quad x \in \partial\Omega. \end{aligned} \right\} \tag{6.297}$$

Since the constant A can be arbitrary, we choose $A = 1$.

Step 3: *Select the type of boundary elements and collocate the BIE*

We now discretize (6.296) by point collocation. First, it is generally necessary to triangulate the boundary $\partial\Omega$ into simpler approximating elements Γ_i:

$$\partial\Omega \sim (\partial\Omega)_h \equiv \bigcup_{i=1}^{m'} \Gamma_i. \tag{6.298}$$

For two-dimensional problems, $\partial\Omega$ is a closed curve. Thus the arclength variable can be used, and an approximation like (6.298) may not be necessary. On $(\partial\Omega)_h$, we assume an approximate solution f_h to the density f in (6.297) as

$$f_h(x) = \sum_{j=1}^{\widehat{m}} \alpha_j \varphi_{j,h}(x), \quad x \in (\partial\Omega)_h, \tag{6.299}$$

where $\varphi_{j,h} \in V_h$, $j = 1, \ldots, \widehat{m}$, can be selected from the various types of finite elements introduced in Chapter 5 (and others). Let $\widetilde{x}_j$, $j = 1, 2, \ldots, \widehat{m}$ be the collocation points corresponding to the finite-element basis $\{\varphi_{j,h} \mid j = 1, \ldots, \widehat{m}\}$. These collocation points are usually chosen to be the *nodal points for odd-degree* and the *internodal midpoints for even-degree* splines when $\partial\Omega$ is one-dimensional. See § 10.3. Then (6.297) is collocated as

$$\left. \begin{aligned} &\int_{(\partial\Omega)_h} f_h(x) d\sigma = A \\ a_h + &\int_{(\partial\Omega)_h} E(\widetilde{x}_j,\xi) f_h(\xi) d\sigma_\xi = g_0(\widetilde{x}_j), \quad j = 1, 2, \ldots, \widehat{m}, \end{aligned} \right\} \tag{6.300}$$

leading to an $(\widehat{m}+1) \times (\widehat{m}+1)$ linear system

$$M_h \vec{\alpha}_h = \vec{\gamma}_h, \tag{6.301}$$

where

$$\vec{\alpha}_h \equiv (a_h, \alpha_1, \ldots, \alpha_{\widehat{m}}), \quad \vec{\gamma}_h \equiv (A, g_0(\widetilde{x}_1), \ldots, g_0(\widetilde{x}_{\widehat{m}})) \in \mathbb{R}^{\widehat{m}+1} \tag{6.302}$$

and the matrix M_h is given by

$$\begin{bmatrix} 0 & \int \varphi_{1,h}(x)\,d\sigma & \int \varphi_{2,h}(x)\,d\sigma & \vdots & \int \varphi_{\widehat{m},h}(x)\,d\sigma \\ 1 & \int E(\widetilde{x}_1,\xi)\varphi_{1,h}(\xi)\,d\sigma & \int E(\widetilde{x}_1,\xi)\varphi_{2,h}(\xi)\,d\sigma & \vdots & \int E(\widetilde{x}_1,\xi)\varphi_{\widehat{m},h}(\xi)\,d\sigma \\ 1 & \int E(\widetilde{x}_2,\xi)\varphi_{1,h}(\xi)\,d\sigma & \int E(\widetilde{x}_2,\xi)\varphi_{2,h}(\xi)\,d\sigma & \vdots & \int E(\widetilde{x}_2,\xi)\varphi_{\widehat{m},h}(\xi)\,d\sigma \\ \vdots & \vdots & \vdots & & \vdots \\ \vdots & \vdots & \vdots & & \vdots \\ 1 & \int E(\widetilde{x}_{\widehat{m}},\xi)\varphi_{1,h}(\xi)\,d\sigma & \int E(\widetilde{x}_{\widehat{m}},\xi)\varphi_{2,h}(\xi)\,d\sigma & \vdots & \int E(\widetilde{x}_{\widehat{m}},\xi)\varphi_{\widehat{m},h}(\xi)\,d\sigma \end{bmatrix},$$

where each integral is evaluated on the support of the corresponding basis function $\varphi_{j,h}(\cdot)$. For h sufficiently small, according to the convergence theorems in Chapter 10, the linear system (6.301) is invertible for all $h,\ 0 < h \leqslant h_0$, for some h_0; thus

$$\vec{\alpha}_h = M_h^{-1}\vec{\gamma}_h \tag{6.303}$$

is obtained. Substituting the coefficients in $\vec{\alpha}_h$ into (6.299) and using $f_h(\cdot)$ and a_h in (6.295), we obtain an approximate solution

$$w_{0,h}(x) \equiv a_h + \int_{(\partial\Omega)_h} E(x,\xi) f_h(\xi)\,d\sigma_\xi, \quad x \in \overline{\Omega}_h. \tag{6.304}$$

Thus numerical solutions to (6.289) are given by

$$w_h(x) \equiv w_{0,h}(x) + v_h(x), \quad x \in \overline{\Omega}_h, \tag{6.305}$$

where $v_h(x)$ corresponds to (6.292) computed by certain quadrature.

In this example, the operator $\mathscr{L}_1'$ maps $\mathbb{R}\oplus H^r(\partial\Omega)$ onto $\mathbb{R}\oplus H^{r+1}(\partial\Omega)$, so it is smoothing by the order of 1. We choose piecewise-constant boundary elements for the approximating space V_n. By Exercise 2.2, each basis element $\varphi_h \in H^{1/2-\varepsilon}(\partial\Omega)$, for any $\varepsilon > 0$. Therefore the approximate solution (6.304) has regularity

$$\left.\begin{aligned} w_{0,h} &\in H^{3/2-\varepsilon}(\partial\Omega) \text{ on } \partial\Omega, \\ w_{0,h} &\in H^{2-\varepsilon}(\Omega) \quad\ \ \text{on } \Omega, \end{aligned}\right\} \tag{6.306}$$

for any $\varepsilon > 0$, which is generally acceptable as satisfactory. *The greatest advantage of using piecewise-constant boundary elements is that numerical quarature is easy to evaluate and computer program coding is almost always the simplest and the most effective.* In this example, as we will see, the rate of convergence is also fast: for $(a_h, f_h(\cdot))$, we have $\mathscr{O}(h)$ convergence in $\mathbb{R}\oplus H^0(\partial\Omega)$. Of course, higher-order splines can always be used and speed up convergence faster still, but the tradeoff is a higher programming cost by the numerical analyst, which is generally more expensive than the cost of CPU time on a super computer or a laptop.

In our numerical calculation of Example 6.2, we divide the circumference $\partial\Omega$ into N equal parts Γ_j, $1 \leqslant j \leqslant N$, such as shown in Fig. 1.3. The midpoints $\widetilde{x}_j$ of Γ_j are the nodal points. Here in terms of the arclength variable on $\partial\Omega$,

$$\widetilde{x}_j = (2j-1)\frac{\pi}{N}, \qquad j = 1, 2, \ldots, N. \tag{6.307}$$

Using $N = 6, 12, 24, 48$ and 96, we have computed $a_h, f_h(\cdot)$ and $w_h(\cdot)$ according to the procedures described above. ($v_h(\cdot)$ is computed by a high accuracy-quadrature rule, and the same $v_h(\cdot)$ is used in (6.305) throughout.)
In Fig. 6.7, the exact solution w as given in (6.290) is plotted.

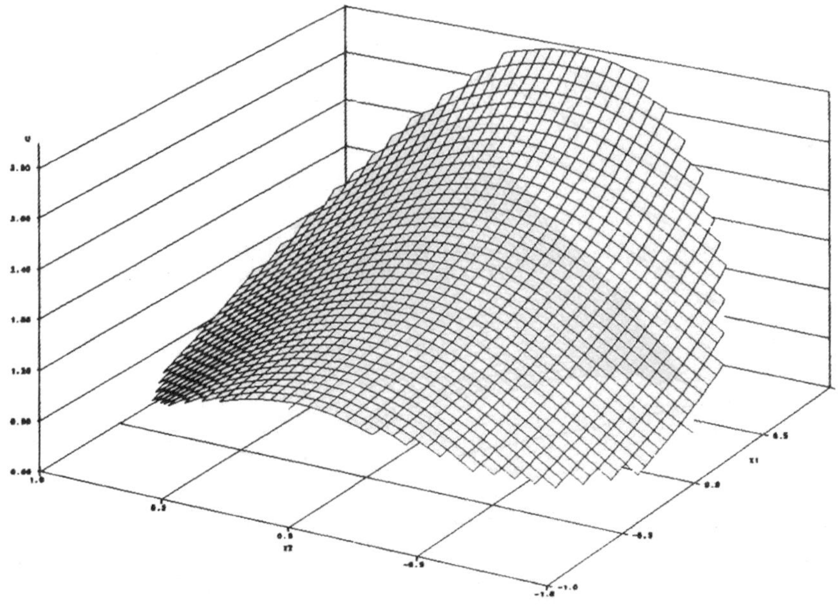

Figure 6.7: The exact solution $w(x)$ of Example 6.3.

In Figs. 6.8, 6.9 and 6.10, numerical solutions $w_h(\cdot)$ (cf. (6.305)) corresponding to $N =$ 6, 12 and 24 are plotted. The reader can see in Fig. 6.8 that even using only six ($N = 6$) piecewise-constant boundary elements, a rather crude discretization by most standards, the numerical solution still seems to have captured the essential features of the profile of the exact solution. Nevertheless, there is considerable ruggedness near the boundary $\partial\Omega$. The same can be said of the numerical solution $w_h(\cdot)$ corresponding to $N = 12$ as shown in Fig. 6.9. When $N = 24$ is used, the profile of the numerical solution $w_h(\cdot)$ in Fig. 6.10 is indistinguishable from that of the exact solution in Fig. 6.7 at the plotting scale used.

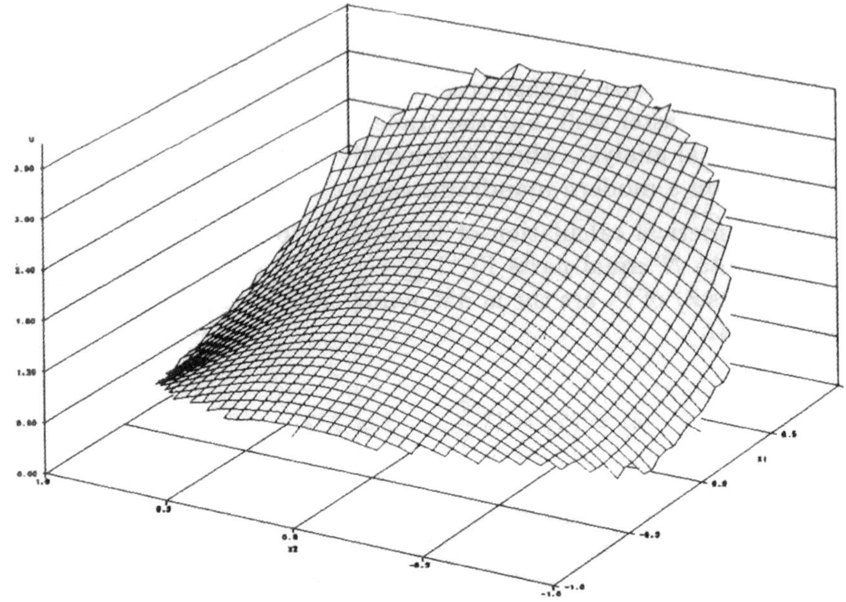

Figure 6.8: The numerical solution $w_h(x)$ of Example 6.3; $N = 6$ piecewise-constant boundary elements are used on $\partial\Omega$.

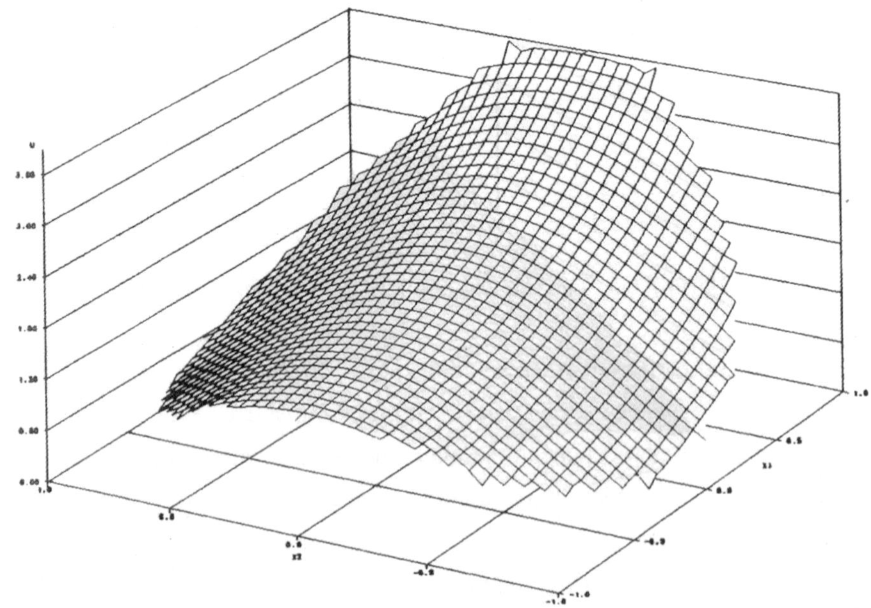

Figure 6.9: The numerical solution $w_h(x)$ of Example 6.3; $N = 12$ piecewise-constant boundary elements are used on $\partial\Omega$.

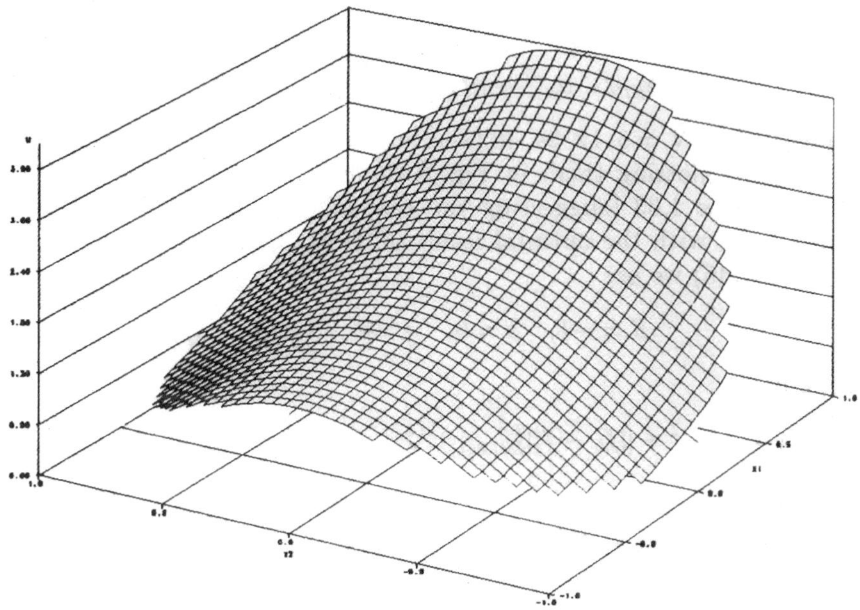

Figure 6.10: The numerical solution $w_h(x)$ of Example 6.3; $N = 24$ piecewise-constant boundary elements are used on $\partial\Omega$.

Numerical solutions corresponding to $N = 48$ and 96 look identical to Fig. 6.10, so their graphics are omitted.

Numerical solutions f_h of the density f are plotted together in Fig. 6.11. The reader can see a clear trend of convergence.

The evaluation of the rate of convergence

$$(|a_h - a|^2 + \|f_h - f\|^2_{L^2(\partial\Omega)})^{1/2} \tag{6.308}$$

to the exact solution $(a, f(\cdot))$ of (6.296) is carried out as follows. Because $(a, f(\cdot))$ is not known in closed form, a direct computation of (6.308) is not possible. Instead, we compute the L^2-norm of two successive solutions

$$(|a_{h/2} - a_h|^2 + \|f_{h/2} - f_h\|^2_{L^2(\partial\Omega)})^{1/2}, \tag{6.309}$$

corresponding to $N = 6,\ 12,\ 24,\ 48$ (and 96). Taking the *negative logarithms* of (6.309) and plotting them in Fig. 6.12, we find that the slope of the connected line is 1.03 (suggesting the rate of convergence $\mathcal{O}(h^1)$). This is in agreement with the optimal result of Theorem 10.14 in Chapter 10, where in (10.146) we use

$$\beta = -1, \quad s = 0, \quad d = 0, \quad t = 1, \tag{6.310}$$

resulting in

$$\left(|a_h - a|^2 + \|f_h - f\|^2_{L^2(\partial\Omega)}\right)^{1/2} \leqslant Ch\|f\|_{H^1(\partial\Omega)}. \tag{6.311}$$

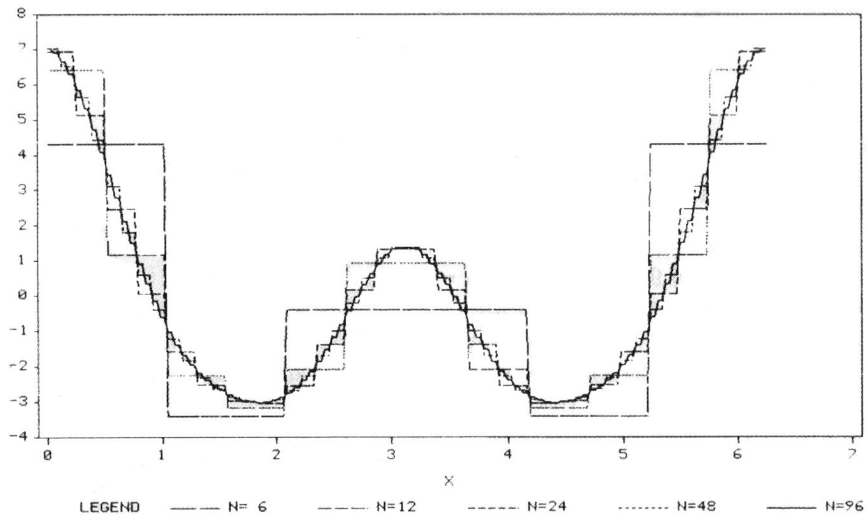

Figure 6.11: Convergence of f_h in Example 6.3 $N = 6,\ 12,\ 24,\ 48$ and 96 piecewise-constant boundary elements are used.

Numerical values of f_h and w_h at sample points and a certain value of a_h can be found in Table 6.1.

Remark 6.12. How do we compute values of partial derivatives by BEM? This is a commonly raised question.

For example, for the 2D interior Dirichlet problem

$$\left.\begin{aligned} \Delta w(x) &= 0 && \text{on } \Omega \subseteq \mathbb{R}^2, \\ w(x) &= g(x), && x \in \partial\Omega, \end{aligned}\right\} \tag{6.312}$$

according to Theorem 6.33, the *ansatz*

$$w(x) = a + \int_{\partial\Omega} E(x,\xi) f(\xi)\, d\sigma_\xi, \quad x \in \overline{\Omega}, \tag{6.313}$$

is applicable. Assume that the Dirichlet data g is sufficiently smooth. We wish to compute some partial derivatives $D^\alpha w(x_0)$ of the solution w at some $x_0 \in \overline{\Omega}$. Consider the following situations.

(i) $x_0 \in \partial\Omega$

In this case, we can compute the tangential derivative by

$$\frac{\partial w(x_0)}{\partial \sigma} = \frac{\partial g(x_0)}{\partial \sigma}, \quad x_0 \in \partial\Omega, \tag{6.314}$$

and the normal derivative by

$$\frac{\partial w(x_0)}{\partial n} = \frac{1}{2} f(x_0) + \int_{\partial\Omega} \frac{\partial E(x,\xi)}{\partial n_x}\Big|_{x=x_0} f(\xi)\,d\sigma_\xi, \quad x_0 \in \partial\Omega. \tag{6.315}$$

Once $\partial w/\partial \sigma$ and $\partial w/\partial n$ are obtained, we use (1.34) and (1.36) to get $\partial w(x_0)/\partial x_1$ and $\partial w(x_0)/\partial x_2$. Thus all first-order derivatives have been obtained. Formula (6.314) can be applied repeatedly to give tangential derivatives of arbitrarily higher order:

$$\frac{\partial^j w(x_0)}{\partial \sigma^j} = \frac{\partial^j g(x_0)}{\partial \sigma^j}.$$

If $\Delta w(x) = 0$ is satisfied on $\partial\Omega$ as well then we can use

$$\begin{aligned} 0 &= \frac{\partial^2 w(x_0)}{\partial x_1^2} + \frac{\partial^2 w(x_0)}{\partial x_2^2} \\ &= \left\{ \left[n_1(x)\frac{\partial}{\partial n} + n_2(x)\frac{\partial}{\partial \sigma}\right]^2 w(x) + \left[n_2(x)\frac{\partial}{\partial n} + n_1(x)\frac{\partial}{\partial \sigma}\right]^2 w(x) \right\}\Bigg|_{x=x_0} \end{aligned}$$

to derive certain information about the second-order mixed derivatives $D^2 w(x_0)$.

(ii) $x_0 \in \Omega$

If x_0 is sufficiently far from the boundary $\partial\Omega$, then we can simply evaluate the boundary integral

$$D^\alpha w(x_0) = \int_{\partial\Omega} D_x^\alpha E(x,\xi) f(\xi)\,d\sigma_\xi. \tag{6.316}$$

Difficulties occur when $x_0 \in \Omega$ is fairly close to $\partial\Omega$. In this circumstance, the singularity in the kernel $D_x^\alpha E(x,\xi)$ will cause a large deviation of the quadrature, rendering (6.316) totally inaccurate. (The larger $|\alpha|$ is or the closer x_0 is to $\partial\Omega$, the worse the accuracy becomes.) According to our experience, the best way to compute $D^\alpha w(x_0)$ is by *finite difference*. Let the coordinates of x_0 be

$$x_0 = (x_{0,1}, x_{0,2}). \tag{6.317}$$

Then we may approximate, say $\partial^2 w(x_0)/\partial x_1^2$, by a central difference

$$\frac{\partial^2 w(x_0)}{\partial x_1^2} \sim \frac{w(x_{0,1}+\widetilde{h}, x_{0,2}) - 2w(x_{0,1}, x_{0,2}) + w(x_{0,1}-\widetilde{h}, x_{0,2})}{\widetilde{h}^2}, \tag{6.318}$$

for example, by choosing a sufficiently small $\widetilde{h} > 0$, where each value of w on the RHS of (6.318) is obtained by the formula (6.313). This usually may produce satisfactory or even very accurate numerical values of $\partial^2 w(x_0)/\partial x_1^2$. A numerical example will be given in Example 6.4.

□

Table 6.1: Numerical values of $f_h(\cdot), w_h(\cdot)$ and $w(\cdot)$ at sample points and the value of a_h in Example 6.3, with $N = 24$.

$f_h(\cdot)$	r	$\theta(2\pi)$	$w_h(\cdot)$	$w(\cdot)$
6.941987	$\frac{1}{4}$	$\frac{1}{8}$	2.134712	2.131513
5.121257	$\frac{1}{4}$	$\frac{3}{8}$	1.877396	1.877583
2.453160	$\frac{1}{4}$	$\frac{3}{8}$	1.774578	1.776115
0.039092	$\frac{1}{4}$	$\frac{4}{8}$	1.777344	1.778801
−1.602749	$\frac{1}{4}$	$\frac{5}{8}$	1.774578	1.776115
−2.535620	$\frac{1}{4}$	$\frac{6}{8}$	1.877396	1.877583
−2.994165	$\frac{1}{4}$	$\frac{7}{8}$	2.134712	2.131513
−3.049866	$\frac{1}{4}$	$\frac{8}{8}$	2.289339	2.284025
−2.555343	$\frac{2}{4}$	$\frac{1}{8}$	2.189698	2.184364
−1.387400	$\frac{2}{4}$	$\frac{2}{8}$	1.537060	1.540302
0.165207	$\frac{2}{4}$	$\frac{3}{8}$	1.458440	1.462433
1.314299	$\frac{2}{4}$	$\frac{4}{8}$	1.604858	1.606531
1.314299	$\frac{2}{4}$	$\frac{5}{8}$	1.458440	1.462433
0.165207	$\frac{2}{4}$	$\frac{6}{8}$	1.537060	1.540302
−1.387400	$\frac{2}{4}$	$\frac{7}{8}$	2.189698	2.184364
−2.555343	$\frac{2}{4}$	$\frac{8}{8}$	2.660820	2.648721
−3.049866	$\frac{3}{4}$	$\frac{1}{8}$	2.194791	2.187789
−2.994165	$\frac{3}{4}$	$\frac{2}{8}$	1.062714	1.070737
−2.535620	$\frac{3}{4}$	$\frac{3}{8}$	1.070082	1.076707
−1.602749	$\frac{3}{4}$	$\frac{4}{8}$	1.472712	1.472367
0.039092	$\frac{3}{4}$	$\frac{5}{8}$	1.070082	1.076707
2.453160	$\frac{3}{4}$	$\frac{6}{8}$	1.062714	1.070737
5.121257	$\frac{3}{4}$	$\frac{7}{8}$	2.194791	2.187789
6.941987	$\frac{3}{4}$	$\frac{8}{8}$	3.138585	3.117000
$a_h = 1.4908$				

6.17 Numerical example (II): an interior Neumann problem in $\mathbb{R}^2$

We now consider the following.

Example 6.3.

$$\left.\begin{aligned}&\Delta w(x)=0 \quad \text{on } \Omega= \text{ unit open disk in } \mathbb{R}^2,\\ &\frac{\partial w(x)}{\partial n}=\tfrac{1}{2}(x_1^3-3x_1x_2^2)\equiv g(x) \quad \text{on } \partial\Omega,\\ &\int_{\partial\Omega} x_1^2 w(x)\,d\sigma_x=-2.\end{aligned}\right\} \tag{6.319}$$

Note that the Neumann boundary condition above satisfies the compatibility condition

$$\int_{\partial\Omega}\frac{1}{2}(x_1^3-3x_1x_2^2)\,d\sigma=0,$$

and the third equation in (6.319) is an accessory condition of the type (4.182). According to Corollary 4.5, the BVP (6.319) has a unique solution w. □

We apply Theorem 6.34. The solution w can be represented as

$$w(x)=a+\int_{\partial\Omega}E(x,\xi)f(\xi)\,d\sigma_\xi, \quad x\in\overline{\Omega}, \tag{6.320}$$

where $(a,f(\cdot))$ satisfies the augmented BIE system

$$\mathscr{L}\left(\begin{bmatrix}a\\ f\end{bmatrix}\right)(x)=\begin{bmatrix}\displaystyle\int_{\partial\Omega}\eta_1^2\left[a+\int_{\partial\Omega}E(\eta,\xi)f(\xi)\,d\sigma_\xi\right]d\sigma_\eta\\ \displaystyle\frac{1}{2}f(x)+\int_{\partial\Omega}E(x,\xi)f(\xi)\,d\sigma_\xi\end{bmatrix}$$
$$=\begin{bmatrix}-2\\ g(x)\end{bmatrix}, \quad x\in\partial\Omega. \tag{6.321}$$

Again, we use piecewise-constant boundary elements with uniform meshes to approximate f:

$$f_h(x)=\sum_{j=1}^{N}\alpha_j\varphi_{j,h}(x), \tag{6.322}$$

just as in Example 6.2. Substituting $(a_h,f_h(\cdot))$ into (6.321) and collocating at nodal points (6.307), we obtain an $(N+1)\times(N+1)$ linear system

$$M_h\vec{\alpha}_h=\vec{\gamma}_h, \tag{6.323}$$

where

$$\vec{\alpha}_h=(a_h,\alpha_1,\alpha_2,\ldots,\alpha_N)^{\mathrm{Tr}},\ \vec{\gamma}_h=(-2,g(\widetilde{x}_1),g(\widetilde{x}_2),\ldots,g(\widetilde{x}_N))^{\mathrm{Tr}}\in\mathbb{R}^{N+1},$$

and $(M_h)_{ij}$, the (i,j)-entry of the matrix M_h, is given by

$$(M_h)_{ij}=\begin{cases}\displaystyle\int_{\partial\Omega}x_1^2\,d\sigma, & (i,j)=(1,1)\\ \displaystyle\int_{\partial\Omega}\int_{\partial\Omega}x_1^2E(x,\xi)\varphi_{j-1,h}(\xi)\,d\sigma_\xi\,d\sigma_x, & i=1,\ j=2,\ldots,N+1,\\ 0, & j=1,\ i=2,\ldots,N+1,\\ \displaystyle\frac{1}{2}\delta_{i-1,j-1}+\int_{\partial\Omega}E(\widetilde{x}_i,\xi)\phi_{j-1,h}(\xi)\,d\sigma_\xi,\ i,j=2,\ldots,N+1. & \end{cases}$$

For N chosen sufficiently large, the matrix equation (6.323) is invertible, so we obtain

$$\vec{\alpha}_h = M_h^{-1}\vec{\gamma}_h.$$

Back-substituting the coefficients in $\vec{\alpha}_h = (a_h, \alpha_1, \alpha_2, \dots, \alpha_N)$ into (6.322) and (6.320), we obtain numerical solutions $w_h(\cdot)$ of (6.319) on $\overline{\Omega}$.

The exact solution w to (6.319) is

$$w(x) = \frac{1}{6}(x_1^3 - 3x_1x_2^2) - \frac{2}{\pi} \quad \text{on } \overline{\Omega}, \tag{6.324}$$

which is plotted in Fig. 6.14.

As before, we use $N = 6,\ 12,\ 24,\ 48$ and 96 in our computations. The numerical solutions $w_h(\cdot)$ corresponding to $N = 24$ and 48 are shown, respectively, in Figs. 6.15 and 6.16.

The approximate density functions $f_h(\cdot)$ corresponding to $N = 6,\ 12, \dots,$

96 are plotted together in Fig. 6.12. Again, convergence of f_h is observed. Evaluating the rate of convergence according to (6.309), we plot the (negative logarithm) points in Fig. 6.17 and find the slope of the line to be 0.97, suggesting that the ideal rate of convergence is $\mathscr{O}(h^1)$. This is again in very good agreement with the optimal result of Theorem 10.14, where in (10.146) we use

$$\beta = 0,\ s = 0,\ d = 0,\ t = 1 \quad 0 = d > -\frac{1}{2} = \beta - \frac{1}{2},$$

resulting in

$$\left(|a_h - a|^2 + \|f_h - f\|^2_{L^2(\partial\Omega)}\right)^{1/2} \leqslant Ch\|f\|_{H^1(\partial\Omega)}.$$

Numerical values of f_h, w_h and w at sample points as well as values of a_h can be found in Table 6.2.

Example 6.4. We use Remark 6.12 to compute a second-order derivative value at an (interior) point near $\partial\Omega$ for Example 6.3. Let the point x_0 in (6.317) be

$$x_0 = (0.98, 0),$$

that is fairly close to $\partial\Omega$. The exact value of $\partial^2 w(x_0)/\partial x_1^2$, from (6.324), is

$$\frac{\partial^2 w(x_0)}{\partial x_1^2} = 0.98. \tag{6.325}$$

If we apply (6.318), with $\widetilde{h} = 1.0 \times 10^{-7}$, we obtain

$$\frac{\partial^2 w_h(x_0)}{\partial x_1^2} = \begin{cases} -0.9381, & N = 24, \\ 0.0882, & N = 48, \\ 1.0159, & N = 96, \end{cases}$$

which becomes a good approximation for (6.325) when $N = 96$. □

Table 6.2: Numerical values of $f_h(\cdot), w_h(\cdot)$ and $w(\cdot)$ at sample points, and the value of a_h, Example 6.3 with $N = 24$.

$f_h(\cdot)$	r	$\theta(2\pi)$	$w_h(\cdot)$	$w(\cdot)$
0.843721	$\frac{1}{4}$	$\frac{1}{8}$	−0.629961	−0.638461
0.349481	$\frac{1}{4}$	$\frac{2}{8}$	−0.628299	−0.636620
−0.349481	$\frac{1}{4}$	$\frac{3}{8}$	−0.626636	−0.634778
−0.843721	$\frac{1}{4}$	$\frac{4}{8}$	−0.630650	−0.639224
−0.843721	$\frac{1}{4}$	$\frac{5}{8}$	−0.626636	−0.634778
−0.349481	$\frac{1}{4}$	$\frac{6}{8}$	−0.628299	−0.636620
0.349481	$\frac{1}{4}$	$\frac{7}{8}$	−0.629961	−0.638461
0.843721	$\frac{1}{4}$	$\frac{8}{8}$	−0.625947	−0.634016
0.843721	$\frac{2}{4}$	$\frac{1}{8}$	−0.641599	−0.651351
0.349481	$\frac{2}{4}$	$\frac{2}{8}$	−0.628299	−0.636620
−0.349481	$\frac{2}{4}$	$\frac{3}{8}$	−0.614998	−0.621888
−0.843721	$\frac{2}{4}$	$\frac{4}{8}$	−0.647108	−0.657453
−0.843721	$\frac{2}{4}$	$\frac{5}{8}$	−0.614998	−0.621888
−0.349481	$\frac{2}{4}$	$\frac{6}{8}$	−0.628299	−0.636620
0.349481	$\frac{2}{4}$	$\frac{7}{8}$	−0.641599	−0.651351
0.843721	$\frac{2}{4}$	$\frac{8}{8}$	−0.609489	−0.615786
0.843721	$\frac{3}{4}$	$\frac{1}{8}$	−0.673180	−0.686338
0.349481	$\frac{3}{4}$	$\frac{2}{8}$	−0.628299	−0.636620
−0.349481	$\frac{3}{4}$	$\frac{3}{8}$	−0.583417	−0.586901
−0.843721	$\frac{3}{4}$	$\frac{4}{8}$	−0.691770	−0.706932
−0.843721	$\frac{3}{4}$	$\frac{5}{8}$	−0.583417	−0.586901
−0.349481	$\frac{3}{4}$	$\frac{6}{8}$	−0.628299	−0.636620
0.349481	$\frac{3}{4}$	$\frac{7}{8}$	−0.673180	−0.686338
0.843721	$\frac{3}{4}$	$\frac{8}{8}$	−0.564827	−0.566307
$a_h = -0.6283$				

6.18 Numerical example (III): an exterior Neumann problem in $\mathbb{R}^2$

Example 6.5. Let Ω^c be the exterior open domain outside the unit disk in $\mathbf{R}^2$. Consider

$$\left.\begin{aligned} &\Delta w(x) = 0 && \text{on } \Omega^c, \\ &\frac{\partial w(x)}{\partial n} = \left(\frac{\partial w(x)}{\partial r} =\right) \frac{1}{2}\cos 3\theta = g(x), && x \in \partial\Omega, \\ &\lim_{|x|\to\infty} w(x) = 0. \end{aligned}\right\} \tag{6.326}$$

In the Neumann boundary condition above, (r,θ) are the polar coordinates of $x \in \partial\Omega$. The exact solution of (6.326) is easily verified to be

$$w(x) = -\frac{x_1^3 - 3x_1x_2^2}{6[(x_1^3 - 3x_1x_2^2)^2 + (3x_1^2x_2 - x_2^3)^2]}. \tag{6.327}$$

Since the data g satisfies

$$\int_{\partial\Omega} g(x)\,d\sigma = 0,$$

by Theorem 6.39, w is $\mathscr{O}(|x|^{-1})$ growth for large $|x|$, so the third equation in (6.326) is satisfied. We have a unique simple-layer representation

$$w(x) = \int_{\partial\Omega} E(x,\xi)f(\xi)\,d\sigma_\xi,$$

where f solves the BIE

$$\mathscr{L}_{2e}(f)(x) \equiv \begin{bmatrix} \int_{\partial\Omega} f(x)\,d\sigma \\ -\frac{1}{2}f(x) + \int_{\partial\Omega} \frac{\partial E(x,\xi)}{\partial n_x} f(\xi)\,d\sigma_\xi \end{bmatrix} = \begin{bmatrix} 0 \\ g(x) \end{bmatrix}.$$

Again, we use piecewise-constant boundary elements to approximate f, with $N =$ 6, 12, 24, 48 and 96. Since the discretization and collocation procedures are identical to Examples 6.2 and 6.3, we omit the details.

The exact solution (6.327) is plotted in Fig. 6.18. It is plotted on a square $[3.0] \times [3.0]$ outside the unit disk. The numerical solution w_h corresponding to $N = 24$ is plotted in Fig. 6.19. The densities f_h are plotted in Fig. 6.20, where convergence can again be observed. Values of f_h, w_h and w at sample points are listed in Table 6.3. □

Table 6.3: Numerical values of $f_h(\cdot), w_h(\cdot)$ and $w(\cdot)$ at sample points in Example 6.5, with $N = 24$.

$f_h(\cdot)$	r	$\theta(2\pi)$	$w_h(\cdot)$	$w(\cdot)$
−0.946722	3	$\frac{1}{8}$	−0.001456	−0.001488
−0.392145	3	$\frac{2}{8}$	0.003514	0.003591
0.392145	3	$\frac{3}{8}$	−0.003514	−0.003591
0.946722	3	$\frac{4}{8}$	0.001456	0.001488
0.946722	3	$\frac{5}{8}$	0.001456	0.001488
0.392145	3	$\frac{6}{8}$	−0.003514	−0.003591
−0.392145	3	$\frac{7}{8}$	0.003514	0.003591
−0.946722	3	$\frac{8}{8}$	−0.001456	−0.001488
−0.946722	5.5	$\frac{1}{8}$	−0.000499	−0.000510
−0.392145	5.5	$\frac{2}{8}$	0.001205	0.001232
0.392145	5.5	$\frac{3}{8}$	−0.001205	−0.001232
0.946722	5.5	$\frac{4}{8}$	0.000499	0.000510
0.946722	5.5	$\frac{5}{8}$	0.000499	0.000510
0.392145	5.5	$\frac{6}{8}$	−0.001205	−0.001232
−0.392145	5.5	$\frac{7}{8}$	0.001205	0.001232
−0.946722	5.5	$\frac{8}{8}$	−0.000499	−0.000510
−0.946722	8	$\frac{1}{8}$	−0.000122	−0.000125
−0.392145	8	$\frac{2}{8}$	0.000294	0.000301
0.392145	8	$\frac{3}{8}$	−0.000294	−0.000301
0.946722	8	$\frac{4}{8}$	0.000122	0.000125
0.946722	8	$\frac{5}{8}$	0.000122	0.000125
0.392145	8	$\frac{6}{8}$	−0.000294	−0.000301
−0.392145	8	$\frac{7}{8}$	0.000294	0.000301
−0.946722	8	$\frac{8}{8}$	−0.000122	−0.000125

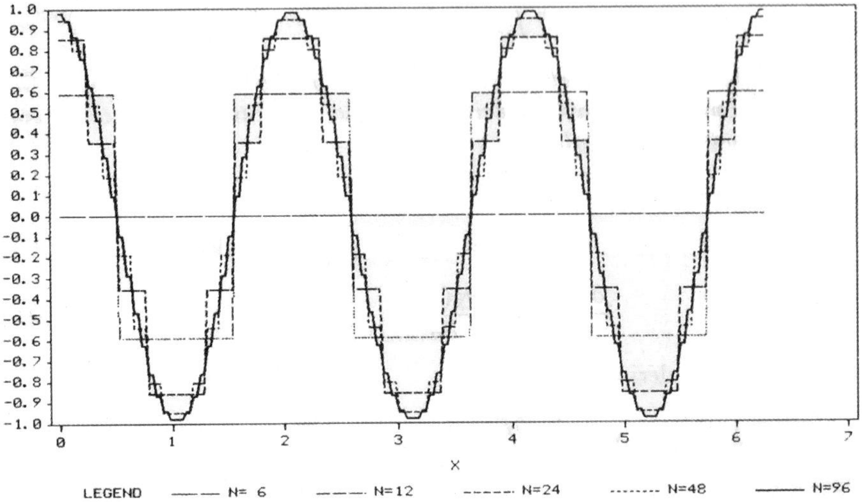

Figure 6.12: Convergence of f_h in Example 6.3, $N = 6,\ 12,\ 24,\ 48$ and 96 piecewise-constant boundary elements are used.

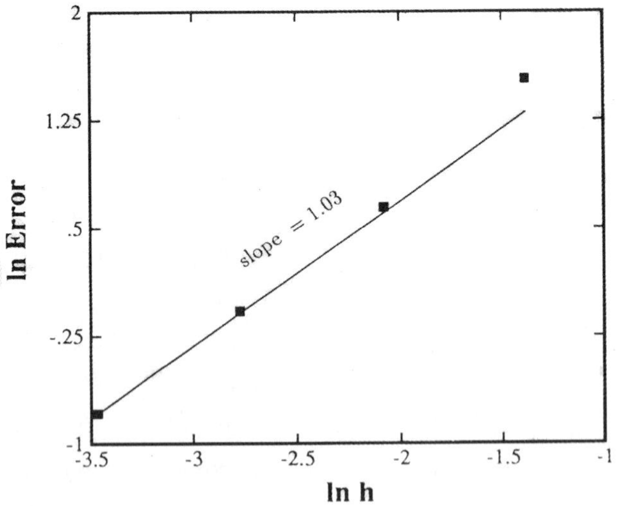

Figure 6.13: The rate of convergence of $(a, f(\cdot))$ by piecewise-constant boundary elements in Example 6.3.

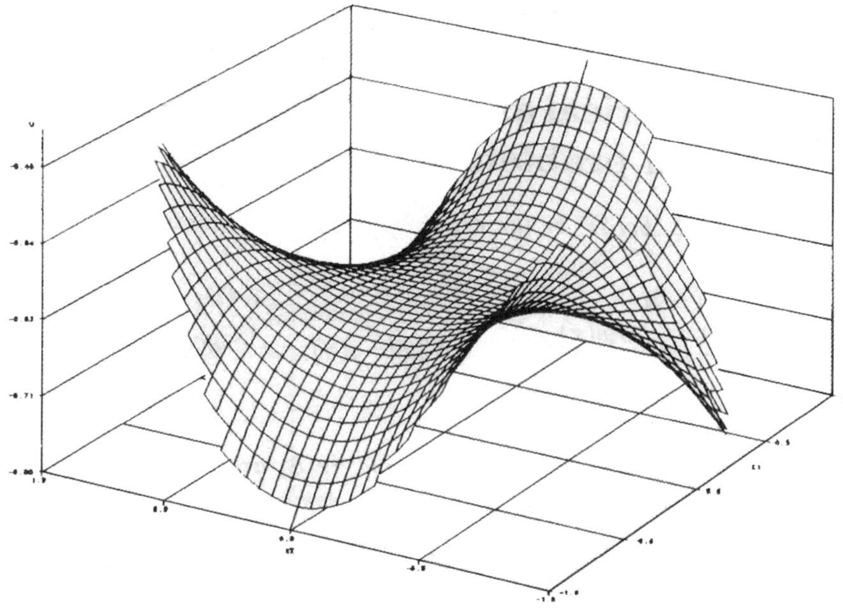

Figure 6.14: The exact solution $w(x)$ of Example 6.3.

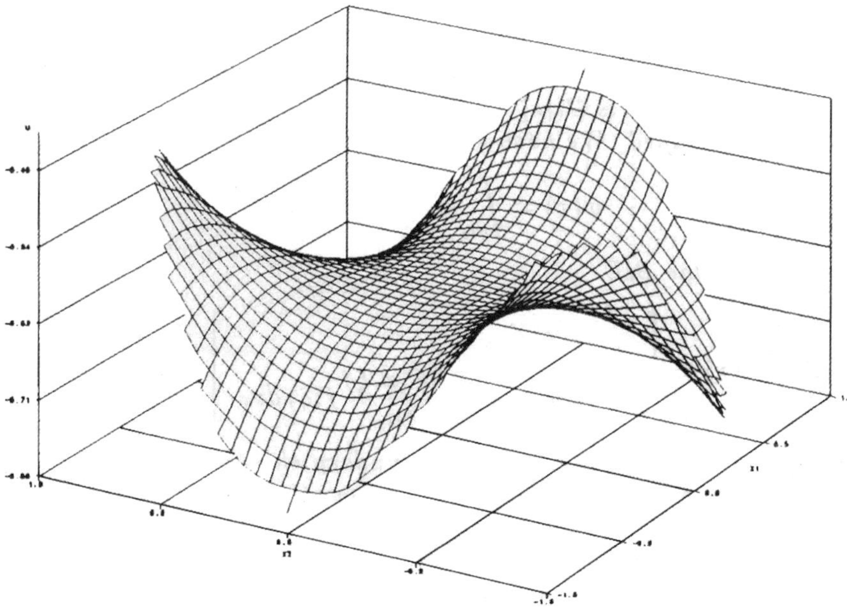

Figure 6.15: The numerical solution $w_h(x)$ of Example 6.3. $N = 24$ piecewise-constant boundary elements are used on $\partial\Omega$.

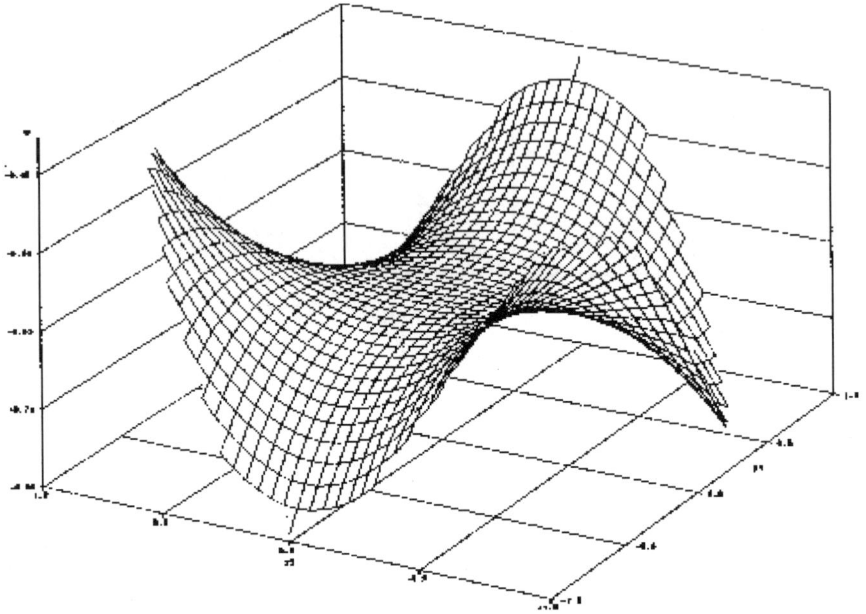

Figure 6.16: The numerical solution $w_h(x)$ of Example 6.3. $N = 48$ piecewise-constant boundary elements are used on $\partial\Omega$.

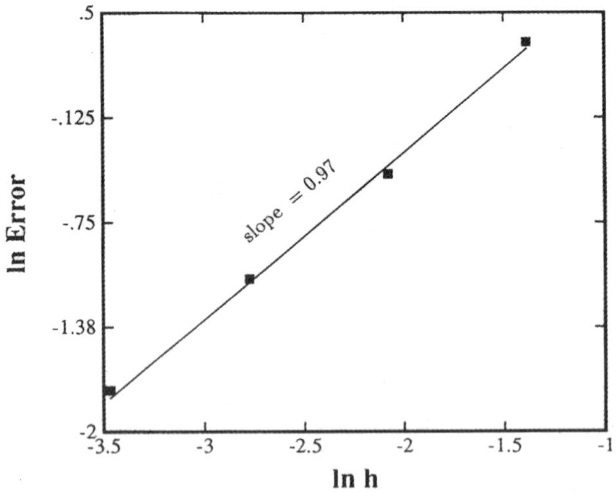

Figure 6.17: The rate of convergence of $(a, f(\cdot))$ by piecewise-constant boundary elements in Example 6.3.

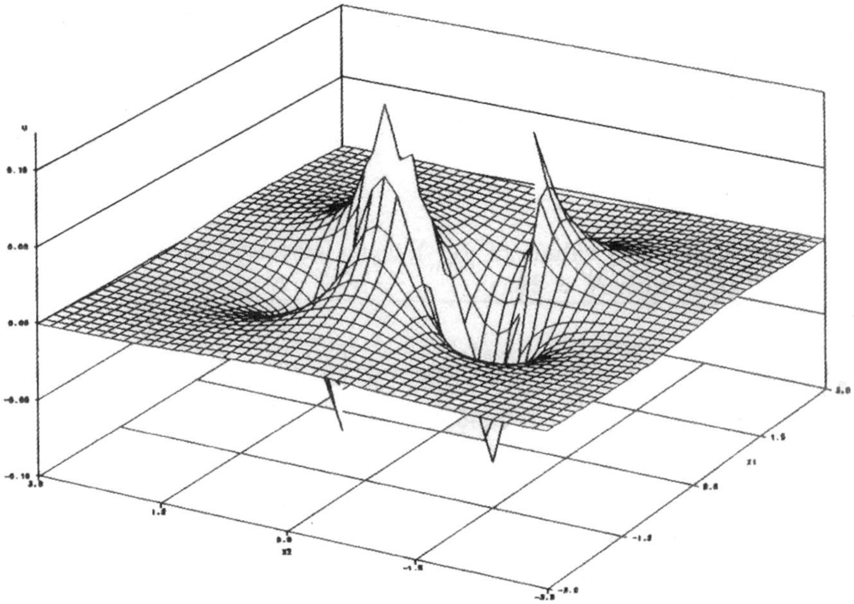

Figure 6.18: The exact solution $w(x)$ of Example 6.5.

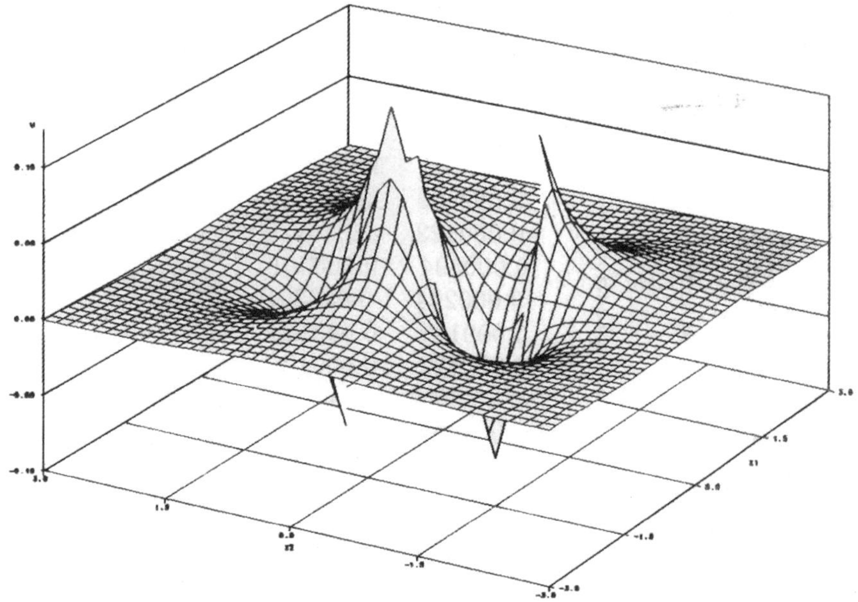

Figure 6.19: The numerical solution $w_h(x)$ of Example 6.5; $N = 24$ piecewise-constant boundary elements are used on $\partial\Omega$.

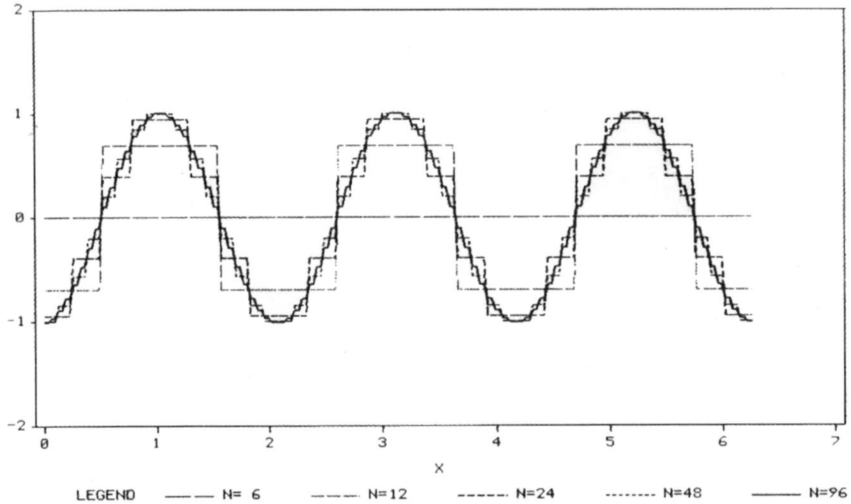

Figure 6.20: Convergence of f_h in Example 6.5; $N = 6,\ 12,\ 24,\ 48$ and 96 piecewise-constant boundary elements are used.

Chapter 7

The Helmholtz Equation

7.1 Background

The Helmholtz equation

$$(\Delta + k^2)w(x) = 0$$

arises naturally in physical applications related to wave propagation and vibration phenomena. We mention several important examples.

(a) Time harmonic acoustic scattering

Consider acoustic wave propagation in an adiabatic gas flow. Assume that the constant equilibrium state is

$$v = (0,0,0), \quad \rho = \rho_0, \quad p = p_0,$$

where v, ρ and p are the velocity, density and pressure of the gas. We consider the isentropic equations

$$\left.\begin{aligned} &\rho_t + v\cdot\nabla\rho + \rho\nabla\cdot v = 0 \text{ (conservation of mass)},\\ &\rho v_t + \rho v\cdot\nabla v + \nabla p = 0 \text{ (conservation of momentum)}.\end{aligned}\right\} \tag{7.1}$$

with the barotropic equation of state

$$p = F(\rho).$$

Introduce the condensation δ, defined by

$$\delta = \frac{\rho - \rho_0}{\rho_0},$$

which is the relative change in density from the constant state ρ_0. We assume that v, δ and all their derivatives are small compared with unity. Then

$$\begin{aligned} p &= F(\rho_0) + F'(\rho_0)(\rho-\rho_0) + \frac{1}{2}F''(\rho_0)(\rho-\rho_0)^2 + \mathscr{O}((\rho-\rho_0)^3)\\ &= F(\rho_0) + \rho_0 F'(\rho_0)\delta + \frac{1}{2}\rho_0^2 F''(\rho_0)\delta^2 + \mathscr{O}(\delta^3),\\ \nabla p &= \rho_0 F'(\rho_0)\nabla\delta + \mathscr{O}(\delta^2),\end{aligned}$$

and (7.1) gives

$$\left.\begin{aligned}&\delta_t + v\cdot\nabla\delta + (1+\delta)\nabla\cdot v = 0,\\&(1+\delta)v_t + (1+\delta)v\cdot\nabla v + F'(\rho_0)\nabla\delta + \mathcal{O}(\delta^2) = 0.\end{aligned}\right\}$$

Linearization yields

$$\left.\begin{aligned}&\delta_t + \nabla\cdot v = 0,\\&v_t + F'(\rho_0)\nabla\delta = 0.\end{aligned}\right\} \tag{7.2}$$

Eliminating v, we obtain

$$\delta_{tt} - c_0^2\Delta\delta = 0, \tag{7.3}$$

the *wave equation*, where

$$c_0^2 \equiv F'(\rho_0),$$

and c_0 corresponds to the propagation speed of acoustic waves. For an ideal gas, $p = F(\rho) = k\rho^\gamma$, so

$$c_0 = \left(\frac{\gamma p_0}{\rho_0}\right)^{1/2}.$$

(It is known that $\gamma = 1.4$ for air and $\gamma = \frac{5}{3}$ for a monatomic gas.) If the source of acoustic disturbance is time-harmonic,

$$\delta(x,t) = e^{-i\omega t}w(x), \quad \omega > 0, \tag{7.4}$$

then substitution of (7.4) into (7.3) gives the Helmholtz equation

$$\Delta w(x) + k^2 w(x) = 0, \quad k^2 \equiv \omega^2/c_0^2.$$

Another derivation is based upon the velocity potential. When the motion of the gas is irrotational, there exists a scalar function ϕ with the property that

$$v = -\nabla\phi.$$

Substituting this into the second equation of (7.2), we get

$$\nabla\left(-\phi_t + c_0^2\delta\right) = 0;$$

thus $-\phi_t + c_0^2\delta$ is constant throughout $\mathbb{R}^3$. We obtain

$$-\phi_{tt} + c_0^2\delta_t = 0. \tag{7.5}$$

Eliminating δ between the first equation of (7.2) and (7.5), we get

$$\phi_{tt} - c_0^2\Delta\phi = 0, \tag{7.6}$$

the wave equation in terms of the velocity potential ϕ.

When there is damping, the medium acts against the velocity of the motion of wave propagation, so we include an additional term $\gamma\phi_t$ in (7.6):

$$\phi_{tt} + \gamma\phi_t - c_0^2\Delta\phi = 0, \ \ \gamma > 0,$$

where γ is called the damping coefficient. A time-harmonic disturbance $\phi(x,t) = e^{-i\omega t}w(x)$ will give the Helmholtz equation

$$\Delta w(x) + k^2 w(x) = 0, \quad k^2 \equiv \frac{\omega^2 + i\omega\gamma}{c_0^2}. \tag{7.7}$$

Several boundary conditions commonly arise for the wave equation (7.6) (in an exterior domain Ω^c):

(i) The Dirichlet condition

$$\phi(x,t) = G(x,t), \quad x \in \partial\Omega.$$

This corresponds physically to the prescription of the pressure for the sound wave on the boundary of the scatterer (called a *sound-soft obstacle*).

(ii) The Neumann condition

$$\frac{\partial\phi(x,t)}{\partial n} = G(x,t), \quad x \in \partial\Omega.$$

This corresponds physically to prescribing the normal component of the velocity of the acoustic wave on the boundary of a *sound-hard obstacle*.

(iii) The impedance boundary condition

$$\frac{\partial\phi(x,t)}{\partial n} - (\alpha + i\beta)\frac{\partial\phi(x,t)}{\partial t} = G(x,t), \quad \alpha > 0, \ \beta \in \mathbb{R}.$$

This corresponds to a linear relation between normal velocity and pressure on the boundary. The impedance boundary condition is useful in modelling sound-absorbing surfaces such as honeycomb or perforated materials [68].

After letting $\phi(x,t) = e^{-i\omega t}w(x)$ and $G(x,t) = g(x)e^{-i\omega t}$ in the boundary conditions (i)–(iii) above, we obtain respectively the Dirichlet-, Neumann- and Robin-type BVP for the Helmholtz equation:

$$\left.\begin{aligned} & w(x) = g(x), \\ & \frac{\partial w(x)}{\partial n} = g(x), \\ & \frac{\partial w(x)}{\partial n} + i\omega(\alpha + i\beta)w(x) = g(x) \end{aligned}\right\} \tag{7.8}$$

on $\partial\Omega$.

(b) Eigenfunctions and eigenvalues of a vibrating membrane

Let $u(x,t)$ denote the vertical displacement of a thin homogeneous membrane at $x \in \Omega \subset \mathbb{R}^2$ at time t. For small displacement, the total energy at time t of the vibrating membrane is

$$E(t) = \frac{1}{2}\int_\Omega \left[\rho |u_t(x,t)|^2 + T|\nabla u(x,t)|^2\right] dx,$$

where ρ and T denote respectively the mass density and tension coefficient of the membrane. If the boundary of the membrane is fixed, we have the homogeneous Dirichlet condition

$$u(x,t) = 0, \quad x \in \partial\Omega, \ t \geqslant 0. \tag{7.9}$$

If the boundary is free (i.e., no force is acting on it), then we have the homogeneous Neumann condition

$$\frac{\partial u(x,t)}{\partial n} = 0, \quad x \in \partial\Omega, \ t \geqslant 0, \tag{7.10}$$

Subject to either (7.9) or (7.10), using the calculus of variations, we easily derive the equation of motion

$$u_{tt}(x,t) - c^2\Delta u(x,t) = 0, \quad x \in \Omega \subset \mathbb{R}^2, \ t > 0, \ c^2 \equiv \frac{T}{\rho},$$

which is a wave equation on a bounded domain Ω. The vibration of the membrane is known to have eigenmodes $u(x,t) = e^{-i\omega t}w(x)$, where w is nontrivial solution to the eigenvalue problem

$$\Delta w(x) + \lambda^2 w(x) = 0, \quad \lambda = \omega^2/c^2, \tag{7.11}$$

a Helmholtz equation, subject to

$$w = 0, \text{ or } \frac{\partial w}{\partial n} = 0,$$

depending on which of (7.9) and (7.10) holds on $\partial\Omega$. The analysis of the eigenmodes w and eigenvalue $-\lambda^2$ in (7.11) is crucial in understanding the vibration behavior of the membrane.

(c) Aeroacoustics

Tam [175] has formulated a linear vortex-sheet shock cell model for a single supersonic jet emitted from a nozzle. See Fig. 7.1.

There is no disturbance outside the jet. The linearized equations of motion inside the vortex sheet is

$$\rho_J \nabla \cdot v + U_J \frac{\partial \rho}{\partial x_1} = 0, \quad \rho_J U_J \frac{\partial}{\partial x_1} v = -\nabla p, \quad p = a_J^2 \rho, \tag{7.12}$$

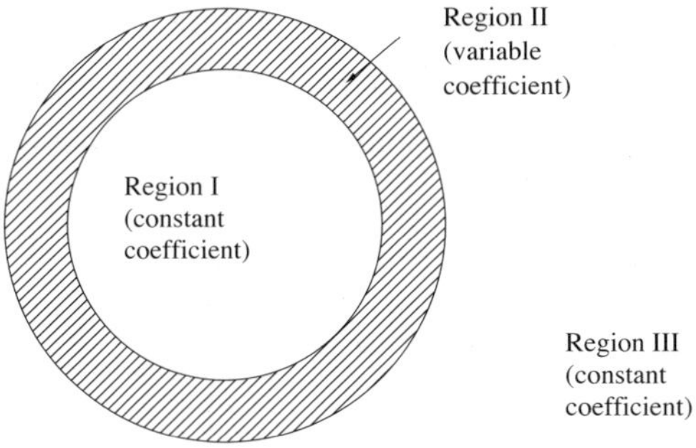

Figure 7.1: A single supersonic jet emitted from a nozzle.

where ρ_J, U_J and a_J are the density, velocity and sound speed of the fully expanded jet, ρ, p and v are the density, pressure and velocity associated with the linear shock cell structure. From (7.12) we eliminate v and ρ to obtain

$$\Delta p - M_J^2 \frac{\partial^2 p}{\partial x_1^2} = 0, \quad M_J \equiv \frac{U_J}{a_J} > 1, \quad M_J \text{ is the Mach number of the jet,}$$

with $p = 0$ on the boundary $\partial\Omega$: $S_0(x_2, x_3) = 0$.
An assumption is made that the shock cell structure is periodic in the x_1 direction so the jet pressure fluctuation is also periodic in the x_1 direction and

$$p(x_1, x_2, x_3) = w(x_2, x_3) \cos kx_1,$$

where k is an as yet unknown axial wavenumber. The equation for $w(x_2, x_3)$ may then be written as

$$\Delta_\perp w + \beta^2 w = 0,$$

a Helmholtz equation, where

$$\Delta_\perp \equiv \frac{\partial^2}{\partial x_2^2} + \frac{\partial^2}{\partial x_3^2} \text{ is the transverse part of the Laplacian,} \tag{7.13}$$
$$\beta^2 \equiv (M_J^2 - 1)k^2 \ (> 0),$$

subject to the Dirichlet boundary condition

$$w = 0 \quad \text{on } \partial\Omega.$$

Once the smallest eigenvalue $-\beta^2$ of the Laplacian Δ_J is determined, one can determine the shock spacing in the jet and the "screech tone" frequencies according to a formula given in [176].

(d) TE and TM waves in electromagnetic wave guides

The propagation of electromagnetic waves in an isotropic, permeable, conducting dielectric absent of sources obeys Maxwell's equations

$$\left.\begin{aligned} &\frac{\partial E}{\partial t} = \frac{c}{\mu\varepsilon}\nabla\times B, \\ &\frac{\partial B}{\partial t} = -c\nabla\times E, \\ &\nabla\cdot E = 0, \\ &\nabla\cdot B = 0, \end{aligned}\right\} \tag{7.14}$$

where $E = (E_1, E_2, E_3)$ and $B = (B_1, B_2, B_3)$ are the electric and magnetic fields, c is the speed of light, and ε, μ are, respectively, the dielectric and permeability constants of the medium. The wave propagates in a cylindrical wave guide with cross section Ω as shown in Fig. 7.2.

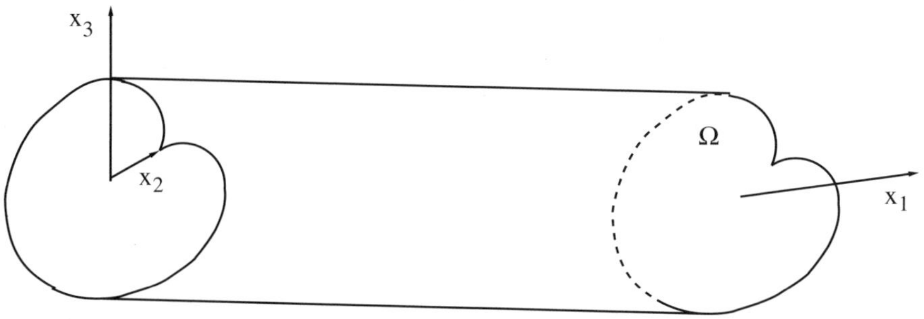

Figure 7.2: Hollow, cylindrical waveguide with cross section Ω.

Assume the time-harmonic dependence $e^{-i\omega t}$. Then Maxwell's equations (7.14) take the form

$$\left.\begin{aligned} &\nabla\times E = i\frac{\omega}{c}B, \\ &\nabla\times B = -i\mu\varepsilon\frac{\omega}{c}E, \\ &\nabla\cdot E = 0, \\ &\nabla\cdot B = 0. \end{aligned}\right\} \tag{7.15}$$

It follows immediately that both E and B satisfy the Helmholtz equation

$$\left(\Delta + \frac{\mu\varepsilon\omega^2}{c^2}\right)\begin{bmatrix} E \\ B \end{bmatrix} = 0. \tag{7.16}$$

Based upon the cylindrical geometry, we assume that the fields are periodic in the x_1 direction and write

$$\begin{bmatrix} E(x_1,x_2,x_3,t) \\ B(x_1,x_2,x_3,t) \end{bmatrix} = \begin{bmatrix} E(x_2,x_3) \\ B(x_2,x_3) \end{bmatrix} e^{i(kx_1-\omega t)}. \tag{7.17}$$

Therefore (components of the new) E and B satisfy the two-dimensional Helmholtz equation

$$\left[\Delta_\perp + \left(\mu_\varepsilon \frac{\omega^2}{c^2} - k^2\right)\right]\begin{bmatrix} E \\ B \end{bmatrix} = 0, \quad \Delta_\perp \text{ as in (7.13).} \tag{7.18}$$

We separate E and B into axial and transverse components:

$$\begin{aligned} E &= E_1 + E_\perp, \quad E_1 = (e_1 \cdot E)e_1, \quad E_\perp = (e_1 \times E) \times e_1, \\ B &= B_1 + B_\perp, \quad B_1 = (e_1 \cdot B)e_1, \quad B_\perp = (e_1 \times B) \times e_1, \end{aligned} \tag{7.19}$$

where e_1 is the unit vector in the x_1 direction. Then from (7.15) and (7.19), we can express $E_\perp$ and $B_\perp$ in terms of E_1 and B_1:

$$\begin{aligned} E_\perp &= \left(\mu\varepsilon\frac{\omega^2}{c^2} - k^2\right)^{-1}\left[\nabla_\perp\left(\frac{\partial E_1}{\partial x_1}\right) - i\frac{\omega}{c} e_1 \times \nabla_\perp B_1\right], \\ B_\perp &= \left(\mu\varepsilon\frac{\omega^2}{c^2} - k^2\right)^{-1}\left[\nabla_\perp\left(\frac{\partial B_1}{\partial x_1}\right) + i\mu\varepsilon\frac{\omega}{c} e_1 \times \nabla_\perp E_1\right], \end{aligned}$$

where

$$\nabla_\perp = \left(0, \frac{\partial}{\partial x_2}, \frac{\partial}{\partial x_3}\right).$$

Thus once E_1 and B_1 are determined, the entire fields are determined.

Assume that the boundary of the waveguide is perfectly conducting:

$$n \times E = 0, \quad n \cdot B = 0 \quad \text{for } (E,B) \text{ in (7.17).} \tag{7.20}$$

Then $n \times E = 0$ on $(-\infty,\infty) \times \partial\Omega$, the boundary of the waveguide, gives

$$E_1 = 0 \quad \text{on } \partial\Omega.$$

Thus $\nabla_\perp(E_1 \cdot e_1)$ points to the direction of n or $-n$ and

$$n \cdot [e_1 \times \nabla_\perp(E_1 \cdot e_1)] = 0. \tag{7.21}$$

Using $n \cdot B = 0$ in (7.20) and (7.19), we obtain

$$\frac{\partial}{\partial n}(B_1 \cdot e_1) = 0.$$

To simplify notation, write $E_1 \cdot e_1$ and $B_1 \cdot e_1$, respectively, as E_1 and B_1. For a given frequency ω, only certain values of the axial wavenumber k will yield nontrivial solutions of

E_1 and B_1. But because the boundary conditions on E_1 and B_1 are different, the eigenvalue problems

$$\left.\begin{aligned}&\left[\Delta_\perp + \left(\mu\varepsilon\frac{\omega^2}{c^2} - k^2\right)\right] E_1 = 0 \qquad \text{(cf. (7.18))},\\ &E_1 = 0 \quad \text{on } \partial\Omega,\end{aligned}\right\} \tag{7.22}$$

$$\left.\begin{aligned}&\left[\Delta_\perp + \left(\mu\varepsilon\frac{\omega^2}{c^2} - k^2\right)\right] B_1 = 0 \qquad \text{(cf. (7.18))},\\ &\frac{\partial B_1}{\partial n} = 0 \quad \text{on } \partial\Omega,\end{aligned}\right\} \tag{7.23}$$

cannot be satisfied simultaneously. Therefore only two possibilities result:

(i) TM (transverse magnetic) wave

$$B_1 \equiv 0 \quad \text{on } \Omega;$$

only (7.22) holds.

(ii) TE (transverse electric) wave

$$E_1 \equiv 0 \quad \text{on } \Omega;$$

only (7.23) holds.

Again, we see that the eigenvalue problems (7.22) and (7.23) are Helmholtz equations.

7.2 The fundamental solution of the Helmholtz equation

We consider the fundamental solution of the Helmholtz equation in N space dimensions:

$$(\Delta + k^2)E(x,\xi;k) = -\delta(x-\xi), \tag{7.24}$$

$$\Delta = \Delta_x = \sum_{i=1}^{N} \frac{\partial^2}{\partial x_i^2}, \quad x,\xi \in \mathbb{R}^N,$$

where (according to (7.7)) k^2 may be complex. We stipulate that the square root k of k^2 is chosen such that

$$k = \mu + i\nu, \qquad \nu = \operatorname{Im} k \geqslant 0. \tag{7.25}$$

We first set $\xi = 0 \in \mathbb{R}^N$ in (7.24) and consider the solution $E(x)$ $(\equiv E(x,0;k))$ of (7.24):

$$(\Delta + k^2)E(x) = -\delta(x). \tag{7.26}$$

Once $E(x)$ is obtained, we can make a simple change of variable $x \longmapsto x-\xi$ and obtain the solution $E(x,\xi;k)$ to (7.24) by

$$E(x,\xi;k) = E(x-\xi) = E(x-\xi,0;k).$$

It is easy to check that (7.24) is invariant under rotations, so the solution E depends only on the radial variable r, i.e., $E(x) = E(r)$, and

$$(\Delta + k^2)E(r) = \frac{1}{r^{N-1}}\frac{\partial}{\partial r}\left[r^{N-1}\frac{\partial E(r)}{\partial r}\right] + k^2 E(r) = 0, \qquad \forall\, r > 0,$$
$$\frac{\partial}{\partial r}\left(r^{N-1}\frac{\partial E}{\partial r}\right) + k^2 r^{N-1} E = 0.$$

We make a change of the (dependent) variable

$$E = r^{1-N/2} F$$

and get

$$\frac{d}{dr}\left(r\frac{dF}{dr}\right) + \left[k^2 r - \frac{(\frac{1}{2}N-1)^2}{r}\right] F = 0.$$

Making another change of the (independent) variable $\rho = kr$, we obtain

$$\frac{d}{d\rho}\left(\rho\frac{dF}{d\rho}\right) + \left[\rho - \frac{(\frac{1}{2}N-1)^2}{\rho}\right] F = 0, \tag{7.27}$$

a Bessel differential equation of order $\frac{1}{2}N-1$. Thus the solution F to (7.27) can be represented as

$$F(r) = C_1 H^{(1)}_{N/2-1}(kr) + C_2 H^{(2)}_{N/2-1}(kr), \tag{7.28}$$

where $H^{(1)}_\alpha$ and $H^{(2)}_\alpha$ are, respectively, Hankel functions of order α of the first and second kinds. They are defined through

$$\left.\begin{aligned} H^{(1)}_\alpha(z) &= J_\alpha(z) + iN_\alpha(z), \\ H^{(2)}_\alpha(z) &= J_\alpha(z) - iN_\alpha(z), \end{aligned}\right\} \tag{7.29}$$

where

$$J_\alpha(z) = \left(\frac{z}{2}\right)^\alpha \sum_{k=0}^{\infty} \frac{(-1)^k}{k!(k+\alpha)!}\left(\frac{z}{2}\right)^{2k}, \quad \alpha \neq -1,-2,-3,\ldots,$$
$$N_\alpha(z) = \frac{1}{\sin\pi\alpha}\left[(\cos\pi\alpha)J_\alpha(z) - J_{-\alpha}(z)\right] \tag{7.30}$$

are, respectively, Bessel and Neumann functions of order α. (When α is an integer, (7.30) is taken in the sense of limit.) Asymptotically,

$$H^{(1),(2)}_\alpha(z) \sim \left(\frac{2}{\pi z}\right)^{1/2} e^{\pm i(z-\alpha\pi/2-\pi/4)}, \quad \text{large } |z|, \tag{7.31}$$

with $+,-$ corresponding to $(1),(2)$, respectively. Recalling the stipulation (7.25), we let $C_2 = 0$ in (7.28) so that $F(r)$ does not grow exponentially when $r \to \infty$. (If it is known that $\operatorname{Im} k = 0$, then alternatively we can set $C_1 = 0$.) Therefore, from (7.28),

$$E(r) = C_1 r^{1-N/2} H^{(1)}_{N/2-1}(kr). \tag{7.32}$$

The next step is to determine C_1. From (7.26), integrate on an ε-ball,

$$\int_{B_\varepsilon} (\Delta E + k^2 E)\, dx = -\int_{B_\varepsilon} \delta(x)\, dx = -1.$$

By the divergence theorem,

$$\int_{\partial B_\varepsilon} \frac{\partial E}{\partial r} d\sigma = -1 - k^2 \int_{B_\varepsilon} E(r)\, dx = -1 + \begin{cases} \mathscr{O}\left(k^2 \varepsilon^N (k\varepsilon)^{\frac{2-N}{2}}\right), & N \geqslant 3, \\ \mathscr{O}\left(k^2 \varepsilon^2 \ln k\varepsilon\right), & N = 2, \end{cases}$$

$$\text{(see (7.35) below)}$$

$$\longrightarrow -1 \quad \text{as } \varepsilon \downarrow 0. \tag{7.33}$$

Since

$$d\sigma = r^{N-1}\, d\omega_N, \text{ with } d\omega_N \text{ the differential element of the N-dimensional unit-ball surface,}$$

from (7.33) we get

$$\lim_{\varepsilon \downarrow 0} \varepsilon^{N-1} |\omega_N| \frac{E}{r}\Big|_{r=\varepsilon} = -1. \tag{7.34}$$

For small values of z,

$$H^{(1)}_\alpha(z) \sim \begin{cases} \dfrac{2i}{\pi} \ln z, & \alpha = 0, \\ -\dfrac{i 2^\alpha (\alpha - 1)!}{\pi} z^{-\alpha}, & \alpha > 0, \end{cases} \tag{7.35}$$

so by (7.32) we obtain

$$C_1 = \frac{i\pi 2^{-N/2} k^{(N-2)/2}}{[(\frac{1}{2}N) - 1]!\, |\omega_N|} = \frac{i}{4} \left(\frac{k}{2\pi}\right)^{(N-2)/2} \quad \left(\text{because } |\omega_N| = \frac{2\pi^{N/2}}{\Gamma(\frac{1}{2}N)}\right).$$

Thus

$$E(x) = \frac{i}{4} \left(\frac{k}{2\pi r}\right)^{(N-2)/2} H^{(1)}_{N/2-1}(kr), \quad N \geqslant 2. \tag{7.36}$$

When $N = 3$, we have

$$E_3(x) = \frac{i}{4} \left(\frac{k}{2\pi r}\right)^{1/2} H^{(1)}_{1/2}(kr) = \frac{e^{ikr}}{4\pi r},$$

because

$$H_{1/2}^{(1)}(z) = \frac{1}{i}\left(\frac{2}{\pi}\right)^{1/2}\frac{e^{iz}}{z^{1/2}}.$$

Therefore, the fundamental solutions to (7.24) are

$$E_2(x,\xi;k) = \frac{1}{4}iH_0^{(1)}(k|x-\xi|), \quad N=2, \tag{7.37}$$

$$E_3(x,\xi;k) = \frac{e^{ik|x-\xi|}}{4\pi|x-\xi|}, \quad N=3. \tag{7.38}$$

If $\operatorname{Im} k = 0$, (for interior boundary value problems), we may also use

$$E_2(x,\xi;k) = -\frac{1}{4}iH_0^{(2)}(k|x-\xi|), \tag{7.39}$$

$$E_3(x,\xi;k) = \frac{e^{-ik|x-\xi|}}{4\pi|x-\xi|}. \tag{7.40}$$

For $k=1$, we have plotted the real and imaginary parts of the fundamental solution E_2 in (7.37) in Figs. 7.3 and 7.4. The reader can see that both graphs exhibit considerable wave behavior.

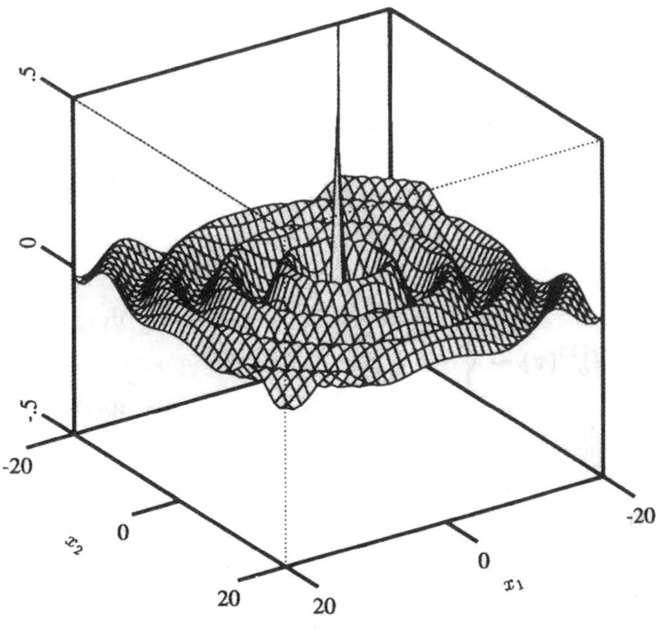

Figure 7.3: Real part of $\frac{1}{4}iH_0^{(1)}(|x|)$.

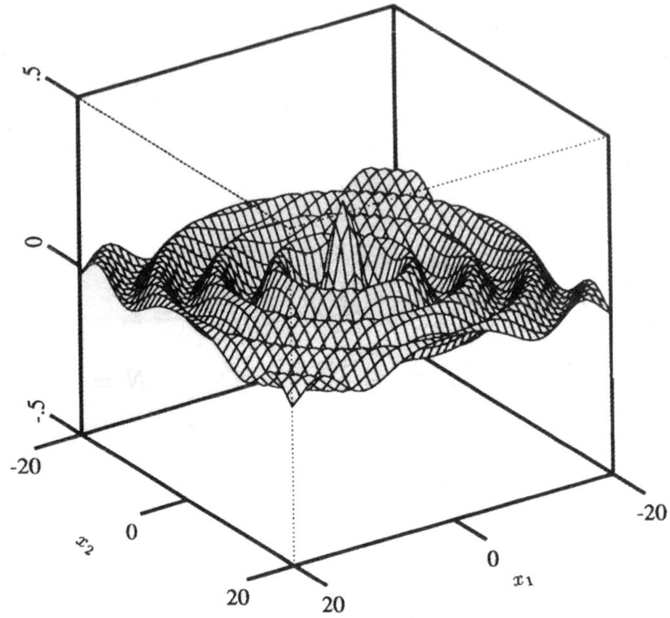

Figure 7.4: Imaginary part of $\frac{1}{4}iH_0^{(1)}(|x|))$.

7.3 Regularity of the layer potentials and jump properties

As in Chapter 6, for a given density function $f(\cdot)$ defined on $\partial\Omega$, we let

$$\left.\begin{aligned} V_1^{\pm}(f)(x) &= \int_{\partial\Omega} E(x,\xi;k) f(\xi)\,d\sigma_\xi \\ V_2^{\pm}(f)(x) &= \int_{\partial\Omega} \frac{\partial E(x,\xi;k)}{\partial n_\xi} f(\xi)\,d\sigma_\xi \end{aligned}\right\}, \quad \text{with} \quad \left\{\begin{aligned} &+ \text{ for } x \in \Omega^c, \\ \\ &- \text{ for } x \in \Omega, \end{aligned}\right. \tag{7.41}$$

be the simple- and double-layer potentials for the Helmholtz equation. When there is no ambiguity about the dependence on k, we simply write $E(x,\xi;k)$ as $E(x-\xi)$ or $E(x,\xi)$. We also write E_N for E in case the space dimension N is important in the discussion.

We observe from (7.35) that in $\mathbb{R}^3$ and $\mathbb{R}^2$, the fundamental solution of the Helmholtz equation has the same type of singularity as that of the Laplace equation, i.e.,

$$\lim_{\xi\to x} \frac{E_N^H(x,\xi;k)}{E_N^L(x,\xi)} = 1, \quad N = 2,\ 3, \tag{7.42}$$

where the superscripts H and L signify, respectively, Helmholtz and Laplace. (As a matter of fact, (7.42) holds for all $N \in \mathbb{Z}^+$, independent of k.) It is therefore not surprising at all that nearly every theorem in § 6.5 also holds for the new kernel

$$K(x,\xi) = E_N(x,\xi;k), \quad N = 2,\ 3, \ \text{ any } k \in \mathbb{C},\ \operatorname{Im} k \geqslant 0,$$

or

$$K(x,\xi) = \frac{\partial}{\partial n_\xi} E_N(x,\xi;k)$$

and for the Helmholtz simple- and double-layer potentials (7.41). We leave it to the reader to verify these properties.

The only adjustment we need to make in § 6.5 is that for the double-layer $V_2(f)$, instead of (6.51) and (6.52) we now have

$$\begin{aligned}\nabla V_2^{\pm}(f)(x) = {} & k^2 \int_{\partial\Omega} E(x-\xi) n_\xi f(\xi)\, d\sigma_\xi \\ & - \int_{\partial\Omega} \nabla_x E(x-\xi) \times \left[\overline{\nabla} f(\xi) \times n_\xi\right] d\sigma_\xi \pm \frac{1}{2}\overline{\nabla} f(\xi), \\ & x \in \partial\Omega, \end{aligned} \tag{7.43}$$

where the first term containing k^2 on the RHS is new. Its appearance is due to the fact that in going from (6.131) to (6.132), we now instead have

$$\begin{aligned}\nabla V_2^{\pm}(f)(x) &= -\nabla\nabla \cdot \int_{\partial\Omega} E(x-\xi) f(\xi) n_\xi\, d\sigma_\xi \\ &= -\int_{\partial\Omega} \Delta_x E(x-\xi) f(\xi) n_\xi\, d\sigma_\xi - \nabla \times \nabla \times \int_{\partial\Omega} E(x-\xi) f(\xi) n_\xi\, d\sigma_\xi \\ &= k^2 \int_{\partial\Omega} E(x-\xi) f(\xi) n_\xi\, d\sigma_\xi \\ &\quad - \nabla \times \nabla \times \int_{\partial\Omega} E(x-\xi) f(\xi) n_\xi\, d\sigma_\xi, \quad \text{for} \quad x \in \mathbb{R}^3/\partial\Omega.\end{aligned}$$

The rest of the derivation follows in the same way as in the proof of Theorem 6.12 in § 6.5. We can actually regard (6.130) in Theorem 6.12 as a special case of (7.43), because it just corresponds to $k = 0$.

7.4 Solution of BVP in scattering theory by layer potentials

As given in (7.7) of § 7.1, the field of the time-harmonic acoustic scattering problem in an exterior domain Ω^c is governed by the Helmholtz equation

$$(\Delta + k^2) w(x) = 0, \quad x \in \Omega^c. \tag{7.44}$$

Assume that w satisfies the homogeneous Dirichlet, Neumann or impedance boundary conditions

$$\left.\begin{aligned} & w(x) = 0, \\ & \frac{\partial w(x)}{\partial n} = 0, \\ & \frac{\partial w(x)}{\partial n} + i\omega(\alpha + i\beta) w(x) = 0, \ \ \alpha, \beta \in \mathbb{R}, \ \ \alpha > 0, \end{aligned}\right\} \tag{7.45}$$

on $\partial\Omega$, where k and ω are related through (7.7). In scattering theory, the total field w can be decomposed into two parts:

$$w(x) = w_I(x) + w_S(x), \tag{7.46}$$

where $w_I(x)$ is the *incident field* and $w_S(x)$ is the *scattered field*. Both w_I and w_S satisfy the Helmholtz equation (7.44). Also, w_I is *given*. Therefore the scattered field w_S satisfies the Helmholtz equation

$$(\Delta + k^2) w_S(x) = 0, \tag{7.47}$$

subject to one of the following boundary conditions:

$$\left.\begin{aligned}
&w_S(x) = -w_I(x) \quad \text{(Dirichlet)},\\
&\frac{\partial w_S(x)}{\partial n} = -\frac{\partial w_I(x)}{\partial n} \quad \text{(Neumann)},\\
&\frac{\partial w_S(x)}{\partial n} + i\omega(\alpha + i\beta) w_S(x) = -\left[\frac{\partial w_I(x)}{\partial n} + i\omega(\alpha + i\beta) w_I(x)\right]\\
&\hspace{20em}\text{(impedance)}.
\end{aligned}\right\} \tag{7.48}$$

It is well known [53] that the solution to such an exterior BVP is *not unique*, unless *the Sommerfeld radiation* conditions are imposed on w for large $|x|$:

(i) in $\mathbb{R}^3$

$$\left.\begin{aligned}
&w(x) = \mathcal{O}\left(|x|^{-1}\right),\\
&\left(\frac{\partial}{\partial r} - ik\right) w(x) = o(|x|^{-1});
\end{aligned}\right\} \tag{7.49}$$

(ii) in $\mathbb{R}^2$

$$\left.\begin{aligned}
&w(x) = \mathcal{O}\left(|x|^{-1/2}\right),\\
&\left(\frac{\partial}{\partial r} - ik\right) w(x) = o\left(|x|^{-1/2}\right).
\end{aligned}\right\} \tag{7.50}$$

We require that the scattered field $w_S(x)$ satisfy the radiation conditions (7.49) or (7.50) so that w_S is genuinely physically admissible. The proof of the uniqueness of exterior BVP satisfying (7.49) or (7.50) will be given in Theorems 7.6 and 7.7.

Our first attempt to solve the BVP subject to (one of) (7.48) is to try a simple- or double-layer representation (7.41), as we have learned from Chapter 6 about the Laplace equation that such an approach is generally superior to the direct approach based upon the Green's formula.

We first note that both the simple- and double-layer Helmholtz (or acoustic) potentials satisfy the Sommerfeld radiation conditions (7.49) and (7.50). To verify this, we need only show that the kernels $E_N(x-\xi)$ and $\partial E_N(x-\xi)/\partial n_\xi$ of the layer potentials (7.41) satisfy the radiation conditions for $N=3$ and 2 when $|x|$ is large. We leave this as an exercise.

Exercise 7.1. Show that for $|x|$ large and $|\xi|$ bounded, $\operatorname{Im} k \geqslant 0$ we have

(i) $N=3$

$$\begin{aligned}
E_3(x-\xi) &= \frac{1}{4\pi}\frac{e^{ik|x-\xi|}}{|x-\xi|} = \mathscr{O}\left(|x|^{-1}\right),\\
\left(\frac{\partial}{\partial r}-ik\right)E_3(x-\xi) &= \frac{x}{|x|}\frac{1}{4\pi}\nabla_x\frac{e^{ik|x-\xi|}}{|x-\xi|} - ik\frac{e^{ik|x-\xi|}}{4\pi|x-\xi|}\\
&= \mathscr{O}\left(|x|^{-2}\right),\\
\frac{\partial E_3(x-\xi)}{\partial n_\xi} &= \mathscr{O}\left(|x|^{-2}\right),\\
\left(\frac{\partial}{\partial r}-ik\right)\frac{\partial E_3(x-\xi)}{\partial n_\xi} &= \mathscr{O}\left(|x|^{-2}\right);
\end{aligned}$$

(ii) $N=2$

$$\begin{aligned}
E_2(x-\xi) &= \frac{1}{4}iH_0^{(1)}(k|x-\xi|)\\
(\text{use } (7.31)) &\sim \frac{i}{4}\left(\frac{2}{\pi k|x-\xi|}\right)^{1/2} e^{i(k|x-\xi|-\pi/4)}\\
&= \mathscr{O}\left(|x|^{-1/2}\right),\\
\left(\frac{\partial}{\partial r}-ik\right)E_2(x-\xi) &= \mathscr{O}\left(|x|^{-3/2}\right),\\
\frac{\partial E_2(x-\xi)}{\partial n_\xi} &= \mathscr{O}\left(|x|^{-1/2}\right),\\
\left(\frac{\partial}{\partial r}-ik\right)\frac{\partial E_2(x-\xi)}{\partial n_\xi} &= \mathscr{O}\left(|x|^{-3/2}\right).
\end{aligned}$$

□

Using the simple-layer potential V_1 in (7.41), e.g., to treat the Dirichlet BVP, we would need to solve the BIE

$$\mathscr{L}_1(f)(x) \equiv \int_{\partial\Omega} E(x-\xi;k)f(\xi)\,d\sigma_\xi = g(x),\quad x\in\partial\Omega, \tag{7.51}$$

$$g(x) = -w_I(x) \text{ is given on } \partial\Omega, \text{as in } (7.48).$$

where f is the simple-layer density representing the solution of (7.44). Unfortunately, the linear integral operator $\mathscr{L}_1$ in (7.51) has a nontrivial kernel, when the wavenumber k makes

$-k^2$ an eigenvalue for the interior Dirichlet problem

$$\left.\begin{array}{l}\Delta u(x) = \lambda u(x), \quad x \in \Omega, \quad \lambda = -k^2, \\ u(x) = 0 \quad \text{on } \partial\Omega,\end{array}\right\} \tag{7.52}$$

because by applying the Green's formula on Ω as in (1.23), we have

$$\begin{aligned} 0 = \frac{1}{2}u(x) &= \int_{\partial\Omega} \left[E(x-\xi;k)\frac{\partial u(\xi)}{\partial n} - \frac{\partial E(x-\xi;k)}{\partial n_\xi}u(\xi) \right] d\sigma \\ &= \int_{\partial\Omega} E(x-\xi;k)\frac{\partial u(\xi)}{\partial n}\, d\sigma_\xi, \quad x \in \partial\Omega, \end{aligned}$$

so

$$\int_{\partial\Omega} E(x-\xi;k)\frac{\partial u(\xi)}{\partial n}\, d\sigma_\xi = 0, \quad x \in \partial\Omega,$$

where the Neumann data $\partial u/\partial n \not\equiv 0$ on $\partial\Omega$ because u as a (nonzero) eigenfunction of (7.52) cannot satisfy the homogeneous Neumann condition at the same time. Similarly, if we use the double-layer potential V_2 in (7.41) to solve the exterior Dirichlet BVP, letting x tend to a boundary point from Ω^c, we have the BIE

$$\mathscr{L}_2(f)(x) = \frac{1}{2}f(x) + \int_{\partial\Omega} \frac{\partial E(x-\xi;k)}{\partial n_\xi} f(\xi)\, d\sigma_\xi = g(x), \tag{7.53}$$
$$x \in \partial\Omega, \quad g(x) = -w_I(x).$$

This operator $\mathscr{L}_2$ again may have a nontrivial cokernel if k makes $\lambda = -k^2$ an eigenvalue for the interior Neumann problem

$$\left.\begin{array}{l}\Delta v(x) = \lambda v(x), \quad x \in \Omega, \ \lambda = -k^2, \\ \dfrac{\partial v(x)}{\partial n} = 0 \quad \text{on } \partial\Omega,\end{array}\right\}$$

because by Green's formula for v, we have

$$\begin{aligned} \frac{1}{2}v(x) &= \int_{\partial\Omega} \left[E(x-\xi;k)\frac{\partial v(\xi)}{\partial n} - \frac{\partial E(x-\xi;k)}{\partial n_\xi}v(\xi) \right] d\sigma_\xi, \quad x \in \partial\Omega \\ &= -\int_{\partial\Omega} \frac{\partial E(x-\xi;k)}{\partial n_\xi} v(\xi)\, d\sigma_\xi, \end{aligned}$$

i.e.,

$$\mathscr{L}_2(v)(x) = 0 \quad \text{on } \partial\Omega,$$

for $v \not\equiv 0$ on $\partial\Omega$. From the foregoing discussion, we see that a straightforward use of the simple- and double-layer potentials for exterior BVP of the Helmholtz equation may not work in general, because we do not know a priori whether a given wavenumber k makes $\lambda = -k^2$ an eigenvalue of the interior Dirichlet or Neumann problem. This difficulty of solvability of the integral equations (7.51) and (7.53) arises from our method of approach rather than from the nature of the (exterior boundary value) problem itself. Note that even if we do know where the interior eigenvalues λ are located and avoid using (7.51) or (7.53) when k makes $-k^2 = \lambda$, we usually still encounter *numerical difficulty*, because if we attempt to discretize (7.51) or (7.53) (or others) by boundary elements then the resulting linear systems will be *ill-conditioned* in a neighborhood of the interior eigenvalues, especially it is also understood that for multidimensional eigenvalues of the Laplacian, the *asymptotic gap* between successive eigenvalues generally is zero, i.e.,

$$\liminf_{0 \leqslant j < \infty} |\lambda_{j+1} - \lambda_j| = 0, \tag{7.54}$$

see [36, 41]. In other applications such as inverse scattering problems [53], the shape of the domain is unknown; thus the interior eigenvalues are also unknown and impossible to avoid a priori [53]. Therefore it is highly desirable to develop methods leading to boundary integral equations that are uniquely solvable for all wavenumbers k. A mixed-layer method will be adopted in § 7.9. Let us first prepare some asymptoticity and uniqueness theorems of Helmholtz BVP in the next section, which are the counterparts of Lemmas 6.6 and 6.5 in Chapter 6 for the Laplace equation.

7.5 Asymptotics and uniqueness of solutions to the Helmholtz equation

Theorem 7.1. *Let $w \in H^1(\Omega^c), \partial w/\partial n \in L^1(\partial\Omega)$, be a weak solution of the Helmholtz equation*

$$\Delta w + k^2 w = 0 \quad \text{on } \Omega^c \subseteq \mathbb{R}^3, \ \operatorname{Im} k \geqslant 0,$$

satisfying

$$\left(\frac{\partial}{\partial r} - ik\right) w(x) = o(|x|^{-1}), \quad \text{large } |x|. \tag{7.55}$$

Then

(i)
$$w(x) = \mathscr{O}(|x|^{-1}); \tag{7.56}$$

(ii)
$$\int_{\partial\Omega} \left[E(x-\xi)\frac{\partial w(\xi)}{\partial n} - \frac{\partial E(x-\xi)}{\partial n_\xi} w(\xi) \right] d\sigma_\xi = \begin{cases} 0, & x \in \Omega, \\ -w(x), & x \in \Omega^c. \end{cases} \tag{7.57}$$

Remark 7.1. This theorem implies that the first condition in the Sommerfeld radiation condition (7.49) is redundant under the integrability assumptions $w \in H^1(\Omega^c)$ and $\partial w/\partial n \in L^1(\partial\Omega)$. □

Proof. By (7.55), we have

$$\begin{aligned}
0 &= \lim_{R\to\infty}\int_{|\xi|=R}\left|\frac{\partial w(\xi)}{\partial r} - ikw(\xi)\right|^2 d\sigma \\
&= \lim_{R\to\infty}\int_{|\xi|=R}\left[\left|\frac{\partial w(\xi)}{\partial n}\right|^2 + |k|^2|w(\xi)|^2 + 2\mathrm{Im}\,(kw\frac{\partial \overline{w}}{\partial r})\right] d\sigma.
\end{aligned} \tag{7.58}$$

But

$$\begin{aligned}
k\int_{|\xi|=R} w\frac{\partial \overline{w}}{\partial r}\,d\sigma - k\int_{\partial\Omega} w\frac{\partial \overline{w}}{\partial n}\,d\sigma &= k\int_{\Omega^c\cap B_R}\nabla\cdot(w\nabla\overline{w})\,d\xi \\
&= k\int_{\Omega^c\cap B_R}|\nabla w|^2\,d\xi \\
&\quad - k\cdot\overline{k}^2\int_{\Omega^c\cap B_R}|w|^2\,d\xi.
\end{aligned} \tag{7.59}$$

Take the imaginary part of the above and substitute it into (7.58), resulting in

$$\begin{aligned}
&\lim_{R\to\infty}\left[\int_{|x|=R}\left(\left|\frac{\partial w}{\partial n}\right|^2 + |k|^2|w|^2\right)d\sigma + 2(\mathrm{Im}\,k)\int_{\Omega^c\cap B_R}\left(|k|^2|w|^2 + |\nabla w|^2\right)dx\right] \\
&= -2\,\mathrm{Im}\left(k\int_{\partial\Omega} w\frac{\partial \overline{w}}{\partial n}\,d\sigma\right).
\end{aligned} \tag{7.60}$$

Since the RHS is finite and the LHS is nonnegative, we see that

$$\int_{|\xi|=R}|w|^2\,d\sigma = \mathscr{O}(1) \quad \text{as } R\to\infty. \tag{7.61}$$

We have

$$\begin{aligned}
&\int_{|\xi|=R}\left[w(\xi)\frac{\partial E(x-\xi)}{\partial n_\xi} - \frac{\partial w(\xi)}{\partial n}E(x-\xi)\right]d\sigma_\xi \\
&= \int_{|\xi|=R} w(\xi)\left[\frac{\partial E(x-\xi)}{\partial n_\xi} - ikE(x-\xi)\right]d\sigma_\xi \\
&\quad - \int_{|\xi|=R} E(x-\xi)\left[\frac{\partial w(\xi)}{\partial r} - ikw(\xi)\right]d\sigma_\xi,
\end{aligned} \tag{7.62}$$

but from (7.55), (7.61) and Exercise 7.1, by applying the Cauchy–Schwarz inequality to the RHS of (7.62), we easily see that

$$\lim_{R\to\infty}\int_{|\xi|=R}\left[w(\xi)\frac{\partial E(x-\xi)}{\partial n_\xi} - \frac{\partial w(\xi)}{\partial n}E(x-\xi)\right]d\sigma = 0. \tag{7.63}$$

Now, for $x\in\Omega^c$ or Ω, we have

$$\begin{aligned}
&\int_{\Omega^c\cap B_R}\{[(\Delta+k^2)w(\xi)]E(x-\xi) - [(\Delta_\xi+k^2)E(x-\xi)]w(\xi)\}\,d\xi \\
&= \begin{cases} 0, & x\in\Omega \\ w(x), & x\in\Omega^c, \end{cases}
\end{aligned}$$

for R large enough containing x in the interior of B_R. But the LHS above is equal to

$$\int_{|\xi|=R}\left[\frac{\partial w(\xi)}{\partial r}E(x-\xi)-\frac{\partial E(x-\xi)}{\partial r_\xi}w(\xi)\right]d\sigma_\xi$$
$$-\int_{\partial\Omega}\left[\frac{\partial w(\xi)}{\partial n}E(x-\xi)-\frac{\partial E(x-\xi)}{\partial n_\xi}w(\xi)\right]d\sigma_\xi,$$

where the first integral tends to 0 as $R\to\infty$, by (7.63). Hence (7.57) is proved. Equation (7.56) follows from

$$\begin{aligned}w(x)&=-\int_{\partial\Omega}\left[E(x-\xi)\frac{\partial w(\xi)}{\partial n}-\frac{\partial E(x-\xi)}{\partial n_\xi}w(\xi)\right]d\sigma_\xi\\&=-\frac{1}{4\pi}\int_{\partial\Omega}\left[\frac{e^{ik|x-\xi|}}{|x-\xi|}\frac{\partial w(\xi)}{\partial n}-\frac{\partial}{\partial n_\xi}\left(\frac{e^{ik|x-\xi|}}{|x-\xi|}\right)w(\xi)\right]d\sigma_\xi\\&=\mathscr{O}\left(\frac{1}{|x|}\right),\end{aligned}$$

because $\partial w/\partial n\in L^1(\partial\Omega)$, $w\in H^{1/2}(\partial\Omega)\subset L^2(\partial\Omega)\subset L^1(\partial\Omega)$, $E(x-\xi)=\mathscr{O}(|x|^{-1})$ and $\partial/\partial n_\xi E(x-\xi)=\mathscr{O}(|x|^{-2})$. □

Exercise 7.2. State and prove the analogue of Theorem 7.1 in $\mathbb{R}^2$. □

The uniqueness of solutions of the interior Dirichlet and Neumann BVP is given next.

Theorem 7.2. *Let* $\operatorname{Im} k\neq 0$. *Assume that* $w\in H^1(\Omega)$ *is a weak solution of either*

$$\begin{aligned}(\Delta+k^2)w(x)&=0\quad\text{on }\Omega,\\w(x)&=0\quad\text{on }\partial\Omega,\end{aligned}$$

or

$$\begin{aligned}(\Delta+k^2)w(x)&=0\quad\text{on }\Omega,\\\frac{\partial w(x)}{\partial n}&=0\quad\text{on }\partial\Omega.\end{aligned}$$

Then w *is identically* 0 *on* Ω.

Proof. Write $k=\alpha+i\beta$, where $\beta\neq 0$. We have

$$\begin{aligned}0&=\int_\Omega(\Delta+k^2)w(x)\cdot\overline{w}(x)\,dx\\&=\int_{\partial\Omega}\frac{\partial w}{\partial n}\overline{w}\,d\sigma+\int_\Omega\left(-|\nabla w|^2+k^2|w|^2\right)dx\\&=-\int_\Omega\left[|\nabla w|^2+(\beta^2-\alpha^2)|w|^2\right]dx+i2\alpha\beta\int_\Omega|w|^2\,dx,\end{aligned}\tag{7.64}$$

because either $w = 0$ or $\partial w/\partial n = 0$ on $\partial\Omega$. (Note that every term in the above is finite by the regularity assumptions.) If $\alpha \neq 0$ then

$$\int_{\Omega} |w|^2\, dx = 0.$$

If $\alpha = 0$ then

$$\int_{\Omega} \left(|\nabla w|^2 + \beta^2 |w|^2\right) dx = 0.$$

In either case, we have $w(x) \equiv 0$ on Ω. □

The existence theorems for interior Dirichlet and Neumann problems with inhomogeneous boundary data will be given in § 7.6.

Remark 7.2. When $\mathrm{Im}\,k = 0$, uniqueness generally does not hold, because k may make $\lambda = -k^2 < 0$ an eigenvalue of the Laplacian subject to homogeneous Dirichlet or Neumann boundary conditions. □

Let us also state the uniqueness theorem for the interior impedance BVP. For simplicity, in (7.7), let

$$c_0 = 1, \quad \gamma = 0; \tag{7.65}$$

so $\omega = k$. We assume that $k \in \mathbb{R}$. The homogeneous impedance boundary condition for an interior Helmholtz problem is now

$$\frac{\partial w(x)}{\partial n} \pm ik(\alpha + i\beta) w(x) = 0, \quad \alpha > 0, \beta \in \mathbb{R}, x \in \partial\Omega. \tag{7.66}$$

Theorem 7.3. *Let $k \in \mathbb{R}$, $k \neq 0$, $\alpha > 0$, $\beta \in \mathbb{R}$. Let $w \in H^1(\Omega)$ be a weak solution of*

$$\left.\begin{aligned} &(\Delta + k^2) w(x) = 0 && \text{on } \Omega, \\ &\frac{\partial w(x)}{\partial n} \pm ik(\alpha + i\beta) w(x) = 0 && \text{on } \partial\Omega. \end{aligned}\right\} \tag{7.67}$$

Then w is identically 0 *on* Ω.

Proof. Similarly to (7.64), we have

$$\begin{aligned} \int_{\Omega} \left(|\nabla w|^2 - k^2 |w|^2\right) dx &= \int_{\partial\Omega} \overline{w} \frac{\partial w}{\partial n}\, d\sigma \\ &= \mp ik(\alpha + i\beta) \int_{\partial\Omega} |w|^2\, d\sigma. \end{aligned} \tag{7.68}$$

Since $k\alpha \neq 0$, taking the imaginary parts on both sides of (7.68), we get

$$\int_{\partial\Omega} |w|^2\, d\sigma = 0.$$

Therefore $w \equiv 0$ on $\partial\Omega$, and so is $\partial w/\partial n = \mp ik(\alpha + i\beta) w = 0$. From integration by parts, we get, for $x \in \Omega$,

$$w(x) = \int_{\partial\Omega} \left[\frac{\partial w(\xi)}{\partial n} E(x - \xi; k) - w(\xi) \frac{\partial E(x - \xi; k)}{\partial n_{\xi}} \right] d\sigma_{\xi} = 0, \quad \forall\, x \in \Omega.$$

□

Next, consider the exterior problems. The field scattered by a bounded obstacle behaves like a spherical wave when the distance to the obstacle is large (see e.g., Figs. 7.19–7.21 below).

In geometrical optics [109], it is known that a spherical wave must be of the form

$$w_S(x) \sim \frac{e^{ikr}}{r} \sum_{m=0}^{\infty} \frac{1}{(ik)^m} \sum_{j=0}^{\infty} f_{jm}(r,\theta) r^{-j}. \tag{7.69}$$

Since w_S satisfies $(\Delta + k^2)w_S = 0$, using spherical coordinates for the Helmholtz operator $\Delta + k^2$, letting it act on the RHS of (7.69) and setting it equal to 0, we obtain the recursion formula

$$f_{jm} = \frac{1}{2j}[j(j-1)+\mathscr{B}] f_{j-1,m-1}, \quad j \neq 0, m \geqslant 1,$$

where $\mathscr{B}$ is the Beltrami operator

$$\mathscr{B} \equiv \frac{1}{\sin\theta}\left[\frac{\partial}{\partial\theta}\left(\sin\theta\frac{\partial}{\partial\theta}\right) + \frac{1}{\sin\theta}\frac{\partial^2}{\partial\phi^2}\right].$$

In the solution, k and r can only appear in the combination kr since the problem contains no characteristic dimension:

$$\begin{aligned} 0 &= (\Delta + k^2)w(x) \\ &= \frac{1}{r^2}\frac{\partial}{\partial r}\left(r^2\frac{\partial w(x)}{\partial r}\right) + k^2 w(x) \quad \text{(cf. (6.10), with } N=3) \\ &= k^2\left\{\frac{1}{(kr)^2}\frac{\partial}{\partial(kr)}\left[(kr)^2\frac{\partial}{\partial(kr)}w(x)\right] + w(x)\right\}. \end{aligned} \tag{7.70}$$

Thus in (7.69), $f_{jm} \equiv 0$ for $j \neq m$, and

$$\begin{aligned} w_S(x) \equiv w_S(r,\theta,\phi) &\sim \frac{e^{ikr}}{kr}\sum_{m=0}^{\infty}\frac{w_m(\theta,\phi)}{(ikr)^m}, \\ w_m(\theta,\phi) &\equiv f_{mm}(\theta,\phi)k. \end{aligned} \tag{7.71}$$

The following theorem establishes rigorously that the scattered field is indeed a spherical wave (see [53, p. 72], [11, 184]).

Theorem 7.4. *Let $w \in H^1(\Omega^c)$, $\partial w/\partial n \in L^1(\partial\Omega)$ be a weak solution of the Helmholtz equation $(\Delta + k^2)w = 0$ on Ω^c satisfying the radiation condition (7.49). Let R_0 be chosen sufficiently large that $\Omega \subset B_{R_0}$. Let (r,θ,ϕ) be the spherical coordinates of x. Then w admits a spherical wave expansion*

$$w(x) = \frac{e^{ikr}}{kr}\sum_{m=0}^{\infty}\frac{w_m(\theta,\phi)}{(ikr)^m}, \quad r = |x|, \tag{7.72}$$

valid for all $r \geqslant R_0$ and convergent absolutely uniformly with respect to (r,θ,ϕ). w_m obeys the recursion relation

$$w_m = \frac{1}{2m}[m(m-1)+\mathscr{B}]w_{m-1}, \quad m \geqslant 1. \tag{7.73}$$

Proof. We apply Theorem 7.1 to $\mathbb{R}^3\backslash\overline{B}_R$, for some large enough $R < R_0$. Then

$$w(x) = \int_{|\xi|=R} \left[w(\xi)\frac{\partial E(x-\xi)}{\partial n_\xi} - \frac{\partial w(\xi)}{\partial n} E(x-\xi) \right] d\sigma_\xi.$$

The above can be written in spherical coordinates as

$$w(x) = \frac{R^2}{4\pi}\int_0^{2\pi}\int_0^{\pi} \left[w(R,\theta',\phi')\left(\frac{\xi}{|\xi|}\frac{\xi-x}{\rho}\right)\frac{d}{d\rho}\left(\frac{e^{ik\rho}}{\rho}\right) \right.$$
$$\left. - \frac{\partial w(R,\theta',\phi')}{\partial n}\frac{e^{ik\rho}}{\rho} \right] \sin\theta'\, d\theta'\, d\phi', \tag{7.74}$$

where

$$\rho = |x-\xi| = (r^2 - 2rR\cos\gamma + R^2)^{1/2},$$
$$\cos\gamma = \cos\theta\cos\theta' + \sin\theta\sin\theta'\cos(\phi-\phi').$$

Since

$$\frac{e^{ik(\rho-r)}}{\rho} = \frac{\zeta\exp\{ikR[(1-2\zeta\cos\gamma+\zeta^2)^{1/2}-1]/\zeta\}}{R(1-2\zeta\cos\gamma+\zeta^2)^{1/2}}, \quad \zeta \equiv \frac{R}{r}, \tag{7.75}$$

(for the square root, take the branch with the value +1 when $\zeta = 0$) and the RHS of (7.75) is an analytic function of ζ for $\zeta < 1$, it admits a power series expansion

$$\begin{aligned}\frac{e^{ik(\rho-r)}}{k\rho} &= \sum_{m=1}^{\infty} a_m(kR,\gamma)\zeta^m \\ &= \sum_{m=1}^{\infty} a_m(kR,\gamma)\left(\frac{R}{r}\right)^m \\ &= \sum_{m=1}^{\infty} \frac{(ikR)^m a_m(kR,\gamma)}{(ikr)^m}.\end{aligned} \tag{7.76}$$

Further, this series is absolutely and uniformly convergent for $r \geqslant R_0$ and $\gamma \in [0, 2\pi]$. But according to (7.70), the LHS of (7.76) has no characteristic dimension, and k and r can only appear in the combination kr; therefore

$$a_m(kR,\gamma) = \frac{1}{(kR)^m} b_m(\gamma), \tag{7.77}$$

where $b_m(\gamma)$ is independent of k and R. Using (7.76) and (7.77) in (7.74), we have shown that w admits the expansion (7.72), where w_m is independent of k. The fact that w_m obeys the recursion relation (7.73) has already been explained in the paragraph between (7.69) and (7.71). □

We write (7.72) as

$$w(x) = \frac{e^{ikr}}{r}F_0(\theta,\phi) + \mathscr{O}\left(\frac{1}{r^2}\right), \quad F_0(\theta,\phi) \equiv \frac{1}{k}w_0(\theta,\phi), \tag{7.78}$$

where $F_0(\theta,\phi)$ is called the *far-field pattern* or *radiation pattern* of w.

Remark 7.3. It easily follows that if a scattered field satisfies the Sommerfeld radiation condition and if its radiation pattern is zero, then by the recursive relation (7.73), the scattered field is zero for $|x| \geqslant R_0$. By the Holmgren uniqueness theorem [92], the entire scattered field is zero on Ω^c. □

We can now state the following uniqueness theorem due to Rellich [158]; see [53, p. 78].

Theorem 7.5. *Let $w \in H^1(\Omega^c)$ be a solution to the Helmholtz equation satisfying the radiation condition and*

$$\operatorname{Im}\left(k \int_{\partial\Omega} w \frac{\partial \overline{w}}{\partial n}\, d\sigma\right) \geqslant 0.$$

Then $w = 0$ in Ω^c.

Proof. There are two possibilities:

(i) $\operatorname{Im} k > 0$: from (7.60), we have

$$\int_{\Omega^c \cap B_R} |w|^2\, dx \to 0 \quad \text{as} \quad R \to \infty;$$

(ii) $\operatorname{Im} k = 0$: from (7.60), we have

$$\int_{|x|=R} |w|^2\, d\sigma \to 0 \quad \text{as} \quad R \to \infty.$$

But from the expansion (7.78), we obtain

$$\int_{|x|=R} |w|^2\, d\sigma = \int_0^{2\pi} \int_0^{\pi} |F_0(\theta,\phi)|^2 \sin\theta\, d\theta\, d\phi + \mathscr{O}\left(\frac{1}{R}\right), \qquad \text{large } R.$$

By Remark 7.3, $w = 0$ on Ω^c. □

Remark 7.4. From part (ii) of the proof of Theorem 7.5 above, we see that if k is positive and

$$\int_{|x|=R} |w|^2\, d\sigma \to 0 \quad \text{as } R \to \infty$$

then $w \equiv 0$ on Ω^c. □

7.6 BIE solutions to the exterior Dirichlet and Neumann BVP

As in Chapter 6, we can find solutions to the exterior Dirichlet and Neumann BVP from candidate solutions that are simple- or double-layer potentials. Consider first the simple-layer representation

$$w(x) = \int_{\partial\Omega} E(x-\xi;k) f(\xi)\, d\sigma_\xi, \quad x \in \Omega^c, \tag{7.79}$$

for the exterior Dirichlet problem

$$\left.\begin{array}{l}(\Delta+k^2)w(x)=0,\ x\in\Omega^c,\\ w(x)\mid_{\partial\Omega}=g(x),\quad x\in\partial\Omega,\ g\in H^{r+1}(\partial\Omega),\ r\in\mathbb{R}.\end{array}\right\}\tag{7.80}$$

If there exists $f\in H^r(\partial\Omega)$ such that

$$(\mathscr{L}_1 f)(x)\equiv\int_{\partial\Omega}E(x-\xi;k)f(\xi)\,d\sigma_\xi=g(x),\quad x\in\partial\Omega,\tag{7.81}$$

then by (an analogous) Theorem 6.16, the regularity of w is in $H^{r+3/2}_{\text{loc}}(\Omega^c)$, w satisfies the radiation condition (7.49), and w is unique by Theorem 7.5. Therefore the simple-layer representation (7.81) provides the unique solution to the exterior Dirichlet BVP (7.80).

We first state some properties of $\mathscr{L}_1$.

Proposition 7.1. *Let* $\operatorname{Im}k>0$. *Then*

$$\mathscr{L}_1:\ H^r(\partial\Omega)\to H^{r+1}(\partial\Omega),\quad r\in\mathbb{R},\tag{7.82}$$

is an isomorphism. Consequently, $\mathscr{L}_1^*$, *the adjoint of* $\mathscr{L}_1$, *mapping* $H^r(\partial\Omega)\to H^{r+1}(\partial\Omega)$, *is also an isomorphism.*

Proof. The regularity property (7.82) is a consequence of Theorem 6.3. We establish the proof as follows. If $\tilde f\in C^\infty(\partial\Omega)$ makes

$$\mathscr{L}_1\tilde f\equiv 0\quad\text{on }\partial\Omega,$$

then

$$\tilde u(x)\equiv\int_{\partial\Omega}E(x-\xi;k)\tilde f(\xi)\,d\sigma_\xi,\quad x\in\Omega,\tag{7.83}$$

would be a solution to

$$\begin{array}{ll}(\Delta+k^2)\tilde u^-(x)=0,\ x\in\Omega,\\ \tilde u^-(x)=0 & \text{on }\partial\Omega,\end{array}$$

i.e., $\tilde u^-(x)$ is an eigenfunction of the Laplacian subject to the Dirichlet boundary condition with a positive or (nonreal) complex eigenvalue $-k^2$. This is possible only if $\tilde u^-(x)\equiv 0$ on Ω. Define $\tilde u^+(x)$ on Ω^c by (7.83) for $x\in\Omega^c$. We see that $\tilde u^+$ satisfies

$$\begin{array}{ll}(\Delta+k^2)\tilde u^+(x)=0,\ x\in\Omega^c,\\ \tilde u^+(x)=0 & \text{on }\partial\Omega.\end{array}$$

By the uniqueness Theorem 7.5, $\tilde u^+(x)\equiv 0$ on Ω^c. Hence

$$\tilde f(x)=\frac{\partial\tilde u^-(x)}{\partial n}-\frac{\partial\tilde u^+(x)}{\partial n}\equiv 0\quad\text{on }\partial\Omega.\tag{7.84}$$

To show that $\mathscr{L}_1\colon\ H^r(\partial\Omega) \to H^{r+1}(\partial\Omega)$ is onto, we prove that $\mathscr{L}_1^*$ defined by

$$(\mathscr{L}_1^*\phi)(x) = \int_{\partial\Omega} \overline{E}(x-\xi;k)\phi(\xi)\,d\sigma_\xi, \quad x \in \partial\Omega,$$
$$\mathscr{L}_1^*\colon\ H^r(\partial\Omega) \to H^{r+1}(\partial\Omega)$$

is also 1–1. Note that

$$\overline{E}(x-\xi;k) = \frac{e^{-i\overline{k}|x-\xi|}}{4\pi|x-\xi|} = E(x-\xi;-\overline{k}),$$

with $\mathrm{Im}(-\overline{k}) = \mathrm{Im}\,k > 0$. We can repeat the proof from (7.83) through (7.84) for $\mathscr{L}_1^*$, by just replacing k everywhere with $-\overline{k}$. Thus $\mathscr{L}_1^*$ is also 1–1. Hence $\mathscr{L}_1$ in (7.81) is 1–1 and onto, provided that $\mathrm{Im}\,k > 0$. □

Therefore in solving the BIE (7.81), we need only be concerned with the case $\mathrm{Im}\,k = 0$, i.e., $k > 0$. (As it later turns out, the simple-layer representation (7.79) is deficient in solving the exterior Dirichlet problem; see Remark 7.6. A double-layer representation is necessary; see Theorem 7.6.)

Next, consider using the simple-layer potential (7.79) to solve the exterior Neumann problem

$$\left.\begin{array}{l}(\Delta+k^2)w(x) = 0,\ x \in \Omega^c \\ \dfrac{\partial w(x)}{\partial n} = g(x), \quad x \in \partial\Omega,\ g \in H^r(\partial\Omega).\end{array}\right\} \tag{7.85}$$

If we can find $f \in H^r(\partial\Omega)$ such that

$$(\mathscr{L}_2 f)(x) \equiv -\frac{1}{2}f(x) + \int_{\partial\Omega} \frac{\partial E(x-\xi;k)}{\partial n_x} f(\xi)\,d\sigma_\xi = g(x),\ x \in \partial\Omega, \tag{7.86}$$

then w in (7.79) will be a solution to the exterior Neumann problem (7.85). The operator $\mathscr{L}_2$ has an adjoint

$$\mathscr{L}_2^*\colon\ H^r(\partial\Omega) \to H^r(\partial\Omega) \quad \forall\, r \in \mathbb{R},$$
$$(\mathscr{L}_2^*\phi)(x) = -\frac{1}{2}\phi(x) + \int_{\partial\Omega} \frac{\partial \overline{E}(x-\xi;k)}{\partial n_\xi}\phi(\xi)\,d\sigma_\xi.$$

We have the relations

$$\mathscr{N}(\mathscr{L}_2) = \mathrm{Coker}\,\mathscr{L}_2^*, \quad \mathscr{N}(\mathscr{L}_2^*) = \mathrm{Coker}\,\mathscr{L}_2.$$

For $\mathscr{L}_2$, we state the following.

Proposition 7.2. *If* $\mathrm{Im}\,k > 0$, *then* $\mathscr{L}_2\colon\ H^r(\partial\Omega) \to H^r(\partial\Omega)$ *is an isomorphism.*

Proof. Let $\operatorname{Im} k > 0$. If $\widetilde{f} \in \mathscr{N}(\mathscr{L}_2), \widetilde{f} \not\equiv 0$, then $\widetilde{f} \in C^\infty(\partial\Omega)$ and

$$\widetilde{w}^+(x) = \int_{\partial\Omega} E(x-\xi;k)\widetilde{f}(\xi)\,d\sigma_\xi, \quad x \in \Omega^c, \tag{7.87}$$

is a solution to (7.85) with $g(x) \equiv 0$. By Theorem 7.5, $\widetilde{w}^+(x) \equiv 0$ on Ω^c. By the continuity of the simple-layer potential across $\partial\Omega$, we have

$$\widetilde{w}^-(x) = \widetilde{w}^+(x) = 0 \quad \text{on } \partial\Omega,$$

and $\widetilde{w}^-$ satisfies

$$(\Delta + k^2)\widetilde{w}^-(x) = 0, \quad x \in \Omega,$$

where $\widetilde{w}^-$ is defined as in (7.87) but with $x \in \Omega$. This is possible only if $\widetilde{w}^-(x) \equiv 0$ on Ω, because $\operatorname{Im} k > 0$, by Theorem 7.2. Hence

$$\widetilde{f}(x) = \frac{\partial \widetilde{w}^-(x)}{\partial n} - \frac{\partial \widetilde{w}^+(x)}{\partial n} = 0 \quad \text{on } \partial\Omega, \text{ by (i) of Corollary 6.1.}$$

Next, we show that $\mathscr{N}(\mathscr{L}_2^*) = \{0\}$. If $\widetilde{\phi} \in \mathscr{N}(\mathscr{L}_2^*)$, $\widetilde{\phi} \not\equiv 0$, then $\widetilde{\phi} \in C^\infty(\partial\Omega)$. Set

$$\begin{aligned} \widetilde{u}^-(x) &\equiv \int_{\partial\Omega} \frac{\partial \overline{E}(x-\xi;k)}{\partial n_\xi}\widetilde{\phi}(\xi)\,d\sigma_\xi \\ &= \int_{\partial\Omega} \frac{\partial E(x-\xi;-\overline{k})}{\partial n_\xi}\widetilde{\phi}(\xi)\,d\sigma_\xi, \quad x \in \Omega. \end{aligned} \tag{7.88}$$

Letting x tend to a boundary point, we see that $\widetilde{u}^-$ satisfies

$$\left.\begin{aligned} &\left[\Delta + (-\overline{k})^2\right]\widetilde{u}^-(x) = 0 \quad \text{on } \Omega, \\ &\widetilde{u}^-(x)\,|_{\partial\Omega} = -\frac{1}{2}\widetilde{\phi}(x) + \int_{\partial\Omega} \frac{\partial E(x-\xi;-\overline{k})}{\partial n_\xi}\widetilde{\phi}(\xi)\,d\sigma_\xi = 0. \end{aligned}\right\}$$

This is possible only if $\widetilde{u}^-(x) \equiv 0$ on Ω, because $\operatorname{Im}(-\overline{k}) > 0$ and by Theorem 7.2. Define $\widetilde{u}^+(x)$ on Ω^c by extending x to Ω^c in (7.88). From the continuity

$$\frac{\partial \widetilde{u}^+(x)}{\partial n} = \frac{\partial \widetilde{u}^-(x)}{\partial n} = 0 \quad \text{on } \partial\Omega,$$

we see that $\widetilde{u}^+$ is a solution to

$$\left[\Delta + (-\overline{k})^2\right]\widetilde{u}^+(x) = 0 \text{ on } \Omega^c,$$

$$\frac{\partial \widetilde{u}^+(x)}{\partial n} = 0 \qquad \text{on } \partial\Omega.$$

By the uniqueness Theorem 7.5, we have $\widetilde{u}^+(x) \equiv 0$ on Ω^c. Hence by (ii) of Corollary 6.1,

$$\widetilde{\phi}(x) = \widetilde{u}^+(x) - \widetilde{u}^-(x) = 0, \quad x \in \partial\Omega.$$

The proof is complete. □

As a consequence of Proposition 7.2, we need only consider the solvability of the BIE (7.86) for $\mathrm{Im}\, k = 0$, i.e., $k > 0$.

The following is a direct consequence of Propositions 7.1 and 7.2.

Corollary 7.1. *Let* $Im\, k > 0$. *Then the exterior Dirichlet (respectively, Neumann) BVP (7.80) (respectively, (7.85)) has the unique solution (7.79) satisfying the radiation condition (7.49). It is represented by a simple-layer potential for a unique* $f \in H^r(\partial\Omega)$ *solving the BIE (7.81) (respectively, (7.86)) for given Dirichlet (respectively, Neumann) data* $g \in H^{1+r}(\partial\Omega)$ *(respectively,* $H^r(\partial\Omega)$*) for any* $r \in \mathbb{R}$. □

Throughout the rest of the section, we assume that $k > 0$. We introduce the double-layer potential

$$w^{\pm}(x) = \int_{\partial\Omega} \frac{\partial E(x-\xi;k)}{\partial n_\xi} f(\xi)\, d\sigma_\xi, \text{ with } \begin{cases} + \text{ for } x \in \Omega^c, \\ - \text{ for } x \in \Omega; \end{cases} \tag{7.89}$$

$w^+(x)$ solves the exterior Dirichlet problem (7.80) if f solves the BIE

$$(\mathscr{L}_3 f)(x) \equiv \frac{1}{2} f(x) + \int_{\partial\Omega} \frac{\partial E(x-\xi;k)}{\partial n_\xi} f(\xi)\, d\sigma_\xi = g(x), \quad x \in \partial\Omega.$$

$$(\mathscr{L}_3 \colon\ H^r(\partial\Omega) \to H^r(\partial\Omega), \quad r \in \mathbb{R}.)$$

With respect to the sesquilinear form $(f,g)_{L^2(\partial\Omega)} = \int_{\partial\Omega} f g\, d\sigma$, $\mathscr{L}_2$ and $\mathscr{L}_3$ have adjoints $\mathscr{L}'_2$ and $\mathscr{L}'_3$ given by

$$\mathscr{L}'_2, \mathscr{L}'_3 \colon\ H^r(\partial\Omega) \to H^r(\partial\Omega), \quad r \in \mathbb{R},$$

$$(\mathscr{L}'_2 \phi)(x) = -\frac{1}{2}\phi(x) + \int_{\partial\Omega} \frac{\partial E(x-\xi;k)}{\partial n_\xi} \phi(\xi)\, d\sigma_\xi,$$

$$(\mathscr{L}'_3 \phi)(x) = \frac{1}{2}\phi(x) + \int_{\partial\Omega} \frac{\partial E(x-\xi;k)}{\partial n_x} \phi(\xi)\, d\sigma_\xi.$$

Define

$$K_1 = \mathscr{N}(\mathscr{L}_1) \ (= \mathscr{N}(\mathscr{L}_1^*)),$$

$$K_j = \mathscr{N}(\mathscr{L}_j), \ j = 2, 3,$$

$$K'_j = \mathscr{N}(\mathscr{L}'_j), \ j = 2, 3.$$

Exercise 7.3. Show that

$$K_1 = K_2 = \left\{ \left. \frac{\partial u^-}{\partial n} \right|_{\partial\Omega} \middle| \ (\Delta + k^2) u^-(x) = 0 \text{ on } \Omega, u^- = 0 \text{ on } \partial\Omega \right\}, \tag{7.90}$$

$$K_3 = \left\{ u^- \big|_{\partial\Omega} \middle| \ (\Delta + k^2) u^-(x) = 0 \ \text{ on } \Omega, \frac{\partial u^-}{\partial n} = 0 \text{ on } \partial\Omega \right\} \tag{7.91}$$

(Hint: First show that K_j, $1 \leqslant j \leqslant 3$, are finite-dimensional. Then argue as in the proof of Proposition 7.2; cf. also [53, Theorems 3.17 and 3.22, pp. 81–84].) □

Proposition 7.3.

$$\dim K_2 = \dim K_2', \quad \dim K_3 = \dim K_3'. \tag{7.92}$$

Proof. We write

$$\mathscr{L}_2 = -\frac{1}{2}(I - 2\boldsymbol{L}_2),$$
$$\mathscr{L}'_2 = -\frac{1}{2}(I - 2\boldsymbol{L}_2'),$$

where

$$(\boldsymbol{L}_2 f)(x) \equiv \int_{\partial\Omega} \frac{\partial E(x-\xi;k)}{\partial n_x} f(\xi)\,d\sigma_\xi, \quad x \in \partial\Omega, \tag{7.93}$$

$$(\boldsymbol{L}_2'\phi)(x) \equiv \int_{\partial\Omega} \frac{\partial E(x-\xi;k)}{\partial n_\xi} \phi(\xi)\,d\sigma_\xi, \quad x \in \partial\Omega. \tag{7.94}$$

By (6.137), $\boldsymbol{L}_2$ and $\boldsymbol{L}_2'$ have the properties

$$\boldsymbol{L}_2, \boldsymbol{L}_2' : \; H^r(\partial\Omega) \to H^{r+1}(\partial\Omega), \quad \forall\, r \in \mathbb{R}.$$

Thus $\boldsymbol{L}_2$ and $\boldsymbol{L}_2'$ are compact operators on $H^r(\partial\Omega)$. Hence $-2\mathscr{L}_2$ and $-2\mathscr{L}'_2$ are equal to the identity operator plus a compact operator. Hence they are Fredholm with index zero, and the Fredholm alternative applies. So

$$\dim\, \mathscr{N}(\mathscr{L}_2) = \dim\, \mathscr{N}(\mathscr{L}'_2)$$

Therefore (7.92) is proved. □

We are now in a position to prove the existence of a unique solution to the exterior Dirichlet problem (for $k > 0$) according to Colton and Kress [53]. We first give the following.

Lemma 7.1. *Let $k > 0$. Let $\phi_1, \ldots, \phi_d$ be a basis of K_3', where $d = \dim K_3' > 0$. Define*

$$u_j^{\pm}(x) = \int_{\partial\Omega} E(x-\xi;k)\phi_j(\xi)\,d\sigma_\xi, \quad 1 \leqslant j \leqslant d, \text{ with } \begin{cases} + \text{ for } x \in \Omega^c, \\ - \text{ for } x \in \Omega. \end{cases} \tag{7.95}$$

Then

(i) $\phi_j = -\dfrac{\partial u_j^+}{\partial n}$ *on* $\partial\Omega, 1 \leqslant j \leqslant d$;

(ii) $\psi_j \equiv -\overline{u}_j^+$ *on* $\partial\Omega$, $1 \leqslant j \leqslant d$,

form a basis for K_3;

(iii) *The $d \times d$ matrix with (j,l)-entries given by*

$$\langle \psi_j, \phi_l \rangle = \int_{\partial\Omega} \overline{u}_j^+(x) \frac{\partial u_l^+(x)}{\partial n}\,d\sigma, \quad 1 \leqslant j,\; l \leqslant d$$

is nonsingular.

Proof. We first note that $\phi_j \in C^\infty(\partial\Omega)$ for $j = 1, 2, \ldots, d$.

(i) From the jump relation (i) in Corollary 6.1,

$$\phi_j = \frac{\partial u_j^-}{\partial n} - \frac{\partial u_j^+}{\partial n}.$$

But

$$\frac{\partial u_j^-}{\partial n} = \mathscr{L}_3'\phi_j = 0, \quad \text{because } \phi_j \in K_3'.$$

So (i) is proved.

(ii) This is a consequence of Exercise 7.3 and Proposition 7.3, because $u_j^+ = u_j^-$ on $\partial\Omega$. (Each u_j^+ can be regarded as real, so the use of the complex conjugate $\overline{u_j^+}$ does not matter.)

(iii) For any given $\alpha_j \in \mathbb{C}$, $1 \leqslant j \leqslant d$, assume

$$\sum_{j=1}^{d} \alpha_j \langle \psi_j, \phi_l \rangle = 0, \quad 1 \leqslant l \leqslant d.$$

Define

$$w^+(x) = \sum_{j=1}^{d} \overline{\alpha}_j u_j^+(x), \quad x \in \Omega^c.$$

Then

$$\int_{\partial\Omega} \overline{w}^+ \frac{\partial w^+}{\partial n} \, d\sigma = \sum_{j,l=1}^{d} \alpha_j \overline{\alpha}_l \int_{\partial\Omega} \overline{u}_j^+(x) \frac{\partial u_l^+}{\partial n} \, d\sigma$$
$$= \sum_{l=1}^{d} \overline{\alpha}_l \left(\sum_{j=1}^{d} \alpha_j \langle \psi_j, \phi_l \rangle \right) = 0.$$

Since w^+ is a solution of the Helmholtz equation satisfying the radiation condition, we conclude that $w^+ \equiv 0$ by Theorem 7.5. Consequently,

$$\frac{\partial w^+}{\partial n} = 0 \quad \text{on } \partial\Omega,$$

i.e.,

$$\sum_{j=1}^{\alpha} \overline{\alpha}_j \frac{\partial u_j^+}{\partial n} = -\sum_{j=1}^{\alpha} \overline{\alpha}_j \phi_j = 0.$$

Hence $\alpha_j = 0$, $1 \leqslant j \leqslant d$, because $\{\phi_j\}$ is a basis. Hence the matrix with entries $\langle \psi_j, \phi_l \rangle$ is nonsingular.

□

Remark 7.5. We can choose ψ_j to be real, because by (ii) in Lemma 7.1 and (7.91), the eigenfunctions for the interior Neumann problem can be chosen real. □

Theorem 7.6. *Let $k > 0$. Then the exterior Dirichlet BVP*

$$\left.\begin{array}{l} (\Delta + k^2)w(x) = 0, \quad x \in \Omega^c \\ w(x) = g(x) \in H^r(\partial\Omega), \quad r \in \mathbb{R}, \\ w \text{ satisfies the radiation condition (7.49)}, \end{array}\right\} \tag{7.96}$$

has a unique solution $w \in H^{r+1/2}_{\text{loc}}(\Omega^c)$.

Proof. If k makes $K_3 = \{0\}$, i.e., $-k^2$ is not an eigenvalue for the interior Neumann problem (cf. (7.91)), then the BIE

$$\mathscr{L}_3 f = g, \quad f \in H^r(\partial\Omega),$$

is uniquely solvable, yielding a solution

$$w(x) = \int_{\partial\Omega} \frac{\partial E(x-\xi;k)}{\partial n_\xi} f(\xi)\, d\sigma_\xi, \quad x \in \Omega^c,$$

to (7.96).

If $-k^2$ is an eigenvalue of the interior Neumann problem then $\dim K_3 \equiv d > 0$. We consider a candidate solution of the form

$$w(x) = \int_{\partial\Omega} \frac{\partial E(x-\xi;k)}{\partial n_\xi} f(\xi)\, d\sigma_\xi + \sum_{j=1}^{d} \alpha_j u_j^+(x), \; x \in \Omega^c, \tag{7.97}$$

cf. (7.95) for u_j^+, $1 \leqslant j \leqslant d$. Letting x tend to a point on $\partial\Omega$, we have

$$w(x) = (\mathscr{L}_3 f)(x) - \sum_{j=1}^{d} \alpha_j \psi_j(x), \quad x \in \partial\Omega.$$

If we can find $f \in H^r(\partial\Omega)$ and $\alpha_j \in \mathbb{C}$, $1 \leqslant j \leqslant d$, such that

$$(\mathscr{L}_3 f)(x) = g(x) + \sum_{j=1}^{d} \alpha_j \psi_j(x), \tag{7.98}$$

then w will be the unique solution to (7.96). But (7.98) is solvable (i.e., nonuniquely) if and only if

$$\left\langle g + \sum_{j=1}^{d} \alpha_j \psi_j, \phi_l \right\rangle = 0, \quad \forall \phi_l \in K_3', 1 \leqslant l \leqslant d,$$

by Lemma 7.1 and the Fredholm alternative. We choose α_j, $1 \leqslant j \leqslant d$, such that

$$\sum_{j=1}^{d} \alpha_j \langle \psi_j, \phi_l \rangle = -\langle g, \phi_l \rangle, \quad l = 1, 2, \ldots, d.$$

This linear system is uniquely solvable, by (iii) of Lemma 7.1. Therefore, there exists f satisfying the BIE (7.98), $f \in H^r(\partial\Omega)$. (Note that ψ_j, $1 \leqslant j \leqslant d$, are C^∞ on $\partial\Omega$.) Therefore w defined by (7.97) is the unique solution to the exterior Dirichlet BVP (7.96). □

Remark 7.6. If we wish to solve the exterior Dirichlet BVP (7.96) by BIE and the simple-layer representation (7.79), it would require seeking the solution f satisfying

$$\mathscr{L}_1 f = g, \tag{7.99}$$

which is not solvable when g is not orthogonal to the cokernel (= kernel) of $\mathscr{L}_1$. No modification of (7.79) would work, either. □

To solve the exterior Neumann problem, we first state the following.

Lemma 7.2. *Let $k > 0$. Let $\eta_1, \ldots, \eta_{d'}$ be a basis of K_2', where $d' \equiv \dim K_2' > 0$. Define*

$$v_j^{\pm}(x) = \int_{\partial\Omega} \frac{\partial E(x-\xi;k)}{\partial n_\xi} \eta_j(\xi) d\sigma_\xi, \text{ with } \begin{cases} + \text{ for } x \in \Omega^c, \\ - \text{ for } x \in \Omega, \end{cases}$$

for $1 \leqslant j \leqslant d'$. Then

(i) $\eta_j = v_j^+$ on $\partial\Omega$, $1 \leqslant j \leqslant d'$;

(ii) $\chi_j^+ \equiv \dfrac{\partial \overline{v}_j^+}{\partial n}$ on $\partial\Omega, 1 \leqslant j \leqslant d'$.
form a basis for K_2;

(iii) *The $d' \times d'$ matrix with (j,l)-entries defined by*

$$\langle \chi_j, \eta_l \rangle = \int_{\partial\Omega} v_l^+ \frac{\partial \overline{v}_j^+}{\partial n} d\sigma, \quad 1 \leqslant j,\ l \leqslant d',$$

is nonsingular.

Proof. This is the same as for Lemma 7.1. □

Theorem 7.7. *Let $k > 0$. Then the exterior Neumann BVP*

$$\left.\begin{array}{l} (\Delta + k^2) w(x) = 0, \quad x \in \Omega^c \\ \dfrac{\partial w(x)}{\partial n} = g(x) \in H^r(\partial\Omega), \quad r \in \mathbb{R}, \\ w \text{ satisfies the radiation condition (7.49)} \end{array}\right\} \tag{7.100}$$

has a unique solution $w \in H^{r+3/2}_{\text{loc}}(\Omega^c)$.

Proof. We consider a candidate solution of the form

$$w(x) = \int_{\partial\Omega} E(x-\xi;k) f(\xi) d\sigma_\xi + \sum_{j=1}^{d'} \alpha_j v_j^+(x), \tag{7.101}$$

$$x \in \Omega^c,\ \alpha_j \in \mathbb{C},\ j = 1, \ldots, d'.$$

Upon taking the normal derivative on $\partial\Omega$, the above leads to the BIE

$$\frac{\partial w}{\partial n} = \mathscr{L}_2 f + \sum_{j=1}^{d'} \alpha_j \chi_j = g. \tag{7.102}$$

Note that we have similarly chosen χ_j to be real according to Remark 7.5. By choosing α_j, $1 \leqslant j \leqslant d'$, to satisfy

$$\sum_{j=1}^{d'} \alpha_j \langle \chi_j, \eta_l \rangle = \langle g, \eta_l \rangle, \quad 1 \leqslant l \leqslant d', \tag{7.103}$$

(see Lemma 7.2(i) for η_l), and by arguing similarly as in the proof of Theorem 7.6, we can find $f \in H^r(\partial\Omega)$ solving the BIE (7.102). So we have solved the exterior Neumann BVP (7.100). □

Remark 7.7. (Solving the exterior Dirichlet and Neumann BVP by boundary integral equations based on Green's formula). Let $w(x)$ be the solution of either the exterior Dirichlet problem (7.96) or the exterior Neumann problem (7.100). Then by an application of the Green's formula and Theorem 7.1, letting x tend to a boundary point, we have

$$\frac{1}{2} w(x) + \int_{\partial\Omega} \left[\frac{\partial w(\xi)}{\partial n} E(x-\xi;k) - \frac{\partial E(x-\xi;k)}{\partial n_\xi} w(\xi) \right] d\sigma_\xi = 0,$$
$$x \in \partial\Omega. \tag{7.104}$$

Therefore for the Dirichlet problem in which

$$w = g_1 \quad \text{on } \partial\Omega$$

is given, we would be solving the Neumann data $\partial w / \partial n$ by the BIE

$$\mathscr{L}_1 \left(\frac{\partial w}{\partial n} \right) = \mathscr{L}'_2(g_1),$$

which is uniquely solvable if $\operatorname{Im} k > 0$ (by Prop. 7.1), but may fail to have a unique solution if $\operatorname{Im} k = 0$, for reasons explained before.

For the Neumann problem in which

$$\frac{\partial w}{\partial n} = g_2 \quad \text{on } \partial\Omega$$

is given, we would be solving the Dirichlet data w by the BIE

$$\mathscr{L}'_2(w) = \mathscr{L}_1(g_2),$$

which is uniquely solvable if $\operatorname{Im} k > 0$ (by Proposition 7.2 and the Fredholm alternative), but may fail to have a unique solution if $\operatorname{Im} k = 0$, because $\mathscr{N}(\mathscr{L}'_2)$ may become nontrivial. □

7.7 Existence and uniqueness of solutions to the exterior impedance BVP ($\mathbf{Im}\,k = \mathbf{0}$)

We consider the impedance boundary condition, the third equation in (7.8). As in (7.65), we assume that $c_0 = 1$ and $\gamma = 0$, so $\omega = k$, and the impedance boundary condition becomes

$$\frac{\partial w(x)}{\partial n} + ik(\alpha + i\beta)w(x) = g(x),\ x \in \partial\Omega,\ \alpha, \beta \in \mathbb{R},\ \alpha > 0, \tag{7.105}$$

where w satisfies the Helmholtz equation and the Sommerfeld radiation condition on the exterior domain Ω^c. We further assume that $k \in \mathbb{R}$, $k \neq 0$.

Remark 7.8. First, one might consider a solution represented by a simple-layer potential

$$w(x) = \int_{\partial\Omega} E(x-\xi;k)f(\xi)\,d\sigma_\xi, \quad x \in \Omega^c.$$

The impedance boundary condition (7.105) is satisfied provided that f satisfies the BIE

$$\begin{aligned} -\frac{1}{2}f(x) + \int_{\partial\Omega} \frac{\partial E(x-\xi;k)}{\partial n_x} f(\xi)\,d\sigma_\xi \\ + ik(\alpha+i\beta)\int_{\partial\Omega} E(x-\xi;k)f(\xi)\,d\sigma_\xi = g(x),\ x \in \partial\Omega. \end{aligned} \tag{7.106}$$

Define

$$\begin{aligned} \mathscr{L}_4\colon\ H^\alpha(\partial\Omega) &\longrightarrow H^\alpha(\partial\Omega), \qquad \alpha \in \mathbb{R}, \\ \mathscr{L}_4(f)(x) &= -\frac{1}{2}f(x) + \int_{\partial\Omega} \frac{\partial E(x-\xi;k)}{\partial n_x} f(\xi)\,d\sigma_\xi \\ &\quad + ik(\alpha+i\beta)\int_{\partial\Omega} E(x-\xi;k)f(\xi)\,d\sigma_\xi. \end{aligned} \tag{7.107}$$

In order for (7.106) to be solvable, we must have

$$\mathscr{N}(\mathscr{L}_4) = \{0\}, \quad \operatorname{Coker}\mathscr{L}_4 = \{0\},$$

which holds provided that $-k^2$ is not an eigenvalue of Δ on Ω with the zero Dirichlet BC, but otherwise may fail in general. □

We now give the following.

Theorem 7.8. *Let $k,\ \alpha,\ \beta \in \mathbb{R}$, $k \neq 0$ and $\alpha > 0$. Then the exterior impedance boundary value problem*

$$\left.\begin{aligned} &(\Delta + k^2)w(x) = 0, \quad x \in \Omega^c \\ &\frac{\partial w(x)}{\partial n} + ik(\alpha+i\beta)w(x) = g(x) \in H^r(\partial\Omega), \quad r \in \mathbb{R}, \end{aligned}\right\} \tag{7.108}$$

has a unique solution represented by a mixed-layer potential

$$w(x) = \int_{\partial\Omega} \left[\frac{\partial E(x-\xi;k)}{\partial n_\xi} + ik(\alpha+i\beta)E(x-\xi;k) \right] f(\xi)\, d\sigma_\xi \in H^{r+3/2}_{\mathrm{loc}}(\Omega^c)$$

where $f \in H^r(\partial\Omega)$ *is the unique solution to the BIE*

$$\mathscr{L}_5 f = g,$$

and where

$$\mathscr{L}_5 \colon\ H^\alpha(\partial\Omega) \to H^{\alpha-1}(\partial\Omega), \quad \alpha \in \mathbb{R},$$

defined by

$$(\mathscr{L}_5 f)(x) = \int_{\partial\Omega} \left\{ \frac{\partial^2 E}{\partial n_x \partial n_\xi} + ik(\alpha+i\beta)\left(\frac{\partial E}{\partial n_x} + \frac{\partial E}{\partial n_\xi} \right) + \left[ik(\alpha+i\beta)\right]^2 E \right\} f(\xi)\, d\sigma_\xi$$

$$(E \equiv E(x-\xi;k)) \tag{7.109}$$

is an isomorphism.

Proof. We only provide a sketch as follows.

(1) Define

$$w^\pm(x) = \int_{\partial\Omega} \left[\frac{\partial E}{\partial n_\xi} + ik(\alpha+i\beta)E \right] f(\xi)\, d\sigma_\xi, \ x \in \Omega^c, \ x \in \Omega. \tag{7.110}$$

Use the jump conditions to show that

$$\frac{\partial w^+}{\partial n} + ik(\alpha+i\beta)w^+ = \frac{\partial w^-}{\partial n} + ik(\alpha+i\beta)w^- \quad \text{on } \partial\Omega. \tag{7.111}$$

(2) Use Theorem 7.5 to show that the homogeneous exterior impedance BVP

$$\begin{aligned} &(\Delta+k^2)w^+(x) = 0 && \text{on } \Omega^c, \\ &\frac{\partial w^+(x)}{\partial n} + ik(\alpha+i\beta)w^+(x) = 0 && \text{on } \partial\Omega, \end{aligned}$$

has only the trivial solution $w^+ \equiv 0$.

(3) By Theorem 7.3, the interior impedance BVP (7.67) also has only the trivial solution.

(4) If

$$\mathscr{L}_5(f) = 0,$$

then $f \in C^\infty(\partial\Omega)$, and w^+ defined in (7.110) satisfies the homogeneous impedance BC on $\partial\Omega$. By (2) and (3),

$$w^\pm(x) \equiv 0 \ \text{ on } \Omega^c \text{ and } \Omega.$$

Now use the jump condition

$$f = w^+|_{\partial\Omega} - w^-|_{\partial\Omega} = 0$$

to conclude that $\mathscr{N}(\mathscr{L}_5) = \{0\}$.

Repeat (1)–(4) for $\mathscr{L}_5^*$ (by adjusting all the $ik(\alpha+i\beta)$ to $-ik(\alpha-i\beta)$ in the above) to conclude that $\mathscr{N}(\mathscr{L}_5^*)=\{0\}$. □

Next, consider a solution of the impedance BVP based upon the Green's formula. From (7.105), using

$$\frac{\partial w(x)}{\partial n}=-ik(\alpha+i\beta)w(x)+g(x), \quad x\in\partial\Omega, \tag{7.112}$$

in (7.104), we obtain the BIE

$$\begin{aligned}-\frac{1}{2}w(x)+\int_{\partial\Omega}\left[\frac{\partial E(x-\xi;k)}{\partial n_\xi}+ik(\alpha+i\beta)E(x-\xi;k)\right]w(\xi)\,d\sigma_\xi \\ =\int_{\partial\Omega}E(x-\xi;k)g(\xi)\,d\sigma_\xi,\end{aligned} \tag{7.113}$$

for the Dirichlet data $w(x)$ on $x\in\partial\Omega$.

Theorem 7.9. *Let $k,\ \alpha,\ \beta\in\mathbb{R}$, $k\neq 0$ and $\alpha>0$. Then the exterior impedance BVP (7.108) has a unique solution*

$$\begin{aligned}w(x)=\int_{\partial\Omega}\left[\frac{\partial E(x-\xi;k)}{\partial n_\xi}+ik(\alpha+i\beta)E(x-\xi;k)\right]w(\xi)\,d\sigma_\xi \\ -\int_{\partial\Omega}E(x-\xi;k)g(\xi)\,d\sigma_\xi\in H_{\text{loc}}^{r+3/2}(\Omega^c),\end{aligned} \tag{7.114}$$

where the Dirichlet data w on $\partial\Omega$ is obtained from (7.113), which admits a unique solution $w\in H^{r+1}(\partial\Omega)$ if and only if $-k^2$ is not an eigenvalue of Δ on Ω with the Dirichlet BC.

Proof. Define the LHS of (7.113) as an integral operator $\mathscr{L}_6$. Then

$$(\mathscr{L}_6 w)(x)=\int_{\partial\Omega}E(x-\xi;k)g(\xi)\,d\sigma_\xi\equiv G(x)\in H^{r+1}(\partial\Omega). \tag{7.115}$$

The adjoint operator $\mathscr{L}_6^*$ of $\mathscr{L}_6$ is then

$$\left(\mathscr{L}_6^*\phi\right)(x)=-\frac{1}{2}\phi(x)+\int_{\partial\Omega}\left[\frac{\partial E(x-\xi;k)}{\partial n_x}-ik(\alpha-i\beta)E(x-\xi;-k)\right]\phi(\xi)\,d\sigma_\xi.$$

But this operator $\mathscr{L}_6^*$ is of the same form as $\mathscr{L}_4$ in (7.107), which is invertible provided that $-k^2$ is not an eigenvalue of Δ on Ω with the zero Dirichlet BC according to Remark 7.8, but otherwise may fail to be invertible in general. □

7.8 Solutions to the interior BVP ($\mathbf{Im}\,k\neq 0$)

Interior boundary value problems governed by the Helmholtz equation occur when there is monochromatic wave propagation in a resonant cavity, which by its very nature is a bounded domain. In this section, we use the simple-layer potentials as candidate solutions to solve these problems.

We give, sequentially, the solutions to the Dirichlet, Neumann and impedance BVP.

Theorem 7.10. *Let* $\operatorname{Im} k \neq 0$. *Consider the interior Dirichlet BVP*

$$\left.\begin{aligned} &(\Delta+k^2)w(x)=0, && x\in\Omega,\\ &w(x)=g(x)\in H^{r+1}(\partial\Omega), && r\in\mathbb{R}. \end{aligned}\right\} \tag{7.116}$$

Then

(i) *for* $\operatorname{Im} k>0$, *the simple-layer BIE*

$$(\mathscr{L}_1 f)(x)=\int_{\partial\Omega}E(x-\xi;k)f(\xi)\,d\sigma_\xi=g(x),\quad x\in\partial\Omega, \tag{7.117}$$

has a unique solution $f\in H^r(\partial\Omega)$, *and the solution* w *of (7.116) is given by*

$$w(x)=\int_{\partial\Omega}E(x-\xi;k)f(\xi)\,d\sigma_\xi\in H^{r+3/2}(\Omega),\ x\in\Omega; \tag{7.118}$$

(ii) *for* $\operatorname{Im} k<0$, *the simple-layer BIE*

$$\int_{\partial\Omega}E(x-\xi;-k)f(\xi)\,d\sigma_\xi=g(x),\quad x\in\partial\Omega, \tag{7.119}$$

has a unique solution $f\in H^r(\partial\Omega)$, *and the solution* w *of (7.116) is given by*

$$w(x)=\int_{\partial\Omega}E(x-\xi;-k)f(\xi)\,d\sigma_\xi\in H^{r+3/2}(\Omega),\ x\in\Omega. \tag{7.120}$$

Proof. Use Proposition 7.1. □

Theorem 7.11. *Let* $\operatorname{Im} k \neq 0$. *Consider the interior Neumann BVP*

$$\left.\begin{aligned} &(\Delta+k^2)w(x)=0, && x\in\Omega,\\ &\frac{\partial w(x)}{\partial n}=g(x)\in H^{r}(\partial\Omega),\ r\in\mathbb{R}. \end{aligned}\right\} \tag{7.121}$$

Then

(i) *for* $\operatorname{Im} k>0$, *the BIE*

$$\frac{1}{2}f(x)+\int_{\partial\Omega}\frac{\partial E(x-\xi;k)}{\partial n_x}f(\xi)d\sigma_\xi=g(x),\quad x\in\partial\Omega,$$

has a unique solution $f\in H^r(\partial\Omega)$, *and the solution* w *of (7.121) is given by the simple-layer potential*

$$w(x)=\int_{\partial\Omega}E(x-\xi;k)f(\xi)d\sigma_\xi\in H^{r+3/2}(\Omega),\quad x\in\partial\Omega;$$

(ii) *for* $\operatorname{Im} k < 0$, *the BIE*

$$\frac{1}{2}f(x) + \int_{\partial\Omega} \frac{\partial E(x-\xi;-k)}{\partial n_x} f(\xi) d\sigma_\xi = g(x), \quad x \in \partial\Omega,$$

has a unique solution $f \in H^r(\partial\Omega)$, *and the solution* w *of (7.121) is given by the simple-layer potential*

$$w(x) = \int_{\partial\Omega} E(x-\xi;-k) f(\xi) d\sigma_\xi \in H^{r+3/2}(\Omega), \quad x \in \Omega.$$

Proof. Use Proposition 7.2. □

Theorem 7.12. *Let* $k \in \mathbb{R}$, $k \neq 0$, $\alpha > 0$, $\beta \in \mathbb{R}$. *Consider the interior impedance BVP*

$$\left.\begin{aligned} &(\Delta + k^2) w(x) = 0 \quad \text{on } \Omega, \\ &\frac{\partial w(x)}{\partial n} + ik(\alpha + i\beta) w(x) = g(x) \in H^r(\partial\Omega), \ x \in \partial\Omega, r \in \mathbb{R}. \end{aligned}\right\} \tag{7.122}$$

Then the associated BIE

$$\frac{1}{2}f(x) + \int_{\partial\Omega} \left[\frac{\partial E(x-\xi;k)}{\partial n_x} + ik(\alpha + i\beta) E(x-\xi;k)\right] f(\xi) d\sigma_\xi = g(x), \quad x \in \partial\Omega \tag{7.123}$$

admits a unique solution $f \in H^r(\partial\Omega)$. *Consequently, the solution* w *to (7.122) is given by*

$$w(x) = \int_{\partial\Omega} E(x-\xi;k) f(\xi d\sigma_\xi \in H^{r+3/2}(\Omega), \ x \in \Omega.$$

Proof. This is left as an exercise. (Or check the proof of Theorem 7.13.) □

For the interior Dirichlet problem (7.116) and Neumann problem (7.121), when $\operatorname{Im} k = 0$, their solutions may or may not exist uniquely, depending on whether $-k^2$ is an eigenvalue of the Laplacian and whether the data g in (7.116) or (7.121) satisfies the compatibility conditions. Certain boundary integral equation theorems still need be proved to cover those cases and to resolve the difficulties involved. However, such theorems would not be useful in applications for reasons explained in the last paragraph of § 7.4 (see particularly (7.54)). Instead, for $\operatorname{Im} k = 0$ the primary interest will be focused on (numerically) finding the eigenvalues (i.e., $-k^2$) and eigenfunctions of the Laplacian subject to the homogeneous Dirichlet, Neumann or other types of boundary conditions on the interior domain Ω. This discussion will be given in § 7.10.

7.9 Modified integral equations approach to exterior scattering problems ($\mathrm{Im}\,k \geqslant 0$)

When $\mathrm{Im}\,k = 0$, the exterior Dirichlet and Neumann BVP were solved according to Theorems 7.6 and 7.7 in § 7.6. In order to apply formulas (7.98) and (7.102), one needs to know what the interior eigenfunctions u_j^+ and v_j^+ are. This is very inconvenient. The complication arises from the method of approach rather than from the nature of the problem itself, because we know that for $\mathrm{Im}\,k = 0$ the exterior scattering problem has a unique solution. It is desirable to develop BIE methods that are uniquely solvable for all wavenumbers k satisfying $\mathrm{Im}\,k \geqslant 0$. Such methods will be particularly significant in the study of the inverse scattering problem [53, Chaps. 6–8], because in that case the geometry of the obstacle is unknown thus the interior eigenfunctions and eigenvalues are not available. In this section, we present the combination layer approach of Leis [118], Brakhage and Werner [19] and Panich [150], who devised the method independently. Other approaches, suggested by Jones, and Kleinman and Roach, can be found in [105, 106, 113].

Theorem 7.13. *Let* $\mathrm{Im}\,k \geqslant 0$. *Define the combined simple and double layer*

$$w(x) = \int_{\partial\Omega} \left[\frac{\partial E(x-\xi;k)}{\partial n_\xi} - i\eta E(x-\xi;k) \right] f(\xi)\, d\sigma_\xi, \quad x \in \Omega^c, \tag{7.124}$$

where $\eta \in \mathbb{R}$, $\eta \neq 0$ *is chosen such that*

$$\eta\, \mathrm{Re}\,k > 0. \tag{7.125}$$

Then the associated BIE

$$\frac{1}{2} f(x) + \int_{\partial\Omega} \left[\frac{\partial E(x-\xi;k)}{\partial n_\xi} - i\eta E(x-\xi;k) \right] f(\xi)\, d\sigma_\xi = g(x) \in H^r(\partial\Omega),$$
$$x \in \partial\Omega, \tag{7.126}$$

has a unique solution $f \in H^r(\partial\Omega)$, *and (7.124) gives a unique solution* $w \in H^{r+1/2}_{\mathrm{loc}}(\Omega^c)$ *to the exterior Dirichlet BVP (7.96).*

Proof. Define the boundary integral operator

$$(K\phi)(x) = 2 \int_{\partial\Omega} \left[\frac{\partial E(x-\xi;k)}{\partial n_\xi} - i\eta E(x-\xi;k) \right] \phi(\xi)\, d\sigma_\xi, \quad x \in \partial\Omega.$$

Then by § 4.4 and (6.137),

$$K\colon\ H^s(\partial\Omega) \to H^{s+1}(\partial\Omega), \quad s \in \mathbb{R},$$

so K mapping from $H^s(\partial\Omega)$ into itself is compact. We need only to show that $\mathscr{N}(I+K) = \{0\}$.

Assume that

$$(I+K)f=0.$$

Then $w^+(x)$ defined by the RHS of (7.124) has zero Dirichlet data. By the exterior uniqueness theorem for the Dirichlet problem, $w^+(x)\equiv 0$ on Ω^c. Now define $w^-(x)$ on Ω by extending the RHS of (7.124) to $x\in\Omega$. By the jump conditions

$$\begin{aligned}
&w^+(x)-w^-(x)=-w^-(x)=f(x),\\
&\frac{\partial w^+(x)}{\partial n}-\frac{\partial w^-(x)}{\partial n}=-\frac{\partial w^-(x)}{\partial n}=i\eta f(x)
\end{aligned}$$

on $\partial\Omega$. From (7.64), we again have

$$i\eta\int_{\partial\Omega}|f(x)|^2\,d\sigma=\int_{\partial\Omega}\overline{w}^-\frac{\partial w^-}{\partial n}d\sigma=\int_{\Omega}\left(|\nabla w^-|^2-k^2|w^-|^2\right)dx.$$

Taking the imaginary parts of the above, we get

$$\eta\int_{\partial\Omega}|f(x)|^2\,d\sigma=-2(\mathrm{Re}\,k)(\mathrm{Im}\,k)\int_{\Omega}|w|^2\,dx.$$

So it follows that $f\equiv 0$ on $\partial\Omega$ because of (7.125) and $\mathrm{Im}\,k\geqslant 0$. □

Theorem 7.14. *Let* $\mathrm{Im}\,k\geqslant 0$. *Define the combined simple and double layer as in (7.124) and (7.125). Then for the exterior Neumann BVP*

$$\begin{aligned}
&(\Delta+k^2)w(x)=0\ \text{ on }\Omega^c,\\
&\frac{\partial w(x)}{\partial n}=g(x)\in H^r(\partial\Omega),\ r\in\mathbb{R},
\end{aligned}$$

the associated BIE

$$\frac{i\eta}{2}f(x)+\int_{\partial\Omega}\left[\frac{\partial^2E(x-\xi;k)}{\partial n_x\partial n_\xi}-i\eta\frac{\partial E(x-\xi;k)}{\partial n_x}\right]f(\xi)\,d\sigma_\xi=g(x),\ \ x\in\partial\Omega,$$

has a unique solution $f\in H^{r+1}(\partial\Omega)$, *yielding the solution* w *in (7.124) with* $w\in H^{r+3/2}_{\mathrm{loc}}(\Omega^c)$.

Proof. The arguments are similar to those in the proofs of Theorems 7.8 and 7.13. We leave it as an exercise. □

Remark 7.9. In an early work by Leis [119] as well as in [53], Theorem 7.14 is proved by a "regularization" procedure. □

7.10 Numerical example (I): computation of eigenfunctions of the Laplacian

A vibrating membrane satisfies the wave equation

$$w_{tt}(x,t) - \Delta w(x,t) = 0, \quad x \in \Omega \subset \mathbb{R}^2, \ t > 0, \tag{7.127}$$

in two space dimensions. For time-harmonic excitations, the equation has a separable solution

$$w(x,t) = e^{i\mu t}\phi(x), \quad \mu \in \mathbb{C}. \tag{7.128}$$

Substituting (7.128) into (7.127), we see that μ and $\phi(\cdot)$ must satisfy

$$\Delta\phi(x) = \lambda\phi(x), \quad x \in \Omega, \quad \lambda \equiv -\mu^2. \tag{7.129}$$

Let the boundary condition associated with (7.127) be

$$w(x,t) = 0, \qquad x \in \partial\Omega, \quad t > 0, \tag{7.130}$$

the Dirichlet condition. Then ϕ satisfies the eigenvalue problem

$$\left.\begin{aligned} \Delta\phi_n(x) &= -\mu_n^2\phi_n(x) \text{ on } \Omega, \\ \phi_n(x) &= 0 \qquad\qquad \text{on } \partial\Omega. \end{aligned}\right\} \tag{7.131}$$

For a bounded domain Ω, it is known that (7.131) has an infinite sequence of negative eigenvalues $\lambda_n = -\mu_n^2$, $n = 1, 2, \ldots$, satisfying

$$0 < \mu_1^2 < \mu_2^2 < \cdots < \mu_n^2 < \cdots \to \infty,$$

where each $\lambda_n = -\mu_n^2$ has finite multiplicity. These eigenfunctions and eigenvalues are essential in the understanding of the vibration behavior of the membrane.
We rewrite (7.131) as

$$\left.\begin{aligned} \Delta\phi_n(x) &= -\mu_n^2\phi_n(x) \text{ on } \Omega, \\ \phi_n(x) &= 0 \qquad\qquad \text{on } \partial\Omega. \end{aligned}\right\} \tag{7.132}$$

We approximate $\phi_n(\cdot)$ and μ_n^2 in (7.132) by piecewise-constant boundary elements. First, represent ϕ_n as a simple-layer potential

$$\phi_n(x) = \int_{\partial\Omega} E(x-\xi;\mu_n) f_n(\xi)\, d\sigma_\xi, \quad x \in \overline{\Omega}, \tag{7.133}$$

where $E(x-\xi;\mu_n) = \frac{1}{4} i H_0^{(1)}(\mu_n|x-\xi|)$ as in (7.37). It is not difficult to show that $(\phi_n(\cdot), \mu_n^2)$ satisfy (7.132) if and only if there exists $f_n \not\equiv 0$ on $\partial\Omega$ such that

$$\int_{\partial\Omega} E(x-\xi;\mu_n) f_n(\xi)\, d\sigma_\xi = 0, \quad \forall x \in \partial\Omega. \tag{7.134}$$

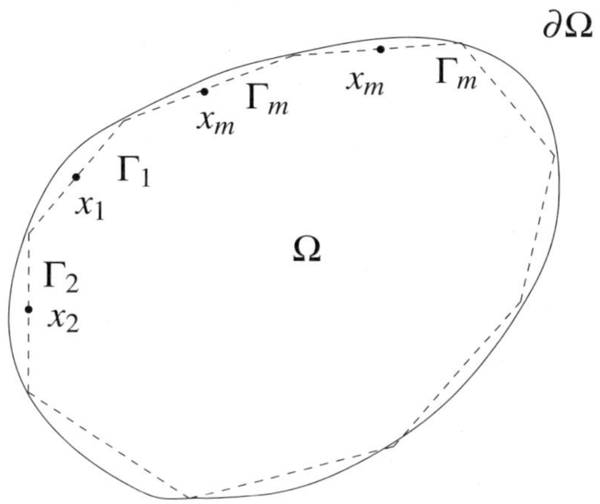

Figure 7.5: Approximation of $\partial\Omega$ by a polygon.

Approximate the boundary $\partial\Omega$ by a polygon $\Gamma \equiv \bigcup_{j=1}^{m} \Gamma_j$, where Γ_j is a line segment with endpoints on $\partial\Omega$ and with midpoint x_j; see Fig. 7.5. Let the density f_n on $\partial\Omega$ be approximated by

$$f_n(\cdot) \approx \sum_{j=1}^{m} f_{nj}\chi_{\Gamma_j}, \quad f_{nj} \in \mathbb{C}, \;\; j = 1, 2, \ldots, m, \tag{7.135}$$

where χ_{Γ_j} is the characteristic function of Γ_j on Γ. The BIE (7.134) is now approximated by

$$\int_{\Gamma} E(x_l - \xi; \mu_n) \sum_{j=1}^{m} f_{nj}\chi_{\Gamma_j}(\xi)\, d\sigma_\xi = 0, \quad l = 1, 2, \ldots, m,$$

$$\sum_{j=1}^{m} f_{nj} \int_{\Gamma_j} E(x_l - \xi; \mu_n)\, d\sigma_\xi = 0, \qquad l = 1, 2, \ldots, m.$$

Thus we obtain a linear system

$$A(\mu_n)\vec{F}_n = 0, \tag{7.136}$$

where

$$A(\mu_n) = [a_{ij}(\mu_n)]_{m\times m},$$
$$a_{ij}(\mu_n) = \int_{\Gamma_j} E(x_i - \xi; \mu_n)\, d\sigma_\xi, \;\; 1 \leqslant i, \, j \leqslant m, \tag{7.137}$$

is an $m \times m$ matrix, and $\vec{F}_n = (f_{n1}, f_{n2}, \ldots, f_{nm})^{\mathrm{Tr}} \in \mathbb{C}^m$, $\vec{F}_n \neq \vec{0}$. Therefore approximate eigenvalues $-\mu_n^2$ of (7.132) can be obtained by computing the roots of the transcendental equation

$$\det A(\mu_n) = 0. \tag{7.138}$$

Numerically, we can approximate μ_n by the Newton–Raphson method as follows:

(1) Pick an initial value $\mu_n^{(0)}$.

(2) Iterate by

$$\mu_n^{(r+1)} = \mu_n^{(r)} - \frac{1}{\mathrm{tr}[A^{-1}(\mu_n^{(r)})A'(\mu_n^{(r)})]}, \quad r = 0, 1, 2, \ldots, \tag{7.139}$$

where tr is the trace of a matrix;

(3) Find an approximate eigenvector $\vec{F}_n$ by the *inverse power method*: Let $\mu_n^{(r)}$ be a "good" approximation to μ_n, but $A(\mu_n^{(r)})$ is still invertible. We iterate by

$$A(\mu_n^{(r)})\vec{F}_n^{(r+1)} = \vec{F}_n^{(r)}, \quad r = 0, 1, 2, \ldots, \tag{7.140}$$

where at each step the vector $\vec{F}_n^{(r)}$ on the RHS above is normalized by

$$\|\vec{F}_n^{(r)}\|_\infty = \max_{1\leqslant j\leqslant m}(|f_{nj}^{(r)}|) = 1. \tag{7.141}$$

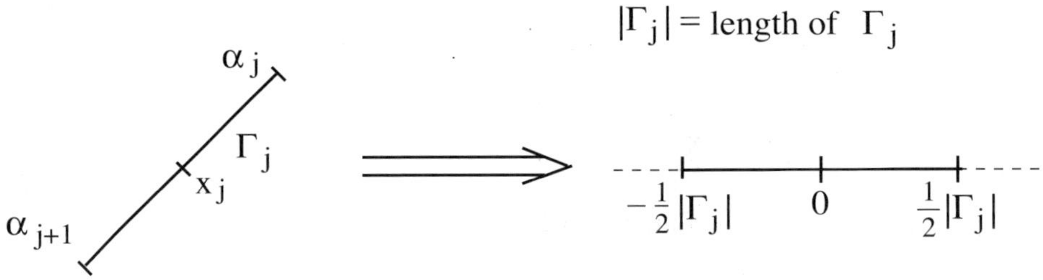

Figure 7.6: Rotation of a boundary segment to a horizontal position.

The accuracy of the above algorithm hinges on the accuracy of the entries $a_{ij}(\mu_n)$ in (7.137). How do we approximate

$$a_{ij}(\mu_n) = \int_{\Gamma_j} E(x_i - \xi; \mu_n)\, d\sigma_\xi ? \tag{7.142}$$

When $i \neq j$, $E(x_i - \xi; \mu_n)$ does not contain a singularity on Γ_j. So we can integrate (7.142) by a standard Gaussian quadrature. When $i = j$, $E(x_j - \xi; \mu_n)$ has a singularity at $\xi = x_j$. More careful handling is required in order to better approximate (7.142). We do it as follows. First rotate Γ_j to a horizontal segment as shown in Fig. 7.6. Then

$$\begin{aligned}\int_{\Gamma_j} E(x_j - \xi; \mu_n)\, d\sigma_\xi &= \frac{i}{4}\int_{-|\Gamma_j|/2}^{|\Gamma_j|/2} H_0^{(1)}(\mu_n|\xi|)\, d\xi \\ &= 2\left(\frac{i}{4}\right)\int_0^{|\Gamma_j|/2} H_0^{(1)}(\mu_n s)\, ds \\ &= \frac{i}{2\mu_n}\int_0^{\mu_n|\Gamma_j|/2} H_0^{(1)}(\tau)\, d\tau. \end{aligned} \tag{7.143}$$

However, by [1, p. 480], for a complex variable z,

$$\int_0^z H_0^{(1)}(\tau)\,d\tau = zH_0^{(1)}(z) + \frac{1}{2}\pi z[\boldsymbol{H}_0(z)H_1^{(1)}(z) - \boldsymbol{H}_1(z)H_0^{(1)}(z)], \tag{7.144}$$

where $\boldsymbol{H}_0$ and $\boldsymbol{H}_1$ are respectively the Struve functions of order 0 and 1, satisfying

$$\boldsymbol{H}_0(z) = \frac{4}{\pi}\sum_{k=0}^{\infty}\frac{1}{2k+1}J_{2k+1}(z), \tag{7.145}$$

$$\boldsymbol{H}_1(z) = \frac{2}{\pi}[1 - J_0(z)] + \frac{4}{\pi}\sum_{k=1}^{\infty}\frac{1}{4k^2-1}J_{2k}(z), \tag{7.146}$$

and where

$$H_1^{(1)}(z) = J_1(z) + iN_1(z) \tag{7.147}$$

is the Hankel function of order 1 of the first kind. (See (7.30) for J_k and N_k.) Note that the recursion relations

$$J_{n+1}(z) = \frac{2n}{z}J_n(z) - J_{n-1}(z), \quad n = 2, 3, 4, 5, \dots, \tag{7.148}$$

can be used in the evaluation of (7.145) and (7.146). (As usual, in the actual evaluations, truncation of the infinite series (7.145) and (7.146) can be made after only a few sums.) The quadrature (7.144) is *exact*. Therefore, from (7.143), we obtain

$$\begin{aligned}\int_{\Gamma_j} E(x_j - \xi;\mu_n)\,d\sigma_\xi = \frac{1}{4}i|\Gamma_j|\Big\{&H_0^{(1)}\left(\frac{1}{2}\mu_n|\Gamma_j|\right)\\ &+\frac{1}{2}\pi\boldsymbol{H}_0\left(\frac{1}{2}\mu_n|\Gamma_j|\right)H_1^{(1)}\left(\frac{1}{2}\mu_n|\Gamma_j|\right)\\ &-\frac{\pi}{2}\boldsymbol{H}_1\left(\frac{\mu_n}{2}|\Gamma_j|\right)H_0^{(1)}\left(\frac{\mu_n}{2}|\Gamma_j|\right)\Big\}.\end{aligned} \tag{7.149}$$

Note that in the above procedures, because of (7.148), we need only know the special functions J_0, N_0 (appearing in (7.39) and (7.29)) and J_1, N_1 (appearing in (7.147)) in order to carry out all the computations. Subroutines of J_0, N_0, J_1 and N_1 are available in Fortran as j_0, y_0, j_1 and y_1 respectively.

Example 7.1. Let Ω be the unit open disk in $\mathbb{R}^2$. We use the above algorithm to compute the numerical solutions of eigenfunctions and eigenvalues of Δ in (7.132). The circumference $\partial\Omega$ is subdivided into 20 equal parts. For $i \neq j$ in (7.142), we use a high-accuracy 48 point Gaussian quadrature formula to compute $a_{ij}(\mu_n)$ while for $i = j$, we use (7.149).
The exact eigenfunctions on the unit disk are known to be

$$J_n(\mu_{nk}r)e^{\pm in\theta}, \quad n = 0, 1, 2, \dots, \tag{7.150}$$

where μ_{nk} is the kth positive root of

$$J_n(\mu) = 0.$$

These eigenfunctions are tabulated in sequential order (according to the magnitudes of μ^2) in the third column of Table 7.1.

The computed numerical values of μ_{nk} are tabulated in the fourth and fifth columns of Table 7.1, where the former represents the real parts of μ_{nk} while the latter represents the imaginary parts. This method of computation yields a small imaginary part for each value of μ_{nk} as part of the numerical error. Thus, the size of $\operatorname{Im}\mu_{nk}$ can serve as an error indicator. For the first 21 eigenvalues (counted up to multiplicity) as tabulated in Table 7.1, we find that the relative error of μ_{nk} is about $1 \sim 2\%$.

The graphs of these eigenfunctions, up to the 17th (that is, $J_5(8.771r)\cdot e^{\pm i5\theta}$), along with their nodal lines (below), are plotted in Figs. 7.7–7.7. Exact solutions and numerical solutions are posted in (a) and (b) for easy comparison. The largest magnitude of each function is normalized to 1. Note that only the real part of $J_n(\mu_{nk}r)e^{\pm in\theta}$ is plotted, because the imaginary part is just a rotation of the real part. Also the numerical solution may differ from the exact solution by an angle of rotation. □

Next, let us consider the eigenvalue problem of the Laplacian with zero Neumann boundary condition

$$\left.\begin{aligned} \Delta\phi_n(x) &= -\mu_n^2\phi_n(x) \quad \text{on } \Omega, \\ \frac{\partial}{\partial n}\phi_n(x) &= 0 \qquad\qquad\quad \text{on } \partial\Omega. \end{aligned}\right\} \tag{7.151}$$

Again we represent the solution as a simple layer (7.133). Taking the normal derivative of (7.133) and using (6.58), we can show that $(\phi_n(\cdot),\mu_n^2)$ satisfies (7.151) if and only if there exists $f_n \in C^\infty(\partial\Omega)$ satisfying

$$0 = \frac{\partial}{\partial n}\phi_n(x) = \frac{1}{2}f_n(x) + \int_{\partial\Omega}\frac{\partial E(x-\xi;\mu_n)}{\partial n_x}f_n(\xi)\,d\sigma_\xi, \quad \forall x \in \partial\Omega. \tag{7.152}$$

The kernel of the above integral operator is

$$\begin{aligned} \frac{\partial E(x-\xi;\mu_n)}{\partial n_x} &= \frac{\partial}{\partial n_x}\left[\frac{1}{4}iH_0^{(1)}(\mu_n|x-\xi|)\right] \\ &= -\frac{1}{4}i\mu_n H_1^{(1)}(\mu_n|x-\xi|)\frac{\langle n_x, x-\xi\rangle}{|x-\xi|}. \end{aligned} \tag{7.153}$$

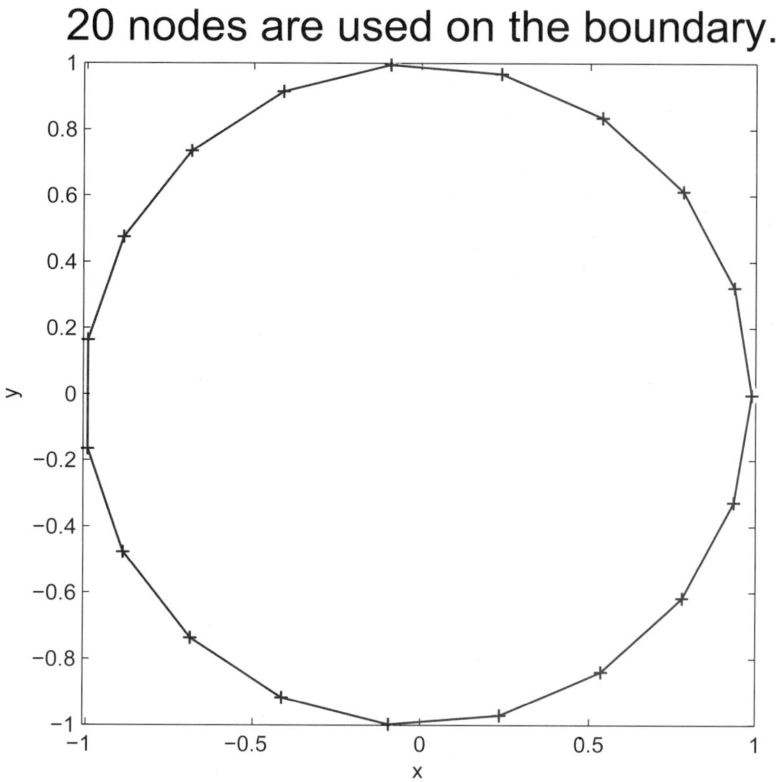

Table 7.1: Exact and computed eigenvalues of Δ with the zero Dirichlet condition on the unit disk.

#	Multiciplity	Exact solution	Numerical eigenvalues $\lambda = -\mu^2$	
			Re μ	Im μ^*
1	1	$J_0(2.405r)$	2.4305	−0.000025
2, 3	2	$J_1(3.832r)e^{\pm i\theta}$	3.8718	−0.000060
4, 5	2	$J_2(5.136r)e^{\pm i2\theta}$	5.1855	−0.000090
6	1	$J_0(5.520r)$	5.5791	−0.000130
7, 8	2	$J_3(6.380r)e^{\pm i3\theta}$	6.4333	−0.000005
9, 10	2	$J_1(7.016r)e^{\pm i\theta}$	7.0888	−0.000208
11, 12	2	$J_4(7.588r)e^{\pm i4\theta}$	7.6340	0.000767
13, 14	2	$J_2(8.417r)e^{\pm i2\theta}$	8.4978	−000258
15	1	$J_0(8.654r)$	8.7464	−0.000318
16, 17	2	$J_5(8.771r)e^{\pm i5\theta}$	8.7909	0.004099
18, 19	2	$J_3(9.761r)e^{\pm i3\theta}$	9.8371	0.000247
20, 21	2	$J_6(9.936r)e^{\pm i6\theta}$	9.8921	0.017548

Error = $1 - 2\%$

*Exact values of μ should satisfy $\text{Im}(\mu) = 0$. Therefore the magnitude of $\text{Im}(\mu)$ serves as an indicator of the size of the numerical error.

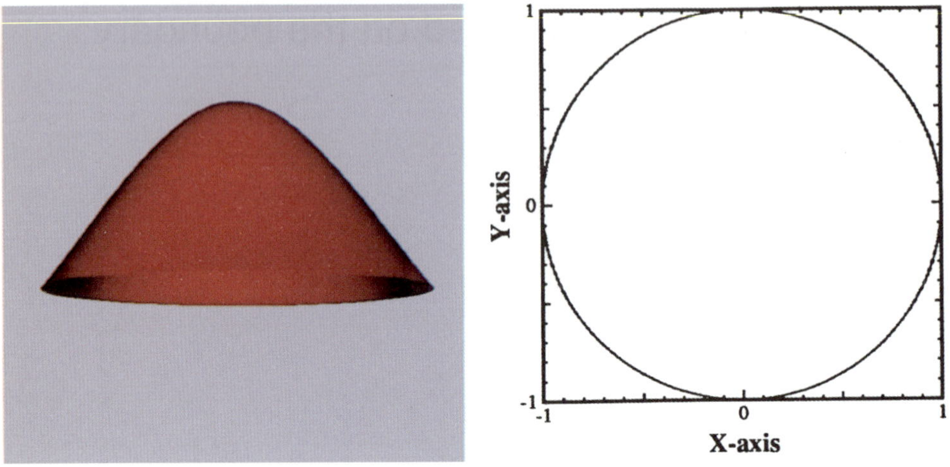

Figure 7.7: (a) Exact eigenfunction $J_0(2.405r)$; multiplicity 1.

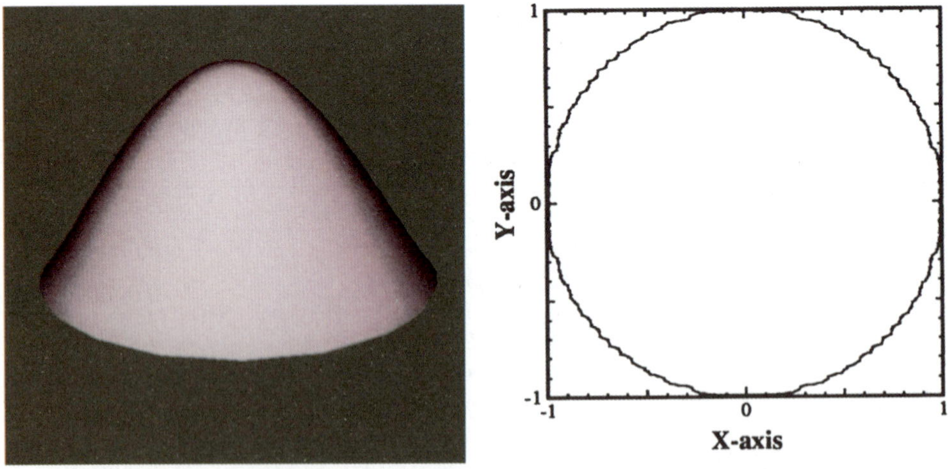

Figure 7.7: (b) BEM solution.

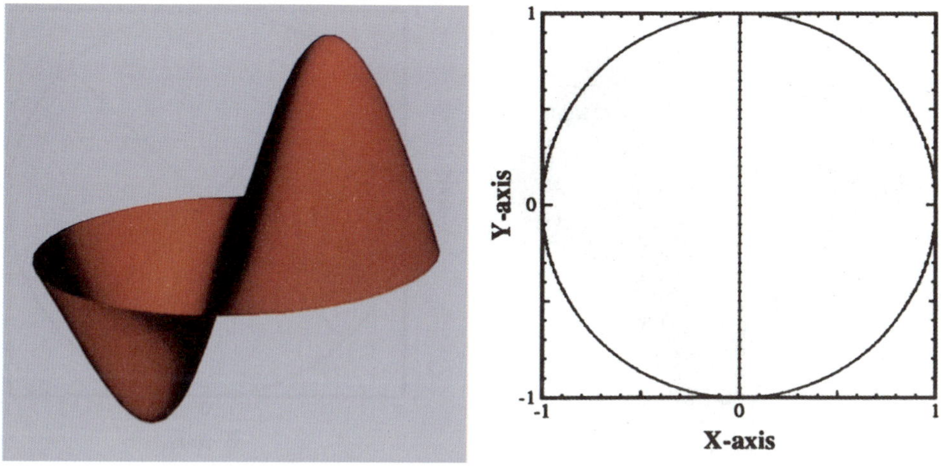

Figure 7.8: (a) Exact eigenfunction $J_1(3.832r)e^{\pm i\theta}$; multiplicity 2, real part.

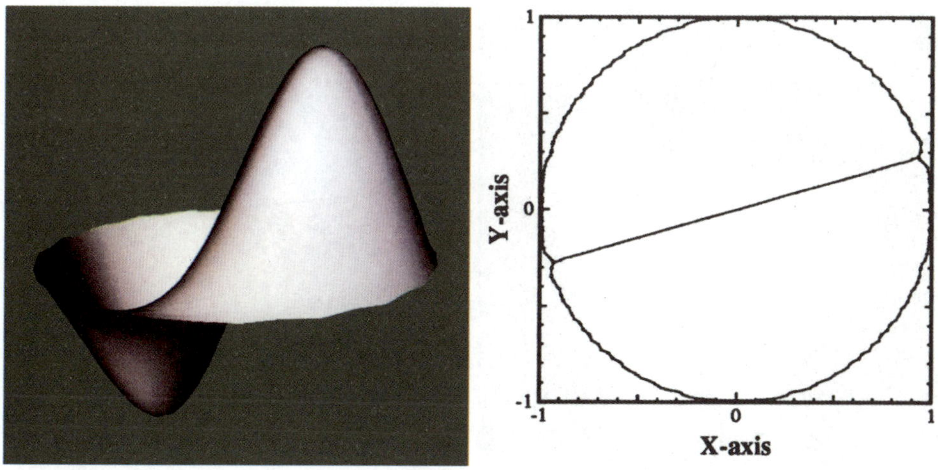

Figure 7.8: (b) BEM solution.

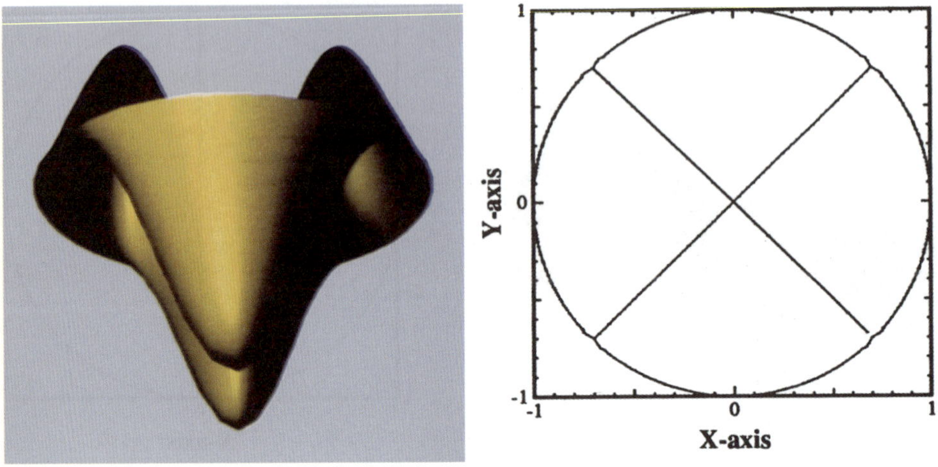

Figure 7.9: (a) Exact eigenfunction $J_2(5.136r)e^{\pm i2\theta}$; multiplicity 2, real part.

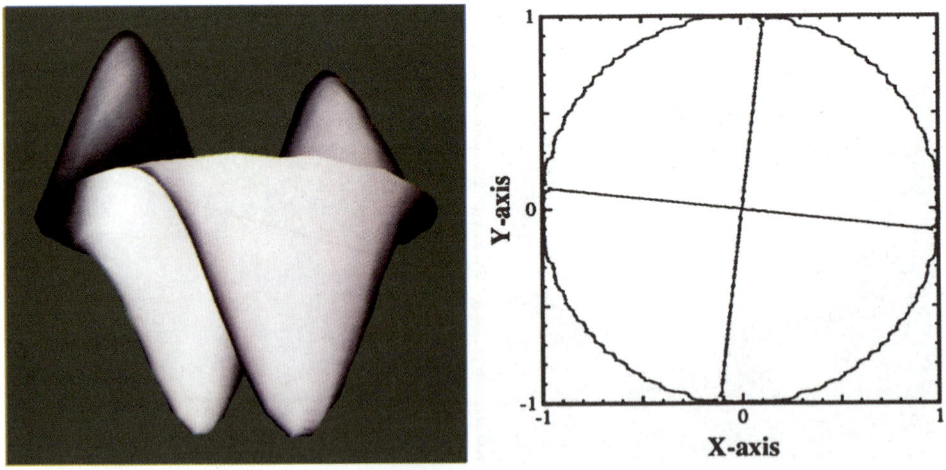

Figure 7.9: (b) BEM solution.

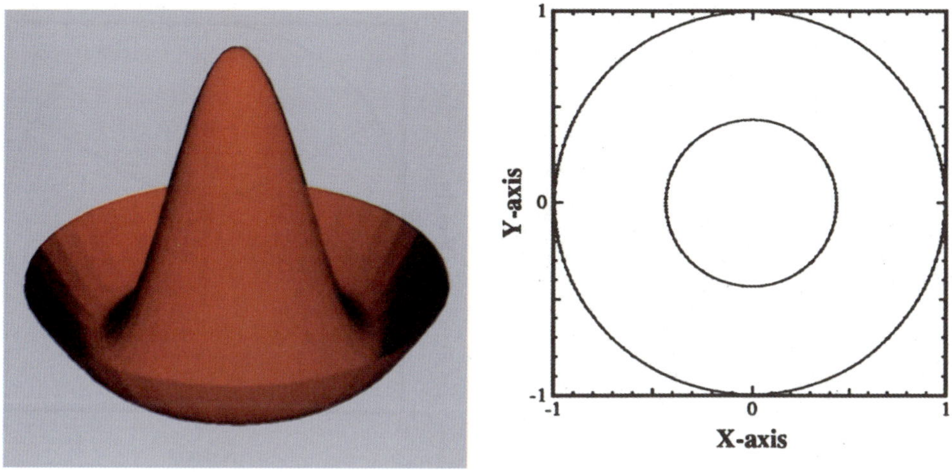

Figure 7.10: (a) Exact eigenfunction $J_0(5.520r)$; multiplicity 1.

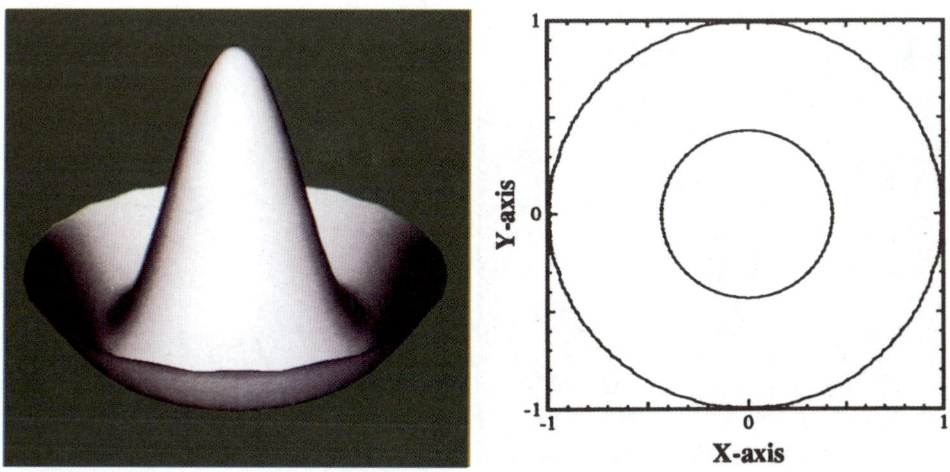

Figure 7.10: (b) BEM solution.

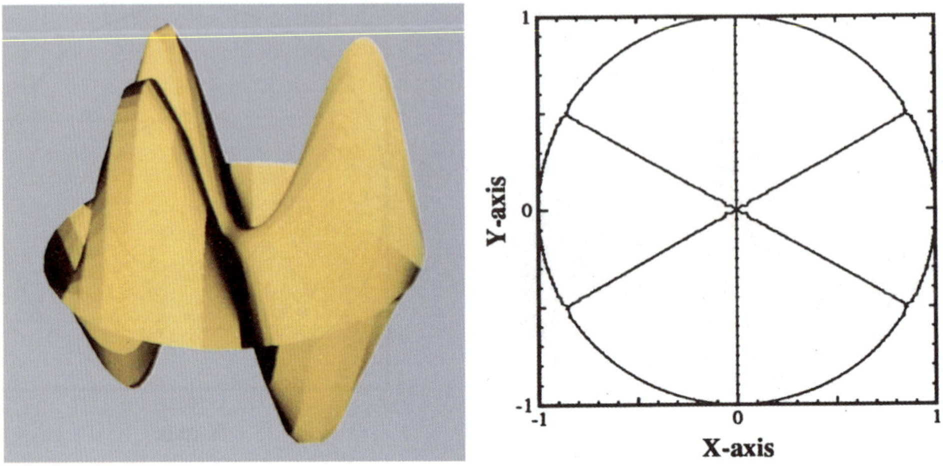

Figure 7.11: (a) Exact eigenfunction $J_3(6.380r)e^{\pm i3\theta}$; multiplicity 2, real part.

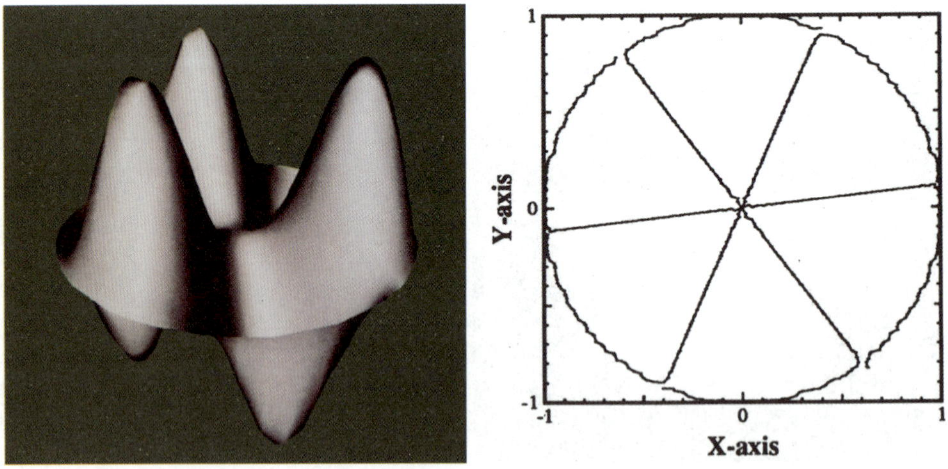

Figure 7.11: (b) BEM solution.

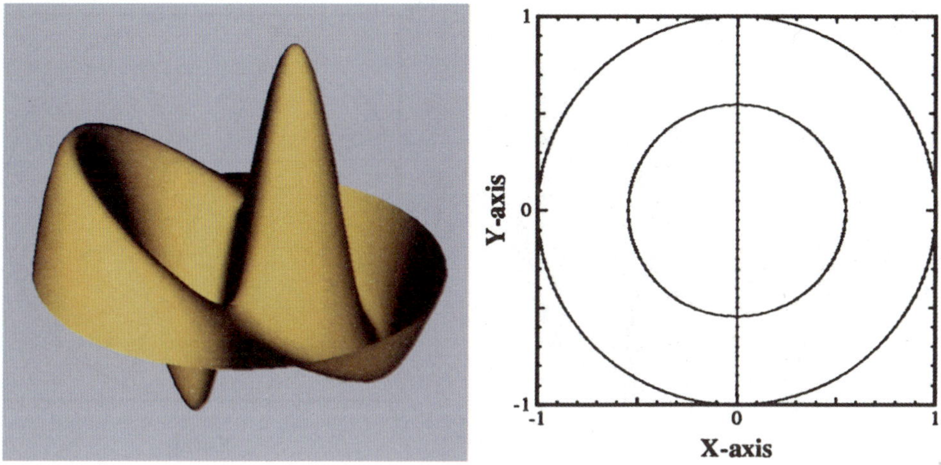

Figure 7.12: (a) Exact eigenfunction $J_1(7.016r)e^{\pm i\theta}$; multiplicity 2, real part.

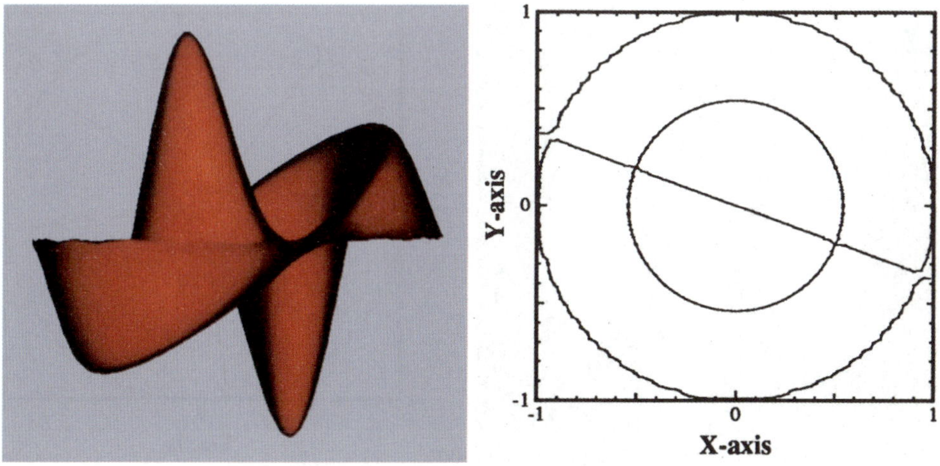

Figure 7.12: (b) BEM solution.

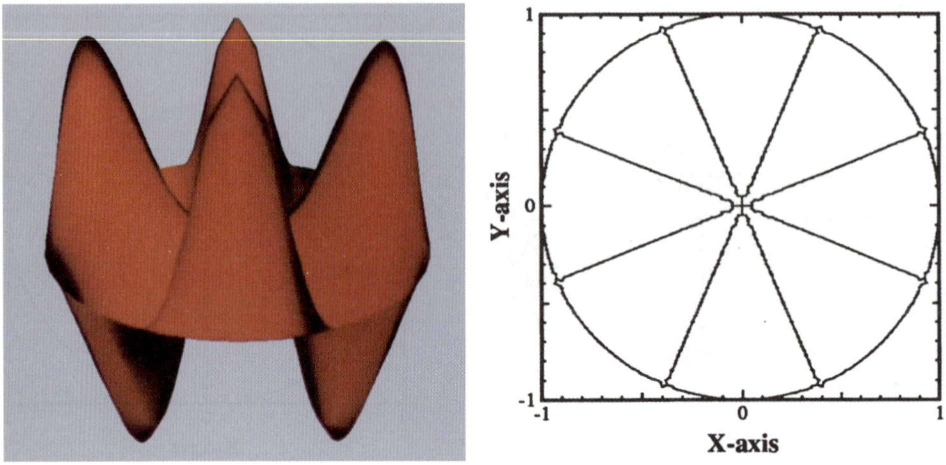

Figure 7.13: (a) Exact eigenfunction $J_4(7.588r)e^{\pm i4\theta}$; multiplicity 2, real part.

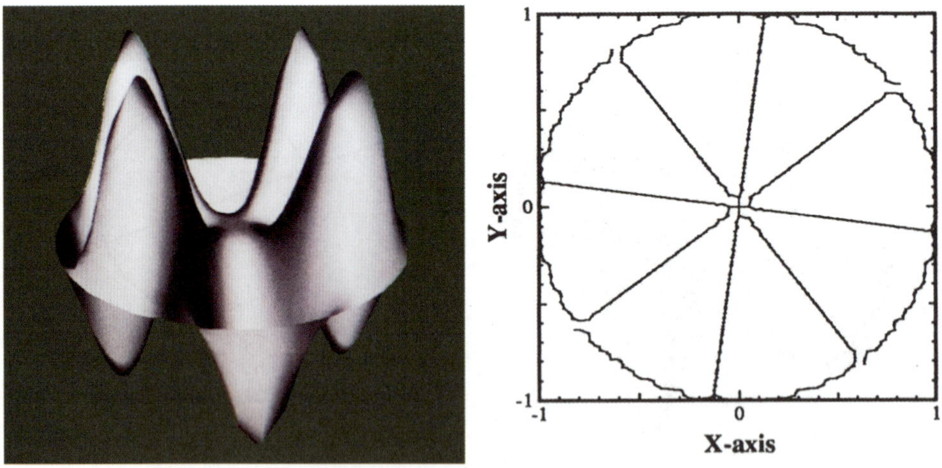

Figure 7.13: (b) BEM solution.

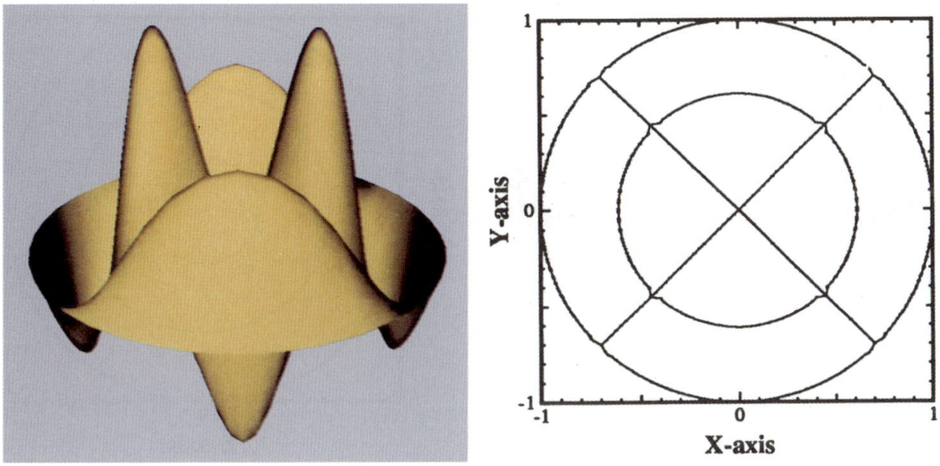

Figure 7.14: (a) Exact eigenfunction $J_2(8.417r)e^{\pm i2\theta}$; multiplicity 2, real part.

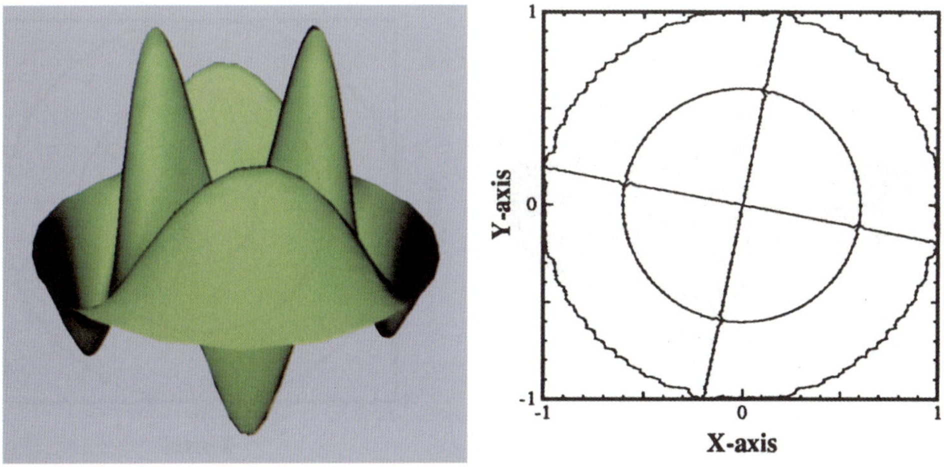

Figure 7.14: (b) BEM solution.

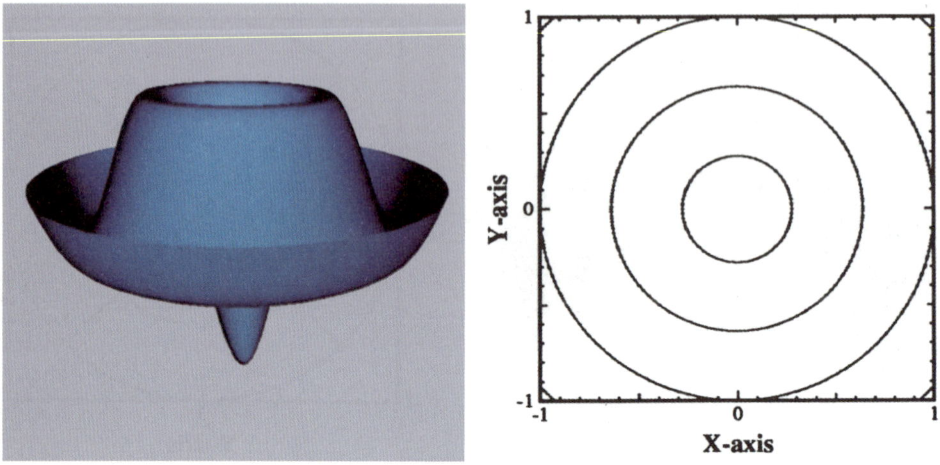

Figure 7.15: (a) Exact eigenfunction $J_0(8.654r)$; multiplicity 1.

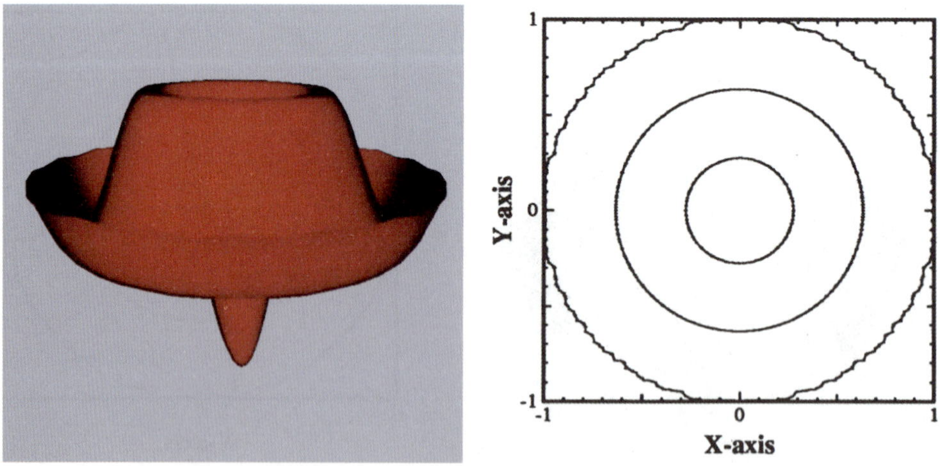

Figure 7.15: (b) BEM solution.

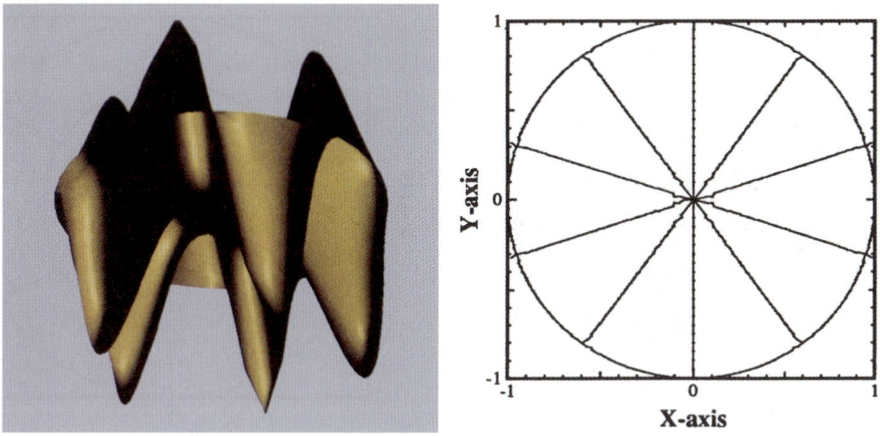

Figure 7.16: (a) Exact eigenfunction $J_5(8.771r)e^{\pm i5\theta}$; multiplicity 2, real part.

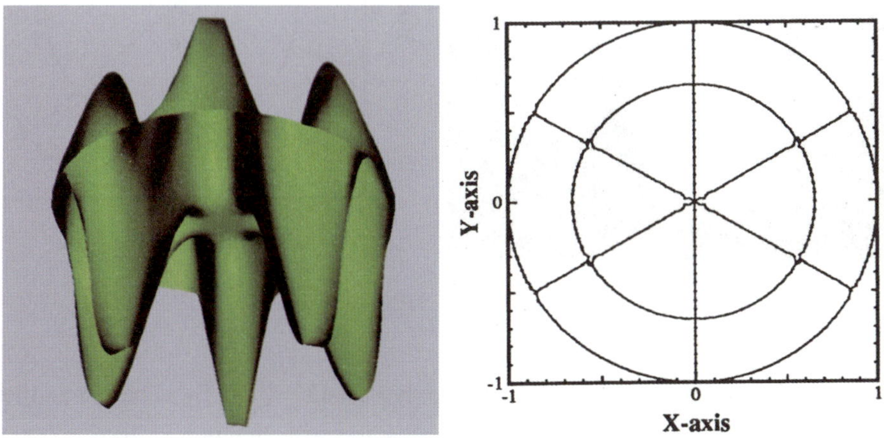

Figure 7.16: (b) BEM solution.

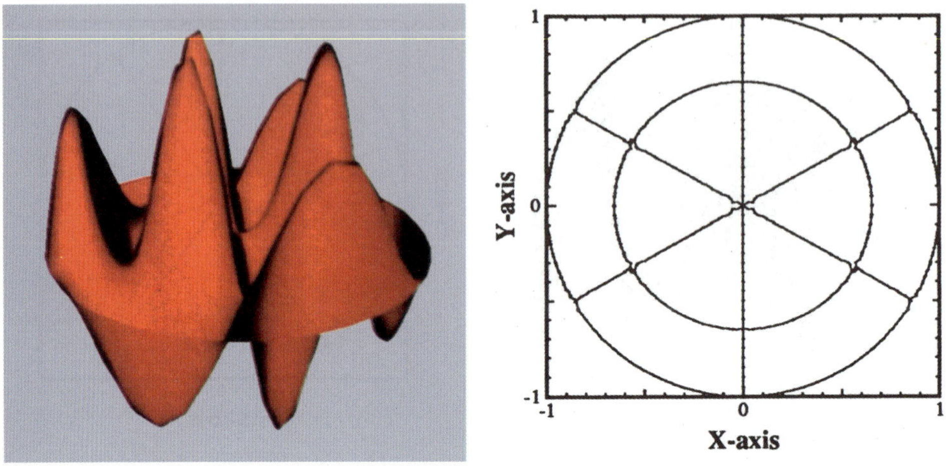

Figure 7.17: (a) Exact eigenfunction $J_3(9.761r)e^{\pm i3\theta}$; multiplicity 2, real part.

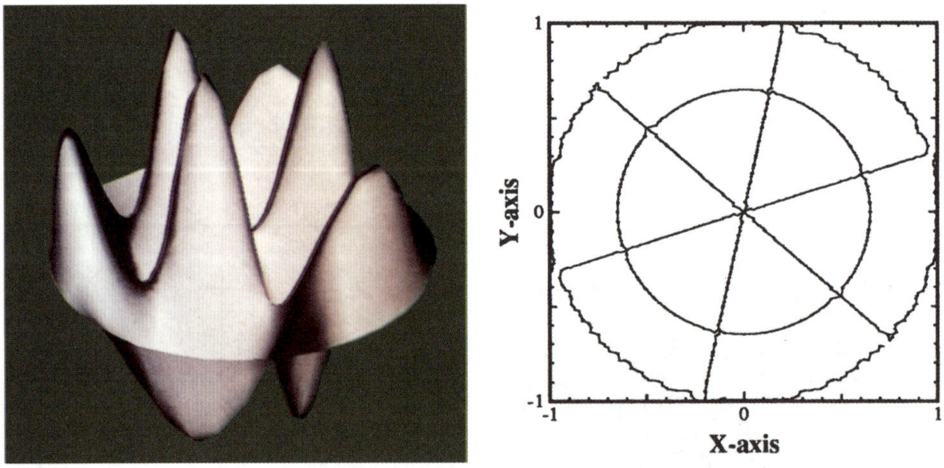

Figure 7.17: (b) BEM solution.

Let the boundary $\partial\Omega$ be approximated by a polygon $\partial\Omega = \bigcup_{j=1}^{m} \Gamma_j$ as before and let the layer density f_n be approximated by a piecewise-constant function as in (7.135). Then (7.152) is discretized as

$$\frac{1}{2} f_{nl} + \sum_{j=1}^{m} f_{nj} \int_{\Gamma_j} \frac{\partial E(x-\xi;\mu_n)}{\partial n_x}\Big|_{x=x_l} d\sigma_\xi = 0, \quad l = 1, 2, \ldots, m. \tag{7.154}$$

The above integral can be approximated as follows:

(i) For $l \neq j$, the integrand is smooth on Γ_j. We use a standard Gaussian quadrature to evaluate it.

(ii) For $l = j$, we have

$$\frac{\partial E(x-\xi;\mu_n)}{\partial n_x}\Big|_{x=x_j} = 0, \quad \forall \xi \in \Gamma_j, \tag{7.155}$$

because from (7.153), $n_x \perp (x-\xi)$ for $x = x_j$ and $\xi \in \Gamma_j$. See Fig. 7.18.

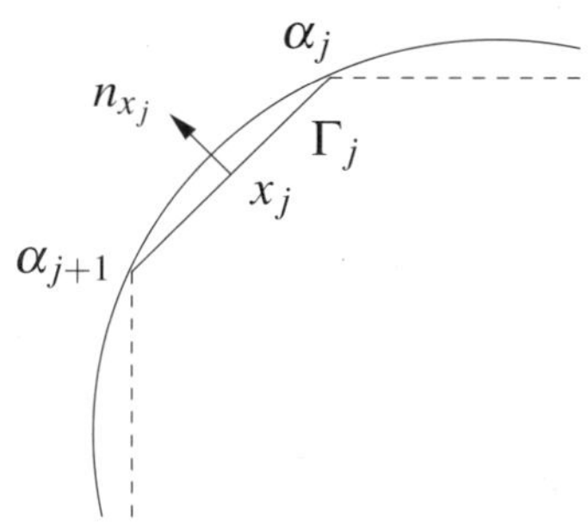

Figure 7.18: n_{x_j} is orthogonal to Γ_j.

Thus we can again assemble the equations in (7.154) into a linear matrix equation of the form

$$B(\mu_n)\vec{F}_n = 0$$

like (7.136), where

$$B(\mu_n) = [b_{ij}(\mu_n)]_{m\times m},$$

$$b_{ij}(\mu_n) = \begin{cases} \displaystyle\int_{\Gamma_j}\left[\frac{\partial E(x-\xi;\mu_n)}{\partial n_x}\right]\Big|_{x=x_i} d\sigma, & i \neq j, \\ \dfrac{1}{2}, & i = j. \end{cases}$$

The remaining procedures are identical to those in (7.134)–(7.141) and are therefore omitted.

Numerical examples of computation of the eigenvalues and eigenfunctions of Δ subject to *impedance boundary conditions* can be found in Zhou [193].

7.11 Numerical example (II): scattering of a plane wave by a 2D elliptic obstacle

Let Ω be an ellipse with major axis length 4 and minor axis length 3:

$$\Omega = \left\{(x_1,x_2) \in \mathbb{R}^2 \,\Big|\, \frac{x_1^2}{4^2} + \frac{x_2^2}{3^2} < 1\right\}. \tag{7.156}$$

Consider the scattering of acoustic waves of a given frequency on the exterior domain Ω^d. The obstacle is assumed to be sound-soft. Therefore, we have a Dirichlet problem

$$\left.\begin{aligned} &\Delta w(x) + k^2 w(x) = 0 \quad \text{on } \Omega^c, \\ &w(x) = 0 \qquad\qquad\qquad\; \text{on } \partial\Omega. \end{aligned}\right\} \tag{7.157}$$

Let the incident wave (cf. (7.46)) be a *plane wave* of the form

$$w_I(x) = e^{-ik(\eta_1 x_1 + \eta_2 x_2)}, \tag{7.158}$$

where $k = 1$ is chosen, and

$$\eta_1 = \cos\alpha, \quad \eta_2 = \sin\alpha. \tag{7.159}$$

α is the *angle of incidence* of the plane wave. Let $w_S(\cdot)$ be the scattered wave. Then from (7.47) and the first equation of (7.48), we see that w_S satisfies the exterior BVP

$$\left.\begin{aligned} &(\Delta + k^2) w_S(x) = 0 \quad \text{on } \Omega^c, \\ &w_S(x) = -e^{-ik(\eta_1 x_1 + \eta_2 x_2)}, \quad x \in \partial\Omega, \\ &w_S \text{ satisfies the Sommerfeld radiation condition (7.50).} \end{aligned}\right\} \tag{7.160}$$

We solve the above by the modified layer method of Theorem 7.13. Choose $\eta = 1$ in (7.124) and write

$$w_S(x) = \int_{\partial\Omega} \left[\frac{\partial E(x-\xi;k)}{\partial n_\xi} - iE(x-\xi;k) \right] f(\xi)\, d\sigma_\xi, \quad x \in \Omega^c, \tag{7.161}$$

where E is given as in (7.37). Thus, f satisfies the BIE (7.126):

$$\begin{aligned} &\frac{1}{2} f(x) + \int_{\partial\Omega} \left[\frac{\partial E(x-\xi;k)}{\partial n_\xi} - iE(x-\xi;k) \right] f(\xi)\, d\sigma_\xi \\ &\qquad = -e^{-ik(\eta_1 x_1 + \eta_2 x_2)}, \quad x \in \partial\Omega. \end{aligned} \tag{7.162}$$

We approximate $\partial\Omega$ by a polygon

$$\partial\Omega \sim \Gamma = \bigcup_{j=1}^{m} \Gamma_j, \tag{7.163}$$

where the vertices of the polygon Γ are located at

$$(x_1, x_2) = \left(4\cos\frac{2j\pi}{m}, 3\sin\frac{2j\pi}{m} \right), \quad j = 0, 1, \ldots, m-1.$$

In other words, each side Γ_j subtends an equal angle $2\pi/m$ from the origin. (This "equal-angle" discretization may not be the way of grid generation to yield the most accurate numerical solution, but it surely is the most convenient one.) The density f is approximated by a piecewise constant function

$$f(\cdot) \approx \sum_{j=1}^{m} f_j \chi_{\Gamma_j} \tag{7.164}$$

just as (7.135). From (7.163) and (7.164), we obtain a system of linear equations

$$\frac{1}{2}f_l + \sum_{j=1}^{m} f_j \int_{\Gamma_j} \left[\frac{\partial E(\widetilde{x}_l - \xi; k)}{\partial n_\xi} - iE(\widetilde{x}_l - \xi; k) \right] d\sigma_\xi$$
$$= -e^{-ik[\eta_1 \widetilde{x}_{l,1} + \eta_2 \widetilde{x}_{l,2}]}, \quad l = 1, 2, \ldots, m, \tag{7.165}$$

where

$$\widetilde{x}_l = (\widetilde{x}_{l,1}, \widetilde{x}_{l,2}) \text{ is the midpoint of } \Gamma_l.$$

In (7.165), the integral

$$\int_{\Gamma_j} E(\widetilde{x}_l - \xi; k)\, d\sigma_\xi \tag{7.166}$$

can be approximated in exactly the same way as described in (7.142)-(7.149). As for the integral

$$\int_{\Gamma_j} \frac{\partial E(\widetilde{x}_l - \xi; k)}{\partial n_\xi} d\sigma_\xi, \tag{7.167}$$

we note that

$$\frac{\partial E(\widetilde{x}_l - \xi; k)}{\partial n_\xi} = \frac{ik}{4} H_1^{(1)}(k|\widetilde{x}_1 - \xi|) \frac{\langle n_\xi, \widetilde{x}_l - \xi \rangle}{|\widetilde{x}_l - \xi|}. \tag{7.168}$$

Thus, if $l \neq j$, the above is smooth on Γ_j and a high-accuracy Gaussian quadrature would suffice for (7.167). If $l = j$, then

$$\langle n_\xi, \widetilde{x}_j - \xi \rangle = 0, \quad \forall \xi \in \Gamma_j; \tag{7.169}$$

thus (7.167) is just 0. Thus, all entries in (7.165) are easily computed and a unique numerical solution of $(f_1, f_2, \ldots, f_m)$ is obtained. Substituting (7.164) into (7.161), we obtain a numerical solution of the scattered field w_S.

In our computations, we use $m = 48$ in (7.163). The number of Gaussian quadrature points in integrals (7.166) and (7.167) (for $l \neq j$) is also chosen to be 48. Two values of incident angles

$$\alpha = \frac{1}{6}\pi \text{ and } \frac{1}{2}\pi$$

are chosen, and two sets of solutions have been plotted in Figs. 7.19–7.20, where the numerical solutions of the density f, graphs of the real and imaginary parts of the incident plane wave, and numerical solutions of the real and imaginary parts of the scattered wave and its intensity are plotted, respectively, in (a), (b), (c), (d) and (e) of each figure. The scattered wave seems to have the strongest intensity (i.e., highest peak) in the direction of reflection corresponding to the angle of incidence; see part (e) of each figure.

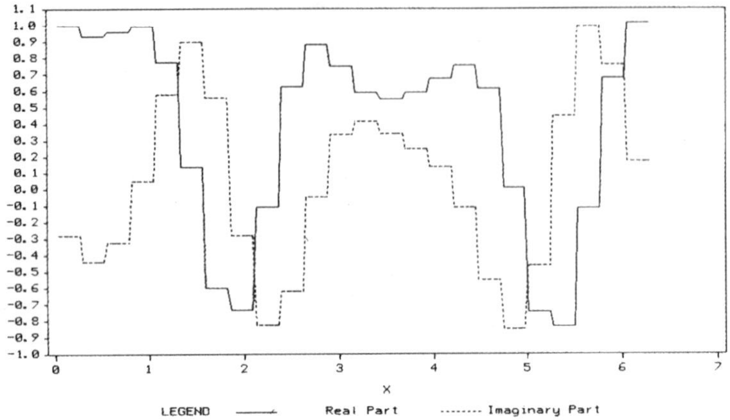

Figure 7.19: (a) Incident angle $\alpha = \frac{1}{6}\pi$: real and imaginary parts of the density f, plotted against the angle θ of polar coordinates.

7.12 Numerical example (III): minimizing the reflection of waves by boundary impedance

The material in this section is based on a modification of our earlier work in [33]. Consider the wave equation

$$y_{tt}(x,t) - \Delta y(x,t) = 0, \quad x \in \Omega^c \subset \mathbb{R}^N, \ t > 0, \tag{7.170}$$

with an impedance boundary condition

$$\frac{\partial y(x,t)}{\partial n} - Z(x)\frac{\partial y(x,t)}{\partial t} = 0, \quad x \in \partial\Omega, \ t > 0, \ Z(x) \in \mathbb{C}, \tag{7.171}$$

where $Z(\cdot)$ satisfying $\operatorname{Re} Z(x) \geqslant 0$ is the boundary impedance. Assume that the initial conditions

$$y(x,0) = y_0(x), \quad y_t(x,0) = y_1(x), \qquad x \in \Omega^c,$$

have finite energy, i.e.,

$$E(0) = \frac{1}{2}\int_{\Omega^c} [|y_1(x)|^2 + |\nabla_x y_0(x)|^2]\,dx < \infty.$$

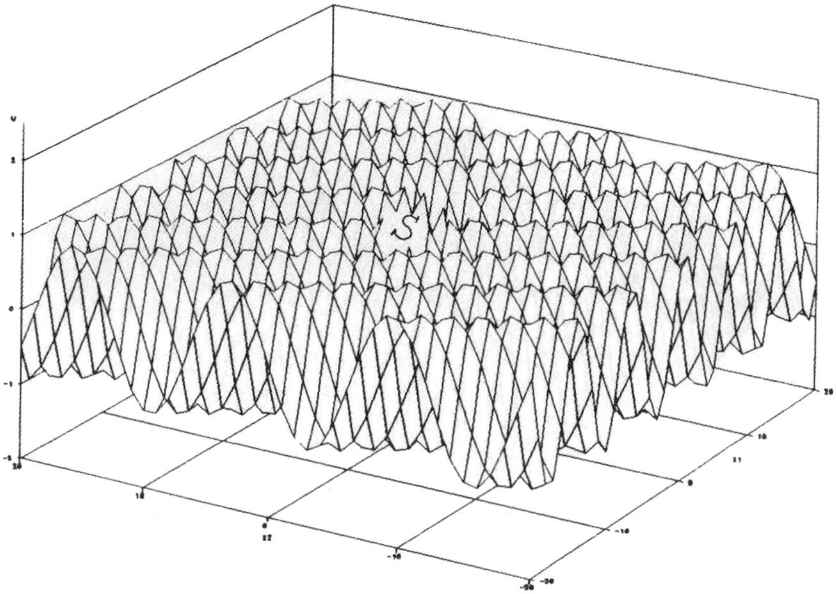

Figure 7.19: (b) Incident angle $\alpha = \frac{1}{6}\pi$: graph of the real part of the incident wave. $\mathscr{S}$ denotes the location of the scatterer.

Then

$$\begin{aligned}\frac{d}{dt}E(t) &= \frac{d}{dt}\frac{1}{2}\int_{\Omega^c}[|y_t(x,t)|^2 + |\nabla_x y(x,t)|^2]\,dx\\ &= \frac{1}{2}\int_{\Omega^c}[y_{tt}(x,t)\overline{y}_t(x,t) + y_t(x,t)\overline{y}_{tt}(x,t)\\ &\quad + \nabla_x y_t(x,t)\cdot\nabla_x\overline{y}(x,t) + \nabla_x y(x,t)\cdot\nabla_x\overline{y}_t(x,t)]\,dx\\ &= -\mathrm{Re}\int_{\partial\Omega} y_t(x,t)\frac{\overline{\partial y(x,t)}}{\partial n}\,d\sigma\\ &= -\int_{\partial\Omega}[\mathrm{Re}\,Z(x)]|y_t(x,t)|^2\,d\sigma \leqslant 0,\end{aligned}$$

so the energy of the total field at time t is decreasing. As we have mentioned in § 7.1 **(iii)**, the impedance boundary condition models wave-absorbing surfaces. For example, in modern "stealth" aircraft technology, the surface coating on the aircraft produces an impedance effect that diminishes the intensity of reflected waves, thereby making the aircraft "invisible" to the radar.

When the incident wave is monochromatic,

$$y_I(x,t) = e^{-ikt}w_I(x), \quad x\in\Omega^c, \quad \mathrm{Im}\,k \geqslant 0,$$

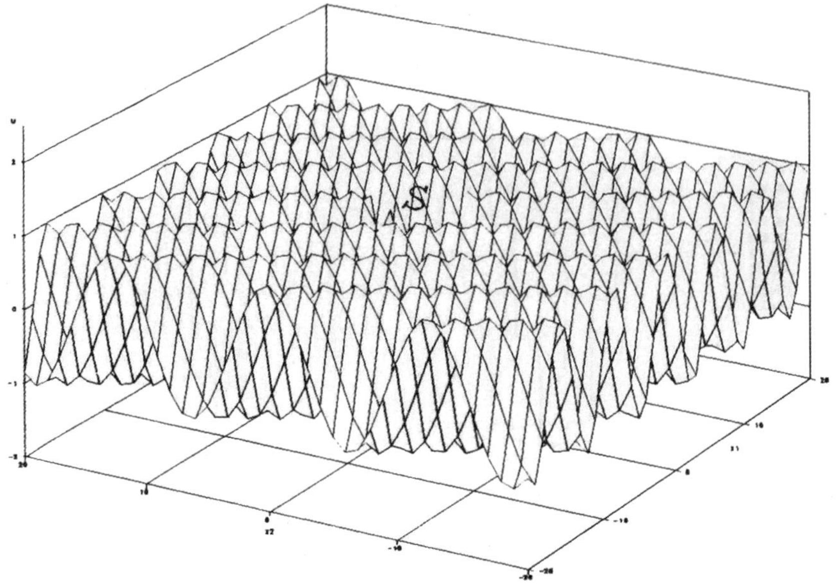

Figure 7.19: (c) Incident angle $\alpha = \frac{1}{6}\pi$: graph of the imaginary part of the incident wave.

the scattered wave is also monochromatic, thus,

$$y_S(x,t) = e^{-ikt} w_S(x).$$

From $y = y_I + y_S$ and (7.170) and (7.171), we obtain the exterior BVP

$$\left.\begin{aligned} &(\Delta + k^2) w_S(x) = 0 \quad \text{on } \Omega^c, \\ &\frac{\partial w_S(x)}{\partial n} + ikZ(x) w_S(x) = -\left[\frac{\partial w_I(x)}{\partial n} + ikZ(x) w_I(x)\right], \quad x \in \partial\Omega, \end{aligned}\right\} \tag{7.172}$$

where the boundary condition is of the third type in (7.8)

We let the boundary impedance $Z(\cdot)$ be the *control variable* and study the minimization of the scattered field at certain specified points for a given incident plane wave. Suppose the cost functional to be minimized is

$$J(Z) = \sum_{j=1}^{3} |w_S(p_i, Z)|, \tag{7.173}$$

where

$$p_1 = (-16, 16), \quad p_2 = (0, 16\sqrt{2}), \quad p_3 = (16, 16) \tag{7.174}$$

are the selected locations. The impedance $Z(\cdot)$ is subject to certain design constraints, such as

$$Z(x) \in U_{\text{ad}} \equiv \{z \in \mathbb{C} \mid 0 \leqslant \text{Re}\, z \leqslant 4, -2 \leqslant \text{Im}\, z \leqslant 2\}. \tag{7.175}$$

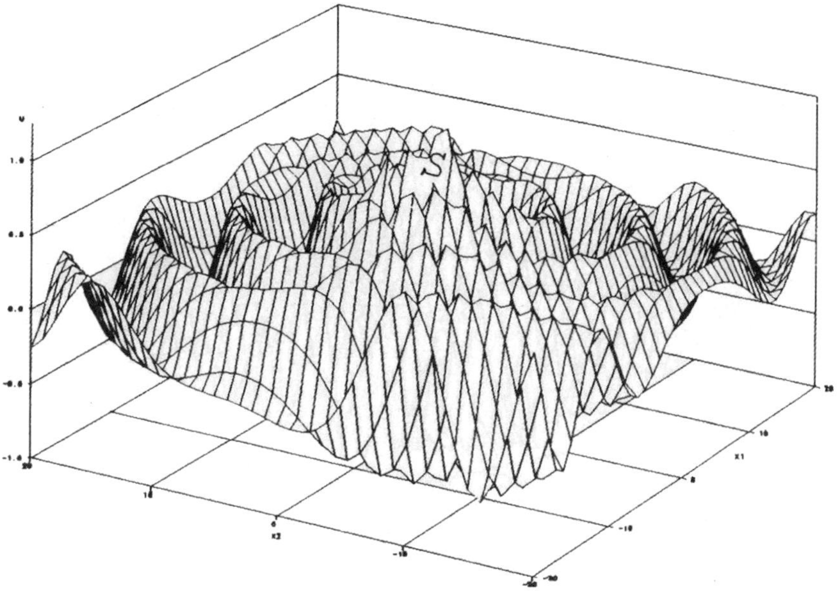

Figure 7.19: (d) Incident angle $\alpha = \frac{1}{6}\pi$: numerical solution of the scattered wave (real part).

Let the incident field be

$$w_I(x) = e^{-i(x_1/2+\sqrt{3}x_2/2)}, \quad x \in \Omega^c, \tag{7.176}$$

a plane wave whose direction of propagation (i.e., angle of incidence) makes an angle $\frac{1}{6}\pi$ with the x_1-axis.

The scatterer Ω considered here is elliptical, the same as (7.156).

We thus have an optimal control/design problem:

$$\left.\begin{array}{l} \inf J(Z), \\ Z(x) \in U_{\mathrm{ad}}, \quad x \in \partial\Omega, \\ \text{subject to (7.172) and (7.176).} \end{array}\right\} \tag{7.177}$$

We consider two cases. In Case I, the impedance $Z(\cdot)$ is required to be constant on the entire $\partial\Omega$, while in Case II, $Z(\cdot)$ can vary on $\partial\Omega$.

Case I: *Optimal constant impedance*

We discretize the boundary ellipse by a polygon Γ with 10 sides, where each side subtends an angle $2\pi/10$ from the origin. See Fig. 7.21.

Let $\widehat{Z} \in \mathbb{C}$ be an optimal solution to (7.177). Since the cost functional J in (7.173) is nonconvex with respect to Z, problem (7.177) may have nonunique solutions. Nevertheless,

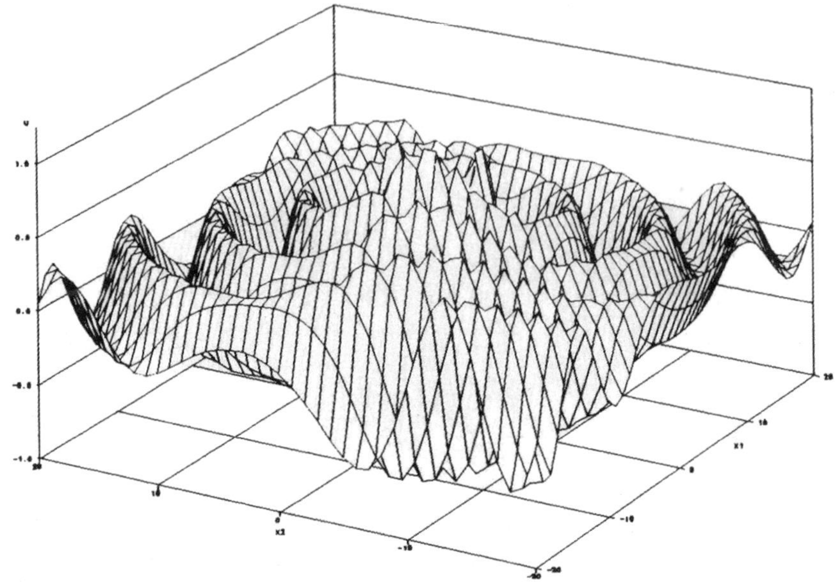

Figure 7.19: (e) Incident angle $\alpha = \frac{1}{6}\pi$: numerical solution of the scattered wave (imaginary part).

each optimal solution $\widehat{Z}$ is expected to satisfy

$$\operatorname{Re}\widehat{Z} > 0, \tag{7.178}$$

since we understand that the impedance effect will reduce wave scattering. From (7.178), we can apply Theorem 7.9, namely the BIE (7.113) based upon the Green's formula:

$$\begin{aligned}
&\frac{1}{2}w_S(x) - \int_{\partial\Omega}\left[\frac{\partial E(x-\xi;k)}{\partial n_\xi} + ikZE(x-\xi;k)\right] w_S(\xi)\, d\sigma_\xi \\
&= \int_{\partial\Omega} E(x-\xi;k)\left[\frac{\partial w_I(x)}{\partial n} + ikZw_I(z)\right] d\sigma_\xi,\ x \in \partial\Omega,
\end{aligned} \tag{7.179}$$

where $k = 1$ and w_I is given by (7.176). (Here we deliberately have $-k^2 = -1$ *not* an eigenvalue of Δ on Ω so that Theorem 7.9 becomes applicable. Otherwise there is no guarantee that the above BIE is invertible.) Again, let $w_S(\cdot)$ on $\partial\Omega$ be approximated by piecewise constants according to $\Gamma = \bigcup_{j=1}^{10}\Gamma_j$ in Fig. 7.21. The procedures are as follows.

(1) Use the discretized form of (7.179) to obtain an approximate solution of $w_S(\cdot)$ on Γ.

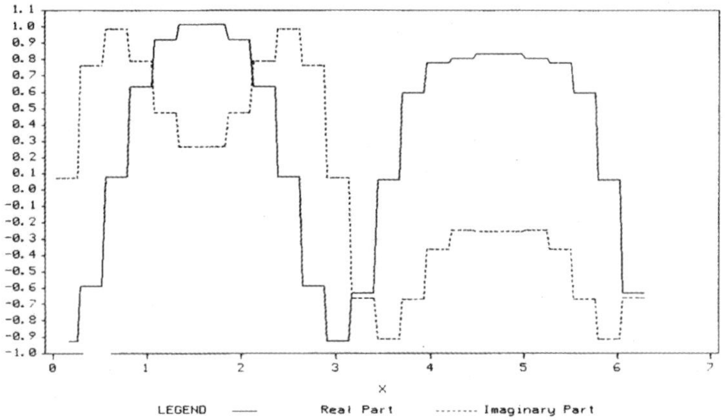

Figure 7.20: (a) Incident angle $\alpha = \frac{1}{2}\pi$: real and imaginary parts of the density f, plotted against the angle θ of polar coordinates.

(2) Substitute $w_S(\cdot)$ into (7.114) to obtain $w_S(x)$ at the points $x = p_1,\ p_2$ and p_3:

$$w_S(p_j, Z) = \int_{\partial\Omega} \left[\frac{\partial E(p_j - \xi; k)}{\partial n_\xi} + ikZE(p_j - \xi; k) \right] w_S(\xi)\, d\sigma_\xi$$
$$+ \int_{\partial\Omega} E(p_j - \xi; k) \left[\frac{\partial w_I(\xi)}{\partial n} + ikZw_I(\xi) \right] d\sigma_\xi, \quad j = 1, 2, 3.$$

(3) Substitute the above values into (7.173) to obtain $J(Z)$.

(4) Minimize $J(Z)$ subject to the constraint $Z \in U_{\text{ad}}$. Because of the lack of convexity and the occurrence of multiple local minima of $J(Z)$, as we mentioned earlier, we have adopted a global optimization algorithm due to Zheng, see [74].

The optimal constant impedance and the corresponding cost are computed to be

$$\widehat{Z} = 0.8848 + i0.001619,$$
$$J(\widehat{Z}) = 0.06721. \tag{7.180}$$

The scattered field w_S is plotted in Figs. 7.22 (a) and (b). Its intensity $|w_S(x; \widehat{Z})|$ is plotted in Fig. 7.23. The reader can see that in the "*illuminated region*" of the incident wave, the reflection has been reduced considerably. Compare the same area in Fig. 7.24 of the scattered field intensity $|w_S(x, 0)|$, where the perfectly reflecting boundary condition $\partial w/\partial n = 0$ (of the total field) corresponding to $Z = 0$ has been used. The corresponding cost is $J(0) = 0.8831$, considerably larger than (7.36). Note that in the "*penumbra*" or "*shadow*" regions (see Fig. 7.25) of the incident wave, there is still considerable wave scattering (i.e., diffraction) where boundary impedance is ineffective in minimizing the scattering intensity.

Case II: *Variable boundary impedance*

In this case, the variable boundary impedance $Z(x)$ is approximated by a piecewise-constant function

$$Z(\cdot) \approx \sum_{j=1}^{10} Z_j \chi_{\Gamma_j}$$

on the approximate polygonal boundary $\Gamma = \bigcup_{j=1}^{10} \Gamma_j$. Since all the procedures are similar to Case I, we omit the details. The optimal variable impedance $\widehat{Z}(x)$ has been computed, and is tabulated in Table 7.2.

Table 7.2: Approximate piecewise-constant solution to the optimal variable boundary impedance $\widehat{Z}(\cdot)$.

j	$\mathrm{Re}\,\widehat{Z}(\widetilde{x}_j)$	$\mathrm{Im}\,\widehat{Z}(\widetilde{x}_j)$
1	1.9674269	0.6844930
2	1.6444479	-0.5589241
3	2.1538300	0.7442251
4	2.0043140	-0.3603203
5	1.8312809	0.3701571
6	1.1527685	0.1014286
7	1.7316518	0.7297065
8	3.6680615·	-0.6052722
9	2.1898028	-07248696
10	2.4092668	0.8712740

The corresponding cost is computed to be

$$J(\widehat{Z}(\cdot)) = 0.0146386,$$

which is even smaller than (7.180), the constant impedance case.

The corresponding scattered wave intensity $|w_S(\cdot;\widehat{Z})|$ is plotted in Fig. 7.26. The reader may find a rather sharp attenuation of the scattered field along three rays passing p_1, p_2 and p_3. Elsewhere in the illuminated region, the scattered field is not as uniformly diminished in intensity as in Case I, however. Compare Fig. 7.26 with Fig. 7.25.

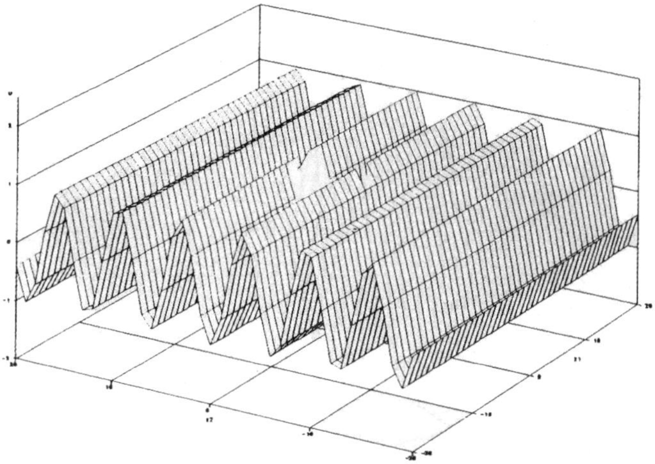

Figure 7.20: (b) Incident angle $\alpha = \frac{1}{2}\pi$: graph of the real part of the incident wave. $\mathscr{S}$ denotes the location of the scatterer.

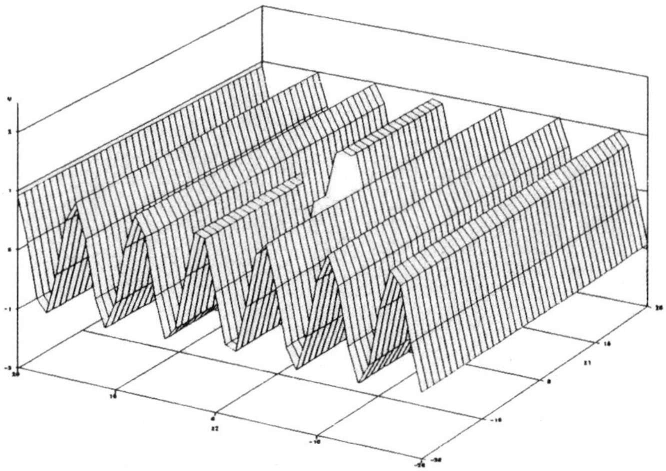

Figure 7.20: (c) Incident angle $\alpha = \frac{1}{2}\pi$: graph of the imaginary part of the incident wave.

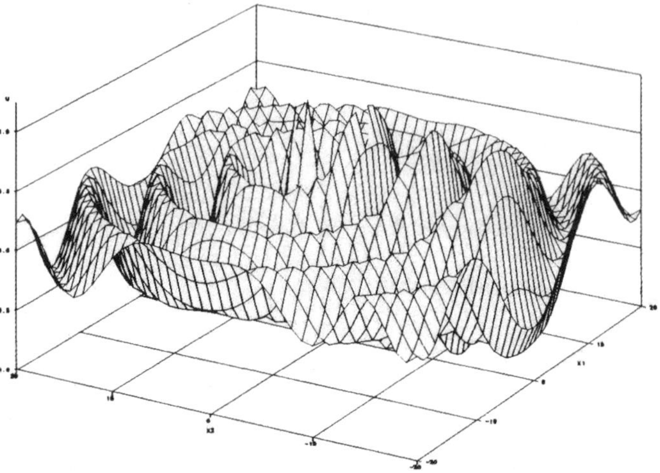

Figure 7.20: (d) Incident angle $\alpha = \frac{1}{2}\pi$: numerical solution of the scattered wave (real part).

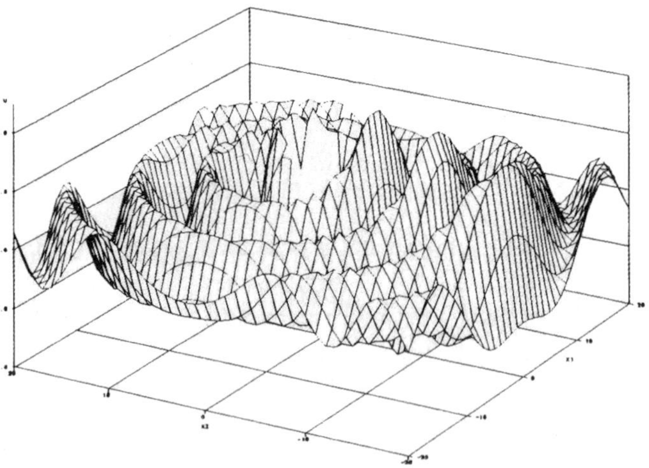

Figure 7.20: (e) Incident angle $\alpha = \frac{1}{2}\pi$: numerical solution of the scattered wave (imaginary part).

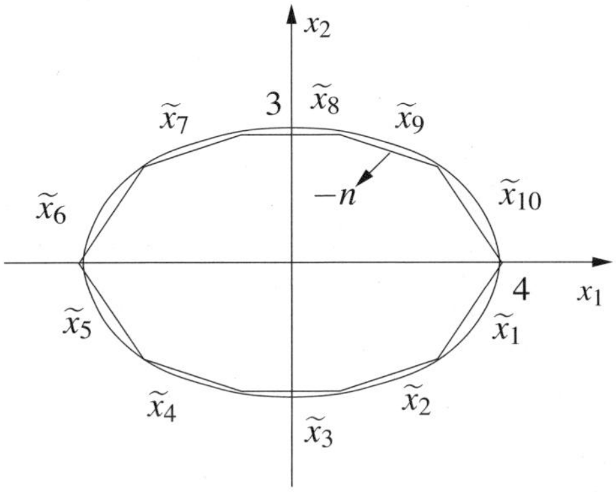

Coordinates of nodal points $\widetilde{x}_i$

i	x_1	x_1
1	3.62	−0.88
2	2.24	2.31
3	0.00	−2.85
4	−2.24	−2.31
5	−3.62	−0.88
6	−3.62	0.88
7	−2.24	2.31
8	0.00	2.85
9	2.24	2.31
10	3.62	0.88

Figure 7.21: Discretization of the boundary of the elliptical scatterer for § 7.12.

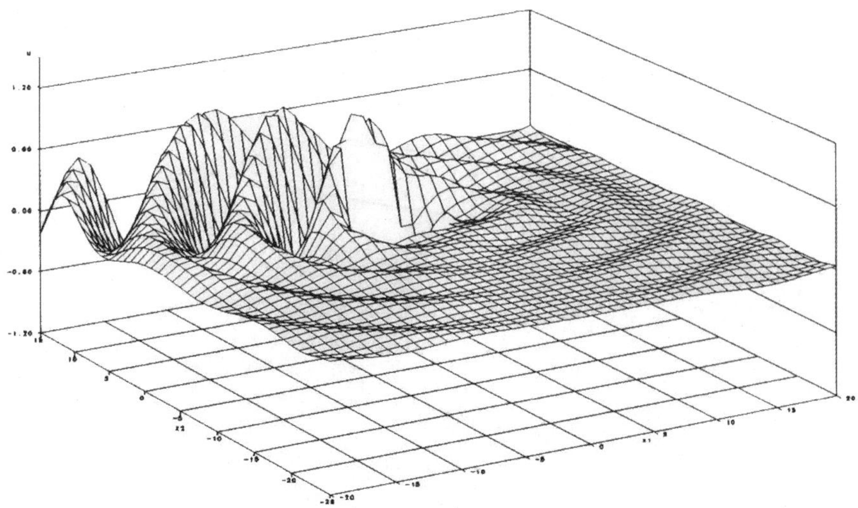

Figure 7.22: (a) The scattered field (real part) corresponding to the optimal constant boundary impedance.

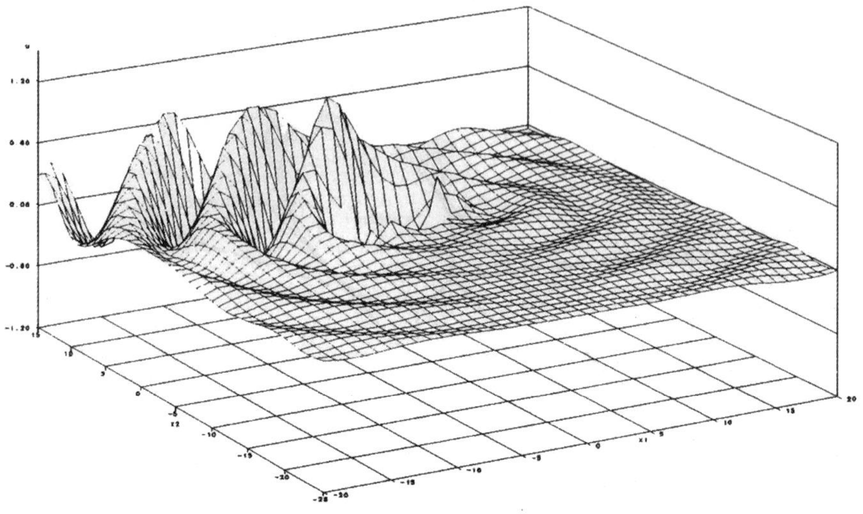

Figure 7.22: (b) The scattered field (imaginary part) corresponding to the optimal constant boundary impedance.

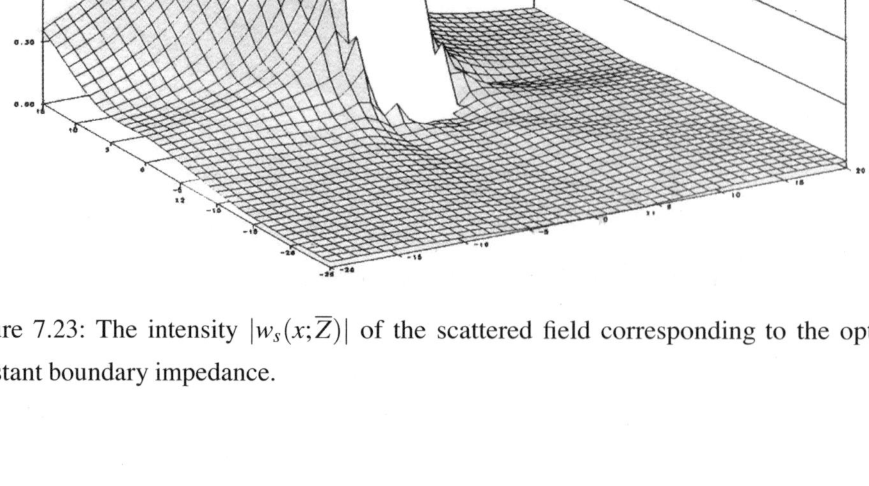

Figure 7.23: The intensity $|w_s(x;\overline{Z})|$ of the scattered field corresponding to the optimal constant boundary impedance.

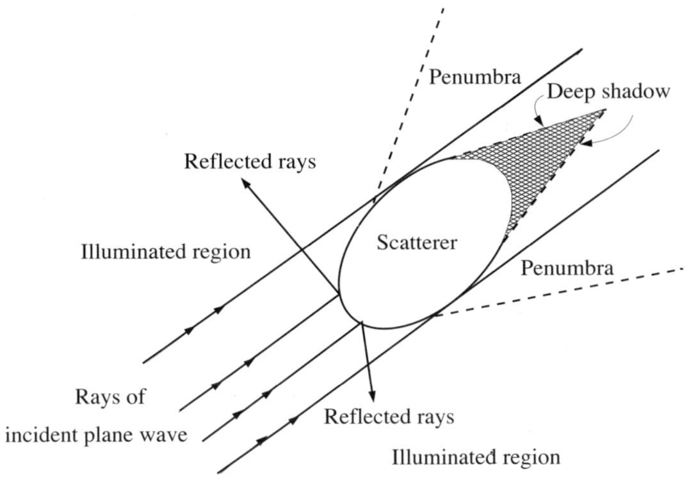

Figure 7.24: The illuminated, penumbra and shadow regions near the scatterer.

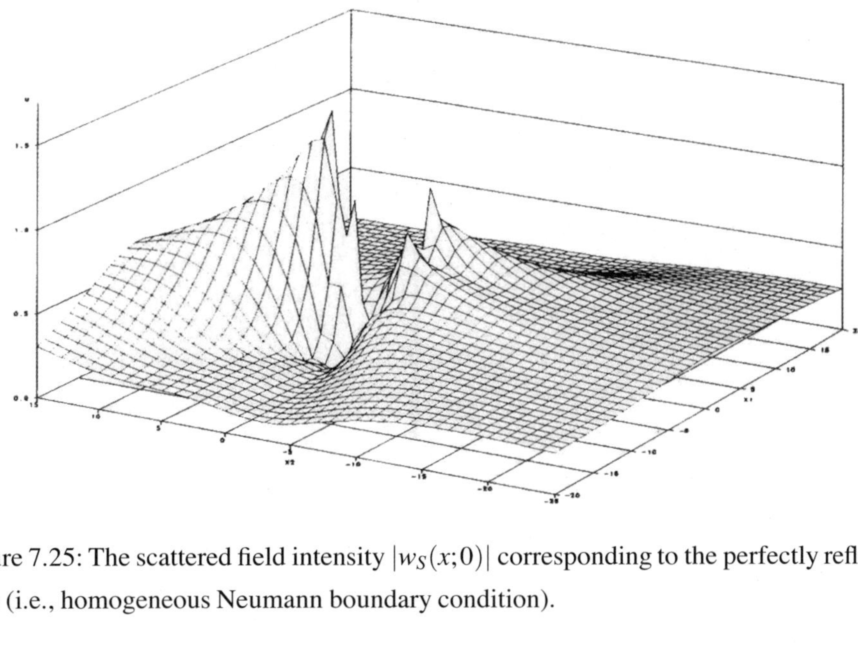

Figure 7.25: The scattered field intensity $|w_S(x;0)|$ corresponding to the perfectly reflecting case (i.e., homogeneous Neumann boundary condition).

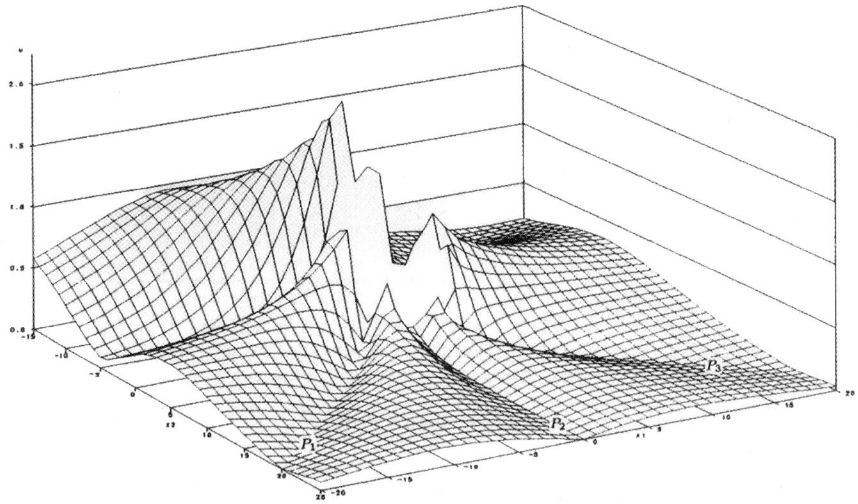

Figure 7.26: The intensity $|w_S(x)|$ of the scattered field corresponding to the optimal variable coefficient boundary impedance, where p_1, p_2 and p_3 are the selected locations (7.174) for the wave to be minimized.

Chapter 8

The Thin Plate Equation

The dynamic thin plate equation

$$\rho w_{tt}(x,t) + D\Delta^2 w(x,t) = 0, \qquad x \in \Omega \subseteq \mathbb{R}^2,\ t > 0, \tag{8.1}$$

where

$$w(x,t) \text{ is the vertical deflection at location } x \text{ at time } t,$$
$$\rho \text{ is mass density per unit area } > 0,$$
$$D \text{ is flexual rigidity } > 0,$$
$$\Delta^2 = \left(\frac{\partial^2}{\partial x_1^2} + \frac{\partial^2}{\partial x_2^2}\right)^2 \text{ is the biharmonic operator,}$$

was first proposed by Kirchhoff [112] to model the vibration of a thin plate subject to *pure bending* with small deflection. It is an important mathematical model in structural mechanics and seismology. When the plate has settled into an elastostatic equilibrium, w is no longer time-dependent; thus

$$\Delta^2 w(x) = 0, \qquad x \in \Omega, \tag{8.2}$$

i.e., w satisfies the *biharmonic equation. Four* types of variational boundary conditions naturally occur for the elastostatic thin plate equation; see § 8.1. We will formulate BIE for each type of those boundary conditions and compute them numerically by collocation.

The biharmonic equation (8.2) is also a standard equation in *two dimensional elasticity* for a homogeneous isotropic elastostatic medium (cf. (9.22) in Chapter 9). Such a medium is also called a plate subject to *stretching* (vis-*à*-vis *pure bending*). The function w in (8.2) normally signifies the Airy stress function, which does not have any concrete physical meaning. To treat such 2D elasticity problems, there exists a vast amount of literature using the *complex variable method* developed by the Russian school [137] (and others). In principle, those integral equations derived in [137] could also be collocated and discretized to yield numerical solutions. Nevertheless, they seem rather unwieldy to manipulate, since the kernels of the integral operators occurring there often appear in complicated forms that are very different from the classical simple- or double-layer expressions.

The biharmonic equation modelling either stretching or pure bending of a plate can also be computed by the standard *finite-element method* (FEM) for any properly posed boundary conditions. Nevertheless, using FEM for the biharmonic equation requires either the use of higher-order *conforming elements* or the check of some *patch test* if lower-order nonconforming elements are used, since the equation is now of higher order than the potential equation. This extra work plus the usual isoparametric transformation for the curved boundary can be quite burdensome for the programmer. The computer memory storage and the number of quadrature evaluations required are also large as the interior of the domain is discretized. In fact, as the reader can clearly determine, the higher-order the PDE gets, the more advantageous and effective BEM becomes over FEM.

Some past work on the BIE/BEM approach for the biharmonic equation includes that of Hsiao and MacCamy [95] on the second biharmonic problem in 2D linear elasticity, Costable, Lusikka and Saranen [54] on the Kirchhoff plate subject to clamped boundary conditions, and Hartmann and Zotemantel [88] on the Rayleigh–Green formula approach and algorithms for the clamped Kirchhoff plate. See also the references therein. However, in the existing literature very little can be found about the BIE/BEM treatment for the thin plate subject to three other types of boundary conditions such as *simply-supported, roller-supported* and *free* boundary conditions. In this chapter, we will provide unified and comprehensive formulations for all types of boundary conditions. Their rigorous mathematical proofs have been only partially completed, but the validity of the stated theorems is evident based upon the analogous nature of these BIE to the potential equation case.

8.1 The Kirchhoff thin static plate model subject to pure bending

Let a thin plate in elastostatic equilibrium have flexual rigidity D and Poisson ratio ν, with $0 < \nu < \frac{1}{2}$. Consider an infinitesimal element of the plate as shown in Fig. 8.1. Let M_{x_1} and M_{x_2} be the bending moments per unit length acting on the edges parallel to the x_2 and x_1 axes respectively.

According to Timoshenko and Woinowsky–Krieger [178, pp. 38–39], we have

$$M_{x_1} = -D\left(\frac{\partial^2 w}{\partial x_1^2} + \nu\frac{\partial^2 w}{\partial x_2^2}\right),$$

$$M_{x_2} = -D\left(\frac{\partial^2 w}{\partial x_2^2} + \nu\frac{\partial^2 w}{\partial x_1^2}\right).$$

The work done by the moments on the infinitesimal plate element is

$$\begin{aligned} dV_1 &= \left(-\frac{1}{2}M_{x_1}\frac{\partial^2 w}{\partial x_1^2}\,dx_1\,dx_2\right) + \left(-\frac{1}{2}M_{x_2}\frac{\partial^2 w}{\partial x_2}\,dx_2\,dx_2\right) \\ &= \frac{1}{2}D\left[\left(\frac{\partial^2 w}{\partial x_1^2}\right)^2 + \left(\frac{\partial^2 w}{\partial x_2^2}\right)^2 + 2\nu\frac{\partial^2 w}{\partial x_1^2}\frac{\partial^2 w}{\partial x_2^2}\right]dx_1\,dx_2. \end{aligned}$$

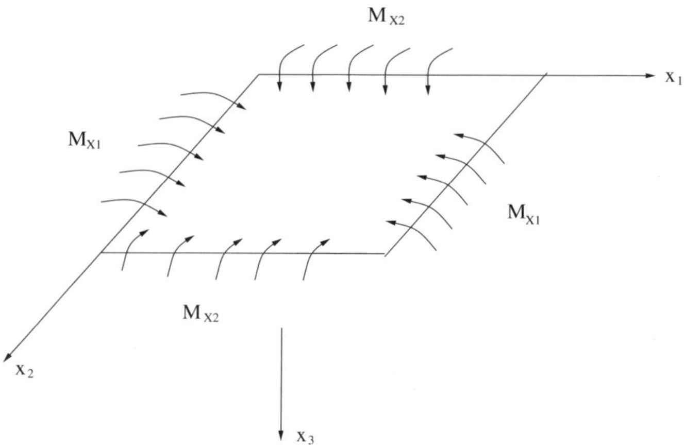

Figure 8.1: Pure bending of an infinitesimal plate element.

In addition to the above strain energy due to bending moments, there is also the contribution due to *twisting moments* $M_{x_1x_2}\,dx_2$ and $M_{x_2x_1}\,dx_1$, where

$$M_{x_1x_2} = M_{x_2x_1} = D(1-\nu)\frac{\partial^2 w}{\partial x_1 \partial x_2}, \qquad \text{(cf. [178, (43), p. 41]).}$$

The sum of their strain energy contribution is [178, p. 47]

$$\begin{aligned} dV_2 &= \frac{1}{2}(M_{x_1x_2}\,dx_2)\left(\frac{\partial^2 w}{\partial x_1\partial x_2}\,dx_1\right) + \frac{1}{2}(M_{x_2x_1}\,dx_1)\left(\frac{\partial^2 w}{\partial x_2\partial x_1}\,dx_2\right) \\ &= D(1-\nu)\left(\frac{\partial^2 w}{\partial x_1\partial x_2}\right)^2 dx_1\,dx_2. \end{aligned}$$

Since the bending moments and twisting moments work independently of each other, the total strain energy of the plate is

$$\begin{aligned} V(w) &= \int_\Omega dV = \int_\Omega (dV_1 + dV_2) \\ &= \frac{D}{2}\int_\Omega \{|\Delta w(x)|^2 + 2(1-\nu)[w_{x_1x_2}^2(x) - w_{x_1x_1}(x)w_{x_2x_2}(x)]\}\,dx. \end{aligned}$$

For convenience, from now on we assume that the flexual rigidity D has been normalized to $D = 1$.

We now apply the principle of virtual work. Let w be the equilibrium state of the plate. Assume that $q(\cdot)$ is the external lateral load distributed over the plate, and there is no boundary stress acting on $\partial\Omega$. Then

$$\delta V(w) = \delta \int_\Omega q(x)w(x)\,dx.$$

We get

$$\begin{aligned}0 &= \delta V(w) - \int_\Omega q(\delta w)\,dx \\ &= \int_\Omega \{\Delta w \Delta(\delta w) + (1-\nu)\left[2w_{x_1x_2}(\delta w)_{x_1x_2} - w_{x_1x_1}(\delta w)_{x_2x_2} - (\delta w)_{x_1x_1}w_{x_2x_2}\right]\}\,dx_1\,dx_2 \\ &\qquad - \int_\Omega q(\delta w)\,dx_1\,dx_2, \end{aligned} \tag{8.3}$$

for all virtual displacement δw.

Lemma 8.1 (Rayleigh–Green formula). *Define a bilinear form*

$$a(w,v) = \int_\Omega [\Delta w \Delta v + (1-\nu)(2w_{x_1x_2}v_{x_1x_2} - w_{x_1x_1}v_{x_2x_2} - w_{x_2x_2}v_{x_1x_1})]\,dx, \qquad 0 < \nu < \frac{1}{2}. \tag{8.4}$$

Then for w and v sufficiently smooth, we have

$$a(w,v) = \int_\Omega (\Delta^2 w)v\,dx - \int_{\partial\Omega}\left[(B_1 w)v - (B_2 w)\frac{\partial v}{\partial n}\right]d\sigma, \tag{8.5}$$

where

$$\begin{aligned} B_1 w &= \frac{\partial \Delta w}{\partial n} - (1-\nu)\frac{\partial}{\partial\sigma}\left[n_1 n_2\left(\frac{\partial^2 w}{\partial x_1^2} - \frac{\partial^2 w}{\partial x_2^2}\right) - (n_1^2 - n_2^2)\frac{\partial^2 w}{\partial x_1 \partial x_2}\right], \\ B_2 w &= \nu\Delta w + (1-\nu)\left[n_1^2\frac{\partial^2 w}{\partial x_1^2} + n_2^2\frac{\partial^2 w}{\partial x_2^2} + 2n_1n_2\frac{\partial^2 w}{\partial x_1\partial x_2}\right]. \end{aligned}$$

$B_1 w$ is the transverse force (consisting of the shear force and twisting moment), $B_2 w$ is the bending moment, $n = (n_1, n_2)$ is the unit outward normal on $\partial\Omega$, and $\partial/\partial\sigma = -n_2\partial/\partial x_1 + n_1\partial/\partial x_2$ is the counterclockwise tangential derivative along $\partial\Omega$.

Proof. Consider the first term on the RHS of (8.4). Integration by parts twice gives

$$\begin{aligned}\int_\Omega \Delta w \Delta v\,dx &= \int_{\partial\Omega} \Delta w \frac{\partial v}{\partial n}\,d\sigma - \int_\Omega \nabla(\Delta w)\cdot\nabla v\,dx \\ &= \int_{\partial\Omega} \Delta w \frac{\partial v}{\partial n}\,d\sigma - \int_{\partial\Omega}\frac{\partial \Delta w}{\partial n} v\,d\sigma + \int_\Omega (\Delta^2 w)v\,dx.\end{aligned}$$

Similarly, for the second term on the right of (8.4), we have

$$\begin{aligned}\int_\Omega w_{x_1x_2}v_{x_1x_2}\,dx &= \int_{\partial\Omega} n_2 w_{x_1x_2}v_{x_1}\,d\sigma - \int_\Omega w_{x_2x_1x_2}v_{x_1}\,dx\\ &= \int_{\partial\Omega} n_2 w_{x_1x_2}v_{x_1}\,d\sigma - \int_{\partial\Omega} n_1 w_{x_2x_2}v_{x_1}\,d\sigma + \int_\Omega w_{x_2x_2}v_{x_1x_1}\,dx,\end{aligned}$$

and

$$\begin{aligned}\int_\Omega w_{x_1x_2}v_{x_1x_2}\,dx &= \int_{\partial\Omega} n_1 w_{x_1x_2}v_{x_2}\,d\sigma - \int_\Omega w_{x_1x_1x_2}v_{x_2}\,dx\\ &= \int_{\partial\Omega} n_1 w_{x_1x_2}v_{x_2}\,d\sigma - \int_{\partial\Omega} n_2 x_{x_1x_1}v_{x_2}\,d\sigma + \int_\Omega w_{x_1x_1}v_{x_2x_2}\,dx.\end{aligned}$$

So we obtain

$$\begin{aligned}a(w,v) = \int_\Omega(\Delta^2 w)v\,dx + \int_{\partial\Omega}\Big[\Delta w\frac{\partial v}{\partial n} - \frac{\partial\Delta w}{\partial n}v + (1-\nu)(n_2 w_{x_1x_2}v_{x_1} - n_1 w_{x_2x_2}v_{x_1}\\ + n_1 w_{x_1x_2}v_{x_2} - n_2 w_{x_1x_1}v_{x_2})\Big]\,d\sigma. \qquad (8.6)\end{aligned}$$

Now, from

$$\begin{aligned}\frac{\partial v}{\partial n} &= n_1\frac{\partial v}{\partial x_1} + n_2\frac{\partial v}{\partial x_2},\\ \frac{\partial v}{\partial \sigma} &= -n_2\frac{\partial v}{\partial x_1} + n_1\frac{\partial v}{\partial x_2},\end{aligned}$$

we get

$$\begin{aligned}\frac{\partial v}{\partial x_1} &= n_1\frac{\partial v}{\partial n} - n_2\frac{\partial v}{\partial \sigma},\\ \frac{\partial v}{\partial x_2} &= n_2\frac{\partial v}{\partial n} + n_1\frac{\partial v}{\partial \sigma}.\end{aligned}$$

Substituting them into (8.6), we get

$$\begin{aligned}a(w,v) = \int_\Omega(\Delta^2 w)v\,dx + \int_{\partial\Omega}\Big\{\Delta w\frac{\partial v}{\partial n} - \frac{\partial\Delta w}{\partial n}v\\ + (1-\nu)\Big[w_{x_1x_2}\Big(n_1n_2\frac{\partial v}{\partial n} - n_2^2\frac{\partial v}{\partial\sigma} + n_1n_2\frac{\partial v}{\partial n} + n_1^2\frac{\partial v}{\partial\sigma}\Big)\\ - w_{x_1x_1}\Big(n_2^2\frac{\partial v}{\partial n} + n_1n_2\frac{\partial v}{\partial\sigma}\Big) - w_{x_2x_2}\Big(n_1^2\frac{\partial v}{\partial n} - n_1n_2\frac{\partial v}{\partial\sigma}\Big)\Big]\Big\}\,d\sigma.\end{aligned}$$

(8.7)

We use the integration by parts formula on $\partial\Omega$

$$\int_{\partial\Omega} f(\sigma)\frac{\partial g(\sigma)}{\partial\sigma}\,d\sigma = -\int_{\partial\Omega}\frac{\partial f(\sigma)}{\partial\sigma}g(\sigma)\,d\sigma,$$

and get

$$\int_{\partial\Omega} \left[w_{x_1x_2}(-n_2^2+n_1^2) - (w_{x_1x_1} - w_{x_2x_2})n_1n_2\right] \frac{\partial v}{\partial \sigma} d\sigma$$
$$= -\int_{\partial\Omega} \frac{\partial}{\partial \sigma} \left[(n_1^2 - n_2^2)w_{x_1x_2} - n_1n_2(w_{x_1x_1} - w_{x_2x_2})\right] v\, d\sigma.$$

So (8.7) becomes

$$a(w,v) = \int_\Omega (\Delta^2 w) v\, dx + \int_{\partial\Omega} \left\{ \Delta w \frac{\partial v}{\partial n} - \left(\frac{\partial}{\partial n}\Delta w\right) v \right.$$
$$+ (1-\nu)\left\{ (2n_1n_2w_{x_1x_2} - n_2^2 w_{x_1x_1} - n_1^2 w_{x_2x_2})\frac{\partial v}{\partial n} \right.$$
$$\left.\left. - \frac{\partial}{\partial \sigma}[(n_1^2 - n_2^2)w_{x_1x_2} - n_1n_2(w_{x_1x_1} - w_{x_2x_2})]v \right\} \right\} d\sigma.$$

Combining terms involving v and $\partial v/\partial n$ and using the relation $n_1^2 + n_2^2 = 1$, we get (8.5). □

From (8.3) and Lemma 8.1, we have

$$0 = \delta V(w) - \int_\Omega q(\delta w)\, dx$$
$$= a(w, \delta w) - \int_\Omega q(\delta w)\, dx$$
$$= \int_\Omega (\Delta^2 w - q)(\delta w)\, dx - \int_{\partial\Omega} \left[(B_1 w)(\delta w) - (B_2 w)\left(\frac{\partial \delta w}{\partial n}\right)\right] d\sigma.$$

Therefore

$$\Delta^2 w(x) = q(x) \quad \text{on} \quad \Omega. \tag{8.8}$$

On $\partial\Omega$, several types of boundary conditions are possible. They include, at least, the following *four homogeneous variational types*:

(i) clamped

$$w = 0,$$
$$\frac{\partial w}{\partial n} = 0;$$

(ii) free

$$B_1 w = 0,$$
$$B_2 w = 0;$$

(iii) hinged (or simply supported)

$$w = 0,$$
$$B_2 w = 0;$$

(iv) roller-supported

$$\frac{\partial w}{\partial n} = 0,$$
$$B_1 w = 0.$$

See Fig. 8.2.

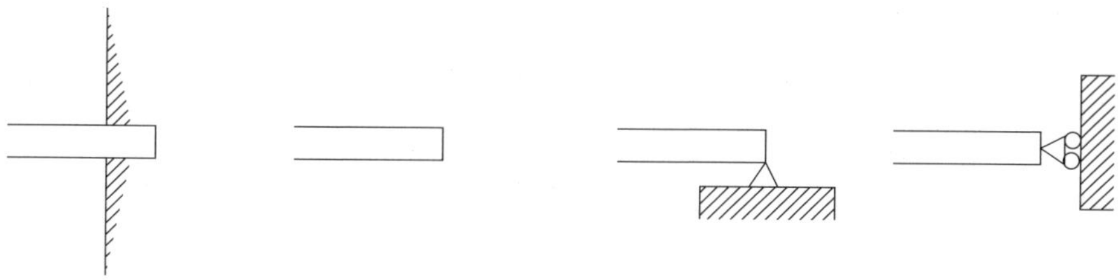

(a) (b) (c) (d)

Figure 8.2: Boundary conditions of a plate (a) clamped; (b) free; (c) hinged; (d) roller supported.

Because the lateral loading $q(x)$ in (8.8) can always be eliminated by subtracting a volume potential as in (6.15)–(6.19), we may simply assume that $q(x) \equiv 0$ in (8.8); thus

$$\Delta^2 w(x) = 0 \quad \text{on} \quad \Omega.$$

After the change of dependent variable through the subtraction of the volume potential, the homogeneous boundary conditions (i)–(iv) above become nonhomogeneous. We obtain the following four types of biharmonic boundary value problems:

Type I (displacement-angle)
$$\begin{cases} \Delta^2 w(x) = 0, & x \in \Omega, \\ w(x) = g_1(x), & x \in \partial\Omega, \\ \dfrac{\partial w(x)}{\partial n} = g_2(x), & x \in \partial\Omega; \end{cases} \tag{8.9}$$

Type II (shear-moment)
$$\begin{cases} \Delta^2 w(x) = 0, & x \in \Omega, \\ B_1 w(x) = g_1(x), & x \in \partial\Omega, \\ B_2 w(x) = g_2(x), & x \in \partial\Omega; \end{cases} \tag{8.10}$$

Type III (displacement-moment)
$$\begin{cases} \Delta^2 w(x) = 0, & x \in \Omega, \\ w(x) = g_1(x), & x \in \partial\Omega, \\ B_2 w(x) = g_2(x), & x \in \partial\Omega; \end{cases} \tag{8.11}$$

Type IV (angle-shear)

$$\begin{cases} \Delta^2 w(x) = 0, & x \in \Omega, \\ \dfrac{\partial w(x)}{\partial n} = g_1(x), & x \in \partial\Omega, \\ B_1 w(x) = g_2(x), & x \in \partial\Omega. \end{cases} \tag{8.12}$$

8.2 Existence, uniqueness and regularity of solutions to the plate equation

Basing on the classical elliptic boundary value problems [127, 64], we state the following existence and uniqueness theorems.

Theorem 8.1 (Biharmonic problem of Type I). *Let* $(g_1, g_2) \in H^{3/2}(\partial\Omega) \oplus H^{1/2}(\partial\Omega)$ *and let*

$$v \in H_0^2(\Omega) = \left\{ v \in H^2(\Omega) \,\middle|\, v = \frac{\partial v}{\partial n} = 0 \quad on \quad \partial\Omega \right\}.$$

Then there exists a unique weak solution $w \in H^2(\Omega)$ *to (8.9) satisfying* $w = g_1$, $\partial w/\partial n = g_2$ *on* $\partial\Omega$ *such that*

$$a(w, v) = 0, \quad \forall v \in H_0^2(\Omega).$$

Generally, if $(g_1, g_2) \in H^{\alpha+1}(\partial\Omega) \oplus H^{\alpha}(\partial\Omega)$, *for some* $a \in \mathbb{R}$, *then there exists a unique generalized solution* $w \in H^{\alpha+3/2}(\Omega)$ *satisfying (8.9).* □

Theorem 8.2 (Biharmonic problem of Type II). *Let* $(g_1, g_2) \in H^{-3/2}(\partial\Omega) \oplus H^{-1/2}(\partial\Omega)$. *Then there exists a weak solution* $w \in H^2(\Omega)$ *to (8.10), unique up to a linear function* $a_0 + a_1 x_1 + a_2 x_2$, a_0, a_1, $a_2 \in \mathbb{R}$, *such that*

$$a(w, v) = -\int_{\partial\Omega} \left[g_1(x) v(x) - g_2(x) \frac{\partial v(x)}{\partial n} \right] d\sigma, \quad \forall v \in H^2(\Omega),$$

provided that

$$\left\langle \begin{bmatrix} g_1 \\ g_2 \end{bmatrix}, \begin{bmatrix} \phi_1 \\ \phi_2 \end{bmatrix} \right\rangle_{[H^{-3/2}(\partial\Omega) \oplus H^{-1/2}(\partial\Omega)], [H^{3/2}(\partial\Omega) \oplus H^{1/2}(\partial\Omega)]} = 0 \tag{8.13}$$

for all

$$\begin{bmatrix} \phi_1 \\ \phi_2 \end{bmatrix} \in \operatorname{span} \left\{ \begin{bmatrix} 1 \\ 0 \end{bmatrix}, \begin{bmatrix} x_1 \\ -n_1(x) \end{bmatrix}, \begin{bmatrix} x_2 \\ -n_2(x) \end{bmatrix} \,\middle|\, x \in \partial\Omega \right\}$$
$$\subseteq C^\infty(\partial\Omega) \oplus C^\infty(\partial\Omega).$$

Note that the inner product in (8.13) is taken with respect to the duality pairing of the two Sobolev spaces in the subscripts. Generally, if (8.13) is satisfied and if $(g_1, g_2) \in H^{\alpha}(\partial\Omega) \oplus H^{\alpha+1}(\partial\Omega)$ *for some* $\alpha \in \mathbb{R}$, *then there exists a generalized solution* $w \in H^{\alpha+7/2}(\Omega)$ *satisfying (8.10); w is unique up to a linear function* $a_0 + a_1 x_1 + a_2 x_2$. □

Theorem 8.3 (Biharmonic problem of Type III). *Let* $(g_1, g_2) \in H^{3/2}(\partial\Omega) \oplus H^{-1/2}(\partial\Omega)$. *Then there exists a unique weak solution* $w \in H^2(\Omega)$ *to (8.11) satisfying* $w = g_1$ *on* $\partial\Omega$ *and*

$$a(w,v) = \int_{\partial\Omega} g_2(x) \frac{\partial v(x)}{\partial n}\, d\sigma, \quad \forall v \in H^2(\Omega), v|_{\partial\Omega} = 0.$$

Generally, if $(g_1, g_2) \in H^{\alpha+2}(\partial\Omega) \oplus H^{\alpha}(\partial\Omega)$ *for some* $\alpha \in \mathbb{R}$, *then there exists a unique generalized solution* $w \in H^{\alpha+5/2}(\Omega)$ *satisfying (8.11).* □

Theorem 8.4 (Biharmonic problem of Type IV). *Let* $(g_1, g_2) \in H^{1/2}(\partial\Omega) \oplus H^{-3/2}(\partial\Omega)$. *Then there exists a weak solution* $w \in H^2(\Omega)$, *unique up to a constant* a_0, *to (8.12) such that*

$$a(w,v) = -\int_{\partial\Omega} g_2(x) v(x)\, d\sigma, \quad \forall v \in H^2(\Omega), \left.\frac{\partial v}{\partial n}\right|_{\partial\Omega} = 0,$$

provided that

$$\left\langle \begin{bmatrix} g_1 \\ g_2 \end{bmatrix}, \begin{bmatrix} 0 \\ 1 \end{bmatrix} \right\rangle_{[H^{1/2}(\partial\Omega)\oplus H^{-3/2}(\partial\Omega)],[H^{-1/2}(\partial\Omega)\oplus H^{3/2}(\partial\Omega)]} = 0. \tag{8.14}$$

Generally, if $(g_1, g_2) \in H^{\alpha+2}(\partial\Omega) \oplus H^{\alpha}(\partial\Omega)$ *for some* $\alpha \in \mathbb{R}$ *and if (8.14) is satisfied, then there exists a generalized solution* $w \in H^{\alpha+7/2}(\partial\Omega)$, *unique up to a constant* a_0, *satisfying (8.12).* □

Exercise 8.1. For $0 < \nu < \frac{1}{2}$, it is clear that $a(w,w)$ defines a seminorm on $H_0^2(\Omega) = \{v \in H^2(\Omega) \mid v = \partial v/\partial n = 0 \text{ on } \partial\Omega\}$. Use the Deny–Lions Lemma 5.2 to show that $a(w,w)$ is a norm on $H_0^2(\Omega)$. (Hint: Define three linear functionals l_1, l_2 and l_3 on $H_0^2(\Omega)$ by

$$\begin{aligned} l_1(f) &= \int_{\partial\Omega} f(x)\, d\sigma, \\ l_2(f) &= \int_{\partial\Omega} n_1(x) \frac{\partial f(x)}{\partial n}\, d\sigma, \\ l_3(f) &= \int_{\partial\Omega} n_2(x) \frac{\partial f(x)}{\partial n}\, d\sigma.) \end{aligned}$$

□

Exercise 8.2. For $0 < \nu < \frac{1}{2}$, show that $a(w,w)$ defines an equivalent $H^2(\Omega)$-norm on $H^2(\Omega) \cap H_0^1(\Omega) = \{v \in H^2(\Omega) \mid v = 0 \text{ on } \partial\Omega\}$. (Hint: Define l_1, l_2 and l_3 on $H^2(\Omega) \cap H_0^1(\Omega)$ by

$$\begin{aligned} l_1(f) &= \int_{\partial\Omega} f(x)\, d\sigma, \\ l_2(f) &= \int_{\partial\Omega} n_2(x) \frac{\partial f(x)}{\partial \sigma}\, d\sigma, \\ l_3(f) &= \int_{\partial\Omega} n_1(x) \frac{\partial f(x)}{\partial \sigma})\, d\sigma.) \end{aligned}$$

□

8.3 Multilayer potentials for the plate BVP

We are now in a position to formulate multilayer potentials as solutions to the biharmonic boundary value problems (8.9)–(8.12). It is straightforward to show that the fundamental solution $E(x,\xi)$ to the biharmonic equation

$$\Delta_x^2 E(x,\xi) = (\Delta_\xi^2 E(x,\xi) =) - \delta(x-\xi), \quad x,\xi \in \mathbb{R}^2, \tag{8.15}$$

is

$$E(x,\xi) = -\frac{1}{8\pi}|x-\xi|^2 \ln|x-\xi|. \tag{8.16}$$

Its graph is shown in Fig. 8.3. Let $x \in \Omega$. Assume that w satisfies $\Delta^2 w(x) = 0$ on Ω. By Lemma 8.1, integration by parts on Ω and $\partial\Omega$ gives

$$\begin{aligned}
w(x) &= \int_\Omega \{[\Delta^2 w(\xi)]E(x,\xi) - [\Delta_\xi^2 E(x,\xi)]w(\xi)]\}\, d\xi \\
&\left(= \int_\Omega \delta(\xi - x) w(\xi)\, d\xi \right) \\
&= \int_{\partial\Omega} \left\{ [B_1 w(\xi)]E(x,\xi) - [B_2 w(\xi)]\frac{\partial E(x,\xi)}{\partial n_\xi} \right\} d\sigma_\xi + a(w, E(x,\cdot)) \\
&\quad - \left\{ \int_{\partial\Omega} \left\{ [B_{1\xi} E(x,\xi)]w(\xi) - [B_{2\xi} E(x,\xi)]\frac{\partial w(\xi)}{\partial n} \right\} d\sigma_\xi - a(E(x,\cdot), w) \right\} \\
&= \int_{\partial\Omega} \left\{ E(x,\xi) B_1 w(\xi) - \frac{\partial E(x,\xi)}{\partial n_\xi} B_2 w(\xi) \right. \\
&\quad \left. - [B_{1\xi} E(x,\xi)]w(\xi) + [B_{2\xi} E(x,\xi)]\frac{\partial w(\xi)}{\partial n} \right\} d\sigma_\xi .
\end{aligned} \tag{8.17}$$

Obviously, $w(x)$ can be determined if we know all four data

$$w, \quad \frac{\partial w}{\partial n}, \quad B_1 w, \quad B_2 w \tag{8.18}$$

on $\partial\Omega$. But throughout (8.9)–(8.12), only two of the four data in (8.18) are prescribed. The remaining two must be *solved* in order to know $w(x)$ on Ω.

The four boundary integrals on the right of (8.17) are layer potentials according to the terminology in [76]:

$$\int_{\partial\Omega} E(x,\xi) B_1 w(\xi)\, d\sigma_\xi \qquad \text{(simple layer);}$$

$$\int_{\partial\Omega} \frac{\partial}{\partial n_\xi} E(x,\xi) B_2 w(\xi)\, d\sigma_\xi \qquad \text{(double layer);}$$

$$\int_{\partial\Omega} B_{2\xi} E(x,\xi) \frac{\partial w(\xi)}{\partial n}\, d\sigma_\xi \qquad \text{(triple layer);}$$

$$\int_{\partial\Omega} B_{1\xi} E(x,\xi) w(\xi)\, d\sigma_\xi \qquad \text{(quadruple layer).}$$

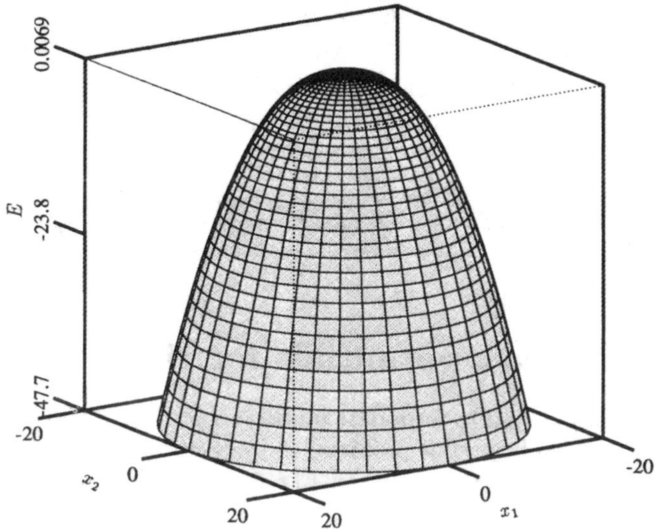

Figure 8.3: Graph of the fundamental solution $E(x,\xi) = -\frac{1}{8\pi}|x-\xi|^2 \ln|x-\xi|$ with $\xi = (0,0)$.

The higher the order of a layer, the more singular the kernel becomes at $x = \xi \in \partial\Omega$. Computationally, it is advantageous to use layer potentials that are not too singular. This motivates us to make an *ansatz* of a combination of simple- and double-layer potentials as our candidate solutions:

$$w(x) = \int_{\partial\Omega} \left[E(x,\xi) f_1(\xi) + \frac{\partial E(x,\xi)}{\partial n_\xi} f_2(\xi) \right] d\sigma_\xi, \quad x \in \Omega, \tag{8.19}$$

where f_1 and f_2 are two potential densities to be solved from the two given boundary conditions in (8.9)–(8.12). This is the basic approach adopted in this chapter. Possible combinations of layer potentials other than (8.19) will be mentioned in § 8.5.

Let us examine the fundamental properties of various layer potentials. For $x \in \mathbb{R}^2$, and for an $L^\infty(\partial\Omega)$ function f, define potentials

$$V_1(f)(x) = \int_{\partial\Omega} E(x,\xi) f(\xi)\, d\sigma_\xi, \tag{8.20}$$

$$V_2(f)(x) = \int_{\partial\Omega} \frac{\partial E(x,\xi)}{\partial n_\xi} f(\xi)\, d\sigma_\xi, \tag{8.21}$$

$$V_3(f)(x) = \int_{\partial\Omega} [B_{2\xi} E(x,\xi)] f(\xi)\, d\sigma_\xi, \tag{8.22}$$

$$V_4(f)(x) = \int_{\partial\Omega} [B_{1\xi} E(x,\xi)] f(\xi)\, d\sigma_\xi. \tag{8.23}$$

Such potentials are all well defined for $x \notin \partial\Omega$. But when $x \in \partial\Omega$, although V_1, V_2 and V_3 evidently remain well defined for each x (indeed, according to the theory in § 6.5, they are at least Hölder-continuous with certain exponents on $\partial\Omega$), V_4 might become divergent. Also, derivatives of V_1, V_2, V_3 and V_4 have higher-order singularities and may correspond to divergent integrals when $x \in \partial\Omega$. To clarify them, we provide a list of some basic properties of these layer potentials below.

Lemma 8.2. *Let $f \in H^{\alpha}(\partial\Omega)$ for $\alpha \in \mathbb{R}$, and $0 < \nu < \frac{1}{2}$. Let $V_i(f)(x)$ be defined as in (8.20)–(8.23). Denote*

$$V_i^+(f)(x) = V_i(f)(x),\ x \in \Omega^c,$$
$$V_i^-(f)(x) = V_i(f)(x),\ x \in \Omega, \quad i = 1, 2, 3, 4.$$

For $x \in \partial\Omega$, define

$$V_i^+(f)(x) = \lim_{\substack{y \to x \\ y \in \Omega^c}} V_i^+(f)(y), \quad V_i^-(f)(x) = \lim_{\substack{y \to x \\ y \in \Omega}} V_i^-(f)(y),$$

where the limits are taken in the sense of approximations of f by $C^{\infty}(\partial\Omega)$ functions when $\alpha \geqslant 0$, and in the sense of duality with respect to $H^{-\alpha}(\partial\Omega)$ when $\alpha < 0$. Then for $f \in H^{\alpha}(\partial\Omega)$, $\alpha \in \mathbb{R}$, and $x \in \partial\Omega$, we have

$$V_1^+(f)(x) = V_1^-(f)(x) = V_1(f)(x) \quad in\ H^{\alpha+3}(\partial\Omega), \tag{8.24}$$
$$V_2^+(f)(x) = V_2^-(f)(x) = V_2(f)(x) \quad in\ H^{\alpha+3}(\partial\Omega), \tag{8.25}$$
$$V_3^+(f)(x) = V_3^-(f)(x) = V_3(f)(x) \quad in\ H^{\alpha+1}(\partial\Omega), \tag{8.26}$$
$$V_4^+(f)(x) = \frac{1}{2}f(x) + V_4(f)(x) \quad in\ H^{\alpha}(\partial\Omega), \tag{8.27}$$
$$V_4^-(f)(x) = -\frac{1}{2}f(x) + V_4(f)(x) \quad in\ H^{\alpha}(\partial\Omega), \tag{8.28}$$

where

$$V_4\colon\ H^{\alpha}(\partial\Omega) \to C^{\infty}(\partial\Omega) \tag{8.29}$$

is continuous for all $\alpha \in \mathbb{R}$,

$$\frac{\partial}{\partial n}V_1^+(f)(x) = \frac{\partial}{\partial n}V_1^-(f)(x)$$
$$= \int_{\partial\Omega} \frac{\partial E(x,\xi)}{\partial n_x} f(\xi)\, d\sigma_\xi \quad in\ H^{\alpha+3}(\partial\Omega), \tag{8.30}$$
$$\frac{\partial}{\partial n}V_2^+(f)(x) = \frac{\partial}{\partial n}V_2^-(f)(x)$$
$$= \int_{\partial\Omega} \frac{\partial^2 E(x,\xi)}{\partial n_x \partial n_\xi} f(\xi)\, d\sigma_\xi \quad in\ H^{\alpha+1}(\partial\Omega), \tag{8.31}$$
$$\frac{\partial}{\partial n}V_3^+(f)(x) = -\frac{1}{2}f(x) + \int_{\partial\Omega} \frac{\partial B_{2\xi}E(x,\xi)}{\partial n_x} f(\xi)\, d\sigma_\xi \quad in\ H^{\alpha}(\partial\Omega), \tag{8.32}$$

$$\frac{\partial}{\partial n}V_3^-(f)(x) = \frac{1}{2}f(x) + \int_{\partial\Omega} \frac{\partial B_{2\xi}E(x,\xi)}{\partial n_x} f(\xi)\,d\sigma_\xi \quad in\ H^\alpha(\partial\Omega), \tag{8.33}$$

where the integral operator

$$f \mapsto \int_{\partial\Omega} \frac{\partial B_{2\xi}E(\cdot,\xi)}{\partial n_x} f(\xi)\,d\sigma_\xi \tag{8.34}$$

is continuous from $H^\alpha(\partial\Omega)$ into $C^\infty(\partial\Omega)$ for any $\alpha \in \mathbb{R}$,

$$\frac{\partial}{\partial n}V_4^+(f)(x) = \frac{\partial}{\partial n}V_4^-(f)(x) = \int_{\partial\Omega} \frac{\partial B_{1\xi}E(x,\xi)}{\partial n_x} f(\xi)\,d\sigma_\xi \quad in\ H^{\alpha-1}(\partial\Omega), \tag{8.35}$$

$$B_2V_1^+(f)(x) = B_2V_1^-(f)(x) = \int_{\partial\Omega} [B_{2x}E(x,\xi)] f(\xi)\,d\sigma_\xi \quad in\ H^{\alpha+1}(\partial\Omega), \tag{8.36}$$

$$B_2V_2^+(f)(x) = \frac{1}{2}f(x) + \int_{\partial\Omega} \left[B_{2x}\frac{\partial E(x,\xi)}{\partial n_\xi}\right] f(\xi)\,d\sigma_\xi \quad in\ H^\alpha(\partial\Omega), \tag{8.37}$$

$$B_2V_2^-(f)(x) = -\frac{1}{2}f(x) + \int_{\partial\Omega} \left[B_{2x}\frac{\partial E(x,\xi)}{\partial n_\xi}\right] f(\xi)\,d\sigma_\xi \quad in\ H^\alpha(\partial\Omega), \tag{8.38}$$

where the integral operator

$$f \mapsto \int_{\partial\Omega} \left[B_{2x}\frac{\partial E(\cdot,\xi)}{\partial n_\xi}\right] f(\xi)\,d\sigma_\xi \tag{8.39}$$

is continuous from $H^\alpha(\partial\Omega)$ into $C^\infty(\partial\Omega)$ for any $\alpha \in \mathbb{R}$,

$$B_2V_3^+(f)(x) = B_2V_3^-(f)(x) = \int_{\partial\Omega} [B_{2x}B_{2\xi}E(x,\xi)] f(\xi)\,d\sigma_\xi \quad in\ H^{\alpha-1}(\partial\Omega), \tag{8.40}$$

$$B_2V_4^+(f)(x) = B_2V_4^-(f)(x) = \int_{\partial\Omega} [B_{2x}B_{1\xi}E(x,\xi)] f(\xi)\,d\sigma_\xi \quad in\ H^{\alpha-1}(\partial\Omega), \tag{8.41}$$

$$B_1V_1^+(f)(x) = -\frac{1}{2}f(x) + \int_{\partial\Omega} [B_{1x}E(x,\xi)] f(\xi)\,d\sigma_\xi \quad in\ H^\alpha(\partial\Omega) \tag{8.42}$$

$$B_1V_1^-(f)(x) = \frac{1}{2}f(x) + \int_{\partial\Omega} [B_{1x}E(x,\xi)] f(\xi)\,d\sigma_\xi \quad in\ H^\alpha(\partial\Omega), \tag{8.43}$$

where the integral operator

$$f \mapsto \int_{\partial\Omega} [B_{1x}E(\cdot,\xi)] f(\xi)\,d\sigma_\xi \tag{8.44}$$

is continuous from $H^{\alpha}(\partial\Omega)$ *into* $C^{\infty}(\partial\Omega)$ *for any* $\alpha \in \mathbb{R}$,

$$B_1V_2^+(f)(x) = B_1V_2^-(f)(x)$$

$$= \int_{\partial\Omega} \left[B_{1x} \frac{\partial E(x,\xi)}{\partial n_\xi} \right] f(\xi)\, d\sigma_\xi \quad in\ H^{\alpha-1}(\partial\Omega), \tag{8.45}$$

$$B_1V_3^+(f)(x) = B_1V_3^-(f)(x)$$

$$= \int_{\partial\Omega} [B_{1x}B_{2\xi}E(x,\xi)] f(\xi)\, d\sigma_\xi \quad in\ H^{\alpha-1}(\partial\Omega), \tag{8.46}$$

$$B_1V_4^+(f)(x) = B_1V_4^-(f)(x)$$

$$= \int_{\partial\Omega} [B_{1x}B_{1\xi}E(x,\xi)] f(\xi)\, d\sigma_\xi \quad in\ H^{\alpha-3}(\partial\Omega), \tag{8.47}$$

where each mapping represented by the corresponding boundary integral operators on the RHS of the equations is continuous with closed range *from* $H^{\alpha}(\partial\Omega)$ *into the Sobolev space where the RHS belongs. (The* C^{∞}*-smooth operators (8.29), (8.34), (8.39) and (8.44) do not have closed ranges in any Sobolev space* $H^{\alpha}(\partial\Omega)$ *for any* $\alpha \in \mathbb{R}$.*)* □

Comments: These regularity properties have all been rigorously verified by us recently, but the work is rather lengthy and cumbersome thus will be omitted. Here we will only make some heuristic observations and arguments of how they come about.

By directly comparing the regularity of the boundary integral operator V_j with the regularity of the solutions of the plate BVP (8.9)–(8.12), in light of (8.17), we observe that V_j should satisfy

$$V_j\colon\ H^{\alpha}(\partial\Omega) \to H^{4+\alpha-j}(\partial\Omega) \text{ continuously}, \quad j = 1,2,3,4.$$

Basically, this is why the regularities in (8.24), (8.26) and (8.27) hold, for $j = 1,\ 3$ and 4. The only exception is (8.25). We note that the kernel of V_2 is

$$\frac{\partial E(x,\xi)}{\partial n_\xi} = -\frac{1}{8\pi}\left(|x-\xi|^2 \frac{\langle n_\xi, \xi - x\rangle}{|x-\xi|^2} + 2\langle n_\xi, \xi - x\rangle \ln|x-\xi| \right)$$

$$= \frac{1}{8\pi}\langle n_\xi, x-\xi\rangle(2\ln|x-\xi|+1).$$

The factor $\langle n_\xi, x-\xi\rangle$ is C^{∞} on $\partial\Omega \times \partial\Omega$, satisfying

$$\langle n_\xi, x-\xi\rangle = \mathcal{O}(|x-\xi|^2) \quad \text{for small } |x-\xi|, \text{ by (6.37).}$$

Thus

$$\frac{\partial E(x,\xi)}{\partial n_\xi} = \mathcal{O}(|x-\xi|^2 \ln|x-\xi|) \tag{8.48}$$

in a neighborhood near the diagonal $\partial\Omega \times \partial\Omega \ni (x,\xi)$. Therefore (8.48) has the same smoothing property as $E(x,\xi) = -(8\pi)^{-1}|x-\xi|^2 \ln|x-\xi|$, the kernel of V_1. Hence

$$V_2\colon\ H^{\alpha}(\partial\Omega) \to H^{\alpha+3}(\partial\Omega) \text{ is continuous.} \tag{8.49}$$

The boundary integral operator on the RHS of (8.30) is the adjoint of V_2. By Theorem 4.2, it has the same smoothing property $H^{\alpha}(\partial\Omega) \to H^{\alpha+3}(\partial\Omega)$. Also note that the boundary integral operators (8.41) and (8.46) have the same smoothing property as (8.40) and (8.45). Our heuristic observation herein is largely based upon the singularity $\mathscr{O}(|\theta-\phi|^{-2})$ of the corresponding kernels on a circle. Refer to (8.68) in Example 8.1.

All the jump properties in (8.24)–(8.47) are obtainable through the use of the Calderòn projector. Let w satisfy

$$\Delta^2 w(x) = 0 \text{ on } \Omega$$

such that on $\partial\Omega$, w has the "Cauchy data"

$$\begin{aligned} B_1 w(x) &= F_1(x), \\ B_2 w(x) &= -F_2(x), \\ \frac{\partial w(x)}{\partial n} &= F_3(x), \\ w(x) &= -F_4(x). \end{aligned}$$

From (8.17), we can write

$$w(x) = V_1^-(F_1)(x) + V_2^-(F_2)(x) + V_3^-(F_3)(x) + V_4^-(F_4)(x), \quad x \in \Omega. \tag{8.50}$$

Letting x tend nontangentially to a boundary point on $\partial\Omega$, we have only half of the contribution of the Dirac delta function $\delta(x-\xi)$ at $\xi = x \in \partial\Omega$, so

$$\frac{1}{2}w(x) = V_1(F_1)(x) + V_2(F_2)(x) + V_3(F_3)(x) + V_4(F_4)(x), \quad x \in \partial\Omega. \tag{8.51}$$

This is possible only if

$$V_4^-(F_4)(x) = -\frac{1}{2}F_4(x) + V_4(F_4)(x)$$

and

$$V_i^-(F_i)(x) = V_i(F_i)(x), \quad i = 1, 2, 3,$$

for $x \in \partial\Omega$.

If $x \in \Omega^c$ instead then it is easy to see that (8.17) is replaced by

$$w(x) = -\sum_{i=1}^{4} V_i^+(F_i)(x). \tag{8.52}$$

Letting x tend nontangentially to a point on $\partial\Omega$, for the same reason as above we have

$$\frac{1}{2}w(x) = -\sum_{i=1}^{4} V_i(F_i)(x), \quad x \in \partial\Omega. \tag{8.53}$$

This is possible only if

$$V_4^+(F_4) = \frac{1}{2}F_4(x) + V_4(F_4)(x)$$

and

$$V_i^+(F_i)(x) = V_i(F_i)(x), \quad i = 1, 2, 3,$$

for $x \in \partial\Omega$.

Therefore (8.24)–(8.28) hold.

To consider (8.30)–(8.35), let $x \in \Omega$. We first use a Gaussian normal extension to extend the boundary normal vector field $\vec{n}(x)$ to a C^∞ field $\widetilde{n}(x)$ on a neighborhood of $\partial\Omega$. Then, from (8.50),

$$\nabla w(x) \cdot \widetilde{n}(x) = \sum_{i=1}^{4} \nabla V_i^-(F_i)(x) \cdot \widetilde{n}(x), \quad x \in \Omega.$$

Letting x tend nontangentially to a boundary point on $\partial\Omega$, as before, we have

$$\frac{1}{2}\frac{\partial w(x)}{\partial n} = \sum_{i=1}^{4} \frac{\partial}{\partial n} V_i^-(F_i)(x). \tag{8.54}$$

This is possible only if

$$\frac{\partial}{\partial n} V_3^-(F_3)(x) = \frac{1}{2}F_3(x) + \int_{\partial\Omega} \frac{\partial B_{2\xi}E(x,\xi)}{\partial n_x} F_3(\xi)\, d\sigma_\xi,$$

and

$$\frac{\partial}{\partial n} V_1^-(F_1)(x) = \int_{\partial\Omega} \frac{\partial E(x,\xi)}{\partial n_x} F_1(\xi)\, d\sigma_\xi,$$
$$\frac{\partial}{\partial n} V_2^-(F_2)(x) = \int_{\partial\Omega} \frac{\partial^2 E(x,\xi)}{\partial n_x \partial n_\xi} F_2(\xi)\, d\sigma_\xi,$$
$$\frac{\partial}{\partial n} V_4^-(F_4)(x) = \int_{\partial\Omega} \frac{\partial B_{1\xi}E(x,\xi)}{\partial n_x} F_4(x)\, d\sigma_\xi$$

for $x \in \partial\Omega$.

If $x \in \Omega^c$, we repeat the same argument, using (8.52) instead. We will then have verified (8.30)–(8.35).

The rest of the properties (8.36)–(8.47) can be verified in a similar manner. The reader can detect that the above arguments are essentially those used in the proof of Theorem 4.6.

Note that ν is required to satisfy $0 < \nu < \frac{1}{2}$ in order for the plate BVP (8.9)–(8.12) to be *elliptic* so that the theory in § 4.5 becomes applicable. If ν does not satisfy $0 < \nu < \frac{1}{2}$, which is *mathematically possible*, it is not clear which of the properties in (8.24)–(8.47) may fail.

In principle, the C^∞ smoothing properties in (8.29), (8.34), (8.39) and (8.44) can be verified just as in (6.146).

To introduce the Calderòn projector, let us first simplify the notations in Lemma 8.2. For $f \in C^\infty(\partial\Omega)$, define boundary integral operators (mapping into $C^\infty(\partial\Omega)$)

$$v_1(f) = \int E(\cdot,\xi) f(\xi)\, d\sigma_\xi \tag{8.55}$$

$$\begin{aligned}
v_2^{(j)}(f) &= \int \frac{\partial E(\cdot,\xi)}{\partial n_j} f(\xi)\, d\sigma_\xi, \quad j = 1,2, \ \frac{\partial}{\partial n_1} \equiv \frac{\partial}{\partial n_x}, \ \frac{\partial}{\partial n_2} \equiv \frac{\partial}{\partial n_\xi}, \\
v_3^{(j)}(f) &= \int [B_{2j}E(\cdot,\xi)] f(\xi)\, d\sigma_\xi, \quad j = 1,2, \ B_{21} \equiv B_{2x}, \ B_{22} \equiv B_{2\xi}, \\
v_4^{(j)}(f) &= \int [B_{1j}E(\cdot,\xi)] f(\xi)\, d\sigma_\xi, \quad j = 1,2, \ B_{11} \equiv B_{1x}, \ B_{12} \equiv B_{1\xi}, \\
v_{2,2}^{(j,l)}(f) &= \int \frac{\partial^2 E(\cdot,\xi)}{\partial n_l \partial n_j} f(\xi)\, d\sigma_\xi, \quad j,l = 1,2, \\
v_{2,3}^{(j,l)}(f) &= \int \left[B_{2l} \frac{\partial E(\cdot,\xi)}{\partial n_j} \right] f(\xi)\, d\sigma_3, \quad j,l = 1,2, \\
v_{2,4}^{(j,l)}(f) &= \int \left[B_{1l} \frac{\partial E(\cdot,\xi)}{\partial n_j} \right] f(\xi)\, d\sigma_\xi, \quad j,l = 1,2, \\
v_{3,2}^{(j,l)}(f) &= \int \left[\frac{\partial B_{2j}E(\cdot,\xi)}{\partial n_l} \right] f(\xi)\, d\sigma_\xi, \quad j,l = 1,2, \\
v_{3,3}^{(j,l)}(f) &= \int [B_{2l}B_{2j}E(\cdot,\xi)] f(\xi)\, d\sigma_\xi, \quad j,l = 1,2, \\
v_{4,2}^{(j,l)}(f) &= \int \left[\frac{\partial B_{1j}E(\cdot,\xi)}{\partial n_l} \right] f(\xi)\, d\sigma_\xi, \quad j,l = 1,2, \\
v_{4,3}^{(j,l)}(f) &= \int [B_{2l}B_{1j}E(\cdot,\xi)] f(\xi)\, d\sigma_\xi, \quad j,l = 1,2, \\
v_{4,4}^{(j,l)}(f) &= \int [B_{1l}B_{1j}E(\cdot,\xi)] f(\xi)\, d\sigma_\xi, \quad j,l = 1,2,
\end{aligned} \tag{8.56}$$

where the integrals are all evaluated on $\partial\Omega$. These operators satisfy $v_{i,k}^{(j,l)} = v_{k,i}^{(l,j)}$. According to the notations in § 4.5, (8.17) and the comments above, we have

$$\mathscr{C}_1 = \begin{bmatrix}
\frac{1}{2} - v_4^{(2)} & v_3^{(2)} & -v_2^{(2)} & v_1 \\
-v_{4,2}^{(2,1)} & \frac{1}{2} + v_{3,2}^{(2,1)} & -v_{2,2}^{(2,1)} & v_2^{(1)} \\
-v_{4,3}^{(2,1)} & v_{3,3}^{(2,1)} & \frac{1}{2} - v_{2,3}^{(2,1)} & v_3^{(1)} \\
-v_{4,4}^{(2,1)} & v_{3,4}^{(2,1)} & -v_{2,4}^{(2,1)} & \frac{1}{2} + v_4^{(1)}
\end{bmatrix}. \tag{8.57}$$

Similarly,

$$\mathscr{C}_2 = \begin{bmatrix} \frac{1}{2} + v_4^{(2)} & -v_3^{(2)} & v_2^{(2)} & -v_1 \\ v_{4,2}^{(2,1)} & \frac{1}{2} - v_{3,2}^{(2,1)} & v_{2,2}^{(2,1)} & -v_2^{(1)} \\ v_{4,3}^{(2,1)} & -v_{3,3}^{(2,1)} & \frac{1}{2} + v_{2,3}^{(2,1)} & -v_3^{(1)} \\ v_{4,4}^{(2,1)} & -v_{3,4}^{(2,1)} & v_{2,4}^{(2,1)} & \frac{1}{2} - v_4^{(1)} \end{bmatrix}. \tag{8.58}$$

Thus

$$\mathscr{C}_1 + \mathscr{C}_2 = I.$$

Also, from $\mathscr{C}_1^2 = \mathscr{C}_1$, many nontrivial identities involving these boundary integral operators can be obtained. All the entries in $\mathscr{C}_i$, $i = 1, 2$, are strongly elliptic ψDO satisfying the Fredholm properties and the Gårding inequality.

Example 8.1. Let Ω be the open disk with radius R centered at the origin in $\mathbb{R}^2$. Represent $x, \xi \in \overline{\Omega}$ by polar coordinates:

$$x = re^{i\theta}, \quad \xi = \rho e^{i\phi}, \quad 0 \leqslant r, \rho \leqslant R, \quad 0 \leqslant \theta, \phi \leqslant 2\pi. \tag{8.59}$$

Then on $\partial\Omega$,

$$\frac{\partial}{\partial n_x} = \frac{\partial}{\partial r}, \quad \frac{\partial}{\partial n_\xi} = \frac{\partial}{\partial \rho}, \quad \frac{\partial}{\partial \sigma_x} = \frac{1}{R}\frac{\partial}{\partial \theta}, \quad \frac{\partial}{\partial \sigma_\xi} = \frac{1}{R}\frac{\partial}{\partial \phi}.$$

We have

$$\begin{aligned} E(x,\xi) &= -\frac{1}{8\pi}\{2R^2[1-\cos(\theta-\phi)]\}\ln\{2R^2[1-\cos(\theta-\phi)]\}^{1/2} \\ &= \mathscr{O}(|\theta-\phi|^2\ln|\theta-\phi|) \quad \text{near the diagonal of } [0,2\pi]\times[0,2\pi], \end{aligned} \tag{8.60}$$

$$\begin{aligned} \frac{\partial E(x,\xi)}{\partial n_\xi} &= -\frac{1}{8\pi}\{R[1-\cos(\theta-\phi)]\}\left[\ln\{2R^2[1-\cos(\theta-\phi)]\}+1\right] \\ &= \frac{\partial E(x,\xi)}{\partial n_x} \\ &= \mathscr{O}(|\theta-\phi|^2\ln|\theta-\phi|), \end{aligned} \tag{8.61}$$

$$\begin{aligned} \frac{\partial^2 E(x,\xi)}{\partial n_x \partial n_\xi} &= -\frac{1}{8\pi}\left\{-\cos(\theta-\phi)\left[\ln\{2R^2[1-\cos(\theta-\phi)]\}+2\right]+1\right\} \\ &= \mathscr{O}(\ln|\theta-\phi|), \end{aligned} \tag{8.62}$$

$$B_{2\xi}E(x,\xi) = -\frac{1}{8\pi}\left[(1+\nu)\ln\{2R^2[1-\cos(\theta-\phi)]\}\right.$$

$$
\begin{aligned}
&\quad + (1+3\nu) - (1-\nu)[1-\cos(\theta-\phi)]] && (8.63)\\
&= B_{2x}E(x,\xi)\\
&= \mathcal{O}(\ln|\theta-\phi|),\\
\frac{\partial B_{2\xi}E(x,\xi)}{\partial n_x} &= -\frac{\nu}{4\pi R} + \frac{1-\nu}{8\pi R}\cos(\theta-\phi)\\
&= \frac{\partial B_{2x}E(x,\xi)}{\partial n_\xi} \in C^\infty([0,2\pi]\times[0,2\pi]), && (8.64)\\
B_{1\xi}E(x,\xi) &= -\frac{1}{4\pi R} - \frac{1-\nu}{8\pi R}\cos(\theta-\phi)\\
&= B_{1x}E(x,\xi) \in C^\infty([0,2\pi]\times[0,2\pi]), && (8.65)\\
\frac{\partial B_{1\xi}E(x,\xi)}{\partial n_x} &= -\frac{1-\nu}{8\pi R^2}\cos(\theta-\phi) + \frac{1+\nu}{8\pi R^2[1-\cos(\theta-\phi)]}\\
&= B_{1x}\frac{\partial E(x,\xi)}{\partial n_\xi}\\
&= \mathcal{O}(|\theta-\phi|^{-2}), && (8.66)\\
B_{2x}B_{2\xi}E(x,\xi) &= \frac{\nu(1-\nu)\cos(\theta-\phi)}{2\pi R^2[1-\cos(\theta-\phi)]} - \frac{(1-\nu)^2}{8\pi R^2}\left[1-4\cos(\theta-\phi)+\frac{3}{1-\cos(\theta-\phi)}\right]\\
&= B_{2\xi}B_{2x}E(x,\xi)\\
&= \mathcal{O}(|\theta-\phi|^{-2}), && (8.67)\\
B_{1x}B_{2\xi}E(x,\xi) &= \frac{(1-\nu)^2[1+\cos(\theta-\phi)]-2(1-\nu)}{4\pi R^3} - \frac{(1-\nu)^2}{8\pi R^3[1-\cos(\theta-\phi)]}\\
&= B_{1\xi}B_{2x}E(x,\xi)\\
&= \mathcal{O}(|\theta-\phi|^{-2}), && (8.68)\\
B_{1x}B_{1\xi}E(x,\xi) &= \frac{(1-\nu)[2+\cos(\theta-\phi)]}{2\pi R^4(1-\cos(\theta-\phi))^2}\\
&\quad + \frac{(1-\nu)^2}{8\pi R^4[1-\cos(\theta-\phi)]}\Bigg[12+22\cos(\theta-\phi)+\cos^2(\theta-\phi)\\
&\quad + \frac{4-2\sin(\theta-\phi)\cos(\theta-\phi)}{1-\cos(\theta-\phi)}\\
&\quad + \frac{16\sin(\theta-\phi)\cos^2(\theta-\phi)-16\cos^3(\theta-\phi)}{1-\cos(\theta-\phi)}\Bigg]\\
&= B_{1\xi}B_{1x}E(x,\xi)\\
&= \mathcal{O}(|\theta-\phi|^{-4}). && (8.69)
\end{aligned}
$$

The above include all the kernels appearing in the Calderòn projectors $\mathscr{C}_1$ and $\mathscr{C}_2$. The reader can clearly see the C^∞-smoothing properties (8.29), (8.34), (8.39) and (8.44) from (8.64) and (8.65), respectively, because these explicit kernel representations are C^∞ on $\partial\Omega \times \partial\Omega$. Also, from the severity of the singularities of the kernels in (8.66)–(8.69), we understand that they are all *hypersingular*, and the regularity properties as stated in (8.40), (8.41) and (8.45)–(8.47) are *optimal.* □

8.4 BIE for interior plate BVP

In this section, we will formulate boundary integral equations for the four types of biharmonic boundary value problems in (8.9)–(8.12) and prove the existence and uniqueness of solutions.

We first prove a fundamental uniqueness theorem for the exterior biharmonic boundary value problem.

Theorem 8.5. *Assume that $w(x)$ is a solution of*

$$\Delta^2 w(x) = 0, \quad x \in \Omega^c$$

subject to one of the following sets of boundary conditions on $\partial\Omega$:

$$\left.\begin{aligned} w(x) &= \alpha_0 + \alpha_1 x_1 + \alpha_2 x_2, \\ \frac{\partial w}{\partial n}(x) &= \alpha_1 n_1(x) + \alpha_2 n_2(x), \end{aligned}\right\} \tag{8.70}$$

$$\left.\begin{aligned} w(x) &= \alpha_0 + \alpha_1 x_1 + \alpha_2 x_2, \\ B_2 w(x) &= 0, \end{aligned}\right\} \tag{8.71}$$

$$\left.\begin{aligned} \frac{\partial w}{\partial n}(x) &= \alpha_1 n_1(x) + \alpha_2 n_2(x), \\ B_1 w(x) &= 0, \end{aligned}\right\} \tag{8.72}$$

$$\left.\begin{aligned} B_1 w(x) &= 0, \\ B_2 w(x) &= 0, \end{aligned}\right\} \tag{8.73}$$

such that

$$|w(x)[B_1(r)w(x)]| = \mathcal{O}(r^{-(1+\delta)}), \tag{8.74}$$

$$\left|\frac{\partial w(x)}{\partial r}[B_2(r)w(x)]\right| = \mathcal{O}(r^{-(1+\delta)}), \tag{8.75}$$

for $r = |x|$ large for some $\delta > 0$, where

$$B_1(r) \equiv -\left\{\frac{\partial}{\partial r}\Delta - (1-\nu)\frac{1}{r}\frac{\partial}{\partial\theta}\left[\frac{x_1x_2}{r^2}\left(\frac{\partial^2}{\partial x_1^2} - \frac{\partial^2}{\partial x_2^2}\right) - \frac{1}{r^2}(x_1^2 - x_2^2)\frac{\partial^2}{\partial x_1 \partial x_2}\right]\right\},$$

$$B_2(r) \equiv -\nu\Delta - \frac{1-\nu}{r^2}\left(x_1^2\frac{\partial^2}{\partial x_1^2} + x_2^2\frac{\partial^2}{\partial x_2^2} + 2x_1x_2\frac{\partial^2}{\partial x_1 \partial x_2}\right)$$

are, respectively, the transverse force and bending moment operators in the radial direction. Then w is a linear function

$$w(x) = \alpha_0 + \alpha_1 x_1 + \alpha_2 x_2 \tag{8.76}$$

throughout $x \in \Omega^c$.

Proof. Since the proofs for (8.70)–(8.73) are similar, we will only prove the case (8.73). Applying the integration by parts procedure as in Lemma 8.1 to the domain $\Omega_R^c \equiv \Omega^c \cap \{x \in \mathbb{R}^2 \mid |x| \leqslant R\}$ for sufficiently large R, we obtain

$$\begin{aligned}
a_R(w,w) &\equiv \int_{\Omega_R^c} [|\Delta w|^2 + 2(1-\nu)(w_{x_1x_2}^2 - w_{x_1x_1}w_{x_2x_2})]\,dx \\
&= -\int_{\partial\Omega} \left\{ [B_1 w(x)]w(x) - [B_2 w(x)]\frac{\partial w(x)}{\partial n} \right\} d\sigma \\
&\quad - \int_{|x|=R} \left\{ [B_1(R)w(x)]w(x) - [B_2(R)w(x)]\frac{\partial w(x)}{\partial n} \right\} d\sigma \\
&\leqslant 0 + \int_0^{2\pi} \frac{C}{R^{1+\alpha}} R\,d\theta \\
&= \frac{2\pi C}{R^\alpha} \to 0 \ \text{ as } R \to \infty.
\end{aligned}$$

Since $a_R(w,w)$ is a norm for w up to a linear function, as $R \to \infty$, we have (8.76). □

The multilayer biharmonic potentials to be used in this section are all based on the simple- and double-layer combination (8.19) slightly augmented by a linear function to account for the existence and/or uniqueness of solutions of BIE. We now formulate BIE for biharmonic problems one by one.

Theorem 8.6. (Simple- and double-layer representations for biharmonic problem of Type I). *Let $\Omega \subseteq \mathbb{R}^2$ be bounded with smooth boundary $\partial\Omega$. The mapping*

$$\mathscr{L}_1 \begin{bmatrix} a_0 \\ a_1 \\ a_2 \\ f_1 \\ f_2 \end{bmatrix}(x) = \begin{bmatrix} \int_{\partial\Omega} f_1(\xi)\,d\sigma \\ \int_{\partial\Omega} [\xi_1 f_1(\xi) + n_1(\xi) f_2(\xi)]\,d\sigma \\ \int_{\partial\Omega} [\xi_2 f_1(\xi) + n_2(\xi) f_2(\xi)]\,d\sigma \\ \int_{\partial\Omega} \left[E(x,\xi) f_1(\xi) + \frac{\partial E(x,\xi)}{\partial n_\xi} f_2(\xi) \right] d\sigma_\xi \\ \quad + a_0 + a_1 x_1 + a_2 x_2 \\ \int_{\partial\Omega} \left[\frac{\partial E(x,\xi)}{\partial n_x} f_1(\xi) + \frac{\partial^2 E(x,\xi)}{\partial n_x \partial n_\xi} f_2(\xi) \right] d\sigma_\xi \\ \quad + a_1 n_1(x) + a_2 n_2(x) \end{bmatrix} \tag{8.77}$$

is an isomorphism from $\mathbb{R}^3 \oplus H^r(\partial\Omega) \oplus H^{r+1}(\partial\Omega)$ *onto* $\mathbb{R}^3 \oplus H^{r+3}(\partial\Omega) \oplus H^{r+2}(\partial\Omega)$ *for any* $r \in \mathbb{R}$, *satisfying*

$$\dim \mathscr{N}(\mathscr{L}_1) = 0, \quad \dim \text{Coker}\, \mathscr{L}_1 = 0. \tag{8.78}$$

Its adjoint operator $\mathscr{L}_1^*$ *mapping* $\mathbb{R}^3 \oplus H^{-(r+3)}(\partial\Omega) \oplus H^{-(r+2)}(\partial\Omega)$ *onto* $\mathbb{R}^3 \oplus H^{-r}(\partial\Omega) \oplus H^{-(r+1)}(\partial\Omega)$ *is given formally by*

$$\mathscr{L}_1^* = \mathscr{L}_1 \quad \text{as in (8.77)}, \tag{8.79}$$

with

$$\dim \mathscr{N}(\mathscr{L}_1^*) = 0, \quad \dim \text{Coker}\, \mathscr{L}_1^* = 0. \tag{8.80}$$

$\mathscr{L}_1^*$ *also maps*

$$\mathscr{L}_1^*:\ \mathbb{R}^3 \oplus H^s(\partial\Omega) \oplus H^{s+1}(\partial\Omega) \to \mathbb{R}^3 \oplus H^{s+3}(\partial\Omega) \oplus H^{s+2}(\partial\Omega) \tag{8.81}$$

isomorphically for any $s \in \mathbb{R}$, *so* $\mathscr{L}_1$ *is self-adjoint.*

Consequently, the solution $w(x)$ *to*

$$\left.\begin{aligned} &\Delta^2 w(x) = 0, && x \in \Omega, \\ &w(x) = g_1(x) \in H^{r+3}(\partial\Omega), && x \in \partial\Omega,\ r \in \mathbb{R}, \\ &\frac{\partial w}{\partial n}(x) = g_2(x) \in H^{r+2}(\partial\Omega), && x \in \partial\Omega, \end{aligned}\right\} \tag{8.82}$$

is uniquely given by

$$\begin{aligned} w(x) = &\int_{\partial\Omega} \left[E(x,\xi)\widetilde{f}_1(\xi) + \frac{\partial E(x,\xi)}{\partial n_\xi}\widetilde{f}_2(\xi) \right] d\sigma_\xi \\ &+ \widetilde{a}_0 + \widetilde{a}_1 x_1 + \widetilde{a}_2 x_2, \quad x \in \Omega, \\ w \in &H^{r+7/2}(\Omega), \end{aligned} \tag{8.83}$$

where $(\widetilde{a}_0, \widetilde{a}_1, \widetilde{a}_2, \widetilde{f}_1, \widetilde{f}_2) \in \mathbb{R}^3 \oplus H^r(\partial\Omega) \oplus H^{r+1}(\partial\Omega)$ *uniquely solves*

$$\mathscr{L}_1(\widetilde{a}_0, \widetilde{a}_1, \widetilde{a}_2, \widetilde{f}_1, \widetilde{f}_2) = (A_0, A_1, A_2, g_1, g_2) \tag{8.84}$$

for given $A_0,\ A_1,\ A_2 \in \mathbb{R}^3$. *(*$A_0,\ A_1,\ A_2 \in \mathbb{R}$ *can be chosen arbitrarily, but* $\widetilde{a}_0,\ \widetilde{a}_1,\ \widetilde{a}_2$ *and* $\widetilde{f}_1,\ \widetilde{f}_2$ *depend on the choice* $A_0,\ A_1,\ A_2$*.)*

Proof. The regularity property of $\mathscr{L}_1$ is a consequence of Lemma 8.2.

We represent $\mathscr{L}_1$ by a matrix kernel

$$K_1(x,\xi) = \begin{bmatrix} 0 & 0 & 0 & 1 & 0 \\ 0 & 0 & 0 & \xi_1 & n_1(\xi) \\ 0 & 0 & 0 & \xi_2 & n_2(\xi) \\ 1 & x_1 & x_2 & E(x,\xi) & \dfrac{\partial E(x,\xi)}{\partial n_\xi} \\ 0 & n_1(x) & n_2(x) & \dfrac{\partial E(x,\xi)}{\partial n_x} & \dfrac{\partial^2 E(x,\xi)}{\partial n_x \partial n_\xi} \end{bmatrix}, \tag{8.85}$$

so that

$$\mathscr{L}_1 \begin{bmatrix} a_0 \\ a_1 \\ a_2 \\ f_1 \\ f_2 \end{bmatrix} = \int_{\partial\Omega} K_1(x,\xi) \begin{bmatrix} a_0 \\ a_1 \\ a_2 \\ f_1 \\ f_2 \end{bmatrix} (\xi)\, d\sigma_\xi,$$

where we have tacitly assumed that $\int_{\partial\Omega} d\sigma = 1$. (This causes no loss of generality.) The matrix $K_1(x,\xi)$ satisfies

$$K_1(x,\xi) = K_1^{\mathrm{Tr}}(\xi,x),$$

so $\mathscr{L}_1^*$ is formally self-adjoint to $\mathscr{L}_1$. This formal adjoint becomes the true adjoint because all the spaces involved are reflexive and the Fredholm property as stated in Lemma 8.2 holds. Hence $\mathscr{L}_1^*$ has the stated smoothing properties.

Since $\mathscr{L}_1$ is (now truly, not just formally) self-adjoint, if we can prove that

$$\dim \mathscr{N}(\mathscr{L}_1) = 0, \tag{8.86}$$

the rest of the proof will be complete, due to the Fredholm alternative. Assume that

$$(\widetilde{a}_0, \widetilde{a}_1, \widetilde{a}_2, \widetilde{f}_1, \widetilde{f}_2) \in \mathscr{N}(\mathscr{L}_1). \tag{8.87}$$

By Corollary 4.4, $(\widetilde{a}_0, \widetilde{a}_1, \widetilde{a}_2, \widetilde{f}_1, \widetilde{f}_2) \in \mathbb{R}^3 \oplus C^\infty(\partial\Omega) \oplus C^\infty(\partial\Omega)$. We want to show that

$$(\widetilde{a}_0, \widetilde{a}_1, \widetilde{a}_2, \widetilde{f}_1, \widetilde{f}_2) = (0,0,0,0,0). \tag{8.88}$$

For $x \in \Omega$, consider

$$\begin{aligned} \widetilde{w}^-(x) \equiv \int_{\partial\Omega} \left[E(x,\xi)\widetilde{f}_1(\xi) + \frac{\partial E(x,\xi)}{\partial n_\xi} \widetilde{f}_2(\xi) \right] d\sigma_\xi \\ + \widetilde{a}_0 + \widetilde{a}_1 x_1 + \widetilde{a}_2 x_2. \end{aligned} \tag{8.89}$$

Then, by (8.87), we have

$$\left. \begin{aligned} \widetilde{w}^-(x) = 0, \\ \frac{\partial \widetilde{w}^-(x)}{\partial n} = 0 \end{aligned} \right\} \quad \text{on} \quad \partial\Omega. \tag{8.90}$$

$\widetilde{w}^-(x)$ satisfies

$$\Delta^2 \widetilde{w}^-(x) = 0, \quad x \in \Omega. \tag{8.91}$$

By Theorem 8.1, the uniqueness of solution implies

$$\widetilde{w}^-(x) \equiv 0 \quad \text{on} \quad \Omega. \tag{8.92}$$

We extend (8.89) to $x \in \Omega^c$ by defining

$$\widetilde{w}^+(x) \equiv \int_{\partial\Omega} \left[E(x,\xi)\widetilde{f}_1(\xi) + \frac{\partial E(x,\xi)}{\partial n_\xi} \widetilde{f}_2(\xi) \right] d\sigma_\xi$$
$$+ \widetilde{a}_0 + \widetilde{a}_1 x_1 + \widetilde{a}_2 x_2, \quad x \in \Omega^c.$$

Then for large $|x|, x \in \Omega^c$, we have the asymptotic expansion

$$\begin{aligned}
\widetilde{w}^+(x) = & -\frac{1}{8\pi}|x|^2 \ln|x| \left\{ \int_{\partial\Omega} \widetilde{f}_1(\xi)\, d\sigma \right\} \\
& + \frac{1}{4\pi} \ln|x| \left\{ x_1 \int_{\partial\Omega} [\xi_1 \widetilde{f}_1(\xi) + n_1(\xi)\widetilde{f}_2(\xi)]\, d\sigma \right. \\
& \left. + x_2 \int_{\partial\Omega} [\xi_2 \widetilde{f}_1(\xi) + n_2(\xi)\widetilde{f}_2(\xi)]\, d\sigma \right\} \\
& + \frac{1}{8\pi} \left\{ x_1 \int_{\partial\Omega} [\xi_1 \widetilde{f}_1(\xi) + n_1(\xi)\widetilde{f}_2(\xi)]\, d\sigma \right. \\
& \left. + x_2 \int_{\partial\Omega} [\xi_2 \widetilde{f}_1(\xi) + n_2(\xi)\widetilde{f}_2(\xi)]\, d\sigma \right\} + \widetilde{a}_1 x_1 + \widetilde{a}_2 x_2 \\
& - \ln|x| \frac{1}{8\pi} \left[\int_{\partial\Omega} |\xi|^2 \widetilde{f}_1(\xi)\, d\sigma + 2 \int_{\partial\Omega} [\xi \cdot n(\xi)] \widetilde{f}_2(\xi)\, d\sigma \right] \\
& + \left\{ \widetilde{a}_0 - \frac{1}{16\pi} \int_{\partial\Omega} \{ |\xi|^2 \widetilde{f}_1(\xi) + 2[\xi \cdot n(\xi)] \widetilde{f}_2(\xi) \}\, d\sigma \right\} \\
& - \frac{1}{8\pi} \frac{1}{|x|^2} \int_{\partial\Omega} \{ (x \cdot \xi)^2 \widetilde{f}_1(\xi) + 2[x \cdot n(\xi)](x \cdot \xi) \widetilde{f}_2(\xi) \}\, d\sigma_\xi + \cdots \\
= & \ \widetilde{a}_1 x_1 + \widetilde{a}_2 x_2 \\
& - \frac{1}{8\pi} \ln|x| \left\{ \int_{\partial\Omega} |\xi|^2 \widetilde{f}_1(\xi)\, d\sigma + 2 \int_{\partial\Omega} [\xi \cdot n(\xi)] \widetilde{f}_2(\xi)\, d\sigma \right\} + \widetilde{R}(x),
\end{aligned} \tag{8.93}$$

where the leading terms involving $|x|^2 \ln|x|$, $x_1 \ln|x|$, $x_2 \ln|x|$ all vanish due to (8.87). The new leading terms

$$v(x) \equiv \widetilde{a}_1 x_1 + \widetilde{a}_2 x_2 - \frac{\ln|x|}{8\pi} \left\{ \int_{\partial\Omega} |\xi|^2 \widetilde{f}_1(\xi)\, d\sigma + 2 \int_{\partial\Omega} [\xi \cdot n(\xi)] \widetilde{f}_2(\xi)\, d\sigma \right\}$$

satisfy

$$\begin{aligned}
|v(x)[B_1(r)v(x)]| &= \mathscr{O}(r^{-2} \ln r), \\
\left| \frac{\partial v(x)}{\partial r} [B_2(r)v(x)] \right| &= \mathscr{O}(r^{-2})
\end{aligned}$$

for $|x| = r$, when r is large. The remainder $\widetilde{R}(x)$ in (8.93) is $\mathscr{O}(1)$, such that

$$B_1(r)\widetilde{R}(x), \quad B_2(r)\widetilde{R}(x)$$

are even more "rapidly" decreasing functions on $|x| = r$ for large r. Therefore,

$$\left.\begin{aligned} |\widetilde{w}^+(x)[B_1(r)\widetilde{w}^+(x)]| &= \mathscr{O}(r^{-2}\ln r), \\ \left|\frac{\partial \widetilde{w}^+(x)}{\partial r}[B_2(x)\widetilde{w}^+(x)]\right| &= \mathscr{O}(r^{-2}), \end{aligned}\right\} \tag{8.94}$$

and hence (8.74) and (8.75) are satisfied.

On $\partial\Omega, \widetilde{w}^+(x)$ satisfies

$$\left.\begin{aligned} \widetilde{w}^+(x) = \widetilde{w}^-(x) = 0, \\ \frac{\partial \widetilde{w}^+(x)}{\partial n} = \frac{\partial \widetilde{w}^-(x)}{\partial n} = 0, \end{aligned}\right\} \tag{8.95}$$

by Lemma 8.2 and by (8.92). Because $\widetilde{w}^+(x)$ satisfies

$$\Delta^2 \widetilde{w}^+(x) = 0 \quad \text{on} \quad \Omega^c$$

as well as (8.95) and (8.94), by Theorem 8.5, we have

$$\widetilde{w}^+(x) \equiv 0 \quad \text{on} \quad \Omega^c. \tag{8.96}$$

By (8.27) and (8.28), we have

$$\widetilde{f}_1(x) = B_1\widetilde{w}^+(x) - B_1\widetilde{w}^-(x) \equiv 0, \text{ by (8.92) and (8.96)}, \tag{8.97}$$

and by (8.32) and (8.33), we have

$$\widetilde{f}_2(x) = -[B_2\widetilde{w}^+(x) - B_2\widetilde{w}^-(x)] \equiv 0 \tag{8.98}$$

on $\partial\Omega$. From (8.97), (8.98), (8.89) and (8.92), we get

$$\widetilde{a}_0 = \widetilde{a}_1 = \widetilde{a}_2 = 0.$$

Hence (8.88) is proved. The proof is complete. □

Solving the biharmonic Type I problem (8.9) requires the introduction of six extra constants a_0, a_1, a_2 and A_0, A_1, A_2 in (8.84). Their appearance is somewhat cumbersome. One wonders whether it is always possible to solve the more direct system of multilayer equations (from (8.19))

$$\left.\begin{aligned} &\int_{\partial\Omega}\left[E(x,\xi)f_1(x) + \frac{\partial E(x,\xi)}{\partial n_\xi}f_2(\xi)\right]d\sigma_\xi = g_1(x),\ x \in \partial\Omega, \\ &\int_{\partial\Omega}\left[\frac{\partial E(x,\xi)}{\partial n_x}f_1(\xi) + \frac{\partial^2 E(x,\xi)}{\partial n_x \partial n_\xi}f_2(\xi)\right]d\sigma_\xi = g_2(x),\ \ x \in \partial\Omega, \end{aligned}\right\} \tag{8.99}$$

for (8.9) without those constants a_0, a_1, a_2, A_0, A_1 and A_2. The following example shows that the answer is negative in general.

Example 8.2. Consider the integral operator

$$K\colon\ \mathscr{H} \equiv H^r(\partial\Omega) \oplus H^{r+1}(\partial\Omega) \to H^{r+3}(\partial\Omega) \oplus H^{r+2}(\partial\Omega),\ \ r \in \mathbb{R},$$

$$K = \begin{bmatrix} K_{11} & K_{12} \\ K_{21} & K_{22} \end{bmatrix},$$

$$\left(K\begin{bmatrix} f_1 \\ f_2 \end{bmatrix}\right)(x) = \begin{bmatrix} (K_{11}f_1)(x) + (K_{12}f_2)(x) \\ (K_{21}f_1)(x) + (K_{22}f_2)(x) \end{bmatrix}$$
$$\equiv \begin{bmatrix} \int_{\partial\Omega} E(x,\xi) f_1(\xi)\, d\sigma_\xi + \int_{\partial\Omega} \frac{\partial E(x,\xi)}{\partial n_\xi} f_2(\xi)\, d\sigma_\xi \\ \int_{\partial\Omega} \frac{\partial E(x,\xi)}{\partial n_\xi} f_1(\xi)\, d\sigma_\xi + \int_{\partial\Omega} \frac{\partial^2 E(x,\xi)}{\partial n_x \partial n_\xi} f_2(\xi)\, d\sigma_\xi \end{bmatrix} \tag{8.100}$$

defined on Ω, an open disk with radius R. □

Let

$$V = \text{span}\left\{ \begin{bmatrix} 1 \\ 0 \end{bmatrix}, \begin{bmatrix} 0 \\ 1 \end{bmatrix}, \begin{bmatrix} x_1 \\ 0 \end{bmatrix}, \begin{bmatrix} 0 \\ x_1 \end{bmatrix}, \begin{bmatrix} x_2 \\ 0 \end{bmatrix}, \begin{bmatrix} 0 \\ x_2 \end{bmatrix} \right\}.$$

Then V is a six-dimensional subspace of $\mathscr{H}$. Note that on $\partial\Omega$,

$$x_1 = R\cos\theta, \quad x_2 = R\sin\theta, \quad 0 \leqslant \theta \leqslant 2\pi.$$

Also, for $x, \xi \in \partial\Omega$, let (8.59) be the polar coordinate representations of x and ξ. Then from (8.60),

$$E(x,\xi) = -\frac{1}{8\pi} R^2[1-\cos(\theta-\phi)]\ln[2R^2(1-\cos(\theta-\phi))].$$

Using

$$\int_0^{2\pi} \cos(\theta-\phi) \begin{bmatrix} 1 \\ \cos\phi \\ \sin\phi \end{bmatrix} d\phi = \begin{bmatrix} 0 \\ \pi\cos\theta \\ \pi\sin\theta \end{bmatrix},$$

$$\int_0^{2\pi} \ln[1-\cos(\theta-\phi)] \begin{bmatrix} 1 \\ \cos\phi \\ \sin\phi \end{bmatrix} d\phi = \begin{bmatrix} -2\pi\ln 2 \\ -2\pi\cos\theta \\ -2\pi\sin\theta \end{bmatrix},$$

$$\int_0^{2\pi} \cos(\theta-\phi)\ln[1-\cos(\theta-\phi)] \begin{bmatrix} 1 \\ \cos\phi \\ \sin\phi \end{bmatrix} d\phi = \begin{bmatrix} -2\pi \\ -\pi\left(\frac{1}{2}+\ln 2\right)\cos\theta \\ -\pi\left(\frac{1}{2}+\ln 2\right)\sin\theta \end{bmatrix},$$

we can easily verify that for

$$\begin{aligned} f_1(\xi) &= \alpha_1 + \frac{1}{R}(\beta_1\xi_1 + \gamma_1\xi_2) \\ &= \alpha_1 + \beta_1\cos\phi + \gamma_1\sin\phi, \\ f_2(\xi) &= \alpha_2 + \frac{1}{R}(\beta_2\xi_1 + \gamma_2\xi_2) \\ &= \alpha_2 + \beta_2\cos\phi + \gamma_2\sin\phi, \end{aligned}$$

we have

$$K\begin{bmatrix} f_1 \\ f_2 \end{bmatrix}(x) = -\frac{1}{8\pi}\begin{bmatrix} 2R^2\pi[R(1+2\ln R)A_2 + 2(1+\ln R)A_2] \\ +R^2\pi\left[R\left(-\frac{3}{2}-2\ln R\right)B_1 + \left(-\frac{5}{2}-2\ln R\right)B_2\right]\cos\theta \\ +R^2\pi\left[R\left(-\frac{3}{2}-2\ln R\right)C_1 + \left(-\frac{5}{2}-2\ln R\right)C_2\right]\sin\theta \\ 2R\pi[R(2+2\ln R)A_1 + A_2] \\ +R\pi\left[R\left(-\frac{5}{2}-2\ln R\right)B_1 + \left(-\frac{3}{2}-2\ln R\right)B_2\right]\cos\theta \\ +R^2\pi\left[R\left(-\frac{5}{2}-2\ln R\right)C_1 + \left(-\frac{3}{2}-2\ln R\right)C_2\right]\sin\theta \end{bmatrix},$$

where A_i, B_i and C_i, $i = 1, 2$, depend only on α_i, β_i and γ_i, $i = 1, 2$. Thus we see that V is an invariant subspace of K in $\mathscr{H}$. Indeed, we further have

$$V = V_1 \oplus V_2 \oplus V_3,$$

where

$$\begin{aligned} V_1 &= \operatorname{span}\left\{\begin{bmatrix} 1 \\ 0 \end{bmatrix}, \begin{bmatrix} 0 \\ 1 \end{bmatrix}\right\}, \\ V_2 &= \operatorname{span}\left\{\begin{bmatrix} x_1 \\ 0 \end{bmatrix}, \begin{bmatrix} 0 \\ x_1 \end{bmatrix}\right\}, \\ V_3 &= \operatorname{span}\left\{\begin{bmatrix} x_2 \\ 0 \end{bmatrix}, \begin{bmatrix} 0 \\ x_2 \end{bmatrix}\right\}, \end{aligned}$$

and each V_i, $i = 1, 2, 3$, is invariant under K. Let

$$\boldsymbol{P}_i\colon\ \mathscr{H} \to V_i, \quad i = 1, 2, 3,$$

be the orthogonal projection of $\mathscr{H}$ onto V_i. Then we have the following.

(i) W.r.t. the basis $\left\{\begin{bmatrix}1\\0\end{bmatrix},\begin{bmatrix}0\\1\end{bmatrix}\right\}$, the matrix representation for $\boldsymbol{P}_1 K \boldsymbol{P}_1$ is

$$M_1 = -\frac{1}{8\pi}\begin{bmatrix} 2R^2\pi[R(1+2\ln R)] & 4R^2\pi(1+\ln R) \\ 4R^2\pi(1+\ln R) & 2R\pi \end{bmatrix},$$

which is *always invertible* on V_1.

(ii) W.r.t. the basis $\left\{\begin{bmatrix}\cos\theta\\0\end{bmatrix},\begin{bmatrix}0\\\cos\theta\end{bmatrix}\right\}$ in V_2, the matrix representation for $\boldsymbol{P}_2 K \boldsymbol{P}_2$ is

$$M_2 = -\frac{R^2}{8}\begin{bmatrix} R\left(-\frac{3}{2}-2\ln R\right) & -\frac{5}{2}-2\ln R \\ -\frac{5}{2}-2\ln R & -\frac{1}{R}\left(\frac{3}{2}+2\ln R\right) \end{bmatrix}, \tag{8.101}$$

which is *singular at* $R = e^{-1}$.

(iii) W.r.t. the basis $\left\{\begin{bmatrix}\sin\theta\\0\end{bmatrix},\begin{bmatrix}0\\\sin\theta\end{bmatrix}\right\}$ in V_3, the matrix representation for $\boldsymbol{P}_3 K \boldsymbol{P}_3$ is the same as in (8.101). So it is again singular at $R = e^{-1}$. Therefore in (8.99), if we let

$$\begin{bmatrix} g_1(x) \\ g_2(x) \end{bmatrix} = \begin{bmatrix} \mu_1 x_1 \\ \mu_2 x_1 \end{bmatrix}, \quad x \in \partial\Omega,$$

such that

$$\begin{bmatrix} \mu_1 \\ \mu_2 \end{bmatrix} \notin \text{range of } M_2 \text{ in (8.101)},$$

then the boundary integral equation system (8.99) will *not be solvable* for such (g_1, g_2). □

The reader can easily test the numerical solvability of (8.99) by discretizing it using point collocation. For a disk Ω with radius R, the discretized linear system is *always invertible* and gives accurate numerical solutions when R is not too close to e^{-1}. But when R is very close (or equal) to e^{-1}, the matrix becomes *ill-conditioned* (albeit always invertible), and the numerical solutions assume excessively large values and become worthless.

Corollary 8.1. *Assume that for any given $\alpha_0,\ \alpha_1,\ \alpha_2 \in \mathbb{R}$, the boundary integral equations*

$$\left.\begin{aligned} &\int_{\partial\Omega}\left[E(x,\xi)\eta_1(\xi) + \frac{\partial E(x,\xi)}{\partial n_\xi}\eta_2(\xi)\right] d\sigma_\xi = \alpha_0 + \alpha_1 x_1 + \alpha_2 x_2, \quad x \in \partial\Omega, \\ &\int_{\partial\Omega}\left[\frac{\partial E(x,\xi)}{\partial n_x}\eta_1(\xi) + \frac{\partial^2 E(x,\xi)}{\partial n_x \partial n_\xi}\eta_2(\xi)\right] d\sigma_\xi = \alpha_1 n_1(x) + \alpha_2 n_2(x), \quad x \in \partial\Omega, \end{aligned}\right\} \tag{8.102}$$

are uniquely solvable for (η_1, η_2). *Then for any given functions* $(g_0, g_1) \in H^{r+3}(\partial\Omega) \oplus H^{r+2}(\partial\Omega)$, $r \in \mathbb{R}$, *the boundary integral equations*

$$\left.\begin{aligned} \int_{\partial\Omega} \left[E(x,\xi) f_1(\xi) + \frac{\partial E(x,\xi)}{\partial n_\xi} f_2(\xi) \right] d\sigma_\xi &= g_1(x),\ x \in \partial\Omega, \\ \int_{\partial\Omega} \left[\frac{\partial E(x,\xi)}{\partial n_x} f_1(\xi) + \frac{\partial^2 E(x,\xi)}{\partial n_x \partial n_\xi} f_2(\xi) \right] d\sigma_\xi &= g_2(x),\ x \in \partial\Omega, \end{aligned}\right\} \tag{8.103}$$

are also uniquely solvable for $(f_1, f_2) \in H^r(\partial\Omega) \oplus H^{r+1}(\partial\Omega)$. *Consequently, any biharmonic function* w *is representable in the simpler form (8.19) (no matter what the boundary conditions are).*

Proof. Choose $A_0 = 0$, $A_1 = 0$, $A_2 = 0$ (or any other real numbers) in (8.84) in Theorem 8.6. They by Theorem 8.6, there exist unique $-\widetilde{a}_0$, $-\widetilde{a}_1$, $-\widetilde{a}_2 \in \mathbb{R}$ and functions $\widetilde{f}_1$, $\widetilde{f}_2$ on $\partial\Omega$ such that for $x \in \partial\Omega$,

$$\begin{aligned} &\int_{\partial\Omega} \widetilde{f}_1(\xi)\, d\sigma = 0, \\ &\int_{\partial\Omega} [\xi_1 \widetilde{f}_1(\xi) + n_1(\xi) \widetilde{f}_2(\xi)]\, d\sigma = 0, \\ &\int_{\partial\Omega} [\xi_2 \widetilde{f}_1(\xi) + n_2(\xi) \widetilde{f}_2(\xi)]\, d\sigma = 0, \\ &\int_{\partial\Omega} \left[E(x,\xi) \widetilde{f}_1(\xi) + \frac{\partial E(x,\xi)}{\partial n_\xi} \widetilde{f}_2(\xi) \right] d\sigma_\xi - (\widetilde{a}_0 + \widetilde{a}_1 x_1 + \widetilde{a}_2 x_2) = g_1(x), \\ &\int_{\partial\Omega} \left[\frac{\partial E(x,\xi)}{\partial n_x} \widetilde{f}_1(\xi) + \frac{\partial^2 E(x,\xi)}{\partial n_x \partial n_\xi} \widetilde{f}_2(\xi) \right] d\sigma_\xi - [\widetilde{a}_1 n_1(x) + \widetilde{a}_2 n_2(x)] = g_2(x). \end{aligned}$$

For such $\widetilde{a}_0$, $\widetilde{a}_1$, $\widetilde{a}_2 \in \mathbb{R}$, let $\widetilde{\eta}_1$, $\widetilde{\eta}_2$ be the unique solutions to

$$\begin{aligned} &\int_{\partial\Omega} \left[E(x,\xi) \widetilde{\eta}_1(\xi) + \frac{\partial E(x,\xi)}{\partial n_\xi} \widetilde{\eta}_2(\xi) \right] d\sigma_\xi = \widetilde{a}_0 + \widetilde{a}_1 x_1 + \widetilde{a}_2 x_2, \qquad x \in \partial\Omega, \\ &\int_{\partial\Omega} \left[\frac{\partial E(x,\xi)}{\partial n_x} \widetilde{\eta}_1(\xi) + \frac{\partial^2 E(x,\xi)}{\partial n_x \partial n_\xi} \widetilde{\eta}_2(\xi) \right] d\sigma_\xi = \widetilde{a}_1 n_1(x) + \widetilde{a}_2 n_2(x),\ x \in \partial\Omega. \end{aligned}$$

Then

$$\left.\begin{aligned} f_1(x) &\equiv \widetilde{f}_1(x) - \widetilde{\eta}_1(x) \\ f_2(x) &\equiv \widetilde{f}_2(x) - \widetilde{\eta}_2(x) \end{aligned}\right\} \quad x \in \partial\Omega,$$

solve (8.103). The solutions f_1 and f_2 are unique because the null space of the integral equations (8.103) contains only zero by assumption. □

Remark 8.1. By Corollary 8.1, we see that the boundary integral equations (8.99) are uniquely solvable for any *disk* Ω whose radius is not equal to e^{-1}.

We will furnish a numerical example based on the integral equations (8.99) discretized by collocation in § 8.7. □

Next, we consider the biharmonic problem of Type II, where shear and moment forces act on the boundary.

Theorem 8.7. (Simple- and double-layer representations for biharmonic problem of Type II). *Let $\Omega \subseteq \mathbb{R}^2$ be bounded with smooth boundary $\partial\Omega$. The mapping*

$$\mathscr{L}_2 \begin{bmatrix} f_1 \\ f_2 \end{bmatrix}(x) = \begin{bmatrix} \int_{\partial\Omega} f_1(\xi)\,d\sigma \\ \int_{\partial\Omega} [\xi_1 f_1(\xi) + n_1(\xi) f_2(\xi)]\,d\sigma \\ \int_{\partial\Omega} [\xi_2 f_1(\xi) + n_2(\xi) f_2(\xi)]\,d\sigma \\ \int_{\partial\Omega} \left[B_{1x}E(x,\xi) f_1(\xi) + B_{1x}\frac{\partial E(x,\xi)}{\partial n_\xi} f_2(\xi) \right] d\sigma_\xi + \frac{1}{2} f_1(x) \\ \int_{\partial\Omega} \left[B_{2x}E(x,\xi)\cdot f_1(\xi) + B_{2x}\frac{\partial E(x,\xi)}{\partial n_\xi} f_2(\xi) \right] d\sigma_\xi - \frac{1}{2} f_2(x) \end{bmatrix} \tag{8.104}$$

is linear continuous:

$$\mathscr{L}_2\colon\ H^r(\partial\Omega) \oplus H^{r+1}(\partial\Omega) \to \mathbb{R}^3 \oplus H^r(\partial\Omega) \oplus H^{r+1}(\partial\Omega) \tag{8.105}$$

for $r \in \mathbb{R}$, with

$$\dim \mathscr{N}(\mathscr{L}_2) = 0,\ \dim \mathrm{Coker}\, \mathscr{L}_2 = 3. \tag{8.106}$$

Its adjoint operator

$$\mathscr{L}_2^*\colon\ \mathbb{R}^3 \oplus H^{-r}(\partial\Omega) \oplus H^{-(r+1)}(\partial\Omega) \to H^{-r}(\partial\Omega) \oplus H^{-(r+1)}(\partial\Omega) \tag{8.107}$$

is given by

$$\mathscr{L}_2^* \begin{bmatrix} b_0 \\ b_1 \\ b_2 \\ \phi_1 \\ \phi_2 \end{bmatrix}(x) = \begin{bmatrix} \int_{\partial\Omega} [B_{1\xi}E(x,\xi)\phi_1(\xi) + B_{2\xi}E(x,\xi)\phi_2(\xi)]\,d\sigma_\xi \\ \qquad + \frac{1}{2}\phi_1(x) + b_0 + b_1 x_1 + b_2 x_2 \\ \int_{\partial\Omega} \left[\frac{\partial B_{1\xi}E(x,\xi)}{\partial n_x}\phi_1(\xi) + \frac{\partial B_{2\xi}E(x,\xi)}{\partial n_x}\phi_2(\xi) \right] d\sigma_\xi \\ \qquad - \frac{1}{2}\phi_2(x) + b_1 n_1(x) + b_2 n_2(x) \end{bmatrix} \tag{8.108}$$

on $\partial\Omega$, with

$$\dim \mathscr{N}(\mathscr{L}_2^*) = 3,\ \dim \mathrm{Coker}\, \mathscr{L}_2^* = 0, \tag{8.109}$$

$$\mathscr{N}(\mathscr{L}_2^*) = \mathrm{span} \left\{ \left. \begin{bmatrix} 0 \\ 0 \\ 0 \\ 1 \\ 0 \end{bmatrix}, \begin{bmatrix} 0 \\ 0 \\ 0 \\ x_1 \\ -n_1(x) \end{bmatrix}, \begin{bmatrix} 0 \\ 0 \\ 0 \\ x_2 \\ -n_2(x) \end{bmatrix} \right| x \in \partial\Omega \right\}. \tag{8.110}$$

$\mathscr{L}_2^*$ maps continuously

$$\mathscr{L}_2^*\colon \mathbb{R}^3 \oplus H^s(\partial\Omega) \oplus H^{s-1}(\partial\Omega) \to H^s(\partial\Omega) \oplus H^{s-1}(\partial\Omega), \tag{8.111}$$

for any $s \in \mathbb{R}$.

Consequently, the solution $w(x)$ to the BVP

$$\left.\begin{aligned} &\Delta^2 w(x) = 0, && x \in \Omega, \\ &B_1 w(x) = g_1(x) \in H^r(\partial\Omega), && r \in \mathbb{R}, \\ &B_2 w(x) = g_2(x) \in H^{r+1}(\partial\Omega), && \end{aligned}\right\} \tag{8.112}$$

is given by

$$w(x) = \int_{\partial\Omega} \left[E(x,\xi)\widetilde{f}_1(\xi) + \frac{\partial}{\partial n_\xi} E(x,\xi) \cdot \widetilde{f}_2(\xi)\right] d\sigma_\xi + c_0 + c_1 x_1 + c_2 x_2,$$

$$x \in \Omega,\ x \in H^{r+7/2}(\Omega), \tag{8.113}$$

for some unique $(\widetilde{f}_1, \widetilde{f}_2) \in H^r(\partial\Omega) \oplus H^{r+1}(\partial\Omega)$ and for arbitrarily given constants c_0, c_1, $c_2 \in \mathbb{R}$, where $(\widetilde{f}_1, \widetilde{f}_2)$ is the unique solution to

$$\mathscr{L}_2(\widetilde{f}_1, \widetilde{f}_2) = (A_0, A_1, A_2, g_1, g_2), \tag{8.114}$$

provided that

$$\begin{aligned} (0,0,0,g_1,g_2) &\perp \operatorname{Coker} \mathscr{L}_2 \\ &= \mathscr{N}(\mathscr{L}_2^*) \text{ in } \mathbb{R}^3 \oplus C^\infty(\partial\Omega) \oplus C^\infty(\partial\Omega). \end{aligned} \tag{8.115}$$

The three constants A_0, A_1, A_2 in (8.114) can be chosen arbitrarily (but the resulting $\widetilde{f}_1$, $\widetilde{f}_2$ in (8.114) depend on the choice of A_0, A_1 and A_2.)

Proof. The regularity properties (8.105), (8.107) of $\mathscr{L}_2$ and $\mathscr{L}_2^*$ are consequences of Lemma 8.2.

We represent $\mathscr{L}_2$ by a matrix kernel

$$K_2(x,\xi) = \begin{bmatrix} 1 & 0 \\ \xi_1 & n_1(\xi) \\ \xi_2 & n_2(\xi) \\ B_{1x}E(x,\xi) + \frac{1}{2}\delta(x-\xi) & B_{1x}\dfrac{\partial E(x,\xi)}{\partial n_\xi} \\ B_{2x}E(x,\xi) & B_{2x}\dfrac{\partial E(x,\xi)}{\partial n_\xi} - \frac{1}{2}\delta(x-\xi) \end{bmatrix},$$

where the delta function is regarded as a distribution on $\mathscr{D}(\partial\Omega)$ so that

$$\mathscr{L}_2 \begin{bmatrix} f_1 \\ f_2 \end{bmatrix}(x) = \int_{\partial\Omega} K_2(x,\xi) \begin{bmatrix} f_1 \\ f_2 \end{bmatrix}(\xi)\, d\sigma_\xi, \quad x \in \partial\Omega.$$

The adjoint operator $\mathscr{L}_2^*$ of $\mathscr{L}_2$ is formally defined through the matrix $K_2^{\mathrm{Tr}}(\xi,x)$, i.e.,

$$\mathscr{L}_2^* \begin{bmatrix} b_0 \\ b_1 \\ b_2 \\ \phi_1 \\ \phi_2 \end{bmatrix}(x) = \int_{\partial\Omega} K_2^{\mathrm{Tr}}(\xi,x) \begin{bmatrix} b_0 \\ b_1 \\ b_2 \\ \phi_1 \\ \phi_2 \end{bmatrix}(\xi)\, d\sigma_\xi, \; x \in \partial\Omega;$$

therefore (8.108) is obtained.

We now show the following.

(i) $\dim \mathscr{N}(\mathscr{L}_2) = 0$

Assume that $(\widetilde{f}_1, \widetilde{f}_2) \in \mathscr{N}(\mathscr{L}_2)$. By Corollary 4.4, we have $(\widetilde{f}_1, \widetilde{f}_2) \in C^\infty(\partial\Omega) \oplus C^\infty(\partial\Omega)$. For $x \in \Omega$, define

$$\widetilde{w}^-(x) = \int_{\partial\Omega} \left[E(x,\xi)\widetilde{f}_1(\xi) + \frac{\partial E(x,\xi)}{\partial n_\xi} \widetilde{f}_2(\xi) \right] d\sigma_\xi. \tag{8.116}$$

Then on $\partial\Omega$

$$B_1 \widetilde{w}^-(x) = \frac{1}{2}\widetilde{f}_1(x) + \int_{\partial\Omega} \left\{ [B_{1x}E(x,\xi)]\widetilde{f}_1(\xi) + \left[B_{1x}\frac{\partial E(x,\xi)}{\partial n_\xi} \right] \widetilde{f}_2(\xi) \right\} d\sigma_\xi = 0$$
$$B_2 \widetilde{w}^-(x) = -\frac{1}{2}\widetilde{f}_2(x) + \int_{\partial\Omega} \left\{ [B_{2x}E(x,\xi)]\widetilde{f}_1(\xi) + \left[B_{2x}\frac{\partial E(x,\xi)}{\partial n_\xi} \right] \widetilde{f}_2(\xi) \right\} d\sigma_\xi = 0,$$

by (8.43), (8.38) and the assumption that $(\widetilde{f}_1, \widetilde{f}_2) \in \mathscr{N}(\mathscr{L}_2)$. Since

$$\Delta^2 \widetilde{w}^-(x) = 0 \quad \text{on} \quad \Omega,$$

by Theorem 8.2 we get

$$\widetilde{w}^-(x) = \alpha_0 + \alpha_1 x_1 + \alpha_2 x_2 \quad \text{on} \quad \Omega, \tag{8.117}$$

i.e., $\widetilde{w}^-$ is a linear polynomial on Ω. Now we extend (8.116) to Ω^c by defining

$$\widetilde{w}^+(x) = \int_{\partial\Omega} \left[E(x,\xi)\widetilde{f}_1(\xi) + \frac{\partial E(x,\xi)}{\partial n_\xi} \widetilde{f}_2(\xi) \right] d\sigma_\xi, \; x \in \Omega^c.$$

Then by Lemma 8.2, we have, on $\partial\Omega$,

$$\left.\begin{aligned} \widetilde{w}^+(x) &= \widetilde{w}^-(x) = \alpha_0 + \alpha_1 x_2 + \alpha_2 x_2 \\ \frac{\partial \widetilde{w}^+(x)}{\partial n} &= \frac{\partial \widetilde{w}^-(x)}{\partial n} = \alpha_1 n_1(x) + \alpha_2 n_2(x). \end{aligned}\right\} \tag{8.118}$$

$\widetilde{w}^+$ is a biharmonic function:

$$\Delta^2 \widetilde{w}^+(x) = 0 \quad \text{on} \quad \Omega^c.$$

Since $(\widetilde{f}_1, \widetilde{f}_2) \in \mathscr{N}(\mathscr{L}_2)$,

$$\int_{\partial\Omega} \widetilde{f}_1(\xi)\, d\sigma = 0,$$

$$\int_{\partial\Omega} [\xi_1 \widetilde{f}_1(\xi) + n_1(\xi)\widetilde{f}_2(\xi)]\, d\sigma = 0,$$

$$\int_{\partial\Omega} [\xi_2 \widetilde{f}_1(\xi) + n_2(\xi)\widetilde{f}_2(\xi)]\, d\sigma = 0,$$

we can do asymptotic estimates as in (8.93) through (8.94) and conclude that

$$\widetilde{w}^+(x) = \alpha_0 + \alpha_1 x_1 + \alpha_2 x_2, \quad x \in \Omega^c, \tag{8.119}$$

by (8.118) and Theorem 8.5. From (8.117) and (8.119), using the jump conditions as in (8.97), (8.98), we conclude that

$$\widetilde{f}_1(x) \equiv 0, \quad \widetilde{f}_2(x) \equiv 0 \quad \text{on} \quad \partial\Omega.$$

(ii) $\dim \mathscr{N}(\mathscr{L}_2^*) \geqslant 3$

Let w be any biharmonic function on $\partial\Omega$:

$$\Delta^2 w(x) = 0 \quad \text{on} \quad \Omega.$$

As in (8.17), (8.51) and (8.54), on $\partial\Omega$, we have

$$\begin{aligned}
\frac{1}{2} w(x) &= -\int_{\partial\Omega} \{[\Delta_\xi^2 E(x,\xi)]w(\xi) - [\Delta^2 w(x)]E(x,\xi)\}\, dx \\
&= -\int_{\partial\Omega} \Big\{ [B_{1\xi} E(x,\xi)]w(\xi) - [B_1 w(\xi)]E(x,\xi) \\
&\quad - [B_{2\xi} E(x,\xi)]\frac{\partial w(\xi)}{\partial n} + [B_2 w(\xi)]\frac{\partial E(x,\xi)}{\partial n_\xi} \Big\}\, d\sigma_\xi, \quad x \in \partial\Omega,
\end{aligned} \tag{8.120}$$

$$\begin{aligned}
\frac{1}{2}\frac{\partial w(x)}{\partial n} &= -\int_{\partial\Omega} \Big\{ \frac{\partial B_{1\xi} E(x,\xi)}{\partial n_x} w(\xi) - [B_1 w(\xi)]\frac{\partial E(x,\xi)}{\partial n_x} \\
&\quad - \frac{\partial B_{2\xi} E(x,\xi)}{\partial n_x}\frac{\partial w(\xi)}{\partial n} + [B_2 w(\xi)]\frac{\partial^2 E(x,\xi)}{\partial n_x \partial n_\xi} \Big\}\, d\sigma_\xi, \quad x \in \partial\Omega.
\end{aligned} \tag{8.121}$$

In (8.120) and (8.121), if we choose

(a) $w(x) = 1$ on Ω, (b) $w(x) = x_1$ on Ω, (c) $w(x) = x_2$ on Ω,

we immediately see that

$$\begin{bmatrix} 0 \\ 0 \\ 0 \\ 1 \\ 0 \end{bmatrix}, \begin{bmatrix} 0 \\ 0 \\ 0 \\ x_1 \\ -n_1(x) \end{bmatrix}, \begin{bmatrix} 0 \\ 0 \\ 0 \\ x_2 \\ -n_2(x) \end{bmatrix} \in \mathscr{N}(\mathscr{L}_2^*).$$

(iii) $\dim \mathscr{N}(\mathscr{L}_2^*) = 3$

Assume that

$$(\widetilde{b}_0, \widetilde{b}_1, \widetilde{b}_2, \widetilde{\phi}_1, \widetilde{\phi}_2) \in \mathscr{N}(\mathscr{L}_2^*). \tag{8.122}$$

Define

$$\left.\begin{array}{l} \widetilde{w}^+(x) \\ \widetilde{w}^-(x) \end{array}\right\} = \int_{\partial\Omega} \{[B_{1\xi}E(x,\xi)]\widetilde{\phi}_1(\xi) + [B_{2\xi}E(x,\xi)]\widetilde{\phi}_2(\xi)\}\, d\sigma_\xi$$
$$+ \widetilde{b}_0 + \widetilde{b}_1 x_1 + \widetilde{b}_2 x_2, \tag{8.123}$$

with

$$\begin{cases} + & \text{for } x \in \Omega^c \\ - & \text{for } x \in \Omega. \end{cases}$$

Then (8.122) and Lemma 8.2 imply that

$$\left.\begin{array}{l} \widetilde{w}^+(x) = 0, \\ \dfrac{\partial \widetilde{w}^+(x)}{\partial n} = 0 \end{array}\right\} \quad \text{on} \quad \partial\Omega. \tag{8.124}$$

It is easy to check that $\widetilde{w}^+$ on Ω^c satisfies (8.74) and (8.75). By Theorem 8.1, we have

$$\widetilde{w}^+(x) \equiv 0 \quad \text{on} \quad \Omega. \tag{8.125}$$

Because $(\widetilde{\phi}_1, \widetilde{\phi}_2) \in C^\infty(\partial\Omega) \oplus C^\infty(\partial\Omega)$, by Lemma 8.2 on $\partial\Omega$, we have

$$\left.\begin{array}{l} B_1\widetilde{w}^+(x) = B_1\widetilde{w}^-(x) = 0, \\ B_2\widetilde{w}^+(x) = B_2\widetilde{w}^-(x) = 0, \end{array}\right\} \tag{8.126}$$

by (8.125). By Theorem 8.2, we have

$$\widetilde{w}^-(x) = \alpha_0 + \alpha_1 x_1 + \alpha_2 x_2 \quad \forall x \in \Omega, \tag{8.127}$$

for some α_0, α_1 and $\alpha_2 \in \mathbb{R}$. Hence from Lemma 8.2, (8.125) and (8.127),

$$\left.\begin{array}{l} \widetilde{\phi}_1(x) = \widetilde{w}^+(x) - \widetilde{w}^-(x) = \alpha_0 + \alpha_1 x_1 + \alpha_2 x_2, \\ \widetilde{\phi}_2(x) = -\left[\dfrac{\partial \widetilde{w}^+(x)}{\partial n} - \dfrac{\partial \widetilde{w}^-(x)}{\partial n}\right] = -\alpha_1 n_1(x) - a_2 n_2(x). \end{array}\right\} \tag{8.128}$$

Substituting (8.128) into (8.123) for $\widetilde{w}^+(x)$, and letting x tend to a boundary point on $\partial\Omega$ from Ω^c, we get

$$\widetilde{b}_0 + \widetilde{b}_1 x_1 + \widetilde{b}_2 x_2 = 0 \quad \text{on } \partial\Omega,$$

by (8.122). Hence

$$\widetilde{b}_0 = \widetilde{b}_1 = \widetilde{b}_2 = 0. \tag{8.129}$$

By (8.128) and (8.129), we conclude that

$$\begin{bmatrix} \widetilde{b}_0 \\ \widetilde{b}_1 \\ \widetilde{b}_2 \\ \widetilde{\phi}_1 \\ \widetilde{\phi}_2 \end{bmatrix} \in \operatorname{span}\left\{ \begin{bmatrix} 0 \\ 0 \\ 0 \\ 1 \\ 0 \end{bmatrix}, \begin{bmatrix} 0 \\ 0 \\ 0 \\ x_1 \\ -n_1(x) \end{bmatrix}, \begin{bmatrix} 0 \\ 0 \\ 0 \\ x_2 \\ -n_2(x) \end{bmatrix} \middle| \, x \in \partial\Omega \right\}. \qquad \square$$

The operator $\mathscr{L}_2$ in Theorem 8.7 has index

$$\begin{aligned} \operatorname{Ind} \mathscr{L}_2 &= \dim \mathscr{N}(\mathscr{L}_2) - \operatorname{codim} \mathscr{R}(\mathscr{L}_2) \\ &= 0 - 3 \;=\; -3. \end{aligned}$$

Sometimes, an operator with zero index is preferable, since Theorem 4.10 is applicable, allowing the flexibility of choosing different sets of accessory conditions such as the examples in § 4.7 show. Thus, we state the following easy corollary from Theorem 8.7.

Corollary 8.2. *Let $\mathscr{L}_2$ denote the operator defined in (8.104). Let $\boldsymbol{P}$ denote the projection operator*

$$\begin{aligned} &\boldsymbol{P}\colon\ \mathbb{R}^3 \oplus H^r(\partial\Omega) \oplus H^{r+1}(\partial\Omega) \to H^r(\partial\Omega) \oplus H^{r+1}(\partial\Omega), \\ &\boldsymbol{P}(a_0, a_1, a_2, f_1, f_2) = (f_1, f_2). \end{aligned}$$

($\boldsymbol{P}$ is independent of $r \in \mathbb{R}$.) Then the operator

$$\widetilde{\mathscr{L}_2} \equiv \boldsymbol{P}\mathscr{L}_2\colon\ H^r(\partial\Omega) \oplus H^{r+1}(\partial\Omega) \to H^r(\partial\Omega) \oplus H^{r+1}(\partial\Omega)$$

is continuous for any $r \in \mathbb{R}$, with

$$\mathscr{N}(\widetilde{\mathscr{L}_2}) = \operatorname{span}\left\{ \boldsymbol{P}\mathscr{L}_1^{-1}k_i \,\middle|\, i = 1, 2, 3,\ k_1 = \begin{bmatrix} 0 \\ 0 \\ 0 \\ 1 \\ 0 \end{bmatrix},\ k_2 = \begin{bmatrix} 0 \\ 0 \\ 0 \\ x_1 \\ n_1(x) \end{bmatrix},\ k_3 = \begin{bmatrix} 0 \\ 0 \\ 0 \\ x_2 \\ n_2(x) \end{bmatrix} \right\}$$

$$\in C^\infty(\partial\Omega) \oplus C^\infty(\partial\Omega), \ \text{ with } \mathscr{L}_1 \textit{ given by (8.77)},$$

$$\operatorname{Coker}(\widetilde{\mathscr{L}_2}) = \textit{span}\left\{ \begin{bmatrix} 1 \\ 0 \end{bmatrix}, \begin{bmatrix} x_1 \\ -n_1(x) \end{bmatrix}, \begin{bmatrix} x_2 \\ -n_2(x) \end{bmatrix} \right\}$$

$$\in C^\infty(\partial\Omega) \oplus C^\infty(\partial\Omega),$$

and hence with index zero. Its adjoint operator $\widetilde{\mathscr{L}_2^*}$ is given by

$$\widetilde{\mathscr{L}_2^*} = \mathscr{L}_2^*\boldsymbol{P}^*\colon\ H^{-r}(\partial\Omega) \oplus H^{-(r+1)}(\partial\Omega) \to H^{-r}(\partial\Omega) \oplus H^{-(r+1)}(\partial\Omega),$$

where

$$\boldsymbol{P}^*:\ H^{-r}(\partial\Omega)\oplus H^{-(r+1)}(\partial\Omega)\to\mathbb{R}^3\oplus H^{-r}(\partial\Omega)\oplus H^{-(r+1)}(\partial\Omega),$$
$$\boldsymbol{P}^*(\phi_1,\phi_2)=(0,0,0,\phi_1,\phi_2),$$

with

$$\mathscr{N}(\widetilde{\mathscr{L}_2^*})=\text{Coker}(\widetilde{\mathscr{L}_2}),\ \text{Coker}(\widetilde{\mathscr{L}_2^*})=\mathscr{N}(\widetilde{\mathscr{L}_2}).$$

The solution w to the BVP (8.112) is given by (8.113), where $(\widetilde{f}_1,\widetilde{f}_2)\in H^r(\partial\Omega)\oplus H^{r+1}(\partial\Omega)$ is a (nonunique) solution to

$$\widetilde{\mathscr{L}_2}(\widetilde{f}_1,\widetilde{f}_2)=(g_1,g_2). \tag{8.130}$$

□

The proofs of Theorems 8.6 and 8.7 above have encompassed two extreme cases (Types I and II) of boundary conditions of plate bending. The boundary integral equation system (8.84) in Theorem 8.6 is basically a Fredholm integral equation of the *first* kind, while (8.114) in Theorem 8.7 is basically a Fredholm integral equation system of the *second* kind.

The remaining two theorems in this section treat the last two biharmonic problems, Types III and IV. The resulting boundary integral equations (8.137) and (8.148) are *Fredholm of mixed first and second kinds*. Their proofs can be emulated from those of Theorems 8.6 and 8.7, and are therefore omitted.

Theorem 8.8. (Simple- and double-layers representation for biharmonic problem of Type III). *Let $\Omega\subseteq\mathbb{R}^2$ be bounded with smooth boundary $\partial\Omega$. The mapping*

$$\mathscr{L}_3\begin{bmatrix}a_0\\a_1\\a_2\\f_1\\f_2\end{bmatrix}(x)=\left[\begin{array}{l}\displaystyle\int_{\partial\Omega}f_1(\xi)\,d\sigma\\ \displaystyle\int_{\partial\Omega}[\xi_1f_1(\xi)+n_1(\xi)f_2(\xi)]\,d\sigma\\ \displaystyle\int_{\partial\Omega}[\xi_2f_1(\xi)+n_2(\xi)f_2(\xi)]\,d\sigma\\ \displaystyle\int_{\partial\Omega}\left[E(x,\xi)f_1(\xi)+\frac{\partial E(x,\xi)}{\partial n_\xi}f_2(\xi)\right]d\sigma_\xi\\ \quad+a_0+a_1x_1+a_2x_2\\ \displaystyle\int_{\partial\Omega}\left\{[B_{2x}E(x,\xi)]f_1(\xi)+\left[B_{2x}\frac{\partial E(x,\xi)}{\partial n_\xi}\right]f_2(\xi)\right]d\sigma_\xi\\ \quad-\dfrac{1}{2}f_2(x)\end{array}\right\} \tag{8.131}$$

is an isomorphism:

$$\mathscr{L}_3:\ \mathbb{R}^3\oplus H^r(\partial\Omega)\oplus H^{r+1}(\partial\Omega)\to\mathbb{R}^3\oplus H^{r+3}(\partial\Omega)\oplus H^{r+1}(\partial\Omega), \tag{8.132}$$

for $r \in \mathbb{R}$. Its adjoint operator

$$\mathscr{L}_3^*:\ \mathbb{R}^3 \oplus H^{-(r+3)}(\partial\Omega) \oplus H^{-(r+1)}(\partial\Omega) \to \mathbb{R}^3 \oplus H^{-r}(\partial\Omega) \oplus H^{-(r+1)}(\partial\Omega),$$

given by

$$\mathscr{L}_3^* \begin{bmatrix} b_0 \\ b_1 \\ b_2 \\ \phi_1 \\ \phi_2 \end{bmatrix}(x) = \begin{bmatrix} \int_{\partial\Omega} \phi_1(\xi)\, d\sigma \\ \int_{\partial\Omega} \xi_1 \phi_1(\xi)\, d\sigma \\ \int_{\partial\Omega} \xi_2 \phi_1(\xi)\, d\sigma \\ \int_{\partial\Omega} \{E(x,\xi)\phi_1(\xi) + [B_{2\xi}E(x,\xi)]\phi_2(\xi)\}\, d\sigma_\xi \\ \quad + b_0 + b_1 x_1 + b_2 x_2 \\ \int_{\partial\Omega} \left[\frac{\partial E(x,\xi)}{\partial n_x}\phi_1(\xi) + \frac{\partial B_{2\xi}E(x,\xi)}{\partial n_x}\phi_2(\xi)\right] d\sigma_\xi \\ \quad -\frac{1}{2}\phi_2(x) + b_1 n_1(x) + b_2 n_2(x) \end{bmatrix}, \tag{8.133}$$

is also an isomorphism. $\mathscr{L}_3^*$ maps isomorphically

$$\mathscr{L}_3^*:\ \mathbb{R}^3 \oplus H^{s}(\partial\Omega) \oplus H^{s+2}(\partial\Omega) \to \mathbb{R}^3 \oplus H^{s+3}(\partial\Omega) \oplus H^{s+2}(\partial\Omega), \tag{8.134}$$

for any $s \in \mathbb{R}$.

Consequently, the solution $w(x)$ to

$$\left.\begin{aligned} &\Delta^2 w(x) = 0, \\ &w(x) = g_1(x) \in H^{r+3}(\partial\Omega), \quad r \in \mathbb{R}, \\ &B_2 w(x) = g_2(x) \in H^{r+1}(\partial\Omega), \end{aligned}\right\} \tag{8.135}$$

is given by

$$\begin{aligned} w(x) = &\int_{\partial\Omega} \left[E(x,\xi)\widetilde{f}_1(\xi) + \frac{\partial E(x,\xi)}{\partial n_\xi}\widetilde{f}_2(\xi)\right] d\sigma_\xi \\ &+ \widetilde{a}_0 + \widetilde{a}_1 x_1 + \widetilde{a}_2 x_2 \in H^{r+7/2}(\Omega), \end{aligned} \tag{8.136}$$

for unique $(\widetilde{a}_0, \widetilde{a}_1, \widetilde{a}_2, \widetilde{f}_1, \widetilde{f}_2) \in \mathbb{R}^3 \oplus H^r(\partial\Omega) \oplus H^{r+1}(\partial\Omega)$, which solves

$$\mathscr{L}_3(\widetilde{a}_0, \widetilde{a}_1, \widetilde{a}_2, \widetilde{f}_1, \widetilde{f}_2) = (A_0, A_1, A_2, g_1, g_2) \in \mathbb{R}^3 \oplus H^{r+3}(\partial\Omega) \oplus H^{r+1}(\partial\Omega), \tag{8.137}$$

for given $A_0,\ A_1,\ A_2 \in \mathbb{R}$. ($A_0,\ A_1,\ A_2 \in \mathbb{R}$ can be chosen arbitrarily, but the resulting $\widetilde{a}_0,\ \widetilde{a}_1,\ \widetilde{a}_2$ and $\widetilde{f}_1,\ \widetilde{f}_2$ depend on the choice of $A_0,\ A_1,\ A_2$.)

Theorem 8.9. (Simple- and doubler-layers representation for biharmonic problem of Type IV). *Let $\Omega \subseteq \mathbb{R}^2$ be bounded with smooth boundary $\partial\Omega$. The mapping*

$$\mathscr{L}_4 \begin{bmatrix} a_1 \\ a_2 \\ f_1 \\ f_2 \end{bmatrix}(x) = \begin{bmatrix} \int_{\partial\Omega} f_1(\xi)\,d\sigma \\ \int_{\partial\Omega} [\xi_1 f_1(\xi) + n_1(\xi) f_2(\xi)]\,d\sigma \\ \int_{\partial\Omega} [\xi_2 f_1(\xi) + n_2(\xi) f_2(\xi)]\,d\sigma \\ \int_{\partial\Omega} \left[\frac{\partial E(x,\xi)}{\partial n_x} f_1(\xi) + \frac{\partial^2 E(x,\xi)}{\partial n_x \partial n_\xi} f_2(\xi)\right] d\sigma_\xi \\ \quad + a_1 n_1(x) + a_2 n_2(x) \\ \int_{\partial\Omega} \left[B_{1x} E(x,\xi) f_1(\xi) + B_{1x} \frac{\partial E(x,\xi)}{\partial n_\xi} f_2(\xi)\right] d\sigma_\xi \\ \quad + \frac{1}{2} f_1(x) \end{bmatrix} \tag{8.138}$$

is linear continuous:

$$\mathscr{L}_4 \colon\ \mathbb{R}^2 \oplus H^r(\partial\Omega) \oplus H^{r+1}(\partial\Omega) \to \mathbb{R}^3 \oplus H^{r+2}(\partial\Omega) \oplus H^r(\partial\Omega), \quad r \in \mathbb{R}, \tag{8.139}$$

with

$$\dim \mathscr{N}(\mathscr{L}_4) = 0, \quad \dim \operatorname{Coker} \mathscr{L}_4 = 1. \tag{8.140}$$

Its adjoint operator

$$\mathscr{L}_4^* \colon\ \mathbb{R}^3 \oplus H^{-(r+2)}(\partial\Omega) \oplus H^{-r}(\partial\Omega) \to \mathbb{R}^2 \oplus H^{-r}(\partial\Omega) \oplus H^{-(r+1)}(\partial\Omega), \tag{8.141}$$

given by

$$\mathscr{L}_4^* \begin{bmatrix} b_0 \\ b_1 \\ b_2 \\ \phi_1 \\ \phi_2 \end{bmatrix}(x) = \begin{bmatrix} \int_{\partial\Omega} n_1(\xi)\phi_1(\xi)\,d\sigma \\ \int_{\partial\Omega} n_2(\xi)\phi_1(\xi)\,d\sigma \\ \int_{\partial\Omega} \left[\frac{\partial}{\partial n_\xi} E(x,\xi)\cdot\phi_1(\xi) + B_{1\xi} E(x,\xi)\cdot\phi_2(\xi)\right] d\sigma_\xi \\ \quad + \frac{1}{2}\phi_2(x) + b_0 + b_1 x_1 + b_2 x_2 \\ \int_{\partial\Omega} \left[\frac{\partial^2 E(x,\xi)}{\partial n_x \partial n_\xi}\phi_1(\xi) + \frac{\partial B_{1\xi} E(x,\xi)}{\partial n_x}\phi_2(\xi)\right] d\sigma_\xi \\ \quad + b_1 n_1(x) + b_2 n_2(x) \end{bmatrix}, \tag{8.142}$$

is also linear continuous, with

$$\dim \mathscr{N}(\mathscr{L}_4^*) = 1, \quad \dim \operatorname{Coker} \mathscr{L}_4^* = 0, \tag{8.143}$$

$$\mathscr{N}(\mathscr{L}_4^*) = \text{Coker}\,\mathscr{L}_4 = \text{span}\left\{\begin{bmatrix}0\\0\\0\\1\\0\end{bmatrix}\right\} \; in \; \mathbb{R}^3 \oplus C^\infty(\partial\Omega) \oplus C^\infty(\partial\Omega), \tag{8.144}$$

$$\begin{aligned}\text{Coker}\,\mathscr{L}_4^* &= \mathscr{N}(\mathscr{L}_4) \\ &= \text{span}\left\{\begin{bmatrix}\alpha_1\\\alpha_2\\\phi_1\\\phi_2\end{bmatrix} \middle| \begin{bmatrix}\alpha_1\\\alpha_2\\\phi_1\\\phi_2\end{bmatrix} \begin{array}{l}\text{are the last four}\\ \text{componemts of}\end{array} \mathscr{L}_1^{-1}\begin{bmatrix}0\\0\\0\\1\\0\end{bmatrix}\right\} \\ &\in \mathbb{R}^2 \oplus C^\infty(\partial\Omega) \oplus C^\infty(\partial\Omega).\end{aligned}$$

$\mathscr{L}_4^*$ maps continuously

$$\mathscr{L}_4^* : \; \mathbb{R}^3 \oplus H^s(\partial\Omega) \oplus H^{s+2}(\partial\Omega) \to \mathbb{R}^2 \oplus H^{s+2}(\partial\Omega) \oplus H^{s+1}(\partial\Omega), \tag{8.145}$$

for any $s \in \mathbb{R}$.

Consequently, the solution $w(x)$ to

$$\left.\begin{aligned} &\Delta^2 w(x) = 0 \quad on\ \Omega, \\ &\frac{\partial w(x)}{\partial n} = g_1(x) \in H^{r+2}(\partial\Omega), \quad r \in \mathbb{R}, \\ &B_1 w(x) = g_2(x) \in H^r(\partial\Omega), \quad g_2 \perp 1, \end{aligned}\right\} \tag{8.146}$$

satisfying (8.14) is given by

$$\begin{aligned} w(x) = &\int_{\partial\Omega} \left[E(x,\xi)\widetilde{f}_1(\xi) + \frac{\partial E(x,\xi)}{\partial n_\xi}\widetilde{f}_2(\xi)\right] d\sigma_\xi \\ &+ \widetilde{a}_0 + \widetilde{a}_1 x_1 + \widetilde{a}_2 x_2 \in H^{r+7/2}(\Omega), \end{aligned} \tag{8.147}$$

for an arbitrary constant $\widetilde{a}_0 \in \mathbb{R}$, where $(\widetilde{a}_1, \widetilde{a}_2, \widetilde{f}_1, \widetilde{f}_2) \in \mathbb{R}^2 \oplus H^r(\partial\Omega) \oplus H^{r+1}(\partial\Omega)$ uniquely solves

$$\mathscr{L}_4(\widetilde{a}_1, \widetilde{a}_2, \widetilde{f}_1, \widetilde{f}_2) = (A_0, A_1, A_2, g_1, g_2) \in \mathbb{R}^3 \oplus H^{r+2}(\partial\Omega) \oplus H^r(\partial\Omega), \tag{8.148}$$

for given A_0, A_1, $A_2 \in \mathbb{R}$. ($A_0, A_1, A_2 \in \mathbb{R}$ can be chosen arbitrarily, but $\widetilde{a}_1$, $\widetilde{a}_2$ and $\widetilde{f}_1$, $\widetilde{f}_2$ depend on the choice of A_0, A_1 and A_2.) □

The operator $\mathscr{L}_4$ in Theorem 8.9 has

$$\operatorname{Ind}\mathscr{L}_4 = \dim\mathscr{N}(\mathscr{L}_4) - \operatorname{codim}\mathscr{R}(\mathscr{L}_4) = -1.$$

As in Corollary 8.1, we can also define a corresponding operator whose index is zero. Here the case is trivial.

Corollary 8.3. *Define*

$$\widetilde{\mathscr{L}_4} : \mathbb{R}^3 \oplus H^r(\partial\Omega) \oplus H^{r+1}(\partial\Omega) \to \mathbb{R}^3 \oplus H^r(\partial\Omega) \oplus H^{r+1}(\partial\Omega)$$

by

$$\widetilde{\mathscr{L}_4}(a_0, a_1, a_2, f_1, f_2) = \mathscr{L}_4(a_1, a_2, f_1, f_2).$$

Then

$$\dim\mathscr{N}(\widetilde{\mathscr{L}_4}) = 1, \ \dim\operatorname{Coker}\widetilde{\mathscr{L}_4} = 1,$$

so $\operatorname{Ind}\widetilde{\mathscr{L}_4} = 0$. □

8.5 Other multilayer representations of biharmonic functions

So far, the formulations of boundary integral equations in the previous section have been based on the combination of simple- and double-layer potentials (8.19). They have the advantage that the solution $w(x)$ has high regularity due to the smoothing properties of V_1 and V_2 in (8.24) and (8.25). Other formulations of BIE are possible using a combination of different (augmented) layer potentials, including

(i) *simple- and triple-layers*

$$w(x) = \int_{\partial\Omega} [E(x,\xi) f_1(\xi) + B_{2\xi} E(x,\xi) f_2(\xi)]\, d\sigma_\xi + a_0 + a_1 x_1 + a_2 x_2; \tag{8.149}$$

(ii) *double- and quadruple-layers*

$$w(x) = \int_{\partial\Omega} \left[\frac{\partial}{\partial n_\xi} E(x,\xi) f_1(\xi) + B_{1\xi} E(x,\xi) f_2(\xi)\right] d\sigma_\xi + a_0 + a_1 x_1 + a_2 x_2; \tag{8.150}$$

(iii) *triple- and quadruple-layers*

$$w(x) = \int_{\partial\Omega} \{[B_{1\xi} E(x,\xi)] f_1(\xi) + [B_{2\xi} E(x,\xi)] f_2(\xi)\}\, d\sigma_\xi + a_0 + a_1 x_1 + a_2 x_2. \tag{8.151}$$

Corresponding to such layer potential representations for the biharmonic boundary value problems (8.9)–(8.12), different BIE formulations and existence and uniqueness theorems can be proved. For example, we give the following.

Theorem 8.10. (Simple- and triple-layers representation for biharmonic problem of Type I). *The boundary integral operator*

$$\mathscr{L}_5\colon\ \mathbb{R}^3 \oplus H^r(\partial\Omega) \oplus H^{r+2}(\partial\Omega) \to \mathbb{R}^3 \oplus H^{r+3}(\partial\Omega) \oplus H^{r+2}(\partial\Omega), \quad r \in \mathbb{R}, \tag{8.152}$$

defined by

$$\mathscr{L}_5 \begin{bmatrix} a_0 \\ a_1 \\ a_2 \\ f_1 \\ f_2 \end{bmatrix} (x) = \begin{bmatrix} \int_{\partial\Omega} f_1(\xi)\,d\sigma \\ \int_{\partial\Omega} \xi_1 f_1(\xi)\,d\sigma \\ \int_{\partial\Omega} \xi_2 f_1(\xi)\,d\sigma \\ \int_{\partial\Omega} \{E(x,\xi) f_1(\xi) + [B_{2\xi}E(x,\xi)] f_2(\xi)\}\,d\sigma_\xi \\ \quad +a_0 + a_1 x_1 + a_2 x_2 \\ \int_{\partial\Omega} \left[\dfrac{\partial E(x,\xi)}{\partial n_x} f_1(\xi) + \dfrac{\partial B_{2\xi}E(x,\xi)}{\partial n_x} f_2(\xi)\right] d\sigma_\xi \\ \quad +\dfrac{1}{2} f_2(x) + a_1 n_1(x) + a_2 n_2(x) \end{bmatrix}, \tag{8.153}$$

is an isomorphism, with adjoint

$$\mathscr{L}_5^*\colon\ \mathbb{R}^3 \oplus H^{-(r+3)}(\partial\Omega) \oplus H^{-(r+2)}(\partial\Omega) \to \mathbb{R}^3 \oplus H^{-r}(\partial\Omega) \oplus H^{-(r+2)}(\partial\Omega), \tag{8.154}$$

given by

$$\mathscr{L}_5^* \begin{bmatrix} b_0 \\ b_1 \\ b_2 \\ \phi_1 \\ \phi_2 \end{bmatrix} (x) = \begin{bmatrix} \int_{\partial\Omega} \phi_1(\xi)\,d\sigma \\ \int_{\partial\Omega} [\xi_1 \phi_1(\xi) + n_1(\xi)\phi_2(\xi)]\,d\sigma \\ \int_{\partial\Omega} [\xi_2 \phi_1(\xi) + n_2(\xi)\phi_2(\xi)]\,d\sigma \\ \int_{\partial\Omega} \left[E(x,\xi)\phi_1(\xi) + \dfrac{\partial E(x,\xi)}{\partial n_\xi}\phi_2(\xi)\right] d\sigma_\xi \\ \quad +b_0 + b_1 x_1 + b_2 x_2 \\ \int_{\partial\Omega} \left\{[B_{2x}E(x,\xi)]\phi_1(\xi) + \left[B_{2x}\dfrac{\partial E(x,\xi)}{\partial n_\xi}\right]\phi_2(\xi)\right\} d\sigma_\xi \\ \quad +\dfrac{1}{2}\phi_2(x) \end{bmatrix}. \tag{8.155}$$

Further,

$$\mathscr{L}_5^*\colon\ \mathbb{R}^3 \oplus H^s(\partial\Omega) \oplus H^{s+1}(\partial\Omega) \to \mathbb{R}^3 \oplus H^{s+3}(\partial\Omega) \oplus H^{s+1}(\partial\Omega) \tag{8.156}$$

is also an isomorphism for any $s \in \mathbb{R}$.

Consequently, the solution $w(x)$ *to*

$$\left.\begin{aligned} &\Delta^2 w(x) = 0 \quad on\ \Omega, \\ &w(x) = g_1(x) \in H^{r+3}(\partial\Omega), \quad r \in \mathbb{R}, \\ &\frac{\partial w(x)}{\partial n} = g_2(x) \in H^{r+2}(\partial\Omega), \end{aligned}\right\} \tag{8.157}$$

is given by

$$\begin{aligned} w(x) = &\int_{\partial\Omega} \{E(x,\xi)\widetilde{f}_1(\xi) + [B_{2\xi}E(x,\xi)]\widetilde{f}_2(\xi)\}\, d\sigma_\xi \\ &+ \widetilde{a}_0 + \widetilde{a}_1 x_1 + \widetilde{a}_2 x_2 \in H^{r+7/2}(\Omega), \end{aligned} \tag{8.158}$$

for unique $(\widetilde{a}_0, \widetilde{a}_1, \widetilde{a}_1, \widetilde{f}_1, \widetilde{f}_2) \in \mathbb{R}^3 \oplus H^r(\partial\Omega) \oplus H^{r+2}(\partial\Omega)$ *as the solution to*

$$\mathscr{L}_5(\widetilde{a}_0, \widetilde{a}_1, \widetilde{a}_2, \widetilde{f}_1, \widetilde{f}_2) = (A_0, A_1, A_2, g_1, g_2) \in \mathbb{R}^3 \oplus H^{r+3}(\partial\Omega) \oplus H^{r+2}(\partial\Omega), \tag{8.159}$$

for given $A_0,\ A_1,\ A_2 \in \mathbb{R}$. *(*$A_0,\ A_1,\ A_2 \in \mathbb{R}$ *can be chosen arbitrarily, but* $\widetilde{a}_0,\ \widetilde{a}_1,\ \widetilde{a}_2$ *and* $\widetilde{f}_1,\ \widetilde{f}_2$ *depend on the choice of* $A_0,\ A_1$ *and* A_2*.)* □

The reader can easily formulate a class of analogous theorems like the above corresponding to (8.149)–(8.151) for biharmonic BVP of Types I–IV.

8.6 BIE for exterior plate BVP

Let us examine BVP of Types I–IV (cf. (8.9)–(8.12)) on an exterior domain Ω^c.

Example 8.3. Let Ω be the exterior of the closed *unit* disk in $\mathbb{R}^2$. Then

$$w(x) = |x|^2 \ln|x| - \ln|x|, \quad x \in \Omega, \tag{8.160}$$

satisfies

$$\left.\begin{aligned} &\Delta^2 w(x) = 0, \\ &\left.\begin{aligned} &w(x) = 0, \\ &\frac{\partial w(x)}{\partial n} = \frac{\partial w(x)}{\partial r} = \left(2r\ln r + r - \frac{1}{r}\right)\Big|_{r=1} = 0 \end{aligned}\right\} \text{ on } \partial\Omega. \end{aligned}\right\} \tag{8.161}$$

Also, $w(x) \equiv 0$ is a trivial solution satisfying the boundary conditions. Therefore, the *exterior* boundary value problem (8.9) (Type I) *does not have uniqueness* of solutions. □

Example 8.4. Let Ω be the exterior of the closed disk with radius R. Let

$$w(x) = |x|^2 \ln|x| + a_0 + a_1 x_1 + a_2 x_2 \tag{8.162}$$

for arbitrary $a_0, a_1, a_2 \in \mathbb{R}$.

Then w satisfies

$$\left.\begin{aligned}&\Delta^2 w(x)=0,\\&\left.\begin{aligned}B_1w(x)&=1/R\equiv g_1(x),\\B_2w(x)&=2(1+\nu)\ln R+3+\nu\equiv g_2(x),\end{aligned}\right\}\text{ on }\partial\Omega.\end{aligned}\right\}\tag{8.163}$$

Then, obviously (g_1,g_2) violates the compatibility condition (8.13) for the bounded domain case. Yet the exterior BVP (8.163) has (nonunique) solutions. □

To fix the uniqueness for a problem like Example 8.3, we need to impose some growth rate restriction on the solution. For a problem like Example 8.4, we not only need to impose the growth rate condition, but also need some additional normalization condition(s).
In [95], the growth rate condition for solutions of the biharmonic equation $\Delta^2 w(x)=0$ on an exterior domain was set as

$$w(x)=(A_1x_1+A_2x_2)\ln|x|+\mathscr{O}(|x|)\ \text{ for large }|x|,\ \ A_1,A_2\in\mathbb{R},\tag{8.164}$$

for a linear elasticity (i.e., plate *stretching*) problem in $\mathbb{R}^2$. Such a growth rate would rule out the pathological solutions (8.160) and (8.162), but is still unsuitable for our plate bending problem because the solution w in (8.164) has *unbounded strain energy* on Ω^c when A_1 or A_2 is nonzero. Here we say that a biharmonic function w satisfies the *linear-logarithmic growth condition* if

$$\left.\begin{aligned}&\text{(a) } w(x)=[a_0+a_1x_1+a_2x_2]+A\ln|x|+o(1)\\&\qquad\text{for some } a_0,\ a_1,\ a_2,\ A\in\mathbb{R},\text{ large }|x|;\\&\text{(b) } B_1(r)w(x)=o(r^{-(1+\delta)}),\\&\qquad B_2(r)w(x)=o(r^{-(1+\delta)}),\\&\qquad \frac{\partial w(x)}{\partial r}=\mathscr{O}(1),\\&\qquad\text{for some }\delta>0\text{ at large }|x|=r,\end{aligned}\right\}\tag{8.165}$$

where

$$\begin{aligned}B_1(r)&\equiv-\left\{\frac{\partial}{\partial r}\Delta-(1-\nu)\frac{1}{r}\frac{\partial}{\partial\theta}\left[\frac{x_1x_2}{r^2}\left(\frac{\partial^2}{\partial x_1^2}-\frac{\partial^2}{\partial x_2^2}\right)-\frac{1}{r^2}(x_1^2-x_2^2)\frac{\partial^2}{\partial x_1\partial x_2}\right]\right\}\\&=\frac{\partial}{\partial r}\Delta+\left(\frac{1-\nu}{r^2}\right)\frac{\partial^3}{\partial r\partial\theta^2}-\frac{1-\nu}{r^3}\frac{\partial^2}{\partial\theta^2},\\B_2(r)&\equiv-\nu\Delta-\frac{1-\nu}{r^2}\left(x_1^2\frac{\partial^2}{\partial x_1^2}+x_2^2\frac{\partial^2}{\partial x_2^2}+2x_1x_2\frac{\partial^2}{\partial x_1\partial x_2}\right)\\&=\Delta-\left(\frac{1-\nu}{r}\right)\frac{\partial}{\partial r}-\frac{1-\nu}{r^2}\frac{\partial^2}{\partial\theta^2},\end{aligned}$$

are respectively the transverse force and bending moment operators in the radial direction as before. It is easy to check that the logarithmic part $A\ln|x|$ of w has *finite strain energy*. An immediate consequence is that if $w \in H^{s+2}_{\text{loc}}(\Omega)$ for $s \geqslant 0$ is a biharmonic function on Ω satisfying (8.165), then w has *finite strain energy:* $[a(w,w)]^{1/2} < \infty$.

Theorem 8.11 (Exterior biharmonic problem of Type I). *Let Ω^c be an exterior domain in $\mathbb{R}^2$ with bounded complement and smooth boundary $\partial\Omega$. Consider the exterior BVP*

$$\left.\begin{aligned} &\Delta^2 w(x) = 0 \quad \text{on } \Omega^c, \\ &w(x) = g_1(x) \in H^{r+3}(\partial\Omega), \quad r \in \mathbb{R}. \\ &\frac{\partial w(x)}{\partial n} = g_2(x) \in H^{r+2}(\partial\Omega), \end{aligned}\right\} \tag{8.166}$$

Then there exists a unique solution $w \in H^{r+7/2}_{\text{loc}}(\Omega^c)$ satisfying the linear-logarithmic growth condition (8.165). The solution can be represented by a combination of simple- and double-layer potentials and a linear function

$$w(x) = \int\limits_{\partial\Omega} \left[E(x,\xi)\widetilde{f}_1(\xi) + \frac{\partial E(x,\xi)}{\partial n_\xi}\widetilde{f}_2(\xi) \right] d\sigma_\xi + \widetilde{a}_0 + \widetilde{a}_1 x_1 + \widetilde{a}_2 x_2$$
$$\in H^{r+7/2}_{\text{loc}}(\Omega^c), \tag{8.167}$$

where $(\widetilde{a}_0, \widetilde{a}_1, \widetilde{a}_2, \widetilde{f}_1, \widetilde{f}_2) \in \mathbb{R}^3 \oplus H^r(\partial\Omega) \oplus H^{r+1}(\partial\Omega)$ is the unique solution to the BIE

$$\mathscr{L}_1 \begin{bmatrix} \widetilde{a}_0 \\ \widetilde{a}_1 \\ \widetilde{a}_2 \\ \widetilde{f}_1 \\ \widetilde{f}_2 \end{bmatrix} = \begin{bmatrix} 0 \\ 0 \\ 0 \\ g_1 \\ g_2 \end{bmatrix}, \tag{8.168}$$

and where

$$\mathscr{L}_1 \colon \ \mathbb{R}^3 \times H^s(\partial\Omega) \times H^{s+1}(\partial\Omega) \to \mathbb{R}^3 \times H^{s+3}(\partial\Omega) \times H^{s+2}(\partial\Omega) \tag{8.169}$$

is an isomorphism for any $r \in \mathbb{R}$, defined by (8.77) as in Theorem 8.6.

Proof. From the *ansatz* (8.167), for large $|x|$, we obtain as the asymptotic expansion the expression on the RHS of (8.93). The terms in the first three sets of curly brackets $\{\cdots\}$ on the RHS of (8.93) vanish because of (8.168). Therefore w satisfies the linear-logarithmic growth condition (8.165). The rest of the proof is obvious from that of Theorem 8.6. □

Next, we state and prove the following.

Theorem 8.12 (Exterior biharmonic problem of Type II). *Let Ω^c be an exterior domain in $\mathbb{R}^2$ with bounded complement and smooth boundary $\partial\Omega$. Let $0 < \nu < \frac{1}{2}$. Consider the exterior BVP*

$$\left.\begin{aligned} &\Delta^2 w(x) = 0 \quad \text{on} \quad \Omega^c, \\ &B_1 w(x) = g_1(x) \in H^r(\partial\Omega), \quad r \in \mathbb{R}, \\ &B_2 w(x) = g_2(x) \in H^{r+1}(\partial\Omega), \end{aligned}\right\} \tag{8.170}$$

where (g_1, g_2) satisfies the orthogonality condition (8.13). Then there exists a weak solution $w \in H^{r+7/2}_{\text{loc}}(\Omega^c)$ satisfying the linear-logarithmic growth condition (8.165), and w is unique up to an arbitrary linear function. The general solution can be represented by a combination of simple- and double-layer potentials plus an arbitrary linear function $c_0 + c_1x_1 + c_2x_2$:

$$w(x) = \int_{\partial\Omega} \left[E(x,\xi)\widetilde{f}_1(\xi) + \frac{\partial E(x,\xi)}{\partial n_\xi}\widetilde{f}_2(\xi) \right] d\sigma_\xi + c_0 + c_1x_1 + c_2x_2,$$

$$x \in \Omega^c, \ c_0, \ c_1, \ c_2 \in \mathbb{R}, \tag{8.171}$$

where $(\widetilde{f}_1, \widetilde{f}_2) \in H^r(\partial\Omega) \oplus H^{r+1}(\partial\Omega)$ corresponds to the unique solution in the boundary integral equation

$$\mathscr{L}_6 \begin{bmatrix} \widetilde{a}_0 \\ \widetilde{a}_1 \\ \widetilde{a}_2 \\ \widetilde{f}_1 \\ \widetilde{f}_2 \end{bmatrix} = \begin{bmatrix} 0 \\ 0 \\ 0 \\ g_1 \\ g_2 \end{bmatrix}, \tag{8.172}$$

and where

$$\mathscr{L}_6 \colon \ \mathbb{R}^3 \oplus H^r(\partial\Omega) \oplus H^{r+1}(\partial\Omega) \to \mathbb{R}^3 \oplus H^r(\partial\Omega) \oplus H^{r+1}(\partial\Omega) \tag{8.173}$$

is an isomorphism for any $r \in \mathbb{R}$, defined by

$$\mathscr{L}_6 \begin{bmatrix} a_0 \\ a_1 \\ a_2 \\ f_1 \\ f_2 \end{bmatrix}(x) = \begin{bmatrix} \displaystyle\int_{\partial\Omega} f_1(\xi)\, d\sigma \\ \displaystyle\int_{\partial\Omega} [f_1(\xi)\xi_1 + f_2(\xi)n_1(\xi)]\, d\sigma \\ \displaystyle\int_{\partial\Omega} [f_1(\xi)\xi_2 + f_2(\xi)n_2(\xi)]\, d\sigma \\ \displaystyle -\frac{1}{2}f_1(x) + \int_{\partial\Omega} \left[B_{1x}E(x,\xi)f_1(\xi) + B_{1x}\frac{\partial E(x,\xi)}{\partial n_\xi}f_2(\xi) \right] d\sigma_\xi \\ \qquad + a_0 + a_1x_1 + a_2x_2 \\ \displaystyle \frac{1}{2}f_2(x) + \int_{\partial\Omega} \left[B_{2x}E(x,\xi)f_1(\xi) + B_{2x}\frac{\partial E(x,\xi)}{\partial n_\xi}f_2(\xi) \right] d\sigma_\xi \\ \qquad - a_1n_1(x) - a_2n_2(x) \end{bmatrix}. \tag{8.174}$$

Furthermore, the solution to (8.170) is unique if we either specify c_0, c_1 *and* c_2 *in (8.171), or if we impose three normalization conditions*

$$\langle w, \eta_j \rangle_{H^{r+3}(\partial\Omega)\times H^{-(r+3)}(\partial\Omega)} = \alpha_j \in \mathbb{R}, \quad j = 1, 2, 3, \tag{8.175}$$

where α_j *are given for some* $\eta_j \in H^{-(r+3)}(\partial\Omega)$ *satisfying*

$$\det \begin{vmatrix} \langle 1, \eta_1 \rangle & \langle x_1, \eta_1 \rangle & \langle x_2, \eta_1 \rangle \\ \langle 1, \eta_2 \rangle & \langle x_1, \eta_2 \rangle & \langle x_2, \eta_2 \rangle \\ \langle 1, \eta_3 \rangle & \langle x_1, \eta_3 \rangle & \langle x_2, \eta_3 \rangle \end{vmatrix} \neq 0. \tag{8.176}$$

Proof. For ν satisfying $0 < \nu < \frac{1}{2}$, by using the same arguments as in § 8.4, it is not difficult to show that the reduced operator

$$\widetilde{\mathscr{L}_6} \begin{bmatrix} f_1 \\ f_2 \end{bmatrix} (x) \equiv \begin{bmatrix} -\frac{1}{2} f_1(x) + \int_{\partial\Omega} [B_{1x} E(x,\xi) f_1(\xi) + B_{1x} \frac{\partial E(x,\xi)}{\partial n_\xi} f_2(\xi)] \, d\sigma_\xi \\ \frac{1}{2} f_2(x) + \int_{\partial\Omega} \left[B_{2x} E(x,\xi) f_1(\xi) + B_{2x} \frac{\partial E(x,\xi)}{\partial n_\xi} f_2(\xi) \right] d\sigma_\xi \end{bmatrix} \tag{8.177}$$

is an elliptic pseudodifferential operator mapping *isomorphically* from $H^r(\partial\Omega) \times H^{r+1}(\partial\Omega)$ into itself. Since $\mathscr{L}_6$ is a *finite-dimensional augmentation* of $\widetilde{\mathscr{L}_6}$, $\mathscr{L}_6$ is a *Fredholm operator of index* 0 [108, Chap. 5]. Therefore, if we can show that $\mathscr{L}_6$ is 1–1, then $\mathscr{L}_6$ will also be onto and be an isomorphism. Let $(a_0, a_1, a_2, f_1, f_2) \in \mathscr{N}(\mathscr{L}_6)$. Then

$$\int_{\partial\Omega} f_1(\xi)\, d\sigma = 0, \tag{8.178}$$

$$\int_{\partial\Omega} [f_1(\xi)\xi_1 + f_2(\xi) n_1(\xi)]\, d\sigma = 0, \tag{8.179}$$

$$\int_{\partial\Omega} [f_1(\xi)\xi_2 + f_2(\xi) n_2(\xi)]\, d\sigma = 0, \tag{8.180}$$

$$-\frac{1}{2} f_1(x) + \int_{\partial\Omega} \left[B_{1x} E(x,\xi) f_1(\xi) + B_{1x} \frac{\partial E(x,\xi)}{\partial n_\xi} f_2(\xi) \right] d\sigma_\xi + a_0 + a_1 x_1 + a_2 x_2 = 0, \tag{8.181}$$

$$\frac{1}{2} f_2(x) + \int_{\partial\Omega} \left[B_{2x} E(x,\xi) f_1(\xi) + B_{2x} \frac{\partial E(x,\xi)}{\partial n_\xi} f_2(\xi) \right] d\sigma_\xi - a_1 n_1(x) - a_2 n_2(x) = 0. \tag{8.182}$$

Multiplying (8.181) by $\phi_1(x)$ and (8.182) by $\phi_2(x)$, adding, integrating over $\partial\Omega$ and interchanging the order of integration, we obtain

$$\begin{aligned}
&-\int_{\partial\Omega}[f_1(x)\phi_1(x)-f_2(x)\phi_2(x)]\,d\sigma\\
&+\int_{\partial\Omega}\Bigg\{f_1(\xi)\int_{\partial\Omega}\{[B_{1x}E(x,\xi)]\phi_1(x)+[B_{2x}E(x,\xi)]\phi_2(x)\}\,d\sigma_x\\
&+f_2(\xi)\frac{\partial}{\partial n_\xi}\int_{\partial\Omega}\{[B_{1x}E(x,\xi)]\phi_1(x)+[B_{2x}E(x,\xi)]\phi_2(x)\}\,d\sigma_x\Bigg\}\,d\sigma_\xi\\
&+\int_{\partial\Omega}\{(a_0+a_1x_1+a_2x_2)\phi_1(x)-[a_1n_1(x)+a_2n_2(x)]\phi_2(x)\}\,d\sigma=0.
\end{aligned}\tag{8.183}$$

Using

$$\begin{bmatrix}\phi_1(x)\\ \phi_2(x)\end{bmatrix}=\begin{bmatrix}1\\0\end{bmatrix},\begin{bmatrix}x_1\\ -n_1(x)\end{bmatrix},\begin{bmatrix}x_2\\ -n_2(x)\end{bmatrix}\tag{8.184}$$

successively in (8.183), and noting (8.51),

$$\int_{\partial\Omega}[B_{1x}E(x,\xi)\cdot 1+B_{2x}E(x,\xi)\cdot 0]\,d\sigma_x=-\frac{1}{2},$$

$$\int_{\partial\Omega}\{[B_{1x}E(x,\xi)]x_j-[B_{2x}E(x,\xi)]n_j(x)\}\,d\sigma_x=-\frac{\xi_j}{2},\quad j=1,2,$$

we obtain, by (8.178)–(8.182),

$$\left\langle a_0\begin{bmatrix}1\\0\end{bmatrix}+a_1\begin{bmatrix}x_1\\ -n_1(x)\end{bmatrix}+a_2\begin{bmatrix}x_2\\ -n_2(x)\end{bmatrix},\begin{bmatrix}\phi_1\\ \phi_2\end{bmatrix}\right\rangle_{L^2(\partial\Omega)\times L^2(\partial\Omega)}=0.\tag{8.185}$$

Since the three functions for $\begin{bmatrix}\phi_1\\ \phi_2\end{bmatrix}$ in (8.184) are linearly independent in $L^2(\partial\Omega)\times L^2(\partial\Omega)$, we see that $a_0=a_1=a_2=0$ in (8.181) and (8.182). Therefore (f_1,f_2) satisfies

$$\widetilde{\mathscr{L}_6}\begin{bmatrix}f_1\\ f_2\end{bmatrix}=\begin{bmatrix}0\\0\end{bmatrix}.$$

But we already showed that $\widetilde{\mathscr{L}_6}$ is an isomorphism. Therefore $(f_1,f_2)=(0,0)$. Hence $(a_0,a_1,a_2,f_1,f_2)=(0,0,0,0,0)$, and $\mathscr{N}(\mathscr{L}_1)$ is trivial.
Now, let $(\widetilde{f}_1,\widetilde{f}_2)\in H^r(\partial\Omega)\oplus H^{r+1}(\partial\Omega)$ along with some $(\widetilde{a}_0,\widetilde{a}_1,\widetilde{a}_2)\in\mathbb{R}^3$ satisfy (8.172). We want to show that

$$w(x)=\int_{\partial\Omega}\left[E(x,\xi)\widetilde{f}_1(\xi)+\frac{\partial E(x,\xi)}{\partial n_\xi}\widetilde{f}_2(\xi)\right]d\sigma_\xi+c_0+c_1x_1+c_2x_2,$$

$$x\in\Omega^c,\quad c_0,c_1,c_2\in\mathbb{R}\text{ are arbitrary,}\tag{8.186}$$

is a solution to (8.170) satisfying the linear-logarithmic growth condition (8.165). We take the boundary values $B_1 w$ and $B_2 w$ on w in (8.186), and by Lemma 8.2, obtain

$$B_1 w(x) = -\frac{1}{2}\widetilde{f}_1(x) + \int_{\partial\Omega}\left[B_{1x}E(x,\xi)\widetilde{f}_1(\xi) + B_{1x}\frac{\partial E(x,\xi)}{\partial n_\xi}\widetilde{f}_2(\xi)\right]d\sigma_\xi, \quad x \in \partial\Omega, \tag{8.187}$$

$$B_2 w(x) = \frac{1}{2}\widetilde{f}_2(x) + \int_{\partial\Omega}\left[B_{2x}E(x,\xi)\widetilde{f}_1(\xi) + B_{2x}\frac{\partial E(x,\xi)}{\partial n_\xi}\widetilde{f}_2(\xi)\right]d\sigma_\xi, \quad x \in \partial\Omega. \tag{8.188}$$

But from the last two components of (8.172), we see that

$$-\frac{1}{2}\widetilde{f}_1(x) + \int_{\partial\Omega}\left[B_{1x}E(x,\xi)\widetilde{f}_1(\xi) + B_{1x}\frac{\partial E(x,\xi)}{\partial n_\xi}\widetilde{f}_2(\xi)\right]d\sigma_\xi + \widetilde{a}_0 + \widetilde{a}_1 x_1 + \widetilde{a}_2 x_2 = g_1(x), \tag{8.189}$$

$$\frac{1}{2}\widetilde{f}_2(x) + \int_{\partial\Omega}\left[B_{2x}E(x,\xi)\widetilde{f}_1(\xi) + B_{2x}\frac{\partial E(x,\xi)}{\partial n_\xi}\widetilde{f}_2(\xi)\right]d\sigma_\xi - \widetilde{a}_1 n_1(x) - \widetilde{a}_2 n_2(x) = g_2(x). \tag{8.190}$$

Therefore w satisfies the boundary conditions $B_1 w = g_1$ and $B_2 w = g_2$ if and only if $\widetilde{a}_0 = \widetilde{a}_1 = \widetilde{a}_2 = 0$ in (8.181) and (8.182). To show this, we again multiply (8.189) by $\phi_1(x)$ and (8.190) by $\phi_2(x)$, using (ϕ_1, ϕ_2) as in (8.184). Repeating similar procedures as before, noting that (g_1, g_2) satisfies (8.13), we again arrive at (8.185), with (a_0, a_1, a_2) replaced by $(\widetilde{a}_0, \widetilde{a}_1, \widetilde{a}_2)$. Therefore $(\widetilde{a}_0, \widetilde{a}_1, \widetilde{a}_2) = (0,0,0)$, and the boundary conditions in (8.170) are satisfied.

To show that $w(x)$ defined by (8.186) satisfies the linear-logarithmic growth condition (8.165), we use the asymptotic expansion (8.93), with $\widetilde{a}_1 = \widetilde{a}_2 = 0$ and $\widetilde{a}_0 = c_0$, and immediately obtain (8.165), because $(\widetilde{f}_1, \widetilde{f}_2)$ satisfies (8.178)–(8.180) by (8.172).

Now we show that any w satisfying (8.170) and (8.165) can differ at most by a constant. Let w_1 and w_2 both satisfy (8.170) and (8.165). Define $w = w_1 - w_2$. Then w satisfies

$$\begin{aligned}
&\Delta^2 w(x) = 0 \quad \text{on } \Omega^c,\\
&\left.\begin{aligned} B_1 w = 0,\\ B_2 w = 0, \end{aligned}\right\} \quad \text{on } \partial\Omega,\\
&\left.\begin{aligned} &w(x) = a_0 + a_1 x_1 + a_2 x_2 + A\ln|x| + o(1),\\ &B_1(r)w(x) = o(r^{-(1+\delta/2)}),\ \ B_2(r)w(x) = o\left(r^{-(1+\delta/2)}\right)\\ &\frac{\partial w}{\partial r} = \mathscr{O}(1), \end{aligned}\right\} \text{ at large } |x| = r,
\end{aligned}$$

for some $\delta > 0$. Then on $\Omega_R^c = \{x \in \Omega^c \,|\, |x| \leqslant R\}$, we have

$$\begin{aligned}
&\int_{\Omega_R^c} [|\Delta w|^2 + 2(1-\nu)(w_{x_1x_2}^2 - w_{x_1x_1}w_{x_2x_2})]\,dx \\
&= -\int_{\partial\Omega} \left[(B_1 w)w - (B_2 w)\frac{\partial w}{\partial n}\right] d\sigma - \int_{|x|=R} \left\{[B_1(R)w]w - [B_2(R)w]\frac{\partial w}{\partial n}\right\} d\sigma \\
&\leqslant 0 + \int_0^{2\pi} \frac{C}{R^{1+\delta/2}} R\,d\theta \\
&= \frac{2\pi C}{R^{\delta/2}} \to 0 \quad \text{as} \quad R \to \infty.
\end{aligned}$$

Since $[a(w,w)]^{1/2}$ is a seminorm, and a norm up to a linear polynomial, we must have

$$w(x) = c_0 + c_1 x_1 + c_2 x_2, \quad x \in \Omega^c.$$

To fix these arbitrary constants c_0, c_1 and c_2, we write a general solution as

$$w(x) = w_p(x) + c_0 + c_1 x_1 + c_2 x_2,$$

where w_p is any particular solution to (8.170) satisfying (8.165). The regularity of solution is $w \in H^{r+3}(\partial\Omega)$ and by (8.175),

$$\begin{aligned}
\alpha_j &= \langle w, \eta_j \rangle_{H^{r+3}(\partial\Omega)\times H^{-(r+3)}(\partial\Omega)} \\
&= \langle w_p + c_0 + c_1 x_1 + c_2 x_2, \eta_j \rangle \\
&= c_0\langle 1, \eta_j \rangle + c_1\langle x_1, \eta_j \rangle + c_2\langle x_2, \eta_j \rangle + \langle w_p, \eta_j \rangle, \quad j = 1, 2, 3.
\end{aligned}$$

By (8.176), c_0, c_1 and c_2 are uniquely determinable from $\langle w_p, \eta_j \rangle$, $j = 1$, 2 and 3. Therefore the solution to (8.170) satisfying (8.165) and (8.175) is unique. □

A more concise version of Theorem 8.12 follows as a consequence.

Corollary 8.4. *Under the same conditions as in Theorem 8.12, there exists a solution $w \in H_{\text{loc}}^{r+7/2}(\Omega^c)$ satisfying the linear-logarithmic growth condition (8.165); w is unique up to a linear polynomial. The general solution can be represented by a combination of simple- and double-layer potentials plus an arbitrary linear polynomial $c_0 + c_1x_1 + c_2x_2$, $c_0, c_1, c_2 \in \mathbb{R}$:*

$$w(x) = \int_{\partial\Omega} \left[E(x,\xi)\widetilde{f}_1(\xi) + \frac{\partial E(x,\xi)}{\partial n_\xi}\widetilde{f}_2(\xi)\right] d\sigma_\xi + c_0 + c_1x_1 + c_2x_2,\ x \in \Omega^c,$$

where $(\widetilde{f}_1, \widetilde{f}_2) \in H^r(\partial\Omega) \oplus H^{r+1}(\partial\Omega)$ corresponds to the unique solution of the BIE

$$\widetilde{\mathscr{L}_6}\begin{bmatrix} \widetilde{f}_1 \\ \widetilde{f}_2 \end{bmatrix} = \begin{bmatrix} g_1 \\ g_2 \end{bmatrix}, \tag{8.191}$$

with $\widetilde{\mathscr{L}_6}$ defined by (8.177).

Proof. We need only show that $(\widetilde{f}_1, \widetilde{f}_2)$ satisfying (8.191) will also satisfy

$$\int_{\partial\Omega} \widetilde{f}_1(\xi)\,d\sigma = 0,$$
$$\int_{\partial\Omega} [\widetilde{f}_1(\xi)\xi_1 + \widetilde{f}_2(\xi)n_2(\xi)]\,d\sigma = 0,$$
$$\int_{\partial\Omega} [\widetilde{f}_1(\xi)\xi_2 + \widetilde{f}_2(\xi)n_2(\xi)]\,d\sigma = 0.$$

But these are immediate consequences of the proof of Theorem 8.12, since (g_1, g_2) satisfies the orthogonality condition (8.13). □

Remark 8.2.

(a) Although Corollary 8.4 is simpler than Theorem 8.12, at the time of computation the BIE (8.172) in Theorem 8.12 is *preferred* because, after discretization, (8.172) allows us to annihilate the fast-growing terms

$$|x|^2 \ln|x|, \quad x_1 \ln|x|, \quad x_2 \ln|x| \tag{8.192}$$

completely. Otherwise they could contribute disastrous errors to the already-growing linear-logarithmic terms

$$x_1, \quad x_2, \quad \ln|x|.$$

The BIE (8.191) in Corollary 8.4 does not have this advantage, because, after discretization from $(\widetilde{f}_1, \widetilde{f}_2)$ to $(\widetilde{f}_{1,h}, \widetilde{f}_{2,h})$, we normally have small residuals ε_1, ε_2 and ε_3:

$$\int_{\partial\Omega} \widetilde{f}_{1,h}(\xi)\,d\sigma = \varepsilon_1,$$
$$\int_{\partial\Omega} [\widetilde{f}_{1,h}(\xi)\xi_1 + \widetilde{f}_{2,h}(\xi)n_1(\xi)]\,d\sigma = \varepsilon_2,$$
$$\int_{\partial\Omega} [\widetilde{f}_{1,h}(\xi)\xi_2 + \widetilde{f}_{2,h}(\xi)n_2(\xi)]\,d\sigma = \varepsilon_2,$$

which are coefficients of (8.192) causing large errors when $|x|$ grows.

(b) As shown in Example 8.4, the data (g_1, g_2) to the exterior BVP (8.170) need not satisfy the compatibility condition (8.13). But after imposing the growth condition (8.165), the compatibility condition (8.13) for the data (g_1, g_2) in Theorem 8.12 naturally follows *as a consequence of* (8.165). □

The next two theorems are of the same nature as Theorems 8.11 and 8.12. We state them without giving proofs.

Theorem 8.13 (Exterior biharmonic problem of Type III). *Let Ω^c be an exterior domain in $\mathbb{R}^2$ with bounded complement and smooth boundary $\partial\Omega$. Consider the exterior BVP*

$$\begin{aligned} &\Delta^2 w(x) = 0 \quad \text{on } \Omega^c, \\ &w(x) = g_1(x) \in H^{r+3}(\partial\Omega), \qquad r \in \mathbb{R}. \\ &B_2 w(x) = g_2(x) \in H^{r+1}(\partial\Omega), \end{aligned}$$

Then there exists a unique solution $w \in H^{r+7/2}_{\mathrm{loc}}(\Omega^c)$ satisfying the linear–logarithmic growth condition (8.165). The solution can be represented by a combination of simple- and double-layer potentials and a linear function:

$$\begin{aligned} w(x) = &\int_{\partial\Omega} \left[E(x,\xi)\widetilde{f}_1(\xi) + \frac{\partial E(x,\xi)}{\partial n_\xi}\widetilde{f}_2(\xi)\right] d\sigma_\xi \\ &+ \widetilde{a}_0 + \widetilde{a}_1 x_1 + \widetilde{a}_2 x_2 \in H^{r+7/2}_{\mathrm{loc}}(\Omega^c), \end{aligned}$$

where $(\widetilde{a}_0, \widetilde{a}_1, \widetilde{a}_2, \widetilde{f}_1, \widetilde{f}_2) \in \mathbb{R}^3 \oplus H^r(\partial\Omega) \oplus H^{r+1}(\partial\Omega)$ is the unique solution to the BIE

$$\mathscr{L}_7 \begin{bmatrix} \widetilde{a}_0 \\ \widetilde{a}_1 \\ \widetilde{a}_2 \\ \widetilde{f}_1 \\ \widetilde{f}_2 \end{bmatrix} = \begin{bmatrix} 0 \\ 0 \\ 0 \\ g_1 \\ g_2 \end{bmatrix},$$

with

$$\mathscr{L}_7 \colon \mathbb{R}^3 \oplus H^s(\partial\Omega) \oplus H^{s+1}(\partial\Omega) \to \mathbb{R}^3 \oplus H^{s+3}(\partial\Omega) \oplus H^{s+1}(\partial\Omega), \text{ continuously,}$$

defined by

$$\mathscr{L}_7 \begin{bmatrix} a_0 \\ a_1 \\ a_2 \\ f_1 \\ f_2 \end{bmatrix}(x) = \begin{bmatrix} \int_{\partial\Omega} f_1(\xi)\, d\sigma \\ \int_{\partial\Omega} [\xi_1 f_1(\xi) + n_1(\xi) f_2(\xi)]\, d\sigma \\ \int_{\partial\Omega} [\xi_2 f_1(\xi) + n_2(\xi) f_2(\xi)]\, d\sigma \\ \int_{\partial\Omega} \left[E(x,\xi) f_1(\xi) + \frac{\partial E(x,\xi)}{\partial n_\xi} f_2(\xi)\right] d\sigma_\xi + a_0 + a_1 x_1 + a_2 x_2 \\ \int_{\partial\Omega} \left[B_{2x} E(x,\xi) f_1(\xi) + B_{2x} \frac{\partial E(x,\xi)}{\partial n_\xi} f_2(\xi)\right] d\sigma_\xi + \frac{1}{2} f_2(x) \end{bmatrix},$$

which is an isomorphism for any $s \in \mathbb{R}$. □

Theorem 8.14 (Exterior biharmonic problem of Type IV). *Let Ω^c be an exterior domain in $\mathbb{R}^2$ with bounded complement and smooth boundary $\partial\Omega$. Consider the exterior BVP*

$$\begin{aligned} &\Delta^2 w(x) = 0 \quad \text{on } \Omega^c, \\ &\frac{\partial w(x)}{\partial n} = g_1(x) \in H^{r+2}(\partial\Omega), \quad r \in \mathbb{R}, \\ &B_1 w(x) = g_2(x) \in H^r(\partial\Omega), \end{aligned}$$

where $g_2 \perp 1$ in $L^2(\partial\Omega)$. Then there exists a solution $w \in H^{r+7/2}_{\text{loc}}(\Omega^c)$, unique up to constant c, satisfying the linear–logarithmic growth condition (8.165). The general solution is represented by a combination of simple- and double-layer potentials and a linear function

$$w(x) = \int_{\partial\Omega} \left[E(x,\xi)\widetilde{f}_1(\xi) + \frac{\partial E(x,\xi)}{\partial n_\xi}\widetilde{f}_2(\xi) \right] d\sigma_\xi + c_0 + \widetilde{a}_1 x_1 + \widetilde{a}_2 x_2,$$

$$x \in \Omega^c,$$

where $(\widetilde{a}_1, \widetilde{a}_2, \widetilde{f}_1, \widetilde{f}_2) \in \mathbb{R}^2 \oplus H^r(\partial\Omega) \oplus H^{r+1}(\partial\Omega)$ is the unique solution to the BIE

$$\mathscr{L}_8 \begin{bmatrix} \widetilde{a}_1 \\ \widetilde{a}_2 \\ \widetilde{f}_2 \\ \widetilde{f}_2 \end{bmatrix} = \begin{bmatrix} 0 \\ 0 \\ 0 \\ g_1 \\ g_2 \end{bmatrix}, \quad \langle 1, g_2 \rangle_{H^{-r}(\partial\Omega), H^r(\partial\Omega)} = 0,$$

with

$$\mathscr{L}_8 \colon \mathbb{R}^2 \oplus H^s(\partial\Omega) \oplus H^{s+1}(\partial\Omega) \to \mathbb{R}^3 \oplus H^{s+2}(\partial\Omega) \oplus H^s(\partial\Omega), \quad \text{1-1 } \textit{and continuously,}$$

defined by

$$\mathscr{L}_8 \begin{bmatrix} a_1 \\ a_2 \\ f_1 \\ f_2 \end{bmatrix}(x) = \begin{bmatrix} \int_{\partial\Omega} f_1(\xi)\,d\sigma_\xi \\ \int_{\partial\Omega} [\xi_1 f_1(\xi) + n_1(\xi) f_2(\xi)]\,d\sigma \\ \int_{\partial\Omega} [\xi_2 f_1(\xi) + n_2(\xi) f_2(\xi)]\,d\sigma \\ \int_{\partial\Omega} \left[\frac{\partial E(x,\xi)}{\partial n_x} f_1(\xi) + \frac{\partial^2 E(x,\xi)}{\partial n_x \partial n_\xi} f_2(\xi) \right] d\sigma_\xi + a_1 n_1(x) + a_2 n_2(x) \\ \int_{\partial\Omega} \left[B_{1x} E(x,\xi) f_1(\xi) + B_{1x} \frac{\partial E(x,\xi)}{\partial n_\xi} f_2(\xi) \right] d\sigma_\xi - \frac{1}{2} f_1(x) \end{bmatrix},$$

for any $s \in \mathbb{R}$. Further, $\mathscr{L}_8$ is an isomorphism from $\mathbb{R}^2 \oplus H^r(\partial\Omega) \oplus H^{r+1}(\partial\Omega)$ onto its range

$$\mathscr{R}(\mathscr{L}_8) = \left\{ (A_0, A_1, A_2, \psi_1, \psi_2) \in \mathbb{R}^3 \oplus H^{s+2}(\partial\Omega) \oplus H^s(\partial\Omega) \mid \langle 1, \psi_2 \rangle_{H^{-s}(\partial\Omega), H^s(\partial\Omega)} = 0 \right\}.$$

□

Exercise 8.3. Prove an analogue of Theorem 6.20 for the plate: Let $\Omega \subset \mathbb{R}^2$ be bounded with smooth boundary $\partial\Omega$. Let $\mathscr{L}_1$ be defined as in (8.77). Then there exists $C > 0$ such that

$$\begin{aligned}
&- \langle \boldsymbol{P}\mathscr{L}_1(\vec{F}), \boldsymbol{P}\vec{F} \rangle_{H^{3/2}(\partial\Omega)\times H^{1/2}(\partial\Omega), H^{-3/2}(\partial\Omega)\times H^{-1/2}(\partial\Omega)} \\
&\quad + |a_0|^2 + |a_1|^2 + |a_2|^2 \geqslant C\|\mathscr{L}_1 \vec{F}\|^2_{R^3\oplus H^{3/2}(\partial\Omega)\oplus H^{1/2}(\partial\Omega)},
\end{aligned}$$

for all $\vec{F} = (a_0, a_1, a_2, f_1, f_2) \in \mathbb{R}^3 \oplus H^{-3/2}(\partial\Omega) \oplus H^{-1/2}(\partial\Omega)$ satisfying

$$\begin{aligned}
&\langle f_1, 1 \rangle_{H^{-3/2}(\partial\Omega), H^{3/2}(\partial\Omega)} = 0, \\
&\langle f_1, x_j \rangle_{H^{-3/2}(\partial\Omega), H^{3/2}(\partial\Omega)} + \langle f_2, n_j(x) \rangle_{H^{-1/2}(\partial\Omega), H^{1/2}(\partial\Omega)} = 0, \quad j = 1, 2,
\end{aligned}$$

where $\boldsymbol{P}$ is the projection operator defined by

$$\boldsymbol{P}(a_0, a_1, a_2, f_1, f_2) = (f_1, f_2). \qquad \square$$

8.7 Numerical computations and examples (I): exterior BVP

In the remaining sections in this chapter, we provide a few examples to illustrate the high accuracy and advantage of BEM for a thin plate. As before, the computational approach is based on the *pointwise collocation* of the various BIE derived earlier.

All of the numerical examples in the rest of this chapter are given on *circular domains*. The *kernels* of the various BIE involved have already been calculated in Example 8.1. The reader can take advantage of formulas (8.60)–(8.69) to write computer programs and compare the numerical results thus obtained with those given here by the authors.

Example 8.5. As an application of Theorem 8.12, we apply it to find the numerical solution of

$$\left.\begin{aligned}
&\Delta^2 w(x) = 0, \quad x \in \Omega^c, \\
&B_1 w(x) = \frac{2(1-\nu)\cos\theta}{[R^2+\alpha_0^2-2R\alpha_0\sin\theta]^2}\left\{1-\frac{4\alpha_0}{R}\sin\theta\right. \\
&\qquad +\frac{4\alpha_0(3R\sin\theta+5\alpha_0\cos^2\theta-3\alpha_0)}{R^2+\alpha_0^2-2R\alpha_0\sin\theta} \\
&\qquad \left.-\frac{24R\alpha_0^2\cos^2\theta(R-\alpha_0\sin\theta)}{[R^2+\alpha_0^2-2R\alpha_0\sin\theta]^2}\right\}, \\
&B_2 w(x) = \frac{2(1-\nu)\cos\theta}{[R^2+\alpha_0^2-2R\alpha_0\sin\theta]^2}\left[2\alpha_0\sin\theta-3R\right. \\
&\qquad \left.+\frac{4R(R-\alpha_0\sin\theta)^2}{R^2+\alpha_0^2-2R\alpha_0\sin\theta}\right], \\
&R=\frac{1}{2}, \alpha_0=\frac{1}{4}, \nu=\frac{1}{8}, \\
&x_1 = R\cos\theta, x_2 = R\sin\theta, 0 \leqslant \theta \leqslant 2\pi,
\end{aligned}\right\} \tag{8.193}$$

where Ω^c is the exterior of the closed disk with radius $\frac{1}{2}$. It is straightforward to check that the data in (8.193) satisfies the compatibility condition (8.13).

The exact solution satisfying (8.193) and the growth condition (8.165) is (a harmonic function)

$$w(x) = \frac{x_1}{x_1^2+\left(x_2-\frac{1}{4}\right)^2}, \quad x=(x_1,x_2)\in\Omega^c. \tag{8.194}$$

We compute w by approximating (8.171) and (8.172), where in (8.171) we have specified $c_j = 0$ for $j = 1$, 2 and 3 using some a priori knowledge of w. The linear system (8.172) is approximated by pointwise-collocating the boundary integral equations. A (3,2)-system of quadratic B-splines (5.61) is used, and there are totally 24 uniformly spaced mesh points. The collocation points are the midpoints between the mesh points. This results in a matrix equation of size 54×54, where

$$\begin{aligned}
&24\times 2\,(\text{one each for } f_1 \text{ and } f_2) + 3\,(\text{compatibility conditions}) \\
&\qquad\qquad + 3\,(\text{coefficients } \widetilde{a}_0,\ \widetilde{a}_1,\ \widetilde{a}_2) = 54.
\end{aligned}$$

Graphics of the exact and numerical solutions have no visible differences from each other, as illustrated in Figs. 8.4 and 8.5. Numerical values also show good agreement at the selected collocation points on the boundary ($R = \frac{1}{2}$), and at points on circles with radius $R = 10$ and 20 in Ω; see Table 8.1. This computation requires very small memory and computer time. □

Note that Example 8.5 is selected to have just $o(1)$ growth at ∞. If a solution has linear–logarithmic growth (8.165) with $A \neq 0$ then the numerical solution is usually accurate only within a small radius of $\partial\Omega$. When $|x|$ becomes large, *the numerical solution quickly loses accuracy with a logarithmic rate*, unless the coefficient A in (8.165) can be computed exactly.

We also remark that the computer program for computing the numerical solution of Example 8.5 (exterior problem) is *essentially identical* to those for computing Examples 8.7 and 8.8 (interior problems) in the next section, because their corresponding BIE formulations (8.174) and (8.104) are virtually the same (except for the difference of coefficients $\pm\frac{1}{2}$ in front of f_1 and f_2). This is a great advantage of BEM.

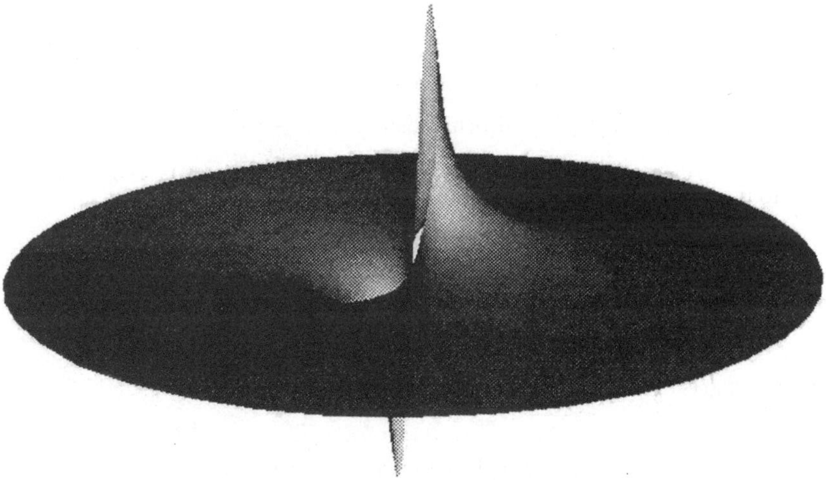

Figure 8.4: The exact solution $w(x) = x_1/[x_1^2 + (x_2 - \frac{1}{4})^2]$, a benchmark for the exterior plate problem, in Example 8.5. The graph is plotted on a disk with radius 6.

8.8 Numerical computations and examples (II): interior BVP

We first consider the following biharmonic boundary value problem of Type I:

$$\begin{aligned} &\Delta^2 w(x) = 0, \quad x \in \Omega, \\ &w(x) = g_1(x) \in H^{r+3}(\partial\Omega), \\ &\frac{\partial w(x)}{\partial n}(x) = g_2(x) \in H^{r+2}(\partial\Omega). \end{aligned}$$

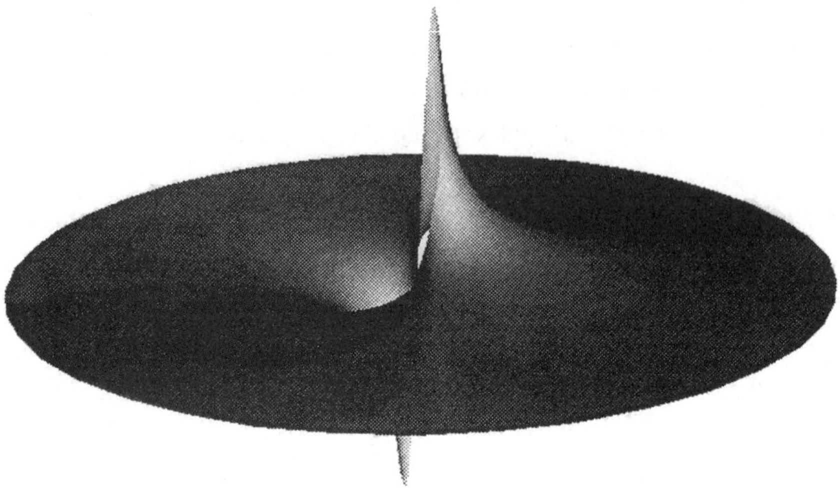

Figure 8.5: Graph of the numerical solution for Example 8.5, obtained by point collocation using quadratic B-splines with 24 uniform meshes ($h = \frac{1}{12}\pi$).

According to Theorem 8.6, the solution $w(x)$ is given by

$$w(x) = \int_{\partial\Omega} \left[E(x,\xi) f_1(\xi) + \frac{\partial E(x,\xi)}{\partial n_\xi} f_2(\xi) \right] d\sigma_\xi + a_0 + a_1 x_1 + a_2 x_2,$$

$$x \in \Omega,$$

where $(a_0, a_1, a_2, f_1, f_2) \in \mathbb{R}^3 \oplus H^r(\partial\Omega) \oplus H^{r+1}(\partial\Omega)$ solves the boundary integral equation (8.84).

The first three components on the RHS of (8.84) are easy to evaluate, so we need only concentrate on the last two components. Let us divide $\partial\Omega$ into meshes $\{\Gamma_1, \Gamma_2, \ldots, \Gamma_{n(h)}\}$, $h = \max_{1 \leqslant i \leqslant n(h)} (\text{length}\,\Gamma_i)$. We assume that the mesh is uniform, i.e., there exists $c > 0$ for all h such that

$$0 < h < c \cdot \min_{1 \leqslant i \leqslant n(h)} (\text{length}\,\Gamma_i).$$

Let $\widehat{S}_h$ be a (t,l)-system on $\partial\Omega$ (see Definition 5.1) with basis $\{\psi_1, \psi_2, \ldots, \psi_{m(h)}\}$. Let $\{z_1, z_2, \ldots, z_{m(h)}\} \subseteq \partial\Omega$ be the collocation points associated with the basis $\{\psi_1, \psi_2, \ldots, \psi_{m(h)}\}$. Let the potential densities f_1 and f_2 be approximated by

$$f_1^h(x) = \sum_{i=1}^{m(h)} \alpha_i \psi_i(x), \quad f_2^h(x) = \sum_{i=1}^{m(h)} \beta_i \psi_i(x), \quad x \in \partial\Omega. \tag{8.195}$$

Note that f_1 and f_2 have different regularity properties, therefore for optimal convergence and error estimates, perhaps splines of different smoothness should be used for f_1 and f_2

respectively. But *such a computational approach would not be convenient in practice*, thus it is not used in (8.195).

The collocated boundary integral equation system is

$$\left.\begin{aligned}
&\int_{\partial\Omega} f_1^h(\xi)\,d\sigma = \sum_{i=1}^{m(h)} \alpha_i \int_{\partial\Omega} \psi_i(\xi)\,d\sigma = A_0,\\
&\int_{\partial\Omega} [\xi_i f_1^h(\xi) + n_1(\xi) f_2^h(\xi)]\,d\sigma = \sum_{i=1}^{m(h)} \alpha_i \int_{\partial\Omega} \xi_1 \psi_i(\xi)\,d\sigma\\
&\quad + \sum_{i=1}^{m(h)} \beta_i \int_{\partial\Omega} n_1(\xi)\psi_i(\xi)\,d\sigma = A_1,\\
&\int_{\partial\Omega} [\xi_2 f_1^h(\xi) + n_2(\xi) f_2^h(\xi)]\,d\sigma = \sum_{i=1}^{m(h)} \alpha_i \int_{\partial\Omega} \xi_2 \psi_i(\xi)\,d\sigma\\
&\quad + \sum_{i=1}^{m(h)} \beta_i \int_{\partial\Omega} n_i(\xi)\psi_i(\xi)\,d\sigma = A_2,\\
&\sum_{i=1}^{m(h)} \alpha_i \int_{\partial\Omega} E(z_j,\xi)\psi_i(\xi)\,d\sigma_\xi + \sum_{i=1}^{m(h)} \beta_i \int_{\partial\Omega} \frac{\partial E(z_j,\xi)}{\partial n_\xi}\psi_i(\xi)\,d\sigma_\xi\\
&\quad + a_0^{(h)} + a_1^{(h)} z_{j,1} + a_2^{(h)} z_{j,2} = g_1(z_j),\ \ j = 1,2,\ldots,m(h),\\
&\sum_{i=1}^{m(h)} \alpha_i \int_{\partial\Omega} \left.\frac{\partial E(x,\xi)}{\partial n_x}\right|_{x=z_j} \psi_i(\xi)\,d\sigma_\xi\\
&\quad + \sum_{i=1}^{m(h)} \beta_i \int_{\partial\Omega} \left.\frac{\partial E(x,\xi)}{\partial n_x \partial n_\xi}\right|_{x=z_j} \psi_i(\xi)\,d\sigma_\xi\\
&\quad + a_1^{(h)} n_1(z_j) + a_2^{(h)} n_2(z_j) = g_2(z_j),\ \ j = 1,2,\ldots,m(h),
\end{aligned}\right\} \tag{8.196}$$

a $[2m(h)+3]\times[2m(h)+3]$ linear system of the form

$$A_h F_h = G_h,$$

where

$$F_h = \begin{bmatrix} a_0^{(h)} \\ a_1^{(h)} \\ a_2^{(h)} \\ \alpha_1 \\ \alpha_2 \\ \vdots \\ \alpha_{m(h)} \\ \beta_1 \\ \beta_2 \\ \vdots \\ \beta_{m(h)} \end{bmatrix}, \quad G_h = \begin{bmatrix} A_0 \\ A_1 \\ A_2 \\ g_1(z_1) \\ g_1(z_2) \\ \vdots \\ g_1(z_{m(h)}) \\ g_2(z_1) \\ g_2(z_2) \\ \vdots \\ g_2(z_{m(h)}), \end{bmatrix}, \tag{8.197}$$

and the $[2m(h)+3]\times[2m(h)+3]$ matrix A_h can be easily assembled from the system of equations (8.196). All the entries of A_h are computable, since the most severe singularity of the kernel is $\partial^2E(x,\xi)/\partial n_x\partial n_\xi=\mathscr{O}(\ln|x-\xi|)$, which is $L^1(\partial\Omega)$. Thus

$$F_h=(A_h)^{-1}G_h, \tag{8.198}$$

yielding an approximate solution $(a_0^{(h)},a_1^{(h)},a_2^{(h)},f_1^h,f_2^h)$.

Example 8.6. Let Ω be the disk with radius $\frac{1}{2}$ centered at the origin in $\mathbb{R}^2$ and

$$\begin{aligned}\zeta(z)=\frac{1}{8\pi}&(10|x-\eta_1|^2\ln|x-\eta_1|+4|x-\eta_2|^2\ln|x-\eta_2|\\&+3|x-\eta_3|^2\ln|x-\eta_3|+2|x-\eta_4|^3\ln|x-\eta_4|),\ \ x\in\mathbb{R}^2,\end{aligned} \tag{8.199}$$

where

$$\eta_1=\left(-\frac{3}{2},\frac{3}{2}\right),\ \ \eta_2=(2,2),\ \ \eta_3=(3,-3),\ \ \eta_4=\left(-\frac{5}{2},-\frac{5}{2}\right).$$

$\zeta(x)$ has a concave mirror shape, as shown in Fig. 8.6. Note that ζ has removable singularities at η_1, η_2, η_3 and η_4 outside Ω.

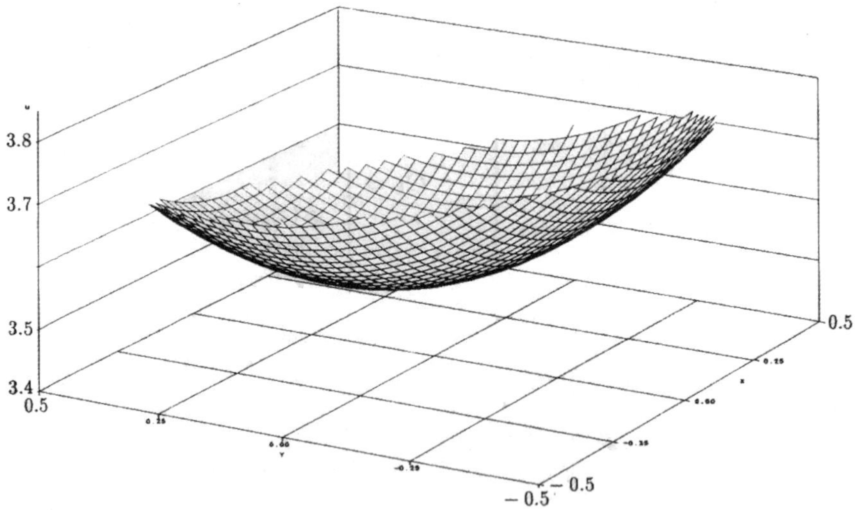

Figure 8.6: The concave mirror shape of $\zeta(x)$ as given by (8.199) in Example 8.6.

It is easy to see that the exact solution to the biharmonic problem

$$\left.\begin{aligned}&\Delta^2w(x)=0, && x\in\Omega,\\&w(x)=\zeta(x), && x\in\partial\Omega,\\&\frac{\partial w(x)}{\partial n}=\frac{\partial}{\partial n}\zeta(x), && x\in\partial\Omega,\end{aligned}\right\} \tag{8.200}$$

is

$$w(x) = \zeta(x), \quad x \in \Omega. \tag{8.201}$$

According to Remark 8.1, we can set up boundary integral equations as

$$\left.\begin{aligned} &\int_{\partial\Omega}\left[E(x,\xi)f_1(\xi)+\frac{\partial E(x,\xi)}{\partial n_\xi}f_2(\xi)\right]d\sigma_\xi = \zeta(x), \qquad x\in\partial\Omega,\\ &\int_{\partial\Omega}\left[\frac{\partial E(x,\xi)}{\partial n_x}f_1(\xi)+\frac{\partial^2 E(x,\xi)}{\partial n_x \partial n_\xi}f_2(\xi)\right]d\sigma_\xi = \frac{\partial\zeta(x)}{\partial n}, \; x\in\partial\Omega, \end{aligned}\right\} \tag{8.202}$$

without the need to introduce three extra constants a_0, a_1, a_2 in Theorem 8.6. Equations (8.202) constitute a bounded invertible mapping from $(f_1, f_2) \in H^r(\partial\Omega) \oplus H^{r+1}(\partial\Omega)$ into $H^{r+3}(\partial\Omega) \oplus H^{r+2}(\partial\Omega)$.

We discretize (8.202) by collocation as in (8.196), using as $\widehat{S^h}$ a $(3,2)$-system of quadratic B-splines (5.61). The collocation points are the midpoints of mesh points. Numerical values of $w(x)$ at sampled nodal points and interior points are tabulated in Table 8.2.

Visually, the graph of $w_h(x)$ has no tangible difference from Fig. 8.6 for $h = \frac{1}{24}\pi$ on the scale of the graph, so we do not plot the graph of $w_h(x)$ here. □

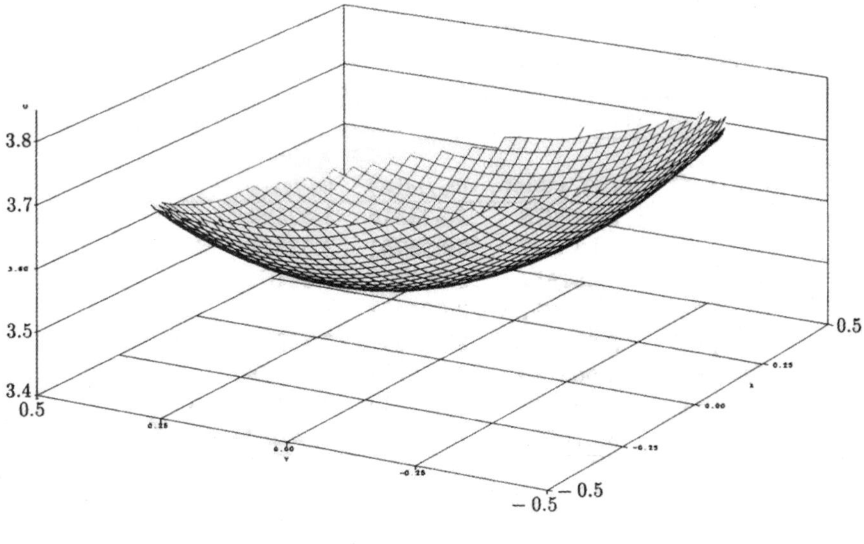

Figure 8.7: The numerical solution $w_h(x)$ of Example 8.6, obtained by point collocation using quadratic B-splines with 48 uniform meshes ($h = \frac{\pi}{24}$).

Next, we consider the biharmonic boundary value problem

$$\left.\begin{aligned}&\Delta^2 w(x) = 0, && x \in \Omega,\\ &B_1 w(x) = g_1(x), && x \in \partial\Omega,\\ &B_2 w(x) = g_2(x), && x \in \partial\Omega, \quad (g_1, g_2) \text{ satisfies (8.115).}\end{aligned}\right\} \tag{8.203}$$

Let $\widehat{S}_h$ be a (t,l)-system as mentioned in § 5.4. We invoke Theorem 8.7. Formally, we can discretize the system of integral equations

$$\mathscr{L}_2 \begin{bmatrix} f_1 \\ f_2 \end{bmatrix} = \begin{bmatrix} A_0 \\ A_1 \\ A_2 \\ g_1 \\ g_2 \end{bmatrix} \tag{8.204}$$

by assuming (8.115), leading to the first three equations in (8.71), and in addition,

$$\begin{aligned}&\sum_{i=1}^{m(h)} \alpha_i \int_{\partial\Omega} [B_{1x} E(x,\xi)|_{x=z_j}] \psi_i(\xi)\, d\sigma_\xi\\ &\quad + \sum_{i=1}^{m(h)} \beta_i \int_{\partial\Omega} \left[B_{1x} \frac{\partial E(x,\xi)}{\partial n_\xi}\bigg|_{x=z_j} \right] \psi_i(\xi)\, d\sigma_\xi + \frac{1}{2} \sum_{i=1}^{m(h)} \alpha_i \psi_i(z_j)\\ &= g_1(z_j), \qquad 1 \leqslant j \leqslant m(h),\end{aligned} \tag{8.205}$$

$$\begin{aligned}&\sum_{i=1}^{m(h)} \alpha_i \int_{\partial\Omega} [B_{2x} E(x,\xi)|_{x=z_j}] \psi_i(\xi)\, d\sigma_\xi\\ &\quad + \sum_{i=1}^{m(h)} \beta_i \int_{\partial\Omega} \left[B_{2x} \frac{\partial E(x,\xi)}{\partial n_\xi}\bigg|_{x=z_j} \right] \psi_i(\xi)\, d\sigma_\xi - \frac{1}{2} \beta_i \psi_i(z_j)\\ &= g_2(z_j), \qquad 1 \leqslant j \leqslant m(h).\end{aligned} \tag{8.206}$$

These equations result in an overdetermined $[2m(h)+3] \times [2m(h)]$ linear system

$$\widetilde{A}_h \widetilde{F}_h = \widetilde{G}_h, \tag{8.207}$$

where

$$\widetilde{F}_h = \begin{bmatrix} \alpha_1 \\ \alpha_2 \\ \vdots \\ \alpha_{m(h)} \\ \beta_1 \\ \beta_2 \\ \vdots \\ \beta_{m(h)} \end{bmatrix}, \quad \widetilde{G}_h = \begin{bmatrix} A_0 \\ A_1 \\ A_2 \\ g_1(z_1) \\ g_1(z_2) \\ \vdots \\ g_1(z_{m(h)}) \\ g_2(z_1) \\ g_2(z_2) \\ \vdots \\ g_1(z_{m(h)}) \end{bmatrix},$$

and $\widetilde{A}_h$ is a $[2m(h)+3] \times [2m(h)]$ matrix assembled from (8.205), (8.206), along with the first three equations in (8.196). As noted earlier in Example 8.1,

$$\begin{aligned} &B_{2x}E(x,\xi) = \mathcal{O}(\ln|x-\xi|), \\ &B_{1x}E(x,\xi) \text{ is } C^\infty\text{-smooth for } x,\xi \in \partial\Omega, \\ &B_{2x}\frac{\partial E(x,\xi)}{\partial n_\xi} \text{ is } C^\infty\text{-smooth for } x,\xi \in \partial\Omega, \\ &B_{1x}\frac{\partial E(x,\xi)}{\partial n_\xi} = \mathcal{O}(|x-\xi|^{-2}). \end{aligned} \tag{8.208}$$

The first kernel is absolutely integrable with respect to ξ for any given x on $\partial\Omega$, so there is no problem. The second and third kernels are C^∞-smooth. The last kernel is hypersingular, which we evaluate by regularization using formula (3.32) since the severity of the singularity in (8.208) is of the type $\mathcal{O}(|\xi|^{-2})$ after making the change of variable $\xi - x \mapsto \xi$. (In the case when Ω is a disk centered at the origin, we can use (8.66) to get

$$\begin{aligned} \int_{\partial\Omega} B_{1x}\frac{\partial E(x,\xi)}{\partial n_\xi}\,d\sigma_\xi &= \int_{\partial\Omega} \frac{\partial B_{1\xi}E(x,\xi)}{\partial n_x}\,d\sigma_x \\ &= \frac{\partial}{\partial n_x}\int_{\partial\Omega} B_{1\xi}E(x,\xi)\,d\sigma_x \\ &= \frac{\partial}{\partial n_x}(-1) \\ &= 0, \end{aligned} \tag{8.209}$$

so

$$\int_{\partial\Omega} B_{1x}\frac{\partial E(x,\xi)}{\partial n_\xi}\psi_i(\xi)\,d\sigma_\xi = \int_{\partial\Omega} B_{1x}\frac{\partial E(x,\xi)}{\partial n_\xi}[\psi_i(\xi)-\psi_i(x)]\,d\sigma_\xi. \tag{8.210}$$

The integrand on the RHS above has singularity $\mathscr{O}((x-\xi)^{-1})$, so we can easily take its *Cauchy principal value* as in (3.33) (or (1.25)), i.e.,

$$\int_{\partial\Omega} B_{1x}\frac{\partial E(x,\xi)}{\partial n_\xi}\psi_i(\xi)\,d\sigma_\xi = \lim_{\varepsilon\downarrow 0}\int_{\partial\Omega\backslash B_\varepsilon(x)}\left[B_{1x}\frac{\partial E(x,\xi)}{\partial n_\xi}\right][\psi_i(\xi)-\psi_i(x)]\,d\sigma_\xi, \quad (8.211)$$

whose evaluation is rather easy.)

We note that (8.207) is an overdetermined linear system; its solvability requires that

$$\widetilde{G}_h \perp [\mathscr{N}(\widetilde{A}_h^{Tr})] \subset \mathbb{R}^{2m(h)+3},$$

which is inconvenient to verify in practice. To compute a solution, instead we will apply the theory in § 4.7 by augmenting three extra accessory conditions.

Example 8.7. Consider

$$\left.\begin{aligned} &\Delta^2 w(x) = 0, \quad x\in\Omega, \text{ an open disk in } \mathbb{R}^2 \text{ with radius } \frac{1}{2}, \\ &B_1 w(x) = B_1\zeta(x), \quad x\in\partial\Omega, \quad \nu=\frac{1}{8}, \\ &B_2 w(x) = B_2\zeta(x), \quad x\in\partial\Omega, \end{aligned}\right\} \quad (8.212)$$

where Ω and $\zeta(x)$ are the same as in Example 8.6. The exact solutions to (8.212) are

$$w(x) = a_0 + a_1x_1 + a_2x_2 + \zeta(x), \quad x\in\Omega, \quad (8.213)$$

for three arbitrary constants $a_0, a_1, a_2 \in \mathbb{R}$. To fix these three constants, we impose three accessory boundary data

$$\begin{aligned} &w(x_0) = \zeta(x_0), \quad \frac{\partial w(x_0)}{\partial\sigma} = \frac{\partial\zeta(x_0)}{\partial\sigma}, \\ &\frac{\partial w(x_0)}{\partial n} = \frac{\partial\zeta(x_0)}{\partial n}, \quad \text{at } x_0 = \left(-\frac{1}{2},0\right), \end{aligned} \quad (8.214)$$

according to Example 4.7. Thus $a_0 = a_1 = a_2 = 0$ in (8.213) and $w(x) = \zeta(z)$ is the solution to (8.212) and (8.214).

We represent the solution as an augmented combination of simple- and double-layers:

$$w(x) = \int_{\partial\Omega}\left[E(x,\xi)f_1(\xi) + \frac{\partial E(x,\xi)}{\partial n_\xi}f_2(\xi)\right]d\sigma_\xi + \mu_0 + \mu_1x_1 + \mu_2x_2, \quad x\in\Omega.$$

The three conditions in (8.214) can be discretized as

$$\sum_{i=1}^{m(h)} \alpha_i \int_{\partial\Omega} E(x_0,\xi)\psi_i(\xi)\,d\sigma + \sum_{i=1}^{m(h)} \beta_i \int_{\partial\Omega} \frac{\partial E(x_0,\xi)}{\partial n_\xi}\psi_i(\xi)\,d\sigma$$
$$+\mu_0+\mu_1 x_{0,1}+\mu_2 x_{0,2} = \zeta(x_0), \tag{8.215}$$

$$\sum_{i=1}^{m(h)} \alpha_i \int_{\partial\Omega} \frac{\partial E(x_0,\xi)}{\partial \sigma_x}\psi_i(\xi)\,d\sigma + \sum_{i=1}^{m(h)} \beta_i \int_{\partial\Omega} \frac{\partial^2 E(x_0,\xi)}{\partial \sigma_x \partial n_\xi}\psi_i(\xi)\,d\sigma$$
$$-\mu_1 n_2(x_0)+\mu_2 n_1(x_0) = \frac{\partial\zeta(x_0)}{\partial\sigma}, \tag{8.216}$$

$$\sum_{i=1}^{m(h)} \alpha_i \int_{\partial\Omega} \frac{\partial E(x_0,\xi)}{\partial n_x}\psi_i(\xi)\,d\sigma + \sum_{i=1}^{m(h)} \beta_i \int_{\partial\Omega} \frac{\partial^2 E(x_0,\xi)}{\partial n_x \partial n_\xi}\psi_i(\xi)\,d\sigma$$
$$+\mu_1 n_1(x_0)+\mu_2 n_2(x_0) = \frac{\partial\zeta(x_0)}{\partial n}. \tag{8.217}$$

The above are used in conjunction with (the discretized form of) the two augmented BIE

$$\frac{1}{2}f_1(x)+\int_{\partial\Omega}\left\{[B_{1x}E(x,\xi)]f_1(\xi)+\left[B_{1x}\frac{\partial E(x,\xi)}{\partial n_\xi}i\right]f_2(\xi)\right\}d\sigma_\xi$$
$$+\mu_0+\mu_1 x_1+\mu_2 x_2 = g_1(x) \qquad (=B_1\zeta(x)),$$
$$-\frac{1}{2}f_2(x)+\int_{\partial\Omega}\left\{[B_{2x}E(x,\xi)]f_1(\xi)+\left[B_{2x}\frac{\partial E(x,\xi)}{\partial n_\xi}\right]f_2(\xi)\right\}d\sigma_\xi$$
$$-\mu_1 n_1(x)-\mu_2 n_2(x) = g_2(x) \qquad (=B_2\zeta(x)).$$

The reader should use the theory in §§ 4.6–4.7 to verify that the overall augmented system is invertible.

We thus have obtained a $[2m(h)+3]\times[2m(h)+3]$ linear system for the unknowns $(\mu_0,\mu_1,\mu_2,\alpha_1,\alpha_2,\ldots,\alpha_{m(h)},\ \beta_1,\beta_2,\ldots,\beta_{m(h)})$. With $\widehat{S}_h$ a (3,2)-system and $h=\frac{1}{24}\pi$, we have computed numerical results, which are tabulated in Table 8.3 for a few sample points. The graph of the numerical solution is provided in Fig. 8.7. The reader can compare it with Fig. 8.6. The rate of convergence is slower than expected, indicating the need to fine tune the hypersingularity quadrature in order to accelerate convergence. □

Example 8.8. Consider

$$\Delta^2 w(x)=0,\quad \text{on}\,\Omega,\ \text{an open disk in } \mathbb{R}^2 \text{ with radius } \tfrac{1}{2},$$
$$\left.\begin{aligned} B_1 w(x) &= 20\cos 3\theta,\\ B_2 w(x) &= 20\sin 4\theta \end{aligned}\right\}\quad \text{on}\,\partial\Omega,\ \ \theta=\arg x,\ \ \nu=\tfrac{1}{4},$$
$$w(x_0)=3.7429,\quad \frac{\partial w(x_0)}{\partial s}=-0.1224,\quad \frac{\partial w(x_0)}{\partial n}=0.811\ \ \text{at}\ x_0=\left(-\frac{1}{2},0\right).$$

For this problem, no closed-form solution is easily obtainable. A numerical solution is illustrated in Fig. 8.8, using $h=\frac{1}{24}\pi$ and a $(3,2)$-system of splines and the same discretization procedures as in Example 8.7. □

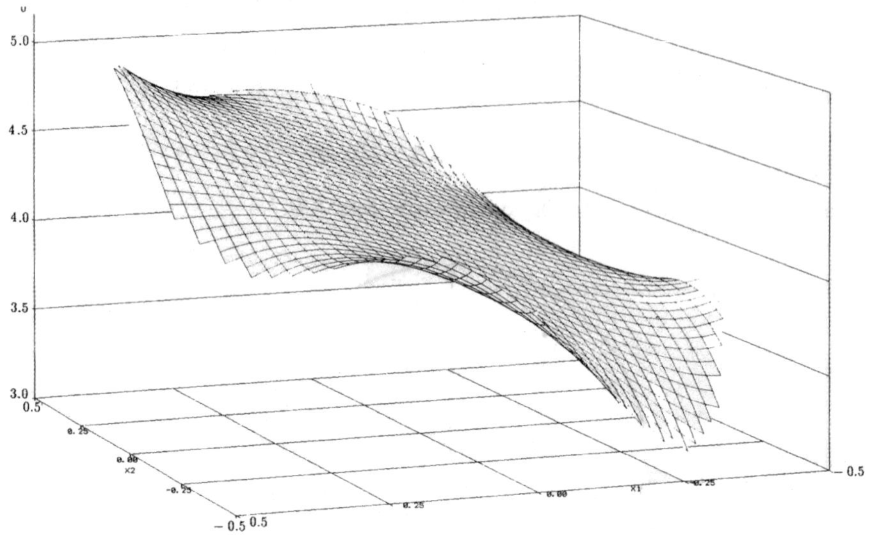

Figure 8.8: Bending and shearing of a circular plate: graph of the numerical solution of Example 8.8, computed by point collocation using quadratic *B*-splines with 48 uniform meshes ($h = \frac{1}{24}\pi$).

Table 8.1: Comparison of exact and numerical values for Example 8.5. The first two columns are the coordinates of $(x = (x_1, x_2) = (x, y))$. Values are tabulated for $r = 0.5$ (the boundary points), and for points on the concentric circles with radii $r = 10$ and 20. The solution is computed by point collocation using quadratic B-splines with 24 uniform meshes.

$r = 0.5$				$r = 20$			
x_1	x_2	Numerical	Exact	x_1	x_2	Numerical	Exact
0.4830	0.1294	1.9363	1.9490	19.3185	5.1764	0.0471	0.0486
0.4330	0.2500	2.2981	2.3084	17.3205	10.0000	0.0415	0.0438
0.3536	0.3536	2.5984	2.6050	14.1421	14.1421	0.0334	0.0360
0.2500	0.4330	2.6071	2.6043	10.0000	17.3205	0.0234	0.0255
0.1294	0.4830	1.8194	1.8222	5.1764	19.3185	0.0120	0.0133
0.0000	0.5000	0.0000	0.0000	0.0000	20.0000	0.0000	0.0000
−0.1294	0.4830	−1.8194	−1.8222	−5.1764	19.3185	−0.0120	−0.0133
−0.2500	0.4330	−2.6071	−2.6043	−10.0000	17.3205	−0.0234	−0.0255
−0.3536	0.3536	−2.5984	−2.6050	−14.1421	14.1421	−0.0334	−0.0360
−0.4330	0.2500	−2.2981	−2.3094	−17.3205	10.0000	−0.0415	−0.0438
−0.4830	0.1294	−1.9363	−1.9490	−19.3185	5.1764	−0.0471	−0.0486
−0.5000	0.0000	−1.5875	−1.6000	−20.0000	0.0000	−0.0496	−0.0500
−0.4830	−0.1294	−1.2691	−1.2804	−19.3185	−5.1764	−0.0489	−0.0480
−0.4330	−0.2500	−0.9803	−0.9897	−17.3205	−10.0000	−0.0446	−0.0428
−0.3536	−0.3536	−0.7154	−0.7226	−14.1421	−14.1421	−0.0370	−0.0347
−0.2500	−0.4330	−0.4677	−0.4726	−10.0000	−17.3205	−0.0265	−0.0245
−0.1294	−0.4830	−0.2312	−0.2336	−5.1764	−19.3185	−0.0138	−0.0126
0.0000	−0.5000	0.0000	0.0000	0.0000	−20.0000	0.0000	0.0000
0.1294	−0.4830	0.2312	0.2336	5.1764	−19.3185	0.0138	0.0126
0.2500	−0.4330	0.4677	0.4726	10.0000	−17.3205	0.0265	0.0245
0.3536	−0.3536	0.7154	0.7226	14.1421	−14.1421	0.0370	0.0347
0.4330	−0.2500	0.9803	0.9897	17.3205	−10.0000	0.0446	0.0428
0.4830	−0.1294	1.2691	1.2804	19.3185	−5.1764	0.0489	0.0480
0.5000	0.0000	1.5875	1.6000	20.0000	0.0000	0.0496	0.0500

$r = 10$			
x_1	x_2	Numerical	Exact
9.6593	2.5882	0.0960	0.0978
8.6603	5.0000	0.0862	0.0888
7.0711	7.0711	0.0706	0.0733
5.0000	8.6603	0.0500	0.0522
2.5882	9.6593	0.0260	0.0272
0.0000	10.0000	0.0000	0.0000
−2.5882	9.6593	−0.0260	−0.0272
−5.0000	8.6603	−0.0500	−0.0522
−7.0711	7.0711	−0.0706	−0.0733
−8.6603	5.0000	−0.0862	−0.0888
−9.6593	2.5882	−0.0960	−0.0978
−10.0000	0.0000	−0.0992	−0.0999
−9.6593	−2.5882	−0.0959	−0.0953
−8.6603	−5.0000	−0.0860	−0.0844
−7.0711	−7.0711	−0.0704	−0.0683
−5.0000	−8.6603	−0.0498	−0.0479
−2.5882	−9.6593	−0.0258	−0.0247
0.0000	−10.0000	0.0000	0.0000
2.5882	−9.6593	0.0258	0.0247
5.0000	−8.6603	0.0498	0.0479
7.0711	−7.0711	0.0704	0.0683
8.6603	−5.0000	0.0860	0.0844
9.6593	−2.5882	0.0959	0.0953
10.0000	0.0000	0.0992	0.0999

Table 8.2: Comparison of numerical and exact solutions at selected points for Example 8.6. The numerical solutions are computed by point collocation using quadratic B-splines with $24, 48$ and 96 uniform meshes ($h = \frac{1}{12}\pi, h = \frac{1}{24}\pi$ and $h = \frac{1}{48}\pi$, respectively).

$x = (r, \theta)$	$w(x)$, exact	$w_{\pi/12}(x)$	$w_{\pi/24}(x)$	$w_{\pi/48}(x)$
$(\frac{1}{3}, \frac{1}{4}\pi)$	3.525278	3.525316	3.525291	3.525286
$(\frac{1}{3}, \frac{1}{2}\pi)$	3.533697	3.533751	3.533711	3.533705
$(\frac{1}{3}, \frac{3}{4}\pi)$	3.581651	3.581712	3.581669	3.581660
$(\frac{1}{3}, \pi)$	3.627217	3.627188	3.627220	3.627222
$(\frac{1}{3}, \frac{5}{4}\pi)$	3.657121	3.657042	3.657117	3.657123
$(\frac{1}{3}, \frac{3}{2}\pi)$	3.663788	3.663738	3.663789	3.663792
$(\frac{1}{3}, \frac{7}{4}\pi)$	3.631413	3.631414	3.631421	3.631419
$(\frac{1}{3}, 2\pi)$	3.570268	3.570301	3.570279	3.570275

Table 8.3: Comparison of numerical and exact solutions at selected points for Example 8.7. The numerical solutions are computed by point collocation using quadratic B-splines with $24, 48$ and 96 uniformmeshes ($h = \frac{1}{12}\pi$, $\frac{1}{24}\pi$ and $\frac{1}{48}\pi$, respectively).

	$h = 12$	$h = 24$	$h = 48$	
$\mu_{0,h}$	0.3584	0.2474	0.2104	
$\mu_{1,h}$	0.0029	0.0015	0.0010	
$\mu_{2,h}$	0.0069	0.0036	0.0024	
$x = (r, \theta)$	$w_{\pi/12}(x)$	$w_{\pi/24}(x)$	$w_{\pi/48}(x)$	$w(x)$, exact
$\left(\frac{1}{3}, \frac{1}{4}\pi\right)$	3.5338	3.5318	3.5308	3.5253
$\left(\frac{1}{3}, \frac{1}{2}\pi\right)$	3.5413	3.5389	3.5378	3.5337
$\left(\frac{1}{3}, \frac{3}{4}\pi\right)$	3.5857	3.5842	3.5835	3.5817
$\left(\frac{1}{3}, \pi\right)$	3.6272	3.6272	3.6273	3.6272
$\left(\frac{1}{3}, \frac{5}{4}\pi\right)$	3.6548	3.6563	3.6568	3.6571
$\left(\frac{1}{3}, \frac{3}{2}\pi\right)$	3.6623	3.6642	3.6648	3.6638
$\left(\frac{1}{3}, \frac{7}{4}\pi\right)$	3.6335	3.6346	3.6346	3.6314
$\left(\frac{1}{3}, 2\pi\right)$	3.5765	3.5759	3.5754	3.5703
$\left(\frac{1}{2}, \frac{1}{3}\pi\right)$	3.5934	3.5899	3.5881	3.5803
$\left(\frac{1}{2}, \frac{2}{3}\pi\right)$	3.6615	3.6584	3.6571	3.6531
$\left(\frac{1}{2}, \pi\right)$	3.7429	3.7429	3.7429	3.7429
$\left(\frac{1}{2}, \frac{4}{3}\pi\right)$	3.7919	3.7946	3.7954	3.7956
$\left(\frac{1}{2}, \frac{5}{3}\pi\right)$	3.7829	3.7850	3.7854	3.7818
$\left(\frac{1}{2}, 2\pi\right)$	3.6711	3.6701	3.6692	3.6616

Chapter 9

Linear Elastostatics

9.1 Derivations of equations in linear elasticity

Let $\vec{w}(x) = (w_1(x), w_2(x), w_3(x))^{\mathrm{Tr}}$ denote the displacement vector at point $x = (x_1, x_2, x_3)$ of an elastic solid originally occupying space Ω in $\mathbb{R}^3$, where x_1, x_2 and x_3 are the Cartesian coordinates of x. (Note that throughout this chapter, $\vec{w}(x)$, as well as many other vectors, generally denotes a *column vector*. But when it is unimportant to differentiate between a column vector and a row vector, we omit the transpose sign "Tr".) In continuum mechanics, the linear strain tensor is given by

$$\varepsilon_{ij}(x) = \frac{1}{2}\left[\frac{\partial w_i(x)}{\partial x_j} + \frac{\partial w_j(x)}{\partial x_i}\right], \quad 1 \leqslant i, j \leqslant 3, \quad x \in \Omega.$$

Let $\sigma_{ij}(x)$, $1 \leqslant i, j \leqslant 3$, denote the stress tensor at x. Assume that the stress (tensor) is zero in the undeformed state, and the solid deforms with a linear stress-strain relationship without any dissipation of energy. (This kind of solid is called a linear elastic solid, or Hookean solid.) Then there are coefficients c_{ijkl} such that

$$\sigma_{ij} \equiv \sum_{k=1}^{3}\sum_{l=1}^{3} c_{ijkl}\varepsilon_{kl}, \quad 1 \leqslant i, j \leqslant 3. \tag{9.1}$$

From now on, we will use the summation convention

$$\sigma_{ij} = c_{ijkl}\varepsilon_{kl} \tag{9.2}$$

to omit the summation signs on the RHS of (9.1) when no ambiguities should occur. Equation (9.1) is called the *generalized Hooke's law* for an elastic solid.

At first sight, there are $3 \times 3 \times 3 \times 3 = 81$ elastic stiffness coefficients c_{ijkl}. But under the additional assumptions that

$$\sigma_{ij} = \sigma_{ji}, \quad \varepsilon_{ij} = \varepsilon_{ji}$$

and that the material is *isotropic* (i.e., the material properties at a given point do not vary with direction or orientation of coordinate axes at that point) and *homogeneous* (i.e., the material properties do not vary with position), it can be further shown that only *two* of the coefficients c_{ijkl} are independent constants:

$$c_{jjjj} = \lambda + 2\mu, \quad c_{jjkk} = \lambda, \quad c_{jkjk} = 2\mu.$$

λ and μ are called the Lamé constants. The remaining constants c_{ijkl} vanish. Therefore, for an isotropic, homogeneous Hookean solid, the stress–strain relations are given by Hooke's law

$$\sigma_{ij} = \lambda(\nabla \cdot \vec{w})\delta_{ij} + 2\mu\varepsilon_{ij}, \quad 1 \leqslant i, j \leqslant 3, \tag{9.3}$$

where ∇ is the divergence operator and δ_{ij} is the Kronecker delta. The equations of motion of an elastic solid under a body force (per unit volume) $F = (F_1, F_2, F_3)$ are then given by

$$\sum_{j=1}^{3} \frac{\partial \sigma_{ij}}{\partial x_j} + F_i = \rho \frac{\partial^2 w_i}{\partial t^2}, \quad 1 \leqslant i \leqslant 3, \tag{9.4}$$

where ρ is the mass density of the solid. Substituting (9.3) into (9.4), we obtain

$$\mu \Delta \vec{w} + (\lambda + \mu)\nabla(\nabla \cdot \vec{w}) + F = \rho \frac{\partial^2 \vec{w}}{\partial t^2}, \tag{9.5}$$

where Δ is the Laplacian. Since

$$\Delta = \nabla(\nabla \cdot) - \nabla \times \nabla \times,$$

(9.5) can also be written as

$$(\lambda + 2\mu)\nabla(\nabla \cdot \vec{w}) - \mu \nabla \times \nabla \times \vec{w} + F = \rho \frac{\partial^2 \vec{w}}{\partial t^2}. \tag{9.6}$$

We will refer to (9.5) (and (9.6)) as the *elastodynamic equation*.

When there is no body force, and when the solid is in static equilibrium,

$$F = 0, \quad \frac{\partial^2 \vec{w}}{\partial t^2} = 0,$$

so (9.6) reduces to the Cauchy–Navier *equation of elastostatics*

$$(\lambda + 2\mu)\nabla(\nabla \cdot \vec{w}) - \mu \nabla \times \nabla \times \vec{w} = 0.$$

Upon using $\nabla \times \nabla \times = \nabla(\nabla \cdot) - \Delta$, we recover

$$\mu \Delta \vec{w} + (\lambda + \mu)\nabla(\nabla \cdot \vec{w}) = 0. \tag{9.7}$$

This equation can also be written as

$$\Delta \vec{w} + \frac{1}{1 - 2\nu}\nabla(\nabla \cdot \vec{w}) = 0, \tag{9.8}$$

where

$$\nu = \frac{\lambda}{2(\lambda + \mu)} \tag{9.9}$$

is the *Poisson ratio*, satisfying $0 < \nu < \frac{1}{2}$.

Let us discuss the physical meanings of various elasticity constants. λ, the first Lamé constant, is not easy to interpret physically. μ, the second Lamé constant, may be identified physically as the *shear modulus* based on the shear stress and strain components; see (9.12). Engineers have introduced the Poisson ratio (9.9) as the ratio of the lateral strain to the longitudinal strain in a uniaxial state of stress. A second constant E, called the modulus of elasticity or *Young's modulus*, is also introduced to express the ratio of stress to strain in a uniaxial state of stress, where

$$E = \frac{\mu(3\lambda + 2\mu)}{\lambda + \mu}; \tag{9.10}$$

see (9.13). Another constant

$$K = \frac{E}{3(1-2\nu)},$$

called the bulk modulus, is also sometimes used; see (9.16). The inverse relations of (9.9) and (9.10) are

$$\lambda = \frac{E\nu}{(1+\nu)(1-2\nu)}, \quad \mu = \frac{E}{2(1+\nu)}. \tag{9.11}$$

The above relationships are described mathematically by

$$\sigma_{ij} = 2\mu\varepsilon_{ij}, \quad i \neq j, \tag{9.12}$$

$$\left.\begin{aligned} \varepsilon_{11} &= \frac{1}{E}\left[\sigma_{11} - \nu(\sigma_{22}+\sigma_{33})\right], \ \varepsilon_{12} = \frac{1+\nu}{E}\sigma_{12} = \frac{1}{2\mu}\sigma_{12}, \\ \varepsilon_{22} &= \frac{1}{E}\left[\sigma_{22} - \nu(\sigma_{33}+\sigma_{11})\right], \ \varepsilon_{13} = \frac{1+\nu}{E}\sigma_{13} = \frac{1}{2\mu}\sigma_{13}, \\ \varepsilon_{33} &= \frac{1}{E}\left[\sigma_{33} - \nu(\sigma_{11}+\sigma_{22})\right], \ \varepsilon_{23} = \frac{1+\nu}{E}\sigma_{23} = \frac{1}{2\mu}\sigma_{23}. \end{aligned}\right\} \tag{9.13}$$

The short form for (9.13) is

$$\varepsilon_{ij} = \frac{1+\nu}{E}\sigma_{ij} - \frac{\nu}{E}\delta_{ij}\sigma_{kk}. \tag{9.14}$$

Contracting the (i, j) indices, we obtain

$$\Theta \equiv \sigma_{kk} = \frac{E}{1-2\nu}\varepsilon_{ii}.$$

Introducing the mean pressure

$$p_M \equiv -\frac{1}{3}\Theta, \tag{9.15}$$

we have

$$p_M = -KD, \tag{9.16}$$

where

$$D \equiv \varepsilon_{ii}$$

is called the cubical dilation. The strain energy of the elastostatic solid satisfying (9.7) is given by

$$\mathscr{E}\left(\vec{w}\right) = \frac{1}{2}\int_{\Omega}\left[\lambda(\nabla\cdot\vec{w})^2 + \frac{\mu}{2}\sum_{j,k=1}^{3}\left(\frac{\partial w_j}{\partial x_k} + \frac{\partial w_k}{\partial x_j}\right)^2\right] dx. \tag{9.17}$$

If the solid is elastodynamic and satisfies (9.5), then its total energy is the strain energy (9.17) plus the kinetic energy

$$\frac{1}{2}\int_{\Omega}\rho\left|\frac{\partial\vec{w}}{\partial t}\right|^2 dx.$$

9.2 Kelvin's fundamental solution to the linear elastostatic equation

There are at least two ways of deriving the fundamental solution $E(x,\xi)$ of linear elastostatics satisfying

$$\begin{aligned}\mu\Delta_\xi E(x,\xi) + (\lambda+\mu)\nabla_\xi[\nabla_\xi\cdot E(x,\xi)] &= \mu\Delta_x E(x,\xi) + (\lambda+\mu)\nabla_x[\nabla_x\cdot E(x,\xi)]\\ &= -\delta(x-\xi)I_N, \quad x,\xi\in\mathbb{R}^N,\end{aligned} \tag{9.18}$$

where I_N is the $N\times N$ identity matrix. The first is due to Lord Kelvin while the second is by the Fourier transform. Kelvin's solution can be obtained from a representation of Papkovich [150] and Neuber [140]. Although the derivations are somewhat lengthier, those procedures are useful in the classical elasticity analysis, so we present them here first. The Fourier transform approach will be given thereafter.

For the time being, let us adopt the equation in the form of (9.8) because it has simpler coefficients. A general solution of the linear elastostatic equation has been provided by Papkovich and Neuber (see also [103, p. 100]):

$$\vec{w}(x) = \vec{h}(x) - \kappa\nabla[x\cdot\vec{h}(x) + f(x)], \tag{9.19}$$

where $\vec{h}(x)$ is a harmonic vector function and $f(x)$ is a harmonic function:

$$\Delta\vec{h}(x) = 0, \quad \Delta f(x) = 0, \tag{9.20}$$

$$\kappa = \frac{1}{4(1-\nu)}. \tag{9.21}$$

Lemma 9.1. *Let $\vec{w}$ be given by (9.19), with $\vec{h}$, f and κ satisfying (9.20) and (9.21). Then $\vec{w}$ satisfies the linear elastostatic equation (9.8).*

Proof. The verification is straightforward. We first have

$$\Delta(x\cdot\vec{h}) = 2\nabla\cdot\vec{h},$$

whence

$$\begin{aligned}\nabla\cdot\vec{w} &= \nabla\cdot\vec{h} - \kappa\Delta(x\cdot\vec{h}+f) = (1-2\kappa)\nabla\cdot\vec{h},\\ \Delta\vec{w} &= \Delta\vec{h} - \kappa\nabla[\Delta(x\cdot\vec{h}+f)],\\ &= 0-\kappa\nabla(2\nabla\cdot\vec{h}) = -2\kappa\nabla(\nabla\cdot\vec{h}).\end{aligned}$$

Therefore the LHS of (9.8) becomes

$$\Delta\vec{w} + \frac{1}{1-2\nu}\nabla(\nabla\cdot\vec{w}) = -2\kappa\nabla(\nabla\cdot\vec{h}) + \frac{1-2\kappa}{1-2\nu}\nabla(\nabla\cdot\vec{h}) = 0,$$

because of (9.21). □

Actually, as pointed out by Mindlin [133], the representation (9.19) is equivalent to another representation

$$\vec{w} = \Delta\vec{\zeta} - 2\kappa\nabla(\nabla\cdot\vec{\zeta}),$$

originally due to Galerkin, where $\vec{\zeta}$ is a biharmonic vector function:

$$\Delta^2\vec{\zeta} = 0. \qquad (9.22)$$

This follows by writing

$$\vec{h} = \Delta\vec{\zeta}$$

and noting that

$$\Delta(\nabla\cdot\vec{\zeta}) = \nabla\cdot\vec{h} = \frac{1}{2}\Delta(x\cdot\vec{h}),$$

from where

$$f \equiv \nabla\cdot\vec{\zeta} - \frac{1}{2}(x\cdot\vec{h})$$

must be a harmonic function; hence the splitting (9.19). Equation (9.22) also points out a natural connection between solutions of linear elastostatics with the biharmonic functions. The generality of the representation (9.19) is shown in the following.

Lemma 9.2. *Let $\vec{w}$ satisfy the linear elastostatic equation (9.8). Then there exist a harmonic vector function $\vec{h}$ and a harmonic function f satisfying (9.20) such that (9.19) holds for κ given by (9.21).*

Proof. Since every sufficiently smooth vector field can be split into the sum of lamellar and solenoidal potentials, we have

$$\vec{w} = \nabla S + \nabla \times \vec{A}, \qquad \nabla \cdot \vec{A} = 0, \tag{9.23}$$

where S is a scalar potential and $\vec{A}$ is a vector potential. Substituting (9.23) into equation (9.8), we have

$$\Delta\left(\nabla S + \nabla \times \vec{A} + \frac{1}{1-2\nu}\nabla S\right) = 0.$$

Thus

$$\left(1+\frac{1}{1-2\nu}\right)\nabla S + \nabla \times \vec{A} = \frac{2(1-\nu)}{1-2\nu}\nabla S + \nabla \times \vec{A} \equiv \vec{h} \tag{9.24}$$

is a harmonic vector function. Taking the divergence of (9.24), we get

$$\Delta S = \frac{1-2\nu}{2(1-\nu)}\nabla \cdot \vec{h} = \frac{1-2\nu}{4(1-\nu)}\Delta(x \cdot \vec{h}).$$

Hence

$$S - \frac{1-2\nu}{4(1-\nu)}(x \cdot \vec{h}) \equiv \frac{1-2\nu}{4(1-\nu)}f \tag{9.25}$$

is a harmonic function. We now substitute $\nabla \times \vec{A}$ from (9.24) into (9.23) to obtain

$$\begin{aligned}\vec{w} &= \nabla S + \vec{h} - \frac{2(1-\nu)}{1-2\nu}\nabla S = \vec{h} - \frac{1}{1-2\nu}\nabla S\\ &= \vec{h} - \frac{1}{1-2\nu}\cdot\frac{1-2\nu}{4(1-\nu)}\nabla[f + (x \cdot \vec{h})] \quad \text{(from (9.25))}\\ &= \vec{h} - \kappa\nabla[x \cdot \vec{h} + f].\end{aligned}$$

□

When $\nu = \frac{1}{2}$, we have $\kappa = \frac{1}{2}$, and the Papkovich–Neuber representation (9.19) becomes

$$\vec{w} = \vec{h} - \frac{1}{2}\nabla(x \cdot \vec{h} + f), \tag{9.26}$$

which satisfies

$$\nabla \cdot \vec{w} = 0, \tag{9.27}$$

namely, the incompressibility of the medium. From (9.15), we have

$$\lim_{\nu\to 1/2}\frac{1}{1-2\nu}(\nabla \cdot \vec{w}) = \lim_{\nu\to 1/2}\frac{1}{1-2\nu}\varepsilon_{ii} = -\frac{3}{E}p_M = -\frac{1}{\mu}p_M.$$

We drop the subscript M and just call p the pressure. Accordingly, as $\nu \to \frac{1}{2}$, the linear elastostatic equation becomes the Stokes flow

$$\left.\begin{aligned}&\mu\Delta\vec{w} - \nabla p = 0,\\ &\nabla \cdot \vec{w} = 0,\end{aligned}\right\} \tag{9.28}$$

where both $\vec{w}$ and p must be solved. By noting from (9.28) that

$$\nabla p = \mu \Delta \vec{w} = \mu \Delta \vec{h} - \frac{1}{2}\mu \nabla \Delta (x \cdot \vec{h} + f) = -\mu \nabla (\nabla \cdot \vec{h}),$$

so we obtain a representation for p:

$$p = -\mu \nabla \cdot \vec{h} + c \quad \text{for some arbitrary constant c.} \tag{9.29}$$

The Stokes equations (9.28) are useful in studying a nonlinear incompressible slow viscous flow with small Reynolds number:

$$\left.\begin{aligned} &\Delta \vec{v} - \nabla p = R\left(\vec{v} \cdot \nabla\right) \vec{v} \text{ on } \Omega^c, \\ &\vec{v} = 0 \text{ on } \partial\Omega; \quad \vec{v} \to e_1 \text{ as } |x| \to \infty \end{aligned}\right\} \tag{9.30}$$

by a singular perturbation approach using a combination of matched asymptotic expansions and strained coordinates. Note that in (9.30), R denotes the Reynolds number and e_1 is the unit vector in the positive x_1 direction.

We now proceed to calculate the fundamental solution tensor $E(x-\xi)$, originally due to Lord Kelvin, of the linear elastostatic equation, satisfying

$$\mu \Delta_\xi E(x-\xi) + (\lambda+\mu)\nabla_\xi [\nabla_\xi \cdot E(x-\xi)] = -\delta(x-\xi) I_3, \quad x, \xi \in \mathbb{R}^3. \tag{9.31}$$

From our understanding of the fundamental solution (6.13) of the Laplacian Δ in $\mathbb{R}^3$, we assume that $E(x-\xi)$ here in (9.31) should also have a singularity of the form $r_{x\xi}^{-1} \equiv |x-\xi|^{-1}$. Therefore we try a Papkovich–Neuber representation (9.19) with

$$\vec{h}(x-\xi) = \left(r_{x\xi}^{-1}, 0, 0\right)^{\mathrm{Tr}}. \tag{9.32}$$

To simplify notation, we translate the location of the singularity to the origin by letting $\xi = 0$, i.e.,

$$\vec{h}(x) = \left(r^{-1}, 0, 0\right)^{\mathrm{Tr}}, \quad r = |x|. \tag{9.33}$$

Corresponding to (9.19), we use (9.33) with $f \equiv 0$ and obtain

$$\vec{w}^*(x) = \left(r^{-1}, 0, 0\right)^{\mathrm{Tr}} - \kappa \nabla (x_1 r^{-1}) \tag{9.34}$$

$$= \left((1-\kappa) r^{-1} + \kappa x_1^2 r^{-3}, \kappa x_1 x_2 r^{-3}, \kappa x_1 x_3 r^{-3}\right)^{\mathrm{Tr}}. \tag{9.35}$$

Thus $\vec{w}$ satisfies (9.8) everywhere except at $x = 0$, where there is a singularity (caused by a concentrated force at $x = 0$ acting in the x_1 direction.) To calculate the magnitude of the concentrated force, we evaluate the integral

$$\int_{B_\varepsilon} \left\{ \mu \Delta \vec{w}^*(x) + (\lambda+\mu) \nabla \left[\nabla \cdot \vec{w}^*(x)\right] \right\} dx \tag{9.36}$$

over a small ball of radius ε centered at the origin. First,

$$(9.36) = \mu \int_{B_\varepsilon} \left\{ \Delta \vec{w}^*(x) + \frac{1}{1-2\nu} \nabla[\nabla \cdot \vec{w}^*(x)] \right\} dx. \tag{9.37}$$

Second,

$$\begin{aligned}\nabla \cdot \vec{w}^*(x) &= \frac{\partial}{\partial x_1}\left[(1-\kappa)r^{-1} + \kappa x_1^2 r^{-3}\right] + \frac{\partial}{\partial x_2}(\kappa x_1 x_2 r^{-3}) + \frac{\partial}{\partial x_3}(\kappa x_1 x_3 r^{-3}) \\ &= (-1+2\kappa)x_1 r^{-3}.\end{aligned}$$

Third, the three components of $\nabla(\nabla \cdot \vec{w}^*)$ are

$$\nabla\left(\nabla \cdot \vec{w}^*(x)\right) = \\ \left(\frac{\partial}{\partial x_1}\left[(-1+2\kappa)x_1 r^{-3}\right], \frac{\partial}{\partial x_2}\left[(-1+2\kappa)x_1 r^{-3}\right], \frac{\partial}{\partial x_3}\left[(-1+2\kappa)x_1 r^{-3}\right]\right)^{\mathrm{Tr}}.$$

Now we can evaluate the three components of the RHS of (9.37):

$$\begin{aligned}&\text{first component} \\ &= \mu \int_{B_\varepsilon} \left\{ \Delta\left[(1-\kappa)r^{-1} + \kappa x_1^2 r^{-3}\right] + \frac{1}{1-2\nu}\frac{\partial}{\partial x_1}\left[(-1+2\kappa)x_1 r^{-3}\right] \right\} dx \\ &= \mu \int_{\partial B_\varepsilon} \left\{ \frac{\partial}{\partial r}\left[(1-\kappa)r^{-1} + \kappa x_1^2 r^{-3}\right] + \frac{-1+2\kappa}{1-2\nu} n_1 x_1 r^{-3} \right\} d\sigma, \end{aligned} \tag{9.38}$$

where we have applied the Green's formula and n_1 is the first component of the unit exterior normal $n = (n_1, n_2, n_3)$. Using the spherical coordinates

$$\left.\begin{aligned} &x_1 = r\sin\theta\cos\phi, \ d\sigma = r^2 \sin\theta\, d\phi\, d\theta, \\ &x_2 = r\sin\theta\sin\phi, \ n_1 = \sin\theta\cos\phi, \\ &x_3 = r\cos\theta, \qquad 0 \leqslant \theta \leqslant \pi, \qquad 0 \leqslant \phi \leqslant 2\pi, \end{aligned}\right\} \tag{9.39}$$

we obtain the RHS of (9.38) as

$$\text{first component} \tag{9.40}$$

$$\begin{aligned} &= \mu \int_0^{2\pi}\int_0^{\pi} \left[-(1-\kappa) - \kappa \sin^2\theta\cos^2\phi\right] \sin\theta\, d\theta\, d\phi \\ &\qquad + \frac{(-1+2\kappa)}{1-2\nu}\int_0^{2\pi}\int_0^{2\pi} \sin^3\theta\cos^2\phi\, d\theta\, d\phi, \end{aligned} \tag{9.41}$$

$$= \mu\left[-4\pi(1-\kappa) - \kappa\frac{4\pi}{3} + \frac{-1+2\kappa}{1-2\nu}\frac{4\pi}{3}\right] \tag{9.42}$$

$$= -4\pi\mu, \ \text{independent of } \varepsilon. \tag{9.43}$$

Similarly, it is easy to evaluate that

second component of (9.37) = 0,

third component of (9.37) = 0.

Therefore, in order to get the magnitude of concentrated force to be -1 (as the RHS of (9.31) shows), we adjust the coefficient of $\vec{w}^*$ in (9.34) by a factor of $1/(4\pi\mu)$ to get the first column vector of the fundamental solution tensor $E(x)$. Similarly, by choosing

$$\vec{h}(x) = (0, r^{-1}, 0)^{\mathrm{Tr}} \text{ and } (0, 0, r^{-1})^{\mathrm{Tr}}$$

successively, we obtain the second and third row vectors of $E(x)$. Assembling them together, we get

$$\begin{aligned} E(x) &= \frac{1}{4\pi\mu}\left[(1-\kappa)r^{-1}\delta_{ij} + \kappa x_i x_j r^{-3}\right]_{1\leqslant i,j\leqslant 3} \\ &= \frac{\lambda+3\mu}{8\pi\mu(\lambda+2\mu)}\frac{1}{r}\left(I_3 + \frac{\lambda+\mu}{\lambda+3\mu}\frac{1}{r^2}xx^{\mathrm{Tr}}\right) \\ &\left(= \frac{1}{16\pi\mu(1-\nu)}\frac{1}{r}\left[(3-4\nu)\delta_{ij} + \frac{\partial r}{\partial x_i}\frac{\partial r}{\partial x_j}\right]_{1\leqslant i,j\leqslant 3}\right). \end{aligned} \tag{9.44}$$

Therefore

$$\begin{aligned} E(x,\xi) &= E(x-\xi) \\ &= \frac{\lambda+3\mu}{8\pi\mu(\lambda+2\mu)}\frac{1}{r_{x\xi}}\left[I_3 + \frac{\lambda+\mu}{\lambda+3\mu}\frac{1}{r_{x\xi}^2}(x-\xi)(x-\xi)^{\mathrm{Tr}}\right] \end{aligned} \tag{9.45}$$

is the fundamental solution tensor satisfying

$$\begin{aligned} \mu\Delta_x E(x,\xi) + (\lambda+\mu)\nabla_x[\nabla_x\cdot E(x,\xi)] &= \mu\Delta_\xi E(x,\xi) + (\lambda+\mu)\nabla_\xi[\nabla_\xi\cdot E(x,\xi)] \\ &= -\delta(x-\xi)I_3. \end{aligned}$$

Next, let us also derive (9.45) by the alternative method of Fourier transform. Let the space dimension be $N = 3$. Let $\xi = 0$ in (9.18) and consider

$$\mu\Delta E(x) + (\lambda+\mu)\nabla[\nabla\cdot E(x)] = -\delta(x)I_3. \tag{9.46}$$

Let $E_k(x)$ denote the kth row of $E(x)$ and denote

$$E_k(x) = (E_{k1}(x), E_{k2}(x), E_{k3}(x)).$$

Then

$$\mu\Delta E_k(x) + (\lambda+\mu)\nabla[\nabla\cdot E_k(x)] = -\delta(x)e_k, \tag{9.47}$$

where e_k is the unit (row) vector in the kth coordinate direction. Taking the Fourier transform of (9.47), we obtain

$$\mu(-4\pi|\xi|^2)\widehat{E}_{kj}(\xi) + (\lambda+\mu)(2\pi i\xi_j)(2\pi i)\left(\sum_{l=1}^{3}\xi_l\widehat{E}_{kl}\right) = -\delta_{kj}, \quad j = 1,2,3, \tag{9.48}$$

where $\xi = (\xi_1, \xi_2, \xi_3)$ is the Fourier transform variable. Multiply (9.48) by ξ_j and sum, resulting in

$$4\pi^2\left[-\mu|\xi|^2\sum_{j=1}^{3}\xi_j\widehat{E}_{kj}-(\lambda+\mu)|\xi|^2\sum_{l=1}^{3}\xi_l\widehat{E}_{kl}\right]=-\xi_k.$$

Thus

$$\sum_{j=1}^{3}\xi_j\widehat{E}_{kj}=\frac{\xi_k}{4\pi^2(\lambda+2\mu)|\xi|^2},\qquad k=1,2,3. \tag{9.49}$$

Substituting (9.49) into (9.48), we get

$$\widehat{E}_{kj}(\xi)=\frac{1}{4\pi^2\mu|\xi|^2}\left(\delta_{kj}-\frac{\lambda+\mu}{\lambda+2\mu}\frac{\xi_j\xi_k}{|\xi|^2}\right). \tag{9.50}$$

Now, take the inverse Fourier transform of the above. From (3.117),

$$\overline{\mathscr{F}}\left(\frac{1}{4\pi^2\mu}|\xi|^{-2}\delta_{kj}\right)=\frac{1}{4\pi\mu}|x|^{-1}\delta_{kj}. \tag{9.51}$$

Also, using $\lambda=-1$ and $N=3$ in (3.115), we obtain

$$\begin{aligned}\overline{\mathscr{F}}(|\xi|^{-4})&=\frac{\Gamma\left(-\frac{1}{2}\right)}{\pi^{-1-3/2}\Gamma\left(\frac{1}{2}(1+3)\right)}|x|\\&=\frac{-2\sqrt{\pi}}{\pi^{-5/2}\cdot 1}|x|\\&=-2\pi^3|x|.\end{aligned}$$

Applying (3.78), we have

$$\begin{aligned}\overline{\mathscr{F}}\left((2\pi i\xi_j)(2\pi i\xi_k)|\xi|^{-4}\right)&=\frac{\partial}{\partial x_j}\frac{\partial}{\partial x_k}(-2\pi^3|x|)\\&=-2\pi^3\left(\frac{\delta_{jk}}{|x|}-\frac{x_jx_k}{|x|^3}\right);\end{aligned}$$

thus

$$\overline{\mathscr{F}}\left(\frac{1}{4\pi^2\mu}\frac{\lambda+\mu}{\lambda+2\mu}\frac{\xi_j\xi_k}{|\xi|^4}\right)=\frac{1}{8\pi\mu}\frac{\lambda+\mu}{\lambda+2\mu}\left(\frac{\delta_{jk}}{|x|}-\frac{x_jx_k}{|x|^3}\right). \tag{9.52}$$

Taking the inverse Fourier transform of (9.50) and using (9.51) and (9.52), we obtain

$$E_{kj}(x)=\frac{1}{8\pi\mu}\left[\left(2-\frac{\lambda+\mu}{\lambda+2\mu}\right)\frac{\delta_{kj}}{|x|}+\frac{\lambda+\mu}{\lambda+2\mu}\frac{x_jx_k}{|x|^3}\right],\qquad 1\leqslant k,\ j\leqslant 3.$$

Making the change of variable $x\mapsto x-\xi$, we obtain (9.45).

Exercise 9.1. Show that in $\mathbb{R}^2$, the fundamental solution tensor $E_2(x,\xi)$ of the linear elastostatic equation, satisfying

$$\mu\Delta_\xi E_2(x,\xi)+(\lambda+\mu)\nabla_\xi[\nabla_\xi\cdot E_2(x,\xi)]=-\delta(x-\xi)I_2,$$

is given by

$$E_2(x,\xi)=\frac{\lambda+3\mu}{4\pi\mu(\lambda+2\mu)}\left[-(\ln r_{x\xi})I_2+\frac{\lambda+\mu}{\lambda+3\mu}\frac{1}{r_{x\xi}^2}(x-\xi)(x-\xi)^{\mathrm{Tr}}\right]. \tag{9.53}$$

□

Exercise 9.2. Show that in $\mathbb{R}^N$, with $N \geqslant 3$, the fundamental solution $E(x-\xi)$ is an $N \times N$ matrix whose (i,j)-entry is given by

$$E_{ij}(x-\xi) = \frac{c_1}{|\omega_N|(N-2)} \frac{\delta_{ij}}{|x-\xi|^{N-2}} + \frac{c_2}{|\omega_N|} \frac{(x_i-\xi_i)(x_j-\xi_j)}{|x-\xi|^N},$$

where $|\omega_N|$ is the surface area of the unit sphere in $\mathbb{R}^N$ and

$$c_1 = \frac{1}{2}\left(\frac{1}{\mu} + \frac{1}{2\mu+\lambda}\right), \quad c_2 = \frac{1}{2}\left(\frac{1}{\mu} - \frac{1}{2\mu+\lambda}\right). \qquad \square$$

9.3 BVP in linear elastostatics

Let an elastic solid occupy a medium Ω in $\mathbb{R}^3$ or $\mathbb{R}^2$. Ω is a bounded open domain with smooth boundary $\partial\Omega$. Let $\vec{w}(x)$ denote the displacement vector at x. For any infinitesimal element dx, the amount of strain energy contained therein (by neglecting a factor of $\frac{1}{2}$) is equal to

$$\begin{aligned}
&\text{stress} \cdot \text{strain} \cdot dx \\
&= \sum_{i,j} \sigma_{ij}(x)\varepsilon_{ij}(x)\,dx \\
&= \sum_{i,j} \{\lambda[\nabla \cdot \vec{w}(x)]\delta_{ij} + 2\mu\varepsilon_{ij}(x)\}\varepsilon_{ij}(x)\,dx \\
&= \left\{\lambda \sum_i [\nabla \cdot \vec{w}(x)]\varepsilon_{ii}(x) + 2\mu \sum_{i,j} \varepsilon_{ij}^2(x)\right\} dx \\
&= \left\{\lambda\left[\nabla \cdot \vec{w}(x)\right]^2 + \frac{\mu}{2}\sum_{i,j}\left[\frac{\partial w_i(x)}{\partial x_j} + \frac{\partial w_j(x)}{\partial x_i}\right]^2\right\} dx.
\end{aligned}$$

Therefore the total potential (strain) energy contained in the solid is equal to

$$\mathscr{E}(\vec{w}) \equiv \int_\Omega \left[\lambda(\nabla \cdot \vec{w})^2 + \frac{\mu}{2}\sum_{i,j=1}^N \left(\frac{\partial w_i}{\partial x_j} + \frac{\partial w_j}{\partial x_i}\right)^2\right] dx, \quad N = 2 \text{ or } 3. \tag{9.54}$$

Lemma 9.3. *Let $N = 3$. If $\mathscr{E}(\vec{w}) = 0$, then*

$$\vec{w}(x) = \sum_{j=1}^6 \alpha_j \vec{m}_j(x) \text{ for some } \alpha_j \in \mathbb{R}, \quad 1 \leqslant j \leqslant 6, \tag{9.55}$$

where $\vec{m}_j(x)$ is the jth column vector of the matrix

$$M(x) = \begin{bmatrix} 1 & 0 & -x_2 & 0 & 0 & x_3 \\ 0 & 1 & x_1 & 0 & -x_3 & 0 \\ 0 & 0 & 0 & 1 & x_2 & -x_1 \end{bmatrix}. \tag{9.56}$$

When $N=2$ and if $\mathscr{E}(\vec{w})=0$, then

$$\vec{w}(x)=\sum_{j=1}^{3}\alpha_j\vec{m}_j(x) \text{ for some } \alpha_j\in\mathbb{R},\ 1\leqslant j\leqslant 3, \tag{9.57}$$

where it is understood that the third components of $\vec{m}_j(x)$, $1\leqslant j\leqslant 3$, are ignored.

Proof. We need only consider $N=3$. Let $\vec{w}\in\mathscr{D}'(\Omega)$ be a distribution such that $\mathscr{E}(\vec{w})=0$. Then for i, j with $1\leqslant i,\ j\leqslant 3$,

$$\frac{\partial w_i}{\partial x_j}+\frac{\partial w_j}{\partial x_i}=0,$$

i.e.,

$$\frac{\partial w_1}{\partial x_1}=0,\quad \frac{\partial w_2}{\partial x_2}=0,\quad \frac{\partial w_3}{\partial x_3}=0, \tag{9.58}$$

$$\frac{\partial w_1}{\partial x_2}+\frac{\partial w_2}{\partial x_1}=0,\quad \frac{\partial w_1}{\partial x_3}+\frac{\partial w_3}{\partial x_1}=0,\quad \frac{\partial w_2}{\partial x_3}+\frac{\partial w_3}{\partial x_2}=0. \tag{9.59}$$

Therefore

$$\nabla w_1=\left(\frac{\partial w_1}{\partial x_1},\frac{\partial w_1}{\partial x_2},\frac{\partial w_1}{\partial x_3}\right)=\left(0,-\frac{\partial w_2}{\partial x_1},-\frac{\partial w_3}{\partial x_1}\right).$$

From

$$\nabla\times(\nabla w_1)=0,$$

we get

$$\frac{\partial^2 w_2}{\partial x_1\partial x_3}-\frac{\partial^2 w_3}{\partial x_1\partial x_2}=0, \tag{9.60}$$

$$\frac{\partial^2 w_3}{\partial x_1^2}=0, \tag{9.61}$$

$$\frac{\partial^2 w_2}{\partial x_1^2}=0. \tag{9.62}$$

From (9.60), we get

$$\frac{\partial w_2}{\partial x_3}-\frac{\partial w_3}{\partial x_2}=F(x_1), \tag{9.63}$$

for some $F\in\mathscr{D}'(\Omega)$ depending on x_1 only. From (9.63) and the third equation in (9.59), we have

$$\frac{\partial w_2}{\partial x_3}=\frac{1}{2}F(x_1),$$

$$\frac{\partial w_3}{\partial x_2}=-\frac{1}{2}F(x_1). \tag{9.64}$$

Combining (9.61), (9.64) and the third equation in (9.59), we easily deduce that

$$w_3(x) = \gamma_0 + \gamma_1 x_1 + \gamma_2 x_2 + \gamma_3 x_1 x_2, \tag{9.65}$$

for some constants $\gamma_j \in \mathbb{R}$, $j = 0, 1, 2, 3$. From symmetry of the indices, we also obtain

$$w_1(x) = \alpha_0 + \alpha_1 x_2 + \alpha_2 x_3 + \alpha_3 x_2 x_3, \tag{9.66}$$

$$w_2(x) = \beta_0 + \beta_1 x_3 + \beta_2 x_1 + \beta_3 x_3 x_1, \tag{9.67}$$

for some constants $\alpha_j, \beta_j \in \mathbb{R}$, $j = 0, 1, 2, 3$. From (9.59) and (9.65)–(9.67), we get

$$\begin{aligned} \alpha_1 + \alpha_3 x_3 + \beta_2 + \beta_3 x_3 &= 0, \\ \alpha_2 + \alpha_3 x_2 + \gamma_1 + \gamma_3 x_2 &= 0, \\ \beta_1 + \beta_3 x_1 + \gamma_2 + \gamma_3 x_1 &= 0. \end{aligned}$$

The solutions to this linear system are characterized by

$$\left. \begin{aligned} &\alpha_3 = \beta_3 = \gamma_3 = 0, \\ &\begin{bmatrix} \alpha_1 \\ \beta_1 \\ \gamma_1 \end{bmatrix} = - \begin{bmatrix} \beta_2 \\ \gamma_2 \\ \alpha_2 \end{bmatrix}. \end{aligned} \right\} \tag{9.68}$$

Also note that in (9.65)–(9.67), the constants α_0, β_0 and γ_0 are arbitrary. By (9.68), we conclude (9.57). □

Remark 9.1. We can also write (9.57) as

$$\vec{w}(x) = \vec{\alpha} + \vec{\beta} \times \vec{x}, \tag{9.69}$$

where $\vec{\alpha}$ and $\vec{\beta}$ are two arbitrary constant vectors in $\mathbb{R}^3$. Lemma 9.3 says that two deformations differing by a rigid body motion (i.e., translation and rotation) have equal potential energy. □

Next, we state without proof, the Korn inequality. The interested reader can refer to [70, p. 110] for a proof.

Theorem 9.1. *Let Ω be a bounded open domain in $\mathbb{R}^N$ with regular boundary $\partial\Omega$. Let $v_j \colon \Omega \to \mathbb{R}$, $1 \leqslant j \leqslant N$ and $\vec{v} = (v_1, \ldots, v_N)$, and $\vec{v}|_{\partial\Omega} = 0$. Then there exists a constant $C > 0$ independent of $\vec{v}$ such that*

$$\sum_{i,j} \int_\Omega \varepsilon_{i,j}^2(v)\,dx + \sum_i \int_\Omega v_i^2\,dx \geqslant C \| \vec{v} \|_{[H^1(\Omega)]^N}^2, \tag{9.70}$$

where

$$\varepsilon_{ij}(v) \equiv \frac{1}{2}\left(\frac{\partial v_i}{\partial x_j} + \frac{\partial v_j}{\partial x_i}\right), \tag{9.71}$$

$$\| \vec{v} \|_{[H^1(\Omega)]^N}^2 \equiv \sum_{i,j} \int_\Omega \frac{\partial v_i^2}{\partial x_j}\,dx + \sum_i \int_\Omega v_i^2\,dx. \tag{9.72}$$

□

Remark 9.2. Korn's inequality (9.70) is not trivial, since the LHS of (9.70) contains only *certain combinations* of first-order derivatives, while the RHS of (9.70) involves *all* first-order derivatives. □

For $\vec{w}, \vec{v} \in [H^1(\Omega)]^N$, $N = 3$ or 2, define a bilinear form

$$a(\vec{w}, \vec{v}) = \int_\Omega \left[\lambda(\nabla \cdot \vec{w})(\nabla \cdot \vec{v}) + \frac{\mu}{2}\sum_{i,j}\left(\frac{\partial w_i}{\partial x_j} + \frac{\partial w_j}{\partial x_i}\right)\left(\frac{\partial v_i}{\partial x_j} + \frac{\partial v_j}{\partial x_i}\right)\right] dx. \tag{9.73}$$

Then $a(\vec{w}, \vec{w}) = \mathscr{E}(\vec{w})$, and a induces only a *seminorm* in $[H^1(\Omega)]^N$. But if we take the algebraic quotient $[H^1(\Omega)]^N$ modulo the linear span of column vectors of $M(x)$, i.e.,

$$H^1_{N,\text{mod}\mathscr{M}}(\Omega) \equiv \left[H^1(\Omega)\right]^N / \mathscr{M}, \ \mathscr{M} = \text{span}\left\{m_j(x) \mid 1 \leqslant j \leqslant 3(N-1)\right\}, \tag{9.74}$$

with the induced norm

$$\|[\vec{v}]\|_{\text{ind}} = \inf_{\vec{m} \in \mathscr{M}} \|\vec{v} + \vec{m}\|_{[H^1(\Omega)]^N}, \tag{9.75}$$

$$\text{for any coset} \quad [\vec{v}] = \vec{v} + \mathscr{M}, \ \vec{v} \in [H^1(\Omega)]^N,$$

then this norm is equivalent to the norm induced by the bilinear form a.

Theorem 9.2. *The bilinear form $\tilde{a}$ defined on $H^1_{N,mod\mathscr{M}}(\Omega)$ by*

$$\tilde{a}([\vec{w}], [\vec{v}]) \equiv a(\vec{w}, \vec{v})$$

for any cosets $[\vec{w}] = \vec{w} + \mathscr{M}$, $[\vec{v}] = \vec{v} + \mathscr{M}$, $\vec{w}, \vec{v} \in [H^1(\Omega)]^N$, is an inner product with which $H^1_{N,mod}(\Omega)$ is a Hilbert space. $\tilde{a}$ defines a norm equivalent to the norm $\| \quad \|_{ind}$ in (9.75).

Proof. It is routine to verify that $H^1_{N,\text{mod}\mathscr{M}(\Omega)}$ is a Hilbert space equipped with the inner product $\tilde{a}$. Now consider the linear transformation

$$\mathscr{L}\colon \left[H^1(\Omega)\right]^N \to H^1_{N,\text{mod}\mathscr{M}}(\Omega),$$

equipped with the inner product $\tilde{a}$,

$$\mathscr{L}(\vec{w}) = [\vec{w}],$$

Because

$$\tilde{a}([\vec{w}], [\vec{w}]) = a(\vec{w}, \vec{w}) \leqslant C(\lambda, \mu)\|\vec{w}\|^2_{[H^1(\Omega)]^N}$$

for some constant C depending on λ and μ only, we see that $\mathscr{L}$ is continuous. It is onto, with

$$\mathscr{N}(\mathscr{L}) = \mathscr{M}.$$

Therefore, $\mathscr{L}$ induces an (algebraic and topological) isomorphism between Banach spaces

$$\widetilde{\mathscr{L}}\colon\ [H^1(\Omega)]^N/\mathscr{M} \longrightarrow H^1_{N,\mathrm{mod}.\mathscr{M}}(\Omega),$$

equipped with the inner product $\widetilde{a}$. Thus there exist constants $C_1(\lambda,\mu)$ and $C_2(\lambda,\mu)$ depending on λ and μ only, such that

$$C_1(\lambda,\mu)[a(\vec{v},\vec{v})]^{1/2} \leqslant \|[\vec{v}]\|_{\mathrm{ind}} \leqslant C_2(\lambda,\mu)[a(\vec{v},\vec{v})]^{1/2}, \quad \forall\, \vec{v} \in \left[H^1(\Omega)\right]^N.$$

□

Therefore, the energy bilinear form a in (9.73) is (essentially) an inner product on $[H^1(\Omega)]^N$, modulo $\mathscr{M}$. From now on, we consider $H^1_{N,\mathrm{mod}.\mathscr{M}}(\Omega)$ the same as $[H^1(\Omega)]^N$, with inner product a defined on $[H^1(\Omega)]^N$.

Let us now consider variational type boundary conditions. For a solid with displacement field $\vec{w}(x)$ on Ω, the strain energy contained therein is $\mathscr{E}(\vec{w})$ (cf. (9.54)). From the principle of virtual work,

$$\begin{aligned} 0 &= \delta\mathscr{E}(\vec{w}) \\ &= 2\int_\Omega \left[\lambda(\nabla\cdot\vec{w})(\nabla\cdot\vec{\delta w}) + \frac{\mu}{2}\sum_{j,k=1}^N \left(\frac{\partial w_j}{\partial x_k}+\frac{\partial w_k}{\partial x_j}\right)\left(\frac{\partial \delta w_j}{\partial x_k}+\frac{\partial \delta w_k}{\partial x_j}\right)\right]dx. \end{aligned} \tag{9.76}$$

We note that

$$\begin{aligned} &\frac{\mu}{2}\sum_{j,k=1}^N \left(\frac{\partial w_j}{\partial x_k}+\frac{\partial w_k}{\partial x_j}\right)\left(\frac{\partial \delta w_j}{\partial x_k}+\frac{\partial \delta w_k}{\partial x_j}\right) \\ &= \frac{\mu}{2}\sum_{j,k=1}^N \left[\left(\frac{\partial w_j}{\partial x_k}-\frac{\partial w_k}{\partial x_j}\right)+2\frac{\partial w_k}{\partial x_j}\right]\left[\left(\frac{\partial \delta w_k}{\partial x_j}-\frac{\partial \delta w_j}{\partial x_k}\right)+2\frac{\partial \delta w_j}{\partial x_k}\right] \\ &= \frac{\mu}{2}\cdot\sum_{j,k}\left(\frac{\partial w_j}{\partial x_k}-\frac{\partial w_k}{\partial x_j}\right)\left(\frac{\partial \delta w_k}{\partial x_j}-\frac{\partial \delta w_j}{\partial x_k}\right) \\ &\quad+\frac{\mu}{2}\cdot 2\cdot\sum_{j,k}\frac{\partial w_k}{\partial x_j}\left(\frac{\partial \delta w_k}{\partial x_j}-\frac{\partial \delta w_j}{\partial x_k}\right)+\frac{\mu}{2}\cdot 2\cdot\sum_{j,k}\frac{\partial \delta w_j}{\partial x_k}\left(\frac{\partial w_j}{\partial x_k}-\frac{\partial w_k}{\partial x_j}\right) \\ &\quad+\frac{\mu}{2}\cdot 4\cdot\sum_{j,k}\frac{\partial w_k}{\partial x_j}\frac{\partial \delta w_j}{\partial x_k} \\ &= -\frac{\mu}{2}\cdot 2(\nabla\times\vec{w})\cdot(\nabla\times\vec{\delta w})+\frac{\mu}{2}\cdot 4\cdot\sum_{j,k}\frac{\partial w_k}{\partial x_j}\frac{\partial \delta w_k}{\partial x_j} \\ &= 2\mu\nabla\vec{w}\cdot\nabla\vec{\delta w}-\mu(\nabla\times\vec{w})\cdot(\nabla\times\delta\vec{w}). \end{aligned}$$

Therefore

$$
\begin{aligned}
2\cdot\frac{\mu}{2}\cdot\int_\Omega&\left[\sum_{j,k}\left(\frac{\partial w_j}{\partial x_k}+\frac{\partial w_k}{\partial x_j}\right)\left(\frac{\partial\delta w_j}{\partial x_k}+\frac{\partial\delta w_k}{\partial x_j}\right)\right]dx\\
&=2\mu\int_\Omega[2\nabla\vec{w}\cdot\nabla\delta\vec{w}-(\nabla\times\vec{w})\cdot(\nabla\times\delta\vec{w})]\,dx\\
&=2\mu\int_{\partial\Omega}\left[2\frac{\partial\vec{w}}{\partial n}\cdot\delta\vec{w}+n\times(\nabla\times\vec{w})\cdot\delta\vec{w}\right]d\sigma\\
&\quad-4\mu\int_\Omega\Delta\vec{w}\cdot\delta\vec{w}\,dx-2\mu\int_\Omega\nabla\times(\nabla\times\vec{w})\cdot\delta\vec{w}\,dx. \qquad (9.77)
\end{aligned}
$$

Also

$$
2\lambda\int_\Omega[(\nabla\cdot\vec{w})(\nabla\cdot\delta\vec{w})]\,dx=2\lambda\int_\Omega(\nabla\cdot\vec{w})n\cdot\delta\vec{w}\,d\sigma-2\lambda\int_\Omega\nabla(\nabla\cdot\vec{w})\cdot\delta\vec{w}\,dx. \qquad (9.78)
$$

Now using

$$
\nabla\times(\nabla\times)=\nabla(\nabla\cdot)-\Delta
$$

in the last term on the RHS of (9.77) and combining (9.76), (9.77) and (9.78), we obtain

$$
\begin{aligned}
0&=\delta\mathscr{E}(\vec{w})\\
&=-2\int_\Omega[\mu\Delta\vec{w}+(\lambda+\mu)\nabla(\nabla\cdot\vec{w})]\cdot\delta\vec{w}\,dx+2\int_{\partial\Omega}\vec{\tau}(\vec{w})\cdot\delta\vec{w}\,d\sigma,
\end{aligned}
\qquad (9.79)
$$

where

$$
\begin{aligned}
\vec{\tau}(\vec{w})&\equiv\lambda(\nabla\cdot\vec{w})n+2\mu\frac{\partial\vec{w}}{\partial n}+\mu n\times(\nabla\times\vec{w})\\
&=\lambda(\nabla\cdot\vec{w})n+\mu(\nabla\vec{w}+\nabla\vec{w}^{\mathrm{Tr}})n, \qquad (9.80)
\end{aligned}
$$

called the *traction vector*, signifies the surface force at a point $x\in\partial\Omega$.

The above derivations also give

$$
\int_\Omega[\mu\Delta\vec{w}+(\lambda+\mu)\nabla(\nabla\cdot\vec{w})]\cdot\vec{v}\,dx=\int_{\partial\Omega}\vec{\tau}(\vec{w})\cdot\vec{v}\,d\sigma-a(\vec{w},\vec{v}) \qquad (9.81)
$$

and

$$
\begin{aligned}
\int_\Omega\Big\{\big[\mu\Delta\vec{w}+(\lambda+\mu)\nabla(\nabla\cdot\vec{w})\big]\cdot\vec{v}-\big[\mu\Delta\vec{v}+(\lambda+\mu)\nabla(\nabla\cdot\vec{v})\big]\cdot\vec{w}\Big\}\,dx\\
=\int_{\partial\Omega}\left[\vec{\tau}(\vec{w})\cdot\vec{v}-\vec{\tau}(\vec{v})\cdot\vec{w}\right]d\sigma, \qquad (9.82)
\end{aligned}
$$

which can be regarded as generalizations of the Gauss–Green integration by parts formulas.

Exercise 9.3. Show that

$$
\vec{\tau}(\vec{w})(x)=(\tau_1(x),\tau_2(x),\tau_3(x))^{\mathrm{Tr}},
$$

the three components of $\vec{\tau}$, satisfy

$$
\tau_j(x)=\sum_{i=1}^{3}\sigma_{ij}(x)n_i(x),\quad j=1,2,3.
$$

□

Remark 9.3. When $N = 2$, the traction vector $\vec{\tau}$ in (9.80) has only two components:

$$\vec{\tau}(\vec{w}) = \left[\lambda\left(\frac{\partial w_1}{\partial x_1} + \frac{\partial w_2}{\partial x_2}\right)^{n_1} + 2\mu\frac{\partial w_1}{\partial n} + \mu n_2\left(\frac{\partial w_2}{\partial x_1} - \frac{\partial w_1}{\partial x_2}\right),\right.$$
$$\left.\lambda\left(\frac{\partial w_1}{\partial x_1} + \frac{\partial w_2}{\partial x_2}\right)^{n_2} + 2\mu\frac{\partial w_2}{\partial n} - \mu n_1\left(\frac{\partial \mu w_2}{\partial x_1} - \frac{\partial w_1}{\partial x_2}\right)\right]^{\mathrm{Tr}}. \qquad \square$$

From the variational equation (9.79), we have therefore recovered the partial differential equation

$$\mu\Delta\vec{w} + (\lambda+\mu)\nabla(\nabla\cdot\vec{w}) = 0 \tag{9.83}$$

(obtained by picking $\overrightarrow{\delta w} \in [\mathscr{D}(\Omega)]^3$). We have also obtained two basic types of variational boundary conditions:

$$\text{(Dirichlet type)} \quad \vec{w}(x) \equiv \vec{g}(x) \quad \text{is given on } \partial\Omega, \tag{9.84}$$

or

$$\text{(Neumann type)} \quad \vec{\tau}(\vec{w}) \equiv \vec{g}(x) \quad \text{is given on } \partial\Omega; \tag{9.85}$$

for $\vec{w}(x)$ satisfying the partial differential equation (9.83). Of course we can have a mixture of both boundary conditions on $\partial\Omega$; namely, (9.84) is prescribed on some part Γ_1 of $\partial\Omega$, and (9.85) is prescribed on the remaining part $\Gamma_2 = \partial\Omega \setminus \Gamma_1$ of $\partial\Omega$.
Also, in other applications such as crack and fracture analysis, it is common to have mixed boundary conditions

$$w_j(x) = g_j(x), \quad j \in I \subseteq \{k \mid 1 \leqslant k \leqslant N,\ k \in \mathbb{Z}\},$$
$$\tau_j(x) = g_j(x), \quad j \in \{k \mid 1 \leqslant k \leqslant N,\ k \in \mathbb{Z}\} \setminus I,$$

prescribed on various parts of the boundary. Such a type of boundary conditions is also variational in nature. But in the subsequent development, our main attention will be directed toward (9.84) and (9.85).

9.4 The Betti–Somigliana formula. Simple- and double-layers

Let $\vec{w}^{(1)}(x)$ and $\vec{w}^{(2)}(x)$ be two elastostatic displacement fields satisfying

$$\mu\Delta\vec{w}^{(k)} + (\lambda+\mu)\nabla(\nabla\cdot\vec{w}^{(k)}) = \vec{f}^{(k)} \quad \text{on } \Omega,\ k = 1, 2, \tag{9.86}$$

with corresponding stress fields $\sigma_{ij}^{(1)}$ and $\sigma_{ij}^{(2)}$ satisfying

$$\sum_j \frac{\partial\sigma_{ij}^{(k)}}{\partial x_j} = f_i^{(k)}, \quad 1 \leqslant i \leqslant N, \tag{9.87}$$

and with corresponding traction vectors $\vec{\tau}(\vec{w}^{(1)})$ and $\vec{\tau}(\vec{w}^{(2)})$ on $\partial\Omega$. Then a simple integration by parts yields

$$0=\sum_i \int_\Omega \left[\left(\sum_j \frac{\partial \sigma_{ij}^{(1)}}{\partial x_j} - f_i^{(1)}\right) w_i^{(2)} - \left(\sum_j \frac{\partial \sigma_{ij}^{(2)}}{\partial x_j} - f_i^{(2)}\right) w_i^{(1)}\right] dx$$
$$= -\int_\Omega \left[\vec{f}^{(1)}\cdot\vec{w}^{(2)} - \vec{f}^{(2)}\cdot\vec{w}^{(1)}\right] dx + \int_{\partial\Omega} \left[\vec{\tau}(\vec{w}^{(1)})\cdot\vec{w}^{(2)} - \vec{\tau}(\vec{w}^{(2)})\cdot\vec{w}^{(1)}\right] d\sigma, \tag{9.88}$$

where we have utilized the fact that

$$\sum_{i,j}\left(\sigma_{ij}^{(1)}\frac{\partial w_i^{(2)}}{\partial x_j} - \sigma_{ij}^{(2)}\frac{\partial w_i^{(1)}}{\partial x_j}\right) = 0$$

from (9.3). Thus (9.88) gives

$$\int_{\partial\Omega} (\vec{\tau}^{(1)}\cdot\vec{w}^{(2)} - \vec{\tau}^{(2)}\cdot\vec{w}^{(1)})\, d\sigma = \int_\Omega (\vec{f}^{(1)}\cdot\vec{w}^{(2)} - \vec{f}^{(2)}\cdot\vec{w}^{(1)})\, dx. \tag{9.89}$$

When the equations (9.87) for $k=1,2$ are source-free (i.e., $\vec{f}^{(k)}\equiv 0$), we get *Betti's reciprocal relation*

$$\int_{\partial\Omega} (\vec{\tau}^{(1)}\cdot\vec{w}^{(2)} - \vec{\tau}^{(2)}\cdot\vec{w}^{(1)})\, d\sigma = 0.$$

If instead we pick

$$\vec{w}^{(1)} = \vec{w}, \quad \text{with } \vec{f}^{(1)} \equiv 0 \text{ in (9.87)},$$

and

$$\vec{w}^{(2)} = \vec{E}^{1}, \quad \text{the first column vector in the fundamental matrix tensor } E(x-\xi) \text{ in (9.45)},$$

then $\vec{E}^{1}(x-\xi)$ satisfies

$$\mu\Delta_\xi \vec{E}^{1}(x-\xi) + (\lambda+\mu)\nabla_\xi[\nabla_\xi\cdot\vec{E}^{1}(x-\xi)] = \begin{bmatrix} -\delta(x-\xi) \\ 0 \\ 0 \end{bmatrix} \left(\equiv \vec{f}^{(2)}(x-\xi)\right).$$

Substituting the above into (9.89), with the understanding that the expression (9.89) remains valid because its derivation from integration by parts can be done through a limiting process, we obtain

$$\int_{\partial\Omega} [\vec{\tau}(\vec{w})(\xi)\cdot\vec{E}^{1}(x-\xi) - \vec{w}(\xi)\cdot\vec{\tau}_\xi(\vec{E}^{1})(x-\xi)]\, d\sigma_\xi = w_1(x), x\in\Omega.$$

Repeating the above procedures for $\vec{E}^2(x-\xi)$ and $\vec{E}^3(x-\xi)$ and assembling them together, we obtain *Betti–Somigliana formula*

$$\vec{w}(x) = \int_{\partial\Omega} [E(x-\xi)\vec{\tau}(\vec{w})(\xi) - \vec{\tau}_{\xi}^{\mathrm{Tr}}(E)(x-\xi)\vec{w}(\xi)]\,d\sigma_{\xi}. \tag{9.90}$$

It is straightforward to verify that the traction field $\vec{\tau}_{\xi}^{\mathrm{Tr}}(E)(x-\xi)$ is given by

$$\begin{aligned}\vec{\tau}_{\xi}^{\mathrm{Tr}}(E)(x-\xi) &= T_{\xi}(x-\xi)\\ &\equiv \frac{\mu}{2(N-1)\pi(\lambda+2\mu)}\left\{\left[I_N + \frac{N(\lambda+\mu)}{\mu|x-\xi|^2}(x-\xi)(x-\xi)^{\mathrm{Tr}}\right]\frac{\partial\gamma(x-\xi)}{\partial n_{\xi}}\right.\\ &\quad\left. - \frac{1}{|x-\xi|^N}\{(x-\xi)n_{\xi}^{\mathrm{Tr}} - [(x-\xi)n_{\xi}^{\mathrm{Tr}}]^{\mathrm{Tr}}\}\right\},\quad N=3 \text{ or } 2,\end{aligned} \tag{9.91}$$

where

$$\gamma(x-\xi) \equiv \begin{cases} |x-\xi|^{-1}, & N=3,\\ -\ln|x-\xi|, & N=2.\end{cases} \tag{9.92}$$

Relation (9.90) parallels the scalar relation (1.6). It is thus natural for us to call expressions

$$\int_{\partial\Omega} E(x-\xi)\vec{f}_1(\xi)\,d\sigma_{\xi} \quad \left(\text{such as } \int_{\partial\Omega} E(x-\xi)\vec{\tau}(\vec{w})(\xi)\,d\sigma_{\xi} \text{ on the RHS of (9.92)}\right),$$

$$\int_{\partial\Omega} T_{\xi}(x-\xi)\vec{f}_2(\xi)\,d\sigma_{\xi} \quad \left(\text{such as } \int_{\partial\Omega} \vec{\tau}_{\xi}^{\mathrm{Tr}}(E)(x-\xi)\vec{w}(\xi)\,d\sigma_{\xi} \text{ on the RHS of (9.91)}\right)$$

simple- and double-layer elastostatic potentials, respectively.

These simple- and double-layer elastostatic potentials behave near the boundary $\partial\Omega$ very much like the ordinary harmonic potentials that have been studied in Chapters 6 and 7, for example. Therefore here we will only restate their main boundary properties without giving proofs. For a given vector valued function $\vec{f}: \partial\Omega \to \mathbb{R}^N$, $N=3$ or 2, define

$$\left.\begin{aligned} V_1^{\pm}(\vec{f})(x) &= \int_{\partial\Omega} E(x-\xi)\vec{f}(\xi)\,d\sigma_{\xi},\\ V_2^{\pm}(\vec{f})(x) &= \int_{\partial\Omega} T_{\xi}(x-\xi)\vec{f}(\xi)\,d\sigma_{\xi},\end{aligned}\right\} \text{ with } \begin{cases} + \text{ for } x \in \Omega^c,\\ - \text{ for } x \in \Omega.\end{cases} \tag{9.93}$$

Then we have the following.

Property 9.1. *Let $\vec{f} \in [C^0(\partial\Omega)]^N$. Then $V_1^+(\vec{f})$ and $V_1^-(\vec{f})$ are extendable to $\overline{\Omega^c}$ and $\overline{\Omega}$ respectively and are continuous across $\partial\Omega$, such that $V_1(\vec{f})$, the combined extended functions, is uniformly Hölder-continuous throughout $\mathbb{R}^N$, satisfying*

$$\|V_1(\vec{f})\|_{[C^{0,\alpha}(\mathbb{R}^N)]^N} \leqslant C_{\alpha}\|\vec{f}\|_{[C^0(\partial\Omega)]^N} \tag{9.94}$$

for all α, $0<\alpha<1$, for some constant C_{α} depending only on $\partial\Omega$ and α.

Property 9.2. *Let $\vec{f} \in [C^0(\partial\Omega)]^N$. Then the double-layer potential(s) $V_2^+(\vec{f})$ and $V_2^-(\vec{f})$ can be continuously extended, respectively, from Ω^c to $\overline{\Omega^c}$ and from Ω to $\overline{\Omega}$, such that on $\partial\Omega$,*

$$V_2^{\pm}(\vec{f})(x) = \pm\frac{1}{2}\vec{f}(x) + \int_{\partial\Omega} T_\xi(x-\xi)\vec{f}(\xi)\,d\sigma_\xi, \quad x \in \partial\Omega, \tag{9.95}$$

where the integral exists as a principal value. Consequently, we have the jump relation

$$V_2^+(\vec{f}) - V_2^-(\vec{f}) = \vec{f} \quad \text{on } \partial\Omega. \tag{9.96}$$

For $x \in \partial\Omega$, define

$$V_2(\vec{f})(x) = \int_{\partial\Omega} T_\xi(x-\xi)\vec{f}(\xi)\,d\sigma. \tag{9.97}$$

Then for $\vec{f} \in [C^{0,\alpha}(\partial\Omega)]^N$, $V_2(\vec{f})$ is uniformly Hölder-continuous on $\partial\Omega$, satisfying

$$\|V_2(\vec{f})\|_{[C^{0,\alpha}(\partial\Omega)]^N} \leqslant C_\alpha \|\vec{f}\|_{[C^{0,\alpha}(\partial\Omega)]^N},$$

for α, $0 < \alpha < 1$, and some constant C_α depending on $\partial\Omega$ and α, and

$$\|V_2(\vec{f})\|_{[C^{0,\beta}(\partial\Omega)]^N} \leqslant qC_\beta \|\vec{f}\|_{[C^{0,1}(\partial\Omega)]^N},$$

for some $C_\beta > 0$ depending only on $\partial\Omega$, if $\vec{f} \in [C^{0,1}(\partial\Omega)]^N$. Further, if the layer density $\vec{f}$ is in $[C^{0,\alpha}(\partial\Omega)]^N$, $0 < \alpha < 1$, then (the extended)
$V_2^+(\vec{f})$ and $V_2^-(\vec{f})$ are uniformly Hölder-continuous on $\overline{\Omega^c}$ and $\overline{\Omega}$, respectively, satisfying

$$\|V_2^+(\vec{f})\|_{[C^{0,\alpha}(\overline{\Omega^c})]^N} \leqslant C_\alpha \|\vec{f}\|_{[C^{0,\alpha}(\partial\Omega)]^N},$$

$$\|V_2^-(\vec{f})\|_{[C^{0,\alpha}(\overline{\Omega})]^N} \leqslant C_\alpha \|\vec{f}\|_{[C^{0,\alpha}(\partial\Omega)]^N},$$

for some $C_\alpha > 0$ depending only on $\partial\Omega$ and α.

In particular, we note, from the principal symbols given in [98, Tables 1 and 2, pp. 46–49], that the integral operator V_2 represents a Cauchy principal value. Consequently, it is not a compact or a smoothing operator as we have had in the similar situations in the preceding three chapters.
Next, let us consider derivatives of layer potentials.

Property 9.3. *Let the density $\vec{f} \in [C^{0,\alpha}(\partial\Omega)]^N, 0 < \alpha < 1$. Then the first derivatives of $V_1^+(\vec{f})$ and $V_1^-(\vec{f})$ can be extended in a uniformly Hölder-continuous fashion, respectively, from Ω^c into $\overline{\Omega^c}$ and from Ω into $\overline{\Omega}$, with limit*

$$\frac{\partial}{\partial x_i}V_{1,j}^{\pm}(\vec{f})(x) = \mp\left[\frac{1}{2\mu}n_{x,i}f_j(x) - \frac{\lambda+\mu}{2\mu(2\mu+\lambda)}n_{x,i}n_{x,j}n_x\cdot\vec{f}(x)\right]$$
$$+\int_{\partial\Omega}\Sigma_k\frac{\partial E_{jk}(x-\xi)}{\partial x_i}f_k(\xi)\,d\sigma_\xi, \quad x \in \partial\Omega,\ N \geqslant 2, \tag{9.98}$$

where $V^{\pm}_{1,j}$ *denotes the jth component of* $V^{\pm}_1$, *and the integral in (9.98) exists as a principal value. Furthermore, we have the estimates*

$$\|\nabla V_1^{+}(\vec{f})\|_{[C^{0,\alpha}(\overline{\Omega^c})]^N} \leqslant C_\alpha \|\vec{f}\|_{[C^{0,\alpha}(\partial\Omega)]^N},$$

$$\|\nabla V_1^{-}(\vec{f})\|_{[C^{0,\alpha}(\overline{\Omega})]^N} \leqslant C_\alpha \|\vec{f}\|_{[C^{0,\alpha}(\partial\Omega)]^N},$$

for some constant $C_\alpha > 0$ *depending only on* α *and* $\partial\Omega$.

As a consequence of (9.98), if we take the traction vector of $V_1^{\pm}$, *we get*

$$\begin{aligned}\vec{\tau}(V_1^{\pm})(x) &= \lambda(\nabla\cdot V_1^{\pm})\vec{n} + \mu\left[\nabla V_1^{\pm} + (\nabla V_1^{\pm})^{Tr}\right]\vec{n} \\ &= \mp\frac{1}{2}\vec{f}(x) + \int_{\partial\Omega} T_x(x-\xi)\vec{f}(\xi)\,d\sigma_\xi,\ \ x\in\partial\Omega,\end{aligned}$$

where

$$\begin{aligned}T_x(x-\xi) &\equiv \vec{\tau}_x(E)(x-\xi) \\ &= \frac{\mu}{2(N-1)\pi(\lambda+2\mu)}\left\{\left[I_N + \frac{N(\lambda+\mu)}{\mu|x-\xi|^2}(x-\xi)(x-\xi)^{Tr}\right]\frac{\partial\gamma(x-\xi)}{\partial n_x}\right. \\ &\quad \left. + \frac{1}{|x-\xi|^N}\left\{(\xi-x)n_x^{Tr} - [(\xi-x)n_x^{Tr}]^{Tr}\right\}\right\},\quad N=3 \text{ or } 2, \qquad (9.99)\end{aligned}$$

(see (9.92) for $\gamma(x-\xi)$*). So the jump relation*

$$\vec{\tau}(V_1^{+})(x) - \vec{\tau}(V_1^{-})(x) = -\vec{f}(x) \qquad (9.100)$$

holds.

Property 9.4. *The first derivatives of a double-layer potential* $V_2^{\pm}(\vec{f})$ *with density* $\vec{f} \in [C^{1,\alpha}(\partial\Omega)]^N, 0<\alpha<1$, *can be extended in a uniformly Hölder-continuous fashion from* Ω^c *into* $\overline{\Omega^c}$ *and from* Ω *into* $\overline{\Omega}$, *respectively, such that*

$$\|\nabla V_2^{+}(\vec{f})\|_{[C^{0,\alpha}(\overline{\Omega^c})]^{N\times N}} \leqslant C_\alpha \|\vec{f}\|_{[C^{1,\alpha}(\partial\Omega)]^N},$$

$$\|\nabla V_2^{-}(\vec{f})\|_{[C^{0,\alpha}(\overline{\Omega})]^{N\times N}} \leqslant C_\alpha \|\vec{f}\|_{[C^{1,\alpha}(\partial\Omega)]^{N\times N}}$$

for some constant $C_\alpha > 0$ *depending on* $\partial\Omega$ *and* α.

The regularity properties of layer potentials in terms of Sobolev space norms can also be discussed. Let the layer density $\vec{f}$ be in $[H^\alpha(\partial\Omega)]^N$ for $\alpha\in\mathbb{R}$. Then for $\alpha \geqslant 0$, $\vec{f}$ can be approximated by a sequence $\{\vec{f}_n\} \in [C^\infty(\partial\Omega)]^N$ such that

$$\lim_{n\to\infty} \|\vec{f}_n - \vec{f}\|_{[H^\alpha(\partial\Omega)]^N} = 0.$$

Let $V_1^{\pm}(\vec{f}), \nabla V_1^{\pm}(\vec{f}), V_2^{\pm}(\vec{f})$ and $\nabla V_2^{\pm}(\vec{f})$ be the limits of $V_1^{\pm}(\vec{f_n}), \nabla V_1^{\pm}(\vec{f_n}), V_2^{\pm}(\vec{f}_n)$ and $\nabla V_2^{\pm}(\vec{f}_n)$, respectively, *in the sense of distributions on* $\partial\Omega$. When $\alpha<0$, a duality transposition argument can be used to interpret $V_1^{\pm}(\vec{f}), \nabla V_1^{\pm}(\vec{f}), V_2^{\pm}(\vec{f})$ and $\nabla V_2^{\pm}(\vec{f})$.

Property 9.5. *Let $\vec{f} \in [H^{\alpha}(\partial\Omega)]^N$ and let $V_1^{\pm}(\vec{f})$, $\nabla V_1^{\pm}(\vec{f})$, $V_2^{\pm}(\vec{f})$ and $\nabla V_2^{\pm}(\vec{f})$ be defined in the sense of distributions as above. Then for $\alpha \in \mathbb{R}$,*

$$V_1^{+}(\vec{f}) = V_1^{-}(\vec{f}) \quad (\equiv V_1(\vec{f})) \text{ in } [H^{\alpha+1}(\partial\Omega)]^N, \tag{9.101}$$

$$V_2^{+}(\vec{f}) = \frac{1}{2}\vec{f} + V_2(\vec{f}) \text{ in } [H^{\alpha}(\partial\Omega)]^N \tag{9.102}$$

(see (9.97) for V_2),

$$V_2^{-}(\vec{f}) = -\frac{1}{2}\vec{f} + V_2(\vec{f}) \text{ in } [H^{\alpha}(\partial\Omega)]^N, \tag{9.103}$$

$$\vec{\tau}(V_1^{+}(\vec{f})) = -\frac{1}{2}\vec{f} + \int_{\partial\Omega} T_x(\cdot - \xi)\vec{f}(\xi)\,d\sigma_\xi \text{ in } [H^{\alpha}(\partial\Omega)]^N \tag{9.104}$$

(see (9.99) for T_x),

$$\vec{\tau}(V_1^{-}(\vec{f})) = \frac{1}{2}\vec{f} + \int_{\partial\Omega} T_x(x - \xi)\vec{f}(\xi)\,d\sigma_\xi \text{ in } [H^{\alpha}(\partial\Omega)]^N \tag{9.105}$$

$$\vec{\tau}(V_2^{+}(\vec{f})) = (F.P.)\int_{\partial\Omega} \vec{\tau}_x\left[\vec{\tau}_\xi^{Tr}(E(x-\xi))\right]\vec{f}(\xi)\,d\sigma_\xi \quad \text{in } [H^{\alpha-1}(\partial\Omega)]^N, \tag{9.106}$$

$$\vec{\tau}(V_2^{-}(\vec{f})) = (F.P.)\int_{\partial\Omega} \vec{\tau}_x\left[\vec{\tau}_\xi^{Tr}(E(x-\xi))\right]\vec{f}(\xi)\,d\sigma_\xi \quad \text{in } [H^{\alpha-1}(\partial\Omega)]^N, \tag{9.107}$$

where in (9.106) and (9.107), $\vec{\tau}_x[\vec{\tau}_\xi^{Tr}(E(x-\xi))]$ is a 3×3 matrix whose (i,j)-entry is given by

$$\begin{aligned}\alpha_{ij}(x,\xi) = r^{-3}\Bigg[&\frac{\partial r}{\partial n_x}\frac{\partial r}{\partial n_\xi}(3\lambda c_1\delta_{ij} - 10\mu c_1 c_2 r_i r_j)\\ &+ \sum_{k=1}^{3} n_{x,k} n_{\xi,k}(-3\lambda c_1 n_{\xi,i} r_j - 2\mu c_1\delta_{ij})\\ &+ \frac{\partial r}{\partial n_x}(-3\lambda c_1 n_{\xi,i} r_j - 6\mu n_{\xi,j} r_i)\\ &+ \frac{\partial r}{\partial n_\xi}(6\mu c_1 n_{x,i} r_j + 3\lambda c_1 n_{x,j} r_i)\\ &+ (\mu - 2\lambda)c_1 n_{x,i} n_{\xi,j} - \mu c_1 n_{x,j} n_{\xi,i}\Bigg],\end{aligned}$$

with $r \equiv |x-\xi|$, $r_i \equiv \partial r/\partial\xi_i$, $c_1 \equiv -\mu/4\pi(\lambda+2\mu)$, $c_2 \equiv 3(\lambda+\mu)/\mu$.

9.5 Solutions of the interior BVP by simple- and double-layer potentials

We first state the existence and uniqueness theorems for boundary value problems in linear elastostatics on a bounded domain. Their proofs can be obtained in the same way as classical elliptic boundary value problems (see e.g., [64, 127]).

Theorem 9.3 (Elastostatic problem with displacement boundary condition). *Let* $\vec{g} \in [H^r(\partial\Omega)]^N$, $r \geqslant \frac{1}{2}$, $N = 3$ *or* 2. *Then the boundary value problem*

$$\left.\begin{array}{l}\mu\nabla\vec{w} + (\lambda+\mu)\nabla(\nabla\cdot\vec{w}) = 0 \ \text{ on } \Omega, \\ \vec{w} = \vec{g} \ \text{ on } \partial\Omega,\end{array}\right\} \tag{9.108}$$

admits a unique generalized (weak) solution $\vec{w} \in [H^{r+1/2}(\Omega)]^N$ *satisfying*

$$\int_\Omega \left[\mu\Delta\vec{v} + (\lambda+\mu)\nabla(\nabla\cdot\vec{v})\right]\cdot\vec{w}\,dx = \int_{\partial\Omega}\vec{\tau}(\vec{v})\cdot\vec{g}\,d\sigma$$

for all $\vec{v} \in [H_0^1(\Omega)]^N$. *Furthermore, if* $\vec{g} \in [H^r(\partial\Omega)]^N$ *for* $r \geqslant \frac{3}{2}$, *then the weak solution* $\vec{w} \in [H^{r+1/2}(\Omega)]^N$ *is a strong solution of (9.108) satisfying*

$$\|\vec{w}\|_{[H^{r+1/2}(\Omega)]^N} \leqslant C\|\vec{g}\|_{[H^r(\partial\Omega)]^N}$$

for a constant $C > 0$ *depending only on* Ω, λ *and* μ. □

Theorem 9.4 (Elastostatic problem with traction boundary condition). *Let* $\vec{g} \in [H^r(\partial\Omega)]^N$, $r \geqslant -\frac{1}{2}$, $N = 3$ *or* 2. *Consider the boundary value problem*

$$\left.\begin{array}{ll}\mu\Delta\vec{w} + (\lambda+\mu)\nabla(\nabla\cdot\vec{w}) = 0 & \text{on } \Omega, \\ \vec{\tau}(\vec{w}) = \vec{g} & \text{on } \partial\Omega.\end{array}\right\} \tag{9.109}$$

If $\vec{g}$ *satisfies the compatibility conditions*

$$\int_{\partial\Omega}\vec{g}\cdot m_i\,d\sigma = 0 \quad \forall\, m_i, 1 \leqslant i \leqslant 3(N-1), \quad \text{cf. (9.56)}, \tag{9.110}$$

then the BVP (9.109) admits a weak solution $\vec{w} \in [H^{r+3/2}(\Omega)]^N$ *satisfying*

$$a(\vec{w},\vec{v}) = \int_{\partial\Omega}\vec{g}\cdot\vec{v}\,d\sigma \qquad \forall v \in [H^1(\Omega)]^N.$$

This solution $\vec{w}$ *is unique modulo* $\mathscr{M}$ *(see (9.74)). Furthermore, if* $r \geqslant \frac{1}{2}$ *then the weak solution* $\vec{w}$ *is a strong solution of (9.109), and we have the estimate*

$$\inf\|\vec{w}\|_{[H^{r+3/2}(\Omega)]^N} \leqslant C\|\vec{g}\|_{[H^r(\partial\Omega)]^N}$$

for some constant $C > 0$ *independent of* $\vec{g}$. □

We are now in a position to solve BVP of the above two types by the simple-layer method. First, we consider three-dimensional problems.

Theorem 9.5 (Simple-layer solution for the displacement elastostatic problem in $\mathbb{R}^3$). *Let Ω be a bounded domain in $\mathbb{R}^3$ with smooth boundary $\partial\Omega$. The linear operator*

$$\mathscr{L}_1\colon [H^\alpha(\partial\Omega)]^3 \to [H^{\alpha+1}(\partial\Omega)]^3, \qquad \alpha \in \mathbb{R},$$

$$\mathscr{L}_1(\vec{f})(x) = \int_{\partial\Omega} E(x-\xi)\vec{f}(\xi)\,d\sigma_\xi \equiv V_1(\vec{f})(x), \quad x \in \partial\Omega,$$

is an isomorphism, with

$$\mathscr{L}_1^*\colon [H^{-(\alpha+1)}(\partial\Omega)]^3 \to [H^{-\alpha}(\partial\Omega)]^3 \text{ isomorphically},$$

and $\mathscr{L}_1^* = \mathscr{L}_1$ on $[H^\alpha(\partial\Omega)]^3$, so $\mathscr{L}_1$ is self-adjoint. $\mathscr{L}_1$ is positive definite, satisfying

$$\langle \mathscr{L}_1\vec{f}, \vec{f}\rangle_{[L^2(\partial\Omega)]^3} \geqslant C\|\mathscr{L}_1\vec{f}\|^2_{[H^{1/2}(\partial\Omega)]^3} \tag{9.111}$$
$$\text{for some } C>0,\ \forall\, \vec{f} \in [L^2(\partial\Omega)]^3.$$

The solution to the displacement boundary value problem

$$\begin{cases} \mu\Delta\vec{w} + (\lambda+\mu)\nabla(\nabla\cdot\vec{w}) = 0 & \text{on } \Omega,\\ \vec{w} = \vec{g} \in [H^{r+1}(\partial\Omega)]^3, & r \in \mathbb{R}, \end{cases}$$

is given uniquely by

$$\vec{w}(x) = \int_{\partial\Omega} E(x-\xi)\vec{f}(\xi)\,d\sigma_\xi \in [H^{r+3/2}(\Omega)]^3, \qquad x \in \Omega, \tag{9.112}$$

where $\vec{f} \in [H^r(\partial\Omega)]^3$ is the unique solution to the BIE

$$\mathscr{L}_1(\vec{f}) = \vec{g} \quad \text{on } \partial\Omega. \tag{9.113}$$

Proof. The arguments are analogous to the proofs of Theorems 6.20 and 6.21. So the details are omitted. □

Theorem 9.6 (Simple-layer solution for the interior traction elastostatic problem in $\mathbb{R}^3$). *Let Ω be a bounded domain in $\mathbb{R}^3$ with smooth boundary $\partial\Omega$. The linear operator*

$$\mathscr{L}_2\colon [H^\alpha(\partial\Omega)]^3 \to [H^\alpha(\partial\Omega)]^3, \quad \alpha \in \mathbb{R},$$

$$\mathscr{L}_2\vec{f}(x) = \frac{1}{2}\vec{f}(x) + \int_{\partial\Omega} T_x(x-\xi)\vec{f}(\xi)\,d\sigma_\xi, \quad x \in \partial\Omega, \text{ cf. (9.99) for } T_x,$$

is bounded, with

$$\left.\begin{aligned} &\dim\mathscr{N}(\mathscr{L}_2) = 6,\ \dim\operatorname{Coker}\mathscr{L}_2 = 6,\\ &\mathscr{N}(\mathscr{L}_2) = \widetilde{\mathscr{M}} = \operatorname{span}\mathscr{L}_1^{-1}\{\vec{m}_k(x)|1 \leqslant k \leqslant 6\},\\ &\operatorname{Coker}\mathscr{L}_2 = \operatorname{span}\{\vec{m}_k(x) \mid 1 \leqslant k \leqslant 6\}; \end{aligned}\right\} \tag{9.114}$$

thus Ind $\mathscr{L}_2$ *is zero.* $\mathscr{L}_2$ *has an adjoint*

$$\mathscr{L}_2^*\colon\ [H^{-\alpha}(\partial\Omega)]^3 \longrightarrow [H^{-\alpha}(\partial\Omega)]^3 \quad \text{continuous},$$
$$\mathscr{L}_2^*(\vec{\phi})(x) = \frac{1}{2}\vec{\phi}(x) + \int_{\partial\Omega} T_\xi(x-\xi)\,\vec{\phi}(\xi)\,d\sigma_\xi, \quad x \in \partial\Omega \ \textit{(cf. (9.91) for } T_\xi)$$

with the property that $\mathscr{L}_2^* : [H^s(\partial\Omega)]^3 \to [H^s(\partial\Omega)]^3$ *is continuous, for any* $s \in \mathbb{R}$, *and*

$$\left.\begin{aligned} &\dim \mathscr{N}(\mathscr{L}_2^*) = 6, \quad \dim\mathrm{Coker}\, \mathscr{L}_2^* = 6,\\ &\mathscr{N}(\mathscr{L}_2^*) = \mathrm{span}\{\vec{m}_k(x) \mid 1 \leqslant k \leqslant 6\}, \ \ \mathrm{Coker}\, \mathscr{L}_2^* = \widetilde{\mathscr{M}}. \end{aligned}\right\} \tag{9.115}$$

The solution to the traction BVP

$$\left.\begin{aligned} &\mu\Delta\vec{w} + (\lambda+\mu)\nabla(\nabla\cdot\vec{w}) = 0 \quad \text{on } \Omega,\\ &\vec{\tau}(\vec{w}) = \vec{g} \in [H^r(\partial\Omega)]^3, \quad \vec{g} \perp \mathscr{M} \ \text{ in } [L^2(\partial\Omega)]^3, \ \ r \in \mathbb{R}, \end{aligned}\right\} \tag{9.116}$$

is given by

$$\vec{w}(x) = \int_{\partial\Omega} E(x-\xi)\,\vec{f}(\xi)\,d\sigma_\xi + \sum_{i=1}^{6} \alpha_i \vec{m}_i(x) \ \in [H^{r+3/2}(\Omega)]^3 \tag{9.117}$$

for arbitrarily given constants $\alpha_i \in \mathbb{R}$, $1 \leqslant i \leqslant 6$, *where* $\vec{f} \in [H^r(\partial\Omega)]^3$ *is a solution to the BIE*

$$\mathscr{L}_2\vec{f} = \vec{g}. \tag{9.118}$$

Proof. The arguments are analogous to Theorem 6.22. So we leave the proof to the reader.

□

Note that $\mathscr{L}_2$ has index zero. We can apply Theorem 4.10 to $\mathscr{L}_2$ and BVP (9.116). In particular, see Example 4.5.
Next, we consider solving boundary value problems by the double-layer method.

Theorem 9.7. (Double-layer solution for the interior displacement elastostatic problem in $\mathbb{R}^3$**).** *Let* Ω *be a bounded domain in* $\mathbb{R}^3$ *with smooth boundary* $\partial\Omega$. *The linear operator*

$$\mathscr{L}_3\colon\ [H^{\alpha}(\partial\Omega)]^3 \to [H^{\alpha}(\partial\Omega)]^3, \quad \alpha \in \mathbb{R},$$
$$\mathscr{L}_3(\vec{f})(x) = -\frac{1}{2}\vec{f}(x) + \int_{\partial\Omega} T_\xi(x-\xi)\,\vec{f}(\xi)\,d\sigma_\xi, \quad x \in \partial\Omega, \tag{9.119}$$

is an isomorphism, with

$$\dim \mathscr{N}(\mathscr{L}_3) = 0, \quad \dim\mathrm{Coker}\, \mathscr{L}_3 = 0.$$

$\mathscr{L}_3$ has an adjoint

$$\mathscr{L}_3^*\colon H^{-\alpha}(\partial\Omega)]^3 \longrightarrow [H^{-\alpha}(\partial\Omega)]^3,$$
$$\mathscr{L}_3^* \vec{\phi}(x) = -\frac{1}{2}\vec{\phi}(x) + \int_{\partial\Omega} T_x(x-\xi)\vec{f}(\xi)\,d\sigma_\xi, \quad x \in \partial\Omega,$$

satisfying $\mathscr{L}_3^*\colon [H^s(\partial\Omega)]^3 \to [H^s(\partial\Omega)]^3$ isomorphically, for any $s \in \mathbb{R}$, and

$$\dim \mathscr{N}(\mathscr{L}_3^*) = 0, \quad \dim \operatorname{Coker} \mathscr{L}_3^* = 0.$$

Consequently, the solution to the displacement BVP

$$\left.\begin{array}{ll} \mu\Delta\vec{w} + (\lambda+\mu)\nabla(\nabla\cdot\vec{w}) = 0 & \text{on } \Omega, \\ \vec{w} = \vec{g} \in [H^r(\partial\Omega)]^3, & r \in \mathbb{R}, \end{array}\right\}$$

is given by

$$\vec{w}(x) = \int_{\partial\Omega} T_\xi(x-\xi)\vec{f}(\xi)\,d\sigma_\xi \in [H^{r+1/2}(\Omega)]^3,$$

where $\vec{f} \in [H^r(\partial\Omega)]^3$ is the unique solution to the BIE

$$\mathscr{L}_3(\vec{f}) = \vec{g}.$$

Proof. Similar to Theorem 6.27. □

Theorem 9.8. (Double-layer solution for the interior traction elastostatic problem in $\mathbb{R}^3$). *Let Ω be a bounded domain in $\mathbb{R}^3$ with smooth boundary $\partial\Omega$. Then the linear operator*

$$\mathscr{L}_4\colon [H^{\alpha+1}(\partial\Omega)]^3 \longrightarrow [H^{\alpha}(\partial\Omega)]^3, \quad \alpha \in \mathbb{R},$$
$$\mathscr{L}_4(\vec{f})(x) = \int_{\partial\Omega} \vec{\tau}_x[\vec{\tau}_\xi^{Tr} E(x-\xi)]\vec{f}(\xi)\,d\sigma_\xi, \quad x \in \partial\Omega,$$

is bounded, where the integral is interpreted as the finite part when $\vec{f} \in [C^\infty(\partial\Omega)]^3$. We have

$$\left.\begin{array}{ll} \dim \mathscr{N}(\mathscr{L}_4) = 6, & \dim \operatorname{Coker} \mathscr{L}_4 = 6, \\ \mathscr{N}(\mathscr{L}_4) = \operatorname{Coker} \mathscr{L}_4 = \mathscr{M}. & \end{array}\right\} \tag{9.120}$$

The operator $\mathscr{L}_4$ satisfies

$$\mathscr{L}_4^*\colon [H^{-\alpha}(\partial\Omega)]^3 \longrightarrow [H^{-(\alpha+1)}(\partial\Omega)]^3 \text{ continuously},$$
$$\mathscr{L}_4^* = \mathscr{L}_4 \text{ formally},$$

and $\mathscr{L}_4^*\colon [H^{s+1}(\partial\Omega)]^3 \to [H^s(\partial\Omega)]^3$ is continuous for any $s \in \mathbb{R}$, so $\mathscr{L}_4$ is self-adjoint, with

$$\left.\begin{array}{l} \dim \mathscr{N}(\mathscr{L}_4^*) = 6, \ \dim \operatorname{Coker} \mathscr{L}_4^* = 6, \\ \mathscr{N}(\mathscr{L}_4^*) = \operatorname{Coker} \mathscr{L}_4^* = \mathscr{M}. \end{array}\right\} \tag{9.121}$$

Consequently, the solution to the traction BVP

$$\begin{cases} \mu\Delta\vec{w} + (\lambda+\mu)\nabla(\nabla\cdot\vec{w}) = 0 & on\ \Omega, \\ \vec{\tau}(\vec{w}) = \vec{g} \in [H^r(\partial\Omega)]^3, & \vec{g} \perp \mathscr{M}\ in\ L^2(\partial\Omega),\ r\in\mathbb{R}, \end{cases}$$

is given by

$$\vec{w}(x) = \int_{\partial\Omega} \vec{\tau}_{\xi}^{Tr}(E(x-\xi))\vec{f}(\xi)\,d\sigma_\xi + \sum_{i=1}^{6}\alpha_i\vec{m}_i(x) \in [H^{r+3/2}(\Omega)]^3,$$

$$\alpha_i \in \mathbb{R},\ 1 \leqslant i \leqslant 6,\ are\ arbitrary, \tag{9.122}$$

where $\vec{f} \in [H^{r+1}(\partial\Omega)]^3$ *is a (nonunique) solution to the BIE*

$$\mathscr{L}_4(\vec{f}) = \vec{g}.$$

Proof. Similar to Theorem 6.28. □

When the domain Ω is two-dimensional, bounded, with smooth boundary $\partial\Omega$, we modify Theorems 9.5–9.8 as follows.

Theorem 9.9. (Simple-layer solution for the interior displacement elastostatic problem in $\mathbb{R}^2$). *Let Ω be a bounded domain in $\mathbb{R}^2$ with smooth boundary $\partial\Omega$. The linear operator*

$$\mathscr{L}_1'\left(\begin{bmatrix} a_1 \\ a_2 \\ \vec{f} \end{bmatrix}\right)(x) = \begin{bmatrix} \int_{\partial\Omega}\vec{f}(\xi)d\xi \\ \int_{\partial\Omega}E(x-\xi)\vec{f}(\xi)\,d\sigma_\xi + \begin{bmatrix} a_1 \\ a_2 \end{bmatrix} \end{bmatrix}, \quad a_1,\ a_2 \in \mathbb{R},$$

$$\mathscr{L}_1' \colon \mathbb{R}^2 \oplus [H^\alpha(\partial\Omega)]^2 \to \mathbb{R}^2 \oplus \left[H^{\alpha+1}(\partial\Omega)\right]^2, \quad \alpha\in\mathbb{R},$$

is an isomorphism, with adjoint

$$\mathscr{L}_1'^* \colon \mathbb{R}^2 \oplus \left[H^{-(\alpha+1)}(\partial\Omega)\right]^2 \to \mathbb{R}^2 \oplus \left[H^{-\alpha}(\partial\Omega)\right]^2, \quad \alpha\in\mathbb{R},$$

$$\mathscr{L}_1'^* = \mathscr{L}_1',$$

also an isomorphism, and $\mathscr{L}_1'$ is self-adjoint.

The solution to the displacement BVP

$$\begin{cases} \mu\Delta\vec{w} + (\lambda+\mu)\nabla(\nabla\cdot\vec{w}) = 0 & \text{on}\ \Omega, \\ \vec{w} = \vec{g} \in [H^{r+1}(\partial\Omega)]^2, & r\in\mathbb{R}, \end{cases}$$

is given uniquely by

$$\vec{w}(x) = \int_{\partial\Omega}E(x-\xi)\vec{f}(\xi)\,d\sigma_\xi + \begin{bmatrix} a_1 \\ a_2 \end{bmatrix} \in [H^{r+3/2}(\Omega)]^2, \quad x\in\Omega,$$

where $\vec{f} \in [H^r(\partial\Omega)]^2$ *is the unique solution to the BIE*

$$\mathscr{L}_1'\left(\begin{bmatrix} a_1 \\ a_2 \\ \vec{f} \end{bmatrix}\right) = \begin{bmatrix} A_1 \\ A_2 \\ \vec{g} \end{bmatrix} \quad \text{on } \partial\Omega.$$

($A_1, A_2 \in \mathbb{R}$ *can be chosen arbitrarily, but* a_1, a_2 *and* $\vec{f}$ *depend on* A_1 *and* A_2.)

Proof. This is left as an exercise. □

The counterpart of Theorem 9.6 in $\mathbb{R}^2$, namely, the simple-layer solution for the interior traction elastostatic problem for $\Omega \subset \mathbb{R}^2$, can be stated essentially the same way, except that (9.114) and (9.115) are modified, respectively, to

$$\left.\begin{aligned} &\dim \mathscr{N}(\mathscr{L}_2) = 3, \quad \dim \operatorname{Coker} \mathscr{L}_2 = 3, \\ &\mathscr{N}(\mathscr{L}_2) = \widetilde{\mathscr{M}} = \{\vec{k}_j \in [C^\infty(\partial\Omega)]^2 \mid \vec{k}_j \neq 0, \\ &\text{either } V_1(\vec{k}_j) = 0 \text{ or } V_1(\vec{k}_j) = \vec{m}_j, \quad j = 1, 2, 3\}, \\ &\operatorname{Coker} \mathscr{L}_2 = \mathscr{M}, \end{aligned}\right\} \tag{9.123}$$

and

$$\left.\begin{aligned} &\dim \mathscr{N}(\mathscr{L}_2)^* = 3, \quad \dim \operatorname{Coker} \mathscr{L}_2^* = 3 \\ &\mathscr{N}(\mathscr{L}_2)^* = \mathscr{M}, \quad \operatorname{Coker} \mathscr{L}_2^* = \widetilde{\mathscr{M}}_2. \end{aligned}\right\} \tag{9.124}$$

The solution (9.117) is now modified to

$$\vec{w}(x) = \int_{\partial\Omega} E(x-\xi)\vec{f}(\xi)\, d\sigma_\xi + \sum_{i=1}^{3} \alpha_i \vec{m}_i(x). \tag{9.125}$$

The counterparts of Theorems 9.7 and 9.8 in $\mathbb{R}^2$ can also be stated essentially the same way, except that (9.120) must be modified by changing "6" to "3", and (9.122) must be also adjusted to (9.125), accordingly.

9.6 BIE for exterior problems in linear elastostatics

Now we consider exterior elastostatic problems in $\mathbb{R}^3$ or $\mathbb{R}^2$.
Let $\vec{w}(x)$ be a solution of (9.7) on Ω^c. Following Kupradze [117, p. 113], we introduce the notion of *regular solutions* $\vec{w}(x)$ by requiring that

$$\left.\begin{aligned} &\vec{w}(x) = o(1), \quad \nabla\vec{w}(x) = o(|x|^{-1}) \text{ as } |x| \to \infty, \\ &x \in \Omega^c \subset \mathbb{R}^N, \quad N = 3 \text{ or } 2. \end{aligned}\right\} \tag{9.126}$$

For the exterior problem with *displacement* boundary condition, however, it is physically reasonable to assume that at infinity the solution is asymptotically a rigid motion

$$\vec{r}_0(x) = \sum_{j=1}^{3(N-1)} \omega_j \vec{m}_j(x) = M(x)\vec{\omega}, \qquad \vec{\omega} \in \mathbb{R}^{3(N-1)}, \quad N = 2 \text{ or } 3, \tag{9.127}$$

i.e., $\vec{w}(x) - \vec{r}_0(x)$ is regular at ∞. According to Hsiao and Wendland [98, p. 34], such functions $\vec{w}(x)$ will be called *generalized regular solutions*. This condition will be appended to the displacement boundary condition

$$\vec{w}(x) = \vec{g}(x), \quad x \in \partial\Omega, \tag{9.128}$$

and is essential to guarantee uniqueness of solutions of the exterior BVP. For the *traction* BVP, we simply require that the solution be regular at ∞, i.e., $\vec{\omega} = 0$ in (9.127).

Among the $3(N-1)$ constants ω_j in (9.127), we can assume that some are given, and the rest are to be determined from the auxiliary physical conditions. We do this as follows. Consider $\mathbb{R}^3$ first. Let

$$\mathscr{I} \equiv \{1,2,3,4,5,6\}, \quad \mathscr{J} \subset \mathscr{I}. \tag{9.129}$$

We let ω_j, $j \in \mathscr{J}$, be *prescribed*. For the remaining ω_j, we impose additional normalization conditions

$$\int_{\partial\Omega} \vec{m}_k(x) \cdot \vec{\tau}(\vec{w})(x)\, d\sigma_x = b_k, \quad b_k \text{ given}, \ k \in \mathscr{I} - \mathscr{J}, \tag{9.130}$$

In $\mathbb{R}^2$, the situation is more restrictive. We will see (Lemma 9.6) that if $\vec{w}(x)$ is a generalized regular solution then its traction on $\partial\Omega$ must satisfy the equilibrium condition

$$\int_{\partial\Omega} \vec{\tau}(\vec{w})(x)\, d\sigma = \vec{0}. \tag{9.131}$$

This limits our consideration to two possibilities

$$\left.\begin{array}{l} (1)\ \omega_3 \text{ is given and } \omega_1, \omega_2 \text{ are unknowns;} \\ (2)\ \omega_1,\ \omega_2 \text{ and } \omega_3 \text{ are all unknowns, and } \vec{w} \\ \quad \text{satisfies the given normalization condition} \\ \quad \displaystyle\int_{\partial\Omega} \vec{m}_3(x) \cdot \vec{\tau}(\vec{w})(x)\, d\sigma_x = b_3 \in \mathbb{R}. \end{array}\right\} \tag{9.132}$$

We present two lemmas due to Hsiao and Wendland [98].

Lemma 9.4. *Let $\vec{w} \in H^1_{loc}(\Omega^c)$ be a generalized regular function. Then if $\vec{w}$ is a weak solution of (9.7), we have the modified Betti–Somigliana formula*

$$\begin{aligned} \vec{w}(x) = &-\int_{\partial\Omega} [E(x-\xi)\vec{\tau}(\vec{w})(\xi) - T_\xi(x-\xi)\vec{w}(\xi)]\, d\sigma_\xi \\ &+ M(x)\vec{\omega}, \quad x \in \Omega^c, \end{aligned} \tag{9.133}$$

where $\vec{r}(x) = M(x)\vec{\omega}$ corresponds to the rigid motion of $\vec{w}$ at ∞.

Proof. Let

$$\vec{v}(x) = \vec{w}(x) - M(x)\vec{\omega}, \quad x \in \Omega^c. \tag{9.134}$$

Then $\vec{v}(x)$ is regular at ∞, i.e., it satisfies (9.126). It can be shown in a similar manner as in (9.90) that

$$\vec{v}(x) = -\int_{\partial\Omega} [E(x-\xi)\vec{\tau}(\vec{v})(\xi) - T_\xi(x-\xi)\vec{v}(\xi)]\, d\sigma_\xi, \quad x \in \Omega^c. \tag{9.135}$$

(In $\mathbb{R}^2$, (9.131) must be imposed first in order to kill the logarithmic growth of $\vec{v}$.) We also have

$$\vec{\tau}(\vec{v})(\xi) = \vec{\tau}(\vec{v} + \vec{r})(\xi), \quad \xi \in \partial\Omega. \tag{9.136}$$

From integration by parts, it is easy to see that

$$\int_{\partial\Omega} T_\xi(x-\xi)\vec{r}(\xi)\, d\sigma_\xi = \int_{\partial\Omega} T_\xi(x-\xi) M(\xi)\vec{\omega}\, d\sigma_\xi = \vec{0}, \quad x \in \Omega^c. \tag{9.137}$$

From (9.134)–(9.137), we have proved (9.133) for a generalized regular solution $\vec{w} \in H^1_{\text{loc}}(\Omega^c)$. □

Remark 9.4. From (9.134) and (9.135), we see that

$$\vec{v}(x) = \vec{w}(x) - M(x)\vec{\omega} = \mathcal{O}\left(\frac{1}{R}\right), \text{ for large } |x|.$$

Thus $\vec{v}(x)$, as the regular part of $\vec{w}(x)$, has a stronger decay property than (9.126). This property will be used in the subsequent proofs. □

Lemma 9.5 ([98, Theorem 2.2, p. 36]). *Define the energy bilinear form $a(\cdot,\cdot)$ on Ω^c just as in (9.73) with Ω^c replacing Ω. Let $\vec{w}$ and $\vec{v}$ be generalized regular functions and be weak solutions of (9.7). Then we have*

$$a(\vec{w},\vec{v}) = -\int_{\partial\Omega} \vec{w}\cdot\vec{\tau}(\vec{v})\, d\sigma + \int_{\partial\Omega} M(x)\vec{\omega}\cdot\vec{\tau}(\vec{v})(x)\, d\sigma, \tag{9.138}$$

where $\vec{r}(x) = M(x)\vec{\omega}$ corresponds to the rigid motion of $\vec{w}$ at ∞. Consequently, the strain energy of $\vec{w}$ is finite and given by

$$a(\vec{w},\vec{w}) = -\int_{\partial\Omega} \vec{w}\cdot\vec{\tau}(\vec{w})\, d\sigma + \int_{\partial\Omega} M(x)\vec{\omega}\cdot\vec{\tau}(\vec{w})(x)\, d\sigma. \tag{9.139}$$

Proof. Let

$$\Omega^c_R = \Omega^c \cap \{x \in \mathbb{R}^N \mid |x| < R, \quad N = 2 \text{ or } 3\}.$$

Integrating by parts on Ω_R^c according to (9.77) and (9.78), we have

$$a_R(\vec{w}, \vec{v}) \equiv \int_{\Omega_R^c} \left[\lambda(\nabla \cdot \vec{w})(\nabla \cdot \vec{v}) + \frac{\mu}{2} \sum_{i,j} \left(\frac{\partial w_i}{\partial x_j} + \frac{\partial w_j}{\partial x_i} \right) \left(\frac{\partial v_i}{\partial x_j} + \frac{\partial v_j}{\partial x_i} \right) \right] dx$$
$$= \int_{|x|=R} \vec{w} \cdot \vec{\tau}(\vec{v})\, d\sigma - \int_{\partial\Omega} \vec{w} \cdot \vec{\tau}(\vec{v})\, d\sigma.$$

By the assumptions on $\vec{v}$, we have $\vec{v}_0(x) \equiv \vec{v}(x) - \vec{r}\,'(x)$ regular at ∞, for some rigid motion $\vec{r}\,'(x)$. By Lemma 9.4, we can represent $\vec{v}_0(x)$ by

$$\vec{v}_0(x) = -\int_{\partial\Omega} E(x-\xi)\, \vec{\tau}(\vec{v})(\xi)\, d\sigma_\xi + \int_{\partial\Omega} T_\xi(x-\xi)\, \vec{v}(\xi)\, d\sigma_\xi.$$

Since $\vec{\tau}(\vec{v}) = \vec{\tau}(\vec{v}_0)$ on $|x| = R$, we have

$$I \equiv \int_{|x|=R} \vec{w}(x) \cdot \vec{\tau}(\vec{v})(x)\, d\sigma$$
$$= -\int_{|x|=R} \int_{\partial\Omega} \vec{w}(x) \cdot \vec{\tau}_x(E)(x,\xi)\, \vec{\tau}(\vec{v})(\xi)\, d\sigma_\xi\, d\sigma_x$$
$$+ \int_{|x|=R} \int_{\partial\Omega} \vec{w}(x) \cdot \left[\vec{\tau}_x \left(\vec{\tau}_\xi(E) \right)(x,\xi) \right]^{\mathrm{Tr}} \vec{v}(\xi)\, d\sigma_\xi\, d\sigma_x$$
$$= \int_{|x|=R} \int_{\partial\Omega} \left[\omega^{\mathrm{Tr}} M^{\mathrm{Tr}}(x) + \mathscr{O}(\frac{1}{R}) \right] \Big[-\tau_x(E)(x,\xi)\, \vec{\tau}(\vec{v})(\xi)$$
$$+ \vec{\tau}_x(\vec{\tau}_\xi E)^{\mathrm{Tr}}(x,\xi)\, \vec{v}(\xi) \Big]\, d\sigma_\xi\, d\sigma_x, \quad \text{by Remark 9.4.}$$

For the rigid motion $\vec{r}(x)$, we use the Betti–Somigliana formula (9.90) inside $|x| < R$:

$$M(\xi)\, \vec{\omega} = -\int_{|x|=R} \vec{\tau}_x^{\,\mathrm{Tr}}(E)(x,\xi) M(x)\, \vec{\omega}\, d\sigma_x, \quad |\xi| < R, \tag{9.140}$$

for sufficiently large R. Substituting (9.140) into I, we simplify by interchanging the order of integrating to obtain

$$I = \int_{\partial\Omega} \omega^{\mathrm{Tr}} M^{\mathrm{Tr}}(\xi)\, \vec{\tau}(\vec{v})(\xi)\, d\sigma_\xi$$
$$+ \int_{\partial\Omega} \left[\int_{|x|=R} \vec{\omega}^{\mathrm{Tr}} M^{\mathrm{Tr}}(x)\, \vec{\tau}_x(\vec{\tau}_\xi E)^{\mathrm{Tr}}(x,\xi)\, d\sigma_x \right] \vec{v}(\xi)\, d\sigma_\xi + \mathscr{O}\left(\frac{1}{R}\right)$$
$$= \int_{\partial\Omega} M(\xi)\, \vec{\omega} \cdot \vec{\tau}(\vec{v})(\xi)\, d\sigma_\xi + \int_{\partial\Omega} \vec{v}(\xi) \cdot \tau_\xi[M(\xi)\omega]\, d\sigma + \mathscr{O}\left(\frac{1}{R}\right)$$
$$= \int_{\partial\Omega} M(\xi)\, \vec{\omega} \cdot \vec{\tau}(\vec{v})(\xi)\, d\sigma_\xi + \mathscr{O}\left(\frac{1}{R}\right).$$

Letting $R \to \infty$, we obtain (9.138) and hence also (9.139) by setting $\vec{v} = \vec{w}$ in (9.138). □

Remark 9.5. For $N = 2$, because of the equilibrium condition as mentioned in (9.131), we see that (9.139) is reduced to

$$a(\vec{w}, \vec{w}) = -\int_{\partial\Omega} \vec{w} \cdot \vec{\tau}(\vec{w})\, d\sigma + \omega_3 \int_{\partial\Omega} \vec{m}_3(x) \cdot \vec{\tau}(\vec{w})(x)\, d\sigma. \tag{9.141}$$

□

Corollary 9.1. ([98, Corollary 2.3, p. 39] (Uniqueness of the generalized regular solution for the exterior displacement BVP)). *Let $\vec{w} \in H^1_{loc}(\Omega^c)$ be a generalized regular solution of (9.7) satisfying*

$$\left.\begin{aligned} &\vec{w} = 0 \quad on\ \partial\Omega, \\ &\omega_j = 0, \quad j \in \mathscr{J}, \\ &\int_{\partial\Omega} \vec{m}_k(x) \cdot \vec{\tau}(\vec{w})(x)\, d\sigma = 0, \quad k \in \mathscr{I} - \mathscr{J}, \end{aligned}\right\} \tag{9.142}$$

cf. (9.129). Then $\vec{w}$ vanishes identically on Ω^c.

Proof. By (9.138) and the assumption (9.142), we have

$$a(\vec{w}, \vec{w}) = 0 \text{ on } \Omega^c.$$

Therefore $\vec{w}$ must be a rigid motion

$$\vec{w}(x) = \sum_{1}^{3(N-1)} \omega_j \vec{m}_j(x), \quad x \in \Omega^c.$$

But (9.142) implies that $\omega_j = 0$ for $j = 1, \ldots, 3(N-1)$. □

9.7 Simple-layer representation for the exterior displacement elastostatic problem

We treat the three-dimensional problem first: find a generalized regular solution $\vec{w}$ satisfying

$$\left.\begin{aligned} &\mu\Delta\vec{w} + (\lambda+\mu)\nabla(\nabla\cdot\vec{w}) = 0 \quad \text{on } \Omega^c, \\ &\vec{w}(x) = \vec{g}(x) \quad \text{on } \partial\Omega, \quad \vec{g} \text{ is given}, \\ &\vec{w}(x) \sim \sum_{i=1}^{6} \omega_i \vec{m}_i(x), \quad \omega_j \text{ is given for } j \in \mathscr{J}, \\ &\int_{\partial\Omega} \vec{m}_k(x) \cdot \vec{\tau}(\vec{w})(x)\, d\sigma = b_k, \ b_k \text{ is given for } k \in \mathscr{I} - \mathscr{J}, \\ &\qquad \text{cf. (9.130).} \end{aligned}\right\} \tag{9.143}$$

According to the modified Betti–Somigliana formula (9.133), it is reasonable to take

$$\vec{w}(x) = \int_{\partial\Omega} E(x-\xi)\vec{f}(\xi)\, d\sigma_\xi + \sum_{i=1}^{6} \omega_i \vec{m}_i(x), \quad x \in \Omega^c, \tag{9.144}$$

as the *ansatz* for the simple-layer solution.

Theorem 9.10. (Generalized regular simple-layer solution for the exterior displacement elastostatic problem in $\mathbb{R}^3$). *Let Ω be a bounded domain in $\mathbb{R}^3$ with* *smooth boundary $\partial\Omega$. The linear operator*

$$\mathscr{L}_{1e}\left(\begin{bmatrix}\omega_{k_1}\\ \vdots \\ \omega_{k_p}\\ \vec{f}\end{bmatrix}\right)(x)=\begin{bmatrix}\int_{\partial\Omega}\vec{m}_{k_1}(x)\cdot\left[-\frac{1}{2}\vec{f}(x)+\int_{\partial\Omega}\vec{\tau}_x(E)(x,\xi)\vec{f}(\xi)\,d\sigma_\xi\right]d\sigma_x\\ \vdots\\ \int_{\partial\Omega}\vec{m}_{k_p}(x)\cdot\left[-\frac{1}{2}\vec{f}(x)+\int_{\partial\Omega}\vec{\tau}_x(E)(x,\xi)\vec{f}(\xi)\,d\sigma_\xi\right]d\sigma_x\\ \int_{\partial\Omega}E(x-\xi)\vec{f}(\xi)\,d\sigma_\xi+\sum_{l=1}^{p}\omega_{k_l}\vec{m}_{k_l}(x)\end{bmatrix},$$

with $k_l \in \mathscr{I}-\mathscr{J}$, $l=1,2,\ldots,p$, $x\in\partial\Omega$,

$$\mathscr{L}_{1e}\colon\ \mathbb{R}^p\oplus[H^\alpha(\partial\Omega)]^3\longrightarrow\mathbb{R}^p\oplus[H^{\alpha+1}(\partial\Omega)]^3,\quad \alpha\in\mathbb{R},$$

is an isomorphism, where $p=\#(\mathscr{I}-\mathscr{J})$, with adjoint

$$\mathscr{L}_{1e}^*\colon\ \mathbb{R}^p\oplus[H^{-(\alpha+1)}(\partial\Omega)]^3\longrightarrow\mathbb{R}^p\oplus[H^{-\alpha}(\partial\Omega)]^3,$$

$$\mathscr{L}_{1e}^*\left(\begin{bmatrix}\nu_{k_1}\\ \vdots \\ \nu_{k_p}\\ \vec{\phi}\end{bmatrix}\right)(x)=\begin{bmatrix}\int_{\partial\Omega}\vec{m}_{k_1}(x)\cdot\vec{\phi}(x)\,d\sigma\\ \vdots\\ \int_{\partial\Omega}\vec{m}_{k_p}(x)\cdot\vec{\phi}(x)\,d\sigma\\ \int_{\partial\Omega}E(x-\xi)\vec{\phi}(\xi)\,d\sigma_\xi-\sum_{l=1}^{p}\nu_{k_l}\vec{m}_{k_l}(x)\end{bmatrix}, \tag{9.145}$$

for $k_l\in\mathscr{I}-\mathscr{J}$, $l=1,\ldots,p$, which is also an isomorphism. $\mathscr{L}_{1e}^*$ satisfies

$$\mathscr{L}_{1e}^*\colon\ \mathbb{R}^p\oplus[H^s(\partial\Omega)]^3\longrightarrow\mathbb{R}^p\oplus[H^{s+1}(\partial\Omega)]^3\ \text{ isomorphically}\ \forall\, s\in\mathbb{R}.$$

The generalized regular solution to the exterior BVP (9.143) is given uniquely by (9.144), where ω_j is given for $j\in\mathscr{J}$ and $(\omega_{k_1},\ldots,\omega_{k_p},\vec{f})$ satisfies

$$\mathscr{L}_{1e}(\omega_{k_1},\ldots,\omega_{k_p},\vec{f})(x)=\left(b_{k_1},\ldots,b_{k_p},\vec{g}(x)-\sum_{j\in\mathscr{J}}\omega_j m_j(x)\right).$$

For $\vec{g}\in[H^{\alpha+1}(\partial\Omega)]^3$, $\alpha\in\mathbb{R}$, we have $\vec{f}\in[H^\alpha(\partial\Omega)]^3$ and $\vec{w}\in[H_{loc}^{\alpha+3/2}(\Omega^c)]^3$.

Proof. Let

$$(\omega_{k_1},\ldots,\omega_{k_p},\vec{f})\in\mathbb{R}^p\oplus[H^\alpha(\partial\Omega)]^3\ \text{ and }\ (\nu_{k_1},\ldots,\nu_{k_p},\vec{\phi})\in\mathbb{R}^p\oplus[H^{-(\alpha+1)}(\partial\Omega)]^3.$$

By formal transposition, it is straightforward to verify that

$$\left(\mathscr{L}_{1e}\left(\begin{bmatrix}\omega_{k_1}\\ \vdots\\ \omega_{k_p}\\ \vec{f}\end{bmatrix}\right),\begin{bmatrix}\nu_{k_1}\\ \vdots\\ \nu_{k_p}\\ \vec{\phi}\end{bmatrix}\right)_{\mathbb{R}^p\oplus[H^{\alpha+1}(\partial\Omega)]^3,\mathbb{R}^p\oplus[H^{-(\alpha+1)}(\partial\Omega)]^3}$$

$$=\left(\begin{bmatrix}\omega_{k_1}\\ \vdots\\ \omega_{k_p}\\ \vec{f}\end{bmatrix},\mathscr{L}_1^*\begin{bmatrix}\nu_{k_1}\\ \vdots\\ \nu_{k_p}\\ \vec{\phi}\end{bmatrix}\right)_{\mathbb{R}^p\oplus[H^{\alpha}(\partial\Omega)]^3,\mathbb{R}^p\oplus[H^{-\alpha}(\partial\Omega)]^3},$$

with $\mathscr{L}_{1e}^*$ given by

$$\mathscr{L}_{1e}^*\left(\begin{bmatrix}\nu_{k_1}\\ \vdots\\ \nu_{k_p}\\ \vec{\phi}\end{bmatrix}\right)(\xi)=\begin{bmatrix}\int_{\partial\Omega}\vec{m}_{k_1}(x)\cdot\vec{\phi}(x)\,d\sigma\\ \vdots\\ \int_{\partial\Omega}\vec{m}_{k_p}(x)\cdot\vec{\phi}(x)\,d\sigma\\ \int_{\partial\Omega}E(\xi-x)\vec{\phi}(x)\,d\sigma_x+\sum_{l=1}^{p}\nu_{k_l}\Big[-\frac{1}{2}\vec{m}_{k_l}(\xi)\\ +\int_{\partial\Omega}\vec{\tau}_x^{\,\mathrm{Tr}}(E)(x-\xi)\vec{m}_{k_l}(x)\,d\sigma_x\Big]\end{bmatrix}.$$

But using the Betti–Somigliana formula (9.90) on $\partial\Omega$, we get

$$\int_{\partial\Omega}\vec{\tau}_x^{\,\mathrm{Tr}}(E)(x-\xi)\vec{m}_{k_l}(x)\,d\sigma_x=-\frac{1}{2}\vec{m}_{k_l}(\xi),\quad x\in\partial\Omega,\ 1\leqslant l\leqslant p.$$

Therefore (9.145) follows. The regularity of the mappings $\mathscr{L}_1$ and $\mathscr{L}_1^*$ is a direct consequence of properties (9.101).

We now proceed to show that the kernels of both $\mathscr{L}_{1e}$ and $\mathscr{L}_{1e}^*$ are trivial.

(i) $\mathscr{N}(\mathscr{L}_{1e})=\{0\}$

Assume that $(\omega_{k_1},\ldots,\omega_{k_p},\vec{f})\in\mathscr{N}(\mathscr{L}_{1e})$. Then $\vec{f}\in[C^\infty(\partial\Omega)]^3$ and

$$\int_{\partial\Omega}E(x-\xi)\vec{f}(\xi)\,d\sigma_\xi+\sum_{l=1}^{p}\omega_{k_l}\vec{m}_{k_l}(x)=0,\quad\forall x\in\partial\Omega.\tag{9.146}$$

Define

$$\vec{w}^{\pm}(x)=\int_{\partial\Omega}E(x-\xi)\vec{f}(\xi)\,d\sigma_\xi,\ \text{with}\ \begin{cases}+\ \text{for}\ x\in\Omega^c,\\ -\ \text{for}\ x\in\Omega.\end{cases}\tag{9.147}$$

Then $\vec{w}^+(x)$ is a regular solution with rigid motion $\vec{r}(x) = \vec{0}$ at ∞. On $\partial\Omega$, $\vec{w}^+$ has displacement and traction boundary data

$$\vec{w}^+(x) = -\sum_{l=1}^{p} \omega_{k_l} \vec{m}_{k_l}(x), \tag{9.148}$$
$$\vec{\tau}(\vec{w}^+)(x) = -\frac{1}{2}\vec{f}(x) + \int_{\partial\Omega} \vec{\tau}_x(E)(x,\xi)\vec{f}(\xi)\,d\sigma_\xi.$$

We now apply (9.139) and obtain

$$a(\vec{w}^+,\vec{w}^+) = -\int_{\partial\Omega}\left[-\sum_{l=1}^{p}\omega_{k_l}\vec{m}_{k_l}(x)\right]\cdot\left[-\frac{1}{2}\vec{f}(x) + \int_{\partial\Omega}\vec{\tau}_x(E)(x,\xi)\vec{f}(\xi)d\sigma_\xi\right]d\sigma_x$$
$$+\int_{\partial\Omega}\vec{0}\cdot\left[-\frac{1}{2}\vec{f}(x) + \int_{\partial\Omega}\vec{\tau}_x(E)(x,\tau)\vec{f}(\xi)\,d\sigma_\xi\right]d\sigma_x = 0.$$

Therefore on Ω^c, $\vec{w}^+$ is a pure rigid motion, which is $\vec{0}$, i.e.,

$$\vec{w}^+(x) = 0 \quad \text{on } \Omega^c.$$

From (9.148), we have $\omega_{k_1} = \cdots = \omega_{k_l} = 0$. From (9.101) and Theorem 9.3, we get

$$\vec{w}^-(x) = \vec{0} \quad \text{on } \Omega.$$

Hence

$$\vec{f}(x) = \vec{\tau}(\vec{w}^-)(x) - \vec{\tau}(\vec{w}^+)(x) = \vec{0}.$$

(ii) $\mathscr{N}(\mathscr{L}_{1e}^*) = \{0\}$

Let $(\nu_{k_1},\ldots,\nu_{k_p},\vec{\phi}) \in \mathscr{N}(\mathscr{L}_{1e}^*)$. Define

$$\vec{w}^\pm(x) = \int_{\partial\Omega} E(x-\xi)\vec{\phi}(\xi)\,d\sigma_\xi - \sum_{l=1}^{p}\nu_{k_l}\vec{m}_{k_l}(x), \tag{9.149}$$

with $+$ for $x\in\Omega^c$, and $-$ for $x\in\Omega$. Then $\vec{w}^+$ is a generalized regular solution on Ω^c, with $\vec{0}$ displacement data on $\partial\Omega$ and a rigid motion $\vec{r} = -\sum_{l=1}^{p}\nu_{k_l}\vec{m}_{k_l}(x)$ at ∞. By the modified Betti–Somigliana formula (9.133), we have

$$\vec{w}^+(x) = -\int_{\partial\Omega}[E(x-\xi)\vec{\tau}(\vec{w}^+)(\xi) - T_\xi(x,\xi)\cdot\vec{0}\,]d\sigma_\xi - \sum_{l=1}^{p}\nu_{k_l}\vec{m}_{k_l}(x), \quad x\in\Omega^c. \tag{9.150}$$

Subtracting (9.150) from (9.149), we get

$$\int_{\partial\Omega} E(x-\xi)[\vec{\phi}(\xi) + \vec{\tau}(\vec{w}^+)(\xi)d\sigma_\xi = \vec{0}.$$

But by Theorem 9.5, the kernel of the simple-layer operator is trivial. Therefore

$$\vec{\phi}(x) = -\vec{\tau}(\vec{w}^+)(x), \quad x\in\partial\Omega. \tag{9.151}$$

From the assumption that $(v_{k_1},\dots,v_{k_p},\vec{\phi}) \in \mathcal{N}(\mathcal{L}_1^*)$ and (9.151), we have

$$0 = -\int_{\partial\Omega} \vec{m}_{k_l}(x)\cdot\vec{\tau}(\vec{w}^+)(x)\,d\sigma_x, \quad l = 1,2,\dots,p.$$

By Corollary 9.1, we conclude that

$$\vec{w}^+(x) \equiv \vec{0} \quad \text{on } \Omega^c, \tag{9.152}$$

and

$$v_{k_1} = v_{k_2} = \cdots = v_{k_p} = 0.$$

From (9.101), we have

$$\vec{w}^-(x) = 0 \quad \text{on } \partial\Omega.$$

By the uniqueness of interior displacement BVP, we have

$$\vec{w}^-(x) \equiv 0 \quad \text{on } \Omega. \tag{9.153}$$

By (9.152), (9.153) and Property 9.5, we have

$$\vec{\phi}(x) = \vec{\tau}(\vec{w}^-)(x) - \vec{\tau}(\vec{w}^+)(x) = 0, \quad \forall x \in \partial\Omega.$$ □

Next, we treat the two-dimensional exterior displacement BVP.

Lemma 9.6. *Let $\vec{w} \in [H_{loc}^{3/2}(\Omega^c)]^2$ be a generalized regular solution of the exterior displacement BVP on $\Omega^c \subset \mathbb{R}^2$. Then the traction $\vec{\tau}(\vec{w})$ on the boundary $\partial\Omega$ satisfies the equilibrium condition*

$$\int_{\partial\Omega} \vec{\tau}(\vec{w})(x)\,d\sigma = 0. \tag{9.154}$$

Proof. Let $\vec{w}$ be asymptotically a rigid motion $\vec{r}(x) = \sum_{i=1}^3 \omega_i \vec{m}_i(x)$ at infinity. According to the modified Betti–Somigliana formula (9.133), we have

$$\vec{w}(x) = -\int_{\partial\Omega} [E(x-\xi)\vec{\tau}(\vec{w})(\xi) - T_\xi(x,\xi)\vec{w}(\xi)]\,d\sigma_\xi + \sum_{i=1}^3 \omega_i \vec{m}_i(x).$$

Thus, $\vec{w}_0(x) \equiv \vec{w}(x) - \vec{r}(x)$ is a regular solution satisfying

$$\vec{w}_0(x) = -\int_{\partial\Omega} [E(x-\xi)\vec{\tau}(\vec{w})(\xi) - T_\xi(x,\xi)\vec{w}(\xi)]\,d\sigma_\xi.$$

Using E given by (9.53), for large $|x|$ we have the asymptotic expansion

$$\vec{w}_0(x) = \frac{\lambda+3\mu}{4\pi\mu(\lambda+2\mu)}\Bigg\{ -\ln|x| \int_{\partial\Omega} \vec{\tau}(\vec{w})(\xi)\,d\sigma_\xi$$
$$+ \frac{\lambda+\mu}{\lambda+3\mu}\frac{1}{|x|^2}\int_{\partial\Omega} \begin{bmatrix} x_1^2 & x_1x_2 \\ x_1x_2 & x_2^2 \end{bmatrix} \vec{\tau}(\vec{w})(\xi)\,d\sigma_\xi + \mathcal{O}(|x|^{-1}) \Bigg\}.$$

We see that $\vec{w}_0$ is a regular solution if and only if $\int_{\partial\Omega} \vec{\tau}(\vec{w})\,d\sigma = \vec{0}$. □

As indicated in (9.132), there are two possible ways of imposing the boundary condition at ∞ for two-dimensional displacement BVP. Let us consider only case (2) in (9.132). Case (1) is left as an exercise after Theorem 9.11.

Theorem 9.11. (Generalized regular simple-layer solution for the exterior displacement elastostatic problem in $\mathbb{R}^2$). *Let Ω be a bounded domain in $\mathbb{R}^2$ with smooth boundary $\partial\Omega$. The linear operator*

$$\mathscr{L}_{1e}\left(\begin{bmatrix}\vec{w}\\ \vec{f}\end{bmatrix}\right)(x)=\begin{bmatrix}\int_{\partial\Omega}\vec{m}_1(\xi)\cdot\vec{f}(\xi)\,d\sigma\\ \int_{\partial\Omega}\vec{m}_2(\xi)\cdot\vec{f}(\xi)\,d\sigma\\ \int_{\partial\Omega}\vec{m}_3(x)\cdot[-\frac{1}{2}\vec{f}(x)+\int_{\partial\Omega}\vec{\tau}_x(E)(x,\xi)\vec{f}(\xi)\,d\sigma_\xi]\,d\sigma_x\\ \int_{\partial\Omega}E(x-\xi)\vec{f}(\xi)\,d\sigma_\xi+\sum_{i=1}^{3}\omega_i\vec{m}_i(x)\end{bmatrix},$$

$$x\in\partial\Omega,$$

$$\mathscr{L}_{1e}\colon\ \mathbb{R}^3\oplus[H^\alpha(\partial\Omega)]^2\longrightarrow\mathbb{R}^3\oplus\left[H^{\alpha+1}(\partial\Omega)\right]^2,\quad \alpha\in\mathbb{R},$$

$$\vec{\omega}=(\omega_1,\omega_2,\omega_3)\in\mathbb{R}^3,\vec{f}=(f_1,f_2)\in[H^\alpha(\partial\Omega)]^2,$$

is an isomorphism, with adjoint

$$\mathscr{L}^*_{1e}\left(\begin{bmatrix}\vec{v}\\ \\ \vec{\phi}\end{bmatrix}\right)(x)=\begin{bmatrix}\int_{\partial\Omega}\vec{m}_1(\xi)\cdot\vec{\phi}(\xi)\,d\sigma\\ \int_{\partial\Omega}\vec{m}_2(\xi)\cdot\vec{\phi}(\xi)\,d\sigma\\ \int_{\partial\Omega}\vec{m}_3(\xi)\cdot\vec{\phi}(\xi)\,d\sigma\\ \int_{\partial\Omega}E(x-\xi)\vec{\phi}(\xi)\,d\sigma_\xi\\ +v_1\vec{m}_1(x)+v_2\vec{m}_2(x)-v_3\vec{m}_3(x)\end{bmatrix},$$

$$x\in\partial\Omega, \tag{9.155}$$

$$\mathscr{L}^*_{1e}\colon\ \mathbb{R}^3\oplus[H^{-(\alpha+1)}(\partial\Omega)]^2\longrightarrow\mathbb{R}^3\oplus[H^{-\alpha}(\partial\Omega)]^2,$$

which is also an isomorphism. $\mathscr{L}^*_{1e}$ satisfies

$$\mathscr{L}^*_{1e}\colon\ \mathbb{R}^3\oplus[H^s(\partial\Omega)]^2\longrightarrow\mathbb{R}^3\oplus[H^{s+1}(\partial\Omega)]^2\ \text{isomorphically}\ \forall s\in R.$$

The generalized regular solution to the exterior BVP

$$\begin{cases}\mu\Delta\vec{w}+(\lambda+\mu)\nabla(\nabla\cdot\vec{w})=0 & \text{on }\Omega^c,\\ \vec{w}(x)=\vec{g}(x)\in[H^{\alpha+1}(\partial\Omega)]^2 & \text{on }\partial\Omega,\ \alpha\in\mathbb{R},\\ \int_{\partial\Omega}\vec{m}_3(x)\cdot\vec{\tau}(\vec{w})(x)\,d\sigma=b\in\mathbb{R},\end{cases}$$

is given uniquely by

$$\vec{w}(x) = \int_{\partial\Omega} E(x-\xi)\vec{f}(\xi)\,d\sigma + \sum_{i=1}^{3} \omega_i \vec{m}_i(x),$$

where

$$\mathscr{L}_1\left(\begin{bmatrix} \vec{\omega} \\ \vec{f} \end{bmatrix}\right) = \begin{bmatrix} 0 \\ 0 \\ b \\ \vec{g} \end{bmatrix},$$

with $\vec{f} \in [H^{\alpha}(\partial\Omega)]^2$ *and* $\vec{w} \in [H^{\alpha+3/2}_{loc}(\Omega^c)]^2$.

Proof. It can be easily verified by transposition that $\mathscr{L}^*_{1e}$ is given by (9.155). We can show that $\mathscr{N}(\mathscr{L}_{1e}) = \{0\}$ and $\mathscr{N}(\mathscr{L}^*_{1e}) = \{0\}$, just as in the proof of Theorem 9.10. The theorem follows. □

Exercise 9.4. Use the fact that

$$\int_{\partial\Omega} \vec{\tau}(\vec{w})(x)\,d\sigma = \begin{bmatrix} 0 \\ 0 \end{bmatrix}$$

and Theorem 9.11 to show that the exterior displacement BVP Case (1) in (9.132) has a unique solution with a simple-layer representation

$$\vec{w}(x) = \int_{\partial\Omega} E(x-\xi)\vec{f}(\xi)\,d\sigma_\xi + \begin{bmatrix} \omega_1 \\ \omega_2 \end{bmatrix} + \omega_3 \vec{m}_3(x),$$

satisfying $\vec{w} = \vec{g} \in [H^{r+1}(\partial\Omega)]^2$, where $\vec{f} \in [H^r(\partial\Omega)]^2$ and $(\omega_1, \omega_2) \in \mathbb{R}^2$ are unique. □

9.8 Simple-layer solution for the exterior traction BVP

Consider

$$\left.\begin{array}{l} \mu\Delta\vec{w} + (\lambda+\mu)\nabla(\nabla\cdot\vec{w}) = 0 \quad \text{on } \Omega^c, \\ \vec{\tau}(\vec{w})(x) = \vec{g}(x) \quad \text{on } \partial\Omega, \vec{g} \text{ is given}, \\ \vec{w} \text{ is a regular solution at } \infty, \end{array}\right\} \tag{9.156}$$

in $\mathbb{R}^3$ or $\mathbb{R}^2$. We want to find a regular solution to this BVP by the simple-layer representation. Note that in $\mathbb{R}^2$, $\vec{\tau}(\vec{w}) = \vec{g}$ must satisfy (9.154).

Theorem 9.12 (Simple-layer regular solution of the exterior traction BVP in $\mathbb{R}^3$). *Let Ω be a bounded domain in $\mathbb{R}^3$ with smooth boundary $\partial\Omega$. The linear operator*

$$\mathscr{L}_{2e}(\vec{f})(x) = -\frac{1}{2}\vec{f}(x) + \int_{\partial\Omega} \vec{\tau}_x(E)(x,\xi)\vec{f}(\xi)\,d\sigma_\xi, \quad x \in \partial\Omega,$$

$$\mathscr{L}_{2e}\colon [H^\alpha(\partial\Omega)]^3 \longrightarrow [H^\alpha(\partial\Omega)]^3, \quad \alpha \in \mathbb{R},$$

is an isomorphism, with adjoint

$$\mathscr{L}^*_{2e}(\vec{\phi})(x) = -\frac{1}{2}\vec{\phi}(x) + \int_{\partial\Omega} T_\xi(x,\xi)\vec{\phi}(\xi)\,d\sigma_\xi, \quad x \in \partial\Omega,$$

$$\mathscr{L}^*_{2e}\colon [H^{-\alpha}(\partial\Omega)]^3 \longrightarrow [H^{-\alpha}(\partial\Omega)]^3,$$

also an isomorphism.

The regular solution to the exterior BVP (9.156) is given uniquely by

$$w(x) = \int_{\partial\Omega} E(x-\xi)\vec{f}(\xi)\,d\sigma_\xi \in [H^{\alpha+3/2}_{\text{loc}}(\Omega^c)]^3,$$

where $\vec{f} \in [H^\alpha(\partial\Omega)]^3$ is the unique solution of the BIE

$$\mathscr{L}_{2e}(\vec{f}) = \vec{g}, \quad \vec{g} \in [H^\alpha(\partial\Omega)]^3.$$

Proof. The arguments are similar to those in the proof of Theorem 9.10. We leave them to the reader. □

For two-dimensional exterior traction BVP, we state the following.

Theorem 9.13 (Simple-layer regular solution of the exterior traction BVP in $\mathbb{R}^2$). *Let Ω be a bounded domain in $\mathbb{R}^2$ with smooth boundary $\partial\Omega$. Define*

$$\mathscr{L}'_{2e}\colon \mathbb{R}^2 \oplus [H^\alpha(\partial\Omega)]^2 \longrightarrow \mathbb{R}^2 \oplus [H^\alpha(\partial\Omega)]^2$$

by

$$\mathscr{L}'_{2e}\left(\begin{bmatrix}\vec{a}\\ \vec{f}\end{bmatrix}\right)(x) = \begin{bmatrix}\int_{\partial\Omega}\vec{f}(\xi)\,d\sigma \\ -\frac{1}{2}\vec{f}(x) + \int_{\partial\Omega}\vec{\tau}_x(E)(x,\xi)\vec{f}(\xi)\,d\sigma_\xi + \vec{a}\end{bmatrix}.$$

Then $\mathscr{L}'_{2e}$ is an isomorphism. $\mathscr{L}'_{2e}$ has an adjoint

$$\mathscr{L}'^*_{2e}\colon \mathbb{R}^2 \oplus [H^{-\alpha}(\partial\Omega)]^2 \longrightarrow \mathbb{R}^2 \oplus [H^{-\alpha}(\partial\Omega)]^2,$$

$$\mathscr{L}'^*_{2e}\left(\begin{bmatrix}\vec{\alpha}\\ \vec{\phi}\end{bmatrix}\right)(x) = \begin{bmatrix}\int_{\partial\Omega}\vec{\phi}(\xi)\,d\sigma \\ -\frac{1}{2}\vec{\phi}(x) + \int_{\partial\Omega}\vec{\tau}^{Tr}_\xi(E)(x,\xi)\vec{\phi}(\xi)\,d\sigma_\xi + \vec{\alpha}\end{bmatrix}.$$

$\mathscr{L}_{2e}^{\prime *}*$ satisfies

$$\mathscr{L}_{2e}^{\prime *}*: \mathbb{R}^2 \oplus [H^s(\partial\Omega)]^2 \longrightarrow \mathbb{R}^2 \oplus [H^s(\partial\Omega)]^2 \text{ isomorphically } \forall\, s \in \mathbb{R}.$$

The regular solution to the exterior traction BVP (9.156) is given uniquely by

$$\vec{w}\,(x) = \int_{\partial\Omega} E(x-\xi)\,\vec{f}\,(\xi)\,d\sigma_\xi, \quad x \in \Omega^c, \quad \vec{w} \in H_{loc}^{\alpha+3/2}(\Omega^c),$$

where $\vec{f} \in [H^\alpha(\partial\Omega)]^2$ is the unique solution to

$$\mathscr{L}_{2e}^{\prime}\left(\begin{bmatrix} \vec{a} \\ \vec{f} \end{bmatrix}\right) = \begin{bmatrix} 0 \\ \vec{g} \end{bmatrix}.$$

for $\vec{g} \in [H^\alpha(\partial\Omega)]^2$ satisfying $\int_{\partial\Omega} \vec{g}\,(\xi)\,d\sigma = \vec{0}$ (implying $\int_{\partial\Omega} \vec{f}\,(\xi)\,d\sigma = \vec{0}$ and $\vec{a} = \vec{0}$).

Proof. The arguments are identical to those in the proof of Theorem 9.10. We leave them to the reader, except that we need only note

$$\begin{aligned}\int_{\partial\Omega}\Big[-\frac{1}{2}\vec{f}\,(x) &+ \int_{\partial\Omega} \vec{\tau}_x(E)(x,\xi)\,\vec{f}\,(\xi)\,d\sigma_\xi + \vec{a}\,\Big]\,d\sigma_x \\ &= \int_{\partial\Omega}[-f(x)+\vec{a}\,]\,d\sigma \\ &= \vec{a}\cdot \text{arclength}\,\partial\Omega = \int_{\partial\Omega} \vec{g}\,(x)\,d\sigma = 0.\end{aligned}$$

□

9.9 Solutions of exterior BVP by double-layer potentials

On an exterior domain $\Omega^c \subseteq \mathbb{R}^3$, the double-layer representation is given formally by

$$V_2(\vec{f}\,)(x) = \int_{\partial\Omega} T_\xi(x,\xi)\,\vec{f}\,(\xi)\,d\sigma_\xi, \quad x \in \Omega^c. \tag{9.157}$$

Its asymptotic behavior in $\mathbb{R}^3$ is

$$V_2(\vec{f}\,)(x) = \mathscr{O}(|x|^{-2}), \quad \text{large } |x|.$$

In order to obtain generalized regular solutions for the exterior displacement BVP based upon the double-layer representation, we may mimic (6.223) and try a *modified double-layer representation*

$$\vec{V}_2(\vec{f}\,)(x) = \int_{\partial\Omega} T_\xi(x,\xi)\,\vec{f}\,(\xi)\,d\sigma_\xi + E(x,0)\,\vec{a} + \sum_{i=1}^{6} \omega_i\,\vec{m}_i(x), \; x \in \Omega^c, \tag{9.158}$$

where $\vec{a} \in \mathbb{R}^3$, and the origin 0 is assumed to be contained in Ω. The term $E(x,0)\,\vec{a}$ is included with the hope of providing the $\mathscr{O}(|x|^{-1})$ decay rate at ∞. Note that using $E(x,\xi_0)\,\vec{a}$

instead of $E(x,0)\vec{a}$ in (9.158) for some $\xi_0 \in \Omega$, $\xi_0 \neq 0$, will not provide extra advantage, since

$$E(x,\xi_0)\vec{a} = E(x,0)\vec{a} + \mathscr{O}(|x|^{-2}), \quad \text{large } |x|,$$

and this $\mathscr{O}(|x|^{-2})$ term would be picked up by the original double-layer (9.157) itself. The use of a combination

$$\sum_{i=1}^{p} E(x,\xi_i)\vec{a}_i, \quad \xi_i \in \Omega,$$

would cause additional complications; thus it also is best avoided. As it turns out, such a modified double-layer potential (9.158) will not encompass all solutions of the displacement BVP with arbitrary momenta. Let (9.158) be a solution of (9.156) on Ω^c. Call it $\vec{w}$. Assume that $\mathscr{J} = \emptyset$ (the empty set) and six momentum data are given:

$$\int_{\partial\Omega} \vec{m}_j(x) \cdot \vec{\tau}(\vec{w})(x)\, d\sigma = b_j \in \mathbb{R}, \quad j = 1,2,\ldots,6. \tag{9.159}$$

Then from (9.158),

$$\begin{aligned} &\int_{\partial\Omega} \vec{m}_j(x) \left\{ \int_{\partial\Omega} \vec{\tau}_x[\vec{\tau}_{\xi}^{\mathrm{Tr}}(E)(x,\xi)]\vec{f}(\xi)\, d\sigma_\xi + \vec{\tau}_x(E)(x,0)\vec{a} \right\} d\sigma_x = b_j, \\ &\int_{\partial\Omega} \vec{f}(\xi)\,\vec{\tau} \int_{\partial\Omega} \vec{\tau}_x^{\mathrm{Tr}}(E)(x,\xi)\vec{m}_j(x)\, d\sigma_x\, d\sigma_\xi - \frac{1}{2}\vec{m}_j(0)\cdot\vec{a} = b_j, \\ &1 \leqslant j \leqslant 6. \end{aligned} \tag{9.160}$$

The first term in (9.160) is 0 by (9.140), so

$$\frac{1}{2}\vec{m}_j(0)\cdot\vec{a} = -b_j, \quad j = 1,2,\ldots,6. \tag{9.161}$$

But

$$\vec{m}_j(0) = \vec{0}, \quad j = 3,5,6,$$

therefore

$$b_j = 0, \quad j = 3,5,6,$$

contradicting the fact that b_j, $1 \leqslant j \leqslant 6$, can be arbitrarily prescribed for exterior displacement BVP. (Again, this appears to be a disadvantage of the double-layer potential.)

Exercise 9.5. Discuss for what given $b_1, b_2, \ldots, b_6$ in (9.159), the exterior displacement BVP is solvable based on the representation

$$\vec{w}(x) = \int_{\partial\Omega} \vec{\tau}_{\xi}^{\mathrm{Tr}}(E)(x,\xi)\vec{f}(\xi)\, d\sigma_\xi + E(x,\xi_0)\vec{a} + \sum_{i=1}^{6} \omega_i \vec{m}_i(x), \quad x \in \Omega^c,$$

where ξ_0 is chosen to be in Ω but $\xi_0 \neq \vec{0}$. □

The difficulty (9.161) associated with (9.158) can be overcome if, instead of (9.158), we make the *ansatz*

$$\widetilde{V}_2(\vec{f})(x) = \int_{\partial\Omega} T_\xi(x,\xi)\,\vec{f}(x)\,d\sigma_\xi + \sum_{i=1}^{6} a_i \int_{\partial\Omega} E(x,\xi)\,\vec{k}_i(\xi)\,d\sigma_\xi$$
$$+ \sum_{i=1}^{6} \omega_i \vec{m}_i(x), \quad x \in \partial\Omega, \tag{9.162}$$

where $\vec{k}_i \in [C^\infty(\partial\Omega)]^3$, $i = 1, 2, \ldots, 6$, are the unique solutions of

$$\mathscr{L}_1(\vec{k}_i) = \vec{m}_i, \quad 1 \leqslant i \leqslant 6,$$

(see Theorem 9.5 for $\mathscr{L}_1$). Then

$$\int_{\partial\Omega} \vec{m}_j(x) \cdot \vec{\tau}(\widetilde{V}_2)(x)\,d\sigma$$
$$= \int_{\partial\Omega} \vec{m}_j(x) \cdot \Big\{ \int_{\partial\Omega} \vec{\tau}_x[\vec{\tau}_\xi^{\,\mathrm{Tr}}(E)(x,\xi)]\,\vec{f}(\xi)\,d\sigma_\xi$$
$$+ \sum_{i=1}^{6} a_j \int_{\partial\Omega} \tau_x(E)(x,\xi)\,\vec{k}_i(\xi)\,d\sigma_\xi \Big\}$$
$$= \int_{\partial\Omega} \vec{f}(\xi) \cdot \vec{\tau} \int_{\partial\Omega} \vec{\tau}_x^{\,\mathrm{Tr}}(E)(x,\xi)\,\vec{m}_j(x)\,d\sigma_x\,d\sigma_\xi$$
$$- \frac{1}{2}\sum_{i=1}^{6} a_i \int_{\partial\Omega} \vec{k}_i(\xi) \cdot \vec{m}_j(\xi)\,d\sigma_\xi$$
$$= 0 - \frac{1}{2}\sum_{i=1}^{6} a_i \langle \mathscr{L}_1 \vec{k}_i, \vec{k}_j \rangle_{L^2(\partial\Omega)} \equiv -A\vec{a}, \tag{9.163}$$

where A is a 6×6 matrix with entries $a_{ij} = \frac{1}{2}\langle \mathscr{L}_1 \vec{k}_i, \vec{k}_j \rangle_{L^2(\partial\Omega)}$, and $\vec{a} = (a_1, a_2, \ldots, a_6) \in \mathbb{R}^6$. From (9.111), the matrix A is positive definite and nonsingular. Therefore, for given b_j, $1 \leqslant j \leqslant 6$,

$$b_j = \int_{\partial\Omega} \vec{m}_j(x) \cdot \vec{\tau}(\widetilde{V}_2)(x)\,d\sigma$$
$$= -\frac{1}{2}\sum_{i=1}^{6} a_i \langle \mathscr{L}_1 \vec{k}_i, \vec{k}_j \rangle, \quad 1 \leqslant j \leqslant 6, \tag{9.164}$$

will hold for some (unique) a_i, $1 \leqslant i \leqslant 6$.

Note that (9.162) is *not a pure (augmented) double-layer potential.*

Theorem 9.14. (Modified double-layer representation for generalized regular solutions of the exterior displacement BVP in $\mathbb{R}^3$). *Let Ω be a bounded domain in $\mathbb{R}^3$ with smooth boundary $\partial\Omega$. Define the operator*

$$\mathscr{L}_{3e} \colon \mathbb{R}^{12} \oplus [H^\alpha(\partial\Omega)]^3 \longrightarrow \mathbb{R}^{12} \oplus [H^\alpha(\partial\Omega)]^3,$$

$$\mathscr{L}_{3e}\left(\begin{bmatrix}\vec{\omega}\\ \vec{a}\\ \vec{f}\end{bmatrix}\right)(x)=\begin{bmatrix}\int_{\partial\Omega}M^{Tr}(\xi)\vec{f}(\xi)\,d\sigma_\xi\\ A\vec{a}\\ \frac{1}{2}\vec{f}(x)+\int_{\partial\Omega}\vec{\tau}_\xi^{Tr}(E)(x,\xi)\vec{f}(\xi)\,d\sigma_\xi\\ \quad+\sum_{k=1}^{6}\left[a_i\int_{\partial\Omega}E(x,\xi)\vec{k}_i(\xi)\,d\sigma_\xi+\omega_i m_i(x)\right]\end{bmatrix},$$

$$x\in\partial\Omega,\quad \vec{\omega}\in\mathbb{R}^6,\quad \vec{a}\in\mathbb{R}^3 \tag{9.165}$$

where A is defined in (9.163). Then $\mathscr{L}_{3e}$ is an isomorphism, with adjoint

$$\mathscr{L}_{3e}^*:\ \mathbb{R}^{12}\oplus[H^{-\alpha}(\partial\Omega)]^3\longrightarrow\mathbb{R}^{12}\oplus[H^{-\alpha}(\partial\Omega)]^3,$$

$$\mathscr{L}_{3e}^*\left(\begin{bmatrix}\vec{\nu}\\ \vec{\alpha}\\ \vec{\phi}\end{bmatrix}\right)(x)=\begin{bmatrix}\int_{\partial\Omega}M^{Tr}(\xi)\vec{\phi}(\xi)\,d\sigma\\ A\vec{\alpha}+\vec{\mu}\\ \frac{1}{2}\vec{\phi}(x)+\int_{\partial\Omega}\vec{\tau}_x(E)(x,\xi)\vec{\phi}(\xi)\,d\sigma_\xi\\ \quad+\sum_{k=1}^{6}\nu_k\vec{m}_k(x)\end{bmatrix},$$

$$x\in\partial\Omega,$$

$$\vec{\mu}=(\mu_1,\ldots,\mu_6)\in\mathbb{R}^6,\quad \mu_j=\langle\vec{m}_j,\vec{\phi}\rangle_{L^2(\partial\Omega)}.$$

$\mathscr{L}_{3e}^*$ is also an isomorphism,

$$\mathscr{L}_{3e}^*:\ \mathbb{R}^{12}\oplus[H^r(\partial\Omega)]^3\longrightarrow\mathbb{R}^{12}\oplus[H^r(\partial\Omega)]^3$$

for any $r\in\mathbb{R}$.

Consequently, for the exterior displacement BVP

$$\left.\begin{aligned}&\mu\Delta\vec{w}+(\lambda+\mu)\nabla(\nabla\cdot\vec{w})=0\ \text{ on }\Omega^c,\\ &\vec{w}(x)=\vec{g}(x)\ \text{ on }\partial\Omega,\ \ \vec{g}\in[H^r(\partial\Omega)]^3,\ r\in\mathbb{R},\\ &\int_{\partial\Omega}\vec{m}_k(x)\cdot\vec{\tau}(\vec{w})(x)\,d\sigma=b_k,\ k=1,2,\ldots,6,\\ &\vec{w}(x)\sim\sum_{k=1}^{6}\omega_k\vec{m}_k(x)\ \text{ at }\infty,\\ &\vec{\omega}=(\omega_1,\omega_2,\ldots,\omega_6)\ \ \textit{unknown},\end{aligned}\right\} \tag{9.166}$$

the solution is given uniquely by

$$\begin{aligned}\vec{w}(x)=&\int_{\partial\Omega}\vec{\tau}_\xi^{Tr}(E)(x,\xi)\vec{f}(\xi)\,d\sigma_\xi\\ &+\sum_{i=1}^{6}\left[a_i\int_{\partial\Omega}E(x,\xi)\vec{k}_i(\xi)\,d\sigma_\xi+\omega_i\vec{m}_i(x)\right],\quad x\in\Omega^c,\end{aligned} \tag{9.167}$$

where $(\vec{\omega}, \vec{a}, \vec{f}) \in \mathbb{R}^{12} \oplus [H^r(\partial\Omega)]^3$ *is the unique solution of*

$$\mathscr{L}_{3e}\left(\begin{bmatrix}\vec{\omega}\\ \vec{a}\\ \vec{f}\end{bmatrix}\right) = \begin{bmatrix}\vec{\eta}\\ \vec{b}\\ \vec{g}\end{bmatrix} \in \mathbb{R}^{12} \oplus [H^r(\partial\Omega)]^3, \tag{9.168}$$

$\vec{\eta} \in \mathbb{R}^6$ *can be given arbitrarily, but* $\vec{\omega}$ *and* $\vec{f}$ *depend on* $\vec{\eta}$. *The solution satisfies* $\vec{w} \in [H^{r+1/2}_{\text{loc}}(\Omega^c)]^3$.

Proof. We easily see that $\mathscr{L}_{3e}$ on $\mathbb{R}^9 \oplus [H^\alpha(\partial\Omega)]^3$ is bounded and invertible iff the reduced operator

$$\widetilde{\mathscr{L}_{3e}}\begin{bmatrix}\vec{\omega}\\ \vec{f}\end{bmatrix}(x) \equiv \begin{bmatrix}\int_{\partial\Omega} M^{\text{Tr}}(\xi)\vec{f}(\xi)\,d\sigma\\ \frac{1}{2}\vec{f}(x) + \int_{\partial\Omega}\vec{\tau}_\xi^{\,\text{Tr}}(E)(x,\xi)\vec{f}(\xi)\,d\sigma_\xi + \sum_{k=1}^{6}\omega_k\vec{m}_k(x)\end{bmatrix} \tag{9.169}$$

is bounded and invertible on $\mathbb{R}^6 \oplus [H^\alpha(\partial\Omega)]^3$. But this can be proved by using the standard arguments.

We now consider the solution of the BVP (9.166) based on the *ansatz* (9.162). We use (9.163): on its RHS,

$$\vec{a} = -A^{-1}\vec{b}, \text{ according to (9.164), for } \vec{a} \in \mathbb{R}^6.$$

Then (9.168) is uniquely solvable, and the solution $\vec{w}$ in (9.167) constructed from $(\vec{\omega}, \vec{a}, \vec{f})$ in (9.168) satisfies (9.166). Therefore the proof is complete. □

Now, consider the two-dimensional exterior displacement BVP, $\Omega^c \subseteq \mathbb{R}^2$. The double-layer representation

$$\vec{V}_2(\vec{f})(x) = \int_{\partial\Omega}\vec{\tau}_\xi^{\,\text{Tr}}(E)(x,\xi)\vec{f}(\xi)\,d\sigma_\xi, \quad x \in \Omega^c,$$

has decay rate $\mathscr{O}(|x|^{-1})$ for large $|x|$. If the solution $\vec{w}$ has rigid motion $\sum_{i=1}^3 \omega_i\vec{m}_i(x)$ at ∞ then its double-layer representation is naturally

$$\vec{w}(x) = \int_{\partial\Omega}\vec{\tau}_\xi^{\,\text{Tr}}(E)(x,\xi)\vec{f}(\xi)\,d\sigma_\xi + \sum_{i=1}^{3}\omega_i\vec{m}_i(x), \ x \in \Omega^c. \tag{9.170}$$

Associated with this representation, let us check the momentum conditions

$$\begin{aligned}&\int_{\partial\Omega}\vec{m}_j(x)\cdot\vec{\tau}(\vec{w})(x)\,d\sigma\\ &= \int_{\partial\Omega}\vec{m}_j(x)\cdot\left\{\int_{\partial\Omega}\vec{\tau}_x[\vec{\tau}_\xi^{\,\text{Tr}}(E)(x,\xi)]\vec{f}(\xi)\,d\sigma_\xi\right\}d\sigma_x\\ &= \int_{\partial\Omega}\vec{f}(x)\cdot\vec{\tau}_\xi\int_{\partial\Omega}\vec{\tau}_x^{\,\text{Tr}}(E)(x,\xi)\vec{m}_j(\xi)\,d\sigma_\xi\,d\sigma_x\\ &= 0, \quad j = 1, 2, 3.\end{aligned}$$

We see that the representation (9.170) only allows us to treat the case of the boundary conditions (9.132), and b_3 *must be zero.*

Theorem 9.15. (Double-layer representation for generalized regular solutions of the exterior displacement BVP with restricted momentum in $\mathbb{R}^2$). *Let Ω be a bounded domain in $\mathbb{R}^2$ with smooth boundary $\partial\Omega$. Define the operator*

$$\mathscr{L}_{3e}\colon \mathbb{R}^3 \oplus [H^\alpha(\partial\Omega)]^2 \longrightarrow \mathbb{R}^3 \oplus [H^\alpha(\partial\Omega)]^2, \quad \alpha \in \mathbb{R},$$

$$\mathscr{L}_{3e}\left(\begin{bmatrix} \vec{\omega} \\ \vec{f} \end{bmatrix}\right)(x) = \begin{bmatrix} \int_{\partial\Omega} M^{Tr}(\xi)\vec{f}(\xi)\,d\sigma \\ \frac{1}{2}\vec{f}(x) + \int_{\partial\Omega} \vec{\tau}_\xi^{Tr}(E)(x,\xi)\vec{f}(\xi)\,d\sigma_\xi + \sum_{k=1}^{3}\omega_k\vec{m}_k(x) \end{bmatrix},$$

$$x \in \partial\Omega.$$

Then $\mathscr{L}_{3e}$ is an isomorphism, with adjoint

$$\mathscr{L}_{3e}^*\colon \mathbb{R}^3 \oplus [H^{-\alpha}(\partial\Omega)]^2 \longrightarrow \mathbb{R}^3 \oplus [H^{-\alpha}(\partial\Omega)]^2,$$

$$\mathscr{L}_{3e}^*\left(\begin{bmatrix} \vec{\nu} \\ \vec{\phi} \end{bmatrix}\right)(x) = \begin{bmatrix} \int_{\partial\Omega} M^{Tr}(\xi)\vec{\phi}(\xi)\,d\sigma \\ \frac{1}{2}\vec{\phi}(x) + \int_{\partial\Omega} \vec{\tau}_\xi^{Tr}(E)(x,\xi)\vec{\phi}(\xi)\,d\sigma_\xi + \sum_{k=1}^{3}\nu_k\vec{m}_k(x) \end{bmatrix},$$

$$x \in \partial\Omega.$$

and $\mathscr{L}_{3e}^*$ satisfies

$$\mathscr{L}_{3e}^*\colon \mathbb{R}^3 \oplus [H^r(\partial\Omega)]^2 \longrightarrow \mathbb{R}^3 \oplus [H^r(\partial\Omega)]^2 \textit{ isomorphically } \forall\, r \in \mathbb{R}.$$

Consequently, for the exterior displacement BVP

$$\left.\begin{aligned} &\mu\Delta\vec{w} + (\lambda+\mu)\nabla(\nabla\cdot\vec{w}) = 0 \text{ on } \Omega^c, \\ &\vec{w}(x) = \vec{g}(x) \text{ on } \partial\Omega, \quad \vec{g} \in [H^r(\partial\Omega)]^2, \ r \in \mathbb{R}, \\ &\int_{\partial\Omega} \vec{m}_j(x)\cdot\vec{\tau}(\vec{w})(x)\,d\sigma = 0, \quad j = 1,2,3, \\ &\vec{w}(x) \sim \sum_{k=1}^{3}\omega_k\vec{m}_k(x) \text{ at } \infty, \quad \vec{\omega} = (\omega_1,\omega_2,\omega_3) \text{ unknown,} \end{aligned}\right\} \tag{9.171}$$

the solution is given uniquely by

$$\vec{w}(x) = \int_{\partial\Omega} \vec{\tau}_\xi^{Tr}(E)(x,\xi)\vec{f}(\xi)\,d\sigma_\xi + \sum_{i=1}^{3}\omega_i\vec{m}_i(x), \quad x \in \Omega^c,$$

where $(\vec{\omega},\vec{f}) \in \mathbb{R}^3 \oplus [H^r(\partial\Omega)]^2$ is the unique solution of

$$\mathscr{L}_{3e}\begin{bmatrix} \vec{\omega} \\ \vec{f} \end{bmatrix} = \begin{bmatrix} \vec{\eta} \\ \vec{g} \end{bmatrix} \in \mathbb{R}^3 \oplus [H^\alpha(\partial\Omega)]^2,$$

$\vec{\eta} \in \mathbb{R}^3$ can be chosen arbitrarily, but $\vec{\omega}$ and $\vec{f}$ depend on $\vec{\eta}$. The solution satisfies $\vec{w} \in [H_{loc}^{r+1/2}(\Omega^c)]^2$.

Proof. This is similar to before. □

Next, let us consider the traction BVP. For an exterior domain Ω^c in $\mathbb{R}^3$, we first tend to try the representation

$$\vec{w}(x) = \int_{\partial\Omega} \vec{\tau}_{\xi}^{\mathrm{Tr}}(E)(x,\xi)\,\vec{f}(\xi)\,d\sigma_{\xi} + E(x,0)\,\vec{a}, \quad x \in \Omega^c. \tag{9.172}$$

where $0 \in \Omega$ and $\vec{a} \in \mathbb{R}^3$. Further analysis again shows that (9.172) is inadequate. It turns out that a representation that works is

$$\vec{w}(x) = \int_{\partial\Omega} \vec{\tau}_{\xi}^{\mathrm{Tr}}(E)(x,\xi)\,\vec{f}(\xi)\,d\sigma_{\xi} + \sum_{i=1}^{6} \omega_i \int_{\partial\Omega} E(x,\xi)\,\vec{F}_i(\xi)\,d\sigma_{\xi}, \quad x \in \Omega^c, \tag{9.173}$$

where $\vec{F}_i \in [C^{\infty}(\partial\Omega)]^3$, $1 \leqslant i \leqslant 6$, are the solutions of

$$\vec{F}_i = \mathscr{L}_{2e}^{-1}(\vec{m}_i), \quad \text{see Theorem 9.12.}$$

Obviously, $\vec{w}$ satisfies the elastostatics equation (9.7), and $\vec{w}$ is a regular function at ∞.

Theorem 9.16. (Modified double-layer representation for regular solution of exterior traction BVP in $\mathbb{R}^3$). *Let Ω be a bounded domain in $\mathbb{R}^3$ with smooth boundary $\partial\Omega$ and containing the origin. Define the operator*

$$\mathscr{L}_{4e}\colon\ \mathbb{R}^6 \oplus [H^{\alpha+1}(\partial\Omega]^3 \longrightarrow \mathbb{R}^6 \oplus [H^{\alpha}(\partial\Omega)]^3, \quad \alpha \in \mathbb{R},$$

$$\mathscr{L}_{4e}\left(\begin{bmatrix} \vec{\omega} \\ \vec{f} \end{bmatrix}\right)(x) = \begin{bmatrix} \displaystyle\int_{\partial\Omega} M^{Tr}(\xi)\,\vec{f}(\xi)\,d\sigma \\ \displaystyle\int_{\partial\Omega} \vec{\tau}_x\left[\vec{\tau}_{\xi}^{Tr}(E)(x,\xi)\right]\vec{f}(\xi)\,d\sigma_{\xi} + M(x)\,\vec{\omega} \end{bmatrix}.$$

Then $\mathscr{L}_{4e}$ is an isomorphism, with an adjoint

$$\mathscr{L}_{4e}^{*}\colon\ \mathbb{R}^6 \oplus [H^{-\alpha}(\partial\Omega)]^3 \longrightarrow \mathbb{R}^6 \oplus [H^{-(\alpha+1)}(\partial\Omega)]^3,$$

$$\mathscr{L}_{4e}^{*} = \mathscr{L}_{4e}, \quad \text{formally.}$$

$\mathscr{L}_{4e}^{*}$ is also an isomorphism,

$$\mathscr{L}_{4e}^{*}\colon\ \mathbb{R}^6 \oplus [H^{r+1}(\partial\Omega)]^3 \longrightarrow \mathbb{R}^6 \oplus [H^{r}(\partial\Omega)]^3,$$

for any $r \in \mathbb{R}$.

Consequently, for the exterior traction BVP

$$\left.\begin{aligned} &\mu\Delta\vec{w} + (\lambda+\mu)\nabla(\nabla\cdot\vec{w}) = 0 \quad \text{on } \Omega^c, \\ &\vec{\tau}(\vec{w})(x) = \vec{g}(x) \quad \text{on } \partial\Omega, \ \vec{g} \in [H^r(\partial\Omega)]^3, \ r \in \mathbb{R}, \\ &\vec{w} \text{ is a regular function at infinity,} \end{aligned}\right\} \tag{9.174}$$

the solution is found uniquely representable by (9.173), where $\vec{f}$ and $\vec{\omega} = (\omega_1, \omega_2, \dots, \omega_6)$ uniquely solve the BIE

$$\mathscr{L}_{4e} \begin{bmatrix} \vec{\omega} \\ \vec{f} \end{bmatrix} = \begin{bmatrix} \vec{\mu} \\ \vec{g} \end{bmatrix}, \tag{9.175}$$

for a given $\vec{\mu} \in \mathbb{R}^6$. The constant vector $\vec{\mu} \in \mathbb{R}^6$ can be chosen arbitrarily, but $\vec{\omega}$ and $\vec{f}$ depend on $\vec{\mu}$. The solution (9.173) has regularity $\vec{w} \in [H^{r+3/2}_{loc}(\Omega^c)]^3$.

Proof. According to Lemmas 9.4 and 9.5, it is easy to verify that $\mathscr{L}_{4e}$ is 1–1. Thus $\mathscr{L}^*_{4e}$ is also 1–1. Since $\mathscr{L}_{4e}$ is Fredholm, it is 1–1 and onto.
If $(\vec{\omega}, \vec{f})$ satisfies the BIE (9.175), then it is easy to see that

$$\vec{w}(x) \equiv \int_{\partial\Omega} \vec{\tau}^{\mathrm{Tr}}_{\xi}(E)(x,\xi)\vec{f}(\xi)d\sigma_\xi + \sum_{i=1}^{6} \omega_i \int_{\partial\Omega} E(x,\xi)\vec{F}_i(\xi)d\sigma_\xi, \quad x \in \Omega^c,$$

satisfies

$$\vec{\tau}(\vec{w})(x) = \int_{\partial\Omega} \vec{\tau}_x \left[\vec{\tau}^{\mathrm{Tr}}_{\xi}(E)\right](x,\xi)\vec{f}(\xi)d\sigma_\xi + \sum_{i=1}^{6} \omega_i \vec{m}_i(x) = \vec{g}(x), \quad x \in \partial\Omega.$$ □

For the two-dimensional case, we need only modify (9.173) to

$$\vec{w}(x) = \int_{\partial\Omega} \vec{\tau}^{\mathrm{Tr}}_{\xi}(E)(x,\xi)\vec{f}(\xi)d\sigma_\xi + \sum_{i=1}^{3} \omega_i \int_{\partial\Omega} E(x,\xi)\vec{F}_i(\xi)d\sigma_\xi,$$

where $\vec{F}_i \in [C^\infty(\partial\Omega)]^2$ is the unique solution in the BIE

$$\mathscr{L}'^{-1}_{2e} \begin{bmatrix} 0 \\ \vec{m}_i \end{bmatrix} = \begin{bmatrix} 0 \\ \vec{F}_i \end{bmatrix}, \quad i = 1, 2, 3,$$

according to Theorem 9.13. The statement of the double-layer representation theorem for exterior traction BVP in two dimensions is then obtained nearly verbatim as that of Theorem 9.16.

9.10 Direct formulations of BIE based upon the Betti–Somigliana formula

Let us consider the interior displacement BVP (9.108) or the traction BVP (9.109). Recall that the Betti–Somigliana formula (9.90) gives

$$\vec{w}(x) = \int_{\partial\Omega} [E(x-\xi)\vec{\tau}(\vec{w})(\xi) - \vec{\tau}^{\mathrm{Tr}}_{\xi}(E)(x,\xi)\vec{w}(\xi)]d\sigma_\xi,\ x \in \Omega, \tag{9.176}$$

$$\frac{1}{2}\vec{w}(x) = \int_{\partial\Omega} [E(x-\xi)\vec{\tau}(\vec{w})(\xi) - \vec{\tau}^{\mathrm{Tr}}_{\xi}(E)(x,\xi)\vec{w}(\xi)]d\sigma_\xi,\ x \in \partial\Omega. \tag{9.177}$$

We divide our discussion into the following cases.

(I) Interior displacement BVP

(I.1) *First approach:* Given the displacement data $\vec{w} = \vec{g}$ on $\partial\Omega$, we wish to find the traction data $\vec{\tau}(\vec{w})$ by solving $\vec{f}$ in the BIE

$$\int_{\partial\Omega} E(x-\xi)\vec{f}(\xi)\,d\sigma_\xi = \frac{1}{2}\vec{g}(x) + \int_{\partial\Omega} \vec{\tau}_\xi^{\,\mathrm{Tr}}(E)(x,\xi)\,\vec{g}(\xi)\,d\sigma_\xi, \quad x \in \partial\Omega, \tag{9.178}$$

where $\vec{f}$ and $\vec{g}$ correspond, respectively, to $\vec{\tau}(\vec{w})$ and $\vec{w}$ in (9.177).

(a) *3D Case:* The above equation is of the form (9.113), and is thus always uniquely solvable by Theorem 9.5. So the traction data $\vec{\tau}\left(\vec{w}\right)$ is obtained.

(b) *2D Case:* The BIE (9.178) itself may not be invertible. We need to quote Theorem 9.9 instead. We can find a vector-valued function $\vec{f}$ and $\vec{a} \in \mathbb{R}^2$ satisfying

$$\int_{\partial\Omega} E(x-\xi)\vec{f}(\xi)\,d\sigma_\xi + \vec{a} = \frac{1}{2}\vec{g}(x) + \int_{\partial\Omega} \vec{\tau}_\xi^{\,\mathrm{Tr}}(E)(x,\xi)\,\vec{g}(\xi)\,d\sigma_\xi. \tag{9.179}$$

Obviously, $\vec{f}$ will not be equal to the traction data $\vec{\tau}(\vec{w})$ if $\vec{a} \neq (0,0)$. Even when $\vec{a} = (0,0)$, $\vec{f}$ still may not correspond to the traction data $\vec{\tau}(\vec{w})$ because $\vec{f}$ and $\vec{f} + \vec{\psi}$ can both satisfy

$$\int_{\partial\Omega} E(x-\xi)\vec{f}(\xi)\,d\sigma_\xi = \int_{\partial\Omega} E(x-\xi)[\vec{f}(\xi) + \vec{\psi}(\xi)]\,d\sigma_\xi, \quad x \in \partial\Omega,$$

provided that

$$\int_{\partial\Omega} E(x-\xi)\vec{\psi}(\xi)\,d\sigma_\xi \equiv 0 \quad \text{on } \partial\Omega \tag{9.180}$$

has a nontrivial solution $\vec{\psi}$. If (9.180) does not admit any nontrivial solution $\vec{\psi}$, then the solution $\vec{f}$ of (9.179) will be exactly the traction data $\vec{\tau}(\vec{w})$ when $\vec{a} = (0,0)$ in (9.179).

(I.2) *Second approach:* Instead of using (9.177) directly, we take the traction boundary data from (9.176), getting

$$\begin{aligned} &\frac{1}{2}\vec{\tau}(w)(x) - \int_{\partial\Omega} \vec{\tau}_x(E)(x,\xi)\,\vec{\tau}(\vec{w})(\xi)\,d\sigma_\xi \\ &= -\vec{\tau}_x \int_{\partial\Omega} \vec{\tau}_\xi^{\,\mathrm{Tr}}(E)(x,\xi)\,\vec{w}(\xi)\,d\sigma_\xi \\ &\left(= -\int_{\partial\Omega} \vec{\tau}_x[\vec{\tau}_\xi^{\,\mathrm{Tr}}(E(x,\xi))][\vec{w}(\xi) - \vec{w}(x)]\,d\sigma_\xi \right) \end{aligned} \tag{9.181}$$

for $x \in \partial\Omega$. Therefore, we consider solving $\vec{f}$ in the BIE

$$\frac{1}{2}\vec{f}(x) - \int_{\partial\Omega} \vec{\tau}_x(E)(x,\xi)\vec{f}(\xi)\,d\sigma_\xi$$
$$= -\vec{\tau}_x \int_{\partial\Omega} \vec{\tau}_\xi^{\,\mathrm{Tr}}(E)(x,\xi)\vec{g}(\xi)\,d\sigma_\xi, \qquad x \in \partial\Omega. \tag{9.182}$$

The operator on the LHS corresponds to the operator $\mathscr{L}_{2e}$ in Theorem 9.12, which is an isomorphism on $[H^\alpha(\partial\Omega)]^N$, for both $N = 2$ and 3. (For $N = 2$, Theorem 9.13 should be quoted instead.) Therefore, if $g \in [H^{r+1}(\partial\Omega)]^N$, the RHS of (9.182) lies in $[H^r(\partial\Omega)]^N$, and (9.182) has a unique solution $\vec{f} \in [H^r(\partial\Omega)]^N$, corresponding to the traction data $\vec{\tau}(\vec{w})$. This approach avoids the difficulty of nonuniqueness mentioned in (I.1)(b) above, namely, the two-dimensional case. However, there is a price to pay: the numerical quadrature of the RHS of (9.182) is evaluated for

$$\int_{\partial\Omega} \vec{\tau}_x[\vec{\tau}_\xi^{\,\mathrm{Tr}}(E(x,\xi))][\vec{g}(\xi) - \vec{g}(x)]\,d\sigma_\xi, \quad x \in \partial\Omega,$$

whose kernel $\vec{\tau}_x[\vec{\tau}_\xi^{\,\mathrm{Tr}}(E(x,\xi))]$ is hypersingular and complicated; thus there is considerably larger amount of work involved.

(II) Interior traction BVP

(II.1) *First approach:* Given the traction data $\vec{\tau}(\vec{w}) = \vec{g} \in [H^\alpha(\partial\Omega)]^N$, $N = 2$ or 3, we wish to find the displacement data $\vec{w}$ by solving $\vec{f}$ in the BIE

$$\frac{1}{2}\vec{f}(x) + \int_{\partial\Omega} \vec{\tau}_\xi^{\,\mathrm{Tr}}(E)(x,\xi)\vec{f}(\xi)\,d\sigma_\xi = \int_{\partial\Omega} E(x,\xi)\vec{g}(\xi)\,d\sigma_\xi, \quad x \in \partial\Omega. \tag{9.183}$$

Thus, this equation corresponds to the direct use of (9.177).
The LHS of (9.183) corresponds to $\mathscr{L}_2^*\vec{f}$ according to Theorem 9.6, such an equation is never uniquely solvable. Instead, we need to quote (9.169) in the proof of Theorem 9.14, which ensures the existence of $\vec{f} \in [H^{\alpha+1}(\partial\Omega)]^N$ and $\vec{b} \in \mathbb{R}^{3(N-1)}$ such that

$$\frac{1}{2}\vec{f}(x) + \int_{\partial\Omega} \vec{\tau}_\xi^{\,\mathrm{Tr}}(E)(x,\xi)\vec{f}(\xi)\,d\sigma_\xi + M(x)\vec{b}$$
$$= \int_{\partial\Omega} E(x-\xi)\vec{g}(\xi)\,d\sigma_\xi, \quad x \in \partial\Omega. \tag{9.184}$$

Any solution $\vec{f}$ of (9.184) will correspond to a displacement data $\vec{w}$, which is nonunique with $3(N-1)$ degrees of freedom.

(II.2) *Second approach:* We use (9.176) instead: Find $\vec{f} = \vec{w}$ such that

$$\int_{\partial\Omega} \vec{\tau}_x[\vec{\tau}_\xi^{\,\mathrm{Tr}}(E(x,\xi))]\vec{f}(\xi)\,d\sigma_\xi$$
$$= -\frac{1}{2}\vec{g}(x) + \int_{\partial\Omega} \vec{\tau}_x(E)(x,\xi)\vec{g}(\xi)\,d\sigma_\xi, \quad x \in \partial\Omega, \tag{9.185}$$

where $\vec{\tau}(\vec{w}) = \vec{g}$ is the given traction data on $\partial\Omega$. The hypersingular operator on the left is not invertible from $[H^{\alpha+1}(\partial\Omega)]^N$ into $[H^{\alpha}(\partial\Omega)]^N$, $N = 2$ or 3. Nevertheless, we can quote Theorem 9.16, which ensures that for $g \in [H^{\alpha}(\partial\Omega)]^N$, there exist $\vec{f} \in [H^{\alpha+1}(\partial\Omega)]^N$ and $\vec{a} \in \mathbb{R}^{3(N-1)}$ such that

$$\begin{aligned}&\int_{\partial\Omega} \vec{\tau}_x[\vec{\tau}_{\xi}^{\mathrm{Tr}}(E)](x,\xi)\vec{f}(\xi)\,d\sigma_{\xi} + M(x)\vec{a}\\ &= -\frac{1}{2}\vec{g}(x) + \int_{\partial\Omega} \vec{\tau}_x(E)(x,\xi)\vec{g}(\xi)\,d\sigma_{\xi}, \quad x \in \partial\Omega. \end{aligned} \tag{9.186}$$

We have the same situation as before: any such a solution $\vec{f}$ corresponds to a displacement data $\vec{w}$, with $3(N-1)$ degrees of freedom.

Next, we consider the exterior displacement BVP or traction BVP. Recall that the Betti–Somigliana formula gives

$$\begin{aligned}\vec{w}(x) = -\int_{\partial\Omega} [E(x-\xi)\vec{\tau}(\vec{w})(\xi) - \vec{\tau}_{\xi}^{\mathrm{Tr}}(E)(x,\xi)\vec{w}(\xi)]\,d\sigma_{\xi} + M(x)\vec{\omega},\\ x \in \Omega^c,\end{aligned} \tag{9.187}$$

and

$$\begin{aligned}\frac{1}{2}\vec{w}(x) = -\int_{\partial\Omega} [E(x-\xi)\vec{\tau}(\vec{w})(\xi) - \vec{\tau}_{\xi}^{\mathrm{Tr}}(E)(x,\xi)\vec{w}(\xi)]\,d\sigma_{\xi} + M(x)\vec{\omega},\\ x \in \partial\Omega.\end{aligned} \tag{9.188}$$

(III) Exterior displacement BVP

(III.1) *First approach:* Given the displacement data $\vec{w} = \vec{g}$ on $\partial\Omega$ as well as $\omega_j \in \mathbb{R}$, $j \in \mathscr{J}$, we wish to find the traction data $\vec{\tau}(\vec{w}) = \vec{f}$ and ω_k, $k \in \mathscr{I} - \mathscr{J}$ from the BIE

$$\left.\begin{aligned}&\int_{\partial\Omega} E(x,\xi)\vec{f}(\xi)\,d\sigma_{\xi} - \sum_{k\in\mathscr{I}-\mathscr{J}} \omega_k \vec{m}_k(x) = -\frac{1}{2}\vec{g}(x)\\ &\quad + \int_{\partial\Omega} \vec{\tau}_{\xi}^{\mathrm{Tr}}(E)(x,\xi)\vec{g}(\xi)\,d\sigma_{\xi} + \sum_{j\in\mathscr{J}} \omega_j \vec{m}_j(x), \quad x \in \partial\Omega,\\ &\int_{\partial\Omega} \vec{m}_k(x)\cdot\vec{f}(x)\,d\sigma = b_k, \quad k \in \mathscr{I} - \mathscr{J}.\end{aligned}\right\} \tag{9.189}$$

(a) *3D Case:* The BIE (9.189) is uniquely solvable, according to Theorem 9.10. For given $\vec{g} \in [H^{\alpha+1}(\partial\Omega)]^3$, $r \geqslant 0$, the density $\vec{f}$ lies in $[H^r(\partial\Omega)]^3$ and $\vec{f}$ corresponds to the traction $\vec{\tau}(\vec{w})$ on $\partial\Omega$.

(b) *2D Case:* According to (9.132), there are two possibilities. The second possibility (in (9.132)) has been directly solved in Theorem 9.11; namely, there uniquely exists $(\vec{\omega}, \vec{f}) \in \mathbb{R}^3 \oplus [H^r(\partial\Omega)]^3$ such that

$$\int_{\partial\Omega} E(x,\xi)\vec{f}(\xi)\,d\sigma_\xi - \sum_{i=1}^{3} \omega_i \vec{m}_i(x) = -\frac{1}{2}\vec{g}(x)$$
$$+\int_{\partial\Omega} \vec{\tau}_\xi^{\,\mathrm{Tr}}(E)(x,\xi)\vec{g}(\xi)\,d\sigma_\xi, \quad x \in \partial\Omega,$$
$$\int_{\partial\Omega} M(x)\cdot\vec{f}(x)\,d\sigma = \begin{bmatrix} 0 \\ 0 \\ b_3 \end{bmatrix},$$

for $\vec{g} \in [H^{r+1}(\partial\Omega)]^2$. This density $\vec{f}$ uniquely corresponds to the traction data $\vec{\tau}(\vec{w})$ on $\partial\Omega$.

The first possibility in (9.132) has been answered in Exercise 9.4, where $\vec{f}$ uniquely corresponds to the traction data $\vec{\tau}(\vec{w})$.

(III.2) *Second approach:* We take the traction boundary data for $\vec{w}$ given by (9.187) and obtain

$$\frac{1}{2}\vec{\tau}(\vec{w})(x) + \int_{\partial\Omega} \vec{\tau}_x(E)(x,\xi)\vec{\tau}(\vec{w})(\xi)\,d\sigma_\xi i$$
$$= \vec{\tau}_x \int_{\partial\Omega} \vec{\tau}_\xi^{\,\mathrm{Tr}}(E)(x,\xi)\vec{w}(\xi)\,d\sigma_\xi, \qquad x \in \partial\Omega, \tag{9.190}$$

plus accessory conditions

$$\int_{\partial\Omega} \vec{m}_k(x)\cdot\vec{\tau}(\vec{w})(x)\,d\sigma = b_k, \quad k \in \mathscr{I} - \mathscr{J}. \tag{9.191}$$

The BIE (9.190) is uniquely solvable for $\vec{\tau}(\vec{w})$ if and only if $\mathscr{J} = \emptyset$, according to Theorem 9.6. (See (9.114).) (For the two-dimensional case, it is further necessary to impose $b_1 = b_2 = 0$.) When $\mathscr{J} \neq \emptyset$, we can again quote Theorem 9.6, which ensures an $(\vec{a}, \vec{f}) \in \mathbb{R}^6 \oplus [H^r(\partial\Omega)]^3$ such that

$$\frac{1}{2}\vec{f}(x) + \int_{\partial\Omega} \vec{\tau}_x(E)(x,\xi)\vec{f}(\xi)\,d\sigma_\xi + M(x)\vec{a}$$
$$= \vec{\tau}_x \int_{\partial\Omega} \vec{\tau}_\xi^{\,\mathrm{Tr}}(E)(x,\xi)\vec{g}(\xi)\,d\sigma_\xi \tag{9.192}$$

if $\vec{w} = \vec{g} \in [H^{r+1}(\partial\Omega)]^3$. It is easy to check that $\vec{a} = \vec{0}$ holds; therefore, $\vec{f}$ will be equal to $\vec{\tau}(\vec{w})$ on $\partial\Omega$ if and only if (9.191) is satisfied. The two-dimensional case can be argued in a similar manner.

(IV) Exterior traction BVP

For the exterior traction BVP, because we do not need to be concerned with the rigid motion at ∞, we set $\vec{\omega} = \vec{0}$ in (9.187) and (9.188).

(IV.1) *First approach:* Given the traction data $\vec{\tau}(\vec{w}) = \vec{g} \in [H^r(\partial\Omega)]^N$, $N = 2$ or 3, we wish to use (9.188) to find the displacement data $\vec{w} = \vec{f}$:

$$\frac{1}{2}\vec{f}(x) - \int_{\partial\Omega} \vec{\tau}_\xi^{\mathrm{Tr}}(E)(x,\xi)\vec{f}(\xi)\,d\sigma_\xi = -\int_{\partial\Omega} E(x-\xi)\vec{g}(\xi)\,d\sigma_\xi, \quad x \in \partial\Omega.$$

According to Theorem 9.7, the operator $\mathscr{L}_3$ representing the LHS is always invertible on $[H^{r+1}(\partial\Omega)]^N$. A unique solution $\vec{f} \in [H^{r+1}(\partial\Omega)]^N$ exists corresponding to the displacement data $\vec{w}$ on $\partial\Omega$.

(IV.2) *Second approach:* We use (9.190): given $\vec{\tau}(\vec{w}) = \vec{g} \in [H^r(\partial\Omega)]^N$, solve $\vec{f} = \vec{w}$ in

$$\int_{\partial\Omega} \vec{\tau}_x(\vec{\tau}_\xi^{\mathrm{Tr}}(E(x,\xi)))\vec{f}(\xi)\,d\sigma_\xi = \frac{1}{2}\vec{g}(x) + \int_{\partial\Omega} \vec{\tau}_x(E)(x,\xi)\vec{g}(\xi)\,d\sigma_\xi, \quad x \in \partial\Omega. \tag{9.193}$$

The situation is similar to that in (II.2) (except that there $\frac{1}{2}\vec{g}(x)$ was $-\frac{1}{2}\vec{g}(x)$ in (9.185), but this has no effect on the solvability of the operator on the LHS of (9.193).) Again, we quote Theorem 9.16, which ensures a unique $(\vec{a}, \vec{f}) \in \mathbb{R}^{3(N-1)} \oplus [H^{r+1}(\partial\Omega)]^N$ such that

$$\int_{\partial\Omega} M^{\mathrm{Tr}}(x)\vec{f}(x)\,d\sigma = \vec{0}, \tag{9.194}$$

$$\int_{\partial\Omega} \vec{\tau}_x[\vec{\tau}_\xi^{\mathrm{Tr}}(E(x,\xi))]\vec{f}(\xi)\,d\sigma_\xi + M(x)\vec{a} = \frac{1}{2}\vec{g}(x) + \int_{\partial\Omega} \vec{\tau}_x(E)(x,\xi)\vec{g}(\xi)\,d\sigma_\xi, \quad x \in \partial\Omega. \tag{9.195}$$

Forming the inner product of (9.195) with $\vec{m}_j(x)$, $1 \leqslant j \leqslant 3(N-1)$ and integrating on $\partial\Omega$, we get $\vec{a} = 0$. Thus there exists a unique $\vec{f}$ satisfying (9.193), which corresponds to a displacement data $\vec{w}$ on $\partial\Omega$. This displacement data is the only data that satisfies the equation (9.194). (Any $\vec{f}(x) + \sum_{j=1}^{3(N-1)} \alpha_j \vec{m}_j(x)$ would also satisfy (9.193) and corresponds to a displacement data, but only one of them satisfies (9.194).)

9.11 Numerical example (I): comparison of direct formulation and simple-layer approaches

In § 9.11–9.13, we treat several numerical examples in $\mathbb{R}^3$. The domain Ω always represents the unit open ball in $\mathbb{R}^3$.

Consider the displacement BVP

$$\left.\begin{aligned}&\mu\Delta\vec{w}+(\lambda+\mu)\nabla(\nabla\cdot\vec{w})=0 \text{ on } \Omega, \quad \lambda=\frac{2}{3}, \quad \mu=1,\\&\vec{w}(x)=E(x,y_0)\,\vec{b}, \quad x\in\partial\Omega,\end{aligned}\right\} \tag{9.196}$$

where $y_0=(3,3,3)\in\Omega^c$, $\vec{b}=(1,0,0)$, and E is the fundamental solution (9.45). Obviously, the unique solution of (9.196) is

$$\vec{w}(x)=E(x,y_0)\,\vec{b} \quad \forall x\in\Omega.$$

This will serve as our benchmark.

Let us first solve (9.196) by the direct formulation (9.178):

$$\left.\begin{aligned}&\text{Find } \vec{f}^{(1)}=\vec{\tau}_x(\vec{w}) \text{ such that}\\&\int_{\partial\Omega}E(x,\xi)\,\vec{f}^{(1)}(\xi)\,d\sigma_\xi=\frac{1}{2}E(x,y_0)\,\vec{b}\\&+\int_{\partial\Omega}\vec{\tau}_\xi^{\,\mathrm{Tr}}(E)(x,\xi)E(\xi,y_0)\,\vec{a}\,d\sigma_\xi, \quad x\in\partial\Omega.\end{aligned}\right\} \tag{9.197}$$

We use piecewise-constant boundary elements to approximate $\vec{f}$:

$$\vec{\tau}_x(\vec{w})=\vec{f}^{(1)}\approx\vec{f}_h^{(1)}=\sum_{i=1}^{n_h}\vec{f}_i^{(1)}\chi_{\Gamma_{h,i}}, \quad \vec{f}_i^{(1)}\in\mathbb{R}^3, \tag{9.198}$$

where

$$\partial\Omega\approx\Gamma_h=\bigcup_{i=1}^{n_h}\Gamma_{h,i}$$

is an approximation of $\partial\Omega$, and $\chi_{\Gamma_{h,i}}$ is the characteristic function on $\Gamma_{h,i}$. Thus (9.197) is discretized into the form

$$\begin{aligned}\sum_{i=1}^{n_h}\int_{\Gamma_{h,i}}E(x^{(j)},\xi)\,d\sigma_\xi\,\vec{f}_i^{(1)}&=\frac{1}{2}E(x^{(j)},y_0)\,\vec{b}\\&+\int_{\Gamma_h}\vec{\tau}_\xi^{\,\mathrm{Tr}}(E)(x^{(j)},\xi)E(\xi,y_0)\,\vec{a}\,d\sigma_\xi, \quad j=1,2,\ldots,n_h,\end{aligned} \tag{9.199}$$

where $x^{(j)}\in\Gamma_{h,j}$ is a representative point on $\Gamma_{h,j}$. $x^{(j)}$ is commonly chosen in the proximity of the "geometric center" of $\Gamma_{h,j}$.

In our computations, we discretize the curved boundary surface $\partial\Omega$ along the latitudinal angle θ and longitudinal angle ϕ (cf. (9.39)). Let (r,θ,ϕ) be the spherical coordinate of a point $x\in\mathbb{R}^3$. Then each piece

$$\left\{(1,\theta,\phi)\mid(p-1)\left(\frac{\pi}{l}\right)<\theta<p\left(\frac{\pi}{l}\right),\ (q-1)\left(\frac{2\pi}{m}\right)<\phi<q\left(\frac{2\pi}{m}\right)\right\}$$

$$p=1,2,\ldots,l, \quad q=1,2,\ldots,m, \tag{9.200}$$

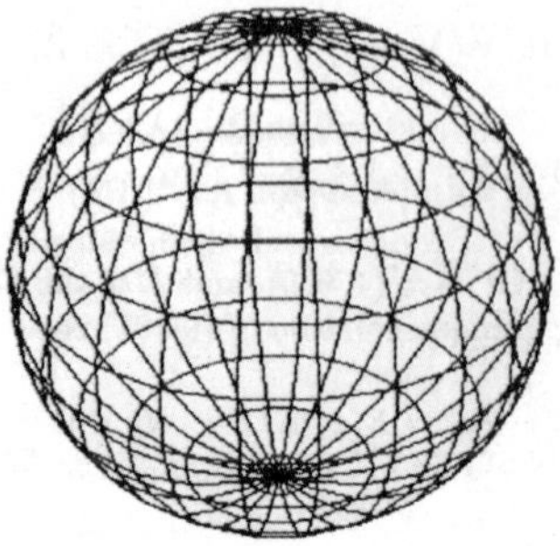

Figure 9.1: Discretization of the unit spherical surface into piecewise-constant boundary elements according to (9.200), using $l = 12$ and $m = 24$.

is an element $\Gamma_{h,j}$. There are a total of $l \times m$ elements. See Fig. 9.1, for example, where $l = 12$ and $m = 24$ are chosen. This "equal-angle" grid generation (9.200) is rather convenient. Nevertheless, those boundary elements with a vertex near the north pole or the south pole will have an angle tending to zero as m tends to infinity. This will lead to a forbidden situation, as mentioned in (iii) after equation (5.88) in § 5.10 of Chapter 5, and may cause additional errors and numerical distortion of the solution. To avoid this, measures can be taken, such as, for example, by combining those boundary elements near the north and south poles into two single elements of the form

$$\left\{(1,\theta,\phi) \mid 0 \leqslant \theta < \frac{\pi}{l}, 0 \leqslant \phi < 2\pi\right\} \text{ (centered at the north pole),}$$
$$\left\{(1,\theta,\phi) \mid \left(\frac{l-1}{l}\right)\pi < \theta \leqslant \pi, 0 \leqslant \phi < 2\pi\right\} \text{ (centered at the south pole).}$$

We have tried this and found some improvement in numerical accuracy. But in our numerical computations in this chapter, we discretize by using only (9.200).

The integrals

$$\int_{\Gamma_{h,i}} E(x^{(j)},\xi)\, d\sigma_\xi, \qquad \int_{\Gamma_h} \vec{\tau}_\xi^{\,\mathrm{Tr}}(E)(x^{(j)},\xi)E(\xi,y_0)\,\vec{a}\, d\sigma_\xi, \tag{9.201}$$

are, respectively, weakly singular and of Cauchy principal type, and are not too difficult to compute, after transforming them from surface integrals into planar integrals using the curvilinear coordinates on the sphere. According to the discretization (9.200) and (9.201), (9.198) becomes a $3lm \times 3lm$ linear system for the unknowns $\vec{f}_i^{(1)}$, $1 \leqslant i \leqslant n_h = lm$, which is solved by matrix inversion.

Once (9.198) is approximated, from (9.176) we use

$$\vec{w}_h^{(1)}(x) = \int_{\Gamma_h} \left[E(x,\xi) \left(\sum_{i=1}^{n_h} \vec{f}_i^{(1)} \chi_{\Gamma_{h,i}} \right) - \vec{\tau}_\xi^{\mathrm{Tr}}(E)(x,\xi) E(\xi, y_0) \vec{a} \right] d\sigma_\xi, \ x \in \Omega, \tag{9.202}$$

to obtain approximate values of $\vec{w}$ on Ω.

Another approach, using the simple-layer potential representation according to Theorem 9.5, is made by writing $\vec{w}$ as (9.111) and by solving (9.113), which is

$$\int_{\partial\Omega} E(x,\xi) \vec{f}^{(2)}(\xi) \, d\sigma_\xi = E(x, y_0) \vec{a}, \quad x \in \partial\Omega. \tag{9.203}$$

Let $\vec{f}^{(2)}$ be approximated by

$$\vec{f}^{(2)} \approx \vec{f}_h^{(2)} = \sum_{i=1}^{n_h} \vec{f}_i^{(2)} \chi_{\Gamma_{h,i}}. \tag{9.204}$$

We discretize (9.203) as

$$\sum_{i=1}^{n_h} \int_{\Gamma_{h,i}} E(x^{(j)}, \xi) \, d\sigma_\xi \, \vec{f}_i^{(2)} = E(x^{(j)}, y_0) \vec{a}, \quad 1 \leqslant j \leqslant n_h. \tag{9.205}$$

Again, (9.205) leads to a $3lm \times 3lm$ linear system for the $3lm$ unknowns in $\vec{f}_i^{(2)}$. According to (9.112), we then use

$$\vec{w}_h^{(2)}(x) = \int_{\Gamma_h} E(x,\xi) \left[\sum_{i=1}^{n_h} \vec{f}_i^{(2)} \chi_{\Gamma_{h,i}}(\xi) \right] d\sigma_\xi, \quad x \in \Omega, \tag{9.206}$$

to obtain an approximate solution over Ω.

Numerical solutions obtained by the two approaches (9.202) and (9.206) at sample points are tabulated in Table 9.1. The reader can find that *the simple-layer approach (9.206) yields more accurate data than those obtained by direct formulation.*

We wish to find the rates of convergence of the simple-layer density $\vec{f}^{(i)}$, $i = 1, 2$. For the approach (9.205), no exact solution of $\vec{f}^{(2)}$ is available. Due to the memory requirement limitation, it is harder to determine $\| \vec{f}^{(2)} - \vec{f}_h^{(2)} \|_{L^2(\partial\Omega)}$. Thus, instead we calculate

$$\| \vec{f}^{(1)} - \vec{f}_h^{(1)} \|_{L^2(\partial\Omega)}. \tag{9.207}$$

The exact solution is

$$\vec{f}^{(1)}(x) = \tau_x^{\mathrm{Tr}}(E)(x, y_0) \vec{a}, \quad x \in \partial\Omega.$$

Using $(l,m) = (4,8), (6,12)$ and (12,24), we have plotted the logarithmic error of (9.196) in Fig. 9.2. The rate of convergence as determined by the slope of the line segment therein is measured to be 1.808, suggesting that piecewise-constant boundary elements ideally may converge with a rate $\mathscr{O}(h_1^2 h_2^2)$, where $h_1 = \pi/l$ and $h_2 = 2\pi/m$, and $h_1/h_2 < c, h_2/h_1 < c$, for some $c > 0$.

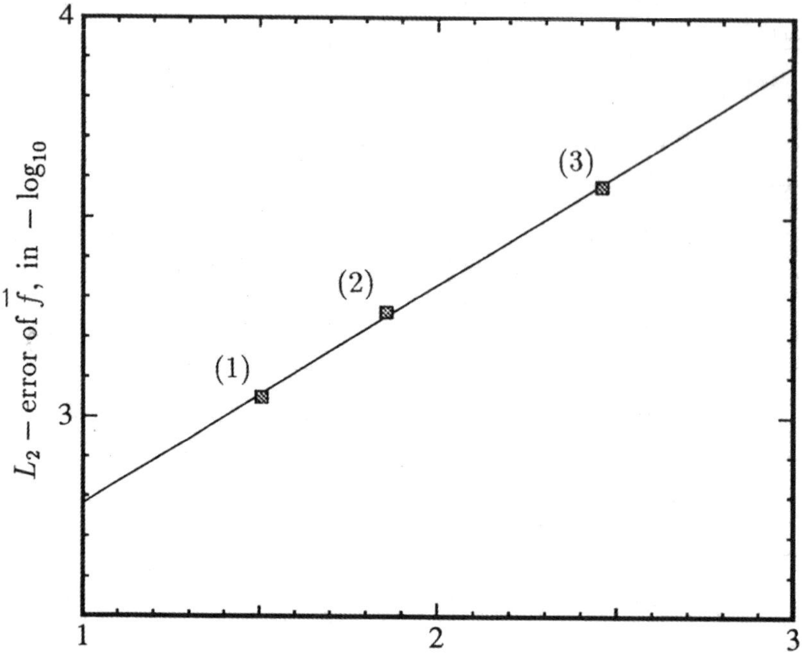

Figure 9.2: The (negative) logarithmic error of $\| \vec{f}^{(1)} - \vec{f}_h^{(1)} \|_{L^2(\partial\Omega)}$ in (9.196). The rate of convergence, corresponding to the slope, of piecewise-constant boundary elements is measured to be 1.808, suggesting $\| \vec{f}^{(1)} - \vec{f}_h^{(1)} \|_{L^2(\partial\Omega)} = \mathcal{O}(h_1^2 h_2^2)$, where $h_1 = \pi/l$ and $h_2 = 2\pi/m$.

Table 9.1: Numerical solutions $\vec{w}_h^{(1)}$, (9.202), and $\vec{w}_h^{(2)}$, (9.206), and the exact solution $\vec{w}$ at sample points.

x (spherical coordinates)	$\vec{w}_h^{(1)}(x)$	$\vec{w}_h^{(2)}(x)$	$\vec{w}(x)$
$(\frac{2}{3},0,0)$	0.13269E-01	0.13271E-01	0.13271E-01
	0.19680E-02	0.19714E-02	0.19716E-02
	0.15372E-02	0.15336E-02	0.15335E-02
$(\frac{2}{3},\frac{1}{4}\pi,0)$	0.13282E-01	0.13284E-01	0.13284E-01
	0.18439E-02	0.18545E-02	0.18549E-02
	0.15571E-02	0.15631E-02	0.15635E-02
	0.13916E-01	0.13920E-01	0.13922E-01

continued on next page ...

Table 9.1: continued ...

x (spherical coordinates)	$\vec{w}_h^{(1)}(x)$	$\vec{w}_h^{(2)}(x)$	$\vec{w}(x)$
$\left(\frac{2}{3},\frac{1}{4}\pi,\frac{1}{2}\pi\right)$	0.18563E-02	0.18545E-02	0.18549E-02
	0.18531E-02	0.18545E-02	0.18549E-02
$\left(\frac{2}{3},\frac{1}{4}\pi,\pi\right)$	0.12528E-01	0.12528E-01	0.12528E-01
	0.18066E-02	0.18012E-02	0.18013E-02
	0.15303E-02	0.15183E-02	0.15182E-02
$\left(\frac{2}{3},\frac{1}{4}\pi,\frac{3}{2}\pi\right)$	0.12002E-01	0.12000E-01	0.12000E-01
	0.17978E-02	0.18013E-02	0.18013E-02
	0.13183E-02	0.13121E-02	0.13121E-02
$\left(\frac{2}{3},\frac{1}{2}\pi,0\right)$	0.12492E-01	0.12492E-01	0.12492E-01
	0.15218E-02	0.15335E-02	0.15335E-02
	0.15205E-02	0.15334E-02	0.15335E-02
$\left(\frac{2}{3},\frac{1}{2}\pi,\frac{1}{2}\pi\right)$	0.13267E-01	0.13270E-01	0.13271E-01
	0.15417E-02	0.15333E-02	0.15335E-02
	0.19661E-02	0.19711E-02	0.19716E-02
$\left(\frac{2}{3},\frac{1}{2}\pi,\frac{3}{2}\pi\right)$	0.11028E-01	0.11026E-01	0.11026E-01
	0.15494E-02	0.15514E-02	0.15514E-02
	0.12735E-02	0.12693E-02	0.12693E-02
$\left(\frac{2}{3},\frac{3}{4}\pi,0\right)$	0.11550E-01	0.11550E-01	0.11549E-01
	0.13093E-02	0.13121E-02	0.13121E-02
	0.15085E-02	0.15183E-02	0.15182E-02
$\left(\frac{2}{3},\frac{3}{4}\pi,\frac{1}{2}\pi\right)$	0.12000E-01	0.12000E-01	0.12000E-01
	0.13190E-02	0.13121E-02	0.13121E-02
	0.17970E-02	0.18012E-02	0.18013E-02
$\left(\frac{2}{3},\frac{3}{4}\pi,\pi\right)$	0.11085E-01	0.11083E-01	0.11083E-01
	0.13676E-02	0.13599E-02	0.13599E-02
	0.15792E-02	0.15735E-02	0.15736E-02
$\left(\frac{2}{3},\frac{3}{4}\pi,\frac{3}{2}\pi\right)$	0.10686E-01	0.10685E-01	0.10684E-01
	0.13607E-02	0.13599E-02	0.13599E-02
	0.13616E-02	0.13599E-02	0.13599E-02
	0.11027E-01	0.11026E-01	0.11026E-01

continued on next page ...

Table 9.1: continued ...

x (spherical coordinates)	$\vec{w}_h^{(1)}(x)$	$\vec{w}_h^{(2)}(x)$	$\vec{w}(x)$
$(\frac{2}{3}, \pi, 0)$	0.12730E-02	0.12693E-02	0.12693E-02
	0.15501E-02	0.15514E-02	0.15514E-02
	0.12695E-01	0.12696E-01	0.12696E-01
$(\frac{1}{3}, 0, 0)$	0.17777E-02	0.17784E-02	0.17786E-02
	0.15828E-02	0.15808E-02	0.15810E-02
	0.12690E-01	0.12690E-01	0.12690E-01
$(\frac{1}{3}, \frac{1}{4}\pi, 0)$	0.17191E-02	0.17232E-02	0.17235E-02
	0.15849E-02	0.15878E-02	0.15881E-02
	0.12971E-01	0.12972E-01	0.12973E-01
$(\frac{1}{3}, \frac{1}{4}\pi, \frac{1}{2}\pi)$	0.17250E-02	0.17231E-02	0.17235E-02
	0.17227E-02	0.17231E-02	0.17235E-02
	0.12352E-01	0.12352E-01	0.12352E-01
$(\frac{1}{3}, \frac{1}{4}\pi, \pi)$	0.17136E-02	0.17098E-02	0.17101E-02
	0.15814E-02	0.15756E-02	0.15757E-02
	0.12093E-01	0.12093E-01	0.12093E-01
$(\frac{1}{3}, \frac{1}{4}\pi, \frac{3}{2}\pi)$	0.17086E-02	0.17099E-02	0.17101E-02
	0.14640E-02	0.14608E-02	0.14609E-02
	0.12323E-01	0.12323E-01	0.12323E-01
$(\frac{1}{3}, \frac{1}{2}\pi, \frac{1}{2}\pi)$	0.15766E-02	0.15808E-02	0.15810E-02
	0.15749E-02	0.15808E-02	0.15810E-02
	0.12695E-01	0.12696E-01	0.12696E-01
$(\frac{1}{3}, \frac{1}{2}\pi, \frac{3}{2}\pi)$	0.15851E-02	0.15807E-02	0.15810E-02
	0.17762E-02	0.17782E-02	0.17786E-02
	0.11566E-01	0.11565E-01	0.11565E-01
$(\frac{1}{3}, \frac{3}{4}\pi, \frac{1}{2}\pi)$	0.15827E-02	0.15832E-02	0.15833E-02
	0.14275E-02	0.14248E-02	0.14249E-02
	0.11854E-01	0.11854E-01	0.11853E-01
$(\frac{1}{3}, \frac{3}{4}\pi, \frac{3}{2}\pi)$	0.14599E-02	0.14608E-02	0.14609E-02
	0.15710E-02	0.15756E-02	0.15757E-02
	0.12093E-01	0.12093E-01	0.12093E-01

continued on next page ...

Table 9.1: continued ...

x (spherical coordinates)	$\vec{w}_h^{(1)}(x)$	$\vec{w}_h^{(2)}(x)$	$\vec{w}(x)$
$(\frac{1}{3}, \pi, \frac{3}{2}\pi)$	0.14652E-02	0.14608E-02	0.14609E-02
	0.17078E-02	0.17098E-02	0.17101E-02
	0.11589E-01	0.11588E-01	0.11588E-01
$(\frac{1}{3}, \frac{1}{4}\pi, \frac{1}{2}\pi)$	0.14790E-02	0.14735E-02	0.14735E-02
	0.15927E-02	0.15891E-02	0.15893E-02
	0.11366E-01	0.11365E-01	0.11365E-01
$(\frac{1}{3}, \frac{1}{4}\pi, \frac{1}{2}\pi)$	0.14745E-02	0.14735E-02	0.14735E-02
	0.14747E-02	0.14734E-02	0.14735E-02
	0.11566E-01	0.11565E-01	0.11565E-01
$(\frac{1}{3}, \frac{1}{4}\pi, \frac{1}{2}\pi)$	0.14278E-02	0.14248E-02	0.14249E-02
	0.15825E-02	0.15831E-02	0.15833E-02
	0.12124E-01	0.12124E-01	0.12124E-01
(0,0,0)	0.15966E-02	0.15951E-02	0.15953E-02
	0.15956E-02	0.15951E-02	0.15953E-02
finished			

9.12 Numerical example (II): a ball subject to vertical compression

Consider the following traction problem on the unit ball Ω, where the traction $\vec{\tau}(\vec{w})(x) = \vec{g}(x) = (g_1(x), g_2(x), g_3(x))^{\mathrm{Tr}}$ (in rectangular coordinates) on $\partial\Omega$ is given by

$$\left\{ \begin{array}{l} \left(\lambda = \dfrac{2}{3}, \mu = 1\right) \\ g_1(x) = g_2(x) = 0, \\ g_3(x) = \begin{cases} -3(1-\sin\theta), & 0 \leqslant \theta \leqslant \dfrac{1}{2}\pi \ \text{(upper hemisphere)}, \\ 3(1-\sin\theta), & \dfrac{1}{2}\pi \leqslant \theta \leqslant \pi \ \text{(lower hemisphere)}, \end{cases} \\ (1,\theta,\phi) \ \text{are the spherical coordinates of } x \in \partial\Omega. \end{array} \right\} \tag{9.208}$$

Thus the ball is subject to a vertically symmetric compression force, as depicted in Fig. 9.3. It is easy to check that the data $\vec{g}$ in (9.208) satisfies the compatibility conditions (9.110). We use Theorem 9.6 to solve the traction problem (9.116), with $\vec{g}$ as given in (9.208). The solution $\vec{w}$ assuming the form (9.117) is nonunique. We need six accessory conditions to be appended to the BIE (9.118) in order to fix the six degrees of freedom $\sum_{i=1}^{6} \alpha_i \vec{m}_i(x)$ in

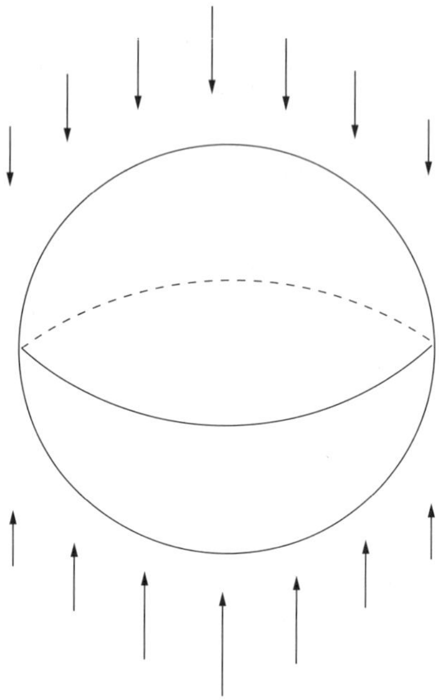

Figure 9.3: A unit ball subject to a vertical compression (traction) force as given by (9.208).

(9.117). This is done as follows. Since the compression force is vertically symmetric, we can expect the vertical displacement of $\vec{w}(x)$ at the equator $\theta = \frac{1}{2}\pi$ to be zero. Thus, we impose six equations

$$\text{the third component of } \vec{w}(x^{(j)}),\ w_3(x^{(j)}) = 0, \quad j = 1,2,\ldots,6,$$
$$x^{(j)} = (r^{(j)}, \theta^{(j)}, \phi^{(j)}) = \left(1, \frac{1}{2}\pi, \frac{1}{6}j2\pi\right). \tag{9.209}$$

The above leads to

$$\left[\int_{\partial\Omega} E(x^{(j)},\xi)\vec{f}(\xi)\,d\sigma_\xi\right]_3 = 0, \quad 1 \leqslant j \leqslant 6, \tag{9.210}$$

where in (9.210), $[\ldots\ldots]_3$ denotes the x_3 component of the three-dimensional vector. The BIE (9.118) is

$$(\mathscr{L}_2 f)(x) = \frac{1}{2}\vec{f}(x) + \int_{\partial\Omega} \vec{\tau}_x(E)(x,\xi)\vec{f}(\xi)\,d\sigma_\xi = \vec{g}(x), \ x \in \partial\Omega, \tag{9.211}$$

which should be augmented as

$$\sum_{i=1}^{6} \alpha_i \vec{m}_i(x) + \left[\frac{1}{2}\vec{f}(x) + \int_{\partial\Omega} \vec{\tau}_x(E)(x,\xi)\vec{f}(\xi)\,d\sigma_\xi\right] = \vec{g}(x), \quad x \in \partial\Omega; \tag{9.212}$$

cf. (4.208), for example. Using piecewise-constant boundary elements $\vec{f}_h$ like (9.204) to approximate $\vec{f}$, the system (9.210) and (9.211) can be discretized pretty much in a similar way as before. Using $(l,m) = (12,24)$, we have computed $\vec{f}_h(\cdot)$ and $\alpha_{j,h}$, $1 \leqslant j \leqslant 6$, and obtained

$$\vec{w}_h(x) = \int_{\Gamma_h} E(x,\xi)\,\vec{f}_h(\xi)\,d\sigma_\xi, \quad x \in \overline{\Omega}, \tag{9.213}$$

which is plotted in Fig. 9.4, showing that the ball has been compressed to the shape of a "tire". Numerical values of $\vec{w}_h(x)$ at sample points can be found in Table 9.2. The reader should use Theorem 4.10 to show that the accessory conditions (9.210) augmenting the operator $\mathscr{L}_2$ in (9.211) lead to an *invertible* operator formed by (9.210) and (9.212).

9.13 Numerical example (III): a ball subject to twisting traction

Again, consider a traction BVP on the unit ball Ω, where the traction

$$\vec{\tau}(\vec{w})(x) = \vec{g}(x) = (g_1(x), g_2(x), g_3(x))^{\mathrm{Tr}} \quad \text{on } \partial\Omega \tag{9.214}$$

is given by

$$\left.\begin{aligned} &(\lambda = \frac{2}{3}, \mu = 1) \\ &g_1(x) = -Ax_3x_2, \\ &g_2(x) = Ax_3x_1, \\ &g_3(x) = 0. \end{aligned}\right\} \tag{9.215}$$

The vector field $\vec{g}$ (extended on an open set G containing $\overline{\Omega}$) satisfies the property

$$\nabla \cdot \vec{g} = 0 \ \text{ on } G.$$

Therefore, $\vec{g}$ represents a solenoidal field. Thus, the ball is subject to a twisting force on the boundary.

Note that $\vec{g}$ satisfies the compatibility conditions

$$\langle \vec{m}_i, \vec{g} \rangle_{L^2(\partial\Omega)} = 0, \quad 1 \leqslant i \leqslant 6.$$

We use Corollary 4.9 to solve this traction BVP. Six accessory conditions (4.206) are used, and the augmented system takes the form (4.219), which is further discretized in the same way as in § 9.11 and 9.12. Using $(l,m) = (12,24)$, we have computed $\vec{f}_h(\cdot)$ and $\alpha_{j,h}$, $1 \leqslant j \leqslant 6$, and obtained the solution (9.213). See the plots in Fig. 9.5. Numerical values of $\vec{w}_h(x)$ at sample points can be found in Table 9.3.

Table 9.2: Numerical solution $\vec{w}_h(x)$ (ball subject to compression) at sample points; $(l,m) = (12,24)$ is used.

x (spherical coordinates)	$\vec{w}_h(x)$		
$(1,0,0)$	-0.757807E-09	0.274607E-13	-0.873107E+00
$(1,\frac{1}{4}\pi,0)$	0.802640E-01	-0.830245E-09	-0.409959E+00
$(1,\frac{1}{4}\pi,\frac{1}{2}\pi)$	0.802640E-01	-0.830257E-09	0.409959E+00
$(1,\frac{1}{4}\pi,\pi)$	-0.802640E-01	0.830285E-09	-0.409959E+00
$(1,\frac{1}{4}\pi,\frac{3}{2}\pi)$	-0.157807E-08	-0.802640E-01	-0.409959E+00
$(1,\frac{1}{2}\pi,0)$	0.152617E+00	0.000000E+00	0.000000E+00
$(1,\frac{1}{2}\pi,\frac{1}{2}\pi)$	0.000000E+00	0.152617E+00	0.000000E+00
$(1,\frac{1}{2}\pi,\pi)$	-0.152617E+00	0.000000E+00	0.000000E+00
$(1,\frac{1}{2}\pi,\frac{3}{2}\pi)$	-0.152617E+00	0.000000E+00	0.000000E+00
$(1,\frac{3}{4}\pi,0)$	0.802640E-01	-0.830257E-09	0.409959E+00
$(1,\frac{3}{4}\pi,\pi)$	0.947985E-10	0.802640E-01	0.409959E+00
$(1,\frac{3}{4}\pi,\frac{3}{2}\pi)$	-0.802640E-01	0.830217E-09	0.409959E+00
$(1,\pi,0)$	-0.157805E-08	-0.802640E-01	0.409959E+00
$(\frac{1}{2},\frac{3}{4}\pi,\frac{3}{2}\pi)$	-0.757807E-09	-0.285709E-13	0.873107E+00
$(\frac{1}{2},\frac{1}{3}\pi,0)$	0.527615E-01	-0.251175E-14	-0.113310E+00
$(\frac{1}{2},\frac{2}{3}\pi,0)$	0.527615E-01	-0.120148E-14	0.113310E+00
$(\frac{1}{2},\pi,0)$	0.794720E-16	0.128555E-14	0.262307E+00
$(\frac{1}{2},\frac{1}{3}\pi,\pi)$	-0.527615E-01	0.112252E-14	-0.113310E+00
$(\frac{1}{2},\frac{2}{3}2\pi,\pi)$	-0.527615E-01	0.244565E-14	0.113310E+00
$(\frac{1}{2},\pi,\pi)$	0.692805E-16	0.128281E-14	0.262307E+00
$(0,0,0)$	-0.114817E-15	-0.384298E-16	-0.962772E-15

$\alpha_1 = 0,\ \alpha_2 = 0,\ \alpha_3 = 0,\ \alpha_4 = 0,\ \alpha_5 = 0,\ \alpha_6 = 0.$

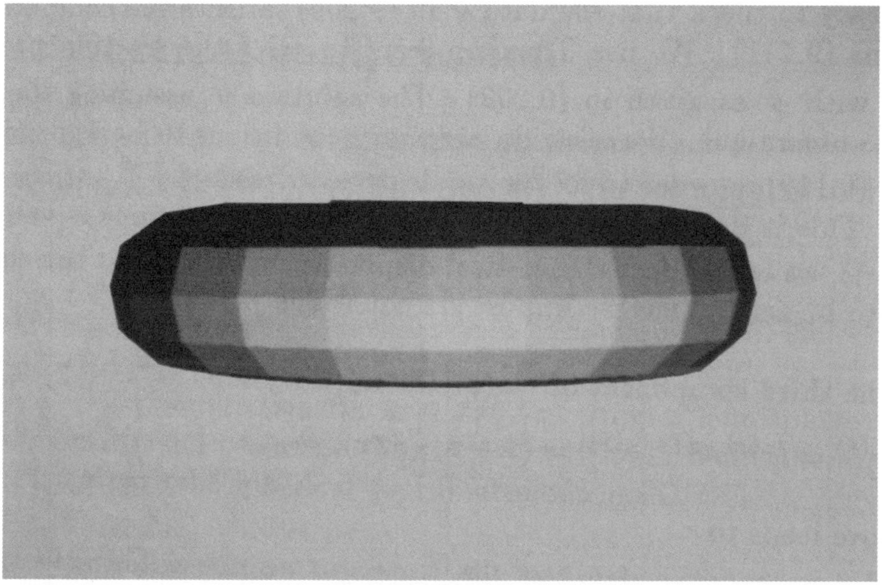

Figure 9.4: (a) front view

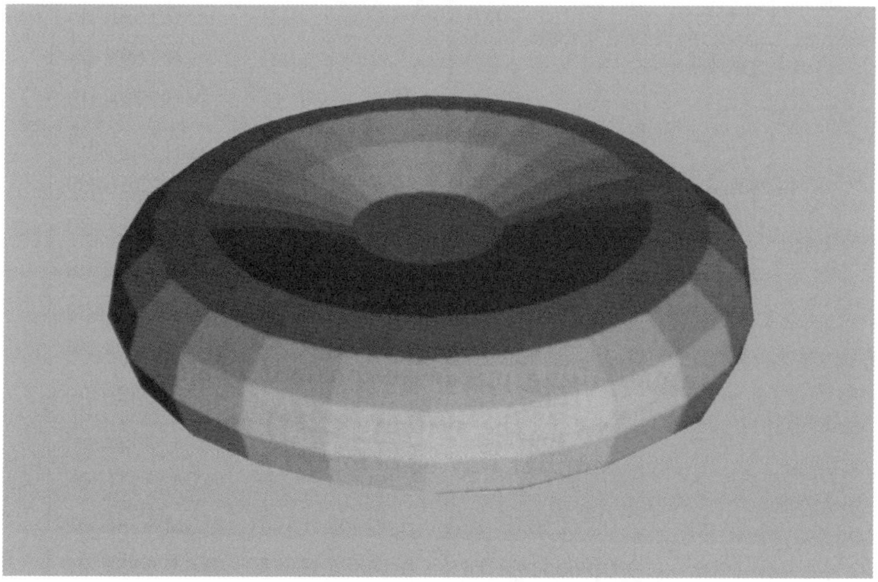

Figure 9.4: (b) top view

Figure 9.4: Plot of the numerical solution $\vec{w}_h$ of the deformed ball due to the vertical compression force (9.208).

Table 9.3: Numerical solution $\vec{w}_h(x)$ (ball subject to twisting) at sample points; $(l,m) = (12,24)$ is used with, $A = -3$ in (9.215).

x (spherical coordinates)	$\vec{w}_h(x)$		
$(1,0,0)$	0.0	0.0	-0.313E-08
$(1,\frac{1}{4}\pi,0)$	0.0	1.33	-0.272E-09
$(1,\frac{1}{4}\pi,\frac{1}{2}\pi)$	-1.33	0.0	-0.272E-09
$(1,\frac{1}{4}\pi,\pi)$	0.0	-1.33	-0.272E-09
$(1,\frac{1}{4}\pi,\frac{3}{2}\pi)$	1.33	0.0	-0.272E-09
$(1,\frac{1}{2}\pi,0)$	0.0	0.0	0.169E-08
$(1,\frac{1}{2}\pi,\frac{1}{2}\pi)$	0.0	0.0	0.169E-08
$(1,\frac{1}{2}\pi,\pi)$	0.0	0.0	0.169E-08
$(1,\frac{1}{2}\pi,\frac{3}{2}\pi)$	0.0	0.0	0.169E-08
$(1,\frac{3}{4}\pi,0)$	0.0	-1.33	-0.272E-09
$(1,\frac{3}{4}\pi,\pi)$	1.33	0.0	-0.272E-09
$(1,\frac{3}{4}\pi,\frac{3}{2}\pi)$	0.0	1.33	-0.272E-09
$(1,\pi,0)$	-1.33	0.0	-0.272E-09
$(\frac{1}{2},\frac{3}{4}\pi,\frac{3}{2}\pi)$	0.0	0.0	-0.313E-08
$(\frac{1}{2},\frac{1}{3}\pi,0)$	-0.453108E-09	0.299619E+00	-0.245543E-08
$(\frac{1}{2},\frac{2}{3}\pi,0)$	0.453110E-09	-0.299619E+00	-0.245543E-08
$(\frac{1}{2},\pi,0)$	0.726787E-14	-0.512228E-14	-0.380938E-08
$(\frac{1}{2},\frac{1}{3}\pi,\pi)$	0.453092E-09	-0.299619E+00	-0.245542E-08
$(\frac{1}{2},\frac{2}{3}\pi,\pi)$	-0.453101E-09	0.299619E+00	-0.245541E-08
$(\frac{1}{2},\pi,\pi)$	0.712565E-14	-0.490024E-14	-0.380938E-08
$(0,0,0)$	-0.160517E-14	0.148273E-15	-0.347613E-08

$\alpha_1 = 0,\ \alpha_2 = 0,\ \alpha_3 = 0,\ \alpha_4 = 0,\ \alpha_5 = 0,\ \alpha_6 = 0.$

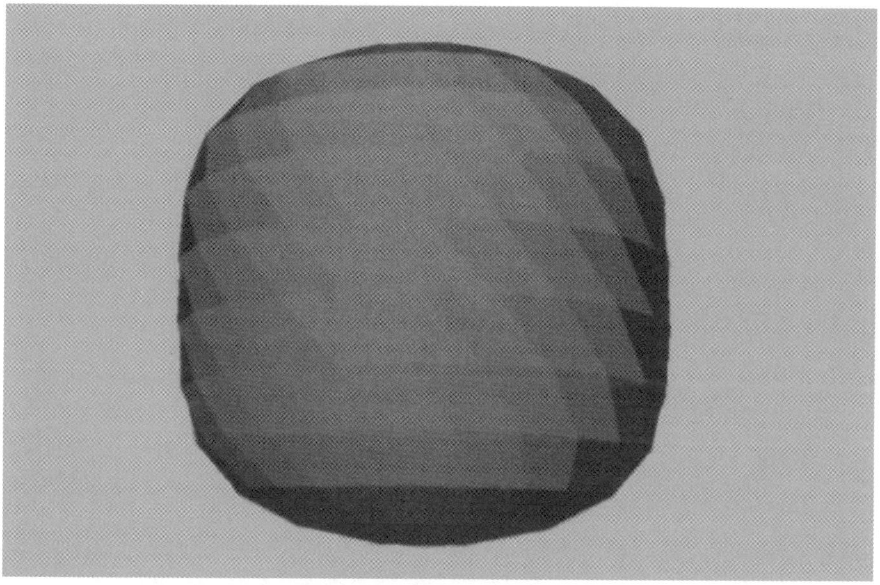

Figure 9.5: (a)

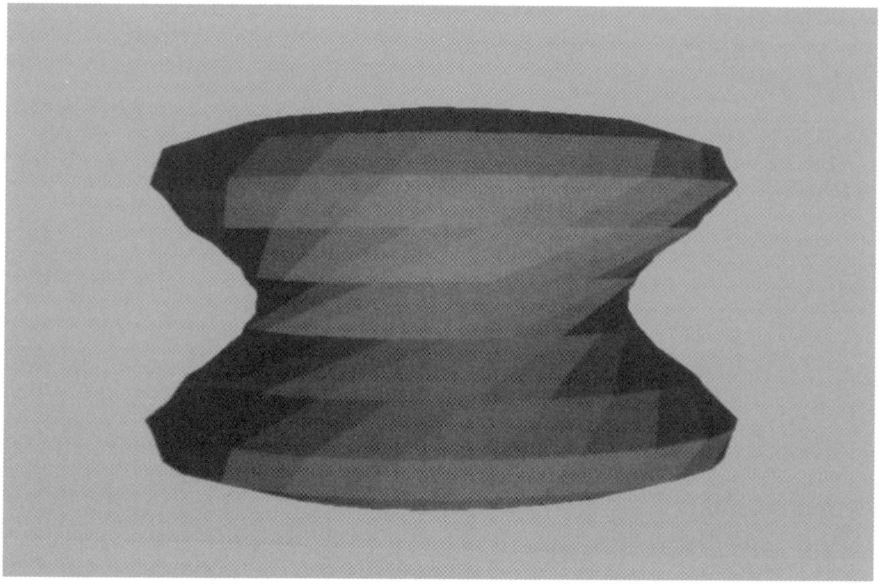

Figure 9.5: (b)

Figure 9.5: Boundary element solutions of a ball subject to a horizontal twisting force (9.215) with (a) $A = 1$ and (b) $A = -3$. Note the different directions of the twisted deformation in (a) and (b), and the severity of the deformation in (b) when $|A|$ becomes large.

Chapter 10

Some Error Estimates for Numerical Solutions of Boundary Integral Equations

Boundary integral equations can be numerically approximated by various methods. Let

$$Au(x) = g(x), \quad x \in \partial\Omega, \tag{10.1}$$

be a boundary integral equation on the boundary $\partial\Omega$ of a domain Ω. Let $\{V_h\}_h$ be a family of finite-element spaces on $\partial\Omega$. We can think of the Galerkin method for approximating (10.1):

$$\text{Find} \quad u_h \in V_h \quad \text{such that} \quad \langle Au_h, v_h\rangle = \langle g, v_h\rangle \ \forall\, v_h \in V_h, \tag{10.2}$$

where in the above, $\langle\cdot,\cdot\rangle$ denotes either the L_2-inner product or duality pairings of certain Sobolev spaces. Sharp error estimates for a Galerkin scheme like (10.2) for a Fredholm integral equation of the first and second kinds can be found in [96, 97]. The theory of convergence and error estimates is rather well developed. We present it in §10.1.

Another possible way of approximating (10.1) is by least-squares approximation. We may consider

$$\text{Find} \quad u_h \in V_h \quad \text{such that} \quad \|Au_h - g\|^2 = \min_{v_h \in V_h} \|Av_h - g\|^2, \tag{10.3}$$

where $\|\cdot\|$ is the norm of a certain Sobolev space (the simplest of which is just the L^2 norm). By the calculus of variations (cf. Theorem 5.1), (10.3) is equivalent to solving

$$\text{Find} \quad u_h \in V_h \quad \text{such that} \quad \langle Au_h - g, Av_h\rangle = 0 \ \forall \ v_h \in V_h, \tag{10.4}$$

which is similar to (10.2). See [23, 24] for some numerical treatment of (10.3). However, it is known that the Galerkin method is superior to the least-squares method whenever the Galerkin method is applicable [148, 149].

In practice, as we have shown in the preceding chapters, most numerical solutions of boundary integral equations are computed by *collocation* procedures:

$$\begin{aligned} &\text{Find} \quad u_h \in V_h \quad \text{such that} \quad (Au_h)(x_i) = g(x_i),\\ &\qquad i = 1, 2, \ldots, n, \quad x_i \in \partial\Omega. \end{aligned} \tag{10.5}$$

For the collocation method, we should point out that the convergence of the approximating operators (for A) and the completeness of the spline collocation function spaces V_h in general may not imply the convergence of approximate solution to the exact solution. See an example given by Collatz in [52, pp. 260–262]. Thus, the question of convergence is very delicate. Indeed, there are still many open questions concerning rigorous convergence of the collocation method. For two-dimensional domains Ω, the BIE (10.1) is one-dimensional. The best convergence results so far were obtained by Arnold and Wendland [10], which have improved earlier results of Arnold and Wendland [9] and Saranen and Wendland [165]. The results in [9] are applicable to nonuniform meshes, while those in [10] are mainly applicable to "smoothly graded partitions" (which are essentially uniformly spaced meshes). In §§10.2 and 10.3, we will follow [9] and [10] closely and present some of the major results.

For the following situations, there are still no collocation convergence results available:

(i) systems of BIE of different orders (such as those arising for plates in Chapter 9);
(ii) BIE arising from BVP on domains with corners;
(iii) BIE arising from mixed BVP.

For three-dimensional domains, so far there are virtually no convergence results available for collocation of boundary integral equations. The latest, most comprehensive results may be found in the book Hsiao and Wendland [99].

10.1 Convergence and error estimates of the Galerkin method

Although the Galerkin method (10.2) has not been used in our earlier numerical examples, the method itself is important in that it essentially leads to the first convergence proofs of the collocation method. The convergence of the Galerkin scheme is valid for a large class of boundary pseudodifferential equations in any space dimension.

Let $\partial\Omega$ be the (C^∞-smooth) boundary of a bounded open domain in $\mathbb{R}^N$. Let A be a strongly elliptic pseudodifferential operator of order 2α such that

$$A\colon\ H^{s+\alpha}(\partial\Omega) \to H^{s-\alpha}(\partial\Omega) \quad \text{isomorphically}, \quad \forall\, s \in \mathbb{R}, \tag{10.6}$$

and A satisfies (4.47) with $m = 2\alpha$. Thus, (10.1) is uniquely solvable for a given $f \in H^{s-\alpha}(\partial\Omega)$. As a consequence of strong ellipticity, A satisfies the Gårding inequality

$$\langle Au, u\rangle_{H^s(\partial\Omega)} \geqslant \gamma \|u\|^2_{H^{s+\alpha}(\partial\Omega)} - \langle Ku, u\rangle_{H^s(\partial\Omega)}, \quad u \in H^{s+\alpha}(\partial\Omega), \tag{10.7}$$

for some $\gamma > 0$ (depending only on s) and

$$K\colon\ H^{s+\alpha}(\partial\Omega) \to H^{s-\alpha}(\partial\Omega) \qquad \text{is compact.} \tag{10.8}$$

For any bounded linear operator $B\colon\ H^{s_1}(\partial\Omega) \to H^{s_2}(\partial\Omega)$, we define its formal t-adjoint $B^*\colon\ H^{2t-s_2}(\partial\Omega) \to H^{2t-s_1}(\partial\Omega)$ by

$$\langle u, B^*v\rangle_{H^t(\partial\Omega)} = \langle Bu, v\rangle_{H^t(\partial\Omega)} \quad \forall u \in H^{s_1}(\partial\Omega),\ \ \forall v \in H^{2t-s_2}(\partial\Omega). \tag{10.9}$$

Let A^* be the formal s-adjoint of A. Then A^* also satisfies the coercivity inequality (10.7). Since A is an isomorphism, from (10.7), A and A^* are both Fredholm operators of zero index; thus A^* is also an isomorphism from $H^{s+\alpha}(\partial\Omega)$ onto $H^{s-\alpha}(\partial\Omega)$. The invertibility of A and A^* gives

$$\|u\|_{H^{s+\alpha}} \leqslant C\|Au\|_{H^{s-\alpha}}, \quad \|u\|_{H^{s+\alpha}} \leqslant C\|A^*u\|_{H^{s-\alpha}} \tag{10.10}$$

for some $C > 0, \forall u \in H^{s+\alpha}(\partial\Omega)$.

Following [14, p. 84], we now introduce an (l,m)-system of approximation spaces $S_h^{l,m}(\partial\Omega) \subset H^m(\partial\Omega)$ (cf. Definition 5.1) satisfying

(i) *convergence property*: let $-l \leqslant s \leqslant r \leqslant l$, $-m \leqslant r$, $s \leqslant m$. Then for any $u \in H^r(\partial\Omega)$, there exists $v_h \in S_h^{l,m}(\partial\Omega)$ such that

$$\|u - v_h\|_{H^s(\partial\Omega)} \leqslant Ch^{r-s}\|u\|_{H^r(\partial\Omega)}, \qquad 0 < h < 1, \tag{10.11}$$

for some $C > 0$ independent of u and h;

(ii) *inverse inequality (cf. §5.13):* for $s \leqslant r$, and $|s|$, $|r| \leqslant m$, there exists a constant $M = M(s,r,m)$ such that

$$\|v_h\|_{H^r(\partial\Omega)} \leqslant Mh^{s-r}\|v_h\|_{H^s(\partial\Omega)} \qquad \forall v_h \in S_h^{l,m}(\partial\Omega). \tag{10.12}$$

We first note that in general the variational problem (10.2) does not necessarily have a solution u_h when h is not small, as the following theorem shows.

Theorem 10.1. *Let V and H be Hilbert spaces such that V is densely imbedded in H, and V' is the dual of V pivotal to H. Let S_h be a finite-dimensional subspace of V and let $A\colon\ V \to V'$ be an isomorphism. Then the problem*

$$(\boldsymbol{P})\ \ \text{find } w_h \in S_h \text{ such that } \langle Aw_h, v_h\rangle_{V'\times V} = \langle Aw, v_h\rangle_{V'\times V}\ \ \forall\, v_h \in S_h$$

for an arbitrarily given $w \in V$ has a unique solution $w_h \in S_h$ if and only if

$$AS_h \cap S_h^{\perp\prime} = \{0\}, \tag{10.13}$$

where

$$S_h^{\perp\prime} \equiv \{z \in V' \mid \langle v_h, z\rangle_{V\times V'} = 0\ \ \forall\, v_h \in S_h.$$

Proof. Since S_h is finite-dimensional, we let $\{e_1, \ldots, e_m\}$ be a basis for S_h. We first show that (**P**) has a unique solution if and only if

$$\det[\langle Ae_j, e_i\rangle_{V'\times V}]_{1\leqslant i,j\leqslant m} \neq 0. \tag{10.14}$$

Let (10.14) be satisfied. For any $w \in V$, let

$$\langle Aw, e_i\rangle = a_i, \qquad 1 \leqslant i \leqslant m.$$

Define

$$w_h = \sum_{j=1}^{m} \beta_j e_j,$$

where $\beta_j, 1 \leqslant j \leqslant m$, satisfy

$$\sum_{j=1}^{m} m_{ij}\beta_j = \alpha_i, \quad 1 \leqslant i \leqslant m,\ m_{ij} \equiv \langle Ae_j, e_i\rangle_{V'\times V}. \tag{10.15}$$

Such $\beta_j, 1 \leqslant j \leqslant m$, and consequently w_h, are unique because of (10.14). Therefore

$$\langle Aw_h, e_i\rangle = \sum_{j=1}^{m} \beta_j\langle Ae_j, e_i\rangle = \sum_{j=1}^{m} m_{ij}\beta_j = \alpha_j = \langle Aw, e_i\rangle,$$

for $1 \leqslant i \leqslant n$. Thus, (***P***) has a unique solution $w_h \in S_h$.

Conversely, if (10.14) is not satisfied, then, because A is an isomorphism, we can choose $w \in V$ and $(\alpha_1, \ldots, \alpha_m) \in \mathbb{R}^m$ such that (10.15) is not solvable for an $(\beta_1, \ldots, \beta_m) \in \mathbb{R}^m$. Thus, (***P***) does not admit a solution $w_h \in S_h$.

But (10.14) is satisfied if and only if

$$\sum_{j=1}^{m} \beta_j\langle Ae_j, e_i\rangle = 0$$

admits only the trivial solution $(\beta_1, \ldots, \beta_m) = (0, \ldots, 0)$, i.e.,

$$\left\langle A\left(\sum_{j=1}^{m} \beta_j e_j\right), e_i\right\rangle_{V'\times V} = 0, \quad 1 \leqslant i \leqslant m,$$

implies $\sum_{j=1}^{m} \beta_j e_j = 0$ in S_h. This is exactly equivalent to (10.13). □

The reader can easily construct an example with $\dim S_h = 1$ to violate (10.13).

To prove the convergence of the Galerkin scheme (10.2), we now show that the variational equation (10.2) induces the Babuska coercivity condition, Theorem 5.5.

Theorem 10.2.

(1) *Let V and H be two Hilbert spaces. V is densely imbedded in H. V' is the dual of V pivotal to H; thus*

$$V \subset H \subset V'. \tag{10.16}$$

(2) *Let $A\colon\ V \to V'$ be an isomorphism and let $K\colon\ V \to V'$ be linear and compact.*

(3) *Let*

$$\langle (A+K)v, v\rangle_{V'\times V} \geqslant \gamma \|v\|_V^2, \quad \gamma > 0 \ \ \forall v \in V.$$

(4) *Let $S_h \subset V$, $0 < h < 1$, be a one-parameter family of subspaces of V satisfying*

$$\inf_{v_h \in S_h} \|w - v_h\|_V \equiv \eta_h(w) \downarrow 0 \ \text{ as } h \downarrow 0 \ \ \forall w \in V. \tag{10.17}$$

Then there exists an $h_0 > 0$ such that

(a)

$$\inf_{\substack{v \in S_h \\ \|v\|_V = 1}} \ \sup_{\substack{w \in S_h \\ \|w\|_V = 1}} \langle Aw, v\rangle_{V'\times V} \geqslant C_1 > 0,$$

$$\text{for some } C_1 \text{ independent of } h, \forall h < h_0. \tag{10.18}$$

(b) *For any given $w \in V$, the variational problem*

$$\text{Find } w_h \in S_h \text{such that } \langle Aw_h, v_h\rangle_{V'\times V} = \langle Aw, v_h\rangle_{V'\times V} \ \forall\, v_h \in S_h, \tag{10.19}$$

has a unique solution $\forall h,\ 0 < h < h_0$, satisfying

$$\|w_h\|_V \leqslant C_2 \|w\|_V$$

and

$$\|w - w_h\|_V \leqslant C_3 \inf_{v_h \in S_h} \|w - v_h\|_V, \tag{10.20}$$

for some $C_2, C_3 > 0$ independent of w and h.

Proof. Since $A+K\colon V \to V'$ is continuous, satisfying (3), for $w, v \in V$, we define

$$a(w,v) = \langle (A+K)w, v\rangle_{V'\times V}. \tag{10.21}$$

Then $a(\cdot,\cdot)$ is a coercive bilinear form.
Since $A\colon\ V \to V'$ is an isomorphism, $A^{-1}\colon\ V' \to V$ is also an isomorphism. By the open mapping theorem (Appendix A.3(i)), there exist $\rho_1, \rho_2 > 0$ such that

$$\left.\begin{array}{l} A(B(1,V)) \supset \rho_1 B(1,V'), \\ A^{-1}(B(1,V')) \supset \rho_2 B(1,V), \end{array}\right\} \tag{10.22}$$

where $B(1,V)$ and $B(1,V')$ are, respectively, the unit balls in V and V'.

Let us define

$$\widetilde{S}_h^{\perp} = \{g \in V | a(g,v) = 0 \; \forall v \in S_h\}. \tag{10.23}$$

Then $\widetilde{S}_h^{\perp}$ is a closed subspace of V. From the coercivity and boundedness of $a(\cdot,\cdot)$, by the Lax–Milgram theorem, we know that for every $u \in V$, there exists a unique $w \in S_h$ such that

$$a(u-w,v) = 0 \qquad \forall v \in S_h.$$

Therefore

$$u = w + g \text{ for some } g \in \widetilde{S}_h^{\perp}, \tag{10.24}$$

i.e.,

$$V = S_h + \widetilde{S}_h^{\perp}. \tag{10.25}$$

Furthermore, the representation (10.24) is unique.

Let $v \in S_h$. By (10.21) and (10.23), we have

$$\sup_{w \in S_h} \langle Aw, v\rangle_{V'\times V} = \sup_{\substack{w \in S_h \\ g \in \widetilde{S}_h^{\perp}}} \left[\langle A(w+g), v\rangle_{V'\times V} + \langle Kg, v\rangle_{V'\times V}\right]. \tag{10.26}$$

Since $K\colon V \to V'$ is compact, $(A+K)^{-1}K\colon V \to V$ is compact. Using Exercise 10.1 below (letting $\widetilde{K} = (A+K)^{-1}K$), we have

$$\sup_{\substack{g \in \widetilde{S}_h^{\perp} \\ \|g\|_V \leqslant 1}} \|(A+K)^{-1}Kg\|_V \equiv \varepsilon_h \longrightarrow 0 \text{ as } h \downarrow 0. \tag{10.27}$$

Since $A+K\colon V \to V'$ is an isomorphism with norm $\rho_0^2 \equiv \|A+K\|$, we have

$$\sup_{\substack{g \in \widetilde{S}_h^{\perp} \\ \|g\|_V \leqslant 1}} \|Kg\|_{V'} \leqslant \rho_0^2 \varepsilon_h \downarrow 0 \text{ as } h \downarrow 0. \tag{10.28}$$

Continuing from (10.26), we obtain

$$\begin{aligned}
\sup_{\substack{w \in S_h \\ \|w\|_V = 1}} \langle Aw, v\rangle_{V'\times V} &\geqslant \sup_{\substack{w \in S_h, \|w\|_V = 1 \\ g \in \widetilde{S}_h^{\perp}, \|g\|_V \leqslant 1}} [\langle A(w+g), v\rangle_{V'\times V}] - \rho_0^2 \varepsilon_h \|v\|_V \quad \text{(by (10.28))} \\
&= \sup_{\substack{w \in S_h, \|w\|_V \leqslant 1 \\ g \in \widetilde{S}_h^{\perp}, \|g\|_V \leqslant 1}} [\langle A(w+g), v\rangle_{V'\times V}] - \rho_0^2 \varepsilon_h \|v\|_V \\
&\geqslant \sup_{q \in V, \|q\|_V \leqslant 1} [\langle Aq, v\rangle_{V'\times V}] - \rho_0^2 \varepsilon_h \|v\|_V \quad \text{(by (10.25))} \\
&\geqslant \rho_1 \sup_{\phi \in V', \|\phi\|_{V'} = 1} [\langle \phi, v\rangle_{V'\times V}] - \rho_0^2 \varepsilon_h \|v\|_V \quad \text{(by(10.22))} \\
&= \rho_1 \|v\|_V - \rho_0^2 \varepsilon_h \|v\|_V \\
&\geqslant \delta \|v\|_V,
\end{aligned}$$

for some $\delta > 0$, for all h sufficiently small such that $\rho_1 - \rho_0^2 \varepsilon_h > \delta$. So (a) is proved.

Since $A^*, K^*: V \to V'$ satisfy the same properties (1), (2) and (3) as A and K, we also have

$$\inf_{\substack{w \in S_h \\ \|w\|_V = 1}} \sup_{\substack{v \in S_h \\ \|v\|_V = 1}} \langle Aw, v \rangle_{V' \times V} \geqslant C_2' > 0,$$

for some C_2', for all sufficiently small h.

Now we can apply Theorem 5.5 and conclude that the variational problem (10.19) has a unique solution $w_h \in S_h$ satisfying

$$\|w - w_h\|_V \leqslant \left(1 + \frac{\|A\|}{C_1}\right) \inf_{\mu \in S_h} \|w - \mu\|_V$$
$$\forall h \leqslant h_0, \quad \text{for some } h_0 > 0,$$

where $\|A\|$ is the norm of A from V into V'. Hence (b) is proved. □

Exercise 10.1. Let V be a Hilbert space and let $\{S_h | 0 < h < 1\}$ be a one-parameter family of subspaces of V satisfying

$$\inf_{v_h \in S_h} \|w - v_h\|_V \equiv \eta_h(w) \downarrow \ 0 \text{ as } h \downarrow 0, \ \forall w \in V.$$

Let $a(\cdot, \cdot)$ be a bounded and coercive bilinear form on V, and let $\widetilde{S}_h^{\perp}$ be defined as in (10.23). Assume that $\widetilde{K}: V \to V$ is compact. Prove that

$$\sup_{g \in \widetilde{S}_h^{\perp}, \|g\| = 1} \|\widetilde{K} g\| \equiv \varepsilon_h \longrightarrow 0 \ \text{ as } h \downarrow 0.$$

(Hint: Approximate $\widetilde{K}$ by a sequence of linear operators of finite rank.) □

Theorem 10.2 is applicable to a strongly elliptic pseudodifferential boundary operator A of order $2\alpha \geqslant 0$. When the order 2α is negative, we reformulate Theorem 10.2 to obtain Theorem 10.3 below. Its proof is identical and thus is omitted.

Theorem 10.3.

(1) *Let $V \subset H \subset V'$ hold as in (1) of Theorem 10.1.*

(2) *Let $A: V' \to V$ be an isomorphism, and let $K: V' \to V$ be linear and compact.*

(3) *Let*

$$\langle (A + K)v, v \rangle_{V \times V'} \geqslant \gamma \|v\|_{V'}, \gamma > 0, \ \forall\, v \in V.$$

(4) *Let $S_h \subset V'$, $0 < h < 1$, be a one-parameter family of subspaces of V' satisfying*

$$\lim_{h \downarrow 0} \inf_{v_h \in S_h} \|w - v_h\|_{V'} = 0, \ \forall\, w \in V'.$$

Then there exists an h_0, $0 < h_0 < 1$ such that

$$\inf_{\substack{v\in S_h\\ \|v\|_{V'}=1}} \sup_{\substack{w\in S_h\\ \|w\|_{V'}=1}} \langle Aw, v\rangle_{V\times V'} \geqslant C_1 > 0$$

for some $C_1 > 0$ independent of h, $\forall h$: $0 < h < h_0$, and for any $w \in V'$, the variational problem

$$\begin{aligned}&\text{Find } w_h \in S_h\\ &\text{such that} \langle Aw_h, v_h\rangle_{V\times V'} = \langle Aw, v_h\rangle_{V\times V'} \qquad \forall v_h \in S_h\end{aligned}$$

has a unique solution $\forall h$, $0 < h < h_0$, satisfying

$$\|w - w_h\|_{V'} \leqslant C_2 \inf_{v\in S_h} \|w - v_h\|_{V'}$$

for some $C_2 > 0$ independent of w and h. □

By slightly adapting the proof of Theorem 10.7, we obtain the following.

Corollary 10.1. *Assume the same conditions as in Theorem 10.2. Let $\widehat{J}_h : V \to V'$ be a family of compact operators such that $\widehat{J}_h v \longrightarrow 0$ as $h \downarrow 0$ for every $v \in V$. Then there exists $h_0 > 0$ such that*

(a)
$$\inf_{\substack{v\in S_h\\ \|v\|_{V}=1}} \sup_{\substack{w\in S_h\\ \|w\|_{V}=1}} \langle (A + \widehat{J}_h)w, v\rangle \geqslant C_1 > 0$$

for some $C_1 > 0$, $\forall h < h_0$, and

(b) *the variational problem*

$$\begin{aligned}&\text{given } v \in V, \text{ find } w_h \in S_h\\ &\text{such that } \langle (A + \widehat{J}_h)w_h, v_h\rangle_{V'\times V} = \langle (A + \widehat{J}_h)w, v_h\rangle \ \ \forall v_h \in S_h\end{aligned}$$

has a unique solution w_h such that

$$\|w_h\|_V \leqslant C_2 \|w\|_V$$

for some $C_2 > 0$, $\forall h$, $0 < h < h_0$. □

The error estimates of the Galerkin scheme (10.2) can now be given.

Theorem 10.4. *Assume that (10.6), (10.7) and (10.8) hold. Let $S_h^{l,m}(\partial\Omega)$ be a family of finite-dimensional approximation subspaces satisfying (10.11). Let $w \in H^\alpha(\partial\Omega)$. Then the Galerkin approximation*

$$\left.\begin{aligned}&\text{Find } w_h \in S_h^{l,m}(\partial\Omega)\\ &\text{such that } \langle Aw_h, v_h\rangle_{H^{-\alpha},H^\alpha} = \langle Aw, v_h\rangle_{H^{-\alpha},H^\alpha},\\ &\qquad\qquad v_h \in S_h^{l,m}(\partial\Omega),\end{aligned}\right\} \tag{10.29}$$

has a unique solution $w_h \in S_h^{l,m}$ *satisfying*

$$\|w - w_h\|_{H^\alpha(\partial\Omega)} \leqslant Ch^{r-\alpha}\|w\|_{H^r(\partial\Omega)}, \quad 0 < h < h_0, \tag{10.30}$$

for some $C > 0$, *and* h_0, $0 < h_0 < 1$, *for all* α *and* r: $-l \leqslant \alpha \leqslant r \leqslant l$, $\alpha \leqslant m$, $-m \leqslant r$, *provided* $w \in H^r(\partial\Omega)$.

Proof. For $\alpha > 0$, apply Theorem 10.2 with $V = H^\alpha(\partial\Omega), H = L^2(\partial\Omega)$ and $V' = H^{-\alpha}(\partial\Omega)$, with $s = \alpha$ in (10.11).
For $\alpha = 0$, use $V = H = V' = L^2(\partial\Omega)$, with $s = \alpha = 0$ in (10.11).
For $\alpha < 0$, apply Theorem 10.3 with $V = H^{-\alpha}(\partial\Omega), H = L^2(\partial\Omega)$ and $V' = H^\alpha(\partial\Omega)$, with $s = \alpha$ in (10.11). □

By adapting the proof of the Aubin–Nitsche lemma (Theorem 5.11), error estimates in $H^s(\partial\Omega)$ for $s \leqslant \alpha$, called the *super-approximation*, property have been obtained by Hsiao and Wendland [97, Theorem 2.1].

Theorem 10.5. *Assume that (10.6)–(10.8) hold. Let* $S_h^{l,m}(\partial\Omega)$ *be a family of finite-dimensional approximation subspaces satisfying the convergence property (10.11). Let* w_h *be the solution of the Galerkin approximation (10.29). Then the super-approximation property*

$$\|w - w_h\|_{H^s(\partial\Omega)} \leqslant Ch^{t-s}\|w\|_{H^t(\partial\Omega)}, \qquad 0 < h < h_0, \tag{10.31}$$

holds for $2\alpha - l \leqslant s \leqslant \alpha \leqslant t \leqslant l$, $-l \leqslant \alpha \leqslant m$, *for some* $C > 0$ *independent of* w *and* h.

Proof. Let

$$e = w - w_h \tag{10.32}$$

denote the error term. From (10.29),

$$\begin{aligned} \langle Ae, v_h\rangle_{H^{-\alpha},H^\alpha} &= \langle Ae, v_h\rangle_{L^2} \\ &= \langle e, A^* v_h\rangle_{L^2} = 0, \qquad \forall v_h \in S_h^{l,m}(\partial\Omega). \end{aligned} \tag{10.33}$$

By assumptions (10.6)–(10.8), the adjoint equation

$$A^* z = \phi \tag{10.34}$$

is uniquely solvable for each $\phi \in H^{-s}(\partial\Omega)$, with $z \in H^{2\alpha-s}$ satisfying

$$\|z\|_{H^{2\alpha-s}} = \|(A^*)^{-1}\phi\|_{H^{2\alpha-s}} \leqslant C_1\|\phi\|_{H^{-s}(\partial\Omega)} \quad \forall \phi \in H^{-s}(\partial\Omega), \tag{10.35}$$

since $A^{*^{-1}}$ is continuous. Therefore, we have

$$\begin{aligned}\|e\|_{H^s(\partial\Omega)} &\leqslant \sup_{\|\phi\|_{H^{-s}}=1} |\langle e,\phi\rangle_{H^s,H^{-s}}| \\ &= \sup|\langle e,A^*z\rangle| \\ &= \sup|\langle e,A^*(z-v_h)\rangle + \langle e,A^*v_h\rangle| \\ &\leqslant \sup[\|e\|_{H^\alpha}\|A^*(z-v_h)\|_{H^{-\alpha}}] \\ &\leqslant C_2\sup(\|e\|_{H^\alpha}\|z-v_h\|_{H^\alpha}).\end{aligned}$$

Using (10.30) and (10.11) in the above, we get

$$\|e\|_{H^s} \leqslant \sup C_3 h^{t-\alpha}\|w\|_{H^t} h^{\alpha-s}\|z\|_{H^{2\alpha-s}},$$

for s satisfying $-m \leqslant 2\alpha - s \leqslant l$, i.e., $2\alpha - l \leqslant s \leqslant m+2\alpha$. Finally, using (10.35) we obtain

$$\|e\|_{H^s} \leqslant \sup_{\|\phi\|_{H^{-s}}=1} (C_4 h^{t-s}\|w\|_{H^t}\|\phi\|_{H^{-s}}) = C_4 h^{t-s}\|w\|_{H^t}. \tag{10.36}$$

For the range $m+2\alpha \leqslant s \leqslant \alpha$, use an interpolation argument between the estimate (10.36) and (10.30) to obtain the final estimate (10.31). □

Corollary 10.2 ([97, Corollary 2.1]). *Let the same assumptions of Theorem 10.5 hold. In addition, assume that the mesh is quasiuniform (cf. (5.133)) such that the inverse inequality (10.12) holds. Then the Galerkin approximation (10.29) satisfies*

$$\|w-w_h\|_{H^s(\partial\Omega)} \leqslant Ch^{t-s}\|u\|_{H^t(\partial\Omega)}, \tag{10.37}$$

for t and s satisfying $2\alpha - l \leqslant s \leqslant t \leqslant l$, $s \leqslant m$, $-l \leqslant \alpha \leqslant m$ and $-m \leqslant \alpha \leqslant s$.

Sketch of Proof. For the cases $\alpha \geqslant t$ and $\alpha \leqslant s$ (which were not covered by Theorem 10.5) using "simultaneous approximation in scales" [21, 89] combined with the inverse inequality (10.12) (see the technique used in the proof of Theorem 10.11 in the next section), we can demonstrate that (10.30) remains valid. Hence (10.37) holds for all t and s in the indicated range. □

Example 10.1. The operator $\mathscr{L}_1$ in Theorem 6.21 satisfies

$$\mathscr{L}_1 : H^{s+\alpha}(\partial\Omega) \to H^{s-\alpha}(\partial\Omega) \text{ isomorphically}, \ \forall s \in \mathbb{R},$$

with $\alpha = -\frac{1}{2}$,

$$\langle \mathscr{L}_1 f, f\rangle_{L^2} \geqslant C_1\|\mathscr{L}_1 f\|_{H^{1/2}} \geqslant C_2\|f\|_{H^{-1/2}} \ \forall f \in H^{1/2}(\partial\Omega);$$

so Theorem 10.5 and Corollary 10.2 are directly applicable. The same is true for the operator $\mathscr{L}_1$ in Theorem 9.3. □

Most of the other boundary integral operators that have been introduced in this book are not invertible but become isomorphisms after the augmentation of proper finite-dimensional spaces; see Theorem 4.10. The augmented system is *strongly elliptic,* and an adapted form of Theorem 10.5 applies to such systems. The treatment of such systems is given in §10.4.

10.2 Convergence and error estimates for collocation by odd degree splines with uniform meshes

In the rest of this chapter, we investigate the convergence problems of the collocation methods in $\mathbb{R}^2$. Let Γ be a smooth Jordan curve in $\mathbb{R}^2$ given by a parametric representation

$$\Gamma: z = (x_1(t), x_2(t)) = x_1(t) + ix_2(t),$$

where $x_1(\cdot)$ and $x_2(\cdot)$ are real-valued periodic functions with period 1, and

$$\left|\frac{dz(t)}{dt}\right| \neq 0 \text{ for } t \in [0,1], \quad j = 1, 2.$$

More generally, if $\Gamma = \bigcup_{j=1}^{n} \Gamma_j$ is a disjoint union of finitely many smooth Jordan curves in $\mathbb{R}^2$, we may parametrize each of them. A scalar function or distribution defined on Γ is then identified as an *n-vector-valued function or distribution*, each component having period 1. Since all 1-periodic distributions can be represented by Fourier series

$$u = \sum_{n\in\mathbb{Z}} \widehat{u}_n e^{i2\pi nt}, \quad t \in \mathbb{R}, \quad \text{with } \widehat{u}_n = \int_0^1 e^{-i2\pi nt} u(t)\, dt, \tag{10.38}$$

all of the results in the rest of this chapter will be presented in terms of the Fourier representation.

Let H^s denote the Sobolev space of order $s \in \mathbb{R}$, i.e., the closure of all smooth real-valued 1-periodic functions with respect to the norm

$$\|u\|_s \equiv \|u\|_{H^s(\Gamma)} \equiv \Big(|\widehat{u}_0|^2 + \sum_{\substack{k\in\mathbb{Z}\\ k\neq 0}} |\widehat{u}_k|^2 |2\pi k|^{2s}\Big)^{1/2}.$$

The inner product in H^s is defined by

$$\langle u, v\rangle_s = \widehat{u}_0 \cdot \overline{\widehat{v}}_0 + \sum_{\substack{k\in\mathbb{Z}\\ k\neq 0}} \widehat{f}_k \cdot \overline{\widehat{g}}_k |2\pi k|^{2s}. \tag{10.39}$$

For any $\alpha \in \mathbb{R}$, this inner product can be extended to a pairing product between $H^{s+\alpha}$ and

$H^{s-\alpha}$:

$$\begin{aligned}
\langle u,v\rangle_s &\equiv \langle u,v\rangle_{s+\alpha,s-\alpha} \\
&\equiv \widehat{u}_0\cdot\overline{\widehat{v}}_0 + \sum_{\substack{k\in\mathbb{Z}\\k\neq 0}} \widehat{f}_k\cdot\overline{\widehat{g}}_k |2\pi k|^{s+\alpha}|2\pi k|^{s-\alpha} \\
&= \widehat{u}_0\cdot\overline{\widehat{v}}_0 + \sum_{\substack{k\in\mathbb{Z}\\k\neq 0}} \widehat{f}_k\cdot\overline{\widehat{g}}_k |2\pi k|^{2s}, \quad u\in H^{s+\alpha}, \quad v\in H^{s-\alpha},
\end{aligned}$$

satisfying

$$\|\langle u,v\rangle_{s+\alpha,s-\alpha}\| < \infty,$$

$$\sup_{v\in H^{s-\alpha}} \frac{\langle u,v\rangle_s}{\|v\|_{s-\alpha}} = \|u\|_{s+\alpha}.$$

We first study the case

$$Au = f,$$

where u and f are (1-periodic) vector-valued functions. As in (10.6), we assume that

$$A\colon H^{j+\alpha} \longrightarrow H^{j-\alpha} \quad \text{is linear and continuous,} \tag{10.40}$$

where $j\in\mathbb{R}$ will be specified later, and 2α is called the *order* of the operator A.

Example 10.2. Consider the operator defined by

$$P^+u = \sum_{n\geqslant o} \widehat{u}_n e^{i2\pi nt}, \quad u\in\mathbb{C}^\infty(\Gamma).$$

Then

$$\begin{aligned}
P^+u(t) &= \sum_{n\geqslant 0}\int_0^1 u(\tau)e^{-i2\pi n\tau}d\tau e^{i2\pi nt} \\
&= \int_0^1 u(\tau)\left(\sum_{n\geqslant 0} e^{i2\pi n(t-\tau)}\right)d\tau \\
&= \int_0^1 \frac{u(\tau)}{1-e^{i2\pi(t-\tau)}}d\tau.
\end{aligned}$$

□

It can be shown [179, pp. 56–58] that P^+ is a pseudodifferential operator of order zero on the class of 1-periodic functions, whose principal symbol is

$$\sigma(P^+,\xi) = \begin{cases} 1, & \xi > 0, \\ 0, & \xi < 0. \end{cases}$$

Thus, the principal symbol of P^+ is the Heaviside function $H(\xi)$ (cf. Example 3.1).

Similarly, we can define

$$P^- u = \sum_{n \leqslant 0} \widehat{u}_n e^{i2\pi nt}.$$

Then P^- is a pseudodifferential operator of order zero with principal symbol $H(-\xi)$. We have

$$P^+ u + P^- u = u + u_0,$$

implying

$$P^+ + P^- \sim I,$$

because the operator $u \longrightarrow u_0$ is obviously regularizing (cf. (4.25)).

Let α and β be two C^∞ 1-periodic functions that do not vanish on $[0,1]$. Then the operator

$$\alpha(t) P^+ + \beta(t) P^-, \quad t \in [0,1],$$

is an *elliptic* operator of order zero on the class of 1-periodic functions.

More generally, we define the operators

$$P_{2\alpha}^+ u = \sum_{n>0} n^{2\alpha} \widehat{u}_n e^{i2\pi nt} + \widehat{u}_0,$$
$$P_{2\alpha}^- u = \sum_{n<0} |n|^{2\alpha} \widehat{u}_n e^{i2\pi nt} + \widehat{u}_0.$$

Then

$$\alpha(t) P_{2\alpha}^+ + \beta(t) P_{2\alpha}^-$$

is an elliptic operator of order 2α on the class of 1-periodic functions.

Example 10.3. On the unit circle, the simple layer potential satisfies

$$\begin{aligned} Vu(t) &= \frac{1}{2\pi} \int_{|\xi|=1} \ln|x-\xi| \widetilde{u}(\xi)\, d\sigma_\xi, \\ &\quad (x = e^{i2\pi t}, \ \xi = e^{i2\pi\tau}, \ \widetilde{u}(\xi) = u(\tau)) \\ &= \frac{1}{2\pi} \int_0^1 \ln|e^{i2\pi t} - e^{i2\pi\tau}| u(\tau) 2\pi\, d\tau \\ &= \sum_{n \neq 0} \frac{1}{|n|} \widehat{u}_n e^{i2\pi nt}, \quad t \in [0,1], \end{aligned} \tag{10.41}$$

thus its principal symbol is $\sigma(V,\xi) = 1/|\xi|$. It is an elliptic operator of order -1. This is consistent with (4.73). □

We now look at the collocation scheme (10.5). We choose an increasing sequence of points

$$\triangle = \{x_i\}_{i=-\infty}^{\infty},$$

satisfying $x_{i+n} = x_i + 1$, for some fixed $n \in \mathbb{N}$, for all $i \in \mathbb{Z}$. Let $S_d(\triangle)$ be the space of all 1-periodic, $d-1$ times continuously differentiable splines of order d subordinate to the partition $\triangle$. For $\phi \in S_d(\triangle), \phi^{(d)}$ is a step function. For $d = 1$, $S_1(\triangle)$ consists of the roof functions in (5.54). For $d = 2$ and 3, $S_d(\triangle)$ consists of the $S^{(3,2)}(x_0, x_N)$ and $S^{(4,3)}(x_0, x_N)$ systems of *B*-spline spaces in § 5.5, etc.

The regularity of elements in $S_d(\triangle)$ can be examined as follows. Since $\phi^{(d)}$ is piecewise continuous, by Exercise 2.2,

$$\phi^{(d)} \in H^{1/2-\varepsilon} \qquad \text{for any} \quad \varepsilon > 0.$$

Thus, $\phi \in S_d(\triangle) \subset H^s$ if and only if

$$d + \frac{1}{2} - \varepsilon = s \tag{10.42}$$

for some $\varepsilon > 0$, i.e., if and only if $s < d + \frac{1}{2}$.

In this section we require that the degree of the splines be *odd*, thus we denote by d a fixed positive odd integer and define

$$j = \frac{1}{2}(d+1), \tag{10.43}$$

a positive integer.

Let $h_\triangle = \max(x_i - x_{i-1})$. Recall from Definition 5.1 the approximation property of $S_d(\triangle)$:

$$\left.\begin{array}{l} \inf\limits_{\phi \in S_d(\triangle)} \|u - \phi\|_r \leqslant C h_\triangle^{s-r} \|u\|_s, \quad 0 \leqslant r \leqslant s \leqslant d+1, \quad r \leqslant d, \\ C > 0 \text{ depending only on } r, s \text{ and } d. \end{array}\right\} \tag{10.44}$$

But, as it turns out, the above can actually be improved to

$$\begin{array}{l} \inf\limits_{\phi \in S_d(\triangle)} \|u - \phi\|_r \leqslant C h_\triangle^{s-r} \|u\|_s, \quad -\infty < r \leqslant s \leqslant d+1, \quad r < d + \dfrac{1}{2}, \\ C > 0 \text{ depending only on } r, s \text{ and } d. \end{array} \tag{10.45}$$

See the comments and references in [9, p. 353]. The proof will not be given here.

We now consider the collocation problem

$$\begin{array}{c} \text{Find } u_\triangle \in S_d(\triangle) \text{ such that } (Au_\triangle)(x_i) = g(x_i), \\ i = 1, 2, \ldots, n, \text{ where } g \text{ satisfies (10.1).} \end{array} \tag{10.46}$$

In order for (10.46) to be well–defined, $Au_\triangle$ must be continuous at nodes x_i. Since $Au_\triangle \in H^{s-2\alpha}$, we assume that $H^{s-2\alpha}$ is imbedded in the space of continuous functions. By the Sobolev imbedding theorem 2.3, we need the assumption

$$s - 2\alpha > \frac{1}{2}. \tag{10.47}$$

By (10.42) and (10.47), this leads to the relation

(A_1)

$$d = 2j - 1 > 2\alpha, \quad \text{i.e., } j - \alpha > \frac{1}{2},$$

between the order of the operator A and the degree of the splines. This relation also implies

$$S_d(\triangle) \subset H^s \subset H^{j+\alpha} \tag{10.48}$$

because of (10.43).

As in (10.6) and (10.7), we make the following additional assumptions on A:

(A_2) $A\colon H^{j+\alpha} \longrightarrow H^{j-\alpha}$ is an isomorphism, i.e., A is 1–1 and onto;

(A_3) A is j-coercive, satisfying the Gårding inequality

$$\begin{aligned} &\langle Au, u\rangle_j \geqslant \gamma \|u\|^2_{j+\alpha} - \langle Ku, u\rangle_j, \quad u \in H^{j+\alpha}, \\ &\text{where } \gamma > 0 \text{ and } K\colon H^{j+\alpha} \to H^{j-\alpha} \text{ is compact.} \end{aligned} \tag{10.49}$$

If A satisfies (A_3) then, as before, A^* will satisfy (10.49). By (A_2), A and A^* are both Fredholm operators of index zero and $A^*: H^{j+\alpha} \to H^{j-\alpha}$ is an isomorphism. Thus, again A and A^* have bounded inverses, so

$$\|u\|_{j+\alpha} \leqslant C\|Au\|_{j-\alpha}, \quad \|u\|_{j+\alpha} \leqslant C\|A^*u\|_{j-\alpha}, \quad u \in H^{j+\alpha}, \tag{10.50}$$

for some $C > 0$ independent of u.

The key idea in [9] is to *reduce the collocation equation (10.46) to equivalent Galerkin equations*. To this end, we first define operators

$$Ju \equiv \int_0^1 u(t)\,dt, \tag{10.51}$$

$$J_\triangle u \equiv \sum_{i=1}^{n} \frac{x_{i-1} - x_i}{2} u(x_i) \quad \text{(the trapezoidal rule).} \tag{10.52}$$

Theorem 10.6. *Let $w \in H^{j-\alpha}$. Then*

$$w(x_i) = 0, \quad i = 1, 2, \ldots, n, \tag{10.53}$$

if and only if

$$\langle w - Jw + J_\triangle w, v\rangle_j = 0 \quad \forall v \in S_d(\triangle). \tag{10.54}$$

Proof. First note that (10.53) is well defined because of (A_1), and so is (10.54) because of (10.48). By (10.39), we have

$$\langle f, g\rangle_j = \int_0^1 f^{(j)}(x)\overline{g}^{(j)}(x)\,dx + \int_0^1 f(x)\,dx \int_0^1 \overline{g}(x)\,dx$$

for $f, g \in H^j$. Integrating by parts once, we have, for $w \in C^\infty$,

$$\langle w - Jw + J_\triangle w, v\rangle_j = [w(x) - Jw + J_\triangle w]^{(j-1)}\overline{v}^{(j)}(x)\Big|_{x=x_0}^{x_n}$$
$$- \int_0^1 w^{(j-1)}(x)\overline{v}^{(j+1)}(x)\,dx + (J_\triangle w)(\overline{Jv})$$
$$= - \int_0^1 w^{(j-1)}(x)\overline{v}^{(j+1)}(x)\,dx + (J_\triangle w)(\overline{Jv}),$$

where we have used the periodicity of the functions $w^{(j-1)}$ and $v^{(j)}$. Repeating this $j-2$ more times and noting that $v^{(p)}$ is continuous for $p = 0, 1, \ldots, 2j-2$, we have

$$\langle w - Jw + J_\triangle w, v\rangle_j = (-1)^{j-1}\int_0^1 w'(x)\overline{v}^{(2j-1)}(x)\,dx + (J_\triangle)(\overline{Jv}). \tag{10.55}$$

We can then extend (10.55) by continuity to $w \in H^{j-\alpha}$.

Note that the operator $(d/dx)^{2j-1}$ maps $V_d \equiv \{v \in S_d(\triangle) | Jv = 0\}$ isomorphically onto $V_0 \equiv \{v \in S_0(\triangle) | Jv = 0\}$, the space of piecewise-constant functions with integral zero. Define $y_i(\cdot) \in V_0$ by

$$y_i(x) = \begin{cases} -h_i^{-1}, & x \in [x_{i-1}, x_i), \\ h_{i+1}^{-1}, & x \in [x_i, x_{i+1}), \\ 0, & x \in [x_{i+1}, x_{i-1}+1), \end{cases} \quad \text{with } h_i = x_i - x_{i-1}.$$

Let $v_i \in V_d \subset S_d(\triangle)$ be the unique function satisfying

$$v_i^{(2j-1)} = y_i, \quad Jv_i = 0.$$

Then

$$\langle w - Jw + J_\triangle w, v_i\rangle = (-1)^{j-1}\left[\frac{w(x_{i+1}) - w(x_i)}{h_{i+1}} - \frac{w(x_i) - w(x_{i-1})}{h_i}\right].$$

If (10.54) holds, there must exist a constant k such that

$$\frac{1}{h_i}[w(x_i) - w(x_{i-1})] = k, \quad i \in \mathbb{Z}. \tag{10.56}$$

Using (10.56) with $i = n, n-1, \ldots, 1$, we get

$$w(x_n) = w(x_0) + k(h_n + h_{n-1} + \ldots + h_1)$$
$$= w(x_0) + k.$$

Since w is periodic, necessarily $k = 0$. Thus (10.54) implies

$$w(x_i) = w(x_n) \quad \forall\, i \in \mathbb{Z}. \tag{10.57}$$

But since $S_d(\triangle)$ is spanned by the v_i and the constant function 1, we see that (10.54) holds if and only if (10.57) holds and

$$\langle w - Jw + J_\triangle w, 1\rangle_j = J_\triangle w = 0. \tag{10.58}$$

The above and (10.57) together are exactly (10.53). The proof is complete. □

Now define

$$A_\triangle = (I - J + J_\triangle)A \colon H^{j+\alpha} \to H^{j-\alpha},$$

which is continuous by (A_1). As a consequence of Theorem 10.6, we have the following.

Theorem 10.7. *A function $u_\triangle \in S_d(\triangle)$ satisfies the collocation equations (10.46) if and only if the Galerkin equations*

$$\langle A_\triangle u_\triangle, v\rangle_j = \langle A_\triangle u, v\rangle_j, \quad \forall v \in S_d(\triangle) \tag{10.59}$$

are satisfied. □

Lemma 10.1. *The operator $A_\triangle$ is invertible with inverse*

$$A_\triangle^{-1} = A^{-1}(I + J - J_\triangle), \tag{10.60}$$

satisfying

$$\|A_\triangle\|_{j+\alpha,j-\alpha} + \|A_\triangle^{-1}\|_{j-\alpha,j+\alpha} \leqslant C$$

for some $C > 0$, for all partitions $\triangle$.

Proof. By (A_1), $j - \alpha > \frac{1}{2}$, so $J_\triangle \colon H^{j+\alpha} \to H^{j-\alpha}$ is uniformly bounded by the uniform boundedness principle (see Appendix A.3(iii)). Therefore $A_\triangle$ is uniformly bounded by the continuity of A. Equation (10.60) follows as a consequence of

$$JJ = J_\triangle J = J, \quad J_\triangle J_\triangle = JJ_\triangle = J_\triangle.$$

The uniform boundedness of $A_\triangle^{-1}$ follows from the continuity of A^{-1}. □

The stability of the Galerkin equations (10.46) is given in the following.

Theorem 10.8 ([9, p. 355]). *There exist positive constants C and h_0 such that if $\triangle$ is a partition for which $h_\triangle = \max(x_i - x_{i-1}) \leqslant h_0$ then*

$$\inf_{\substack{v \in S_d(\triangle) \\ \|v\|_{j+\alpha}=1}} \; \sup_{\substack{z \in S_d(\triangle) \\ \|z\|_{j+\alpha}=1}} \langle A_\triangle v, z >_j \geqslant C > 0. \tag{10.61}$$

Proof. We apply Corollary 10.1 by letting

$$V = H^{j+\alpha}, \quad H = H^j, \quad V' = H^{j-\alpha}, \tag{10.62}$$

$$\widehat{J}_h = (J_\triangle - J)A, \quad A + \widehat{J}_h = A_\triangle, \quad S_h = S_d(\triangle). \tag{10.63}$$

Because of (A_1) and (10.45), using $r = j + \alpha$ and $s = d + 1$ in (10.45), we have

$$\inf_{\phi \in S_d(\triangle)} \|u - \phi\|_r \leqslant C h^{d+1-(j+\alpha)} \|u\|_s \equiv \eta_h(u) \downarrow 0 \text{ as } h \downarrow 0,$$

for

$$d+1-(j+\alpha)=\frac{1}{2}+(\frac{1}{2}d-\alpha)>\frac{1}{2},$$

for $u\in H^s$. But H^s is dense in $H^{j+\alpha}$, so (10.17) is satisfied. So Corollary 10.1 is applicable and the proof is complete. □

As a consequence of (10.45) and Theorems 10.8 and 5.5, we obtain the error estimates in the natural space $H^{j+\alpha}$.

Theorem 10.9. *There exist positive constants C and h_0 such that for any partition $\triangle$ with $h_\triangle \leqslant h_0$, there is a unique solution $u_\triangle \in S_d(\triangle)$ of the collocation equations (10.46) (equivalently (10.59)). We have*

$$\|u-u_\triangle\|_{j+\alpha}\leqslant C\inf_{v\in S_d(\triangle)}\|u-v\|_{j+\alpha}.$$

In addition, if $u\in H^s$ for some $s\in[j+\alpha,2j]$, then

$$\|u-u_\triangle\|_{j+\alpha}\leqslant Ch_\triangle^{s-j-\alpha}\|u\|_s \tag{10.64}$$

for $C>0$ independent of u, $u_\triangle$ and $h_\triangle$. □

Next, we use the Aubin–Nitsche duality argument (see Theorem 5.11) to find *error estimates in Sobolev spaces of lower order.*

Let $s>j+\alpha$, and let $B\colon H^{j+\alpha}\to H^{j-\alpha}$ be an isomorphism. We say that B is *s-regular* if B^{-1} maps $H^{s-2\alpha}$ boundedly into H^s.

Theorem 10.10. *Let $\tau\in[2\alpha,j+\alpha]$ and assume that A^* is $(2(j+\alpha)-\tau)$-regular. In the case $\tau\leqslant 2\alpha+\frac{1}{2}$ and $j>\alpha+2$, assume in addition that A maps $H^{2+2\alpha}$ boundedly into H^2. Then*

$$\|u-u_\triangle\|_\tau\leqslant Ch_\triangle^{j+\alpha-\tau}\|u-u_\triangle\|_{j+\alpha}. \tag{10.65}$$

Proof. We consider the cases $\tau\in(2\alpha+\frac{1}{2},j+\alpha]$ and $\tau\in[2\alpha,2\alpha+\frac{1}{2}]$ separately below.

(i) $\tau\in(2\alpha+1/2,j+\alpha]$

Because $\tau-2\alpha>\frac{1}{2}$,

$$J_\triangle\colon H^{\tau-2\alpha}\to H^{\tau-2\alpha} \tag{10.66}$$

is uniformly bounded with respect to $\triangle$, so the j-adjoint operators

$$J_\triangle^*\colon H^{2j-(\tau-2\alpha)}\to H^{2j-(\tau-2\alpha)} \quad (=H^{2(j+\alpha)-\tau})$$

are uniformly bounded for fixed τ. From the regularity assumption on A^*, it follows that

$$A^{*^{-1}}_{\triangle} = (I+J-J^*_{\triangle})A^{*-1} : H^{2j-\tau} \to H^{2(j+\alpha)-\tau} \tag{10.67}$$

are uniformly bounded with respect to $\triangle$.

By duality, we can find $v_{\triangle} \in H^{2j-\tau}$ such that

$$\langle u-u_{\triangle}, v_{\triangle}\rangle_j = \|u-u_{\triangle}\|^2_{\tau} = \|v_{\triangle}\|^2_{2j-\tau}. \tag{10.68}$$

Let $y_{\triangle} = (A^*_{\triangle})^{-1}v_{\triangle}$. Then the uniform boundedness of $(A^*_{\triangle})_{-1}$ and (10.68) imply

$$\|y_{\triangle}\|_{2(j+\alpha)-\tau} \leqslant C\|v_{\triangle}\|_{2j-\tau} = C\|u-u_{\triangle}\|_{\tau} \tag{10.69}$$

for some $C > 0$ independent of $\triangle$ and u. From (10.68) and (10.69), we have

$$\begin{aligned}
\|u-u_{\triangle}\|^2_{\tau} = \langle u-u_{\triangle}, A^*_{\triangle}y_{\triangle}\rangle_j &= \langle A_{\triangle}(u-u_{\triangle}), y_{\triangle}\rangle_j \\
&= \inf_{\phi\in S_d(\triangle)} \langle A_{\triangle}(u-u_{\triangle}), y_{\triangle}-\phi\rangle_j \\
&\leqslant C\|A_{\triangle}(u-u_{\triangle})\|_{j-\alpha} \inf_{\phi\in S_{d(\triangle)}} \|y-\phi\|_{j+\alpha} \\
&\leqslant C\|u-u_{\triangle}\|_{j+\alpha} h^{j+\alpha-\tau}_{\triangle} \|y_{\triangle}\|_{2(j+\alpha)-\tau}.
\end{aligned} \tag{10.70}$$

The inequalities (10.69) and (10.70) lead to the estimate (10.65).

(ii) $\tau \in [2\alpha, 2\alpha+\frac{1}{2}]$

Again we define $v_{\triangle}$ by (10.68). However, (10.66) is no longer well defined, so $A^*_{\triangle}$ is not $(2(j+\alpha)-\tau)$-regular, i.e., (10.67) is not continuous. We let $y_{\triangle} = A^{*^{-1}}v_{\triangle}$. From (10.69), we get

$$\begin{aligned}
\|u-u_{\triangle}\|^2_{\tau} &= \langle u-u_{\triangle}, A^*y_{\triangle}\rangle_j \\
&= \langle A_{\triangle}(u-u_{\triangle}), y_{\triangle}\rangle_j + \langle (A-A_{\triangle})(u-u_{\triangle}), y_{\triangle}\rangle_j.
\end{aligned} \tag{10.71}$$

The first term above can be estimated in the same way as (10.70). For the second term we have

$$\begin{aligned}
|\langle (A-A_{\triangle})(u-u_{\triangle}), y_{\triangle}\rangle_j| &= |(J-J_{\triangle})[A(u-u_{\triangle})]|\,|Jy_{\triangle}| \\
&\leqslant C|(J-J_{\triangle})[A(u-u_{\triangle})]|\|y_{\triangle}\|_{2(j+\alpha)-\tau} \\
&\leqslant C|(J-J_{\triangle})[A(u-u_{\triangle})]|\|u-u_{\triangle}\|_{\tau}.
\end{aligned} \tag{10.72}$$

If $j-\alpha \leqslant 2$, we use (A_1) and simple error estimates for the trapezoidal rule (see (10.52)) to get

$$\begin{aligned} |(J-J_\triangle)[A(u-u_\triangle)]| &\leqslant Ch_\triangle^{j-\alpha}\|A(u-u_\triangle)\|_{j-\alpha} \\ &\leqslant Ch_\triangle^{j-\alpha}\|u-u_\triangle\|_{j+\alpha} \\ &\leqslant Ch_\triangle^{j+\alpha-\tau}\|u-u_\triangle\|_{j+\alpha}, \\ &\quad \text{because } 2\alpha-\tau \leqslant 0. \end{aligned}$$

We substitute these estimates into (10.71), yielding (10.65). If $j-\alpha > 2$, we invoke the additional boundedness assumption on A to get

$$\begin{aligned} |(J-J_\triangle)[A(u-u_\triangle)]| &\leqslant Ch_\triangle^{2}\|A(u-u_\triangle)\|_{2} \\ &\leqslant Ch_\triangle^{2}\|u-u_\triangle\|_{2\alpha+2}. \end{aligned}$$

Since $2\alpha+2 \in (2\alpha+\frac{1}{2}, j+\alpha)$, we may apply (10.65) with τ replaced by $2\alpha+2$ in the previous case, yielding

$$\begin{aligned} |(J-J_\triangle)[A(u-u_\triangle)]| &\leqslant Ch_\triangle^{j-\alpha}\|u-u_\triangle\|_{j+\alpha} \\ &\leqslant Ch_\triangle^{j+\alpha-\tau}\|u-u_\triangle\|_{j+\alpha}. \end{aligned} \tag{10.73}$$

Combining (10.72) and (10.73) and substituting into (10.71), we obtain (10.65). □

Combining (10.64) and (10.65), we see that if $u \in H^s$ for some $s \in [j+\alpha, 2j]$, then

$$\|u-u_\triangle\|_\tau \leqslant Ch_\triangle^{s-\tau}\|u\|_s \quad \text{for} \quad \tau \in [2\alpha, j+\alpha]. \tag{10.74}$$

Thus the highest-order estimate afforded by the above is

$$\|u-u_\triangle\|_{2\alpha} = \mathscr{O}(h_\triangle^{2(j-\alpha)}).$$

According to a remark in [9, p. 358], this estimate is the best possible.

We now further examine error estimates in higher-order Sobolev spaces H^τ for $\tau > j+\alpha$. To achieve this, we need inverse properties of finite elements like (10.12). Thus we must use a family of quasiuniform meshes (cf. (5.133) in Chapter 5). We say that a mesh $\triangle$ is ρ-quasiuniform for $\rho > 0$ if

$$h_\triangle \leqslant \rho \inf(x_i - x_{i-1}).$$

The inverse properties we will need here can be stated as follows:

$$\begin{aligned} &\text{for any } \rho, \eta, \mu \in \mathbb{R}, \rho > 0, \mu \leqslant \eta < d + \frac{1}{2}, \\ &\text{there exists a } C > 0 \text{ such that} \\ &\|v\|_\eta \leqslant Ch_\triangle^{\mu-\eta}\|v\|_\mu \;\; \forall\, v \in S_d(\triangle) \\ &\text{for all } \rho\text{-quasiuniform meshes } \triangle. \end{aligned} \tag{10.75}$$

Theorem 10.11. *Let s, $\tau \in \mathbb{R}$ satisfy $j+\alpha \leqslant \tau \leqslant \sigma \leqslant 2j$, $\tau < 2j-\frac{1}{2}$ and assume that $u \in H^s$. Also assume that the inverse inequality (10.75) holds for $S_d(\triangle)$ for all ρ-quasiuniform meshes $\triangle$, with $\mu = j+\alpha$ and $\eta = \tau$. Then*

$$\|u-u_\triangle\|_\tau \leqslant Ch_\triangle^{s-\tau}\|u\|_s.$$

Proof. By "simultaneous approximation in scales" [21, 89], to each ρ-quasiuniform mesh $\triangle$ there corresponds an operator $\boldsymbol{P}_\triangle : H^\tau \to S_d(\triangle)$ providing optimal order of approximation in H^τ and $H^{j+\alpha}$ and satisfying

$$h_\triangle^\tau\|u-\boldsymbol{P}_\triangle u\|_\tau + h_\triangle^{j+\alpha}\|u-\boldsymbol{P}_\triangle u\|_{j+\alpha} \leqslant Ch_\triangle^s\|u\|_s. \tag{10.76}$$

Consequently,

$$\begin{aligned}
\|u-u_\triangle\|_\tau &\leqslant \|u-\boldsymbol{P}_\triangle u\|_\tau + \|u_\triangle - \boldsymbol{P}\triangle u\|_\tau \\
&\leqslant \|u-\boldsymbol{P}_\triangle u\|_\tau + Ch_\triangle^{j+\alpha-\tau}\|u_\triangle - \boldsymbol{P}_\triangle u\|_{j+\alpha}, \quad \text{by (10.75)}, \\
&\leqslant \|u-\boldsymbol{P}_\triangle u\|_\tau + Ch_\triangle^{j+\alpha-\tau}(\|u-u_\triangle\|_{j+\alpha} \\
&\quad + \|u-\boldsymbol{P}_\triangle u\|_{j+\alpha}) \\
&\leqslant Ch_\triangle^{s-\tau}\|u\|_s, \quad \text{by (10.65) and (10.76)}.
\end{aligned}$$

□

10.3 Collocation estimates by Fourier analysis techniques for both even- and odd-degree splines

The method of analysis in the previous section appears difficult to extend to the case of even-degree splines. In [165], Saranen and Wendland used Fourier analysis techniques to treat even-degree spline collocations based upon the equivalence of the collocation equations with certain Galerkin–Petrov equations. (By a Galerkin–Bubnov method, we mean a Galerkin method with the same test and trial function spaces such as in §10.1; whereas by a Galerkin–Petrov method as in [165] we mean distinct test and trial functions are used.) In [10], such techniques have been further strengthened to treat *both* odd- and even-degree spline collocations while relaxing some of the restrictions in [165] for the even-degree case, namely that in [165] the strongly elliptic boundary integral operator must have a leading symbol with constant coefficients, and that the spacing of the knots of the splines must be uniform. (In the odd-degree case, the results in [10] are actually somewhat weaker than those in [9] as presented in §10.2, because the meshes in [10] must be essentially uniform.) We have two objectives in this section. The first is to introduce the Fourier analysis techniques. The second is to cover the case of even-degree spline collocations (and in fact, a unified treatment of both even- and odd-degree splines.) The major reference for the material in this section is [10].

Let $d \in \mathbb{Z}^+$ and $n \in \mathbb{N}, h = 1/n$. Denote

$$\triangle = \{jh \mid j \in \mathbb{Z}\} \tag{10.77}$$

$$\Xi = \begin{cases} \triangle, & d \text{ odd}, \\ \{(j+\frac{1}{2})h \mid j \in \mathbb{Z}\}, & d \text{ even}. \end{cases} \tag{10.78}$$

Let S_h^d denote the space of smooth splines of degree d on the uniform mesh Ξ. For $\beta \in \mathbb{R}$, define operators Q_β^+ and Q_β^- on the class of 1-periodic distributions u (10.38) by

$$Q_\beta^+ u(t) = \sum_{n \neq 0} |n|^\beta \widehat{u}_n e^{i2\pi nt},$$

$$Q_\beta^- u(t) = \sum_{n \neq 0} \operatorname{sgn}(n)\, |n|^\beta \widehat{u}_n e^{i2\pi nt}.$$

Then $Q_\beta^\pm$ map H^s boundedly into $H^{s-\beta}$ for all $s \in \mathbb{R}$. The operators Q_β^+ and Q_β^- have respectively symbols

$$|\xi|^\beta, \quad \xi|\xi|^{\beta-1}$$

and therefore are *elliptic* pseudodifferential of order β. Define an operator

$$A_0 = b_+ Q_\beta^+ + b_- Q_\beta^- + J, \tag{10.79}$$

where $Ju = \int_0^1 u(t)\,dt$ is the same as in (10.51), and b_+ and b_- are complex constants satisfying

$$\gamma \equiv \min[\operatorname{Re}(b_+ + b_-), \operatorname{Re}(b_+ - b_-)] > 0. \tag{10.80}$$

As a consequence, the leading symbol of A_0 satisfies

$$\operatorname{Re}[b_+|\xi|^\beta + b_- \operatorname{sgn}(\xi)\,|\xi|^\beta] \geqslant \gamma|\xi|^\beta,$$

so by (4.47), A_0 is *strongly elliptic*.

We have

$$\widehat{A_0 u}(m) = \widehat{A_0 u}_m = \int_0^1 e^{-i2\pi mt} A_0 u(t)\,dt \tag{10.81}$$

$$= \begin{cases} [b_+ + \operatorname{sgn}(m) a b_-]|m|^\beta \widehat{u}_m, & m \neq 0, \\ \widehat{u}_0, & m = 0. \end{cases} \tag{10.82}$$

It follows easily from the above and Parseval's identity that

$$\min\left(1, \frac{\gamma}{(2\pi)^\beta}\right) \|v\|_{s+\beta} \leqslant \|A_0 v\|_s$$

$$\leqslant \max\left(1, \frac{|b_+| + |b_-|}{(2\pi)^\beta}\right) \|v\|_{s+\beta}. \tag{10.83}$$

As in [10], let us use the notation $\equiv$ to indicate congruence modulo n, and let $\Lambda_n = \{p \in \mathbb{Z} \mid -\frac{1}{2}n < p \leqslant \frac{1}{2}n\}$. Also denote

$$\sum_m = \lim_{l\to\infty} \sum_{m=-l}^{l},$$

$$\sum_{m\equiv p} = \lim_{l\to\infty} \sum_{\substack{m=-l\\ m\equiv p}}^{l}.$$

The essence of the approach in [10] is the analysis of the collocation method using Fourier series. We first find an equivalence between the collocation equations and the Fourier coefficients as given in the following key lemma.

Lemma 10.2 ([10, Lemma 2.1]). *Let $\phi \in L^1$ and assume that ϕ is Hölder-continuous in some neighborhood of $\triangle$. Then $\sum_{m\equiv p} \widehat{\phi}_m < \infty$ for all p. Moreover,*

$$\sum_{m\equiv p} \widehat{\phi}_m = 0, \quad \forall\, p \in \Lambda_n$$

if and only if

$$\phi = 0 \ \text{on} \ \triangle.$$

Proof. $\phi \in L^1$ and ϕ being Hölder continuous in a neighborhood of $\triangle$ imply the convergence of the Fourier series of ϕ at $x = q/n$:

$$\phi\left(\frac{q}{n}\right) = \sum_m \widehat{\phi}_m e^{i2\pi mq/n}, \quad \forall q \in \mathbb{Z}.$$

Hence

$$\begin{aligned}\frac{1}{n}\sum_{q\in\Lambda_n} \phi\left(\frac{q}{n}\right) e^{-i2\pi pq/n} &= \frac{1}{n}\sum_{q\in\Lambda_n}\sum_m \widehat{\phi}_m e^{i2\pi(m-p)q/n}\\ &= \sum_m \widehat{\phi}_m \left(\frac{1}{n}\sum_{q\in\Lambda_n} e^{i2\pi(m-p)q/n}\right).\end{aligned} \tag{10.84}$$

It is straightforward to check that

$$\sum_{q\in\Lambda_n} e^{i2\pi(m-p)q/n} = \begin{cases} 0, & m \not\equiv p \,(\mathrm{mod}\, n),\\ n, & m \equiv p \,(\mathrm{mod}\, n).\end{cases} \tag{10.85}$$

Therefore, (10.84) and (10.85) give

$$\frac{1}{n}\sum_{q\in\Lambda_n} \phi\left(\frac{q}{n}\right) e^{-i2\pi pq/n} = \sum_{m\equiv p} \widehat{\phi}_m \quad \forall p \in \Lambda_n.$$

Since the Vandermonde matrix $[e^{-i2\pi pq/n}]_{p,q\in\Lambda_n}$ is nonsingular, the lemma follows. □

Proposition 10.1 ([10, Proposition 2.2]). *Let $u \in H^\tau$ for some $\tau > \beta + \frac{1}{2}$ and $u_n \in S_h^d$, where $d \in \mathbb{Z}^+$ and*

$$\left.\begin{array}{ll} d > \beta & \textit{odd } d, \\ d > \beta - \dfrac{1}{2} & \textit{even } d. \end{array}\right\} \tag{10.86}$$

Then the collocation equations

$$A_0 u_h(x) = A_0 u(x), \quad x \in \triangle, \tag{10.87}$$

are satisfied if and only if

$$\sum_{m \equiv p} \widehat{A_0 u}(m) = \sum_{m \equiv p} \widehat{A_0 u_h}(m) \quad \forall p \in \Lambda_n. \tag{10.88}$$

Proof. Since $A_0 u \in H^{\tau-\beta}$ and $\tau - \beta > \frac{1}{2}$, by Theorem 2.3, $A_0 u$ is Hölder-continuous. Also $S_h^d \subset H^s$ for all $s < d + \frac{1}{2}$, so the hypotheses imply that $A_0 u_h \in L^2 \subset L^1$ and $A_0 u_h$ is Hölder-continuous if d is odd. If d is even, u_h is smooth in a neighborhood of $\triangle$; thus $A_0 u_h$ is also Hölder-continuous in a neighborhood of $\triangle$. Hence Lemma 10.2 is applicable to $\phi = A_0 u - A_0 u_h$ and the proof is complete. □

From (10.79), for any u with $A_0 u$ absolutely integrable and Hölder-continuous in a neighborhood of $\triangle$, we have

$$\sum_{m \equiv p} \widehat{A_0 u}(m) = b_0(p)\widehat{u}_p + \sum_{k \in \mathbb{Z}_{\widehat{0}}} [b_+ + \operatorname{sgn}(k) b_-] |p + kn|^\beta \widehat{u}_{p+kn}, \quad p \in \Lambda_n, \tag{10.89}$$

where

$$b_0(p) \equiv \begin{cases} [b_+ + \operatorname{sgn}(p) b_-] |p|^\beta, & p \neq 0, \\ 1, & p = 0, \end{cases} \tag{10.90}$$

and

$$\mathbb{Z}_{\widehat{0}} = \mathbb{Z} \setminus \{0\}.$$

Lemma 10.3 ([8]). *For $\phi \in S_h^d$, we have the recursion relation for the Fourier coefficients*

$$\widehat{\phi}_{p+kn}(p + kn)^{d+1} = (-1)^{k(d+1)} \widehat{\phi}_p p^{d+1}, \quad \forall p, k \in \mathbb{Z}. \tag{10.91}$$

Proof. The elements in S_h^d are B-splines (Remark 5.4), which are obtainable as convolutions defined below. Let

$$\chi_n^1(x) = \begin{cases} n, & x \in \left[m - \dfrac{1}{2n}, m + \dfrac{1}{2n}\right], \quad m \in \mathbb{Z}, \\ 0, & \text{elsewhere}, \end{cases}$$

i.e., χ_n^1 is equal to n times a characteristic function. Define

$$\chi_n^{d+1}(x) = \int_0^1 \chi_n^d(x-y)\chi_n^1(y)\,dy, \;\; x \in [0,1].$$

Then the B-splines $\chi_n^{d+1}(\cdot - j/n), j \in \Lambda_n$, with their 1-periodic extensions, form a basis for S_h^d, with dimension n.

Take $\phi(x) = \chi_n^{d+1}(x - j/n)$, $j \in \mathbb{Z}$. Then its Fourier coefficient is easily found to satisfy

$$\begin{aligned}\widehat{\phi}_p p^{d+1} &= e^{-i2\pi pj/n}(\widehat{\chi_n^{d+1}})_p p^{d+1} \\ &= e^{-i2\pi pj/n}[(\widehat{\chi_n^1})_p p]^{d+1} = e^{-i2\pi pj/n}\left[\frac{n}{\pi}\sin\left(\frac{p\pi}{n}\right)\right]^{d+1}.\end{aligned}$$

Similarly,

$$\begin{aligned}\widehat{\phi}_{p+kn}(p+kn)^{d+1} &= e^{-i2\pi(p+kn)j/n}\left\{\frac{n}{\pi}\sin\left[\frac{(p+kn)\pi}{n}\right]\right\}^{d+1} \\ &= e^{-i2\pi pj/n}\left[\frac{n}{\pi}\sin\left(\frac{p\pi}{n}\right)\right]^{d+1}(-1)^{k(d+1)}.\end{aligned}$$

So (10.91) holds for any basis element $\phi \in S_h^d$. The lemma is proved. □

From (10.89) and (10.91), for $u_h \in S_h^d$, $p \in \Lambda_n$, we have

$$\begin{aligned}\sum_{m\equiv p}\widehat{A_0u_h}(m) &= b_0(p)\widehat{u}_h(p) + p^{d+1}\sum_{k\in\mathbb{Z}_{\widehat{0}}}[b_+ + \operatorname{sgn}(k)\,b_-] \\ &\quad\times(-1)^{k(d+1)}(\operatorname{sgn}k)^{d+1}|p+kn|^{\beta-d-1}\widehat{u}_h(p) \\ &= \Bigg\{b_0(p) + (\operatorname{sgn}p)^{d+1}|p|^{\beta}\left|\frac{2p}{n}\right|^{d+1-\beta} \\ &\quad\times\sum_{k\in\mathbb{Z}_{\widehat{0}}}[b_+ + \operatorname{sgn}(k)\,b_-](-1)^{k(d+1)}(\operatorname{sgn}k)^{d+1} \\ &\quad\times\left|\frac{2p}{n}+2k\right|^{\beta-d-1}\Bigg\}\widehat{u}_h(p).\end{aligned} \tag{10.92}$$

For $y \in [-1,1]$, define

$$\left.\begin{aligned} f(y) &= (\operatorname{sgn}y)^{d+1}|y|^{d+1-\beta}\textstyle\sum_{k\in\mathbb{Z}_{\widehat{0}}}(-1)^{k(d+1)}(\operatorname{sgn}k)^{d+1} \\ &\quad\times|y+2k|^{\beta-d-1}, \\ g(y) &= -(\operatorname{sgn}y)^{d}|y|^{d+1-\beta}\textstyle\sum_{k\in\mathbb{Z}_{\widehat{0}}}(-1)^{k(d+1)}(\operatorname{sgn}k)^{d} \\ &\quad\times|y+2k|^{\beta-d-1}.\end{aligned}\right\} \tag{10.93}$$

Then (10.92) can be written more compactly as

$$\sum_{m\equiv p}\widehat{A_0u_h}(m) = \begin{cases} |p|^{\beta}\left\{b_+\left[1+f\left(\dfrac{2p}{n}\right)\right]\right. \\ \left.+\operatorname{sgn}(p)\,b_-\left[1-g\left(\dfrac{2p}{n}\right)\right]\right\}\widehat{u}_h(p), & p\in\Lambda_n, p\neq 0, \\ \widehat{u}_h(0), & p=0.\end{cases} \tag{10.94}$$

The elementary properties of f and g are given below.

Lemma 10.4 ([10, Lemma 2.3]). *Assume that $d \in \mathbb{Z}^+$ and*

$$\left.\begin{aligned} &\beta < d, && \text{odd } d, \\ &\beta < d+1, && \text{even } d. \end{aligned}\right\} \tag{10.95}$$

Then f and g defined by (10.93) are continuous, even, nonnegative functions on $[-1,1]$ and are strictly increasing on $[0,1]$. Moreover, $g(1)=1$, and there exists a constant C depending only on β and d such that

$$|f(y)| + |g(y)| \leqslant C|y|^{d+1-\beta}. \tag{10.96}$$

Proof. First consider the case when d is odd. Then (10.93) may be rewritten as

$$f(y) = |y|^{d+1-\beta} \sum_{k=1}^{\infty} [(2k-y)^{\beta-d-1} + (2k+y)^{\beta-d-1}],$$

$$g(y) = \operatorname{sgn}(y)\, |y|^{d+1-\beta} \sum_{k=1}^{\infty} [(2k-y)^{\beta-d-1} - (2k+y)^{\beta-d-1}].$$

Since $\beta < d$, both series above converge absolutely and uniformly for $y \in [-1,1]$. We can easily see that both f and g are even functions bounded by $|y|^{d+1-\beta}$ and they satisfy $f(0) = g(0) = 0$ and $g(1) = 1$. Thus, we need only show that both functions are strictly increasing on [0,1]. Since term-by-term differentiation of both series is admissible, for $y \in (0,1)$ we have

$$\begin{aligned} f'(y) &= (d+1-\beta)y^{d-\beta} \sum_{k=1}^{\infty} \Big\{[(2k-y)^{\beta-d-1} + (2k+y)^{\beta-d-1}] \\ &\quad + y[(2k-y)^{\beta-d-2} - (2k+y)^{\beta-d-2}]\Big\} \\ &= (d+1-\beta)y^{d-\beta} \sum_{k=1}^{\infty} \Big\{[(2k-y)+y](2k-y)^{\beta-d-2} \\ &\quad + [(2k+y)-y](2k+y)^{\beta-d-2}\Big\} \\ &= 2(d+1-\beta)y^{d-\beta} \sum_{k=1}^{\infty} k[(2k-y)^{\beta-d-2} - (2k+y)^{\beta-d-2}] > 0, \end{aligned}$$

and similarly

$$g'(y) = 2(d+1-\beta)y^{d-\beta} \sum_{k=1}^{\infty} k[(2k-y)^{\beta-d-2} - (2k+y)^{\beta-d-2}] > 0.$$

Hence the lemma is proved for odd d.

When d is even, (10.93) becomes

$$f(y) = \operatorname{sgn}(y)\,|y|^{d+1-\beta}\sum_{k=1}^{\infty}(-1)^{k+1}[(2k-y)^{\beta-d-1}-(2k+y)^{\beta-d-1}],$$

$$g(y) = |y|^{d+1-\beta}\sum_{k=1}^{\infty}(-1)^{k+1}[(2k-y)^{\beta-d-1}+(2k+y)^{\beta-d-1}].$$

Now since $\beta < d+1$, the series $\sum_{k=1}^{\infty}(-1)^{k+1}(2k\pm y)^{\beta-d-1}$ converges uniformly for $y\in[-1,1]$ and admits termwise differentiation. Therefore, f and g define continuous even functions on $[-1,1]$ that are $\mathscr{O}(|y|^{d+1-\beta})$, $f(0)=g(0)=0$, $g(1)=1$ and, for $y\in(0,1)$,

$$\begin{aligned} f'(y) &= 2(d+1-\beta)y^{d-\beta}\sum_{k=1}^{\infty}(-1)^{k+1}k[(2k-y)^{\beta-d-2}-(2k+y)^{\beta-d-2}] \\ &\geqslant 2(d+1-\beta)y^{d-\beta}\sum_{k=1}^{2}(-1)^{k+1}[k(2k-y)^{\beta-d-2}-(2k+y)^{\beta-d-2}] \\ &> 0, \\ g'(y) &= 2(d+1-\beta)y^{d-\beta}\sum_{k=1}^{\infty}(-1)^{k+1}k[(2k-y)^{\beta-d-2}+(2k+y)^{\beta-d-2}] \\ &\geqslant 2(d+1-\beta)y^{d-\beta}\sum_{k=1}^{2}(-1)^{k+1}k[(2k-y)^{\beta-d-2}+(2k+y)^{\beta-d-2}] \\ &> 0. \end{aligned}$$

So f and g are increasing on [0,1], and the proof is complete. □

We are now in a position to state and prove the first major theorem in this section.

Theorem 10.12 ([10, § 2]). *Let A_0 be given as in (10.79). Let $d\in\mathbb{Z}^+$ satisfy (10.95). Then there exists $h_0>0$ such that for $0<h\leqslant h_0$ and any continuous function $\widetilde{g}$ the collocation equations*

$$A_0u_h = \widetilde{g} \quad \text{on } \triangle \tag{10.97}$$

for

$$A_0u(x) = \widetilde{g}(x), \quad \forall x\in[0,1] \tag{10.98}$$

are uniquely solvable for $u_h\in S_h^d$. Moreover, if $s,\tau\in\mathbb{R}$ satisfy

$$\beta\leqslant s\leqslant\tau\leqslant d+1,\ s<d+\frac{1}{2},\ \beta+\frac{1}{2}<\tau, \tag{10.99}$$

and the solution u to (10.98) is in H^τ, then the error estimate

$$\|u-u_h\|_s\leqslant Ch^{\tau-s}\|u\|_\tau \tag{10.100}$$

holds for some $C>0$ independent of h and u.

Proof. Define

$$D(y) = [1+f(y)]b_+ + \operatorname{sgn}(y)\,[1-g(y)]b_-, \quad y\in[-1,1], \tag{10.101}$$

where f and g are defined as in (10.93). By Proposition 10.1 and (10.94), A_0u_h collocates A_0u, i.e., (10.97) holds, if and only if

$$\left.\begin{aligned} &|p|^\beta D(2p/n)\widehat{u}_h(p) = \sum_{m\equiv p}\widehat{A_0u}(m), \quad \forall p\in\Lambda_n, p\neq 0,\\ &\widehat{u}_h(0) = \sum_{m\equiv 0}\widehat{A_0u}(m). \end{aligned}\right\} \tag{10.102}$$

Since

$$0\leqslant 1-g(y)\leqslant 1\leqslant 1+f(y)<\infty, \; y\in[-1,1]$$

by Lemma 10.4 and by the strong ellipticity (10.79) and (10.80), we have

$$\operatorname{Re}D(y)\geqslant\gamma>0, \quad \forall y\in[-1,1]; \tag{10.103}$$

hence the Fourier coefficients $\widehat{u}_h(p)$, $p\in\Lambda_n$, in (10.102) are uniquely solvable:

$$\left.\begin{aligned} &\widehat{u}_h(p) = D(\tfrac{2p}{n})^{-1}|p|^{-\beta}\Big\{b_0(p)\widehat{u}_p + \sum_{k\in\mathbb{Z}_{\widehat{0}}}[b_+ + \operatorname{sgn}(k)\,b_-]|p+kn|^\beta\widehat{u}_{p+kn}\Big\},\\ &\widehat{u}_h(0) = \widehat{u}_0 + \sum_{k\in\mathbb{Z}_{\widehat{0}}}[b_+ + \operatorname{sgn}(k)\,b_-]|kn|^\beta\widehat{u}_{kn}. \end{aligned}\right\} \tag{10.104}$$

By Lemma 10.3, these coefficients determine u_h uniquely, therefore there exists a unique solution u_h to the collocation equations (10.97).

We now proceed to prove the error estimates

$$\|u-u_h\|_s \leqslant Ch^{\tau-s}\|u\|_\tau = Cn^{s-\tau}\|u\|_\tau.$$

It suffices to show that

$$\begin{aligned} \|u-u_h\|_s^2 &\leqslant C\Big\{[\widehat{u}_0-\widehat{u}_h(0)]^2 + \Big[\sum_{m\in\mathbb{Z}\setminus\Lambda_n}|m|^{2s}|\widehat{u}_m|^2\Big]\\ &\quad + \Big[\sum_{m\in\mathbb{Z}\setminus\Lambda_n}|m|^{2s}|\widehat{u}_h(m)|^2\Big] + \Big[\sum_{\substack{p\in\Lambda_n\\ p\neq 0}}|p|^{2s}|\widehat{u}_p-\widehat{u}_h(p)|^2\Big]\Big\}\\ &\equiv C\{T_1+T_2+T_3+T_4\} \leqslant Cn^{2(s-\tau)}\|u\|_\tau^2. \end{aligned}$$

Note that the generic constant C in the proof depends only on β, d, s, τ and γ. Also, we will use several times in the proof that

$$\sum_{k\in\mathbb{Z}_{\widehat{0}}}|y+2k|^r < C(r), \quad r<-1, \; \forall y\in[-1,1]. \tag{10.105}$$

To bound T_1, we use the second equation in (10.104), the Cauchy–Schwarz inequality, (10.105) with $r = 2(\beta - \tau) < -1$ and $s \geqslant \beta$ in (10.99):

$$\begin{aligned} T_1 &\leqslant C\Big[\sum_{k\in\mathbb{Z}_{\widehat{0}}} |kn|^{\beta}|\widehat{u}_{kn}|\Big]^2 \\ &\leqslant Cn^{2(\beta-\tau)}\Big[\sum_{k\in\mathbb{Z}_{\widehat{0}}} |k|^{2(\beta-\tau)}\Big]\Big[\sum_{k\in\mathbb{Z}_{\widehat{0}}} |kn|^{2\tau}|\widehat{u}_{kn}|^2\Big] \\ &\leqslant Cn^{2(s-\tau)}\|u\|_{\tau}^2. \end{aligned}$$

Next, note that $|m| \geqslant \frac{1}{2}n$ for $m \in \mathbb{Z} \setminus \Lambda_n$, so

$$T_2 = \sum_{m\in\mathbb{Z}\setminus\Lambda_n} (|m|^{2\tau}|\widehat{u}_m|^2|m|^{2s-2\tau}) \leqslant Cn^{2(s-\tau)}\|u\|_{\tau}^2.$$

To bound T_3, first apply Lemma 10.3 and the first equation in (10.104) to get

$$\begin{aligned} |\widehat{u}_h(p+ln)| = \left|\frac{2p}{n}\right|^{d+1}\left|\frac{2p}{n}+2l\right|^{-d-1} &|D(2p/n)^{-1}\Big\{[b_+ + \operatorname{sgn}(p)\,b_-]\widehat{u}_p \\ &+ |p|^{-\beta}\sum_{k\in\mathbb{Z}_{\widehat{0}}}[b_+ + \operatorname{sgn}(k)\,b_-]|p+kn|^{\beta}\widehat{u}_{p+kn}\Big\}| \\ &p\in\Lambda_n,\quad p\neq 0,\quad l\in\mathbb{Z}_{\widehat{0}}. \end{aligned} \tag{10.106}$$

Now, use (10.105) with $r = 2(s-d-1) < -1$, by $s < d+\frac{1}{2}$ and $\tau \leqslant d+1$ in (10.99)

$$\begin{aligned} &\sum_{\substack{p\in\Lambda_n\\ p\neq 0}}\sum_{l\in\mathbb{Z}_{\widehat{0}}} |\widehat{u}_p|^2\left|\frac{2p}{n}\right|^{2d+2}\left|\frac{2p}{n}+2l\right|^{-2d-2}|p+ln|^{2s} \\ &= \left|\frac{n}{2}\right|^{2s-2\tau}\sum_{\substack{p\in\Lambda_n\\ p\neq 0}}\left|\frac{2p}{n}\right|^{2d+2-2\tau}|\widehat{u}_p|^2|p|^{2\tau}\sum_{l\in\mathbb{Z}_{\widehat{0}}}\left|\frac{2p}{n}+2l\right|^{2s-2d-2} \\ &\leqslant Cn^{2(s-\tau)}\|u\|_{\tau}^2, \end{aligned} \tag{10.107}$$

because $2d+2-2\tau > 0$ and $|2p/n| \leqslant 1$. Also, by the Cauchy–Schwarz inequality, (10.105) with $r = 2s-2d-2 < -1$, (10.105) with $r = 2(\beta-\tau) < -1$ and $\tau > \beta + \frac{1}{2}$ and $\tau \leqslant d+1$

in (10.99), we get

$$\begin{aligned}
&\sum_{\substack{p\in\Lambda_n\\ p\neq 0}}\sum_{l\in\mathbb{Z}_{\widehat{0}}}\left|\frac{2p}{n}\right|^{2d+2}\left|\frac{2p}{n}+2l\right|^{-2d-2}|p|^{-2\beta}\\
&\qquad\times\Big(\sum_{k\in\mathbb{Z}_{\widehat{0}}}|p+kn|^{\beta}|\widehat{u}_{p+kn}|\Big)^2|p+ln|^{2s}\\
&\leqslant Cn^{2(s-\tau)}\sum_{\substack{p\in\Lambda_n\\ p\neq 0}}\Big[\left|\frac{2p}{n}\right|^{-2\beta+2d+2}\Big(\sum_{l\in\mathbb{Z}_{\widehat{0}}}\left|\frac{2p}{n}+2l\right|^{2s-2d-2}\Big)\\
&\qquad\times\Big(\sum_{l\in\mathbb{Z}_{\widehat{0}}}\left|\frac{2p}{n}+2l\right|^{2(\beta-\tau)}\Big)\Big(\sum_{k\in\mathbb{Z}_{\widehat{0}}}|p+kn|^{2\tau}|\widehat{u}_{p+kn}|^2\Big)\Big]\\
&\leqslant Cn^{2(s-\tau)}\sum_{\substack{p\in\Lambda_n\\ p\neq 0}}\sum_{k\in\mathbb{Z}_{\widehat{0}}}|p+kn|^{2\tau}|\widehat{u}_{p+kn}|^2\leqslant Cn^{2(s-\tau)}\|u\|_\tau^2.
\end{aligned}\tag{10.108}$$

Since $\widehat{u}_h(ln)=0$ for $l\in\mathbb{Z}_{\widehat{0}}$ by Lemma 10.3, using (10.106)–(10.108) and (10.103), we obtain

$$T_3=\sum_{\substack{p\in\Lambda_n\\ p\neq 0}}\sum_{l\in\mathbb{Z}_{\widehat{0}}}|p+ln|^{2s}|\widehat{u}_h(p+ln)|^2\leqslant Ch^{2(s-\tau)}\|u\|_\tau^2.\tag{10.109}$$

It remains to bound T_4. From the first equation in (10.104), (10.89) and the definitions of b_0 and D in (10.90) and (10.101), we have for $p\in\Lambda_n$, $p\neq 0$, that

$$\begin{aligned}
\widehat{u}_p-\widehat{u}_h(p)&=\widehat{u}_p-D\left(\frac{2p}{n}\right)^{-1}|p|^{-\beta}\\
&\qquad\times\Big\{b_0(p)\widehat{u}_p+\sum_{k\in\mathbb{Z}_{\widehat{0}}}[b_++\operatorname{sgn}(k)\,b_-]|p+kn|^{\beta}\widehat{u}_{p+kn}\Big\}\\
&=D\left(\frac{2p}{n}\right)^{-1}\Big\{\Big[f\left(\frac{2p}{n}\right)b_+-\operatorname{sgn}(p)\,g\left(\frac{2p}{n}\right)b_-\Big]\widehat{u}_p\\
&\qquad-|p|^{-\beta}\sum_{k\in\mathbb{Z}_{\widehat{0}}}[b_++\operatorname{sgn}(k)\,b_-]|p+kn|^{\beta}\widehat{u}_{p+kn}\Big\}.
\end{aligned}\tag{10.110}$$

By (10.103), Lemma 10.4 and $s\geqslant\beta,\tau\leqslant d+1$ in (10.99), we get

$$\begin{aligned}
&\left|D\left(\frac{2p}{n}\right)^{-1}\Big[f\left(\frac{2p}{n}\right)b_+-\operatorname{sgn}(p)\,g\left(\frac{2p}{n}\right)b_-\Big]\widehat{u}_p\right|^2|p|^{2s}\\
&\leqslant Cn^{2(s-\tau)}\left|\frac{2p}{n}\right|^{2(d+1-\beta+s-\tau)}|\widehat{u}_p|^2|p|^{2\tau}\leqslant Cn^{2(s-\tau)}|p|^{2\tau}|\widehat{u}_p|^2.
\end{aligned}\tag{10.111}$$

For the remaining terms in (10.110), we use (10.103), the Cauchy–Schwarz inequality, $s \geqslant \beta$ and $\tau > \beta + \frac{1}{2}$ in (10.99), and (10.105) to get

$$\begin{aligned}
&\left\{D\left(\frac{2p}{na}\right)^{-1} |p|^{-\beta} \sum_{k\in\mathbb{Z}_{\widehat{0}}} [b_+ + \operatorname{sgn}(k)\, b_-] |p+kn|^{\beta} |\widehat{u}_{p+kn}|\right\}^2 |p|^{2s} \\
&\leqslant Cn^{2(s-\tau)} \left|\frac{2p}{n}\right|^{2(s-\beta)} \sum_{j\in\mathbb{Z}_{\widehat{0}}} \left|\frac{2p}{n} + 2j\right|^{2(\beta-\tau)} \sum_{k\in\mathbb{Z}_{\widehat{0}}} |p+kn|^{2\tau} |\widehat{u}_{p+kn}|^2 \\
&\leqslant Cn^{2(s-\tau)} \sum_{k\in\mathbb{Z}_{\widehat{0}}} |p+kn|^{2\tau} |\widehat{u}_{p+kn}|^2. \qquad (10.112)
\end{aligned}$$

Collecting (10.110)–(10.112), we have

$$T_4 \leqslant Cn^{2(s-\tau)} \|u\|_\tau^2.$$

The proof is complete. □

For many boundary integral equations derived from two-dimensional elliptic boundary value problems, the principal parts of the corresponding pseudodifferential operators are either a pure convolution (BIE of the first kind based upon the simple-layer potential representation) or $\pm\frac{1}{2}I$ (BIE of the second kind). In the framework of pseudodifferential operators, this means that the principal symbol $\sigma = \sigma(\xi)$ of many such operators depends only on the Fourier-transformed variable ξ. In addition, because such pseudifferential operators are strongly elliptic, they yield BIE of the form

$$Au = \widetilde{g}, \qquad (10.113)$$

where

$$\left.\begin{aligned}
&A = b_+ Q_\beta^+ + b_- Q_\beta^- + K, \\
&K:\ H^r \to H^{r-\beta+\delta} \text{ is continuous for some} \delta > 0,\ \ \forall r \in \mathbb{R}.
\end{aligned}\right\} \qquad (10.114)$$

Nevertheless, for hypersingular BIE of the first kind, the principal part of the corresponding pseudodifferential operator is *not* a pure convolution; therefore the principal symbols may have variable coefficients depending on the parametrization. More generally, as has been shown in § 4.4, although many pseudodifferential operators have local principal symbols $\sigma(\xi)$ depending on ξ only, their global symbols will appear with spatially varying coefficients such as

$$\left.\begin{aligned}
&A = b_+(x) Q_\beta^+ + b_-(x) Q_\beta^- + K, \\
&K:\ H^r \to H^{r-\beta+\delta} \text{ is continuous for some } \delta > 0,
\end{aligned}\right\} \qquad (10.115)$$

because we cannot "flatten out" the boundary simultaneously. In [10, § 3], error estimates for the collocation of (10.113) corresponding to A in the form (10.115) have been obtained.

Here we will abridge the proofs in [10] to obtain the error estimates for the simpler operator (10.114).

Let S_h^d be the space of smoothest splines with quasiuniform mesh of degree d as before. Then the inverse inequality (10.45) holds. Here we will use

$$\|v\|_t \leqslant Ch^{s-t}\|v\|_s \quad \forall v \in S_h^d, \tag{10.116}$$

for $s \leqslant t < d+\frac{1}{2}$, for some $C>0$ independent of v and h.

The proof of collocation error estimates hinges on the existence of a family of approximation operators $\boldsymbol{P}_h^d$ stated below.

Lemma 10.5 ([10, Lemma 3.1]). *Let $r < d+\frac{1}{2}$. Then there exists a family of operators $\boldsymbol{P}_h^d\colon\ H^r \to S_h^d$ and a constant $C>0$ independent of h and u such that*

$$\|u-\boldsymbol{P}_h^d u\|_s \leqslant Ch^{t-s}\|u\|_t, \quad \forall u \in H^t,\ \ s \leqslant r \leqslant t \leqslant d+1. \tag{10.117}$$

The operators $\boldsymbol{P}_h^d$ satisfy the additional property that

$$\|u_h-\boldsymbol{P}_h^d u_h\|_r \leqslant Ch^{\delta}\|u_h\|_r \text{ for some } C>0, \forall u_h \in S_h^d, \tag{10.118}$$

for r lying in the range of (10.117). Moreover, for $s \leqslant t < d+\frac{1}{2}$, there exists a constant $C>0$ independent of h and u such that

$$\|u-v\|_t \leqslant C(h^{s-t}\|u-v\|_s+\|u\|_t), \quad \forall u \in H^t, v \in S_h^d. \tag{10.119}$$

Proof. (10.117) is a familiar inequality, cf. [49] and, partly, § 5.11. (When the mesh is uniform, $\boldsymbol{P}_h^d$ may be constructed independently of r.) The inequality (10.119) follows from (10.116) and (10.117):

$$\begin{aligned}
\|u-v\|_t &\leqslant \|u-\boldsymbol{P}_h^d u\|_t+\|v-\boldsymbol{P}_h^d u\|_t \\
&\leqslant C\|u\|_t+Ch^{s-t}\|v-\boldsymbol{P}_h^d u\|_s \\
&\leqslant C\|u\|_t+Ch^{s-t}[\|v-u\|_s+\|u-\boldsymbol{P}_h^d u\|_s] \\
&\leqslant C\|u\|_t+Ch^{s-t}[\|v-u\|_s+Ch^{t-s}\|u\|_t] \\
&= C\|u\|_t+Ch^{s-t}\|u-v\|_s.
\end{aligned}$$

To prove (10.118), we use (10.117):

$$\|u_h-\boldsymbol{P}_h^d u_h\|_r \leqslant Ch^{d+1-r}\|u_h\|_{d+1}. \tag{10.120}$$

Since $u_h^{(d+1)}=0$, we have

$$\|u_h\|_{d+1}=\|u_h\|_d.$$

Now apply the inverse inequality (10.12)

$$\|u_h\|_d \leqslant Ch^{\min(r-d,0)}\|u_h\|_r$$

and combine it with (10.120) to get (10.118). □

We are now ready to prove the main collocation theorem of this chapter.

Theorem 10.13. *Let A be a strongly elliptic pseudodifferential operator of order β with constant coefficient leading symbol of the form (10.114). Assume that*

$$A\colon\ H^r \to H^{r-\beta} \text{ is an isomorphism, } \forall r \in \mathbb{R}. \tag{10.121}$$

Let $u \in H^t$ and $u_h \in S_h^d$ satisfy the collocation equations

$$Au = Au_h \text{ on } \triangle. \tag{10.122}$$

Then there exist a constant $C > 0$ independent of h, $0 < h \leqslant h_0$, and u_h such that

$$\|u - u_h\|_s \leqslant Ch^{t-s}\|u\|_t \tag{10.123}$$

for all β, d, s, $t \in \mathbb{R}$ satisfying the assumptions of Theorem 10.12 (with $\tau = t$), for sufficiently small $h_0 > 0$.

Proof. Assume that $h_0 > 0$ is sufficiently small and that the collocation solution u_h satisfying (10.122) also satisfies the inequality

$$\|u - u_h\|_s \leqslant Ch^{t-s}\|u\|_t + Ch^{\delta}\|u - u_h\|_s + \begin{cases} 0, & s < t, \\ C\|u - u_h\|_{s-\delta}, & s = t, \end{cases} \tag{10.124}$$

for $t, s \in \mathbb{R}$ satisfying $\beta \leqslant s \leqslant t < d + \frac{1}{2}$, $t \neq \beta$, and for some $\delta > 0$.
We discuss three cases separately.

(i) If $\beta \leqslant s < t < d + \frac{1}{2}$, $t > \beta + \frac{1}{2}$, then (10.123) follows directly from (10.123) and (10.100) with $\tau = t$:

$$\frac{1}{2}\|u - u_h\|_s \leqslant (1 - Ch^{\delta})\|u - u_h\|_s \leqslant Ch^{t-s}\|u\|_t, \tag{10.125}$$

for $h \leqslant h_0$ and h_0 sufficiently small.

(ii) If $\beta + \frac{1}{2} < s = t < d + \frac{1}{2}$, we again use (10.124) but replace δ by $\min(\delta, s - \beta)$ so that $\delta \leqslant s - \beta$ and the second inequality in (10.125) remains valid for the new δ. Note that now $\beta \leqslant s - \delta$. We apply case (i), with (s,t) replaced by $(s - \delta, s)$, so $\beta \leqslant s - \delta < s < d + \frac{1}{2}$ and $s > \beta + \frac{1}{2}$, and we get

$$\|u - u_h\|_{s-\delta} \leqslant Ch^{\delta}\|u\|_s. \tag{10.126}$$

Substituting the above into (10.124) and choosing h_0 sufficiently small we again obtain (10.123).

(iii) Finally, we let $\beta \leqslant s < d+\frac{1}{2} \leqslant t \leqslant d+1$. We let E_h be the error operator

$$E_h(u) = u - u_h.$$

From cases (i) and (ii) and (10.123), we know that u_h exists (for s and t in the given ranges in (i) and (ii)) and is unique because the homogeneous collocation equations admit only the trivial solution. Thus, E_h is linear and

$$E_h(\chi) = 0 \qquad \forall \chi \in S_d^h,$$

since

$$A\chi_h = A\chi \quad \text{on } \triangle, \ \forall \chi \in S_d^h.$$

Consequently,

$$E_h(u-\chi) = E_h(u) \qquad \forall u \in H^t.$$

Choose $t_0 \in \left[s, d+\frac{1}{2}\right)$, $t_0 > \beta + \frac{1}{2}$. Then, using cases **(i)** and **(ii)** and (10.117) in Lemma 10.5 with $r = s$, we have

$$\begin{aligned}
\|E_h(u)\|_s &= \inf_{\chi \in S_h^d} \|E_h(u-\chi)\|_s \leqslant \inf_{\chi \in S_h^d} [Ch^{t_0-s}\|u-\chi\|_{t_0}] \\
&= Ch^{t_0-s} \inf_{\chi \in S_h^d} \|u-\chi\|_{t_0} \\
&\leqslant Ch^{t_0-s} \cdot Ch^{t-t_0}\|u\|_t \\
&= Ch^{t-s}\|u\|_t.
\end{aligned}$$

Therefore, it only remains to prove (10.124). We do it in the following. □

Proposition 10.2 ([10, Proposition 3.4]). *Let $s, t \in \mathbb{R}$ satisfy $\beta \leqslant s \leqslant t < d+\frac{1}{2}$ and let $u \in H^t$ and $u_h \in S_h^d$ satisfy the collocation equations (10.122). Then there exist constants $C > 0$ and $\delta > 0$ independent of u such that*

$$\begin{aligned}
\|u-u_h\|_s \leqslant\ & Ch^{t-s}\|u\|_t + Ch^{\delta}\|u-u_h\|_s \\
&+ \begin{cases} 0, & s < t, \\ C\|u-u_h\|_{s-\delta}, & s = t. \end{cases}
\end{aligned}$$

Proof. Define A_0 as in (10.79). From (10.83), we have

$$C_1^{-1}\|v\|_{s+\beta} \leqslant \|A_0 v\|_s \leqslant C_1\|v\|_{s+\beta} \quad \forall v \in H^{s+\beta}, s \in \mathbb{R}. \tag{10.127}$$

If $w \in H^t$ and $w_h \in S_h^d$ are related by

$$A_0 w = A_0 w_h \qquad \text{on } \triangle, \tag{10.128}$$

then, by Theorem 10.12, we have

$$\|w-w_h\|_s \leqslant C_2 h^{t-s}\|w\|_t \tag{10.129}$$

for $\beta \leqslant s < d+\frac{1}{2}, s \leqslant t \leqslant d+\frac{1}{2}$ and $\beta+\frac{1}{2} < t$.

Let u and u_h satisfy (10.122). From the identities

$$A_0 \boldsymbol{P}_h^d u_h = A_0 u + A_0(u_h - u) + A_0(\boldsymbol{P}_h^d u_h - u_h),$$

$$A_0(u_h - u) = (A_0 - A)(u_h - u) + A(u_h - u),$$

we get, by (10.121),

$$\begin{aligned} A_0 \boldsymbol{P}_h^d u_h &= A_0 u + (A_0 - A)(u_h - u) + A(u_h - u) + A_0(\boldsymbol{P}_h^d u_h - u_h) \\ &= A_0 u + (A_0 - A)(u_h - u) + A_0(\boldsymbol{P}_h^d u_h - u_h) \\ &= A_0[u + A_0^{-1}(A_0 - A)(u_h - u) + \boldsymbol{P}_h^d u_h - u_h)], \text{ on } \triangle. \end{aligned} \tag{10.130}$$

Let

$$\left.\begin{aligned} w &\equiv u + A_0^{-1}(A_0 - A)(u_h - u) + (\boldsymbol{P}_h^d u_h - u_h), \\ w_h &\equiv \boldsymbol{P}_h^d u_h. \end{aligned}\right\} \tag{10.131}$$

Then by (10.130), Theorem 10.12 is applicable to $u = w$ and $u_h = w_h$; so we use them in (10.100) to get

$$\begin{aligned} \|w - w_h\|_s &= \|[u + (u_h - u) - A_0^{-1}A(u_h - u) + (\boldsymbol{P}_h^d u_h - u_h)] - \boldsymbol{P}_h^d u_h\|_s \\ &= \|A_0^{-1}A(u_h - u)\|_s \\ &\leqslant C_2 h^{t-s}\|w\|_t. \end{aligned} \tag{10.132}$$

We now estimate $\|w\|_t$ on the RHS above by bounding the three terms on the RHS of the first equation in (10.130). Obviously,

$$\|u\|_t \leqslant \|u\|_t. \tag{10.133}$$

Also,

$$\begin{aligned} \|A_0^{-1}(A_0 - A)(u_h - u)\|_t &= \|A_0^{-1}(K - J)(u_h - u)\|_t \\ &\leqslant C_3\|(K - J)(u_h - u)\|_{t-\beta} \\ &\leqslant C_4\|u - u_h\|_{t-\delta}, \end{aligned} \tag{10.134}$$

by the second inequality in (10.127) and the smoothing property of J. By (10.118), the triangle inequality and (10.119), we have

$$\begin{aligned} \|\boldsymbol{P}_h^d u_h - u_h\|_t &\leqslant C_5 h^\delta \|u_h\|_t \\ &\leqslant C_6(h^{s-t+\delta}\|u - u_h\|_s + h^\delta\|u\|_t). \end{aligned} \tag{10.135}$$

Combining (10.133)–(10.135) and using (10.132), we get

$$\begin{aligned}\|A_0^{-1}A(u_h-u)\|_s &\leqslant C_2h^{t-s}[\|u\|_t+C_4\|u-u_h\|_{t-\delta}\\ &\quad +C_6h^{s-t+\delta}\|u-u_h\|_s+C_6h^{\delta}\|u\|_t]\\ &\leqslant C_7[h^{t-s}\|u\|_t+h^{t-s}\|u-u_h\|_{t-\delta}+h^{\delta}\|u-u_h\|_s].\end{aligned}$$

By (10.121), (10.127) and the above, we have

$$\begin{aligned}\|u_h-u\|_s &\leqslant C_8\|A(u_h-u)\|_{s-\beta}\leqslant C_9\|A_0^{-1}A(u_h-u)\|_s\\ &\leqslant C_{10}[h^{t-s}\|u\|_t+h^{t-s}\|u-u_h\|_{t-\delta}+h^{\delta}\|u-u_h\|_s].\end{aligned} \tag{10.136}$$

In case $s=t$, we have completed the proof of (10.124). If $s<t$, we decrease δ in(10.136) so that $0<\delta\leqslant s-t$ and use (10.121) (with $t-\delta$ in place of t) to complete the proof. □

Remark 10.1.

(a) Using the ideas of *partition of unity* and *freezing the variable coefficients locally*, Arnold and Wendland [10, § 3] proved the collocation error estimates (10.123) for strongly elliptic pseudodifferential operators of the general form (10.115). The interested reader should consult the original paper [10] for the complete proof.

(b) When the pseudodifferential operator A appears in matrix form (as in a *system*), it is understood that b_+ and b_- are (say) $p\times p$ constant matrices and $Q_\beta^{\pm}$ represents an entry in the operator matrices $\boldsymbol{Q}_\beta^+$ and $\boldsymbol{Q}_\beta^-$. In this case, if we adapt the strong ellipticity condition (10.80) to

$$\gamma\equiv\min_{\substack{|\zeta|=1\\ \zeta\in\mathbb{C}^p}}\{\mathrm{Re}[\zeta^{\mathrm{Tr}}(b_++b_-)\overline{\zeta}],\mathrm{Re}[\zeta^{\mathrm{Tr}}(b_+-b_-)\overline{\zeta}]\}>0, \tag{10.137}$$

then Theorem 10.13 remains valid. Furthermore, we can treat augmented systems such as in (4.141). This is sketched in the next section. □

10.4 Convergence and error estimates for collocation of augmented systems of BIE

Let us discuss augmented systems of the form (4.141). Let

$$\mathscr{H}^s=\mathbb{R}^m\oplus(H^s)^p,\quad p\in\mathbb{N},m\in\mathbb{Z}^+.$$

Since R in (4.141) is generally set to zero, we assume that

$$\mathscr{A}=\begin{bmatrix}0 & T\\ S & A\end{bmatrix}:\ \mathscr{H}^r\rightarrow\mathscr{H}^{r-\beta} \tag{10.138}$$

for all $r > r_0$ for some $r_0 \in [-\infty, \infty)$. (For example, in §4.7, for the accessory condition (4.173), we need to take $r_0 = 0$.) The operator A is a $p \times p$ matrix of strongly elliptic pseudodifferential operators satisfying (10.137). By Gårding's inequality, we have

$$\langle Au, u \rangle_r \geqslant \gamma \|u\|^2_{r+\beta/2} - \langle Ku, u \rangle \quad \forall u \in (H^{r+\beta/2})^p, \tag{10.139}$$

for some $\gamma > 0$ and $K\colon\ (H^{r+\beta/2})^p \to (H^{r-\beta/2})^p$ compact. (Here, naturally, $\langle \cdot, \cdot \rangle_r$ and $\| \cdot \|_r$ denote the inner product and norm in $(H^r)^p$.) Since the operators S and T are of finite rank and hence compact, as a consequence of (10.139), we have

$$\begin{aligned} &\langle \mathscr{A} \vec{u}, \vec{u} \rangle_{\mathscr{H}^r} \geqslant \widetilde{\gamma} \| \vec{u} \|^2_{\mathscr{H}^{r+\beta/2}} - \langle \mathscr{K} \vec{u}, \vec{u} \rangle_{\mathscr{H}^r}, \\ &\forall \vec{u} = (a, u) \in \mathbb{R}^m \oplus (H^{r+\beta/2})^p, \ \ r \geqslant r_0, \end{aligned} \tag{10.140}$$

for some $\widetilde{\gamma} > 0$ and $\mathscr{K} :\ \mathscr{H}^{r+\beta/2} \to \mathscr{H}^{r-\beta/2}$ compact. We further assume

$$\mathscr{A} :\ \mathscr{H}^r \to \mathscr{H}^{r-\beta} \text{ is an isomorphism}, \ \ \forall r > r_0. \tag{10.141}$$

Given $\vec{G} = (a, G) \in \mathscr{H}^{r-\beta}$, we wish to approximate the solution $\vec{u} = (a, u) \in \mathscr{H}^r$ of

$$\mathscr{A} \vec{u} = \vec{G}$$

by collocation. Let $\vec{S}^d_h \equiv \mathbb{R}^m \oplus (S^d_h)^p$, where S^d_h satisfies the same properties as in §10.3. The collocation solution $\vec{u}_\triangle = (a_\triangle, u_\triangle) \in \vec{S}^d_h$ is defined by

$$\left.\begin{aligned} &Tu_\triangle = \alpha \in \mathbb{R}^m, \\ &Sa_\triangle + Au_\triangle = G \qquad \text{on } (\triangle)^p, \end{aligned}\right\} \tag{10.142}$$

where $(\triangle)^p$ is the p-fold Cartesian product of the partition $\triangle$ in (10.77).

Define

$$\mathscr{A}_0 = \begin{bmatrix} 0 & T \\ S & A_0 \end{bmatrix}, \tag{10.143}$$

where

$$A_0 = b_+ \boldsymbol{Q}^+_\beta + b_- \boldsymbol{Q}^-_\beta + J, \tag{10.144}$$

$\boldsymbol{Q}^+_\beta$, $\boldsymbol{Q}^-_\beta$, b_+ and b_- satisfy the properties as indicated in Remark 10.1(b), and J is understood to apply componentwise. Then the following analogue of Theorem 10.12 can be proved.

Theorem 10.14. *Let A_0 be given as in (10.144), satisfying 10.137). Let $d \in \mathbb{Z}^+$ satisfy (10.95). Then there exists $h_0 > 0$ such that for $0 < h \leqslant h_0$ and any continuous function G, the collocation equations (10.142) for*

$$\mathscr{A}_0 \vec{u} = \vec{G} \tag{10.145}$$

are uniquely solvable for $\vec{u}_\triangle \in \vec{S}^d_h$. *Moreover, if* $s,\tau \in \mathbb{R}$ *satisfy*

$$\beta \leqslant s \leqslant \tau \leqslant d+1, r_0 < s < d+\frac{1}{2}, \beta+\frac{1}{2} < \tau, \tau > r_0, \tag{10.146}$$

and the solution $\vec{u} = (a,u)$ *of (10.145) is in* $\mathscr{H}^T$, *then*

$$|a-a_\triangle| + \|u-u_\triangle\|_s \leqslant Ch^{\tau-s}\|u\|_T. \qquad \square$$

The inverse inequality (10.116) remains valid for $v \in (S^d_h)^p$, for $s \leqslant t < d+\frac{1}{2}$, for a (quasi) uniform mesh $\triangle$. Also, Lemma 10.5 holds for $u \in (H^t)^p$, for $r < d+\frac{1}{2}$ and $s \leqslant r \leqslant t \leqslant d+\frac{1}{2}$. Therefore, we mimic the work in the proof of Theorem 10.13 to obtain the following.

Theorem 10.15. *Let A be a* $p \times p$ *matrix of (scalar) strongly elliptic pseudodifferential operators of order* β *of the form*

$$A = b_+\boldsymbol{Q}^+_\beta + b_-\boldsymbol{Q}^-_\beta + K,$$

satisfying (10.137). Assume that (10.141) holds. Let $\vec{u} = (a,u) \in \mathscr{H}^t$ *and* $\vec{u}_\triangle = (a_\triangle, u_\triangle) \in \vec{S}^d_h$ *satisfy the collocation equations*

$$\begin{cases} Tu_\triangle = Tu & \text{in} \quad \mathbb{R}^m, \\ Sa_\triangle + Au_\triangle = Sa + Au & \text{on} \quad (\triangle)^p. \end{cases}$$

Then there exists a constant $C > 0$ *independent of* h: $0 < h \leqslant h_0$ *for some* $h_0 > 0$ *sufficiently small and a solution* $\vec{u}_\triangle$ *such that*

$$|a-a_\triangle| + \|u-u_\triangle\|_s \leqslant Ch^{t-s}\|u\|_t$$

for all $\beta, d, s, t \in \mathbb{R}$ *satisfying (10.146).* $\square$

Empirically, the sharpness of Theorems 10.13 and 10.15 has been confirmed in the numerical computations of Examples 6.3 and 6.5 of Chapter 6, where piecewise-constant boundary elements were used. Other types of smoother splines have also been experimented with and have yielded the sharpest convergence rates exactly as indicated in these two theorems.

Chapter 11

Boundary Element Methods for Semilinear Elliptic Partial Differential Equations (I): The Monotone Iteration Scheme and Error Estimates

11.1 Introduction

Numerical solutions of nonlinear partial differential equations (PDEs) are important in applications. Historically, such work is done primarily by the finite difference methods (FDM) and finite element methods (FEM). We note a few of such papers containing FDM and FEM numerical analysis of semilinear and quasilinear elliptic BVPs:

Bers [17], Parter [155], Greenspan and Parter [83], for FDM;
Ciarlet, Schultz and Varga [50], Dupont and Douglas [67], Brenner and Scott [28], for FEM.

At present, even though the results therein are not directly applicable to the problems considered here because of the difficulties related to the often lack of stability of the solution as well as the possible inappropriate iteration algorithms (to be explained shortly), the error estimates and algorithms given by those authors above and elsewhere (such as Eydeland and Spruck [71], for example) are quite delicate, and many ideas remain useful. We believe they will eventually help the analysis of convergence and errors for nonlinear problems in both Chapters 11 and 12.

BEM is as competitive as FDM and FEM in solving nonlinear elliptic BVPs. Progress has been made by several researchers to apply BEM to encompass nonlinearities in elliptic BVPs with some success. For example,

(i) when the governing equation itself is linear, with nonlinearities occurring only in the boundary conditions, such as

$$\begin{cases} \Delta u(x) = f(x) & \text{on } \Omega, \\ \dfrac{\partial u(x)}{\partial n} = g_1(x) + g_2(x,u) & \text{on } \partial\Omega, \end{cases}$$

then the applicability of BEM is quite ready. A nonlinear integral equation can be formulated on the boundary; an essential feature and an advantage of BEM are retained. Error analysis for such problems has been done by Ruotsalainen and Wendland [163], e.g.;

(ii) when the governing equation is nonlinear, with the nonlinearities occurring in the lowest-order terms (leading to a semilinear PDE), Brebbia and Walker [26], and Sakakihara [164] have considered iteration schemes for such problems. Accurate numerical results for certain examples have been obtained. But no error analysis and convergence for the discretized scheme were established in [26] and [164]. See more details on these below.

The semilinear elliptic PDEs considered in [26] and [164] are of the form

$$\begin{cases} \Delta u(x) = F(x, u(x)) & \text{on } \Omega, \\ Bu(x) = g(x) & \text{on } \partial\Omega, \end{cases} \tag{11.1}$$

where Ω is a simply-connected bounded domain in $\mathbb{R}^N$ with C^∞ smooth boundary $\partial\Omega$, $F\colon\ \overline{\Omega}\times\mathbb{R}\to\mathbb{R}$ is smooth such that $F(x,u)$ is nonlinear in u, B is a linear boundary operator of the form

$$Bu = u \quad \text{or} \quad Bu = \frac{\partial u}{\partial n} + \alpha(\cdot)u, \text{ with } \alpha(x) \geqslant 0 \quad \forall x \in \partial\Omega, \tag{11.2}$$

and n is the unit outward normal on $\partial\Omega$. A *straightforward* iteration scheme suggested by Brebbia and Walker in [26] goes as follows:

$$\begin{cases} \text{Initialize: Take an initial guess } u_0(x); \\ \text{Iterate: Solve, for } n = 0,\ 1,\ 2, \ldots \\ \quad \begin{cases} \Delta u_{n+1}(x) = F(x, u_n(x)), & x \in \Omega, \\ (Bu_{n+1})(x) = g(x), & x \in \partial\Omega, \end{cases} \end{cases} \tag{11.3}$$

where at each iteration $u_{n+1}(\cdot)$ is solved by a boundary element method. Although this scheme appears quite natural and sometimes even produces nice numerical results, unfortunately, for "strong" nonlinearities, it fails to converge in general, as will be shown in § 11.2.

On the other hand, the iteration scheme used by Sakakihara in [164] is the *monotone iteration scheme*:

$$\begin{cases} (1) \text{ Choose a } \lambda > 0 \text{ "sufficiently large"; } \quad \text{(cf. (11.13))} \\ (2) \text{ Initialize: Take an initial guess } u_0(x); \\ (3) \text{ Iterate: Solve, for } n = 0, 1, 2, \ldots \\ \quad \begin{cases} \Delta u_{n+1}(x) - \lambda u_{n+1}(x) = -\lambda u_n(x) + F(x, u_n(x)), & x \in \Omega, \\ (Bu_{n+1})(x) = g(x), & x \in \partial\Omega. \end{cases} \\ \text{(Here } u_{n+1} \text{ is also solved by a boundary element method, but} \\ \text{other general numerical methods may also be used.)} \end{cases} \tag{11.4}$$

The monotone iteration scheme is a well-known *constructive* method for solving semilinear elliptic PDEs of the type (11.1). When some *supersolutions* and *subsolutions* (also known as *barriers*) are available, monotone convergence of the iterates $u_{n+1}(x)$ (of the undiscretized scheme) can be established [4, 141, 167]. In [164], Sakakihara used the Green's formula, i.e., the commonly called *direct approach*, to formulate boundary integral equations (BIEs) to solve each $u_{n+1}(x)$. The convergence of boundary layer densities in those BIEs can be readily established, yielding the convergence of $u_n(x)$ to some $u(x)$ satisfying (11.1), i.e., a solution of (11.1). Nevertheless, in practical BEM computations, a BIE formulation of (11.4) must be *discretized*. Analysis of convergence and error for the discretized problem requires the knowledge of regularities of several singular integral operators involved in the BIEs. Such work is missing in [164].

In boundary element computations of PDEs, *indirect formulations* using only simple- or double-layer potentials to represent the solution are known to have several important advantages, one among which as we have noted in preceding chapters is a significant reduction of the amount of quadrature involved. Here, in our experience, using the simple-layer potential representation (plus a volume potential to account for the right-hand side $-\lambda u_n(x) + F(x, u_n(x))$ in (11.4), see the second integral in (11.29)) is preferable to the direct formulation adopted in [164]. This will be the basic approach taken in this chapter. We will undertake the task of establishing convergence and error estimates for the discretized Galerkin boundary-element scheme corresponding to this approach. First, we need to emphasize some fundamental issues in the study of nonlinear PDEs: nonexistence of solutions, or the existence of multiple solutions, some of which are known to be *unstable* [141, 167]. (Unstable solutions may be briefly explained as a saddle-point type critical point on the solution manifold, which must be obtained through a Mountain–Pass lemma [1] kind of argument and are thus unobtainable as solutions of a maximization/minimization variational problem. Or, unstable solutions may be regarded as an unstable equilibrium of a corresponding dynamic problem, say, the parabolic equation $u_t - \Delta u + f(\cdot, u) = 0$ with the same boundary condition and certain initial conditions.) Only after these questions are properly understood, can one talk about which solution the numerical iterates are converging to. Indeed, this may help explain why earlier workers had not been totally successful in deriving any error estimates or convergence for the boundary element monotone iteration scheme. Not surprisingly, a certain additional working assumption (see ([H]) in § 11.4) is needed in order for the Galerkin boundary element scheme to converge. This assumption, although looking somewhat restrictive at first glance, is actually natural because of the needs to be able to estimate errors, especially after seeing several types of concrete examples of semilinear PDEs analyzed and computed in § 11.7. The monotone iteration scheme studied in this chapter *can only capture stable solutions*. The computation of unstable solutions must be done by other numerical schemes, to be developed in Chapter 12.

This chapter is based primarily on our work in [34, 62][1]. It is organized as follows:

(i) In § 11.2, we provide some further background material and discussion.

(ii) In § 11.3, we prove the *antimonotone* convergence property of simple-layer densities of the (undiscretized) boundary integral equations according to our formulation.

(iii) In § 11.4, we establish convergence and error estimates for a Galerkin boundary element scheme, subject to the Dirichlet boundary condition. By comparing various intermediate approximate solutions, error estimates $\|u_h^n - u_\infty\|_{H^r(\Omega)}$ are obtained which are sharp for the $H^r(\Omega)$ Sobolev space norms, $0 \leqslant r \leqslant 2$.

(iv) A *distinctive feature* of BEM for semilinear PDEs is that we are able to obtain "higher than optimal order" error estimates with respect to h, provided that the nonlinearity is *separable* (cf. (11.132)). This is presented in § 11.5.

(v) Error estimates for the semilinear PDE subject to the Neumann or Robin boundary conditions are discussed in § 11.6.

(vi) In § 11.7, some typical cases of semilinear elliptic PDEs are discussed and computed, with numerical solutions illustrated.

(vii) § 11.8 studies 2×2 systems of coupled semilinear elliptic PDEs, their error estimates and BEM computations.

---

[1]The authors wish to thank the American Mathematical Society for copyright permissions regarding [34, 62] published in Mathematics of Computation.

The *collocation scheme* is computationally more convenient than the Galerkin one. Our numerical results in § 11.7 and Subsection 11.8.4 are obtained via the former rather than the latter.

11.2 A straightforward iteration scheme and the monotone iteration scheme

We first review the straightforward iteration scheme (11.3). We show through an ODE example that (11.3) does not converge in general.

Example 11.1. Consider

$$\begin{cases} \frac{d^2}{dx^2}u(x) - u^3(x) = 0, \quad 0 < x < 1, \\ u(0) = u(1) = 0. \end{cases} \tag{11.5}$$

The differential operator $\frac{d^2}{dx^2}$ represents the simplest elliptic operator in one dimension. Multiplying the equation in (11.5) by u and integrating by parts, we obtain

$$\int_0^1 [|u'(x)|^2 + u^4(x)]dx = 0.$$

Therefore, (11.5) admits a unique solution that is trivial. (This trivial solution is, of course, stable.)

Let us find iterative solutions of (11.5) by (11.3):

$$\begin{cases} \text{Set } u_0(x) = C > 0; \\ \text{Solve, for } n = 0, 1, 2, \ldots \\ \qquad \begin{cases} u''_{n+1}(x) = u_n^3(x), \quad x \in [0,1], \\ u_{n+1}(0) = u_{n+1}(1) = 0. \end{cases} \end{cases} \tag{11.6}$$

For $n = 0$, for example, we get

$$u_1''(x) = C^3,$$

$$u_1(x) = \frac{C^3}{2}x^2 + \alpha_1 x + \beta_1,$$

$$u_1(0) = u_1(1) = 0 \Rightarrow \beta_1 = 0, \quad \alpha_1 = -\frac{C^3}{2},$$

hence

$$u_1(x) = \frac{C^3}{2}(x^2 - x). \tag{11.7}$$

In general, we have

$$u_n(x) = \alpha_n x^{m_n} + \cdots, \tag{11.8}$$

where the leading coefficient α_n and the power m_n can be uniquely determined recursively:

$$\begin{cases} \alpha_0 = C, \quad m_0 = 0, \\ \alpha_{n+1} = \dfrac{\alpha_n^3}{(3m_n+2)(3m_n+1)}, \quad m_{n+1} = 3m_n + 2, \quad n = 0, 1, 2, \ldots . \end{cases} \tag{11.9}$$

From (11.9), it is not difficult to show that if we choose $C > \frac{2}{3}k_0$ for any $k_0 \geqslant 6$, then

$$\alpha_n > k_0 \left(m_n + \frac{2}{3} \right) \longrightarrow \infty \text{ as } n \longrightarrow \infty.$$

Consequently, the iterates $u_n(x)$ in (11.8) cannot converge to the unique solution $u(x) \equiv 0$ of (11.5). □

Thus, the direct iteration scheme (11.3) does not work for Example 11.1 if the initial state $u_0(x) \equiv C$ is not small. In contrast, the monotone iteration scheme (11.4) works for Example 11.1 for any $C \in \mathbb{R}$ if $u_0(x) \equiv C_i$, $i = 1, 2$, are properly chosen, because

(i) if $C_1 > 0$, then $\overline{u}(x) \equiv C_1$ satisfies

$$\begin{cases} \overline{u}''(x) - \overline{u}(x)^3 = -C_1^3 < 0 \quad \text{on } (0,1), \\ \overline{u}(x) = C_1 > 0 \quad \text{at} \quad x = 0, 1. \end{cases}$$

Therefore, $\overline{u}(x)$ is a supersolution, cf. (11.11).

(ii) If $C_2 < 0$, then $\overline{v}(x) \equiv C_2$ satisfies

$$\begin{cases} \overline{v}''(x) - \overline{v}(x)^3 = -C_2^3 > 0 \quad \text{on } (0,1), \\ \overline{v}(x) = C_2 < 0 \quad \text{at} \quad x = 0, 1. \end{cases}$$

Therefore $\overline{v}(x)$ is a subsolution, cf. (11.12).

If we choose $\lambda > 0$ and $C \in \mathbb{R}$ such that

$$\lambda > \max(3C_1^2, 3C_2^2), \quad C \in (C_2, C_1),$$

then for the nonlinearity $F(x,u) = u^3$ in Example 11.1, we have

$$\lambda - \frac{\partial F}{\partial u} = \lambda - 3u^2 > 0 \qquad \forall u \in (C_2, C_1).$$

The monotone iteration scheme (11.4) will work for Example 11.1, by Theorem 11.1 below. We now state a version of the monotone iteration theorem. Since only its utilization is needed, we refer the interested reader to [4, 141, 167] for a proof.

Theorem 11.1. *Let $F(x,u)$ be C^1 with respect to $(x,u) \in \overline{\Omega} \times \mathbb{R}$. Consider the boundary value problem*

$$\begin{cases} \Delta u(x) - F(x, u(x)) = 0 & \text{on } \Omega, \\ Bu(x) = g(x) & \text{on } \partial\Omega, \end{cases} \tag{11.10}$$

where B is given by (11.2) *with* $\alpha \in C^\infty(\partial\Omega)$, $\alpha(x) \geqslant 0$, *and* $\alpha(x) \not\equiv 0$ *if* $Bu \neq u$ *on* $\partial\Omega$, *and* $g \in C^2(\partial\Omega)$. *Let* $\overline{u}, \overline{v} \in C^2(\overline{\Omega})$ *satisfy* $\overline{u} \geqslant \overline{v}$ *as well as*

$$\begin{cases} \Delta\overline{u}(x) - F(x,\overline{u}(x)) \leqslant 0, & x \in \Omega, \\ B\overline{u}(x) \geqslant g(x), & x \in \partial\Omega, \end{cases} \tag{11.11}$$

$$\begin{cases} \Delta\overline{v}(x) - F(x,\overline{v}(x)) \geqslant 0, & x \in \Omega, \\ B\overline{v}(x) \leqslant g(x), & x \in \partial\Omega. \end{cases} \tag{11.12}$$

(We call $\overline{u}$ and $\overline{v}$, respectively, a supersolution and a subsolution for satisfying (11.11) *and* (11.12).*) Choose a number* $\lambda > 0$ *such that*

$$\lambda - \frac{\partial F(x,u)}{\partial u} > 0 \quad \forall (x,u) \in \overline{\Omega} \times [a,b], \quad a = \inf\overline{v}, \quad b = \sup\overline{u}, \tag{11.13}$$

and such that the operator $(\Delta - \lambda, B|_{\partial\Omega} = 0)$ *has its spectrum strictly contained in the open left-half complex plane. Then the mapping*

$$T\colon\ \phi \mapsto w, \quad w = T\phi, \tag{11.14}$$

$$\phi \in C^2(\overline{\Omega}), \quad \phi(x) \in [a,b] \quad \forall x \in \overline{\Omega}, \tag{11.15}$$

where $w(x)$ is the unique solution of the BVP

$$\begin{cases} \Delta w(x) - \lambda w(x) = -\lambda\phi(x) + F(x,\phi(x)), & x \in \Omega, \\ Bw(x) = g(x), \quad x \in \partial\Omega, \end{cases}$$

is monotone, i.e., for any ϕ_1, ϕ_2 *satisfying* (11.15) *and* $\phi_1 \leqslant \phi_2$, *we have*

$$T\phi_1, T\phi_2 \text{ satisfy (11.15) and } T\phi_1 \leqslant T\phi_2 \text{ on } \Omega. \tag{11.16}$$

Consequently, by letting $F_\lambda(x,u) \equiv -\lambda u + F(x,u)$, *the iterations*

$$\begin{cases} u_0(x) = \overline{u}(x), \\ (\Delta - \lambda)u_{n+1}(x) = F_\lambda(x,u_n(x)), \quad x \in \Omega, \quad n = 0, 1, 2, \ldots, \\ Bu_{n+1} = g \quad \text{on } \partial\Omega, \end{cases} \tag{11.17}$$

and

$$\begin{cases} v_0(x) = \overline{v}(x), \\ (\Delta - \lambda)v_{n+1}(x) = F_\lambda(x,v_n(x)), \quad x \in \Omega, \quad n = 0, 1, 2, \ldots, \\ Bv_{n+1} = g \quad \text{on } \partial\Omega, \end{cases} \tag{11.18}$$

yield iterates u_n and v_n satisfying

$$\overline{v} = v_0 \leqslant v_1 \leqslant \cdots \leqslant v_n \leqslant \cdots \leqslant u_n \leqslant \cdots \leqslant u_1 \leqslant u_0 = \overline{u}, \tag{11.19}$$

so that the limits

$$u_\infty(x) \equiv \lim_{n\to\infty} u_n(x) \tag{11.20}$$

and

$$v_\infty(x) \equiv \lim_{n\to\infty} v_n(x) \tag{11.21}$$

exist in $C^2(\overline{\Omega})$. *We have*

(i) $v_\infty \leqslant u_\infty$ on $\overline{\Omega}$;

(ii) v_∞ and u_∞ are both stable solutions of (11.10);

(iii) If $v_\infty \not\equiv u_\infty$, then there exists an unstable solution $\phi \in C^2(\overline{\Omega})$ of (11.10) such that

$$v_\infty \leqslant \phi \leqslant u_\infty.$$

□

Remark 11.1. The rate of convergence in (11.20) and (11.21) is not available in general. To make the numerical analysis possible in our subsequent development, obviously some additional working assumption is needed. Otherwise, numerical discretizations cause errors and make the computed solutions deviate outside the range of monotone interations (11.13). This will be formulated in the important ([H]) condition in § 11.4. □

Remark 11.2.

(a) We note that there is a useful further formulation of the above theorem in the weak (i.e., distributional) sense.

(b) For simplicity of presentation, from now on, we will assume that F, g, $\overline{u}$, $\overline{v}$ and $\partial\Omega$ (consequently, also u_∞ and v_∞) are all $C^\infty(\overline{\Omega})$, considering that this is not the main issue here. □

11.3 Formulation of boundary integral equations based on the simple-layer representation

For given $\lambda > 0$, let $E(x,\lambda)$ be the fundamental solution satisfying

$$(\Delta - \lambda)E(x;\lambda) = -\delta(x), \quad x \in \mathbb{R}^N, \tag{11.22}$$

where $\delta(x)$ is the Dirac delta function. It is known from §7.2 of Chapter 7, that

$$E(x;\lambda) = \frac{1}{4}(i)^{\frac{N}{2}} \left(\frac{\sqrt{\lambda}}{2\pi|x|} \right)^{\frac{N-2}{2}} H^{(1)}_{\frac{N}{2}-1}(i\sqrt{\lambda}|x|), \quad N \geqslant 2, \tag{11.23}$$

where $H^{(1)}_{\frac{N}{2}-1}(\cdot)$ is the Hankel function of order $\frac{N}{2}-1$ of the first kind. In particular, we have

$$E(x;\lambda) = \frac{1}{2\pi} K_0(\sqrt{\lambda}|x|), \quad N = 2, \tag{11.24}$$

where K_0 is the Macdonald function of order 0, and

$$E(x;\lambda) = \frac{e^{-\sqrt{\lambda}|x|}}{4\pi|x|}, \quad N = 3. \tag{11.25}$$

Let the assumptions of Theorem 11.1 be satisfied. We first consider the case when the boundary condition is Dirichlet:

$$Bu(x) \equiv u(x) = g(x), \qquad \text{on } \partial\Omega, \tag{11.26}$$

and defer the discussion of the Neumann and Robin boundary conditions to § 11.6. For the boundary integral formulation of the monotone iteration

$$\begin{cases} (\Delta-\lambda)u_{n+1}(x) = F_\lambda(x,u_n(x)), & x\in\Omega, \\ u_{n+1}(x) = g(x), \quad x\in\partial\Omega, \end{cases} \tag{11.27}$$

the boundary integral equation by the standard *direct formulation* for the unknown $\partial u_{n+1}/\partial n$ is

$$\int_{\partial\Omega} E(x-y;\lambda)\frac{\partial u_{n+1}(y)}{\partial n}d\sigma_y = \frac{1}{2}g(x) + \int_{\partial\Omega}\frac{\partial E(x-y;\lambda)}{\partial n_y}g(y)d\sigma_y \tag{11.28}$$
$$-\int_\Omega E(x-y;\lambda)F_\lambda(y,u_n(y))dy, \qquad x\in\partial\Omega,$$

as given in Sakakihara [164], (11.24). Once $\partial u_{n+1}/\partial n$ is solved in (11.28), u_{n+1} on the entire Ω can be obtained by quadratures on Ω and $\partial\Omega$.

The *indirect formulation* is based upon the following ansatz ([72], **??**])

$$u_{n+1}(x) = (\widetilde{S}_\lambda\eta_{n+1})(x) - \widetilde{V}_\lambda(u_n)(x) \tag{11.29}$$
$$\equiv \int_{\partial\Omega} E(x-y;\lambda)\eta_{n+1}(y)d\sigma_y - \int_\Omega E(x-y;\lambda)F_\lambda(y,u_n(y))dy \quad x\in\overline{\Omega},$$

where $\eta_{n+1}(\cdot)$ is an unknown boundary layer density defined on $\partial\Omega$ to be solved by the BIE

$$\int_{\partial\Omega} E(x-y;\lambda)\eta_{n+1}(y)d\sigma_y = g(x) + \int_\Omega E(x-y;\lambda)F_\lambda(y,u_n(y))dy, \qquad x\in\partial\Omega. \tag{11.30}$$

Comparing (11.30) with (11.28), we see that a large amount of numerical quadrature (involved in the evaluation of $\int_{\partial\Omega}\frac{\partial E}{\partial n_y}g d\sigma_y$) is bypassed. This approach is more advantageous.

Let $a, b\in\mathbb{R}$, $\lambda>0$ be chosen such that for any $\phi\in C^\infty(\overline{\Omega})$,

$$\frac{\partial F(x,\phi(x))}{\partial u} - \lambda < 0, \qquad \forall\phi\colon\ a\leqslant\phi(x)\leqslant b, \quad x\in\overline{\Omega}. \tag{11.31}$$

For any $\phi\in C^\infty(\overline{\Omega})$, the volume potential in (11.29) defines a continuous mapping

$$\widetilde{V}_\lambda\colon\ C^\infty(\overline{\Omega})\to C^\infty(\overline{\Omega}), \quad \widetilde{V}_\lambda(\phi)(x) \equiv \int_\Omega E(x-y;\lambda)F_\lambda(y,\phi(y))dy, \qquad x\in\overline{\Omega},$$

cf. §6.5 of Chapter 6, e.g. By the trace theorem, we have

$$\widetilde{V}_{\lambda,\Gamma}(\phi) \equiv \widetilde{V}_\lambda(\phi)|_{\partial\Omega} \in C^\infty(\partial\Omega). \tag{11.32}$$

The trace of the simple-layer potential $\widetilde{S}_\lambda$ in (11.29) is denoted by S_λ:

$$S_\lambda\eta \equiv \widetilde{S}_\lambda\eta|_{\partial\Omega}. \tag{11.33}$$

It is well known that S_λ has the following regularity:

$$\begin{cases} S_\lambda\colon\ H^r(\partial\Omega)\to H^{r+1}(\partial\Omega), \text{ isomorphically}, \forall r\in\mathbb{R}, \\ S_\lambda \text{ is a strongly elliptic pseudodifferential operator of order } -1 \text{ on } \partial\Omega. \end{cases} \tag{11.34}$$

Now define a mapping $\mathscr{T}$ on $C^\infty_{a,b}(\overline{\Omega}) \equiv \{\phi \in C^\infty(\overline{\Omega}) \mid a \leqslant \phi(x) \leqslant b \text{ on } \overline{\Omega}\}$:

$$\eta = \mathscr{T}\phi, \tag{11.35}$$

where η is a function defined on $\partial\Omega$ satisfying the BIE

$$\int_{\partial\Omega} E(x-y;\lambda)\eta(y)d\sigma_y = \int_\Omega E(x-y;\lambda)F_\lambda(y,\phi(y))dy, \qquad x \in \partial\Omega. \tag{11.36}$$

By (11.32), and (11.34)–(11.36), we have $\eta \in C^\infty(\partial\Omega)$, and that

$$\mathscr{T}:\ C^\infty_{a,b}(\overline{\Omega}) \to C^\infty(\partial\Omega) \tag{11.37}$$

is continuous.

We now establish the main theorem in this section.

Theorem 11.2. *The mapping* $\mathscr{T}$ *in* (11.35) *is antimonotone, i.e.,*

$$\phi_1,\phi_2 \in C^\infty_{a,b}(\overline{\Omega}), \qquad \phi_1(x) \leqslant \phi_2(x) \qquad \forall x \in \overline{\Omega}$$

implies

$$\mathscr{T}\phi_1 \geqslant \mathscr{T}\phi_2 \qquad \text{on } \partial\Omega. \tag{11.38}$$

Proof. Let

$$\eta_i = \mathscr{T}\phi_i, \qquad i=1,2, \quad \text{on } \partial\Omega. \tag{11.39}$$

Extend ϕ_1 and ϕ_2 to $\mathbb{R}^N$ by

$$\widetilde{\phi}_i(x) = \begin{cases} 0, & x \in \Omega^c \equiv \mathbb{R}^N \backslash \overline{\Omega}, \\ \phi_i(x), & x \in \overline{\Omega}, \end{cases} \qquad i=1,2. \tag{11.40}$$

Then through a localization procedure in §6.5 of Chapter 6, it is not difficult to show that

$$\widetilde{\phi}_i \in H^{\frac{1}{2}-\varepsilon}(\mathbb{R}^N), \quad \text{for any } \varepsilon > 0, i=1,2. \tag{11.41}$$

From the definition of $\widetilde{S}_\lambda\eta$ in (11.29), by abuse of notation, extend $\widetilde{S}_\lambda\eta$ to all $\mathbb{R}^N$, i.e.,

$$(\widetilde{S}_\lambda\eta)(x) = \int_{\partial\Omega} E(x-y;\lambda)\eta(y)d\sigma_y, \qquad x \in \mathbb{R}^N.$$

We also extend $\widetilde{V}_\lambda$ to $\mathbb{R}^N$ by defining

$$(\widetilde{V}_\lambda\widetilde{\phi})(x) = \int_\Omega E(x-y;\lambda)\widetilde{F}_\lambda(y,\phi(y))dy, \qquad x \in \mathbb{R}^N, \tag{11.42}$$

where

$$\widetilde{F}_\lambda(x,\phi(x)) \equiv \begin{cases} 0, & \text{if } x \in \Omega^c, \\ F_\lambda(x,\phi(x)), & \text{if } x \in \Omega. \end{cases} \tag{11.43}$$

From (11.37), (11.40), (11.41), (11.43) and § 6.5, Chapter 6, we have the regularities

$$\eta_i \in C^\infty(\partial\Omega), \quad \widetilde{V}_\lambda \widetilde{\phi}_i \in H^{\frac{5}{2}-\varepsilon}_{\text{loc}}(\mathbb{R}^N), \quad \text{for any} \quad \varepsilon > 0. \tag{11.44}$$

Now define

$$w_i^\pm(x) = (\widetilde{S}_\lambda \eta_i)(x) - (\widetilde{V}_\lambda \phi_i)(x), \qquad i = 1,2, \tag{11.45}$$
$$\text{“}+\text{” if } x \in \Omega^c, \quad \text{“}-\text{” if } x \in \Omega.$$

By (11.35) and (11.36), we have $w_i^-|_{\partial\Omega} \equiv 0$, $i = 1,2$. Also, on Ω, we have

$$(\Delta - \lambda)w_i^-(x) = F_\lambda(x, \phi_i(x)) = -\lambda\phi_i(x) + F(x, \phi_i(x)), \qquad i = 1,2.$$

Therefore, w_i^- is the (unique) solution of the BVP

$$\left.\begin{cases} (\Delta - \lambda)w_i^-(x) = -\lambda\phi_i(x) + F(x, \phi_i(x)) & \text{on } \Omega, \\ w_i^-(x) = 0 & \text{on } \partial\Omega, \end{cases}\right\} \qquad i = 1,2. \tag{11.46}$$

Similarly, from (11.36) and (11.43)–(11.45), w_i^+ satisfies the exterior BVP

$$\begin{cases} (\Delta - \lambda)w_i^+(x) = 0 & \text{on } \Omega^c, \\ w_i^+(x) = 0 & \text{on } \partial\Omega, \end{cases} \qquad i = 1,2. \tag{11.47}$$

We now show that w_i^+, $i = 1,2$, satisfying (11.45) and (11.47), are trivial. For $x \in \Omega^c$, $|x|$ large, and $y \in \Omega$, we have

$$\begin{aligned} |x - y| &= \sqrt{|x|^2 - 2\langle x,y\rangle + |y|^2} \\ &= |x|\left[1 - \frac{1}{2}\left(\frac{2\langle x,y\rangle - |y|^2}{|x|^2}\right) + \frac{\frac{1}{2}(-\frac{1}{2})}{1\cdot 2}\left(\frac{-2\langle x,y\rangle + |y|^2}{|x|^2}\right)^2 \pm \cdots\right]. \end{aligned} \tag{11.48}$$

The fundamental solution $E(x;\lambda)$ in (11.23) is known to have fast decay for $|x|$ large when $\lambda > 0$. For example, when $N = 3$, this is obvious from (11.25). For $N = 2$, we have [1]

$$\begin{aligned} E(x,\lambda) &= \frac{1}{2\pi} K_0(\sqrt{\lambda}|x|) \\ &\sim \frac{1}{2}\sqrt{\frac{1}{2\sqrt{\lambda}|x|}}\, e^{-\sqrt{\lambda}|x|}\left\{1 - \frac{1}{8\sqrt{\lambda}|x|} \pm \cdots\right\}. \end{aligned} \tag{11.49}$$

Using (11.48) and (11.49) in (11.45), we obtain

$$w_i^+(x) = \begin{cases} \frac{1}{2}\lambda^{-\frac{1}{4}}\sqrt{\frac{1}{2|x|}}[1 + \mathscr{O}(|x|^{-\frac{1}{2}})]\left\{\int_{\partial\Omega} e^{-\sqrt{\lambda}|x-y|}\eta(y)d\sigma_y \right. \\ \qquad \left. -\int_\Omega e^{-\sqrt{\lambda}|x-y|}F_\lambda(y, \phi_i(y))dy\right\}, & N = 2; \\ \frac{1}{4\pi}\frac{1}{|x|}[1 + \mathscr{O}(|x|^{-1})]\left\{\int_{\partial\Omega} e^{-\sqrt{\lambda}|x-y|}\eta(y)d\sigma_y \right. \\ \qquad \left. -\int_\Omega e^{-\sqrt{\lambda}|x-y|}F_\lambda(y, \phi_i(y))dy\right\}, & N = 3. \end{cases} \tag{11.50}$$

For space dimension $N > 3$, similar expressions can be obtained. Note that each integral inside the parentheses above decay with an exponential rate $e^{-\sqrt{\lambda}|x|}$ for $|x|$ large.
Let

$$\Omega_R^c = \Omega^c \cap \{x \in \mathbb{R}^N \;\big|\; |x| < R\}, \quad \text{for large} \quad R > 0.$$

From (11.47), we obtain

$$\begin{aligned}0 &= \int_{\Omega^c} [(\Delta - \lambda)w_i^+(x)]w_i^+(x)dx \\ &= \lim_{R\to\infty} \left[\int_{\partial\Omega_R^c} \frac{\partial w_i^+(x)}{\partial n} w_i^+(x)d\sigma - \int_{\Omega_R^c} (|\nabla w_i^+(x)|^2 + \lambda |w_i^+(x)|^2)dx \right] \\ &= -\int_{\Omega^c} (|\nabla w_i^+(x)|^2 + \lambda |w_i^+(x)|^2)dx,\end{aligned}$$

where the limit of the boundary integral over $\partial\Omega_R^c$ tends to 0 as $R \to \infty$ because

$$\begin{aligned}\int_{\partial\Omega_R^c} \frac{\partial w_i^+(x)}{\partial n} w_i^+(x)d\sigma &= \int_{\partial\Omega^c} \frac{\partial w_i^+(x)}{\partial n} w_i^+(x)d\sigma + \int_{|x|=R} \frac{\partial w_i^+(x)}{\partial r} w_i^+(x)d\sigma \\ &= \int_{|x|=R} \frac{\partial w_i^+(x)}{\partial r} w_i^+(x)d\sigma \qquad (r = |x|),\end{aligned}$$

and the above integral tends to zero as $R \to \infty$ by (11.50). Hence $w_i^+(x) \equiv 0$ on Ω^c, $i = 1,2$.
Define

$$w(x) = \begin{cases} w_2^+(x) - w_1^+(x), & x \in \Omega^c, \\ w_2^-(x) - w_1^-(x), & x \in \Omega. \end{cases} \tag{11.51}$$

Then

$$w(x) \equiv 0 \quad \text{for} \quad x \in \Omega^c. \tag{11.52}$$

On Ω, the function w satisfies

$$\begin{aligned}(\Delta - \lambda)w(x) &= -\lambda[\phi_2(x) - \phi_1(x)] + [F(x,\phi_2(x)) - F(x,\phi_1(x))] \\ &= [\phi_2(x) - \phi_1(x)] \cdot \left[-\lambda + \frac{F(x,\phi_2(x)) - F(x,\phi_1(x))}{\phi_2(x) - \phi_1(x)} \right] \\ &\leqslant 0 \qquad \text{by (11.33)},\end{aligned} \tag{11.53}$$

and on $\partial\Omega$, it satisfies $w \equiv 0$. Thus,

$$\begin{cases} (\Delta - \lambda)w \leqslant 0 & \text{on} \quad \Omega, \\ w|_{\partial\Omega} = 0 & \text{on } \partial\Omega. \end{cases} \tag{11.54}$$

If $w(x) < 0$ for some $x \in \Omega$, then

$$\min_{x\in\overline{\Omega}} w(x) < 0. \tag{11.55}$$

Assume that

$$w(x_0) = \min_{x\in\overline{\Omega}} w(x).$$

Then $x_0 \notin \Omega$ by Hopf's boundary point lemma [156]. Thus $x_0 \in \partial\Omega$. But

$$0 > w(x_0) = 0, \quad \text{by the boundary condition in (11.54), a contradiction.} \tag{11.56}$$

Therefore,

$$w(x) \geqslant 0 \quad \text{on } \Omega. \tag{11.57}$$

Let $\widetilde{V}_\lambda \widetilde{\phi}_i$ be the volume potential in (11.44). Since $\widetilde{F}_\lambda(\cdot,\phi_i(\cdot)) \in H^{\frac{1}{2}-\varepsilon}(\mathbb{R}^N)$ (just as in (11.41)), by elliptic regularity we have

$$\widetilde{V}_\lambda \widetilde{\phi}_i \in H^{\frac{5}{2}-\varepsilon}(\mathbb{R}^N), \qquad i = 1,2, \text{ for any } \varepsilon > 0. \tag{11.58}$$

Further, let

$$\begin{aligned}(\widetilde{V}_\lambda \widetilde{\phi}_i)^+ &= \widetilde{V}_\lambda \widetilde{\phi}_i(x) \quad \text{if } x \in \Omega^c,\\ (\widetilde{V}_\lambda \widetilde{\phi}_i)^-(x) &= \widetilde{V}_\lambda \widetilde{\phi}_i(x) \quad \text{if } x \in \Omega.\end{aligned}$$

Then the trace theorem gives

$$\frac{\partial}{\partial n}(\widetilde{V}_\lambda \widetilde{\phi}_i)^\pm \in H^{1-\varepsilon}(\partial\Omega), \quad \text{for any} \quad 0 < \varepsilon < 1.$$

Choose any ε: $0 < \varepsilon < 1$. Then by the continuity of the normal trace operation we have

$$\frac{\partial}{\partial n}(\widetilde{V}_\lambda \widetilde{\phi}_i)^+ - \frac{\partial}{\partial n}(\widetilde{V}_\lambda \widetilde{\phi}_i)^- = \frac{\partial}{\partial n}\widetilde{V}_\lambda \widetilde{\phi}_i - \frac{\partial}{\partial n}\widetilde{V}_\lambda \widetilde{\phi}_i = 0 \quad \text{on } \partial\Omega, i = 1,2. \tag{11.59}$$

Also, let

$$(\widetilde{S}_\lambda \eta_i)^\pm(x) = \int_{\partial\Omega} E(x-\xi;\lambda)\eta_i(\xi)d\sigma_\xi, \qquad i = 1,2, \tag{11.60}$$

$$\text{"+" if } x \in \Omega^c, \quad \text{"−" if } x \in \Omega.$$

Then, on $\partial\Omega$,

$$\begin{aligned}\frac{\partial}{\partial n}[w^+(x) - w^-(x)] &= -\frac{\partial}{\partial n}w^-(x) \qquad \text{(by (11.52))} \\ &= \lim_{h\downarrow 0}\frac{w^-(x-nh) - w^-(x)}{h} \\ &= \lim_{h\downarrow 0}\frac{w^-(x-nh)}{h} \qquad (x \in \partial\Omega) \\ &\geqslant 0 \quad \text{by (11.57).}\end{aligned} \tag{11.61}$$

But from (11.59), we also have

$$
\begin{aligned}
&\frac{\partial}{\partial n}[w^+(x)-w^-(x)] \\
&=\frac{\partial}{\partial n}\{[((\widetilde{S}_\lambda\eta_2)^+(x)-((\widetilde{V}_\lambda\widetilde{\phi}_2)^+(x))-((\widetilde{S}_\lambda\eta_1)^+(x)-(\widetilde{V}_\lambda\widetilde{\phi}_1)^+(x))] \\
&\quad-[((\widetilde{S}_\lambda\eta_2)^-(x)-(\widetilde{V}_\lambda\widetilde{\phi}_2)^-(x))-((\widetilde{S}_\lambda\eta_1)^-(x)-(\widetilde{V}_\lambda\phi_1)^-(x))]\} \\
&=\frac{\partial}{\partial n}\{[(\widetilde{S}_\lambda\eta_2)^+(x)-(\widetilde{S}_\lambda\eta_1)^+(x)]-[(\widetilde{S}_\lambda\eta_2)^-(x)-(\widetilde{S}_\lambda\eta_1)^-(x)]\} \\
&=-\eta_2(x)-(-\eta_1(x)) \\
&=\eta_1(x)-\eta_2(x)\geqslant 0 \qquad \text{(by (11.61))}.
\end{aligned}
$$

Hence the proof is complete. □

Some numerical evidence of Theorem 11.2 is provided in Fig. 11.2.

Corollary 11.1. *Let $g\in C^\infty(\partial\Omega)$ be given. Let (11.31) be satisfied. Define*

$$
\mathscr{T}_g:\ C^\infty_{a,b}(\overline{\Omega})\to C^\infty(\partial\Omega),\quad \mathscr{T}_g\phi=\eta, \tag{11.62}
$$

where η is the solution of the BIE

$$
\int_{\partial\Omega}E(x-y;\lambda)\eta_{n+1}(y)d\sigma_y=g(x)+\int_\Omega E(x-y;\lambda)F_\lambda(y,\phi(y))dy,\quad x\in\partial\Omega. \tag{11.63}
$$

Then $\mathscr{T}_g$ is also antimonotone.

Corollary 11.2. *Let the assumptions of Theorem 11.1 be satisfied. Let the monotone iterations*

$$
\begin{cases}
u_0(x)=\overline{u}(x)\in C^\infty(\overline{\Omega}), \\
(\Delta-\lambda)u_{n+1}(x)=F_\lambda(x,u_n(x)),\ x\in\Omega, \\
u_{n+1}(x)=g(x), \qquad x\in\partial\Omega,
\end{cases} \tag{11.64}
$$

and

$$
\begin{cases}
v_0(x)=\overline{v}(x)\in C^\infty(\overline{\Omega}), \\
(\Delta-\lambda)v_{n+1}(x)=F_\lambda(x,v_n(x)),\ x\in\Omega,\ , \\
v_{n+1}(x)=g(x), \qquad x\in\partial\Omega
\end{cases} \tag{11.65}
$$

have iterative solutions u_{n+1} and v_{n+1} represented, for $n=0,1,2,\ldots$, as

$$
u_{n+1}(x)=\int_{\partial\Omega}E(x-y;\lambda)\mu_{n+1}(y)d\sigma_y-\int_\Omega E(x-y;\lambda)F_\lambda(y,u_n(y))dy,\ x\in\overline{\Omega}, \tag{11.66}
$$

$$
v_{n+1}(x)=\int_{\partial\Omega}E(x-y;\lambda)v_{n+1}(y)d\sigma_y-\int_\Omega E(x-y;\lambda)F_\lambda(y,v_n(y))dy,\ x\in\overline{\Omega}, \tag{11.67}
$$

where the simple-layer densities $\mu_{n+1}(\cdot)$ and $\nu_{n+1}(\cdot)$ are determined from

$$\mu_{n+1} = \mathscr{T}_g u_n, \quad \nu_{n+1} = \mathscr{T}_g v_n, \qquad n = 0,1,2,\dots . \tag{11.68}$$

Then

$$\mu_\infty = \lim_{n\to\infty} \mu_n, \qquad \nu_\infty = \lim_{n\to\infty} \nu_n \tag{11.69}$$

exist in $C^\infty(\partial\Omega)$ such that

$$\mu_1 \leqslant \mu_2 \leqslant \cdots \leqslant \mu_n \leqslant \cdots \to \mu_\infty \leqslant \nu_\infty \leftarrow\leqslant \nu_n \leqslant \cdots \leqslant \nu_2 \leqslant \nu_1. \tag{11.70}$$

As in Remark 11.2, we again note that the rate of convergence of (11.69) is not available in general.

11.4 Error estimates for the Galerkin boundary element scheme

We first recall some basic regularity properties of the operators corresponding to the simple-layer and volume potentials. For $\phi \in C^\infty(\overline{\Omega})$, let $\mathscr{V}_\lambda \phi$ denote the volume potential defined by

$$(\mathscr{V}_\lambda \phi)(x) = \int_\Omega E(x-y;\lambda)\phi(y)dy, \quad x \in \overline{\Omega}. \tag{11.71}$$

Then by Theorem 6.1, we know that $\mathscr{V}_\lambda$ can be extended to be an operator such that

$$\begin{cases} \mathscr{V}_\lambda : H^r(\Omega) \to H^{r+2}(\Omega) \quad \text{continuous, } \forall r \geqslant -1. \\ \|\mathscr{V}_\lambda \phi\|_{H^{r+2}(\Omega)} \leqslant \overline{C}_r \|\phi\|_{H^r(\Omega)}, \qquad \forall \phi \in H^r(\Omega), \quad r \geqslant -1, \end{cases} \tag{11.72}$$

for a constant $\overline{C}_r > 0$ independent of ϕ. The trace of (11.71) induces an operator $\mathscr{V}_{\lambda,\Gamma}$ defined by

$$(\mathscr{V}_{\lambda,\Gamma}\phi)(x) \equiv (\mathscr{V}_\lambda \phi)(x), \qquad x \in \partial\Omega; \tag{11.73}$$

it satisfies

$$\begin{cases} \mathscr{V}_{\lambda,\Gamma}\colon H^r(\Omega) \to H^{r+\frac{3}{2}}(\partial\Omega) \quad \text{continuous, } \forall r \geqslant -1, \\ \text{there exists } \overline{\overline{C}}_r > 0 \text{ such that} \\ \|\mathscr{V}_{\lambda,\Gamma}\phi\|_{H^{r+\frac{3}{2}}(\partial\Omega)} \leqslant \overline{\overline{C}}_r \|\phi\|_{H^r(\Omega)}, \qquad \forall \phi \in H^r(\Omega), r \geqslant -1. \end{cases} \tag{11.74}$$

Note the relationship

$$\widetilde{V}_\lambda \phi = \mathscr{V}_\lambda F(\cdot, \phi(\cdot)) \tag{11.75}$$

from (11.29).

Let the simple-layer potential $\widetilde{S}_\lambda$ be as defined in (11.29), and let S_λ be the corresponding trace as given in (11.33). It is known from §6.5 of Chapter 6 that

$$\begin{cases} \widetilde{S}_\lambda:\ H^r(\partial\Omega) \to H^{r+\frac{3}{2}}(\Omega) \quad \text{continuous, } \forall r \in \mathbb{R}, \\ \text{there exists } \widetilde{C}_r > 0, \text{ such that} \\ \|\widetilde{S}_\lambda \eta\|_{H^{r+\frac{3}{2}}(\Omega)} \leqslant \widetilde{C}_r \|\eta\|_{H^r(\partial\Omega)}, \qquad \forall \eta \in H^r(\partial\Omega). \end{cases} \tag{11.76}$$

Also, from (11.34), for each $r \in \mathbb{R}$, there exists a $\widetilde{\widetilde{C}}_r > 0$ such that

$$\widetilde{\widetilde{C}}_r^{-1} \|\eta\|_{H^r(\partial\Omega)} \leqslant \|S_\lambda \eta\|_{H^{r+1}(\partial\Omega)} \leqslant \widetilde{\widetilde{C}}_r \|\eta\|_{H^r(\partial\Omega)}, \qquad \forall \eta \in H^r(\partial\Omega). \tag{11.77}$$

Let $\{\mathscr{S}_h \mid 0 < h \leqslant h_0\}$ be a 1-parameter family of finite-dimensional boundary element approximation spaces that form an (ℓ,m)-system $S_h^{\ell,m}(\partial\Omega)$, with $\ell, m \in \mathbb{Z}^+ \equiv \{0,1,2,\ldots\ldots\}$, $\ell \geqslant m+1$, on $\partial\Omega$, in the sense of Babuška and Aziz [14], cf. §5.13 of Chapter 5, satisfying

(1) **Approximation property:**

For each $\phi \in H^t(\partial\Omega)$, there exists a $\phi_h \in \mathscr{S}_h$ such that

$$\|\phi - \phi_h\|_{H^s(\partial\Omega)} \leqslant C_{t,s} h^{t-s} \|\phi\|_{H^t(\partial\Omega)}, \qquad \forall h:\ 0 < h \leqslant h_0, \tag{11.78}$$

where $-\ell \leqslant s \leqslant t \leqslant \ell; |s|, |t| \leqslant m$, and $C_{t,s}$ is a constant independent of h and ϕ.

(2) **Inverse property:**

There exists a constant $M_{s,t} > 0$ such that

$$\|\phi_h\|_{H^t(\partial\Omega)} \leqslant M_{s,t} h^{s-t} \|\phi_h\|_{H^s(\partial\Omega)}, \quad \forall \phi_h \in \mathscr{S}_h, \quad \forall h:\ 0 < h \leqslant h_0, \tag{11.79}$$

where $s \leqslant t$, and $|s|, |t| \leqslant m$.

These two properties, (11.78) and (11.79), provide the foundation for error estimations in this Chapter. Further estimations on elements in $\mathscr{S}_h$ after being noted by boundary integral operators (which are pseudodifferential operators) can be found mostly from the study in Chapter 10. Here, throughout the rest of this Chapter, we also include the original sources of references on such estimates as an additional aid for the reader.

We now implement the monotone iteration scheme according to Corollary 11.2. For definiteness, we will only discretize the (supersolution) systems (11.68) and (11.70). The subsolution systems (11.64) and (11.67) are analogous and thus can be handled in a nearly identical way. The Galerkin boundary element method can be formulated as follows:

(i) Set

$$u_h^0(x) = \overline{u}(x), x \in \overline{\Omega}, \text{ cf. the first equation in (11.64).} \tag{11.80}$$

(ii) For each $n = 0,1,2,\ldots$, and $h:\ 0 < h \leqslant h_0$, find $\eta_h^{n+1} \in \mathscr{S}_h$ such that

$$\langle S_\lambda \eta_h^{n+1}, \phi_h \rangle = \langle g + \mathscr{V}_{\lambda,\Gamma} F(\cdot, u_h^n), \phi_h \rangle, \qquad \forall \phi_h \in \mathscr{S}_h, \tag{11.81}$$

where $\langle\ ,\ \rangle$ signifies the $L^2(\partial\Omega)$ inner product.

(iii) Define

$$u_h^{n+1}(x) = (\widetilde{S}_\lambda \eta_h^{n+1})(x) - (\mathscr{V}_\lambda F(\cdot, u_h^n))(x), \qquad x \in \overline{\Omega}, \quad n = 0,1,2,\ldots . \tag{11.82}$$

Note that (11.81) and (11.82) correspond, respectively, to the discretization of (11.63) and (11.66). We also see that in (11.81) and (11.82), quadrature evaluations are involved in the determination of $\mathscr{V}_\lambda F(\cdot, u_h^n)$ and $\mathscr{V}_{\lambda,\Gamma} F(\cdot, u_h^n)$. Since, in principle, these quadrature evaluations may be carried out to as high an accuracy as desired, we assume that $\mathscr{V}_\lambda F(\cdot, u_h^n)$ and $\mathscr{V}_{\lambda,\Gamma} F(\cdot, u_h^n)$ are *exact* (although in practice, these quadrature evaluations are rather tedious because the integration is carried out on the domain, and numerical errors are involved.) From now on, without further mention, we also maintain all the assumptions and notation used in Theorem 11.1.

Our main objective is to estimate $\|u_\infty - u_h^n\|_{H^r(\Omega)}$, where u_∞ is a stable solution as announced in Theorem 11.1(ii). The following working assumption is crucial; it enables us to obtain rates of convergence:

$$\left.\begin{array}{l} \text{There exist constants } \gamma_1 \text{ and } \gamma_2 \text{ such that} \\ -\lambda_1 < \gamma_1 \leqslant \dfrac{\partial F(x,u)}{\partial u} \leqslant \gamma_2, \qquad \forall (x,u) \in \overline{Q}, \qquad \overline{Q} \equiv \overline{\Omega} \times [a,b], \\ \text{where } \lambda_1 > 0 \text{ is the smallest eigenvalue of } -\Delta \text{ on } \Omega \text{ subject to} \\ \text{the homogeneous Dirichlet boundary condition.} \end{array}\right\} \tag{H}$$

Because $\frac{\partial F}{\partial u}$ is continuous, it easily follows from ([H]) that for a sufficiently small $\delta > 0$, there exist $\widetilde{\gamma}_1$ and $\widetilde{\gamma}_2$ such that

$$-\lambda_1 < \widetilde{\gamma}_1 \leqslant \frac{\partial F(x,u)}{\partial u} \leqslant \widetilde{\gamma}_2, \qquad \forall (x,u) \in \overline{Q}_\delta, \qquad \overline{Q}_\delta \equiv \overline{\Omega} \times [a-\delta, b+\delta]. \tag{11.83}$$

This small δ will provide the "breathing room" to accommodate the errors in the numerical operation for the iterates to stay within the range of monotone iteration.

Lemma 11.1. *Let u_{n+1} be iterated according to* (11.64) *and let u_∞ be the limit as in* (11.69). *Denote $e^n = u_\infty - u_n$. Then we have*

$$\|e^{n+1}\|_{L^2(\Omega)} \leqslant \alpha_\lambda^{n+1} \|e^0\|_{L^2(\Omega)}, \qquad n = 0,1,2,\ldots, \tag{11.84}$$

where

$$0 < \alpha_\lambda \equiv \frac{\lambda - \widetilde{\gamma}_1}{\lambda + \lambda_1} < 1. \tag{11.85}$$

Proof. Since u_∞ and u_{n+1} satisfy, respectively, (11.10) (with $Bu \equiv u$) and (11.64), we obtain by subtraction

$$\left.\begin{aligned}(\Delta-\lambda)e^{n+1}(x) &= F_\lambda(x,u_\infty(x)) - F_\lambda(x,u_n(x)), \; x\in\Omega,\\ e^{n+1}(x) &= 0, \qquad\qquad\qquad\qquad\qquad\qquad x\in\partial\Omega.\end{aligned}\right\} \tag{11.86}$$

Multiplying the above by $-e^{n+1}$, integrating by parts and using $\lambda_1\|e^{n+1}\|^2 \leqslant \|\nabla e^{n+1}\|^2$, we get

$$\|e^{n+1}\| \leqslant \frac{1}{\lambda+\lambda_1}\|F_\lambda(\cdot,u_\infty) - F_\lambda(\cdot,u_n)\|, \tag{11.87}$$

where all the norms without subscripts are $L^2(\Omega)$. Let

$$M_\lambda \equiv \sup_{(x,u)\in\overline{Q}_\delta}\left[\lambda - \frac{\partial F(x,u)}{\partial u}\right]. \tag{11.88}$$

Then

$$0 < M_\lambda < \lambda - \widetilde{\gamma}_1. \tag{11.89}$$

(Actually, $\widetilde{\gamma}_1$ can be taken to be γ_1 here.) We get

$$\begin{aligned}\|F_\lambda(\cdot,u_\infty) - F_\lambda(\cdot,u_n)\| &= \left\|-\left[\lambda - \frac{F(\cdot,u_\infty)-F(\cdot,u_n)}{u_\infty-u_n}\right](u_\infty-u_n)\right\| \\ &\leqslant M_\lambda\|u_\infty-u_n\|.\end{aligned} \tag{11.90}$$

Substituting (11.90) into (11.87), we obtain

$$\|e^{n+1}\| \leqslant \frac{M_\lambda}{\lambda+\lambda_1}\|e^n\| \leqslant \alpha_\lambda\|e^n\|, \quad \alpha_\lambda \equiv \frac{\lambda-\widetilde{\gamma}_1}{\lambda+\lambda_1} < 1.$$

Hence (11.84) follows. □

Lemma 11.2. *Let η_h^{n+1} and u_h^{n+1} be defined as in (11.81) and (11.82) for $n = 0,1,2,\ldots$. Let ζ_h^{n+1}, defined on $\partial\Omega$, be the (unique) solution of the BIE*

$$(S_\lambda\zeta_h^{n+1})(x) = g(x) + \mathscr{V}_{\lambda,\Gamma}F(x,u_h^n(x)), \qquad x\in\partial\Omega. \tag{11.91}$$

Then there exists $\widetilde{C}_{t,s} > 0$, depending only on s and t, such that

$$\|\eta_h^{n+1} - \zeta_h^{n+1}\|_{H^s(\partial\Omega)} \leqslant \widetilde{C}_{t,s}h^{t-s}\|\zeta_h^{n+1}\|_{H^t(\partial\Omega)}, \tag{11.92}$$

where $m \leqslant \ell-1$, $-(\ell+1) \leqslant s \leqslant t \leqslant \ell$, $s \leqslant m$, $-\ell \leqslant -\frac{1}{2} \leqslant m$ and $-m \leqslant -\frac{1}{2} \leqslant t$.

Proof. Since S_λ satisfies (11.77), according to Hsiao and Wendland [97], the order of S_λ is $2\alpha = -1$, i.e., $\alpha = -\frac{1}{2}$. Invoking [97, Corollary 2.1], see also Chapter 10, we obtain the conclusion. □

Lemma 11.3. *Let $N = 2$ or 3, and let*

$$\varepsilon_0(N) = \begin{cases} 0, & \text{if } N = 2, \\ \text{an arbitrarily small positive number,} & \text{if } N = 3. \end{cases} \tag{11.93}$$

Then for $\ell \geqslant 2$ and $m \geqslant 1$, there exist positive constants $K_{\varepsilon_0(N)}$ and K such that

$$\begin{aligned} \|\eta_h^{n+1} - \zeta_h^{n+1}\|_{H^{\varepsilon_0(N)}(\partial\Omega)} &\leqslant K_{\varepsilon_0(N)} h^{\frac{1}{2}-\varepsilon_0(N)} [\|u_\infty\|_{H^2(\Omega)} + M_\lambda \alpha_\lambda^n \|e^0\|_{L^2(\Omega)} \\ &\quad + \|F_\lambda(\cdot, u_h^n) - F_\lambda(\cdot, u_n)\|_{L^2(\Omega)}], \end{aligned} \tag{11.94}$$

$$\begin{aligned} \|\eta_h^{n+1} - \zeta_h^{n+1}\|_{H^{-\frac{3}{2}}(\partial\Omega)} &\leqslant K h^2 [\|u_\infty\|_{H^2(\Omega)} + M_\lambda \alpha_\lambda^n \|e^0\|_{L^2(\Omega)} \\ &\quad + \|F_\lambda(\cdot, u_h^n) - F_\lambda(\cdot, u_n)\|_{L^2(\Omega)}], \\ &\quad \forall n = 0, 1, 2, \ldots, 0 < h \leqslant h_0. \end{aligned} \tag{11.95}$$

Proof. We have from (11.92) and property (11.77) of S_λ,

$$\begin{aligned} &\|\eta_h^{n+1} - \zeta_h^{n+1}\|_{H^{\varepsilon_0(N)}(\partial\Omega)} \leqslant \widetilde{C}_{\frac{1}{2},\varepsilon_0(N)} h^{\frac{1}{2}-\varepsilon_0(N)} \cdot \widetilde{\widetilde{C}}_{\frac{1}{2}} \|S_\lambda \zeta_h^{n+1}\|_{H^{\frac{3}{2}}(\partial\Omega)} \\ &\leqslant (\widetilde{C}_{\frac{1}{2},\varepsilon_0(N)} \widetilde{\widetilde{C}}_{\frac{1}{2}}) h^{\frac{1}{2}-\varepsilon_0(N)} [\|g\|_{H^{\frac{3}{2}}(\partial\Omega)} + \|\mathscr{V}_{\lambda,\Gamma} F_\lambda(\cdot, u_h^n)\|_{H^{\frac{3}{2}}(\partial\Omega)}] \text{ (by (11.91))} \\ &\leqslant (\widetilde{C}_{\frac{1}{2},\varepsilon_0(N)} \widetilde{\widetilde{C}}_{\frac{1}{2}}) h^{\frac{1}{2}-\varepsilon_0(N)} [\|g\|_{H^{\frac{3}{2}}(\partial\Omega)} + \overline{\overline{C}}_0 \|F_\lambda(\cdot, u_h^n)\|_{L^2(\Omega)}] \text{ (by (11.74))} \\ &\leqslant (\widetilde{C}_{\frac{1}{2},\varepsilon_0(N)} \widetilde{\widetilde{C}}_{\frac{1}{2}}) h^{\frac{1}{2}-\varepsilon_0(N)} [\|g\|_{H^{\frac{3}{2}}(\partial\Omega)} + \overline{\overline{C}}_0 (\|F_\lambda(\cdot, u_h^n) - F_\lambda(\cdot, u_n)\|_{L^2(\Omega)} \\ &\quad + \|F_\lambda(\cdot, u_n) - F_\lambda(\cdot, u_\infty)\|_{L^2(\Omega)} + \|F_\lambda(\cdot, u_\infty)\|_{L^2(\Omega)}] \\ &\leqslant (\widetilde{C}_{\frac{1}{2},\varepsilon_0(N)} \widetilde{\widetilde{C}}_{\frac{1}{2}}) h^{\frac{1}{2}-\varepsilon_0(N)} [C \|u_\infty\|_{H^2(\Omega)} + \overline{\overline{C}}_0 M_\lambda \alpha_\lambda^n \|e^0\|_{L^2(\Omega)} \\ &\quad + \overline{\overline{C}}_0 \|F_\lambda(\cdot, u_h^n) - F_\lambda(\cdot, u_n)\|_{L^2(\Omega)}], \end{aligned} \tag{11.96}$$

where in the last inequality we have utilized the properties that u_∞ satisfies (11.10), the trace theorem, (11.84) and (11.90). Therefore (11.94) has been established. From (11.92), we have

$$\begin{aligned} \|\eta_h^{n+1} - \zeta_h^{n+1}\|_{H^{-\frac{3}{2}}(\partial\Omega)} &\leqslant \widetilde{C}_{\frac{1}{2},-\frac{3}{2}} h^2 \|\zeta_h^{n+1}\|_{H^{\frac{1}{2}}(\partial\Omega)} \\ &\leqslant \widetilde{C}_{\frac{1}{2},-\frac{3}{2}} h^2 \cdot \widetilde{C}_{\frac{1}{2}} \|S_\lambda \zeta_h^{n+1}\|_{H^{\frac{3}{2}}(\partial\Omega)}. \end{aligned}$$

The rest is the same as in (11.96). □

Remark 11.3. Lemma 11.3 remains valid even if $\varepsilon_0(N) = 0$ when $N = 3$, where the constant $K_{\varepsilon_0(N)}$ remains bounded as $\varepsilon_0(N) \downarrow 0$. But in the discussions henceforth we will only utilize the part when $\varepsilon_0(N)$ is positive yet nonvanishingly small. □

From (11.91), we now define

$$w_h^{n+1}(x) = (\widetilde{S}_\lambda \zeta_h^{n+1})(x) - (\mathscr{V}_\lambda F(\cdot, u_h^n))(x), \qquad x \in \Omega, n = 0, 1, 2, \ldots, 0 < h \leqslant h_0. \quad (11.97)$$

It is readily seen that w_h^{n+1} satisfies

$$\left.\begin{aligned} (\Delta - \lambda) w_h^{n+1} &= F_\lambda(\cdot, u_h^n) \text{ on } \Omega, \\ w_h^{n+1} &= g \qquad\qquad \text{ on } \partial\Omega. \end{aligned}\right\} \quad (11.98)$$

Lemma 11.4. *Let $N = 2$ or 3. Let $C^{0,\beta}(\overline{\Omega})$ be the Hölder space of continuous functions with exponent β, $0 < \beta < 1$, on $\overline{\Omega}$. Then for $\varepsilon_0(N)$ satisfying* (11.93)*, and for β sufficiently small, we have a positive constant $K'_{\varepsilon_0(N)}$ such that*

$$\begin{aligned} \|u_h^{n+1} - w_h^{n+1}\|_{C^{0,\beta}(\overline{\Omega})} \leqslant K'_{\varepsilon_0(N)} h^{\frac{1}{2} - \varepsilon_0(N)} [\|u_\infty\|_{H^2(\Omega)} + M_\lambda \alpha_\lambda^n \|e^0\|_{L^2(\Omega)} \\ + \|F_\lambda(\cdot, u_h^n) - F_\lambda(\cdot, u^n)\|_{L^2(\Omega)}], \end{aligned} \quad (11.99)$$

for $n = 0, 1, 2, \ldots, 0 < h \leqslant h_0$, where $K'_{\varepsilon_0(3)}$ may grow unbounded as $\varepsilon_0(3) \downarrow 0$ when $N = 3$.

Proof. We use the Sobolev imbedding theorem, Theorem 2.2:

$$\|f\|_{C^{0,\beta}(\overline{\Omega})} \leqslant C \|f\|_{H^s(\Omega)}, \quad \text{for} \quad 0 < \beta = s - \frac{N}{2} < 1, \quad (11.100)$$

to get

$$\|u_h^{n+1} - w_h^{n+1}\|_{C^{0,\beta}(\overline{\Omega})} \leqslant c_{\varepsilon_0} \|u_h^{n+1} - w_h^{n+1}\|_{H^{\frac{3}{2} + \varepsilon_0(N)}(\Omega)}, \quad (11.101)$$

where

$$\left.\begin{aligned} 0 < \beta < \frac{1}{2} \\ \beta = \varepsilon_0(3) \end{aligned}\right\} \quad \text{if} \quad \left\{\begin{aligned} N = 2, \\ N = 3. \end{aligned}\right. \quad (11.102)$$

Note that $c_{\varepsilon_0} \uparrow \infty$ as $\varepsilon_0(3) \downarrow 0$. Continuing from the right-hand side of (11.101), we have

$$\begin{aligned} \|u_h^{n+1} - w_h^{n+1}\|_{C^{0,\beta}(\overline{\Omega})} &\leqslant c_{\varepsilon_0} \|\widetilde{S}_\lambda (\eta_h^{n+1} - \zeta_h^{n+1})\|_{H^{\frac{3}{2} + \varepsilon_0(N)}(\Omega)} \\ &\qquad \text{(by (11.82) and (11.97))} \\ &\leqslant c_{\varepsilon_0} \cdot \widetilde{C}_{\varepsilon_0(N)} \|\eta_h^{n+1} - \zeta_h^{n+1}\|_{H^{\varepsilon_0(N)}} \quad \text{(by (11.76))}. \end{aligned}$$

The rest of the proof follows from Lemma 11.3. □

Lemma 11.5. *Let $N = 2$ or 3. Then there exists a constant $K > 0$ such that*

$$\begin{aligned} &\|w_h^{n+1} - u_{n+1}\|_{C^{0,\beta}(\overline{\Omega})} \leqslant K \|F_\lambda(\cdot, u_h^n) - F_\lambda(\cdot, u_n)\|_{L^2(\Omega)} \\ &\forall n = 0, 1, 2, \ldots, 0 < h \leqslant h_0, \text{ where } 0 < \beta < 1/2. \end{aligned} \quad (11.103)$$

Proof. From (11.66), (11.97), (11.100) and (11.102), we get

$$
\begin{aligned}
&\|w_h^{n+1} - u_{n+1}\|_{C^{0,\beta}(\overline{\Omega})} \\
&\leqslant c\|w_h^{n+1} - u_{n+1}\|_{H^2(\Omega)} \quad (\text{for } 0<\beta<\tfrac{1}{2}, \text{ by } (11.100)) \\
&\leqslant c[\|\widetilde{S}_\lambda(\zeta_h^{n+1} - \mu_{n+1})\|_{H^2(\Omega)} + \|\mathscr{V}_\lambda(F_\lambda(\cdot,u_n^h) - F_\lambda(\cdot,u_n))\|_{H^2(\Omega)}] \\
&\leqslant c[\widetilde{C}_{\frac{1}{2}}\widetilde{\widetilde{C}}_{\frac{1}{2}}\|S_\lambda(\zeta_h^{n+1} - \mu_{n+1})\|_{H^{\frac{3}{2}}(\partial\Omega)} \\
&\quad + \|\mathscr{V}_\lambda(F_\lambda(\cdot,u_n^h) - F_\lambda(\cdot,u_n))\|_{H^2(\Omega)}] \quad (\text{by } (11.76), (11.77)) \\
&\leqslant c[\widetilde{C}_{\varepsilon_0(N)}\widetilde{\widetilde{C}}_{\varepsilon_0(N)}\|\mathscr{V}_{\lambda,\Gamma}(F_\lambda(\cdot,u_h^n) - F_\lambda(\cdot,u_n))\|_{H^{\frac{3}{2}}(\partial\Omega)} \\
&\quad + \|\mathscr{V}_\lambda(F_\lambda(\cdot,u_h^n) - F_\lambda(\cdot,u_n))\|_{H^2(\Omega)}] \\
&\leqslant c'\|F_\lambda(\cdot,u_h^n) - F_\lambda(\cdot,u_n)\|_{L^2(\Omega)} \quad (\text{by } (11.72) \text{ and } (11.74)). \qquad \square
\end{aligned}
$$

Similar to Lemmas 11.4 and 11.5, we give the parallel estimates in the $L^2(\Omega)$ norms in Lemma 11.6.

Lemma 11.6. *We have*

$$
\begin{aligned}
\|u_h^{n+1} - w_h^{n+1}\|_{L^2(\Omega)} &\leqslant Kh^2[\|u_\infty\|_{H^2(\Omega)} + M_\lambda\alpha_\lambda^n\|e^0\|_{L^2(\Omega)} \\
&\quad + \|F_\lambda(\cdot,u_h^n) - F_\lambda(\cdot,u_n)\|_{L^2(\Omega)}], \textit{ for some } K>0, \qquad (11.104)\\
\|w_h^{n+1} - u_{n+1}\|_{L^2(\Omega)} &\leqslant \frac{1}{\lambda+\lambda_1}\|F_\lambda(\cdot,u_h^n) - F_\lambda(\cdot,u_n)\|_{L^2(\Omega)}, \qquad (11.105)\\
&\forall n = 0,1,2,3,\ldots,\quad 0<h\leqslant h_0.
\end{aligned}
$$

Proof. From (11.82) and (11.97),

$$
\begin{aligned}
\|u_h^{n+1} - w_h^{n+1}\|_{L^2(\Omega)} &= \|\widetilde{S}_\lambda(\eta_h^{n+1} - \zeta_h^{n+1})\|_{L^2(\Omega)} \\
&\leqslant \widetilde{C}_{\frac{3}{2}}\|\eta_h^{n+1} - \zeta_h^{n+1}\|_{H^{-\frac{3}{2}}(\partial\Omega)} \qquad (\text{by } (11.76)).
\end{aligned}
$$

The rest follows from (11.102).

To show (11.105), we use that u_{n+1} and w_h^{n+1} satisfy, respectively, (11.64) and (11.98). Subtracting (11.98) from (11.64) and estimating as in (11.86) and (11.87), we obtain (11.105). □

Corollary 11.3. (i) *Let $N = 2$ or 3. Then for $\varepsilon_0(N)$ satisfying* (11.93) *and for $\beta>0$ sufficiently small, with β being related to $\varepsilon_0(N)$ through* (11.102)*, we have a constant $K'_{\varepsilon_0(N)}>0$ such that*

$$
\begin{aligned}
\|u_h^{n+1} - u_{n+1}\|_{C^{0,\beta}(\overline{\Omega})} &\leqslant K'_{\varepsilon_0(N)}h^{\frac{1}{2}-\varepsilon_0(N)}[\|u_\infty\|_{H^2(\Omega)} + M_\lambda\alpha_\lambda^n\|e^0\|_{L^2(\Omega)}] \\
&\quad + \left[\frac{1}{\lambda+\lambda_1} + K'_{\varepsilon_0(N)}h^{\frac{1}{2}-\varepsilon_0(N)}\right]\|F_\lambda(\cdot,u_h^n) - F_\lambda(\cdot,u_n)\|_{L^2(\Omega)}, \qquad (11.106)
\end{aligned}
$$

$\forall n = 0,1,2,\ldots,0 < h \leqslant h_0$, where $K'_{\varepsilon_0(N)}$ may grow unbounded as $\varepsilon_0(3) \downarrow 0$.

(ii) *There exists a constant $K > 0$ such that*

$$\begin{aligned}\|u_h^{n+1} - u_{n+1}\|_{L^2(\Omega)} \leqslant Kh^2[\|u_\infty\|_{H^2(\Omega)} + M_\lambda \alpha_\lambda^n \|e^0\|_{L^2(\Omega)}] \\ + \left(\frac{1}{\lambda+\lambda_1} + Kh^2\right) \cdot \|F_\lambda(\cdot,u_h^n) - F_\lambda(\cdot,u_n)\|_{L^2(\Omega)}, \\ \forall n = 0,1,2,\ldots,0 < h \leqslant h_0.\end{aligned} \tag{11.107}$$

Proof. The inequality (11.106) follows from an application of the triangle inequality to (11.99) and (11.103), while (11.107) follows likewise from (11.104) and (11.105). □

Theorem 11.3. *Let $N = 2$ or 3 and let $\varepsilon_0(N)$ be given by (11.93) and $\beta > 0$ be sufficiently small, with β being related to $\varepsilon_0(N)$ through (11.102). Then there exist $\overline{h}_0 : 0 < \overline{h}_0 \leqslant h_0$ and $\rho_\lambda : 0 < \rho_\lambda < 1$, independent of h and n, such that*

$$\|u_h^{n+1} - u_{n+1}\|_{C^{0,\beta}(\overline{\Omega})} \leqslant K''_{\varepsilon_0(N)} h^{\frac{1}{2}-\varepsilon_0(N)} [\|u_\infty\|_{H^2(\Omega)} + \rho_\lambda^{n+1} \|e^0\|_{L^2(\Omega)}], \tag{11.108}$$

$$\|u_h^{n+1} - u_{n+1}\|_{L^2(\Omega)} \leqslant \overline{K} h^2 [\|u_\infty\|_{H^2(\Omega)} + \rho_\lambda^{n+1} \|e^0\|_{L^2(\Omega)}], \tag{11.109}$$

and

$$a - \delta \leqslant u_h^{n+1}(x) \leqslant b + \delta, \qquad \forall n = 0,1,2,\ldots,0 < h \leqslant \overline{h}_0, \tag{11.110}$$

where $K''_{\varepsilon_0(N)}$ and $\overline{K}$ are constants independent of h and n.

Proof. Since $0 < \alpha_\lambda < 1$, we can choose $\overline{h}_0 > 0$ so small that

$$0 < \alpha_\lambda + KM_\lambda \overline{h}_0^2 \leqslant \rho_\lambda < 1, \tag{11.111}$$

$$\begin{aligned}[K'_{\varepsilon_0(N)} \overline{h}_0^{\frac{1}{2}-\varepsilon_0(N)} + (\alpha_\lambda + K'_{\varepsilon_0(N)} M_\lambda \overline{h}_0^{\frac{1}{2}-\varepsilon_0(N)}) \left(\frac{K}{1-\rho_\lambda}\right) \overline{h}_0^2] \\ \cdot [\|u_\infty\|_{H^2(\Omega)} + M_\lambda \|e^0\|_{L^2(\Omega)}] \leqslant \delta,\end{aligned} \tag{11.112}$$

for some $\rho_\lambda :\ 0 < \rho_\lambda < 1$. By induction, we may assume

$$u_h^k(x) \in [a-\delta, b+\delta],\ \forall x \in \overline{\Omega},$$
$$k = 0,1,2,\ldots,n, \quad 0 < h \leqslant \overline{h}_0.$$

Hence, similar to (11.88)–(11.90), we have

$$\|F(\cdot,u_h^k) - F(\cdot,u_k)\|_{L^2(\Omega)} \leqslant M_\lambda \|u_h^k - u_k\|_{L^2(\Omega)}, \quad k = 0,1,2,\ldots,n, \quad 0 < h \leqslant \overline{h}_0. \tag{11.113}$$

Using (11.111) and (11.113) in (11.107), we obtain

$$\begin{aligned}
&\|u_h^{n+1} - u_{n+1}\|_{L^2(\Omega)} \qquad (11.114)\\
&\quad \leqslant Kh^2[\|u_\infty\|_{H^2(\Omega)} + M_\lambda \alpha_\lambda^n \|e^0\|_{L^2(\Omega)}]\\
&\qquad + \rho_\lambda \{Kh^2[\|u_\infty\|_{H^2(\Omega)} + M_\lambda \alpha_\lambda^{n-1}\|e^0\|_{L^2(\Omega)}] + \rho_\lambda \|u_h^{n-1} - u_{n-1}\|_{L^2(\Omega)}\}\\
&\quad \leqslant \cdots\\
&\quad \leqslant Kh^2(1+\rho_\lambda + \cdots + \rho_\lambda^n)\|u_\infty\|_{H^2(\Omega)}\\
&\qquad + KM_\lambda h^2(\alpha_\lambda^n + \rho_\lambda \alpha_\lambda^{n-1} + \cdots + \rho_\lambda^{n-1}\alpha_\lambda + \rho_\lambda^n)\|e^0\|_{L^2(\Omega)},
\end{aligned}$$

where in the last inequality, we have used $u_h^0 = u_0 = \overline{u}$. Since $\alpha_\lambda < 1$, we derive

$$\|u_h^{n+1} - u_{n+1}\|_{L^2(\Omega)} \leqslant \left(\frac{K}{1-\rho_\lambda}\right) h^2[\|u_\infty\|_{H^2(\Omega)} + M_\lambda \|e^0\|_{L^2(\Omega)}],$$
$$\forall n = 0,1,2,\ldots, \quad 0 < h \leqslant \overline{h}_0. \qquad (11.115)$$

By applying (11.113) and (11.115) to (11.106), we then obtain

$$\|u_h^{n+1} - u_{n+1}\|_{C^{0,\beta}(\overline{\Omega})} \leqslant \overline{K}_{\varepsilon_0(N)} h^{\frac{1}{2}-\varepsilon_0(N)}[\|u_\infty\|_{H^2(\Omega)} + M_\lambda \|e^0\|_{L^2(\Omega)}],$$
$$\forall n = 0,1,2,\ldots, \quad 0 < h \leqslant \overline{h}_0, \qquad (11.116)$$

where

$$\overline{K}_{\varepsilon_0(N)} = K'_{\varepsilon_0(N)} + (\alpha_\lambda + K'_{\varepsilon_0(N)} M_\lambda \overline{h}_0^{\frac{1}{2}-\varepsilon_0(N)}) \left(\frac{K}{1-\rho_\lambda}\right) \overline{h}_0^{\frac{3}{2}+\varepsilon_0(N)}. \qquad (11.117)$$

Thus, (11.110) is obtained from (11.112), (11.116), (11.117) and the induction hypothesis. To get (11.109), we note that $\alpha_\lambda < \rho_\lambda$ from (11.111). Then, instead of (11.115), we obtain from (11.114)

$$\begin{aligned}
\|u_h^{n+1} - u_{n+1}\|_{L^2(\Omega)} &\leqslant \left(\frac{K}{1-\rho_\lambda}\right) h^2 \|u_\infty\|_{H^2(\Omega)} + KM_\lambda \left(\frac{\rho_\lambda^{n+1}}{\rho_\lambda - \alpha_\lambda}\right) h^2 \|e^0\|_{L^2(\Omega)},\\
&\leqslant \left(\frac{K}{1-\rho_\lambda}\right) h^2 \|u_\infty\|_{H^2(\Omega)} + \left(\frac{1}{\overline{h}_0^2}\right) \rho_\lambda^{n+1} h^2 \|e^0\|_{L^2(\Omega)}\\
&\forall n = 0,1,2,\ldots, \quad 0 < h \leqslant \overline{h}_0, \qquad (11.118)
\end{aligned}$$

where we used (11.111). Thus, (11.109) follows from (11.118) by setting $\overline{K} = \max\{\frac{K}{1-\rho_\lambda}, \frac{1}{\overline{h}_0^2}\}$.

Finally, by using (11.110), (11.111) and then (11.118), we obtain (11.108) from (11.106). The proof is complete. □

The inequality (11.110) is important because it guarantees that each numerical iterate $u_h^n(x)$ does not fall out of the range of validity, $\Omega \times [a-\delta, b+\delta]$, of the monotone iteration, and for the applicability of (11.88)–(11.90). It also partially explains why we need an assumption like ([H]).

Theorem 11.4. *Let $\{\mathscr{S}_h \mid 0 < h \leqslant h_0\}$ be a family of (ℓ, m)-systems on $\partial\Omega$ with $\ell \geqslant 2$ and $m \geqslant 1$, and let $N = 2$ or 3. Assume* ([H]), *so that there exists a small number $\delta > 0$ for which* (11.83) *holds. Then there exist constants $K_1 > 0$, $K_2 > 0$, and an $\overline{h}_0$: $0 < \overline{h}_0 \leqslant h_0$, such that for any r: $0 \leqslant r \leqslant 2$, we have*

$$\|u_h^n - u_\infty\|_{H^r(\Omega)} \leqslant K_1 h^{2-r} \|u_\infty\|_{H^2(\Omega)} + K_2 \rho_\lambda^n \|e^0\|_{L^2(\Omega)} \tag{11.119}$$

and

$$a - \delta \leqslant u_h^n(x) \leqslant b + \delta,$$

for all $n = 0, 1, 2, \ldots, h$: $0 < h \leqslant \overline{h}_0$, where $e^0 = u_\infty - \overline{u}$ and $0 < \rho_\lambda < 1$. Therefore, u_h^n converges to u_∞ as n tends to ∞ and h tends to 0.

Proof. Similar to the procedures in (11.96), by using (11.109) and (11.113), we have

$$\begin{aligned}
\|\eta_h^{n+1} - \zeta_h^{n+1}\|_{H^s(\partial\Omega)} &\leqslant \widetilde{C}_{\frac{1}{2},s} h^{\frac{1}{2}-s} \|\zeta_h^{n+1}\|_{H^{\frac{1}{2}}(\partial\Omega)} \\
&\leqslant (\widetilde{C}_{\frac{1}{2},s} \widetilde{C}_{\frac{1}{2}}) h^{\frac{1}{2}-s} [C\|u_\infty\|_{H^2(\Omega)} + \overline{\overline{C}}_0 M_\lambda \alpha_\lambda^n \|e^0\|_{L^2(\Omega)} \\
&\quad + \overline{\overline{C}}_0 \|F_\lambda(\cdot, u_h^n) - F_\lambda(\cdot, u_n)\|_{L^2(\Omega)}] \\
&\leqslant (\widetilde{C}_{\frac{1}{2},s} \widetilde{C}_{\frac{1}{2}}) h^{\frac{1}{2}-s} [C\|u_\infty\|_{H^2(\Omega)} + \overline{\overline{C}}_0 M_\lambda \alpha_\lambda^n \|e^0\|_{L^2(\Omega)} \\
&\quad + \overline{\overline{C}}_0 M_\lambda \overline{K} h^2 (\|u_\infty\|_{H^2(\Omega)} + \rho_\lambda^n \|e^0\|_{L^2(\Omega)}] \\
&\leqslant C_1 h^{\frac{1}{2}-s} [\|u_\infty\|_{H^2(\Omega)} + \rho_\lambda^n \|e^0\|_{L^2(\Omega)}],
\end{aligned} \tag{11.120}$$

where $-(\ell+1) \leqslant s \leqslant 1/2$. From here we then have, cf. the proof of Lemma 11.6,

$$\begin{aligned}
\|u_h^{n+1} - w_h^{n+1}\|_{H^{s+\frac{3}{2}}(\Omega)} &= \|\widetilde{S}_\lambda(\eta_h^{n+1} - \zeta_h^{n+1})\|_{H^{s+\frac{3}{2}}(\Omega)} \leqslant \widetilde{C}_s \|\eta_h^{n+1} - \zeta_h^{n+1}\|_{H^s(\partial\Omega)} \\
&\leqslant C_2 h^{\frac{1}{2}-s} [\|u_\infty\|_{H^2(\Omega)} + \rho_\lambda^n \|e^0\|_{L^2(\Omega)}] \quad \text{(by (11.120))}.
\end{aligned} \tag{11.121}$$

Let $r = s + \frac{3}{2}$ in (11.121). Then

$$\|u_h^{n+1} - w_h^{n+1}\|_{H^r(\Omega)} \leqslant C_2 h^{2-r} [\|u_\infty\|_{H^2(\Omega)} + \rho_\lambda^n \|e^0\|_{L^2(\Omega)}], \tag{11.122}$$

where $-\ell + \frac{1}{2} \leqslant r \leqslant 2$.

For $0 \leqslant r \leqslant 2$, we have

$$\begin{aligned}\|w_h^{n+1} - u_{n+1}\|_{H^r(\Omega)} &\leqslant \|w_h^{n+1} - u_{n+1}\|_{H^2(\Omega)} \\ &\leqslant \overline{C}_0 \|F_\lambda(\cdot, u_h^n) - F_\lambda(\cdot, u_n)\|_{L^2(\Omega)} \\ &\quad \text{(by (11.17) and (11.98))} \\ &\leqslant \overline{C}_0 M_\lambda \|u_h^n - u_n\|_{L^2(\Omega)} \qquad \text{(by (11.113))} \\ &\leqslant (\overline{C}_0 M_\lambda K) h^2 [\|u_\infty\|_{H^2(\Omega)} + \rho_\lambda^n \|e^0\|_{L^2(\Omega)}] \quad \text{(by (11.109)).}\end{aligned} \tag{11.123}$$

Combining (11.122) and (11.123), we obtain

$$\|u_h^{n+1} - u_{n+1}\|_{H^r(\Omega)} \leqslant C_3 h^{2-r} [\|u_\infty\|_{H^2(\Omega)} + \rho_\lambda^n \|e^0\|_{L^2(\Omega)}], \qquad 0 \leqslant r \leqslant 2, \\ \forall n = 0, 1, 2, \ldots, 0 < h \leqslant \overline{h}_0. \tag{11.124}$$

Also,

$$\begin{aligned}\|u_{n+1} - u_\infty\|_{H^r(\Omega)} &\leqslant \|u_{n+1} - u_\infty\|_{H^2(\Omega)} \\ &\leqslant C_\lambda \|F_\lambda(\cdot, u_h) - F_\lambda(\cdot, u_\infty)\|_{L^2(\Omega)} \quad \text{(by (11.86))} \\ &\leqslant C_\lambda M_\lambda \|u_n - u_\infty\|_{L^2(\Omega)} \\ &\leqslant C_\lambda M_\lambda \alpha_\lambda^n \|e^0\|_{L^2(\Omega)} \qquad \text{(by (11.84)).}\end{aligned} \tag{11.125}$$

Then, upon noting $\alpha_\lambda < \rho_\lambda$, we see that (11.119) follows from (11.124) and (11.125). □

Let μ_∞ be the limit of the simple-layer densities $\{\mu_n\}_1^\infty$ given in Corollary 11.2. We give the error estimate $\|\eta_h^n - \mu_\infty\|$ in the following theorem.

Theorem 11.5. *Let $N = 2$ or 3. Let $\{\mathscr{S}_h \mid 0 < h \leqslant h_0\}$ be a family of (ℓ, m)-systems with $\ell \geqslant 2$ and $m \geqslant 1$. Then for s: $-\frac{3}{2} \leqslant s \leqslant \frac{1}{2}$, there are constants $K_1 > 0$ and $K_2 > 0$ such that*

$$\|\eta_h^n - \mu_\infty\|_{H^s(\partial\Omega)} \leqslant K_1 h^{\frac{1}{2}-s} \|u_\infty\|_{H^2(\Omega)} + K_2 \rho_\lambda^n \|e^0\|_{L^2(\Omega)}, \\ \forall n = 0, 1, 2, \ldots, \quad h\colon\ 0 < h \leqslant \overline{h}_0. \tag{11.126}$$

Proof. First, from (11.91) and (11.68), we have that for $s \leqslant 1/2$,

$$\begin{aligned}\|\zeta_h^{n+1} - \mu_n\|_{H^s(\partial\Omega)} &= \|S_\lambda^{-1}[\mathscr{V}_{\lambda,\Gamma}(F_\lambda(\cdot, u_h^n) - F_\lambda(\cdot, u_n))]\|_{H^s(\partial\Omega)} \\ &\leqslant \widetilde{\widetilde{C}}_{s+1} \|\mathscr{V}_{\lambda,\Gamma}[F_\lambda(\cdot, u_h^n) - F_\lambda(\cdot, u_n)]\|_{H^{\frac{3}{2}}(\partial\Omega)} \\ &\quad \text{(for any } s \in \mathbb{R}\text{, by (11.77))} \\ &\leqslant K \|u_h^n - u_n\|_{L^2(\Omega)} \qquad \text{(by (11.72) and (11.113))} \\ &\leqslant K' h^2 [\|u_\infty\|_{H^2(\Omega)} + \rho_\lambda^n \|e^0\|_{L^2(\Omega)}], \text{ from (11.109).}\end{aligned} \tag{11.127}$$

Also, from (11.120),

$$\|\eta_h^{n+1} - \zeta_h^{n+1}\|_{H^s(\partial\Omega)} \leqslant K'' h^{\frac{1}{2}-s}[\|u_\infty\|_{H^2(\Omega)} + \rho_\lambda^n \|e^0\|_{L^2(\Omega)}], \quad -(\ell+1) \leqslant s \leqslant \frac{1}{2}. \tag{11.128}$$

Similarly, from (11.68), (11.69), and the proofs of (11.127) and Lemma 4.1, we have

$$\|\mu_n - \mu_\infty\|_{H^s(\partial\Omega)} \leqslant K''' \alpha_\lambda^n \|e^0\|_{L^2(\Omega)} \quad \text{for} \quad s \leqslant \frac{1}{2}. \tag{11.129}$$

From (11.127)–(11.129) and the triangle inequality, (11.126) follows. □

In our proofs above, for the $S_h^{\ell,m}(\partial\Omega)$-systems we have used, because of the need of the inverse inequality the requirement that $|\alpha| \leqslant m$ (cf. the proof of Lemma 11.2 for α), $\alpha = -\frac{1}{2}$, leading to $m \geqslant 1$. Thus piecewise constant boundary element spaces $S_h^{1,0}(\partial\Omega)$ are excluded. However, this restriction may be relaxed for $N = 2$ if we use S_h^d, the space of smoothest splines of degree d with respect to a quasi-uniform mesh on $\partial\Omega$, which is a smooth Jordan curve in $\mathbb{R}^2$. In this case the approximation property (11.78) holds for all $s \leqslant t \leqslant d+1, s < d+\frac{1}{2}$, and the inverse property (11.79) holds for $s \leqslant t < d+\frac{1}{2}$ [141]. By [162, Corollary 4], in (11.94) we now have

$$\|\eta_h^{n+1} - \zeta_h^{n+1}\|_{H^s(\partial\Omega)} \leqslant C h^{t-s} \|\zeta_h^{n+1}\|_{H^t(\partial\Omega)} - (d+2) \leqslant s \leqslant t \leqslant d+1,$$
$$s < d+\frac{1}{2}, \quad t > -\left(d+\frac{3}{2}\right), \quad d \geqslant 0. \tag{11.130}$$

Therefore, all of the lemmas and theorems after Lemma 11.2 in this section remain valid for these S_h^d systems with a quasiuniform mesh on $\partial\Omega$ when $N = 2$. In particular, we state the following.

Theorem 11.6. *Let $N = 2$ and let $\{\mathscr{S}_h \mid 0 < h \leqslant h_0\}$ be a family of S_h^d spaces of smoothest splines of degree d ($\in \mathbb{Z}^+$) with respect to a quasiuniform mesh on $\partial\Omega$. Under otherwise the same assumptions as in Theorems 11.4 and 11.5, we have $K_i > 0$, $i = 1,2,3,4$, and $\overline{h}_0$: $0 < \overline{h}_0 \leqslant h_0, 0 < \rho_\lambda < 1$, such that*

$$\|u_h^n - u_\infty\|_{H^2(\Omega)} \leqslant K_1 h^{2-r} \|u_\infty\|_{H^2(\Omega)} + K_2 \rho_\lambda^n \|e^0\|_{L^2(\Omega)},$$
$$\|\eta_h^n - u_\infty\|_{H^s(\partial\Omega)} \leqslant K_3 h^{\frac{1}{2}-s} \|u_\infty\|_{H^2(\Omega)} + K_4 \rho_\lambda^n \|e^0\|_{L^2(\Omega)}$$

(K_3 and K_4 depend on s),

$\forall r$: $0 \leqslant r \leqslant 2$, $n = 0, 1, 2, \ldots$, h: $0 < h \leqslant \overline{h}_0$; s: $-(d+2) \leqslant s < 1/2$. □

Remark 11.4. The error estimates obtained in Theorems 11.4–11.6 are all of the separable form

$$K_1 h^\gamma \|u_\infty\|_{H^2(\Omega)} + K_2 \rho_\lambda^n \|e^0\|_{L^2(\Omega)}, \qquad \gamma > 0.$$

It is obvious that these estimates are asymptotically tight. □

Remark 11.5. Error estimates of $\|u_h^n - u_\infty\|_{H^r(\Omega)}$ with respect to higher-order Sobolev space norms $H^r(\Omega)$, with $r > 2$, and for space dimension $N > 3$, are possible if we make an assumption like

$$\|F_\lambda(\cdot,u_1(\cdot)) - F_\lambda(\cdot,u_2(\cdot))\|_{H^m(\Omega)} \leqslant \widetilde{M}_\lambda \|u_1 - u_2\|_{H^m(\Omega)},$$
$$m \geqslant 1,\ \forall u_1,u_2 \in H^m(\Omega), \tag{11.131}$$

for some sufficiently small $\widetilde{M}_\lambda$, to supersede (11.90). But (11.131) appears unnatural, and leads to cumbersome notation. So we do not go into this. □

11.5 Higher than regular-order error estimates for nonlinearities that are separable

A special feature of boundary elements for nonlinear PDEs is that for an important class of nonlinearities, one can obtain higher than "regular order" error estimates with respect to the h-parameter. Let us assume that $F(x,u)$ in (11.1) is separable, of the form

$$F(x,u) = a_0(x)f(u) + f_0(x), \tag{11.132}$$

where $a_0, f_0 \in C^\infty(\overline{\Omega})$, and $f \in C^\infty(\mathbb{R})$, and that $g(x) \equiv 0$ on $\partial\Omega$.

Lemma 11.7. *Let* $\{\mathscr{S}_h \mid 0 < h \leqslant h_0\}$ *be* $S_h^{\ell,m}(\partial\Omega)$*-systems with* $\ell \geqslant 2$ *and* $m \geqslant 1$ *when* $N = 2$ *or 3, or be* S_h^d*-systems of smoothest splines of degree d with quasi-uniform mesh when* $N = 2$. *Then there are* $\overline{h}_0\colon\ 0 < \overline{h}_0 \leqslant h_0$, $\rho_\lambda\colon\ 0 < \rho_\lambda < 1$, *and constants* $K_1 > 0$, $K_2 > 0$ *such that*

$$\|\eta_h^{n+1} - \zeta_h^{n+1}\|_{H^{-3/2}(\partial\Omega)} \leqslant K_1 h^{2+\sigma}[\|u\|_{H^2(\Omega)} + \rho_\lambda^n \|e^0\|_{L^2(\Omega)} + K_2],$$
$$\forall \sigma\colon\ 0 < \sigma < 1, \tag{11.133}$$

for all $0 < h \leqslant \overline{h}_0, n = 0,1,2,\ldots$, *provided that* $F(x,u)$ *is of the form* (11.132), *and* $g \equiv 0$. *In particular,* $K_2 = 0$ *if* $a_0(x)$ *and* $f_0(x)$ *in* (11.132) *are constant functions.*

Proof. We first recall the Sloboditskii norm $\|u\|_\sigma^2 \equiv |u|_\sigma^2 + \|u\|_{L^2(\Omega)}^2$, where

$$|u|_\sigma^2 = \iint\limits_{\Omega\times\Omega} \frac{|u(x)-u(y)|^2}{|x-y|^{N+2\sigma}} dxdy$$

for $0 < \sigma < 1$. It is known that the Sloboditskii norm is equivalent to the Sobolev space norm $\|\cdot\|_{H^\sigma(\Omega)}^2$. From (11.132), we have, for $a - \delta \leqslant u(x) \leqslant b + \delta, x \in \Omega$,

$$\begin{aligned}|F_\lambda(x,u(x)) - F_\lambda(y,u(y))| = |&\lambda[u(x)-u(y)] - [a_0(x)-a_0(y)]f(u(x)) \\ &- a_0(y)[f(u(x)) - f(u(y))] + [f_0(x) - f_0(y)]|,\end{aligned}$$

from which it follows that

$$|F_\lambda(\cdot,u)|_\sigma \leqslant \lambda|u|_\sigma + \|f\|_{C[a-\delta,b+\delta]}|a_0|_\sigma + \|a_0\|_{C^0(\overline{\Omega})}\|f'\|_{C^0[a-\delta,b+\delta]}|u|_\sigma + |f_0|_\sigma$$
$$\equiv C_1|u|_\sigma + C_2, \qquad 0<\sigma<1. \tag{11.134}$$

Because of (11.118), we can get

$$\|F_\lambda(\cdot,u_h^n) - F_\lambda(\cdot,u_\infty)\|_{L^2(\Omega)} \leqslant M_\lambda\|u_h^n - u_\infty\|_{L^2(\Omega)}. \tag{11.135}$$

Similarly as in the proof of Lemma 11.4, we obtain

$$\begin{aligned}
&\|\eta_h^{n+1} - \zeta_h^{n+1}\|_{H^{-3/2}(\partial\Omega)} \leqslant C_3h^{2+\sigma}\|\zeta_h^{n+1}\|_{H^{\frac{1}{2}+\sigma}(\partial\Omega)}\\
&\leqslant C_4h^{2+\sigma}\|F_\lambda(\cdot,u_h^n)\|_{H^\sigma(\Omega)} \qquad \text{(because } g=0 \text{ in (11.91))}\\
&\leqslant C_4h^{2+\sigma}[|F_\lambda(\cdot,u_h^n)|_\sigma + \|F_\lambda(\cdot,u_h^n) - F_\lambda(\cdot,u_\infty)\|_{L^2(\Omega)} + \|F_\lambda(\cdot,u_\infty)\|_{L^2(\Omega)}]\\
&\leqslant C_4h^{2+\sigma}[C_1|u_h^n|_\sigma + C_2 + M_\lambda\|u_h^n - u_\infty\|_{L^2(\Omega)} + C_5\|u_\infty\|_{H^2(\Omega)}]\\
&\quad \text{(by (11.134)) and (11.135))}\\
&\leqslant C_4h^{2+\sigma}[C_6(\|u_h^n - u_\infty\|_{H^\sigma(\Omega)} + C_7\|u_\infty\|_{H^2(\Omega)}) + C_2]\\
&\leqslant C_8h^{2+\sigma}[\|u_\infty\|_{H^2(\Omega)} + \rho_\lambda^n\|e^0\|_{L^2(\Omega)} + C_9] \quad \text{(by Theorem 11.4).}
\end{aligned}$$

□

Now, the following is obvious.

Theorem 11.7. *Assume the same conditions as in Lemma 11.7. Then there are three constants $C_i>0$, $i=1,2,3$, such that*

$$\|u_h^n - u_\infty\|_{H^r(\Omega)} \leqslant C_1\rho_\lambda^n\|e^0\|_{L^2(\Omega)} + C_2h^{2-r+\sigma}(\|u_\infty\|_{H^2(\Omega)} + C_3)$$
$$\forall h:\ 0<h\leqslant \overline{h}_0, \quad n=0,1,2,\ldots, \quad 0\leqslant\sigma<1, \quad 0\leqslant r\leqslant 2.$$

In particular, $C_3=0$ if $a_0(x)$ is a constant function and $f_0(x)\equiv 0$ in (11.132). □

11.6 Neumann and Robin boundary conditions

The treatments in § 11.4 and § 11.5 deal with the Dirichlet boundary condition. We now consider the case when the boundary condition is

$$\frac{\partial}{\partial n}u(x) + \alpha(x)u(x) = g(x), \qquad x\in\partial\Omega, \tag{11.136}$$

where $\alpha(x)\geqslant 0$, $\alpha\in C^\infty(\partial\Omega)$. The condition (11.136) corresponds to a Neumann or a Robin boundary value problem depending on whether $\alpha(x)$ is or is not identically equal to zero.

For (11.136), we formulate a corresponding assumption similar to ([H]). Let λ_1' be the smallest eigenvalue of the operator $\big(-\Delta, [\frac{\partial}{\partial n}+\alpha(\cdot)]\big|_\Gamma\big)$. Then $\lambda_1'=0$ if $\alpha(x)\equiv 0$ and $\lambda_1'>0$ if $\alpha(x)\geqslant 0$ but $\alpha(x)\not\equiv 0$. We assume:

([H])$'$ There exist constants γ_1' and γ_2' such that

$$-\lambda_1' < \gamma_1' \leqslant \frac{\partial F(x,u)}{\partial u} \leqslant \gamma_2', \qquad \forall (x,u)\in \overline{Q}\equiv \overline{\Omega}\times[a,b].$$

Similar to (11.83), we now have that for a sufficiently small $\delta>0$, there exist $\widetilde{\gamma}_1'$ and $\widetilde{\gamma}_2'$ such that

$$-\lambda_1' < \widetilde{\gamma}_1' \leqslant \frac{\partial F(x,u)}{\partial u} \leqslant \widetilde{\gamma}_2', \qquad \forall (x,u)\in \overline{Q}_\delta\equiv \overline{\Omega}\times[a-\delta,b+\delta]. \tag{11.137}$$

Let $\lambda>0$ be chosen sufficiently large such that

$$M_\lambda' \equiv \sup_{(x,u)\in\overline{Q}_\delta}\left[\lambda-\frac{\partial F(x,u)}{\partial u}\right] > 0. \tag{11.138}$$

Then

$$0 < M_\lambda' < \lambda - \widetilde{\gamma}_1'. \tag{11.139}$$

Lemma 11.8. *Let the assumptions in Theorem 2.9 and* (11.136)–(11.139) *hold. Let* u_{n+1} *be iterated according to*

$$\begin{cases} (\Delta-\lambda)u_{n+1}(x) = F_\lambda(x,u_n(x)), \qquad x\in\Omega, \\ \dfrac{\partial}{\partial n}u_{n+1}(x)+\alpha(x)u_{n+1}(x) = g(x),\ x\in\partial\Omega,\ g\in C^\infty(\partial\Omega), \end{cases} \tag{11.140}$$

and let u_∞ *be the limit as in* (11.20). *Denote* $e^n=u_\infty-u_n$. *Then we have*

$$\|e^{n+1}\|_{L^2(\Omega)} \leqslant \alpha_\lambda'^{\,n+1}\|e^0\|_{L^2(\Omega)}, \qquad n=0,1,2,\ldots,$$

where

$$0<\alpha_\lambda' \equiv \frac{\lambda-\widetilde{\gamma}_1'}{\lambda+\lambda_1'}<1.$$

Proof. Refer to the proof of Lemma 11.1. Instead of (11.86), we now have

$$\begin{cases} (\Delta-\lambda)e^{n+1}(x) = F_\lambda(x,u_\infty(x))-F_\lambda(x,u_n(x)),\ x\in\Omega, \\ \dfrac{\partial}{\partial n}e^{n+1}(x)+\alpha(x)e^{n+1}(x)=0, \qquad x\in\partial\Omega. \end{cases}$$

Multiplying the above by $-e^{n+1}$ and integrating by parts, we get

$$\int_{\partial\Omega}\alpha|e^{n+1}|^2 d\sigma + \int_\Omega |\nabla e^{n+1}|^2 dx + \lambda\int_\Omega |e^{n+1}|^2 dx \\ \leqslant \|F_\lambda(\cdot,u_\infty)-F_\lambda(\cdot,u_n)\|_{L^2(\Omega)}\|e^{n+1}\|_{L^2(\Omega)}.$$

But

$$\inf_{v\in H^1(\Omega)} \frac{\left[\int_{\partial\Omega}\alpha|v|^2 d\sigma + \int_{\Omega}|\nabla v|^2 dx\right]}{\|v\|^2_{L^2(\Omega)}} = \lambda_1'.$$

Thus,

$$(\lambda+\lambda_1')\|e^{n+1}\|^2_{L^2(\Omega)} \leqslant \|F_\lambda(\cdot,u_\infty)-F_\lambda(\cdot,u_n)\|_{L^2(\Omega)}\|e^{n+1}\|_{L^2(\Omega)}.$$

The rest of the arguments is the same as in the proof of Lemma 11.1. □

Proceeding as in § 11.4, we now note where the places are that need to be adjusted to accommodate the boundary condition (11.136). The iterate $u_{n+1}(\cdot)$ satisfying (11.140) is still represented as in (11.29), but η_{n+1} is determined from the BIE

$$\begin{aligned}\mathscr{B}_\lambda u_{n+1}(x) &\equiv \frac{1}{2}\eta_{n+1}(x) + \int_{\partial\Omega}\left[\frac{\partial}{\partial n_x}E(x-y;\lambda)+\alpha(x)E(x-y;\lambda)\right]\eta_{n+1}(y)d\sigma_y \\ &= g(x)+\alpha(x)(\mathscr{V}_{\lambda,\Gamma}F(\cdot,u_n))(x)+\frac{\partial}{\partial n}(\mathscr{V}_\lambda F(\cdot,u_n))(x), \quad x\in\partial\Omega, \qquad (11.141)\end{aligned}$$

instead of (11.68). Note that $\mathscr{B}_\lambda$ is a strongly elliptic pseudodifferential operator of order 0 ($=2\alpha$ according to the notation in [97]) with principal symbol $\frac{1}{2}$ such that

$$\mathscr{B}_\lambda:\ H^s(\partial\Omega)\longrightarrow H^s(\partial\Omega) \quad \text{isomorphically}, \quad \forall s\in\mathbb{R}. \qquad (11.142)$$

Therefore, (11.141) is uniquely solvable.

The Galerkin boundary element method for (11.140) is almost the same as in (11.80)–(11.82), except that (11.81) is replaced by

(ii)′ For each $n=0,1,2,\ldots$, and $h:\ 0<h\leqslant h_0$, find $\eta_h^{n+1}\in\mathscr{S}_h$ such that

$$\langle\mathscr{B}_\lambda\eta_h^{n+1},\phi_h\rangle_{L^2(\partial\Omega)} = \langle *,\phi_h\rangle_{L^2(\partial\Omega)}, \qquad \forall\phi_h\in\mathscr{S}_h,$$

where $*$ denotes the sum in the last equality of (11.141), but with u_h^n taking the place of u_h. Next, we let ζ_h^{n+1} be the unique solutions of the BIE

$$(\mathscr{B}_\lambda\zeta_h^{n+1})(x) = g(x)+\alpha(x)(\mathscr{V}_{\lambda,\Gamma}F(\cdot,u_h^n))(x)+\frac{\partial}{\partial n}(\mathscr{V}_\lambda F(\cdot,u_h^n))(x), \quad x\in\partial\Omega,$$

(instead of (11.91))

from which we define $w_h^{n+1}(x)$ just as in (11.97).

It is now routine to check that beginning from (11.91) in Lemma 11.2 and ending at (11.130), all the estimates remain valid. Indeed, we obtain the same results under a *less restrictive assumption* on $S_h^{\ell,m}(\partial\Omega)$ – here we only require that $\ell\geqslant 1$ (and $m\geqslant 0$) because 2α, the order of $\mathscr{B}_\lambda$, is zero. Thus the requirement that $|\alpha|\leqslant m$ made in [97, Corollary 2.1] and [162, Corollary 3] is now automatically satisfied with $m\geqslant 0$.

11.7 Numerical examples

The theory developed so far was molded by computations of concrete examples of semi-linear PDEs. In the following, we will present several such examples featuring various situations, with numerical solutions illustrated by computer graphics.

The steps that we take are indicated by a flowchart next. For definiteness, we only consider the Dirichlet boundary condition for $\Omega \equiv D_1$, the unit open disk in $\mathbb{R}^2$. The boundary value problem is of the form

FLOW-CHART

The Monotone Iteration Algorithm (MIA)

(1) Initialize: Take an initial guess $w_0(x), x \in \Omega$.

Set $j = 0$. Select $\lambda > 0$ to satisfy (11.13).

(2) Compute the volume potential

$$V_{\lambda,j}(x) \equiv \int_\Omega E(x-y)[\lambda w_j(y) - F(y, w_j(y))]dy, \qquad x \in \overline{\Omega}.$$

(3) Solve the unknown simple-layer density $\eta_{j+1}(\cdot)$ from the boundary integral equation

$$\int_{\partial\Omega} E(x-y)\eta_{j+1}(y)d\sigma_y = g(x) - \int_\Omega E(x-y)[\lambda w_j(y) - F(y, w_j(y))]dy, x \in \partial\Omega.$$

(4) Compute the solution

$$w_{j+1}(x) = \int_{\partial\Omega} E(x-y)\eta_{j+1}(y)d\sigma_y + \int_\Omega E(x-y)[\lambda w_j(y) - F(y, w_j(y))]dy, x \in \overline{\Omega}.$$

(5) If $\|w_{j+1} - w_j\|_{L^2(\Omega)} < \varepsilon$, exit. Output and stop.

(6) Else $j := j+1$;

continue.

$$\begin{cases} \Delta u(x) - F(x, u(x)) = 0, & x \in D_1, \\ u(x) = g(x), & x \in \partial D_1. \end{cases} \tag{11.143}$$

Note that at Steps 2 and 4 of the flowchart, the volume potential $V_{\lambda,j}(x)$ and the iterate $w_{j+1}(x)$ are computed by quadratures of high accuracy with 864 Gauss points placed on the unit disk. The Gauss points are selected by using the Cartesian Gauss rule for the polar coordinate variables (r,θ): dividing the unit disk into $4_r \times 24_\theta$ elements and using $3_r \times 3_\theta$ Gauss points in each element. The key step is the solution of the boundary integral equation for the simple-layer density η_{j+1} in Step 3. In our numerical work, η_{j+1} is solved *by point collocation* using piecewise constant boundary elements (i.e., $S_h^{1,0}(\partial\Omega)$-systems) with uniform mesh on $\partial\Omega$, where $\partial\Omega$ is divided into 64 equal pieces of arc (i.e., $h = 2\pi/64$). Here collocation is taken as a measure of convenience, because the Galerkin boundary element method as stated in (11.80) and (11.82) involve significantly more programming work. It is also for this reason that in this section we have not attempted to confirm the error estimates given in § 11.5 and § 11.6 experimentally. We nevertheless hope that the examples furnished here will serve as useful benchmarks for researchers doing similar work, and that in the future we will also be able to derive and numerically verify error estimates for the boundary element point-collocation scheme.

In all our numerical computations, $\varepsilon = 10^{-4}$ has been used as the relative L_2 error in solutions to terminate the iteration, see Flow-chart Step 5.

We first test the accuracy of our computer program this way. Choose a known (smooth) function $y(x), x \in \overline{\Omega}$, and for a somewhat arbitrary function F, where F is nonlinear in u, we compute the solution of

$$\begin{cases} \Delta u(x) - F(x,u(x)) = 0, & x \in \Omega, \\ u(x) = y(x), & x \in \partial\Omega, \end{cases}$$

where

$$F(x,u(x)) = F(u(x)) + [\Delta y(x) - F(y(x))], \quad x \in \Omega.$$

Obviously, $y(x)$ is an *exact solution* of the (artificially set example) above. We then use $y(x)$ as a benchmark to test whether we can numerically recover $u(x) = y(x)$. The computer results are affirmative: the output data have shown high accuracy.

In the following, we provide numerical examples which do not have exact solutions available. The accompanying theoretical discussions are to a large extent motivated by visualization of the numerical results.

Example 11.2. Multiple solutions – two stable solutions bounding an unstable one.

Consider

$$\begin{cases} \Delta u(x) - \gamma u(x)[u(x) - a][u(x) - 1] = 0, & x \in D_1, \\ u(x) = 0, & x \in \partial D_1, \end{cases} \tag{11.144}$$

where $\gamma > 0$, $0 < a < 1/2$. It is easy to see that $u(x) = \overline{v}(x) \equiv 0$ is a solution. One can further establish that $\overline{v}(x) \equiv 0$ is *stable*. This solution is of less interest because it is *trivial*. Substituting $\overline{u}(x) \equiv 1$ for u into (11.144), we also easily verify that $\overline{u}(x)$ is a supersolution. (Actually, $\overline{u}(x) \equiv C$ for any $C \geqslant 1$ will do.)

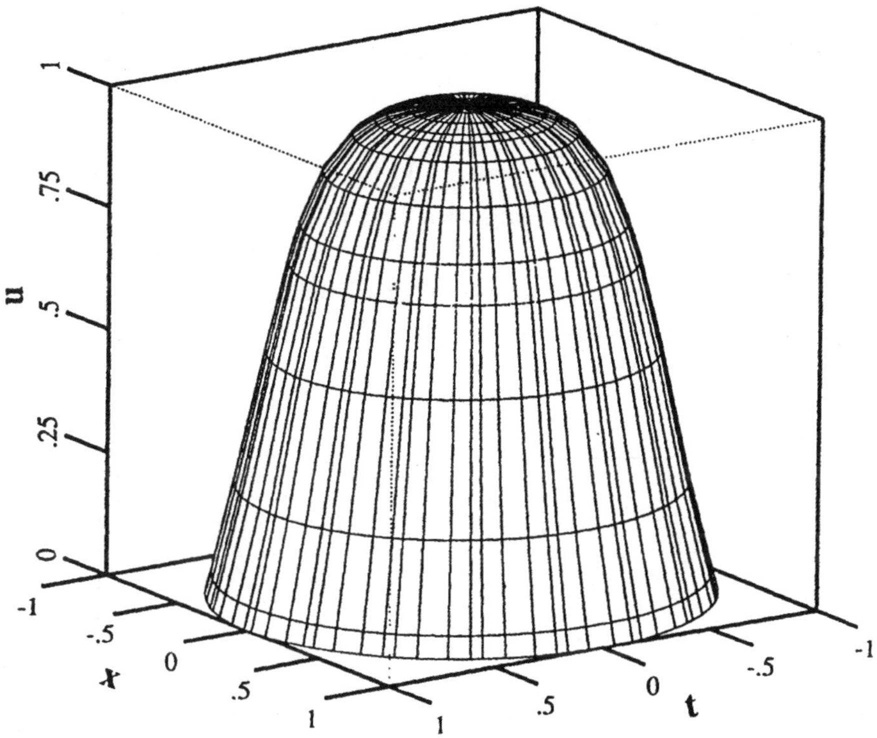

Figure 11.1: A bell-shaped stable positive boundary element numerical solution $u(x)$ satisfying $\Delta u - u(u - \frac{1}{4})(u-1) = 0$ and the zero Dirichlet boundary condition for Example 11.2.

In (11.144), we have

$$F(x,u) \equiv F(u) = \gamma u(u-a)(u-1), \tag{11.145}$$

$$\frac{\partial F}{\partial u} = \gamma[(u-a)(u-1) + u(u-1) + u(u-\alpha)]$$

$$= \gamma[3u^2 - 2(a+1)u + a].$$

In our computation, we choose

$$\gamma = 100, \quad a = \frac{1}{4}, \quad \text{yielding} \quad \max \frac{\partial F}{\partial u} = 75, \quad u \in [0,1].$$

Thus, if we choose $\overline{u}(x) \equiv 1$, $\lambda = 75.69$, then (11.31) is satisfied.

The monotone scheme beginning from $u_0(x) = \overline{u}(x) \equiv 1$ takes 30 iterations, yielding a bell-shaped function $u_\infty(x)$ as shown in Fig. 11.1. According to Theorem 11.1, this is a stable solution. Since $u_\infty(x) \not\equiv 0 \equiv \overline{v}(x) = v_\infty(x)$, we know from Theorem 11.1 that there is at least one unstable solution of (11.144), called a "spike-layer" solution, sandwiched between v_∞ and u_∞.

Generally, it is known that for (11.144), the following are true:

(i) Any solution u of (11.144) must satisfy $0 \leqslant u(x) \leqslant 1$;

(ii) For $\gamma > 0$ sufficiently small, (11.144) has the trivial solution as its only solution;

(iii) The spike-layer solution exists only when $\gamma > 0$ is large.

Meanwhile, the antimonotone convergence of the simple-layer densities, guaranteed by Theorem 11.2, is reflected in Fig. 11.2. □

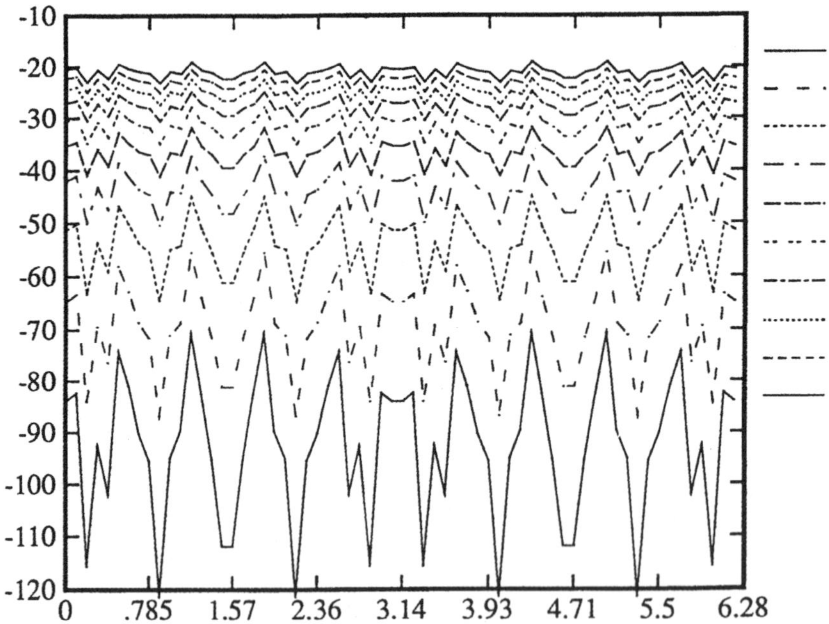

Figure 11.2: The antimonotone convergence of the simple- layer densities for Example 11.2. While the solutions $u_h^n(x)$ decrease monotonically on the unit disk, the simple layer densities $\eta_h^n(x)$ increase monotonically for $n = 1$ (bottom curve) to $n = 10$ (top curve).

Example 11.3. A monotone dissipative operator equation.

Consider

$$\begin{cases} \Delta u(x) - 4u^3(x) = \gamma x_1^2, & x = (x_1, x_2) \in D_1, \quad \gamma > 0, \\ u(x) = 0, & x \in \partial D_1. \end{cases} \tag{11.146}$$

The nonlinear operator $\mathcal{N}$ on the left-hand side of the PDE in (11.143) is known to be a *monotone dissipative operator* as it satisfies

$$\begin{aligned} \langle \mathcal{N} u_1 - \mathcal{N} u_2, u_1 - u_2 \rangle_{L^2(D_1)} &= \langle (\Delta u_1 - 4u_1^3) - (\Delta u_2 - 4u_2^3), u_1 - u_2 \rangle \\ &= \int_{D_1} [\Delta(u_1 - u_2) - 4(u_1^3 - u_2^3)](u_1 - u_2) dx \\ &= -\int [|\nabla(u_1 - u_2)|^2 + 4(u_1^3 - u_2^3)(u_1 - u_2)] dx \\ &\leqslant 0, \end{aligned}$$

for $u_1, u_2 \in H_0^2(D_1)$. From the theory of monotone dissipative operators it is straightforward ([126], e.g.) to establish that (11.146) has a *unique* solution $u \in C^\infty(\overline{D}_1)$.

From the way (11.146) is given, we can further prove that the solution satisfies $u(x) < 0$, $x \in D_1$. Assume the contrary. Then by the usual maximum principle [156] $G \equiv \{x \in D_1 \mid u(x) > 0\}$ is nonempty. Thus, on G, u satisfies

$$\begin{cases} \Delta u(x) = 4u^3(x) + \gamma x_1^2 > 0, & x \in G, \\ u(x) > 0, & x \in G, \\ u(x) = 0, & x \in \partial G. \end{cases}$$

But for u satisfying $\Delta u \geqslant 0$ on G and $u = 0$ on ∂G, the maximum principle implies that $u(x) \leqslant 0$ on G, a contradiction.

For (11.146), we have $F(x,u) = 4u^3 + \gamma x_1^2$. For D_1, it is known that λ_1, the smallest eigenvalue of $-\Delta$ corresponding to the homogeneous Dirichlet condition, is (cf. [41, p. 188], e.g.)

$$\lambda_1 \approx (2.4048)^2 \approx 5.783063.$$

Therefore,

$$\frac{\partial F}{\partial u} = 12u^2 \geqslant 0 > -\lambda_1 \approx -5.783063. \tag{11.147}$$

We may take $\overline{u}(x) \equiv 0$ as a supersolution, and $\overline{v}(x) \equiv -C$, $C > 0$, as a subsolution, if C is chosen such that $4C^3 - \gamma x_1^2 \geqslant 0$, for all $x \in \overline{D}_1$.

Because of (11.147), we easily see that ([H]) and (11.85) are satisfied for $\widetilde{\gamma}_1 < 0$ if $|\widetilde{\gamma}_1|$ is small.

In our numerical computations, we choose $\gamma = 10$ and $\lambda = 4$. The computation takes 8 iterations to converge. The numerical solution is plotted in Fig. 11.3. The reader can easily see that the solution $u(x)$ is nonpositive. □

Note that Theorem 11.7 applies to Examples 11.2 and 11.3.

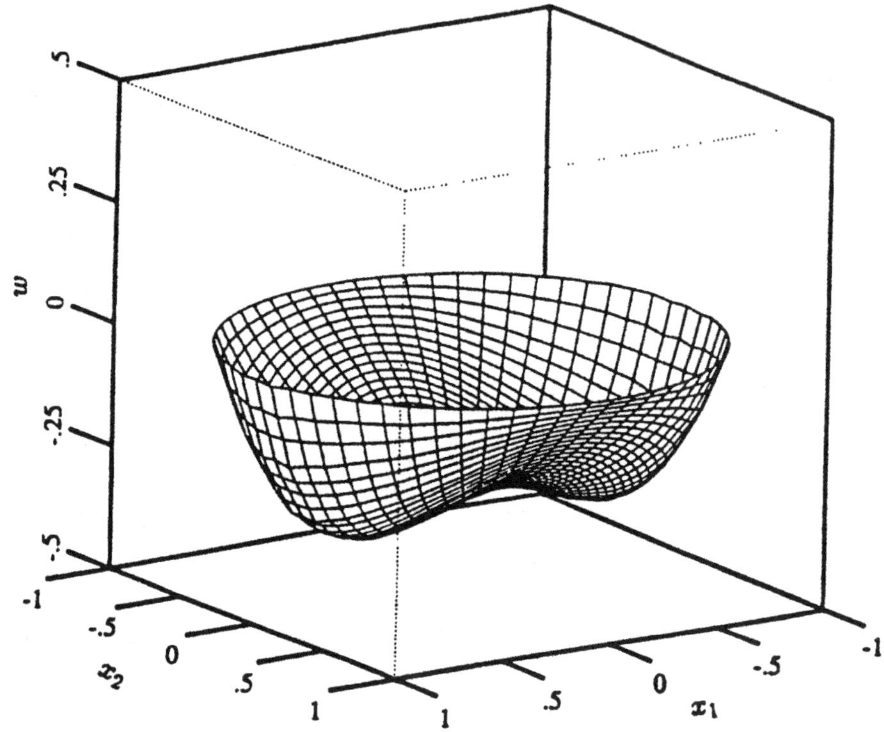

Figure 11.3: Boundary element numerical solution of $\Delta u - 4u^3 \equiv 10x_1^2$ subject to the zero Dirichlet boundary condition, Example 11.3.

Example 11.4. A nonlinear PDE of nonmonotone type whose existence or nonexistence of stable solutions of one sign depends on the size of forcing.

Consider

$$\begin{cases} \Delta u(x) + 4u^3(x) = \gamma x_1^2, & x \in D_1, \quad \gamma > 0, \\ u(x) = 0, & x \in \partial D_1. \end{cases} \tag{11.148}$$

Suppose we are interested in finding nonpositive solutions

$$u(x) \leqslant 0 \quad \text{on} \quad D_1. \tag{11.149}$$

This condition is now appended to (11.148). Later on, we will be able to remove (11.149) provided that $\gamma > 0$ is sufficiently small, and that proper supersolutions and subsolutions are chosen; see (iii) below.

We now show the following.

(i) *The system* (11.148) *and* (11.149) *has no solution if* γ *is large.*

Let ϕ_1 be the first eigenfunction of $-\Delta$ satisfying

$$\begin{cases} \Delta\phi_1 + \lambda_1\phi_1 = 0 \text{ on } D_1 (\lambda_1 \approx 5.783063), \\ \phi_1(x) > 0 \qquad \text{on } D_1, \\ \phi_1|_{\partial D_1} = 0. \end{cases} \tag{11.150}$$

Here, $\phi_1(x) > 0$ because we know that ϕ_1 is of one sign on D_1 (cf. § 7.10, Chapter 7, for some graphics of eigenfunctions). From (11.148)–(11.150), we get

$$\begin{aligned} \gamma\int_{D_1} x_1^2\phi_1(x)dx &= \int_{D_1} [\Delta u(x) + 4u^3(x)]\phi_1(x)dx \\ &= \int_{D_1} [-\lambda_1 u(x)\phi_1(x) + 4u^3(x)]\phi_1(x)dx \\ &= \int_{D_1} u(x)\phi_1(x)[4u^2(x) - \lambda_1]dx \\ &= \int_{D_{1,1}} u(x)\phi_1(x)[4u^2(x) - \lambda_1]dx \\ &\quad + \int_{D_{1,2}} u(x)\phi_1(x)[4u^2(x) - \lambda_1]dx, \end{aligned}$$

where

$$D_{1,1} = \{x \in D_1 \mid -\sqrt{\lambda_1}/2 \leqslant u(x) \leqslant 0\}, \quad D_{1,2} = \{x \in D_1 \mid u(x) < -\sqrt{\lambda_1}/2\}.$$

Hence,

$$0 < \gamma\int_{D_1} x_1^2\phi_1(x)dx \leqslant \int_{D_{1,1}} u(x)\phi_1(x)[4u^2(x) - \lambda_1]dx. \tag{11.151}$$

On $D_{1,1}$, we have $\sqrt{\lambda_1}/2 \geqslant -u(x)$, and $\lambda_1 \geqslant \lambda_1 - 4u^2(x)$. Using the above in (11.151), we obtain

$$\frac{1}{2}\int_{D_{1,1}} \sqrt{\lambda_1}\phi_1(x)\lambda_1 dx \geqslant \gamma\int_{D_1} x_1^2\phi_1(x)dx,$$

$$\gamma \leqslant \frac{\lambda_1^{3/2}\int_{D_{1,1}}\phi_1(x)dx}{2\int_{D_1} x_1^2\phi_1(x)dx} \leqslant \frac{\lambda_1^{3/2}\int_{D_1}\phi_1(x)dx}{2\int_{D_1} x_1^2\phi_1(x)dx}. \tag{11.152}$$

Thus, if $\gamma > 0$ is chosen large enough to violate (11.152), the system (11.148) and (11.149) will not have a solution.

(ii) *If the system* (11.148) *and* (11.149) *has a solution for* $\gamma = \gamma_1 > 0$, *then for all* γ: $0 < \gamma < \gamma_1$, *the system* (11.148) *and* (11.149) *also has a solution.*

Let $\widetilde{u}$ be the solution corresponding to γ_1:

$$\begin{cases} \Delta\widetilde{u}(x) + 4\widetilde{u}^3(x) = \gamma_1 x_1^2, & x \in D_1, \\ \widetilde{u}(x) \leqslant 0 & \text{on } D_1, \\ \widetilde{u}(x) = 0 & \text{on } \partial D_1. \end{cases} \tag{11.153}$$

Consider u, satisfying

$$\begin{cases} \Delta u(x) + 4u^3(x) = \gamma x_1^2, & x \in D_1, \quad 0 < \gamma < \gamma_1, \\ u(x) \leqslant 0 & \text{on } D_1, \\ u(x) = 0 & \text{on } \partial D_1. \end{cases} \tag{11.154}$$

Thus, it is easy to check that $\overline{u}(x) \equiv 0$ is a supersolution for (11.154). We also claim that $\overline{v}(x) = \widetilde{u}(x)$ is a subsolution, since

$$\begin{aligned} \Delta \overline{v}(x) + 4\overline{v}^3(x) - \gamma x_1^2 &= [\Delta \widetilde{u}(x) + 4\widetilde{u}^3(x) - \gamma_1 x_1^2] + (\gamma_1 - \gamma)x_1^2 \\ &= (\gamma_1 - \gamma)x_1^2 \geqslant 0 \quad \text{on } D_1, \end{aligned}$$

and $\overline{v}(x) = \widetilde{u}(x) = 0$ on ∂D_1. By Theorem 11.1, a solution $u(x)$ of (11.154) exists, satisfying $\widetilde{u}(x) \leqslant u(x) \leqslant 0$ on D_1. □

(iii) *The system* (11.148) *and* (11.149) *possesses a stable solution* $\widetilde{u_1}: \ \widetilde{u_1}(x) \leqslant 0$ *as well as a second solution* $\widetilde{u_2}: \ \widetilde{u_2}(x) < \widetilde{u_1}(x), x \in \Omega$, *if* $\gamma > 0$ *is sufficiently small.*
We already know that $\overline{u}(x) \equiv 0$ is a supersolution of (11.148) and (11.149). We now construct a subsolution. Let $w(x)$ satisfy the Lane-Emden equation

$$\begin{cases} \Delta w(x) + 4w^3(x) = 0, & \text{on } D_1, \\ w(x) > 0, & \text{on } D_1, \\ w(x) = 0, & \text{on } \partial D_1. \end{cases} \tag{11.155}$$

It is known (cf. [141], e.g., and see many examples in § 12.3 of Chapter 12) that (11.156) has a unique solution (that is, also radially symmetric and monotonically decreasing in $|x|$). Thus, we write $w(x) = w(r)$, $r = |x|$, $x \in D_1$. Define

$$w_\alpha(x) = \alpha w(\alpha r), \quad 0 < \alpha < 1. \tag{11.156}$$

Then it is routine to verify that $w_\alpha(x)$ satisfies

$$\Delta w_\alpha(x) + 4w_\alpha^3(x) = 0, \quad w_\alpha(x) \geqslant \varepsilon > 0, \text{ for } \varepsilon = \alpha w(\alpha) \text{ and } x \in D_1.$$

Let

$$\overline{v}(x) = -\beta w_\alpha(x), \quad 1 > \beta > 0, \text{ for some fixed } \beta. \tag{11.157}$$

Then

$$\begin{aligned} \Delta \overline{v}(x) + 4\overline{v}^3(x) - \gamma x_1^2 &= -\beta \Delta w_\alpha(x) - 4\beta^3 w_\alpha^3(x) - \gamma x_1^2 \\ &= 4\beta w_\alpha^3(x) - 4\beta^3 w_\alpha^3(x) - \gamma x_1^2 \\ &= 4\beta(1-\beta^2) w_\alpha^3(x) - \gamma x_1^2 \\ &\geqslant 4\beta(1-\beta^2)\varepsilon^3 - \gamma x_1^2 \geqslant 0, \end{aligned}$$

if $\gamma > 0$ is sufficiently small. Also,

$$\overline{v}(x) = -\beta w_\alpha(x) = -\beta\alpha w(\alpha) \leqslant -\beta\alpha\varepsilon, \quad \text{if} \quad x \in \partial D_1.$$

Thus $\overline{v}(x)$ is a subsolution, if $\gamma > 0$ is sufficiently small. Hence the system (11.148) and (11.149) admits a stable solution for $\gamma > 0$ sufficiently small. Call this solution $\widetilde{u}_1$. From (ii), we further see that $\widetilde{u}_1 \uparrow 0$ uniformly as $\gamma \downarrow 0$.

To establish the existence of the second solution $\widetilde{u_2} < \widetilde{u_1}$ (on Ω), we use a variational approach. For ease of discussion, set $z = -u$ in (11.148) and (11.149), giving

$$\begin{cases} \Delta z(x) + 4z^3(x) + \gamma {x_1}^2 = 0, & \text{on } D_1, \\ z(x) > 0, & \text{on } D_1, \\ z(x) = 0, & \text{on } \partial D_1. \end{cases} \tag{11.158}$$

Note that $z_1 = -\widetilde{u_1}$ is a solution of (11.158). Setting $w = z - z_1$ in (11.158), we obtain

$$\begin{cases} \Delta w(x) + [12{z_1}^2(x)]w(x) + [12z_1(x)]w^2(x) + 4w^3(x) = 0, & x \in D_1, \\ w(x) = 0, & x \in \partial D_1. \end{cases} \tag{11.159}$$

We now look for a solution of (11.159) that further satisfies

$$w(x) > 0, \quad x \in D_1. \tag{11.160}$$

When $\gamma > 0$ is sufficiently small, $\|z_1\|_{L^\infty(\Omega)}$ is small. Thus $\|12{z_1}^2\|_{L^\infty(\Omega)} < \lambda_1$ (cf. (11.150) for λ_1) is satisfied for small $\gamma > 0$. Then an application of the usual Mountain–Pass Lemma (see the forthcoming Theorem 12.1 in Chapter 12 and [6, 141]) yields a classical solution w of (11.159) and (11.160). Hence, $\widetilde{u}_2 \equiv -(z_1 + w) < \widetilde{u}_1 < 0$ on Ω is a second solution.

For (11.148), we have

$$F(x,u) = -4u^3 + \gamma x_1^2, \quad \frac{\partial F}{\partial u} = -12u^2.$$

Although the subsolution as given in (11.156) and (11.157) is *not known explicitly*, it can be verified that the conditions ([H]) and (11.85) are satisfied if $\lambda \geqslant 0$.

In our numerical experiments, we have chosen $\gamma = 10$ and $\lambda = 4$. The numerical iterates converge after 16 iterations. The graph of the numerical solution $\widetilde{u}_1$ can be found in Fig. 11.4.

Example 11.5. A nonlinear PDE of nonmonotone type with a unique stable nonpositive solution.

Consider

$$\begin{cases} \Delta u(x) + 4u^2(x) = \gamma x_1^2, & x \in D_1, \quad \gamma > 0, \\ u(x) = 0, & x \in \partial D_1. \end{cases} \tag{11.161}$$

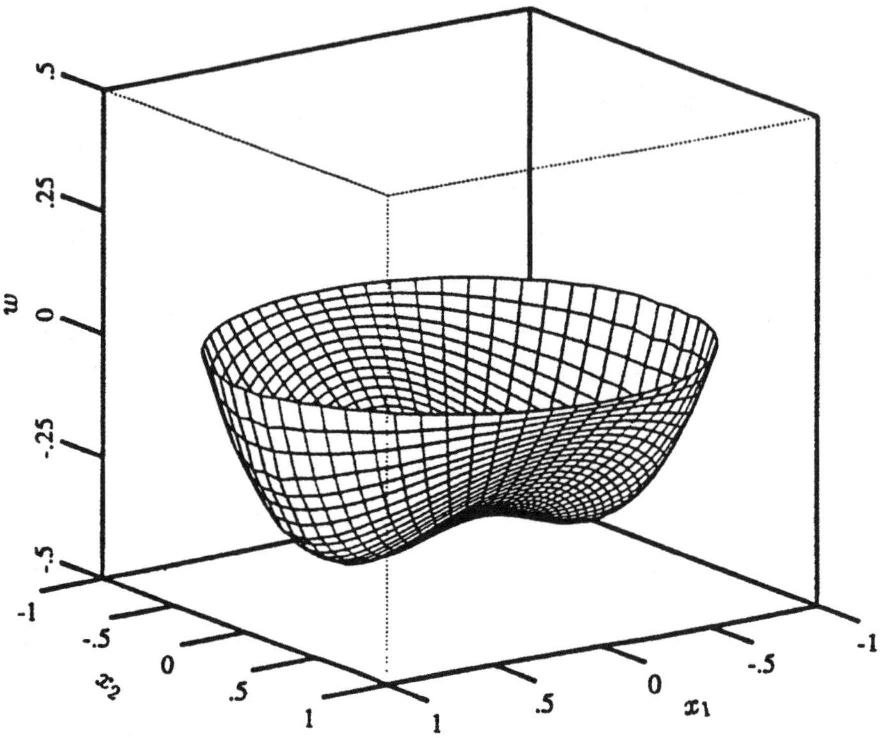

Figure 11.4: Boundary element numerical solution of $\Delta u + 4u^3 = 10x_1^2$ subject to the zero Dirichlet boundary condition, Example 11.4.

It is easy to see that $\overline{u}(x) \equiv 0$ is a supersolution, and $\overline{v}(x) \equiv -C, C > 0$, is a subsolution if C is chosen such that $4C^2 - \gamma x_1^2 \geqslant 0, \forall x_1 : \ |x_1| \leqslant 1$. Thus, by Theorem 11.1, there exists a stable solution $u(x)$ such that $-C \leqslant u(x) \leqslant 0, \forall x \in D_1$. We now show that *nonpositive* solutions of (11.161) are unique. Suppose $u^{(1)}$ and $u^{(2)}$ both are nonpositive and satisfy (11.161). We may assume that

$$u^{(1)}(x) \geqslant u^{(2)}(x) \quad \forall x \in D_1, \quad u^{(1)} \not\equiv u^{(2)}. \tag{11.162}$$

Otherwise, by Theorem 11.1 and Remark 11.2(a), $\overline{v}(x) = \min\{u^{(1)}(x), u^{(2)}(x)\}, x \in D_1$, is a subsolution of (11.161) and monotone iteration with $\overline{v}(x)$ as the initial state will yield a limit solution $u_\infty(x)$ such that $u^{(1)}(x) \geqslant u_\infty(x)$, for all $x \in D_1$, $u^{(1)} \not\equiv u_\infty$. From (11.161) and

(11.162), we have

$$\begin{aligned} 0 &= \int_{\partial D_1} \left[u^{(2)} \frac{\partial u^{(1)}}{\partial n} - u^{(1)} \frac{\partial u^{(2)}}{\partial n} \right] d\sigma = \int_{D_1} [u^{(2)} \Delta u^{(1)} - u^{(1)} \Delta u^{(2)}] dx \\ &= -\int_{D_1} [u^{(2)}(4u^{(1)^2} - \gamma x_1^2) - u^{(1)}(4u^{(2)^2} - \gamma x_1^2)] dx \\ &= -\int_{D_1} [4u^{(1)}u^{(2)}(u^{(1)} - u^{(2)}) + \gamma x_1^2 (u^{(1)} - u^{(2)})] dx < 0, \end{aligned}$$

a contradiction. □

Thus, in contrast to Example 11.4, the size of γ here does not affect the existence or nonexistence of the unique stable nonpositive solution. (Nevertheless, we do not rule out that some *other stable solutions not of one sign may still exist.*)

For (11.161), we have

$$F(x,u) = -4u^2 + \gamma x_1^2, \quad \frac{\partial F}{\partial u} = -8u. \tag{11.163}$$

Note that we need only consider $u \leqslant 0$. Here, to ensure that ([H]) and (11.85) are satisfied, we need only use $\lambda = 0$.

In our numerical computation, we choose $\gamma = 10$ in (11.161) and use $\lambda = 4$. The numerical solutions converge after 9 iterations. The limiting solution is plotted in Fig. 11.5.

Although the PDEs in Examples 11.3, 11.4 and 11.5 seem to have vastly different nonlinearities in a qualitative sense, we nevertheless notice that the solution graphics in Figs. 11.3, 11.4 and 11.5 seem to display similar profiles.

Example 11.6. A nonlinear PDE of nonmonotone type, whose stable solutions, not of one sign, may be obtained through symmetry.

Consider

$$\begin{cases} \Delta u(x) + 4u^3(x) = \gamma x_1^3, & x \in D_1, \quad \gamma > 0, \\ u(x) = 0, & x \in \partial D_1. \end{cases} \tag{11.164}$$

Here the forcing term γx_1^3 changes sign as x moves from $D_{1,-}$ into $D_{1,+}$:

$$D_{1,-} \equiv \{x \in D_1 \mid x_1 < 0\}, \quad D_{1,+} \equiv \{x \in D_1 \mid x_1 > 0\}.$$

Note that, if $u(x)$ is a solution of (11.164), then so is $-u(-x_1,x_2)$, i.e., u is antisymmetric with respect to the x_2-axis, and thus we deduce that $u(x) = 0$, $x \in \overline{D}_{1,-} \cap \overline{D}_{1,+}$. So we need only consider the boundary value problem

$$\begin{cases} \Delta \widetilde{u}(x) + 4\widetilde{u}^3(x) = \gamma x_1^3, & x \in D_{1,+}, \\ \widetilde{u}(x) = 0, & x \in \partial D_{1,+}, \end{cases} \tag{11.165}$$

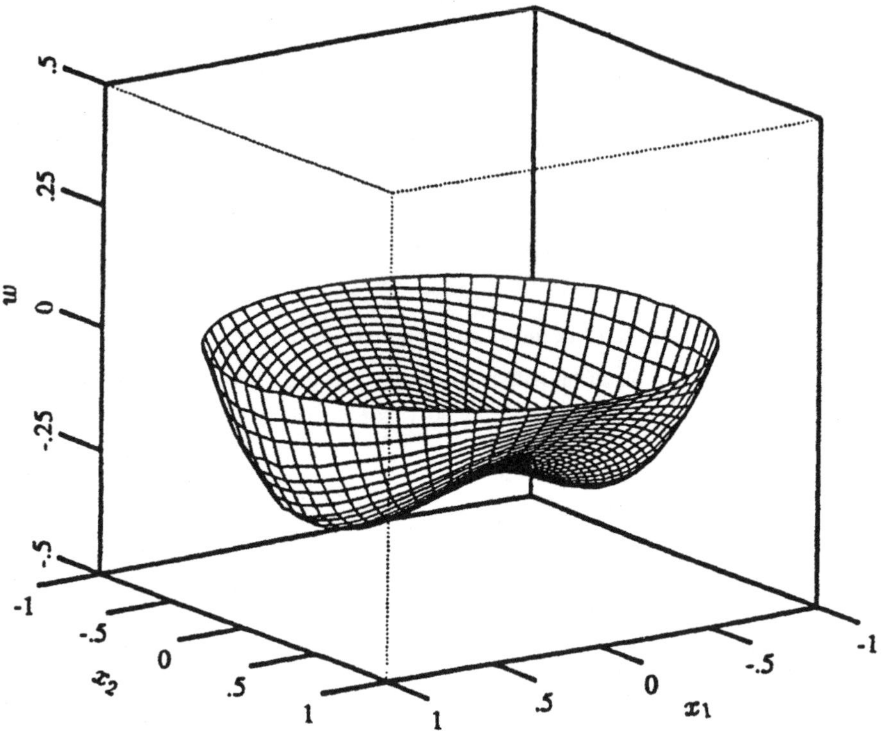

Figure 11.5: Boundary element numerical solution of $\Delta u + 4u^2 = 10x_1^2$ subject to the zero Dirichlet boundary condition, Example 11.5.

because on $D_{1,-}$ we simply have $u(x_1,u_2) = -\widetilde{u}(-x_1,x_2)$ for $x \in D_{1,-}$. The nonlinear boundary value problem (11.165) is similar to the one considered in (11.148) because the forcing terms γx_1^3 and γx_1^2 are all nonnegative on each domain. So essentially all the arguments in parts (i)–(iii) of Example 11.4 apply. The only concern one might have for (11.165) may be that $\partial D_{1,+}$ is not a C^∞ curve. Nevertheless, $\partial D_{1,+}$ is Lipschitz continuous. With suitable modifications, the arguments still go through and there will be no problems. In our numerical computations, we again choose $\gamma = 10$ in (11.164) and use $\lambda = 4$. The numerical solutions converge after 13 iterations, using $u_0(x) \equiv 0$. The graphics is given in Fig. 11.6. The reader will find that the solution changes sign across the x_2-axis. □

Remark 11.6. How restrictive is the working assumption ([H]) made in § 11.4? From Examples 11.2–11.6, we see that for the monotone equation (11.146) in Example 11.3, ([H]) is always satisfied. The BVP (11.146) also has the least trouble with existence and unique-

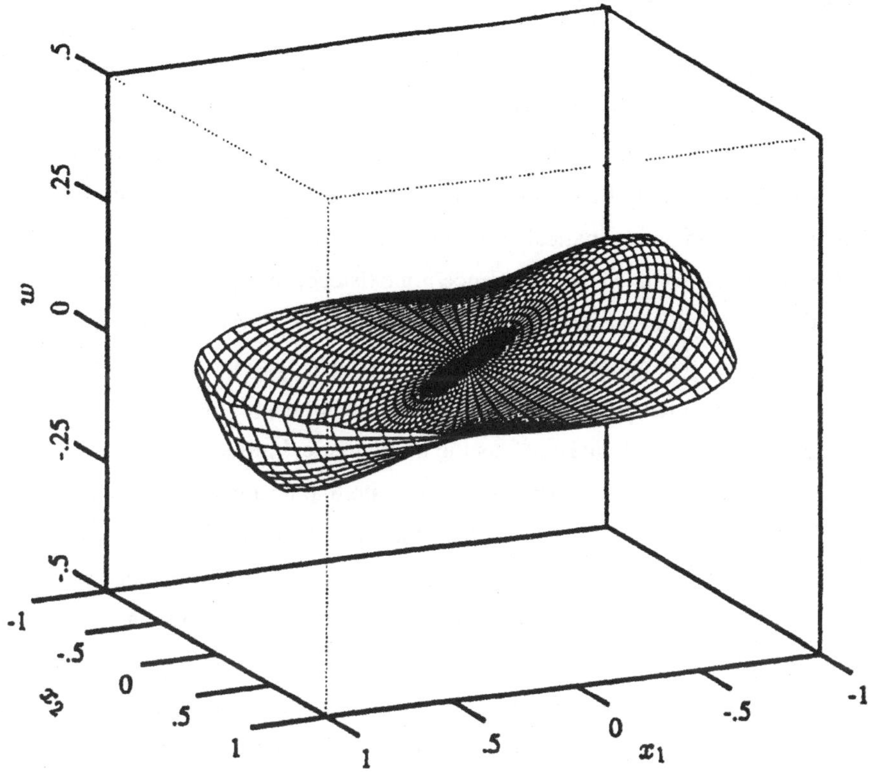

Figure 11.6: Boundary element numerical solution of $\Delta u + 4u^3 = 10x_1^3$ subject to the zero Dirichlet boundary condition, Example 11.6.

ness. For the BVP in Example 11.2, when γ is sufficiently small, condition ([H]) holds. However, ([H]) fails when γ is large, because the system (11.143) has multiple solutions that can be separated only by nonconstant sup- and subsolutions. So it appears that an alternative assumption to ([H]) is required. Nevertheless, computationally we are still able to obtain convergent numerical solutions.

For Examples 11.4 and 11.5, even though the existence/nonexistence and uniqueness/multiplicity of solutions are harder to determine, condition ([H]) can still largely be confirmed. In any case, regardless of whether ([H]) can be confirmed a priori or a posteriori, numerical solutions can always be generated by the computer and motivate us to establish further theoretical properties of solutions as given in this section. This is an indication of the usefulness of the development of a combination of theory (i.e., monotone iterations) and numerical methods (i.e., BEM). □

11.8 Quasimonotone iteration for coupled 2×2 semilinear elliptic systems

In the rest of this chapter, we extend our previous work to cover 2×2 systems of coupled semilinear elliptic boundary value problems.

Systems of coupled nonlinear elliptic equations arise frequently in applications. Solutions of such systems may be regarded as steady states of certain reaction-diffusion systems or problems in mathematical biology. For nonlinear problems, we face the ever present difficulties associated with existence/nonexistence, uniqueness/multiplicity and stability/instability of solutions. Since the nonlinearity is now further coupled and may appear somewhat arbitrary, the treatment is naturally expected to be more complex and challenging than the case of a single scalar semilinear elliptic equation and, thus, many more problems remain open [141]. Though inasmuch as the mathematical intractability for coupled semilinear elliptic systems is more severe, it is fortunate that a satisfactory method for finding *stable solutions* for a certain class of such systems has been developed. This class consists of coupled systems whose nonlinearities are of the *quasimonotone nondecreasing, quasimonotone nonincreasing,* or *mixed quasimonotone* types. The method is a generalization of the monotone iteration scheme that we employed in the preceding sections. A good reference can be found in Pao [152]. Since the method is *constructive*, it can be discretized and implemented numerically by the established methods of finite difference, finite element and boundary element. Obviously, researchers in science and engineering have computed many examples of coupled semilinear elliptic (or parabolic) systems arising from applications, with or without the knowledge that the quasimonotone iteration scheme can be incorporated in their algorithms. In Pao [151], e.g., a finite difference-monotone iteration method is used to solve some coupled elliptic systems numerically. Nevertheless, not many *global error analysis* results are available in the literature. In particular, when BEM is used, rates of convergence with respect to the discretization parameter have never been analyzed for coupled elliptic systems, except in [34]. The crux of the rest of this chapter, therefore, is to carry out the boundary element numerical analysis and computation for coupled semilinear elliptic systems using the quasimonotone iteration scheme.

A coupled semilinear elliptic system of boundary value problems may be likened to a 2×2 system of nonlinear matrix equations in an infinite-dimensional space. Here we exploit this 2×2 matrix structure of the PDE system, and formalize it to an infinite-dimensional setting. This is carried out in Subsections 11.8.1 and 11.8.3. A critical assumption, $[H^2_{2\times 2}]$, is stated, which is a generalization of our assumption $[H]$ in [62]. This $[H^2_{2\times 2}]$ condition is *directly verifiable* and provides the basis of the local contraction argument in the permissible range of (quasi) monotone iteration. Once $[H^2_{2\times 2}]$ is well formulated, the rest of the error estimation procedures are pretty much the same as in the preceding sections. We state the error estimates for the Galerkin boundary-element iterates $u_h^n(x)$ in the $H^r(\Omega)$, $0 \leqslant r < 2$, Sobolev space norms in Subsection 11.8.2 but leave out the proof. In Subsection 11.8.4, we provide a concrete example of a mixed quasimonotone 2×2 elliptic system for computation and discussion.

11.8.1 *Monotone iteration scheme for nonlinearities of quasimonotone nonincreasing, quasimonotone nondecreasing, and mixed quasimonotone types*

We consider a coupled 2×2 system of semilinear elliptic partial differential equations of the form

$$(S) \qquad \begin{cases} \Delta u_i(x) = f_i(x, u_1(x), u_2(x)), & x \in \Omega \in \mathbb{R}^N,\ N = 2 \text{ or } 3, \quad (11.166) \\ B_i u_i(x) = g_i(x), & x \in \partial\Omega, \quad (11.167) \\ \quad i = 1, 2, \end{cases}$$

where B_i, $i = 1, 2$, are linear boundary operators given by

$$B_i u_i = u_i \quad \text{(Dirichlet type)} \tag{11.168}$$

or

$$B_i u_i = \frac{\partial u_i}{\partial n} + \alpha(\cdot) u_i, \quad \alpha(x) \geqslant 0 \text{ on } \partial\Omega,\ \alpha \in C^\infty(\partial\Omega), \tag{11.169}$$
$$\text{(Neumann or Robin types)}$$

and $f_i \in C^\infty(\Omega \times \mathbb{R} \times \mathbb{R})$, $g_i \in C^\infty(\partial\Omega)$, for $i = 1, 2$.
Throughout the rest of the chapter, for $\Sigma = \Omega$ or $\partial\Omega$, we use $\|u\|_{s,\Sigma}$ to denote the norm of a scalar function u in the Sobolev space $H^s(\Sigma)$, and use $\|(u_1, u_2)\|_{s,\Sigma}$ to denote the norm of a vector-valued function (u_1, u_2) in the product Sobolev space $H^s(\Sigma) \times H^s(\Sigma)$:

$$\|(u_1, u_2)\|^2_{s,\Sigma} \equiv \|u_1\|^2_{s,\Sigma} + \|u_2\|^2_{s,\Sigma}. \tag{11.170}$$

It is well understood that the monotone iteration scheme for scalar semilinear elliptic boundary value problems has been generalized to system (S); see Pao [152], for example. For (S), the construction of monotone iterates depends on the so-called *quasimonotone properties* of the *reaction functions* f_1 and f_2 in (S). Let $J = [a_1, b_1] \times [a_2, b_2] \subset \mathbb{R}^2$, for some a_i, b_i, $i = 1, 2$, satisfying $a_i < b_i$. A vector-valued function $(f_1(x, u_1, u_2), f_2(x, u_1, u_2))$ is called *quasimonotone nonincreasing* in $(x, u_1, u_2) \in \Omega \times J$, if

$$\frac{\partial f_1}{\partial u_2}(x, u_1, u_2) \leqslant 0, \quad \frac{\partial f_2}{\partial u_1}(x, u_1, u_2) \leqslant 0, \quad (x, u_1, u_2) \in \Omega \times J; \tag{11.171}$$

quasimonotone nondecreasing in $(x,u_1,u_2) \in \Omega \times J$, if

$$\frac{\partial f_1}{\partial u_2}(x,u_1,u_2) \geqslant 0, \quad \frac{\partial f_2}{\partial u_1}(x,u_1,u_2) \geqslant 0, \quad (x,u_1,u_2) \in \Omega \times J; \tag{11.172}$$

and *mixed quasimonotone* in $(x,u_1,u_2) \in \Omega \times J$, if

$$\frac{\partial f_1}{\partial u_2}(x,u_1,u_2) \geqslant 0, \quad \frac{\partial f_2}{\partial u_1}(x,u_1,u_2) \leqslant 0, \quad (x,u_1,u_2) \in \Omega \times J. \tag{11.173}$$

A pair of vector-valued functions $(\underline{u}_1,\underline{u}_2)$ and $(\overline{u}_1,\overline{u}_2)$ with $\underline{u}_i,\overline{u}_i \in C(\overline{\Omega}) \cap C^2(\Omega)$, $i=1,2$, are called *ordered subsolution* and *supersolution*, respectively, if they satisfy $\underline{u}_i \leqslant \overline{u}_i$, $i=1,2$, and

$$B_i\underline{u}_i - g_i \leqslant 0, \qquad B_i\overline{u}_i - g_i \geqslant 0, \quad \text{on } \partial\Omega, \quad i=1,2, \tag{11.174}$$

plus

(i) for quasimonotone nonincreasing (f_1,f_2),

$$\begin{cases} \Delta\underline{u}_1 - f_1(x,\underline{u}_1,\underline{u}_2) \geqslant 0, & \Delta\underline{u}_2 - f_2(x,\underline{u}_1,\underline{u}_2) \geqslant 0, \\ \Delta\overline{u}_1 - f_1(x,\overline{u}_1,\overline{u}_2) \leqslant 0, & \Delta\overline{u}_2 - f_2(x,\overline{u}_1,\overline{u}_2) \leqslant 0; \end{cases} \tag{11.175}$$

(ii) for quasimonotone nondecreasing (f_1,f_2),

$$\begin{cases} \Delta\underline{u}_1 - f_1(x,\underline{u}_1,\overline{u}_2) \geqslant 0, & \Delta\underline{u}_2 - f_2(x,\overline{u}_1,\underline{u}_2) \geqslant 0, \\ \Delta\overline{u}_1 - f_1(x,\overline{u}_1,\underline{u}_2) \leqslant 0, & \Delta\overline{u}_2 - f_2(x,\underline{u}_1,\overline{u}_2) \leqslant 0; \end{cases} \tag{11.176}$$

(iii) for mixed quasimonotone (f_1,f_2),

$$\begin{cases} \Delta\underline{u}_1 - f_1(x,\underline{u}_1,\overline{u}_2) \geqslant 0, & \Delta\underline{u}_2 - f_2(x,\underline{u}_1,\underline{u}_2) \geqslant 0, \\ \Delta\overline{u}_1 - f_1(x,\overline{u}_1,\underline{u}_2) \leqslant 0, & \Delta\overline{u}_2 - f_2(x,\overline{u}_1,\overline{u}_2) \leqslant 0. \end{cases} \tag{11.177}$$

Henceforth, we assume that λ_i, $i=1,2$, are chosen sufficiently large such that

$$\begin{cases} \text{(i)} \ \dfrac{\partial f_i(x,u_1,u_2)}{\partial u_i} \leqslant \lambda_i, \quad \forall (x,u_1,u_2) \in \Omega \times J, \quad i=1,2; \ \lambda_i \geqslant 0; \\ \text{(ii) the spectrum of the operators } (\lambda_i - \Delta, B_i), \ i=1,2, \\ \qquad \text{entirely lie on the open right half plane.} \end{cases} \tag*{$[H^1_{2\times 2}]$}$$

Let

$$\lambda_i^* = \text{the smallest eigenvalue of the operator } (-\Delta, B_i), \ i=1,2. \tag{11.178}$$

Then obviously

$$\begin{cases} \lambda_i^* \geqslant 0, \quad i=1,2; \\ \lambda_i^* = 0 \text{ iff } B_i \text{ corresponds to the homogeneous} \\ \text{Neumann boundary condition.} \end{cases} \tag{11.179}$$

From $([H^1_{2\times 2}]$(i)) and $([H^1_{2\times 2}]$(ii)), we see that, the following iteration scheme is well-defined: for any initial state $(u_1^{(0)}, u_2^{(0)})$ with $u_i^{(0)} \in C(\overline{\Omega}) \cap C^2(\Omega)$, $i = 1, 2$,

$$\begin{cases} (\lambda_i - \Delta) u_i^{(n+1)} = F_{i,\lambda_i}(x, u_1^{(n)}, u_2^{(n)}) \quad \text{in } \Omega, \\ B_i u_i^{(n+1)} = g_i \quad \text{on } \partial\Omega, \\ i = 1, 2, \ \text{ and } \ n = 0, 1, 2, \ldots, \end{cases} \tag{11.180}$$

where

$$F_{i,\lambda_i}(x, u_1, u_2) \equiv \lambda_i u_i - f_i(x, u_1, u_2), \quad i = 1, 2. \tag{11.181}$$

The convergence of the scheme (11.180) and the monotonicity properties of the iterates depend on the choice of the initial state $(u_1^{(0)}, u_2^{(0)})$, which in turn depends on the quasi-monotone type of (f_1, f_2). Let us proceed according to the recipe given in [152], § 8.4:

(i) For quasimonotone nonincreasing (f_1, f_2) in $\Omega \times J$, let

$$(u_1^{(0)}, u_2^{(0)}) = (\underline{u}_1, \underline{u}_2) \quad (\text{resp. } (\overline{u}_1, \overline{u}_2)). \tag{11.182}$$

We denote by $(\underline{u}_1^{(n)}, \underline{u}_2^{(n)})$ (resp. $(\overline{u}_1^{(n)}, \overline{u}_2^{(n)})$), $n = 1, 2, \ldots$, the corresponding iterates from (11.180).

(ii) For quasimonotone nondecreasing (f_1, f_2) in $\Omega \times J$, let

$$(u_1^{(0)}, u_2^{(0)}) = (\overline{u}_1, \underline{u}_2) \quad (\text{resp. } (\underline{u}_1, \overline{u}_2)). \tag{11.183}$$

We denote by $(\overline{u}_1^{(n)}, \underline{u}_2^{(n)})$ (resp. $(\underline{u}_1^{(n)}, \overline{u}_2^{(n)})$), $n = 1, 2, \ldots$, the corresponding iterates.

(iii) For mixed quasimonotone (f_1, f_2) in $\Omega \times J$, the initial state is chosen as in (11.182), but the iteration procedure (11.180) is modified to

$$\begin{cases} \left.\begin{array}{l} (\lambda_1 - \Delta) \underline{u}_1^{(n+1)} = F_{1,\lambda_1}(\cdot, \underline{u}_1^{(n)}, \overline{u}_2^{(n)}), \\ (\lambda_2 - \Delta) \underline{u}_2^{(n+1)} = F_{2,\lambda_2}(\cdot, \underline{u}_1^{(n)}, \underline{u}_2^{(n)}), \end{array}\right\} \quad \text{on} \quad \Omega \\ B_i \underline{u}_i^{(n+1)} = g_i \quad \text{on } \partial\Omega, \quad i = 1, 2; \quad n = 0, 1, 2, \ldots, \end{cases} \tag{11.184}$$

$$\begin{cases} \left.\begin{array}{l} (\lambda_1 - \Delta) \overline{u}_1^{(n+1)} = F_{1,\lambda_1}(\cdot, \overline{u}_1^{(n)}, \underline{u}_2^{(n)}), \\ (\lambda_2 - \Delta) \overline{u}_2^{(n+1)} = F_{2,\lambda_2}(\cdot, \overline{u}_1^{(n)}, \overline{u}_2^{(n)}), \end{array}\right\} \quad \text{on } \Omega, \\ B_i \overline{u}_i^{(n+1)} = g_i \quad \text{on } \partial\Omega, \quad i = 1, 2; \quad n = 0, 1, 2, \ldots . \end{cases} \tag{11.185}$$

Following [152, (4.7), p. 403], we define the *sector*

$$\langle \underline{\mathbf{u}}, \overline{\mathbf{u}} \rangle = \{(u_1, u_2) \mid u_i \in C(\overline{\Omega}), \quad \underline{u}_i \leqslant u_i \leqslant \overline{u}_i, \quad i = 1, 2\}. \tag{11.186}$$

We now have the following convergence theorem.

Theorem 11.8. *For each type of monotonicity, the sequences $\underline{u}_i^{(n)}$, $\overline{u}_i^{(n)}$, described in (11.180)–(11.185) satisfy the monotone property*

$$\underline{u}_i^{(n)} \leqslant \underline{u}_i^{(n+1)} \leqslant \overline{u}_i^{(n+1)} \leqslant \overline{u}_i^{(n)} \quad on\, \overline{\Omega}, \quad i=1,2, \quad n=0,1,2,\ldots,$$

and there exist functions $\underline{u}_i^{(\infty)}$ and $\overline{u}_i^{(\infty)}$, $i=1,2$, satisfying $\underline{u}_i^{(\infty)} \leqslant \overline{u}_i^{(\infty)}$ on $\overline{\Omega}$ such that $\underline{u}_i^{(n)}$ and $\overline{u}_i^{(n)}$ converge monotonically to $\underline{u}_i^{(\infty)}$ and $\overline{u}_i^{(\infty)}$, respectively, for $i=1,2$. We have

(i) *if (f_1,f_2) is quasimonotone nonincreasing, then $(\underline{u}_1^{(\infty)},\underline{u}_2^{(\infty)})$ and $(\overline{u}_1^{(\infty)},\overline{u}_2^{(\infty)})$ are solutions of (S);*

(ii) *if (f_1,f_2) is quasimonotone nondecreasing, then $(\underline{u}_1^{(\infty)},\overline{u}_2^{(\infty)})$ and $(\overline{u}_1^{(\infty)},\underline{u}_2^{(\infty)})$ are solutions of (S);*

(iii) *if (f_1,f_2) is mixed quasimonotone and satisfies*

$$\max\left\{\frac{\partial f_i(x,u_1,u_2)}{\partial u_i} \mid (x,u_1,u_2) \in \Omega \times \langle \underline{\mathbf{u}}, \overline{\mathbf{u}} \rangle \right\} < \lambda_i^* \tag{11.187}$$

for either $i=1$ or $i=2$, then $(\underline{u}_1^{(\infty)},\underline{u}_2^{(\infty)})$ and $(\overline{u}_1^{(\infty)},\overline{u}_2^{(\infty)})$ are solutions of (S).

Moreover, any solution (u_1,u_2) in $\langle \underline{\mathbf{u}}, \overline{\mathbf{u}} \rangle$ satisfies $\underline{u}_i^{(\infty)} \leqslant u_i \leqslant \overline{u}_i^{(\infty)}$ on $\overline{\Omega}$ for $i=1,2$.

Proof. See Pao [152, pp. 403–409]. □

It is easy to see that if all the data in (S), namely, $\partial\Omega$, f_i, g_i, $i=1,2$, are C^∞-smooth, then the limits of monotone convergence $\underline{u}_i^{(\infty)}$ and $\overline{u}_i^{(\infty)}$, $i=1,2$, have regularity $C^\infty(\overline{\Omega})$ because the pairs $(\underline{u}_1^{(\infty)},\underline{u}_2^{(\infty)})$, $(\overline{u}_1^{(\infty)},\overline{u}_2^{(\infty)})$, $(\underline{u}_1^{(\infty)},\overline{u}_2^{(\infty)})$, or $(\overline{u}_1^{(\infty)},\underline{u}_2^{(\infty)})$ are classical solutions of (S) for respective quasimonotone types.

With the existence of an ordered subsolution and supersolution, by Theorem 11.8, any iterate $(\underline{u}_1^{(n)},\underline{u}_2^{(n)})$, $(\underline{u}_1^{(n)},\overline{u}_2^{(n)})$, $(\overline{u}_1^{(n)},\underline{u}_2^{(n)})$ and $(\overline{u}_1^{(n)},\overline{u}_2^{(n)})$ will lie in $\langle \underline{\mathbf{u}}, \overline{\mathbf{u}} \rangle$. Thus, we may replace the product interval $J=[a_1,b_1]\times[a_2,b_2]$ with $\langle \underline{\mathbf{u}}, \overline{\mathbf{u}} \rangle$ throughout all of the preceding discussions.

We now need to introduce the following notations. Given any sufficiently smooth function $f(x,u_1,u_2)$ of three variables (x,u_1,u_2), we define two associated functions $\mathscr{D}_1 f$ and $\mathscr{D}_2 f$

of five variables (x,u_1,u_2,v_1,v_2) by

$$(\mathscr{D}_1 f)(x,u_1,u_2,v_1,v_2) = \begin{cases} \dfrac{1}{2(u_1-v_1)}[f(x,u_1,u_2)-f(x,v_1,u_2) \\ \qquad +f(x,u_1,v_2)-f(x,v_1,v_2)], & \text{if } u_1 \neq v_1, \\ \dfrac{1}{2}\left[\dfrac{\partial f(x,u_1,u_2)}{\partial u_1}+\dfrac{\partial f(x,u_1,v_2)}{\partial u_1}\right], & \text{if } u_1 = v_1; \end{cases} \tag{11.188}$$

$$(\mathscr{D}_2 f)(x,u_1,u_2,v_1,v_2) = \begin{cases} \dfrac{1}{2(u_2-v_2)}[f(x,u_1,u_2)-f(x,u_1,v_2) \\ \qquad +f(x,v_1,u_2)-f(x,v_1,v_2)], & \text{if } u_2 \neq v_2, \\ \dfrac{1}{2}\left[\dfrac{\partial f(x,u_1,u_2)}{\partial u_2}+\dfrac{\partial f(x,v_1,v_2)}{\partial u_2}\right], & \text{if } u_2 = v_2. \end{cases} \tag{11.189}$$

Then $\mathscr{D}_1 f$ and $\mathscr{D}_2 f$ are continuous functions of the five variables (x,u_1,u_2,v_1,v_2), and we have

$$\begin{aligned} f(x,u_1,u_2)-f(x,v_1,v_2) &= (\mathscr{D}_1 f)(x,u_1,u_2,v_1,v_2)\cdot(u_1-v_1) \\ &\quad + (\mathscr{D}_2 f)(x,u_1,u_2,v_1,v_2)\cdot(u_2-v_2). \end{aligned} \tag{11.190}$$

Although convergence is guaranteed in Theorem 11.8, the rates are not known in general. In order to obtain such explicit rates of convergence, as in the preceding sections of this chapter, some additional assumption is required. We now formulate a condition that is sufficient to guarantee an algebraically fast rate of convergence. It is a generalization of assumption [H] in § 11.4 for the scalar semilinear elliptic equation. It is based upon estimations in Subsection 11.8.3 for a 2×2 linear matrix problem in $\mathbb{R}^2$, but now generalized to an infinite dimensional setting. From the arguments that follow in Subsection 11.8.3, the reader may agree that the assumption is rather "generically tight":

$$\left.\begin{aligned} \min[\lambda_1^* + \mathscr{D}_1 f_1(x,u_1,u_2,v_1,v_2)] > 0, \\ \min[\lambda_2^* + \mathscr{D}_2 f_2(x,u_1,u_2,v_1,v_2)] > 0, \end{aligned}\right\} \tag*{$[H^2_{2\times 2}]$}$$

for $x \in \overline{\Omega}$, $(u_1,u_2) \in \langle \underline{\mathbf{u}}, \overline{\mathbf{u}} \rangle$, $(v_1,v_2) \in \langle \underline{\mathbf{u}}, \overline{\mathbf{u}} \rangle$.

Theorem 11.9. *Let the semilinear system* (S) *be quasimonotone nonincreasing, quasimonotone nondecreasing, or mixed quasimonotone with pairs of ordered subsolutions* $(\underline{u}_1,\underline{u}_2)$ *and* $(\overline{u}_1,\overline{u}_2)$. *Let assumptions* $[H^i_{2\times 2}]$ *hold for* $i=1,2$, *let the iteration be performed as in (11.180)–(11.185) for some* $\overline{\lambda} = (\overline{\lambda}_1,\overline{\lambda}_2)$, $\overline{\lambda}_1 \geqslant 0$, $\overline{\lambda}_2 \geqslant 0$, *satisfying* $[H^1_{2\times 2}]$, *and let*

$$u_i^{(n)} \equiv \underline{u}_i^{(n)} \quad (\textit{resp. } u_i^{(n)} \equiv \overline{u}_i^{(n)}), \quad i=1,2; \quad n=0,1,2,\ldots,$$

and

$$u_i^{(\infty)} \equiv \underline{u}_i^{(\infty)} \quad (resp.\ u_i^{(\infty)} \equiv \overline{u}_i^{(\infty)}), \quad i = 1, 2,$$

where $\underline{u}_i^{(\infty)}$ *and* $\overline{u}_i^{(\infty)}$, $i = 1, 2$, *are promised by Theorem 11.8. Denote* $e_i^{(m)} = u_i^{(m)} - u_i^{(\infty)}$, $i = 1, 2$. *Then there exist a* $\overline{\lambda}$ *and a* $\theta_{\overline{\lambda}}$: $0 < \theta_{\overline{\lambda}} < 1$, *such that*

$$\|(e_1^{(n)}, e_2^{(n)})\|_{0,\Omega} \leqslant \theta_{\overline{\lambda}}^{n} \|e_1^{(0)}, e_2^{(0)})\|_{0,\Omega}, \quad n = 1, 2, \ldots . \tag{11.191}$$

Proof. See [34, pp. 634–636]. □

By the intermediate value theorem and the continuity of $\partial f_i/\partial u_i$, $i = 1, 2$, we easily see from (11.189) that $[H_{2\times 2}^2]$ is equivalent to

$$[H_{2\times 2}^2]\colon \min\left[\lambda_1^* + \frac{\partial f_1(x, u_1, u_2)}{\partial u_1}\right] > 0, \quad \min\left[\lambda_2^* + \frac{\partial f_2(x, u_1, u_2)}{\partial u_2}\right] > 0, \tag{11.192}$$
$$\text{for } x \in \overline{\Omega}, (u_1, u_2) \in \langle \underline{\mathbf{u}} - \delta, \overline{\mathbf{u}} + \delta \rangle, \text{ for some small } \delta > 0.$$

By Remark 11.7 in Subsection 11.8.3, we see that, generically, $[H_{2\times 2}^2]$ is about the tightest we can get in order for (11.191) to hold for some $\theta_{\overline{\lambda}}$: $0 < \theta_{\overline{\lambda}} < 1$.

11.8.2 *Error analysis for a Galerkin boundary element monotone iteration scheme*

We now introduce a Galerkin boundary element scheme for computing (11.180), (11.184) or (11.185), and analyze the errors. Let $E(x - y; \lambda)$ be the fundamental solution of the Helmholtz operator $\Delta - \lambda$ satisfying

$$(\Delta - \lambda) E(x - y; \lambda) = -\delta(x - y), \quad \lambda > 0, \quad x, y \in \mathbb{R}^N.$$

As before, denote the volume potential by $\widetilde{V}_\lambda$:

$$(\widetilde{V}_\lambda \phi)(x) = \int_\Omega E(x - y; \lambda) \phi(y) dy, \quad x \in \Omega,$$

and the simple-layer potential by $\widetilde{S}_\lambda$:

$$(\widetilde{S}_\lambda \eta)(x) = \int_{\partial\Omega} E(x - y; \lambda) \eta(y) d\sigma_y, \quad x \in \overline{\Omega}.$$

Also, let S_λ and V_λ be, respectively, the trace of $\widetilde{S}_\lambda$ and $\widetilde{V}_\lambda$:

$$S_\lambda \eta \equiv (\widetilde{S}_\lambda \eta)|_{\partial\Omega}, \quad V_\lambda \phi \equiv (V_\lambda \phi)|_{\partial\Omega}.$$

The normal derivative of the volume potential is denoted by

$$(V_\lambda' \phi)(x) = \int_\Omega \frac{\partial E(x - y; \lambda)}{\partial n_x} \phi(y) dy, \quad x \in \partial\Omega.$$

In addition, we define boundary operator K_λ:

$$(K_\lambda \eta)(x) = \int_{\partial\Omega} \frac{\partial E(x-y;\lambda)}{\partial n_x} \eta(y) d\sigma_y, \quad x \in \partial\Omega.$$

Recall the regularity properties of these integral operators from Chapters 4 and 6 in the following.

Theorem 11.10. *Let $\lambda > 0$. Then*

(i) *$\widetilde{V}_\lambda$ maps continuously from $H^s(\Omega)$ into $H^{s+2}(\Omega)$, for all $s \geqslant -1$;*
(ii) *V_λ maps continuously from $H^s(\Omega)$ into $H^{s+\frac{3}{2}}(\partial\Omega)$, for all $s \geqslant -1$;*
(iii) *V'_λ maps continuously from $H^s(\Omega)$ into $H^{s+\frac{1}{2}}(\partial\Omega)$, for all $s \geqslant -1$;*
(iv) *$\widetilde{S}_\lambda$ maps continuously from $H^s(\partial\Omega)$ into $H^{s+\frac{3}{2}}(\Omega)$, for all $s \in \mathbb{R}$;*
(v) *S_λ maps isomorphically from $H^s(\partial\Omega)$ onto $H^{s+1}(\partial\Omega)$, and is positive definite, for all $s \in \mathbb{R}$; S_λ is a strongly elliptic pseudodifferential operator of order -1;*
(vi) *K_λ maps continuously from $H^s(\partial\Omega)$ into*

$$\begin{cases} H^{s+1}(\partial\Omega), & \forall s \in \mathbb{R}, \text{ if} N \geqslant 3, \\ C^\infty(\partial\Omega), & \forall s \in \mathbb{R}, \text{ if} N = 2. \end{cases}$$

Furthermore, $\frac{1}{2}I + K_\lambda$ maps $H^s(\partial\Omega)$ onto $H^s(\partial\Omega)$ isomorphically for every $s \in \mathbb{R}$ such that $\frac{1}{2}I + K_\lambda = \left(\frac{\partial}{\partial n}\widetilde{S}_\lambda\right)\big|_{\partial\Omega}$, and $\frac{1}{2}I + K_\lambda$ is a strongly elliptic pseudodifferential operator of order 0.

(vii) *For $\alpha(\cdot)$ given as in (11.169), the operator*

$$T_\lambda \equiv \frac{1}{2}I + K_\lambda + \alpha(\cdot)S_\lambda \tag{11.193}$$

is a strongly elliptic pseudodifferential operator of order 0 mapping isomorphically from $H^s(\partial\Omega)$ onto $H^s(\partial\Omega)$ for all $s \in \mathbb{R}$. □

Return to the iteration (11.180)–(11.185). In the two boundary conditions $B_i u_i = g_i$ for $i = 1, 2$, we have several different possible combinations for (B_1, B_2), say, for example, (Dirichlet, Dirichlet), (Dirichlet, Neumann) or (Dirichlet, Robin) type. In the subsequent error analysis, the arguments for the Neumann type boundary condition are identical to those for the Robin type. Therefore, for definiteness, let us treat the most representative case, $(B_1, B_2) =$ (Dirichlet, Robin), i.e.,

$$B_1 u_1 \equiv u_1|_{\partial\Omega} = g_1; \quad B_2 u_2 = \left(\frac{\partial u_2}{\partial n} + \alpha(\cdot)u\right)\bigg|_{\partial\Omega} = g_2, \quad \text{(cf. (11.169))}. \tag{11.194}$$

Noting from the preceding sections that the *indirect formulation* is more advantageous, we continue to adopt it here by using the ansatz that the solution is a sum of a simple-layer potential plus a volume potential

$$u_i^{(n+1)}(x) = (\widetilde{S}_{\lambda_i}\eta_i^{(n+1)})(x) - (\widetilde{V}_{\lambda_i}F_{i,\lambda_i}(\cdot, u_1^{(n)}, u_2^{(n)}))(x)$$
$$= \int_{\partial\Omega} E(x-y;\lambda_i)\eta_i^{(n+1)}(y)d\sigma_y - \int_{\Omega} E(x-y;\lambda_i)F_{i,\lambda_i}(y, u_1^{(n)}(y), u_2^{(n)}(y))dy,$$
$$x \in \overline{\Omega}, \quad i = 1, 2, \tag{11.195}$$

where $\eta_i^{(n+1)}$, $i = 1, 2$, are the unknown simple layer densities to be solved, *on $\partial\Omega$ only*. Applying the boundary conditions in (11.167) to (11.195), we get the following two boundary integral equations for the determination of $\eta_i^{(n+1)}$, $i = 1, 2$:

$$\left.\begin{aligned} S_{\lambda_1}\eta_1^{(n+1)} &= g_1 + V_{\lambda_1}F_{1,\lambda_1}^{(n)}, \quad &(11.196)\\ T_{\lambda_2}\eta_2^{(n+1)} &= g_2 + (V'_{\lambda_2} + \alpha V_{\lambda_2})F_{2,\lambda_2}^{(n)} \quad &(11.197) \end{aligned}\right\} \text{ on } \partial\Omega,$$

where

$$F_{i,\lambda_i}^{(n)} \equiv F_{i,\lambda_i}\left(\cdot, u_1^{(n)}, u_2^{(n)}\right). \tag{11.198}$$

Lemma 11.9. *Let $g_i \in C^\infty(\partial\Omega)$ and $F_{i,\lambda_i}^{(n)} \in C^\infty(\overline{\Omega})$ for $i = 1, 2$ in (11.167) and (11.198). Then there exist unique $\eta_1^{(n+1)}$, $\eta_2^{(n+1)} \in C^\infty(\partial\Omega)$ which are, respectively, the solution of the BIEs (11.196) and (11.197).*

Proof. The BIE (11.196) is solvable with a unique solution $\eta_1^{(n+1)}$ because the property (v) in Theorem 11.10.

The BIE (11.197) is also solvable with a unique solution $\eta_2^{(n+1)}$ because of property (vii) in Theorem 11.10. □

The following theorem is a generalization and combination of Theorem 11.2 and Corollary 11.2. The indicated antimonotone property is useful and has also been confirmed numerically.

Theorem 11.11. *Let the assumptions of Theorem 11.8 hold, including $[H^1_{2\times 2}]$ but not $[H^2_{2\times 2}]$. Let the boundary conditions be given as (11.194), and let $\underline{\eta}_1^{(n+1)}$, $\overline{\eta}_1^{(n+1)}$, $\underline{\eta}_2^{(n+1)}$, $\overline{\eta}_2^{(n+1)}$ be the simple layer densities in (11.196) and (11.197) corresponding, respectively, to $\underline{u}_1^{(n+1)}$, $\overline{u}_1^{(n+1)}$, $\underline{u}_2^{(n+1)}$, $\overline{u}_2^{(n+2)}$ in the quasimonotone iteration. Then we have the antimonotone property*

$$\underline{\eta}_i^{(n)} \geqslant \underline{\eta}_i^{(n+1)} \geqslant \overline{\eta}_i^{(n+1)} \geqslant \overline{\eta}_i^{(n)} \quad \text{on } \partial\Omega, \quad i = 1, 2, n = 0, 1, 2, \ldots.$$

Consequently, $\underline{\eta}_i^{(n)}$ and $\overline{\eta}_i^{(n)}$, converge monotonically and respectively, to some $\underline{\eta}_i^{(\infty)}$ and $\overline{\eta}_i^{(\infty)}$, naturally corresponding to $\underline{u}_i^{(\infty)}, \overline{u}_i^{(\infty)}$, respectively, for $i = 1, 2$ in Theorem 11.8.

Proof. The boundary operator S_{λ_1} in (11.196) is the same simple-layer boundary operator as in (11.33), so we can apply Theorem 11.2 directly. The boundary operator T_{λ_2} in (11.197) corresponds to the Robin boundary B_2 in (11.194), and therefore, we need to establish an analogue of Theorem 11.2 for the operator T_{λ_2} based upon the representation (11.195). Let us go to the proof of Theorem 11.2 we need only modify certain segments therein. First, inherit all the work there, except that in $(11.46)_2$ and $(11.47)_2$, we change the homogeneous Dirichlet boundary conditions to the homogeneous Robin conditions. But note that $(11.54)_2$ remains unchanged. Then the rest of all the arguments in the proof of Theorem 11.2 remain the same, except that near the end of the proof of Theorem 11.2, we now write instead

$$\begin{aligned}
&\frac{\partial}{\partial n}[w^+(x)-w^-(x)]+\alpha(x)[w^+(x)-w^-(x)] = \frac{\partial}{\partial n}[w^+(x)-w^-(x)] \\
&\quad = \ldots\ldots \text{ (same as in the proof of Theorem 11.2)} \\
&\quad = \eta_1(x)-\eta_2(x) \geqslant 0.
\end{aligned}$$

Therefore the proof is complete. □

Throughout the discussions in the rest of this Subsection, without further mention, we have assumed that $[H^i_{2\times 2}]$, $i=1,2$.

For later use, we define, for given $\lambda=(\lambda_1,\lambda_2)$,

$$\begin{aligned}
M_\lambda = \max\Big\{ &\big[(\lambda_1-\mathscr{D}_1 f_1(x,u_1,u_2,v_1,v_2))^2+(\mathscr{D}_2 f_1(x,u_1,u_2,v_1,v_2)^2\big]^{1/2}, \\
&\big[(\lambda_2-\mathscr{D}_2 f_2(x,u_1,u_2,v_1,v_2))^2+(\mathscr{D}_1 f_2(x,u_1,u_2,v_1,v_2))^2\big]^{1/2}\Big\} \\
&x\in\overline{\Omega};\ (u_1,u_2),(v_1,v_2)\in\langle\underline{\mathbf{u}}-\delta,\overline{\mathbf{u}}+\delta\rangle.
\end{aligned} \tag{11.199}$$

As in § 11.4, let $\mathscr{S}_h$: $0<h\leqslant h_0$, be a 1-parameter family of finite-dimensional boundary element approximation spaces that form an (ℓ,m)-system $\mathscr{S}_h^{\ell,m}$, with ℓ, $m\in\mathbb{Z}^+\equiv\{0,1,2,\ldots\}$, $\ell\geqslant m+1$, on $\partial\Omega$, in the sense of Babuška and Aziz [14]. The Galerkin boundary element scheme to solve (11.196) and (11.197) is:

(i) Initialization: Set $u^{(0)}_{i,h}(x)=\underline{u}_i(x)$ or $\overline{u}_i(x)$, $i=1,2$, according to whether the semilinear system is quasimonotone nonincreasing, nondecreasing, or mixed, and according to (11.180)–(11.187)

(ii) Choose $\lambda_i\geqslant 0$, $i=1,2$, satisfying $[H^1_{2\times 2}]$ and $[H^2_{2\times 2}]$.

(iii) For each $n=0,1,2,\ldots$, and h, $0<h\leqslant h_0$, find $\eta^{(n+1)}_{i,h}\in\mathscr{S}_h$ such that

$$\left.\begin{aligned}
\langle S_{\lambda_1}\eta^{(n+1)}_{1,h},\phi_h\rangle &= \langle g_1+V_{\lambda_1}F^{(n)}_{1,\lambda_1,h},\phi_h\rangle, \\
\langle T_{\lambda_2}\eta^{(n+1)}_{2,h},\phi_h\rangle &= \langle g_2+(V'_{\lambda_2}+\alpha V_{\lambda_2})F^{(n)}_{2,\lambda_2,h},\phi_h\rangle
\end{aligned}\right\}\quad \forall\phi_h\in\mathscr{S}_h, \tag{11.200}$$

where $\langle\ ,\ \rangle$ signifies the $L^2(\partial\Omega)$ inner product, and

$$F^{(n)}_{i,\lambda_i,h} \equiv F_{i,\lambda_i}(\cdot, u^{(n)}_{1,h}, u^{(n)}_{2,h}).$$

(iv) Define

$$u^{(n+1)}_{i,h}(x) = (\widetilde{S}_{\lambda_i}\eta^{(n+1)}_{i,h})(x) - (\widetilde{V}_{\lambda_i}F^{(n)}_{i,\lambda_i,h})(x), \quad x \in \overline{\Omega}, \quad n = 0, 1, 2, \ldots. \tag{11.201}$$

Note that the two equations in (11.200) can be computed *separately*. Let us state the main theorem of error estimates.

Theorem 11.12. *Let $\{\mathscr{S}_h \mid 0 < h \leqslant h_0\}$ be a family of (ℓ,m)-systems on $\partial\Omega$ with $\ell \geqslant 2$, $m \geqslant 1$, and let $N = 2$ or 3. Assume $[H^i_{2\times 2}]$, $i = 1, 2$. Then there exist $K_1 > 0$, $K_2 > 0$ (the same as stated in (E4) and (E5) below), ρ_λ, $0 < \rho_\lambda < 1$, and a small $\overline{h}_0 > 0$ satisfying (11.213) such that for any r, $0 \leqslant r < 2$, we have*

$$\left\| \left(u^{(n)}_{1,h} - u^{(\infty)}_1, u^{(n)}_{2,h} - u^{(\infty)}_2 \right) \right\|_{r,\Omega} \leqslant K_1 h^{2-r} \left\| \left(u^{(\infty)}_1, u^{(\infty)}_2 \right) \right\|_{2,\Omega} + K_2 \rho^n_\lambda \left\| \left(e^{(0)}_1, e^{(0)}_2 \right) \right\|_{0,\Omega}, \tag{11.202}$$

for all $n = 1, 2, \ldots$, $0 < h \leqslant \overline{h}_0$. Therefore $(u^{(n)}_{1,h}, u^{(n)}_{2,h})$ converges to $(u^{(\infty)}_1, u^{(\infty)}_2)$ as $n \to \infty$ and $h \downarrow 0$. Also, for $-\frac{3}{2} \leqslant s < \frac{1}{2}$, there are constants $K_{3,s} > 0$ and $K_{4,s} > 0$ such that

$$\left\| \left(\eta^{(n)}_{1,h} - \eta^{(\infty)}_1, \eta^{(n)}_{2,h} - \eta^{(\infty)}_2 \right) \right\|_{s,\partial\Omega} \leqslant K_{3,s} h^{\frac{1}{2}-s} \left\| \left(u^{(\infty)}_1, u^{(\infty)}_2 \right) \right\|_{2,\Omega} + K_{4,s} \rho^n_\lambda \left\| \left(e^{(0)}_1, e^{(0)}_2 \right) \right\|_{2,\Omega}, \tag{11.203}$$

for all $n = 1, 2, \ldots$, $0 < h \leqslant \overline{h}_0$.
In particular, if $N = 2$ and if $\{\mathscr{S}_h \mid 0 < h \leqslant h_0\}$ is a family of S^d_h spaces of smoothest splines of degree $d \geqslant 0$ with respect to a quasiuniform mesh on $\partial\Omega$, then (11.202) holds for $0 \leqslant r < 2$, and (11.203) holds for s, $-(d+2) \leqslant s < 1/2$.

Proof. See (E1)–(E6) below and Theorems 11.4 and 11.6. □

We will be concise in deriving and stating error estimates, since much of the procedure is similar to the work in the preceding sections. The estimation is done through a string of comparisons. Note that

(I) The exact simple-layer densities $(\eta^{(n+1)}_1, \eta^{(n+1)}_2)$ for the iterative solution $(u^{(n+1)}_1, u^{(n+1)}_2)$ (cf. (11.180)) satisfy the BIEs (11.196) and (11.197);

(II) The Galerkin boundary element approximation

$$\left(\eta^{(n+1)}_{1,h}, \eta^{(n+1)}_{2,h} \right)$$

for $(\eta^{(n+1)}_1, \eta^{(n+1)}_2)$ satisfies (11.200);

(III) The numerical solution

$$\left(u_{1,h}^{(n+1)}, u_{2,h}^{(n+1)}\right)$$

approximating $(u_1^{(n+1)}, u_2^{(n+1)})$ at each step of the iteration is defined from $(\eta_{1,h}^{(n+1)}, \eta_{2,h}^{(n+1)})$ through (11.201). The errors are defined to be

$$e_{i,h}^{(n+1)} = u_{i,h}^{(n+1)} - u_i^{(n+1)}, \quad i = 1, 2; \quad n = 0, 1, 2, \ldots; \tag{11.204}$$

(IV) An intermediate simple-layer density pair $(\zeta_{1,h}^{(n+1)}, \zeta_{2,h}^{(n+1)})$ on $\partial\Omega$ for the comparison purpose is defined to be the unique solutions of the BIEs

$$\left\{\begin{array}{l} S_{\lambda_1}\zeta_{1,h}^{(n+1)} = g_1 - V_{\lambda_1}F_{1,\lambda_1,h}^{(n)}, \\ T_{\lambda_2}\zeta_{2,h}^{(n+1)} = g_2 - (V_{\lambda_2}' + \alpha V_{\lambda_2})F_{2,\lambda_2,h}^{(n)}, \end{array}\right. \quad \text{on } \partial\Omega, \quad n = 0, 1, 2, \ldots,$$

(as guaranteed in the same way by Lemma 11.9);

(V) An intermediate solution pair $(w_{1,h}^{(n+1)}, w_{2,h}^{(n+1)})$ on $\overline{\Omega}$ for the comparison purpose is defined by

$$w_{i,h}^{(n+1)}(x) = \left(\widetilde{S}_{\lambda_i}\zeta_{i,h}^{(n+1)}\right)(x) - \left(\widetilde{V}_{\lambda_i}F_{i,\lambda_i,h}^{(n)}\right)(x), \tag{11.205}$$
$$x \in \overline{\Omega}, \quad i = 1, 2; \ n = 0, 1, 2, \ldots .$$

It is obvious that $w_{i,h}^{(n+1)}$, $i = 1, 2$, satisfy

$$\left.\begin{array}{ll} (\lambda_i - \Delta)w_{i,h}^{(n+1)} = F_{i,\lambda_i,h}^{(n)} & \text{on } \Omega, \\ B_i w_{i,h}^{(n+1)} = g_i & \text{on } \partial\Omega, \\ i = 1, 2; \ n = 0, 1, 2, \ldots . & \end{array}\right\} \tag{11.206}$$

(VI) The limit of $(u_1^{(n+1)}, u_2^{(n+1)})$ as $n \to \infty$ is $(u_1^{(\infty)}, u_2^{(\infty)})$, as guaranteed by Theorem 11.8. Note that $(u_1^\infty, u_2^\infty) = (\underline{u}_1^{(\infty)}, \underline{u}_2^{(\infty)})$, or $(\overline{u}_1^{(\infty)}, \overline{u}_2^{(\infty)})$, or $(\underline{u}_1^{(\infty)}, \overline{u}_2^{(\infty)})$, or $(\overline{u}_1^{(\infty)}, \underline{u}_2^{(\infty)})$ according to Theorem 11.8. The pair $(u_1^{(\infty)}, u_2^{(\infty)})$ is an exact solution of (S). Corresponding to $(u_1^{(\infty)}, u_2^{(\infty)})$, we have simple-layer densities $(\eta_1^{(\infty)}, \eta_2^{(\infty)})$ which are solutions of the BIEs

$$\left.\begin{array}{l} S_{\lambda_1}\eta_1^{(\infty)} = g_1 + V_{\lambda_1}F_{1,\lambda_1}\left(\cdot, u_1^{(\infty)}, u_2^{(\infty)}\right) \\ T_{\lambda_2}\eta_2^{(\infty)} = g_2 + (V_{\lambda_2}' + \alpha V_{\lambda_2})F_{2,\lambda_2}\left(\cdot, u_1^{(\infty)} u_2^{(\infty)}\right) \end{array}\right\} \quad \text{on } \partial\Omega.$$

The initial error $(e_1^{(0)}, e_2^{(0)})$ is defined to be

$$e_i^{(0)} = u_i^{(0)} - u_i^{(\infty)}. \tag{11.207}$$

The glossary of error estimates can now be stated sequentially as follows:

(E1)

$$\left\|\eta_{i,h}^{(n+1)} - \zeta_{i,h}^{(n+1)}\right\|_{t,\partial\Omega} \leqslant C_{i,s,t} h^{t-s} \left\|\zeta_{i,h}^{(n+1)}\right\|_{s,\partial\Omega}, \qquad i = 1, 2; \quad n = 0, 1, 2, \ldots, \tag{11.208}$$

where $C_{i,s,t}$, $i = 1, 2$, are positive constants independent of n and h, and

$$2\omega - \ell \leqslant t \leqslant s \leqslant \ell, \quad t \leqslant m, \quad -\ell \leqslant \omega \leqslant m, \quad -m \leqslant \omega \leqslant t,$$

with

$$\omega \equiv \begin{cases} -\dfrac{1}{2}, & i = 1, \\ 0, & i = 2. \end{cases}$$

Proof. Cf. Lemma 11.2 and § 11.6, Hsiao and Wendland [97, Cor. 2.1] and Ruotsalainen and Saranen [162, Cor. 3]. □

(E2) Let $N = 2$ or 3, and

$$\varepsilon_0(N) = \begin{cases} 0, & N = 2, \\ \text{an arbitrarily small positive number, } 0 < \varepsilon_0(3) < 1/2, & N = 3. \end{cases} \tag{11.209}$$

Then for $\ell \geqslant 2$, $m \geqslant 1$, and some sufficiently small h_0, there are positive constants K_0 and $K_{\varepsilon_0(N)}$ such that

$$\begin{aligned} \left\|\eta_{i,h}^{(n+1)} - \zeta_{i,h}^{(n+1)}\right\|_{\varepsilon_0(N),\partial\Omega} &\leqslant K_{\varepsilon_0(N)} h^{\frac{1}{2}-\varepsilon_0(N)} \cdot \Big[\left\|u_i^{(\infty)}\right\|_{2,\Omega} + M_\lambda \theta_\lambda^n \left\|(e_1^{(0)}, e_2^{(0)})\right\|_{0,\Omega} \\ &\qquad + \left\|F_{i,\lambda_i,h}^{(n)} - F_{i,\lambda_i}^{(n)}\right\|_{0,\Omega}\Big], \end{aligned} \tag{11.210}$$

$$\begin{aligned} \left\|\eta_{i,h}^{(n+1)} - \zeta_{i,h}^{(n+1)}\right\|_{-\frac{3}{2},\partial\Omega} &\leqslant K_0 h^2 \Big[\left\|u_i^{(\infty)}\right\|_{2,\Omega} + M_\lambda \theta_\lambda^n \left\|(e_1^{(0)}, e_2^{(0)})\right\|_{0,\Omega} \\ &\qquad + \left\|F_{i,\lambda_i,h}^{(n)} - F_{i,\lambda_i}^{(n)}\right\|_{0,\Omega}\Big], \end{aligned}$$

for $i = 1, 2$, $n = 0, 1, 2, \ldots$, $0 < h \leqslant h_0$.

Proof. Cf. Lemma 11.3 and § 11.6. □

(E3) Let $N = 2$ or 3. For $\varepsilon_0(N)$ given by (11.209) and for β with

$$\begin{cases} 0 < \beta < 1/2, & \text{if } N = 2, \\ \beta = \varepsilon_0(3), & \text{if } N = 3, \end{cases} \tag{11.211}$$

then there exist positive constants K_0' and $K_{\varepsilon_0(N)}'$ such that

$$\begin{aligned}\left\|u_{i,h}^{(n+1)}-w_{i,h}^{(n+1)}\right\|_{C^{0,\beta}(\overline{\Omega})} &\leqslant K_{\varepsilon_0(N)}' h^{\frac{1}{2}-\varepsilon_0(N)}\Big[\left\|u_i^{(\infty)}\right\|_{2,\Omega}+M_\lambda\theta_\lambda^n\left\|(e_1^{(0)},e_2^{(0)})\right\|_{0,\Omega}\\&\qquad+\left\|F_{i,\lambda_i,h}^{(n)}-F_{i,\lambda_i}^{(n)}\right\|_{0,\Omega}\Big],\\ \left\|u_{i,h}^{(n+1)}-w_{i,h}^{(n+1)}\right\|_{0,\Omega} &\leqslant K_0'h^2\Big[\left\|u_i^{(\infty)}\right\|_{2,\Omega}+M_\lambda\theta_\lambda^n\left\|(e_1^{(0)},e_2^{(0)})\right\|_{0,\Omega}+\left\|F_{i,\lambda_i,h}^{(n)}-F_{i,\lambda_i}^{(n)}\right\|_{0,\Omega}\Big],\end{aligned}$$

for $i=1,2$, $n=0,1,2,\ldots$, $0<h\leqslant h_0$, where $K_{\varepsilon_0(N)}'$ may grow unbounded as $\varepsilon_0(3)\downarrow 0$ when $N=3$.

Proof. Cf. Lemma 11.4. □

(E4) Let $N=2$ or 3, and let β satisfy (11.211). Then

$$\begin{aligned}\|w_{i,h}^{(n+1)}-u_i^{(n+1)}\|_{C^{0,\beta}(\overline{\Omega})} &\leqslant K_1\|F_{i,\lambda_i,h}^{(n)}-F_{i,\lambda_i}^{(n)}\|_{0,\Omega},\\ \|w_{i,h}^{(n+1)}-u_i^{(n+1)}\|_{0,\Omega} &\leqslant \frac{1}{\lambda_i+\lambda_i^*}\|F_{i,\lambda_i,h}^{(n)}-F_{i,\lambda_i}^{(n)}\|_{0,\Omega}\end{aligned}$$

for some $K_1>0$ independent of n,h and β, and for $i=1,2$, $n=0,1,2,\ldots$, $0<h\leqslant h_0$.

Proof. Cf. Lemmas 11.5 and 11.6. □

(E5) Let $N=2$ or 3, and let β be related to $\varepsilon_0(N)$ through (11.211). If we have $(u_{1,h}^{(n)}(x),u_{2,h}^{(n)}(x))\in\langle\underline{\mathbf{u}}-\delta,\overline{\mathbf{u}}+\delta\rangle$ for all $x\in\overline{\Omega}$, then

$$\begin{aligned}\|e_{i,h}^{(n+1)}\|_{C^{0,\beta}(\overline{\Omega})} &\leqslant K_{\varepsilon_0(N)}'h^{\frac{1}{2}-\varepsilon_0(N)}\big[\|(u_1^{(\infty)},u_2^{(\infty)})\|_{2,\Omega}+M_\lambda\theta_\lambda^n\|(e_1^{(0)},e_2^{(0)})\|_{0,\Omega}\big]\\&\quad+\big(K_1'M_\lambda+K_{\varepsilon_0(N)}'h^{\frac{1}{2}-\varepsilon_0(N)}\big)\|(e_{1,h}^{(n)},e_{2,h}^{(n)})\|_{0,\Omega}, \qquad (11.212)\\ \|(e_{1,h}^{(n+1)},e_{2,h}^{(n+1)})\|_{0,\Omega} &\leqslant K_2h^2\big[\|(u_1^{(\infty)},u_2^{(\infty)})\|_{2,\Omega}+M_\lambda\theta_\lambda^n\|(e_1^{(0)},e_2^{(0)})\|_{0,\Omega}\big]\\&\quad+(\theta_\lambda+K_0'M_\lambda h^2)\|(e_{1,h}^{(n)},e_{2,h}^{(n)})\|_{0,\Omega},\end{aligned}$$

for $i=1,2$, $0<h\leqslant h_0$, for some $K_1'>0$, $K_2>0$ independent of n and h.

Proof. Cf. Corollary 11.3. □

(E6) Let $N=2$ or 3. Choose $\overline{h}_0>0$ so small that

$$0<\theta_\lambda+K_2M_\lambda\overline{h}_0^2\leqslant\rho_\lambda<1, \qquad (11.213)$$

for some $\rho_\lambda=\rho_{\lambda_1,\lambda_2}$, $0<\rho_\lambda<1$. Then

$$\begin{aligned}&\big(u_{1,h}^{(n)}(x),u_{2,h}^{(n)}(x)\big)\in\langle\underline{\mathbf{u}}-\delta,\overline{\mathbf{u}}+\delta\rangle, \quad \forall x\in\overline{\Omega}, \qquad (11.214)\\ &i=1,2, \quad n=1,2,\ldots, \quad 0<h\leqslant\overline{h}_0,\end{aligned}$$

and there exist $\overline{K}_{\varepsilon_0(N)} > 0, \overline{K}_0 > 0$ such that

$$\|e_{i,h}^{(n+1)}\|_{C^{0,\beta}(\overline{\Omega})} \leqslant \overline{K}_{\varepsilon_0(N)} h^{\frac{1}{2}-\varepsilon_0(N)} \big[\|(u_1^{(\infty)}, u_2^{(\infty)})\|_{2,\Omega} + \rho_\lambda^{n+1} h^2 \|(e_1^{(0)}, e_2^{(0)})\|_{0,\Omega}\big],$$
$$\|(e_{1,h}^{(n+1)}, e_{2,h}^{(n+1)})\|_{0,\Omega} \leqslant \overline{K}_0 h^2 \big[\|(u_1^{(\infty)}, u_2^{(\infty)})\|_{2,\Omega} + \rho_\lambda^{n+1} h^2 \|(e_1^{(0)}, e_2^{(0)})\|_{0,\Omega}\big],$$

for $n = 0, 1, 2, \ldots, 0 < h \leqslant \overline{h}_0$.

Proof. A direct consequence of (11.213), (11.214) and (E5). □

11.8.3 *Comparison of eigenvalues and ℓ^2-norms of 2×2 matrices*

We now provide the rudiments of 2×2 matrices which form the basis in the formulation of the critical condition $[H^2_{2\times 2}]$ in § 11.2.

Let

$$A = \begin{bmatrix} a & b \\ c & d \end{bmatrix} \tag{11.215}$$

be a real 2×2 constant matrix. Then the ℓ^2-norm of A is defined by

$$\|A\| = \sup_{\substack{x \in \mathbb{R}^2 \\ x \neq 0}} \frac{|Ax|}{|x|}, \quad |x| = \sqrt{x_1^2 + x_2^2}. \tag{11.216}$$

The next theorem gives an explicit characterization of $\|A\|$ in terms of its entries a, b, c, d.

Theorem 11.13. *Let A be given as in (11.215). Then*

$$\|A\| = \frac{1}{2}\left[\sqrt{(a+d)^2 + (b-c)^2} + \sqrt{(a-d)^2 + (b+c)^2}\right]. \tag{11.217}$$

Proof. It is well known that (see, e.g., [192, p. 766])

$$\|A\| = \max\{\sqrt{\mu_1}, \sqrt{\mu_2}\}, \text{ where } \mu_1, \mu_2 = \text{ eigenvalues of } A^T A. \tag{11.218}$$

Since

$$A^T A = \begin{bmatrix} a^2 + c^2 & ab + cd \\ ab + cd & b^2 + d^2 \end{bmatrix},$$

we get the characteristic polynomial of $A^T A$:

$$\mu^2 - (a^2 + b^2 + c^2 + d^2)\mu + \big[(a^2 + c^2)(b^2 + d^2) - (ab + cd)^2\big] = 0.$$

Therefore

$$\begin{aligned} \mu_1, \mu_2 &= \frac{1}{2}\Big\{(a^2 + b^2 + c^2 + d^2) \\ &\qquad \pm \sqrt{(a^2 + b^2 + c^2 + d^2)^2 - 4[(a^2 + c^2)(b^2 + d^2) - (ab + cd)^2]}\Big\} \\ &= \frac{1}{4}\left[\sqrt{(a+d)^2 + (b-c)^2} \pm \sqrt{(a-d)^2 + (b+c)^2}\right]^2, \end{aligned}$$

from which we obtain

$$\max\{\sqrt{\mu_1},\sqrt{\mu_2}\} = \frac{1}{2}[\sqrt{(a+d)^2+(b-c)^2}+\sqrt{(a-d)^2+(b+d)^2}].$$

By (11.218), we have proved (11.217). □

Theorem 11.14. *Let*

$$D = \begin{bmatrix} \eta_1 & 0 \\ 0 & \eta_2 \end{bmatrix}, \quad \eta_i > 0, \quad i = 1,2, \tag{11.219}$$

be a diagonal matrix, and let $x, y \in \mathbb{R}^2$ *satisfy*

$$Dx = Ay, \quad A \text{ given by (11.215)}. \tag{11.220}$$

Then

$$|x| \leqslant \theta|y|, \tag{11.221}$$

where

$$\theta \equiv \frac{1}{2}\left[\sqrt{\left(\frac{a}{\eta_1}+\frac{d}{\eta_2}\right)^2+\left(\frac{b}{\eta_1}-\frac{c}{\eta_2}\right)^2}+\sqrt{\left(\frac{a}{\eta_1}-\frac{d}{\eta_2}\right)^2+\left(\frac{b}{\eta_1}+\frac{c}{\eta_2}\right)^2}\right].$$

Proof. From (11.220), we get

$$x = D^{-1}Ay = \begin{bmatrix} a/\eta_1 & b/\eta_1 \\ c/\eta_2 & d/\eta_2 \end{bmatrix} y \equiv \widetilde{A}y. \tag{11.222}$$

Applying Theorem 11.13 to $\widetilde{A}$, we get (11.221). □

We now consider a special form of the relation (11.220), where

$$D = \begin{bmatrix} \lambda_1+\lambda_1^* & 0 \\ 0 & \lambda_2+\lambda_2^* \end{bmatrix}, \tag{11.223}$$

$$A = \begin{bmatrix} \lambda_1 - a & -b \\ -c & \lambda_2 - d \end{bmatrix}, \tag{11.224}$$

where

$$\left.\begin{array}{llll} \lambda_1^* \geqslant 0, & \lambda_2^* \geqslant 0; & a,b,c,d \in \mathbb{R}; & \lambda_1^*, \lambda_2^*, a, b, c \text{ and } d \text{ are given;} \\ \lambda_1 \geqslant 0, & \lambda_2 \geqslant 0; & \lambda_1+\lambda_1^* > 0, & \lambda_2+\lambda_2^* > 0. \end{array}\right\} \tag{11.225}$$

Let $x, y \in \mathbb{R}^2$ satisfy

$$Dx = Ay. \tag{11.226}$$

Then by Theorem 11.14, we get

$$|x| \leqslant \|D^{-1}A\|\,|y| \equiv \theta_\lambda\,|y|, \tag{11.227}$$

where

$$\theta_\lambda = \frac{1}{2}\left\{\left[\left(\frac{\lambda_1 - a}{\lambda_1+\lambda_1^*} + \frac{\lambda_2 - d}{\lambda_2+\lambda_2^*}\right)^2 + \left(-\frac{b}{\lambda_1+\lambda_1^*} + \frac{c}{\lambda_2+\lambda_2^*}\right)^2\right]^{1/2} \right.$$
$$\left. + \left[\left(\frac{\lambda_1 - a}{\lambda_1+\lambda_1^*} - \frac{\lambda_2 - d}{\lambda_2+\lambda_2^*}\right)^2 + \left(\frac{b}{\lambda_1+\lambda_1^*} + \frac{c}{\lambda_2+\lambda_2^*}\right)^2\right]^{1/2}\right\}. \quad (11.228)$$

We want to investigate whether there exist any $\lambda_1 \geqslant 0$, $\lambda_2 \geqslant 0$, such that $0 < \theta_\lambda < 1$, so that (11.227) becomes a contraction.

Theorem 11.15. *Assume that (11.223)–(11.226) hold. If*

$$\lambda_1^* + a > 0, \quad \lambda_2^* + d > 0, \quad (11.229)$$

then there exist some $\lambda_1 > 0$, $\lambda_2 > 0$, sufficiently large, such that $0 < \theta_\lambda < 1$, and (11.227) is a strict contraction.

Proof. Write θ_λ in (11.228) into the following form:

$$\theta_\lambda = \frac{1}{2}\left\{[(\alpha+\delta)^2 + (-\beta+\gamma)^2]^{1/2} + [(\alpha-\delta)^2 + (\beta+\gamma)^2]^{1/2}\right\},$$

where

$$\alpha = \frac{\lambda_1 - a}{\lambda_1+\lambda_1^*}, \quad \delta = \frac{\lambda_2 - d}{\lambda_2+\lambda_2^*}, \quad \beta = \frac{b}{\lambda_1+\lambda_1^*}, \quad \gamma = \frac{c}{\lambda_2+\lambda_2^*}. \quad (11.230)$$

Note that β and γ can be made arbitrarily small by taking λ_1 and λ_2 large. Therefore for large $\lambda_1 > 0$ and $\lambda_2 > 0$, we have $\theta_\lambda < 1$ if and only if

$$\frac{1}{2}\left\{[(\alpha+\delta)^2]^{1/2} + [(\alpha-\delta)^2]^{1/2}\right\} < 1,$$

i.e.,

$$\frac{1}{2}[|\alpha+\delta| + |\alpha-\delta|] < 1.$$

But for $\lambda_1 > 0$ and $\lambda_2 > 0$ sufficiently large, α and δ are positive, and

$$\frac{1}{2}[|\alpha+\delta| + |\alpha-\delta|] = \begin{cases} \alpha & \text{if } \alpha \geqslant \delta, \\ \delta & \text{if } \alpha < \delta. \end{cases}$$

Hence for large λ_1 and λ_2, if both $\alpha < 1$ and $\delta < 1$, then $\theta_\lambda < 1$. But from (11.230) $\alpha < 1$ and $\delta < 1$ mean that

$$\frac{\lambda_1 - a}{\lambda_1+\lambda_1^*} < 1, \quad \frac{\lambda_2 - d}{\lambda_2+\lambda_2^*} < 1,$$

which hold if and only if

$$\lambda_1^* + a > 0 \quad \text{and} \quad \lambda_2^* + d > 0.$$

□

We now consider the converse of Theorem 11.15.

Theorem 11.16. *Assume that (11.223)–(11.225) hold. If*

$$\lambda_1^* + a < 0, \quad \lambda_2^* + d < 0, \tag{11.231}$$

then

$$\theta_\lambda \equiv \|D^{-1}A\| > 1.$$

Proof. From (11.231), we get

$$-a > \lambda_1^*, \quad -d > \lambda_2^*.$$

Therefore, for $\lambda_1 \geqslant 0$, $\lambda_2 \geqslant 0$,

$$\frac{\lambda_1 - a}{\lambda_1 + \lambda_1^*} > 1, \quad \frac{\lambda_2 - d}{\lambda_2 + \lambda_2^*} < 1.$$

From (11.228), we easily see that $\theta_\lambda > 1$. □

Is it still possible to have $0 < \theta_\lambda < 1$ if $(\lambda_1^* + a)(\lambda_2^* + d) \leqslant 0$? The answer is also negative, as given by the following.

Theorem 11.17. *Assume (11.223)–(11.225), and*

$$(\lambda_1^* + a)(\lambda_2^* + d) \leqslant 0. \tag{11.232}$$

Then

$$\theta_\lambda \equiv \|D^{-1}A\| \geqslant 1.$$

Proof. Condition (11.232) implies one of the following four possibilities:

(i) $\lambda_1^* + a = 0$;
(ii) $\lambda_2^* + d = 0$;
(iii) $\lambda_1^* + a < 0$, $\lambda_2^* + d > 0$;
(iv) $\lambda_1^* + a > 0$, $\lambda_2^* + d < 0$.

Cases (i) and (ii) are similar, as are (iii) and (iv). Here we only prove (iii). From $\lambda_1^* + a < 0$, utilizing the facts that $\lambda_1 \geqslant 0$, $\lambda_1^* \geqslant 0$ and $\lambda_1 + \lambda_1^* > 0$, we have

$$\frac{\lambda_1 - a}{\lambda_1 + \lambda_1^*} > 1.$$

From (11.228), we get

$$\theta_\lambda \geqslant \frac{1}{2}\left\{\left|\frac{\lambda_1 - a}{\lambda_1 + \lambda_1^*} + \frac{\lambda_2 - d}{\lambda_2 + \lambda_2^*}\right| + \left|\frac{\lambda_1 - a}{\lambda_1 + \lambda_1^*} - \frac{\lambda_2 - d}{\lambda_2 + \lambda_2^*}\right|\right\} > 1.$$ □

Remark 11.7. Theorems 11.15–11.17 show that (11.229) is a *necessary and sufficient condition* for $0 < \theta_\lambda < 1$. It is also interesting to note that the strict contraction property (11.227) does not depend on the off-diagonal entries $-b$ and $-c$ in the matrix A in (11.224). Since our critical condition $[H^2_{2\times 2}]$ in § 11.2 was formulated on the basis of (11.229), we see that the strict contraction property (11.191) in Theorem 11.9 is independent of the "off-diagonal nonlinearities" in $\partial f_1/\partial u_2$ and $\partial f_2/\partial u_1$, for f_1 and f_2 given in (11.166). □

11.8.4 *Numerical example*

We construct an annular domain Ω as shown in Fig. 11.7, wherein $\partial\Omega$ consists of an exterior bounding circle C_1 an an interior nonconcentric bounding circle C_2. C_1 and C_2 in Fig. 11.7 are discretized uniformly into, respectively, 96 and 48 panels which are piecewise constant boundary elements, as indicated in Fig. 11.7. Thus, what we have is an $S_h^{1,0}(\partial\Omega)$-system, which is also an S_h^d-system ($d = 0$) of smoothest splines of degree 0 with a quasiuniform mesh. So Theorem 11.12 is applicable. For the volume potential (i.e., the domain integral) in (11.195), the numerical integration is achieved by placing 1620 Gaussian quadrature points on $\overline{\Omega}$, which are rather evenly distributed on $\overline{\Omega}$. Note that Ω is a nonconvex, non-starshaped, doubly connected domain. The algorithmic flowchart is nearly the same as the Flow-Chart in § 11.7. The computer programs on semilinear 2×2 system were first tested with *synthetic data*, and produced numerical solutions of high accuracy with strong trends of convergence.

We include a simple numerical example below.

Example 11.7. A nonlinear monotone dissipative operator system that is mixed quasimonotone. Consider

$$\begin{cases} \Delta u_1(x) = 3[3u_1^3(x) + 2u_2(x)] + \dfrac{15}{4} & \text{on } \Omega, \\ \Delta u_2(x) = 3[-2u_1(x) + 3u_2^3(x)] - 6 & \text{on } \Omega, \\ u_1(x) = u_2(x) = 0 & \text{on } \partial\Omega. \end{cases} \tag{11.233}$$

Then

$$\begin{cases} f_1(x,u_1,u_2) = f_1(u_1,u_2) = 3(3u_1^3 + 2u_2) + \dfrac{15}{4}, \\ f_2(x,u_1,u_2) = f_2(u_1,u_2) = 3(-2u_1 + 3u_2^3) - 6, \end{cases} \tag{11.234}$$

satisfy

$$\frac{\partial f_1}{\partial u_2} = 6 > 0, \quad \frac{\partial f_2}{\partial u_1} = -6 < 0.$$

Therefore (11.233) is mixed quasimonotone. □

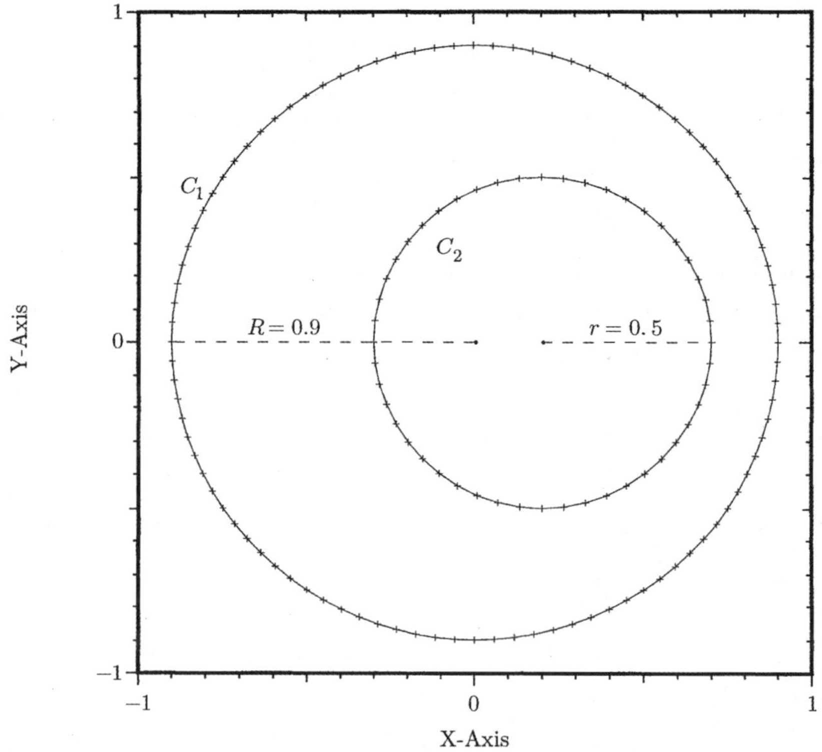

Figure 11.7: The domain Ω for Example 11.7; it is bounded by an exterior circle C_1 and an interior circle C_2. C_1 and C_2 are uniformly discretized, respectively, into 96 and 48 piecewise constant boundary elements.

Define a nonlinear operator

$$\mathscr{N}\left(\begin{bmatrix} u_1 \\ u_2 \end{bmatrix}\right) = \begin{bmatrix} \Delta u_1 - 3(3u_1^3 + 2u_2) \\ \Delta u_2 - 3(-2u_1 + 3u_2^3) \end{bmatrix}. \tag{11.235}$$

Then

$$\begin{aligned} &\left\langle \mathscr{N}\left(\begin{bmatrix} u_1 \\ u_2 \end{bmatrix}\right) - \mathscr{N}\left(\begin{bmatrix} v_1 \\ v_2 \end{bmatrix}\right), \begin{bmatrix} u_1 \\ u_2 \end{bmatrix} - \begin{bmatrix} v_1 \\ v_2 \end{bmatrix} \right\rangle \\ = & -\sum_{i=1}^{2} \int_{\Omega} [|\nabla(u_i - v_i)|^2 + 9(u_i^3 - v_i^3)(u_i - v_i)]dx \leqslant 0, \end{aligned} \tag{11.236}$$

for u_i, $v_i \in H_0^2(\Omega)$. Therefore $\mathscr{N}$ is a monotone dissipative operator, and (11.233) admits a *unique* solution pair (u_1, u_2) [126].

It is straightforward to check that

$$(\underline{u}_1,\overline{u}_1) = (-1,0), \quad (\underline{u}_2,\overline{u}_2) = \left(0, \left(\frac{2}{3}\right)^{\frac{1}{3}}\right)$$

form ordered pairs of constant subsolutions and supersolutions for u_1 and u_2, respectively. Also, since

$$\frac{\partial f_1}{\partial u_1} = 27u_1^2 \geqslant 0, \quad \frac{\partial f_2}{\partial u_2} = 27u_2^2 \geqslant 0, \tag{11.237}$$

we see that (11.192), and consequently $[H^2_{2\times 2}]$, is trivially satisfied because $\lambda_1^* > 0$. Although condition (11.187) in Theorem 11.8 is *violated*, still, according to the monotone dissipative property of $\mathscr{N}$ stated after (11.233), it is not hard to modify the proof in [152, p. 409] to establish that the quasimonotone iterations (11.184) and (11.185) converge to the unique solution. Since $[H^1_{2\times 2}]$ is also easily verifiable, Theorem 11.12 holds with $d = 0$, and $0 \leqslant r < 2$.

As with § 11.7, the Galerkin approximation (11.200) is replaced by a point collocation scheme:

$$\begin{cases} (S_\lambda \eta^{(n+1)}_{j,h})(x_i) = g_j(x_i) + (V_\lambda F^{(n)}_{1,\lambda,h})(x_i), \quad j = 1, 2, \\ x_i \text{ are the midpoints of the piecewise constant boundary elements,} \\ \text{for } 1 \leqslant i \leqslant M, \text{ for some } M, \end{cases} \tag{11.238}$$

where only $(11.200)_1$ is used because the boundary conditions for both u_1 and u_2 are Dirichlet. We choose $\lambda (= \lambda_1 = \lambda_2) = (5.4)^2$. Actually, any small to moderate size positive λ will work. The point collocation scheme (11.235) should have the same type of accuracy as the Galerkin scheme (11.200), according to Arnold and Wendland [10].

The error estimate (11.201) contains two parameters, h and n. We verify its convergence rates by the computer as follows:

(i) Fix $n = 2$. Vary h by doubling the number of meshes as follows:

$$\begin{aligned} &h_0: 48 \text{ (on } C_1) + 24 \quad \text{(on } C_2); \\ &h_1: 96 \text{ (on } C_1) + 48 \quad \text{(on } C_2); \\ &h_2: 192 \text{ (on } C_1) + 96 \quad \text{(on } C_2); \\ &h_3: 384 \text{ (on } C_1) + 192 \text{ (on } C_2). \end{aligned} \tag{11.239}$$

We then compute the logarithmic relative error

$$\ln \left\| \left(\eta^{(2)}_{1,h_j} - \eta^{(2)}_{1,h_{j-1}}, \eta^{(2)}_{2,h_j} - \eta^{(2)}_{2,h_{j-1}} \right) \right\|_{0,\partial\Omega}, \quad j = 1, 2, 3,$$

and plot them against $\ln h_j$. We obtain a line in Fig. 11.8, whose slope is measured to be 0.496. This is totally consistent with the rate $\mathscr{O}(h^{\frac{1}{2}})$ (setting $s = 0$ in (11.201)) given in Theorem 11.12.

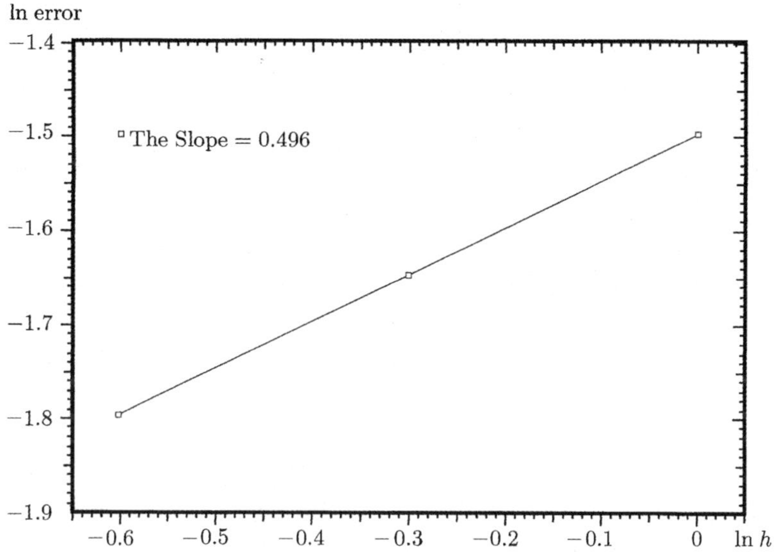

Figure 11.8: The relative convergence rate of $\ln\|(\eta^{(2)}_{1,h}-\eta^{(2)}_{1,0.5h},\eta^{(2)}_{2,h}-\eta^{(2)}_{2,0.5h})\|_{L^2(\partial\Omega)}$ for Example 11.7, plotted against $\ln h$. The slope of the line, 0.496, is close to the theoretical estimate 0.5, i.e., $\mathcal{O}(h^{\frac{1}{2}})$.

(ii) Fix $h=h_2$; see (11.239). We compute the logarithmic relative errors

$$\ln\left\|\left(\eta^{(n)}_{1,h_2}-\eta^{(n-1)}_{1,h_2},\eta^{(n)}_{2,h_2}-\eta^{(n-1)}_{2,h_2}\right)\right\|_{0,\partial\Omega},\quad n=2,3,4,5,6,$$

and plot them against n, yielding Fig. 11.9. Again we have found that these five points lie neatly on a line, with a slope equal to -0.39. This is totally consistent with the rate $\mathcal{O}(\rho^n_\lambda)$ in (11.201). In fact, it also tells that

$$\ln\rho_\lambda\approx-0.39,\quad \text{i.e.,}\quad \rho_\lambda\approx 0.677.$$

Finally, using $h=h_3$, see (11.239), and iterate 21 times, we obtain the approximate profiles of $u_1^{(\infty)}$ and $u_2^{(\infty)}$, given, respectively, in Figs. 11.10 and 11.11.

For this example, any negative constant $u_1^{(0)}$, $-1<u_1^{(0)}<0$, can be used as an initial iterate for $u_1(\cdot)$, and any positive constant constant $u_2^{(0)}$, $0<u_2^{(0)}<(\frac{2}{3})^{\frac{1}{3}}$, can be used as an initial iterate for $u_2(\cdot)$.

Many other examples, such as the steady state Lotka–Volterra system of predator-prey model in Pao [152, Chap. 8] and the population migration pattern as studied in Matano and Mimura [130], in principle can also be computed by quasimonotone iterations. However, we have not yet obtained totally satisfactory numerical results. The primary reason

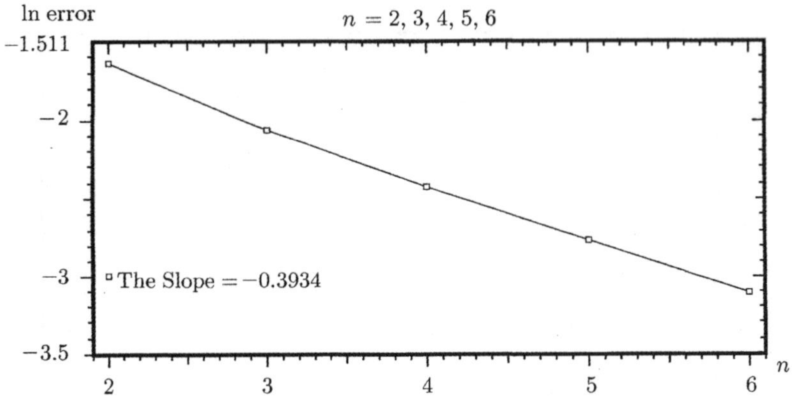

Figure 11.9: The relative convergence rate of $\ln \|(\eta_{1,h}^{(n)} - \eta_{1,h}^{(n+1)}, \eta_{2,h}^{(n)} - \eta_{2,h}^{(n+1)})\|_{L^2(\partial\Omega)}$ for Example 11.7, plotted against n with h being fixed. The points lie neatly on a line, consistent with the theoretical estimate $\mathcal{O}(\rho_\lambda^n)$.

for this is due to the difficulty in choosing the specific upper and lower solutions associated with such problems (unlike Example 11.7, where such choices as constants can be easily made). Some further numerical experiments and improvements are needed.

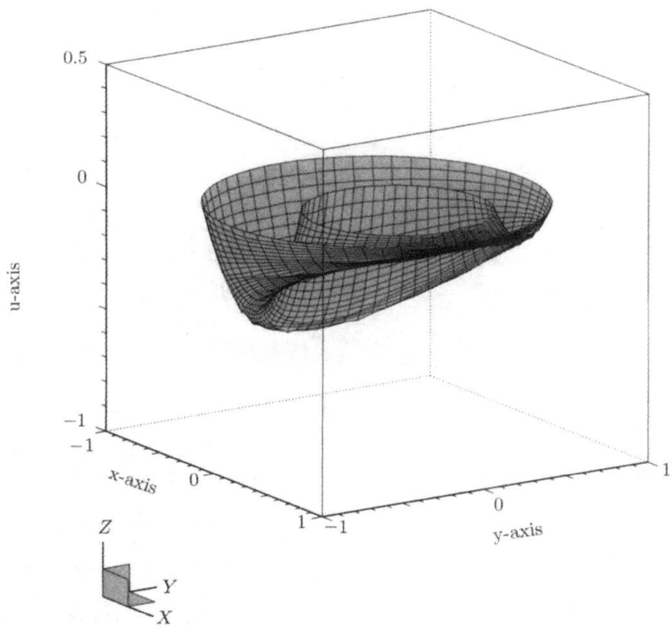

Figure 11.10: The profile of u_1 for Example 11.7.

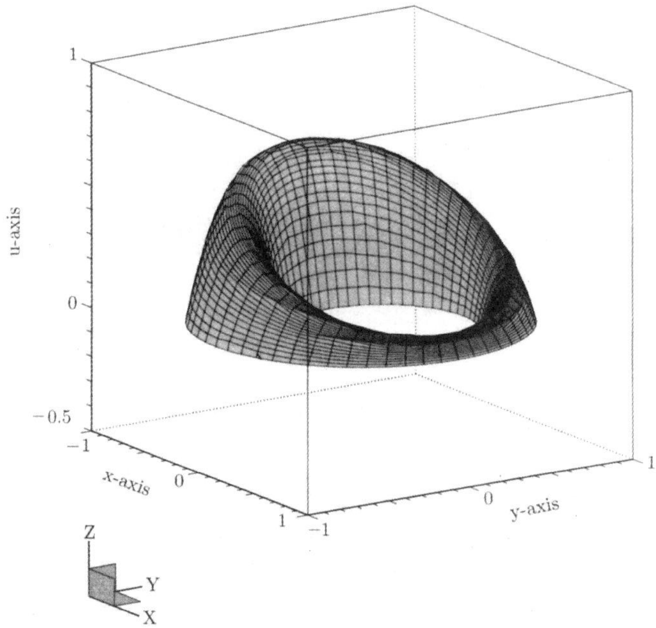

Figure 11.11: The profile of u_2 for Example 11.7.

Chapter 12

Boundary Element Methods for Semilinear Elliptic Partial Differential Equations (II): Algorithms and Computations for Unstable Solutions from Various Models

12.1 Introduction

We continue the study of semilinear elliptic BVPs of the form

$$\begin{cases} \Delta u(x) + f(x,u(x)) = 0, & x \in \Omega, \\ u(x) = 0, & x \in \partial\Omega, \end{cases} \tag{12.1}$$

where Ω is a bounded open domain in $\mathbb{R}^N$, $N = 2$, and f is a nonlinear function of x and u. We will deal with $f \equiv u^p$, $-u + u^p$, or variants thereof. We wish to compute numerical solutions of (12.1) by BEM and plot their graphics for visualization. on domains with various geometries and topologies. We also hope to survey existing algorithms and to introduce new ones, set certain numerical benchmarks, and explore singular perturbation cases. This chapter is based mainly on our prior work [37][1].

For "unstable" solutions consided in this chapter, the simplest, yet most basic form of nonlinearity is the power type. In this connection, we may first mention the Lane–Emden (–Fowler) equation in astrophysics ([31, Chap. 3] and Fowler [73])

$$\Delta u + u^p = 0, \quad u > 0, \quad \text{on } \Omega, \quad p > 1, \quad u|_{\partial\Omega} = 0, \tag{12.2}$$

where u^p is proportional to the density of the gaseous star [31]. From the point of view of analysis, this equation is interesting and challenging because the Laplace operator Δ is "negative" while the nonlinear operator $u \mapsto u^p$ is "positive" in appropriate function spaces, e.g., in $H_0^1(\Omega)$. There is a "competition" between these linear and nonlinear operators. In order for a solution to exist, a certain "balance" is required. There arises the prospect of

[1]The authors wish to thank World Scientific, Singapore, for copyright permissions regarding [37] published in the International Journal for Bifurcation and Chaos.

either nonexistence of solutions, or of the existence of possibly multiple solutions, for the general semilinear equation (12.1). Any nontrivial solution of (12.2) is an *unstable* solution. In contrast, if problem (12.1) is of the form

$$\left\{\begin{array}{l}\Delta u(x)-|u(x)|^{p-1}u(x)=g(x), \quad p>1,\\ u=0 \text{ on } \partial\Omega,\end{array}\right\} \quad g \text{ is given on } \Omega, \tag{12.3}$$

i.e., the sign in front of the nonlinearity $|u|^{p-1}u$ in (12.2) is adjusted from "+" to "−", then both the linear and nonlinear operators on the left of the first equation in (12.3) are "negative" and, as we can expect, a rather different theory applies. Indeed, the existence and uniqueness of solutions of (12.3) are well established; see the monotone dissipative operator theory in Lions [126], and Example 11.3 in Chapter 11.

Let us return to (12.2) and to astrophysics; the domain Ω with the most physical interest is B_R, the open ball of radius R in $\mathbb{R}^3$ centered at the origin. When Henon [90] studied rotating stellar structures in 1973, he proposed a variant of (12.2) as follows:

$$\Delta u+|x|^{\ell}u^p=0, \quad u>0, \text{ on } \Omega, \quad p>1; \ \ell>0; \quad u|_{\partial\Omega}=0. \tag{12.4}$$

(For his choice of nonlinearity, Henon commented: "This choice, although arbitrary, has the advantages of simplicity and convenience.") In 1987, Lieb and Yau [123] considered Chandrasekhar's theory of stellar collapse. They showed that the Chandrasekhar equation for the white dwarf problem without the general relativistic effect is equivalent to the following equation

$$\Delta u+4\pi(2u+u^2)^{3/2}=0, \text{ in } B_R. \tag{12.5}$$

Thus this equation is referred to as the Chandrasekhar equation in this paper. In both Chandrasekhar [31] and Lieb and Yau [123], the primary interest is in *radial solutions* $u \geqslant 0$. These equations provide some of the physical background for the model (12.1). Certain equations in *catalysis* theory also can be described in the form (12.1), but they often contain a varying parameter (Aris [7]) and thus manifest bifurcation phenomena. Other reaction-diffusion models such as Gierer and Meinhardt's system in biology (see Gierer and Meinhardt [75] and Ni [143]), have a "strong relationship," in a certain asymptotic sense to equations of the form (12.1).

The motivation for the model (12.1), along with its significance in other applications, will be given in due course, below.

Concerning the algorithmic development for unstable solutions of semilinear elliptic BVPs, we note the all-important theoretical method by the *Mountain–Pass Lemma* (MPL), published by Ambrosetti and Rabinowitz [6] in 1973. It provides a minimax variational formulation of critical points of functionals that are neither bounded from above nor below, such as those which correspond to (12.2). MPL is a powerful method for semilinear elliptic BVPs, as well as for a host of other nonlinear PDEs.

The mathematical proof of MPL as given in [6] contains both constructive and not-so-constructive elements and, therefore algorithmic realization of MPL is by no means straightforward. However, Choi and McKenna [47, 48] were able to devise ingenious algorithms by fusing the finite element method (FEM) with the method of steepest descent, and obtain multiple solutions of nonlinear elliptic PDEs on a rectangle. They and their collaborators have also successfully applied MPL-based algorithms to suspension bridge and traveling wave problems [100, 46] and to water waves [91]. In our opinion, a more appropriate term for the algorithm devised by Choi and McKenna is the "Mini-Max Algorithm" rather than the "Mountain–Pass Algorithm". An explanation will be given in Remark 12.2 in § 12.2.

Numerical analysis and results in this direction using FDM may be found in [101, 151, 153, 152], etc.

Rigorous justification of convergent and error estimates for unstable solutions is generally much more challenging.

As we can see from Subsection 12.3.4 below, for example, it is possible that a problem (12.2) can have a continuous one-parameter family of uncountably many solutions, each of which is infinitely close to neighboring solutions. There is no point in trying to give a convergence proof for such a case, since one never knows which solution the iterates are converging to. The best hope one can shoot for is to establish convergence for *isolated critical points* by assuming that such critical points are *nondegenerate*; see Remark 12.6. We will make more comments on numerical analysis and convergence in the survey below.

The organization of this chapter is as follows. In § 12.2, we introduce numerical algorithms, methods, error estimators, and some theoretical foundation. In § 12.3, we compute and illustrate solutions of the Lane–Emden equation with power $p = 3$, for various geometries, in great length. In § 12.4, we study a singularly perturbed equation and display the spike-layer pattern of solutions. In § 12.5, we compute the equations of Henon, Chandrasekhar and Lane–Emden (with "large" powers) for selected geometries. Finally, in § 12.6, we study the equation with sublinear growth, based on DIA, along with certain monotonicity properties of solutions observed by us. Commentaries, whenever available and appropriate, are given alongside the graphics to aid in our understanding of these nonlinear PDEs.

12.2 Iterative algorithms and numerical methods

In principle, any constructive proof should be realizable into a useful computational scheme for numerical solutions. Therefore, the true issue is: How efficient is that computational scheme in comparison with other viable ones? If the coding work is too involved, or if the requirement of CPU memory and time is beyond capacity, then that scheme becomes unattractive or even unworthwhile. Nowadays, the CPU time in a high performance computer is virtually costless. The choice of an algorithmic development and the actual programming task, thus, in our opinion, *depends mostly on the level of ease or difficulty of coding by the programmer and analyst.* The following two examples provide some condensed arguments, alternative to MPL, as to why the solutions of semilinear elliptic PDEs in § 12.3–12.5 *exist.* The arguments therein are all constructive. However, we will explain why such constructive proofs are not suitable or compatible with our numerical approach (based on BEM) here and, therefore, are not converted into algorithms by us for practical computational purposes.

Example 12.1. A constrained minimization method for the Lane–Emden equation (12.2). The following argument is now standard; see Ni [141, pp. 18–20]. We include it here for the benefit of those who are not nonlinear PDE specialists. Let p satisfy $1 < p < (N+2)/(N-2)$, where N is the space dimension of Ω. Consider the minimization problem

$$\inf_{v\in\mathscr{C}} \int_\Omega |\nabla v|^2\, dx, \tag{12.6}$$

where $\mathscr{C}$, the constraint set, is defined to be

$$\mathscr{C} = \left\{ v \in H_0^1(\Omega) \;\middle|\; \int_\Omega |v|^{p+1}\, dx = 1 \right\}. \tag{12.7}$$

$\mathscr{C}$ is well-defined by the Sobolev Imbedding Theorem. Let $\{v_k\}$ be a minimizing sequence for (12.6):

$$\int_\Omega |\nabla v_k|^2\, dx \to \lambda \equiv \inf_{v\in\mathscr{C}} \int_\Omega |\nabla v|^2\, dx, \quad \int_\Omega |v_k|^{p+1}\, dx = 1, \quad k = 1, 2.$$

Since $(\int_\Omega |\nabla v|^2\, dx)^{1/2}$ defines an equivalent H^1-norm in $H_0^1(\Omega)$, the sequence $\{v_k\}$ is bounded in $H_0^1(\Omega)$. Therefore it contains a subsequence $\{\widetilde{v}_k\}$ weakly converging to some $\widetilde{u}$ in $H_0^1(\Omega)$. By the compact imbedding of $H_0^1(\Omega)$ in $L^{p+1}(\Omega)$, we have strong convergence in $L^{p+1}(\Omega)$ since $p+1 < 2N/(N-2)$. Thus $\int_\Omega |\widetilde{u}|^{p+1}\, dx = 1$. On the other hand, since $\widetilde{v}_k \to \widetilde{u}$ weakly in $H_0^1(\Omega)$,

$$\int_\Omega |\nabla \widetilde{u}|^2\, dx \leqslant \varliminf_{k\to\infty} \int_\Omega |\nabla \widetilde{v}_k|^2\, dx \leqslant \lambda. \tag{12.8}$$

Therefore $\widetilde{u}$ is a minimizer for (12.6) in $\mathscr{C}$. We claim that $\widetilde{u} \geqslant 0$, since otherwise we simply replace $\widetilde{u}$ by $|\widetilde{u}|$. Also, $\nabla\widetilde{u} \not\equiv 0$, and thus $\lambda > 0$, since $\widetilde{u} = 0$ on $\partial\Omega$ and $\int_\Omega |\widetilde{u}|^{p+1}\,dx = 1$. Now, from standard arguments in calculus of variations one easily concludes that $\widetilde{u}$ is a *weak* solution of

$$\Delta\widetilde{u} + \alpha\widetilde{u}^p = 0 \text{ on } \Omega, \quad \widetilde{u}|_{\partial\Omega} = 0, \tag{12.9}$$

where α is the Lagrange multiplier. From elliptic regularity estimates [79] one sees that $\widetilde{u}$ is a *classical* solution of (12.9). We conclude that $\widetilde{u} > 0$ by the usual maximum principle [156]. Further, multiplying $(12.9)_1$ by $\widetilde{u}$ and integrating by parts, we obtain

$$\lambda = \int_\Omega |\nabla\widetilde{u}|^2\,dx = \alpha\int_\Omega |\widetilde{u}|^{p+1}\,dx = \alpha.$$

Setting $u = \lambda^{1/(p-1)}\widetilde{u}$, we have solved (12.2). □

Example 12.2. Let $f \in C^1(\mathbb{R})$ such that

$$f'(t) > \frac{f(t)}{t}, \quad \forall t > 0. \tag{12.10}$$

Then Ding and Ni [142] shows that a solution of the more general problem

$$\Delta u + f(u) = 0, \quad u > 0 \quad \text{on } \Omega, \quad u|_{\partial\Omega} = 0, \tag{12.11}$$

exists, which is a solution of the following constrained minimization problem

$$\inf_{v\in\mathscr{M}} \frac{1}{2}\int_\Omega [|\nabla v|^2 - F(v)]\,dx, \tag{12.12}$$

where

$$\mathscr{M} \equiv \left\{ v \in H_0^1(\Omega) \mid v \not\equiv 0, \quad \int_\Omega [|\nabla v|^2 - vf(v)]\,dx = 0 \right\}, \quad F(t) = \int_0^t f(s)ds; \tag{12.13}$$

$\mathscr{M}$ is the solution manifold, an idea first due to Nehari [139]. □

Let us attempt to find numerical solutions of (12.2) or (12.11) by directly using the arguments suggested in Examples 12.1 and 12.2. Because of the variational formulations given there, FEM becomes the natural method to be used. A family of finite dimensional approximating spaces $V_n = \text{span}\{\phi_i^h \mid 1 \leqslant i \leqslant n(h)\}$, $0 < h \leqslant h_0$, of subspaces of $H_0^1(\Omega)$ is then chosen such that $V_h \to H_0^1(\Omega)$ as the mesh size h decreases to zero. In proceeding to do the minimization problem (12.6) or (12.12), one must first choose functions

$$v_h = \sum_{j=1}^{n(h)} c_j\phi_j^h, \quad v_h \in V_h,$$

to be admissible in $\mathscr{C}$ (resp., $\mathscr{M}$):

$$\int_\Omega \left|\sum c_j \phi_j^h\right|^{p+1} dx = 1,$$
$$\left(\text{resp., } \int_\Omega \left[\left|\nabla \sum c_j \phi_j^h\right|^2 - \sum c_j \phi_j^h f\left(\sum c_j \phi_j^h\right)\right]^2 dx = 0\right). \tag{12.14}$$

Resolving the constraint relations in (12.14) *explicitly* requires triangulation of the entire domain and extensive quadratures. The workload involved in such software programming, along with the CPU time, are essentially of the same order of magnitude as that for computing the entire problems of (12.2) or (12.11). If one chooses instead to treat (12.14) *implicitly* by using the Lagrange multiplier method to handle the constraint(s), and follow up with local minimization (for *inf* from (12.6) and (12.12)) using Newton's algorithm, for example, extensive domain quadratures must still be evaluated for various linear and nonlinear terms. This becomes overburdensome, especially when the domain has curvilinear boundary (because of the "daunting" work of triangulation, as we have mentioned in §12.1). Thus, these approaches are not compatible with BEM, our choice of elliptic solver, to be described in Subsection 12.2.4, which does not require extensive domain triangulation, and is easily adaptable to change of geometry.

Remark 12.1. The approach as suggested in Examples 12.1 and 12.2 may still have some virtue: consequent FEM numerical error analysis seems to be more amenable, because the problem is now a *minimization* problem (see (12.6) or (12.12)) over a solution manifold $\mathscr{C}$ or $\mathscr{M}$ (see (12.7) and (12.13)). Thus, there should be more numerical stability. However, the minimization problem still may have multiple solutions in general. See Remark 12.2 below. □

In the remainder of this section, we describe the iterative algorithms and numerical methods for computing solutions of (12.1).

12.2.1 *The mountain–pass algorithm (MPA)*

Let E be a Banach space with norm $\| \ \|$, and let J be a C^1 functional on E, with Fréchet derivative J'. We say that J satisfies the Palais–Smale (PS) condition if for any sequence $\{x_n\} \subseteq E$ such that $\{J(x_n)\}$ bounded and $J'(x_n) \to 0$ strongly in E', the dual of E, then there exists a convergent subsequence in E. The PS condition is basically a *compactness* condition.

We now state the famous Mountain–Pass Lemma (MPL).

Theorem 12.1 (The Mountain–Pass Lemma [6]). *Let E be a Banach space and let $J \in C^1(E,\mathbb{R})$ satisfy the PS condition. If there exist an $e \in E$ and $\delta, r > 0$ such that*

(i) $J(0) = J(e) = 0$,

(ii) $r < \|e\|$ *and* $J(x) \geqslant \delta > 0, \quad \forall x \in S_r \equiv \{x \in E \mid \|x\| = r\}$,

then

$$c = \inf_{h\in\Gamma} \max_{t\in[0,1]} J(h(t)) \geqslant \delta \tag{12.15}$$

is a critical value of J, where

$$\Gamma \equiv \{h \in C([0,1],E) \mid h(0) = 0, h(1) = e\}. \qquad \square$$

By using MPL for J given in (12.16) below, it is not difficult to show that (12.2), e.g., has a solution for $\Omega \in \mathbb{R}^N$ when $1 < p < p^* = (N+2)/(N-2)$; p^* is the so-called critical Sobolev exponent.

The proof of MPL contains ingredients such as the contrapositive argument from the Deformation Lemma [157], which defies straightforward numerical implementation. It appears that a numerical algorithm realizing the *full extent* of the proof of MPL for general nonlinear elliptic BVPs is quite involved and difficult, if not impossible, because the saddle point stated in MPL lies in an infinite dimensional space. Some kind of adaptation is required in order to devise a viable algorithm. Choi and McKenna [47] utilize a constructive form of MPL, an idea from Aubin and Ekeland [13]. We reword their algorithm below:

Mountain–Pass Alogrithm

Step 1. Take an initial guess $w_0 \in E$ such that $w_0 \neq 0$ and $J(w_0) \leqslant 0$, under the assumption that 0 is a local minimum of J;

Step 2. Find $t^* \in (0,1)$ such that $J(t^* w_0) = \max_{t\in[0,1]} J(tw_0)$, and set $w_1 = t^* w_0$;

Step 3. Find the steepest descent direction $\widehat{v} \in E$ such that $[J(w_1 + \varepsilon v) - J(w_1)]/\varepsilon$ is as negative as possible as $\varepsilon \downarrow 0$, obtaining $\widehat{v} = -J'(w_1)$ [see (12.17)] below. If $\|\widehat{v}\| < \varepsilon$, then output and stop. Else go to the next step;

Step 4. Let $\lambda > 0$ be such that $J(w_1 + \alpha\widehat{v})$ attains its minimum at $\alpha = \lambda$, $\forall \alpha > 0$;

Step 5. Redefine $w_0 =: w_1 + \lambda\widehat{v}$. Go to Step 2.

For a semilinear elliptic BVP of the form (12.1), the corresponding functional J is of the form

$$J(v) = \int_\Omega \left[\frac{1}{2}|\nabla v|^2 - F(x,v)\right] dx; \quad F(x,t) \equiv \int_0^t f(x,s)ds, \tag{12.16}$$

with $E \equiv H_0^1(\Omega)$. As noted in Choi and McKenna [47], in Step 3 above, the steepest descent $\widehat{v} = -J'(w_1)$ corresponds to solving a *linear* elliptic BVP

$$\Delta\widehat{v} = -\Delta w_1 - f(x,w_1), \quad \text{on } \Omega, \quad v|_{\partial\Omega} = 0. \tag{12.17}$$

Assume further that $J \in C^2(E,\mathbb{R})$. The *Morse index* of J at a critical point $w \in E$ (cf. Chang [32], e.g.) is defined to be the dimension of the maximal negative definite subspace of $J''(w)$. Choi and McKenna's algorithm applies mainly to solutions that have Morse index 1 of the canonical functional J because the obtained critical point of J is the maximum in only one direction. In certain cases when the underlying domain Ω of the PDE has symmetry, their algorithm may generate solutions with Morse index 2 or higher.
An adapted version of Steps 1–5 above may be found in Ding, Costa and Chen [65]. In this paper, our adapted algorithm is given as follows:

(Adapted) Mountain–Pass Algorithm (MPA)

Step 1. Choose an initial state $w_0 \in H_0^1(\Omega)$; set $w_1 = w_0$.

Step 2. If

$$\|\Delta w_1 + f(w_1)\|_{L^2(\Omega)} \leqslant \varepsilon, \tag{12.18}$$

stop and exit. Otherwise from w_1, solve $\widetilde{v}$:

$$\Delta\widetilde{v} = -f(w_0) \text{ on } \Omega, \widetilde{v}|_{\partial\Omega} = 0. \tag{12.19}$$

Set

$$\widehat{v} = \widetilde{v} - w_1. \tag{12.20}$$

Then $\Delta\widehat{v} = \Delta\widetilde{v} - \Delta w_1 = -[\Delta w_1 + f(w_1)]$.

Step 3. For $t : T > t > 0$, let $\lambda(t)$ be such that[1]

$$J(\lambda(t)(w_1 + t\widehat{v})) = \max_{\lambda\in[0,1]} J(\lambda(w_1 + t\widehat{v})).$$

Find $\widehat{t} : T \geqslant \widehat{t} \geqslant 0$ such that

$$J(\lambda(\widehat{t})(w_1 + \widehat{t}\widehat{v})) = \min_{T\geqslant t\geqslant 0} J(\lambda(t)(w_1 + t\widehat{v})).$$

Step 4. Update: $w_1 := \lambda(\widehat{t})(w_1 + \widehat{t}\widehat{v}), \Delta w_1 := \lambda(\widehat{t})(\Delta w_1 + \widehat{t}\Delta\widehat{v})$. Go to Step 2.

Note that solving (12.17) in the mountain–pass algorithm of Choi and McKenna [47, (13)] is identical to solving (12.19), due to (12.20).

If the DO LOOP in (MPA) above stops after n iterations, (12.18) actually say that[2]

$$\|\Delta w_{n+1} + f(w_{n+1})\|_{L^2(\Omega)} \leqslant \varepsilon.$$

The quantity

$$\varepsilon_{n+1} \equiv \|\Delta w_{n+1} + f(w_{n+1})\|_{L^2(\Omega)} \tag{12.21}$$

serves as an excellent indicator of how closely w_{n+1} satisfies $\Delta u + f(u) = 0$. We may thus regard ε_n as an *absolute convergence* error indicator.

Remark 12.2. Return to Example 12.2. Ding, Ni [66] and Ni [142] have shown that under (12.10), we have

$$c = \inf_{v \in \mathscr{M}} J(v) = \inf_{\substack{v>0 \\ v \in H_0^1(\Omega)}} \max_{t \geqslant 0} J(tv), \tag{12.22}$$

where c is the smallest possible positive critical value as given in the MPL. Furthermore, the two infima in (12.22) are both attained, and c is independent of the choice e in Theorem 12.1.

Our adapted algorithm MPA actually has considerable feature more similar to the mini-maximization problem on the right-hand side of (12.22), than to the proof of the MPL itself. Thus, perhaps it should be called MMA (Mini-Max Algorithm) instead.

The critical point as guaranteed by (12.22) has the lowest energy value J. Borrowing terminology from quantum mechanics, we call this critical point (i.e., solution) a (or sometimes, the) *ground state*, or a least-energy solution. Other critical points, which are local minima of $J(v)$ on the solution manifold $\mathscr{M}$, are called *local ground states* in this chapter. □

The very original algorithm due to Choi and McKenna [47] was not devised according to (12.15) in the sense that they did not determine the critical point by testing the inf on all paths Γ in (12.15). Thus, rigorously speaking, their algorithm, as well as our adapted version, cannot be called a mountain–pass algorithm. Nevertheless, MPA here is effective in finding local ground states, with multiplicity generated through varying the initial states.

The strongest result concerning the uniqueness of the ground state so far may be found in Lin [124], where he showed that on a convex domain, the ground state is unique.

More recently, Li and Zhou [121, 122] developed and implemented *minimax methods* to compute multiple solutions of semilinear elliptic PDEs of higher Morse indices without any assumption on domain symmetry.

To aid in the visual understanding we include Figs. 12.1 and 12.2 to illustrate MPA.

[2] In our computer programs, we begin by choosing $T = 1$ for the first three iterations, and gradually increase T to 10 (or larger, if necessary).

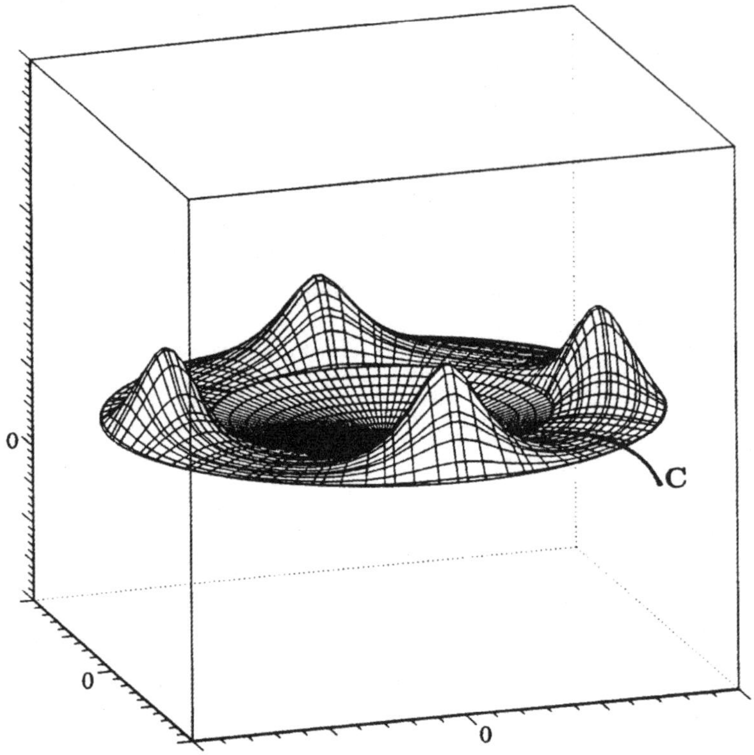

Figure 12.1: The origin is located at the center of a basin, with altitude 0. The protruding dark curve C on the right signifies an optimal path through a mountain pass to get out of the basin to another point, also with altitude 0. There are several other mountain passes overlooking the depression, but their altitudes are higher than that of the mountain pass which is crossed by the path C.

12.2.2 *The scaling iterative algorithm (SIA)*

To explain how this algorithm works, let us use the following problem as a model:

$$\Delta u - au + bu^p = 0, \quad u > 0 \quad \text{on } \Omega, \quad u|_{\partial\Omega} = 0, \tag{12.23}$$

where $a \geqslant 0$ and $b > 0$ are given constants, and $p > 1$. By Example 12.2 or MPL, we know that (12.23) has at least one solution. Choose a sequence of numbers $\beta_n > 0$, $n = 1, 2, \ldots$ and define $v_n(x) = u(x)/\beta_n$. Then each v_n is a scaling of u; v_{n+1} satisfies

$$\left.\begin{aligned} &\Delta v_{n+1}(x) - av_{n+1}(x) + \alpha_{n+1} b v_n^p(x) = 0 \text{ on } \Omega, \alpha_{n+1} \equiv \frac{\beta_n^p}{\beta_{n+1}}, \\ &v_{n+1}(x) > 0 \qquad\qquad\qquad\qquad\qquad\quad \text{on } \Omega, \\ &v_{n+1}|_{\partial\Omega} = 0. \end{aligned}\right\} \tag{12.24}$$

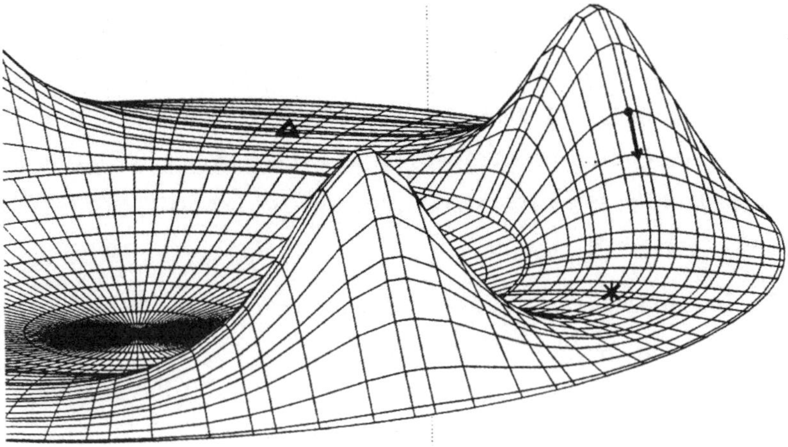

Figure 12.2: This picture is a zoom of the right portion of Fig. 12.1. Note that "$*$" signifies the location of the mountain pass with the lowest altitude. This is the critical point established in (12.22). We call it a (or, sometimes, the) (global) ground state. The location marked with "Δ" is another mountain pass whose altitude is higher than "$*$"'s. We call "Δ" a local ground state. The arrow indicates a steepest descent direction. It takes several changes of descent directions in order to reach a close vicinity of "$*$".

The equations in (12.24) suggest the following iteration algorithm:

Step 1. Choose any $u_0(x) \geqslant 0$ on Ω, u_0 sufficiently smooth, $u_0 \not\equiv 0$;

Step 2. Let

$$\beta_0 = \|u_0\|_{L^\infty(\Omega)} \quad \text{and} \quad v_0 = \frac{u_0}{\beta_0},$$

and solve $v_{n+1}(\cdot)$ and $\alpha_{n+1} > 0$ satisfying

$$\begin{cases} \Delta v_{n+1}(x) - a v_{n+1}(x) = -\alpha_{n+1} b v_n^p(x) \text{ on } \Omega, \\ \|v_{n+1}\|_{L^\infty(\Omega)} = 1, \\ v_{n+1}|_{\partial\Omega} = 0; \qquad n = 0, 1, 2, \ldots, ; \end{cases} \tag{12.25}$$

and let

$$\beta_{n+1} = \frac{\beta_n^p}{\alpha_{n+1}}, \qquad n = 0, 1, 2, \ldots . \tag{12.26}$$

Step 3. If $\delta_{n+1} = \|v_{n+1} - v_n\|_{H_0^1(\Omega)} < \varepsilon$, let

$$u_{n+1}(\cdot) = \beta_{n+1} v_{n+1}(\cdot), \tag{12.27}$$

output and stop. Else go to Step 2.

The usefulness of the above algorithm is shown in the following.

Theorem 12.2. *Let Ω be a sufficiently smooth bounded open domain in $\mathbb{R}^2$. Assume that $a \geqslant 0, b > 0, p > 0$ and $p \neq 1$. Choose any sufficiently smooth $v_0(x) \geqslant 0$ on Ω, $v_0 \not\equiv 0$, and let α_{n+1} and $v_{n+1}(\cdot)$ satisfy (12.25) and (12.26). If $\{\alpha_n \mid n = 1, 2, \ldots\}$ is bounded, then $(v_{n+1}(\cdot), \alpha_{n+1})$ has a convergent subsequence $(\widetilde{v}_{n+1}(\cdot), \widetilde{\alpha}_{n+1}) \to (v_\infty(\cdot), \alpha_\infty)$ in $C^{1,\gamma}(\overline{\Omega}) \times \mathbb{R}$ for any γ: $0 < \gamma < 1$.*

Proof. The iterations in (12.25) and (12.26) can be rewritten into two steps

$$\left\{\begin{array}{l} \Delta w_{n+1} - a w_{n+1} = -b v_n^p, \\ w_{n+1}|_{\partial\Omega} = 0, \end{array}\right\} \quad n = 0, 1, 2, \ldots, \tag{12.28}$$

and

$$\left\{\begin{array}{l} \Delta v_{n+1} - a v_{n+1} = -\alpha_{n+1} b v_n^p, \\ v_{n+1}|_{\partial\Omega} = 0, \end{array}\right\} \quad n = 0, 1, 2, \ldots,$$

where

$$v_{n+1} = \frac{w_{n+1}}{\|w_{n+1}\|_{L^\infty(\Omega)}}, \qquad \alpha_{n+1} = \frac{1}{\|w_{n+1}\|_{L^\infty(\Omega)}}. \tag{12.29}$$

Since $\|v_n\|_{L^\infty(\Omega)} = 1$, by the standard elliptic estimates on (12.28) Gilbarg and Trudinger [79], we have some $C > 0$ such that

$$\|w_{n+1}\|_{W^{2,q}(\Omega)} \leqslant C, \quad \forall q > 1, \quad n = 0, 1, 2, \ldots$$

Therefore, the sequence $\{\alpha_{n+1}\}$ is bounded away from 0, and, by the Sobolev Imbedding Theorem, w_{n+1} has a bounded convergent subsequence $\widetilde{w}_{n+1}$ such that $\widetilde{w}_{n+1} \to w_\infty$ in $C^{1,\gamma}(\overline{\Omega})$ for any γ: $0 < \gamma < 1$. By assumption, $\{\alpha_{n+1} \mid n = 1, 2, \ldots\}$ is bounded, the sequence $\{\overline{\alpha}_{n+1} \mid n = 0, 1, 2, \ldots\}$ is also bounded. Since $w_{n+1} > 0$ in Ω by the standard maximum principle, it follows from $(12.29)_2$ that $w_\infty > 0$ in Ω. Choose a convergent subsequence of $\{\widetilde{\alpha}_{n+1}\}$ and still call it $\{\overline{\alpha}_{n+1}\}$, such that $\widetilde{\alpha}_n \to \alpha_\infty > 0$. From (12.29), we conclude that $\overline{v}_{n+1} = \widetilde{\alpha}_{n+1}\widetilde{w}_{n+1} \to \alpha_\infty w_\infty \equiv v_\infty$ in $C^{1,\gamma}(\Omega)$. □

Remark 12.3.

(1) We suspect that if $p > 1$, then the assumption that $\{\alpha_n \mid n = 1, 2, \ldots\}$ be bounded is unnecessary. (However, if $0 < p < 1$, there is some experimental indication that this assumption is required.)

(2) The iterations in (12.28)–(12.29) above define a map T: $w_{n+1} = T w_n$. In order to claim that the limit w_∞ of the subsequence $\widetilde{w}_n$ in the proof of Theorem 12.2 yields a solution of (12.23), we must establish that w_∞ is a fixed point of the map T. We will have not been able to achieve this so far.

(3) By restricting p to $0 < p < p^* \equiv (N+2)/(N-2)$, one can establish a version of Theorem 12.2 for $\Omega \subset \mathbb{R}^N, N \geqslant 3$. □

Equation $(12.25)_2$, or equivalently, $(12.29)_1$ requires the search of max w_{n+1} on the entire Ω in order to determine

$$\alpha_{n+1} = \frac{1}{\|w_{n+1}\|_{L^\infty(\Omega)}} = \frac{1}{\max w_{n+1}}.$$

This is computationally feasible but inconvenient as far as implementation is concerned. Based upon our understanding of the problem (12.23), as we will see from the graphics in §12.3 and 12.4, we know that its solutions display the pattern of a *spike-layer*, and thus the maxima of successive iterates often occur just near the "center" x_0 of some subdomain G of Ω where "Ω has the most open space"; see Theorem 12.5. This suggests the choice of $\beta_n = u_n(x_0)$ in (12.24) for $n = 1, 2, \ldots$. Therefore $(12.25)_2$ now becomes $v_{n+1}(x_0) = u_{n+1}(x_0)/u_{n+1}(x_0) = 1$, and we obtain the following simpler algorithm.

Scaling Iterative Algorithm (SIA)

Step 1. Choose any $v_0(x) \geqslant 0$ on $\Omega, v_0 \not\equiv 0; v_0$ sufficiently smooth;

Step 2. Find $\alpha_{n+1} > 0$ and $v_{n+1}(\cdot)$ such that

$$\begin{cases} \Delta v_{n+1}(x) - a v_{n+1}(x) = -\alpha_{n+1} b v_n^p(x) \text{ on } \Omega, \\ v_{n+1}(x_0) = 1, \\ v_{n+1}|_{\partial\Omega} = 0; \end{cases} \tag{12.30}$$

Step 3. If

$$\widetilde{\varepsilon}_n \equiv \|v_{n+1} - v_n\|_{H_0^1(\Omega)} < \varepsilon, \tag{12.31}$$

output and stop. Else go to Step 2.

Note that the $\widetilde{\varepsilon}_n$ in (12.31) provides a *relative convergence error indicator*. Unlike the ε_n in (12.21), $\widetilde{\varepsilon}_n$ here does not provide information as to how closely v_{n+1} satisfy the equation $\Delta u + f(u) = 0$.

The key idea in SIA lies in condition $(12.30)_2$. By requiring $v_{n+1}(x_0) = 1$, we hope to avoid, as much as possible, the possibility that $v_n \to 0$, i.e., the convergence to the trivial solution 0. This *pointwise normalization condition* $(12.30)_2$ is an outgrowth of our empirical pursuit of iteration algorithms for (12.21) because, in our experimental trials of various possible designs of iterative algorithms for (12.23), we have observed that the number one cause of algorithm failures is $v_n \to 0$ or $v_n \to \infty$.

The following theorem is obvious.

Theorem 12.3. *Consider the iterative algorithm (12.30). Let* $v_0(\cdot)$ *be sufficiently regular and* $v_0(x) \geqslant 0$, $v_0(x) \underset{\not\equiv}{>} 0$ *on* Ω. *Assume that* $(v_n(\cdot), \alpha_n)$ *converges to* $(v_\infty(\cdot), \alpha_\infty)$ *in* $H_0^1(\Omega) \oplus \mathbb{R}$ *with* $\alpha_\infty \neq 0$. *Then* $u \equiv \alpha_\infty^{\frac{1}{p-1}} v_\infty$ *is a solution of (12.30).*

Remark 12.4. According to our numerical experience, the choice of the location of $x_0 \in \Omega$ rarely has affected the limit of the converging subsequence. However, at this point we are still unable to offer any rigorous proof concerning the convergence (of a subsequence) of SIA. □

Example 12.3. Here we provide some data for the comparison of MPA and SIA. Let us leap ahead to Subsection 12.3.5 concerning the computation of positive solutions of Lane–Emden's equation $\Delta u + u^3 = 0$ with zero Dirichlet condition on the dumbbell-shaped domain Ω_7.

(i) Choosing $u_0(x) =$ RHS of (12.63) on Ω_7 and using (12.64), we obtain an initial state $w_0 \in H_0^1(\Omega)$ for MPA iterations. We get a sequence which is numerically convergent to the ground state u, with

$$\max u = 3.562, \quad J = \text{ the energy } (16) = 10.90, \quad \varepsilon_n = 10^{-4}, \quad n = 7.$$

(ii) Choosing the very same initial state $u_0(x)$ for SIA iterations, we also get a numerically convergent sequence with the same limit u, with

$$\max u = 3.562, \quad J = 10.90, \quad \widetilde{\varepsilon}_n = 10^{-6},$$
$$\alpha_n = 12.69, \quad n = 9, \quad x_0 = (2,0).$$

We see that MPA and SIA provide about the same numerical efficiency and accuracy. The number of iterations required here are, respectively, 7 and 9. In general, for most cases, only some twenty iterations will suffice by either MPA and SIA. □

12.2.3 *The direct iteration algorithm (DIA) and the monotone iteration algorithm (MIA)*

A straightforward iteration algorithm for a semilinear elliptic BVP (12.1) can be stated as follows:

Direct Iteration Algorithm (DIA) for (12.1)

Step 1. Choose a sufficiently smooth initial state $v_0(\cdot)$;

Step 2. Find $v_{n+1}(\cdot)$ such that

$$\begin{cases} \Delta v_{n+1}(x) = -f(x, v_n(x)),\ x \in \Omega, \\ v_{n+1}|_{\partial\Omega} = 0; \end{cases} \tag{12.32}$$

Step 3. If $\|v_{n+1} - v_n\|_{H_0^1(\Omega)} < \varepsilon$, output and stop. Else go to Step 2.

DIA lacks sophistication . Even for a "very stable" nonlinear ODE like

$$u''(x) - u^3(x) = 0 \quad \text{on} \quad x \in [0,1]; \quad u(0) = u(1) = 0, \tag{12.33}$$

we already showed in Example 11.1 that DIA may not produce convergent solutions in general. For nearly all the nonlinear equations studied in this chapter, DIA either diverges quickly or converges to the trivial solution 0. The only exception is the case with sublinear nonlinearity; see § 12.6. In that case, DIA actually provides an algorithm more convenient and efficient than the others. In general, we must use the monotone iteration scheme (MIA), see the Flow-Chart in § 11.7.

12.2.4 *A boundary element numerical elliptic solver based on the simple-layer and volume potentials*

Each iterative algorithm MPA, SIA, DIA or MIA requires a numerical elliptic PDE solver. In principle, the three basic numerical methods FDM, FEM and BEM should all be viable. The basis of BEM solutions of inhomogeneous linear elliptic BVPs is given in the following theorem, wherein the solution is represented as a sum of a volume (or synonymously, Newtonian) potential and a simple-layer potential.

Theorem 12.4. *Let Ω be a bounded open domain in $\mathbb{R}^2$ with C^∞-smooth boundary $\partial\Omega$. Then the solution of*

$$\begin{cases} \Delta u = f \in H^{s_1}(\Omega), & s_1 \geqslant -1, \\ u|_{\partial\Omega} = g \in H^{s_2}(\partial\Omega), & s_2 \in \mathbb{R}, \end{cases} \tag{12.34}$$

can be uniquely represented as

$$u(x) = \int_\Omega E(x-y)f(y)dy + \int_{\partial\Omega} E(x-y)\eta(y)d\sigma_y + \mathrm{a} \in H^{r_1}(\Omega), \tag{12.35}$$

where $E(x-y) = -(1/2\pi)\ln|x-y|$, $r_1 = \min(s_1+2, w_2+3/2)$, and η, the unknown simple-layer density, and a, *an unknown constant, can be uniquely solved by the bound-*

ary integral equations (BIEs)

$$\int_{\partial\Omega} \eta(y)d\sigma_y = 0, \quad \int_{\partial\Omega} E(x-y)\eta(y)d\sigma_y + a$$
$$= -\int_{\Omega} E(x-y)f(y)dy + g(x), \quad \forall x \in \partial\Omega, \tag{12.36}$$

with $\eta \in H^{r_2}(\partial\Omega)$, $r_2 = \min(s_1+1, s_2-1)$.

Proof. See Chapter 6. □

Remark 12.5. If somehow it is known that the simple-layer equation

$$\int_{\partial\Omega} E(x-y)\eta(y)d\sigma_y = 0, \quad x \in \partial\Omega,$$

does not have a nontrivial solution η, then instead of solving the two BIEs in (12.36), one can set a $= 0$ in (12.35) and (12.36), and solve the simpler BIE

$$\int_{\partial\Omega} E(x-y)\eta(y)d\sigma_y = -\int_{\Omega} E(x-y)f(y)dy + g(x), \quad \forall x \in \partial\Omega, \tag{12.37}$$

and represent the solution u of (12.34) as

$$u(x) = \int_{\Omega} E(x-y)f(y)dy + \int_{\partial\Omega} E(x-y)\eta(y)d\sigma_y, \quad x \in \overline{\Omega}. \tag{12.38}$$

See Remark 6.7. Indeed, for all domains Ω_i, $i = 1, 2, \ldots, 9$, used in the remaining sections for computations (excluding Ω_4 because it is "pathological"), only Ω_1, the unit disk, requires the use of (12.35) and (12.36) with $a \neq 0$. For the rest Ω_i, $i \neq 1, 4$, we can just use (12.37) and (12.38).

Note that if $N = 3$, then the representation (12.38) is unique without requiring any $a \in \mathbb{R}$ as in (12.35), with the fundamental solution $E(x) = [4\pi|x|]^{-1}$ being used in (12.37) and (12.38). □

The way to use BEM to solve the elliptic BVPs (12.17) and (12.32) in MPA and DIA, respectively, is now clear form Theorem 12.4 and Remark 12.1. The only case that is not totally clear is that for (12.30) in SIA. Actually, this requires just a tiny amount of extra work. Let us use (12.37) and (12.38). Return to (12.30), and consider for the time being $a = 0$ therein. Then we first solve

$$\Delta v^0_{n+1}(x) = -bv^p_n(x) \quad \text{on } \Omega, \quad v^0_{n+1}|_{\partial\Omega} = 0,$$

by writing

$$v^0_{n+1}(x) = -b\int_{\Omega} E(x-y)v^p_n(y)dy + \int_{\partial\Omega} E(x-y)\eta^0(y)d\sigma_y, \quad x \in \overline{\Omega}, \tag{12.39}$$

according to (12.37) with $E(x-y) = -\ln|x-y|/2\pi$, where the unknown simple-layer density η^0 is the unique solution of the BIE

$$\int_{\partial\Omega} E(x-y)\eta^0(y)d\sigma_y = b\int_\Omega E(x-y)v_n^p(y)dy, \quad x\in\partial\Omega. \tag{12.40}$$

If $v_0(x) > 0$, then by the maximum principle $v_{n+1}^0(x) > 0$ on Ω. Therefore $v_{n+1}^0(x_0) > 0$. Define

$$v_{n+1}(x) = \frac{v_{n+1}^0(x)}{v_{n+1}^0(x_0)}, \quad \alpha_{n+1} = [v_{n+1}^0(x_0)]^{-1}, \quad \eta(\cdot) = \eta^0(\cdot)[v_{n+1}^0(x_0)]^{-1}. \tag{12.41}$$

Then v_{n+1} satisfies

$$\begin{cases} \Delta v_{n+1}(x) = -\alpha_{n+1} b v_n^p(x) & \text{on } \Omega, \\ v_{n+1}(x_0) = 1, \quad v_{n+1}|_{\partial\Omega} = 0. \end{cases}$$

If $a \gneqq 0$ in (12.30), we repeat (12.40) and (12.41), except that now we use

$$E(x-y;a) = \begin{cases} \dfrac{1}{2\pi}K_0(\sqrt{a}|x-y|), & N=2 \\ \dfrac{e^{-\sqrt{a}|x-y}}{4\pi|x-y|}, & N=3, \end{cases} \tag{12.42}$$

therein. Then v_{n+1} satisfies (12.30).

We now determine the enumeration of successive errors for the BEM approach. We use (12.30) as an exemplar case, because the others are virtually the same. By (12.38)–(12.42), we have the representation

$$v_{n+1}(x) = -\alpha_{n+1} b\int_\Omega E(x-y;a)v_n^p(y)dy + \int_{\partial\Omega} E(x-y)\eta(y)d\sigma_y, \qquad x\in\overline{\Omega}.$$

Then $\Delta v_{n+1} - a v_{n+1} = -\alpha_{n+1} b v_n^p$. Similarly, $\Delta v_n - a v_n = -\alpha_n b v_{n-1}^p$. Therefore the error $\widetilde{\varepsilon}_n$ in (12.31) can be obtained by the following:

$$\begin{aligned} \widetilde{\varepsilon}_n^2 &= \int_\Omega [|\nabla(v_{n+1}-v_n)|^2 + a|v_{n+1}-v_n|^2]\,dx = \|v_{n+1}-v_n\|^2_{H_0^1(\Omega)} \\ &= -\int_\Omega [\Delta(v_{n+1}-v_n) - a(v_{n+1}-v_n)](v_{n+1}-v_n)\,dx \\ &= b\int_\Omega (\alpha_{n+1}v_n^p - \alpha_n v_{n-1}^p)(v_{n+1}-v_n)\,dx. \end{aligned} \tag{12.43}$$

In the evaluation of the volume potential in (12.39), v_n^p and v_{n-1}^p have already been computed and put into storage in the computer. We just fetch those data, substitute them into (12.43) and thus obtain $\widetilde{\varepsilon}_{n+1}$.

The key step in the BEM approach is the numerical solution of the BIE (12.12). It is solved by discretization using an $S_h^{\ell,m}(\partial\Omega)$ system as noted in (11.78) and (11.79).

A *collocation scheme* is used for a basis $\{\phi_{h,i}, 1 \leqslant i \leqslant n(h)\} \subseteq S_h(\partial\Omega)$, write

$$\eta_h = \sum_{i=1}^{n(k)} a_i \phi_{h,i}. \tag{12.44}$$

One chooses a set of collocation points $\{x_i \mid 1 \leqslant i \leqslant n(h)\} \subseteq \partial\Omega$ and solves the unknown coefficients a_i in (12.44) from

$$(\mathscr{L}\eta_h)(x_i) = G(x_i), \quad i = 1, 2, \ldots, n(h), \tag{12.45}$$

i.e.,

$$\sum_{j=1}^{n(h)} a_j \int_{\partial\Omega} E(x_i - y)\phi_{h,i}(y) d\sigma_y = G(x_i), \quad i = 1, 2, \ldots, n(h). \tag{12.46}$$

All domains $\Omega_i, i = 1, 2, \ldots, 9$, in subsequent sections are C^∞, except for Ω_4 in Subsection 12.3.3, which is "pathological," and Ω_j, $j = 7, 8, 9$ in Subsections 12.3.5–12.3.7 which have two or four obtuse angular corner points. For such nonsmooth domains, loss of regularity of solutions may occur from elliptic BVPs; see Grisvard [84]. However, these domains Ω_j, $j = 7, 8, 9$, can be modified to be C^∞ by a minuscule local smoothing of $\partial\Omega$ near the corner points in an obvious way. We have found no discernible difference of accuracy at all between Ω_j, $j = 7, 8, 9$ and the modified domains where all the corner points have been smoothed out by minuscule local refining using piecewise quadratic curve segments near the corner points. Thus, for all practical purposes, we may regard Ω_j, $j = 7, 8, 9$, as C^∞ domains.

Also occasional in use in this chapter is FDM, *when the domain and the solution has radial symmetry*; see (12.60), Remarks 12.7 and and 12.9, etc. FDM solutions also provide corroborations for, and comparison of accuracy with, the BEM solutions. But FDM applies to very few limited cases in this paper.

12.3 Graphics for visualization of the Dirichlet problem of $\Delta u + u^3 = 0$

The main equation, whose solutions are to be computed by BEM and visualized in great detail in this section, is the Lane–Emden equation $\Delta u + u^3 = 0$. Other variant equations, to be studied in subsequent sections, have solutions with strikingly similar profiles; their graphics will be displayed only for selected geometries.

We have chosen four types of domains for the computation of numerical solutions:

(i) the unit open disk;
(ii) concentric on nonconcentric annuli;
(iii) dumbbell-shaped domains with varying corridor width;
(iv) dumbbell-shaped domains with cavities. (12.47)

The rationale for choosing (i)–(iv) is based on the *special geometrical and topological features* offered by each type of domain(s); the disk, type (i), has the strongest symmetry, whereupon the analytic information about the solution is also best known from the work of Gidas, Ni and Nirenberg [77, 78]. We compute solutions on (i) also for the purpose of *setting a benchmark for other analysts and researchers*. By changing from a disk to an annulus, i.e., to type (ii), the topology of the domain has lost simple connectedness and becomes multiconnected with "genus 1," i.e., with 1-connectivity. Topologically, its fundamental group is isomorphic to $\mathbb{Z}$ [69]. Rotational symmetry is destroyed after the internal and external bounding circles of the annulus are made nonconcentric. Furthermore, an annular domain is never convex.

Dumbbell-shaped domains were mentioned much earlier on by a few researchers (who should be credited with priority, however, is unclear) as an interesting type of domain in the study of behavior of solutions of BVPs. It has particular significance in mathematical biology and population dynamics due to the *compartmental feature and effect* such domains have [130]. Our dumbbell-shaped domains are constructed by connecting two nonidentical disks through a straight (symmetric) corridor. When the width of the corridor is small, the dumbbell-shaped domain is *non-starshaped*. As the width increases and eventually equals the diameter of the smaller disk, the domain then becomes starshaped. Therefore, the reader can see the *degeneration of dumbbell-shaped domains from being non-starshaped into a starshaped one*.

Lastly, we choose dumbbell-shaped domains with two cavities, i.e., of type (iv), for the desirable special features that they are non-starshaped, lack any global symmetry while still maintaining some local symmetry, and have two-connectivity. All told, nine different, mostly dissimilar domains Ω_i, $i = 1, 2\ldots, 9$, have been selected for use here and in subsequent sections. A total of 21 cases will be computed and visualized in this section.

We are now in a position to treat the problem

$$\begin{cases} \Delta u + u^3 = 0 & \text{on } \Omega, \\ u > 0 & \text{on } \Omega, \\ u|_{\partial\Omega} = 0. \end{cases} \tag{12.48}$$

The corresponding energy functional of (12.48) is

$$J(v) = \int_{\Omega} \left[\frac{1}{2} |\nabla v|^2 - \frac{1}{4} v^4 \right] dx, \quad v \in H_0^1(\Omega). \tag{12.49}$$

It is easy to check that for any solution u of (12.48), we have the energy level $J(u) > 0$.

12.3.1 *The unit disk*

Here $\Omega_1 = \{x \in \mathbb{R}^2 \mid |x| < 1\}$. The boundary $\partial\Omega_1$ is divided into 384 uniform panels, and the number of Gaussian quadrature points for the domain integral (i.e., the first integral on the RHS) in (12.35) is 1537.

Case 12.3.1.a The unique, radially symmetric positive solution [77] on a disk. This is the ground state of (12.48), displayed in Fig. 12.3, with

$$\text{(SIA)} \qquad \max u = 3.5741, \quad J = 10.99, \quad \widetilde{\varepsilon}_{13} = 10^{-6}, \alpha_{13} = 12.77, \ x_0 = (0,0).$$

Here and in the following, (SIA) denotes the algorithm used, $\max u$ denotes the global approximate maximum value of the solution, $J = J(u)$ is the energy level of the solution, $\widetilde{\varepsilon}_{13}$ denotes the relative convergence error from (12.31) and (12.43), α_{13} and x_0 denote those used in (12.30). The choice of the initial state v_0 in Step 1 of SIA is unimportant here so we do not need to describe it.

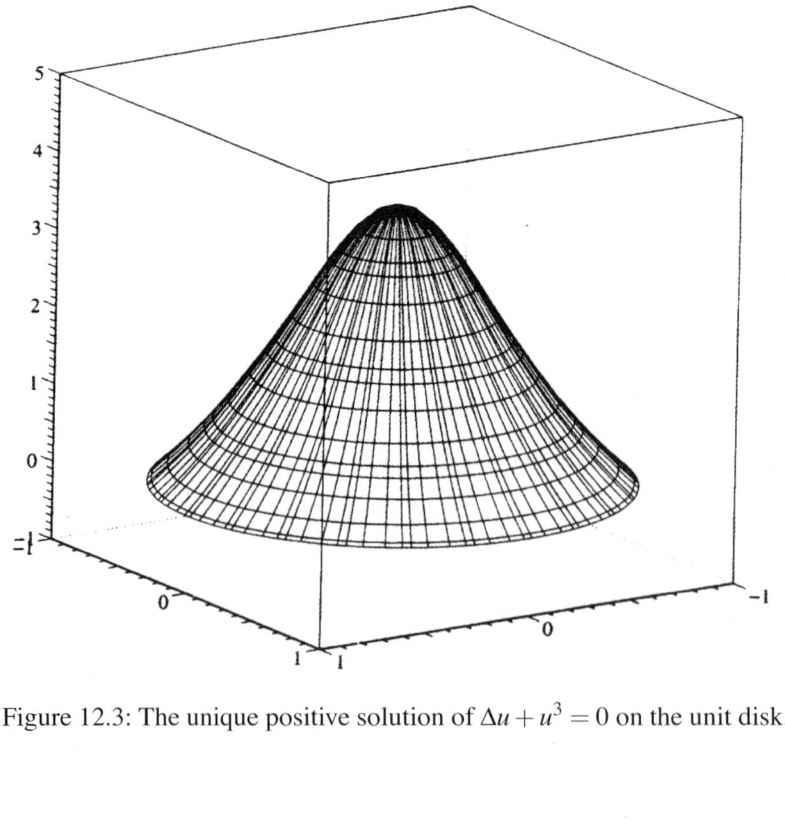

Figure 12.3: The unique positive solution of $\Delta u + u^3 = 0$ on the unit disk.

12.3.2 *Nonconcentric annular domains*

First, let

$$\Omega_2 = \{x \in \mathbb{R}^2 \mid |x| < 0.9,\ |x-(0.2,0)| < 0.5\}. \tag{12.50}$$

On the boundary $\partial\Omega_2$, $384+192=576$ uniform panels are used, with 192 of them placed on the inner circle of Ω_2.

Case 12.3.2.a The ground state on the nonconcentric annular domain Ω_2. The ground state and its contours are displayed in Figs. 12.4(a) and 12.6(b). We have

(SIA) $\quad \max u = 9.12, \quad J = 72.09, \quad \widetilde{\varepsilon}_{16} = 10^{-5}, \quad \alpha_{16} = 80.19, \quad x_0 = (-0.60, 0).$

Next, we change Ω_2 in (12.50) by moving its inner boundary to be farther off-center. Let

$$\Omega_3 = \{x \in \mathbb{R}^2 \mid |x| < 0.9,\ |x-(0.35,0)| < 0.5\}. \tag{12.51}$$

Case 12.3.2.b The ground state on the nonconcentric annular domain Ω_3. The ground state and its contours are displayed in Figs. 12.5(a) and 12.7(b). We have

(SIA) $\quad \max u = 7.38, \quad J = 47.43, \quad \widetilde{\varepsilon}_{11} = 10^{-4}, \quad \alpha_{11} = 51.99, \quad x_0 = (-0.525, 0).$

We have not been able to find any other positive solutions on domains (12.50) and (12.51).

12.3.3 *A "pathological" annulus, with boundary formed by two tangent circles*

If we push the inner bounding circle in (12.50) and (12.51) to the extreme, we obtain a domain defined by

$$\Omega_4 = \{x \in \mathbb{R}^2 \mid |x| < 0.9,\ |x-(0.4,0)| < 0.5\}. \tag{12.52}$$

In contrast to Ω_2 and Ω_3, Ω_4 returns to being *simply connected*. Once again, 576 panels are used to discretize $\partial\Omega_4$. It is "pathological" in the sense that, at the point $\widehat{x} = (0.9, 0)$ on $\partial\Omega_4$, the unit exterior normal is not well defined. There are two *cusps* with vertex at $\widehat{x}$. If a domain contains cusps, then the *cone condition fails* (at $\widehat{x}$), and many classical elliptic estimates may not hold. Nevertheless, numerical computations of (12.48) by (12.35)–(12.38) using BEM can proceed without any trouble at all because, in the representation (12.35) or (12.38), we never need to use the normal derivative $\partial/\partial n$. This is certainly an advantage of BEM.

Case 12.3.3.a The ground state on the pathological annular domain Ω_4. The ground state is displayed in Figs. 12.6(a) and 12.6(b). We have

(SIA) $\quad \max u = 6.95, \quad J = 42.14, \quad \widetilde{\varepsilon}_{12} = 10^{-4}, \quad \alpha_{12} = 47.21, \quad x_0 = (-0.5, 0).$

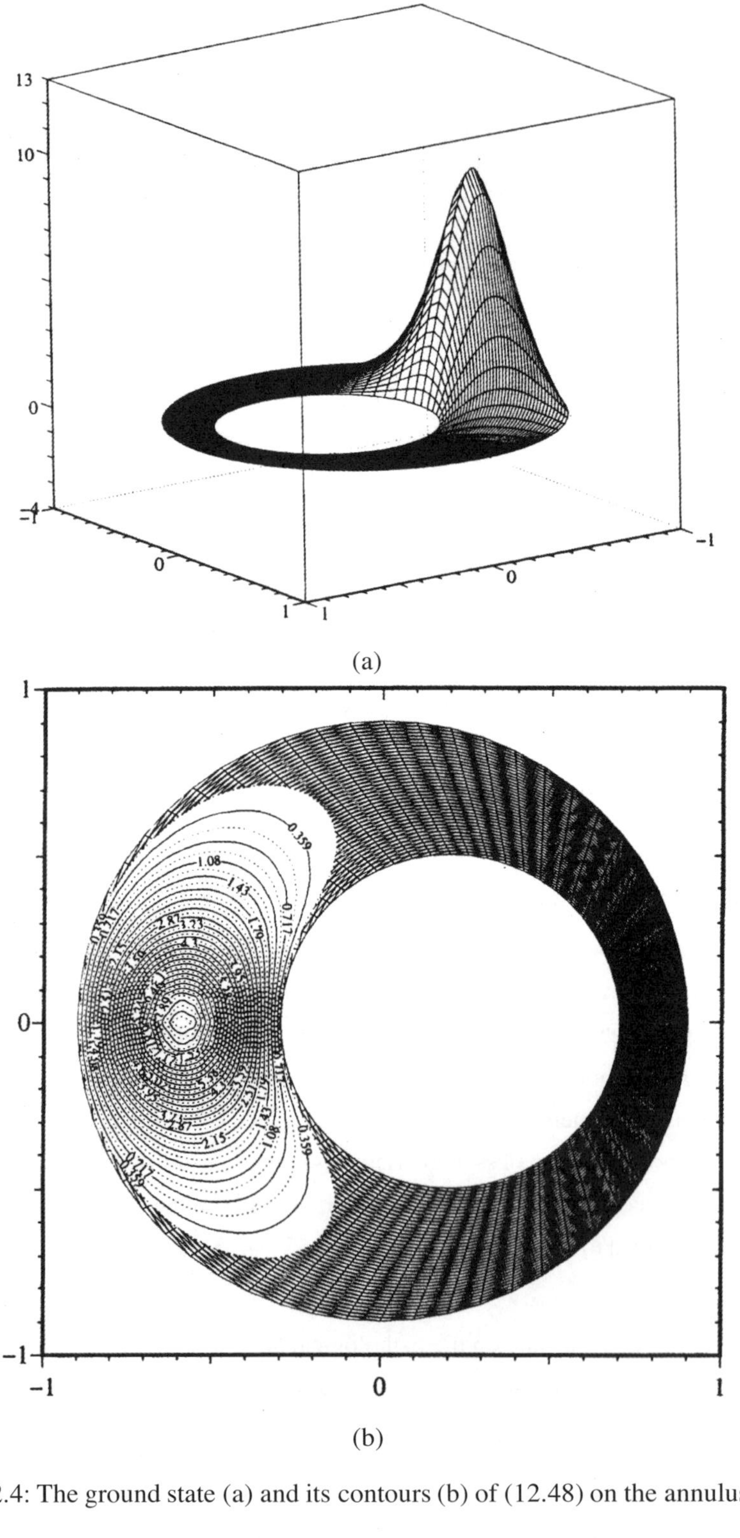

Figure 12.4: The ground state (a) and its contours (b) of (12.48) on the annulus (12.50).

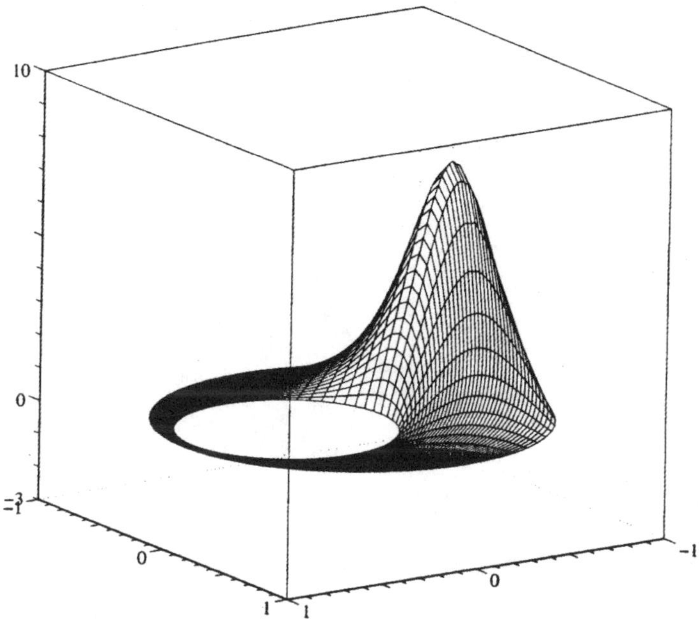

(a)

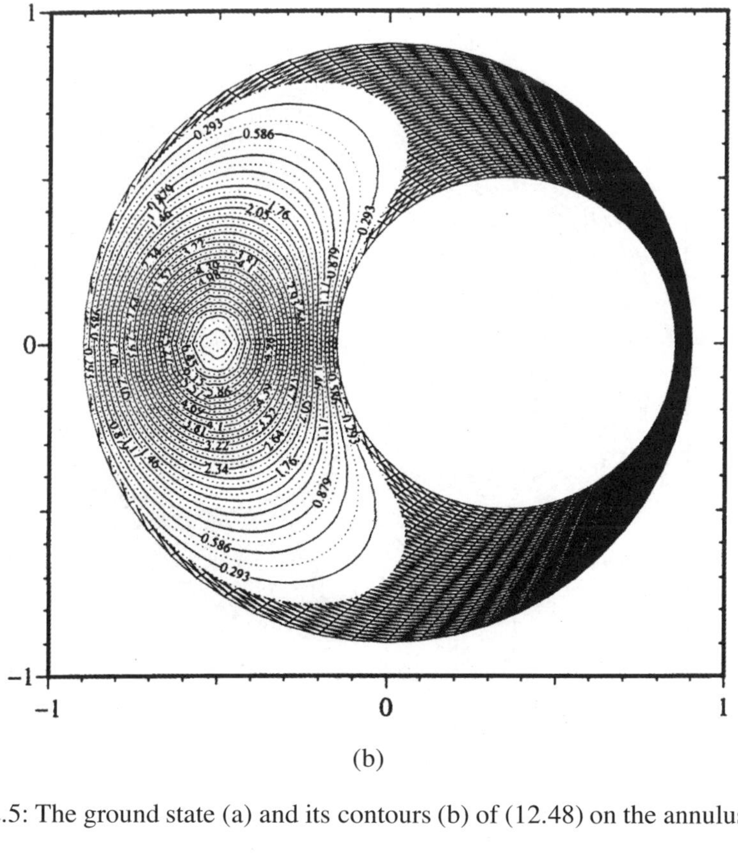

(b)

Figure 12.5: The ground state (a) and its contours (b) of (12.48) on the annulus (12.51).

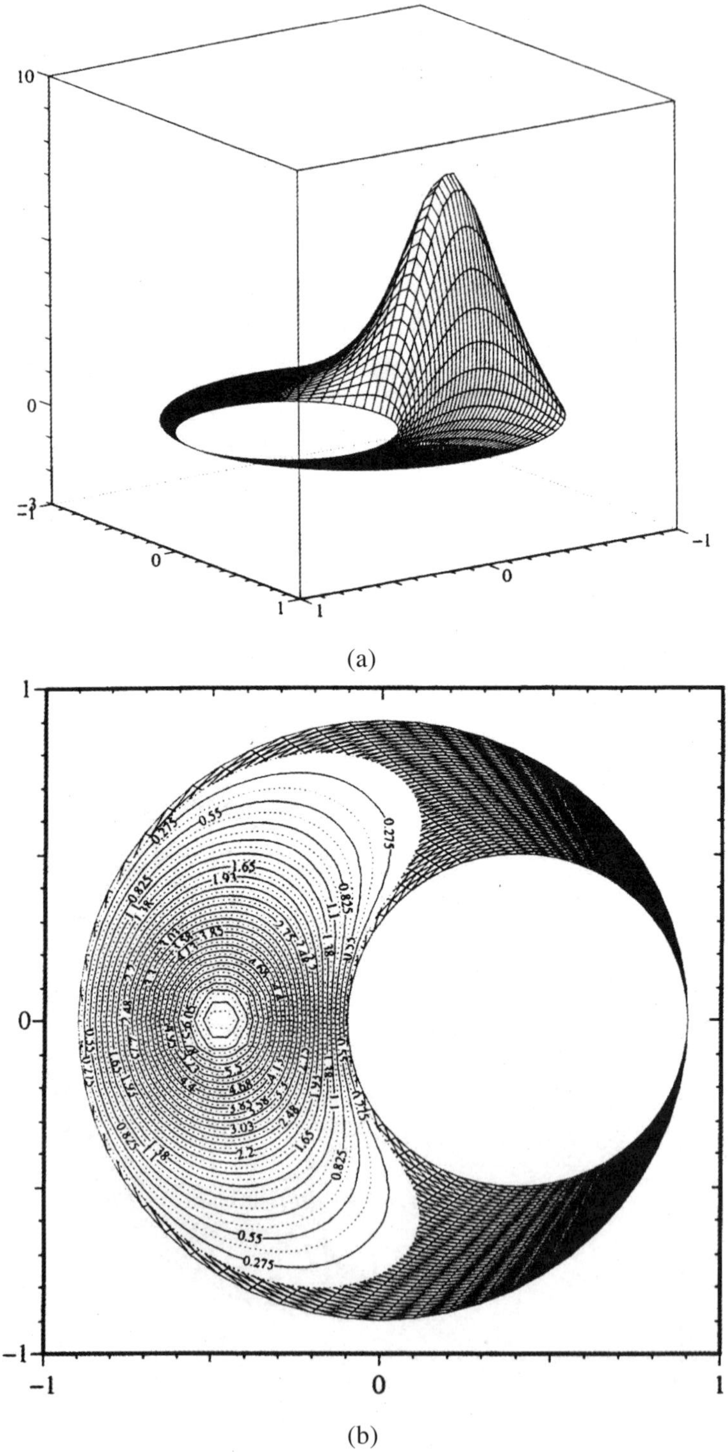

Figure 12.6: The ground state (a) and its contours (b) of (12.48) on the pathological annulus.

12.3.4 *The radially symmetric annulus*

First, let

$$\Omega_5 = \{x \in \mathbb{R}^2 \mid 0.5 < |x| < 0.9\}. \tag{12.53}$$

A ground state of (12.48) on this radially symmetric annulus Ω_5 is known (see Coffman [51] and Li [120]) to be *nonradially symmetric*, if the inner radius of the annulus increases and passes a certain positive number. Here we will actually see that the ground state looks like *a hill with a single peak*. Because of the symmetry which Ω_5 has, any rotation of a given solution is again a solution. Therefore, we have a *continuous one-parameter family of ground states on* Ω_5, whereupon each solution is infinitely close to a continuum of neighboring solutions. This unusual richness of ground states causes difficult circumstances for numerical computation. Actually, among all the cases computed in this chapter, this geometry constitutes the *most difficult* one to deal with, with respect to numerical work. With the use of *MPA*, what we have experienced is that the numerical iterates at first appear to be converging from whatever chosen initial state, but then the trend of convergence slows down because the iterates begin to "get confused" as to where they should "settle down" with respect to rotational symmetry. Small fluctuations of ε_n [cf. (12.21)] last for quite a few iterations (about 10 or so); the programmer must then realize that something is going on, ask the computer to spit out some iterates, make comparisons and then make the decision to terminate. Otherwise the said fluctuations may persist indefinitely. However, with the use of *SIA*, this situation will not occur because the choice of x_0 in $(12.30)_2$ has a symmetry-breaking effect. Numerical solutions obtained by SIA converge fairly fast. Their profiles show that the peaks happen at points close to x_0.

Case 12.3.4.a A ground state on the symmetric annulus Ω_5. It is displayed in Fig. 12.7, with

$$\text{(SIA)} \qquad \begin{aligned} &\max u = 13.41, \quad J = 162.3, \quad \varepsilon_{12} = 10^{-4}, \\ &\alpha_{12} = 179.9, \quad x_0 = (0.7, 0). \end{aligned} \tag{12.54}$$

Remark 12.6. Let u be a solution of (12.2). It is established in Ni [142] that the linearized operator

$$\mathscr{L} = -[\Delta + pu^{p-1}\cdot]: \; v \longmapsto -[\Delta v + (pu^{p-1})v]$$

has the smallest eigenvalue $\lambda_1 < 0$. The next smallest eigenvalue λ_2 of $\mathscr{L}$ satisfies $\lambda_2 \geqslant 0$. Note that $\mathscr{L}$ corresponds to the second derivative of J in (12.49). In general, if $\mathscr{L}$ is

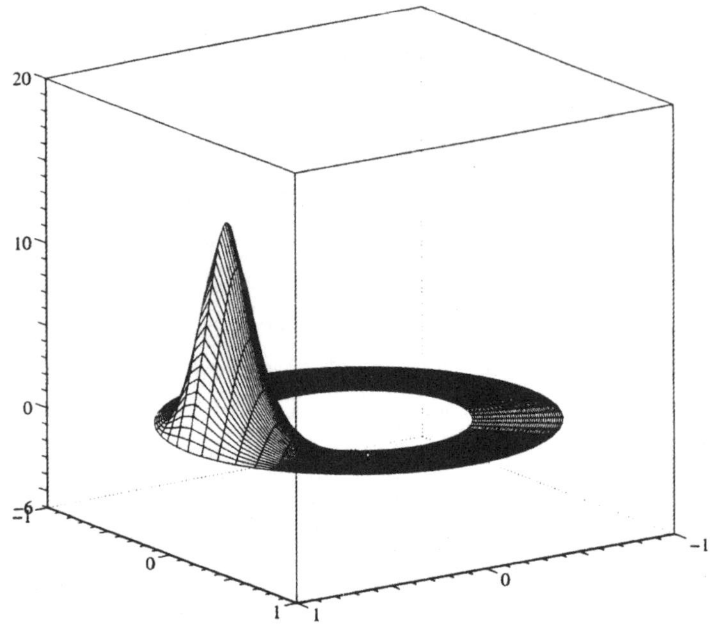

(a)

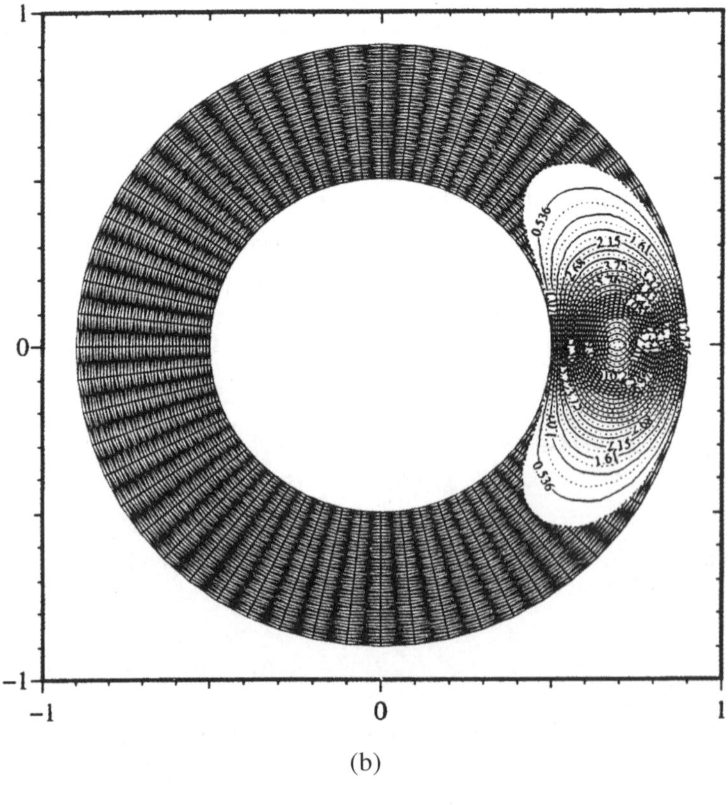

(b)

Figure 12.7: A ground state (a) and its contours (b) of (12.48) on the concentric annulus (12.53). Any rotation of this solution is again a solution with minimal energy.

invertible on the appropriate functional space, we say that u is a *nondegenerate critical point*.

Now, consider an annular domain

$$\Omega = \{x \in \mathbb{R}^2 \mid 0 < a < |x| < b\}.$$

Then as Li [120] has shown, when a is sufficiently close to b, nonradial ground states as shown in Fig. 12.7 occur. Let $u = u(v, \theta)$ be such a ground state. Differentiating (12.2) with respect to θ, and using the commutativity $(\partial/\partial\theta)\Delta = \Delta(\partial/\partial\theta)$ in polar coordinates, we get

$$\Delta\frac{\partial u}{\partial\theta} + pu^{p-1}\frac{\partial u}{\partial\theta} = 0, \quad \text{on } \Omega, \quad \left.\frac{\partial u}{\partial\theta}\right|_{\partial\Omega} = 0.$$

Note that $\partial u/\partial\theta \not\equiv 0$. This implies that $\mathscr{L}$ has an eigenfunction $\partial u/\partial\theta$ with 0 as the eigenvalue. Therefore $\mathscr{L}$ is not invertible and u is a *degenerate critical point* of J. At a degenerate critical point, the applicability of the Morse Lemma [32, 185] is not clear.

It is easy to see from the Implicit Function Theorem that any nondegenerate critical point is *isolated*. Case 12.3.4.a provides an example that the ground states are all degenerate critical points and form a continuum. □

Next, let us search for *multipeak* positive solutions of (12.48), which are again known to exist if the thickness of the symmetric annulus is sufficiently small. Such multipeak positive solutions are known to be the mountain pass solutions corresponding to the functional J on certain invariant subspaces of the rotational symmetry group on Ω_5. (For example, a two-peak solution would be a ground state of J on the space of functions which are invariant with respect to rotation by 180° on Ω.) Therefore, the initial state should be chosen in those invariant subspaces in the hope that successive iterates will also stay in that same invariant subspace. Obviously, this is a "numerically unstable" situation, since discretization and roundoff errors can accumulate and damage the rotational symmetry after a large number of iterations is performed. What we have observed numerically is that, after a large number of iterations, whether by MPA or SIA, the iterates always converge to a single-peak solution that is a global ground state.

On the other hand, if we perform only *a small number of iterations* using an initial state with rotational symmetry of angle $2\pi/n$, $n \in \mathbb{Z}^+$, and, say, observe a trend of numerical convergence, we let the numerical solution data exit and terminate the iterations. Then the solution data is *expected to be close* to the n-peak positive solution (because the numerical error accumulated after a small number of iterations has not damaged the rotational symmetry of angle $2\pi/n$). In the following (Cases 12.3.4b–12.3.4d), we use SIA to perform a small number of iterations with initial states as indicated. The errors $\widetilde{\varepsilon}_n$ are comparatively larger than in the preceding cases. This behavior elicits suspicion as to whether or not the numerical solution is close or not to the true solution. Our *safeguard* here is that we take several different initial states with that same rotational symmetry and iterate from them. If those output multipeak solution data are all close to each other (implying the independence of the initial states), we accept such multipeak numerical solutions as authentic. Otherwise, we reject them.

For later use let us introduce the "mound" function

$$M_{r_0,x_0}(x) = \cos\left(\frac{\pi}{2}\frac{|x-x_0|}{r_0}\right) \tag{12.55}$$

for given $r_0 > 0$ and $x_0 \in \mathbb{R}^2$. Note that $M(r_0,x_0,x) = 0$ for x: $|x-x_0| = r_0$. Let us also introduce a rotation operator R_θ by

$$(R_\theta f)(x) = f(e^{-i\theta}x); \qquad e^{-i\theta}x = (x_1\cos\theta + x_2\sin\theta, x_2\cos\theta - x_1\sin\theta).$$

Case 12.3.4.b A two-peak positive solution on the symmetric annulus Ω_5. First, define

$$\widetilde{v}_0(x) = \begin{cases} M_{r_0,x_0}(x), & r_0 = 0.2, \quad x_0 = (0.7,0), \quad |x-x_0| \leqslant r_0, \\ 0, & \text{elsewhere} \quad x \in \Omega_5, \end{cases} \tag{12.56}$$

and let $v_0(x) = \widetilde{v}_0(x) + (R_\pi \widetilde{v}_0)(x)$ be a period-π initial state for SIA, resulting in

$$\text{(SIA)} \qquad \begin{aligned} \max u = 13.60, \quad J = 327.0, \quad \widetilde{\varepsilon}_2 = 10^{-2}, \\ \alpha_2 = 185.0, \quad x_0 = (0.7,0). \end{aligned} \tag{12.57}$$

The two-peak solution and its contours are displayed in Figs. 12.8(a) and 12.8(b). By comparing Fig. 12.7 with Fig. 12.8 and the data in (12.54) with those in (12.57), we see that a two-peak solution is virtually a linear superposition of the one-peak solution with itself but rotated 180°.

Case 12.3.4.c A three-peak positive solution on the symmetric annulus Ω_5. Continuing from Case 12.3.4b, but using a period-$2\pi/3$ initial state u_0:

$$v_0(x) = \widetilde{v}_0(x) + (R_{\frac{2\pi}{3}}\widetilde{v}_0)(x) + (R_{\frac{4\pi}{3}}\widetilde{v}_0)(x); \qquad \text{cf. } \widetilde{v}_0 \text{ in (12.56)};$$

we get

$$\text{(SIA)} \qquad \begin{aligned} \max u = 13.60, \quad J = 488.6, \quad \widetilde{\varepsilon}_2 = 10^{-2}, \\ \alpha_2 = 185.0, \quad x_0 = (0.7,0); \end{aligned} \tag{12.58}$$

see Fig. 12.9.

Case 12.3.4.d A four-peak positive solution on the symmetric annulus Ω_5. We use a period-$\pi/2$ initial state

$$u_0(x) = \sum_{k=0}^{3} (R_{\frac{2k\pi}{4}} \widetilde{v}_0)(x),$$

resulting in Fig. 12.10

$$\text{(SIA)} \qquad \begin{aligned} &\max u = 13.59, \quad J = 650.3, \quad \widetilde{\varepsilon}_2 = 10^{-2}, \\ &\alpha_2 = 184.6, \quad x_0 = (0.7, 0). \end{aligned} \tag{12.59}$$

Four is the largest number of peaks that we are able to obtain for positive positions of (12.48) on Ω_5.

Note that the values of J in $(12.54)_2$–$(12.59)_2$ appear to grow linearly with respect to the number of peaks.

Case 12.3.4.e A radially symmetric positive solution on Ω_5. A radially symmetric positive solution of (12.48) does not seem to be obtainable by MPA or SIA. Iterations of arbitrarily chosen radially symmetric initial states u_0 quickly converge to a single-peak ground state in Case 12.3.4.a. In this connection, a radially symmetric positive solution has *infinity as its Morse index* because it has a descent direction to any single-peak ground state.

Here, we first assume the radial symmetry of (12.48) and then write it in polar coordinates:

$$\frac{1}{r}\frac{d}{dr}\left(r\frac{du}{dr}\right) + u^3 = 0, \quad u(r)|_{r=0.9} = u(r)|_{r=0.5} = 0; \quad u(r) > 0, \quad 0.5 < r < 0.9.$$

The nonlinear ODE above is replaced by a centered finite difference scheme:

$$\left.\begin{aligned} &u_0 = u_n = 0, \\ &\frac{u_{i+1} - 2u_i + u_{i-1}}{h^2} + \frac{1}{r_i}\frac{u_{i+1} - u_{i-1}}{2h} + u_i^3 = 0, u_i = u(0.5 + i \cdot h), \\ &r_i = 0.5 + ih, \quad h = \tfrac{0.4}{50}, \quad i = 1, 2, \ldots, n, \quad n = 50. \end{aligned}\right\} \tag{12.60}$$

SIA can be easily adapted for the above finite difference approach. We get Fig. 12.11 and

$$\text{(SIA)} \qquad \begin{aligned} &\max u = 9.2492, \quad J = 1557.19, \quad \widetilde{\varepsilon}_8 = 10^{-8}, \\ &\alpha_8 = 85.1481, \quad x_0 = (0.7, 0). \end{aligned} \tag{12.61}$$

By comparing $(12.54)_3$–$(12.61)_3$, we see that the radially symmetric solution has an energy level J much higher than multipeak solutions in Cases 12.3.4(a)–12.3.4(d).

Remark 12.7. On the unit disk Ω_1 or the radially symmetric annulus Ω_5, a finite difference scheme like (12.60) affords us a different method for validating whether our BEM solutions are numerically accurate, if the solutions have radial symmetry. FDM is certainly the easiest to program among the various numerical PDE schemes. We have taken advantage of this to confirm the accuracy of most of the radially symmetric solutions below in this chapter; see Case 12.4.1a, for example. □

Remark 12.8. Examining the data of J in (12.54), (12.57), (12.58) and (12.59), we have found that

$$J_k \approx k \cdot (163), \qquad k = 1,2,3,4,$$

i.e., the energy J_k of the k-peak solution grows approximately linearly with respect to k. This looks very reasonable. Extrapolating, we thus obtain

$$J_9 \approx 9 \cdot 163 = 1467, \quad J_{10} \approx 10 \cdot 163 = 1630.$$

From $(12.61)_2$, we see that the energy of the radially symmetric solution is $J = 1557.19$. Since $J_9 < 1557.19 < J_{10}$, we suspect that for the annulus Ω_5, the largest number of peaks a solution of (12.48) has is 9. Theoretical estimates in Coffman [51] and Li [120] do not furnish any information about the shape or quantity of such solutions. This is a situation where numerical computation shows its value. □

Now, let us reduce the thickness of the annulus Ω_5. We define a thinner concentric annulus

$$\Omega_6 = \{x \in \mathbb{R}^2 \mid 0.7 < |x| < 0.9\}. \tag{12.62}$$

Case 12.3.4.f A single-peak solution on the thin symmetric annulus Ω_6. This solution is displayed in Fig. 12.12, with

$$\text{(SIA)} \qquad \begin{aligned} \max u = 27.11, \quad J = 635.2, \quad \widetilde{\varepsilon}_7 = 10^{-4}, \\ \alpha_4 = 735.0, \quad x_0 = (0.8,0). \end{aligned}$$

Since the annulus Ω_6 is thinner than Ω_5, according to Li [120], there will be positive solutions with more peaks.

Case 12.3.4.g A positive solution, with eight peaks, on the thin annulus Ω_6. Computed by using

$$u_0(x) = \sum_{j=0}^{7} R_{\frac{2\pi j}{8}} M_{r_0,x_0}(x), \quad r_0 = 0.1, \quad x_0 = (0.8,0).$$

It has

$$\text{(SIA)} \qquad \begin{aligned} \max u = 29.15, \quad J = 5571, \quad \widetilde{\varepsilon}_4 = 10^{-2}, \\ \alpha_4 = 247.3, \quad x_0 = (0.8,0). \end{aligned}$$

and is displayed in Fig. 12.13. Eight is the largest number of peaks we are able to produce numerically.

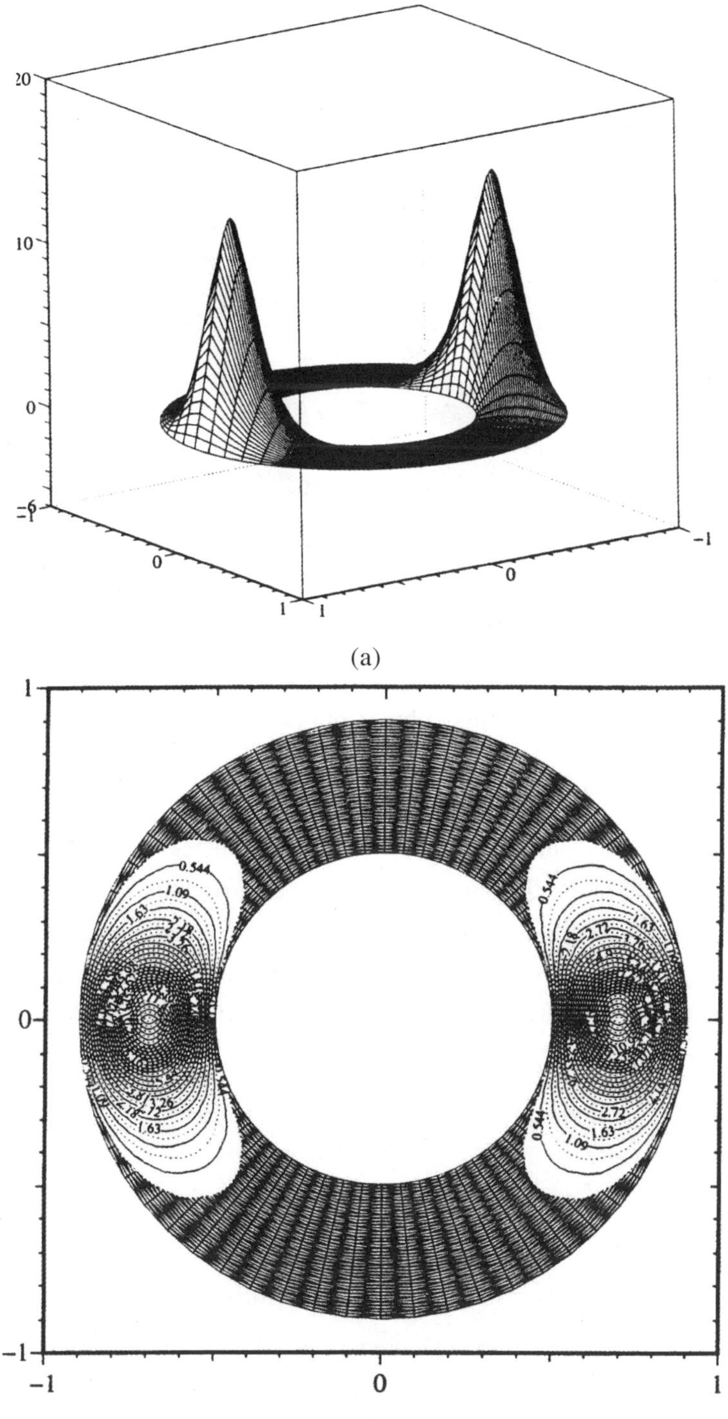

(a)

(b)

Figure 12.8: A two-peak positive solution of (12.48) on the concentric annulus Ω_5.

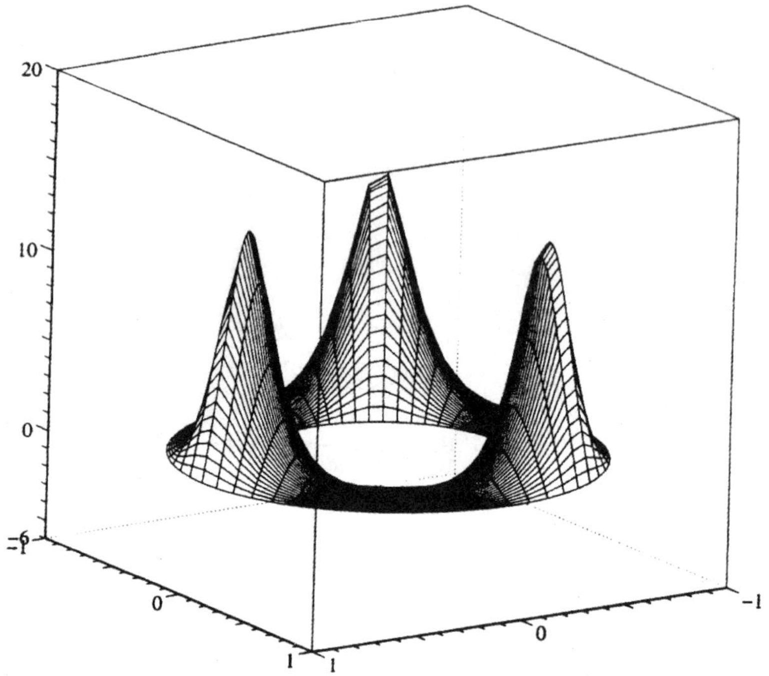

Figure 12.9: A three-peak positive solution of (12.48) on the concentric annulus Ω_5.

12.3.5 *A dumbbell-shaped domain*

We consider a dumbbell-shaped domain Ω_7 as shown in Fig. 12.13. It contains a left, smaller disk D_L with radius 0.5, and a right, larger disk D_R with radius 1, whose centers are separated by a distance equal to 3 along the x_1-axis. A horizontal corridor, symmetric with respect to the x_1-axis, of width $W = 0.4$, is constructed to link the two disks. In our BEM computations, $\partial\Omega$ is discretized into 408 panels, and on Ω, 992 Gaussian quadrature points are used for integrating the volume potentials.

Case 12.3.5.a The ground state on the dumbbell Ω_7. Choose

$$u_0(x) = \begin{cases} -10, & x \in D_R, \\ 0, & x \in \Omega_7 \setminus D_R \end{cases} \tag{12.63}$$

and solve the elliptic BVP

$$\Delta w_0(x) = u_0(x), \quad \text{on } \Omega, \quad w_0|_{\partial\Omega} = 0. \tag{12.64}$$

Obviously, $w_0 \in H_0^1(\Omega)$. This w_0 will be used as the initial state for MPA. Iterate by MPA. We then obtain a single-peak positive solution as shown in Fig. 12.14(a), with

$$\text{(MPA)} \qquad \max u = 3.562, \quad J = 10.90, \quad \varepsilon_7 = 10^{-4}. \tag{12.65}$$

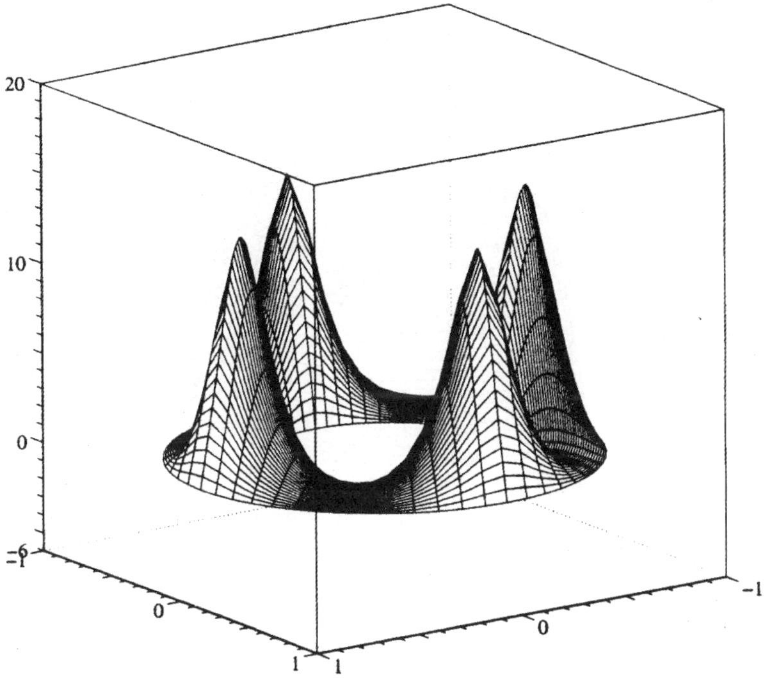

Figure 12.10: A four-peak positive solution of (12.48) on the concentric annulus Ω_5.

Its contours are plotted in Fig. 12.14(b). A careful examination of the contours shows that the peak point of the contours has moved slightly leftward (from the original center of the right disk D_R), toward the corridor. This agrees with our intuitive understanding of solutions of (12.48) that they prefer "open space" [147].

This solution has the lowest energy. Its support is concentrated on the right disk (i.e., the larger of the two disks) and near the right end of the corridor. Elsewhere, u is exponentially small.

Case 12.3.5.b A local ground state concentrated on the small disk. Choose the initial state

$$u_0(x) = \begin{cases} -10, \; x \in D_L, \\ \quad 0, \; x \in \Omega_7 \setminus D_L, \end{cases} \tag{12.66}$$

obtain w_0 as in (12.64) and iterate by MPA. We get a positive solution concentrated mainly on and near the small left disk D_L, as displayed in Fig. 12.15(a), with contours shown in Fig. 12.15(b). This solution has

$$\text{(MPA)} \qquad \max u = 7.037, \quad J(u) = 42.22, \quad \varepsilon_{12} = 10^{-4}. \tag{12.67}$$

Actually, if we choose any positive u_0 [instead of the u_0 in (12.66)] supported on D_L, MPA will yield convergence to the one in Fig. 12.15(a).

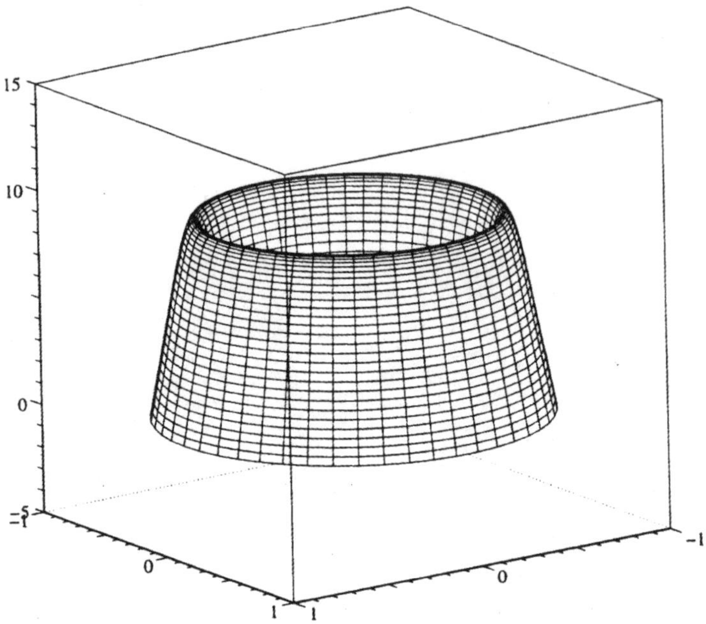

Figure 12.11: A radially symmetric positive solution of (12.48) on the concentric annulus Ω_5.

Case 12.3.5.c A local ground state concentrated near the center of the corridor. Choose the initial state

$$u_0(x) = \begin{cases} 0, & x \in D_L \cup D_R, \\ -10, & x \in \Omega_7 \setminus (D_L \cup D_R), \end{cases} \tag{12.68}$$

and obtain the initial state w_0 from u_0 by (12.64), and iterate by MPA. We obtain a local ground state concentrated nearly on the center of the corridor, given in Figs. 12.17(a) and 12.17(b), with

$$\text{(MPA)} \qquad \max u = 13.63, \quad J(u) = 159.0, \quad \varepsilon_{31} = 10^{-4}. \tag{12.69}$$

This "solution" has generated a considerable debate among some researchers as to whether such a solution could possibly be obtained analytically by the Mountain–Pass approach. The same comment also applies to similar situations below in Subsections 12.3.7, 12.4.3 and 12.5.2.

Case 12.3.5.d A two-peak positive solution. Adapting an algorithm with Morse index 2 first developed in Ding, Costa and Chen [65]; see also Subsection 12.3.8, we are able to obtain a two-peak positive solution of (12.48) concentrated on both the left and right disks

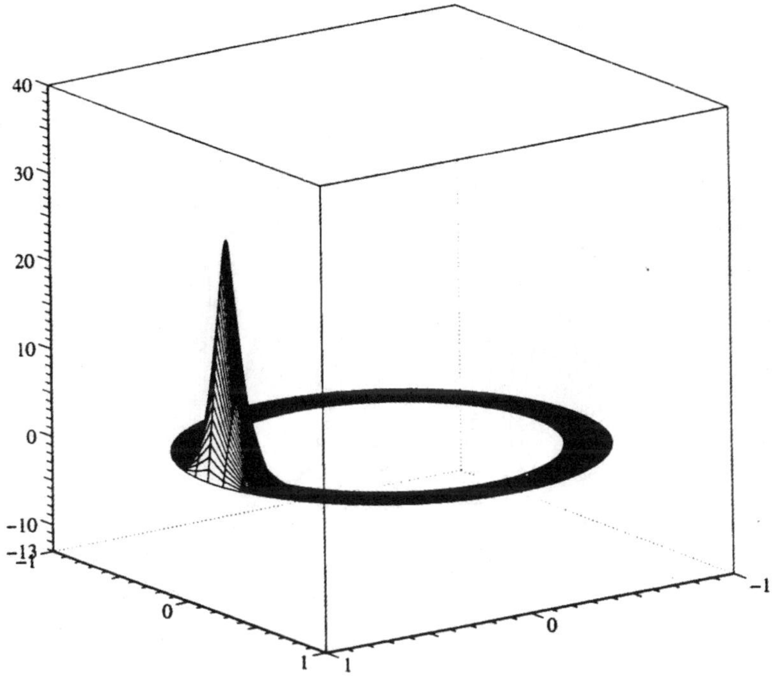

Figure 12.12: A single peak, ground-state solution of (12.48) on the concentric annulus Ω_6.

of Ω_7, with

$$\max u = 7.037, \quad J(u) = 53.12, \quad \varepsilon_{34} = 10^{-x4};$$

see Fig. 12.16. From the data above and Figs. 12.17–12.20, we see that this solution u essentially is a combination of the two solutions in Case 12.3.5.a and 12.3.5.b. It has Morse index 2. As with Subsection 12.3.8, the purpose of including Fig. 12.16 is to satisfy the reader's curiosity about the existence of multipeak positive solutions.

12.3.6 *A starshaped domain degenerated from a dumbbell*

We expand the width W of the corridor in Ω_7 from $W = 0.4$ to $W = 1$. Then we get a new domain Ω_8 as shown in Fig. 12.18. Even though Ω_8 does not look exactly like a star, it satisfies the starshapedness condition, and it is no longer dumbbell-shaped. We discretize $\partial\Omega_8$ into 408 panels, and place 992 Gaussian quadrature points on Ω.

Case 12.3.6.a The ground state of (12.48) on the starshaped domain Ω_8. We choose

$$u_0(x) = \begin{cases} -10, \; x \in D_R, & (D_R \text{ is the disk with radius 1} \\ 0, \quad x \in \Omega_8 \setminus D_R, & \text{on the right of } \Omega_8), \end{cases}$$

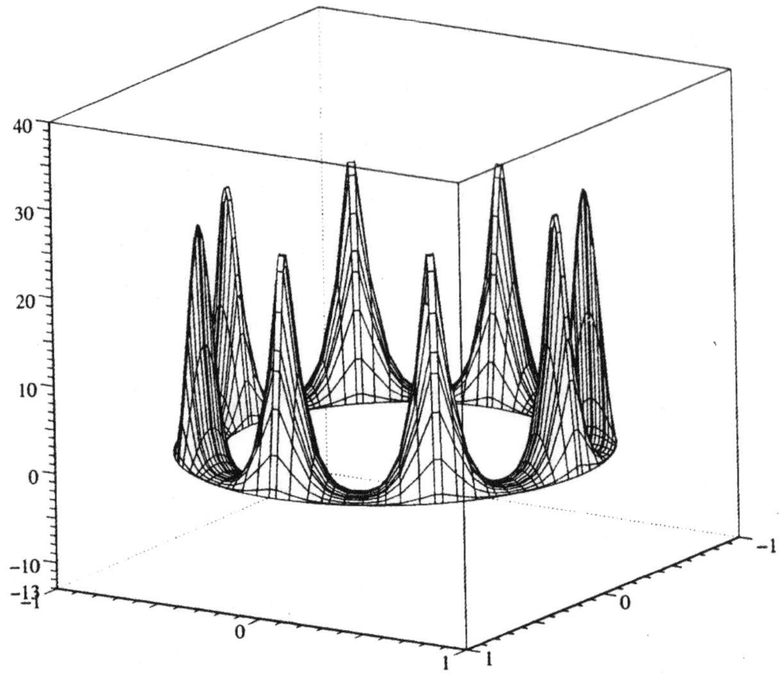

Figure 12.13: A positive solution of (12.48) with eight peaks on the concentric annulus Ω_6 (12.62).

and use (12.64) to get the initial state w_0. Iterating with MPA we obtain the ground state u of (12.48) as shown in Figs. 12.19(a) and 12.19(b), with

$$\text{(MPA)} \qquad \max u = 3.4599, \quad J(u) = 10.43, \quad \varepsilon_{13} = 10^{-1}. \qquad (12.70)$$

The reader may compare Fig. 12.19 against Fig. 12.14, and (12.70) against (12.65).

Case 12.3.6.b A local ground state of (12.48) on the starshaped domain Ω_8. Choose

$$u_0(x) = \begin{cases} -10,\ x = (x_1, x_2) \in \Omega_8, & x_1 < 2 - \dfrac{\sqrt{3}}{2}, \\ 0, \quad \text{elsewhere}, \end{cases}$$

similarly as before and iterate by (MPA). We obtain a positive local mountain pass solution as shown in Figs. 12.20(a) and 12.20(b), with

$$\text{(MPA)} \qquad \max u = 5.362, \quad J(u) = 26.18, \quad \varepsilon_{13} = 10^{-4}. \qquad (12.71)$$

Here, we see that the two local ground states in Case 12.3.5.b and 12.3.5.c coalesce into one.

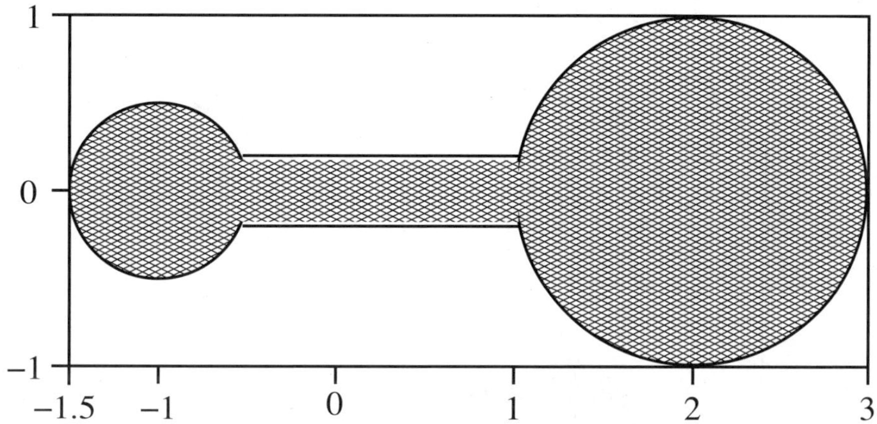

Figure 12.14: Dumbbell-shaped domain Ω_7; the left disk D_L has radius 0.5 and the right D_R has 1. The distance between the two centers of the disks is 3. The corridor has width $W = 0.4$. The dots on $\partial\Omega$ represent collocation points; 408 of them are placed.

From Figs. 12.19(b) and 12.20(b), we can understand that the positive local ground state as shown in Fig. 12.20(a) may be "swallowed" by the global ground state in Fig. 12.19(a) and, thus, disappear, if the corridor portion is not long enough.

12.3.7 *Dumbbell-shaped domains with cavities lacking symmetry*

We remove two circular holes from the dumbbell-shaped domain Ω_7, and obtain a dumbbell-shaped domain with cavities, Ω_9, as shown in Fig. 12.21. Because of the placement of the hole inside the right disk D_R, the new domain Ω_9 lacks any symmetry, of a two-connectivity or with "genus 2". Such topological effects are of interest to analysts in nonlinear PDEs, see e.g., Benci and Cerami [16]. We discretize $\partial\Omega_9$ into 536 panels, and place 847 Gaussian quadrature points on Ω for volume potentials.

Case 12.3.7.a The ground state (12.48) on Ω_9. We use

$$u_0(x) = \begin{cases} -10, & \text{for } x \text{ on the right disk with cavity,} \\ 0, & \text{elsewhere,} \end{cases}$$

to get an initial state $w_0 \in H_0^1(\Omega)$ by (12.64), and iterate with MPA, obtaining

$$\text{(MPA)} \qquad \max u = 6.003, \quad J = 33.40, \quad \varepsilon_{19} = 10^{-3},$$

and Fig. 12.22.

Case 12.3.7.b Local ground states of (12.48) on Ω_9. By making various choices of u_0

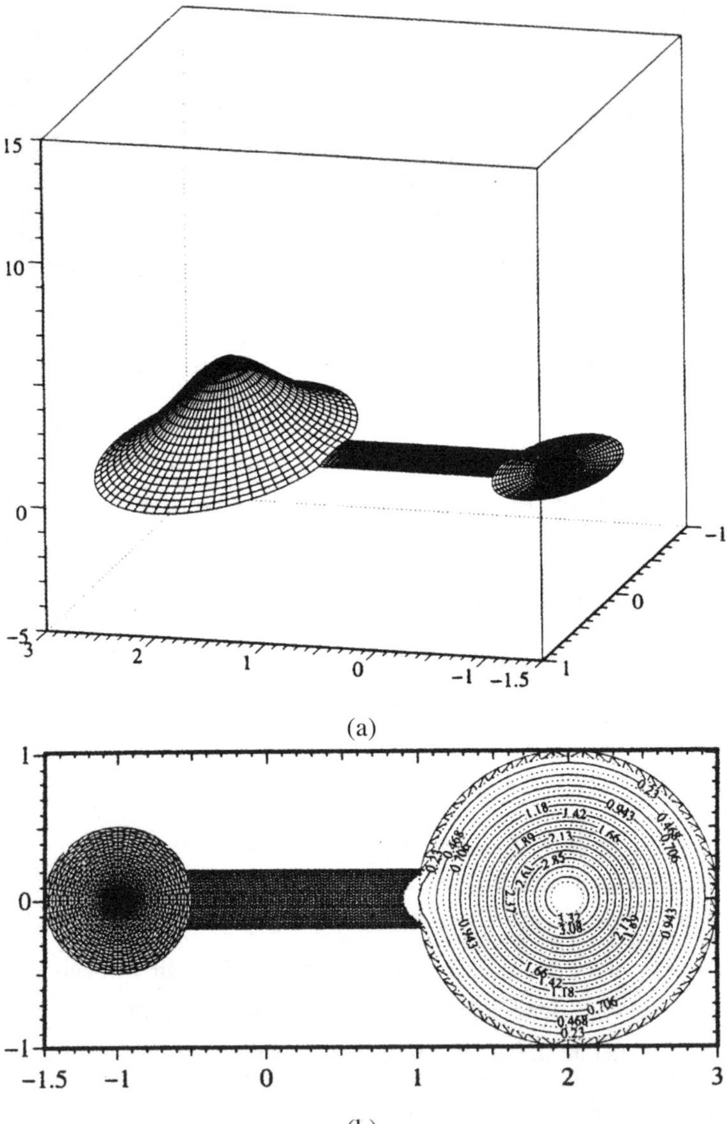

Figure 12.15: The ground state (a) and its contours (b) on (12.48) on the dumbbell-shaped domain Ω_7.

(with MPA) and v_0 (with SIA), we have obtained three other local ground states of (12.48) arranged by increasing energy level J, given in Figs. 12.23–12.25.
Obviously, there appear to exist quite a few other positive solutions of (12.48) on Ω_9. But we have decided to stop at this point, and leave the pursuit to other interested readers.

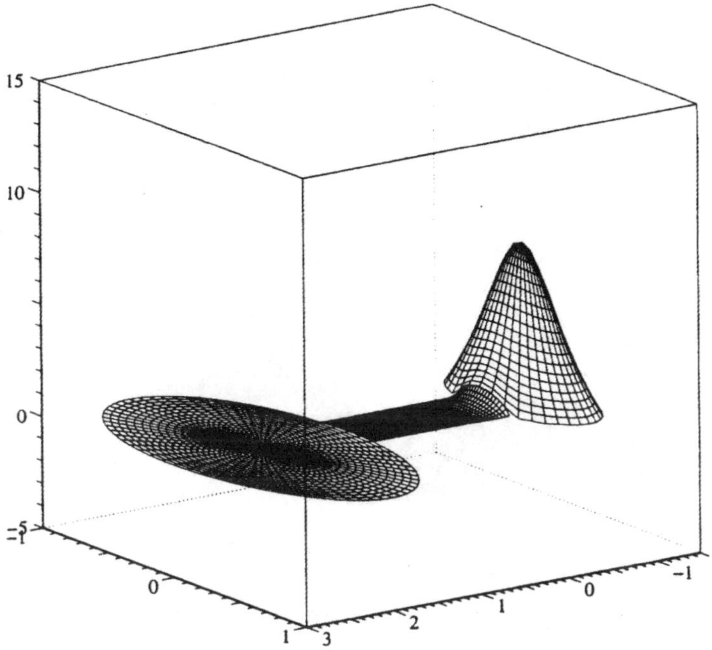

(a)

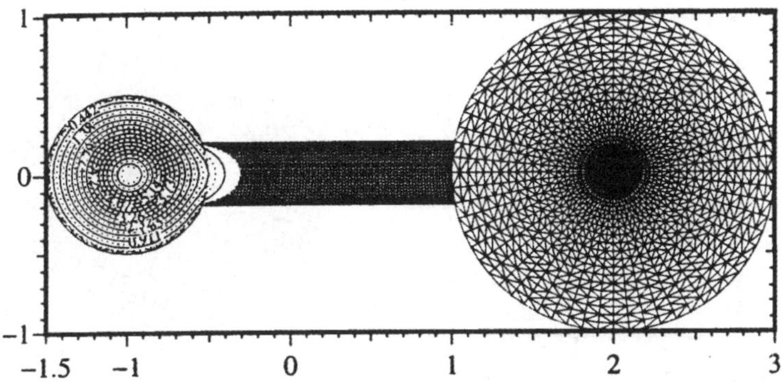

(b)

Figure 12.16: A positive solution (a) and its contours of (12.48) concentrated on and near the small disk D_L of Ω_7.

12.3.8 *Sign-changing solutions*

We return to Subsection 12.3.5, and use the dumbbell-shaped domain Ω_7 for the study in this subsection. We want to consider solutions that are *not necessarily positive*, i.e., find

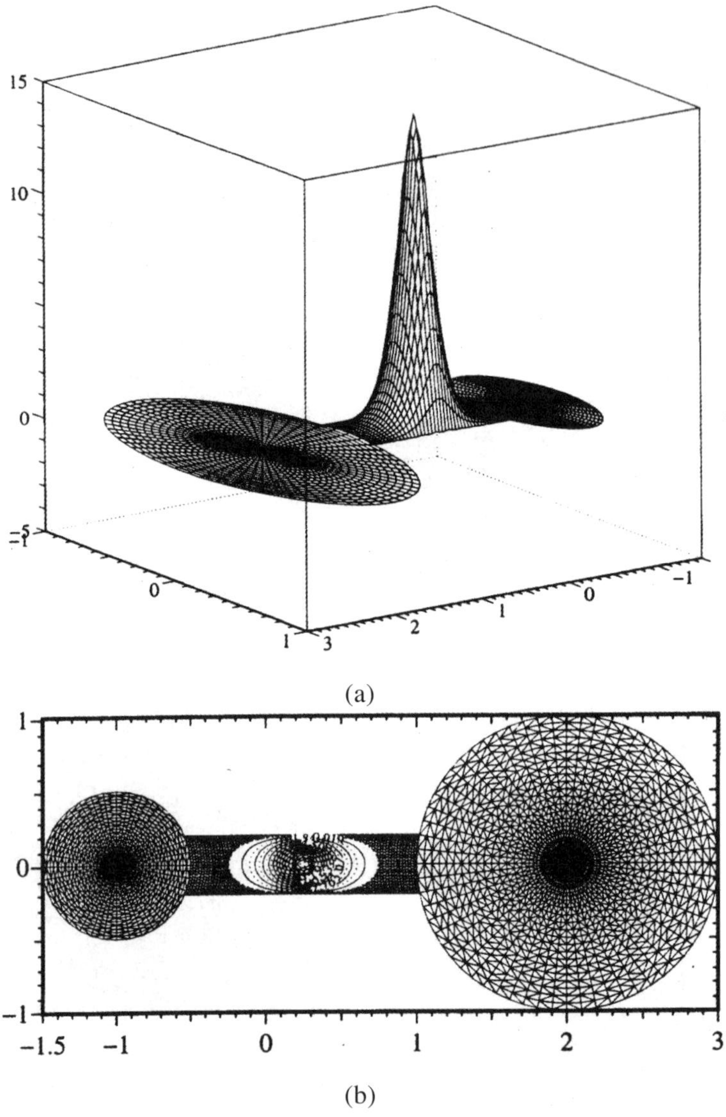

Figure 12.17: A positive solution (a) and its countours (b) of (12.48) concentrated near the center of the corridor of the dumbbell-shaped domain Ω_7.

some u such that

$$\begin{cases} \Delta u + u^p = 0 \text{ on } \Omega_7, \\ u|_{\partial\Omega_7} = 0. \end{cases} \tag{12.72}$$

Even though Choi and McKenna [47] have displayed some sign-changing solutions of (12.72) on a square, we know that in general MPA is not capable of producing sign-

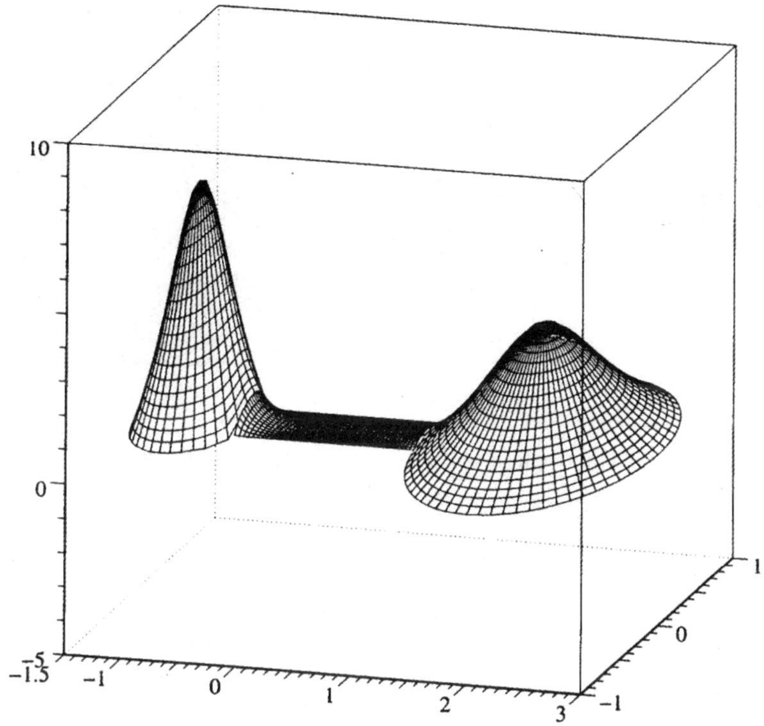

Figure 12.18: A two-peak positive solution on the dumbbell-shaped domain Ω_7.

changing solutions when the domain Ω is not symmetric with respect to some hyperplane in $\mathbb{R}^N$; see Bartsch and Wang [15], Castro, Cossio and Neuberger [30], Wang [181] and Willem [185]. (Here $N = 2$ and a hyperplane in $\mathbb{R}^2$ is just a line.) A sign-changing solution usually has Morse index 2 or higher. Thus, more elaborate MPA need to be developed in order to manage the higher Morse index. A useful numerical algorithm for sign-changing solutions of semilinear elliptic problems with Morse index 2 was developed by Ding, Costa and Chen [65]; it was incorporated with FEM and yielded exemplar sign-changing solutions on triangular domains. Actually, at least from an algorithmic point of view, one can see that it is possible to generalize the ideas in [65] so that an elaborate minimax method can even produce solutions of semilinear elliptic BVP with Morse index $3, 4, \ldots$. But such work requires a certain time duration for theoretical development and numerical testing and, therefore, will be presented elsewhere, after the study has matured. Here, for the sake of comprehensiveness, we include two examples of sign-changing solutions of (12.72) for $p = 3$. Each of them is computed by a variant of the algorithm in [65] for Morse index 2 coupled with BEM.

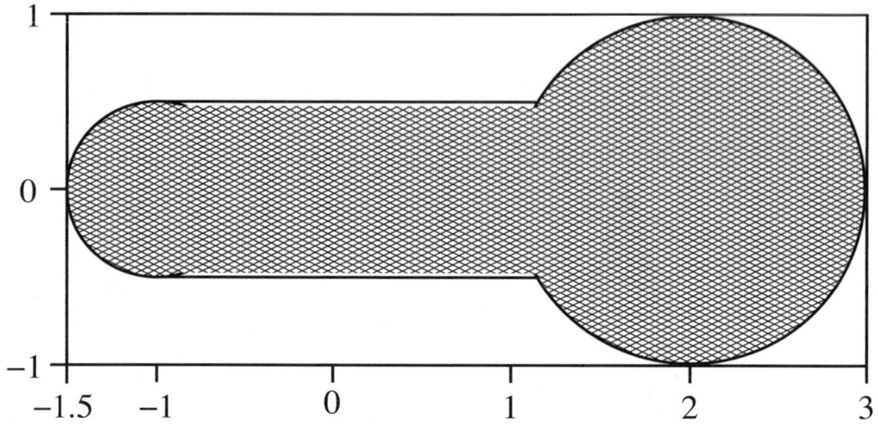

Figure 12.19: A starshaped domain Ω_8. It is a degenerate dumbbell. Each dot on $\partial\Omega$ represents a collocation point; there are 408 of them.

Case 12.3.8.a A sign-changing solution on the unit disk. We get

$$\max u = 5.850, \quad J = 60.03, \quad \varepsilon_{10} = 10^{-3}.$$

This solution is displayed in Fig. 12.27. Note that any rotation of this solution is again a solution.

Dancer [60, p. 140] has pointed out that, for a dumbbell-shaped domain like Ω_7, there should be a solution which is positive on the right disk D_R and negative on the left disk D_L, and with the signs reversed on D_L and D_R too, because, if u is a solution of (12.72) with $p = 3$, then so is $-u$.

Case 12.3.8.b A sign-changing solution on the dumbbell-shaped domain Ω_7. A sign-changing solution has been obtained as in Fig. 12.27; it has

$$\max u = 3.562, \quad \min u = -7.035, \quad J = 53.18, \quad \varepsilon_{20} = 10^{-4}. \tag{12.73}$$

By comparing the data in (12.65), (12.67) and (12.73), we see that this sign-changing solution in Fig. 12.27 is essentially a combination of the two solutions displayed in Figs. 12.14(a) and 12.15(a), except that the part from Fig. 12.14(a) has reversed its sign. The energy value J in (12.73) is nearly equal to the sum of the values of J in (12.65) and (12.67).

Other profiles of sign-changing solutions (which are "combinatorial combinations" of sorts of pairs of Figs. 12.14(a), 12.15(a) and 12.17(a), with just one member in the pair having sign reversed) have also been obtained. There are five such solutions (not counting Fig. 12.27). Their graphics are omitted.

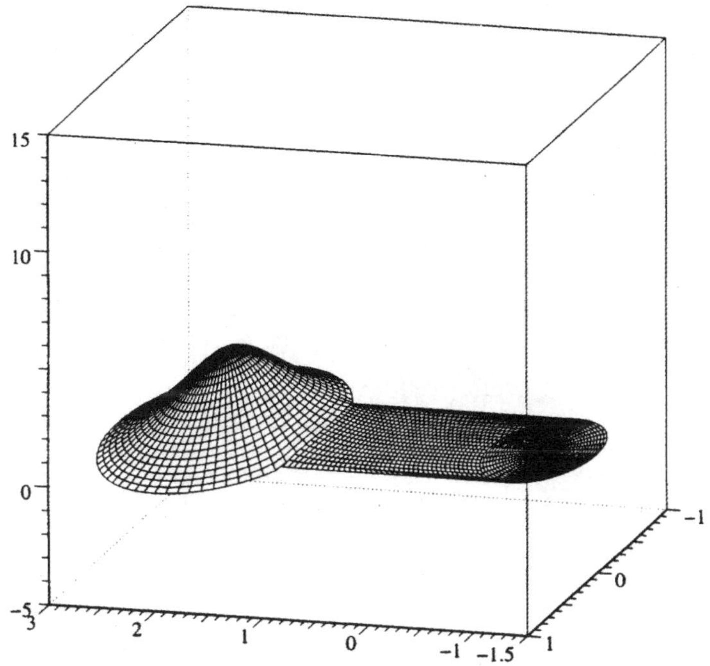

(a)

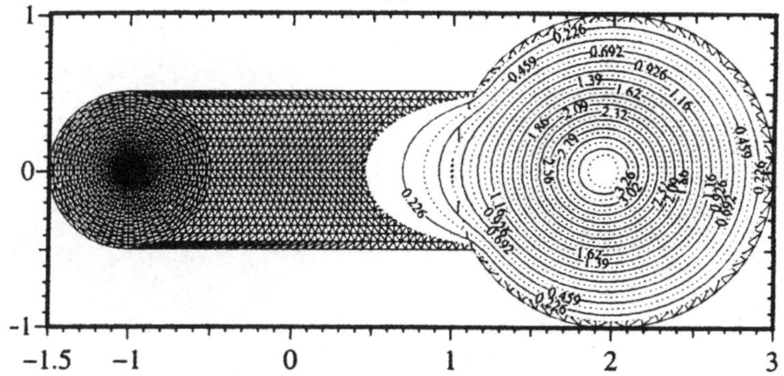

(b)

Figure 12.20: The ground state (a) and its contours (b) on the starshaped domain Ω_8.

After seeing so many graphics in this section, let us now make some conclusive and inferential comments. Obviously, for the Lane–Emden equation here and the other semilinear equations to be addressed in the following sections, the geometry (i.e., shape), symmetry and topology of the domain all have strong bearing on the multiplicity of positive solutions. The influence of symmetry is somewhat easier to understand. How about geometry

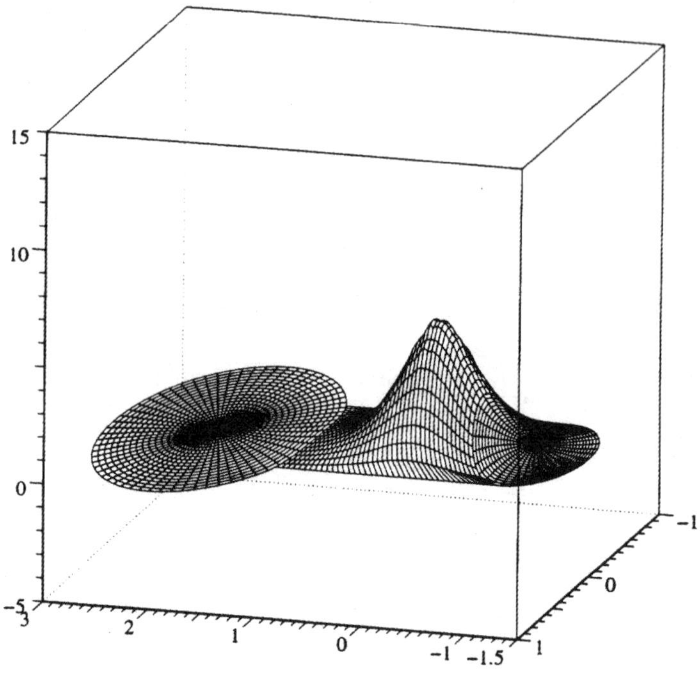

(a)

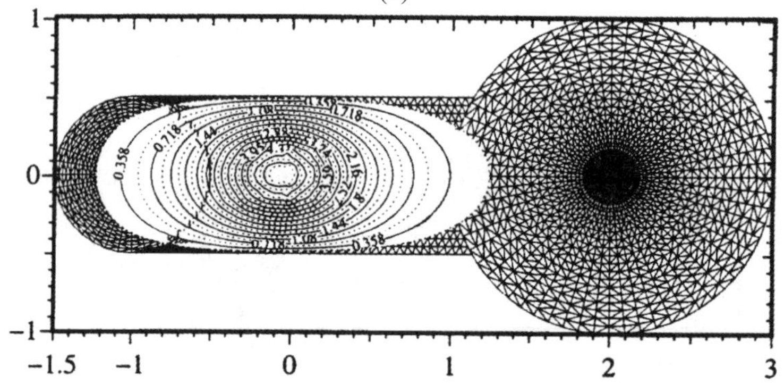

(b)

Figure 12.21: A second positive solution (a) and its contours (b) of (12.48) on the star-shaped domain Ω_8.

and topology? We regard the former as a more decisive factor in generating multiple solutions than the latter. In an irregularly shaped domain such as the one shown in Fig. 12.28, there are many "pockets" (or "small compartments") and "corridors" where local ground states of the Lane–Emden equation thrive, such as Subsection 12.3.5 has obviously led us to believe. The drilling of holes (i.e., change of topology) on a dumbbell-shaped domain in

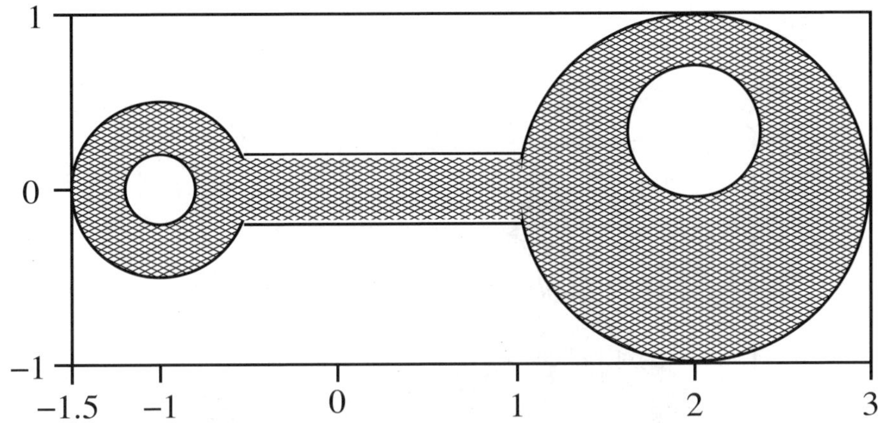

Figure 12.22: A dumbbell-shaped domain with two cavities, Ω_9. Two circular holes are drilled on the previous dumbbell-shaped domain Ω_7. The left hole is centered at $(x_1,x_2) = (-1,0)$ with radius 0.2; it maintains some local symmetry. The right hold is centered at $(x_1,x_2) = (2,0.3)$ with radius 0.4; it destroys any global symmetry.

Subsection 12.3.7 leads to extra local ground states. However, the actual effect of drilling is not necessarily in the change of topology itself but rather, in the formation of new pockets for the extra solutions to live on. This is our observation.

12.4 The singularly perturbed Dirichlet problem $\varepsilon^2\Delta u - u + u^3 = 0$

The semilinear elliptic *Neumann* boundary value problem

$$\begin{cases} d\Delta u - u + u^p = 0, \quad u > 0, \quad \text{on } \Omega, \quad d > 0, \\ \left.\dfrac{\partial u}{\partial n}\right|_{\partial\Omega} = 0, \end{cases} \tag{12.74}$$

is known to be a so-called *shadow system* for the asymptotic state of the following model of the chemotactic aggregation stage of cellular slime molds (amoebae) [111] in mathematical biology:

$$\left.\begin{array}{l} \left.\begin{array}{l} \dfrac{\partial\phi}{\partial t} = D_1\Delta\phi - \chi\nabla\cdot(\phi\nabla\ln\psi), \\ \dfrac{\partial\psi}{\partial t} = D_2\Delta\psi - a\psi + b\phi, \end{array}\right\} \quad \text{on } \Omega, t > 0; D_1, D_2, a, b > 0; \\ \dfrac{\partial\phi}{\partial n} = \dfrac{\partial\psi}{\partial n} = 0 \quad \text{on } \partial\Omega, t > 0; \phi(x,0) = \phi_0(x), \psi(x,0) = \psi_0(x), \quad x \in \Omega. \end{array}\right\} \tag{12.75}$$

See Lin, Ni and Takagi [125], Ni and Takagi [145, 146]. The case of interest is when *d is small* in (12.74), which becomes a singular perturbation problem. Because of the close

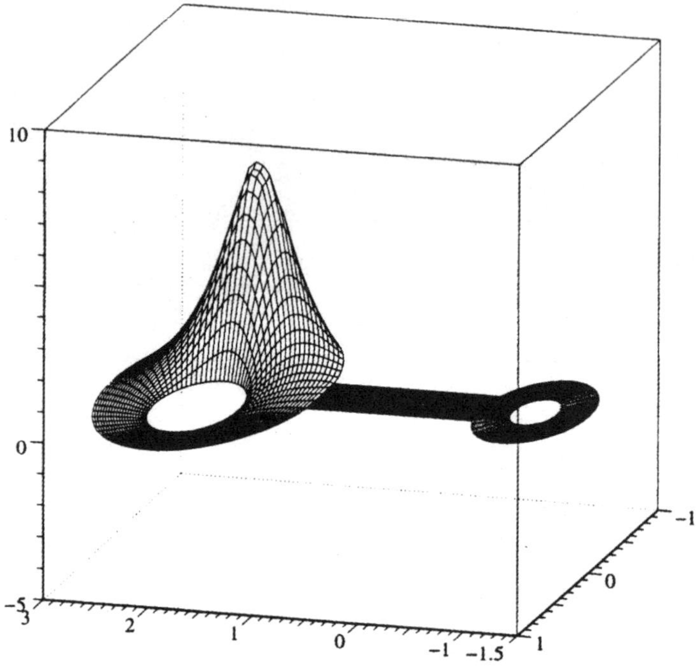

(a)

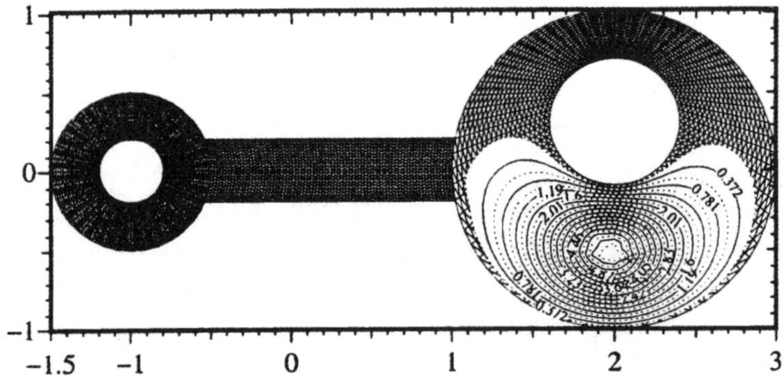

(b)

Figure 12.23: The ground state (a) and its contours (b) of (12.48) on the dumbbell-shaped domain with cavities, Ω_9.

relationship between the Dirichlet problem and the Neumann problem (12.74), Ni and Wei [147] studied the singularly perturbed semilinear BVP

$$\left.\begin{aligned} &\varepsilon^2 \Delta u - u + u^p = 0, \quad u > 0 \quad \text{on } \Omega, \quad \varepsilon \downarrow 0, \\ &u|_{\partial\Omega} = 0, \end{aligned}\right\} \tag{12.76}$$

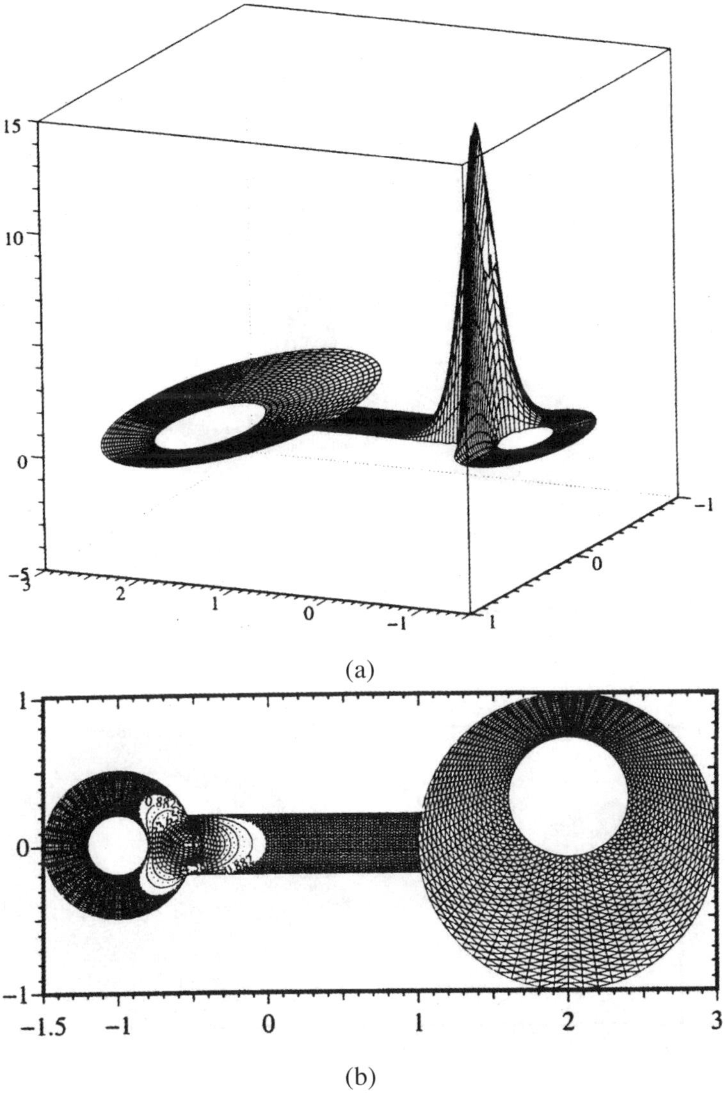

Figure 12.24: A first local ground state (a) and its contours (b) of (12.48) on Ω_9. It is obtained by SIA, with $\max u = 12.83$, $J = 164.4$, $\widetilde{\varepsilon}_{20} = 10^{-4}$, $\alpha_{20} = 79.69$, $x_0 = (1.5, 0)$.

in Ni and Wei [147], where ε^2 corresponds to d in (12.74). Variants of (12.76) have also been discussed in Benci and Cerami [16] and Dancer [60], for example. In this section, we will visualize solutions of (12.76) primarily for the case $p = 3$ on a few selected domains Ω_i, $i = 1, 2, \ldots, 9$, used in §12.3. The visualization of the singularly perturbed Neumann problem (12.74) has been computed by FEM in [39].

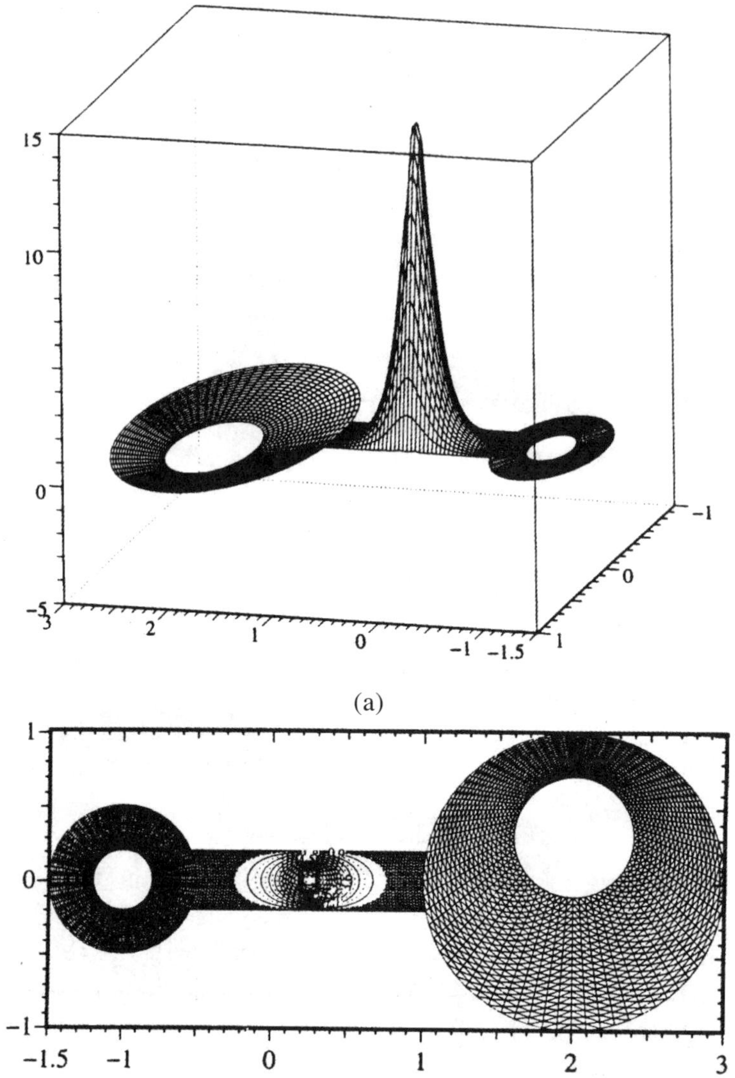

Figure 12.25: A second local ground state (a) and its contours (b) of (12.48) on Ω_9. It is obtained by SIA, with $\max u = 13,747$, $J = 162.6$, $\widetilde{\varepsilon}_{27} = 189.0$, $x_0 = (0,25,0)$.

Corresponding to (12.76), the natural energy functional is defined to be

$$J_\varepsilon(u) = \frac{1}{2}\int_\Omega (\varepsilon^2|\nabla u|^2 + u^2)\,dx - \frac{1}{p+1}\int_\Omega u^{p+1}\,dx.$$

Let u_ε be a critical point of J_ε corresponding to the Mountain–Pass Lemma (i.e., $J'_\varepsilon(u_\varepsilon) = 0$

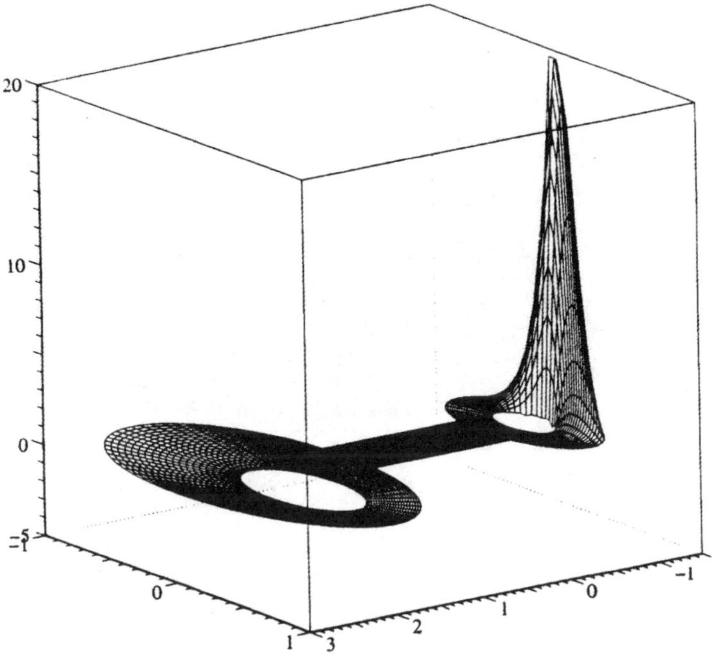

(a)

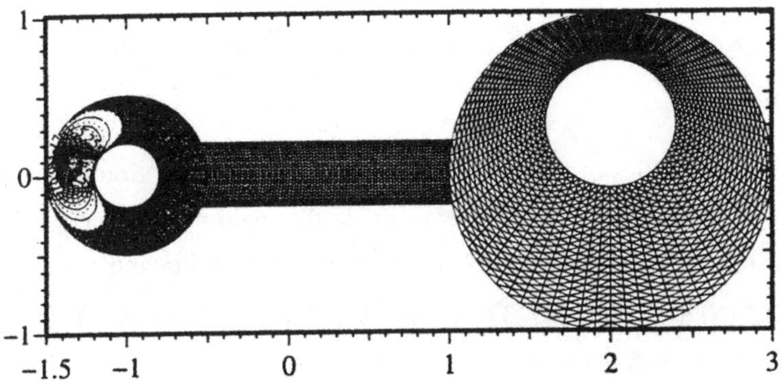

(b)

Figure 12.26: A third local ground state (a) and its contours (b) of (12.48) on Ω_9. It is obtained by SIA, with $\max u = 17.005, J = 361.7, \widetilde{\varepsilon}_{100} = 10^{-4}$, $\alpha_{100} = 32.1, x_0 = (2.5,0)$.

and $J_\varepsilon(u_\varepsilon) = c_\varepsilon$), where

$$c_\varepsilon = \inf_{h\in\Gamma} \max_{0\leqslant t\leqslant 1} J_\varepsilon(h(t)),$$

and where Γ is the set of all continuous paths joining the origin and a fixed nonzero element e in $H_0^1(\Omega)$ with $e \geqslant 0$ and $J_\varepsilon(e) = 0$, cf. c in (12.15). Then Ni and Wei [147, Prop. 2.1, p. 734] $c_\varepsilon > 0$ and c_ε is independent of the choice of ε.

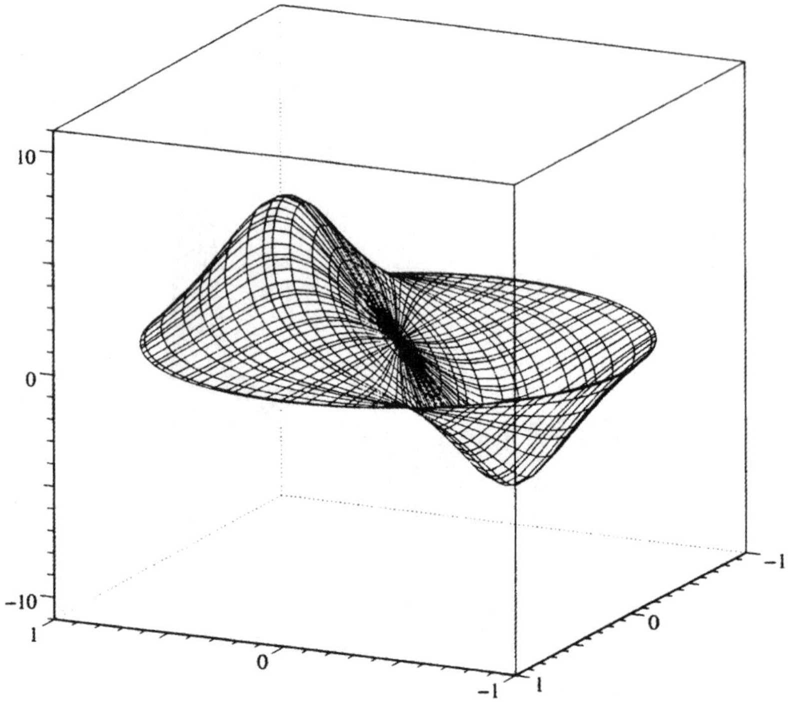

Figure 12.27: A sign-changing solution of $\Delta u + u^3 = 0$ with zero boundary condition on the unit disk.

The theoretical properties of a singularly perturbed sequence of ground states u_ε (12.76) have been studied at length; see [147]. We quote the main result from these and state it (restricted just to the special case (12.76) in $\mathbb{R}^2$ here) in the following.

Theorem 12.5 (Ni and Wei [147, Theorem 2.2, p. 734]). *Let u_ε be a ground state of (12.76). Then, for ε sufficiently small, we have*

(i) *u_ε has at most one local maximum and it is achieved at exactly one point P_ε in Ω. Moreover, $u_\varepsilon(\cdot + P_\varepsilon) \to 0$ in $C^1_{\text{loc}}(\Omega - P_\varepsilon \setminus \{0\})$, where $\Omega - P_\varepsilon = \{x - P_\varepsilon \mid x \in \Omega\}$;*

(ii) *$d(P_\varepsilon, \partial\Omega) \to \max_{P\in\Omega}(P, \partial\Omega)$ as $\varepsilon \to 0$, where $d(P_\varepsilon, \partial\Omega)$ is the distance from P_ε to $\partial\Omega$.*

Theorem 12.5 indicates that u_ε has exactly one peak at some point $P_\varepsilon \in \Omega$. As $\varepsilon \to 0$, $u_\varepsilon \to 0$ except at the peak P_ε, thereby exhibiting a single "spike-layer". Property (ii) says that $\{P_\varepsilon\}$ will tend to a point P_0 satisfying $d(P_0, \partial\Omega) = \max_{P\in\Omega} d(P, \partial\Omega)$, i.e., P_0 is located near the center of some subdomain G of Ω where "Ω has the most open space".

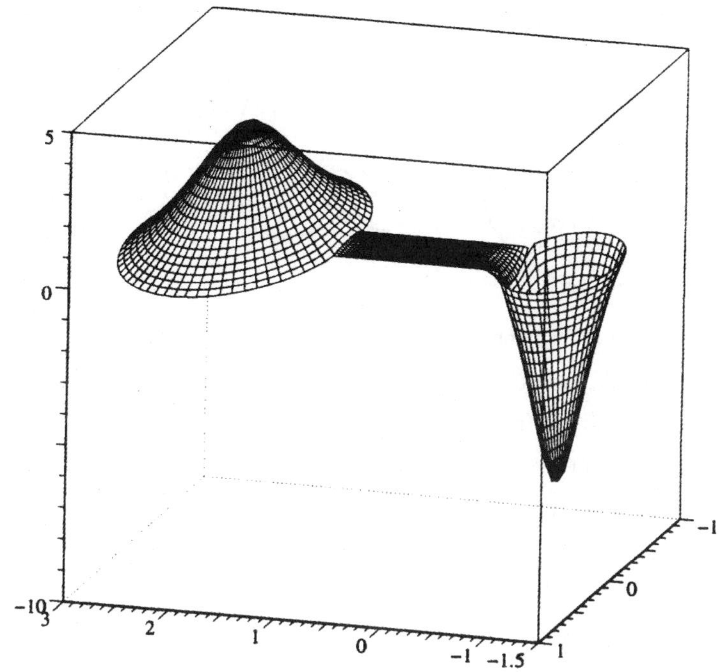

Figure 12.28: A sign-changing solution of $\Delta u + u^3 = 0$ with zero boundary condition on the dumbbell-shaped domain Ω_7.

Remark 12.9. (a) According to Theorem 12.5, and as confirmed by visualization from the graphics below, when ε in (12.76) becomes small, solutions of (12.76) display "spikes". Our boundary element numerical method seems to capture the spike feature better because the numerical solution is $\underline{C^\infty}$ on the *interior* of the domain Ω, where the spikes occur. If, instead, say the finite difference or finite element method were used, then one likely needs to take adaptive measures (domain decomposition, multi-grids, etc.) near the spike locations in order to capture the special feature of their profiles.

(b) The error measure ε_n of the nth iterate u_n of (12.76), by (12.21), is

$$\varepsilon_n = \left\{ \int_\Omega \left| \Delta u_n - \frac{1}{\varepsilon^2}(u_n - u_n^p) \right|^2 dx \right\}^{1/2}. \tag{12.77}$$

Our numerical experience has indicated that, when ε is not small, then both MPA and SIA produce virtually identical, equally accurate numerical solutions. But as ε becomes small, MPA begins to lose accuracy. We believe that this fact may be attributed to the numerical

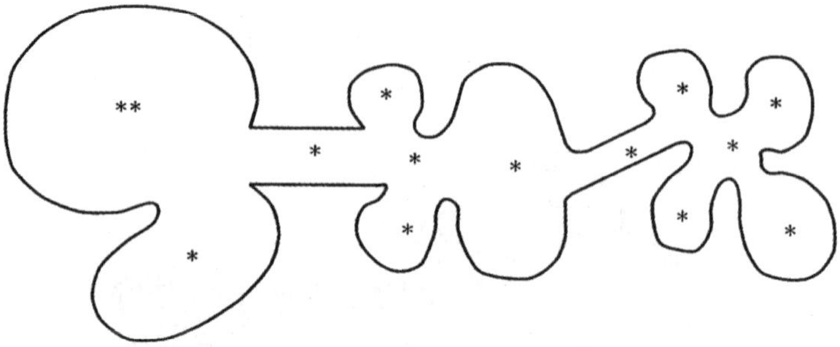

Figure 12.29: An irregular shaped domain with many compartments and corridors. Each "$*$" indicates the likely location of the *numerical solution* of a local ground state, while "$**$" indicates ("beyond any reasonable doubt") the location of the global ground state because that compartment is the largest. There may exist additional local ground states that we are either not aware of, or not able to capture numerically.

ill-conditionedness of the Dirichlet problem

$$\begin{cases} \Delta v = -\Delta w_1 - \dfrac{1}{\varepsilon^2}(-w_1 + w_1^p), \text{ on } \Omega \\ v|_{\partial\Omega} = 0, \end{cases} \tag{12.78}$$

which was required in (12.17) of MPA. In contrast, SIA does not seem to suffer any loss of accuracy, according to our numerical experiments. We attribute this advantage to the scaling condition $(12.30)_2$; it may have helped "normalize" the profile of the solution surface, at least at the point x_0. It is for this reason that most of the numerical solutions in this section are obtained *by SIA*. □

A total of 11 graphics will be presented for visualization in this section.

12.4.1 *The unit disk*

We use the domain Ω_1 from Subsection 12.3.1.

Case 12.4.1.a The unique solution of (12.76) on the unit disk, with $p = 3$, and $\varepsilon^2 = 1, 10, 100, 1000$. We have obtained the following data and graphics, as indicated in Table 12.1.

We may add that, when $\varepsilon^2 = 1$ and $\varepsilon^2 = 10^{-1}$ in Table 12.1, both MPA and SIA work and produce nearly equal results. However, when $\varepsilon^2 = 10^{-2}$ and 10^{-3}, MPA fails to work. The numerical results in Table 12.1 and Figs. 12.32 and 12.33 for $\varepsilon^2 = 10^{-2}$ and 10^{-3} can only be obtained by SIA.

Table 12.1: The data corresponding to the unique positive solution of (12.76) on the unit disk, with $p = 3$, MPA* at two entries mean that the numerical solutions obtained by MPA and SIA agree. But for $\varepsilon^2 = 10^{-2}$ and 10^{-3}, MPA does not yield convergence.

Value of ε^2	$\max u$	J_ε	ε_n or $\widetilde{\varepsilon}_n$	n	x_0	Graphics	Algorithm
1	3.951	14.71	10^{-4}	7		Fig. 12.30	MPA*
10^{-1}	2.242	5.972	10^{-4}	15		Fig. 12.31	MPA*
10^{-2}	2.422	5.795	10^{-5}	46	(0,0)	Fig. 12.32	SIA, $\alpha_n = 5.867$
10^{-3}	1.396	3.790	10^{-6}	10	(0,0)	Fig. 12.33	SIA, $\alpha_n = 1.948$

Since the solution is radially symmetric, we can also apply an FDM to (12.76). Let us choose the finite difference step size $h = 1/50$ in the following [which is analogous to (12.60)]:

$$\begin{cases} \dfrac{1}{\varepsilon^2}\left(\dfrac{u_{i+1} - 2u_i + u_{i-1}}{h^2} + \dfrac{1}{r_i}\dfrac{u_{i+1} - u_{i-1}}{2h}\right) - u_i + u_i^3 = 0, \quad i = 1, 2, \ldots, n-1, \\ u_0 = u_1, \quad u_n = 0, \quad h = \dfrac{1}{n}. \end{cases}$$

We only list the values of $\max u$ here, obtained from finite difference, for the purpose of comparison with Table 12.1:

$$\begin{aligned} &\varepsilon^2 = 1: \ \max u = 3.9477; \\ &\varepsilon^2 = 10^{-1}: \ \max u = 2.2479; \\ &\varepsilon^2 = 10^{-2}: \ \max u = 2.1546; \\ &\varepsilon^2 = 10^{-3}: \ \max u = 0.0001. \end{aligned} \tag{12.79}$$

We see that as $\varepsilon^2 \downarrow 0$, the difference between the data in Table 12.1 from those in (12.79) has widened, indicating that measures (such as finer mesh) must be taken to account for the singular perturbation effect. The singular perturbation parameter ε^2 and the mesh size h are somehow coupled in the error bounds, as the work of Adjerid, Aiffa and Flaherty [3] has shown. Despite the fact that our numerical scheme has not properly adjusted the mesh size with respect to decreasing ε^2, the graphics in this section have captured the essence of the spike-layer feature. However, we hope to be able to address the issue of coupling between h and ε^2 elsewhere in the future.

Case 12.4.1.b The unique solution of (12.76) on the unit disk, $p = 9$, and $\varepsilon^2 = 1, 10, 100, 1000$. To see how different powers p work, we choose a medium value: $p = 9$.

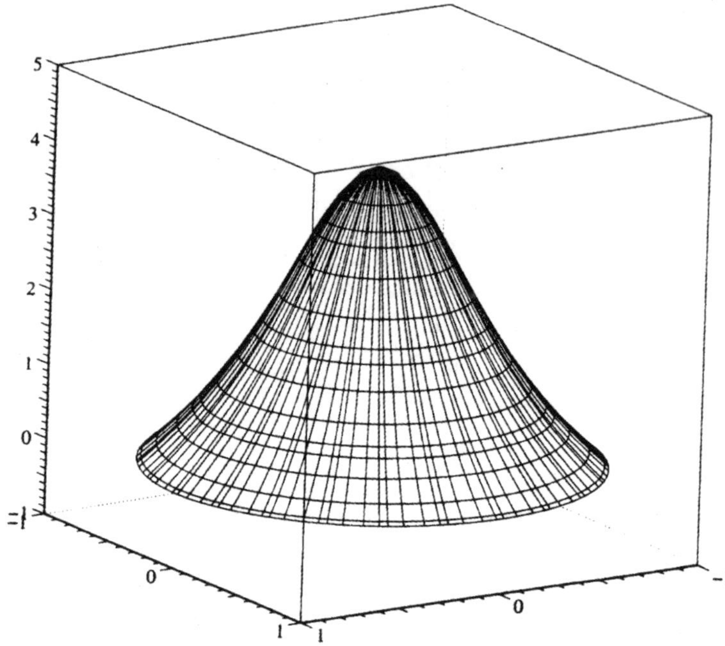

Figure 12.30: The ground state and the unique positive solution of the singularly perturbed problem (12.76), with $p = 3$, $\varepsilon^2 = 1$, on the unit disk Ω_1.

Table 12.2: The data obtained by SIA, corresponding to the unique positive solutions of (12.76) on the unit disk, with $p = 9$. Note that we are not able to compute the case when $\varepsilon^2 = 10^{-3}$.

Value of ε^2	$\max u$	J_ε	$\widetilde{\varepsilon}_n$	α_n	n	x_0	Graphics
1	2.902	2.902	10^{-6}	366.6	10	(0,0)	Fig. 12.34
10	1.657	2.279	4×10^{-6}	56.93	28	(0,0)	Fig. 12.35
10^{-2}	1.489	2.204	4×10^{-6}	3.379	10	(0,0)	Fig. 12.36

The data in this case may also serve as benchmarks for other researchers. We have tabulated them in Table 12.2.

Note that all the solutions in Figs. 12.30–12.36 are the *unique* mountain–pass solutions.

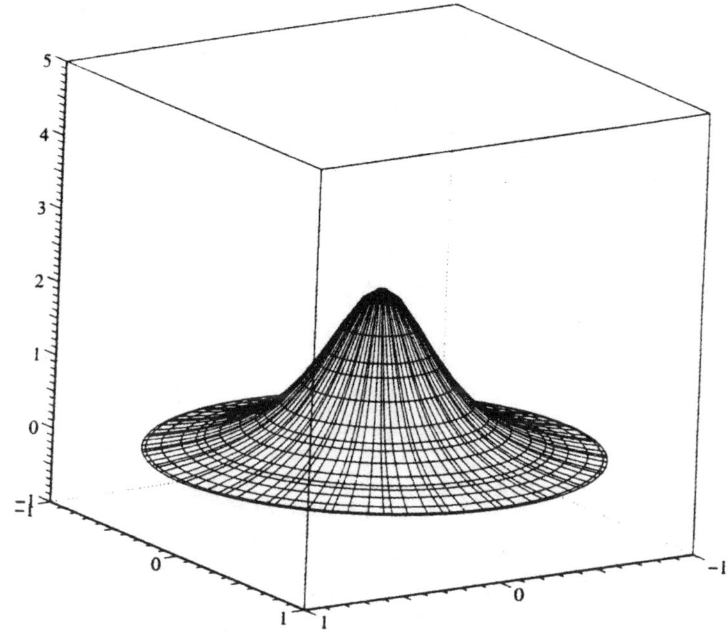

Figure 12.31: Same as Fig. 12.30, but with $\varepsilon^2 = 10^{-1}$.

12.4.2 *The radially symmetric annulus* Ω_6

The existence of nonradially symmetric positive solutions of (12.76) has been established in Coffman [51] and Li [120]. Graphically, we have found them to be multipeak.

Case 12.4.2.a A single-peak ground state of (12.76) ($p = 3$). Let $p = 3$, $\varepsilon^2 = 10^{-2}$ in (12.76), with the initial iterate $v_0(x) = M_{r_0,x_0}(x)$, $r_0 = 0.2, x_0 = (0.7, 0)$. We obtain

$$\text{(SIA)} \qquad \begin{aligned} \max u = 2.3797, \quad J_\varepsilon = 6.320, \quad \widetilde{\varepsilon}_{55} = 10^{-4}, \\ \alpha_{55} = 5.663, \quad x_0 = (0.7, 0), \end{aligned}$$

as displayed in Fig. 12.37. Note that, as above, each rotation of u is again a solution. This solution is a ground state.

We have not been able to obtain multipeak positive solutions for Case 12.4.2.a so far.

12.4.3 *The dumbbell-shaped domain* Ω_7

Case 12.4.3.a Three single-peak, positive solutions concentrated, respectively, on the large disk, small disk, and the corridor. Let $p = 3, \varepsilon^2 = 1/900$ in (12.76). We choose three mound functions $M_{r_0 x_0}(x)$ for the initial state $v_0(x)$ in SIA iterations, and obtain the data in Table 12.3.

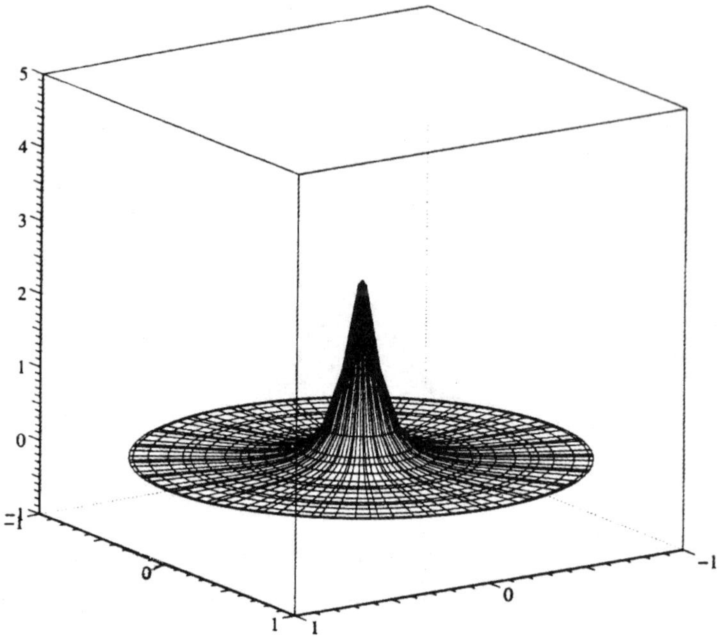

Figure 12.32: Same as Fig. 12.30, but with $\varepsilon^2 = 10^{-2}$.

Table 12.3: Data for three single-peak solutions of (12.76), with $p = 3$, $\varepsilon^2 = 1/900$

Location of the Single Peak	max u	J_ε	$\widetilde{\varepsilon}_n$	$v_0 = M_{r_0,x_0}$		α_n	n	x_0	Graphics
				r_0	x_0				
D_R (large disk)	1.251	4.01	10^{-6}	1	(2,0)	1.564	14	(2,0)	Fig. 12.37
D_L (small disk)	1.801	4.145	10^{-6}	0.5	(−1,0)	3.242	14	(−1,0)	Fig. 12.38
Corridor	1.416	4.133	10^{-6}	0.2	(0,25,0)	4.854	17	(0.25,0)	Fig. 12.39

However, if we choose $\varepsilon^2 = 1/1000$, with the same initial states v_0 as given in Table 12.3 and with everything else unchanged, we obtain the data in Table 12.4.

The data entries for $J_\varepsilon = 4.060$ are *inconsistent* with the theory [147] that the energy of the global ground state on D_R should be lower than that of the local ground state on D_L because "D_R has more open space than D_L". The reason for this inconsistency is easy to explain: *As* $\varepsilon^2 \downarrow 0$, *one needs to make finer discretizations of* $\partial\Omega$ *and* Ω *in order to have correspondingly higher resolution for the singularly perturbed problem. Otherwise numerical inconsistencies may occur.*

Note that Fig. 12.37 corresponds to the ground state of (12.76), while Fig. 12.38 and 12.39

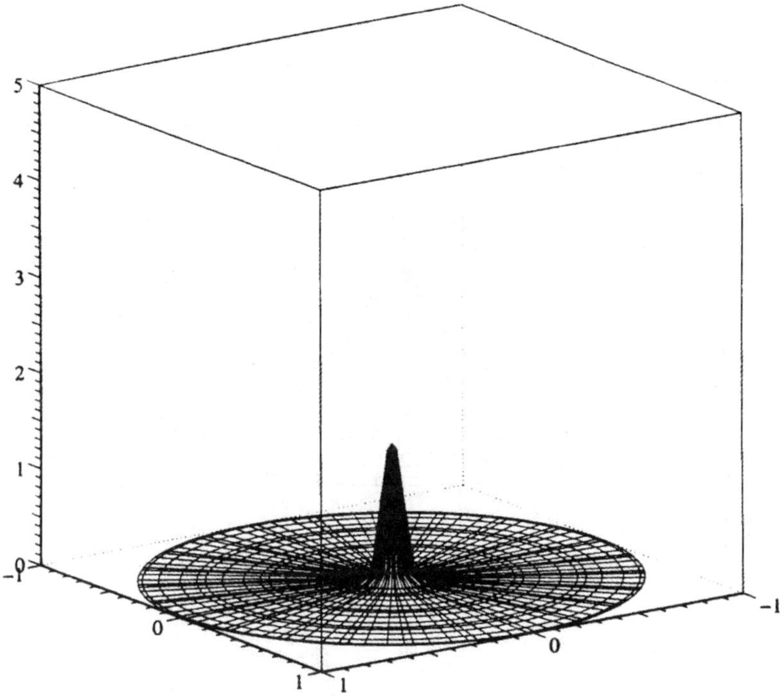

Figure 12.33: Same as Fig. 12.30, but with $\varepsilon^2 = 10^{-3}$.

Table 12.4: Data for three single-peak solutions of (12.76), with $p = 3$, $\varepsilon^2 = 10^{-3}$.

Location of the Single Peak	$\max u$	J_ε	$\widetilde{\varepsilon}_n$	α_n	n
D_R (large disk)	1.2233	4.121	10^{-6}	1.497	14
D_L (small disk)	1.7456	4.060	10^{-6}	3.047	13
Corridor	1.3825	4.163	10^{-6}	0.805	16

correspond to local ground states. The location of each peak for each of the positive solutions in Figs. 12.37–12.39 has fallen almost exactly at, respectively, the geometrical center of the large disk, the small disk and the corridor. This serves as a visual confirmation of Ni and Wei [147, Theorem 4.1].

One can also construct sign-changing solutions from these three single-peak solutions just as in Subsection 12.3.8.

Remark 12.10. The limiting peak value, max $u_{\varepsilon,p}$, of a ground state $u_{\varepsilon,p}$ in (12.76), as

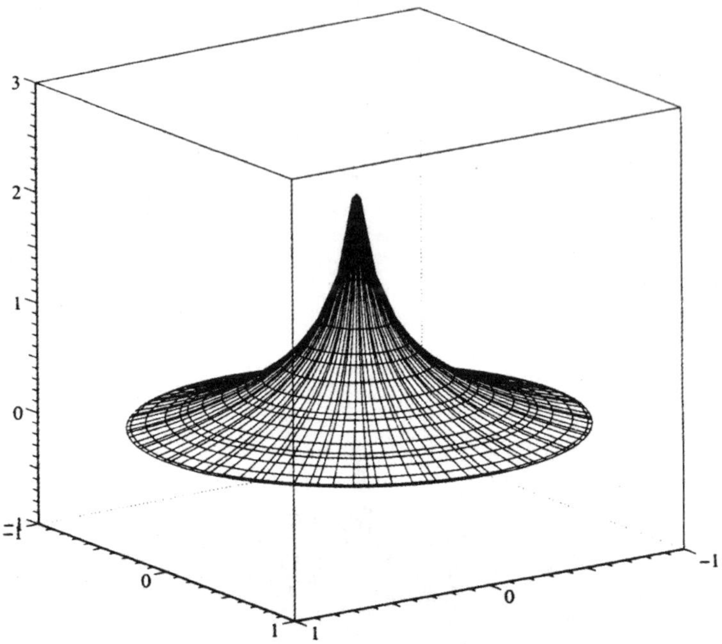

Figure 12.34: The ground state and the unique positive solution of the singularly perturbed problem (12.76), with $p = 9$, $\varepsilon^2 = 1$, on the unit disk Ω_1.

$\varepsilon^2 \downarrow 0$, of this section may be obtained as follows. (The arguments were given implicitly in Ni and Wei [147].) Consider the following ODE

$$\left.\begin{aligned} & w_p''(r) + \frac{1}{r}w_p'(r) - w_p(r) + w_p^p(r) = 0, \quad 0 < r < \infty, \\ & w_p'(0) = 0, \\ & w_p > 0 \quad \text{on} \quad [0,\infty), \quad \lim_{r\to\infty} w_p(r) = 0. \end{aligned}\right\} \tag{12.80}$$

(If the domain Ω is in $\mathbb{R}^N$, then the term $(1/r)w'(r)$, in (12.80) above should be replaced by $((N-1)/r)w'(r)$.) Equation (12.80) is the radially symmetric version of the PDE $\varepsilon^{-2}\Delta u - u + u^p = 0$ in polar coordinates, where ε^{-2} has been "scaled out" and the angular dependence is omitted. It is known that (12.80) has a unique solution satisfying $w_p(0) = \alpha_p$ such that

$$\lim_{\varepsilon\downarrow 0} \max u_{\varepsilon,p} = \alpha_p. \tag{12.81}$$

To obtain α_p, we consider the initial value problem

$$\left.\begin{aligned} & w_p'' + \frac{1}{r}w_p' - w_p + w_p^p = 0, \quad \text{on} \quad [0,\infty), \\ & w_p(0) = \beta, \quad w_p'(0) = 0. \end{aligned}\right\} \tag{12.82}$$

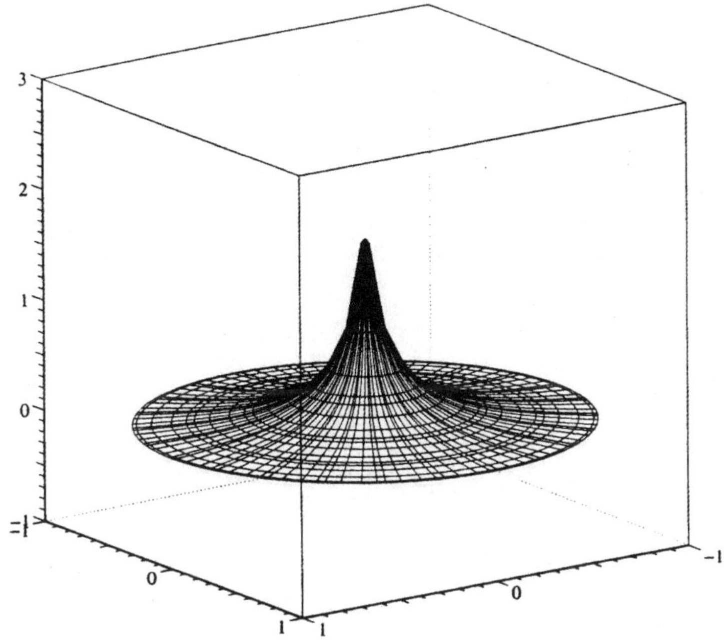

Figure 12.35: Same as Fig. 12.34, but with $\varepsilon^2 = 10$.

This problem has a solution $w_p(r;\beta)$ such that

(i) if $\beta > \alpha_p$, then there exists $r_0 > 0$ such that $w(r_0;\beta) = 0$;
(ii) if $0 < \beta < \alpha_p$, then $\lim\limits_{r \to \infty} w(r;\beta) = 1$.

Using a method of bisection on the parameter β for the ODE $(12.82)_1$ discretized by a finite difference scheme like (12.60) with step size h, we obtain
(1) for $p = 3$,

$$\alpha_{p,h} = \begin{cases} 2.205636, \; h = 10^{-2}, \\ 2.206171, \; h = 5 \times 10^{-3}, \\ 2.206193, \; h = 10^{-3}; \end{cases} \tag{12.83}$$

(2) for $p = 9$,

$$\alpha_{p,h} = \begin{cases} 1.810892, \; h = 10^{-2}, \\ 1.819922, \; h = 5 \times 10^{-3}, \\ 1.820333, \; h = 10^{-3}. \end{cases} \tag{12.84}$$

The dependence of $\alpha_{p,h}$ on h causes the loss of accuracy of α_p. To obtain an accurate approximation of α_p, we use a quadratic *extrapolation* by writing

$$\alpha_{p,h} = \alpha_p + c_1(p)h + c_2(p)h^2, \tag{12.85}$$

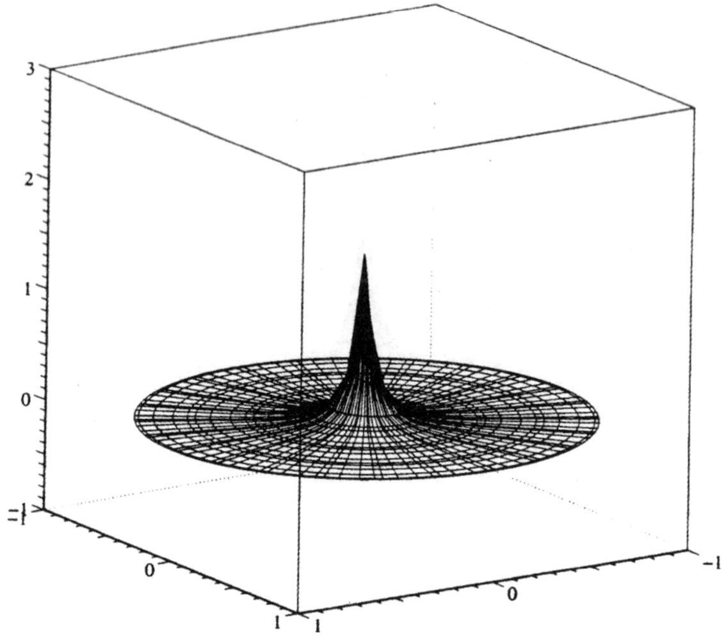

Figure 12.36: Same as Fig. 12.34, but with $\varepsilon^2 = 10^{-2}$.

with three unknowns $\alpha_p, c_1(p)$ and $c_2(p)$ to be determined from (12.83) and (12.84), respectively. We thus obtain

$$\alpha_p|_{p=3} \approx 2.206205, \tag{12.86}$$

$$\alpha_p|_{p=9} \approx 1.820585. \tag{12.87}$$

The reader may use (12.86) and (12.87) to compare the values of $\max u$ in, respectively, Tables 12.1 and 12.2 to see how close $\max u$ is to the "asymptotic regime." □

12.5 Other variant semilinear elliptic Dirichlet problems

In this section, we consider three different PDEs: Henon's equation (12.4), Chandrasekhar's (12.5), and the Lane–Emden equation (12.99) with $p \neq 3$.

12.5.1 *Henon's equation*

Consider equation $(12.4)_1$ on a general, bounded domain Ω in $\mathbb{R}^2$:

$$\Delta u + |x|^{\ell} u^p = 0, \quad u > 0, \quad \text{on } \Omega, \quad u = 0 \quad \text{on } \partial\Omega, \quad k, \ell > 0. \tag{12.88}$$

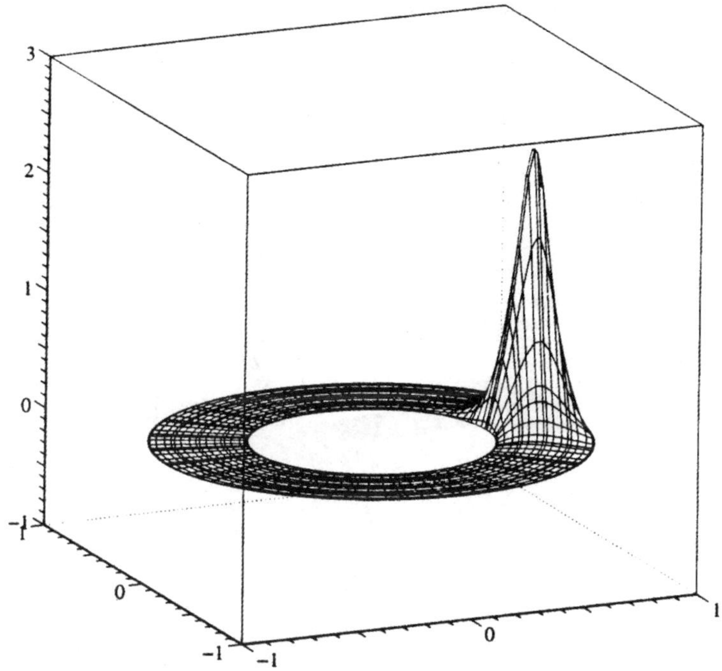

Figure 12.37: A single-peak, global mountain pass solution of (12.76), with $p = 3$, $\varepsilon^2 = 10^{-2}$, on the thin concentric annulus Ω_6.

Since we consider only bounded domains Ω here, the growth factor $|x|^\ell$ should not matter very much. One might expect that solutions of (12.88) behave like those of $\Delta u + u^p = 0$, other conditions being identical. However, this is not true.

Case 12.5.1.a Henon's equation (12.88) on the unit disk Ω_1, with $\ell = 1$, $p = 3$. Even though the governing equation in (12.88) is radially symmetric on the disk Ω_1, the main result in Gidas, Ni and Nirenberg [77, p. 221, Theorem 1′], does not apply because equation (12.88) has explicit x-dependence. As it turns out, some symmetry breaking occurs and a ground state of (12.88) is *not radially symmetric* [144]; see Figs. 12.41(a) and 12.41(b). This seems sort of a surprise to the novice but can be explained roughly in just a few paragraphs given in Remark 12.11 below. We have obtained, for $\ell = 1$, $p = 3$, $k = 0.81$,

$$\text{(SIA)} \qquad \begin{array}{c} \max u = 6.075, \quad J = 35.47, \quad \widetilde{\varepsilon}_{133} = 10^{-6}, \\ \alpha_{133} = 17.83, \quad x_0 = (0,0). \end{array} \qquad (12.89)$$

We also have found a radially symmetric positive solution by MPA, with Morse index 2. It

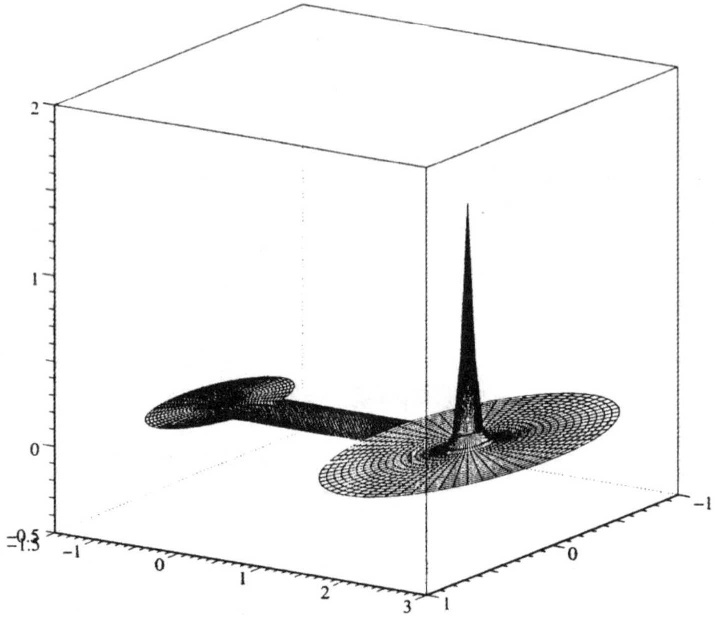

Figure 12.38: A single-peak, positive solution of (12.76), with $p = 3$, $\varepsilon^2 = 1/900$, concentrated on the large disk of the dumbbell-shaped domain Ω_7.

is displayed in Fig. 12.42, with

$$\text{(SIA)} \qquad \begin{aligned} \max u = 5.361, \quad J = 37.09, \quad \widetilde{\varepsilon}_{11} = 2 \times 10^{-5}, \\ \alpha_{11} = 28.74, \quad x_0 = (0,0). \end{aligned}$$

Thus it seems that, in this case, the energy of a radially symmetric positive solution is just slightly higher than that of a ground state.

The following is a case with a medium large power, $\ell = 9$. Note that $|x| < 1$ for all $x \in \Omega_1$ and, thus, $|x|^9$ becomes a small number unless x is very near to $\partial\Omega_1$.

Case 12.5.1.b Henon's equation (12.88) on the unit disk Ω_1, with $\ell = 9$, and $p = 3$. Using $u_0(x) = -10$ and w_0 in (12.64) for the initial state of MPA, we have obtained a ground state with the following data:

$$\text{(SIA)} \qquad \begin{aligned} \max u = 25.23, \quad J = 584, \quad \widetilde{\varepsilon}_{51} = 10^{-6}, \\ \alpha_{51} = 13.91, \quad x_0 = (0,0), \end{aligned} \tag{12.90}$$

and a radially symmetric solution with the following data

$$\text{(SIA)} \qquad \begin{aligned} \max u = 19.76, \quad J = 1843, \quad \widetilde{\varepsilon}_6 = 3 \times 10^{-3}, \\ \alpha_3 = 388.3, \quad x_0 = (0,0). \end{aligned} \tag{12.91}$$

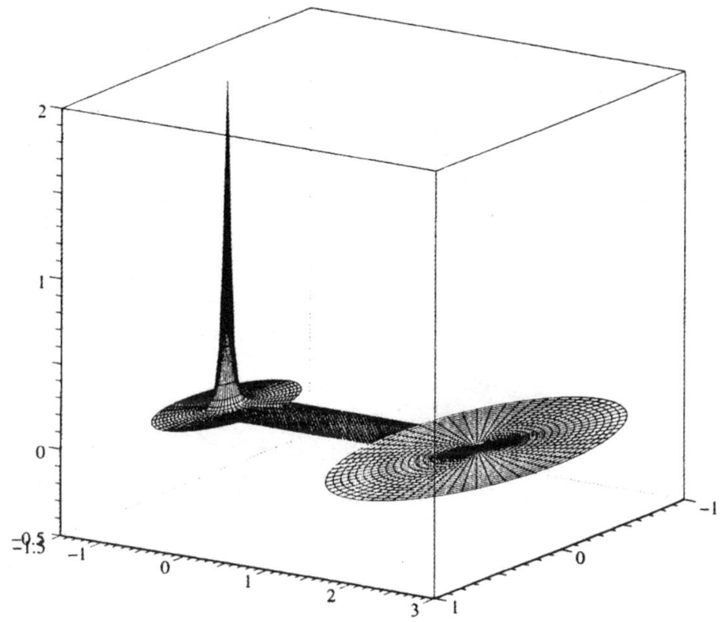

Figure 12.39: A single-peak, positive solution of (12.76), with $p = 3$, $\varepsilon^2 = 1/900$, concentrated on the small disk of the dumbbell-shaped domain Ω_7.

as shown in Figs. 12.43(a) and 12.43(b). One can see that the peak of the ground state is quite far off center, and the difference in J between the ground state and the radially symmetric solution is much larger than that in Case 12.5.1.a.

The data in both (12.89) and (12.90) may serve as benchmarks for other researchers.

Remark 12.11. Let us explain briefly why the ground state of Henon's equation may lack radial symmetry on the unit disk Ω_1. If a semilinear equation takes the form

$$\Delta u + g(r,u) = 0, \quad u > 0 \quad \text{on} \quad B_R; \quad u|_{\partial B_R} = 0, \tag{12.92}$$

where $r = |x|$, B_R is the open ball with radius R, and g, $(\partial g/\partial u)$ are continuous with g nonincreasing in r, then the method in Gidas, Ni and Nirenberg [77] applies which shows that any solution of (12.92) must be *radially symmetric*. For Henon's equation, $g(r,u) = r^\ell u^p$, the property of being nonincreasing in r is clearly violated.

In a way just similar to Example 12.1, one may consider the following two constrained variational problems:

$$M_p \equiv \sup_{u \in \mathscr{C}} \int_\Omega r^\ell |u|^{p+1}\, dx, \quad \mathscr{C} \equiv \left\{ u \in H_0^1(B_R) \,\middle|\, \int_\Omega |\nabla u|^2\, dx = 1 \right\}, \tag{12.93}$$

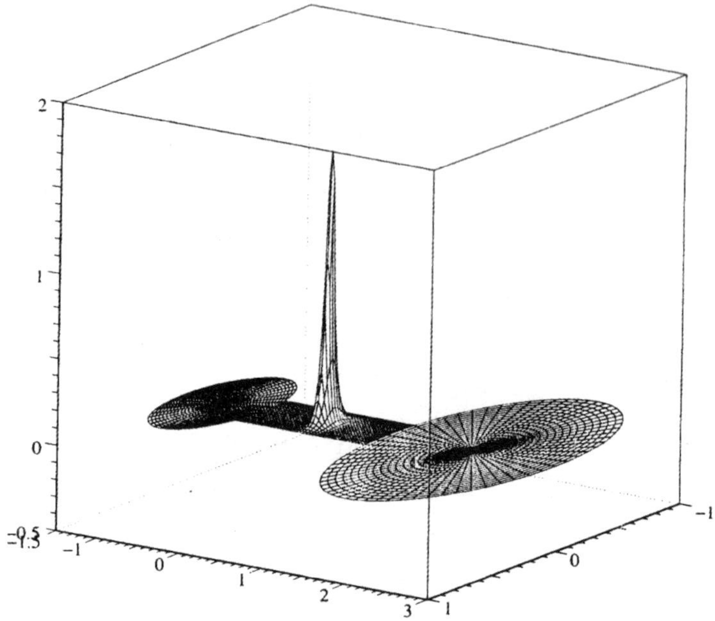

Figure 12.40: A single-peak, positive solution of (12.76), with $p = 3$, $\varepsilon^2 = 1/900$, concentrated on the corridor of the dumbbell-shaped domain Ω_7.

and

$$M_{p,r} \equiv \sup_{u\in\mathscr{C}_r} \int_\Omega r^\ell |u|^{p+1}\,dx, \tag{12.94}$$

$$\mathscr{C}_r \equiv \left\{ u \in H_0^1(B_R) \,\middle|\, \int_\Omega |\nabla u|^2\,dx = 1, u \text{ is radially symmetric} \right\}.$$

Solutions to (12.93) and (12.94) will yield solutions of Henon's equation after rescaling. In particular, a solution to (12.94) will be radially symmetric. However, the distribution of the weight r^ℓ, $\ell > 0$, heavily "favors" the part of the domain consisting of those points x such that $|x| = r$ is large, such as the annular strip bordering the boundary of a disk. This, therefore, creates an effect similar to that of the annulus case as in Subsection 12.3.4. Thus nonradially symmetric ground states are expected for (12.93).

Also, a consequence of this is that $M_p > M_{p,r}$ (because $\mathscr{C} \supset \mathscr{C}_r$). □

12.5.2 *Chandrasekhar's equation*

From (12.5), we consider

$$\Delta u + 4\pi(u^2 + 2u)^{3/2} = 0, \quad u > 0, \quad \text{on } \Omega, \quad u = 0 \quad \text{on } \partial\Omega. \tag{12.95}$$

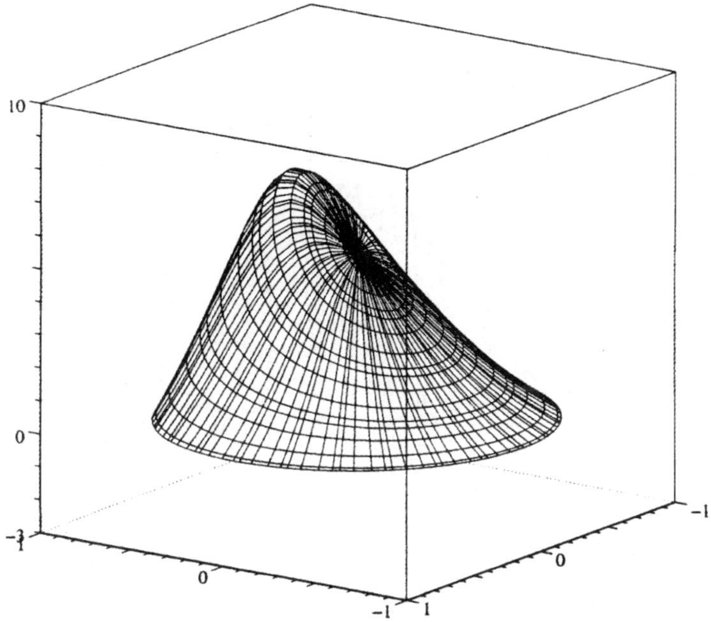

(a)

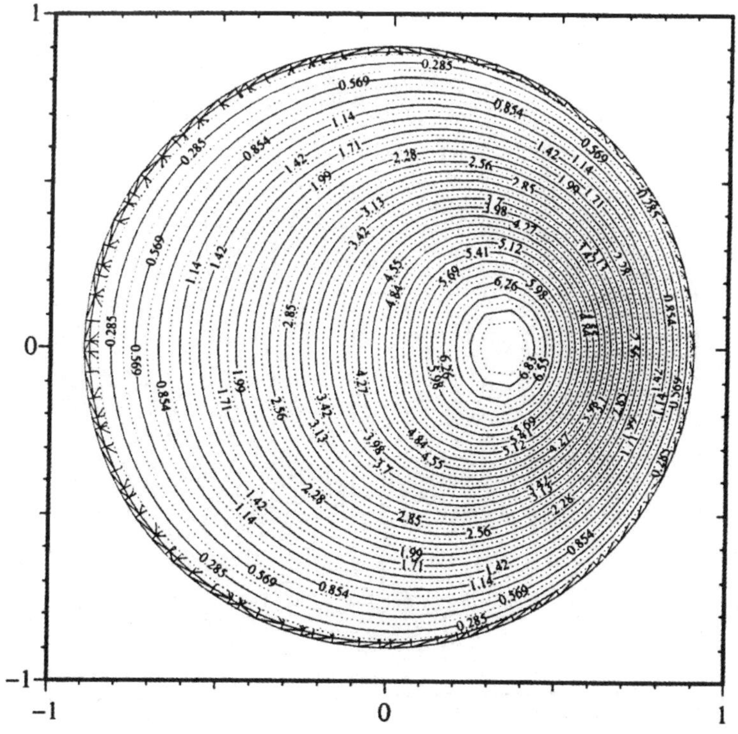

(b)

Figure 12.41: A ground state (a) and its contours (b) of Henon's equation (12.88) on the unit disk Ω_1, with $\ell = 1$, $p = 3$, and $k = 0.81$.

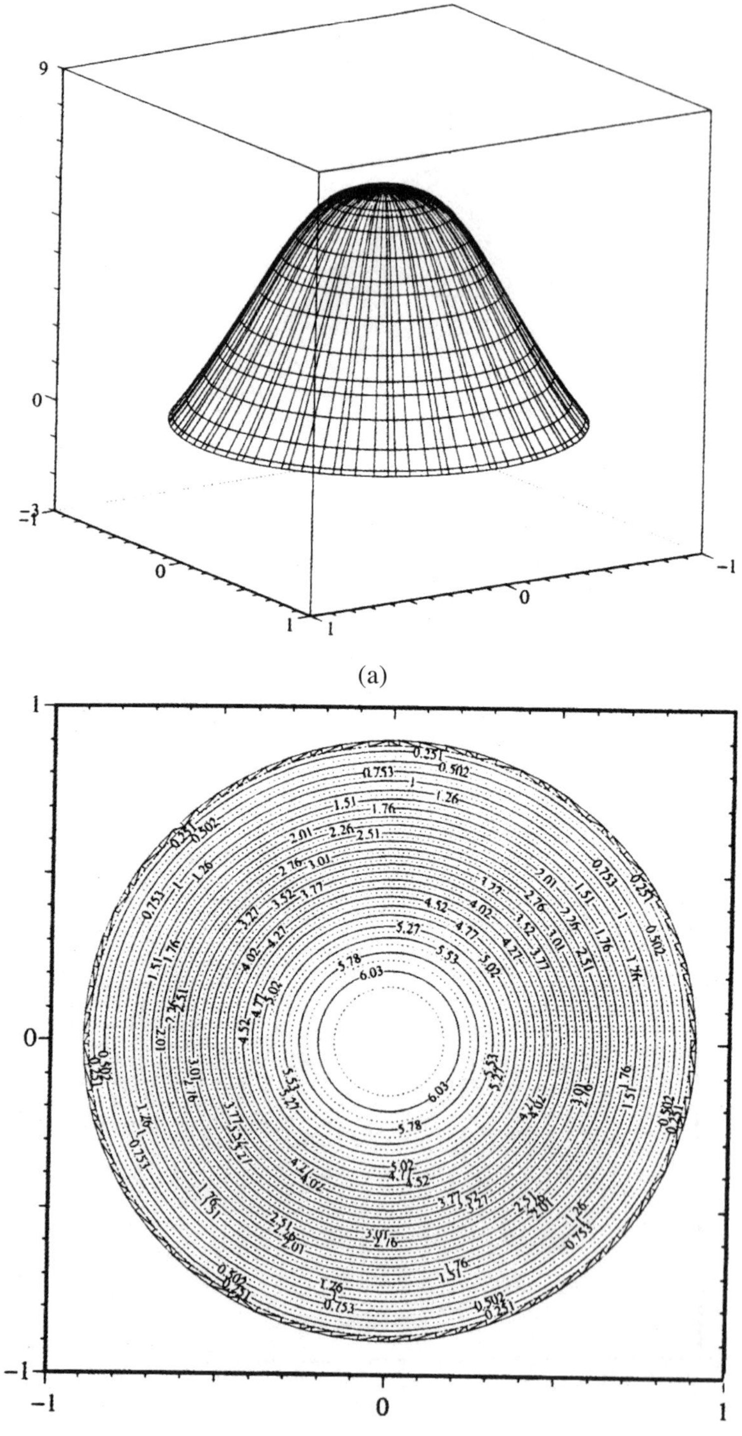

Figure 12.42: A radially symmetric positive solution (a) and its contours (b) of Henon's equation (12.88) on the unit disk, with $\ell = 1$, and $p = 3$. This solution has an energy level slightly higher than that of the ground state displayed in Fig. 12.41.

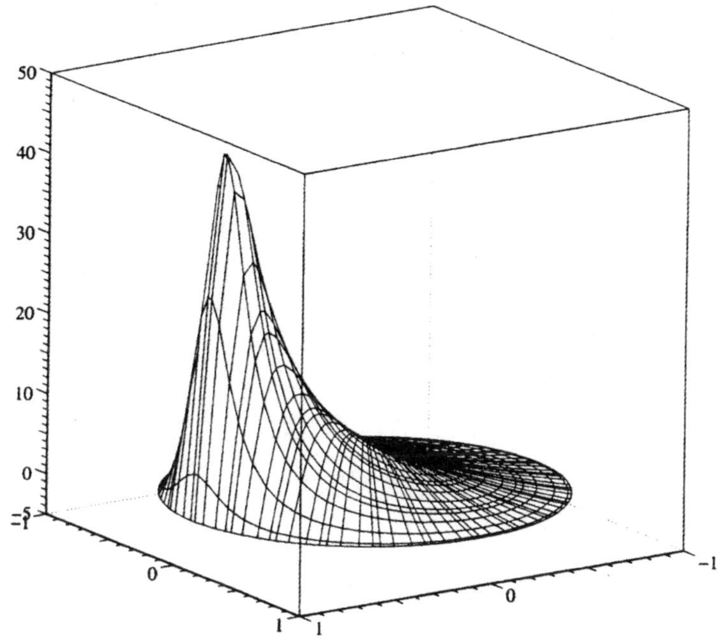

(a)

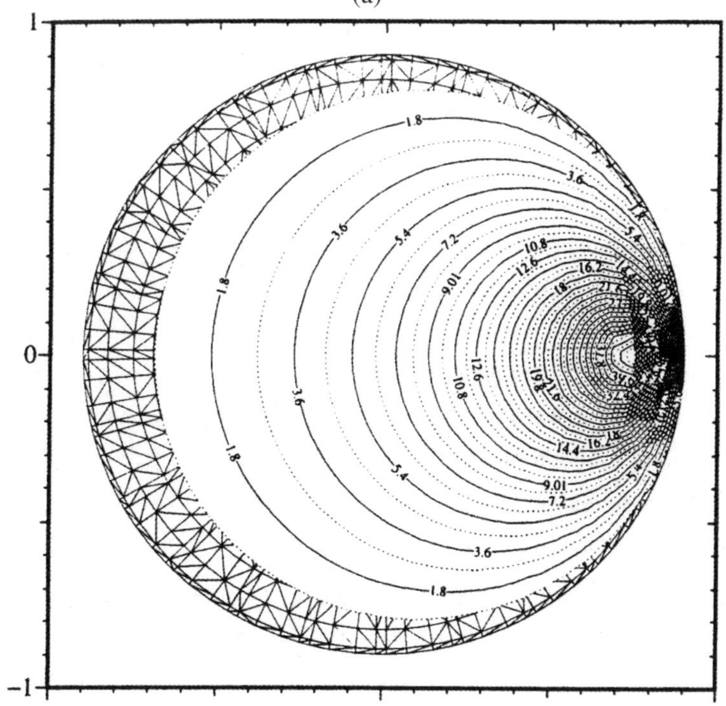

(b)

Figure 12.43: A ground state (a) and its contours (b) of Henon's equation (12.88) on the unit disk Ω_1, with $\ell = 9$, and $p = 3$.

The energy functional of (12.95) is given by

$$J(u) = \int_\Omega \left\{ \frac{1}{2}|\nabla u|^2 - \pi \left[(u+1)(u^2+2u)^{3/2} - \frac{3}{2}(u+1)(u^2+2u)^{1/2} \right.\right.$$
$$\left.\left. + \frac{3}{2}\ln(u+1+\sqrt{u^2+2u}) \right] \right\} dx. \qquad (12.96)$$

Solutions of (12.95) can be computed by MPA. However, because of the more involved appearance of J in (12.96), we also have correspondingly much more work to do at Steps 3 and 4 of MPA. In view of this, let us consider the alternative method, SIA. Mimicking the procedure in (12.30) and (12.31), we would have derived the following iterative algorithm, with Step 2 therein replaced by

Step 2′: For $n = 0, 1, 2, \ldots$, find $\alpha_{n+1} \geqslant 0$ and $v_{n+1}(x)$ such that

$$\begin{cases} \text{find } \alpha_{n+1} \geqslant 0 \text{ and } v_{n+1}(x) \text{ such that} \\ \Delta v_{n+1}(x) = -\alpha_{n+1}^{1/2} \cdot 4\pi[\alpha_{n+1} v_n^2(x) + 2v_n(x)]^{3/2}, \quad x \in \Omega, \\ v_{n+1}(x_0) = 1, \\ v_{n+1}|_{\partial\Omega} = 0. \end{cases} \qquad (12.97)$$

As it turns out from numerical experiments we see that the adapted SIA containing (12.97) is *divergent*. After a few cases of trial-and-error, we have found that a *relaxation* of the unknown parameter α_n by the following:

1. set $\alpha_0 = 1$, and compute α_1 by (12.97);
2. for $n = 1, 2, 3, \ldots$, compute $\alpha_{n+1} \geqslant 0$ and $v_{n+1}(x)$ by solving (12.98)

$$\Delta v_{n+1} = -\alpha_{n+1}^{1/2} \cdot 4\pi[\overline{\alpha}_{n+1} v_n^2(x) + 2v_n(x)]^{3/2}, x \in \Omega; \overline{\alpha}_{n+1} = \frac{\alpha_{n-1}+\alpha_n}{2},$$

$$v_{n+1}(x_0) = 1, v_{n+1}|_{\partial\Omega} = 0,$$

leads to numerical convergence of both α_n and $v_n(\cdot)$ with $\lim_{n\to\infty} \alpha_n \equiv \alpha_\infty > 0$, $v_\infty(\cdot) = \lim_{n\to\infty} v_n(\cdot)$. Consequently, a solution u of (12.95) is found numerically, given by $u(\cdot) = \alpha_\infty v_\infty(\cdot)$.

We call the above ASIA, the *adapted scaling iterative algorithm.*

We have found that many of the profiles of the Chandrasekhar equation are similar to those of the Lane–Emden equation on Ω_i, $i = 1, 2, \ldots, 9$, in § 12.3. Note that it now makes no sense to talk about sign-changing solutions, such as in Subsection 12.3.8, for the Chandrasekhar equation, because of the power 3/2 appearing in the nonlinearity.

Case 12.5.2.a Single-peak positive solutions of the Chandrasekhar equation (12.95) on a dumbbell-shaped domain, Ω_7. The profiles of three single-peak positive solutions have been obtained by ASIA. Their data and graphics are indicated in Table 12.5.

Fig. 12.44 represents the ground state, while Figs. 12.45 and 12.46 represent local ground states.

Table 12.5: The data corresponding to the three single-peak positive solutions of the Chandrasekhar equation (12.95) on dumbbell-shaped domain Ω_7

Initial State $v_0(x) = M_{r_0 x_0}(x), r_0 =$	$\max u$	J	$\widetilde{\varepsilon}_n$	α_n	n	x_0	Graphics
$1(D_R)$	0.03678	0.0006	10^{-6}	0.1918	16	(2,0)	Fig. 12.44
$0.5(D_L)$	0.3913	0.07954	10^{-6}	0.6242	17	$(-1,0)$	Fig. 12.45
0.2 (corridor)	1.6375	2.061	2×10^{-5}	1.286	23	(0.25,0)	Fig. 12.46

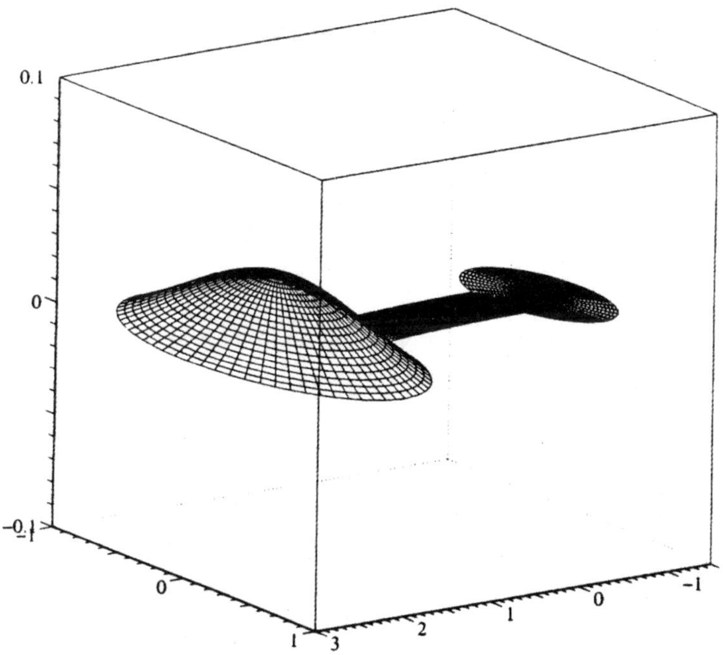

Figure 12.44: A single-peak ground state of the Chandrasekhar equation (12.95) concentrated mainly on the larger disk D_R. This solution can reasonably be expected to be the global mountain–pass solution.

12.5.3 *The Lane–Emden equation* $\Delta u + u^p = 0, p \neq 3$

We consider the Lane–Emden equation with powers $p \neq 3$:

$$\Delta u + u^p = 0, \quad u > 0 \quad \text{on } \Omega, \quad u|_{\partial\Omega} = 0. \tag{12.99}$$

Case 12.5.3.a The unique positive solutions of (12.99) on the unit disk Ω_1, for $p = 5$ and $p = 15$. For $p = 5$ and 15, using SIA we have obtained the data and graphics indicated in Table 12.6.

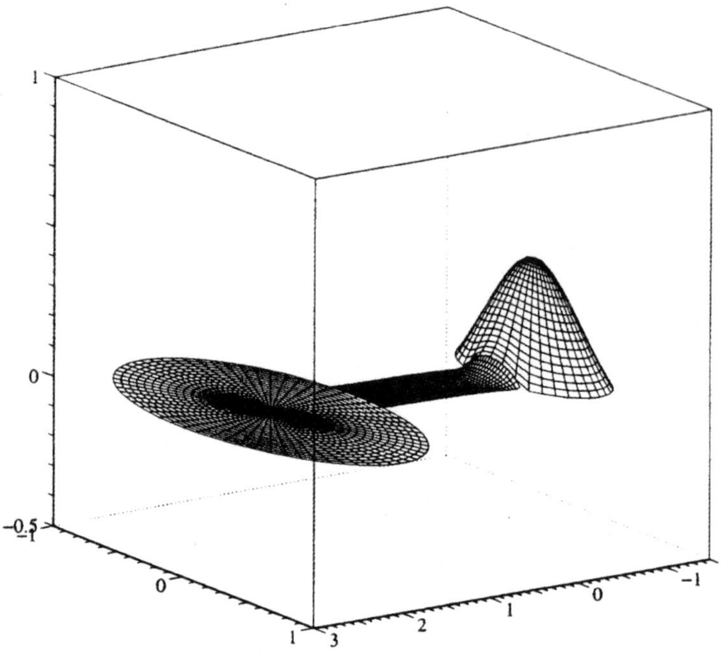

Figure 12.45: A single-peak positive solution of the Chandrasekhar equation (12.95) concentrated mainly on the smaller disk D_L.

Table 12.6: The data corresponding to the unique positive solutions of the Lane–Emden equation (12.99) on the unit disk Ω_1, for $p = 5$ and 15.

p	$\max u$	J	$\widetilde{\varepsilon}_n$	α_n	n	x_0	Graphics
5	1.3307	4.831	10^{-6}	29.51	21	(0,0)	Fig. 12.47
15	1.5469	1.617	2.5×10^{-5}	449.1	8	(0,0)	Fig. 12.48

Remark 12.12. For the unit disk Ω_1, we have found that, for different p's, the unique positive solution in each case satisfies a monotone decreasing property:

$$u_{p_1}(x) \leqslant u_{p_2}(x), \quad x \in \Omega_1, \quad \text{if} \quad p_1 > p_2 > 1. \tag{12.100}$$

We still have not been able to prove (12.100). However, we must remark that (12.100) relies on the fact that $\Omega_1 \subset \mathbb{R}^2$; it will *no longer hold* for the unit ball in $\mathbb{R}^3$, due to the presence of the critical exponent $p^* = (N+2)/(N-2)$. □

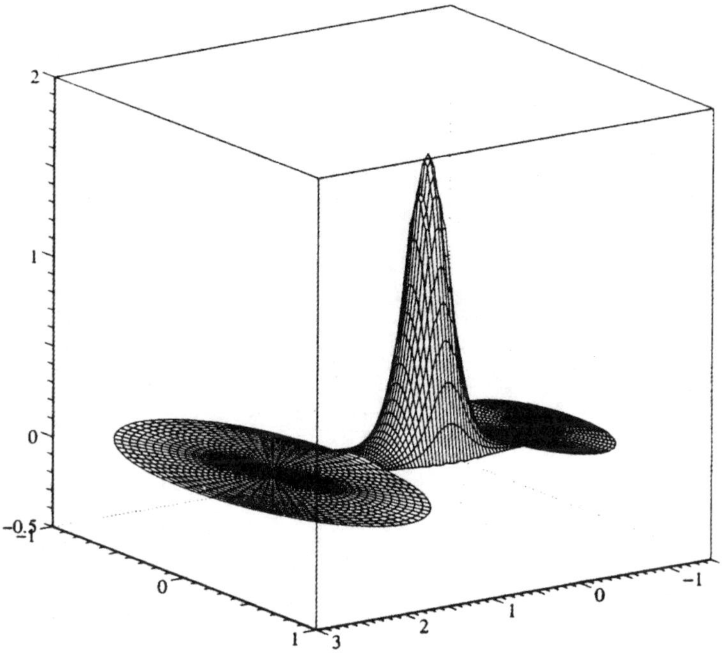

Figure 12.46: A single-peak positive solution of the Chandrasekhar equation (12.95) concentrated mainly on the corridor.

Remark 12.13. A recent paper by Ren and Wei [159] gives a sharp characterization of the asymptotic behavior of the ground states of the Lane–Emden equation on a 2D smooth domain Ω when the exponent p grows large. Let u_p be a ground state of (12.99). Then [159, Theorem 1.4] shows that

$$1 \leqslant \underline{\lim}_{p\to\infty} \|u_p\|_{L^\infty(\Omega)} \leqslant \overline{\lim}_{p\to\infty} \|u_p\|_{L^\infty(\Omega)} \leqslant \sqrt{e}\,. \tag{12.101}$$

In Table 12.6, we have $\max u_{15} = 1.5469$, which differs from $\sqrt{e} \approx 1.6487$ with about 6% of relative deviation. (This deviation is not necessarily an error because it is not known so far whether (12.101) gives the tightest estimates.) As p increases past 15, we have found that $\max u_p$ decreases in value, somehow indicating that a finer discretization is called for in order to improve accuracy. After p passes 20, computer arithmetic *overflow* occurs, preventing further computations.

Ren and Wei [159, Theorem 1.3] have further shown that as $p \to \infty$, for a subsequence p_n of p, we have

$$\frac{u_{p_n}^{p_n}}{\int_\Omega u_{p_n}^{p_n}(x)dx} \to \delta(x_0) \tag{12.102}$$

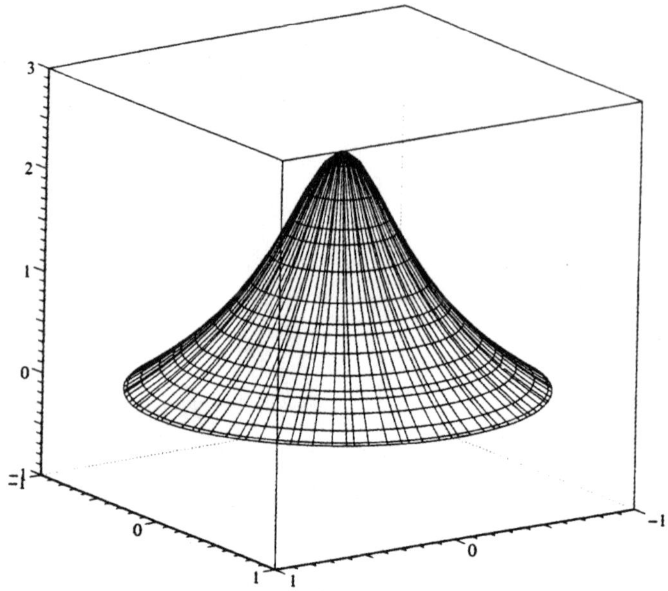

Figure 12.47: The unique ground state of (12.99) on the unit disk Ω_1, with $p = 5$.

in the sense of distribution for some unique point $x_0 \in \Omega$, called the "blow-up" or "condensation" point, where x_0 can be characterized as a critical point of the function $\phi(x) \equiv g(x,x)$, where $g(x,y)$ is the "regular part" of the Green's function $G(x,y)$ for the domain Ω:

$$\Delta_x G(x,y) = -\delta(x-y), \qquad x,y \in \Omega; \quad G(x,y)|_{x\in\partial\Omega} = 0, \quad y \in \Omega.$$

Furthermore, if Ω is convex, then (12.102) holds for the entire sequence p. Thus, as p grows large, the ground states "look more and more like a single spike".

We nevertheless wish to reiterate that for the Lane–Emden equation (12.2), even though it is believed to be true that its ground states all have a single peak, there seems to be no rigorous proof so far. □

12.6 The sublinear Dirichlet problem $\Delta u + u^p = 0, 0 < p < 1$

Let us consider

$$\Delta u + u^p = 0, \quad u > 0 \quad \text{on } \Omega, \quad u|_{\partial\Omega} = 0, \quad 0 < p < 1. \tag{12.103}$$

With the *sublinear* growth of nonlinearity u^p in (12.103), we have a complete grasp of existence and uniqueness of the positive solution.

Theorem 12.6. *Equation (12.103) has a unique solution.* □

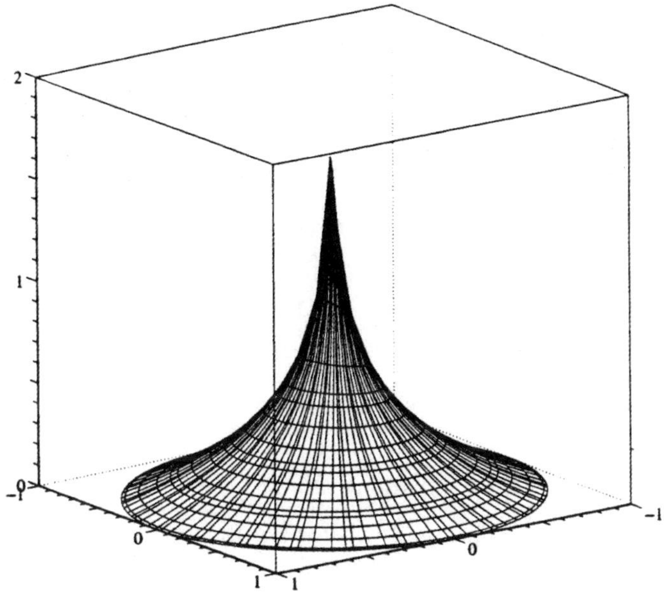

Figure 12.48: The unique ground state of (12.99) on the unit disk Ω_1, with $p = 15$.

Its proof, involving the Monotone Iteration Scheme, Theorem 12.4, may be found in Ni [141, pp. 69–70], for example.

12.6.1 *Solutions of (12.103) by direct iteration*

Obviously, MIA appears to be "the algorithm" to be used for (12.103), since the proof of Theorem 12.6 was carried out that way. Our numerical experience has easily confirmed this. MPA does not work for (12.103) because $p < 1$. We have experimented with SIA, which also converges nicely, and the numerical solutions coincide with those obtained by MIA. In our numerical study, however, we have also found that the use of DIA for (12.103) suffices. This fact is given in the following.

Theorem 12.7 (Convergence of the Direct Iteration Algorithm for (12.103)). *Let Ω be a bounded open domain in $\mathbb{R}^N$, with $C^{2,\delta}$-smooth boundary $\partial\Omega$ for some $0 < \delta < 1$. Let an initial state $u_0(\cdot)$ be sufficiently smooth, $u_0(x) \geqslant 0$ on Ω, and $u_0 \not\equiv 0$. Let u_{n+1} be the solution of*

$$\begin{cases} \Delta u_{n+1} + u_n^p = 0, & \text{on } \Omega, \quad \text{for given } p\colon\ 0 < p < 1, \\ u_{n+1}|_{\partial\Omega} = 0. \end{cases} \tag{12.104}$$

Then u_n converges to the unique solution u of (12.103) in $C^2(\overline{\Omega})$.

Proof. Let $G(x,y)$ be the Green's function satisfying

$$\begin{cases} \Delta G(x,y) = -\delta(x-y), \ \forall x,y \in \Omega, \\ G(x,y) = 0, \qquad\qquad \forall y \in \partial\Omega, \ \text{for each } x \in \overline{\Omega}, x \neq y. \end{cases}$$

Then it is known that $G(x,y) \geqslant 0$ almost everywhere on $\overline{\Omega} \times \overline{\Omega}$. We have the representation

$$u_{n+1}(x) = \int_\Omega G(x,y) u_n^p(y) dy, \qquad \forall x \in \overline{\Omega}. \tag{12.105}$$

Let $\rho(\cdot)$ be the solution of

$$\Delta\rho + 1 = 0, \quad \text{on } \Omega, \quad \rho|_{\partial\Omega} = 0. \tag{12.106}$$

Then from the fact that $G \geqslant 0$, along with (12.105) and (12.106), we get

$$\begin{aligned} \|u_{n+1}\|_{C^0} &\leqslant \|u_n^p\|_{C^0} \cdot \max_{x\in\overline{\Omega}} \left| \int_\Omega G(x,y) dy \right| \\ &= \|u_n\|_{C^0}^p \|\rho\|_{C^0}, \quad \text{where } C^0 \text{ is the } C(\overline{\Omega})\text{-norm}. \end{aligned} \tag{12.107}$$

From the maximum principle, we have $u_n(x) > 0$ on Ω for $n \geqslant 1$. Choose any open subdomain Ω' such that $\overline{\Omega}' \subset \Omega$. Then

$$\begin{aligned} u_{n+1}(x) &\geqslant \int_{\Omega'} G(x,y) u_n^p(y) dy \geqslant (\min_{\Omega'} u_n)^p \int_{\omega'} G(x,y) dy \\ &\geqslant (\min_{\Omega'} u_n)^p D(\Omega'), \qquad \forall x \in \Omega', \end{aligned} \tag{12.108}$$

where $C(\Omega') \equiv \min_{x\in\overline{\Omega}'} \int_{\Omega'} G(x,y) dy > 0$. From (12.107) and (12.108), therefore, we have

$$C(\Omega') \cdot (\min_{\Omega'} u_n)^p \leqslant \min_{\Omega'} u_{n+1} \leqslant \|u_{n+1}\|_{C^0} \leqslant \|u_n\|_{C^0}^p \|\rho\|_{C^0}. \tag{12.109}$$

Since $0 < p < 1$, $1 - p > 0$, we can choose a positive number $M > 1$ so large that $\|u_0\|_{L^\infty(\Omega)} \leqslant M$ and $\|\rho\|_{C^0} \leqslant M^{1-p}$. Then, from (12.109),

$$\|u_1\|_{C^0} \leqslant \|u_0\|_{L^\infty(\Omega)}^p \|\rho\|_{C^0} \leqslant M^p M^{1-p} = M.$$

Similarly,

$$\|u_i\|_{C^0} \leqslant M, \quad \text{for} \quad i = 2, 3, \ldots . \tag{12.110}$$

From the standard elliptic estimate for (12.104) [79, Chap. 7], we have

$$\|u_{n+1}\|_{W^{2,q}(\Omega)} \leqslant C_1 \|u_n\|_{L^q(\Omega)}^p \leqslant C_2 \|u_n\|_{L^\infty(\Omega)}^p \leqslant C_3 M^p, \text{ by (12.110)}.$$

Choose q sufficiently large. By the Sobolev Imbedding Theorem, we have, for some α: $0 < \alpha < 1$,

$$\|u_{n+1}\|_{C^{1,\alpha}(\overline{\Omega})} \leqslant C_4 M^p \leqslant C_4 M,$$

and thus, by integrating $d(u_{n+1}^p)$, the differential of u_{n+1}^p, from a point on $\partial\Omega$, we get

$$\|u_{n+1}^p\|_{C^\alpha(\overline{\Omega})} \leqslant C_5(M), \quad \text{independent of } n.$$

By the Schauder estimates,

$$\|u_{n+2}\|_{C^{2,\alpha}(\overline{\Omega})} \leqslant C_6(M), \quad \text{independent of } n, \text{ by [79, Chap. 6]}.$$

Therefore, the sequence $\{u_n\}$ is bounded in $C^{2,\alpha}(\overline{\Omega})$ and, therefore it contains a convergent subsequence in $C^2(\overline{\Omega})$ with limit $\widetilde{u}$. This $\widetilde{u}$ satisfies $\widetilde{u} \geqslant 0$ and

$$\Delta\widetilde{u} + \widetilde{u}^p = 0, \quad \text{on } \Omega, \quad \widetilde{u}|_{\partial\Omega} = 0.$$

We now show that $\widetilde{u} \not\equiv 0$.
Choose $m > 0$ so small that $m = C(\Omega')^{\frac{1}{1-p}}$, cf. (12.109). We then have, if $a_n = \min_{\Omega'} u_n \leqslant m$ for some n, then, from (12.108),

$$\begin{aligned} a_{n+1} &\equiv \min_{\Omega'} u_{n+1} \geqslant C(\Omega')(\min_{\Omega'} u_n)^p = C(\Omega')a_n^p = m^{1-p}a_n^p \\ &\geqslant a_n^{1-p}a_n^p = a_n. \end{aligned}$$

This implies that the sequence $\{a_n\}$ must be increasing below the level m. On the other hand, if $a_n \geqslant m$, then

$$\begin{aligned} a_{n=1} &\geqslant C(\Omega')a_n^p = m^{1-p} \cdot a_n^p \\ &\geqslant m^{1-p}m^p = m, \end{aligned}$$

i.e., once the sequence $\{a_n\}$ gets above the level m, it must stay above the level m.
The above guarantees that

$$\min_{\Omega'} u_n \equiv a_n \geqslant \min\{a_0, m\} > 0,$$

for a subdomain Ω' such that $v_0 > a_0 > 0$ almost everywhere on Ω'. Hence $\widetilde{u}$, as the limit of $\{u_n\}$, cannot be identically 0. Therefore, $\widetilde{u}$ must be the unique solution of u of (12.103). Note that the entire sequence $\{u_n\}$ converges to u, because every subsequence of $\{u_n\}$ does. □

We now provide some examples obtained by DIA.

Case 12.6.1.a The unique solution of (12.103) on the unit disk Ω_1, for p=1/3 and 2/3. Using DIA, we have obtained the following data and graphics in Table 12.7.

Case 12.6.1.b The unique solution of (12.103) on the dumbbell-shaped domain Ω_7. The data are given in Table 12.8.

Table 12.7: The data corresponding to the unique solution of (12.103) on the unit disk Ω_1, for p=1/3, 2/3.

p	$\max u$	J	ε_n	n	Graphics
1/3	0.1046	-1.562×10^{-2}	10^{-6}	11	Fig. 12.49
2/3	0.00756	-2.998×10^{-5}	10^{-6}	24	Fig. 12.50

Table 12.8: The data corresponding to the unique solution of (12.103) on the dumbbell-shaped domain Ω_7.

p	$\max u$	J	ε_n	n	Graphics
1/3	1.058×10^{-1}	-1.641×10^{-2}	10^{-6}	12	Fig. 12.51
2/3	7.735×10^{-3}	-3.162×10^{-5}	10^{-6}	24	Fig. 12.52

12.6.2 *A consequence of visualization: monotonicity of solutions of (12.103) with respect to* p

In comparing the data for the solutions of (12.103), we have found an interesting phenomenon: *solutions of (12.103) decrease pointwise with respect to the power* p, i.e., let u_p denote the solution of (12.103) corresponding to p, $0<p<1$. Then $u_{p_1}(x)<u_{p_2}(x)$ if $0<p_2<p_1<1$, for all $x\in\Omega$. The telltale sign of this can be visualized from Figs. 12.49–12.52. Can we establish a rigorous proof of this? The following is a theoretical outcome of visualization.

Theorem 12.8. *Let* $\Omega\subseteq S$, *where* S *is a strip with width at most* $\sqrt{2}$: $S=\{(x_1,x_2)\in\mathbb{R}^2 \mid |x_2|<\sqrt{2}\}$. *Then* $u_p\downarrow$ *pointwise on* Ω *as* $p\uparrow$, *for* $0<p<1$. □

We first show that the following lemma holds.

Lemma 12.1. *We have* $0<u_p(x)<1$ *on* Ω, *if* $\Omega\subseteq S$.

Proof. Set $\phi(x)=1-(1/2)x_2^2$. We have $0\leqslant\varphi\leqslant1$ on S, and

$$\Delta\phi+\phi^p=-1+\phi^p\leqslant0 \quad \text{on} \quad S.$$

Since $\phi\geqslant0$ on $\partial\Omega,\phi$ is a supersolution of (12.103), for all p: $0<p<1$. A subsolution ψ with $\psi\leqslant\phi$ can be constructed (in the "usual" way) as follows. Set $\psi=\varepsilon\psi_1$ for some

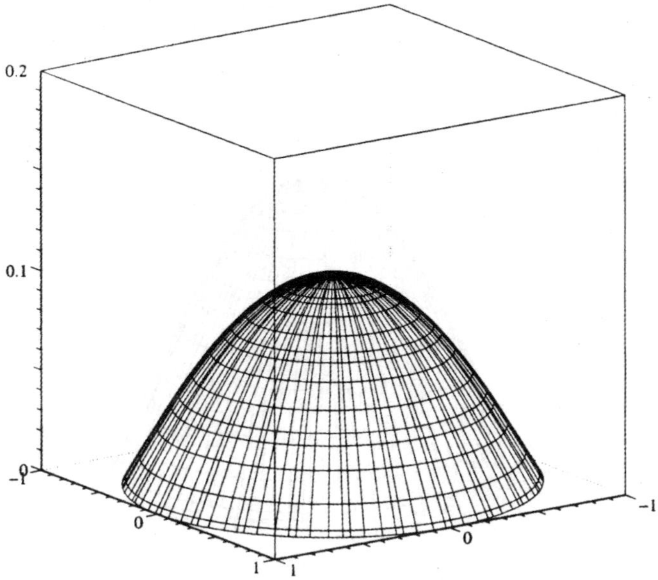

Figure 12.49: The unique solution of (12.103) on the unit disk, $p = 1/3$.

$\varepsilon > 0$, where ψ_1 is the first eigenfunction of Δ, characterized by the following properties:

$$\Delta\psi_1 + \lambda_1\psi_1 = 0, \quad \psi_1 > 0 \quad \text{on } \Omega, \quad \psi_1| - \partial\Omega = 0, \quad \lambda_1 > 0.$$

(We may normalize ψ_1 by $\|\psi_1\|_{L^2(\Omega)} = 1$). We then have

$$\begin{aligned}\Delta\psi + \psi^p &= \varepsilon\Delta\psi_1 + \varepsilon^p\psi_1^p = \varepsilon^p\psi_1^p - \varepsilon\lambda_1\psi_1 \\ &= \varepsilon\psi_1(\varepsilon^{p-1}\psi_1^{p-1} - \lambda_1) > 0\end{aligned}$$

if and only if

$$(\varepsilon\psi_1)^{p-1} > \lambda_1, \quad \text{or} \quad \frac{1}{\lambda_1} > \varepsilon^{1-p}\psi_1^{1-p}. \tag{12.111}$$

Since $1 - p > 0$ and ψ_1 is bounded, (12.111) can always be achieved by taking ε small. Now, choosing ε still smaller if necessary, we have $\psi < \phi$. By the Monotone Iteration Scheme, Theorem 11.1, we conclude that there exists a solution v_p of (12.103) for this p, $0 < p < 1$, such that $\psi \leqslant v_p \leqslant \phi$. Since (12.103) has u_p as the unique solution, we must have $u_p \equiv v_p$. Therefore $0 \leqslant \psi \leqslant u_p \leqslant \phi \leqslant 1$. We then have $u_p < 1$ on Ω since ϕ is a strict supersolution. □

We can now complete the proof.

Proof. [Proof of Theorem 12.8.] First, we note that, if $q > p > 0$, then u_p is a supersolution for equation (12.103) corresponding to q:

$$\Delta u_p + u_p^q = -u_p^p + u_p^q = u_p^p(u_p^{q-p} - 1) < 0,$$

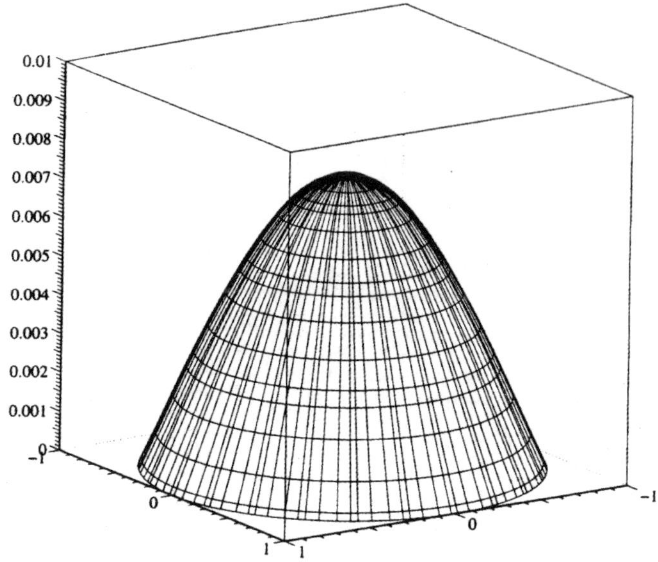

Figure 12.50: The unique solution of (12.103) on the unit disk, $p = 2/3$.

by Lemma 12.1, since $q - p > 0$. Let ψ_1 be the first eigenfunction of Δ as in the proof of Lemma 12.1 and choose $\varepsilon > 0$ so small that $\psi = \varepsilon\psi_1$ is again a subsolution of (12.103) corresponding to q. Then $\psi = \varepsilon\psi_1 < u_p$ on Ω by the Hopf Boundary Point Lemma. Again, by the uniqueness of u_q, we have $0 < \psi = \varepsilon\psi_1 < u_q < u_p < 1$ on Ω, for all $1 > q > p > 0$. The proof is complete. □

All the domains Ω_i, $i = 1, 2, \ldots, 9$, in this paper satisfies the strip-width condition $\Omega_i \subseteq S$. We believe Theorem 12.8 holds for any bounded open domain Ω, but so far we have not been able to provide a proof.

For readers who are interested in computational methods, algorithms and convergence analysis for unstable and/or saddle-type solutions of nonlinear elliptic BVPs and systems, we refer to the more recent developments in [44, 45, 182, 183, 187, 186, 188, 189, 190, 194, 196, 195].

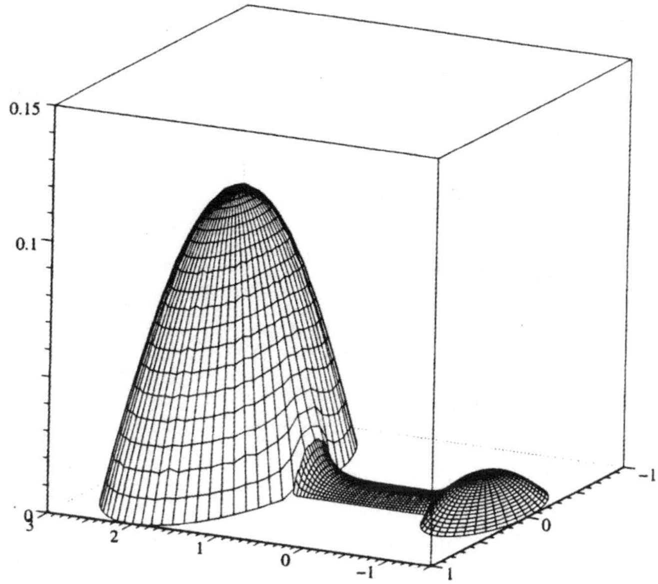

Figure 12.51: The unique solution of (12.103) on the dumbbell-shaped domain Ω_7, $p = 1/3$.

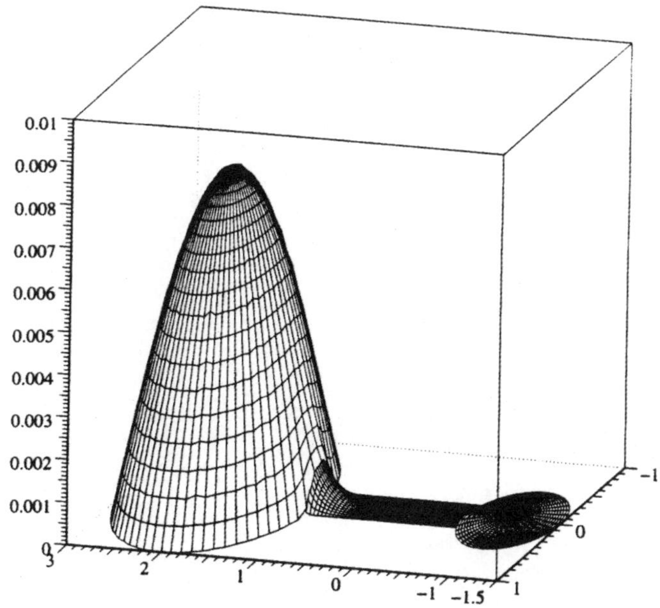

Figure 12.52: The unique solution of (12.103) on the dumbbell-shaped domain Ω_7, $p = 2/3$.

Appendix A

A.1 Integration by parts and the Gauss–Green formulas

Let Ω be a bounded domain in $\mathbb{R}^N$ with piecewise-smooth boundary $\partial\Omega$. Let $n = (n_1, n_2, \ldots, n_N)$ denote the outward unit normal vector field on $\partial\Omega$, which exists everywhere except at finitely many nonsmooth points of $\partial\Omega$. Let f and g be sufficiently regular functions defined on Ω (e.g., $f, g \in H^1(\Omega)$). Then

$$\int_\Omega \frac{\partial f(x)}{\partial x_i} g(x)\,dx = \int_{\partial\Omega} f(x)g(x)n_i(x)\,d\sigma - \int_\Omega f(x)\frac{\partial g(x)}{\partial x_i}\,dx$$

holds. This is the integration-by-parts formula in multi-dimensional space.

Let $\vec{F}(x) = (f_1(x), \ldots, f_N(x))$ be a sufficiently smooth vector-valued function defined on Ω. It follows that

$$\int_\Omega (\nabla \cdot \vec{F}(x))g(x)\,dx = \int_{\partial\Omega} [n(x) \cdot \vec{F}(x)]g(x)\,d\sigma - \int_\Omega \vec{F}(x) \cdot \nabla g(x)\,dx,$$

the first Gauss–Green formula, and

$$\int_\Omega [f(x)\Delta g(x) - g(x)\Delta f(x)]\,dx = \int_{\partial\Omega} \left[f(x)\frac{\partial g(x)}{\partial n} - g(x)\frac{\partial f(x)}{\partial n} \right] d\sigma,$$

the second Gauss–Green formula.

The material in the rest of this Appendix can be found in [161] and [191].

A.2 Banach spaces. Linear operators and linear functionals. Reflexivity

Let X be a real or complex vector space. A *norm* $\|\cdot\|$ on X is a function mapping from X to $\mathbb{R}$ such that

(i) $\|0\| = 0$ and $\|x\| > 0$ if $x \neq 0$;

(ii) $\|\alpha x\| = |\alpha|\,\|x\|, \;\forall x \in X$ and $\alpha \in \mathbb{R}$ (or $\mathbb{C}$);

(iii) $\|x+y\| \leqslant \|x\| + \|y\|, \ \forall x, y \in X$.

X equipped with norm $\|\cdot\|$ is said to be a *normed linear (vector) space*.

A sequence of elements $\{x_1, x_2, \ldots, x_n, \ldots\} \subset X$ is said to be a *Cauchy sequence* in X if for every given $\varepsilon > 0$, there exists $N(\varepsilon) \in \mathbb{N}$ such that

$$\|x_m - x_n\| < \varepsilon, \qquad \forall m, n \geqslant N(\varepsilon).$$

For an arbitrary sequence $\{x_1, x_2, \ldots, x_n, \ldots\}$ in a normed linear space X, we say that this sequence converges to a limit x_0 if for every $\varepsilon > 0$, there exists $N(\varepsilon) \in \mathbb{N}$ such that

$$\|x_m - x_0\| < \varepsilon, \qquad \forall m \geqslant N(\varepsilon).$$

It is easy to see that every convergent sequence is a Cauchy sequence.

A Banach space is a *complete* normed linear space. By completeness here, we mean that every Cauchy sequence in X converges to a *limit in* X.

Let X and Y be two Banach spaces, and let $L\colon X \to Y$ be a linear transformation (or, mapping, operator), i.e., L satisfies

$$L(\alpha x_1 + x_2) = \alpha L(x_1) + L(x_2), \ \forall \alpha \in \mathbb{R} \text{ (or } \mathbb{C}), \ \forall x_1, x_2 \in X.$$

We say that L is (uniformly) *continuous* if for every $\varepsilon > 0$, there exists $\delta > 0$ such that for all $x_1, x_2 \in X$, $\|x_1 - x_2\|_X < \delta$,

$$\|L(x_1) - (x_2)\|_Y < \varepsilon$$

is satisfied. It can be shown that L is continuous iff L is *bounded*, i.e., there exists $B > 0$ such that

$$\|L(x)\|_Y \leqslant B\|x\|_X, \qquad \forall x \in X.$$

Let $L\colon X \to Y$ be a bounded linear transformation. We denote the operator norm of L by

$$\|L\| \equiv \sup_{\substack{x \neq 0 \\ x \in X}} \frac{\|Lx\|_Y}{\|x\|_X}.$$

The set of all bounded linear transformations from X into Y is denoted by $\mathscr{L}(X,Y)$. $\mathscr{L}(X,Y)$ is a Banach space with the operator norm.

When $Y = \mathbb{R}$ or $\mathbb{C}$, depending on whether X is a real or a complex Banach space, elements in $\mathscr{L}(X,\mathbb{R})$ (or $\mathscr{L}(X,\mathbb{C})$) are said to be *linear functionals*. The Banach space $\mathscr{L}(X,\mathbb{R})$ (or $\mathscr{L}(X,\mathbb{C})$) is called the *dual* space of X and is denoted by X'. The dual space of X' is called the *second dual* of X and is denoted by X''. It can be shown that any Banach space X is (algebraically and topologically) isomorphic to a subspace of X''. When

$$X \cong X'' \text{ (canonical isomorphism)},$$

we say that X is a *reflexive* Banach space.

Let Ω be a (bounded or unbounded) open domain in $\mathbb{R}^N$. Let $p \in \mathbb{R}$ satisfy $1 \leqslant p < \infty$. Then $L^p(\Omega)$, the space of all p-th power Lebesgue-integrable functions on Ω, is a Banach space with norm

$$\|f\|_{L^p(\Omega)} \equiv \left(\int_\Omega |f(x)|^p \, dx \right)^{1/p}.$$

We have

$$[L^p(\Omega)]' = L^q(\Omega) \text{ if } 1 < p < \infty, \quad \frac{1}{p} + \frac{1}{q} = 1$$
$$\left[L^1(\Omega)\right]' = L^\infty(\Omega),$$

where $L^\infty(\Omega)$ is the space of all essentially bounded functions on Ω. $L^p(\Omega)$ is reflexive if $1 < p < \infty$.

A.3 The basic principles of linear analysis

(i) *Banach's open mapping theorem*: Let X be a Banach space. A subset U is said to be *open* in X if U is a union of sets of the form (of small open balls)

$$B(x_\alpha, \varepsilon_\alpha) \equiv \{x \in X \mid \|x - x_\alpha\| < \varepsilon_\alpha, \ \varepsilon_\alpha > 0\}.$$

A subset C of X is said to be *closed* if the complement of C is open in X. Then Banach's open mapping theorem can be stated as follows:

> Let X and Y be Banach spaces and let $L \in \mathscr{L}(X,Y)$ map X *onto* Y. Then $L(U)$ is open in Y for each open subset U of X.

(ii) *Closed graph theorem*: Let X and Y be Banach spaces. Then the product space $X \times Y$ is also a Banach space normed with

$$\|(x,y)\|_{X \times Y} \equiv \|x\|_X + \|y\|_Y, \quad \forall (x,y) \in X \times Y.$$

Let L be a linear transformation from X into Y. The graph of L is denoted as

$$\text{Graph}(L) = \{(x, L(x)) \in X \times Y\}.$$

The closed graph theorem can be stated as follows:

> Let X and Y be Banach spaces and let L be a linear transformation from X into Y such that L has a closed graph. Then L is continuous.

(iii) *Uniform boundedness principle*: This can be stated as follows:

> Let X and Y be Banach spaces and let $\{L_\alpha \mid \alpha \in I\}$ be a nonempty subset of $\mathscr{L}(X,Y)$ such that
> $$\sup\{\|L_\alpha(x)\| \mid \alpha \in I\} < \infty$$
> for each $x \in X$. Then
> $$\sup\{\|L_\alpha\| \mid \alpha \in I\} < \infty.$$

(iv) *Hahn–Banach theorem*: Let X be a real linear vector space. A real valued function $p\colon X \to \mathbb{R}$ is said to be a *sublinear functional* if

$$p(x_1 + x_2) \leqslant p(x_1) + p(x_2)$$

and

$$p(rx) \leqslant rp(x)$$

hold for every $x, x_1, x_2 \in X$ and every $r > 0$. A mapping $f\colon X \to \mathbb{R}$ is linear if

$$f(\alpha x_1 + x_2) = \alpha f(x_1) + f(x_2), \quad \forall \alpha \in \mathbb{R}, \quad \forall x_1, x_2 \in X.$$

Then the Hahn–Banach theorem can be stated as follows:

> Let X be a real linear vector space and let M be a linear subspace of X. Assume that p is a sublinear functional defined on X and that $f\colon\ M \to \mathbb{R}$ is linear such that $f(x) \leqslant p(x)$ for all $x \in M$. Then there exists an extension F of f defined on X such that
>
> $$F(x) = f(x), \quad \forall x \in M$$
>
> and
>
> $$F(x) \leqslant p(x), \quad \forall x \in X.$$

A.4 Hilbert spaces. The Riesz representation theorem

Let H be a linear vector space. A mapping $\langle \cdot\,, \cdot \rangle$ from $H \times H$ into $\mathbb{C}$ is said to be an *inner product* if for all $x, y, z \in H$ and $\alpha \in \mathbb{C}$,

$$\begin{aligned}
\langle x + y, z \rangle &= \langle x, z \rangle + \langle y, z \rangle, \\
\langle \alpha x, y \rangle &= \alpha \langle x, y \rangle, \\
\overline{\langle x, y \rangle} &= \langle y, x \rangle, \\
\langle x, x \rangle &> 0 \ \text{ if } \ x \neq 0.
\end{aligned}$$

If a linear vector space H is equipped with an inner product $\langle \cdot\,, \cdot \rangle$, we say that H is an inner product space or a pre-Hilbert space. For $x \in H$, define

$$\|x\| = (\langle x, x \rangle)^{1/2}.$$

If an inner product space H is *complete* with respect to this norm, we say that H is a *Hilbert space*.

Every Hilbert space H, as a Banach space, has a dual H'. Elements in H' are bounded linear functionals on H. Then the Riesz representation theorem can be stated as follows:

Let H be a Hilbert space and let $f \in H'$ be a bounded linear functional. Then there exists a unique $y_f \in H$ such that

$$f(x) = \langle x, y_f \rangle, \quad \forall x \in H,$$

and

$$\|f\|_{H'} = \|y_f\|_H.$$

Conversely, any vector $y \in H$ defines a bounded linear functional $f_y \in H'$ by

$$f_y(x) = \langle x, y \rangle, \quad \forall x \in H,$$

satisfying

$$\|f_y\|_{H'} = \|y\|_H.$$

A.5 Compactness. Completely continuous operators

A topological space is a space endowed with a class of open sets, called the topological structure. A Banach space is a topological space with open sets defined as in Appendix A.3. A collection of open sets $\{\mathscr{O}_\alpha \mid \alpha \in I\}$ in a Banach space X is said to be an *open covering* of a set $S \subset X$ if $S \subset \bigcup_{\alpha \in I} \mathscr{O}_\alpha$. A set S is said to be *compact* if every open covering of S has a finite subsystem also covering S. A compact set S is necessarily closed.

A set $S \subset X$ is said to be *precompact* if its closure (i.e., the smallest set containing S) is compact.

A linear transformation L from a Banach space X into a Banach space Y is said to be *compact* or *completely continuous* if L maps every bounded set in X into a relatively compact set in Y. A compact operator is necessarily bounded.

For a Banach space X, we also define its *weak topology* to be the topological structure formed by basic open sets of the form

$$\{x \in X \mid |f(x)| < \varepsilon\}$$

for some $\varepsilon > 0$ and $f \in X'$, for all $f \in X'$ and $\varepsilon > 0$. Then the weakly compact subsets and weakly precompact subsets can be defined in the same fashion.

In a finite-dimensional space $\mathbb{R}^N$ or $\mathbb{C}^N$, a set is compact if and only if it is closed and bounded. (This is the well-known Bolzano–Weierstrass theorem.) This is not true in infinite-dimensional spaces.

A Hilbert space is said to be *separable* if it contains a countable dense subset. In a *separable Hilbert space, every bounded subset is weakly precompact.*

A.6 Quotient spaces

Let X be a normed vector space, with norm $\|\cdot\|$, and let M be a *closed* linear subspace of X. Define the *quotient* space $X/M = \{x+M \mid x \in X\}$, where the linear operations are defined by

$$(x+M)+(y+M) = (x+y)+M$$

and

$$\alpha(x+M) = (\alpha x)+M,$$

for any scalar α and any vectors x and y. The *quotient norm* on X/M is defined by

$$\|x+M\| = \inf\{\|x+m\| \mid m \in M\}.$$

Then X/M becomes a normed vector space equipped with the quotient norm. Further, if X is a Banach space then so is X/M, and the natural mapping (i.e., projection) ϕ from X onto X/M,

$$\phi(x) = x+M,$$

is a bounded linear transformation satisfying $\|\phi\| \leqslant 1$. By the open mapping theorem, ϕ maps open sets onto open sets.

A.7 Direct sums. Projection operators

Let H be a Hilbert space and let M be a closed linear subspace of H. Let $M^\perp$ be the orthogonal complement of M in H, i.e.,

$$M^\perp \equiv \{y \in H \mid \langle x,y\rangle = 0, \ \forall x \in M\}.$$

Then every vector $z \in H$ can be uniquely represented as a sum

$$z = x+y, \ x \in M, \ y \in M^\perp \text{ and } \langle x,y\rangle = 0.$$

Accordingly, we write

$$H = M \oplus M^\perp$$

and say that H is the *orthogonal direct sum* of M with $M^{\perp}$. For any $z \in H$, let

$$z = x + y,\ \ x \in M,\ y \in M^{\perp} \text{ and } \langle x, y \rangle = 0.$$

Define a linear operator

$$\boldsymbol{P}\colon H \to M$$

by

$$\boldsymbol{P}z = x.$$

Then $\boldsymbol{P}$ satisfies

(i) $\|\boldsymbol{P}\| = 1$;
(ii) $\boldsymbol{P}^2 = \boldsymbol{P}, \boldsymbol{P}(I - \boldsymbol{P}) = 0$;
(iii) $\boldsymbol{P}x = x$ iff $x \in M$.

$\boldsymbol{P}$ is said to be the (orthogonal) projection operator on M.

A.8 The Cauchy–Schwarz inequality and the Hölder–Young inequality

Let $a, b \in \mathbb{C}$. Then

$$|ab| \leqslant |a|\,|b| \leqslant \frac{1}{2}\left(|a|^2 + |b|^2\right).$$

This is the Cauchy–Schwarz inequality for scalars. In practice, we often use a weighted form

$$|ab| \leqslant \frac{\varepsilon}{2}\,|a|^2 + \frac{1}{2\varepsilon}\,|b|^2,$$

where $\varepsilon > 0$ is arbitrary (so that a small weight is placed on $|a|^2$ while a large weight is placed on $|b|^2$, if ε is small).

The scalar form of the Hölder–Young inequality is

$$|ab| \leqslant |a|\,|b| \leqslant \frac{|a|^p}{p} + \frac{|b|^q}{q},$$

for $1 < p < \infty$, $1 < q < \infty$, and $1/p + 1/q = 1$.

Let $f \in L^p(\Omega)$ and $g \in L^q(\Omega)$, where $1 < p < \infty$ and $1 < q < \infty$. Then $fg \in L^1(\Omega)$,

$$\left|\int_{\Omega} fg\,dx\right| \leqslant \int_{\Omega} |fg|\,dx \leqslant \|f\|_{L^p(\Omega)}\|g\|_{L^q(\Omega)}$$

and

$$\left|\int_{\Omega} fg\,dx\right| \leqslant \int_{\Omega} |fg|\,dx \leqslant \frac{1}{p}\,\|f\|^p_{L^p(\Omega)} + \frac{1}{q}\,\|g\|^q_{L^q(\Omega)}.$$

These are the Hölder–Young inequalities. When $p = q = 2$, they are called the *Cauchy–Schwarz inequalities*.

Bibliography

[1] M. Abramowitz and I.A. Stegun, *Handbook of Mathematical Functions*, Dover, New York, 1965.

[2] R. Adams, *Sobolev Spaces*, Academic Press, New York, 1975.

[3] S. Adjerid, M. Aiffa and J.E. Flaherty, High-order finite element methods for singular perturbed elliptic and parabolic problems, *SIAM J. Appl. Math.* **55** (1995), 520–543.

[4] H. Amann, Supersolution, monotone iteration and stability, *J. Diff. Eq.* **21** (1976), 367–377.

[5] H. Amann and M.G. Crandall, On some existence theorems for semilinear elliptic equations, *Indiana Univ. Math. J.* **27** (1978), 779–790.

[6] A. Ambrosetti and P. Rabinowitz, Dual variational methods in critical point theory and applications, *J. Funct. Anal.* **14** (1973), 349–381.

[7] R. Aris, *The Mathematical Theory of Diffusion and Reaction in Permeable Catalysts*, Vol. I, Oxford Univ. Press, London, 1975.

[8] D.N. Arnold, A spline-trigonometric Galerkin method and an exponentially convergent boundary integral method, *Math. Comp.* **41** (1983), 383–397.

[9] D.N. Arnold and W.L. Wendland, On the asymptotic convergence of collocation methods, *Math. Comp.* **41** (1983), 349–381.

[10] D.N. Arnold and W.L. Wendland, The convergence of spline collocation for strongly elliptic equations on curves, *Numer. Math.* **47** (1985), 317–341.

[11] F.V. Atkinson, On Sommerfeld's "Radiation Condition", *Phil. Mag.* **40** (1949), 645–651.

[12] J.P. Aubin, *Approximation of Elliptic Boundary Value Problems*, Wiley Interscience, New York, 1972.

[13] J.P. Aubin and I. Ekeland, *Applied Nonlinear Analysis*, Wiley, New York, 1984.

[14] I. Babuska and A.K. Aziz, *The Mathematical Foundation of the Finite Element Method with Applications to Partial Differential Equations* A.K. Aziz (ed.), Academic Press, New York, 1972.

[15] T. Bartsch and Z.Q. Wang, On the existence of sign-changing solutions for semilinear Dirichlet problems, *Topol. Methods Nonlinear Anal.* **7** (1996), 115–131.

[16] V. Benci and G. Cerami, The effect of the domain topology on the number of positive solutions of nonlinear elliptic problems, *Arch. Rational Mech. Anal.* **114** (1991), 79–94.

[17] L. Bers, On mildly nonlinear partial differential equations of elliptic type, *J. Res. Nat. Bureau Standards* **51** (1953), 229–236.

[18] L. Bieberbach, $\Delta u = e^u$ und die automorphen Functionen, *Math. Ann.* **77** (1916), 173–212.

[19] H. Brakhage and P. Werner, Über das Dirichletsche Aussenraumproblem für die Helmholtzsche Schwingungsgleichung, *Arch. Math.* **16** (1965), 325–329.

[20] J.H. Bramble, *Multigrid Methods*, Pitman Research Notes in Mathematics, Vol. 294, Wiley, New York, 1993.

[21] J.H. Bramble and R. Scott, Simultaneous approximation in scales of Banach spaces, *Math. Comp.* **32** (1978), 947–954.

[22] C.A. Brebbia, *The Boundary Element Method for Engineers*, Pentech Press, London, 1978.

[23] C.A. Brebbia, *Boundary Element Techniques in Engineering*, Wiley, New York, 1978.

[24] C.A. Brebbia, *Boundary Element Methods*, Springer-Verlag, New York, 1981.

[25] C.A. Brebbia, *Boundary Element Techniques, Methods in Engineering*, Springer-Verlag, New York, 1983.

[26] C.A. Brebbia and S. Walker, *Boundary Element Techniques in Engineering*, Newnes-Butterworths, London, 1980.

[27] H. Bremermann, *Distributions, Complex Variables and Fourier Transforms*, Addison-Wesley, Reading, Massachusetts, 1965.

[28] S.C. Brenner and R.L. Scott, *The Mathematical Theory of Finite Element Methods*, Springer-Verlag, New York, 1994.

[29] A.P. Calderòn, Uniqueness in the Cauchy problem of partial differential equations, *Amer. J. Math.* **80** (1958), 16–36.

[30] A. Castro, J. Cossio and J.M. Neuberger, A sign-changing solution for a superlinear Dirichlet problem, *Rocky Mountain J. Math.* **351** (1997), 1919–1945.

[31] S. Chandrasekhar, *An Introduction to the Study of Stellar Structure*, University of Chicago Press, 1939.

[32] K.C. Chang, *Infinite Dimensional Morse Theory and Multiple Solution Problems*, Birkhäuser, Boston, 1993.

[33] G. Chen, T.J. Bridges and J. Zhou, Minimizing the reflection of waves by surface impedance using boundary elements and global optimization, *Wave Motion*, **10** (1988), 239–255.

[34] G. Chen, Y. Deng, W.-M. Ni, and J. Zhou, Boundary element monotone iteration scheme for semilinear elliptic partial differential equations, Part II: quasimonotone iteration for coupled 2×2 systems, *Math. Comp.* **69** (1999), 629–652.

[35] G. Chen, Y. Deng, W.-M. Ni, and J. Zhou, Boundary element monotone iteration scheme for semilinear elliptic partial differential equations, regular oblique derivative nonlinear boundary conditions, in *Variational Methods for Nonlinear Elliptic Partial Differential Equations: Open Problems, Recent Progress and Numerical Algorithms*, J.M. Neuberger (ed.), Amer. Math. Soc. Cont. Math. **357** (2004), Providence, R.I., U.S.A., 17–48.

[36] G. Chen, S.A. Fulling, F.J. Narcowich, and S. Sun, Exponential decay of energy of evolution equations with locally distributed damping, *SIAM J. Appl. Math.* **51** (1991), 266–301.

[37] G. Chen, J. Zhou and W.-M. Ni, Algorithms and visualization for solutions of nonlinear elliptic equations, *Internat. J. Bifur. Chaos* **7** (2000), 1565–1612.

[38] G. Chen, P.J. Morris and J. Zhou, Visualization of special eigenmode shapes of a vibrating elliptical membrane, *SIAM Rev.* **36** (1994), 453–469.

[39] G. Chen, W.-M. Ni, A. Perronnet, and J. Zhou, Algorithms and visualization for solutions of nonlinear elliptic equations, Part II, Dirichlet, Neumann and Robin boundary conditions and problems in 3D, *Internat. J. Bifur. Chaos* **11** (2001), 1781–1977.

[40] G. Chen and S. Sun, Augmenting a Fredholm operator of zero index to achieve invertibility for elliptic boundary value problems, *J. Math. Anal. Appl.* **176** (1993), 24–48.

[41] G. Chen and J. Zhou, *Vibration and Damping in Distributed Systems, Vol. II: WKB and Wave Methods, Visualization and Experimentation*, CRC Press, Boca Raton, Florida, 1993.

[42] G. Chen and J. Zhou, Boundary integral equations and boundary element computations for the biharmonic plate equation with applications to boundary shape control, unpublished manuscript.

[43] G. Chen and J. Zhou, Some boundary control problems and computation for the linear elastostatic Kirchhoff plate on an exterior domain, in *Boundary Control and Variations*, Proc.

of IFIP Workshop, *Springer Lecture Notes on Control and Information Science #178, 82–117*, J.P. Zolesio (ed.), Springer, New York, 1992.

[44] X. Chen and J. Zhou, On homotopy continuation method for computing multiple solutions to the Henon equation, *Numer. Methods Partial Differential Equations* **24** (2008), 728–748.

[45] X. Chen, J. Zhou and X. Yao, A numerical method for finding multiple co-existing solutions to nonlinear cooperative systems, *Appl. Numer. Math.* **58** (2008), 1614–1627.

[46] Y. Chen and P.J. McKenna, Numerical variational methods for approximating travelling waves in a nonlinearly suspended beam, *J. Diff. Eq.* **136** (1997), 325–355.

[47] Y.S. Choi and P.J. McKenna, A mountain pass method for the numerical solution of semilinear elliptic problems, *Nonlinear Analysis, Theory, Methods Appl.* **20** (1993), 417–437.

[48] Y.S. Choi and P.J. McKenna, Numerical mountain pass methods for nonlinear boundary value problems, in *Nonlinear Problems in Applied Mathematics*, Angell, Cook, Kleinman, and Olmstead (eds.), SIAM, Philadelphia, 1996, pp. 86–95.

[49] P.B. Ciarlet, *The Finite Element Method for Elliptic Problems*, North-Holland, Amsterdam, 1978.

[50] P.G. Ciarlet, M.H. Schultz and R.S. Varga, Numerical methods of high-accuracy for nonlinear boundary value problems, *Numer. Math.* **9** (1967), 394–430.

[51] C.V. Coffman, A nonlinear boundary value problem with many positive solutions, *J. Diff. Eq.* **54** (1984), 429–437.

[52] L. Collatz, *The Numerical Treatment of Differential Equations*, Springer-Verlag, New York, 1966.

[53] D. Colton and R. Kress, *Integral Equation Methods in Scattering Theory*, Wiley Interscience, New York, 1983.

[54] M. Costable, I. Lusikka and J. Saranen, Comparison of three boundary element approaches for the solution of the clamped plate problem, *Boundary Elements IX*, C.A. Brebbia (ed.), Springer-Verlag, New York, 1989, pp. 19–34.

[55] M. Costable and W. Wendland, Strong ellipticity of boundary integral operators, *J. Reine Angew. Math.* **372** (1986), 39–63.

[56] R. Courant, Variational methods for the solutions of problems of equilibrium and vibrations, *Bull. Amer. Math. Soc.* **39** (1943), 1–23.

[57] G.R. Cowper, E. Kosko, G.M. Lindberg, and M.D. Olson, Static and dynamic applications of a high-precision triangular plate bending element, *AIAA J.* **7** (1969), 1957–1965.

[58] T.A. Cruse, Numerical solutions in three dimensional elastostatics, *Int. J. Solids & Structures* **5** (1969), 1259–1274.

[59] T.A. Cruse and F.J. Rizzo (eds.), *Boundary Integral Equation Method: Computational Applications in Applied Mechanics*, ASME, New York, 1975.

[60] E.N. Dancer, The effect of domain shape on the number of positive solutions of certain nonlinear equations, *J. Diff. Eq.* **74** (1988), 120–156.

[61] Y. Deng, Boundary element methods for nonlinear partial differential equations, Ph.D. dissertation, Math. Dept., Texas A&M Univ., College Station, Texas, August 1994.

[62] Y. Deng, G. Chen, W.M. Ni, and J. Zhou, Boundary element monotone iteration scheme for semilinear elliptic partial differential equations, *Math. Comp.* **65** (1996), 943–982.

[63] J. Deny and J.L. Lions, Les espaces du type de Beppo Levi, *Ann. Institut Fourier (Grenoble)* **5** (1953-54), 305–370.

[64] J. Dieudonne, *Eléments d'Analyse*, Vol. 8, Gauthier-Villars, Paris, 1978.

[65] Z. Ding, D. Costa and G. Chen, A high linking method for sign changing solutions of semilinear elliptic equations, *Nonlinear Anal.* **38** (1999), 151–172.

[66] W.Y. Ding and W.M. Ni, On the existence of positive entire solutions of a semilinear elliptic equation, *Arch. Rational Mech. Anal.* **91** (1986), 283–308.

[67] J. Douglas and T. Dupont, A Galerkin method for a nonlinear Dirichlet problem, *Math. Comp.* **29** (1975), 689–696.

[68] A.P. Dowling and J.E. Ffowcs- Williams, *Sound and Sources of Sound*, Ellis Horwood, Chichester, England, 1983.

[69] J. Dugundji, *Topology*, Allyn and Bacon, Boston, 1966.

[70] G. Duvaut and J.L. Lions, *Inequalities in Mechanics and Physics*, Springer-Verlag, New York, 1976.

[71] A. Eydeland and J. Spruck, The inverse power method for semilinear elliptic equations, in *Nonlinear Diffusion Equations and Their Equilibrium States I*, MSRI Publications Vol. 12, W.M. Ni, et al. (eds.), Springer-Verlag, New York, 1988.

[72] G. Fichera, Linear elliptic equations of higher order in two independent variables and singular integral equations with applications to anisotropic inhomogeneous elasticity, *Proceedings Symposium on Partial Differential Equations and Continuous Mechanics*, R.E. Langer (ed.), University of Wisconsin Press, Madison, Wisconsin, 1961, pp. 55–80.

[73] R.H. Fowler, Further studies on Emden's and similar differential equations, *Quart. J. Math.* **2** (1931), 259–288.

[74] E. Galperin and Q. Zheng, *Global Solutions in Optimal Control and Games*, NP Research Publication, Presses de l'Université du Quebec à Montréal, 1991.

[75] A. Geirer and H. Meinhardt, A theory of biological pattern formation, *Kybernelic* **12** (1972), 30–39.

[76] I.M. Gelfand and G.E. Shilov, *Generalized Functions*, Vol. 1, Academic Press, New York, 1963.

[77] B. Gidas, W.M. Ni and L. Nirenberg, Symmetry and related properties via the maximum principle, *Comm. Math. Phys.* **68** (1979), 209–243.

[78] B. Gidas, W.M. Ni and L. Nirenberg, Symmetry of positive solutions of nonlinear elliptic equations in $\mathbb{R}^n$, *Adv. Math. Supplementary Studies* **7A** (1981), 369–402.

[79] D. Gilbarg and N.S. Trudinger, *Elliptic Partial Differential Equations of Second Order* 2nd ed., Springer-Verlag, Berlin, 1983.

[80] P. Gilkey, *The Index Theorem and the Heat Equation*, Publish or Perish, Boston, 1974.

[81] G. Giraud, Equations à intégrales principales, étude suivie d'une application, *Ann. Ecole Norm. (3)*, **51**, fasc. 3, (1934) and fasc. 4, (1936).

[82] D. Gottlieb and S. Orszag, *Numerical Analysis of Spectral Methods, Theory and Applications*, SIAM Philadelphia, 1977.

[83] D. Greenspan and S.V. Parter, Mildly nonlinear elliptic partial differential equations and their numerical solution, II, *Numer. Math.* **7** (1965), 129–146.

[84] P. Grisvard, *Elliptic Problems in Nonsmooth Domains*, Pitman Advanced Publishing Program, Boston, 1985.

[85] G. Grubb, *Functional Calculus of Pseudo-Differential Boundary Problems*, Birkhäuser, Boston, 1986.

[86] G. Grubb, On coerciveness and semiboundedness of general boundary problems, *Israel J. Math.* **10** (1971), 32–95.

[87] N.M. Günter, *Die Potential theorie und Ihre Anwendungen auf Grundaufgaben der Mathematischen Physik*, Teubner-Verlag, Leipzig, 1957.

[88] F. Hartmann and R. Zotemantel, The direct boundary element method in plate bending, *Internat. J. Numer. Methods Engrg.* **23** (1986), 2049–2069.

[89] H.P. Helfrich, Simultaneous approximation in negative norms of arbitrary order, *RAIRO Numer. Anal.* **15** (1981), 231–235.

[90] M. Henon, Numerical experiments on the stability of spherical stellar systems, *Astronom. Astrophys.* **24** (1973), 229–238.

[91] S. Hill, Numerical and theoretical investigation of the variational formulation of a water wave

problem, Ph.D. Dissertation, Math. Dept., Univ. of Connecticut, Storrs, Connecticut, 1997.

[92] L. Hörmander, *Linear Partial Differential Operators*, Springer-Verlag, New York, 1963.

[93] G. Hsiao, Singular perturbations of boundary value problems for a class of nonlinear differential equations with a small parameter, Ph.D. Thesis, Department of Mathematics, Carnegie-Mellon University, Pittsburgh, 1969.

[94] G. Hsiao, On the stability of integral equations of the first kind with logarithmic kernels, *Arch. Rat. Mech. Anal.* **94** (1986), 179–192.

[95] G. Hsiao and R.C. MacCamy, Solutions of boundary value problems by integral equations of the first kind, *SIAM Rev.* **15** (1973), 687–705.

[96] G. Hsiao and W.L. Wendland, A finite element method for some integral equations of the first kind, *J. Math. Anal. Appl.* **58** (1977), 449–481.

[97] G. Hsiao and W.L. Wendland, The Aubin–Nitsche lemma for integral equations, *J. Integral Equations* **3** (1981), 299–315.

[98] G. Hsiao and W.L. Wendland, On a boundary integral equation method for some exterior problems in elasticity, *Proc. of Tbiliski University*, Tbiliski University Press, 1985, pp. 30–60.

[99] G. Hsiao and W.L. Wendland, *Boundary Integral Equations*, Springer-Verlag, Berlin, 2008.

[100] L.D. Humphreys and P.J. McKenna, Multiple periodic solutions for a nonlinear suspension bridge equations, *IMA J. Appl. Math.* **63** (1999), 37–49.

[101] C.U. Huy, P.J. McKenna and W. Walter, Finite difference approximations to the Dirichlet problem for elliptic systems, *Numer. Math.* **49** (1986), 227–237.

[102] K. Ishihara, Monotone explicit iterations of the finite element approximations for the nonlinear boundary value problems, *Numer. Math.* **45** (1984), 419–437.

[103] M.A. Jaswon and G.T. Symm, *Integral Equation Methods in Potential Theory and Elastostatics*, Academic Press, London, 1977.

[104] L. Ji, Boundary element methods and supercomputer computations for some distributed parameter control problems, Doctoral Dissertation, Department of Mathematics, Texas A&M University, College Station, Texas, 1991.

[105] D.S. Jones, *Methods in Electromagnetic Wave Propagation*, Clarendon Press, Oxford, 1979.

[106] D.S. Jones, Integral equations for the exterior acoustic problem, *Q.J. Mech. Appl. Math.* **27** (1974), 129–142.

[107] R. Kanwal, *Generalized Functions: Theory and Technique*, Academic Press, New York, 1983.

[108] T. Kato, *Perturbation Theory for Linear Operators*, SpringerVerlag, New York, 1966.

[109] J.B. Keller, R.M. Lewis and B.D. Seckler, Asymptotic solution of some diffraction problems, *Commun. Pure Appl. Math.* **9** (1956), 207–265.

[110] H.B. Keller, *Numerical Solutions for Two-Point Boundary-Value Problems*, Blaisdell, Waltham, Massachusetts, 1968.

[111] E.F. Keller and S.A. Segel, Initiation of shine mode aggregation viewed as an instability, *J. Theo. Biol.* **26** (1970), 399–415.

[112] G. Kirchhoff, Über das Gleichgewicht und die Bewegung einer elastischen Scheibe, *J. Reine Angew. Math.* **401** (1850), 51–58.

[113] R. Kleinman and G.F. Roach, On modified Green's functions in exterior problems for the Helmholtz equation, *Proc. R. Soc. Lond.* **A383** (1982), 313–332.

[114] J.J. Kohn and L. Nirenberg, An algebra of pseudodifferential operators, *Commun. Pure Appl. Math.* **18** (1965), 269–305.

[115] R. Kress, *Linear Integral Equations*, Springer–Verlag, New York, 1989.

[116] H. Kumano-go, *Pseudodifferential Operators*, MIT Press, Cambridge, Massachusetts, 1981.

[117] V. D. Kupradze, *Potential Methods in the Theory of Elasticity*, Israel Program of Scientific Translations, Jerusalem, 1965.

[118] R. Leis, Zur Dirichletschen Rantwertanfgabe des Aussenraums der Schwingungsgleichung, *Math. Z.*, **90** (1965), 205–211.

[119] R. Leis, *Vorlesungen über Partielle Differentialgleichungen Zweiter Ordnung*, Bibliographisches Institut, Mannheim, Germany, 1967.

[120] Y.Y. Li, Existence of many positive solutions of semilinear elliptic equations on annulus, *J. Diff. Eq.* **83** (1990), 348–367.

[121] Y. Li and J. Zhou, A minimax method for finding multiple critical points and its applications to semilinear PDE, *SIAM J. Sci. Comp.* **23** (2001), 840–856.

[122] Y. Li and J. Zhou, Convergence results of a local minimax method for finding multiple critical points, *SIAM J. Sci. Comp.* **24** (2002), 865–885.

[123] E. Lieb and H.-T. Yau, The Chandrasekhar theory of stellar collapse as the limit of quantum mechanics, *Comm. Math. Phys.* **112** (1987), 147–174.

[124] C.S. Lin, Uniqueness of solutions minimizing the functional $\int_\Omega |\nabla u|^2 / \left(\int_\Omega |u|^{p+1}\right)^{2/(p+1)}$ in $\mathbb{R}^n$, *Manuscripta Math.* **84** (1994), 13–19.

[125] C.S. Lin, W.M. Ni and I. Takagi, Large amplitude solutions to a chemotaxis system, *J. Diff. Eq.* **72** (1988), 1–27.

[126] J.L. Lions, *Quelques Méthodes de Résolution des Problèmes aux Limites Non Lineaires*, Dunod, Paris, 1969.

[127] J.L. Lions and E. Magenes, *Nonhomogeneous Boundary Value Problems and Applications*, Springer-Verlag, New York, 1970.

[128] E. Martensen, *Potentialtheorie*, Teubner, Stuttgart, 1968.

[129] J.T. Marti, *Introduction to Sobolev Spaces and Finite Elements Solution of Elliptic Boundary Value Problems*, Academic Press, London, 1986.

[130] H. Matano and M. Mimura, Pattern formation in competition–diffusion systems in nonconvex domain, *Publ. RIMS*, Kyoto Univ., Kyoto, Japan, **19** (1983), 1049–1079.

[131] J. Mawhin and M. Willem, *Critical Point Theory and Hamiltonian System*, Springer-Verlag, New York, 1989.

[132] N. Meyers and J. Serrin, The exterior Dirichlet problem for second order elliptic partial differential equations, *J. Math. Mech.*, **9** (1960), 513–538.

[133] R.D. Mindlin, Notes on the Galerkin and Papkovich stress functions, *Bull. Amer. Math. Soc.* **42** (1936), 373–379.

[134] A.R. Mitchell and R. Wait, *The Finite Element Method in Partial Differential Equations*, Wiley, New York, 1977.

[135] P.J. Morris, T.R.S. Bhat and G. Chen, A linear shock cell model for jets of arbitrary exit geometry, *J. Sound Vib.* **132** (1989), 199–211.

[136] C. Müller, *Grund probleme der Mathematischen theorie Electromagnetischer Schwingungen*, Springer-Verlag, Berlin, 1957.

[137] N.I. Muskhelishvili, *Singular Integral Equations*, Noordhoff, Groningen, 1953.

[138] J. Necas, *Les Methodes Directes dans la Theorie des Equations Elliptiques*, Academie, Tchécoslovaque des Sciences, Prague, 1967.

[139] Z. Nehari, On a class of nonlinear second-order differential equations, *Trans. Amer. Math. Soc.* **95** (1960), 101–123.

[140] H. Neuber, Ein neuer Ansatz zür lösung räumlicher Probleme der Elastizitätstheorie, *Z. argew. Math. Mech.* **14** (1934), 203–212.

[141] W.M. Ni, Some aspects of semilinear elliptic equations, Lecture Notes published by Institute of Mathematics, National Tsing Hua Univ., Hsinchu, Taiwan, Rep. of China, May 1987.

[142] W.M. Ni, Recent progress in semilinear elliptic equations, in *RIMS Kokyuroku* 679, Kyoto

University, Kyoto, Japan, 1989, 1–39.

[143] W.M. Ni, Diffusion, cross diffusion and their spike-layer steady states, *Notices Amer. Math. Soc.* **45** (1998), 9–18.

[144] W.M. Ni and R. Nussbaum, Uniqueness and nonuniqueness for positive radial solutions of $\Delta u + f(u,r) = 0$, *Comm. Pure Appl. Math.* **38** (1985), 67–108.

[145] W.M. Ni and I. Takagi, On the shape of least energy solutions to a semilinear Neumann problem, *Comm. Pure Appl. Math.* **44** (1991), 819–851.

[146] W.M. Ni and I. Takagi, Locating the peaks of least energy solutions to a semilinear Neumann problem, *Duke Math. J.* **70** (1983), 247–281.

[147] W.M. Ni and J. Wei, On the location and profile of spike-layer solutions to singularly perturbed semilinear Dirichlet problems, *Comm. Pure Appl. Math.* **48** (1995), 731–768.

[148] J.A. Nitsche, Zur Convergenz des Ritzschen Verfahrens und der Fehler-quadratmethode I, *Numerische Mathematik, Differentialgleichungen, Approximationstheorie*, Birkhäuser, Basel, 1968, 97–103.

[149] J.A. Nitsche, Vergleich der Konvergenzgeschwindigkeit des Ritzschen Verfahrens und der Fehlerquadratmethode, *Z. angew. Math. Mech.* **49** (1969), 591–596.

[150] O.I. Panich, On the question of the solvability of the exterior boundary value problem for the wave equation and Maxwell's equations, *Russ. Math. Surv.* **20** (1965), 221–226.

[151] C.V. Pao, Numerical solutions for some coupled systems of nonlinear boundary value problems, *Numer. Math.* **51** (1987), 381–394.

[152] C.V. Pao, *Nonlinear Parabolic and Elliptic Equations*, Plenum Press, New York, 1992.

[153] C.V. Pao, Block monotone iterative methods for numerical solutions of nonlinear elliptic equations, *Numer. Math.* **72** (1995), 239–262.

[154] P.F. Papkovich, Solution générale des equations differentielles fondamentales d'élasticité expremée par trois fonctions harmoniques, *C. R. Acad. Sci. Paris*, **195** (1932), 513–515.

[155] S.V. Parter, Mildly nonlinear elliptic partial differential equations and their numerical solutions I, *Numer. Math.* **7** (1965), 113–128.

[156] M. Protter and H. Weinberger, *Maximum Principles in Differential Equations*, Prentice-Hall, Englewood Cliffs, New Jersey, 1967.

[157] P. Rabinowitz, *Minimax Method in Critical Point Theory with Applications to Differential Equations*, CBMS Regional Conf. Series in Math., No. 65, AMS, Providence, 1986.

[158] F. Rellich, Über das asymptotische Verhalten der Lösungen von $\Delta u + \lambda u = 0$ in umendlichen Gebieten, *Jber. Deutsch. Math. Verei.* **53** (1943), 57–65.

[159] X. Ren and J. Wei, Single-point condensation and least energy solutions, *Proc. Amer. Math. Soc.* **124** (1996), 111–120.

[160] F.J. Rizzo, An integral equation approach to boundary value problems of classical elastostatics, *Q. Appl. Math.* **25** (1967), 83–95.

[161] W. Rudin, *Real and Complex Analysis*, McGraw-Hill, New York, 1966.

[162] K. Ruotsalainen and J. Saranen, On the convergence of the Galerkin method for nonsmooth solutions of integral equations, *Num. Math.* **54** (1988), 295–302.

[163] K. Ruotsalainen and W.L. Wendland, On the boundary element method for a nonlinear boundary value problem, in *Boundary Elements IX*, C.A. Brebbia (ed.), Springer-Verlag, New York, 1987, pp. 385–393.

[164] M. Sakakihara, An iterative boundary integral equation method for mildly nonlinear elliptic partial differential equations, in *Boundary Elements VII, Vol. II*, C.A. Brebbia and G. Maier (eds.), Springer-Verlag, Berlin-Heidelberg, 1985, pp. 13.49–13.58.

[165] J. Saranen and W.L. Wendland, On the asymptotic convergence of collocatiòn methods with spline functions of even degree, *Math. Comp.* **45** (1985), 91–108.

[166] D. Sattinger, Monotone methods in nonlinear elliptic and parabolic boundary value problems, *Indiana Univ. Math. J.* **21** (1972/72), 979–1000.

[167] D. Sattinger, *Topics in Stability and Bifurcation Theory*, Lecture Notes in Mathematics, Vol. 309, Springer-Verlag, New York, 1973.

[168] L. Schwartz, *Théorie des Distributions*, Hermann, Paris, 1966.

[169] R.T. Seeley, Singular integrals and boundary value problems, *Amer. J. Math.* **88** (1966), 781–809.

[170] R.T. Seeley, Topics in pseudo-differential operators, *Psuedo-Differential Operators*, L. Nirenberg (ed.), Roma, Edizioni Cremonese, Rome, pp. 169–305.

[171] I. Sneddon, *Partial Differential Equations*, McGraw-Hill, New York, 1958.

[172] I. Stakgold, *Boundary Value Problems of Mathematical Physics*, Vol. II, Macmillan, New York, 1967.

[173] G. Strang and G.J. Fix, *An Analysis of the Finite Element Method*, Prentice-Hall, Englewood Cliffs, New Jersey, 1973.

[174] A.H. Stroud, *Approximate Calculation of Multiple Integrals*, Prentice-Hall, Englewood Cliffs, New Jersey, 1971.

[175] C.K.W. Tam, On the screech tones of supersonic rectangular jets, *American Institute of Aeronautics and Astronautics Paper* No. 86–1866 (1986).

[176] C.K.W. Tam, J.M. Seiner and J.C. Yu, Proposed relationship between broadband shock associated noise and screech tones, *J. Sound Vib.* **110** (1986), 309–321.

[177] M. Taylor, *Pseudodifferential Operators*, Princeton University Press, Princeton, 1981.

[178] S. Timoshenko and S. Woinowsky-Krieger, *Theory of Plates and Shells*, 2nd ed., McGraw-Hill, New York, 1959.

[179] F. Treves, *Introduction to Pseudodifferential and Fourier Integral Operators*, Vol. 1, Plenum Press, New York, 1980.

[180] F. Treves, *Introduction to Pseudodifferential and Fourier Integral Operators*, Vol. 2, Plenum Press, New York, 1980.

[181] Z.-Q. Wang, On a superlinear elliptic equation, *Ann. Inst. Henri Poincaré* **8** (1991), 43–57.

[182] Z.-Q. Wang and J. Zhou, A local minimax-Newton method for finding multiple saddle points with symmetries, *SIAM J. Numer. Anal.* **42** (2004), 1745–1759.

[183] Z.-Q. Wang and J. Zhou, An efficient and stable method for computing multiple saddle points with symmetries, *SIAM J. Numer. Anal.* **43** (2005),891–907.

[184] C. Wilcox, A generalization of theorems of Rellich and Atkinson, *Proc. Amer. Math. Soc.* **7** (1956), 271–276.

[185] M. Willem, *Minimax Theorems*, Birkhäuser, Boston, 1996.

[186] X. Yao and J. Zhou, A local minimax characterization for computing multiple nonsmooth saddle critical points, *Math. Program.* **104** (2005), Ser. B, 749–760.

[187] X. Yao and J. Zhou, A minimax method for finding multiple critical points in Banach spaces and its application to quasi-linear elliptic PDE, *SIAM J. Sci. Comput.* **26** (2005), 1796–1809.

[188] X. Yao and J. Zhou, Unified convergence results on a minimax algorithm for finding multiple critical points in Banach spaces, *SIAM J. Numer. Anal.* **45** (2007), 1330–1347.

[189] X. Yao and J. Zhou, Numerical methods for computing nonlinear eigenpairs, I. Iso-homogeneous cases, *SIAM J. Sci. Comput.* **29** (2007), 1355–1374.

[190] X. Yao and J. Zhou, Numerical methods for computing nonlinear eigenpairs, II. Non-iso-homogeneous cases, *SIAM J. Sci. Comput.* **30** (2008), no. 2, 937–956.

[191] K. Yosida, *Functional Analysis*, Springer-Verlag, New York, 1966.

[192] D.M. Young and R.T. Gregory, *A Survey of Numerical Mathematics*, Vol. II, Addison-Wesley, Reading, MA, 1973.

[193] J. Zhou, Computations of eigenfunctions and eigenfrequencies of two dimensional vibrating structures by boundary element method, *Proceedings of 28th IEEE–CDC*, Tampa, Florida, December 1989, 2045–2049.

[194] J. Zhou, Instability analysis of saddle points by a local minimax method, *Math. Comp.* **74** (2005), 1391–1411.

[195] J. Zhou, Saddle critical point analysis and computation, *Nonlinear Anal.* **63** (2005), 1000–1009.

[196] J. Zhou and X. Yao, Stabilization in computing saddle points, in *Control and Boundary Analysis, 281–292, Lect. Notes Pure Appl. Math.*, 240, Chapman & Hall/CRC, Boca Raton, FL, 2005.

Subject Index